Classical Numerical Analysis

Numerical analysis is a broad field, and coming to grips with all of it may seem like a daunting task. This text provides a thorough and comprehensive exposition of all the topics contained in a classical graduate sequence in numerical analysis. With an emphasis on theory and connections with linear algebra and analysis, the book shows all the rigor of numerical analysis. Its high level and exhaustive coverage will prepare students for research in the field and will become a valuable reference as they continue their career. Students will appreciate the simple notation and clear assumptions and arguments, as well as the many examples and classroom-tested exercises ranging from simple verification to qualifying exam-level problems. In addition to the many examples with hand calculations, readers will also be able to translate theory into practical computational codes by running sample MATLAB codes as they try out new concepts.

Abner J. Salgado is Professor of Mathematics at the University of Tennessee, Knoxville. He obtained his PhD in Mathematics in 2010 from Texas A&M University. His main area of research is the numerical analysis of nonlinear partial differential equations, and related questions.

Steven M. Wise is Professor of Mathematics at the University of Tennessee, Knoxville. He obtained his PhD in 2003 from the University of Virginia. His main area of research interest is the numerical analysis of partial differential equations that describe physical phenomena, and the efficient solution of the ensuing nonlinear systems. He has authored more than 80 publications.

Classical Numerical Analysis

A Comprehensive Course

ABNER J. SALGADO
University of Tennessee, Knoxville

STEVEN M. WISE
University of Tennessee, Knoxville

CAMBRIDGE
UNIVERSITY PRESS

CAMBRIDGE
UNIVERSITY PRESS

Shaftesbury Road, Cambridge CB2 8EA, United Kingdom

One Liberty Plaza, 20th Floor, New York, NY 10006, USA

477 Williamstown Road, Port Melbourne, VIC 3207, Australia

314–321, 3rd Floor, Plot 3, Splendor Forum, Jasola District Centre,
New Delhi – 110025, India

103 Penang Road, #05–06/07, Visioncrest Commercial, Singapore 238467

Cambridge University Press is part of Cambridge University Press & Assessment,
a department of the University of Cambridge.

We share the University's mission to contribute to society through the pursuit of
education, learning and research at the highest international levels of excellence.

www.cambridge.org
Information on this title: www.cambridge.org/9781108837705

DOI: 10.1017/9781108942607

First published 2023

Printed in the United Kingdom by TJ Books Limited, Padstow Cornwall

A catalogue record for this publication is available from the British Library

Library of Congress Cataloging-in-Publication Data
Names: Salgado, Abner J., author. | Wise, Steven M. (Mathematician), author.
Title: Classical numerical analysis : a comprehensive course / Abner J. Salgado,
University of Tennessee, Knoxville, Steven M. Wise, University of Tennessee, Knoxville.
Description: Cambridge, United Kingdom ; New York, NY : Cambridge University Press, 2023. |
Includes bibliographical references and index.
Identifiers: LCCN 2022022842 (print) | LCCN 2022022843 (ebook) |
ISBN 9781108837705 (hardback) | ISBN 9781108942607 (epub)
Subjects: LCSH: Numerical analysis–Textbooks. |
BISAC: MATHEMATICS / Mathematical Analysis
Classification: LCC QA297 .S25 2023 (print) | LCC QA297 (ebook) |
DDC 518–dc23/eng20220823
LC record available at https://lccn.loc.gov/2022022842
LC ebook record available at https://lccn.loc.gov/2022022843

ISBN 978-1-108-83770-5 Hardback

Contents

Preface

This book on numerical analysis grew out of an ever expanding set of lecture notes that, over the years, the authors developed, corrected, used, and misused, while teaching the year-long sequence on this topic at the introductory graduate level at the University of Tennessee, Knoxville (UTK). The purpose of this sequence can be simply stated: prepare students for the PhD subject examination in numerical analysis and equip them for research in a rich, active, and expanding field.

The prerequisites for the book are (i) a solid understanding of linear algebra, at the level of Horn and Johnson [44], for example, and (ii) a working knowledge of advanced calculus, at the level of Rudin [76] or Bartel and Sherbert [6]. Both of these topics are thoroughly reviewed in the appendices. Those comfortable with the material in Appendices A and B should be well prepared for the book. Some important topics from differential equations, functional analysis, and measure and integration theory are also used. But, these are reviewed, as needed, and are not treated as prerequisites.

Our mission while writing this book was to present a spartan, but thoroughly understandable text in numerical analysis that students can use to pass PhD exams and get quickly started on research, similar to the philosophy behind the well-received graduate-level analysis text by Bass [7]. As a competing goal, we have designed the text so that no important topic should be left out; the student using this book should have all the details at their fingertips, to the extent possible. We make a concerted effort to use good notation; make definitions clear and concise; make hypotheses of theorems, lemmas, etc., apparent, perhaps to the point of being too pedantic; use the simplest versions of proofs; and keep proofs of theorems, lemmas, etc., where they are most natural for quick reference, after the respective results are stated. Important facts that students will need later are never buried deep in the exercises; they are front and center in the presentation. We may have fallen short in some or all of these goals, but they were our goals, nonetheless.

The end result is, we believe, and hope, a text that has a very broad coverage, but with a simple, modern, and easy-to-read style, and, importantly, with notation made as clear as possible, even if that means breaking with tradition. We deliberately choose not to be overly expository or conversational with our readers. The reader will find few long paragraphs of explanation. While some instructors will lament the the loss of those paragraphs, to be honest, in our experience, most students do not read them. They want core principles, easily locatable facts, and clear proofs. Thus, we focus on illustrative examples, good problems, and clean presentation.

Some other texts do a much better job of using eloquent language to present the ideas and their historical significance. Specifically, the books by Trefethen [95], Scott [83], and Süli and Mayers [89] include fascinating historical accounts of the development of numerical analysis and are must-reads. In our book, whenever a named concept or result is first introduced, we have included, as a footnote, the name and some minimal background on this person. This is done to convey that mathematics in general, and numerical analysis in particular, is a lively subject, made by and *for* people. We have relied on `Wikipedia` [101] and the `MacTutor History of Mathematics archive` [66] as references.

We emphasize theory over implementation in the book, and there are many good, classroom-tested problems that involve proofs and theoretical insight. This is a book that uses a lot of linear algebra and analysis, subjects that are near and dear to our hearts. The disciplined student will sharpen their theoretical abilities in those subjects with this book. It is our realization, however, that, at least in universities in the United States, advanced undergraduate or beginning graduate students sometimes do not have an adequate preparation in linear algebra and basic analysis. For this reason, we have included rather substantial reviews of the necessary results from linear algebra and analysis, in Appendices A and B, respectively. In addition, many of the ideas and techniques of Part V rely on the theory of partial differential equations. Chapter 23 and Appendix D provide a cooking-recipe list of facts. Some other background material is recalled as it is needed, and references to the literature are provided. To the extent possible, the presentation was designed to be self-contained.

We believe that the subject of numerical analysis does not need to be hard — or harder than it actually is — but, on the other hand, there is deep, beautiful, and, yes, sometimes difficult mathematics under the hood. One of our goals is to dispel the myth that numerical analysis is an *ad hoc* mixture of whatever seems to work, requiring little to no theory, proofs, or analysis. Our subject is a fundamental discipline of mathematics, a core sub-field of analysis, and many of the great mathematicians, from antiquity to the present — including Erdős and von Neumann, for example — have contributed to its advance. The biographical footnotes that we have added reflect this fact.

This book is not merely a catalog of numerical methods, though most of the important ones are contained herein. In almost every case, methods are developed and supported by rigorous theory. We give plenty of examples with hand calculations and output from numerical simulations. We include several sample codes, so that students can get their hands dirty. The codes are written in `MATLAB`® and were tested in its open source counterpart `GNU Octave` for compatibility. These codes are designed to help students learn the subject. To this end, we emphasize clarity over efficiency in our programming. Still, we adopt the point of view, shared by many others, that performing computational experiments can be both fun and deeply enlightening. Students are encouraged to code the methods and try them out. The interested student could use this text as a starting point for learning implementation issues which, we admit, are highly nontrivial and almost completely ignored in this text. The active researcher could use it to learn how to design a better algorithm or to understand why a particular algorithm works, or does not work. The codes listed in the text — in addition to those not listed but used to generate examples and figures — can be obtained from GitHub:

`https://github.com/stevenmwise/ClassicalNumericalAnalysis`

While this text covers more than what can be presented in a year-long course, we have undoubtedly omitted several topics that can be found in other books on the subject. Most of these omissions were deliberate, made either because we believed that the topics were more advanced or because we deemed the topics to be nonessential. For instance, many numerical analysis books begin the discussion by addressing rounding errors and floating point arithmetic. We chose not to discuss these points, as we believe that, while important, they skew the student's perception about what is numerical analysis. This, then, begs the question: *What is numerical analysis?* For an answer, we defer to one of the classics [41]:

> ... we shall mean by numerical analysis the *theory of constructive methods of mathematical analysis.* The emphasis is on the word "constructive."

We urge the reader to examine the Introduction to this reference for a very insightful definition of what numerical analysis is, and *what it is not*, one that was established in the early days of our beloved discipline. For an update on this viewpoint, which only reinforces our beliefs, we refer to the appendix of [96].

Regarding usage, the graduate-level numerical analysis sequence at UTK focuses heavily on numerical linear algebra, nonlinear equations, and numerical differential equations. These topics are well represented in the text, specifically in Parts I, III, IV, and V, and students using this book will get a thorough and modern introduction in those areas. However, we are well aware that other universities choose to emphasize other topics of numerical analysis. While the subjects of interpolation, constructive approximation, and quadrature theory, the focus of Part II, are not specifically in the catalog description of the introductory graduate sequence at UTK, these are included to support the theory needed to analyze numerical methods in the other parts of the book, to broaden the background knowledge of our students when they study these subjects in other classes, and for audiences outside of our university. We have taught the material from Part II in a separate, single-semester, graduate course on classical approximation theory.

We envision, and hope, that this text will be used in several year-long numerical analysis sequences. For this reason the dependency of all chapters is linear, with the exception of the appendices. Some sample course plans and ways we have used this text to teach are as follows:

- *UTK, numerical algebra:* (1 semester) Part I and Chapter 15.
- *UTK, numerical differential equations:* (1 semester) Part IV and Chapters 23 — 25, 28 and 29, referencing results from Part II, as needed.
- *UTK, classical approximation theory:* (1 semester) Part II, with Chapters 26 and 27 as applications.

- *Classical numerical analysis:* (1 semester) Chapters 3, 4, 6, 7, 9, 12, 14, 15, and Part IV.
- *Topics in numerical analysis:* (1 semester) Part II, Chapters 2 and 5, Part IV, Chapters 6 and 7 (using Chapter 24 as motivation), and Chapters 15 and 16.

Acknowledgments

The authors thank the many Numerical Analysis students at the University of Tennessee who endured countless versions of this text. They provided valuable feedback on the stylistic structure of the book and also found numerous typos, errors, and blunders. We hope that most of the inaccuracies have been amended. For those that remain, we are solely responsible.

We acknowledge support from the National Science Foundation, which supported us in our research during the writing of the book. This project represents an attempt on our part to make the subject of numerical analysis more accessible, from the theoretical point of view, and more interesting to a broader mathematics audience, especially those who might have thought that the subject lacks rigor and beauty. At the same time, much of what we have learned at the frontiers of research in numerical partial differential equations and scientific computing has made it into the textbook.

AJS: I wish to thank all the numerical analysis instructors that I have had throughout the years. In particular, and in chronological order, my MSc advisor, V.G. Korneev; my PhD advisor J.-L. Guermond; V. Girault; and my postdoc mentor, R.H. Nochetto. All the good things I contributed to this book are because of the ideas and "way of doing business" that they taught me. All the bad ideas that remain are due only to the fact that, as everyone knows, I am very stubborn. I also wish to thank all the other numerical analysts who had the misfortune to cross paths with me at some point. I thank also Steve M. Wise for bringing me along on this journey, and for his patience. I am glad we were able to stay friends after completing this project. Finally, I thank my parents for supporting me in every step of my professional formation. Without their support I would not have made it to where I am today.

SMW: I must thank my numerical analysis teachers and advisors, Chris Beattie and Layne Watson, at Virginia Tech. I got a really great foundation from them. From my PhD advisor, Bill Johnson, a materials scientist by training, I learned to appreciate the power of computing in scientific exploration. The equations that I learned from Bill still motivate and drive my numerical analysis research 20 years later. My postdoctoral advisor, long-time collaborator, and friend, John Lowengrub, taught me a lot about the intersection of practical scientific computing, modeling, and numerical analysis. I fortunately learned a valuable lesson early in my career; namely, that I should not be afraid to admit that I do not know something. This enabled me to learn from those around me and helped to fill in the gaps in my education, which are still considerable. Specifically, my colleagues and collaborators, Xiaobing Feng, Ohannes Karakashian, Cheng Wang, and, of course, Abner J. Salgado, have given me a great on-the-job education. I am so grateful that Abner agreed to go on this journey with me. He is the smartest person I know, and I have learned so much from him. It has been a pleasure. I thank my wife, Nicole, and children, Jude and Cece, for giving up a lot of family time while I worked on this project over the last few years. I love you so much. Finally, I dedicate my work on this book to the memory of my dear mother, Mary Ann Wise. I love you, miss you, and think of you every day.

Symbols

This list describes several symbols that will be commonly used within the text. The page number indicates where its definition, or first appearance, is found.

$(\cdot,\cdot)_2$	The Euclidean inner product, page 849
$(\cdot,\cdot)_{L^2_h}$	The discrete inner product on $\mathcal{V}_0(\bar{\Omega}_h)$, page 673
$[\boldsymbol{x}]_i$	The ith component of the n-vector $\boldsymbol{x}$, page 840
$[\cdot,\cdot]_{L^2_h}$	The discrete inner product on $\mathcal{V}(\bar{\Omega}_h)$, page 673
$[\mathsf{A}]_{i,j}$	The element in the ith row and jth column of the matrix A, page 4
$\bar{z}$	The complex conjugate of $z \in \mathbb{C}$, page 838
$\bar{\delta}_h$	Backward difference operator, page 669
$\bar{\Omega}_h$	$[0,1]^d \cap \mathbb{Z}^d_h$, with $d \in \mathbb{N}$, page 666
$\boldsymbol{u} \otimes \boldsymbol{v}$	The exterior product of $\boldsymbol{u}$ and $\boldsymbol{v}$, also denoted $\boldsymbol{u}\boldsymbol{v}^{\mathsf{H}}$, page 27
$\mathcal{C}$	In the context of parabolic equations, this is the space–time cylinder, $\Omega \times (0,T)$, page 646
$\mathcal{D}^j_h v$	For a periodic grid function v and $j \in \{1,2\}$, this denotes its pseudo-spectral derivative of order j, page 732
$\mathcal{F}_n[\cdot]$	For $n \in \mathbb{N}$, this denotes the Discrete Fourier Transform (DFT), page 350
$\mathcal{G}^B_h$	For a grid domain $\mathcal{G}_h$, these are the boundary points (with respect to a finite difference operator), page 670
$\mathcal{G}^I_h$	For a grid domain $\mathcal{G}_h$, these are the interior points (with respect to a finite difference operator), page 670
$\mathcal{I}_X$	For a nodal set $X \subset [a,b] \subset \mathbb{R}$, this denotes the interpolation operator subordinate to X, page 234
$\mathcal{K}_m(\mathsf{A},\boldsymbol{q})$	The Krylov subspace of the matrix A of degree m, page 169
$\mathcal{L}_X$	For a nodal set X, this denotes the Lagrange nodal basis, page 235
$\mathcal{R}(a,b)$	The same as $\mathcal{R}([a,b])$, page 862
$\mathcal{R}(a,b;\mathbb{C})$	The collection of complex-valued Riemann integrable functions, page 865
$\mathcal{R}(I)$	For I a finite interval, this denotes the collection of functions that are Riemann integrable, page 862
$\mathcal{V}(\mathcal{G}_h)$	For a grid domain $\mathcal{G}_h \subseteq \mathbb{Z}^d_h$ with $h = 1/(N+1)$, this is the collection of grid functions, page 666
$\mathcal{V}(\mathbb{C})$	The space of functions $\mathbb{Z} \to \mathbb{C}$, page 348
$\mathcal{V}_0(\bar{\Omega}_h)$	The collection of functions in $\mathcal{V}(\bar{\Omega}_h)$ that vanish on the discrete boundary $\partial\Omega_h$, page 667

$\mathcal{V}_{M,\mathrm{p}}(\mathbb{C})$	For $M \in \mathbb{N}$, this denotes the space of complex-valued grid functions that are, in addition, periodic, page 722
$\mathcal{Z}[\cdot]$	The Fourier-$\mathbb{Z}$, or Discrete Fourier, transform on grid functions in $L^2_h(\mathbb{Z}_h;\mathbb{C})$, page 794
χ_A	The characteristic polynomial of the matrix A, page 12
$\mathrm{clo}_s(K)$	For a subset K of a Hilbert space, this denotes its closure, page 455
$\mathrm{clo}_w(K)$	For a subset K of a Hilbert space, this denotes its weak closure, page 455
col(A)	The column space of the matrix A, page 6
$\mathbb{C}$	The set of complex numbers, page 838
$\mathbb{C}^n$	The vector space of complex n-vectors, page 840
$\mathbb{C}^n_\star$	The collection of nonzero vectors in $\mathbb{C}^n$, page 9
$\mathbb{C}^{n\times n}_{\mathrm{Her}}$	The space of Hermitian matrices of size n, page 57
$\Delta\phi$	For a smooth scalar-valued function ϕ, this denotes its Laplacian, page 618
$\delta^{n,\mathrm{p}}$	The n-periodic grid delta function, page 349
Δ_h	The discrete Laplacian, page 669
δ_h	Forward difference operator, page 669
$\delta_h^\diamond$	The discrete mixed derivative, page 669
$\Delta_h^\square$	The two-dimensional skew Laplacian, page 670
$\delta_{i,j}$	The Kronecker delta, page 842
$\ell^2(\mathbb{Z};\mathbb{C})$	The collection of all sequences $\{a_j\}_{j\in\mathbb{Z}} \subset \mathbb{C}$ that are square summable, page 330
$\mathfrak{B}(\mathbb{V})$	The same as $\mathfrak{B}(\mathbb{V},\mathbb{V})$, page 457
$\mathfrak{B}(\mathbb{V},\mathbb{W})$	For normed spaces $\mathbb{V}$ and $\mathbb{W}$, this denotes the vector space of bounded linear operators $\mathbb{V}\to\mathbb{W}$, page 457
$\mathfrak{L}(\mathbb{V})$	The same as $\mathfrak{L}(\mathbb{V},\mathbb{V})$, page 3
$\mathfrak{L}(\mathbb{V},\mathbb{W})$	The set of linear operators from $\mathbb{V}$ to $\mathbb{W}$, page 3
$\Im z$	The imaginary part of the complex number z, i.e., $\Im z = b$, if $z = a + \mathrm{i}b$, page 838
im(A)	The image (or range) of the matrix A, also denoted $\mathcal{R}(\mathsf{A})$, page 6
i	The imaginary unit, page 838
$\kappa(\mathsf{A})$	The condition number of the matrix A, page 80
$\kappa_2(\mathsf{A})$	The spectral condition number of the matrix A, page 80
ker(A)	The kernel (or null space) of the matrix A, also denoted $\mathcal{N}(\mathsf{A})$, page 6
$\langle\cdot,\cdot\rangle$	The duality pairing between a Hilbert space $\mathcal{H}$ and its dual $\mathcal{H}'$, page 464
$[u\star v]^{n,\mathrm{p}}$	For $u, v \in \mathcal{V}_{n,\mathrm{p}}(\mathbb{C})$, this denotes their discrete periodic convolution, page 352
$\leq$	For a vector space $\mathbb{V}$ and $\mathbb{W}\subseteq\mathbb{V}$, $\mathbb{W}\leq\mathbb{V}$ denotes that $\mathbb{W}$ is a subspace of $\mathbb{V}$. If $\mathbb{W}\neq\mathbb{V}$, then we denote $\mathbb{W}<\mathbb{V}$, page 841
$\mathcal{B}_\alpha(-1,1)$	For α the Chebyshev weight function, this denotes the subspace of $C^1([-1,1])$ of functions such that $\alpha(x)g(x)\to 0$ as $x\to\pm1$, page 743
$\mathcal{F}^m(S)$	For $m\in\mathbb{N}$, this is $F^1(S)\cap C^m(S;\mathbb{R}^d)$, page 517

$\mathcal{V}_{n,\mathrm{p}}(\mathbb{C})$	For $n \in \mathbb{N}$, this denotes the space of n-periodic grid functions, page 348
$\mathring{\mathcal{V}}_{M,\mathrm{p}}(\mathbb{C})$	For $M \in \mathbb{N}$, this denotes the space of mean-zero, complex-valued, periodic grid functions, page 723
$\mathring{\delta}_h$	Centered difference operator, page 669
$\mathring{\mathscr{S}}_N(0,1;\mathbb{C})$	For $N \in \mathbb{N}$, this is the space of complex-valued, mean-zero, trigonometric polynomials of degree at most N, page 730
$\mathring{C}^\infty_\mathrm{p}(0,1;\mathbb{C})$	The space infinitely differentiable, complex-valued, periodic functions that have mean-zero, page 884
$\mathring{C}^m_\mathrm{p}(0,1;\mathbb{C})$	For $m \in \mathbb{N}_0$, this denotes the space functions in $C^m_\mathrm{p}(0,1;\mathbb{C})$ that have mean-zero, page 884
$\mathring{H}^m_\mathrm{p}(0,L;\mathbb{C})$	For $m \in \mathbb{N}_0$, this is the space of functions in $H^m_\mathrm{p}(0,L;\mathbb{C})$ that have mean-zero, page 892
$\mathsf{A} \asymp \mathsf{B}$	The matrix A is similar to the matrix B, page 13
$\mathsf{A}(S)$	For $\mathsf{A} \in \mathbb{C}^{n\times n}$ and $S \subseteq \{1,\ldots,n\}$, this denotes the sub-matrix obtained by deleting the rows and columns whose indices are not in S, page 36
A^H	The conjugate transpose of the matrix A, page 7
$\mathsf{A}^\dagger$	The Moore–Penrose pseudo-inverse of A, page 30
$\mathsf{A}^\intercal$	The transpose of the matrix A, page 7
A^{-1}	The inverse of the matrix A, page 8
∇v	For a smooth scalar-valued function, this denotes its gradient, page 612
$\nabla \cdot \boldsymbol{u}$	For a smooth vector-valued function $\boldsymbol{u}$, this denotes its divergence, page 612
$\Vert\mathsf{A}\Vert_{\max}$	The matrix max-norm of the matrix A, page 9
$\Vert\cdot\Vert_{H^1_h}$	The discrete H^1_h-norm on the space of grid functions on $\mathcal{V}(\bar{\Omega}_h)$, page 686
$\Vert\cdot\Vert_{L^p_h}$	For $p \in [1,\infty]$, this denotes the discrete L^p_h-norm on the spaces of grid functions on $(0,1)^d$, page 671
$\Vert\mathsf{A}\Vert_F$	The Frobenius norm of the matrix A, page 9
$\Vert\mathsf{A}\Vert_p$	The induced p-norm of the matrix A, page 10
$\Vert f\Vert_{L^p(\Omega;\mathbb{C})}$	For a function $f\colon \Omega \to \mathbb{C}$, this denotes its L^p-norm, $p \in [1,\infty]$, page 881
$\Vert f\Vert_{L^p_w(a,b;\mathbb{C})}$	For a function $f\colon [a,b] \to \mathbb{C}$, this denotes its weighted L^p-norm, $p \in [1,\infty)$, with weight w, page 882
Ω_h	$(0,1)^d \cap \mathbb{Z}^d_h$, with $d \in \mathbb{N}$, page 666
$\partial\Omega_h$	$\bar{\Omega}_h \backslash \Omega_h$, page 666
$\partial_p\mathcal{C}$	The parabolic boundary of $\mathcal{C}$, page 646
$\mathbb{H}^*$	For a complex Hilbert space $\mathbb{H}$, this denotes the anti-dual, page 736
$\mathbb{K}^n$	The vector space of n-vectors, page 840
$\mathbb{K}^{m\times n}$	The set of matrices with m rows and n columns with coefficients in $\mathbb{K}$, page 4
$\mathbb{P}_n$	This is, typically, $\mathbb{P}_n(\mathbb{R})$ or $\mathbb{P}_n(\mathbb{C})$, depending upon the context, page 840

$\mathbb{P}_n(\mathbb{K})$	The vector space of polynomials of degree no larger than n with coefficients in $\mathbb{K}$, page 840
$\mathbb{P}_{m/n}$	For $m, n \in \mathbb{N}_0$, this is the set of rational polynomials whose numerator and denominator lie in $\mathbb{P}_m$ and $\mathbb{P}_n$, respectively, page 588
$\mathbb{Q}$	The set of rational numbers, page 838
$\mathbb{T}_n$	For $n \in \mathbb{N}_0$, this denotes the space of all one-periodic trigonometric polynomials, page 321
$\mathbb{Z}_h^d$	The collection of vectors in $\mathbb{R}^d$ of the form hz with $\boldsymbol{z} \in \mathbb{Z}^d$, page 665
$\Re z$	The real part of the complex number z, i.e., $\Re z = a$, if $z = a + \mathrm{i}b$, page 838
$\mathbb{R}$	The set of real numbers, page 838
$\mathbb{R}^n$	The vector space of real n-vectors, page 840
$\mathbb{R}^n_\star$	The collection of nonzero vectors in $\mathbb{R}^n$, page 9
$\mathbb{R}^{n\times n}_{\mathrm{sym}}$	The space of real symmetric matrices of size n, page 57
$\rho(\mathsf{A})$	The spectral radius of matrix $\mathsf{A} \in \mathbb{C}^{n\times n}$, page 73
$\mathrm{row}(\mathsf{A})$	The row space of the matrix A, page 6
$\sigma(\mathsf{A})$	The spectrum of the square matrix A, page 12
$\sigma(A)$	The spectrum of the linear operator A, page 15
$\mathscr{S}^{1,0}(\mathscr{T}_h)$	The space of continuous piecewise linear functions subject to the triangulation $\mathscr{T}_h$, page 705
$\mathscr{S}_0^{1,0}(\mathscr{T}_h)$	This is $\mathscr{S}^{1,0}(\mathscr{T}_h) \cap H_0^1(\Omega)$, page 705
$\mathscr{S}^{\boldsymbol{p},0}(\mathscr{T}_h)$	For a one-dimensional mesh $\mathscr{T}_h$, with $\#\mathscr{T}_h = N$ and $\boldsymbol{p} \in \mathbb{N}^{N+1}$, this is the space of functions that are continuous, and for every $I_i \in \mathscr{T}_h$ their restriction to I_i is a polynomial of degree p_{i+1}, page 710
$\mathscr{S}_0^{\boldsymbol{p},0}(\mathscr{T}_h)$	This is $\mathscr{S}^{\boldsymbol{p},0}(\mathscr{T}_h) \cap H_0^1(0,1)$, page 710
$\mathscr{S}^{p,-1}(\boldsymbol{\tau};\mathcal{H})$	For a Hilbert space $\mathcal{H}$, this is the space of $\mathcal{H}$-valued piecewise polynomials of degree at most p over the partition $\boldsymbol{\tau}$, page 599
$\mathscr{S}^{p,0}(\boldsymbol{\tau};\mathcal{H})$	This is $\mathscr{S}^{p,-1}(\boldsymbol{\tau};\mathcal{H}) \cap C([0,T];\mathcal{H})$, page 605
$\mathscr{S}^{p,0}(\mathscr{T}_h)$	For $p \in \mathbb{N}$, this is the space of functions that are continuous and piecewise polynomials, of degree p, subject to the triangulation $\mathscr{T}_h$, page 710
$\mathscr{S}_0^{p,0}(\mathscr{T}_h)$	This is $\mathscr{S}^{p,0}(\mathscr{T}_h) \cap H_0^1(\Omega)$, page 710
$\mathscr{S}^{p,r}(\boldsymbol{\tau};\mathcal{H})$	This is $\mathscr{S}^{p,-1}(\boldsymbol{\tau};\mathcal{H}) \cap C^r([0,T];\mathcal{H})$, page 605
$\mathscr{S}_{N,0}(-1,1)$	For $N \in \mathbb{N}$, this denotes the set of polynomials of degree at most N that vanish at $x = \pm 1$, page 747
$\mathrm{span}(S)$	The span of the set S, also denoted $\langle S \rangle$, page 840
$\mathrm{supp}\, g$	The support of the function g, page 705
$\mathrm{supp}(\phi)$	For a function ϕ, this denotes its support, page 887
$\tilde{\delta}^{n,\mathrm{p}}$	The singular n-periodic grid delta function, page 353
$\tilde{\mathsf{E}}_h^\tau(\xi)$	The symbol of a two-layer, matrix-valued, finite difference method, page 828
$\tilde{E}_h^\tau(\xi)$	The symbol of a two-layer finite difference method, page 795
$\mathscr{T}_h$	A mesh with mesh size $h > 0$, page 705
$\{\boldsymbol{x}_k\}_{k=1}^\infty$	A sequence of vectors in either $\mathbb{C}^d$ or $\mathbb{R}^d$, page 854
A^*	The adjoint of the linear operator A, page 7

$C(A;B)$	The vector space of continuous functions with domain A and range in B, page 842
$C(I)$	For I and interval this denotes the set of functions $f\colon I\to\mathbb{R}$ that are continuous, page 858
$C^0(I)$	The same as $C(I)$, page 859
$C^m(I)$	For $m\in\mathbb{N}$ and I an interval, this denotes the collection of functions $f\colon I\to\mathbb{R}$ whose derivatives up to and including mth order exist and are continuous on I, page 859
$C^m_{\mathrm{p}}(0,1;\mathbb{C})$	For $m\in\mathbb{N}_0$, this denotes the space of complex-valued, m-times continuously differentiable periodic functions, page 884
$C^{0,1}(I)$	For I an interval, this denotes the collection of functions $f\colon I\to\mathbb{R}$ that are Lipschitz continuous, page 859
$C^{0,\alpha}([0,1])$	For $\alpha>0$, this denotes the set of functions $v\colon[0,1]\to\mathbb{R}$ that are Hölder continuous of order α, page 895
$C^{0,\alpha}(I)$	For I an interval and $\alpha\in(0,1]$, this denotes the collection of functions $f\colon I\to\mathbb{R}$ that are Hölder continuous of order α, page 859
$C_b(\mathbb{R}^d)$	The space of continuous functions $\mathbb{R}^d\to\mathbb{R}$ that, in addition, are bounded on $\mathbb{R}^d$, page 639
$C^m_b(I)$	For $m\in\mathbb{N}$ and I an interval, this denotes the collection of functions in $C^m(I)$ such that, in addition, the function and all its derivatives up to and including order m are bounded on I, page 860
$f=\mathcal{O}(g)$	The Landau symbol. Whenever f and g are two related quantities, this is used to denote that f is, asymptotically, of the order of g, page 856
$F^1(S)$	The class of slope functions that are continuously differentiable on S and whose partial $\boldsymbol{u}$-derivatives are bounded, page 517
$H^1(\Omega)$	For a bounded domain $\Omega\in\mathbb{R}^d$, with $d\in\mathbb{N}$ this denotes the Sobolev space of functions $v\in L^2(\Omega)$ such that $\nabla v\in L^2(\Omega;\mathbb{R}^d)$, page 888
$H^1_0(\Omega)$	The subspace of $H^1(\Omega)$ of functions that vanish on the boundary, page 888
$H^1_{\alpha,0}(-1,1)$	The subspace of $H^1_\alpha(-1,1)$ of functions that vanish at $x=\pm1$, page 744
$H^m_\alpha(-1,1)$	For $m\in\mathbb{N}_0$ and α the Chebyshev weight function, this denotes the Chebyshev weighted Sobolev space of order m, page 743
$H^m_{\mathrm{p}}(0,L;\mathbb{C})$	For $L>0$ and $m\in\mathbb{N}_0$, this denotes the space of L-periodic Sobolev functions, page 892
$L^2_h(\mathbb{Z}_h)$	The collection of grid functions $\mathcal{V}(\mathbb{Z}_h)$ that are square summable, page 793
$L^2_{\mathrm{p}}(0,1;\mathbb{C})$	The set of all one-periodic, locally square integrable functions, page 886
L_ℓ	For a nodal set X of size $n+1$ and $0\le\ell\le n$, this denotes the ℓth element of the Lagrange nodal basis, page 235
S_1+S_2	For $S_1,S_2\le\mathbb{C}^n$, this denotes their sum, page 94
$S_1\oplus S_2$	For $S_1,S_2\le\mathbb{C}^n$, this means that they are complementary subspaces, i.e., $S_1+S_2=\mathbb{C}^n$, page 94

$S_1 \overset{\perp}{\oplus} S_2$	For $S_1, S_2 \leq \mathbb{C}^n$, this means that they are complementary, and orthogonal, subspaces, i.e., $S_1 + S_2 = \mathbb{C}^n$ and $\boldsymbol{s}_1 \in S_1$ $\boldsymbol{s}_2 \in S_2$ implies $\boldsymbol{s}_2^{\mathsf{H}} \boldsymbol{s}_1 = 0$, page 95
$W^{\perp}$	The orthogonal complement of the set W, page 849
$x \perp y$	The vector x is orthogonal to y, page 849
$X \hookrightarrow Y$	For normed spaces X and Y, this means that X is continuously embedded in Y, page 706
$\mathsf{A}^{(k)}$	The leading principal sub-matrix of order k of A, page 36
$\\|\boldsymbol{x}\\|_p$	The p-norm of a complex n-vector $\boldsymbol{x}$. Also denoted $\\|\boldsymbol{x}\\|_{\ell^p(\mathbb{C}^n)}$, page 844
$H^m(\Omega)$	For $m \in \mathbb{N}$, this denotes the collection of functions $v \in L^2(\Omega)$ whose weak derivatives up to order m belong to $L^2(\Omega)$ as well, page 890
$L^p(0,T;\mathbb{V})$	For a Banach space $\mathbb{V}$, this denotes the space of functions such that the mapping $t \mapsto \\|v(t)\\|_{\mathbb{V}}$ belongs to $L^p(0,T)$, page 644
$\#S$	The cardinality of the set S, page 841
$\|z\|$	The modulus of the complex number z, page 839

Part I

Numerical Linear Algebra

1 Linear Operators and Matrices

We begin our disussion by presenting several facts about the natural transformations between vector spaces and their representations, i.e., matrices. These form the foundation for our study of *numerical* linear algebra. Herein, we will set in place much of our notation, especially for matrices, that will be used not just for Part I but for the entirety of the book. Every student using this text should master the material from Appendix A and this chapter before moving on. The book by Horn and Johnson [44] is an excellent external reference.

Why is linear algebra so important to numerical analysis? That is a fair question. The answer is that many algorithms in numerical analysis — for a broad range of problem types, interpolation, approximation of functions, approximating solutions to differential or integral equations — require, at some stage in the algorithm, the investigation of a system of linear equations:

$$\begin{cases} a_{1,1}x_1 + a_{1,2}x_2 + \cdots + a_{1,n}x_n = f_1, \\ a_{2,1}x_1 + a_{2,2}x_2 + \cdots + a_{2,n}x_n = f_2, \\ \qquad\qquad\qquad\qquad\vdots \\ a_{m,1}x_1 + a_{m,2}x_2 + \cdots + a_{m,n}x_n = f_m. \end{cases}$$

Many algorithms will require the solution of such systems. Others, by contrast, may need some or all of the eigenvalues or singular values of the associated coefficient matrix for the system.

Before we jump into the topic of how to practically solve such a system of equations, which we will cover in Chapter 3 — or how to compute singular values (Chapter 2) and/or eigenvalues (Chapter 8) of the coefficient matrix — we need to understand the properties of such systems. This will be the topic of this chapter. Let us get started.

1.1 Linear Operators and Matrices

We study the *natural* mappings between vector spaces, i.e., those that preserve the vector space structure.

Definition 1.1 (linear operator). Let $\mathbb{V}$ and $\mathbb{W}$ be complex vector spaces. The mapping $A\colon \mathbb{V} \to \mathbb{W}$ is called a **linear operator** if and only if

$$A(\alpha x + \beta y) = \alpha Ax + \beta Ay, \quad \forall \alpha, \beta \in \mathbb{C},\ \forall x, y \in \mathbb{V}.$$

The set of all linear operators from $\mathbb{V}$ to $\mathbb{W}$ is denoted by $\mathfrak{L}(\mathbb{V},\mathbb{W})$. For simplicity, we denote by $\mathfrak{L}(\mathbb{V})$ the set of linear operators from $\mathbb{V}$ to itself. Suppose that $A, B \in \mathfrak{L}(\mathbb{V},\mathbb{W})$ and $\alpha, \beta \in \mathbb{C}$ are arbitrary. We define, in a natural way, the object $\alpha A + \beta B$ via

$$(\alpha A + \beta B)x = \alpha Ax + \beta Bx, \quad \forall x \in \mathbb{V}.$$

It is straightforward to prove that $\alpha A + \beta B$ is a linear operator and we get the following result.

Proposition 1.2 (properties of $\mathfrak{L}(\mathbb{V},\mathbb{W})$). *Let $\mathbb{V}$ and $\mathbb{W}$ be complex vector spaces. The set $\mathfrak{L}(\mathbb{V},\mathbb{W})$ is a vector space using the natural definitions of addition and scalar multiplication given in the last definition. If $\dim(\mathbb{V}) = m$ and $\dim(\mathbb{W}) = n$, then $\dim(\mathfrak{L}(\mathbb{V},\mathbb{W})) = mn$.*

Proof. See Problem 1.2. □

Definition 1.3 ($m \times n$ matrices). Let $\mathbb{K}$ be a field. We define, for any $m, n \in \mathbb{N}$,

$$\mathbb{K}^{m\times n} = \{\mathsf{A} = [a_{i,j}] \mid a_{i,j} \in \mathbb{K}, \ i = 1, \dots, m, \ j = 1, \dots, n\}.$$

The object A is called a **matrix** and the elements $a_{i,j} \in \mathbb{K}$ are called its **components** or **entries**. We call $\mathbb{C}^{m\times n}$ the set of **complex** $m \times n$ **matrices** and $\mathbb{R}^{m\times n}$ the set of **real** $m \times n$ **matrices**.

To extract the entry in the ith row and jth column of the $m\times n$ matrix $\mathsf{A} \in \mathbb{K}^{m\times n}$, we use the notation

$$[\mathsf{A}]_{i,j} = a_{i,j} \in \mathbb{K}.$$

The convention is that the entries of a matrix are denoted by the respective lowercase roman symbol. For example, the matrix C has entries $c_{i,j}$. We often make this identification explicit, as in writing $\mathsf{A} = [a_{i,j}] \in \mathbb{C}^{m\times n}$. We say that there are m rows and n columns in an $m \times n$ matrix A. We naturally define $m \times n$ matrix addition and scalar multiplication component-wise via

$$[\mathsf{A} + \mathsf{B}]_{i,j} = a_{i,j} + b_{i,j}, \quad [\alpha \mathsf{A}]_{i,j} = \alpha a_{i,j}, \quad i = 1, \dots, m, \quad j = 1, \dots, n,$$

where $\mathsf{A}, \mathsf{B} \in \mathbb{K}^{m\times n}$ are arbitrary $m \times n$ matrices and $\alpha \in \mathbb{K}$ is an arbitrary scalar.

Proposition 1.4 ($\mathbb{K}^{m\times n}$ is a vector space). *With addition and scalar multiplication defined as above, $\mathbb{K}^{m\times n}$ is a vector space over $\mathbb{K}$ and $\dim(\mathbb{K}^{m\times n}) = m \cdot n$.*

Proof. See Problem 1.3. □

Of course, the reader will remember that matrices can be combined in more exotic ways.

Definition 1.5 (matrix product). Let $\mathsf{A} = [a_{i,k}] \in \mathbb{K}^{m\times p}$ and $\mathsf{B} = [b_{k,j}] \in \mathbb{K}^{p\times n}$. The **matrix product** $\mathsf{C} = \mathsf{AB}$ is a matrix in $\mathbb{K}^{m\times n}$ whose entries are computed according to the formula

$$[\mathsf{C}]_{i,j} = c_{i,j} = \sum_{k=1}^{p} a_{i,k} b_{k,j}, \quad i = 1, \dots, m, \quad j = 1, \dots, n.$$

Next, we define a matrix–vector product, which, the reader will see, is similar to the last definition.

Definition 1.6 (matrix–vector product). Suppose that $\boldsymbol{x} = [x_s] \in \mathbb{K}^n$ and $\mathsf{A} = [a_{k,s}] \in \mathbb{K}^{m\times n}$. Then the **matrix–vector product** $\boldsymbol{y} = \mathsf{A}\boldsymbol{x}$ is a vector in $\mathbb{K}^m$ whose components are computed via the formula

$$[\boldsymbol{y}]_k = y_k = \sum_{s=1}^{n} a_{k,s}x_s, \quad k = 1, \dots, m.$$

Remark 1.7 (identification). Suppose that $\mathsf{A} \in \mathbb{C}^{m\times n}$. Then the (canonical) mapping $\mathsf{A}\colon \mathbb{C}^n \to \mathbb{C}^m$ defined by $\boldsymbol{y} = \mathsf{A}\boldsymbol{x}$ — where $\boldsymbol{x} \in \mathbb{C}^n$, so that the matrix–vector product $\boldsymbol{y}$ is in $\mathbb{C}^m$ — is linear. Mimicking the identification process outlined in Theorem A.24, we can also identify $\mathfrak{L}(\mathbb{C}^n, \mathbb{C}^m)$ with the space $\mathbb{C}^{m\times n}$ of matrices having m rows and n columns of complex entries. This says that all linear mappings from $\mathbb{C}^n$ to $\mathbb{C}^m$ are, essentially, matrices. This result can be generalized to identify $\mathfrak{L}(\mathbb{K}^n, \mathbb{K}^m)$ with $\mathbb{K}^{m\times n}$ for a generic field $\mathbb{K}$.

Remark 1.8 (notation). It will be helpful from this point on to always view $\mathbb{C}^k$ as a vector space of column k-vectors, i.e., $\mathbb{C}^{k\times 1}$. When we consider $\boldsymbol{x} \in \mathbb{C}^k$, we think

$$\boldsymbol{x} = \begin{bmatrix} | \\ x \\ | \end{bmatrix} = \begin{bmatrix} x_1 \\ \vdots \\ x_k \end{bmatrix}.$$

Upon introducing the *transpose* operation $\cdot^{\mathsf{T}}\colon \mathbb{C}^{k\times 1} \to \mathbb{C}^{1\times k}$ as mapping column k-vectors to row k-vectors, we will often express $\boldsymbol{x} \in \mathbb{C}^k$ inline as $\boldsymbol{x} = [x_1, \dots, x_k]^{\mathsf{T}}$, i.e., as the transpose of a row vector. In a related way, given a matrix $\mathsf{A} \in \mathbb{C}^{m\times n}$ we commonly wish to represent it in a column-wise format (as a collection of column vectors) via

$$\mathsf{A} = \begin{bmatrix} | & & | \\ \boldsymbol{c}_1 & \cdots & \boldsymbol{c}_n \\ | & & | \end{bmatrix}, \quad \boldsymbol{c}_j \in \mathbb{C}^m, \ j = 1, \dots, n,$$

or in a row-wise format (as a collection of row vectors) via

$$\mathsf{A} = \begin{bmatrix} - & \boldsymbol{r}_1^{\mathsf{T}} & - \\ & \vdots & \\ - & \boldsymbol{r}_m^{\mathsf{T}} & - \end{bmatrix}, \quad \boldsymbol{r}_i \in \mathbb{C}^n, \ i = 1, \dots, m.$$

As a further shorthand, we will often write (inline) $\mathsf{A} = [\boldsymbol{c}_1, \dots, \boldsymbol{c}_n]$ and $\mathsf{A} = [\boldsymbol{r}_1, \dots, \boldsymbol{r}_m]^{\mathsf{T}}$. It is important to notice that if we view the matrix A in column-wise format, then the matrix–vector product $\boldsymbol{y} = \mathsf{A}\boldsymbol{x} \in \mathbb{C}^m$ is precisely

$$\boldsymbol{y} = \sum_{k=1}^{n} x_k \boldsymbol{c}_k.$$

In other words, the column vector $\boldsymbol{y}$ is a linear combination of the columns of A.

Thinking about $A \in \mathbb{C}^{m\times n}$ as a mapping from $\mathbb{C}^n$ to $\mathbb{C}^m$, the following definitions are natural.

Definition 1.9 (range and kernel)**.** Let $A \in \mathbb{C}^{m\times n}$. The **image** (or **range**) of A is defined as

$$\mathrm{im}(A) = \mathcal{R}(A) = \{\boldsymbol{y} \in \mathbb{C}^m \mid \exists \boldsymbol{x} \in \mathbb{C}^n,\ \boldsymbol{y} = A\boldsymbol{x}\} \subseteq \mathbb{C}^m.$$

The **kernel** (or **null space**) of A is

$$\ker(A) = \mathcal{N}(A) = \{\boldsymbol{x} \in \mathbb{C}^n \mid A\boldsymbol{x} = \boldsymbol{0}\} \subseteq \mathbb{C}^n.$$

Definition 1.10 (row and column space)**.** Suppose that the matrix $A \in \mathbb{C}^{m\times n}$ is expressed column-wise as $A = [\boldsymbol{c}_1, \ldots, \boldsymbol{c}_n]$ and row-wise as $A = [\boldsymbol{r}_1, \ldots, \boldsymbol{r}_m]^\intercal$. The **row space** of A is

$$\mathrm{row}(A) = \mathrm{span}(\{\boldsymbol{r}_1, \ldots, \boldsymbol{r}_m\}) \leq \mathbb{C}^n$$

and the **column space** of A is

$$\mathrm{col}(A) = \mathrm{span}(\{\boldsymbol{c}_1, \ldots, \boldsymbol{c}_n\}) \leq \mathbb{C}^m.$$

The **row rank** of A is the dimension of row(A); similarly, the **column rank** is the dimension of col(A).

A very important result in linear algebra states that the row and column ranks coincide. For a proof, see, for example, [44].

Theorem 1.11 (row and column rank)**.** *Suppose that* $A \in \mathbb{C}^{m\times n}$. *The row and column ranks of* A *are equal.*

Since this is an important invariant between the domain and range of an operator, we give it a name.

Definition 1.12 (rank)**.** The **rank** of a matrix $A \in \mathbb{C}^{m\times n}$ is the dimension of its row/column space. We denote it by the symbol rank(A).

Theorem 1.13 (range and column space)**.** *Let* $A \in \mathbb{C}^{m\times n}$ *be represented column-wise as* $A = [\boldsymbol{c}_1, \ldots, \boldsymbol{c}_n]$. *Then*

$$\mathrm{im}(A) = \mathrm{span}(\{\boldsymbol{c}_1, \ldots, \boldsymbol{c}_n\}) = \mathrm{col}(A).$$

In other words, the range of A *coincides with its column space.*

Proof. ($\subseteq$) Let $\boldsymbol{y} \in \mathrm{im}(A) \subseteq \mathbb{C}^m$. Then, by definition, there is an $\boldsymbol{x} \in \mathbb{C}^n$ for which $\boldsymbol{y} = A\boldsymbol{x}$, or

$$\boldsymbol{y} = \sum_{k=1}^{n} x_k \boldsymbol{c}_k,$$

which implies that $\boldsymbol{y} \in \mathrm{col}(A)$.

($\supseteq$) On the other hand, if $\boldsymbol{y} \in \mathrm{col}(A)$, this implies that there are $\alpha_i \in \mathbb{C}$, $i = 1, \ldots, n$ such that

$$\boldsymbol{y} = \sum_{i=1}^{n} \alpha_i \boldsymbol{c}_i.$$

Define $\boldsymbol{x} = [\alpha_1, \ldots, \alpha_n]^\intercal \in \mathbb{C}^n$. The previous identity shows that $\boldsymbol{y} = \mathsf{A}\boldsymbol{x}$, so that $\boldsymbol{y} \in \text{im}(\mathsf{A})$. □

Corollary 1.14 (range and rank). *For any* $\mathsf{A} \in \mathbb{C}^{m\times n}$,

$$\dim(\text{im}(\mathsf{A})) = \text{rank}(\mathsf{A}).$$

Definition 1.15 (nullity). Suppose that $\mathsf{A} \in \mathbb{C}^{m\times n}$. The **nullity** of A is the dimension of $\ker(\mathsf{A})$:

$$\text{nullity}(\mathsf{A}) = \dim(\ker(\mathsf{A})).$$

Theorem 1.16 (properties of the rank). *Let* $\mathsf{A} \in \mathbb{C}^{m\times n}$. *Then*

1. $\text{rank}(\mathsf{A}) \le \min\{m, n\}$.
2. $\text{rank}(\mathsf{A}) + \text{nullity}(\mathsf{A}) = n$.
3. *For any* $\mathsf{B} \in \mathbb{C}^{n\times p}$, *we have* $\text{rank}(\mathsf{AB}) \ge \text{rank}(\mathsf{A}) + \text{rank}(\mathsf{B}) - n$.
4. *For any* $\mathsf{C} \in \mathbb{C}^{m\times m}$ *with* $\text{rank}(\mathsf{C}) = m$ *and any* $\mathsf{B} \in \mathbb{C}^{n\times n}$ *with* $\text{rank}(\mathsf{B}) = n$, *it holds that*

$$\text{rank}(\mathsf{CA}) = \text{rank}(\mathsf{A}) = \text{rank}(\mathsf{AB}).$$

5. $\text{rank}(\mathsf{AB}) \le \min\{\text{rank}(\mathsf{A}), \text{rank}(\mathsf{B})\}$.
6. $\text{rank}(\mathsf{A} + \mathsf{B}) \le \text{rank}(\mathsf{A}) + \text{rank}(\mathsf{B})$.

Proof. Some of these are given as exercises. Otherwise, see, for example, [44]. □

Definition 1.17 (adjoint). Suppose that $(\mathbb{V}, (\cdot, \cdot)_{\mathbb{V}})$ and $(\mathbb{W}, (\cdot, \cdot)_{\mathbb{W}})$ are inner product spaces over $\mathbb{C}$. Let $A \in \mathfrak{L}(\mathbb{V}, \mathbb{W})$. The **adjoint** of A is a linear operator $A^* \in \mathfrak{L}(\mathbb{W}, \mathbb{V})$ that satisfies

$$(Ax, y)_{\mathbb{W}} = (x, A^*y)_{\mathbb{V}}, \quad \forall x \in \mathbb{V},\ y \in \mathbb{W}.$$

A linear operator $A \in \mathfrak{L}(\mathbb{V}) = \mathfrak{L}(\mathbb{V}, \mathbb{V})$ is called **self-adjoint** if and only if $A = A^*$.

For matrices, the adjoint has a familiar definition.

Definition 1.18 (matrix adjoint, conjugate transpose). Let $\mathsf{A} = [a_{i,j}] \in \mathbb{C}^{m\times n}$. The **matrix adjoint** (or **conjugate transpose**) of A is the matrix $\mathsf{A}^\mathsf{H} \in \mathbb{C}^{n\times m}$ with entries

$$\left[\mathsf{A}^\mathsf{H}\right]_{i,j} = \bar{a}_{j,i}.$$

The **transpose** of A is the matrix $\mathsf{A}^\intercal \in \mathbb{C}^{n\times m}$ with entries

$$[\mathsf{A}^\intercal]_{i,j} = a_{j,i}.$$

A matrix $\mathsf{A} \in \mathbb{C}^{n\times n}$ is called **Hermitian**[1] if and only if $\mathsf{A} = \mathsf{A}^\mathsf{H}$. A is called **skew-Hermitian** if and only if $\mathsf{A} = -\mathsf{A}^\mathsf{H}$. A matrix $\mathsf{A} \in \mathbb{R}^{n\times n}$ is called **symmetric** if and only if $\mathsf{A} = \mathsf{A}^\intercal$ and **skew-symmetric** if $\mathsf{A} = -\mathsf{A}^\intercal$.

Simple calculations yield the following results.

[1] Named in honor of the French mathematician Charles Hermite (1822–1901).

Proposition 1.19 (properties of matrix adjoints). *Let* $\mathsf{A} \in \mathbb{C}^{m\times p}$ *and* $\mathsf{B} \in \mathbb{C}^{p\times n}$. *Then* $(\mathsf{AB})^{\mathsf{H}} = \mathsf{B}^{\mathsf{H}}\mathsf{A}^{\mathsf{H}}$ *and* $(\mathsf{A}^{\mathsf{H}})^{\mathsf{H}} = \mathsf{A}$.

Proof. See Problem 1.8. □

Remark 1.20 (notation). Observe that, above, we have naturally extended the domain of definition of the operator $\cdot^{\intercal}$. Let $\boldsymbol{x} = [x_1, \dots, x_n]^{\intercal} \in \mathbb{C}^n$. The *conjugate transpose* of $\boldsymbol{x}$ is defined as the row vector $\boldsymbol{x}^{\mathsf{H}} = [\bar{x}_1, \dots, \bar{x}_n]$. This conforms to the definition above, provided that we view any column n-vector as a matrix with n rows and one column. A direct computation shows that $(\boldsymbol{x}^{\mathsf{H}})^{\mathsf{H}} = \boldsymbol{x}$ for all $\boldsymbol{x} \in \mathbb{C}^n$. Moreover, upon identifying $\mathbb{C}^{1\times 1}$ with $\mathbb{C}$, if $\boldsymbol{x}, \boldsymbol{y} \in \mathbb{C}^m$,

$$(\boldsymbol{x}, \boldsymbol{y})_{\ell^2(\mathbb{C}^m)} = (\boldsymbol{x}, \boldsymbol{y})_2 = \boldsymbol{y}^{\mathsf{H}}\boldsymbol{x} = \overline{\boldsymbol{x}^{\mathsf{H}}\boldsymbol{y}} = \overline{(\boldsymbol{y}, \boldsymbol{x})_2} = \overline{(\boldsymbol{y}, \boldsymbol{x})_{\ell^2(\mathbb{C}^m)}}.$$

Furthermore, if $\mathsf{A} \in \mathbb{C}^{m\times n}$, $\boldsymbol{x} \in \mathbb{C}^n$ and $\boldsymbol{y} \in \mathbb{C}^m$, then it follows that

$$(\mathsf{A}\boldsymbol{x}, \boldsymbol{y})_{\ell^2(\mathbb{C}^m)} = \boldsymbol{y}^{\mathsf{H}}\mathsf{A}\boldsymbol{x} = \left(\mathsf{A}^{\mathsf{H}}\boldsymbol{y}\right)^{\mathsf{H}}\boldsymbol{x} = (\boldsymbol{x}, \mathsf{A}^{\mathsf{H}}\boldsymbol{y})_{\ell^2(\mathbb{C}^n)},$$

where $(\,\cdot\,,\,\cdot\,)_{\ell^2(\mathbb{C}^m)}$ is the Euclidean inner product on $\mathbb{C}^m$. For any $\boldsymbol{x} \in \mathbb{R}^n$, $\boldsymbol{x}^{\mathsf{H}} = \boldsymbol{x}^{\intercal}$, and for $\mathsf{A} \in \mathbb{R}^{m\times n}$, the conjugate transpose coincides with the transpose, $\mathsf{A}^{\intercal}$.

Theorem 1.21 (properties of the conjugate transpose). *Let* $\mathsf{A} \in \mathbb{C}^{m\times n}$. *Then*

1. $\operatorname{rank}(\mathsf{A}) = \operatorname{rank}(\mathsf{A}^{\mathsf{H}}) = \operatorname{rank}(\mathsf{A}^{\intercal})$.
2. $\ker(\mathsf{A}) = \operatorname{im}(\mathsf{A}^{\mathsf{H}})^{\perp}$.
3. $\operatorname{im}(\mathsf{A})^{\perp} = \ker(\mathsf{A}^{\mathsf{H}})$.

Proof. We prove the second result and leave the first and last to exercises; see Problem 1.10.

($\subseteq$) Let $\boldsymbol{x} \in \ker(\mathsf{A})$. By definition, $\mathsf{A}\boldsymbol{x} = \boldsymbol{0} \in \mathbb{C}^m$. Let $\boldsymbol{z} \in \operatorname{im}(\mathsf{A}^{\mathsf{H}})$, i.e., $\exists \boldsymbol{y} \in \mathbb{C}^m$ for which $\boldsymbol{z} = \mathsf{A}^{\mathsf{H}}\boldsymbol{y}$. Now compute

$$(\boldsymbol{z}, \boldsymbol{x})_2 = (\mathsf{A}^{\mathsf{H}}\boldsymbol{y}, \boldsymbol{x})_2 = (\boldsymbol{y}, \mathsf{A}\boldsymbol{x})_2 = 0,$$

which shows that $\boldsymbol{x} \in \operatorname{im}(\mathsf{A}^{\mathsf{H}})^{\perp}$.

($\supseteq$) Conversely, if $\boldsymbol{x} \in \operatorname{im}(\mathsf{A}^{\mathsf{H}})^{\perp}$, then $0 = (\boldsymbol{x}, \mathsf{A}^{\mathsf{H}}\boldsymbol{y})_2 = (\mathsf{A}\boldsymbol{x}, \boldsymbol{y})_2$ for every $\boldsymbol{y} \in \mathbb{C}^m$. Thus, $\mathsf{A}\boldsymbol{x} = \boldsymbol{0}$. □

Definition 1.22 (identity). The matrix $\mathsf{I}_n \in \mathbb{C}^{n\times n}$, defined by

$$[\mathsf{I}_n]_{i,j} = \delta_{i,j},$$

is known as the **matrix identity** of order n.

Definition 1.23 (inverse). Let $\mathsf{A} \in \mathbb{C}^{n\times n}$. If there is $\mathsf{B} \in \mathbb{C}^{n\times n}$ such that $\mathsf{AB} = \mathsf{BA} = \mathsf{I}_n$, then we say that A is **invertible** and call the matrix B an inverse of A.

In light of Problem 1.13, we denote the inverse of A by A^{-1}.

Theorem 1.24 (properties of the inverse). *Let* $\mathsf{A} \in \mathbb{C}^{n\times n}$. *Then* A *is invertible if and only if* $\operatorname{rank}(\mathsf{A}) = n$. *Moreover, if* A *is invertible,*

1. A^{-1} *is invertible and* $(\mathsf{A}^{-1})^{-1} = \mathsf{A}$.

2. A^{H} *is invertible and* $(\mathsf{A}^{\mathsf{H}})^{-1} = (\mathsf{A}^{-1})^{\mathsf{H}}$. *In this case, we write*
$$\mathsf{A}^{-\mathsf{H}} = (\mathsf{A}^{\mathsf{H}})^{-1}.$$
3. $\mathsf{A}^{\intercal}$ *is invertible and* $(\mathsf{A}^{\intercal})^{-1} = (\mathsf{A}^{-1})^{\intercal}$. *In this case, we write*
$$\mathsf{A}^{-\intercal} = (\mathsf{A}^{\intercal})^{-1}.$$
4. *For all* $\alpha \in \mathbb{C}_\star = \mathbb{C}\backslash\{0\}$, $\alpha\mathsf{A}$ *is invertible and* $(\alpha\mathsf{A})^{-1} = \frac{1}{\alpha}\mathsf{A}^{-1}$.
5. *If* $\mathsf{B} \in \mathbb{C}^{n\times n}$ *is also invertible, then the product* AB *is invertible and* $(\mathsf{AB})^{-1} = \mathsf{B}^{-1}\mathsf{A}^{-1}$.

Proof. See Problem 1.14. □

Definition 1.25 (unitary matrices)**.** Let $\mathsf{A} \in \mathbb{R}^{m\times m}$. We say that A is **orthogonal** if and only if $\mathsf{A}^{-1} = \mathsf{A}^{\intercal}$. Similarly, for $\mathsf{A} \in \mathbb{C}^{m\times m}$, we say that A is **unitary** if and only if $\mathsf{A}^{\mathsf{H}} = \mathsf{A}^{-1}$.

1.2 Matrix Norms

Since, for any two vector spaces $\mathbb{V}$ and $\mathbb{W}$, the set $\mathfrak{L}(\mathbb{V}, \mathbb{W})$ is a vector space itself, we can think of ways of norming it. An immediate way of doing so is by simply considering elements of $\mathbb{C}^{m\times n}$ as a collection of mn numbers, i.e., by identifying $\mathbb{C}^{m\times n}$ with $\mathbb{C}^{m\cdot n}$.

Definition 1.26 (Frobenius norm[2])**.** Let $\mathsf{A} = [a_{i,j}] \in \mathbb{C}^{m\times n}$. The **Frobenius norm** is defined via
$$\|\mathsf{A}\|_F^2 = \sum_{i=1}^{m}\sum_{j=1}^{n} |a_{i,j}|^2.$$

Definition 1.27 (max norm)**.** The **matrix max norm** is defined via
$$\|\mathsf{A}\|_{\max} = \max_{\substack{1\le i\le m \\ 1\le j\le n}} |a_{i,j}|$$
for all $\mathsf{A} = [a_{i,j}] \in \mathbb{C}^{m\times n}$.

However, it turns out that it is often more useful when the norms on $\mathfrak{L}(\mathbb{V}, \mathbb{W})$ are, in a sense, compatible with those of $\mathbb{V}$ and $\mathbb{W}$.

Definition 1.28 (induced norm)**.** Let $(\mathbb{V}, \|\cdot\|_{\mathbb{V}})$ and $(\mathbb{W}, \|\cdot\|_{\mathbb{W}})$ be complex, finite-dimensional normed vector spaces. The **induced norm** on $\mathfrak{L}(\mathbb{V}, \mathbb{W})$ is
$$\|A\|_{\mathfrak{L}(\mathbb{V},\mathbb{W})} = \sup_{x\in\mathbb{V}_\star} \frac{\|Ax\|_{\mathbb{W}}}{\|x\|_{\mathbb{V}}}, \quad \forall A \in \mathfrak{L}(\mathbb{V}, \mathbb{W}),$$
where $\mathbb{V}_\star = \mathbb{V}\backslash\{0\}$. When $\mathbb{V} = \mathbb{W}$ it is understood that $\|\cdot\|_{\mathbb{V}} = \|\cdot\|_{\mathbb{W}}$ as well.

[2] Named in honor of the German mathematician Ferdinand Georg Frobenius (1849–1917).

Remark 1.29 (convention)**.** Regarding the last point, in our presentation, the following object would *not* define an induced matrix norm:

$$\|A\|_{\mathfrak{L}(\ell^p(\mathbb{C}^n),\ell^q(\mathbb{C}^n))} = \sup_{x\in\mathbb{C}^n_\star} \frac{\|Ax\|_{\ell^q(\mathbb{C}^n)}}{\|x\|_{\ell^p(\mathbb{C}^n)}}, \qquad \forall A \in \mathbb{C}^{n\times n}$$

for $p \neq q$. While this definition is meaningful for every $p, q \in [1, \infty]$, and it indeed defines a norm, we will only consider it to be an induced norm for $p = q$.

Definition 1.30 (matrix p-norm)**.** Let $A \in \mathbb{C}^{m\times n}$ be given and $p \in [1, \infty]$. The induced $\mathfrak{L}(\ell^p(\mathbb{C}^n), \ell^p(\mathbb{C}^m))$ norm, called simply the **induced matrix p-norm**, is denoted $\|A\|_p$ and is defined as

$$\|A\|_p = \sup_{x\in\mathbb{C}^n_\star} \frac{\|Ax\|_{\ell^p(\mathbb{C}^m)}}{\|x\|_{\ell^p(\mathbb{C}^n)}}.$$

Proposition 1.31 (matrix 1-norm)**.** *Let* $A = [a_{i,j}] = [a_{i,j}] = [\boldsymbol{c}_1, \ldots, \boldsymbol{c}_n] \in \mathbb{C}^{m\times n}$ *be arbitrary. The induced matrix 1-norm, which is, by definition,*

$$\|A\|_1 = \sup_{x\in\mathbb{C}^n_\star} \frac{\|Ax\|_{\ell^1(\mathbb{C}^m)}}{\|x\|_{\ell^1(\mathbb{C}^n)}},$$

may be calculated via the following formula:

$$\|A\|_1 = \max_{j=1}^{n} \left(\sum_{i=1}^{m} |a_{i,j}| \right).$$

Proof. Given any $\boldsymbol{x} = [x_1, \ldots, x_n]^\intercal \in \mathbb{C}^n$,

$$\begin{aligned}
\|A\boldsymbol{x}\|_{\ell^1(\mathbb{C}^m)} &= \left\| \sum_{j=1}^{n} x_j \boldsymbol{c}_j \right\|_{\ell^1(\mathbb{C}^m)} \\
&\leq \sum_{j=1}^{n} |x_j| \|\boldsymbol{c}_j\|_{\ell^1(\mathbb{C}^m)} \\
&\leq \max_{j=1}^{n} \|\boldsymbol{c}_j\|_{\ell_1(\mathbb{C}^m)} \sum_{j=1}^{n} |x_j| \\
&= \max_{j=1}^{n} \|\boldsymbol{c}_j\|_{\ell_1(\mathbb{C}^m)} \|\boldsymbol{x}\|_{\ell^1(\mathbb{C}^n)}.
\end{aligned}$$

This shows that

$$\|A\|_1 \leq \max_{j=1}^{n} \|\boldsymbol{c}_j\|_{\ell_1(\mathbb{C}^m)} = \max_{j=1}^{n} \left(\sum_{i=1}^{m} |a_{i,j}| \right).$$

On the other hand, there must be an index j_0 where the maximum in the previous inequality is attained. Choose $\boldsymbol{x} = \boldsymbol{e}_{j_0}$, the j_0th canonical basis vector, and notice then that

$$\|A\boldsymbol{x}\|_{\ell^1(\mathbb{C}^m)} = \|\boldsymbol{c}_{j_0}\|_{\ell^1(\mathbb{C}^m)}.$$

It is not difficult to see that the supremum in the definition of induced norm is attained at this vector. This implies that the norm is the maximum absolute column sum, i.e.,

$$\|A\|_1 = \max_{j=1}^{n} \left(\sum_{i=1}^{m} |a_{i,j}| \right). \qquad \square$$

Definition 1.32 (sub-multiplicativity)**.** Suppose that $\|\cdot\| : \mathbb{C}^{n\times n} \to \mathbb{R}$ is a matrix norm, i.e., a norm on the vector space $\mathfrak{L}(\mathbb{C}^n)$. We say that the norm is **sub-multiplicative** if and only if

$$\|AB\| \le \|A\| \, \|B\|, \quad \forall A, B \in \mathbb{C}^{n\times n}.$$

Definition 1.33 (consistency)**.** Suppose that $\|\cdot\|_{\mathbb{C}^n} : \mathbb{C}^n \to \mathbb{R}$ and $\|\cdot\|_{\mathbb{C}^m} : \mathbb{C}^m \to \mathbb{R}$ are norms, and $\|\cdot\| : \mathbb{C}^{m\times n} \to \mathbb{R}$ is a matrix norm. We say that $\|\cdot\|$ is **consistent** with respect to the norms $\|\cdot\|_{\mathbb{C}^n}$ and $\|\cdot\|_{\mathbb{C}^m}$ if and only if

$$\|A\boldsymbol{x}\|_{\mathbb{C}^m} \le \|A\| \, \|\boldsymbol{x}\|_{\mathbb{C}^n}$$

for all $A \in \mathbb{C}^{m\times n}$ and $\boldsymbol{x} \in \mathbb{C}^n$.

Proposition 1.34 (property of induced norms)**.** *Suppose that $\|\cdot\|_{\mathbb{C}^n} : \mathbb{C}^n \to \mathbb{R}$ is a norm on $\mathbb{C}^n$ and $\|\cdot\| : \mathbb{C}^{n\times n} \to \mathbb{R}$ is the induced matrix norm*

$$\|A\| = \sup_{\boldsymbol{x}\in\mathbb{C}^n_\star} \frac{\|A\boldsymbol{x}\|_{\mathbb{C}^n}}{\|\boldsymbol{x}\|_{\mathbb{C}^n}}, \quad \forall A \in \mathbb{C}^{n\times n}.$$

Then $\|\cdot\|$ is a sub-multiplicative norm, and it is consistent with respect to $\|\cdot\|_{\mathbb{C}^n}$.

Proof. See Problem 1.27. $\square$

Example 1.1 Let $A \in \mathbb{C}^{1\times n}$, i.e., $A = \boldsymbol{a}^{\mathsf{H}}$ for some $\boldsymbol{a} \in \mathbb{C}^n$. Then $A\boldsymbol{x} = (\boldsymbol{x}, \boldsymbol{a})_2$, so that

$$|A\boldsymbol{x}| = |(\boldsymbol{x}, \boldsymbol{a})_2| \le \|\boldsymbol{x}\|_2 \|\boldsymbol{a}\|_2.$$

In addition,

$$|A\boldsymbol{a}| = |(\boldsymbol{a}, \boldsymbol{a})_2| = \|\boldsymbol{a}\|_2^2,$$

from which we may conclude that $\|A\|_2 = \|\boldsymbol{a}\|_2$. This matrix $A\colon \mathbb{C}^n \to \mathbb{C}$ is a prototype of an object called a *linear functional.*

Proposition 1.35 (norm of a unitary matrix)**.** *Let $A \in \mathbb{C}^{m\times n}$ be arbitrary and $Q \in \mathbb{C}^{m\times m}$ be unitary. Then we have*

$$\|QA\|_2 = \|A\|_2.$$

Proof. Recall that, owing to Problem 1.16, for any unitary matrix we have $\|Q\boldsymbol{x}\|_2 = \|\boldsymbol{x}\|_2$. The result follows from this fact. $\square$

1.3 Eigenvalues and Spectral Decomposition

As a final topic in this chapter we discuss eigenvalues and spectral decomposition of square matrices. We begin with a definition.

Definition 1.36 (spectrum). Let $\mathsf{A} \in \mathbb{C}^{n\times n}$. We say that $\lambda \in \mathbb{C}$ is an **eigenvalue** of A if and only if there exists a vector $\boldsymbol{x} \in \mathbb{C}^n_\star = \mathbb{C}^n \backslash \{\mathbf{0}\}$ such that

$$\mathsf{A}\boldsymbol{x} = \lambda \boldsymbol{x}.$$

This vector is called an **eigenvector** of A associated with λ. The **spectrum** of A, denoted by $\sigma(\mathsf{A})$, is the collection of all eigenvalues of A. The pair $(\lambda, \boldsymbol{x})$ is called an **eigenpair** of A.

Theorem 1.37 (properties of the spectrum). *Let $\mathsf{A} \in \mathbb{C}^{n\times n}$. Then*

1. *$\lambda \in \sigma(\mathsf{A})$ if and only if $\bar{\lambda} \in \sigma(\mathsf{A}^\mathsf{H})$.*
2. *A is invertible if and only if $0 \notin \sigma(\mathsf{A})$.*
3. *The eigenvectors corresponding to distinct eigenvalues are linearly independent.*
4. *$\lambda \in \sigma(\mathsf{A})$ if and only if $\chi_\mathsf{A}(\lambda) = 0$, where χ_A is a polynomial of degree n, defined via*
$$\chi_\mathsf{A}(\lambda) = \det(\lambda \mathsf{I}_n - \mathsf{A}).$$
χ_A is called the characteristic polynomial.
5. *There are at most n distinct complex-valued eigenvalues of* A.

Proof. See Problem 1.28. □

Since we are dealing with matrices with complex entries, the fundamental theorem of algebra (see [18, Section 2.8]) implies that the characteristic polynomial can be written as a product of factors, i.e.,

$$\chi_\mathsf{A}(\lambda) = \prod_{i=1}^{L} (\lambda - \lambda_i)^{m_i} \tag{1.1}$$

with $n = \sum_{i=1}^{L} m_i$.

Definition 1.38 (algebraic multiplicity). Let $\mathsf{A} \in \mathbb{C}^{n\times n}$ be given. The number m_i in (1.1) is called the **algebraic multiplicity** of the eigenvalue λ_i.

Definition 1.39 (geometric multiplicity). Let $\mathsf{A} \in \mathbb{C}^{n\times n}$ and $\lambda \in \sigma(\mathsf{A})$. Define the eigenspace

$$E(\lambda, \mathsf{A}) = \{\boldsymbol{x} \in \mathbb{C}^n \mid \mathsf{A}\boldsymbol{x} = \lambda \boldsymbol{x}\}.$$

This is a vector subspace of $\mathbb{C}^n$; its dimension $\dim(E(\lambda, \mathsf{A}))$ is called the **geometric multiplicity** of λ.

The following result gives a relation between the algebraic and geometric multiplicities of an eigenvalue. For a proof of this result, see [44].

Theorem 1.40 (relation between multiplicities). *Let $\mathsf{A} \in \mathbb{C}^{n\times n}$ and $\lambda \in \sigma(\mathsf{A})$. The geometric multiplicity of λ is not larger than the algebraic multiplicity of λ.*

Definition 1.41 (triangular matrices)**.** The square matrix $\mathsf{A} = [a_{i,j}] \in \mathbb{C}^{n\times n}$ is called **upper triangular** if and only if $a_{i,j} = 0$ for all $i > j$. A is called **lower triangular** if and only if $a_{i,j} = 0$ for all $i < j$. A matrix is called **triangular** if and only if it is either upper or lower triangular. A is called **diagonal** if and only if $a_{i,j} = 0$ for all $i \neq j$. A matrix $\mathsf{A} = [a_{i,j}] \in \mathbb{C}^{n\times n}$ is called **unit lower triangular** (**unit upper triangular**) if and only if it is lower (upper) triangular and $a_{i,i} = 1$, $i = 1, \dots, n$.

Definition 1.42 (similarity)**.** Let $\mathsf{A}, \mathsf{B} \in \mathbb{C}^{n\times n}$. We say that A and B are **similar**, denoted by $\mathsf{A} \asymp \mathsf{B}$, if and only if there is an invertible matrix S such that

$$\mathsf{A} = \mathsf{S}^{-1}\mathsf{B}\mathsf{S}.$$

We say that matrix A is **diagonalizable** if it is similar to a diagonal matrix.

Proposition 1.43 (spectrum of similar matrices)**.** *Let* $\mathsf{A}, \mathsf{B} \in \mathbb{C}^{n\times n}$ *be such that* $\mathsf{A} \asymp \mathsf{B}$. *Then* $\chi_\mathsf{A} = \chi_\mathsf{B}$ *and, consequently,* $\sigma(\mathsf{A}) = \sigma(\mathsf{B})$. *Furthermore,* $\det(\mathsf{A}) = \det(\mathsf{B})$ *and* $\operatorname{tr}(\mathsf{A}) = \operatorname{tr}(\mathsf{B})$.

Proof. See Problem 1.34. □

Definition 1.44 (defective matrix)**.** A matrix $\mathsf{A} \in \mathbb{C}^{n\times n}$ is called **defective** if and only if there is an eigenvalue λ_k with geometric multiplicity strictly smaller than the algebraic multiplicity. Otherwise, the matrix is called **nondefective**.

One of the main results in the spectral theory of matrices is the following.

Theorem 1.45 (diagonalizability criterion)**.** *Let* $\mathsf{A} \in \mathbb{C}^{n\times n}$ *be nondefective. Then it is diagonalizable.*

Proof. Let $\sigma(\mathsf{A}) = \{\lambda_k\}_{k=1}^{L}$, where $\lambda_k \neq \lambda_j$, $k \neq j$. For each k,

$$E(\lambda_k, \mathsf{A}) = \operatorname{span}(\{\boldsymbol{x}_1^{(k)}, \dots, \boldsymbol{x}_{m_k}^{(k)}\}) = \operatorname{span}(S_k),$$

where the set $S_k = \{\boldsymbol{x}_1^{(k)}, \dots, \boldsymbol{x}_{m_k}^{(k)}\}$ is linearly independent. Then $S = \cup_{k=1}^{L} S_k$ is a basis of $\mathbb{C}^n$. Indeed, item 3 of Theorem 1.37 shows that the set S is linearly independent. Moreover, $\#(S) = \sum_{k=1}^{L} m_k = n$, since the matrix A is nondefective.

Now set $\mathsf{D} = \operatorname{diag}(\lambda_1, \dots, \lambda_1, \dots, \lambda_L, \dots, \lambda_L)$ and

$$\mathsf{X} = \begin{bmatrix} | & & | & & | & & | \\ \boldsymbol{x}_1^{(1)} & \cdots & \boldsymbol{x}_{m_1}^{(1)} & \cdots & \boldsymbol{x}_1^{(L)} & \cdots & \boldsymbol{x}_{m_L}^{(L)} \\ | & & | & & | & & | \end{bmatrix},$$

where in D each eigenvalue λ_k appears exactly m_k times. Notice now that, since all the columns of X are linearly independent, we have $\operatorname{rank}(\mathsf{X}) = n$ and this implies that X is invertible.

Since, for all $j = 1, \dots, m_k$, we have $\mathsf{A}\boldsymbol{x}_j^{(k)} = \lambda_k \boldsymbol{x}_j^{(k)}$, we see that

$$\mathsf{A}\mathsf{X} = \mathsf{A}[\boldsymbol{x}_1, \dots, \boldsymbol{x}_n] = [\mathsf{A}\boldsymbol{x}_1, \dots, \mathsf{A}\boldsymbol{x}_n] \quad \text{and} \quad \mathsf{X}\mathsf{D} = [\lambda_1\boldsymbol{x}_1, \dots, \lambda_n\boldsymbol{x}_n].$$

This implies that $\mathsf{AX} = \mathsf{XD}$, or, since X is invertible, $\mathsf{A} = \mathsf{X}\mathsf{D}\mathsf{X}^{-1}$. In conclusion, A is diagonalizable. □

An important class of nondefective matrices are those that are self-adjoint, or Hermitian. To investigate these, we use the Schur factorization. For a proof, again, we refer to [44].

Lemma 1.46 (Schur normal form[3]). *Let* $\mathsf{A} \in \mathbb{C}^{n\times n}$. *There are, not necessarily unique, matrices* $\mathsf{U}, \mathsf{R} \in \mathbb{C}^{n\times n}$, *with* U *unitary and* R *upper triangular, such that*

$$\mathsf{A} = \mathsf{U}\mathsf{R}\mathsf{U}^{\mathsf{H}}.$$

Notice that, in the setting of Lemma 1.46, we have that $\mathsf{A} \asymp \mathsf{R}$ and that, since R is upper triangular, its diagonal entries coincide with its spectrum.

Proposition 1.47 (Spectral Decomposition Theorem). *Let* $\mathsf{A} \in \mathbb{C}^{n\times n}$ *be self-adjoint (Hermitian), i.e.,* $\mathsf{A}^{\mathsf{H}} = \mathsf{A}$. *Then* $\sigma(\mathsf{A}) \subseteq \mathbb{R}$ *and there is a unitary* $\mathsf{U} \in \mathbb{C}^{n\times n}$ *such that*

$$\mathsf{A} = \mathsf{U}\mathsf{D}\mathsf{U}^{\mathsf{H}},$$

where the matrix $\mathsf{D} = \operatorname{diag}(\lambda_1, \ldots, \lambda_n)$. *Furthermore, there exists an orthonormal basis* $B = \{\boldsymbol{u}_1, \ldots, \boldsymbol{u}_n\}$ *of eigenvectors of* A *for the space* $\mathbb{C}^n$ *and* $\mathsf{A}\boldsymbol{u}_i = \lambda_i \boldsymbol{u}_i$, $i = 1, \ldots, n$.

Proof. From Lemma 1.46 we are guaranteed that there is a unitary matrix $\mathsf{U} \in \mathbb{C}^{n\times n}$ and an upper triangular matrix $\mathsf{D} \in \mathbb{C}^{n\times n}$ such that

$$\mathsf{A} = \mathsf{U}\mathsf{D}\mathsf{U}^{\mathsf{H}}.$$

But, since A is self-adjoint,

$$\mathsf{A}^{\mathsf{H}} = \mathsf{U}\mathsf{D}^{\mathsf{H}}\mathsf{U}^{\mathsf{H}} = \mathsf{U}\mathsf{D}\mathsf{U}^{\mathsf{H}} = \mathsf{A}.$$

This implies that $\mathsf{D}^{\mathsf{H}} = \mathsf{D}$, i.e., D is self-adjoint. Since D is triangular, it must be diagonal. Furthermore, the diagonal elements of D must be real. Otherwise, D could not be self-adjoint. Therefore, we have the desired factorization.

Now the eigenvalues of a diagonal matrix are precisely its diagonal entries. Since A is similar to the diagonal matrix D, the eigenvalues of A are precisely $\lambda_i = d_{i,i} \in \mathbb{R}$, $i = 1, \ldots, n$.

Finally, observe that the columns of U form an orthonormal basis for $\mathbb{C}^n$. Indeed, suppose that the kth column of U is denoted $\boldsymbol{u}_k$. Then $\mathsf{A}\mathsf{U} = \mathsf{U}\mathsf{D}$ if and only if

$$\mathsf{A}\boldsymbol{u}_k = d_{k,k}\boldsymbol{u}_k = \lambda_k \boldsymbol{u}_k.$$

Thus, the eigenvectors of A, namely $\boldsymbol{u}_k$, $k = 1, \ldots, n$, form an orthonormal basis for $\mathbb{C}^n$: $(\boldsymbol{u}_k, \boldsymbol{u}_j)_2 = \boldsymbol{u}_j^{\mathsf{H}}\boldsymbol{u}_k = \delta_{k,j}$, $k, j = 1, \ldots, n$. □

Notice that the previous result shows that, for A self-adjoint, there exists an orthonormal basis of $\mathbb{C}^n$ consisting of eigenvectors of A. This is a result that is used countless times in the text.

There are numerous generalizations of the last theorem. We will be interested in one that is rather straightforward to establish. First, we need what is perhaps an obvious definition.

[3] Named in honor of the Russian-born German–Israeli mathematician Issai Schur (1875–1941).

Definition 1.48 (eigenvalue). Suppose that $\mathbb{V}$ is a complex vector space and $A \in \mathfrak{L}(\mathbb{V})$. The scalar $\lambda \in \mathbb{C}$ for which there is $w \in \mathbb{V} \setminus \{0\}$ such that

$$Aw = \lambda w,$$

is called an **eigenvalue** of A and w is a corresponding **eigenvector**. The spectrum of A, $\sigma(A)$, is the set of all eigenvalues of A. The pair (λ, w) is called an **eigenpair** of A.

For self-adjoint operators we have the following general result.

Theorem 1.49 (Spectral Decomposition Theorem). *Suppose that $(\mathbb{V}, (\cdot, \cdot))$ is an n-dimensional complex inner product space and $A \in \mathfrak{L}(\mathbb{V})$ is self-adjoint. Then there are precisely n eigenvalues, counting multiplicities, and $\sigma(A) \subseteq \mathbb{R}$. Moreover, there is an orthonormal basis $B = \{w_1, \ldots, w_n\}$ of eigenvectors of A for the space $\mathbb{V}$: $(w_i, w_j) = \delta_{i,j}$, $i, j = 1, \ldots, n$.*

Proof. A proof for this is, for instance, furnished by the theory developed in Chapter 7. □

Finally, the class of normal matrices, which contains as a proper subset the class of Hermitian matrices, is sometimes important.

Definition 1.50 (normal matrix). The square matrix $\mathsf{A} \in \mathbb{C}^{n \times n}$ is called **normal** if and only if $\mathsf{A}^{\mathsf{H}}\mathsf{A} = \mathsf{A}\mathsf{A}^{\mathsf{H}}$.

We will need the following technical lemma.

Lemma 1.51 (normal and triangular). *Suppose that $\mathsf{A} \in \mathbb{C}^{n \times n}$ is normal and upper triangular. Then it must be diagonal.*

Proof. See Problem 1.45. □

Theorem 1.52 (diagonalization of normal matrices). *Suppose that $\mathsf{A} \in \mathbb{C}^{n \times n}$ is normal. Then A is unitarily diagonalizable, i.e., there is a unitary matrix $\mathsf{U} \in \mathbb{C}^{n \times n}$ and a diagonal matrix $\mathsf{D} \in \mathbb{C}^{n \times n}$ such that*

$$\mathsf{A} = \mathsf{U}\mathsf{D}\mathsf{U}^{\mathsf{H}}.$$

Proof. Use the Schur factorization and Lemma 1.51. See Problem 1.46. □

Corollary 1.53 (orthonormal basis). *Suppose that $\mathsf{A} \in \mathbb{C}^{n \times n}$ is normal. There is an orthonormal basis of eigenvectors of A for $\mathbb{C}^n$.*

Proof. Repeat the construction of Proposition 1.47. □

Problems

1.1 Let $(\mathbb{V}, \|\cdot\|)$ be a finite-dimensional normed space and $A \in \mathfrak{L}(\mathbb{V})$. Does

$$\|\cdot\|_A = \|A \cdot\| : \mathbb{V} \to \mathbb{R}$$

define a norm? Why or why not?

1.2 Prove Proposition 1.2.

1.3 Prove Proposition 1.4.
1.4 For $A \in \mathbb{C}^{m\times n}$, prove that $\text{im}(A) \leq \mathbb{C}^m$ (i.e., $\text{im}(A)$ is a vector subspace of $\mathbb{C}^m$) and $\ker(A) \leq \mathbb{C}^n$ (i.e., $\ker(A)$ is a vector subspace of $\mathbb{C}^n$).
1.5 Suppose that $A \in \mathbb{C}^{m\times n}$. Prove that $\text{im}(A) = C_A$, where $\text{im}(A)$ is the range of A and C_A is its column space.
1.6 Suppose that $A \in \mathbb{C}^{m\times n}$ with $m \geq n$. Prove that the following are equivalent:
a) $\text{rank}(A) = n$.
b) A maps no two distinct vectors in $\mathbb{C}^n$ to the same vector in $\mathbb{C}^m$.
c) $\ker(A) = \{\mathbf{0}\}$.
1.7 Let $A \in \mathbb{C}^{m\times n}$. Prove that $\text{im}(A)^\perp = \ker(A^H)$.
1.8 Prove Proposition 1.19.
1.9 Show that the definitions "adjoint" and the "conjugate transpose" coincide for matrices when we use the canonical inner product

$$(\boldsymbol{x}, \boldsymbol{y})_{\ell^2(\mathbb{C}^m)} = (\boldsymbol{x}, \boldsymbol{y})_2 = \boldsymbol{y}^H\boldsymbol{x}$$

for $\mathbb{C}^m$.
1.10 Complete the proof of Theorem 1.21.
1.11 Show that $I_n \in \mathbb{C}^{n\times n}$ acts as multiplicative identity with respect to matrix multiplication. In other words, for every $A \in \mathbb{C}^{n\times n}$, we have

$$AI_n = I_nA = A.$$

1.12 Suppose that $C \in \mathbb{C}^{n\times n}$ is invertible and the set $S = \{\boldsymbol{w}_1, \ldots, \boldsymbol{w}_k\} \subseteq \mathbb{C}^n$ is linearly independent. Prove that $CS = \{C\boldsymbol{w}_1, \ldots, C\boldsymbol{w}_k\} \subseteq \mathbb{C}^n$ is linearly independent.
1.13 Suppose that $A \in \mathbb{C}^{n\times n}$ is invertible. Prove that its inverse must be unique.
1.14 Prove Theorem 1.24.
1.15 Let $A \in \mathbb{C}^{m\times n}$. Prove that $\text{rank}(A) = \text{rank}(AB)$ for any $B \in \mathbb{C}^{n\times n}$ that is invertible.
1.16 Let $U \in \mathbb{C}^{n\times n}$ be unitary. Show that, for any $\boldsymbol{x}, \boldsymbol{y} \in \mathbb{C}^n$, we have $(U\boldsymbol{x}, U\boldsymbol{y})_2 = (\boldsymbol{x}, \boldsymbol{y})_2$, so that $\|U\boldsymbol{x}\|_2 = \|\boldsymbol{x}\|_2$.
1.17 Show that the Frobenius and matrix max norms are indeed norms on the vector space $\mathfrak{L}(\mathbb{C}^n, \mathbb{C}^m)$.
1.18 Show that

$$\|A\|_F^2 = \text{tr}(A^HA) = \text{tr}(AA^H),$$

where, for any square matrix, $M = [m_{i,j}] \in \mathbb{C}^{n\times n}$, $\text{tr}(M) = \sum_i^n m_{i,i}$ denotes its *trace*.
1.19 Let $\mathbb{V}$ and $\mathbb{W}$ be finite-dimensional complex-normed vector spaces. Show that the induced norm is indeed a norm on the vector space $\mathfrak{L}(\mathbb{V}, \mathbb{W})$. Prove that

$$\|A\|_{\mathfrak{L}(\mathbb{V},\mathbb{W})} = \sup\left\{\|Ax\|_{\mathbb{W}} \mid x \in \mathbb{V},\ \|x\|_{\mathbb{V}} = 1\right\}.$$

1.20 Let, for $a, b \in \mathbb{R}$,

$$A = \begin{bmatrix} a & b \\ b & a \end{bmatrix}.$$

Show that $\|A\|_1 = \|A\|_2 = \|A\|_\infty$.

1.21 Let, for $a, b \in \mathbb{R}$,

$$\mathsf{A} = \begin{bmatrix} a & b \\ b & -a \end{bmatrix}.$$

Show that $\|\mathsf{A}\|_2 = (a^2 + b^2)^{1/2}$.

1.22 Show that

$$\|\mathsf{A}\|_\infty = \max_{1 \le i \le n} \sum_{j=1}^{n} |a_{ij}|, \quad \forall \mathsf{A} \in \mathbb{C}^{n \times n},$$

and also that $\|\mathsf{A}\|_1 = \left\|\mathsf{A}^{\mathsf{H}}\right\|_\infty$.

1.23 Show that, for every $\mathsf{A} \in \mathbb{C}^{n \times n}$,

$$\frac{1}{\sqrt{n}}\|\mathsf{A}\|_2 \le \|\mathsf{A}\|_\infty \le \sqrt{n}\|\mathsf{A}\|_2.$$

1.24 Show that

$$\|\mathsf{A}\|_2^2 \le \|\mathsf{A}\|_1 \|\mathsf{A}\|_\infty, \quad \forall \mathsf{A} \in \mathbb{C}^{n \times n}.$$

1.25 Show that, for every $\mathsf{A} \in \mathbb{C}^{n \times n}$,

$$\|\mathsf{A}\|_{\max} \le \|\mathsf{A}\|_\infty \le n \|\mathsf{A}\|_{\max}, \tag{1.2}$$

where $\|\cdot\|_\infty$ is the induced matrix ∞-norm, and recall that

$$\|\mathsf{A}\|_{\max} = \max_{1 \le i,j \le n} |a_{i,j}|$$

is the matrix max norm.

1.26 Let $\mathsf{A} \in \mathbb{R}^{n \times n}$ be such that $\mathsf{A}^\intercal = \mathsf{A}$ and $\operatorname{tr} \mathsf{A} = 0$. Show that

$$\|\mathsf{A}\|_2^2 \le \frac{n-1}{n}\|\mathsf{A}\|_F^2.$$

Is the assumption that $\operatorname{tr} \mathsf{A} = 0$ essential? You may justify your answer with an example or counterexample.

1.27 Prove Proposition 1.34.

1.28 Prove Theorem 1.37.

1.29 Show that, for every $\mathsf{A} \in \mathbb{C}^{n \times n}$,

$$\|\mathsf{A}\|_2 = \max_{\lambda \in \sigma(\mathsf{A}^{\mathsf{H}}\mathsf{A})} \sqrt{\lambda}.$$

Hint: You need some facts about the eigenvalues and eigenvectors of Hermitian matrices.

1.30 Suppose that $\|\cdot\| : \mathbb{C}^{m \times n} \to \mathbb{R}$ is the induced norm with respect to the vector norms $\|\cdot\|_{\mathbb{C}^m}$ and $\|\cdot\|_{\mathbb{C}^n}$ and that $\mathsf{A} \in \mathbb{C}^{m \times n}$. Prove that the function $\|\mathsf{A}(\cdot)\|_{\mathbb{C}^m} : \mathbb{C}^n \to \mathbb{R}$ is uniformly continuous. Use this fact to prove that there is a vector $\boldsymbol{x} \in S_{\mathbb{C}^n}^{n-1}$ such that

$$\|\mathsf{A}\| = \|\mathsf{A}\boldsymbol{x}\|_{\mathbb{C}^m}.$$

1.31 Suppose that $\|\cdot\| : \mathbb{C}^{n\times n} \to \mathbb{R}$ is the induced norm with respect to the vector norm $\|\cdot\| : \mathbb{C}^n \to \mathbb{R}$. Let $A \in \mathbb{C}^{n\times n}$ be invertible. Prove that

$$\frac{1}{\|A^{-1}\|} = \min_{y\in\mathbb{C}^n_\star} \frac{\|Ay\|}{\|y\|} .$$

1.32 Let $T_k, T \in \mathbb{C}^{n\times n}$, $k = 1, 2$, be lower triangular matrices.

a) Show that T_1T_2 is lower triangular.
b) If T_1 and T_2 are unit lower triangular, show that T_1T_2 is unit lower triangular.
c) If $[T]_{i,i} \neq 0$, show that T is invertible and T^{-1} is lower triangular.
d) If T is unit lower triangular, prove that it is invertible and T^{-1} is unit lower triangular.
e) If $[T]_{i,i} > 0$, show that $\left[T^{-1}\right]_{i,i} = \frac{1}{[T]_{i,i}} > 0$.

1.33 Show that if $A \in \mathbb{C}^{n\times n}$ is both unitary (i.e., $AA^H = A^HA = I_n$) and triangular, then it is diagonal.

Hint: You need a fact about the inverse of a triangular matrix.

1.34 Prove Proposition 1.43.

1.35 Let $A \in \mathbb{C}^{n\times n}$. Suppose that $\lambda_1, \dots, \lambda_k$ are distinct eigenvalues of A and suppose that $\boldsymbol{x}_1, \dots, \boldsymbol{x}_k$ are eigenvectors associated with the respective eigenvalues. Prove that $\{\boldsymbol{x}_1, \dots, \boldsymbol{x}_k\}$ is linearly independent.

1.36 Let $A \in \mathbb{C}^{n\times n}$. Prove that if A has n distinct eigenvalues, then A is diagonalizable.

1.37 Let $A \in \mathbb{C}^{n\times n}$ be Hermitian, i.e., $A^H = A$.

a) Prove directly that all eigenvalues of A are real.
b) Prove that if $\boldsymbol{x}$ and $\boldsymbol{y}$ are eigenvectors associated with distinct eigenvalues, then they are orthogonal, i.e., $\boldsymbol{x}^H\boldsymbol{y} = 0$.

1.38 Let $A \in \mathbb{C}^{m\times n}$ and $B \in \mathbb{C}^{n\times m}$. Show that $\sigma(AB)\backslash\{0\} = \sigma(BA)\backslash\{0\}$, i.e., the nonzero eigenvalues of AB and BA coincide.

1.39 Let $A \in \mathbb{C}^{m\times n}$. Show that $\sigma(A^HA) \cup \sigma(AA^H) \subseteq [0, \infty)$.

1.40 Let $A \in \mathbb{C}^{n\times n}$ and $\lambda \in \sigma(A)$. The vector $\boldsymbol{y} \in \mathbb{C}^n_\star$ is called a left eigenvector associated with λ if and only if $\boldsymbol{y}^HA = \lambda\boldsymbol{y}^H$. Now suppose that λ, $\mu \in \sigma(A)$ are distinct. Let $\boldsymbol{y}$ be a left eigenvector associated with λ and $\boldsymbol{x}$ be a right (usual) eigenvector associated with μ. Prove that $\boldsymbol{y}^H\boldsymbol{x} = 0$.

1.41 Let $A \in \mathbb{C}^{n\times n}$ be skew-Hermitian.

a) Prove directly that the eigenvalues of A are purely imaginary.
b) Prove that if $\boldsymbol{x}$ and $\boldsymbol{y}$ are eigenvectors associated with distinct eigenvalues, then they are orthogonal, i.e., $\boldsymbol{x}^H\boldsymbol{y} = 0$.
c) Show that $I - A$ is nonsingular.
d) Prove that $Q = (I - A)^{-1}(I + A)$ is unitary.

1.42 Let $\boldsymbol{u}, \boldsymbol{v} \in \mathbb{C}^n$. Set $A = I_n + \boldsymbol{u}\boldsymbol{v}^H \in \mathbb{C}^{n\times n}$.

a) Suppose that A is invertible. Prove that $A^{-1} = I_n + \alpha\boldsymbol{u}\boldsymbol{v}^H$ for some $\alpha \in \mathbb{C}$. Give an expression for α.
b) For what $\boldsymbol{u}$ and $\boldsymbol{v}$ is A singular, i.e., not invertible?
c) Suppose that A is singular. What is the kernel space of A, ker(A), in this case?

1.43 Suppose that $\boldsymbol{q} \in \mathbb{C}^n$, $\|\boldsymbol{q}\|_2 = 1$. Set $P = I - \boldsymbol{q}\boldsymbol{q}^H$.

a) Find im(P).

b) Find ker(P).
c) Find the eigenvalues of P.

1.44 Characterize the eigenvalues of a unitary matrix.

1.45 Prove Lemma 1.51.

1.46 Prove Theorem 1.52.

2 The Singular Value Decomposition

In Chapter 1, we saw that if $A \in \mathbb{C}^{n\times n}$ is Hermitian, then it is unitarily diagonalizable, i.e., there is a diagonal matrix $D \in \mathbb{R}^{n\times n}$ and a unitary matrix $U \in \mathbb{C}^{n\times n}$ such that

$$A = UDU^H.$$

This gives us, at least for this class of matrices, a nice geometric interpretation of the action of a matrix on a vector.

1. Since U^H is unitary and $\left\|U^H x\right\|_2 = \|x\|_2$, the action of U^H is *essentially* that of a rotation/reflection (it does not change the magnitude of the vector).
2. The matrix D is diagonal; its action is a (signed) dilation in each coordinate direction.
3. Finally, $U = \left(U^H\right)^{-1}$ reverses the rotation/reflection implemented by U^H.

Now suppose that the elements of $D = \operatorname{diag}(\lambda_1, \ldots, \lambda_n)$, i.e., the eigenvalues of A, are ordered by magnitude:

$$|\lambda_1| \geq |\lambda_2| \geq \cdots \geq |\lambda_n| \geq 0.$$

It is not hard to show that A is the sum of rank-one matrices (matrices of the type $\mu \boldsymbol{u}\boldsymbol{v}^H$):

$$A = \sum_{i=1}^{n} \lambda_i \boldsymbol{u}_i \boldsymbol{u}_i^H,$$

where $U = [\boldsymbol{u}_1, \ldots, \boldsymbol{u}_n]$. Thus, the action of A in matrix–vector multiplication is quite simple: for any $\boldsymbol{x} \in \mathbb{C}^n$,

$$A\boldsymbol{x} = \sum_{i=1}^{n} \lambda_i (\boldsymbol{x}, \boldsymbol{u}_i)_2 \boldsymbol{u}_i.$$

Next, suppose that, after some index $r \in \{1, 2, \ldots, n-1\}$, the eigenvalues are either very small in magnitude relative to $|\lambda_r| > 0$ or zero. Then it seems reasonable that

$$A \approx \sum_{i=1}^{r} \lambda_i \boldsymbol{u}_i \boldsymbol{u}_i^H \quad \Longrightarrow \quad A\boldsymbol{x} \approx \sum_{i=1}^{r} \lambda_i (\boldsymbol{x}, \boldsymbol{u}_i)_2 \boldsymbol{u}_i.$$

In some special circumstances, we have $r \ll n$, so that the last approximations are very inexpensive to assemble (once we have the spectral decomposition, of course). This is the idea of data compression and low-rank approximation, which are related to principal component analysis (PCA). One tries to reduce matrices to their primary (or principal) components. For some matrices with special structure, r can be quite small, and the information of that matrix can be significantly compressed into just a handful of numbers and vectors. We will make these arguments above rigorous, and, in fact, generalize them, with the proof of the Eckart–Young Theorem (2.15) later in the chapter.

Of course, not all matrices are Hermitian. We would like to have an analogue of unitary diagonalization for generic, nonsquare matrices. This is the purpose of the so-called *singular value decomposition* (SVD).

2.1 Reduced and Full Singular Value Decompositions

Definition 2.1 (SVD). Let $\mathsf{A} \in \mathbb{C}^{m\times n}$. A **singular value decomposition** (SVD) of the matrix A is a factorization of the form

$$\mathsf{A} = \mathsf{U}\Sigma\mathsf{V}^{\mathsf{H}},$$

where $\mathsf{U} \in \mathbb{C}^{m\times m}$ and $\mathsf{V} \in \mathbb{C}^{n\times n}$ are unitary, $\Sigma \in \mathbb{R}^{m\times n}$ is diagonal — meaning that $[\Sigma]_{i,j} = 0$, for $i \neq j$ — and the diagonal entries $[\Sigma]_{i,i} = \sigma_i$ are nonnegative and in nonincreasing order: $\sigma_1 \geq \sigma_2 \geq \cdots \geq \sigma_p \geq 0$ with $p = \min(m, n)$. The elements of the Σ are called the **singular values of** A. The columns of U and V are called the **left** and **right singular vectors**, respectively.

Remark 2.2 (reduced SVD). Let us, for the sake of definiteness, assume that $m \geq n$. If $\mathsf{A} \in \mathbb{C}^{m\times n}$ has an SVD, then we can write

$$\mathsf{A}\boldsymbol{v}_j = \sigma_j \boldsymbol{u}_j, \quad j = 1, \ldots, n,$$

or, equivalently,

$$\mathsf{A}[\boldsymbol{v}_1, \ldots, \boldsymbol{v}_n] = [\boldsymbol{u}_1, \ldots, \boldsymbol{u}_n] \begin{bmatrix} \sigma_1 & & \\ & \ddots & \\ & & \sigma_n \end{bmatrix}.$$

In other words, we have obtained the representation

$$\mathsf{A}\mathsf{V} = \hat{\mathsf{U}}\hat{\Sigma},$$

where:

1. $\hat{\Sigma} \in \mathbb{R}^{n\times n}$ is square and diagonal with nonnegative diagonal entries.
2. $\hat{\mathsf{U}} \in \mathbb{C}^{m\times n}$ has orthonormal columns.
3. $\mathsf{V} \in \mathbb{C}^{n\times n}$ is unitary.

Writing this another way, we have

$$\mathsf{A} = \hat{\mathsf{U}}\hat{\Sigma}\mathsf{V}^{\mathsf{H}}.$$

This is the so-called *reduced SVD* of a matrix. The standard SVD implies that the existence of the reduced SVD. Conversely, if one wants to obtain the *full SVD* from the *reduced SVD*, we observe that the columns of $\hat{\mathsf{U}}$ can be completed to a full orthonormal basis of $\mathbb{C}^m$ (using the Gram–Schmidt process described in Section A.5) to obtain

$$\mathsf{U} = [\boldsymbol{u}_1, \ldots, \boldsymbol{u}_n, \boldsymbol{u}_{n+1}, \ldots, \boldsymbol{u}_m] \in \mathbb{C}^{m\times m},$$

which is square and unitary. Next, we define

$$\Sigma = \begin{bmatrix} \hat{\Sigma} \\ \mathsf{O} \end{bmatrix},$$

where $\mathsf{O} \in \mathbb{C}^{(m-n)\times n}$ is a matrix of zeros. It is easy to see that the reduced SVD is equivalent to the representation

$$\mathsf{A} = \mathsf{U}\Sigma\mathsf{V}^{\mathsf{H}}.$$

A similar conclusion can be reached if $m < n$, where in this case the zero padding on the matrix Σ occurs to the right of $\hat{\Sigma}$.

2.2 Existence and Uniqueness of the SVD

Let us now show that *every* matrix has an SVD.

Theorem 2.3 (existence of SVD). *Every matrix* $\mathsf{A} \in \mathbb{C}^{m\times n}$ *has a singular value decomposition. The singular values are unique and*

$$\{\sigma_i^2\}_{i=1}^p = \begin{cases} \sigma(\mathsf{A}^{\mathsf{H}}\mathsf{A}) & \text{if } \; m \geq n, \\ \sigma(\mathsf{A}\mathsf{A}^{\mathsf{H}}) & \text{if } \; m \leq n, \end{cases}$$

where $p = \min(m, n)$. *Recall that the symbol* $\sigma(\mathsf{B})$ *stands for the spectrum of the square matrix* B.

Proof. (existence) Let us set

$$\sigma_1 = \|\mathsf{A}\|_2 = \sup_{\|\boldsymbol{x}\|_{\ell^2(\mathbb{C}^n)}=1} \|\mathsf{A}\boldsymbol{x}\|_{\ell^2(\mathbb{C}^m)}.$$

Arguing by compactness, and using Theorem B.47, there is a vector $\boldsymbol{v}_1 \in \mathbb{C}^n$ with $\|\boldsymbol{v}_1\|_{\ell^2(\mathbb{C}^n)} = 1$ such that $\|\mathsf{A}\|_2 = \|\mathsf{A}\boldsymbol{v}_1\|_{\ell^2(\mathbb{C}^m)} = \sigma_1$. Define $\boldsymbol{u}_1 = \|\mathsf{A}\|_2^{-1}\mathsf{A}\boldsymbol{v}_1$. Then

$$\mathsf{A}\boldsymbol{v}_1 = \sigma_1\boldsymbol{u}_1, \qquad \|\boldsymbol{u}_1\|_{\ell^2(\mathbb{C}^m)} = 1.$$

Using the Gram–Schmidt orthogonalization process described in Section A.5, we can extend $\{\boldsymbol{v}_1\}$ to an orthonormal basis of $\mathbb{C}^n$ and $\{\boldsymbol{u}_1\}$ to an orthonormal basis of $\mathbb{C}^m$. In doing so, we obtain matrices

$$\mathsf{U}_1 = [\boldsymbol{u}_1, \ldots, \boldsymbol{u}_m] \in \mathbb{C}^{m\times m}, \quad \mathsf{V}_1 = [\boldsymbol{v}_1, \ldots, \boldsymbol{v}_n] \in \mathbb{C}^{n\times n},$$

which are unitary and, more importantly, satisfy

$$\mathsf{A}\mathsf{V}_1 = \mathsf{U}_1 \begin{bmatrix} \sigma_1 & \boldsymbol{w}^{\mathsf{H}} \\ \mathbf{0} & \mathsf{B} \end{bmatrix} = \mathsf{U}_1\mathsf{S}$$

for some $\boldsymbol{w} \in \mathbb{C}^{n-1}$ and $\mathsf{B} \in \mathbb{C}^{(m-1)\times(n-1)}$.

Notice that

$$\mathsf{S}\begin{bmatrix}\sigma_1\\ \boldsymbol{w}\end{bmatrix}=\begin{bmatrix}\sigma_1 & \boldsymbol{w}^{\mathsf{H}}\\ \mathbf{0} & \mathsf{B}\end{bmatrix}\begin{bmatrix}\sigma_1\\ \boldsymbol{w}\end{bmatrix}=\begin{bmatrix}\sigma_1^2+\boldsymbol{w}^{\mathsf{H}}\boldsymbol{w}\\ \mathsf{B}\boldsymbol{w}\end{bmatrix}.$$

Therefore,

$$\begin{aligned}\left\|\mathsf{S}\begin{bmatrix}\sigma_1\\ \boldsymbol{w}\end{bmatrix}\right\|_{\ell^2(\mathbb{C}^m)} &= \sqrt{(\sigma_1^2+\boldsymbol{w}^{\mathsf{H}}\boldsymbol{w})^2+\|\mathsf{B}\boldsymbol{w}\|^2_{\ell^2(\mathbb{C}^{m-1})}}\\ &\geq \sigma_1^2+\boldsymbol{w}^{\mathsf{H}}\boldsymbol{w}\\ &= \left(\sigma_1^2+\boldsymbol{w}^{\mathsf{H}}\boldsymbol{w}\right)^{1/2}\left\|\begin{bmatrix}\sigma_1\\ \boldsymbol{w}\end{bmatrix}\right\|_{\ell^2(\mathbb{C}^m)},\end{aligned}$$

which shows the lower bound $\|\mathsf{S}\|_2 \geq \left(\sigma_1^2+\boldsymbol{w}^{\mathsf{H}}\boldsymbol{w}\right)^{1/2}$. On the other hand, since $\mathsf{U}_1^{\mathsf{H}}\mathsf{A}\mathsf{V}_1=\mathsf{S}$, and U_1 and V_1 are unitary, then we must have that $\sigma_1=\|\mathsf{A}\|_2=\|\mathsf{S}\|_2$, which forces us to conclude that $\boldsymbol{w}=\mathbf{0}$. This shows the result if $n=1$ or $m=1$. Otherwise, we proceed by induction.

For the induction step, we factorize $\mathsf{A}\in\mathbb{C}^{m\times n}$ as above,

$$\mathsf{A}\mathsf{V}_1=\mathsf{U}_1\begin{bmatrix}\sigma_1 & \mathbf{0}\\ \mathbf{0} & \mathsf{B}\end{bmatrix},$$

and assume that B has an SVD, say $\mathsf{B}=\mathsf{U}_2\Sigma_2\mathsf{V}_2^{\mathsf{H}}$. Therefore, we have

$$\mathsf{A}=\mathsf{U}_1\begin{bmatrix}1 & \mathsf{O}\\ \mathsf{O} & \mathsf{U}_2\end{bmatrix}\begin{bmatrix}\sigma_1 & 0\\ 0 & \Sigma_2\end{bmatrix}\begin{bmatrix}1 & \mathsf{O}\\ \mathsf{O} & \mathsf{V}_2^{\mathsf{H}}\end{bmatrix}\mathsf{V}_1^{\mathsf{H}},$$

which is the sought-after SVD for A with

$$\mathsf{U}=\mathsf{U}_1\begin{bmatrix}1 & \mathsf{O}\\ \mathsf{O} & \mathsf{U}_2\end{bmatrix},\quad \Sigma=\begin{bmatrix}\sigma_1 & \mathsf{O}\\ \mathsf{O} & \Sigma_2\end{bmatrix},\quad \mathsf{V}=\mathsf{V}_1\begin{bmatrix}1 & \mathsf{O}\\ \mathsf{O} & \mathsf{V}_2\end{bmatrix}.$$

The existence part is finished by induction.
(uniqueness) Next, notice that

$$\mathsf{A}^{\mathsf{H}}\mathsf{A}=\left(\mathsf{U}\Sigma\mathsf{V}^{\mathsf{H}}\right)^{\mathsf{H}}\mathsf{U}\Sigma\mathsf{V}^{\mathsf{H}}=\mathsf{V}\Sigma^{\intercal}\mathsf{U}^{\mathsf{H}}\mathsf{U}\Sigma\mathsf{V}^{\mathsf{H}}=\mathsf{V}\Sigma^{\intercal}\Sigma\mathsf{V}^{\mathsf{H}}$$

and

$$\mathsf{A}\mathsf{A}^{\mathsf{H}}=\mathsf{U}\Sigma\mathsf{V}^{\mathsf{H}}\left(\mathsf{U}\Sigma\mathsf{V}^{\mathsf{H}}\right)^{\mathsf{H}}=\mathsf{U}\Sigma\Sigma^{\intercal}\mathsf{U}^{\mathsf{H}},$$

where $\Sigma^{\intercal}\Sigma\in\mathbb{R}^{n\times n}$ and $\Sigma\Sigma^{\intercal}\in\mathbb{R}^{m\times m}$ are diagonal matrices with the diagonal entries $\sigma_1^2,\ldots,\sigma_p^2$, plus zeros for padding, as needed. Thus, $\mathsf{A}^{\mathsf{H}}\mathsf{A}\asymp\Sigma^{\intercal}\Sigma$ and $\mathsf{A}\mathsf{A}^{\mathsf{H}}\asymp\Sigma\Sigma^{\intercal}$. Using this fact and the fact that eigenvalues are uniquely determined proves the result. □

The uniqueness results for the right and left singular vectors are a little more subtle. We have for instance the following result.

Theorem 2.4 (uniqueness of singular vectors). *Suppose that* $\mathsf{A}\in\mathbb{C}^{m\times n}$ *with* $m\geq n$. *If*

$$\mathsf{A}=\mathsf{U}_1\Sigma_1\mathsf{V}_1^{\mathsf{H}}=\mathsf{U}_2\Sigma_2\mathsf{V}_2^{\mathsf{H}}$$

are two SVDs for A, *then* $\Sigma_1 = \Sigma_2$, *the columns of* V_1 *and* V_2 *form an orthonormal basis of eigenvectors of* $A^H A$, *and if* $A^H A$ *has n distinct eigenvalues, then*

$$V_1 = V_2 D$$

for some $D = \text{diag}[e^{i\theta_1}, \ldots, e^{i\theta_n}]$ *with angles* $\theta_i \in \mathbb{R}$, $i = 1, \ldots, n$. *Finally, if* $\text{rank}(A) = n$ *(*A *has full rank) and*

$$A = U_1 \Sigma V^H = U_2 \Sigma V^H$$

are two SVDs for A, *then the first n columns of* U_1 *and* U_2 *are equal.*

Proof. The fact that $\Sigma_1 = \Sigma_2$ was already proved in Theorem 2.3. Suppose now that $A^H A$ has n distinct eigenvalues and

$$A = U_1 \Sigma V_1^H = U_2 \Sigma V_2^H,$$

where we have used the fact that the singular values are uniquely determined. Also from Theorem 2.3,

$$A^H A = V_1 \Sigma^T \Sigma V_1^H = V_2 \Sigma^T \Sigma V_2^H,$$

which proves that the columns of V_1 and V_2 are eigenvectors of $A^H A$.

Now, if the eigenvalues $\lambda_i = \sigma_i^2$ of $A^H A$ are all simple, the corresponding eigenspaces $E(\lambda_i, A^H A)$ are all one dimensional and

$$E(\lambda_i, A^H A) = \text{span}\{\boldsymbol{v}_{1,i}\} = \text{span}\{\boldsymbol{v}_{2,i}\}, \quad i = 1, \ldots, n,$$

with

$$\|\boldsymbol{v}_{1,i}\|_{\ell^2(\mathbb{C}^n)} = \|\boldsymbol{v}_{2,i}\|_{\ell^2(\mathbb{C}^n)} = 1, \quad i = 1, \ldots, n,$$

where $V_k = [\boldsymbol{v}_{k,1}, \ldots, \boldsymbol{v}_{k,n}]$, $k = 1, 2$. The only possibility is that $\boldsymbol{v}_{1,i} = \gamma_i \boldsymbol{v}_{2,i}$, where $|\gamma_i| = 1$, for $i = 1, \ldots, n$. This proves that $V_1 = V_2 D$, where D is a diagonal matrix with the required structure.

Finally, if

$$A = U_1 \Sigma V^H = U_2 \Sigma V^H$$

are two SVDs for A, then we have the following family of equations:

$$A\boldsymbol{v}_i = \sigma_i \boldsymbol{u}_{k,i}, \quad k = 1, 2, \quad i = 1, \ldots, n,$$

where $U_k = [\boldsymbol{u}_{k,1}, \ldots, \boldsymbol{u}_{k,m}]$, $k = 1, 2$. Since A has full rank and $\sigma_i > 0$, $\boldsymbol{u}_{1,i} = \boldsymbol{u}_{2,i}$, $i = 1, \ldots, n$ is the only possibility. □

Remark 2.5 (geometric interpretation of the SVD)**.** Let $\boldsymbol{b} \in \mathbb{C}^m$ and expand it in the basis of left singular vectors U. This gives the coordinate vector $\boldsymbol{b}' = U^H \boldsymbol{b}$. Do the same for $\boldsymbol{x} \in \mathbb{C}^n$ to obtain its coordinate vector $\boldsymbol{x}' = V^H \boldsymbol{x}$. Once we have this, notice that, if $\boldsymbol{b} = A\boldsymbol{x}$, we can proceed as follows:

$$\boldsymbol{b}' = U^H \boldsymbol{b} = U^H A \boldsymbol{x} = U^H U \Sigma V^H \boldsymbol{x} = \Sigma V^H \boldsymbol{x} = \Sigma \boldsymbol{x}'.$$

In other words, the SVD is essentially saying that every matrix, once proper bases for the domain and target spaces are chosen, may be viewed as a diagonal matrix.

2.3 Further Properties of the SVD

The main motivation for the SVD was to try to construct an analogue of the spectral decomposition, which we know is only valid for square, nondefective matrices. Let us study now the relation between these two constructions, which in principle are not related to each other.

Theorem 2.6 (SVD and rank). *Let* $\mathsf{A} \in \mathbb{C}^{m\times n}$. *Then* $\operatorname{rank}(\mathsf{A})$ *coincides with the number of nonzero singular values.*

Proof. We write the SVD: $\mathsf{A} = \mathsf{U}\Sigma\mathsf{V}^{\mathsf{H}}$. Since U and V are unitary, they are full rank. By Theorem 1.16, $\operatorname{rank}(\mathsf{A}) = \operatorname{rank}(\Sigma)$ and since Σ is diagonal, the assertion follows. □

Theorem 2.7 (range and kernel through SVD). *Let* $\mathsf{A} \in \mathbb{C}^{m\times n}$ *with* $\operatorname{rank}(\mathsf{A}) = r$. *Suppose that an SVD for* A *is given by* $\mathsf{A} = \mathsf{U}\Sigma\mathsf{V}^{\mathsf{H}}$, *where* $\boldsymbol{u}_1, \dots, \boldsymbol{u}_m$ *denote the columns of* U *and* $\boldsymbol{v}_1, \dots, \boldsymbol{v}_n$ *denote the columns of* V. *Then*

$$\langle \boldsymbol{u}_1, \dots, \boldsymbol{u}_r \rangle = \operatorname{im}(\mathsf{A}) \quad \textit{and} \quad \langle \boldsymbol{v}_{r+1}, \dots, \boldsymbol{v}_n \rangle = \ker(\mathsf{A}).$$

Proof. ($\subseteq$) From the SVD one can easily write $\mathsf{A}\boldsymbol{v}_i = \sigma_i \boldsymbol{u}_i$, for $i = 1, \dots, r$. This proves immediately that $\boldsymbol{u}_i \in \operatorname{im}(\mathsf{A})$, for $i = 1, \dots, r$. Since $\operatorname{im}(\mathsf{A})$ is a subspace of $\mathbb{C}^m$, any linear combination of $\boldsymbol{u}_1, \dots, \boldsymbol{u}_r$ is in $\operatorname{im}(\mathsf{A})$. Hence, $\langle \boldsymbol{u}_1, \dots, \boldsymbol{u}_r \rangle \subseteq \operatorname{im}(\mathsf{A})$.

($\supseteq$) Let $\boldsymbol{y} \in \operatorname{im}(\mathsf{A})$. Then there exists $\boldsymbol{x} \in \mathbb{C}^n$ such that $\mathsf{A}\boldsymbol{x} = \boldsymbol{y}$. This implies that $\mathsf{U}\Sigma\mathsf{V}^{\mathsf{H}}\boldsymbol{x} = \boldsymbol{y}$ for some $\boldsymbol{x}$. Let $\boldsymbol{x}' = \mathsf{V}^{\mathsf{H}}\boldsymbol{x}$. Then, for some $\boldsymbol{x}' \in \mathbb{C}^n$, $\mathsf{U}\Sigma\boldsymbol{x}' = \boldsymbol{y}$. Set $\boldsymbol{x}'' = \Sigma\boldsymbol{x}'$. Note that $\boldsymbol{x}'' \in \mathbb{C}^m$ and $x''_{r+1} = \cdots = x''_m = 0$. Hence, for some $\boldsymbol{x}'' \in \mathbb{C}^m$, $\mathsf{U}\boldsymbol{x}'' = \boldsymbol{y}$. Now we write

$$\boldsymbol{y} = \mathsf{U}\boldsymbol{x}'' = \sum_{j=1}^{m} x''_j \boldsymbol{u}_j = \sum_{j=1}^{r} x''_j \boldsymbol{u}_j \in \langle \boldsymbol{u}_1, \dots, \boldsymbol{u}_r \rangle .$$

This proves that $\langle \boldsymbol{u}_1, \dots, \boldsymbol{u}_r \rangle \supseteq \operatorname{im}(\mathsf{A})$, and we are done.

($\subseteq$) From the SVD one can easily write $\mathsf{A}\boldsymbol{v}_i = \boldsymbol{0}$, for $i = r+1, \dots, n$. This proves immediately that $\boldsymbol{v}_i \in \ker(\mathsf{A})$, for $i = r+1, \dots, n$. Since $\ker(\mathsf{A})$ is subspace of $\mathbb{C}^n$, any linear combination of $\boldsymbol{v}_{r+1}, \dots, \boldsymbol{v}_n$ is in $\ker(\mathsf{A})$. Hence, $\langle \boldsymbol{v}_{r+1}, \dots, \boldsymbol{v}_n \rangle \subseteq \ker(\mathsf{A})$.

($\supseteq$) Let $\boldsymbol{x} \in \ker(\mathsf{A})$. Then $\mathsf{A}\boldsymbol{x} = \boldsymbol{0}$. This implies that $\mathsf{U}\Sigma\mathsf{V}^{\mathsf{H}}\boldsymbol{x} = \boldsymbol{0}$. Let $\boldsymbol{x}' = \mathsf{V}^{\mathsf{H}}\boldsymbol{x}$. This implies that $\boldsymbol{x} = \mathsf{V}\boldsymbol{x}'$ and $\mathsf{U}\Sigma\boldsymbol{x}' = \boldsymbol{0}$. Since U is invertible, this implies that $\Sigma\boldsymbol{x}' = \boldsymbol{0}$. This homogeneous system always has a solution of the form

$$\boldsymbol{x}' = \begin{bmatrix} x'_1 = 0 \\ \vdots \\ x'_r = 0 \\ x'_{r+1} = \alpha_{r+1} \\ \vdots \\ x'_n = \alpha_n \end{bmatrix},$$

where $\alpha_{r+1}, \ldots, \alpha_n$ are arbitrary. But this shows that

$$x = Vx' = \sum_{j=1}^{n} x'_j \boldsymbol{v}_j = \sum_{j=r+1}^{n} \alpha_j \boldsymbol{v}_j \in \langle \boldsymbol{v}_{r+1}, \ldots, \boldsymbol{v}_n \rangle .$$

This proves that $\langle \boldsymbol{v}_{r+1}, \ldots, \boldsymbol{v}_n \rangle \supseteq \ker(A)$, and we are done. □

The last result gives another proof of the famous *Rank-Plus-Nullity Theorem*; see Theorem 1.16.2.

Corollary 2.8 (Rank-Plus-Nullity Theorem). *Suppose that* $A \in \mathbb{C}^{m \times n}$. *Then*

$$\operatorname{rank}(A) + \operatorname{nullity}(A) = \dim(\operatorname{im}(A)) + \dim(\ker(A)) = n.$$

Theorem 2.9 ($A^H A$ is nonsingular). *Suppose that* $A \in \mathbb{C}^{m \times n}$ *with* $m \geq n$. $A^H A$ *is nonsingular if and only if* $\operatorname{rank}(A) = n$.

Proof. Let $A = U\Sigma V^H$ be an SVD for A. Then $A^H A = V\Sigma^\intercal \Sigma V^H$ yields a unitary diagonalization of $A^H A$. Note that

$$\Sigma^\intercal \Sigma = \operatorname{diag}\left[\sigma_1^2, \ldots, \sigma_n^2\right],$$

where $\sigma_1, \ldots, \sigma_n$ are the singular values of A, some of which may be zero.

It is clear that $\operatorname{rank}(A) = r$, where r is the number of nonzero singular values. Of course, it must be that $r \leq n$. Likewise, $\operatorname{rank}(A^H A)$ is the number of nonzero elements on the diagonal of $\Sigma^H \Sigma$. This number must also be r. In other words,

$$\operatorname{rank}(A) = r = \operatorname{rank}(A^H A),$$

which proves the result. □

Theorem 2.10 (SVD and norms). *Let* $A \in \mathbb{C}^{m \times n}$. *Then* $\|A\|_2 = \sigma_1$ *and* $\|A\|_F^2 = \sum_{i=1}^{r} \sigma_i^2$.

Proof. The first statement is by construction. The second follows from Problem 1.18 and the fact that U and V are unitary. In this case, $\left\|U\Sigma V^H\right\|_F = \left\|\Sigma V^H\right\|_F = \|\Sigma\|_F$. □

Theorem 2.11 (SVD and self-adjoint matrices). *If* A *is Hermitian, i.e.,* $A^H = A$, *then the singular values of* A *are the absolute values of its eigenvalues. If* A *is Hermitian with nonnegative eigenvalues,*[1] *then the eigenvalues and the singular values coincide.*

Proof. Since A is self-adjoint, it is orthogonally diagonalizable and $\sigma(A) \subset \mathbb{R}$, i.e.,

$$A = Q\Lambda Q^H = Q|\Lambda| \operatorname{sgn}(\Lambda) Q^H,$$

where the notation has the obvious meaning. Define $U = Q$ and $V^H = \operatorname{sgn}(\Lambda) Q^H$ to obtain the SVD of A. Conclude by uniqueness. □

[1] We will see in Chapter 3 that this is a special class of matrices called Hermitian positive semi-definite (HPSD) matrices.

Theorem 2.12 (SVD and determinants). *Let* $\mathsf{A} \in \mathbb{C}^{n \times n}$. *Then*

$$|\det(\mathsf{A})| = \prod_{i=1}^{n} \sigma_i.$$

Proof. By the usual rules for determinants

$$|\det(\mathsf{A})| = |\det(\mathsf{U}\Sigma\mathsf{V}^{\mathsf{H}})| = |\det(\mathsf{U})||\det(\Sigma)||\det(\mathsf{V}^{\mathsf{H}})| = |\det(\Sigma)| = \prod_{i=1}^{n} \sigma_i,$$

where we have used the facts that the determinant of a unitary matrix has modulus one, the determinant of a diagonal matrix is the product of its diagonals, and Σ is diagonal with nonnegative entries. □

2.4 Low Rank Approximations

Now this is where the SVD really shines. If we can find an SVD for a matrix A, we can analyze the singular values and use the SVD to compress the information in A in an optimal way.

Definition 2.13 (rank-one matrix). Given $\boldsymbol{u} \in \mathbb{C}^m_\star$, $\boldsymbol{v} \in \mathbb{C}^n_\star$, and $\sigma \in \mathbb{C}_\star$, the matrix $\mathsf{A} \in \mathbb{C}^{m \times n}$ defined via

$$\mathsf{A} = \sigma \boldsymbol{u}\boldsymbol{v}^{\mathsf{H}} = \sigma \boldsymbol{u} \otimes \boldsymbol{v}$$

is called a **rank-one matrix**.

For every $\boldsymbol{x} \in \mathbb{C}^n$, the rank-one matrix A, defined above, acts as

$$\mathsf{A}\boldsymbol{x} = \sigma(\boldsymbol{x}, \boldsymbol{v})_2 \boldsymbol{u} \in \operatorname{span}\{\boldsymbol{u}\}.$$

This shows, as a consequence of Corollary 1.14, that the rank of this matrix is exactly equal to one. This justifies the name. The question we want to address now is whether *every* matrix can be represented (or at least approximated) by linear combinations of rank-one matrices.

Theorem 2.14 (Rank-One Decomposition Theorem). *Let* $\mathsf{A} \in \mathbb{C}^{m \times n}$ *be such that* $r = \operatorname{rank}(\mathsf{A})$ *and* $\mathsf{A} = \mathsf{U}\Sigma\mathsf{V}^{\mathsf{H}}$ *is an SVD. Then* A *is a linear combination of* r *rank-one matrices*

$$\mathsf{A} = \sum_{j=1}^{r} \sigma_j \boldsymbol{u}_j \otimes \boldsymbol{v}_j. \tag{2.1}$$

Proof. It suffices to write Σ as the sum of matrices of the form

$$\Sigma_j = \operatorname{diag}[0, \dots, 0, \sigma_j, 0, \dots, 0],$$

where the element σ_j is in the jth entry. The rest of the details are left to the reader as an exercise; see Problem 2.6. □

Let us now prove the so-called Eckart–Young low rank approximation theorem, which states that truncating the SVD of a matrix gives, in a sense, *the best* low rank approximation to it.

Theorem 2.15 (Eckart–Young Theorem[2]). *Let* $\mathsf{A} \in \mathbb{C}^{m\times n}$ *be such that* $r = \operatorname{rank}(\mathsf{A})$. *Let* $\mathsf{A} = \mathsf{U}\Sigma\mathsf{V}^{\mathsf{H}}$ *be an SVD of* A. *For* $k < r$ *define* $\mathsf{A}_k = \sum_{j=1}^{k} \sigma_j \boldsymbol{u}_j \otimes \boldsymbol{v}_j$. *Let us denote by* $\mathcal{C}_k$ *the collection of all matrices* $\mathsf{B} \in \mathbb{C}^{m\times n}$ *such that* $\operatorname{rank}(\mathsf{B}) \leq k$. *Then* A_k *is of rank* k *and*

$$\|\mathsf{A} - \mathsf{A}_k\|_2 = \sigma_{k+1} = \inf_{\mathsf{B}\in\mathcal{C}_k} \|\mathsf{A} - \mathsf{B}\|_2.$$

Furthermore,

$$\|\mathsf{A} - \mathsf{A}_k\|_F = \sqrt{\sum_{j=k+1}^{r} \sigma_j^2} = \inf_{\mathsf{B}\in\mathcal{C}_k} \|\mathsf{A} - \mathsf{B}\|_F.$$

Proof. The values of the norms of the difference $\mathsf{A}-\mathsf{A}_k$ follow from the representation (2.1) and Theorem 2.10. Next, it is straightforward to see that $\operatorname{rank}(\mathsf{A}_k) = k$.

To show that the first infimum is attained at A_k, we use a contradiction argument. Namely, let us assume that there is a matrix $\mathsf{B} \in \mathcal{C}_k$ such that

$$\|\mathsf{A} - \mathsf{B}\|_2 < \|\mathsf{A} - \mathsf{A}_k\|_2 = \sigma_{k+1}.$$

Since $\operatorname{rank}(\mathsf{B}) \leq k$, by Theorem 1.16, $\dim\ker(\mathsf{B}) \geq n - k$. There is a subspace $W_1 \leq \ker(\mathsf{B})$ such that $\dim(W_1) = n - k$ and, for every $\boldsymbol{w} \in W_1$, $\mathsf{B}\boldsymbol{w} = \boldsymbol{0}$.

For any $\boldsymbol{w} \in W_1$, it follows that $\mathsf{A}\boldsymbol{w} = (\mathsf{A} - \mathsf{B})\boldsymbol{w}$ and

$$\|\mathsf{A}\boldsymbol{w}\|_2 = \|(\mathsf{A} - \mathsf{B})\boldsymbol{w}\|_2 \leq \|\mathsf{A} - \mathsf{B}\|_2\|\boldsymbol{w}\|_2 < \sigma_{k+1}\|\boldsymbol{w}\|_2.$$

Next, define $W_2 = \operatorname{span}\{\boldsymbol{v}_1, \ldots, \boldsymbol{v}_{k+1}\}$. We claim that, for every $\boldsymbol{w} \in W_2$, $\|\mathsf{A}\boldsymbol{w}\|_2 \geq \sigma_{k+1}\|\boldsymbol{w}\|_2$. To see this, observe that $\boldsymbol{w} \in W_2$ can be written as $\boldsymbol{w} = \sum_{i=1}^{k+1} \beta_i \boldsymbol{v}_i$, for some $\beta_1, \ldots, \beta_{k+1} \in \mathbb{C}$. By orthonormality, it follows that

$$\|\boldsymbol{w}\|_2 = \sqrt{\sum_{i=1}^{k+1} |\beta_i|^2}.$$

Using orthonormality again, we see that

$$\|\mathsf{A}\boldsymbol{w}\|_2 = \left\|\sum_{i=1}^{k+1} \sigma_i\beta_i\boldsymbol{u}_i\right\|_2 = \sqrt{\sum_{i=1}^{k+1} \sigma_i^2|\beta_i|^2} \geq \sigma_{k+1}\sqrt{\sum_{i=1}^{k+1} |\beta_i|^2} = \sigma_{k+1}\,\|\boldsymbol{w}\|_2\,.$$

Thus, $W_1 \leq \mathbb{C}^n$ and $W_2 \leq \mathbb{C}^n$, and the sum of the dimensions of these subspaces exceeds n. Therefore, there must be a nonzero vector in their intersection. But this yields a contradiction, because such a vector $\boldsymbol{z} \in W_1 \cap W_2\setminus\{\boldsymbol{0}\}$ would satisfy

$$\sigma_{k+1}\,\|\boldsymbol{z}\|_2 \leq \|\mathsf{A}\boldsymbol{z}\|_2 < \sigma_{k+1}\|\boldsymbol{z}\|_2.$$

The proof for the second case involving the Frobenius norm is left to the reader as an exercise; see Problem 2.7. □

[2] Named in honor of the American physicist Carl Henry Eckart (1902–1973) and the American engineer Gale J. Young (1912–1990).

Example 2.1 Consider the matrix

$$A = \begin{bmatrix} 2.9057 & 6.5457 & 3.8587 & 4.6737 & 2.4171 & 3.8703 & 0.9922 \\ 2.6460 & 0.8889 & 3.2542 & 1.7574 & 6.8099 & 1.4357 & 6.2114 \\ 2.1268 & 3.4270 & 2.6255 & 2.6030 & 3.0228 & 2.1093 & 2.2459 \\ 2.1686 & 2.6778 & 3.8653 & 3.6416 & 3.6917 & 3.3553 & 2.2689 \\ 3.8927 & 6.6301 & 6.4863 & 6.8192 & 5.0281 & 6.0910 & 2.5521 \\ 4.0461 & 6.2798 & 3.1713 & 2.7880 & 6.1632 & 1.6456 & 5.8120 \\ 2.9612 & 1.9820 & 6.1018 & 5.1754 & 6.4646 & 5.0316 & 4.2031 \end{bmatrix}.$$

The singular values are, to 13 decimal digits of precision,

$$\begin{aligned} \sigma_1 &= 27.775\,450\,511\,276\,4, \\ \sigma_2 &= 08.024\,842\,310\,514\,9, \\ \sigma_3 &= 05.224\,556\,262\,211\,5, \\ \sigma_4 &= 00.000\,196\,585\,865\,6, \\ \sigma_5 &= 00.000\,085\,666\,006\,1, \\ \sigma_6 &= 00.000\,062\,891\,962\,9, \\ \sigma_7 &= 00.000\,007\,169\,799\,2. \end{aligned}$$

This matrix is very nearly singular and is well approximated by the rank-three (compressed) matrix

$$A_3 = \sum_{i=1}^{3} \sigma_i \boldsymbol{u}_i \boldsymbol{v}_i^{\mathsf{T}},$$

where $\boldsymbol{u}_i, \boldsymbol{v} \in \mathbb{R}^7$ are the singular vectors, which are suppressed for brevity. In other words, to a good approximation, there are really only three important components of A — namely, $\sigma_i \boldsymbol{u}_i \boldsymbol{v}_i^{\mathsf{T}}$, $i = 1, 2, 3$ — that express its action. In particular, according to the Eckart–Young Theorem 2.15, the relative error in the compressed matrix is relatively small,

$$\frac{\|A - A_3\|_2}{\|A\|_2} = \frac{\sigma_4}{\sigma_1} = 7.077\,68 \times 10^{-6}.$$

Problems

2.1 Let $A, B \in \mathbb{C}^{n\times n}$ be unitarily equivalent. Prove that they have the same singular values.

2.2 Show that if A is real, then it has a real SVD ($U \in \mathbb{R}^{m\times m}$ and $V \in \mathbb{R}^{n\times n}$).

2.3 Show that any matrix in $\mathbb{C}^{m\times n}$ is the limit of a sequence of matrices of full rank.

2.4 Suppose that $A \in \mathbb{C}^{m\times m}$ has an SVD $A = U\Sigma V^{H}$. Find an eigenvalue decomposition of the matrix

$$\begin{bmatrix} O & A^{H} \\ A & O \end{bmatrix}.$$

2.5 Let $\mathsf{A} \in \mathbb{C}^{m\times n}$ with $\text{rank}(\mathsf{A}) = r$. If A has the SVD $\mathsf{A} = \mathsf{U}\Sigma\mathsf{V}^{\mathsf{H}}$, the *Moore–Penrose pseudo-inverse* of A is defined by

$$\mathsf{A}^{\dagger} = \mathsf{V}\Sigma^{\dagger}\mathsf{U}^{\mathsf{H}},$$

where $\Sigma^{\dagger} = \text{diag}[\sigma_1^{-1}, \ldots, \sigma_r^{-1}, 0, \ldots, 0] \in \mathbb{R}^{n\times m}$. Show the following:

a) If A is square and A^{-1} exists, then $\mathsf{A}^{\dagger} = \mathsf{A}^{-1}$.
b) If $m \geq n$ and A has full rank, then $\mathsf{A}^{\dagger} = (\mathsf{A}^{\mathsf{H}}\mathsf{A})^{-1}\mathsf{A}^{\mathsf{H}}$.
c) $\mathsf{A}\mathsf{A}^{\dagger}\mathsf{A} = \mathsf{A}$.
d) $\mathsf{A}^{\dagger}\mathsf{A}\mathsf{A}^{\dagger} = \mathsf{A}^{\dagger}$.

2.6 Complete the proof of Theorem 2.14.

2.7 Complete the proof of Theorem 2.15.

2.8 Let $\mathsf{A} \in \mathbb{C}^{m\times n}$. Use the SVD to prove the following:

a) $\text{rank}(\mathsf{A}^{\mathsf{H}}\mathsf{A}) = \text{rank}(\mathsf{A}\mathsf{A}^{\mathsf{H}}) = \text{rank}(\mathsf{A}) = \text{rank}(\mathsf{A}^{\mathsf{H}})$.
b) $\mathsf{A}^{\mathsf{H}}\mathsf{A}$ and $\mathsf{A}\mathsf{A}^{\mathsf{H}}$ have the same nonzero eigenvalues.
c) If the eigenvectors $\mathbf{w}_1$ and $\mathbf{w}_2$ of $\mathsf{A}^{\mathsf{H}}\mathsf{A}$ are orthogonal, then $\mathsf{A}\mathbf{w}_1$ and $\mathsf{A}\mathbf{w}_2$ are orthogonal.

2.9 The purpose of this problem is to provide another proof of the existence of an SVD for any matrix $\mathsf{A} \in \mathbb{C}^{m\times n}$. Assume, for the sake of definiteness, that $m \geq n$.

a) Show that $\mathsf{A}\mathsf{A}^{\mathsf{H}} = \mathsf{U}\Sigma\Sigma^{\mathsf{T}}\mathsf{U}^{\mathsf{H}}$, where $\mathsf{U} \in \mathbb{C}^{m\times m}$ is unitary and $\Sigma \in \mathbb{R}^{m\times n}$ is diagonal with nonnegative diagonal entries.
Hint: Recall Problem 1.39.
b) Show that $\mathsf{A}^{\mathsf{H}}\mathsf{A} = \mathsf{V}\Sigma^{\mathsf{T}}\Sigma\mathsf{V}^{\mathsf{H}}$, where $\mathsf{V} \in \mathbb{C}^{n\times n}$ is unitary and Σ is the same matrix from the previous item.
Hint: Recall Problem 1.38.
c) Show that $\mathsf{A} = \mathsf{U}\Sigma\mathsf{V}^{\mathsf{H}}$.
Hint: Take $\mathsf{V} = [\mathbf{v}_1, \ldots, \mathbf{v}_n]$ as given in part (b). Define $\hat{\mathbf{u}}_i = \mathsf{A}\mathbf{v}_i/\sigma_i$, $i = 1, \ldots, r = \text{rank}(\mathsf{A})$. Show that these newly defined vectors can be identified as the first r vectors of the matrix U from part (a).

3 Systems of Linear Equations

In this chapter, we will be concerned with the following problem: Given the matrix $\mathsf{A} = [a_{i,j}] \in \mathbb{C}^{n\times n}$ and the vector $\boldsymbol{f} = [f_i] \in \mathbb{C}^n$, find $\boldsymbol{x} = [x_i] \in \mathbb{C}^n$ such that

$$\mathsf{A}\boldsymbol{x} = \boldsymbol{f}. \tag{3.1}$$

Of course, this is shorthand for the following system of linear equations:

$$\begin{cases} a_{1,1}x_1 + a_{1,2}x_2 + \cdots + a_{1,n}x_n = f_1, \\ a_{2,1}x_1 + a_{2,2}x_2 + \cdots + a_{2,n}x_n = f_2, \\ \qquad\qquad\qquad\vdots \\ a_{n,1}x_1 + a_{n,2}x_2 + \cdots + a_{n,n}x_n = f_n. \end{cases}$$

We call A the *coefficient matrix*. First of all, we need to make sure that a solution exists and is unique. The following result is nothing but a recapitulation of statements that the reader will have encountered before.

Theorem. *The system of linear equations (3.1) has a unique solution if and only if* $\det(\mathsf{A}) \neq 0$ *if and only if* $\mathsf{A}\boldsymbol{x} = \boldsymbol{0}$ *has only the trivial solution if and only if* A^{-1} *exists.*

This theorem gives necessary and sufficient conditions for the inverse of the coefficient matrix, A^{-1}, to exist. If it does, then the solution is $\boldsymbol{x} = \mathsf{A}^{-1}\boldsymbol{f}$, but it is not usually computationally tractable to calculate the inverse, as we discuss later. Thus, we will look for ways of computing $\boldsymbol{x}$ without first explicitly finding A^{-1}. Most of us, in a course on linear algebra, learned of a method called Gaussian elimination. This simple, powerful, and sometimes mysterious technique will be the basis of most of what we do in this chapter. In particular, Gaussian elimination is the foundation for some well-known factorization techniques, such as the LU decomposition method and the Cholesky factorization method for positive definite matrices.

Why is Gaussian elimination mysterious? The reason for this is that one of the big open questions of numerical linear algebra is

Why is Gaussian elimination (with partial pivoting) usually so numerically stable in practice?

Except for certain classes of truly pathological matrices, our best generic estimates for the growth of roundoff error in the algorithm tend to be overly pessimistic. In other words, this simple algorithm usually performs much better than the worst-case scenario for the average matrix. Gaussian elimination is much more reliable than numerical analysts would expect. A discussion of this topic is beyond the scope of our text, but see [34, 96].

3.1 Solution of Simple Systems

Before we describe the general case, in this section, we develop some algorithms to find the solution to (3.1) for some simple cases, all of which avoid the direct construction of A^{-1}.

3.1.1 Diagonal Matrices

If the coefficient matrix A is diagonal, i.e., $A = \text{diag}(a_1, \dots, a_n)$ with $a_k \neq 0$, $k = 1, \dots, n$, then the solution can be easily found by $x_k = f_k / a_k$.

3.1.2 Triangular Matrices

Let us, to be definite, consider the case when A is upper triangular. The system of equations reads

$$\begin{cases} a_{1,1}x_1 + a_{1,2}x_2 + \cdots + a_{1,n}x_n = f_1, \\ \qquad\quad a_{2,2}x_2 + \cdots + a_{2,n}x_n = f_2, \\ \qquad\qquad\qquad \vdots \\ \qquad\qquad\qquad a_{n,n}x_n = f_n. \end{cases}$$

A unique solution exists if and only if $a_{i,i} \neq 0$ for all $i = 1, \dots, n$. In this case, the solution can be easily found by first computing the value of the last variable,

$$x_n = f_n / a_{n,n},$$

and, after that, recursively computing

$$x_k = \frac{1}{a_{k,k}} \left(f_k - \sum_{j=k+1}^{n} a_{k,j} x_j \right), \qquad k = n-1, n-2, \dots, 2, 1.$$

The order of execution of this algorithm is vital: one must start with $k = n - 1$ and proceed in reverse order, finishing with $k = 1$. This algorithm is known as *back substitution*.

Remark 3.1 (forward substitution). A similar procedure, known as *forward substitution*, can be applied to lower triangular matrices.

3.1.3 Tridiagonal Matrices

We begin with a definition.

Definition 3.2 (tridiagonal matrix). Let $A = [a_{i,j}] \in \mathbb{C}^{n\times n}$. We say that A is **tridiagonal** if and only if when $i, j \in \{1, \ldots, n\}$ and $|i-j| > 1$ then $a_{i,j} = 0$.

A generic system of equations with a tridiagonal coefficient matrix can be conveniently expressed as

$$a_k x_{k-1} + b_k x_k + c_k x_{k+1} = f_k, \quad k = 1, \ldots, n, \tag{3.2}$$

with $a_1 = c_n = 0$. This can be visualized as

$$\begin{bmatrix} b_1 & c_1 & 0 & \cdots & 0 & 0 \\ a_2 & b_2 & c_2 & 0 & & 0 \\ 0 & a_3 & b_3 & \ddots & \ddots & \vdots \\ \vdots & 0 & \ddots & \ddots & \ddots & 0 \\ 0 & & \ddots & \ddots & b_{n-1} & c_{n-1} \\ 0 & 0 & \cdots & 0 & a_n & b_n \end{bmatrix} \begin{bmatrix} x_1 \\ x_2 \\ x_3 \\ \vdots \\ x_{n-2} \\ x_{n-1} \\ x_n \end{bmatrix} = \begin{bmatrix} f_1 \\ f_2 \\ f_3 \\ \vdots \\ f_{n-2} \\ f_{n-1} \\ f_n \end{bmatrix}.$$

To find the solution — assuming that a unique solution exists — we begin by assuming that it has the following form:

$$x_k = \alpha_k x_{k+1} + \beta_k.$$

This seems reasonable since, for $k = 1$, we have

$$x_1 = -\frac{c_1}{b_1} x_2 + \frac{f_1}{b_1},$$

which conforms to our solution ansatz with

$$\alpha_1 = -\frac{c_1}{b_1}, \quad \beta_1 = \frac{f_1}{b_1}.$$

Substituting our solution expression into the general form of the equations gives

$$a_k(\alpha_{k-1} x_k + \beta_{k-1}) + b_k x_k + c_k x_{k+1} = f_k,$$

from which we get

$$x_k = -\frac{c_k}{a_k \alpha_{k-1} + b_k} x_{k+1} + \frac{f_k - a_k \beta_{k-1}}{a_k \alpha_{k-1} + b_k} = \alpha_k x_{k+1} + \beta_k. \tag{3.3}$$

Then, since $c_n = 0$,

$$x_n = \frac{f_n - a_n \beta_{n-1}}{b_n + a_n \alpha_{n-1}}.$$

Then, for $k = n-1, \ldots, 1$, we can use (3.3) to find the remaining components of the solution.

An implementation of the just described algorithm is presented in Listing 3.1. The reader can easily verify that the obtained $\boldsymbol{x}$ is indeed a solution to system (3.2). In the literature, this algorithm is sometimes called the *Thomas algorithm.* [1]

[1] Named in honor of the British physicist and applied mathematician Llewellyn Hilleth Thomas (1903–1992).

Remark 3.3 (structure)**.** The reader may wonder how useful the algorithm that we just devised may be, as it requires a very special structure on the system matrix, namely that it is tridiagonal. Later, in Chapters 24 and 28, we will see that many schemes for the solution of one-dimensional boundary and initial boundary problems entail solving (a sequence of) systems of linear equations with tridiagonal matrices.

3.1.4 Cyclically Tridiagonal Matrices

Consider a system of the form

$$\begin{cases} b_1 x_1 + c_1 x_2 + a_1 x_n = f_1, \\ a_k x_{k-1} + b_k x_k + c_k x_{k+1} = f_k, \quad k = 2, \dots, n-1, \\ c_n x_1 + a_n x_{n-1} + b_n x_n = f_n. \end{cases} \tag{3.4}$$

The coefficient matrix for this system is said to be *cyclically tridiagonal.* Equation (3.4) can also be visualized as

$$\begin{bmatrix} b_1 & c_1 & 0 & \cdots & 0 & a_1 \\ a_2 & b_2 & c_2 & 0 & & 0 \\ 0 & a_3 & b_3 & \ddots & \ddots & \vdots \\ \vdots & 0 & \ddots & \ddots & \ddots & 0 \\ 0 & & \ddots & \ddots & b_{n-1} & c_{n-1} \\ c_n & 0 & \cdots & 0 & a_n & b_n \end{bmatrix} \begin{bmatrix} x_1 \\ x_2 \\ x_3 \\ \vdots \\ x_{n-2} \\ x_{n-1} \\ x_n \end{bmatrix} = \begin{bmatrix} f_1 \\ f_2 \\ f_3 \\ \vdots \\ f_{n-2} \\ f_{n-1} \\ f_n \end{bmatrix}.$$

Notice that the coefficient matrix differs from a tridiagonal one, only in the $(1, n)$ and $(n, 1)$ entries. This hints at the fact that, to solve (3.4), we will make use of the solution of systems with tridiagonal matrices.

Indeed, to find the solution of (3.4), we will first solve the tridiagonal systems

$$\begin{cases} b_2 u_2 + c_2 u_3 = f_2, \\ a_3 u_2 + b_3 u_3 + c_3 u_4 = f_3, \\ \vdots \\ a_n u_{n-1} + b_n u_n = f_n \end{cases} \tag{3.5}$$

and

$$\begin{cases} b_2 v_2 + c_2 v_3 = -a_2, \\ a_3 v_2 + b_3 v_3 + c_3 v_4 = 0, \\ \vdots \\ a_n v_{n-1} + b_n v_n = -c_n. \end{cases} \tag{3.6}$$

Then we set

$$x_k = u_k + x_1 v_k, \quad k = 1, \dots, n, \tag{3.7}$$

with $u_1 = 0$ and $v_1 = 1$. Substituting this representation in the first equation yields

$$b_1 x_1 + c_1(u_2 + x_1 v_2) + a_1(u_n + x_1 v_n) = f_1,$$

which implies that

$$x_1 = \frac{f_1 - c_1 u_2 - a_1 u_n}{b_1 + c_1 v_2 + a_1 v_n}. \tag{3.8}$$

Let us verify that (3.7) and (3.8) are indeed the solution to (3.4). To do so, multiply the first equation of system (3.6) by x_1 and add it to the first equation of system (3.5) to obtain

$$a_2 x_1 + b_2(u_2 + x_1 v_2) + c_2(u_3 + x_1 v_3) = f_2,$$

which implies that

$$a_2 x_1 + b_2 x_2 + c_2 x_3 = f_2.$$

A similar calculation can be made for the remaining equations, up until the last one, where we get

$$c_n x_1 + a_n(u_{n-1} + x_1 v_{n-1}) + b_n(u_n + x_1 v_n) = f_n,$$

which implies that

$$c_n x_1 + a_n x_{n-1} + b_n x_n = f_n.$$

An implementation of this procedure is presented in Listing 3.2. Once again, the reader may wonder how often one encounters cyclically tridiagonal matrices. Many discretization schemes for one-dimensional boundary and initial boundary value problems with periodic boundary conditions entail the solution of (a collection of) systems of linear equations with cyclically tridiagonal matrices; see Chapters 24–28 for more details.

3.2 LU Factorization

In this section, we give a practical and theoretical description of the method of LU factorization for solving a square system of linear equations. The idea is based upon the very familiar concept of Gaussian elimination. We need the following preliminary results.

Theorem 3.4 (properties of triangular matrices). *Let the matrices* $\mathsf{T}, \mathsf{T}_k \in \mathbb{C}^{n\times n}$, *for* $k = 1, 2$, *be lower (upper) triangular. Then the following are true.*

1. *The product* $\mathsf{T}_1\mathsf{T}_2$ *is lower (upper) triangular.*
2. *If, in addition* $[\mathsf{T}_k]_{i,i} = 1$, *for* $k = 1, 2$ *and* $i = 1, \dots, n$ *— i.e.,* $\mathsf{T}_1, \mathsf{T}_2 \in \mathbb{C}^{n\times n}$ *are unit lower (unit upper) triangular — then the product* $\mathsf{T}_1\mathsf{T}_2$ *is unit lower (unit upper) triangular.*
3. *The matrix* T *is nonsingular if and only if* $[\mathsf{T}]_{i,i} \neq 0$ *for all* $i = 1, \dots, n$.
4. *If* T *is nonsingular,* $\mathsf{T}^{-1} \in \mathbb{C}^{n\times n}$ *is lower (upper) triangular.*
5. *If* T *is unit lower (unit upper) triangular, then it is invertible and* T^{-1} *is unit lower (unit upper) triangular.*
6. *If* $[\mathsf{T}]_{i,i} > 0$, *then* $[\mathsf{T}_k^{-1}]_{i,i} = \frac{1}{[\mathsf{T}]_{i,i}} > 0$.

Proof. See Problem 1.32. □

Definition 3.5 (sub-matrix). Suppose that $\mathsf{A} \in \mathbb{C}^{n \times n}$ and $S \subseteq \{1, 2, \ldots, n\}$ is nonempty with cardinality $k = \#(S) > 0$. The **sub-matrix** $\mathsf{A}(S) \in \mathbb{C}^{k \times k}$ is that matrix obtained by deleting the columns and rows of A whose indices are not in S. In symbols,

$$[\mathsf{A}(S)]_{i,j} = [\mathsf{A}]_{m_i, m_j}, \quad i, j = 1, \ldots, k,$$

where

$$S = \{m_1, \ldots, m_k\} \quad \text{and} \quad 1 \leq m_1 < m_2 < \cdots < m_k \leq n.$$

Example 3.1 Suppose that

$$\mathsf{A} = \begin{bmatrix} 1 & -7 & 12 & 4 \\ 6 & 9 & -3 & -4 \\ 1 & -6 & 8 & 9 \\ 4 & 4 & -11 & 17 \end{bmatrix}, \quad S = \{2, 4\}.$$

Then $m_1 = 2$ and $m_2 = 4$ and

$$\mathsf{A}(S) = \begin{bmatrix} 9 & -4 \\ 4 & 17 \end{bmatrix}.$$

Definition 3.6 (leading principal sub-matrix). Let $\mathsf{A} \in \mathbb{C}^{n \times n}$ and $S = \{1, 2, \ldots, k\}$ with $k \leq n$. Then we define

$$\mathsf{A}^{(k)} = \mathsf{A}(S) \in \mathbb{C}^{k \times k}$$

and we call $\mathsf{A}^{(k)}$ the **leading principal sub-matrix** of A of order k.

Typically, the LU factorization is produced via Gaussian elimination. But the proof of the following theorem obscures this fact and guarantees, independently of Gaussian elimination, the existence and uniqueness of the LU factorization.

Theorem 3.7 (LU factorization). *Let $n \geq 2$ and $\mathsf{A} \in \mathbb{C}^{n \times n}$. Suppose that all the leading principal sub-matrices of A are nonsingular, i.e., $\det(\mathsf{A}^{(k)}) \neq 0$ for all $k = 1, \ldots, n-1$. Then there exists a unit lower triangular matrix $\mathsf{L} \in \mathbb{C}^{n \times n}$ and an upper triangular matrix $\mathsf{U} \in \mathbb{C}^{n \times n}$ such that*

$$\mathsf{A} = \mathsf{LU}.$$

Proof. The proof is by induction on n, the size of the matrix.
($n = 2$) Consider

$$\mathsf{A} = \begin{bmatrix} a & b \\ c & d \end{bmatrix},$$

with $a \neq 0$, by assumption. Define

$$\mathsf{L} = \begin{bmatrix} 1 & 0 \\ m & 1 \end{bmatrix}, \qquad \mathsf{U} = \begin{bmatrix} u & v \\ 0 & \eta \end{bmatrix},$$

where

$$u = a, \quad v = b, \quad m = \frac{c}{a}, \quad \eta = d - b\frac{c}{a}.$$

Then

$$mu = c, \qquad mv + \eta = d,$$

and consequently $\mathsf{A} = \mathsf{LU}$, as is easily confirmed.
($n = m$) The induction hypothesis is as follows: suppose that the result is valid for any $\mathsf{A} \in \mathbb{C}^{m\times m}$, provided that $\mathsf{A}^{(k)}$ is nonsingular for all $k = 1, \ldots, m-1$.
($n = m+1$) Suppose that $\mathsf{A}^{(k)}$ is nonsingular for $k = 1, \ldots, m$. Set

$$\mathsf{A} = \begin{bmatrix} \mathsf{A}^{(m)} & \boldsymbol{b} \\ \boldsymbol{c}^\mathsf{T} & d \end{bmatrix} \in \mathbb{C}^{(m+1)\times(m+1)}.$$

From the induction hypothesis, there is a unit lower triangular matrix $\mathsf{L}^{(m)}$ and an upper triangular matrix $\mathsf{U}^{(m)}$ such that $\mathsf{A}^{(m)} = \mathsf{L}^{(m)}\mathsf{U}^{(m)}$, where $\mathsf{A}^{(m)}$ is the leading principal sub-matrix of A of order m. Define

$$\mathsf{L} = \begin{bmatrix} \mathsf{L}^{(m)} & \boldsymbol{0} \\ \boldsymbol{m}^\mathsf{T} & 1 \end{bmatrix}, \quad \mathsf{U} = \begin{bmatrix} \mathsf{U}^{(m)} & \boldsymbol{v} \\ \boldsymbol{0}^\mathsf{T} & \eta \end{bmatrix},$$

where $\boldsymbol{b}, \boldsymbol{c}, \boldsymbol{m}, \boldsymbol{v}, \boldsymbol{0} \in \mathbb{C}^m$. Then

$$\mathsf{LU} = \begin{bmatrix} \mathsf{L}^{(m)}\mathsf{U}^{(m)} & \mathsf{L}^{(m)}\boldsymbol{v} \\ \boldsymbol{m}^\mathsf{T}\mathsf{U}^{(m)} & \boldsymbol{m}^\mathsf{T}\boldsymbol{v} + \eta \end{bmatrix}.$$

Let us set this equal to A and determine whether or not the resulting equations are solvable. It is easy to see that $\mathsf{A} = \mathsf{LU}$ if and only if

$$\mathsf{L}^{(m)}\mathsf{U}^{(m)} = \mathsf{A}^{(m)}, \qquad \mathsf{L}^{(m)}\boldsymbol{v} = \boldsymbol{b},$$
$$\boldsymbol{m}^\mathsf{T}\mathsf{U}^{(m)} = \boldsymbol{c}^\mathsf{T}, \qquad \boldsymbol{m}^\mathsf{T}\boldsymbol{v} + \eta = d.$$

The last three equations are uniquely solvable, as we now show: since $\mathsf{L}^{(m)}$ is invertible,

$$\boldsymbol{v} = \left(\mathsf{L}^{(m)}\right)^{-1}\boldsymbol{b}.$$

The matrix $\mathsf{U}^{(m)}$ is invertible since

$$0 \neq \det(\mathsf{A}^{(m)}) = \det(\mathsf{L}^{(m)}\mathsf{U}^{(m)}) = \det(\mathsf{U}^{(m)}).$$

Hence,

$$\boldsymbol{m}^\mathsf{T} = \boldsymbol{c}^\mathsf{T}\left(\mathsf{U}^{(m)}\right)^{-1} \quad \text{or} \quad \boldsymbol{m} = \left(\mathsf{U}^{(m)}\right)^{-\mathsf{T}}\boldsymbol{c}.$$

Finally,

$$\eta = d - \boldsymbol{m}^\mathsf{T}\boldsymbol{v}.$$

The proof by induction is complete. □

Before we go any further, we ought to say why it is that an LU factorization of a matrix is useful. Suppose that we want to solve the indexed family of problems

$$\mathsf{A}\boldsymbol{x}^{(k)} = \boldsymbol{f}^{(k)}, \quad k = 1, \dots, K,$$

and that there exists a unit lower triangular matrix $\mathsf{L} \in \mathbb{C}^{n\times n}$ and an upper triangular matrix $\mathsf{U} \in \mathbb{C}^{n\times n}$ such that $\mathsf{A} = \mathsf{LU}$. To find the solutions $\boldsymbol{x}^{(k)}$, we solve the following equivalent family:

$$\mathsf{L}\boldsymbol{y}^{(k)} = \boldsymbol{f}^{(k)}, \qquad \mathsf{U}\boldsymbol{x}^{(k)} = \boldsymbol{y}^{(k)}, \quad k = 1, \dots, K.$$

The vector $\boldsymbol{y}^{(k)}$ can be obtained easily and cheaply via forward substitution. Subsequently, the vector $\boldsymbol{x}^{(k)}$ can be obtained by back substitution; see Section 3.1.2.

Now we show a practical connection between the LU factorization and what is commonly called Gaussian elimination. We use an example to motivate our discussion.

Example 3.2 Consider the following system of linear equations:

$$\begin{cases} x_1 + & x_2 + & x_3 = & 6, \\ 2x_1 + & 4x_2 + & 2x_3 = & 16, \\ -x_1 + & 5x_2 - & 4x_3 = & -3. \end{cases}$$

Of course, we can represent this as a matrix–vector equation $\mathsf{A}\boldsymbol{x} = \boldsymbol{f}$. We write this as an augmented matrix and perform Gaussian elimination to put the system into so-callled row echelon form,

$$[\mathsf{A}|\boldsymbol{f}] = \left[\begin{array}{ccc|c} \boxed{1} & 1 & 1 & 6 \\ 2 & 4 & 2 & 16 \\ -1 & 5 & -4 & -3 \end{array}\right] \xrightarrow[1R_1+R_3\to R_3]{-2R_1+R_2\to R_2} \left[\begin{array}{ccc|c} 1 & 1 & 1 & 6 \\ 0 & \boxed{2} & 0 & 4 \\ 0 & 6 & -3 & 3 \end{array}\right]$$

$$\xrightarrow{-3R_2+R_3\to R_3} \left[\begin{array}{ccc|c} 1 & 1 & 1 & 6 \\ 0 & 2 & 0 & 4 \\ 0 & 0 & -3 & -9 \end{array}\right].$$

The boxed entries indicate the so-called pivot elements. The values of the pivot elements help to determine the row multipliers in the algorithm. As long as these are nonzero, the algorithm can run to completion. Gaussian elimination uses elementary row operations to produce an equivalent upper triangular system of linear equations, $\mathsf{U}\boldsymbol{x} = \boldsymbol{b}$. By *equivalent*, we mean that the solution sets are exactly the same, even though the coefficient matrix and right-hand-side vector are changed.

Let us focus on the left-hand side of the augmented system, as this will be the important part with respect to the LU factorization. We have

$$\mathsf{L}^{(3,2)}\mathsf{L}^{(3,1)}\mathsf{L}^{(2,1)}\mathsf{A} = \mathsf{U} = \begin{bmatrix} 1 & 1 & 1 \\ 0 & 2 & 0 \\ 0 & 0 & -3 \end{bmatrix},$$

where $\mathsf{L}^{(2,1)}, \mathsf{L}^{(3,1)}, \mathsf{L}^{(3,2)}$ are elementary matrices encoding the elementary row operations performed in our Gaussian elimination process. To produce the matrix representations of these operations, recall that we need only to apply the corresponding elementary row operations on the identity matrix:

$$\begin{bmatrix} 1 & 0 & 0 \\ 0 & 1 & 0 \\ 0 & 0 & 1 \end{bmatrix} \xrightarrow{-2R_1+R_2 \to R_2} \begin{bmatrix} 1 & 0 & 0 \\ -2 & 1 & 0 \\ 0 & 0 & 1 \end{bmatrix} = \mathsf{L}^{(2,1)}.$$

Likewise,

$$\begin{bmatrix} 1 & 0 & 0 \\ 0 & 1 & 0 \\ 0 & 0 & 1 \end{bmatrix} \xrightarrow{1R_1+R_3 \to R_3} \begin{bmatrix} 1 & 0 & 0 \\ 0 & 1 & 0 \\ 1 & 0 & 1 \end{bmatrix} = \mathsf{L}^{(3,1)}$$

and

$$\begin{bmatrix} 1 & 0 & 0 \\ 0 & 1 & 0 \\ 0 & 0 & 1 \end{bmatrix} \xrightarrow{-3R_2+R_3 \to R_3} \begin{bmatrix} 1 & 0 & 0 \\ 0 & 1 & 0 \\ 0 & -3 & 1 \end{bmatrix} = \mathsf{L}^{(3,2)}.$$

Then it is easy to see that

$$\mathsf{L}^{(3,1)}\mathsf{L}^{(2,1)}\mathsf{A} = \begin{bmatrix} 1 & 1 & 1 \\ 0 & 2 & 0 \\ 0 & 6 & -3 \end{bmatrix} \quad \text{and} \quad \mathsf{L}^{(3,2)}\mathsf{L}^{(3,1)}\mathsf{L}^{(2,1)}\mathsf{A} = \begin{bmatrix} 1 & 1 & 1 \\ 0 & 2 & 0 \\ 0 & 0 & -3 \end{bmatrix}.$$

Observe that the order of application of $\mathsf{L}^{(2,1)}$ and $\mathsf{L}^{(3,1)}$ does not matter:

$$\mathsf{L}^{(3,1)}\mathsf{L}^{(2,1)} = \begin{bmatrix} 1 & 0 & 0 \\ -2 & 1 & 0 \\ 1 & 0 & 1 \end{bmatrix} = \mathsf{L}^{(2,1)}\mathsf{L}^{(3,1)}.$$

This will be shown to be true in general.

Suppose that $\mathsf{A} \in \mathbb{C}^{n \times n}$, where $n \geq 2$. If Gaussian elimination for A proceeds to completion without encountering any zero pivots, one gets

$$\mathsf{L}^{(n,n-1)} \cdots \mathsf{L}^{(n,2)} \cdots \mathsf{L}^{(3,2)}\mathsf{L}^{(n,1)} \cdots \mathsf{L}^{(2,1)}\mathsf{A} = \mathsf{U},$$

where U is square and upper triangular. Moreover, since we assume that no zero pivot entries are encountered, $[\mathsf{U}]_{i,i} \neq 0$, for $i = 1, \cdots, n-1$. However, it is possible that $[\mathsf{U}]_{n,n} = 0$. We can group the elementary operations into column operations as follows:

$$\begin{aligned} &(\text{column } 1): && \mathsf{L}_1 = \mathsf{L}^{(n,1)} \cdots \mathsf{L}^{(2,1)}, \\ &(\text{column } 2): && \mathsf{L}_2 = \mathsf{L}^{(n,2)} \cdots \mathsf{L}^{(3,2)}, \\ & && \vdots \\ &(\text{column } n-2): && \mathsf{L}_{n-2} = \mathsf{L}^{(n,n-2)}\mathsf{L}^{(n-1,n-2)}, \\ &(\text{column } n-1): && \mathsf{L}_{n-1} = \mathsf{L}^{(n,n-1)}, \end{aligned}$$

so that

$$\mathsf{L}_{n-1}\mathsf{L}_{n-2} \cdots \mathsf{L}_2\, \mathsf{L}_1\mathsf{A} = \mathsf{U}.$$

Furthermore, we can prove that the matrices defining the L_i matrices commute and can be multiplied in any order, as we will show. The matrices L_i are examples of what we will call column-i complete elementary matrices.

In any case,

$$A = L_1^{-1} \cdots L_{n-1}^{-1} U = LU.$$

It only remains to show that L is unit lower triangular. If so, we have connected Gaussian elimination to the LU factorization.

Now consider the implications of the following example.

Example 3.3 What is the inverse of an elementary matrix? Suppose that

$$L = \begin{bmatrix} 1 & 0 & 0 \\ -\frac{1}{2} & 1 & 0 \\ 0 & 0 & 1 \end{bmatrix} \quad \text{and} \quad L' = \begin{bmatrix} 1 & 0 & 0 \\ \frac{1}{2} & 1 & 0 \\ 0 & 0 & 1 \end{bmatrix}.$$

Then it is easy to see that $L'L = I_3$. Or, in other words, $L' = L^{-1}$.

Definition 3.8 (elementary matrix). A matrix $E \in \mathbb{C}^{n \times n}$ is called **elementary** if and only if $E = I + \mu_{r,s} M^{(r,s)}$ for some $\mu_{r,s} \in \mathbb{C}$ and for some $1 \le s < r \le n$, where

$$M^{(r,s)} = \boldsymbol{e}_r \boldsymbol{e}_s^{\mathsf{T}},$$

i.e.,

$$\left[M^{(r,s)} \right]_{i,j} = \delta_{i,r} \delta_{j,s}.$$

Remark 3.9 (generality). We remark that our last definition is more restrictive than might be found in some other references, specifically in our requirement that r is strictly greater than s. But this is all that is needed for our purposes.

Proposition 3.10 (properties of elementary matrices). *Suppose that*

$$E_k = I + \mu_{r_k,s} M^{(r_k,s)}, \quad k = 1, 2$$

are two elementary matrices with $r_1 \neq r_2$. Then both matrices are invertible, the inverses are elementary, and the matrices commute. Furthermore,

$$E_k^{-1} = I - \mu_{r_k,s} M^{(r_k,s)},$$

$$(E_1 E_2)^{-1} = E_2^{-1} E_1^{-1} = E_1^{-1} E_2^{-1} = I - \mu_{r_1,s} M^{(r_1,s)} - \mu_{r_2,s} M^{(r_2,s)},$$

and

$$E_1 E_2 = E_2 E_1 = I + \mu_{r_1,s} M^{(r_1,s)} + \mu_{r_2,s} M^{(r_2,s)}.$$

Proof. See Problem 3.5. □

Definition 3.11 (complete elementary matrix)**.** Suppose that $n \geq 2$. Let the index $s \in \{1, 2, \ldots, n-1\}$ be given. The matrix $\mathsf{F} \in \mathbb{C}^{n\times n}$ is called a **column-s complete elementary matrix** if and only if

$$\mathsf{F} = \mathsf{I} + \sum_{r=s+1}^{n} \mu_{r,s}\mathsf{M}^{(r,s)}$$

for some scalars $\mu_{r,s} \in \mathbb{C}$, $r = s+1, \ldots, n$. In other words, F is a unit lower triangular matrix of the form

$$\mathsf{F} = \begin{bmatrix} 1 & & & & & \\ & \ddots & & & & \\ & & 1 & & & \\ & & \mu_{s+1,s} & 1 & & \\ & & \vdots & & \ddots & \\ & & \mu_{n,s} & & & 1 \end{bmatrix}.$$

Definition 3.12 (Gaussian elimination[2])**.** Let $\mathsf{A} \in \mathbb{C}^{n\times n}$ be given with $n \geq 2$. We define the **Gaussian elimination** algorithm recursively as follows. Suppose that k stages of Gaussian elimination have been completed, where $k \in \{0, \ldots, n-1\}$, such that no zero pivots have been encountered, producing the matrix factorization

$$\mathsf{L}_k \cdots \mathsf{L}_1\mathsf{A} = \mathsf{A}^{(k)}, \quad k = 1, \ldots, n-1,$$

where $\mathsf{A}^{(0)} = \mathsf{A}$ and, for $k = 1, \ldots, n-1$,

$$\mathsf{A}^{(k)} = \begin{bmatrix} a_{1,1}^{(0)} & a_{1,2}^{(0)} & a_{1,3}^{(0)} & a_{1,4}^{(0)} & \cdots & a_{1,k+1}^{(0)} & \cdots & a_{1,n}^{(0)} \\ 0 & a_{2,2}^{(1)} & a_{2,3}^{(1)} & a_{2,4}^{(1)} & \cdots & a_{2,k+1}^{(1)} & \cdots & a_{2,n}^{(1)} \\ 0 & 0 & a_{3,3}^{(2)} & a_{3,4}^{(2)} & \cdots & a_{3,k+1}^{(2)} & \cdots & a_{3,n}^{(2)} \\ \vdots & & \ddots & \ddots & \ddots & \vdots & & \vdots \\ 0 & & & 0 & a_{k,k}^{(k-1)} & a_{k,k+1}^{(k-1)} & \cdots & a_{k,n}^{(k-1)} \\ 0 & & & 0 & 0 & a_{k+1,k+1}^{(k)} & \cdots & a_{k+1,n}^{(k)} \\ \vdots & & & \vdots & \vdots & \vdots & & \vdots \\ 0 & 0 & \cdots & 0 & 0 & a_{n,k+1}^{(k)} & \cdots & a_{n,n}^{(k)} \end{bmatrix}.$$

If $k = n-1$, we are done and we set $\mathsf{U} = \mathsf{A}^{(n-1)}$. Otherwise, if the $(k+1)$st pivot entry, $a_{k+1,k+1}^{(k)}$, is not equal to zero, the algorithm may proceed. Construct the column-$(k+1)$ complete elementary matrix

$$\mathsf{L}_{k+1} = \mathsf{I} + \sum_{r=k+2}^{n} \mu_{r,k+1}\mathsf{M}^{(r,k+1)},$$

where

$$\mu_{r,k+1} = -\frac{a_{r,k+1}^{(k)}}{a_{k+1,k+1}^{(k)}}, \quad r = k+2, \ldots, n.$$

[2] Named in honor of the German mathematician and physicist Johann Carl Friedrich Gauss (1777–1855).

Then set

$$L_{k+1}A^{(k)} = A^{(k+1)},$$

obtaining

$$A^{(k+1)} = \begin{bmatrix} a_{1,1}^{(0)} & a_{1,2}^{(0)} & a_{1,3}^{(0)} & a_{1,4}^{(0)} & \cdots & a_{1,k+1}^{(0)} & a_{1,k+2}^{(0)} & \cdots & a_{1,n}^{(0)} \\ 0 & a_{2,2}^{(1)} & a_{2,3}^{(1)} & a_{2,4}^{(1)} & \cdots & a_{2,k+1}^{(1)} & a_{2,k+2}^{(1)} & \cdots & a_{2,n}^{(1)} \\ 0 & 0 & a_{3,3}^{(2)} & a_{3,4}^{(2)} & \cdots & a_{3,k+1}^{(2)} & a_{3,k+2}^{(2)} & \cdots & a_{3,n}^{(2)} \\ \vdots & & \ddots & \ddots & \ddots & \vdots & \vdots & & \vdots \\ 0 & & & 0 & a_{k,k}^{(k-1)} & a_{k,k+1}^{(k-1)} & a_{k,k+2}^{(k-1)} & \cdots & a_{k,n}^{(k-1)} \\ 0 & & & 0 & 0 & a_{k+1,k+1}^{(k)} & a_{k+1,k+2}^{(k)} & \cdots & a_{k+1,n}^{(k)} \\ 0 & & & 0 & 0 & 0 & a_{k+2,k+2}^{(k+1)} & \cdots & a_{k+2,n}^{(k+1)} \\ \vdots & & & \vdots & \vdots & \vdots & \vdots & & \vdots \\ 0 & 0 & \cdots & 0 & 0 & 0 & a_{n,k+2}^{(k+1)} & \cdots & a_{n,n}^{(k+1)} \end{bmatrix}.$$

Based on our previous computations, the following result should be clear.

Theorem 3.13 (Gaussian elimination). *Let $A \in \mathbb{C}^{n\times n}$. If Gaussian elimination proceeds to completion without encountering any zero pivots, then there are column-k complete elementary matrices $L_k \in \mathbb{C}^{n\times n}$, for $k = 1, \ldots, n-1$, such that*

$$L_{n-1} \cdots L_2 L_1 A = U,$$

where $U \in \mathbb{C}^{n\times n}$ is upper triangular and

$$[U]_{i,i} \neq 0, \quad i = 1, \ldots, n-1,$$

since no zero pivots are encountered. Furthermore,

$$A = L_1^{-1} \cdots L_{n-1}^{-1} U = LU,$$

where L is unit lower triangular. Writing

$$L_k = I + \sum_{r=k+1}^{n} \mu_{r,s} M^{(r,k)},$$

it follows that

$$L = I - \sum_{k=1}^{n-1} \sum_{r=k+1}^{n} \mu_{r,k} M^{(r,k)}.$$

In other words,

$$L = \begin{bmatrix} 1 & & & & \\ -\mu_{2,1} & 1 & & & \\ -\mu_{3,1} & -\mu_{3,2} & 1 & & \\ \vdots & \vdots & \ddots & \ddots & \\ -\mu_{n,1} & -\mu_{n,2} & \cdots & -\mu_{n,n-1} & 1 \end{bmatrix}.$$

Proof. See Problem 3.7. □

Theorem 3.14 (uniqueness). *Suppose that $n \geq 2$ and $A \in \mathbb{C}^{n\times n}$ is invertible. Suppose that there is a unit lower triangular matrix $L \in \mathbb{C}^{n\times n}$ and an upper triangular matrix $U \in \mathbb{C}^{n\times n}$ such that $A = LU$. Then this LU factorization is unique.*

Proof. Suppose that there are two factorizations with the desired properties:

$$L_1U_1 = A = L_2U_2.$$

Since A is invertible, U_1 and U_2 must be invertible, i.e., there are no zeros on their diagonals. Furthermore,

$$L_2^{-1}L_1 = U_2U_1^{-1} = D,$$

where D is by necessity diagonal. Therefore,

$$L_1 = L_2D.$$

But it must be that $D = I_n$, since the diagonal elements of L_1 and L_2 are all ones. □

Listing 3.3 provides a more streamlined, computable version of the LU factorization algorithm. The algorithm proceeds to completion provided that no zero pivots are encountered. This listing can also be used to estimate the complexity of the LU factorization algorithm.

Proposition 3.15 (complexity of LU). *Let $A \in \mathbb{C}^{n\times n}$. Then the LU factorization algorithm requires, to leading order, $\frac{2}{3}n^3$ operations.*

Proof. We only care about the leading order of operations, which, from Listing 3.3, can easily be seen to be roughly

$$\begin{aligned}\sum_{k=1}^{n-1}\sum_{j=k}^{n}\sum_{t=k}^{n} 2 &= 2\sum_{k=1}^{n-1}(n-k)\sum_{j=k+1}^{n} 1\\ &\approx 2\sum_{k=1}^{n-1}(n-k)^2\\ &= \frac{1}{3}(n-1)n(2n-1)\\ &\approx \frac{2}{3}n^3.\end{aligned}$$

□

3.3 Gaussian Elimination with Column Pivoting

In the last section, we did not consider what one should do if a zero pivot is encountered. Let us examine a simple situation where zero pivots appear.

Example 3.4 Suppose that $\mathsf{A} \in \mathbb{C}^{3\times 3}$ is given by

$$\mathsf{A} = \begin{bmatrix} 0 & 1 & 5 \\ -2 & 1 & 1 \\ 4 & -2 & 6 \end{bmatrix}.$$

We want to use Gaussian elimination to obtain an LU factorization of A. However, we notice from the beginning that there is a zero in the first pivot location. But a simple row interchange operation will fix this situation. In the following algorithm, let us agree to interchange rows, so that the element with largest modulus in the column at or below the pivot position moves into the pivot position. This is called Gaussian elimination with maximal column pivoting:

$$\mathsf{A} = \begin{bmatrix} \boxed{0} & 1 & 5 \\ -2 & 1 & 1 \\ 4 & -2 & 6 \end{bmatrix} \xrightarrow{R_1 \leftrightarrows R_3} \begin{bmatrix} \boxed{4} & -2 & 6 \\ -2 & 1 & 1 \\ 0 & 1 & 5 \end{bmatrix}$$

$$\xrightarrow[0R_1+R_3 \to R_3]{\frac{1}{2}R_1+R_2 \to R_2} \begin{bmatrix} 4 & -2 & 6 \\ 0 & \boxed{0} & 4 \\ 0 & 1 & 5 \end{bmatrix}$$

$$\xrightarrow{R_2 \leftrightarrows R_3} \begin{bmatrix} 4 & -2 & 6 \\ 0 & \boxed{1} & 5 \\ 0 & 0 & 4 \end{bmatrix}$$

$$\xrightarrow{0R_2+R_3 \to R_3} \begin{bmatrix} 4 & -2 & 6 \\ 0 & 1 & 5 \\ 0 & 0 & 4 \end{bmatrix}.$$

Our procedure may be expressed as

$$\mathsf{L}_2\mathsf{P}_2\mathsf{L}_1\mathsf{P}_1\mathsf{A} = \mathsf{U},$$

where P_1 and P_2 are simple permutation matrices and L_1 and L_2 are column-1 and column-2 complete elementary matrices, respectively.

Definition 3.16 (permutation). A matrix $\mathsf{P} \in \mathbb{C}^{n\times n}$ is called a **simple permutation** matrix if and only if it is obtained from the $n \times n$ identity matrix I by interchanging exactly two rows of I. P is called a **regular permutation** (or just a **permutation**) matrix if and only if P is the product of simple permutation matrices.

Proposition 3.17 (action of permutations). *Let $n \geq 2$. Suppose that $\mathsf{A} \in \mathbb{C}^{n\times n}$ is any matrix and $\mathsf{P} \in \mathbb{C}^{n\times n}$ is a simple permutation matrix obtained by interchanging rows r and s of the identity matrix with $1 \leq r < s \leq n$. Then* PA *is identical to* A*, except with rows r and s interchanged. Furthermore,* AP *is identical to* A*, except with columns r and s interchanged.*

Proof. See Problem 3.8. □

Lemma 3.18 (properties of permutations). *Suppose that* $\mathsf{P}, \mathsf{Q} \in \mathbb{C}^{n\times n}$, *with* $n \geq 2$, *are permutation matrices. Then*

1. *The product* PQ *is a permutation matrix.*
2. $\det(\mathsf{P}) = \pm 1$ *according to whether* P *is the product of an even* ($\det(\mathsf{P}) = 1$) *or an odd* ($\det(\mathsf{P}) = -1$) *number of simple permutation matrices.*
3. *The inverse of a simple permutation matrix is itself. Any regular permutation matrix* P *is invertible, and, if*

$$\mathsf{P} = \mathsf{P}_1\mathsf{P}_2\cdots\mathsf{P}_k,$$

where P_i *is a simple permutation matrix, for* $1 \leq i \leq k$, *then*

$$\mathsf{P}^{-1} = \mathsf{P}_k\cdots\mathsf{P}_2\mathsf{P}_1 = \mathsf{P}^{\intercal}.$$

Proof. See Problem 3.9. □

Example 3.5 Let us continue with our 3×3 example, but in general terms. We have

$$\mathsf{L}_2\mathsf{P}_2\mathsf{L}_1\mathsf{P}_1\mathsf{A} = \mathsf{U},$$

where P_j, $j = 1, 2$ are simple permutation matrices or the identity matrix (in the case that no row interchange took place) and L_j are column-j complete elementary matrices, $j = 1, 2$. Now observe that

$$\mathsf{L}_2\mathsf{P}_2\mathsf{L}_1\mathsf{P}_2\mathsf{P}_2\mathsf{P}_1\mathsf{A} = \mathsf{U}.$$

Therefore,

$$\hat{\mathsf{L}}_2\hat{\mathsf{L}}_1\mathsf{P}\mathsf{A} = \mathsf{U},$$

where

$$\hat{\mathsf{L}}_2 = \mathsf{L}_2, \quad \hat{\mathsf{L}}_1 = \mathsf{P}_2\mathsf{L}_1\mathsf{P}_2, \quad \mathsf{P} = \mathsf{P}_2\mathsf{P}_1.$$

Example 3.6 Suppose that Gaussian elimination with maximal column pivoting is applied to $\mathsf{A} \in \mathbb{C}^{4\times 4}$. Then it should be clear that one obtains

$$\mathsf{L}_3\mathsf{P}_3\mathsf{L}_2\mathsf{P}_2\mathsf{L}_1\mathsf{P}_1\mathsf{A} = \mathsf{U},$$

which can be rewritten as

$$\hat{\mathsf{L}}_3\hat{\mathsf{L}}_2\hat{\mathsf{L}}_1\mathsf{P}\mathsf{A} = \mathsf{U},$$

where

$$\hat{\mathsf{L}}_3 = \mathsf{L}_3, \quad \hat{\mathsf{L}}_2 = \mathsf{P}_3\mathsf{L}_2\mathsf{P}_3, \quad \hat{\mathsf{L}}_1 = \mathsf{P}_3\mathsf{P}_2\mathsf{L}_1\mathsf{P}_2\mathsf{P}_3, \quad \mathsf{P} = \mathsf{P}_3\mathsf{P}_2\mathsf{P}_1.$$

It turns out — and it is probably not so hard to see — that the $\hat{\mathsf{L}}_k$ matrices constructed above are still column-k complete elementary matrices.

Proposition 3.19 (permutations and elementary matrices). *Suppose that* $\mathsf{L}_k \in \mathbb{C}^{n\times n}$ *is a column-k complete elementary matrix,*

$$\mathsf{L}_k = \mathsf{I}_n + \sum_{r=k+1}^{n} \mu_{r,k}\mathsf{M}^{(r,k)},$$

for some constants $\mu_{r,k} \in \mathbb{C}$, *for* $k = k+1, \ldots, n$. *Assume that* $\mathsf{Q} \in \mathbb{C}^{n\times n}$ *is a simple permutation matrix encoding the interchange of rows* r' *and* s', *where* $k < r' < s' \le n$. *Then the matrix* $\mathsf{Q}\mathsf{L}_k\mathsf{Q}$ *is a column-k complete matrix. In particular,* $\mathsf{Q}\mathsf{L}_k\mathsf{Q}$ *is identical to* L_k, *except that entries* $\mu_{r',k}$ *and* $\mu_{s',k}$ *are interchanged.*

Proof. It follows that

$$\mathsf{Q}\mathsf{L}_k\mathsf{Q} = \mathsf{Q}\mathsf{I}_n\mathsf{Q} + \sum_{r=k+1}^{n} \mu_{r,k}\mathsf{Q}\boldsymbol{e}_r\boldsymbol{e}_k^{\mathsf{T}}\mathsf{Q}.$$

But observe that $\boldsymbol{e}_k^{\mathsf{T}}\mathsf{Q} = \boldsymbol{e}_k^{\mathsf{T}}$ and

$$\mathsf{Q}\boldsymbol{e}_r = \begin{cases} \boldsymbol{e}_{s'}, & r = r', \\ \boldsymbol{e}_{r'}, & r = s', \\ \boldsymbol{e}_r, & r \in \{1, \ldots, n\}\backslash\{r', s'\}. \end{cases}$$

Therefore,

$$\mathsf{Q}\mathsf{L}_k\mathsf{Q} = \mathsf{I}_n + \sum_{\substack{r=k+1 \\ r\ne r',s'}}^{n} \mu_{r,k}\boldsymbol{e}_r\boldsymbol{e}_k^{\mathsf{T}} + \mu_{s',k}\boldsymbol{e}_{r'}\boldsymbol{e}_k^{\mathsf{T}} + \mu_{r',k}\boldsymbol{e}_{s'}\boldsymbol{e}_k^{\mathsf{T}}.$$

In other words, $\mathsf{Q}\mathsf{L}_k\mathsf{Q}$ is a column-k complete elementary matrix that is identical to L_k, except that the positions of $\mu_{r',k}$ and $\mu_{s',k}$ are swapped. □

Definition 3.20 (Gaussian elimination with maximal column pivoting). Let $\mathsf{A} \in \mathbb{C}^{n\times n}$ be given with $n \ge 2$. We define the **Gaussian elimination with maximal column pivoting** algorithm recursively as follows. Suppose that k stages of Gaussian elimination with maximal column pivoting have been completed, where $k = 0, \ldots, n-1$, producing the matrix decomposition

$$\mathsf{L}_k\mathsf{P}_k \cdots \mathsf{L}_1\mathsf{P}_1\mathsf{A} = \mathsf{A}^{(k)}, \quad k = 1, \ldots, n-1,$$

where $\mathsf{A}^{(0)} = \mathsf{A}$ and, for $k = 1, \ldots, n-1$,

$$\mathsf{A}^{(k)} = \begin{bmatrix} a_{1,1}^{(0,1)} & a_{1,2}^{(0,1)} & a_{1,3}^{(0,1)} & a_{1,4}^{(0,1)} & \cdots & a_{1,k+1}^{(0,1)} & \cdots & a_{1,n}^{(0,1)} \\ 0 & a_{2,2}^{(1,1)} & a_{2,3}^{(1,1)} & a_{2,4}^{(1,1)} & \cdots & a_{2,k+1}^{(1,1)} & \cdots & a_{2,n}^{(1,1)} \\ 0 & 0 & a_{3,3}^{(2,1)} & a_{3,4}^{(2,1)} & \cdots & a_{3,k+1}^{(2,1)} & \cdots & a_{3,n}^{(2,1)} \\ \vdots & & \ddots & \ddots & \ddots & \vdots & & \vdots \\ 0 & & & 0 & a_{k,k}^{(k-1,1)} & a_{k,k+1}^{(k-1,1)} & \cdots & a_{k,n}^{(k-1,1)} \\ 0 & & & 0 & 0 & a_{k+1,k+1}^{(k)} & \cdots & a_{k+1,n}^{(k)} \\ \vdots & & & \vdots & \vdots & \vdots & & \vdots \\ 0 & 0 & \cdots & 0 & 0 & a_{n,k+1}^{(k)} & \cdots & a_{n,n}^{(k)} \end{bmatrix},$$

If $k = n-1$, we are done, and we set $\mathsf{U} = \mathsf{A}^{(n-1)}$. Otherwise, use a simple permutation matrix to interchange rows $k+1$ and r, with $r \geq k+1$ and

$$\left|a_{r,k+1}^{(k)}\right| \geq \left|a_{j,k+1}^{(k)}\right|, \quad j = k+1, \ldots, n,$$

with the understanding that the simple permutation is the identity matrix if $r = k+1$, obtaining

$$\mathsf{P}_{k+1}\mathsf{A}^{(k)} = \mathsf{A}^{(k,1)},$$

where

$$\mathsf{A}^{(k,1)} = \begin{bmatrix} a_{1,1}^{(0,1)} & a_{1,2}^{(0,1)} & a_{1,3}^{(0,1)} & a_{1,4}^{(0,1)} & \cdots & a_{1,k+1}^{(0,1)} & \cdots & a_{1,n}^{(0,1)} \\ 0 & a_{2,2}^{(1,1)} & a_{2,3}^{(1,1)} & a_{2,4}^{(1,1)} & \cdots & a_{2,k+1}^{(1,1)} & \cdots & a_{2,n}^{(1,1)} \\ 0 & 0 & a_{3,3}^{(2,1)} & a_{3,4}^{(2,1)} & \cdots & a_{3,k+1}^{(2,1)} & \cdots & a_{3,n}^{(2,1)} \\ \vdots & & \ddots & \ddots & \ddots & \vdots & & \vdots \\ 0 & & & 0 & a_{k,k}^{(k-1,1)} & a_{k,k+1}^{(k-1,1)} & \cdots & a_{k,n}^{(k-1,1)} \\ 0 & & & 0 & 0 & a_{k+1,k+1}^{(k,1)} & \cdots & a_{k+1,n}^{(k,1)} \\ \vdots & & & \vdots & \vdots & \vdots & & \vdots \\ 0 & 0 & \cdots & 0 & 0 & a_{n,k+1}^{(k,1)} & \cdots & a_{n,n}^{(k,1)} \end{bmatrix}.$$

If the updated $(k+1)$st pivot entry, $a_{k+1,k+1}^{(k,1)}$, is equal to zero, we set $\mathsf{L}_{k+1} = \mathsf{I}_n$. Otherwise, construct the column-$(k+1)$ complete elementary matrix

$$\mathsf{L}_{k+1} = \mathsf{I}_n + \sum_{r=k+2}^{n} \mu_{r,k+1}\mathsf{M}^{(r,k+1)},$$

where

$$\mu_{r,k+1} = -\frac{a_{r,k+1}^{(k,1)}}{a_{k+1,k+1}^{(k,1)}}.$$

Then set

$$\mathsf{L}_{k+1}\mathsf{A}^{(k,1)} = \mathsf{A}^{(k+1)},$$

obtaining

$$\mathsf{A}^{(k+1)} = \begin{bmatrix} a_{1,1}^{(0,1)} & a_{1,2}^{(0,1)} & a_{1,3}^{(0,1)} & a_{1,4}^{(0,1)} & \cdots & a_{1,k+1}^{(0,1)} & a_{1,k+2}^{(0,1)} & \cdots & a_{1,n}^{(0,1)} \\ 0 & a_{2,2}^{(1,1)} & a_{2,3}^{(1,1)} & a_{2,4}^{(1,1)} & \cdots & a_{2,k+1}^{(1,1)} & a_{2,k+2}^{(1,1)} & \cdots & a_{2,n}^{(1,1)} \\ 0 & 0 & a_{3,3}^{(2,1)} & a_{3,4}^{(2,1)} & \cdots & a_{3,k+1}^{(2,1)} & a_{3,k+2}^{(2,1)} & \cdots & a_{3,n}^{(2,1)} \\ \vdots & & \ddots & \ddots & \ddots & \vdots & \vdots & & \vdots \\ 0 & & & 0 & a_{k,k}^{(k-1,1)} & a_{k,k+1}^{(k-1,1)} & a_{k,k+2}^{(k-1,1)} & \cdots & a_{k,n}^{(k-1,1)} \\ 0 & & & 0 & 0 & a_{k+1,k+1}^{(k,1)} & a_{k+1,k+2}^{(k,1)} & \cdots & a_{k+1,n}^{(k,1)} \\ 0 & & & 0 & 0 & 0 & a_{k+2,k+2}^{(k+1)} & \cdots & a_{k+2,n}^{(k+1)} \\ \vdots & & & \vdots & \vdots & \vdots & \vdots & & \vdots \\ 0 & 0 & \cdots & 0 & 0 & 0 & a_{n,k+2}^{(k+1)} & \cdots & a_{n,n}^{(k+1)} \end{bmatrix}.$$

Theorem 3.21 (LU factorization with pivoting). *Suppose that $n \geq 2$ and $\mathsf{A} \in \mathbb{C}^{n\times n}$. The Gaussian elimination with maximal column pivoting algorithm always proceeds to completion to yield an upper triangular matrix U. In particular, there are matrices $\mathsf{L}_j, \mathsf{P}_j \in \mathbb{C}^{n\times n}$, $j = 1, \ldots, n-1$ such that*

$$\mathsf{L}_{n-1}\mathsf{P}_{n-1}\cdots\mathsf{L}_2\mathsf{P}_2\mathsf{L}_1\mathsf{P}_1\mathsf{A} = \mathsf{U},$$

where L_j is a column-j complete elementary matrix and P_j is either the $n\times n$ identity or a simple permutation matrix. Furthermore, there are column-j complete elementary matrices $\hat{\mathsf{L}}_j$, for $j = 1, \ldots, n-1$, and a permutation matrix P such that

$$\hat{\mathsf{L}}_{n-1}\cdots\hat{\mathsf{L}}_1\mathsf{P}\mathsf{A} = \mathsf{U},$$

where

$$\mathsf{P} = \mathsf{P}_{n-1}\cdots\mathsf{P}_1,$$

$$\hat{\mathsf{L}}_j = \mathsf{P}_{n-1}\cdots\mathsf{P}_{j+1}\mathsf{L}_j\mathsf{P}_{j+1}\cdots\mathsf{P}_{n-1}, \quad j = 1, \ldots, n-2,$$

and

$$\hat{\mathsf{L}}_{n-1} = \mathsf{L}_{n-1}.$$

Finally, there is a unit lower triangular matrix L such that

$$\mathsf{P}\mathsf{A} = \mathsf{L}\mathsf{U}.$$

Proof. This is nothing but an exercise in applying our previous results and definitions; see Problem 3.10. □

Listing 3.4 computes the LU factorization with pivoting. As before, and for a different perspective, it is also possible to prove the existence of the factorization $\mathsf{A} = \mathsf{P}^\intercal\mathsf{L}\mathsf{U}$ independent of the consideration of the Gaussian elimination algorithm. Here, we follow the presentation in [89].

Theorem 3.22 (LU factorization with pivoting). *Suppose that $n \geq 2$. Let $\mathsf{A} \in \mathbb{C}^{n\times n}$ be given. There exists a permutation matrix $\mathsf{P} \in \mathbb{R}^{n\times n}$, a unit lower triangular matrix $\mathsf{L} \in \mathbb{C}^{n\times n}$, and an upper triangular matrix $\mathsf{U} \in \mathbb{C}^{n\times n}$ such that*

$$\mathsf{P}\mathsf{A} = \mathsf{L}\mathsf{U}.$$

Proof. We proceed by induction.

($n = 2$) Consider

$$\mathsf{A} = \begin{bmatrix} a & b \\ c & d \end{bmatrix}.$$

If $a \neq 0$, then set $\mathsf{P} = \mathsf{I}_2$,

$$\mathsf{L} = \begin{bmatrix} 1 & 0 \\ \frac{c}{a} & 1 \end{bmatrix}, \qquad \mathsf{U} = \begin{bmatrix} a & b \\ 0 & d - \frac{c}{a}b \end{bmatrix}.$$

Clearly, $\mathsf{PA} = \mathsf{LU}$. On the other hand, if $a = 0$, but $c \neq 0$, set

$$\mathsf{P} = \begin{bmatrix} 0 & 1 \\ 1 & 0 \end{bmatrix}, \qquad \mathsf{L} = \mathsf{I}_2, \qquad \mathsf{U} = \begin{bmatrix} c & d \\ 0 & b \end{bmatrix},$$

and again observe that $\mathsf{PA} = \mathsf{LU}$. Finally, if $a = 0 = c$, there is not much to do: set

$$\mathsf{P} = \mathsf{I}_2 = \mathsf{L}, \qquad \mathsf{U} = \begin{bmatrix} 0 & b \\ 0 & d \end{bmatrix},$$

and conclude the result.

($n = m$) The induction hypothesis is to suppose that the result is true for every matrix $\mathsf{A} \in \mathbb{C}^{k \times k}$ for all $k = 2, \ldots, m$.

($n = m+1$) Suppose that $\mathsf{A} \in \mathbb{C}^{(m+1)\times(m+1)}$ is arbitrary. Suppose that, in the first column of A, the largest element by modulus is contained in row r. Set P_1 as the simple permutation matrix interchanging rows 1 and r. Then the result is, for some $\boldsymbol{p}, \boldsymbol{w} \in \mathbb{C}^m$ and $\mathsf{B} \in \mathbb{C}^{m\times m}$,

$$\mathsf{P}_1\mathsf{A} = \begin{bmatrix} \alpha & \boldsymbol{w}^\intercal \\ \boldsymbol{p} & \mathsf{B} \end{bmatrix},$$

where

$$|\alpha| \geq |[\boldsymbol{p}]_i|, \quad i = 1, \ldots, m.$$

Observe that it is possible that $\alpha = 0$, in which case $\boldsymbol{p} = \boldsymbol{0}$. Next, we seek a solution, if possible, to the following intermediate problem:

$$\mathsf{P}_1\mathsf{A} = \begin{bmatrix} \alpha & \boldsymbol{w}^\intercal \\ \boldsymbol{p} & \mathsf{B} \end{bmatrix} = \begin{bmatrix} 1 & \boldsymbol{0}^\intercal \\ \boldsymbol{m} & \mathsf{I}_m \end{bmatrix} \begin{bmatrix} \alpha & \boldsymbol{v}^\intercal \\ \boldsymbol{0} & \mathsf{C} \end{bmatrix},$$

where

$$\mathsf{C} \in \mathbb{C}^{m\times m}, \qquad \boldsymbol{m}, \boldsymbol{v}, \boldsymbol{0} \in \mathbb{C}^m.$$

The block matrix equation is satisfied if and only if

$$\boldsymbol{v} = \boldsymbol{w}, \quad \alpha\boldsymbol{m} = \boldsymbol{p}, \quad \mathsf{C} = \mathsf{B} - \boldsymbol{m}\boldsymbol{v}^\intercal.$$

Recall that, if $\alpha = 0$, $\boldsymbol{p} = \boldsymbol{0}$. In this case,

$$\boldsymbol{m} = \boldsymbol{0}, \quad \boldsymbol{v} = \boldsymbol{w}, \quad \mathsf{C} = \mathsf{B}$$

is one possible solution. If $\alpha \neq 0$, then

$$\boldsymbol{m} = \frac{1}{\alpha}\boldsymbol{p}, \quad \boldsymbol{v} = \boldsymbol{w}, \quad \mathsf{C} = \mathsf{B} - \frac{1}{\alpha}\boldsymbol{p}\boldsymbol{w}^\intercal.$$

Observe that

$$|[\boldsymbol{m}]_i| \leq 1, \quad i = 1, \ldots, m.$$

Now, from the induction hypothesis, there is a permutation matrix $\tilde{\mathsf{P}} \in \mathbb{C}^{m\times m}$, a unit upper triangular matrix $\tilde{\mathsf{L}} \in \mathbb{C}^{m\times m}$, and an upper triangular matrix $\tilde{\mathsf{U}} \in \mathbb{C}^{m\times m}$ such that

$$\tilde{\mathsf{P}}\mathsf{C} = \tilde{\mathsf{L}}\tilde{\mathsf{U}}.$$

Finally, the reader will observe that

$$\begin{bmatrix} 1 & \mathbf{0}^\intercal \\ \mathbf{0} & \tilde{\mathsf{P}}^\intercal \end{bmatrix} \begin{bmatrix} 1 & \mathbf{0}^\intercal \\ \tilde{\mathsf{P}}\boldsymbol{m} & \tilde{\mathsf{L}} \end{bmatrix} \begin{bmatrix} \alpha & \boldsymbol{v}^\intercal \\ \mathbf{0} & \tilde{\mathsf{U}} \end{bmatrix} = \mathsf{P}_1\mathsf{A},$$

using the fact that $\tilde{\mathsf{P}}^\intercal = \tilde{\mathsf{P}}^{-1}$. Therefore,

$$\mathsf{LU} = \mathsf{PA},$$

where

$$\mathsf{P} = \begin{bmatrix} 1 & \mathbf{0}^\intercal \\ \mathbf{0} & \tilde{\mathsf{P}} \end{bmatrix} \mathsf{P}_1, \qquad \mathsf{L} = \begin{bmatrix} 1 & \mathbf{0}^\intercal \\ \tilde{\mathsf{P}}\boldsymbol{m} & \tilde{\mathsf{L}} \end{bmatrix}, \qquad \mathsf{U} = \begin{bmatrix} \alpha & \boldsymbol{v}^\intercal \\ \mathbf{0} & \tilde{\mathsf{U}} \end{bmatrix},$$

and the matrices are of the required types. □

The LU factorization can be used to efficiently compute the solution to (3.1). Let us suppose that A is nonsingular. The algorithm is as follows: if PA = LU, then we have the equivalent system

$$\mathsf{L}\boldsymbol{y} = \mathsf{P}\boldsymbol{f} = \boldsymbol{q}, \qquad \mathsf{U}\boldsymbol{x} = \boldsymbol{y}. \tag{3.9}$$

We use the forward substitution algorithm to solve $\mathsf{L}\boldsymbol{y} = \boldsymbol{q}$ for $\boldsymbol{y}$ and, then, we use the backward substitution algorithm to solve $\mathsf{U}\boldsymbol{x} = \boldsymbol{y}$ for $\boldsymbol{x}$. These algorithms were covered in Section 3.1.2.

3.4 Implementation of the LU Factorization

We conclude the discussion of Gaussian elimination with some practical considerations. First of all, one does not need to store the permutation matrix P. Instead one only needs to remember which rows were swapped. This means that we only need to store a vector $\boldsymbol{s} \in \mathbb{N}^n$ of indices which is used to indirectly reference the entries of the matrix. Second, since the matrices L and U are lower and upper triangular, respectively, and the diagonal entries of L are always equal to one, both of these matrices can be conveniently stored in the already allocated array for A. This is a convenient and efficient way of storing the LU factorization of the system matrix A. Listing 3.5 provides an implementation of this idea.

Once this factorization has taken place, we can use it, together with the swap vector $\boldsymbol{s}$, to perform a back substitution and find the solution to system (3.1), as described by (3.9). Listing 3.6 provides the implementation details for this. It is important to note that the LU factorization, the most expensive part of this procedure, needs to only be called once. After that, solving several systems of the form (3.1) where the system matrix A does not change, but the right-hand-side $\boldsymbol{f}$ does, is rather efficient, as it only requires n^2 operations. See Problem 3.2. This is a situation that is very common in practice. See, for instance, Chapters 24–28.

3.5 Special Matrices

While Theorems 3.21 and 3.22 provide, in general, the existence and uniqueness of an LU factorization with pivoting, it is of interest to study this process for some special kinds of matrix that often appear in applications.

3.5.1 Diagonally Dominant Matrices

Definition 3.23 (diagonal dominance). A matrix $\mathsf{A} = [a_{i,j}] \in \mathbb{C}^{n \times n}$ is called **diagonally dominant** if and only if

$$|a_{i,i}| \geq \sum_{\substack{k=1 \\ k \neq i}}^{n} |a_{i,k}|, \quad \forall i = 1, \dots, n.$$

A is **strictly diagonally dominant** (SDD) if and only if

$$|a_{i,i}| > \sum_{\substack{k=1 \\ k \neq i}}^{n} |a_{i,k}|, \quad \forall i = 1, \dots, n.$$

A is called **strictly diagonally dominant of dominance** δ if and only if there is a $\delta > 0$ such that

$$|a_{i,i}| \geq \delta + \sum_{\substack{k=1 \\ k \neq i}}^{n} |a_{i,k}|, \quad \forall i = 1, \dots, n.$$

The reader should verify that, essentially, the last two definitions are equivalent. It turns out that SDD matrices are always invertible, and we have an easy bound on the norm of its inverse.

Theorem 3.24 (properties of an SDD matrix). *If* $\mathsf{A} \in \mathbb{C}^{n \times n}$ *is SDD, then* A *is invertible. If* A *is SDD of dominance* $\delta > 0$*, then*

$$\|\mathsf{A}^{-1}\|_{\infty} < \frac{1}{\delta}.$$

Proof. Suppose that A is singular. If that is the case there is an $\boldsymbol{x} = [x_i] \in \mathbb{C}^n_\star$ such that $\mathsf{A}\boldsymbol{x} = \boldsymbol{0}$. Suppose that $k \in \{1, \dots, n\}$ is an index for which $|x_k| = \|\boldsymbol{x}\|_\infty$. Since $\mathsf{A}\boldsymbol{x} = \boldsymbol{0}$, we must have that, for each $i = 1, \dots, n$,

$$\sum_{j=1}^{n} a_{i,j} x_j = 0.$$

In particular, $\sum_{j=1}^{n} a_{k,j} x_j = 0$. Then, from the triangle inequality,

$$|a_{k,k}| \cdot \|\boldsymbol{x}\|_\infty = |a_{k,k} x_k| = \left| - \sum_{\substack{j=1 \\ j \neq k}}^{n} a_{k,j} x_j \right| \leq \sum_{\substack{j=1 \\ j \neq k}}^{n} |a_{k,j}| \cdot |x_j| \leq \|\boldsymbol{x}\|_\infty \sum_{\substack{j=1 \\ j \neq k}}^{n} |a_{k,j}|.$$

Since $\|\boldsymbol{x}\|_\infty > 0$, we have

$$|a_{k,k}| \le \sum_{\substack{j=1\\j\ne k}}^{n} |a_{k,j}|.$$

This proves that A is not SDD, a contradiction.

Next, suppose that A has dominance $\delta > 0$. Let $\boldsymbol{x}$ be arbitrary. Set $\mathsf{A}\boldsymbol{x} = \boldsymbol{f}$. Assume that $\|\boldsymbol{x}\|_\infty = |x_k|$, for some $k = 1, \ldots, n$. Then

$$a_{k,1}x_1 + \cdots + a_{k,k}x_k + \cdots + a_{k,n}x_n = f_k$$

and, using the reverse triangle inequality,

$$|f_k| \ge |a_{k,k}||x_k| - \sum_{\substack{j=1\\j\ne k}}^{n} |a_{j,k}||x_j| \ge \left(|a_{k,k}| - \sum_{\substack{j=1\\j\ne k}}^{n} |a_{j,k}| \right) |x_k| \ge \delta\|\boldsymbol{x}\|_\infty.$$

This shows that

$$\frac{\|\mathsf{A}\boldsymbol{x}\|_\infty}{\|\boldsymbol{x}\|_\infty} \ge \delta, \quad \forall \boldsymbol{x} \in \mathbb{C}^n,$$

which is equivalent to

$$\frac{1}{\delta} \ge \frac{\|\mathsf{A}^{-1}\boldsymbol{w}\|_\infty}{\|\boldsymbol{w}\|_\infty}, \quad \forall \boldsymbol{w} \in \mathbb{C}^n_\star.$$

This, in turn, implies that

$$\frac{1}{\delta} \ge \sup_{\boldsymbol{w}\in\mathbb{C}^n_\star} \frac{\|\mathsf{A}^{-1}\boldsymbol{w}\|_\infty}{\|\boldsymbol{w}\|_\infty} = \|\mathsf{A}^{-1}\|_\infty,$$

as we intended to show. □

Theorem 3.25 (Gaussian elimination and SDD). *Let* $\mathsf{A} = [a_{i,j}] \in \mathbb{C}^{n\times n}$ *be SDD and assume that it is represented as*

$$\mathsf{A} = \begin{bmatrix} \alpha & \boldsymbol{v}^\intercal \\ \boldsymbol{p} & \hat{\mathsf{A}} \end{bmatrix},$$

where $\alpha \in \mathbb{C}$, $\boldsymbol{p}, \boldsymbol{v} \in \mathbb{C}^{n-1}$, *and* $\hat{\mathsf{A}} = [\hat{a}_{i,j}] \in \mathbb{C}^{(n-1)\times(n-1)}$. *After one step of Gaussian elimination (without pivoting),* A *will be reduced to the matrix*

$$\begin{bmatrix} \alpha & \boldsymbol{v}^\intercal \\ \boldsymbol{0} & \mathsf{B} \end{bmatrix},$$

where $\mathsf{B} = [b_{i,j}] \in \mathbb{C}^{(n-1)\times(n-1)}$ *is SDD.*

Proof. Let us construct a matrix $\mathsf{L} \in \mathbb{C}^{n\times n}$ such that

$$\mathsf{LA} = \begin{bmatrix} \alpha & \boldsymbol{v}^\intercal \\ \boldsymbol{0} & \mathsf{B} \end{bmatrix},$$

if possible. Consider

$$\mathsf{L} = \begin{bmatrix} 1 & \mathbf{0}^\intercal \\ \boldsymbol{m} & \mathsf{I}_{n-1} \end{bmatrix}.$$

Then

$$\mathsf{LA} = \begin{bmatrix} 1 & \mathbf{0}^\intercal \\ \boldsymbol{m} & \mathsf{I}_{n-1} \end{bmatrix} \begin{bmatrix} \alpha & \boldsymbol{v}^\intercal \\ \boldsymbol{p} & \hat{\mathsf{A}} \end{bmatrix} = \begin{bmatrix} \alpha & \boldsymbol{v}^\intercal \\ \alpha\boldsymbol{m} + \boldsymbol{p} & \boldsymbol{m}\boldsymbol{v}^\intercal + \hat{\mathsf{A}} \end{bmatrix}.$$

Note that, since A is SDD, $\alpha \neq 0$ and $\hat{\mathsf{A}}$ is SDD. Choosing $\boldsymbol{m} = -\alpha^{-1}\boldsymbol{p}$, we have

$$\mathsf{LA} = \begin{bmatrix} \alpha & \boldsymbol{v}^\intercal \\ \mathbf{0} & \hat{\mathsf{A}} - \alpha^{-1}\boldsymbol{p}\boldsymbol{v}^\intercal \end{bmatrix}.$$

Having successfully constructed L, we find $\mathsf{B} = \hat{\mathsf{A}} - \alpha^{-1}\boldsymbol{p}\boldsymbol{v}^\intercal$.

All that remains is to show that B is SDD. To see that this is the case, consider for row i of B,

$$\begin{aligned}
\sum_{\substack{j=1 \\ j\neq i}}^{n-1} |b_{i,j}| &= \sum_{\substack{j=1 \\ j\neq i}}^{n-1} \left|\hat{a}_{i,j} - \alpha^{-1} p_i v_j\right| \\
&\leq \sum_{\substack{j=1 \\ j\neq i}}^{n-1} |\hat{a}_{i,j}| + \sum_{\substack{j=1 \\ j\neq i}}^{n-1} \left|\alpha^{-1} p_i v_j\right| \\
&= \sum_{\substack{j=1 \\ j\neq i}}^{n-1} |\hat{a}_{i,j}| + \frac{|p_i|}{|\alpha|} \sum_{\substack{j=1 \\ j\neq i}}^{n-1} |v_j| \\
&= \sum_{\substack{j=1 \\ j\neq i}}^{n-1} |a_{i+1,j+1}| + \frac{|a_{i+1,1}|}{|a_{1,1}|} \sum_{\substack{j=1 \\ j\neq i}}^{n-1} |a_{1,j+1}| \\
&= \sum_{\substack{j=2 \\ j\neq i+1}}^{n} |a_{i+1,j}| + \frac{|a_{i+1,1}|}{|a_{1,1}|} \sum_{\substack{j=2 \\ j\neq i+1}}^{n} |a_{1,j}| \\
&= \sum_{\substack{j=1 \\ j\neq i+1}}^{n} |a_{i+1,j}| - |a_{i+1,1}| + \frac{|a_{i+1,1}|}{|a_{1,1}|} \sum_{j=2}^{n} |a_{1,j}| - \frac{|a_{i+1,1}| \cdot |a_{1,i+1}|}{|a_{1,1}|}.
\end{aligned}$$

Now, since A is SDD, we can continue this string of inequalities to obtain

$$\begin{aligned}
\sum_{\substack{j=1 \\ j\neq i}}^{n-1} |b_{i,j}| &< |a_{i+1,i+1}| - |a_{i+1,1}| + |a_{i+1,1}| - \frac{|a_{i+1,1}| \cdot |a_{1,i+1}|}{|a_{1,1}|} \\
&= |a_{i+1,i+1}| - \frac{|a_{i+1,1}| \cdot |a_{1,i+1}|}{|a_{1,1}|} \\
&\leq \left| a_{i+1,i+1} - \frac{a_{i+1,1} a_{1,i+1}}{a_{1,1}} \right| \\
&= |b_{i,i}|,
\end{aligned}$$

where we used the reverse triangle inequality. Thus, as claimed, B is SDD. □

Corollary 3.26 (SDD of magnitude δ). *If* $\mathsf{A} \in \mathbb{C}^{n\times n}$ *is SDD of magnitude* $\delta > 0$, *then* $\mathsf{B} \in \mathbb{C}^{(n-1)\times(n-1)}$, *introduced above, is SDD of magnitude* δ.

Proof. The details of our last proof still hold up to the point where we apply the fact that A is SDD. Thus,

$$
\begin{aligned}
\sum_{\substack{j=1\\ j\neq i}}^{n-1} |b_{i,j}| &\leq \sum_{\substack{j=1\\ j\neq i+1}}^{n} |a_{i+1,j}| - |a_{i+1,1}| + \frac{|a_{i+1,1}|}{|a_{1,1}|}\sum_{j=2}^{n} |a_{1,j}| - \frac{|a_{i+1,1}|\cdot|a_{1,i+1}|}{|a_{1,1}|} \\
&\leq |a_{i+1,i+1}| - \delta - |a_{i+1,1}| + \frac{|a_{i+1,1}|}{|a_{1,1}|}(|a_{1,1}| - \delta) - \frac{|a_{i+1,1}|\cdot|a_{1,i+1}|}{|a_{1,1}|} \\
&\leq |a_{i+1,i+1}| - \delta - |a_{i+1,1}| + |a_{i+1,1}| - \frac{|a_{i+1,1}|\cdot|a_{1,i+1}|}{|a_{1,1}|} \\
&= |a_{i+1,i+1}| - \frac{|a_{i+1,1}|\cdot|a_{1,i+1}|}{|a_{1,1}|} - \delta \\
&\leq \left| a_{i+1,i+1} - \frac{a_{i+1,1}a_{1,i+1}}{a_{1,1}} \right| - \delta \\
&= |b_{i,i}| - \delta.
\end{aligned}
$$

Thus, B is SDD of magnitude δ. □

Corollary 3.27 (Gaussian elimination and SDD). *If* $\mathsf{A} \in \mathbb{C}^{n\times n}$ *is SDD, then Gaussian elimination without pivoting applied to* A *proceeds to completion without encountering any zero pivot elements.*

Proof. We only need to proceed recursively using the previous two results. □

Remark 3.28 (modified Gaussian elimination). In a variant of Gaussian elimination, the pivot entry is normalized to one. Namely, given the system of linear equations

$$
\left\{
\begin{aligned}
a_{1,1}x_1 + a_{1,2}x_2 + \cdots + a_{1,n}x_n &= f_1, \\
a_{2,1}x_1 + a_{2,2}x_2 + \cdots + a_{2,n}x_n &= f_2, \\
&\vdots \\
a_{n,1}x_1 + a_{n,2}x_2 + \cdots + a_{n,n}x_n &= f_n,
\end{aligned}
\right.
$$

we could proceed as follows.

1. Using the first equation, express x_1 in terms of all the other variables to obtain
$$x_1 = a_{1,2}^{(1)}x_2 + \cdots + a_{1,n}^{(1)}x_n + f_1^{(1)}.$$
Eliminate x_1 from equations indexed 2 through n.
2. Using the second equation, express x_2 only in terms of $x_3, \ldots, x_n$:
$$x_2 = a_{2,3}^{(2)}x_3 + \cdots + a_{2,n}^{(2)}x_n + f_2^{(2)}.$$
Eliminate x_2 from equations indexed 3 through n.

3. Using the third equation, express x_3 only in terms of $x_4, \dots, x_n$:

$$x_3 = a_{3,4}^{(3)} x_4 + \cdots + a_{3,n}^{(3)} x_n + f_3^{(3)}.$$

Eliminate x_3 from equations indexed 4 through n.

i. For $i = 4, \dots, n-1$, using the ith equation, express x_i only in terms of the variables $x_{i+1}, \dots, x_n$:

$$x_i = a_{i,i+1}^{(i)} x_{i+1} + \cdots + a_{i,n}^{(i)} x_n + f_i^{(i)}. \tag{3.10}$$

Eliminate x_i from equations indexed $i+1$ through n.

n. Using the last remaining equation, express x_n as

$$x_n = f_n^{(n)}.$$

$n+1$. Use back substitution to solve for $\mathbf{x}$.

We will call this the *modified Gaussian elimination* process. It works as long as no zero pivots are encountered. For systems whose coefficient matrix is SDD, this process yields an interesting and desirable property. The following two results address this case and give a new perspective to our methodology.

Lemma 3.29 (modified Gaussian elimination). *Let $n \geq 2$ and $\mathsf{A}\mathbf{x} = \mathbf{f}$ be a system of n equations with n unknowns, where the coefficient matrix $\mathsf{A} = [a_{i,j}] \in \mathbb{C}^{n \times n}$ is SDD of magnitude $\delta > 0$. Then one may reduce the first equation to the form*

$$x_1 = a_{1,2}^{(1)} x_2 + \cdots + a_{1,n}^{(1)} x_n + f_1^{(1)} \tag{3.11}$$

for some coefficients $a_{1,j}^{(1)}$, $j = 2, \dots, n$, and $f_1^{(1)}$, with no division by zero. Moreover,

$$\sum_{j=2}^{n} |a_{1,j}^{(1)}| < 1.$$

Finally, x_1 can be eliminated from equations indexed 2 through n to obtain a subsystem, $\mathsf{A}^{(1)}\mathbf{x}^{(1)} = \mathbf{f}^{(1)}$, of $n-1$ equations with $n-1$ unknowns. The coefficient matrix $\mathsf{A}^{(1)}$ has diagonal dominance of magnitude $\delta > 0$.

Proof. Since the matrix A has diagonal dominance of magnitude $\delta > 0$, we observe that

$$|a_{1,1}| \geq \delta + |a_{1,2}| + \cdots + |a_{1,n}| > 0,$$

so that $a_{1,1} \neq 0$, and we can define $a_{1,j}^{(1)} = -a_{1,j}/a_{1,1}$, for $j = 2, \dots, n$, and $f_1^{(1)} = f_1/a_{1,1}$. Notice also that

$$\sum_{j=2}^{n} |a_{1,j}^{(1)}| = \frac{|a_{1,2}| + \cdots + |a_{1,n}|}{|a_{1,1}|} < 1.$$

Now, to eliminate x_1 from the system, we substitute (3.11) into all the remaining equations. The ith equation, for $i = 2, \dots, n$, reads

$$a_{i,1} x_1 + a_{i,2} x_2 + \cdots + a_{i,n} x_n = f_i,$$

and, when we substitute for x_1, we find, for $i = 2, \dots, n$,

$$(a_{i,2} + a_{i,1}a_{1,2}^{(1)})x_2 + (a_{i,3} + a_{i,1}a_{1,3}^{(1)})x_3 + \cdots + (a_{i,n} + a_{i,1}a_{1,n}^{(1)})x_n = f_i^{(1)},$$

where

$$f_i^{(1)} = f_i - a_{i,1}f_1^{(1)}, \quad i = 2, \dots, n.$$

It is convenient to index the matrix $\mathsf{A}^{(1)}$ starting at row and column 2 and ending at row and column n. In this case, the entries of $\mathsf{A}^{(1)}$ are

$$\left[\mathsf{A}^{(1)}\right]_{i,j} = a_{i,j} + a_{i,1}a_{1,j}^{(1)}, \quad i,j = 2, \dots, n.$$

The proof that $\mathsf{A}^{(1)}$ is SDD of magnitude $\delta > 0$ follows as before. □

We can now provide a sufficient condition for our modified Gaussian elimination process to proceed to completion and include some further results.

Theorem 3.30 (modified Gaussian elimination, SDD case)**.** *Let $n \geq 2$. Suppose that $\mathsf{A} \in \mathbb{C}^{n\times n}$ has diagonal dominance of magnitude $\delta > 0$ and $\boldsymbol{f} \in \mathbb{C}^n$ is given. Then the modified Gaussian elimination process (without pivoting) used to solve $\mathsf{A}x = \boldsymbol{f}$ does not fail. Moreover, we have that, at every step,*

$$\sum_{j=i+1}^{n} |a_{i,j}^{(i)}| < 1, \quad i = 1, \dots, n-1,$$

and

$$|f_i^{(i)}| \leq \frac{2}{\delta}\|\boldsymbol{f}\|_\infty, \quad i = 1, \dots, n.$$

Proof. To prove the first part, proceed by induction, using the previous lemma to show that the process does not fail. Invoking the fact that $\mathsf{A}^{(i-1)}$ is SDD of magnitude $\delta > 0$, we can always conclude that

$$\sum_{j=i+1}^{n} |a_{i,j}^{(i)}| < 1, \quad i = 1, \dots, n-1.$$

For the second part, notice that, since A has diagonal dominance of magnitude $\delta > 0$, we can recall Theorem 3.24, which indicates

$$\|\boldsymbol{x}\|_\infty = \|\mathsf{A}^{-1}\boldsymbol{f}\|_\infty \leq \|\mathsf{A}^{-1}\|_\infty\|\boldsymbol{f}\|_\infty \leq \frac{1}{\delta}\|\boldsymbol{f}\|_\infty.$$

Now, from (3.10), we see that

$$
\begin{aligned}
|f_i^{(i)}| &= \left| x_i - \sum_{j=i+1}^{n} a_{i,j}^{(i)} x_j \right| \\
&\leq |x_i| + \sum_{j=i+1}^{n} |a_{i,j}^{(i)}||x_j| \\
&\leq \left(1 + \sum_{j=i+1}^{n} |a_{i,j}^{(i)}| \right) \|\boldsymbol{x}\|_\infty \\
&\leq 2\|\boldsymbol{x}\|_\infty \\
&\leq \frac{2}{\delta}\|\boldsymbol{f}\|_\infty,
\end{aligned}
$$

as we intended to show. □

3.5.2 Positive Definite Matrices

Definition 3.31 (HPD matrices)**.** A matrix $\mathsf{A} \in \mathbb{C}^{n\times n}$ is called **Hermitian positive semi-definite** (HPSD) if and only if $\mathsf{A} = \mathsf{A}^{\mathsf{H}}$ and

$$\boldsymbol{x}^{\mathsf{H}}\mathsf{A}\boldsymbol{x} \geq 0, \quad \forall \boldsymbol{x} \in \mathbb{C}^n.$$

A is called **Hermitian positive definite** (HPD) if and only if $\mathsf{A} = \mathsf{A}^{\mathsf{H}}$ and

$$\boldsymbol{x}^{\mathsf{H}}\mathsf{A}\boldsymbol{x} > 0, \quad \forall \boldsymbol{x} \in \mathbb{C}^n_\star.$$

Remark 3.32 (notation)**.** We often use the symbol $\mathbb{C}^{n\times n}_{\text{Her}}$ to denote the vector space of Hermitian matrices. For real, symmetric matrices we use the notation $\mathbb{R}^{n\times n}_{\text{sym}}$.

Theorem 3.33 (properties of HPD matrices)**.** *Suppose that* $\mathsf{A} = [a_{i,j}] \in \mathbb{C}^{n\times n}$ *is HPD. Then*

1. $a_{i,i} > 0$ *for all* $i = 1, \dots, n$.
2. $\sigma(\mathsf{A}) \subset (0, \infty)$.
3. $\det(\mathsf{A}) > 0$ *and* $\operatorname{tr}(\mathsf{A}) = \sum_{i=1}^n a_{i,i} > 0$.
4. *For all* $\emptyset \neq S \subseteq \{1, \dots, n\}$, *we have that* $\mathsf{A}(S)$ *is HPD.*
5. $|a_{i,j}|^2 \leq a_{i,i}a_{j,j}$ *for all* $i \neq j$.
6. $\max_{1\leq i,j\leq n} |a_{i,j}| \leq \max_{1\leq i\leq n} |a_{i,i}|$.

Proof. We will prove statements 4–6 and leave the others for exercises; see Problem 3.13.

4: Given $S \subseteq \{1, \dots, n\}$ nonempty. Suppose that $\boldsymbol{y} \in \mathbb{C}^{\#(S)}_\star$ is arbitrary. Define $\boldsymbol{x} \in \mathbb{C}^n_\star$ such that $x_i = 0$ for all $i \notin S$; otherwise,

$$x_{i_k} = y_k, \quad k = 1, \dots, \#(S),$$

where

$$S = \{i_1, \ldots, i_{\#(S)}\}$$

and

$$i_k < i_{k+1}, \quad k = 1, \ldots, \#(S) - 1.$$

Then $\mathsf{A}(S) = \mathsf{A}(S)^{\mathsf{H}}$ and

$$\boldsymbol{y}^{\mathsf{H}}\mathsf{A}(S)\boldsymbol{y} = \boldsymbol{x}^{\mathsf{H}}\mathsf{A}\boldsymbol{x} > 0, \quad \boldsymbol{y} \in \mathbb{C}_{\star}^{\#(S)}.$$

5: Let $S = \{r, s\}$, $r < s$. Then

$$\mathsf{A}(S) = \begin{bmatrix} a_{r,r} & a_{r,s} \\ a_{s,r} & a_{s,s} \end{bmatrix};$$

consequently,

$$0 < \det(\mathsf{A}(S)) = a_{r,r}a_{s,s} - |a_{r,s}|^2.$$

6: We argue by contradiction. Suppose that the element with largest modulus is off-diagonal in, say, row r and column s, $r \neq s$. Then

$$|a_{r,s}| > a_{r,r}$$

and

$$|a_{r,s}| > a_{s,s}.$$

Thus,

$$|a_{r,s}|^2 > a_{r,r}a_{s,s},$$

which is a contradiction. □

Theorem 3.34 (factorization of HPD matrices)**.** *Let $n \geq 2$. Suppose that $\mathsf{A} \in \mathbb{C}^{n\times n}$ is HPD. There exists a unit lower triangular matrix $\mathsf{L} \in \mathbb{C}^{n\times n}$ and a diagonal matrix $\mathsf{D} \in \mathbb{R}^{n\times n}$ with positive diagonal entries such that*

$$\mathsf{A} = \mathsf{L}\mathsf{D}\mathsf{L}^{\mathsf{H}}.$$

Proof. Since A is HPD, for $k = 1, \ldots, n-1$ the principal sub-matrices $\mathsf{A}^{(k)} \in \mathbb{C}^{k\times k}$ are invertible. By Theorem 3.7 there exists a unit lower triangular matrix L and an upper triangular matrix U such that $\mathsf{A} = \mathsf{L}\mathsf{U}$.

Next, we claim that all of the diagonal elements of U are real and positive. This follows by an induction argument, as in the proof of Theorem 3.7, and the fact that $\det(\mathsf{A}) = \det(\mathsf{U})$. The details are left to the reader see Problem 3.14.

Now set $\mathsf{D} = \operatorname{diag}(u_{1,1}, \ldots, u_{n,n})$ and $\tilde{\mathsf{U}} = \mathsf{D}^{-1}\mathsf{U}$. Then

$$\mathsf{A} = \mathsf{L}\mathsf{D}\mathsf{D}^{-1}\mathsf{U} = \mathsf{L}\mathsf{D}\tilde{\mathsf{U}}.$$

It follows that, for $i = 1, \ldots, n$, $\tilde{u}_{i,i} = 1$. Taking the conjugate transpose, we have

$$\tilde{\mathsf{U}}^{\mathsf{H}}\mathsf{D}\mathsf{L}^{\mathsf{H}} = \mathsf{A}^{\mathsf{H}} = \mathsf{A} = \mathsf{L}\mathsf{D}\tilde{\mathsf{U}}.$$

It follows that

$$\mathsf{L}^{-1}\tilde{\mathsf{U}}^{\mathsf{H}}\mathsf{D} = \mathsf{D}\tilde{\mathsf{U}}\mathsf{L}^{-\mathsf{H}}. \tag{3.12}$$

Recall that, by Theorem 3.4, $\mathsf{L}^{-1}\tilde{\mathsf{U}}^{\mathsf{H}}$ must be unit lower triangular and $\tilde{\mathsf{U}}\mathsf{L}^{-\mathsf{H}}$ must be unit upper triangular. The only way for (3.12) to hold is for $\mathsf{L}^{-1}\tilde{\mathsf{U}}^{\mathsf{H}}$ and $\tilde{\mathsf{U}}\mathsf{L}^{-\mathsf{H}}$ to be diagonal. But, as these products must be unit triangular, they are both equal to the identity. In other words,

$$\mathsf{L} = \tilde{\mathsf{U}}^{\mathsf{H}},$$

and we have proven that

$$\mathsf{A} = \mathsf{L}\mathsf{D}\mathsf{L}^{\mathsf{H}}.$$

□

Corollary 3.35 (Cholesky factorization[3]). *Let $n \geq 2$. Suppose that $\mathsf{A} \in \mathbb{C}^{n\times n}$ is HPD. Then there is a lower triangular matrix $\mathsf{L} \in \mathbb{C}^{n\times n}$ such that*

$$\mathsf{A} = \mathsf{L}\mathsf{L}^{\mathsf{H}}.$$

This is known as the Cholesky factorization.

Proof. From the last theorem, there is a unit lower triangular matrix $\tilde{\mathsf{L}}$ and a diagonal matrix D with positive real diagonal entries such that

$$\mathsf{A} = \tilde{\mathsf{L}}\mathsf{D}\tilde{\mathsf{L}}^{\mathsf{H}}.$$

Suppose that $\mathsf{D} = \operatorname{diag}(d_1, \ldots, d_n)$. Define $\tilde{\mathsf{D}} = \operatorname{diag}(\sqrt{d_1}, \ldots, \sqrt{d_n})$. Then, setting $\mathsf{L} = \tilde{\mathsf{L}}\tilde{\mathsf{D}}$, we see that

$$\mathsf{A} = \mathsf{L}\mathsf{L}^{\mathsf{H}}.$$

The proof is complete. □

Theorem 3.36 (uniqueness of Cholesky factorization). *Let $n \geq 2$. Suppose that $\mathsf{A} \in \mathbb{C}^{n\times n}$ is HPD. Then there is a unique lower triangular matrix L such that the diagonal entries of L are positive real numbers and*

$$\mathsf{A} = \mathsf{L}\mathsf{L}^{\mathsf{H}}.$$

In other words, the Cholesky factorization is unique.

Proof. Suppose that there are two lower triangular matrices L_1 and L_2 with positive real diagonal entries and

$$\mathsf{L}_1\mathsf{L}_1^{\mathsf{H}} = \mathsf{A} = \mathsf{L}_2\mathsf{L}_2^{\mathsf{H}}.$$

Then

$$\mathsf{L}_2^{-1}\mathsf{L}_1 = \mathsf{L}_2^{\mathsf{H}}\mathsf{L}_1^{-\mathsf{H}};$$

by Theorem 3.4, $\mathsf{L}_2^{-1}\mathsf{L}_1$ is lower triangular and $\mathsf{L}_2^{\mathsf{H}}\mathsf{L}_1^{-\mathsf{H}}$ is upper triangular. Thus, there is a diagonal matrix D such that

$$\mathsf{L}_2^{-1}\mathsf{L}_1 = \mathsf{D} = \mathsf{L}_2^{\mathsf{H}}\mathsf{L}_1^{-\mathsf{H}}.$$

Therefore,

$$\mathsf{L}_1 = \mathsf{L}_2\mathsf{D} \tag{3.13}$$

[3] Named in honor of the French military officer and mathematician André-Louis Cholesky (1875–1918).

and

$$\mathsf{D}\mathsf{L}_1^{\mathsf{H}} = \mathsf{L}_2^{\mathsf{H}},$$

or, equivalently,

$$\mathsf{L}_2 = \mathsf{L}_1\mathsf{D}^{\mathsf{H}}. \tag{3.14}$$

Combining (3.13) and (3.14), we have

$$\mathsf{L}_1 = \mathsf{L}_2\mathsf{D} = \mathsf{L}_1\mathsf{D}^{\mathsf{H}}\mathsf{D}.$$

Since L_1 is invertible, the cancellation property holds and $\mathsf{D}^{\mathsf{H}}\mathsf{D} = \mathsf{I}_n$. But since L_1 and L_2 have positive diagonal entries, so must D have positive diagonal entries. It follows that $\mathsf{D} = \mathsf{I}_n$, which implies that $\mathsf{L}_1 = \mathsf{L}_2$. □

Theorem 3.37 (HPD and spectrum). *Suppose that $\mathsf{A} \in \mathbb{C}^{n\times n}$ is Hermitian. Then A is HPD if and only if $\sigma(\mathsf{A}) \subset (0, \infty)$.*

Proof. We only prove one direction here, as the other has already been proven. Since $\mathsf{A} \in \mathbb{C}^{n\times n}_{\mathrm{Her}}$, there exists a unitary matrix U and a diagonal matrix $\mathsf{D} = \mathrm{diag}(\lambda_1, \ldots, \lambda_n)$, where $\lambda_i \in \sigma(\mathsf{A})$, for $i = 1, \ldots, n$, such that $\mathsf{A} = \mathsf{U}^{\mathsf{H}}\mathsf{D}\mathsf{U}$. Let $\boldsymbol{x} \in \mathbb{C}^n_\star$. Set $\boldsymbol{y} = \mathsf{U}\boldsymbol{x}$ and note that $\boldsymbol{y} \neq \boldsymbol{0}$. Then

$$\boldsymbol{x}^{\mathsf{H}}\mathsf{A}\boldsymbol{x} = (\mathsf{U}\boldsymbol{x})^{\mathsf{H}}\mathsf{D}\mathsf{U}\boldsymbol{x} = \boldsymbol{y}^{\mathsf{H}}\mathsf{D}\boldsymbol{y} = \sum_{i=1}^{n} \lambda_i |y_i|^2 \geq \min_{1\leq i\leq n} \lambda_i \boldsymbol{y}^{\mathsf{H}}\boldsymbol{y} = \min_{1\leq i\leq n} \lambda_i \|\boldsymbol{y}\|_2^2 > 0.$$

This proves that A is HPD. □

The matrix encountered in the next result comes up repeatedly in the text. It is an example of an HPD matrix, but with all real entries. Such matrices are called symmetric positive definite (SPD). This matrix is also an example of a *Toeplitz symmetric tridiagonal*[4] (TST) matrix.

Theorem 3.38 (TST matrix). *Define $\mathsf{A} \in \mathbb{R}^{(n-1)\times(n-1)}$ via*

$$\mathsf{A} = \begin{bmatrix} 2 & -1 & 0 & \cdots & 0 \\ -1 & 2 & \ddots & \ddots & \vdots \\ 0 & \ddots & \ddots & -1 & 0 \\ \vdots & \ddots & -1 & 2 & -1 \\ 0 & \cdots & 0 & -1 & 2 \end{bmatrix}.$$

Then A is SPD. Let $h = 1/n$. The set

$$S = \{\boldsymbol{w}_1, \boldsymbol{w}_2, \ldots, \boldsymbol{w}_{n-1}\},$$

where the ith component of $\boldsymbol{w}_k$ is defined via

$$[\boldsymbol{w}_k]_i = \sin(k\pi i h),$$

is an orthogonal set of eigenvectors of A.

[4] Named in honor of the German mathematician Otto Toeplitz (1881–1940).

Proof. Note that for $k = 1, \ldots, n-1$,

$$\begin{aligned} [\mathsf{A}\boldsymbol{w}_k]_i &= -\sin(k\pi(i-1)h) + 2\sin(k\pi ih) - \sin(k\pi(i+1)h) \\ &= 2\sin(k\pi ih) - 2\cos(k\pi h)\sin(k\pi ih) \\ &= (2 - 2\cos(k\pi h))\sin(k\pi ih) \\ &= 2(1 - \cos(k\pi h))[\boldsymbol{w}_k]_i. \end{aligned}$$

Hence, the distinct eigenvalues are $\lambda_k = 2 - 2\cos(k\pi h)$. To see that these are strictly positive for $k = 1, \ldots, n-1$, note that

$$1 > \cos(k\pi h) > -1,$$

which implies that

$$-2 < -2\cos(k\pi h) < 2,$$

which implies that

$$0 = 2 - 2 < 2 - 2\cos(k\pi h) < 2 + 2 = 4.$$

Since A is symmetric, the eigenvectors associated with distinct eigenvalues are orthogonal. A is SPD since its eigenvalues are strictly positive. □

Proposition 3.39 (HPD and similarity transformations). *Let $\mathsf{A} \in \mathbb{C}^{m\times m}$ be HPD and $\mathsf{X} \in \mathbb{C}^{m\times n}$ with $m \geq n$ have full rank. Then $\mathsf{X}^{\mathsf{H}}\mathsf{A}\mathsf{X} \in \mathbb{C}^{n\times n}$ is HPD.*

Proof. Notice that

$$(\mathsf{X}^{\mathsf{H}}\mathsf{A}\mathsf{X})^{\mathsf{H}} = \mathsf{X}^{\mathsf{H}}\mathsf{A}^{\mathsf{H}}\mathsf{X} = \mathsf{X}^{\mathsf{H}}\mathsf{A}\mathsf{X}.$$

Suppose that $\boldsymbol{x} \in \mathbb{C}^n_\star$ is arbitrary. Since X is full rank, it follows that $\boldsymbol{y} = \mathsf{X}\boldsymbol{x} \neq \mathbf{0}$, i.e., $\boldsymbol{y} \in \mathbb{C}^m_\star$. Then

$$\boldsymbol{x}^{\mathsf{H}}\mathsf{X}^{\mathsf{H}}\mathsf{A}\mathsf{X}\boldsymbol{x} = \boldsymbol{y}^{\mathsf{H}}\mathsf{A}\boldsymbol{y} > 0,$$

since A is HPD. □

Theorem 3.40 (HPD and Gaussian elimination). *Let $\mathsf{A} \in \mathbb{C}^{n\times n}$ be HPD and represented as*

$$\mathsf{A} = \begin{bmatrix} \alpha & \boldsymbol{p}^{\mathsf{H}} \\ \boldsymbol{p} & \hat{\mathsf{A}} \end{bmatrix},$$

where $\alpha \in \mathbb{C}$, $\boldsymbol{p} \in \mathbb{C}^{n-1}$, and $\hat{\mathsf{A}} \in \mathbb{C}^{(n-1)\times(n-1)}$. After one step of Gaussian elimination (without pivoting), A will be reduced to the matrix

$$\begin{bmatrix} \alpha & \boldsymbol{p}^{\mathsf{H}} \\ \mathbf{0} & \mathsf{B} \end{bmatrix},$$

where $\mathsf{B} \in \mathbb{C}^{(n-1)\times(n-1)}$. Then B is HPD and the corresponding diagonal elements of B are smaller than those of $\hat{\mathsf{A}}$.

Proof. Let us construct a matrix $\mathsf{L} \in \mathbb{C}^{n\times n}$ such that

$$\mathsf{LA} = \begin{bmatrix} \alpha & \boldsymbol{p}^{\mathsf{H}} \\ \mathbf{0} & \mathsf{B} \end{bmatrix},$$

if possible. Consider

$$\mathsf{L} = \begin{bmatrix} 1 & \mathbf{0}^{\mathsf{T}} \\ \boldsymbol{m} & \mathsf{I}_{n-1} \end{bmatrix}.$$

Then

$$\mathsf{LA} = \begin{bmatrix} 1 & \mathbf{0}^{\mathsf{T}} \\ \boldsymbol{m} & \mathsf{I}_{n-1} \end{bmatrix} \begin{bmatrix} \alpha & \boldsymbol{p}^{\mathsf{H}} \\ \boldsymbol{p} & \hat{\mathsf{A}} \end{bmatrix} = \begin{bmatrix} \alpha & \boldsymbol{p}^{\mathsf{H}} \\ \alpha\boldsymbol{m} + \boldsymbol{p} & \boldsymbol{m}\boldsymbol{p}^{\mathsf{H}} + \hat{\mathsf{A}} \end{bmatrix}.$$

Note that, since A is HPD, $\alpha > 0$ and $\hat{\mathsf{A}}$ is HPD. Choosing $\boldsymbol{m} = -\alpha^{-1}\boldsymbol{p}$, we have

$$\mathsf{LA} = \begin{bmatrix} \alpha & \boldsymbol{p}^{\mathsf{H}} \\ \mathbf{0} & \hat{\mathsf{A}} - \alpha^{-1}\boldsymbol{p}\boldsymbol{p}^{\mathsf{H}} \end{bmatrix}.$$

Having successfully constructed L, we find $\mathsf{B} = \hat{\mathsf{A}} - \alpha^{-1}\boldsymbol{p}\boldsymbol{p}^{\mathsf{H}}$. Notice that this is not the only way to find the matrix B.

Now let $\boldsymbol{x} \in \mathbb{C}^{n-1}_{\star}$ be arbitrary. Define $\boldsymbol{y} \in \mathbb{C}^{n}_{\star}$ via

$$\boldsymbol{y} = \begin{bmatrix} \gamma \\ \boldsymbol{x} \end{bmatrix},$$

where $\gamma \in \mathbb{C}$ is arbitrary. Then, since A is HPD,

$$\begin{aligned} 0 &< \boldsymbol{y}^{\mathsf{H}}\mathsf{A}\boldsymbol{y} \\ &= \begin{bmatrix} \bar{\gamma} & \boldsymbol{x}^{\mathsf{H}} \end{bmatrix} \begin{bmatrix} \alpha & \boldsymbol{p}^{\mathsf{H}} \\ \boldsymbol{p} & \hat{\mathsf{A}} \end{bmatrix} \begin{bmatrix} \gamma \\ \boldsymbol{x} \end{bmatrix} \\ &= \begin{bmatrix} \bar{\gamma} & \boldsymbol{x}^{\mathsf{H}} \end{bmatrix} \begin{bmatrix} \alpha\gamma + \boldsymbol{p}^{\mathsf{H}}\boldsymbol{x} \\ \gamma\boldsymbol{p} + \hat{\mathsf{A}}\boldsymbol{x} \end{bmatrix} \\ &= \alpha|\gamma|^2 + \bar{\gamma}\boldsymbol{p}^{\mathsf{H}}\boldsymbol{x} + \gamma\boldsymbol{x}^{\mathsf{H}}\boldsymbol{p} + \boldsymbol{x}^{\mathsf{H}}\hat{\mathsf{A}}\boldsymbol{x}. \end{aligned}$$

Now we set $\gamma = -\alpha^{-1}\boldsymbol{p}^{\mathsf{H}}\boldsymbol{x}$. From the last calculation

$$\begin{aligned} 0 &< \boldsymbol{y}^{\mathsf{H}}\mathsf{A}\boldsymbol{y} \\ &= \alpha^{-1}|\boldsymbol{p}^{\mathsf{H}}\boldsymbol{x}|^2 - \alpha^{-1}|\boldsymbol{p}^{\mathsf{H}}\boldsymbol{x}|^2 - \alpha^{-1}|\boldsymbol{p}^{\mathsf{H}}\boldsymbol{x}|^2 + \boldsymbol{x}^{\mathsf{H}}\hat{\mathsf{A}}\boldsymbol{x} \\ &= \boldsymbol{x}^{\mathsf{H}}\hat{\mathsf{A}}\boldsymbol{x} - \alpha^{-1}|\boldsymbol{p}^{\mathsf{H}}\boldsymbol{x}|^2 \\ &= \boldsymbol{x}^{\mathsf{H}}\mathsf{B}\boldsymbol{x}. \end{aligned}$$

This proves that B is HPD.

Now the diagonal elements of B, which must be positive since B is HPD, are precisely $[\mathsf{B}]_{i,i} = [\hat{\mathsf{A}}]_{i,i} - \alpha^{-1}|[\boldsymbol{p}]_i|^2$. Hence, $0 < [\mathsf{B}]_{i,i} \le [\hat{\mathsf{A}}]_{ii}$, since $\alpha^{-1}|[\boldsymbol{p}]_i|^2 \ge 0$. □

Corollary 3.41 (HPD and Gaussian elimination)**.** *Suppose that* $\mathsf{A} \in \mathbb{C}^{n\times n}$ *is HPD. Then Gaussian elimination without pivoting proceeds to completion to produce a unit lower triangular matrix* $\mathsf{L} \in \mathbb{C}^{n\times n}$ *and an upper triangular matrix* $\mathsf{U} \in \mathbb{C}^{n\times n}$ *with positive diagonal elements such that* $\mathsf{A} = \mathsf{LU}$.

Proof. Apply recursively the previous result. □

Theorem 3.42 (HPD criterion). *$A \in \mathbb{C}^{n\times n}$ is HPD if and only if $A = LL^H$, where $L \in \mathbb{C}^{n\times n}$ is invertible.*

Proof. Suppose that $A = LL^H$, where L is invertible. Let $\boldsymbol{x} \in \mathbb{C}^n$ be arbitrary. Set $\boldsymbol{y} = L^H\boldsymbol{x}$. Since L is invertible, L^H is invertible; and $\boldsymbol{y} = \boldsymbol{0}$ if and only if $\boldsymbol{x} = \boldsymbol{0}$. Then

$$\boldsymbol{x}^H A\boldsymbol{x} = \boldsymbol{x}^H LL^H\boldsymbol{x} = \left(L^H\boldsymbol{x}\right)^H L^H\boldsymbol{x} = \boldsymbol{y}^H\boldsymbol{y} = \|\boldsymbol{y}\|_2^2 \geq 0,$$

with equality if and only if $\boldsymbol{x} = \boldsymbol{0}$. This proves that A is HPD.

The converse direction follows from the Cholesky factorization provided in Corollary 3.35. □

Theorem 3.43 (block matrices). *Let $k, m \in \mathbb{N}$. Set $n = k + m$. Suppose that $A \in \mathbb{C}^{n\times n}$ has the decomposition*

$$A = \begin{bmatrix} B & C^H \\ C & D \end{bmatrix},$$

where $B \in \mathbb{C}^{k\times k}$, $C \in \mathbb{C}^{m\times k}$, and $D \in \mathbb{C}^{m\times m}$.

1. *If A is HPD, then B, D, and $S = D - CB^{-1}C^H$ are HPD. S is called the* Schur complement *of B in A.*
2. *If A is HPD, the Cholesky factorization of A may be expressed in terms of the matrix C and the Cholesky factorizations of B and S.*

Proof. We prove each statement separately.

1. Since A is HPD, $\boldsymbol{x}^H A\boldsymbol{x} > 0$ for any $\boldsymbol{x} \in \mathbb{C}^n_\star$. Let $\boldsymbol{y} \in \mathbb{C}^k_\star$ and $\boldsymbol{w} \in \mathbb{C}^m_\star$ be arbitrary. Setting $\boldsymbol{x} = \left[\boldsymbol{y}^H, \boldsymbol{0}^H\right]^H \in \mathbb{C}^n_\star$, we have

$$0 < \boldsymbol{x}^H A\boldsymbol{x} = \boldsymbol{y}^H B\boldsymbol{y};$$

on the other hand, setting $\boldsymbol{x} = \left[\boldsymbol{0}^H, \boldsymbol{w}^H\right]^H \in \mathbb{C}^n_\star$, we have

$$0 < \boldsymbol{x}^H A\boldsymbol{x} = \boldsymbol{w}^H D\boldsymbol{w}.$$

Thus, B and D are HPD. More generally, suppose that $\boldsymbol{x} = \left[\boldsymbol{x}_1^H, \boldsymbol{x}_2^H\right]^H \in \mathbb{C}^n_\star$. Then

$$0 < \boldsymbol{x}^H A\boldsymbol{x} = \boldsymbol{x}_1^H B\boldsymbol{x}_1 + \boldsymbol{x}_1^H C^H\boldsymbol{x}_2 + \boldsymbol{x}_2^H C\boldsymbol{x}_1 + \boldsymbol{x}_2^H D\boldsymbol{x}_2.$$

Now pick $\boldsymbol{x}_1 = -B^{-1}C^H\boldsymbol{x}_2$ and suppose that $\boldsymbol{x}_2 \neq \boldsymbol{0}$. Then

$$\begin{aligned} 0 &< \boldsymbol{x}^H A\boldsymbol{x} \\ &= \boldsymbol{x}_2^H CB^{-H}BB^{-1}C^H\boldsymbol{x}_2 - \boldsymbol{x}_2^H CB^{-H}C^H\boldsymbol{x}_2 - \boldsymbol{x}_2^H CB^{-1}C^H\boldsymbol{x}_2 + \boldsymbol{x}_2^H D\boldsymbol{x}_2 \\ &= \boldsymbol{x}_2^H \left(D - CB^{-1}C^H\right)\boldsymbol{x}_2, \end{aligned}$$

where we used the fact that $B^{-1} = B^{-H}$ — since B is HPD — on the last step. It follows that S is HPD.

2. Since A is HPD, there is a unique lower triangular matrix $\mathsf{L}_\mathsf{A} \in \mathbb{C}^{n\times n}$ with positive diagonal entries, such that $\mathsf{A} = \mathsf{L}_\mathsf{A}\mathsf{L}_\mathsf{A}^\mathsf{H}$. Likewise there are unique lower triangular matrices $\mathsf{L}_\mathsf{B} \in \mathbb{C}^{k\times k}$ and $\mathsf{L}_\mathsf{S} \in \mathbb{C}^{m\times m}$, both with positive diagonal entries, such that $\mathsf{B} = \mathsf{L}_\mathsf{B}\mathsf{L}_\mathsf{B}^\mathsf{H}$ and $\mathsf{S} = \mathsf{L}_\mathsf{S}\mathsf{L}_\mathsf{S}^\mathsf{H}$. Suppose that

$$\mathsf{L}_\mathsf{A} = \begin{bmatrix} \mathsf{L}_1 & \mathsf{O} \\ \mathsf{M} & \mathsf{L}_2 \end{bmatrix},$$

where $\mathsf{L}_1 \in \mathbb{C}^{k\times k}$ is lower triangular with positive diagonal entries; $\mathsf{M} \in \mathbb{C}^{m\times k}$; and $\mathsf{L}_2 \in \mathbb{C}^{m\times m}$ is lower triangular with positive diagonal entries. Then

$$\mathsf{L}_\mathsf{A}\mathsf{L}_\mathsf{A}^\mathsf{H} = \begin{bmatrix} \mathsf{L}_1 & \mathsf{O} \\ \mathsf{M} & \mathsf{L}_2 \end{bmatrix} \begin{bmatrix} \mathsf{L}_1^\mathsf{H} & \mathsf{M}^\mathsf{H} \\ \mathsf{O}^\mathsf{H} & \mathsf{L}_2^\mathsf{H} \end{bmatrix} = \begin{bmatrix} \mathsf{L}_1\mathsf{L}_1^\mathsf{H} & \mathsf{L}_1\mathsf{M}^\mathsf{H} \\ \mathsf{M}\mathsf{L}_1^\mathsf{H} & \mathsf{M}\mathsf{M}^\mathsf{H} + \mathsf{L}_2\mathsf{L}_2^\mathsf{H} \end{bmatrix} = \begin{bmatrix} \mathsf{B} & \mathsf{C}^\mathsf{H} \\ \mathsf{C} & \mathsf{D} \end{bmatrix}.$$

Comparing entries we conclude that

$$\begin{aligned} \mathsf{L}_1\mathsf{L}_1^\mathsf{H} = \mathsf{B} \quad &\Longrightarrow \quad \mathsf{L}_1 = \mathsf{L}_\mathsf{B}, \\ \mathsf{M}\mathsf{L}_1^\mathsf{H} = \mathsf{C} \quad &\Longrightarrow \quad \mathsf{M} = \mathsf{C}\mathsf{L}_\mathsf{B}^{-\mathsf{H}}, \\ \mathsf{M}\mathsf{M}^\mathsf{H} + \mathsf{L}_2\mathsf{L}_2^\mathsf{H} = \mathsf{D} \quad &\Longrightarrow \quad \mathsf{L}_2\mathsf{L}_2^\mathsf{H} = \mathsf{D} - \mathsf{C}\mathsf{L}_\mathsf{B}^{-\mathsf{H}}\mathsf{L}_\mathsf{B}^{-1}\mathsf{C}^\mathsf{H} \\ &\Longrightarrow \quad \mathsf{L}_2\mathsf{L}_2^\mathsf{H} = \mathsf{S} \\ &\Longrightarrow \quad \mathsf{L}_2 = \mathsf{L}_\mathsf{S}, \end{aligned}$$

and the result is proven. □

3.5.3 Cholesky Factorization, Revisited

In this section, our aim is to produce a practical, computable algorithm for the Cholesky factorization. Let us first recall how Gaussian elimination acts on an HPD matrix. Let us consider a simplified HPD matrix $\mathsf{A} \in \mathbb{C}^{n\times n}$:

$$\mathsf{A} = \begin{bmatrix} 1 & \boldsymbol{w}^\mathsf{H} \\ \boldsymbol{w} & \mathsf{K} \end{bmatrix} \to \mathsf{L}_1\mathsf{A} = \begin{bmatrix} 1 & \boldsymbol{w}^\mathsf{H} \\ \mathbf{0} & \hat{\mathsf{K}} \end{bmatrix}$$

with

$$\mathsf{L}_1 = \begin{bmatrix} 1 & \mathbf{0}^\intercal \\ \boldsymbol{m} & \mathsf{I}_{n-1} \end{bmatrix},$$

so that

$$\mathsf{L}_1\mathsf{A} = \begin{bmatrix} 1 & \mathbf{0}^\intercal \\ \boldsymbol{m} & \mathsf{I}_{n-1} \end{bmatrix} \begin{bmatrix} 1 & \boldsymbol{w}^\mathsf{H} \\ \boldsymbol{w} & \mathsf{K} \end{bmatrix} = \begin{bmatrix} 1 & \boldsymbol{w}^\mathsf{H} \\ \boldsymbol{m} + \boldsymbol{w} & \mathsf{K} + \boldsymbol{m}\boldsymbol{w}^\mathsf{H} \end{bmatrix}.$$

In other words, we must have $\boldsymbol{m} = -\boldsymbol{w}$ and $\hat{\mathsf{K}} = \mathsf{K} - \boldsymbol{w}\boldsymbol{w}^\mathsf{H}$. Therefore,

$$\mathsf{A} = \mathsf{L}_1^{-1} \begin{bmatrix} 1 & \boldsymbol{w}^\mathsf{H} \\ \mathbf{0} & \hat{\mathsf{K}} \end{bmatrix} = \begin{bmatrix} 1 & \mathbf{0}^\intercal \\ \boldsymbol{w} & \mathsf{I}_{n-1} \end{bmatrix} \begin{bmatrix} 1 & \boldsymbol{w}^\mathsf{H} \\ \mathbf{0} & \mathsf{K} - \boldsymbol{w}\boldsymbol{w}^\mathsf{H} \end{bmatrix},$$

so that, in the process, we lost that the matrix is Hermitian.

But being Hermitian is an important feature, and it would be desirable to keep it throughout the Gaussian elimination process. To do so, we will introduce zeros on the first row to match the ones we introduced on the first column. To achieve this, notice that

$$\begin{bmatrix} 1 & \boldsymbol{w}^{\mathsf{H}} \\ \mathbf{0} & \mathsf{K} - \boldsymbol{w}\boldsymbol{w}^{\mathsf{H}} \end{bmatrix} = \begin{bmatrix} 1 & \mathbf{0}^{\mathsf{T}} \\ \mathbf{0} & \mathsf{K} - \boldsymbol{w}\boldsymbol{w}^{\mathsf{H}} \end{bmatrix} \begin{bmatrix} 1 & \boldsymbol{w}^{\mathsf{H}} \\ \mathbf{0} & \mathsf{I}_{n-1} \end{bmatrix}.$$

Notice that the factor on the right is nothing but $\mathsf{L}_1^{-\mathsf{H}}$. In combining these two operations, we get

$$\mathsf{A} = \begin{bmatrix} 1 & \mathbf{0}^{\mathsf{T}} \\ \boldsymbol{w} & \mathsf{I}_{n-1} \end{bmatrix} \begin{bmatrix} 1 & \mathbf{0}^{\mathsf{T}} \\ \mathbf{0} & \mathsf{K} - \boldsymbol{w}\boldsymbol{w}^{\mathsf{H}} \end{bmatrix} \begin{bmatrix} 1 & \boldsymbol{w}^{\mathsf{H}} \\ \mathbf{0} & \mathsf{I}_{n-1} \end{bmatrix}.$$

This is the main idea behind the algorithmic version of the Cholesky factorization. In general, since $\alpha^2 = a_{1,1} > 0$, we can proceed as follows:

$$\mathsf{A} = \begin{bmatrix} \alpha^2 & \boldsymbol{w}^{\mathsf{H}} \\ \boldsymbol{w} & \mathsf{K} \end{bmatrix} = \begin{bmatrix} \alpha & \mathbf{0}^{\mathsf{T}} \\ \frac{1}{\alpha}\boldsymbol{w} & \mathsf{I}_{n-1} \end{bmatrix} \begin{bmatrix} 1 & \mathbf{0}^{\mathsf{T}} \\ \mathbf{0} & \mathsf{K} - \frac{1}{a_{1,1}}\boldsymbol{w}\boldsymbol{w}^{\mathsf{H}} \end{bmatrix} \begin{bmatrix} \alpha & \frac{1}{\alpha}\boldsymbol{w}^{\mathsf{H}} \\ \mathbf{0} & \mathsf{I}_{n-1} \end{bmatrix} = \mathsf{R}_1^{\mathsf{H}}\mathsf{A}_1\mathsf{R}_1,$$

where R_1 is an upper triangular matrix. Applying this repeatedly we obtain a computable version of the Cholesky factorization, yielding an upper triangular matrix R with positive diagonal entries such that

$$\mathsf{A} = \mathsf{R}^{\mathsf{H}}\mathsf{I}_n\mathsf{R} = \mathsf{R}^{\mathsf{H}}\mathsf{R}.$$

Remark 3.44 (terminology). Since $\alpha_k = \sqrt{a_{k,k}}$ uniquely determines the algorithm, this Cholesky factorization is sometimes also referred to as the *square root method.*

Listing 3.7 describes how to compute this factorization. Once this factorization is computed, system (3.1), in the case that A is HPD, can be solved rather efficiently.

Proposition 3.45 (complexity of Cholesky factorization). *Let $\mathsf{A} \in \mathbb{C}^{n\times n}$ be HPD. The Cholesky factorization algorithm requires of the order of $\frac{1}{3}n^3$ operations.*

Proof. We just need to notice that the innermost loop of Listing 3.7 requires two (2) operations. Thus, to leading order, the total complexity is

$$\sum_{i=2}^{n}\sum_{j=1}^{i-1}\sum_{k=1}^{j-1} 2 \approx \frac{1}{3}n^3.$$

□

Problems

3.1 Write the computational formulas used to solve a system of equations with a lower triangular coefficient matrix. In other words, describe the forward substitution algorithm.

3.2 Show that when the generic back (forward) substitution algorithm is applied to solve a system of $n \in \mathbb{N}$ unknowns, there are n divisions, $\frac{n^2-n}{2}$ multiplications, and $\frac{n^2-n}{2}$ additions/subtractions. Use this to show that *complexity*, i.e., the number of operations of this algorithm, is

$$n + \frac{n^2-n}{2} + \frac{n^2-n}{2} = n^2.$$

3.3 Find the complexity of the algorithm presented in Listing 3.1 to solve a system with a tridiagonal coefficient matrix.

3.4 Given a nonsingular matrix $A \in \mathbb{C}^{n \times n}$ propose an algorithm, based on Gaussian elimination, to find A^{-1}.

3.5 Prove Proposition 3.10.

3.6 Show that a column-s complete elementary matrix is invertible and find its inverse.

3.7 Prove Theorem 3.13.

3.8 Prove Proposition 3.17.

3.9 Prove Lemma 3.18.

3.10 Complete the proof of Theorem 3.21.

3.11 Suppose that $A \in \mathbb{R}^{n \times n}$ is a nonsingular matrix whose leading principal sub-matrices are all nonsingular. Partition A as

$$A = \begin{bmatrix} A_{11} & A_{12} \\ A_{21} & A_{22} \end{bmatrix},$$

where $A_{11} \in \mathbb{R}^{k \times k}$.

a) Show that there is a matrix M such that

$$\begin{bmatrix} I & O \\ -M & I \end{bmatrix} \begin{bmatrix} A_{11} & A_{12} \\ A_{21} & A_{22} \end{bmatrix} = \begin{bmatrix} A_{11} & A_{12} \\ O & \widetilde{A}_{22} \end{bmatrix}$$

and write out the explicit expresssions for M and $\widetilde{A}_{22}$.

b) Show that

$$\begin{bmatrix} A_{11} & A_{12} \\ A_{21} & A_{22} \end{bmatrix} = \begin{bmatrix} I & O \\ M & I \end{bmatrix} \begin{bmatrix} A_{11} & A_{12} \\ O & \widetilde{A}_{22} \end{bmatrix}.$$

c) The leading principal sub-matrices of A_{11} are, of course, all nonsingular. Prove that $\widetilde{A}_{22}$ is also nonsingular.

d) By the previous statement, both A_{11} and $\widetilde{A}_{22}$ have LU decompositions, say $A_{11} = L_1 U_1$ and $\widetilde{A}_{22} = L_2 U_2$. Show that

$$A = \begin{bmatrix} L_1 & O \\ ML_1 & L_2 \end{bmatrix} \begin{bmatrix} U_1 & L_1^{-1} A_{12} \\ O & U_2 \end{bmatrix},$$

which is the LU decomposition of A.

3.12 Suppose that $A \in \mathbb{R}^{n \times n}$ is SDD with positive diagonal entries and nonpositive off-diagonal entries. Show that, in this case, A^{-1} exists and contains only nonnegative elements.

Hint: Use the procedure for inverting a matrix using Gaussian elimination.

3.13 Complete the proof of Theorem 3.33.

3.14 Complete the proof of Theorem 3.34.

3.15 Suppose that $A \in \mathbb{C}^{n \times n}$ is Hermitian, has nonnegative (real) diagonal elements, and is SDD. Prove that it is HPD.

Listings

```
function [x, err] = TriDiagonal( a, b, c, f )
% Solution of a linear system of equations with a tridiagonal
% matrix.
%
%   a(k) x(k-1) + b(k) x(k) + c(k) x(k+1) = f(k)
%
% with a(1) = c(n) = 0.
%
% Input
%   a(1:n), b(1:n), c(1:n) : the coefficients of the system
%                            matrix
%   f(1:n) : the right hand side vector
%
% Output
%   x(1:n) : the solution to the linear system of equations, if
%            no division by zero occurs
%   err : = 0, if no division by zero occurs
%         = 1, if division by zero is encountered
  n = length(f);
  alpha = zeros(n,1);
  beta = zeros(n,1);
  err = 0;
  if abs(b(1)) > eps( b(1) )
    alpha(1) = -c(1)/b(1);
    beta(1) = f(1)/b(1);
  else
    err = 1;
    return;
  end
  for k=2:n
    denominator = a(k)*alpha(k-1) + b(k);
    if abs(denominator) > eps( denominator )
      alpha(k) = - c(k)/denominator;
      beta(k) = ( f(k) - a(k)*beta(k-1) )/denominator;
    else
      err = 1;
      return;
    end
  end
  if abs(a(n)*alpha(n-1) + b(n)) > eps( b(n) )
     x(n) = ( f(n) - a(n)*beta(n-1) )/( a(n)*alpha(n-1) + b(n) );
  else
    err = 1;
    return;
  end
  for k=n-1:-1:1
    x(k) = alpha(k)*x(k+1) + beta(k);
  end
end
```

Listing 3.1 Solution of a system of equations with a tridiagonal coefficient matrix.

```
function [x, err] = CyclicallyTriDiagonal( a, b, c, f )
% Solution of a cyclically tridiagonal linear system of equations:
%
%   b(1) x(  1) + c(1) x(  2) + a(1) x(  n) = f(1),
%   a(k) x(k-1) + b(k) x(  k) + c(k) x(k+1) = f(k),
%   c(n) x(  1) + a(n) x(n-1) + b(n) x(  n) = f(n).
%
% Input
%   a(1:n), b(1:n), c(1:n) : the coefficients of the system
%                            matrix
%   f(1:n) : the right hand side vector
%
% Output
%   x(1:n) : the solution to the linear system of equations, if
%            no division by zero is encountered
%   err : = 0, if division by zero does not occur
%         = 1, if division by zero occurs
  n = length(f);
  err = 0;
  newa = a(2:n);
  newa(1) = 0;
  newb = b(2:n);
  newc = c(2:n);
  newc(n-1) = 0;
  newf = f(2:n);
  [u, err] = TriDiagonal( newa, newb, newc, newf );
  if err == 1
    return;
  end
  newf = zeros(n-1, 1);
  newf(1) = - a(2);
  newf(n-1) = -c(n);
  [v, err] = TriDiagonal( newa, newb, newc, newf );
  if err == 1
    return;
  end
  x = zeros(n,1);
  denmoniator = b(1) + c(1)*v(1) + a(1)*v(n-1);
  if abs(denominator) > eps( denorminator )
    x(1) =  ( f(1) - c(1)*u(1) - a(1)*u(n-1) )/denominator;
  else
    err = 1;
    return;
  end
  for k=2:n
    x(k) = u(k-1) + x(1)*v(k-1);
  end
end
```

Listing 3.2 Solution of a system of equations with a cyclically tridiagonal coefficient matrix.

```
function [L, U, err] = LUFactSimple( A )
% LU factorization of a square matrix.
%
% Input
%   A(1:n,1:n) : the matrix to be factorized
%
% Output
%   L(1:n,1:n), U(1:n,1:n) : the factors in A = LU, if Gaussian
%                            elimination proceeds to completion
%   err : = 0 if Gaussian elimination proceeds to completion
%         = 1 if a zero pivot is encountered
  n = size(A,1);
  U = A;
  L = eye(n);
  err = 0;
  for k=1:n-1
    for j=k+1:n
      if abs( U(k,k) ) > eps( U(k,k) )
        L(j,k) = U(j,k)/U(k,k);
      else
        err = 1;
        return;
      end
      for t = k:n
        U(j,t) = U(j,t)-L(j,k)*U(k,t);
      end
    end
  end
end
```

Listing 3.3 LU factorization.

```
function [L, U, P, err] = LUFactPivot( A )
% LU factorization with pivoting of a square matrix.
%
% Input
%   A(1:n,1:n) : the matrix to be factorized
%
% Output
%   L(1:n,1:n), U(1:n,1:n), P(1:n,1:n) : the factors in A = P'LU
%   err : = 0 if no error was encountered
%         = 1 if a division by zero occurred
  n = size(A,1);
  U = A;
  L = eye(n);
  P = eye(n);
  err = 0;
  for k=1:n-1
    i=k;
    for t=k:n
      if abs( U(t,k) ) > abs( U(i,k) )
        i=t;
      end
    end
```

```
      for t=k:n
        temp = U(k,t);
        U(k,t) = U(i,t);
        U(i,t) = temp;
      end
      for t=1:k-1
        temp = L(k,t);
        L(k,t) = L(i,t);
        L(i,t) = temp;
      end
      for t=1:n
        temp = P(k,t);
        P(k,t) = P(i,t);
        P(i,t) = temp;
      end
      if abs( U(k,k) ) > eps( U(k,k) )
        for j=k+1:n
          L(j,k) = U(j,k)/U(k,k);
          for t=k:n
            U(j,t) = U(j,t) - L(j,k)*U(k,t);
          end
        end
      else
        err = 1;
        return;
      end
    end
  end
```

Listing 3.4 LU factorization with pivoting.

```
function [Afact, swaps, err] = LUFactEfficient( A )
% Efficient implementation of LU factorization with partial
% pivoting.
%
% Input
%   A(1:n,1:n) : the matrix to be factorized into A = P'LU
%
% Output
%   Afact(1:n,1:n) : a matrix containing the matrices L and U
%                    below and above the diagonal, respectively
%   swaps : the indices that indicate the permutations
%           (row swaps)
%   err : = 0 if no the factorization finished successfully
%         = 1 if a division by zero was encountered
  n=size(A,1);
  swaps = 1:n;
  err = 0;
  for k=1:n-1
    i = k;
    for t=k:n
      if abs( A(t,k) ) > abs( A(i,k) )
        i=t;
      end
    end
```

```
    tt = swaps(k);
    swaps(k) = swaps(i);
    swaps(i) = tt;
    if abs( A( swaps(k), k ) <= eps( A( swaps(k), k) )
      err = 1;
      return;
    end
    for i=k+1:n
      xmult = A(swaps(i),k)/A(swaps(k),k);
      A( swaps(i), k ) = xmult;
      for j=k+1:n
        A(swaps(i),j) = A(swaps(i),j) - xmult*A(swaps(k),j);
      end
    end
  end
  Afact = A;
end
```

Listing 3.5 Efficient implementation of LU factorization with pivoting.

```
function [x, err] = Solve( A, f )
% Solves the system Ax = f.
%
% Input
%   A(1:n,1:n) : the coefficient matrix
%   f(1:n) : the RHS vector
%
% Output:
%   x(1:n) : the solution vector
%   err : = 0 if the solution was found successfully
%         = 1 if an error occurred during the process
  n = length(f);
  err = 0;
  [AA, swaps, err] = LUFactEfficient( A );
  if err == 1
    return
  end
  for k=1:n-1
    for i=k+1:n
      f( swaps(i) ) = f( swaps(i) ) - AA( swaps(i), k ) ...
        *f( swaps(k) );
    end
  end
  if abs( AA( swaps(n), n ) ) > eps( AA(swaps(n),n) )
    x(n) = f( swaps(n) )/AA( swaps(n), n );
  else
    err = 1;
    return;
  end
  for i=n-1:-1:1
    xsum = f( swaps(i) );
    for j=i+1:n
      xsum = xsum - AA( swaps(i), j )*x(j);
    end
    if abs( AA( swaps(i), i ) ) > eps( AA( swaps(i), i ) )
```

```
      x(i) = xsum/AA( swaps(i), i );
    else
      err = 1;
      return;
    end
  end
end
```

Listing 3.6 Solution of (3.1) after LU factorization.

```
function [R, err] = CholeskyDecomposition( A )
% Choleksy factorization of the HPD matrix A:
%    A = R'*R
%
% Input:
%    A(1:n,1:n) : an HPD matrix
%
% Output:
%    R(1:n,1:n) : The upper triangular matrix satisfying A = R'*R
%    err : = 0 if the decomposition finished successfully
%          = 1 if an error occurred
  err = 0;
  n = size(A,1);
  R = zeros(n,n);
  R(1,1) = sqrt( A(1,1) );
  for i=2:n
    for j=1:i-1
      sum = A(i,j);
      for k=1:j-1
        sum = sum - R(k,i)* conj( R(k,j) );
      end
      if abs( R(j,j) ) > eps( R(j,j) )
        R(j,i) = sum/R(j,j);
      else
        err = 1;
        return;
      end
    end
    sum = A(i,i);
    for k=1:i-1
      sum = sum - R(k,i)*conj( R(k,i) );
    end
    R(i,i) = sqrt( sum );
  end
end
```

Listing 3.7 Computation of the Cholesky factorization for an HPD matrix.

4 Norms and Matrix Conditioning

In this chapter, we look at some qualitative aspects of the solution to (3.1), which we studied in Chapter 3. While it is usually assumed that the coefficient matrix $\mathsf{A} \in \mathbb{C}^{n\times n}$ and the right-hand-side vector $\boldsymbol{f} \in \mathbb{C}^n$ are known, fixed, and perfectly represented in our computational device, here we are interested to see what happens to the solution to our linear system if these are somehow perturbed.

To fix ideas, suppose, for example, that A is invertible and that $\boldsymbol{x} \in \mathbb{C}^n$ is the solution to the (ideal) system

$$\mathsf{A}\boldsymbol{x} = \boldsymbol{f}. \tag{4.1}$$

Now suppose that, in some hypothetical computing device, A and $\boldsymbol{f}$ are perturbed in storage: $\mathsf{A} \to \mathsf{A} + \delta\mathsf{A}$ and $\boldsymbol{f} \to \boldsymbol{f} + \delta\boldsymbol{f}$, where $\delta\mathsf{A} \in \mathbb{C}^{n\times n}$ and $\delta\boldsymbol{f} \in \mathbb{C}^n$. Assuming that $\mathsf{A} + \delta\mathsf{A}$ is invertible, there is some $\delta\boldsymbol{x} \in \mathbb{C}^n$ such that $\boldsymbol{x} + \delta\boldsymbol{x} \in \mathbb{C}^n$ is the solution to the perturbed system

$$(\mathsf{A} + \delta\mathsf{A})\,(\boldsymbol{x} + \delta\boldsymbol{x}) = \boldsymbol{f} + \delta\boldsymbol{f}. \tag{4.2}$$

Clearly, $\delta\boldsymbol{x}$ measures the error resulting from the perturbations to our data. How large is this error vector? How large is the relative error, $\frac{\|\delta\boldsymbol{x}\|}{\|\boldsymbol{x}\|}$? How do the error vector and relative error relate to the sizes of the perturbations? It turns out that the answers to our questions depend upon the so-called condition number of the matrix A, defined as

$$\kappa(\mathsf{A}) = \|\mathsf{A}\| \left\|\mathsf{A}^{-1}\right\|.$$

Of course, it is not practical to compute $\kappa(\mathsf{A})$ according to the formula above, as it involves the inverse of the coefficient matrix. But, often, the condition number can be accurately estimated. We will see that, if the condition number is large, the relative error can be quite large, even when other measures of error are actually small.

Now, since all the ideas in this chapter depend heavily upon the notions of vector norms on $\mathbb{C}^n$ and induced matrix norms on $\mathbb{C}^{n\times n}$, we urge the reader to, if necessary, review these concepts in Sections A.3 and 1.2.

4.1 The Spectral Radius

We begin with a definition.

Definition 4.1 (spectral radius)**.** Suppose that $A \in \mathfrak{L}(\mathbb{V})$, where $\mathbb{V}$ is a complex n-dimensional vector space. The **spectral radius** of A is

$$\rho(A) = \max\{|\lambda| \mid \lambda \in \sigma(A)\}.$$

An analogous definition is made for any matrix $\mathsf{A} \in \mathbb{C}^{n\times n}$.

For self-adjoint operators, the spectral radius has a very precise meaning.

Theorem 4.2 (self-adjoint operator)**.** *Let $\mathbb{V}$ be an n-dimensional complex inner product space. Suppose that $\|x\| = (x,x)^{1/2}$ is the Euclidean norm. Let $A \in \mathfrak{L}(\mathbb{V})$ be self-adjoint. Then the induced norm satisfies*

$$\|A\| = \rho(A).$$

Proof. Since $A\colon \mathbb{V} \to \mathbb{V}$ is self-adjoint, there exists an orthonormal basis of eigenvectors $S = \{e_1, \dots, e_n\}$, i.e., $(e_i, e_j) = \delta_{i,j}$, $\mathbb{V} = \operatorname{span}(S)$, and $Ae_i = \lambda_i e_i$. Expanding $x \in \mathbb{V}$ in this basis, i.e., $x = \sum_{i=1}^n x_i e_i$ with $x_i \in \mathbb{C}$, we see that

$$Ax = \sum_{i=1}^n \lambda_i x_i e_i.$$

Since this basis is orthonormal,

$$\|x\|^2 = \sum_{i=1}^n |x_i|^2 \quad \text{and} \quad \|Ax\|^2 = \sum_{i=1}^n |\lambda_i|^2 |x_i|^2.$$

With this at hand we notice that

$$\|Ax\| \le \max\{|\lambda| \; |\lambda \in \sigma(A)\}\, \|x\|,$$

which implies that

$$\|A\| \le \rho(A).$$

Problem 4.2 gives the reverse inequality, and this concludes the proof. □

For more general operators and norms, all that can be established is the following.

Theorem 4.3 (norms and spectral radius)**.** *Suppose that $\|\cdot\| : \mathbb{C}^{n\times n} \to \mathbb{R}$ is any induced matrix norm. Then there is a constant $C > 0$ such that*

$$\rho(\mathsf{A}) \le \|\mathsf{A}\| \le C\sqrt{\rho(\mathsf{A}^{\mathsf{H}}\mathsf{A})}, \quad \forall \mathsf{A} \in \mathbb{C}^{n\times n}.$$

Proof. Let $\|\cdot\|_{\mathbb{C}^n}$ be the vector norm that induces $\|\cdot\|$. Owing to Problem A.12, this norm is equivalent to $\|\cdot\|_2$, i.e., there are constants $0 < C_1 \le C_2$ for which

$$C_1 \|\boldsymbol{x}\|_{\mathbb{C}^n} \le \|\boldsymbol{x}\|_2 \le C_2 \|\boldsymbol{x}\|_{\mathbb{C}^n}, \quad \forall \boldsymbol{x} \in \mathbb{C}^n.$$

This, in turn, implies that, if $\boldsymbol{x} \in \mathbb{C}^n_\star$,

$$\frac{\|\mathsf{A}\boldsymbol{x}\|_{\mathbb{C}^n}}{\|\boldsymbol{x}\|_{\mathbb{C}^n}} \le \frac{C_2}{C_1}\frac{\|\mathsf{A}\boldsymbol{x}\|_2}{\|\boldsymbol{x}\|_2} \le C\,\|\mathsf{A}\|_2 \le C\sqrt{\rho(\mathsf{A}^{\mathsf{H}}\mathsf{A})},$$

where we denoted $C = C_2/C_1$ and used Problem 1.29. Taking supremum over $\boldsymbol{x} \in \mathbb{C}^n_\star$ implies the upper bound. The lower bound is an exercise; see Problem 4.2. □

While the spectral radius may not necessarily be a norm it is almost one, as the following, rather technical, result shows.

Theorem 4.4 (spectral radius and norms). *For every matrix* $\mathsf{A} \in \mathbb{C}^{n\times n}$ *and any* $\varepsilon > 0$, *there is a norm* $\|\cdot\|_{\mathsf{A},\varepsilon} : \mathbb{C}^n \to \mathbb{R}$ *such that the induced matrix norm*

$$\|\mathsf{M}\|_{\mathsf{A},\varepsilon} = \sup_{x\in\mathbb{C}^n_\star} \frac{\|\mathsf{M}\boldsymbol{x}\|_{\mathsf{A},\varepsilon}}{\|\boldsymbol{x}\|_{\mathsf{A},\varepsilon}} = \sup_{\|x\|_{\mathsf{A},\varepsilon}=1} \|\mathsf{M}\boldsymbol{x}\|_{\mathsf{A},\varepsilon}, \quad \forall \mathsf{M} \in \mathbb{C}^{n\times n},$$

satisfies

$$\|\mathsf{A}\|_{\mathsf{A},\varepsilon} \le \rho(\mathsf{A}) + \varepsilon.$$

Proof. Appealing to the Schur factorization, Lemma 1.46, there is a unitary matrix $\mathsf{P} \in \mathbb{C}^{n\times n}$ and an upper triangular matrix $\mathsf{B} \in \mathbb{C}^{n\times n}$ such that

$$\mathsf{A} = \mathsf{P}^{\mathsf{H}}\mathsf{B}\mathsf{P}.$$

The diagonal elements of B are the eigenvalues of A. Let us write

$$\mathsf{B} = \Lambda + \mathsf{U},$$

where $\Lambda = \operatorname{diag}(\lambda_1, \dots, \lambda_n)$ with

$$\sigma(\mathsf{A}) = \{\lambda_1, \dots, \lambda_n\} = \sigma(\mathsf{B}),$$

and $\mathsf{U} = [u_{i,j}] \in \mathbb{C}^{n\times n}$ is strictly upper triangular. Let $\delta > 0$ be arbitrary. Define

$$\mathsf{D} = \operatorname{diag}\left(1, \delta^{-1}, \dots, \delta^{1-n}\right).$$

Next, define

$$\mathsf{C} = \mathsf{D}\mathsf{B}\mathsf{D}^{-1} = \Lambda + \mathsf{E},$$

where

$$\mathsf{E} = \mathsf{D}\mathsf{U}\mathsf{D}^{-1}.$$

Now observe that E, like U, must be strictly upper triangular, and the elements of $\mathsf{E} = [e_{i,j}]$ must satisfy

$$e_{i,j} = \begin{cases} 0, & j \le i, \\ \delta^{j-i} u_{i,j}, & j > i. \end{cases}$$

Consequently, the elements of E can be made arbitrarily small in modulus, depending on our choice of δ.

Now notice that

$$\mathsf{A} = \mathsf{P}^{-1}\mathsf{D}^{-1}\mathsf{C}\mathsf{D}\mathsf{P},$$

and, since DP is nonsingular, the following defines a norm and an induced matrix norm: for any $\boldsymbol{x} \in \mathbb{C}^n$,

$$\|\boldsymbol{x}\|_\star = \|\mathsf{D}\mathsf{P}\boldsymbol{x}\|_2,$$

and, for any $\mathsf{M} \in \mathbb{C}^{n\times n}$,

$$\|\mathsf{M}\|_\star = \sup_{\|x\|_\star=1} \|\mathsf{M}\boldsymbol{x}\|_\star.$$

Observe that

$$\|\mathsf{A}\boldsymbol{y}\|_\star = \|\mathsf{DPA}\boldsymbol{y}\|_2 = \|\mathsf{CDP}\boldsymbol{y}\|_2 .$$

Define $\boldsymbol{z} = \mathsf{DP}\boldsymbol{y}$. Then

$$\|\mathsf{A}\boldsymbol{y}\|_\star = \|\mathsf{C}\boldsymbol{z}\|_2 = \sqrt{\boldsymbol{z}^{\mathsf{H}}\mathsf{C}^{\mathsf{H}}\mathsf{C}\boldsymbol{z}}.$$

But

$$\mathsf{C}^{\mathsf{H}}\mathsf{C} = \left(\Lambda^{\mathsf{H}} + \mathsf{E}^{\mathsf{H}}\right)(\Lambda + \mathsf{E}) = \Lambda^{\mathsf{H}}\Lambda + \mathsf{M}(\delta),$$

where

$$\mathsf{M}(\delta) = \mathsf{E}^{\mathsf{H}}\Lambda + \Lambda^{\mathsf{H}}\mathsf{E} + \mathsf{E}^{\mathsf{H}}\mathsf{E}.$$

As an exercise, the reader should prove that, for a given matrix A, there is a constant $K_1 > 0$ such that

$$\|\mathsf{M}(\delta)\|_2 \le K_1\delta$$

for all $0 < \delta \le 1$. Thus, using the definition of the spectral radius, the Cauchy–Schwarz inequality, and induced norm consistency,

$$\begin{aligned}
\boldsymbol{z}^{\mathsf{H}}\mathsf{C}^{\mathsf{H}}\mathsf{C}\boldsymbol{z} &= \boldsymbol{z}^{\mathsf{H}}\Lambda^{\mathsf{H}}\Lambda\boldsymbol{z} + \boldsymbol{z}^{\mathsf{H}}\mathsf{M}(\delta)\boldsymbol{z} \\
&\le \max_{k=1,\dots,n} |\lambda_k|^2\boldsymbol{z}^{\mathsf{H}}\boldsymbol{z} + \|\boldsymbol{z}\|_2 \|\mathsf{M}(\delta)\boldsymbol{z}\|_2 \\
&\le \left(\rho(\mathsf{A})^2 + \|\mathsf{M}(\delta)\|_2\right)\|\boldsymbol{z}\|_2^2 \\
&\le \left(\rho(\mathsf{A})^2 + K_1\delta\right)\|\boldsymbol{z}\|_2^2 .
\end{aligned}$$

To finish up, note that

$$\|\boldsymbol{y}\|_\star = \left\|(\mathsf{DP})^{-1}\boldsymbol{z}\right\|_\star = \|\boldsymbol{z}\|_2 .$$

Hence, $\|\boldsymbol{y}\|_\star = 1$ if and only if $\|\boldsymbol{z}\|_2 = 1$ and

$$\left\{\|\mathsf{A}\boldsymbol{y}\|_\star \mid \|\boldsymbol{y}\|_\star = 1\right\} = \left\{\|\mathsf{C}\boldsymbol{z}\|_\star \mid \|\boldsymbol{z}\|_2 = 1\right\}.$$

Consequently,

$$\begin{aligned}
\|\mathsf{A}\|_\star &= \sup_{\|\boldsymbol{y}\|_\star=1} \|\mathsf{A}\boldsymbol{y}\|_\star \\
&= \sup_{\|\boldsymbol{z}\|_2=1} \|\mathsf{C}\boldsymbol{z}\|_2 \\
&\le \sup_{\|\boldsymbol{z}\|_2=1} \sqrt{\rho(\mathsf{A})^2 + K_1\delta}\,\|\boldsymbol{z}\|_2 \\
&= \sqrt{\rho(\mathsf{A})^2 + K_1\delta} \le \rho(\mathsf{A}) + K_2\delta,
\end{aligned}$$

for some $K_2 > 0$, for all $0 < \delta \le 1$. The result follows on choosing

$$\delta \le \min(1, \varepsilon/K_2).$$

□

As a consequence of this result we can provide an extension of Theorem 4.2 for a broader class of matrices.

Corollary 4.5 (equality). *Suppose that* $A \in \mathbb{C}^{n \times n}$ *is diagonalizable. Then there exists a norm* $\|\cdot\|_\star : \mathbb{C}^n \to \mathbb{R}$ *such that the induced matrix norm satisfies*

$$\|A\|_\star = \rho(A).$$

Proof. See Problem 4.3. □

We provide now a notion of convergence for matrices and, with the aid of the spectral radius, provide necessary and sufficient conditions for convergence.

Definition 4.6 (convergence). We say that the square matrix $A \in \mathbb{C}^{n \times n}$ is **convergent to zero** if and only if $A^k \to O \in \mathbb{C}^{n \times n}$, i.e., if and only if

$$\lim_{k \to \infty} \|A^k\| \to 0$$

for any matrix norm $\|\cdot\| : \mathbb{C}^{n \times n} \to \mathbb{R}$.

Remark 4.7 (norm equivalence). We recall that, owing to Theorem A.29, all norms on $\mathbb{C}^{m \times n}$, whether induced or not, are equivalent. For this reason, the norm in this last definition does not matter.

Theorem 4.8 (convergence criteria). *Let* $A \in \mathbb{C}^{n \times n}$. *The following are equivalent.*

1. *A is convergent to zero.*
2. $\rho(A) < 1$.
3. *For all* $\mathbf{x} \in \mathbb{C}^n$,

$$\lim_{k \to \infty} A^k \mathbf{x} = \mathbf{0}.$$

Proof. ($1 \implies 2$) We recall two facts. First, if $\lambda \in \sigma(A)$, then $\lambda^k \in \sigma(A^k)$. This follows from the Schur factorization: if $A = UTU^H$, where T is upper triangular and U is unitary, then

$$A^k = UT^kU^H.$$

Second, $\rho(A) \le \|A\|$, for any induced matrix norm. Therefore,

$$0 \le \rho^k(A) = \rho(A^k) \le \left\|A^k\right\|.$$

Thus, if $\left\|A^k\right\| \to 0$, it follows that

$$\rho^k(A) \to 0.$$

This implies that $\rho(A) < 1$.

($2 \implies 1$) By Theorem 4.4, there is an induced matrix norm $\|\cdot\|_\star$ such that

$$\|A\|_\star \le \rho(A) + \varepsilon$$

for any $\varepsilon > 0$. Recall that the choice of $\|\cdot\|_\star$ depends upon A and $\varepsilon > 0$. Since, by assumption, $\rho(A) < 1$, there is an $\varepsilon > 0$ such that $\rho(A) + \varepsilon < 1$, and, therefore, an induced norm $\|\cdot\|_\star$ such that

$$\|A\|_\star \le \rho(A) + \varepsilon < 1.$$

Then, using sub-multiplicativity,

$$\left\|A^k\right\|_\star \leq \|A\|_\star^k \leq (\rho(A)+\varepsilon)^k \to 0.$$

Consequently,

$$\lim_{k\to\infty} \left\|A^k\right\|_\star = 0.$$

($1 \implies 3$) Suppose that $\lim_{k\to\infty} \left\|A^k\right\|_\infty = 0$. Let $\boldsymbol{x} \in \mathbb{C}^n$ be arbitrary. Then

$$\left\|A^k\boldsymbol{x}\right\|_\infty \leq \left\|A^k\right\|_\infty \|\boldsymbol{x}\|_\infty \to 0$$

since $\left\|A^k\right\|_\infty \to 0$. Hence, $\left\|A^k\boldsymbol{x}\right\|_\infty \to 0$. This implies that

$$\lim_{k\to\infty} A^k\boldsymbol{x} = \boldsymbol{0}.$$

($3 \implies 1$) Suppose that, for any $\boldsymbol{x} \in \mathbb{C}^n$,

$$\lim_{k\to\infty} A^k\boldsymbol{x} = \boldsymbol{0}.$$

Then it follows that, for all $\boldsymbol{x}, \boldsymbol{y} \in \mathbb{C}^n$,

$$\boldsymbol{y}^H A^k \boldsymbol{x} \to 0.$$

Now suppose that $\boldsymbol{y} = \boldsymbol{e}_i$ and $\boldsymbol{x} = \boldsymbol{e}_j$, then, since

$$\boldsymbol{y}^H A^k \boldsymbol{x} = \boldsymbol{e}_i^H A^k \boldsymbol{e}_j = \left[A^k\right]_{i,j},$$

it follows that

$$\lim_{k\to\infty} \left[A^k\right]_{i,j} = 0.$$

This implies that

$$\lim_{k\to\infty} \left\|A^k\right\|_{\max} = 0.$$

Hence, A is convergent to zero. □

As an easy consequence we obtain a sufficient criterion for convergence to zero.

Corollary 4.9 (convergence condition)**.** *Let* $M \in \mathbb{C}^{n\times n}$*. Assume that, for some induced matrix norm* $\|\cdot\| : \mathbb{C}^{n\times n} \to \mathbb{R}$,

$$\|M\| < 1,$$

then M *is convergent to zero.*

Proof. Recall that, for each and every induced matrix norm $\|\cdot\| : \mathbb{C}^{n\times n} \to \mathbb{R}$,

$$\rho(M) \leq \|M\|,$$

where $\rho(M)$ is the spectral radius of M. □

The spectral radius of a matrix can also be estimated via powers of this matrix.

Proposition 4.10 (upper bound). *Suppose that* $\|\cdot\| : \mathbb{C}^{n\times n} \to \mathbb{R}$ *is an induced matrix norm. Then, for all* $A \in \mathbb{C}^{n\times n}$ *and* $k \in \mathbb{N}$, *we have*

$$\rho(A) \leq \|A^k\|^{1/k}.$$

Proof. See Problem 4.4. □

In fact, for large values of k the previous upper bound is tight.

Theorem 4.11 (Gelfand[1]). *Suppose that* $\|\cdot\| : \mathbb{C}^{n\times n} \to \mathbb{R}$ *is an induced matrix norm. Then, for all* $A \in \mathbb{C}^{n\times n}$, *we have*

$$\rho(A) = \lim_{k\to\infty} \|A^k\|^{1/k}.$$

Proof. Let $0 < \varepsilon < \rho(A)/2$. We define two matrices

$$A_{\pm} = \frac{1}{\rho(A) \pm \varepsilon} A.$$

Clearly,

$$\rho(A_{\pm}) = \frac{\rho(A)}{\rho(A) \pm \varepsilon},$$

which implies that

$$\rho(A_+) < 1 < \rho(A_-).$$

Therefore, A_+ is convergent to zero; consequently, there is a number $K_+ \in \mathbb{N}$ such that, for $k \geq K_+$, we have

$$\|A_+^k\| < 1 \quad \Longrightarrow \quad \frac{1}{(\rho(A)+\varepsilon)^k}\|A^k\| < 1 \quad \Longrightarrow \quad \|A^k\| < (\rho(A)+\varepsilon)^k.$$

On the other hand, the bound of Proposition 4.10 implies that

$$1 < \rho(A_-)^k \leq \|A_-^k\| = \frac{1}{(\rho(A)-\varepsilon)^k}\|A^k\|.$$

In conclusion, for sufficiently large k, we have shown that

$$\rho(A) - \varepsilon < \|A^k\|^{1/k} < \rho(A) + \varepsilon$$

and the result follows. □

Corollary 4.12 (product of matrices). *Suppose that* $\|\cdot\| : \mathbb{C}^{n\times n} \to \mathbb{R}$ *is an induced matrix norm. Let* $\{A_i\}_{i=1}^k \subset \mathbb{C}^{n\times n}$ *be a family of matrices that commute, i.e.,*

$$A_i A_j = A_j A_i, \qquad \forall i, j = 1, \ldots, k.$$

Then

$$\rho\left(\prod_{i=1}^k A_i\right) \leq \prod_{i=1}^k \rho(A_i).$$

Proof. See Problem 4.6. □

[1] This result is due to the Ukrainian–American mathematician Izrail Moiseevic Gelfand (1913–2009).

4.2 Condition Number

We can now introduce the notion of the condition number of a matrix.

Definition 4.13 (condition number)**.** Suppose that $\mathsf{A} \in \mathbb{C}^{n\times n}$ is invertible. The **condition number** of A with respect to the matrix norm $\|\cdot\| : \mathbb{C}^{n\times n} \to \mathbb{R}$ is

$$\kappa(\mathsf{A}) = \|\mathsf{A}\| \left\|\mathsf{A}^{-1}\right\|.$$

Before we get on to the meaning and utility of the condition number, let us present some elementary properties of this quantity.

Proposition 4.14 (properties of κ)**.** *Suppose that $\|\cdot\| : \mathbb{C}^{n\times n} \to \mathbb{R}$ is an induced matrix norm and that $\mathsf{A} \in \mathbb{C}^{n\times n}$ is invertible. Then*

$$\kappa(\mathsf{A}) = \|\mathsf{A}\| \left\|\mathsf{A}^{-1}\right\| \geq 1.$$

Furthermore,

$$\frac{1}{\left\|\mathsf{A}^{-1}\right\|} \leq \|\mathsf{A} - \mathsf{B}\|$$

for any $\mathsf{B} \in \mathbb{C}^{n\times n}$ that is singular. Consequently,

$$\frac{1}{\kappa(\mathsf{A})} \leq \inf_{\det(\mathsf{B})=0} \frac{\|\mathsf{A} - \mathsf{B}\|}{\|\mathsf{A}\|}. \tag{4.3}$$

Proof. See Problem 4.10. □

Remark 4.15 (interpretation of $\kappa(\mathsf{A})$)**.** Estimate (4.3) is useful in a couple of ways. First, it says that if A is close in norm to a singular matrix B, then $\kappa(\mathsf{A})$ will be very large. Thus, nearly singular matrices are ill-conditioned. Second, this formula gives an upper bound on $\kappa(\mathsf{A})^{-1}$.

There are some nice formulas for and estimates of the condition number with respect to the induced matrix 2-norm, which is usually called the *spectral condition number* and denoted κ_2.

Proposition 4.16 (spectral condition number)**.** *Suppose that $\mathsf{A} \in \mathbb{C}^{n\times n}$ is invertible and $\|\cdot\|_2 : \mathbb{C}^{n\times n} \to \mathbb{R}$ is the induced matrix 2-norm.*

1. *If the singular values of A are $\sigma_1 \geq \sigma_2 \geq \cdots \geq \sigma_n > 0$,*

$$\kappa_2(\mathsf{A}) = \|\mathsf{A}\|_2 \left\|\mathsf{A}^{-1}\right\|_2 = \frac{\sigma_1}{\sigma_n}.$$

2. *If the eigenvalues of $\mathsf{B} = \mathsf{A}^{\mathsf{H}}\mathsf{A}$ are $0 < \mu_1 \leq \mu_2 \leq \cdots \leq \mu_n$, then*

$$\kappa_2(\mathsf{A}) = \sqrt{\frac{\mu_n}{\mu_1}}. \tag{4.4}$$

3. *Let, for $p \in [1, \infty]$, $\kappa_p(\mathsf{A}) = \|\mathsf{A}\|_p \cdot \left\|\mathsf{A}^{-1}\right\|_p$, where $\|\cdot\|_p$ is the induced matrix norm with respect to the p-norm. We have*

$$\kappa_2(\mathsf{A}) \leq \sqrt{\kappa_1(\mathsf{A})\kappa_\infty(\mathsf{A})}.$$

4.

$$\frac{1}{\kappa_2(\mathsf{A})} = \inf_{\det(\mathsf{B})=0} \frac{\|\mathsf{A}-\mathsf{B}\|_2}{\|\mathsf{A}\|_2}.$$

5. *If* A *is Hermitian, then*

$$\kappa_2(\mathsf{A}) = \frac{\max_{\lambda\in\sigma(\mathsf{A})}|\lambda|}{\min_{\lambda\in\sigma(\mathsf{A})}|\lambda|}.$$

6. *If* A *is Hermitian positive definite with eigenvalues* $0 < \lambda_1 \le \lambda_2 \le \cdots \le \lambda_n$, *then*

$$\kappa_2(\mathsf{A}) = \frac{\lambda_n}{\lambda_1}.$$

Proof. See Problem 4.12. □

The first and last items in Proposition 4.16 give an easy geometric interpretation of the spectral condition number. It is the ratio of the *maximal stretching* to the *minimal stretching* under the action of the matrix A.

Example 4.1 It is well known that $\mathsf{A} \in \mathbb{C}^{n\times n}$ is singular if and only if $\det(\mathsf{A}) = 0$. Thus, it may be thought that, similar to item 4 of Proposition 4.16, the quantity $|\det(\mathsf{A})|$ may also be used to quantify how close to singular a matrix can be. The following example shows that this is not necessarily the case.

Let $\mathsf{A} \in \mathbb{R}^{n\times n}$ have the singular value decomposition

$$\mathsf{A} = \mathsf{U}\Sigma\mathsf{V}^{\mathsf{H}}, \qquad \sigma_j = \frac{1}{j}, \quad j = 1, \dots, n.$$

Then

$$|\det(\mathsf{A})| = \prod_{j=1}^{n} \frac{1}{j} = \frac{1}{n!},$$

but, owing to the first item in Proposition 4.16,

$$\kappa_2(\mathsf{A}) = \frac{\sigma_1}{\sigma_n} = \frac{1}{1/n} = n.$$

Definition 4.17 (error and residual)**.** Given a matrix $\mathsf{A} \in \mathbb{C}^{n\times n}$ and a vector $\boldsymbol{f} \in \mathbb{C}^n$ with A nonsingular, let $\boldsymbol{x} \in \mathbb{C}^n$ solve (3.1). The **residual vector** with respect to $\boldsymbol{x}' \in \mathbb{C}^n$ is defined as

$$\boldsymbol{r} = \boldsymbol{r}(\boldsymbol{x}') = \boldsymbol{f} - \mathsf{A}\boldsymbol{x}' = \mathsf{A}(\boldsymbol{x} - \boldsymbol{x}').$$

The **error vector** with respect to $\boldsymbol{x}'$ is defined as

$$\boldsymbol{e} = \boldsymbol{e}(\boldsymbol{x}') = \boldsymbol{x} - \boldsymbol{x}'.$$

Consequently,

$$\mathsf{A}\boldsymbol{e} = \boldsymbol{r}.$$

It often happens that we have obtained an approximate solution $\boldsymbol{x}' \in \mathbb{C}^n$. We would like to have some measure of the error, but a direct measurement of the error would require the exact solution $\boldsymbol{x}$. The next best thing is the residual, which is an indirect measurement of the error, as the last definition suggests. The next theorem tells us how useful the residual is in determining the relative size of the error.

Theorem 4.18 (relative error estimate). *Let* $\mathsf{A} \in \mathbb{C}^{n\times n}$ *be invertible,* $\boldsymbol{f} \in \mathbb{C}^n_\star$*, and* $\boldsymbol{x}$ *solves* (3.1)*. Assume that* $\|\cdot\| : \mathbb{C}^{n\times n} \to \mathbb{R}$ *is the induced matrix norm with respect to the vector norm* $\|\cdot\| : \mathbb{C}^n \to \mathbb{R}$*. Then*

$$\frac{1}{\kappa(\mathsf{A})}\frac{\|\boldsymbol{r}\|}{\|\boldsymbol{f}\|} \leq \frac{\|\boldsymbol{e}\|}{\|\boldsymbol{x}\|} \leq \kappa(\mathsf{A})\frac{\|\boldsymbol{r}\|}{\|\boldsymbol{f}\|}.$$

Proof. Since $\boldsymbol{e} = \mathsf{A}^{-1}\boldsymbol{r}$, using consistency of the induced norm

$$\|\boldsymbol{e}\| = \left\|\mathsf{A}^{-1}\boldsymbol{r}\right\| \leq \left\|\mathsf{A}^{-1}\right\| \|\boldsymbol{r}\|.$$

Likewise,

$$\|\boldsymbol{f}\| = \|\mathsf{A}\boldsymbol{x}\| \leq \|\mathsf{A}\| \|\boldsymbol{x}\|,$$

which implies that

$$\frac{1}{\|\boldsymbol{x}\|} \leq \|\mathsf{A}\| \frac{1}{\|\boldsymbol{f}\|}.$$

Combining the first and third inequalities, we obtain the claimed upper bound,

$$\frac{\|\boldsymbol{e}\|}{\|\boldsymbol{x}\|} \leq \|\mathsf{A}\| \left\|\mathsf{A}^{-1}\right\| \frac{\|\boldsymbol{r}\|}{\|\boldsymbol{f}\|} = \kappa(\mathsf{A})\frac{\|\boldsymbol{r}\|}{\|\boldsymbol{f}\|}.$$

The rest of the proof is left to the reader as an exercise; see Problem 4.13. □

4.3 Perturbations and Matrix Conditioning

Let us now return to the motivating problem with which we began the chapter, i.e., trying to estimate how much the solution to (3.1) changes under perturbations to the data A and $\boldsymbol{f}$. It is not difficult to imagine a scenario where the data are perturbed. Perturbations may come from measurement errors, and so they are not exactly known. Or, perhaps, perturbations may be introduced when numbers are stored in finite precision in the computer. Let $\delta\mathsf{A} \in \mathbb{C}^{n\times n}$ and $\delta\boldsymbol{f} \in \mathbb{C}^n$ be known (or estimable) perturbations of the data. The problem that is actually solved then is

$$(\mathsf{A} + \delta\mathsf{A})(\boldsymbol{x} + \delta\boldsymbol{x}) = \boldsymbol{f} + \delta\boldsymbol{f}.$$

We then wish to provide an estimate for how large is the relative error $\frac{\|\delta\boldsymbol{x}\|}{\|\boldsymbol{x}\|}$. Formally, the perturbation $\delta\boldsymbol{x} \in \mathbb{C}^n$ is

$$\delta\boldsymbol{x} = (\mathsf{A} + \delta\mathsf{A})^{-1}(\boldsymbol{f} + \delta\boldsymbol{f}) - \boldsymbol{x},$$

provided that $A + \delta A$ is invertible. Observe that, since A is invertible, we have

$$(A + \delta A)^{-1} = (A(I_n + A^{-1}\delta A))^{-1} = (I_n + A^{-1}\delta A)^{-1}A^{-1}.$$

Therefore, we have reduced the question of the invertibility of $A + \delta A$ to a more general question: Given $M \in \mathbb{C}^{n\times n}$, when is $I_n \pm M$ invertible?

Theorem 4.19 (Neumann series). *Suppose that $\|\cdot\| : \mathbb{C}^{n\times n} \to \mathbb{R}$ is an induced matrix norm with respect to the vector norm $\|\cdot\| : \mathbb{C}^n \to \mathbb{R}$. Let $M \in \mathbb{C}^{n\times n}$ with $\|M\| < 1$. Then $I_n - M$ is invertible,*

$$\left\|(I_n - M)^{-1}\right\| \leq \frac{1}{1 - \|M\|}$$

and

$$(I_n - M)^{-1} = \sum_{k=0}^{\infty} M^k.$$

The series $\sum_{k=0}^{\infty} M^k$ is known as the Neumann series.[2]

Proof. Using the reverse triangle inequality and consistency, since $\|M\| < 1$, for any $\boldsymbol{x} \in \mathbb{C}^n$,

$$\|(I_n - M)\boldsymbol{x}\| \geq \big|\|\boldsymbol{x}\| - \|M\boldsymbol{x}\|\big| \geq (1 - \|M\|)\|\boldsymbol{x}\|.$$

This inequality implies that if $(I_n - M)\boldsymbol{x} = \boldsymbol{0}$, then $\boldsymbol{x} = \boldsymbol{0}$. Therefore, $I_n - M$ is invertible.

To obtain the norm estimate, notice that

$$\begin{aligned}
1 &= \|I_n\| \\
&= \left\|(I_n - M)(I_n - M)^{-1}\right\| \\
&= \left\|(I_n - M)^{-1} - M(I_n - M)^{-1}\right\| \\
&\geq \left\|(I_n - M)^{-1}\right\| - \|M\| \left\|(I_n - M)^{-1}\right\|,
\end{aligned}$$

where we have used the reverse triangle inequality and sub-multiplicativity. The upper bound of the quantity $\left\|(I_n - M)^{-1}\right\|$ now follows.

Finally, for $N \in \mathbb{N}$, define

$$R_N = \sum_{k=0}^{N} M^k.$$

Let us show that $R_N(I_n - M) \to I_n$ as $N \to \infty$. Indeed,

$$R_N(I_n - M) = \sum_{k=0}^{N} M^k(I_n - M) = \sum_{k=0}^{N} M^k - \sum_{k=0}^{N} M^{k+1} = I_n - M^{N+1},$$

which shows that, as $N \to \infty$,

$$\|R_N(I_n - M) - I_n\| = \left\|M^{N+1}\right\| \leq \|M\|^{N+1} \to 0,$$

using the sub-multiplicativity of the induced norm and the fact that $\|M\| < 1$. □

[2] Named in honor of the German mathematician Carl Gottfried Neumann (1832–1925).

A consequence of Theorem 4.19 is that the set of invertible matrices is *open*. In this context, this means that any matrix that is sufficiently close to an invertible one will also be invertible.

Corollary 4.20 (inverse of a perturbation). *Suppose that* $\|\cdot\| : \mathbb{C}^{n\times n} \to \mathbb{R}$ *is an induced matrix norm with respect to the vector norm* $\|\cdot\| : \mathbb{C}^n \to \mathbb{R}$. *If* $\mathsf{R} \in \mathbb{C}^{n\times n}$ *is invertible and* $\mathsf{T} \in \mathbb{C}^{n\times n}$ *satisfies*

$$\left\|\mathsf{R}^{-1}\right\| \|\mathsf{R} - \mathsf{T}\| < 1,$$

then T *is invertible.*

Proof. Notice that

$$\mathsf{T} = \mathsf{R}(\mathsf{I}_n - (\mathsf{I}_n - \mathsf{R}^{-1}\mathsf{T}));$$

therefore, T will be invertible provided that $\mathsf{I}_n - (\mathsf{I}_n - \mathsf{R}^{-1}\mathsf{T})$ is invertible. Define $\mathsf{M} = \mathsf{I}_n - \mathsf{R}^{-1}\mathsf{T}$ to conclude that, according to Theorem 4.19, we need $\|\mathsf{M}\| < 1$. Observe that

$$\|\mathsf{M}\| = \left\|\mathsf{I}_n - \mathsf{R}^{-1}\mathsf{T}\right\| = \left\|\mathsf{R}^{-1}(\mathsf{R} - \mathsf{T})\right\| \le \left\|\mathsf{R}^{-1}\right\| \|\mathsf{R} - \mathsf{T}\| < 1,$$

and so T is invertible. □

With these results at hand we can give an estimate for the relative size of the error in the problem we were originally interested in. Let us first begin by assuming that $\delta \boldsymbol{f} = \mathbf{0}$.

Theorem 4.21 (relative error estimate, case $\delta \boldsymbol{f} = \mathbf{0}$). *Let* $\mathsf{A} \in \mathbb{C}^{n\times n}$ *be invertible,* $\boldsymbol{f} \in \mathbb{C}^n$, *and* $\boldsymbol{x} \in \mathbb{C}^n$ *solves* (3.1). *Suppose that* $\|\cdot\| : \mathbb{C}^{n\times n} \to \mathbb{R}$ *is the induced matrix norm with respect to the vector norm* $\|\cdot\| : \mathbb{C}^n \to \mathbb{R}$. *Assume that* $\delta\mathsf{A} \in \mathbb{C}^{n\times n}$ *satisfies* $\|\mathsf{A}^{-1}\delta\mathsf{A}\| < 1$ *and that* $\boldsymbol{x} + \delta\boldsymbol{x} \in \mathbb{C}^n$ *solves the perturbed problem*

$$(\mathsf{A} + \delta\mathsf{A})(\boldsymbol{x} + \delta\boldsymbol{x}) = \boldsymbol{f}.$$

Then $\delta\boldsymbol{x}$ *is uniquely determined and*

$$\frac{\|\delta\boldsymbol{x}\|}{\|\boldsymbol{x}\|} \le \frac{\kappa(\mathsf{A})}{1 - \kappa(\mathsf{A})\frac{\|\delta\mathsf{A}\|}{\|\mathsf{A}\|}} \frac{\|\delta\mathsf{A}\|}{\|\mathsf{A}\|}.$$

Proof. Let us begin by repeating a previous computation. Since A is invertible, we can write $\mathsf{A} + \delta\mathsf{A} = \mathsf{A}(\mathsf{I}_n + \mathsf{A}^{-1}\delta\mathsf{A})$. Define $\mathsf{M} = -\mathsf{A}^{-1}\delta\mathsf{A}$, which satisfies $\|\mathsf{M}\| < 1$. Invoking Theorem 4.19 we conclude that $\mathsf{A} + \delta\mathsf{A}$ is invertible. Therefore, $\delta\boldsymbol{x}$ exists and is unique. In addition, we have

$$(\mathsf{A} + \delta\mathsf{A})^{-1} = (\mathsf{I}_n - \mathsf{M})^{-1}\mathsf{A}^{-1}$$

and $\left\|(\mathsf{I}_n - \mathsf{M})^{-1}\right\| \le \frac{1}{1 - \|\mathsf{M}\|}$. Moreover, the obvious estimate

$$\|\mathsf{M}\| \le \left\|\mathsf{A}^{-1}\right\| \|\delta\mathsf{A}\|$$

implies that

$$\frac{1}{1 - \|\mathsf{M}\|} \le \frac{1}{1 - \left\|\mathsf{A}^{-1}\right\| \|\delta\mathsf{A}\|}.$$

Now

$$\begin{aligned}\delta \boldsymbol{x} &= (\mathsf{A} + \delta\mathsf{A})^{-1}\boldsymbol{f} - \mathsf{A}^{-1}\boldsymbol{f} \\ &= (\mathsf{I}_n - \mathsf{M})^{-1}\mathsf{A}^{-1}\boldsymbol{f} - \mathsf{A}^{-1}\boldsymbol{f} \\ &= (\mathsf{I}_n - \mathsf{M})^{-1}(\mathsf{A}^{-1}\boldsymbol{f} - (\mathsf{I}_n - \mathsf{M})\mathsf{A}^{-1}\boldsymbol{f}) \\ &= (\mathsf{I}_n - \mathsf{M})^{-1}\mathsf{M}\mathsf{A}^{-1}\boldsymbol{f} \\ &= (\mathsf{I}_n - \mathsf{M})^{-1}\mathsf{M}\boldsymbol{x}.\end{aligned}$$

Consequently,

$$\|\delta\boldsymbol{x}\| \le \left\|(\mathsf{I}_n - \mathsf{M})^{-1}\right\| \; \|\mathsf{M}\| \; \|\boldsymbol{x}\| \le \frac{\left\|\mathsf{A}^{-1}\right\| \|\delta\mathsf{A}\|}{1 - \left\|\mathsf{A}^{-1}\right\| \|\delta\mathsf{A}\|}\|\boldsymbol{x}\| = \frac{\kappa(\mathsf{A})}{1 - \kappa(\mathsf{A})\frac{\|\delta\mathsf{A}\|}{\|\mathsf{A}\|}} \frac{\|\delta\mathsf{A}\|}{\|\mathsf{A}\|}\|\boldsymbol{x}\|.$$

The result follows. □

To conclude our discussion, let us see what happens when we perturb both A and $\boldsymbol{f}$.

Theorem 4.22 (relative error estimate, general case)**.** *Let* $\mathsf{A} \in \mathbb{C}^{n\times n}$ *be invertible,* $\boldsymbol{f} \in \mathbb{C}^n$, *and* $\boldsymbol{x} \in \mathbb{C}^n$ *solves* (3.1). *Suppose that* $\|\cdot\| : \mathbb{C}^{n\times n} \to \mathbb{R}$ *is an induced matrix norm with respect to the vector norm* $\|\cdot\| : \mathbb{C}^n \to \mathbb{R}$. *Assume that* $\delta\mathsf{A} \in \mathbb{C}^{n\times n}$ *satisfies* $\left\|\mathsf{A}^{-1}\delta\mathsf{A}\right\| < 1$, $\delta\boldsymbol{f} \in \mathbb{C}^n$ *is given, and* $\boldsymbol{x} + \delta\boldsymbol{x} \in \mathbb{C}^n$ *satisfies the perturbed problem*

$$(\mathsf{A} + \delta\mathsf{A})(\boldsymbol{x} + \delta\boldsymbol{x}) = \boldsymbol{f} + \delta\boldsymbol{f}.$$

Then $\delta\boldsymbol{x}$ *is uniquely determined and*

$$\frac{\|\delta\boldsymbol{x}\|}{\|\boldsymbol{x}\|} \le \frac{\kappa(\mathsf{A})}{1 - \kappa(\mathsf{A})\frac{\|\delta\mathsf{A}\|}{\|\mathsf{A}\|}}\left(\frac{\|\delta\boldsymbol{f}\|}{\|\boldsymbol{f}\|} + \frac{\|\delta\mathsf{A}\|}{\|\mathsf{A}\|}\right).$$

Proof. Let $\mathsf{M} = -\mathsf{A}^{-1}\delta\mathsf{A}$. We then have that $\boldsymbol{x} = \mathsf{A}^{-1}\boldsymbol{f}$ and $\boldsymbol{x} + \delta\boldsymbol{x} = (\mathsf{I}_n - \mathsf{M})^{-1}\mathsf{A}^{-1}(\boldsymbol{f} + \delta\boldsymbol{f})$. Therefore,

$$\begin{aligned}\delta\boldsymbol{x} &= (\mathsf{I}_n - \mathsf{M})^{-1}\mathsf{A}^{-1}(\boldsymbol{f} + \delta\boldsymbol{f}) - \mathsf{A}^{-1}\boldsymbol{f} \\ &= (\mathsf{I}_n - \mathsf{M})^{-1}\left(\mathsf{A}^{-1}\boldsymbol{f} + \mathsf{A}^{-1}\delta\boldsymbol{f} - (\mathsf{I}_n - \mathsf{M})\mathsf{A}^{-1}\boldsymbol{f}\right) \\ &= (\mathsf{I}_n - \mathsf{M})^{-1}(\mathsf{A}^{-1}\delta\boldsymbol{f} + \mathsf{M}\mathsf{A}^{-1}\boldsymbol{f}).\end{aligned}$$

This shows that

$$\|\delta\boldsymbol{x}\| \le \frac{1}{1 - \kappa(\mathsf{A})\frac{\|\delta\mathsf{A}\|}{\|\mathsf{A}\|}}\left(\left\|\mathsf{A}^{-1}\delta\boldsymbol{f}\right\| + \left\|\mathsf{M}\mathsf{A}^{-1}\boldsymbol{f}\right\|\right).$$

Notice also that

$$\left\|\mathsf{M}\mathsf{A}^{-1}\boldsymbol{f}\right\| = \|\mathsf{M}\boldsymbol{x}\| \le \|\mathsf{M}\|\|\boldsymbol{x}\| \le \left\|\mathsf{A}^{-1}\right\| \; \|\delta\mathsf{A}\| \; \|\boldsymbol{x}\| = \kappa(\mathsf{A})\frac{\|\delta\mathsf{A}\|}{\|\mathsf{A}\|}\|\boldsymbol{x}\|$$

and

$$\left\|\mathsf{A}^{-1}\delta\boldsymbol{f}\right\| \le \left\|\mathsf{A}^{-1}\right\| \|\delta\boldsymbol{f}\|\frac{\|\mathsf{A}\boldsymbol{x}\|}{\|\mathsf{A}\boldsymbol{x}\|} \le \kappa(\mathsf{A})\frac{\|\delta\boldsymbol{f}\|}{\|\boldsymbol{f}\|}\|\boldsymbol{x}\|.$$

The previous three inequalities, when combined, yield

$$\|\delta \boldsymbol{x}\| \leq \frac{\kappa(\mathsf{A})}{1-\kappa(\mathsf{A})\frac{\|\delta \mathsf{A}\|}{\|\mathsf{A}\|}}\left(\frac{\|\delta \boldsymbol{f}\|}{\|\boldsymbol{f}\|}+\frac{\|\delta \mathsf{A}\|}{\|\mathsf{A}\|}\right)\|\boldsymbol{x}\|,$$

as we intended to show. □

Problems

4.1 Does the spectral radius, introduced in Definition 4.1, define a norm?

4.2 Suppose that $\mathbb{V}$ is a finite-dimensional complex normed vector space. Show that if $\|\cdot\|$ is any induced operator norm, then $\rho(A) \leq \|A\|$ for all $A \in \mathcal{L}(\mathbb{V})$.

4.3 Prove Corollary 4.5.

4.4 Prove Proposition 4.10.

4.5 Prove, using induction on $n \in \mathbb{N}$, that if $\mathsf{A}, \mathsf{B} \in \mathbb{C}^{n\times n}$ commute, then they are simultaneously triangularizable, i.e., there is a nonsingular $\mathsf{P} \in \mathbb{C}^{n\times n}$ for which $\mathsf{P}^{-1}\mathsf{A}\mathsf{P}$ and $\mathsf{P}^{-1}\mathsf{B}\mathsf{P}$ are upper triangular.

Hint: Show that if $(\lambda, \boldsymbol{x})$ is an eigenpair of A, then so is $(\lambda, \mathsf{B}\boldsymbol{x})$.

4.6 Prove Corollary 4.12.

Hint: See the previous problem.

4.7 Let $\mathsf{A} \in \mathbb{C}^{n\times n}$ and $\mu \in \mathbb{C}$ be such that $|\mu| > \rho(\mathsf{A})$. Show that the series

$$\sum_{k=0}^{\infty} \frac{1}{\mu^k}\mathsf{A}^k$$

converges to $(\mathsf{I}_n - \mu^{-1}\mathsf{A})^{-1}$.

4.8 Let $\mathsf{A} \in \mathbb{C}^{n\times n}$. Define

$$\mathsf{S}_k = \mathsf{I}_n + \mathsf{A} + \cdots + \mathsf{A}^k.$$

a) Prove that the sequence $\{\mathsf{S}_k\}_{k=0}^{\infty}$ converges if and only if A is convergent to zero.

b) Prove that if A is convergent to zero, then $\mathsf{I} - \mathsf{A}$ is nonsingular and

$$\lim_{k\to\infty} \mathsf{S}_k = (\mathsf{I} - \mathsf{A})^{-1} .$$

4.9 Show that if $\|\mathsf{A}\| < 1$ for some induced matrix norm, then $\mathsf{I} - \mathsf{A}$ is nonsingular and

$$\frac{1}{1+\|\mathsf{A}\|} \leq \left\|(\mathsf{I}-\mathsf{A})^{-1}\right\| \leq \frac{1}{1-\|\mathsf{A}\|}.$$

4.10 Prove Proposition 4.14.

4.11 Suppose that $\|\cdot\| : \mathbb{C}^{n\times n} \to \mathbb{R}$ is the induced norm with respect to the vector norm $\|\cdot\| : \mathbb{C}^n \to \mathbb{R}$. Show that if λ is an eigenvalue of $\mathsf{A}^{\mathsf{H}}\mathsf{A}$, where $\mathsf{A} \in \mathbb{C}^{n\times n}$, then

$$0 \leq \lambda \leq \|\mathsf{A}^{\mathsf{H}}\|\,\|\mathsf{A}\| .$$

4.12 Prove Proposition 4.16.

4.13 Complete the proof of Theorem 4.18.

4.14 Let $\mathsf{A} \in \mathbb{C}^{n\times n}$ be nonsingular. Show that the condition numbers $\kappa_\infty(\mathsf{A})$ and $\kappa_1(\mathsf{A})$ will not change after permutation of rows or columns.

4.15 Suppose that $\|\cdot\| : \mathbb{C}^{n\times n} \to \mathbb{R}$ is a matrix norm and κ is the condition number defined with respect to it. Let $\mathsf{A} \in \mathbb{C}^{n\times n}$ be nonsingular and $0 \neq \alpha \in \mathbb{C}$. Show that $\kappa(\alpha\mathsf{A}) = \kappa(\mathsf{A})$.

4.16 Show that if $\mathsf{Q} \in \mathbb{C}^n$ is unitary, then $\kappa_2(\mathsf{Q}) = 1$.

4.17 Suppose that $\|\cdot\| : \mathbb{C}^{n\times n} \to \mathbb{R}$ is the induced norm with respect to the vector norm $\|\cdot\| : \mathbb{C}^n \to \mathbb{R}$ and κ is the condition number defined with respect to this norm. Let $\mathsf{A} \in \mathbb{C}^n$ be invertible. Show that

$$\kappa(\mathsf{A}) \geq \frac{\max_{\lambda\in\sigma(\mathsf{A})} |\lambda|}{\min_{\lambda\in\sigma(\mathsf{A})} |\lambda|}.$$

4.18 Let $\mathsf{A} = \mathsf{R}^{\mathsf{H}}\mathsf{R}$ with $\mathsf{R} \in \mathbb{C}^{n\times n}$ nonsingular. Give an expression for $\kappa_2(\mathsf{A})$ in terms of $\kappa_2(\mathsf{R})$.

4.19 Let $\mathsf{A} \in \mathbb{C}^{n\times n}$ be invertible, $\boldsymbol{f} \in \mathbb{C}^n_\star$, and $\boldsymbol{x} \in \mathbb{C}^n$ solves (3.1). Suppose that $\|\cdot\| : \mathbb{C}^{n\times n} \to \mathbb{R}$ is the induced norm with respect to the vector norm $\|\cdot\| : \mathbb{C}^n \to \mathbb{R}$. Let the perturbations $\delta\boldsymbol{x}, \delta\boldsymbol{f} \in \mathbb{C}^n$ satisfy $\mathsf{A}\delta\boldsymbol{x} = \delta\boldsymbol{f}$, so that $\mathsf{A}(\boldsymbol{x} + \delta\boldsymbol{x}) = \boldsymbol{f} + \delta\boldsymbol{f}$.

a) Prove the error (or perturbation) estimate

$$\frac{1}{\kappa(\mathsf{A})}\frac{\|\delta\boldsymbol{f}\|}{\|\boldsymbol{f}\|} \leq \frac{\|\delta\boldsymbol{x}\|}{\|\boldsymbol{x}\|} \leq \kappa(\mathsf{A})\frac{\|\delta\boldsymbol{f}\|}{\|\boldsymbol{f}\|}.$$

b) Show that, for any invertible matrix A, the upper bound for $\frac{\|\delta\boldsymbol{x}\|}{\|\boldsymbol{x}\|}$ above can be attained for suitable choices of $\boldsymbol{f}$ and $\delta\boldsymbol{f}$.

4.20 Show that, for every nonsingular $\mathsf{A} \in \mathbb{C}^{n\times n}$, we have

$$\frac{1}{n} \leq \frac{\kappa_\infty(\mathsf{A})}{\kappa_2(\mathsf{A})} \leq n.$$

4.21 Let

$$\mathsf{A} = \begin{bmatrix} 1.0000 & 2.0000 \\ 1.0001 & 2.0000 \end{bmatrix}.$$

a) Calculate $\kappa_1(\mathsf{A})$ and $\kappa_\infty(\mathsf{A})$.

b) Use (4.3) to obtain upper bounds on $\kappa_1(\mathsf{A})^{-1}$ and $\kappa_\infty(\mathsf{A})^{-1}$.

c) Suppose that you wish to solve $\mathsf{A}\boldsymbol{x} = \boldsymbol{f}$, where $\boldsymbol{f} = \begin{bmatrix} 3.0000 \\ 3.0001 \end{bmatrix}$. Instead of $\boldsymbol{x}$ you obtain the approximation $\boldsymbol{x}' = \boldsymbol{x} + \delta\boldsymbol{x} = \begin{bmatrix} 0.0000 \\ 1.5000 \end{bmatrix}$. For this approximation you discover $\boldsymbol{f}' = \boldsymbol{f} + \delta\boldsymbol{f} = \begin{bmatrix} 3.0000 \\ 3.0000 \end{bmatrix}$, where $\mathsf{A}\boldsymbol{x}' = \boldsymbol{f}'$. Calculate $\|\delta\boldsymbol{x}\|_1/\|\boldsymbol{x}\|_1$ exactly. (You will need the exact solution, of course.) Then use the general estimate

$$\frac{\|\delta\boldsymbol{x}\|}{\|\boldsymbol{x}\|} \leq \kappa(\mathsf{A})\frac{\|\delta\boldsymbol{f}\|}{\|\boldsymbol{f}\|}$$

to obtain an upper bound for $\|\delta\boldsymbol{x}\|_1/\|\boldsymbol{x}\|_1$. How good is $\|\delta\boldsymbol{f}\|_1/\|\boldsymbol{f}\|_1$ as indicator of the size of $\|\delta\boldsymbol{x}\|_1 / \|\boldsymbol{x}\|_1$?

5 Linear Least Squares Problem

We begin by providing some motivation for what we want to accomplish in this chapter — namely, the solution of problems like the following. Suppose that we are given a table of values

$$(x_k, y_k), \quad k = 1, \ldots, n,$$

that is obtained, say, by a series of measurements. We wish to find a function $y = y(x)$ that, in some sense, best represents the data of this table. In particular, we may have some reason to believe that, apart from some measurement error, there really is a such a function y — for example, $y(x) = \sin(2\pi x)$ — that would otherwise generate the observed data in the table.

We immediately see two possible solutions.

1. Find a function, say a polynomial p, that *interpolates* the points (x_k, y_k), i.e., $y_k = p(x_k)$ for all $k = 1, \ldots, n$. There are sufficient conditions that indicate when such a polynomial exists and is unique, and there are algorithms to efficiently find it. (We shall examine these later in Chapter 9.) However, this procedure can be *unstable*. But more importantly, since we are dealing with measurements, this approach does not take into account the fact that we might have errors and/or redundancy in the data.
2. Find a simple function — a linear function, for example $y = c_1 x + c_0$, where $c_0, c_1 \in \mathbb{C}$ — that *fits* the data in some exact or approximate sense. If we demand that it matches the data exactly, then

   $$y_k = c_1 x_k + c_0, \quad k = 1, \ldots, n.$$

   This can also be expressed in vector form as

   $$\begin{bmatrix} y_1 \\ \vdots \\ y_n \end{bmatrix} = \begin{bmatrix} x_1 & 1 \\ \vdots & \vdots \\ x_n & 1 \end{bmatrix} \begin{bmatrix} c_1 \\ c_0 \end{bmatrix},$$

   so that we obtain a system of linear equations of the form $\mathsf{A}\boldsymbol{c} = \boldsymbol{y}$ with

   $$\mathsf{A} = \begin{bmatrix} x_1 & 1 \\ \vdots & \vdots \\ x_n & 1 \end{bmatrix} \in \mathbb{C}^{n\times 2}, \quad \boldsymbol{c} = \begin{bmatrix} c_1 \\ c_0 \end{bmatrix} \in \mathbb{C}^2, \quad \boldsymbol{y} = \begin{bmatrix} y_1 \\ \vdots \\ y_n \end{bmatrix} \in \mathbb{C}^n.$$

Usually, n is much larger than 2, so that we end up with an *overdetermined* system of equations, i.e., there are more equations than unknowns. There is no solution in general.

On the other hand, if we only approximately enforce the matching conditions

$$y_k \approx c_1 x_k + c_0, \quad k = 1, \dots, n,$$

then it is not clear how to proceed. There may be an infinite number of ways that we can reasonably satisfy the approximation. The purpose of this chapter is to provide a way to tackle this problem, i.e., we will give a precise way to enforce the matching conditions and show how the numbers c_0, c_1 may be computed.

5.1 Linear Least Squares: Full Rank Setting

In general, it will be important to consider systems of the form $\mathsf{A}\boldsymbol{x} = \boldsymbol{f}$ with a *rectangular* matrix $\mathsf{A} \in \mathbb{C}^{m\times n}$, $m > n$. The question is this: *When can we expect that this system has a solution?* The answer is *almost never*!

The following result is standard in the theory of linear systems of equations.

Theorem 5.1 (existence). *Let* $\mathsf{A} \in \mathbb{C}^{m\times n}$ *and* $\boldsymbol{f} \in \mathbb{C}^m$. *If* $\operatorname{rank}(\mathsf{A})$ *coincides with the rank of the augmented matrix* $[\mathsf{A}|\boldsymbol{f}]$, *then the system of equations* $\mathsf{A}\boldsymbol{x} = \boldsymbol{f}$ *has at least one solution.*

Proof. Let $\mathsf{A} = [\boldsymbol{c}_1, \dots, \boldsymbol{c}_n]$ for $\boldsymbol{c}_i \in \mathbb{C}^m$. If $\operatorname{rank}(\mathsf{A}) = \operatorname{rank}([\mathsf{A}|\boldsymbol{f}])$, then $\boldsymbol{f} \in \operatorname{col}(\mathsf{A})$. Thus, there are $x_i \in \mathbb{C}$, for $i = 1, \dots, n$, such that

$$\sum_{i=1}^{n} x_i \boldsymbol{c}_i = \boldsymbol{f}.$$

Setting $\boldsymbol{x} = [x_1, \dots, x_n]^\intercal \in \mathbb{C}^n$, we immediately see that $\mathsf{A}\boldsymbol{x} = \boldsymbol{f}$. □

Systems of equations $\mathsf{A}\boldsymbol{x} = \boldsymbol{f}$ that have solutions are called *consistent*, and the previous theorem, essentially, is telling us that a system is consistent *provided* that $\boldsymbol{f} \in \operatorname{im}(\mathsf{A})$. But, in general, there is no reason to expect that the data $\boldsymbol{f}$ are in the range of the coefficient matrix A. What do we do in this case? For a given vector $\boldsymbol{z} \in \mathbb{C}^n$, we can always compute its *residual vector*

$$\boldsymbol{r}(\boldsymbol{z}) = \boldsymbol{f} - \mathsf{A}\boldsymbol{z}.$$

The vector $\boldsymbol{z} \in \mathbb{C}^n$ is a solution of $\mathsf{A}\boldsymbol{x} = \boldsymbol{f}$ if and only if $\boldsymbol{r}(\boldsymbol{z}) = \boldsymbol{0}$. If it is not possible to find a solution, the next best thing may be to find $\boldsymbol{z}$ that gives the smallest residual vector, measured in some norm.

Definition 5.2 (generalized solution). Suppose that $\|\cdot\| : \mathbb{C}^m \to \mathbb{R}$ is a norm. Given $\mathsf{A} \in \mathbb{C}^{m\times n}$, $m \geq n$, and $\boldsymbol{f} \in \mathbb{C}^m$, we say that $\boldsymbol{x} \in \mathbb{C}^n$ is a **weak** or **generalized** solution of the system $\mathsf{A}\boldsymbol{x} = \boldsymbol{f}$ if and only if

$$\boldsymbol{x} \in \operatorname*{argmin}_{\boldsymbol{w}\in\mathbb{C}^n} \|\boldsymbol{r}(\boldsymbol{w})\| = \operatorname*{argmin}_{\boldsymbol{w}\in\mathbb{C}^n} \|\boldsymbol{f} - \mathsf{A}\boldsymbol{w}\|.$$

We say that $\boldsymbol{x} \in \mathbb{C}^n$ is a **least squares** solution of the system $\mathsf{A}\boldsymbol{x} = \boldsymbol{f}$ if and only if

$$\boldsymbol{x} \in \underset{\boldsymbol{w}\in\mathbb{C}^n}{\operatorname{argmin}} \|\boldsymbol{r}(\boldsymbol{w})\|_{\ell^2(\mathbb{C}^m)}^2 = \underset{\boldsymbol{w}\in\mathbb{C}^n}{\operatorname{argmin}} \|\boldsymbol{f} - \mathsf{A}\boldsymbol{w}\|_{\ell^2(\mathbb{C}^m)}^2.$$

When these minima exist and are unique, we replace $\in$ with $=$.

Remark 5.3 (squaring)**.** Observe that, for the special case of the 2-norm, we square the norm. This is a cosmetic alteration, since $\boldsymbol{x}$ minimizes $\|\boldsymbol{r}(\boldsymbol{w})\|_{\ell^2(\mathbb{C}^m)}^2$ if and only if it minimizes $\|\boldsymbol{r}(\boldsymbol{w})\|_{\ell^2(\mathbb{C}^m)}$ over all $\boldsymbol{w} \in \mathbb{C}^n$.

Using the 2-norm leads to a relatively simple theory, and deriving the equations that determine the approximations we seek is straightforward. To this end, let us define

$$\Phi(\boldsymbol{z}) = \|\boldsymbol{r}(\boldsymbol{z})\|_{\ell^2(\mathbb{C}^m)}^2, \quad \boldsymbol{r}(\boldsymbol{z}) = \boldsymbol{f} - \mathsf{A}\boldsymbol{z}, \quad \forall \boldsymbol{z} \in \mathbb{C}^n. \tag{5.1}$$

We wish to find conditions that lead to a minimum of $\Phi(\boldsymbol{x})$. Let $\delta\boldsymbol{x} \in \mathbb{C}^n$ be arbitrary. The variation of Φ can be computed via

$$\begin{aligned}
\Phi(\boldsymbol{x} + \delta\boldsymbol{x}) - \Phi(\boldsymbol{x}) &= (\boldsymbol{f} - \mathsf{A}(\boldsymbol{x} + \delta\boldsymbol{x}), \boldsymbol{f} - \mathsf{A}(\boldsymbol{x} + \delta\boldsymbol{x}))_2 - (\boldsymbol{f} - \mathsf{A}\boldsymbol{x}, \boldsymbol{f} - \mathsf{A}\boldsymbol{x})_2 \\
&= (\boldsymbol{r} - \mathsf{A}\delta\boldsymbol{x}, \boldsymbol{r} - \mathsf{A}\delta\boldsymbol{x})_2 - (\boldsymbol{r}, \boldsymbol{r})_2 \\
&= (\boldsymbol{r}, \boldsymbol{r})_2 - (\mathsf{A}\delta\boldsymbol{x}, \boldsymbol{r})_2 - (\boldsymbol{r}, \mathsf{A}\delta\boldsymbol{x})_2 + (\mathsf{A}\delta\boldsymbol{x}, \mathsf{A}\delta\boldsymbol{x})_2 - (\boldsymbol{r}, \boldsymbol{r})_2 \\
&= -\overline{(\boldsymbol{r}, \mathsf{A}\delta\boldsymbol{x})_2} - (\boldsymbol{r}, \mathsf{A}\delta\boldsymbol{x})_2 + (\mathsf{A}\delta\boldsymbol{x}, \mathsf{A}\delta\boldsymbol{x})_2 \\
&= -2\Re\left((\mathsf{A}^{\mathsf{H}}\boldsymbol{r}, \delta\boldsymbol{x})_2\right) + (\mathsf{A}\delta\boldsymbol{x}, \mathsf{A}\delta\boldsymbol{x})_2.
\end{aligned}$$

Thus, if $\boldsymbol{x} \in \mathbb{C}^n$ satisfies

$$\boldsymbol{0} = \mathsf{A}^{\mathsf{H}}\boldsymbol{r} = \mathsf{A}^{\mathsf{H}}(\boldsymbol{f} - \mathsf{A}\boldsymbol{x}), \tag{5.2}$$

then $\boldsymbol{x}$ is a least squares solution to $\mathsf{A}\boldsymbol{x} = \boldsymbol{f}$ (and a minimizer of $\Phi(\boldsymbol{x})$). Note that (5.2) is equivalent to

$$\mathsf{A}^{\mathsf{H}}\mathsf{A}\boldsymbol{x} = \mathsf{A}^{\mathsf{H}}\boldsymbol{f},$$

which is called the *normal equation.* The following result provides necessary and sufficient conditions for when it has a unique solution.

Lemma 5.4 (full rank)**.** *Let* $\mathsf{A} \in \mathbb{C}^{m\times n}$ *with* $m \geq n$. *Then* $\mathsf{A}^{\mathsf{H}}\mathsf{A}$ *is Hermitian positive definite (HPD) if and only if* A *is full rank, i.e.,* $\operatorname{rank}(\mathsf{A}) = n$.

Proof. Clearly the matrix $\mathsf{A}^{\mathsf{H}}\mathsf{A}$ is Hermitian. By construction, this matrix is also nonnegative definite since, for any $\boldsymbol{x} \in \mathbb{C}^n$,

$$(\mathsf{A}^{\mathsf{H}}\mathsf{A}\boldsymbol{x}, \boldsymbol{x})_2 = (\mathsf{A}\boldsymbol{x}, \mathsf{A}\boldsymbol{x})_2 = \|\mathsf{A}\boldsymbol{x}\|_2 \geq 0.$$

($\Longrightarrow$) Suppose that $\mathsf{A}^{\mathsf{H}}\mathsf{A}$ is HPD and, to reach a contradiction, that $\operatorname{rank}(A) < n$. This is equivalent to saying that there is a nonzero $\boldsymbol{x} \in \mathbb{C}^n$ such that $\mathsf{A}\boldsymbol{x} = \boldsymbol{0}$. But then we must have $\mathsf{A}^{\mathsf{H}}\mathsf{A}\boldsymbol{x} = \boldsymbol{0}$ and $(\mathsf{A}^{\mathsf{H}}\mathsf{A}\boldsymbol{x}, \boldsymbol{x})_2 = 0$, contradicting the assumption that $\mathsf{A}^{\mathsf{H}}\mathsf{A}$ is positive definite.

($\Longleftarrow$) Let us now assume that A is of full rank. To reach a contradiction, suppose that $\mathsf{A}^{\mathsf{H}}\mathsf{A}$ is not positive definite. There must be a nonzero $\boldsymbol{x} \in \mathbb{C}_\star^n$ for which

$$0 = (\mathsf{A}^{\mathsf{H}}\mathsf{A}\boldsymbol{x}, \boldsymbol{x})_2 = (\mathsf{A}\boldsymbol{x}, \mathsf{A}\boldsymbol{x})_2 = \|\mathsf{A}\boldsymbol{x}\|_2^2,$$

which implies that

$$\mathsf{A}x = \mathbf{0} \implies x \in \ker(\mathsf{A}).$$

This contradicts the assumption that A has full rank. □

Before we tackle the general case, let us consider the case that the data are composed of real numbers.

Theorem 5.5 (least squares: real case)**.** *Let* $\mathsf{A} \in \mathbb{R}^{m\times n}$ *with* $m \geq n$ *and* $\operatorname{rank}(\mathsf{A}) = n$ *and let* $\boldsymbol{f} \in \mathbb{R}^m$*. The vector* $\boldsymbol{x} \in \mathbb{R}^n$ *is the unique least squares solution to* $\mathsf{A}\boldsymbol{x} = \boldsymbol{f}$*, i.e.,*

$$\boldsymbol{x} = \operatorname*{argmin}_{\boldsymbol{w}\in\mathbb{R}^n} \|\boldsymbol{f} - \mathsf{A}\boldsymbol{w}\|^2_{\ell^2(\mathbb{R}^m)}, \tag{5.3}$$

if and only if $\boldsymbol{x}$ *is the unique solution to the normal equation*

$$\mathsf{A}^\intercal\mathsf{A}\boldsymbol{x} = \mathsf{A}^\intercal\boldsymbol{f}. \tag{5.4}$$

Proof. ($\Longrightarrow$) Suppose that $\boldsymbol{x} \in \mathbb{R}^n$ solves the least squares problem. Set $\boldsymbol{r} = \boldsymbol{f} - \mathsf{A}\boldsymbol{x}$. Now fix $\boldsymbol{y} \in \mathbb{R}^n$ and define

$$g(s) = \Phi(\boldsymbol{x} + s\boldsymbol{y}) = \|\boldsymbol{r} - s\mathsf{A}\boldsymbol{y}\|_2^2$$

for all $s \in \mathbb{R}$, where $\Phi(\,\cdot\,)$ is as defined in (5.1). Then, for any $s \in \mathbb{R}$,

$$g(0) = \Phi(\boldsymbol{x}) \leq \Phi(\boldsymbol{x} + s\boldsymbol{y}) = g(s),$$

since $\boldsymbol{x}$ is a least squares solution. Calculating, we find

$$\begin{aligned} g(s) &= (\boldsymbol{r} - s\mathsf{A}\boldsymbol{y})^\intercal (\boldsymbol{r} - s\mathsf{A}\boldsymbol{y}) \\ &= \boldsymbol{r}^\intercal\boldsymbol{r} - s\boldsymbol{y}^\intercal\mathsf{A}^\intercal\boldsymbol{r} - s\boldsymbol{r}^\intercal\mathsf{A}\boldsymbol{y} + s^2\boldsymbol{y}^\intercal\mathsf{A}^\intercal\mathsf{A}\boldsymbol{y} \\ &= g(0) - 2s\,\boldsymbol{r}^\intercal\mathsf{A}\boldsymbol{y} + s^2\boldsymbol{y}^\intercal\mathsf{A}^\intercal\mathsf{A}\boldsymbol{y}. \end{aligned}$$

Since $\mathsf{A}^\intercal\mathsf{A}$ is symmetric positive definite (SPD), g is a positive quadratic function of one variable with a global minimum at $s = 0$. Hence,

$$0 = \left.\frac{dg}{ds}\right|_{s=0} = -2\boldsymbol{r}^\intercal\mathsf{A}\boldsymbol{y}$$

for arbitrary $\boldsymbol{y} \in \mathbb{R}^n$. From this condition, we conclude that $\boldsymbol{r}^\intercal\mathsf{A} = \mathbf{0}^\intercal$. This is equivalent to (5.4).

($\Longleftarrow$) Now suppose that $\boldsymbol{x} \in \mathbb{R}^n$ solves the normal equation (5.4). Set $\boldsymbol{r} = \boldsymbol{f} - \mathsf{A}\boldsymbol{x}$. This implies that $\boldsymbol{r}^\intercal\mathsf{A}\boldsymbol{y} = 0$ for all $\boldsymbol{y} \in \mathbb{R}^n$, or, equivalently, $\boldsymbol{r} \in \operatorname{im}(\mathsf{A})^\perp$. Then

$$\begin{aligned} \Phi(\boldsymbol{x} + \boldsymbol{y}) &= (\boldsymbol{r} - \mathsf{A}\boldsymbol{y})^\intercal (\boldsymbol{r} - \mathsf{A}\boldsymbol{y}) \\ &= \Phi(\boldsymbol{x}) - \boldsymbol{r}^\intercal\mathsf{A}\boldsymbol{y} - \boldsymbol{y}\mathsf{A}^\intercal\boldsymbol{r} + \boldsymbol{y}^\intercal\mathsf{A}^\intercal\mathsf{A}\boldsymbol{y} \\ &= \Phi(\boldsymbol{x}) + \boldsymbol{y}^\intercal\mathsf{A}^\intercal\mathsf{A}\boldsymbol{y} \\ &\geq \Phi(\boldsymbol{x}), \end{aligned}$$

since $\boldsymbol{y}^\intercal\mathsf{A}^\intercal\mathsf{A}\boldsymbol{y} \geq 0$ for any $\boldsymbol{y}$. More importantly, since $\mathsf{A}^\intercal\mathsf{A}$ is SPD,

$$\Phi(\boldsymbol{x} + \boldsymbol{y}) > \Phi(\boldsymbol{x}), \quad \forall\, \boldsymbol{y} \in \mathbb{R}^n_\star.$$

Thus, $\boldsymbol{x} \in \mathbb{R}^n$ is a least squares solution. □

Now let us treat the complex case. Note the slight difference in the approach.

Theorem 5.6 (least squares: complex case)**.** *Let* $\mathsf{A} \in \mathbb{C}^{m\times n}$ *with* $m \geq n$ *and* $\operatorname{rank}(\mathsf{A}) = n$ *and let* $\boldsymbol{f} \in \mathbb{C}^m$*. The vector* $\boldsymbol{x} \in \mathbb{C}^n$ *is the unique least squares solution to* $\mathsf{A}\boldsymbol{x} = \boldsymbol{f}$*, i.e.,*

$$\boldsymbol{x} = \operatorname*{argmin}_{\boldsymbol{w}\in\mathbb{C}^n} \|\boldsymbol{f} - \mathsf{A}\boldsymbol{w}\|_{\ell^2(\mathbb{C}^m)}^2, \tag{5.5}$$

if and only if $\boldsymbol{x}$ *is the unique solution to the normal equation*

$$\mathsf{A}^{\mathsf{H}}\mathsf{A}\boldsymbol{x} = \mathsf{A}^{\mathsf{H}}\boldsymbol{f}. \tag{5.6}$$

Proof. ($\Longrightarrow$) Suppose that $\boldsymbol{x} \in \mathbb{C}^n$ solves the least squares problem. Set $\boldsymbol{r} = \boldsymbol{f} - \mathsf{A}\boldsymbol{x}$. Now fix $\boldsymbol{y} \in \mathbb{C}^n$ and define

$$g(s,t) = \Phi(\boldsymbol{x} + z\boldsymbol{y}) = \|\boldsymbol{r} - z\mathsf{A}\boldsymbol{y}\|_2^2$$

for all $z = s + \mathrm{i}t \in \mathbb{C}$ with $s, t \in \mathbb{R}$, $\mathrm{i} = \sqrt{-1}$. Then, for any $s, t \in \mathbb{R}$,

$$g(0,0) = \Phi(\boldsymbol{x}) \leq \Phi(\boldsymbol{x} + z\boldsymbol{y}) = g(s,t),$$

since $\boldsymbol{x}$ is a least squares solution. Calculating, we find

$$\begin{aligned} g(s,t) &= (\boldsymbol{r} - z\mathsf{A}\boldsymbol{y})^{\mathsf{H}} (\boldsymbol{r} - z\mathsf{A}\boldsymbol{y}) \\ &= \boldsymbol{r}^{\mathsf{H}}\boldsymbol{r} - \bar{z}\boldsymbol{y}^{\mathsf{H}}\mathsf{A}^{\mathsf{H}}\boldsymbol{r} - z\boldsymbol{r}^{\mathsf{H}}\mathsf{A}\boldsymbol{y} + |z|^2\boldsymbol{y}^{\mathsf{H}}\mathsf{A}^{\mathsf{H}}\mathsf{A}\boldsymbol{y} \\ &= g(0,0) - 2\Re\left(z\boldsymbol{r}^{\mathsf{H}}\mathsf{A}\boldsymbol{y}\right) + \left(s^2 + t^2\right)\boldsymbol{y}^{\mathsf{H}}\mathsf{A}^{\mathsf{H}}\mathsf{A}\boldsymbol{y} \\ &= g(0,0) - 2s\Re\left(\boldsymbol{r}^{\mathsf{H}}\mathsf{A}\boldsymbol{y}\right) + 2t\Im\left(\boldsymbol{r}^{\mathsf{H}}\mathsf{A}\boldsymbol{y}\right) + \left(s^2 + t^2\right)\boldsymbol{y}^{\mathsf{H}}\mathsf{A}^{\mathsf{H}}\mathsf{A}\boldsymbol{y}. \end{aligned}$$

Since $\mathsf{A}^{\mathsf{H}}\mathsf{A}$ is HPD, g is a positive quadratic function of two variables. Hence,

$$0 = \left.\frac{\partial g}{\partial s}\right|_{s,t=0} = -2\Re\left(\boldsymbol{r}^{\mathsf{H}}\mathsf{A}\boldsymbol{y}\right), \quad 0 = \left.\frac{\partial g}{\partial t}\right|_{s,t=0} = 2\Im\left(\boldsymbol{r}^{\mathsf{H}}\mathsf{A}\boldsymbol{y}\right)$$

for arbitrary $\boldsymbol{y} \in \mathbb{C}^n$. From these two conditions, we conclude that $\boldsymbol{r}^{\mathsf{H}}\mathsf{A} = \boldsymbol{0}^{\intercal}$. Hence, $\boldsymbol{r} \in \operatorname{im}(\mathsf{A})^{\perp}$, as desired.

($\Longleftarrow$) This step is more or less the same as in the real case. Suppose that $\boldsymbol{x} \in \mathbb{C}^n$ satisfies $\boldsymbol{r} \in \operatorname{im}(\mathsf{A})^{\perp}$. Set $\boldsymbol{r} = \boldsymbol{f} - \mathsf{A}\boldsymbol{x}$. Then

$$\begin{aligned} \Phi(\boldsymbol{x} + \boldsymbol{y}) &= (\boldsymbol{r} + \mathsf{A}\boldsymbol{y})^{\mathsf{H}} (\boldsymbol{r} + \mathsf{A}\boldsymbol{y}) \\ &= \Phi(\boldsymbol{x}) + \boldsymbol{r}^{\mathsf{H}}\mathsf{A}\boldsymbol{y} + \boldsymbol{y}^{\mathsf{H}}\mathsf{A}^{\mathsf{H}}\boldsymbol{r} + \boldsymbol{y}^{\mathsf{H}}\mathsf{A}^{\mathsf{H}}\mathsf{A}\boldsymbol{y} \\ &= \Phi(\boldsymbol{x}) + \boldsymbol{y}^{\mathsf{H}}\mathsf{A}^{\mathsf{H}}\mathsf{A}\boldsymbol{y} \\ &\geq \Phi(\boldsymbol{x}), \end{aligned}$$

since $\boldsymbol{y}^{\mathsf{H}}\mathsf{A}^{\mathsf{H}}\mathsf{A}\boldsymbol{y} \geq 0$ for any $\boldsymbol{y}$. And, since $\mathsf{A}^{\mathsf{H}}\mathsf{A}$ is HPD,

$$\Phi(\boldsymbol{x} + \boldsymbol{y}) > \Phi(\boldsymbol{x}), \quad \forall\, \boldsymbol{y} \in \mathbb{C}^n_{\star}.$$

Thus, $\boldsymbol{x} \in \mathbb{C}^n$ is a least squares solution. $\square$

Remark 5.7 (weighted least squares)**.** Instead of doing our least squares computations using the norm $\|\cdot\|_{\ell^2(\mathbb{C}^m)}$, we could, in fact, use any norm on $\mathbb{C}^n$ that arises from an inner product. For example, suppose that $(\cdot,\cdot)_w\colon \mathbb{C}^n\times\mathbb{C}^n\to\mathbb{C}$ is an inner product and $\|\cdot\|_w$ is the norm induced by this inner product. Then we can use the inner product $(\cdot,\cdot)_w$ to construct a slightly different least squares theory, a kind of *w-weighted least squares theory*, if you like. Instead of the normal equation, we will get an interesting analogue. For instance, suppose that $\mathsf{B}\in\mathbb{C}^{n\times n}$ is HPD. Consider the inner product $(\cdot,\cdot)_{\mathsf{B}}$, defined by

$$(\boldsymbol{x},\boldsymbol{y})_{\mathsf{B}}=\boldsymbol{y}^{\mathsf{H}}\mathsf{B}\boldsymbol{x},\quad \forall \boldsymbol{x},\boldsymbol{y}\in\mathbb{C}^n.$$

Define

$$\Phi_{\mathsf{B}}(\boldsymbol{z})=\|\boldsymbol{r}(\boldsymbol{z})\|_{\mathsf{B}}^2,\quad \boldsymbol{r}(\boldsymbol{z})=\boldsymbol{f}-\mathsf{A}\boldsymbol{z},\quad \forall\boldsymbol{z}\in\mathbb{C}^n. \tag{5.7}$$

What is the analogue of the normal equations for this case? See Problem 5.5.

5.2 Projection Matrices

The previous section presented the basic theory for the linear least squares problem for the full rank case. To get a more detailed picture, and, in particular, to analyze the rank-deficient case, we need to develop some more tools. The first one of these is projection matrices.

Definition 5.8 (projection matrix)**.** The square matrix $\mathsf{P}\in\mathbb{C}^{n\times n}$ is called a **projection matrix** if and only if it is idempotent, i.e.,

$$\mathsf{P}^2=\mathsf{P}.$$

Proposition 5.9 (image of P)**.** *Let* $\mathsf{P}\in\mathbb{C}^{n\times n}$ *be a projection matrix and* $\boldsymbol{v}\in\operatorname{im}(\mathsf{P})$. *Then*

$$\mathsf{P}\boldsymbol{v}=\boldsymbol{v}.$$

Proof. If $\boldsymbol{v}\in\operatorname{im}(\mathsf{P})$, then there is a vector $\boldsymbol{w}\in\mathbb{C}^n$ such that $\mathsf{P}\boldsymbol{w}=\boldsymbol{v}$. Then

$$\mathsf{P}\boldsymbol{v}=\mathsf{P}(\mathsf{P}\boldsymbol{w})=\mathsf{P}^2\boldsymbol{w}=\mathsf{P}\boldsymbol{w}=\boldsymbol{v}.$$ □

Theorem 5.10 (properties of a projection matrix)**.** *Suppose that* $\mathsf{P}\in\mathbb{C}^{n\times n}$ *is a projection matrix. Then* $\mathsf{I}_n-\mathsf{P}$ *is also a projection matrix and*

1. $\operatorname{im}(\mathsf{I}_n-\mathsf{P})=\ker(\mathsf{P})$.
2. $\ker(\mathsf{I}_n-\mathsf{P})=\operatorname{im}(\mathsf{P})$.
3. $\operatorname{im}(\mathsf{P})\cap\ker(\mathsf{P})=\{\mathbf{0}\}$.
4. $\operatorname{im}(\mathsf{I}_n-\mathsf{P})\cap\ker(\mathsf{I}_n-\mathsf{P})=\{\mathbf{0}\}$.

Proof. We will prove the first property and leave the remaining ones as an exercise; see Problem 5.2. Suppose that $\boldsymbol{x}\in\operatorname{im}(\mathsf{I}_n-\mathsf{P})$. Then there is a vector $\boldsymbol{y}\in\mathbb{C}^n$ such that $(\mathsf{I}_n-\mathsf{P})\boldsymbol{y}=\boldsymbol{x}$. So,

$$\mathsf{P}\boldsymbol{x}=\mathsf{P}(\mathsf{I}_n-\mathsf{P})\boldsymbol{y}=\mathsf{P}\boldsymbol{y}-\mathsf{P}^2\boldsymbol{y}=\mathbf{0}.$$

Consequently $\mathsf{P}\boldsymbol{x} = \mathbf{0}$ and $\boldsymbol{x} \in \ker(\mathsf{P})$.

Suppose now that $\boldsymbol{x} \in \ker(\mathsf{P})$. Then $\mathsf{P}\boldsymbol{x} = \mathbf{0}$. Therefore,

$$(\mathsf{I}_n - \mathsf{P})\boldsymbol{x} = \boldsymbol{x},$$

which implies that $\boldsymbol{x} \in \mathrm{im}(\mathsf{I}_n - \mathsf{P})$. □

Definition 5.11 (sum of subspaces)**.** Let $S_1, S_2 \subseteq \mathbb{C}^n$ be subspaces. Recall that we write $S_1, S_2 \leq \mathbb{C}^n$ for short. Then

$$S_1 + S_2 = \{\boldsymbol{w} \in \mathbb{C}^n \mid \boldsymbol{w} = \boldsymbol{v}_1 + \boldsymbol{v}_2,\ \exists\, \boldsymbol{v}_i \in S_i,\ i = 1, 2\}.$$

Proposition 5.12 (property of the sum)**.** *Let $S_1, S_2 \leq \mathbb{C}^n$. Then $S_1 + S_2 \leq \mathbb{C}^n$.*

Proof. See Problem 5.6. □

Definition 5.13 (complementary subspaces)**.** Suppose that $S_1, S_2 \leq \mathbb{C}^n$. If $S_1 + S_2 = \mathbb{C}^n$ and $S_1 \cap S_2 = \{\mathbf{0}\}$, then we call S_1 and S_2 **complementary** subspaces and we write $S_1 \oplus S_2 = \mathbb{C}^n$.

Theorem 5.14 (decomposition)**.** *Suppose that $\mathsf{P} \in \mathbb{C}^{n\times n}$ is a projection matrix. Then $\mathrm{im}(\mathsf{P}) \oplus \ker(\mathsf{P}) = \mathbb{C}^n$, i.e., $\mathrm{im}(\mathsf{P})$ and $\ker(\mathsf{P})$ are complementary.*

Proof. This follows from Theorem 5.10 and the simple decomposition

$$\boldsymbol{v} = \mathsf{P}\boldsymbol{v} + \boldsymbol{v} - \mathsf{P}\boldsymbol{v}.$$ □

Theorem 5.15 (existence)**.** *Let $S_1, S_2 \leq \mathbb{C}^n$ be complementary subspaces. Then there is a projection matrix $\mathsf{P} \in \mathbb{C}^{n\times n}$ such that*

$$S_1 = \mathrm{im}(\mathsf{P}) \quad \text{and} \quad S_2 = \ker(\mathsf{P}).$$

Proof. Suppose that

$$B_i = \{\boldsymbol{w}_1^{(i)}, \ldots, \boldsymbol{w}_{k_i}^{(i)}\} \subset S_i$$

is a basis for S_i, $i = 1, 2$. Then it is left to the reader to prove that the set

$$B = B_1 \cup B_2$$

is a basis for $\mathbb{C}^n = S_1 \oplus S_2$. Now define a mapping $\mathsf{P}\colon \mathbb{C}^n \to \mathbb{C}^n$ such that

$$\mathsf{P}(\boldsymbol{w}_j^{(1)}) = \boldsymbol{w}_j^{(1)}, \quad j = 1, \ldots, k_1,$$

and

$$\mathsf{P}(\boldsymbol{w}_j^{(2)}) = \mathbf{0}, \quad j = 1, \ldots, k_2.$$

We require P to be linear, so that

$$\begin{aligned}
\mathsf{P}\left(\sum_{j=1}^{k_1} c_j^{(1)}\boldsymbol{w}_j^{(1)} + \sum_{j=1}^{k_2} c_j^{(2)}\boldsymbol{w}_j^{(2)}\right) &= \sum_{j=1}^{k_1} c_j^{(1)}\mathsf{P}(\boldsymbol{w}_j^{(1)}) + \sum_{j=1}^{k_2} c_j^{(2)}\mathsf{P}(\boldsymbol{w}_j^{(2)}) \\
&= \sum_{j=1}^{k_1} c_j^{(1)}\boldsymbol{w}_j^{(1)}.
\end{aligned}$$

It is now straightforward to prove that $\mathsf{P}^2 = \mathsf{P}$ and $\text{im}(\mathsf{P}) = S_1$ and $\ker(\mathsf{P}) = S_2$, as required. □

Definition 5.16 (orthogonal subspaces). Two subspaces $S_1, S_2 \leq \mathbb{C}^n$ are called **orthogonal** if and only if

$$(\boldsymbol{v}_1, \boldsymbol{v}_2)_{\ell^2(\mathbb{C}^n)} = (\boldsymbol{v}_1, \boldsymbol{v}_2)_2 = \boldsymbol{v}_2^{\mathsf{H}} \boldsymbol{v}_1 = 0$$

for all $\boldsymbol{v}_1 \in S_1$ and $\boldsymbol{v}_2 \in S_2$.

Proposition 5.17 (orthogonality). *If $S_1, S_2 \leq \mathbb{C}^n$ are orthogonal subspaces, then $S_1 \cap S_2 = \{\boldsymbol{0}\}$.*

Proof. See Problem 5.7. □

Proposition 5.18 (orthogonal decomposition). *Suppose that $L \leq \mathbb{C}^n$ has dimension $1 \leq k < n$. Then $L^\perp$ is a complementary subspace of dimension $n - k$,*

$$L \oplus L^\perp = \mathbb{C}^n.$$

Furthermore, the decomposition of any $\boldsymbol{w} \in \mathbb{C}^n$ into

$$\boldsymbol{w} = \boldsymbol{x} + \boldsymbol{y}, \quad \boldsymbol{x} \in L, \quad \boldsymbol{y} \in L^\perp$$

is unique.

Proof. See Problem 5.8. □

Remark 5.19 (notation). In light of Propositions 5.17 and 5.18, we see that whenever $S_1, S_2 \leq \mathbb{C}^n$ are orthogonal, then they are complementary. Moreover, whenever $\boldsymbol{s}_1 \in S_1$ and $\boldsymbol{s}_2 \in S_2$ we must necessarily have that $\boldsymbol{s}_2^{\mathsf{H}} \boldsymbol{s}_1 = 0$. We will indicate this fact by

$$\mathbb{C}^n = S_1 \overset{\perp}{\oplus} S_2.$$

As the concepts of complementary and orthogonal subspaces can be naturally extended to any vector space that has an inner product, a similar notation will be used in this case.

Definition 5.20 (orthogonal projection). The matrix $\mathsf{P} \in \mathbb{C}^{n \times n}$ is called an **orthogonal projection** if and only if $\mathsf{P} = \mathsf{P}^2$ and $\text{im}(\mathsf{P})$ and $\ker(\mathsf{P})$ are orthogonal subspaces.

Theorem 5.21 (characterization of orthogonal projection). *Let $\mathsf{P} \in \mathbb{C}^{n \times n}$ be a projection matrix. Then P is an orthogonal projection if and only if $\mathsf{P} = \mathsf{P}^{\mathsf{H}}$.*

Proof. Suppose that $\mathsf{P}^{\mathsf{H}} = \mathsf{P}$. Let $\boldsymbol{v}_1 \in \ker(\mathsf{P})$ and $\boldsymbol{v}_2 \in \text{im}(\mathsf{P})$ be arbitrary. Then

$$\mathsf{P}\boldsymbol{v}_2 = \boldsymbol{v}_2, \qquad (\mathsf{I}_n - \mathsf{P})\boldsymbol{v}_1 = \boldsymbol{v}_1,$$

and

$$\boldsymbol{v}_2^{\mathsf{H}} \boldsymbol{v}_1 = (\mathsf{P}\boldsymbol{v}_2)^{\mathsf{H}} (\mathsf{I}_n - \mathsf{P})\boldsymbol{v}_1 = \boldsymbol{v}_2^{\mathsf{H}} \mathsf{P}^{\mathsf{H}} (\mathsf{I}_n - \mathsf{P})\boldsymbol{v}_1 = \boldsymbol{v}_2^{\mathsf{H}} (\mathsf{P} - \mathsf{P}^2)\boldsymbol{v}_1 = 0.$$

Thus, $\text{im}(\mathsf{P})$ and $\ker(\mathsf{P})$ are orthogonal subspaces, which implies that P is an orthogonal projection.

Suppose now that P is an orthogonal projection. Set

$$S_1 = \operatorname{im}(\mathsf{P}), \qquad S_2 = \ker(\mathsf{P})$$

with

$$\dim(S_1) = k < n, \qquad \dim(S_2) = n - k.$$

We want to prove that $\mathsf{P}^{\mathsf{H}} = \mathsf{P}$, using the fact that S_1 and S_2 are orthogonal subspaces of $\mathbb{C}^n$. Let

$$B_1 = \{\boldsymbol{q}_1, \ldots, \boldsymbol{q}_k\} \subset S_1, \qquad B_2 = \{\boldsymbol{q}_{k+1}, \ldots, \boldsymbol{q}_n\} \subset S_2$$

be orthonormal bases for the respective spaces. This is always possible owing to the Gram–Schmidt algorithm, as described in Theorem A.38. We leave it as an exercise for the reader to prove that $B = B_1 \cup B_2$ is an orthonormal basis for $\mathbb{C}^n$. (Use the fact that S_1 and S_2 are complementary orthogonal subspaces.)

By our construction,

$$\mathsf{P}\boldsymbol{q}_j = \begin{cases} \boldsymbol{q}_j, & j = 1, \ldots, k, \\ \mathbf{0}, & j = k+1, \ldots, n. \end{cases}$$

Set

$$\mathsf{Q} = \begin{bmatrix} | & & | \\ \boldsymbol{q}_1 & \cdots & \boldsymbol{q}_n \\ | & & | \end{bmatrix} \in \mathbb{C}^{n \times n}.$$

Then

$$\mathsf{PQ} = \begin{bmatrix} | & & | & | & & | \\ \boldsymbol{q}_1 & \cdots & \boldsymbol{q}_k & \mathbf{0} & \cdots & \mathbf{0} \\ | & & | & | & & | \end{bmatrix} \in \mathbb{C}^{n \times n}$$

and

$$\mathsf{Q}^{\mathsf{H}}\mathsf{PQ} = \begin{bmatrix} \mathsf{I}_k & \mathsf{O} \\ \mathsf{O} & \mathsf{O}_{n-k} \end{bmatrix} = \Sigma \in \mathbb{C}^{n \times n}.$$

Consequently,

$$\mathsf{P} = \mathsf{Q}\Sigma\mathsf{Q}^{\mathsf{H}}$$

and $\mathsf{P}^{\mathsf{H}} = \mathsf{P}$. □

Remark 5.22 (representation)**.** In the notation of the proof of Theorem 5.21, let us observe that $\mathsf{P} = \hat{\mathsf{Q}}\hat{\mathsf{Q}}^{\mathsf{H}}$, where

$$\hat{\mathsf{Q}} = \begin{bmatrix} | & & | \\ \boldsymbol{q}_1 & \cdots & \boldsymbol{q}_k \\ | & & | \end{bmatrix} \in \mathbb{C}^{n \times k}, \quad n > k.$$

Theorem 5.23 (special projectors)**.** *Let $k \in \{1, \ldots, n\}$. Suppose that the collection of vectors $\{\boldsymbol{q}_1, \ldots, \boldsymbol{q}_k\} \subset \mathbb{C}^n$ is orthonormal. Define*

$$\hat{\mathsf{Q}} = \begin{bmatrix} | & & | \\ \boldsymbol{q}_1 & \cdots & \boldsymbol{q}_k \\ | & & | \end{bmatrix} \in \mathbb{C}^{n \times k},$$

then the matrices

$$\mathsf{P} = \hat{\mathsf{Q}}\hat{\mathsf{Q}}^{\mathsf{H}}, \quad \mathsf{I}_n - \mathsf{P} = \mathsf{I}_n - \hat{\mathsf{Q}}\hat{\mathsf{Q}}^{\mathsf{H}}$$

are orthogonal projectors.

Definition 5.24 (rank-one projection)**.** Suppose that $\boldsymbol{q} \in \mathbb{C}^n$, $\|\boldsymbol{q}\|_2 = 1$. The matrix $\mathsf{P} = \boldsymbol{q}\boldsymbol{q}^{\mathsf{H}}$ is called a **rank-one orthogonal projection**. The complement, $\mathsf{I}_n - \boldsymbol{q}\boldsymbol{q}^{\mathsf{H}}$, is called a **rank-$(n-1)$ orthogonal projection**.

Theorem 5.25 (norm of a projection)**.** *Let $\mathsf{P} \in \mathbb{C}^{n\times n}$ be a nonzero projection matrix. Then $\|\mathsf{P}\|_2 \geq 1$. Moreover, $\|\mathsf{P}\|_2 = 1$ if and only if P is an orthogonal projection, i.e., $\mathsf{P}^{\mathsf{H}} = \mathsf{P}$.*

Proof. Suppose that P is a projection matrix, i.e., $\mathsf{P}^2 = \mathsf{P}$. Then, using the sub-multiplicativity of the 2-norm,

$$\|\mathsf{P}\|_2 = \left\|\mathsf{P}^2\right\|_2 \leq \|\mathsf{P}\|_2 \|\mathsf{P}\|_2 .$$

Since P is not the zero matrix, $\|\mathsf{P}\|_2 > 0$ and $\|\mathsf{P}\|_2 \geq 1$.

Now, for the second part, we have two directions to prove. Assume first that $\mathsf{P}^2 = \mathsf{P}$ and $\mathsf{P}^{\mathsf{H}} = \mathsf{P}$. In the proof of Theorem 5.21, we showed that

$$\mathsf{P} = \mathsf{Q}\Sigma\mathsf{Q}^{\mathsf{H}},$$

where $\mathsf{Q} = [\boldsymbol{q}_1, \ldots, \boldsymbol{q}_n] \in \mathbb{C}^{n\times n}$ is unitary and

$$\Sigma = \begin{bmatrix} \mathsf{I}_k & \mathsf{O} \\ \mathsf{O} & \mathsf{O}_{n-k} \end{bmatrix}$$

for some $1 \leq k < n$. In other words, $\Sigma = \operatorname{diag}[\sigma_1, \ldots, \sigma_k, 0, \ldots, 0]$, where $\sigma_i = 1$, $1 \leq i \leq k$. Recall that $\{\boldsymbol{q}_1, \ldots, \boldsymbol{q}_k\}$ is a basis for $\operatorname{im}(\mathsf{P})$ and $\{\boldsymbol{q}_{k+1}, \ldots, \boldsymbol{q}_n\}$ is a basis for $\ker(\mathsf{P})$. In any case, P is Hermitian positive semi-definite (HPSD), $\mathsf{P} = \mathsf{Q}\Sigma\mathsf{Q}^{\mathsf{H}}$ is a unitary diagonalization of P, and

$$\|\mathsf{P}\|_2 = \sigma_1 = \rho(\mathsf{P}) = 1.$$

The other implication is proved by the contrapositive. Let us assume that $\mathsf{P}^2 = \mathsf{P}$, but $\mathsf{P}^{\mathsf{H}} \neq \mathsf{P}$. We want to show that this implies that

$$\|\mathsf{P}\|_2 > 1.$$

Since $\mathsf{P}^{\mathsf{H}} \neq \mathsf{P}$, $\ker(\mathsf{P}) \cap \operatorname{im}(\mathsf{P}) = \{\boldsymbol{0}\}$, but

$$\ker(\mathsf{P}) \not\perp \operatorname{im}(\mathsf{P}).$$

So, there is some nonzero vector $\boldsymbol{v}_1 \in \operatorname{im}(\mathsf{P})$ and some nonzero vector $\boldsymbol{v}_2 \in \ker(\mathsf{P})$ such that

$$(\boldsymbol{v}_1, \boldsymbol{v}_2)_2 = \boldsymbol{v}_2^{\mathsf{H}}\boldsymbol{v}_1 \neq 0.$$

Set

$$\boldsymbol{v} = \boldsymbol{v}_1 + \alpha\boldsymbol{v}_2, \quad \alpha \in \mathbb{C}.$$

Then $\mathsf{P}\boldsymbol{v} = \boldsymbol{v}_1$. Now we want to choose $\alpha \in \mathbb{C}$, so that

$$\|\mathsf{P}\boldsymbol{v}\|_2 > \|\boldsymbol{v}\|_2 > 0.$$

Indeed, if such an $\alpha \in \mathbb{C}$ exists, then

$$\|\mathsf{P}\|_2 = \sup_{x \in \mathbb{C}^n_\star} \frac{\|\mathsf{P}\boldsymbol{x}\|_2}{\|\boldsymbol{x}\|_2} \geq \frac{\|\mathsf{P}\boldsymbol{v}\|_2}{\|\boldsymbol{v}\|_2} > 1.$$

Since

$$\mathsf{P}\boldsymbol{v} = \mathsf{P}(\boldsymbol{v}_1 + \alpha \boldsymbol{v}_2) = \mathsf{P}\boldsymbol{v}_1 + \alpha \mathsf{P}\boldsymbol{v}_2 = \boldsymbol{v}_1,$$

it follows that

$$\|\boldsymbol{v}_1\|_2^2 = \|\mathsf{P}\boldsymbol{v}\|_2^2$$

and

$$\|\boldsymbol{v}\|_2^2 = \|\boldsymbol{v}_1 + \alpha \boldsymbol{v}_2\|_2^2 = \|\mathsf{P}\boldsymbol{v}\|_2^2 + 2\Re\left(\alpha \boldsymbol{v}_1^{\mathsf{H}} \boldsymbol{v}_2\right) + |\alpha|^2 \|\boldsymbol{v}_2\|_2^2 = \|\mathsf{P}\boldsymbol{v}\|_2^2 + T,$$

where

$$T = 2\Re\left(\alpha \boldsymbol{v}_1^{\mathsf{H}} \boldsymbol{v}_2\right) + |\alpha|^2 \|\boldsymbol{v}_2\|_2^2.$$

Therefore, it suffices to choose $\alpha \in \mathbb{C}$, so that $T < 0$, for, in that case,

$$\|\boldsymbol{v}\|_2^2 < \|\mathsf{P}\boldsymbol{v}\|_2^2.$$

The key is to choose $\alpha \in \mathbb{C}$ such that $\boldsymbol{v} \perp \boldsymbol{v}_2$, i.e., $\boldsymbol{v}_2^{\mathsf{H}}\boldsymbol{v} = 0$. This is equivalent to

$$\begin{aligned}
& \boldsymbol{v}_2^{\mathsf{H}}(\boldsymbol{v}_1 + \alpha \boldsymbol{v}_2) = 0 \\
\iff \quad & \boldsymbol{v}_2^{\mathsf{H}}\boldsymbol{v}_1 = -\alpha \boldsymbol{v}_2^{\mathsf{H}}\boldsymbol{v}_2 \\
\iff \quad & \bar{\alpha}\boldsymbol{v}_2^{\mathsf{H}}\boldsymbol{v}_1 = -|\alpha|^2 \|\boldsymbol{v}_2\|_2^2 \in \mathbb{R} \\
\iff \quad & \overline{\bar{\alpha}\boldsymbol{v}_1^{\mathsf{H}}\boldsymbol{v}_2} = -|\alpha|^2 \|\boldsymbol{v}_2\|_2^2 \in \mathbb{R} \\
\iff \quad & \alpha\boldsymbol{v}_1^{\mathsf{H}}\boldsymbol{v}_2 = -|\alpha|^2 \|\boldsymbol{v}_2\|_2^2 \in \mathbb{R}.
\end{aligned}$$

In this case,

$$T = -|\alpha|^2 \|\boldsymbol{v}_2\|_2^2 \in \mathbb{R}.$$

Thus, the result follows upon choosing

$$\alpha = -\frac{\boldsymbol{v}_2^{\mathsf{H}}\boldsymbol{v}_1}{\|\boldsymbol{v}_2\|_2^2}. \qquad \square$$

5.3 Linear Least Squares: The Rank-Deficient Case

In this section, we address the the rank-deficient case, i.e., the case when $\mathsf{A} \in \mathbb{C}^{m\times n}$, $m > n$, and $\operatorname{rank}(\mathsf{A}) < n$. In this setting, it is not clear, at first glance, that the normal equations will have a solution.

Theorem 5.26 (general least squares). *Suppose that $m \geq n$ and the matrix $\mathsf{A} \in \mathbb{C}^{m\times n}$ is such that* $\operatorname{rank}(\mathsf{A}) \leq n$*, i.e.,* A *may be rank deficient. Let $\boldsymbol{f} \in \mathbb{C}^m$ be given. Then the normal equations,*

$$\mathsf{A}^{\mathsf{H}}\mathsf{A}\boldsymbol{x} = \mathsf{A}^{\mathsf{H}}\boldsymbol{f}, \tag{5.8}$$

always have at least one solution; for any two solutions, $\boldsymbol{x}_1, \boldsymbol{x}_2 \in \mathbb{C}^n$ — in the case that there are multiple solutions — we find

$$\boldsymbol{r}(\boldsymbol{x}_1) = \boldsymbol{r}(\boldsymbol{x}_2),$$

where, for $\boldsymbol{w} \in \mathbb{C}^n$, we defined

$$\boldsymbol{r}(\boldsymbol{w}) = \boldsymbol{f} - \mathsf{A}\boldsymbol{w}.$$

In other words, the residual is always unique. Furthermore, the following are equivalent.

1. *$\boldsymbol{x}_o \in \mathbb{C}^n$ is a solution to*

$$\boldsymbol{x} \in \operatorname*{argmin}_{\boldsymbol{w}\in\mathbb{C}^n} \Phi(\boldsymbol{w}), \quad \Phi(\boldsymbol{w}) = \|\boldsymbol{r}(\boldsymbol{w})\|_2^2. \tag{5.9}$$

2. *$\boldsymbol{x}_o \in \mathbb{C}^n$ is a solution to the normal equations (5.8).*
3. *$\boldsymbol{x}_o \in \mathbb{C}^n$ has the property that*

$$\boldsymbol{r}(\boldsymbol{x}_o) \perp \operatorname{im}(\mathsf{A}).$$

Proof. First, let us prove that the normal equations (5.8) have a solution. Set $L = \operatorname{im}(\mathsf{A})$. Then L and $L^\perp$ are complementary, orthogonal subspaces of $\mathbb{C}^m$:

$$\mathbb{C}^m = L \oplus L^\perp.$$

Therefore, the decomposition

$$\boldsymbol{f} = \boldsymbol{s} + \boldsymbol{r}, \quad \boldsymbol{s} \in L = \operatorname{im}(\mathsf{A}), \quad \boldsymbol{r} \in L^\perp$$

is unique. Since $\boldsymbol{s} \in \operatorname{im}(\mathsf{A})$, there is at least one vector $\boldsymbol{x}_o \in \mathbb{C}^n$ such that

$$\mathsf{A}\boldsymbol{x}_o = \boldsymbol{s}.$$

Since $\boldsymbol{r} \in L^\perp$, $\boldsymbol{r}^{\mathsf{H}}\mathsf{A}\boldsymbol{x} = 0$ for all $\boldsymbol{x} \in \mathbb{C}^n$. This implies that

$$\mathsf{A}^{\mathsf{H}}\boldsymbol{r} = \boldsymbol{0} \in \mathbb{C}^n.$$

Recall that

$$\boldsymbol{f} = \mathsf{A}\boldsymbol{x}_o + \boldsymbol{r},$$

which implies that

$$\boldsymbol{r} = \boldsymbol{f} - \mathsf{A}\boldsymbol{x}_o = \boldsymbol{r}(\boldsymbol{x}_o).$$

Hence,

$$\mathsf{A}^{\mathsf{H}}\boldsymbol{r}(\boldsymbol{x}_o) = \boldsymbol{0} \in \mathbb{C}^n,$$

which is equivalent to the normal equations. Thus, the normal equations have at least one solution. This solution is not necessarily unique. However, since the decomposition $\boldsymbol{f} = \boldsymbol{s} + \boldsymbol{r}$ is unique, this is enough to prove that the residual is uniquely determined. Nevertheless, we show this explicitly.

Suppose that $\boldsymbol{x}_1, \boldsymbol{x}_2 \in \mathbb{C}^n$ are solutions to the normal equations (5.8). Recall that the decomposition

$$\boldsymbol{f} = \boldsymbol{s} + \boldsymbol{r}, \quad \boldsymbol{s} \in L = \operatorname{im}(\mathsf{A}), \quad \boldsymbol{r} \in L^{\perp}$$

is unique. But

$$\boldsymbol{f} = \mathsf{A}\boldsymbol{x}_1 + \boldsymbol{r}(\boldsymbol{x}_1) = \mathsf{A}\boldsymbol{x}_2 + \boldsymbol{r}(\boldsymbol{x}_2),$$

so $\boldsymbol{s} = \mathsf{A}\boldsymbol{x}_1 = \mathsf{A}\boldsymbol{x}_2$ and $\boldsymbol{r} = \boldsymbol{r}(\boldsymbol{x}_1) = \boldsymbol{r}(\boldsymbol{x}_2)$.

Let us now prove the equivalences.

(2 $\Longleftrightarrow$ 3) This follows from the calculations we carried out above.

(2 $\Longrightarrow$ 1) This argument is similar to previous ones. Suppose that $\boldsymbol{x}_o \in \mathbb{C}^n$ is a solution to the normal equation (5.8). Let $\boldsymbol{w} \in \mathbb{C}^n$ be arbitrary. Then

$$\begin{aligned}\Phi(\boldsymbol{x}_o + \boldsymbol{w}) &= (\boldsymbol{r}(\boldsymbol{x}_o) - \mathsf{A}\boldsymbol{w})^{\mathsf{H}} (\boldsymbol{r}(\boldsymbol{x}_o) - \mathsf{A}\boldsymbol{w}) \\ &= \Phi(\boldsymbol{x}_o) - (\boldsymbol{r}(\boldsymbol{x}_o))^{\mathsf{H}}\mathsf{A}\boldsymbol{w} - \boldsymbol{w}^{\mathsf{H}}\mathsf{A}^{\mathsf{H}}\boldsymbol{r}(\boldsymbol{x}_o) + \boldsymbol{w}^{\mathsf{H}}\mathsf{A}^{\mathsf{H}}\mathsf{A}\boldsymbol{w} \\ &= \Phi(\boldsymbol{x}_o) + \boldsymbol{w}^{\mathsf{H}}\mathsf{A}^{\mathsf{H}}\mathsf{A}\boldsymbol{w} \geq \Phi(\boldsymbol{x}_o),\end{aligned}$$

since $\mathsf{A}^{\mathsf{H}}\mathsf{A}$ is HPSD. Note that we cannot claim that $\Phi(\boldsymbol{x}_o + \boldsymbol{w}) > \Phi(\boldsymbol{x}_o)$ for all $\boldsymbol{w} \in \mathbb{C}^n_\star$, since we do not know that $\mathsf{A}^{\mathsf{H}}\mathsf{A}$ is HPD. However, we can still assert that $\boldsymbol{x}_o$ is a minimizer, though it might not be unique.

(1 $\Longrightarrow$ 3) Suppose that $\boldsymbol{x}_o \in \mathbb{C}^n$ is a solution to (5.9). We want to show that $\boldsymbol{r}(\boldsymbol{x}_o) \perp \operatorname{im}(\mathsf{A})$. To get a contradiction, suppose that $\boldsymbol{r}(\boldsymbol{x}_o) \not\perp \operatorname{im}(\mathsf{A})$. If this is the case, there is some $\boldsymbol{q} \in \operatorname{im}(\mathsf{A})$, $\boldsymbol{q} \neq \boldsymbol{0}$, such that $\boldsymbol{q}^{\mathsf{H}}\boldsymbol{r}(\boldsymbol{x}_o) \neq 0$. Since $\boldsymbol{q} \in \operatorname{im}(\mathsf{A})$, there is a vector $\boldsymbol{w} \in \mathbb{C}^n$ such that $\mathsf{A}\boldsymbol{w} = \boldsymbol{q}$. Since $\boldsymbol{x}_o$ is a minimizer of Φ, for any $\alpha \in \mathbb{C}$,

$$\begin{aligned}\|\boldsymbol{r}(\boldsymbol{x}_o)\|_2^2 &= \Phi(\boldsymbol{x}_o) \\ &\leq \Phi(\boldsymbol{x}_o + \alpha\boldsymbol{w}) \\ &= \Phi(\boldsymbol{x}_o) - \alpha(\boldsymbol{r}(\boldsymbol{x}_o))^{\mathsf{H}}\mathsf{A}\boldsymbol{w} - \bar{\alpha}\boldsymbol{w}^{\mathsf{H}}\mathsf{A}^{\mathsf{H}}\boldsymbol{r}(\boldsymbol{x}_o) + |\alpha|^2\boldsymbol{w}^{\mathsf{H}}\mathsf{A}^{\mathsf{H}}\mathsf{A}\boldsymbol{w} \\ &= \|\boldsymbol{r}(\boldsymbol{x}_o)\|_2^2 - 2\Re\left(\bar{\alpha}\boldsymbol{q}^{\mathsf{H}}\boldsymbol{r}(\boldsymbol{x}_o)\right) + |\alpha|^2\boldsymbol{q}^{\mathsf{H}}\boldsymbol{q}.\end{aligned}$$

Thus, for all $\alpha \in \mathbb{C}$,

$$2\Re\left(\bar{\alpha}\boldsymbol{q}^{\mathsf{H}}\boldsymbol{r}(\boldsymbol{x}_o)\right) \leq |\alpha|^2\boldsymbol{q}^{\mathsf{H}}\boldsymbol{q}.$$

Now set

$$\alpha = \frac{\boldsymbol{q}^{\mathsf{H}}\boldsymbol{r}(\boldsymbol{x}_o)}{\boldsymbol{q}^{\mathsf{H}}\boldsymbol{q}}$$

to get a contradiction. □

5.4 The QR Factorization and the Gram–Schmidt Algorithm

In this section, we construct a new factorization for the coefficient matrix A that will aid us in the practical solution of the least squares problem.

Theorem 5.27 (QR factorization). *Suppose that* $\mathsf{A} \in \mathbb{C}^{m\times n}$, $m \geq n$. *There is an upper triangular matrix* $\hat{\mathsf{R}} = [\hat{r}_{i,j}] \in \mathbb{C}^{n\times n}$ *with nonnegative (real) diagonal elements and a matrix* $\hat{\mathsf{Q}} \in \mathbb{C}^{m\times n}$ *satisfying* $\hat{\mathsf{Q}}^{\mathsf{H}}\hat{\mathsf{Q}} = \mathsf{I}_n$ *such that*

$$\mathsf{A} = \hat{\mathsf{Q}}\hat{\mathsf{R}}.$$

If $\operatorname{rank}(\mathsf{A}) = n$, *then* $\hat{r}_{i,i} > 0$ *for all* $i = 1, \ldots, n$.

Proof. The proof is by induction on the number of columns of A, $n \geq 1$.

($n = 1$) $\mathsf{A} = \boldsymbol{a} \in \mathbb{C}^m$. If $\boldsymbol{a} \neq \boldsymbol{0}$, then

$$\hat{\mathsf{Q}} = \frac{1}{\|\boldsymbol{a}\|_2}\boldsymbol{a}, \quad \hat{\mathsf{R}} = [\|\boldsymbol{a}\|_2]$$

gives the result. If $\boldsymbol{a} = \boldsymbol{0}$, pick any $\boldsymbol{q} \in \mathbb{C}^m$ with the property that $\|\boldsymbol{q}\|_2 = 1$. Set

$$\hat{\mathsf{Q}} = \boldsymbol{q}, \quad \hat{\mathsf{R}} = [0]$$

and observe that the result is still satisfied.

($n = k < m$) Suppose that the result is true for any $\mathsf{A} \in \mathbb{C}^{m\times k}$, i.e., there is a matrix $\hat{\mathsf{Q}} \in \mathbb{C}^{m\times k}$ with $\hat{\mathsf{Q}}^{\mathsf{H}}\hat{\mathsf{Q}} = \mathsf{I}_k$ and an upper triangular matrix $\hat{\mathsf{R}} \in \mathbb{C}^{k\times k}$ such that

$$\mathsf{A} = \hat{\mathsf{Q}}\hat{\mathsf{R}}.$$

($n = k + 1 \leq m$) Suppose that $\mathsf{A} \in \mathbb{C}^{m\times(k+1)}$. Write

$$\mathsf{A} = [\mathsf{A}_k \ \boldsymbol{a}], \quad \mathsf{A}_k \in \mathbb{C}^{m\times k}, \quad \boldsymbol{a} \in \mathbb{C}^m.$$

There exist $\hat{\mathsf{Q}}_k \in \mathbb{C}^{m\times k}$ and $\hat{\mathsf{R}}_k \in \mathbb{C}^{k\times k}$, as above, such that $\mathsf{A}_k = \hat{\mathsf{Q}}_k\hat{\mathsf{R}}_k$. Define

$$\hat{\mathsf{R}} = \begin{bmatrix} \hat{\mathsf{R}}_k & \boldsymbol{r} \\ \boldsymbol{0}^{\mathsf{T}} & \alpha \end{bmatrix}, \quad \hat{\mathsf{Q}} = \begin{bmatrix} \hat{\mathsf{Q}}_k & \boldsymbol{q} \end{bmatrix},$$

where $\boldsymbol{r} \in \mathbb{C}^k$, $\boldsymbol{q} \in \mathbb{C}^m$, and $\alpha \in \mathbb{R}$ are to be determined. Now we observe that

$$\mathsf{A} = \hat{\mathsf{Q}}\hat{\mathsf{R}}, \quad \hat{\mathsf{Q}}^{\mathsf{H}}\hat{\mathsf{Q}} = \mathsf{I}_{k+1} \tag{5.10}$$

has a solution if and only if the following are satisfied:

$$\mathsf{A}_k = \hat{\mathsf{Q}}_k\hat{\mathsf{R}}_k, \tag{5.11}$$

$$\boldsymbol{a} = \hat{\mathsf{Q}}_k\boldsymbol{r} + \alpha\boldsymbol{q}, \tag{5.12}$$

$$\hat{\mathsf{Q}}_k^{\mathsf{H}}\hat{\mathsf{Q}}_k = \mathsf{I}_k, \tag{5.13}$$

$$\boldsymbol{q}^{\mathsf{H}}\hat{\mathsf{Q}}_k = \boldsymbol{0}^{\mathsf{T}}, \tag{5.14}$$

$$\boldsymbol{q}^{\mathsf{H}}\boldsymbol{q} = 1. \tag{5.15}$$

Equations (5.11) and (5.13) are satisfied as part of the induction hypothesis. Next, set

$$\alpha = \left\| a - \hat{Q}_k \hat{Q}_k^H a \right\|_2 .$$

If $\alpha > 0$, (5.11)–(5.15) have a solution, namely,

$$r = \hat{Q}_k^H a \in \mathbb{C}^k,$$

$$q = \frac{1}{\alpha}\left(a - \hat{Q}_k \hat{Q}_k^H a\right).$$

It is easy to see that $q^H q = 1$; also, notice that

$$q^H \hat{Q}_k = \frac{1}{\alpha}\left(a - \hat{Q}_k \hat{Q}_k^H a\right)^H \hat{Q}_k = \frac{1}{\alpha}\left(a^H \hat{Q}_k - a^H \hat{Q}_k \hat{Q}_k^H \hat{Q}_k\right) = \mathbf{0}^\top.$$

If $\alpha = 0$, which is possible, the construction fails. In this case, pick any $q \in \mathbb{C}_\star^m$ that satisfies (5.14) and (5.15). Then set $\alpha = 0$ and

$$r = \hat{Q}_k^H a.$$

The proof of the existence of the factorization is completed.

Finally, suppose that $\operatorname{rank}(A) = n$. We want to prove that the diagonal elements of $\hat{R}$ must all be positive. To get a contradiction, suppose that $\hat{R}$ has a zero diagonal element and is, therefore, singular. In this case, there is a vector $x \in \mathbb{C}_\star^n$ such that

$$\hat{R}x = \mathbf{0} \in \mathbb{C}^n.$$

This implies that

$$Ax = \hat{Q}\hat{R}x = \mathbf{0} \in \mathbb{C}^m,$$

which, in turn, implies that A is rank deficient. This is a contradiction. □

Definition 5.28 (reduced QR factorization)**.** Suppose that $A \in \mathbb{C}^{m\times n}$, $m \geq n$. A factorization of the form

$$A = \hat{Q}\hat{R},$$

where $\hat{R} = [\hat{r}_{i,j}] \in \mathbb{C}^{n\times n}$ is an upper triangular matrix with nonnegative (real) diagonal elements and $\hat{Q} \in \mathbb{C}^{m\times n}$ is a matrix satisfying $\hat{Q}^H\hat{Q} = I_n$, is called a **reduced QR factorization**, or sometimes a **reduced nonnegative QR factorization** to emphasize the fact that the diagonal entries are nonnegative.

Theorem 5.29 (uniqueness)**.** *Suppose that* $A \in \mathbb{C}^{m\times n}$, $m \geq n$, *and* $\operatorname{rank}(A) = n$. *The reduced nonnegative QR factorization is unique.*

Proof. See Problem 5.14. □

Recall the Gram–Schmidt process that was introduced in Section A.5. We will now show that this process is related to the reduced QR factorization; in so doing, we give an alternate proof of the existence of the reduced QR factorization in the full rank case.

Lemma 5.30 (reduced QR factorization via Gram–Schmidt). *Let* $\mathsf{A} \in \mathbb{C}^{m\times n}$ *with* $m \geq n = \operatorname{rank}(\mathsf{A})$. *Then there exists a unique factorization*

$$\mathsf{A} = \hat{\mathsf{Q}}\hat{\mathsf{R}},$$

where $\hat{\mathsf{Q}} \in \mathbb{C}^{m\times n}$ *has orthonormal columns and* $\hat{\mathsf{R}} = [\hat{r}_{i,j}] \in \mathbb{C}^{n\times n}$ *is upper triangular with positive real diagonal entries.*

Proof. Let $\mathsf{A} = [\boldsymbol{a}_1, \ldots, \boldsymbol{a}_n]$, where $\boldsymbol{a}_i \in \mathbb{C}^m$ are the columns of A. We will inductively construct the columns of $\hat{\mathsf{Q}} = [\boldsymbol{q}_1, \ldots, \boldsymbol{q}_n]$ with $\boldsymbol{q}_i \in \mathbb{C}^m$ from the columns of A as follows.

1. Since $\operatorname{rank}(\mathsf{A}) = n$, we know that the columns of A are linearly independent. This, in particular, implies that $\boldsymbol{a}_1 \neq 0$. Define

$$\boldsymbol{q}_1 = \frac{1}{\|\boldsymbol{a}_1\|_2}\boldsymbol{a}_1$$

and

$$\hat{r}_{j,1} = \begin{cases} \|\boldsymbol{a}_1\|_2 > 0, & j = 1, \\ 0, & j = 2, \ldots, n. \end{cases}$$

2. Assume that, for $k \leq n-1$, we have found $\boldsymbol{q}_1, \ldots, \boldsymbol{q}_k$ that are orthonormal and, moreover, that

$$L_k = \operatorname{span}\{\boldsymbol{a}_1, \ldots, \boldsymbol{a}_k\} = \operatorname{span}\{\boldsymbol{q}_1, \ldots, \boldsymbol{q}_k\}.$$

3. Since A is full rank, we know that $\boldsymbol{a}_{k+1} \neq \boldsymbol{0}$ and $\boldsymbol{a}_{k+1} \notin L_k$. Therefore, the vector

$$\boldsymbol{v}_{k+1} = \boldsymbol{a}_{k+1} - \sum_{j=1}^{k} (\boldsymbol{a}_{k+1}, \boldsymbol{q}_j)_2 \boldsymbol{q}_j \neq \boldsymbol{0}.$$

We define

$$\boldsymbol{q}_{k+1} = \frac{1}{\|\boldsymbol{v}_{k+1}\|_2}\boldsymbol{v}_{k+1}$$

and

$$\hat{r}_{j,k+1} = \begin{cases} (\boldsymbol{a}_{k+1}, \boldsymbol{q}_j)_2, & j = 1, \ldots, k, \\ \|\boldsymbol{v}_{k+1}\|_2 > 0, & j = k+1, \\ 0, & j = k+2, \ldots, n. \end{cases}$$

By construction, $\hat{\mathsf{Q}}\hat{\mathsf{R}} = \mathsf{A}$, as the reader should confirm. Furthermore, the matrix $\hat{\mathsf{Q}}$ has orthonormal columns and

$$\operatorname{span}(\{\boldsymbol{a}_1, \ldots, \boldsymbol{a}_k\}) = \operatorname{span}(\{\boldsymbol{q}_1, \ldots, \boldsymbol{q}_k\}), \quad k = 1, \ldots, n.$$

The matrix $\hat{\mathsf{R}}$ is upper triangular with positive diagonal entries. The uniqueness of this decomposition follows from Theorem 5.29. □

Remark 5.31 (instability of Gram–Schmidt). As mentioned in Section A.5, the *classical* Gram–Schmidt process used in the construction of the reduced QR factorization in the proof of Lemma 5.30 suffers from numerical instabilities. For this reason, it is never used in practice. There exists a variant that is numerically stable, which we discuss in Section 5.6.

The existence of the so-called QR factorization, given in Lemma 5.30, allows us to provide a solution for the normal equations.

Theorem 5.32 (solution of the normal equations). *Let* $\mathsf{A} \in \mathbb{C}^{m\times n}$ *be such that* $m \geq n = \operatorname{rank}(\mathsf{A})$. *Suppose that* $\boldsymbol{f} \in \mathbb{C}^m$ *is given. The vector* $\tilde{\boldsymbol{x}} \in \mathbb{C}^n$ *is the unique least squares solution to* $\mathsf{A}\boldsymbol{x} = \boldsymbol{f}$ *if and only if*

$$\tilde{\boldsymbol{x}} = \hat{\mathsf{R}}^{-1}\hat{\mathsf{Q}}^{\mathsf{H}}\boldsymbol{f},$$

where $\hat{\mathsf{Q}}$ *and* $\hat{\mathsf{R}}$ *are the matrices from Lemma 5.30.*

Proof. See Problem 5.21. □

Oftentimes it is convenient to have what is known as the full QR factorization, as opposed to the reduced version.

Theorem 5.33 (full QR factorization). *Let* $\mathsf{A} \in \mathbb{C}^{m\times n}$, $m > n$. *There exists a unitary matrix* $\mathsf{Q} \in \mathbb{C}^{m\times m}$ *and an upper triangular matrix* $\mathsf{R} = [r_{i,j}] \in \mathbb{C}^{m\times n}$ *(i.e.,* $r_{i,j} = 0$, *if* $i > j$*) with nonnegative diagonal entries (*$r_{i,i} \geq 0$, *for each* $i = 1, \ldots, n$*) such that*

$$\mathsf{A} = \mathsf{QR}.$$

Such a factorization is known as a full QR factorization.

Proof. By Theorem 5.27, there is an upper triangular matrix $\hat{\mathsf{R}} = [\hat{r}_{i,j}] \in \mathbb{C}^{n\times n}$ with nonnegative (real) diagonal elements and a matrix $\hat{\mathsf{Q}} \in \mathbb{C}^{m\times n}$ satisfying $\hat{\mathsf{Q}}^{\mathsf{H}}\hat{\mathsf{Q}} = \mathsf{I}_n$ such that

$$\mathsf{A} = \hat{\mathsf{Q}}\hat{\mathsf{R}}.$$

Suppose that $\hat{\mathsf{Q}} = [\boldsymbol{q}_1, \ldots, \boldsymbol{q}_n]$. The columns of $\hat{\mathsf{Q}}$ form an orthonormal set. We create the square matrix

$$\mathsf{Q} = \left[\hat{\mathsf{Q}}\ \tilde{\mathsf{Q}}\right] \in \mathbb{C}^{m\times m},$$

where the columns of $\tilde{\mathsf{Q}} = [\boldsymbol{q}_{n+1}, \ldots, \boldsymbol{q}_m] \in \mathbb{C}^{m\times m-n}$ are orthonormal and chosen, so that the set $\{\boldsymbol{q}_1, \ldots, \boldsymbol{q}_m\}$, i.e., the set of columns of Q, is an orthonormal basis for $\mathbb{C}^n$. This can always be accomplished by a combination of the basis extension (Theorem A.22) and the Gram–Schmidt process. Next, we define

$$\mathsf{R} = \begin{bmatrix} \hat{\mathsf{R}} \\ \mathsf{O} \end{bmatrix} \in \mathbb{C}^{m\times n},$$

where $\mathsf{O} \in \mathbb{C}^{m-n\times n}$ is a zero matrix used for padding. Then just note that $\mathsf{A} = \mathsf{QR}$ and the proof is complete. □

The following is a simple consequence of the full QR factorization of a square matrix.

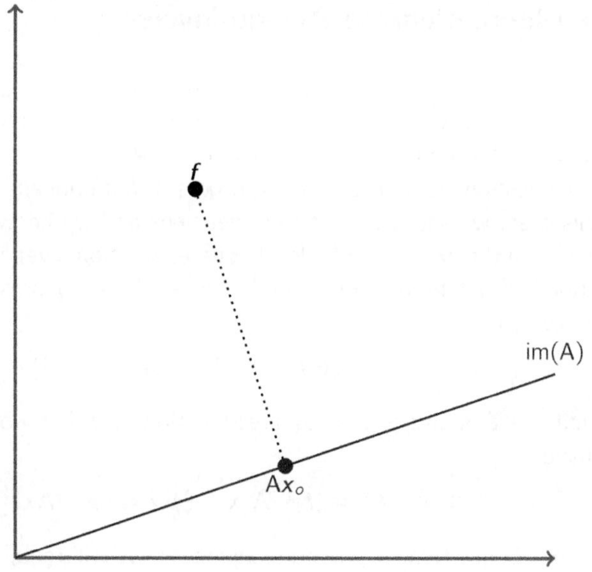

Figure 5.1 Finding the least squares solution to $\mathsf{A}\boldsymbol{x} = \boldsymbol{f}$ amounts to finding the orthogonal projection of $\boldsymbol{f} \in \mathbb{C}^m$ onto the subspace $\operatorname{im}(\mathsf{A})$.

Theorem 5.34 (Hadamard inequality[1]). *Suppose that* $\mathsf{A} \in \mathbb{C}^{n\times n}$. *Denote the columns of* A *by* $\mathbf{a}_j$, $j = 1, \ldots, n$. *Then*

$$|\det(\mathsf{A})| \le \prod_{j=1}^{n} \|\mathbf{a}_j\|_2 .$$

Proof. See Problem 5.23. □

Remark 5.35 (range of A). Let $\mathsf{A} \in \mathbb{C}^{m\times n}$, $m \ge n$ have full rank, i.e., $\operatorname{rank}(\mathsf{A}) = n$. The orthogonal projection onto the range of A is completely determined by the reduced QR factorization. Problem 5.22 shows that this projection is precisely $\mathsf{P} = \hat{\mathsf{Q}}\hat{\mathsf{Q}}^{\mathsf{H}}$. If A is rank deficient, the orthogonal projection onto the range of A is still well defined, but its form is not as simple.

In the next result, we use the orthogonal projection onto the range of A, but we will not assume that A is of full rank. An illustration of this result is given in Figure 5.1.

Theorem 5.36 (least squares and projection). *Let* $\mathsf{A} \in \mathbb{C}^{m\times n}$, $m \ge n$, *and* $\boldsymbol{f} \in \mathbb{C}^m$. *The vector* $\boldsymbol{x}_o \in \mathbb{C}^n$ *is a least squares solution to* $\mathsf{A}\boldsymbol{x} = \boldsymbol{f}$ *if and only if* $\mathsf{A}\boldsymbol{x}_o = \mathsf{P}\boldsymbol{f}$, *where* $\mathsf{P} \in \mathbb{C}^{m\times m}$ *is the orthogonal projection onto* $\operatorname{im}(\mathsf{A})$.

Proof. See Problem 5.24. □

[1] Named in honor of the French mathematician Jacques Salomon Hadamard (1865–1963).

5.5 The Moore–Penrose Pseudo-inverse

If $A \in \mathbb{C}^{m\times n}$ with $m \geq n = \text{rank}(A)$, then there is a unique least squares solution to the system $Ax = f$, which can be found by solving the normal equations or by computing the reduced QR factorization of A.

The question we want to address now is: *What happens if $m \geq n > r = \text{rank}(A)$?* In this case, we say that A is rank deficient and we know that $\ker(A) \neq \{\mathbf{0}\}$. We have shown the existence of a least squares solution, even in this case. However, the solution will not be unique. Indeed, if x is a least squares solution and $\boldsymbol{\xi} \in \ker(A)$ is nonzero, then

$$A^H f = A^H A x = A^H A(x + \boldsymbol{\xi}),$$

so that $x + \boldsymbol{\xi}$ is also a least squares solution. In other words, for every $\boldsymbol{\xi} \in \ker(A)$ we have

$$\Phi(x + \boldsymbol{\xi}) = \|f - A(x + \boldsymbol{\xi})\|_2^2 = \|f - Ax\|_2^2 = \Phi(x).$$

To be able to remove the nonuniqueness, we will require the solution to be, in a sense, the smallest.

Definition 5.37 (minimum norm least squares solution). Let $A \in \mathbb{C}^{m\times n}$ with $m \geq n > r = \text{rank}(A)$ and $f \in \mathbb{C}^m$. Define

$$\Phi(x) = \|f - Ax\|_2^2.$$

The **minimum norm least squares solution** of $Ax = f$ is $\hat{x} \in \mathbb{C}^n$ that satisfies:

1. $\hat{x}$ is a least squares solution, i.e., $\Phi(\hat{x}) \leq \Phi(x)$ for all $x \in \mathbb{C}^n$.
2. If $\Phi(\hat{x}) = \Phi(x)$, then $\|\hat{x}\|_2 \leq \|x\|_2$.

We find this minimum norm solution using the so-called pseudo-inverse of A, which was introduced in Problem 2.5. We recall here its definition and basic properties for convenience.

Definition 5.38 (Moore–Penrose pseudo-inverse[2]). Let the matrix $A \in \mathbb{C}^{m\times n}$ with $\min(m, n) \geq r = \text{rank}(A)$. Assume that A has the SVD $A = U\Sigma V^H$ with

$$\Sigma = \underset{m\times n}{\text{diag}}\,(\sigma_1, \ldots, \sigma_r, 0, \ldots, 0),$$

where $\sigma_1 \geq \cdots \geq \sigma_r > 0$. Define

$$\Sigma^\dagger = \underset{n\times m}{\text{diag}}\left(\sigma_1^{-1}, \ldots, \sigma_r^{-1}, 0, \ldots, 0\right).$$

Then the matrix

$$A^\dagger = V\Sigma^\dagger U^H$$

is called the **Moore–Penrose pseudo-inverse** of A.

[2] Named in honor of the American mathematician Eliakim Hastings Moore (1862–1932) and the British mathematician and physicist Sir Roger Penrose (1931–).

Theorem 5.39 (properties of $A^\dagger$). *In the notation of Definition 5.38, the Moore–Penrose pseudo-inverse satisfies:*

1. *If* $m = n$ *and* A^{-1} *exists, then* $A^\dagger = A^{-1}$.
2. *If* $m \geq n$ *and* A *has full rank, then* $A^\dagger = (A^H A)^{-1} A^H$.
3. $AA^\dagger A = A$.
4. $A^\dagger A A^\dagger = A^\dagger$.

Using the Moore–Penrose pseudo-inverse we can show the existence and uniqueness of a minimal norm least squares solution.

Theorem 5.40 (minimal norm least squares solution). *Let* $A \in \mathbb{C}^{m\times n}$ *with* $m \geq n > r = \text{rank}(A)$. *Then there is a unique minimum norm least squares solution* $\hat{\boldsymbol{x}}$, *which is given by* $\hat{\boldsymbol{x}} = A^\dagger \boldsymbol{f}$.

Proof. Let $A = U\Sigma V^H$. For a fixed $\boldsymbol{x} \in \mathbb{C}^n$, set $\boldsymbol{w}_x = V^H \boldsymbol{x}$. Then

$$\Phi(\boldsymbol{x}) = \|\boldsymbol{f} - U\Sigma V^H \boldsymbol{x}\|_2^2 = \|\boldsymbol{f} - U\Sigma \boldsymbol{w}_x\|_2^2 = \|U^H \boldsymbol{f} - \Sigma \boldsymbol{w}_x\|_2^2.$$

In other words, finding a minimal norm least squares solution is equivalent to finding $\hat{\boldsymbol{w}} \in \mathbb{C}^n$ such that

$$\|U^H \boldsymbol{f} - \Sigma \hat{\boldsymbol{w}}\|_2^2 \leq \|U^H \boldsymbol{f} - \Sigma \boldsymbol{w}\|_2^2, \quad \forall \boldsymbol{w} \in \mathbb{C}^n,$$

which, since $\|\hat{\boldsymbol{x}}\|_2 = \|\hat{\boldsymbol{w}}\|_2$, also has a minimal norm.

Given that $r = \text{rank}(A)$, there are exactly r nonzero diagonal entries in Σ. Therefore,

$$\|U^H \boldsymbol{f} - \Sigma \boldsymbol{w}\|_2^2 = \sum_{i=1}^{r} |\sigma_i w_i - (U^H \boldsymbol{f})_i|^2 + \sum_{j=r+1}^{m} |(U^H \boldsymbol{f})_i|^2.$$

The second sum does not depend on $\boldsymbol{w}$, so to minimize the norm of $\boldsymbol{w} = [w_i]$ we will set $w_i = 0$ for $i = r+1, \ldots, n$. In addition, since $\sigma_i > 0$ for $i \leq r$, we can set

$$w_i = \frac{1}{\sigma_i}(U^H \boldsymbol{f})_i, \quad i = 1, \ldots, r$$

to make the first sum vanish. We have thus constructed the vector $\hat{\boldsymbol{w}} = \Sigma^\dagger U^H \boldsymbol{f}$ that minimizes Φ and has a minimal norm. As a consequence,

$$\hat{\boldsymbol{x}} = V\Sigma^\dagger U^H \boldsymbol{f}$$

is the minimal norm least squares solution. □

5.6 The Modified Gram–Schmidt Process

Let us redefine the Gram–Schmidt process using the language of orthogonal projection matrices. To do so, for $k = 1, \ldots, n$, we define the matrix $P_k \in \mathbb{C}^{n\times n}$ by its action on a vector $\boldsymbol{w} \in \mathbb{C}^n$,

$$P_k \boldsymbol{w} = \boldsymbol{w} - \sum_{j=1}^{k-1} (\boldsymbol{w}, \boldsymbol{q}_j)_2 \boldsymbol{q}_j, \tag{5.16}$$

where $\{\boldsymbol{q}_1, \ldots, \boldsymbol{q}_{k-1}\}$ is an orthonormal set.

Definition 5.41 (modified Gram–Schmidt). Suppose that $S = \{\boldsymbol{a}_1, \ldots, \boldsymbol{a}_k\} \subset \mathbb{C}^n_\star$ with $k \leq n$. The **modified Gram–Schmidt process** is an algorithm for generating the set of vectors $Q = \{\boldsymbol{q}_1, \ldots, \boldsymbol{q}_k\}$ recursively as follows: for $m = 1$,

$$\boldsymbol{q}_1 = \frac{1}{\|\boldsymbol{a}_1\|_2}\boldsymbol{a}_1.$$

For $2 \leq m \leq k$, suppose that $\{\boldsymbol{q}_1, \ldots, \boldsymbol{q}_{m-1}\}$ have been computed. Set

$$\begin{aligned} \boldsymbol{v}_m^1 &= \boldsymbol{a}_m, \\ \boldsymbol{v}_m^2 &= \mathsf{P}_{\boldsymbol{q}_1^\perp}\boldsymbol{v}_m^1, \\ &\vdots \\ \boldsymbol{v}_m = \boldsymbol{v}_m^m &= \mathsf{P}_{\boldsymbol{q}_{m-1}^\perp}\boldsymbol{v}_m^{m-1}. \end{aligned}$$

If $\boldsymbol{v}_m = \boldsymbol{0}$, the process terminates. Otherwise, the process continues with

$$\boldsymbol{q}_m = \frac{1}{\|\boldsymbol{v}_m\|_2}\boldsymbol{v}_m.$$

As noted in Remark 5.31 the classical Gram–Schmidt algorithm is unstable. But it turns out that the modified Gram–Schmidt process is stable and is the one that is used in practical computations. Since it results from only cosmetic changes to the definition of the original Gram–Schmidt process, we have the following result.

Proposition 5.42 (modified Gram–Schmidt). *Let $\{\boldsymbol{a}_k\}_{k=1}^n \subset \mathbb{C}^m$ with $m \geq n$ be linearly independent. The sequence $\{\boldsymbol{q}_k\}_{k=1}^n \subset \mathbb{C}^m$ obtained by the modified Gram–Schmidt process is orthonormal and, for every $k \in \{1, \ldots, n\}$, it holds that*

$$\operatorname{span}(\{\boldsymbol{a}_1, \ldots, \boldsymbol{a}_k\}) = \operatorname{span}(\{\boldsymbol{q}_1, \ldots, \boldsymbol{q}_k\}).$$

Proof. See Problem 5.25. □

We conclude the discussion on the modified Gram–Schmidt process by addressing its complexity. The algorithm is described in Listing 5.1

When looking at Listing 5.1, we notice that every step in the outermost loop can be realized as the right multiplication by a square upper triangular matrix. Indeed, what we are effectively doing is multiplying the first column of $\mathsf{A} = [\boldsymbol{a}_1, \ldots, \boldsymbol{a}_n] = [\boldsymbol{v}_1, \ldots, \boldsymbol{v}_n]$ by $\frac{1}{r_{1,1}}$ and subtracting this $r_{1,j}$ times from the other columns. This is

$$\begin{bmatrix} | & & | \\ \boldsymbol{v}_1 & \cdots & \boldsymbol{v}_n \\ | & & | \end{bmatrix} \begin{bmatrix} \frac{1}{r_{1,1}} & -\frac{r_{1,2}}{r_{1,1}} & -\frac{r_{1,3}}{r_{1,1}} & \cdots & -\frac{r_{1,n}}{r_{1,1}} \\ 0 & 1 & 0 & \cdots & 0 \\ 0 & 0 & \ddots & \ddots & \vdots \\ \vdots & \vdots & \ddots & 1 & 0 \\ 0 & 0 & \cdots & 0 & 1 \end{bmatrix} = \begin{bmatrix} | & | & & | \\ \boldsymbol{q}_1 & \boldsymbol{v}_2^2 & \cdots & \boldsymbol{v}_n^2 \\ | & | & & | \end{bmatrix}.$$

This observation lends itself to the following generalization: at step i of the modified Gram–Schmidt process we subtract $r_{i,j}/r_{i,i}$ times the column i from the columns $j > i$, which is then replaced by $r_{i,i}^{-1}$ times the column itself. This operation can be encoded in the following upper triangular matrix:

$$
R_i = \begin{bmatrix}
1 & 0 & \cdots & 0 & 0 & 0 & 0 & \cdots & 0 \\
0 & 1 & \ddots & \vdots & \vdots & \vdots & & & \vdots \\
\vdots & \ddots & \ddots & 0 & 0 & 0 & 0 & \cdots & 0 \\
0 & \cdots & 0 & 1 & 0 & 0 & 0 & \cdots & 0 \\
0 & \cdots & 0 & 0 & \frac{1}{r_{i,i}} & -\frac{r_{i,i+1}}{r_{i,i}} & -\frac{r_{i,i+2}}{r_{i,i}} & \cdots & -\frac{r_{i,n}}{r_{i,i}} \\
0 & \cdots & 0 & 0 & 0 & 1 & 0 & \cdots & 0 \\
0 & & 0 & 0 & 0 & 0 & \ddots & \ddots & \vdots \\
\vdots & & \vdots & \vdots & \vdots & \vdots & \ddots & 1 & 0 \\
0 & \cdots & 0 & 0 & 0 & 0 & \cdots & 0 & 1
\end{bmatrix},
$$

where the highlighted row occupies the ith position.

With these matrices, we find

$$
AR_1R_2\cdots R_n = \hat{Q} \iff A\hat{R}^{-1} = \hat{Q} \tag{5.17}
$$

with

$$
\hat{R}^{-1} = R_1R_2\cdots R_n,
$$

where each R_i is *upper triangular*.

This shows that the QR factorization can be understood as a *triangular orthogonalization* process.

Theorem 5.43 (complexity of triangular orthogonalization). *Let* $A \in \mathbb{C}^{m\times n}$ *with* $m \geq n = \text{rank}(A)$. *Then the modified Gram–Schmidt process requires* $\mathcal{O}(2mn^2)$ *floating point operations to compute the reduced QR factorization of* A.

Proof. We begin by noticing that, in the innermost part of the loop, we must compute $r_{ij} = (\boldsymbol{v}_j, \boldsymbol{q}_i)_2$ and this requires m multiplications and $m-1$ additions (as $\boldsymbol{v}_j, \boldsymbol{q}_i \in \mathbb{C}^m$). Once this is done we need to compute $\boldsymbol{v}_j = \boldsymbol{v}_j - r_{ij}\boldsymbol{q}_j$, which needs m multiplications and m subtractions. In total, this means that we need

$$
m + (m-1) + m + m \approx 4m
$$

operations.

This loop is performed for $m = i+1, \ldots, n$ inside a loop of size $i = 1, \ldots, n$, so the total number of operations must be about

$$
\sum_{i=1}^{n}\sum_{m=i+1}^{n} 4m \approx 4m\sum_{i=1}^{n} i \approx 2mn^2,
$$

as we intended to show. □

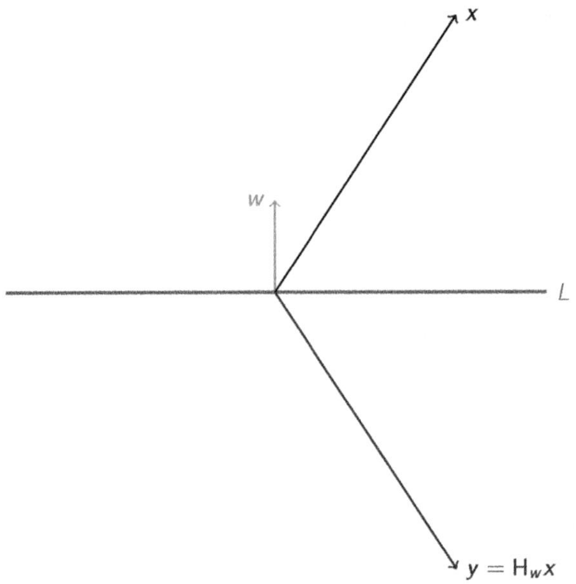

Figure 5.2 The reflection of the point $\boldsymbol{x}$ about the subspace $L = \text{span}\{\boldsymbol{w}\}^{\perp}$.

5.7 Householder Reflectors

Let us develop now a dual idea to triangular orthogonalization, i.e., we will construct a sequence $Q_1, \ldots, Q_n$ of unitary matrices such that

$$Q_n \cdots Q_1 A = R \tag{5.18}$$

with R upper triangular. If we construct this, the QR factorization of the matrix A is given by $Q^H = Q_n \cdots Q_1$ and the R matrix above. This process will be obtained with the help of the so-called *Householder reflectors*.

Remark 5.44 (full QR)**.** Notice that in (5.18) we obtain a matrix Q that is unitary, i.e., we are computing a *full* QR factorization.

Definition 5.45 (Householder reflector[3])**.** Suppose that $\boldsymbol{w} \in \mathbb{C}^n$ with $\|\boldsymbol{w}\|_2 = 1$. The **Householder reflector** with respect to $\text{span}\{\boldsymbol{w}\}^{\perp}$ is the matrix

$$H_{\boldsymbol{w}} = I_n - 2\boldsymbol{w}\boldsymbol{w}^H.$$

As illustrated in Figure 5.2, the action of $H_{\boldsymbol{w}}$ on a point $\boldsymbol{x} \notin \text{span}\{\boldsymbol{w}\}^{\perp}$ is the mirror reflection of $\boldsymbol{x}$ about $\text{span}\{\boldsymbol{w}\}^{\perp}$.

Proposition 5.46 (properties of $H_{\boldsymbol{w}}$)**.** *Suppose that $\boldsymbol{w} \in \mathbb{C}^n$ with $\|\boldsymbol{w}\|_2 = 1$. The reflector $H_{\boldsymbol{w}}$ satisfies the following properties:*

[3] Named in honor of the American mathematician Alston Scott Householder (1904–1993). Householder was a former professor at the University of Tennessee and a researcher at Oak Ridge National Laboratory in Oak Ridge, Tennessee.

1. $\mathsf{H}_{\boldsymbol{w}}$ *is Hermitian.*
2. $\mathsf{H}_{\boldsymbol{w}}$ *is unitary.*
3. $\mathsf{H}_{\boldsymbol{w}}$ *is involutory, i.e.,* $\mathsf{H}_{\boldsymbol{w}}^2 = \mathsf{I}_n$.
4. *For any* $\boldsymbol{x} \in \mathbb{C}^n$, $\|\mathsf{H}_{\boldsymbol{w}}\boldsymbol{x}\|_2 = \|\boldsymbol{x}\|_2$.

Proof. Clearly,

$$\mathsf{H}_{\boldsymbol{w}}^{\mathsf{H}} = \mathsf{I}_n^{\mathsf{H}} - 2\left(\boldsymbol{w}\boldsymbol{w}^{\mathsf{H}}\right)^{\mathsf{H}} = \mathsf{I}_n - 2\boldsymbol{w}\boldsymbol{w}^{\mathsf{H}} = \mathsf{H}_{\boldsymbol{w}},$$

which shows that $\mathsf{H}_{\boldsymbol{w}}$ is Hermitian. To see the second and third properties, observe that

$$\mathsf{H}_{\boldsymbol{w}}^{\mathsf{H}}\mathsf{H}_{\boldsymbol{w}} = \mathsf{H}_{\boldsymbol{w}}^2 = \left(\mathsf{I}_n - 2\boldsymbol{w}\boldsymbol{w}^{\mathsf{H}}\right)\left(\mathsf{I}_n - 2\boldsymbol{w}\boldsymbol{w}^{\mathsf{H}}\right) = \mathsf{I}_n - 4\boldsymbol{w}\boldsymbol{w}^{\mathsf{H}} + 4\boldsymbol{w}\boldsymbol{w}^{\mathsf{H}}\boldsymbol{w}\boldsymbol{w}^{\mathsf{H}} = \mathsf{I}_n.$$

The fourth property follows from the fact that $\mathsf{H}_{\boldsymbol{w}}$ is unitary:

$$\|\mathsf{H}_{\boldsymbol{w}}\boldsymbol{x}\|_2^2 = \boldsymbol{x}^{\mathsf{H}}\mathsf{H}_{\boldsymbol{w}}^{\mathsf{H}}\mathsf{H}_{\boldsymbol{w}}\boldsymbol{x} = \boldsymbol{x}^{\mathsf{H}}\boldsymbol{x} = \|\boldsymbol{x}\|_2^2. \qquad \square$$

Let us now see how Householder reflectors can be used to obtain the full QR factorization. The following two results essentially show that, for any vector with a nonzero jth entry, there is a reflection that sends it to (a multiple of) the canonical vector $\boldsymbol{e}_j$.

Lemma 5.47 (action of a reflector). *Let* $\boldsymbol{x} = [x_j] \in \mathbb{C}_\star^n$ *with*

$$x_j = |x_j|e^{\mathrm{i}\alpha_j}, \quad \alpha_j \in [0, 2\pi).$$

Suppose that $\boldsymbol{w} \in \mathbb{C}^n$, $\|\boldsymbol{w}\|_2 = 1$, *has the property that*

$$\mathsf{H}_{\boldsymbol{w}}\boldsymbol{x} = k\,\|\boldsymbol{x}\|_2\,\boldsymbol{e}_j,$$

where $k \in \mathbb{C}$ *and* $\boldsymbol{e}_j$ *is the* j*th canonical basis vector. Then it must be that*

$$k = \pm e^{\mathrm{i}\alpha_j}.$$

Proof. From the properties of the Householder reflector $\mathsf{H}_{\boldsymbol{w}}$, we have that

$$\|\boldsymbol{x}\|_2 = \|\mathsf{H}_{\boldsymbol{w}}\boldsymbol{x}\|_2 = |k|\,\|\boldsymbol{x}\|_2\,\|\boldsymbol{e}_j\|,$$

which necessarily implies that $|k| = 1$. In other words, there is $\beta \in \mathbb{R}$ such that

$$k = e^{\mathrm{i}\beta}.$$

Next, observe that, since $\mathsf{H}_{\boldsymbol{w}}$ is Hermitian,

$$\boldsymbol{x}^{\mathsf{H}}\mathsf{H}_{\boldsymbol{w}}\boldsymbol{x} = \boldsymbol{x}^{\mathsf{H}}k\,\|\boldsymbol{x}\|_2\,\boldsymbol{e}_j = k\bar{x}_j\,\|\boldsymbol{x}\|_2 \in \mathbb{R}.$$

As a consequence, we find

$$k\bar{x}_j\,\|\boldsymbol{x}\|_2 \in \mathbb{R} \iff k\bar{x}_j \in \mathbb{R} \iff e^{\mathrm{i}\beta}|x_j|e^{-\mathrm{i}\alpha_j} \in \mathbb{R} \iff e^{\mathrm{i}(\alpha_j - \beta)} \in \mathbb{R},$$

but this is only possible if there is $m \in \mathbb{Z}$ for which

$$\beta = \alpha_j + m\pi$$

or, equivalently, $k = \pm e^{\mathrm{i}\alpha_j}$ as we had claimed. $\square$

Theorem 5.48 (existence of reflector). *Let $\boldsymbol{x} = [x_j] \in \mathbb{C}^n_\star$ with*

$$x_j = |x_j| e^{\mathrm{i}\alpha_j}, \quad \alpha_j \in [0, 2\pi).$$

Then, provided that

$$\boldsymbol{x} \neq \pm e^{\mathrm{i}\alpha_j} \|\boldsymbol{x}\|_2 \, \boldsymbol{e}_j,$$

there is a vector $\boldsymbol{w} \in \mathbb{C}^n$, $\|\boldsymbol{w}\|_2 = 1$ such that

$$\mathsf{H}_{\boldsymbol{w}} \boldsymbol{x} = \pm e^{\mathrm{i}\alpha_j} \|\boldsymbol{x}\|_2 \, \boldsymbol{e}_j.$$

Proof. Let $\sigma = \pm 1$. Define

$$\boldsymbol{v} = \boldsymbol{x} - \sigma e^{\mathrm{i}\alpha_j} \|\boldsymbol{x}\|_2 \, \boldsymbol{e}_j;$$

since by assumption $\boldsymbol{v} \neq \boldsymbol{0}$, we can define

$$\boldsymbol{w} = \frac{1}{\|\boldsymbol{v}\|_2} \boldsymbol{v}.$$

Let us now show that

$$\mathsf{H}_{\boldsymbol{w}} \boldsymbol{x} = \sigma e^{\mathrm{i}\alpha_j} \|\boldsymbol{x}\|_2 \, \boldsymbol{e}_j.$$

A straightforward computation reveals that

$$\boldsymbol{v}^{\mathsf{H}} \boldsymbol{x} = \|\boldsymbol{x}\|_2^2 - \sigma e^{-\mathrm{i}\alpha_j} \|\boldsymbol{x}\|_2 \, x_j = \|\boldsymbol{x}\|_2^2 - \sigma |x_j| \, \|\boldsymbol{x}\|_2 \in \mathbb{R}$$

and

$$\boldsymbol{v}^{\mathsf{H}} \boldsymbol{v} = \|\boldsymbol{x}\|_2^2 - 2\sigma |x_j| \, \|\boldsymbol{x}\|_2 + \sigma^2 \|\boldsymbol{x}\|_2^2 = 2\left(\|\boldsymbol{x}\|_2^2 - \sigma |x_j| \, \|\boldsymbol{x}\|_2 \right) = 2 \boldsymbol{v}^{\mathsf{H}} \boldsymbol{x}.$$

With these computations, it follows that

$$\mathsf{H}_{\boldsymbol{w}} \boldsymbol{x} = \boldsymbol{x} - 2\boldsymbol{v} \frac{\boldsymbol{v}^{\mathsf{H}} \boldsymbol{x}}{\boldsymbol{v}^{\mathsf{H}} \boldsymbol{v}} = \boldsymbol{x} - \boldsymbol{v} = \sigma e^{\mathrm{i}\alpha_j} \|\boldsymbol{x}\|_2 \, \boldsymbol{e}_j,$$

as claimed. □

Remark 5.49 (real coordinates). Let $\boldsymbol{x} = [x_j] \in \mathbb{C}^n_\star$ with

$$x_j = |x_j| e^{\mathrm{i}\alpha_j}, \quad \exists \alpha_j \in [0, 2\pi).$$

We can find a vector $\boldsymbol{w} \in \mathbb{C}^n$, $\|\boldsymbol{w}\|_2 = 1$ such that

$$\mathsf{H}_{\boldsymbol{w}} \boldsymbol{x} = \pm \|\boldsymbol{x}\|_2 \, \boldsymbol{e}_j \in \mathbb{R}^n$$

only if $x_j \in \mathbb{R}$, i.e., $\alpha_j = 0$ or π.

Lemma 5.50 (construction of reflectors). *Assume that $m > k$. Suppose that $\boldsymbol{x} = [x_j] \in \mathbb{C}^k_\star$ with*

$$x_1 = |x_1| e^{\mathrm{i}\alpha_1}, \quad \exists \alpha_1 \in [0, 2\pi)$$

satisfies the property that

$$\boldsymbol{x} \neq \pm e^{\mathrm{i}\alpha_1} \|\boldsymbol{x}\|_{\ell^2(\mathbb{C}^k)} \, \widehat{\boldsymbol{e}}_1,$$

where $\widehat{e}_1 \in \mathbb{C}^k$ *is the first canonical basis vector. Let* $\mathsf{H}_{w} \in \mathbb{C}^{k\times k}$ *be the Householder reflector that satisfies*

$$\mathsf{H}_{w}x = \sigma e^{i\alpha_1} \|x\|_{\ell^2(\mathbb{C}^k)} \widehat{e}_1,$$

where $\sigma = \pm 1$*. Then the matrix*

$$\mathsf{H} = \begin{bmatrix} \mathsf{I}_{m-k} & \mathsf{O} \\ \mathsf{O}^\intercal & \mathsf{H}_{w} \end{bmatrix} \in \mathbb{C}^{m\times m}$$

is a Householder reflector in the following sense: there is a vector $z \in \mathbb{C}^m$ *with* $\|z\|_{\ell^2(\mathbb{C}^m)} = 1$ *such that*

$$\mathsf{H} = \mathsf{I}_n - 2zz^{\mathsf{H}}.$$

Furthermore, if $c \in \mathbb{C}^{m-k}$ *is any arbitrary vector, then*

$$\mathsf{H}\begin{bmatrix} c \\ x \end{bmatrix} = \begin{bmatrix} \mathsf{I}_{m-k} & \mathsf{O} \\ \mathsf{O}^\intercal & \mathsf{H}_{w} \end{bmatrix}\begin{bmatrix} c \\ x \end{bmatrix} = \begin{bmatrix} c \\ r\widehat{e}_1 \end{bmatrix}, \quad \textit{where} \quad r = \sigma e^{i\alpha_1} \|x\|_{\ell^2(\mathbb{C}^k)}.$$

Proof. See Problem 5.30. □

The previous construction suggests, given a matrix A, that a triangular matrix R may be obtained from it using reflections by first multiplying by a matrix that sends the first column of A to a multiple of e_1 and then proceeding similarly on the sub-matrix obtained by dropping the first row and column of A. This algorithm is called the *Householder triangularization* and is described below.

Definition 5.51 (Householder triangularization)**.** Suppose that

$$\mathsf{A} = \mathsf{A}^{(0)} = \left[a_1^{(0)}, \dots, a_n^{(0)}\right] \in \mathbb{C}^{m\times n}$$

with $m \geq n$ is given. The **Householder triangularization process** is a recursive algorithm for converting A into an upper triangular matrix $\mathsf{R} \in \mathbb{C}^{m\times n}$ by applying a sequence of Householder reflections; it is defined as follows: for $s = 1$, suppose that $a_1^{(0)} \in \mathbb{C}^m$ has as its first element

$$a_{1,1}^{(0)} = \left|a_{1,1}^{(0)}\right| e^{i\alpha_1}, \quad \alpha_1 \in [0, 2\pi).$$

If

$$a_1^{(0)} \neq \pm e^{i\alpha_1} \left\|a_1^{(0)}\right\|_{s^2(\mathbb{C}^m)} \widehat{e}_1, \tag{5.19}$$

where $\widehat{e}_1 \in \mathbb{C}^m$ is the first canonical basis vector, there is a vector $w_0 \in \mathbb{C}^m$ with $\|w_0\|_{\ell^2(\mathbb{C}^m)} = 1$ such that

$$\mathsf{H}_{w_0} a_1^{(0)} = r_{1,1}\widehat{e}_1,$$

where $|r_{1,1}| = \left\|a_1^{(0)}\right\|_{\ell^2(\mathbb{C}^m)}$. Set $\mathsf{H}_1 = \mathsf{H}_{w_0}$. If one or both of the conditions in (5.19) fails, set $\mathsf{H}_1 = \mathsf{I}_m$. In either case, we have

$$\mathsf{H}_1\mathsf{A}^{(0)} = \begin{bmatrix} r_{1,1} & f^\intercal \\ \mathbf{0} & \mathsf{A}^{(1)} \end{bmatrix},$$

where $\mathsf{A}^{(1)} = \left[a_1^{(1)}, \dots, a_{n-1}^{(1)}\right] \in \mathbb{C}^{(m-1)\times(n-1)}$.

For step $s = k + 1$ with $1 \le k \le n - 1$, suppose that k Householder reflections have been applied resulting in the decomposition

$$\mathsf{H}_k \cdots \mathsf{H}_1 \mathsf{A} = \begin{bmatrix} \mathsf{R}^{(k)} & \mathsf{B}^{(k)} \\ \mathsf{O}^{(k)} & \mathsf{A}^{(k)} \end{bmatrix},$$

where $\mathsf{R}^{(k)} \in \mathbb{C}^{k\times k}$ is an upper triangular matrix with diagonal elements

$$r_{1,1}, \ldots, r_{k,k} \in \mathbb{C},$$

$\mathsf{B}^{(k)} \in \mathbb{C}^{k\times(m-k)}$, $\mathsf{O}^{(k)} \in \mathbb{C}^{(m-k)\times k}$ is a zero matrix, and

$$\mathsf{A}^{(k)} = \left[\boldsymbol{a}_1^{(k)}, \ldots, \boldsymbol{a}_{n-k}^{(k)}\right] \in \mathbb{C}^{(m-k)\times(n-k)}.$$

Suppose that $\boldsymbol{a}_1^{(k)} \in \mathbb{C}^{m-k}$ has as its first element

$$a_{1,1}^{(k)} = \left|a_{1,1}^{(k)}\right| e^{\mathrm{i}\alpha_1}, \quad \alpha_1 \in [0, 2\pi).$$

If

$$\boldsymbol{a}_1^{(k)} \neq \pm e^{\mathrm{i}\alpha_1} \left\|\boldsymbol{a}_1^{(k)}\right\|_{\ell^2(\mathbb{C}^{m-k})} \widehat{\boldsymbol{e}}_1, \tag{5.20}$$

where $\widehat{\boldsymbol{e}}_1 \in \mathbb{C}^{m-k}$ is the first canonical basis vector, there is a vector $\boldsymbol{w}_k \in \mathbb{C}^{m-k}$ with $\|\boldsymbol{w}_k\|_{\ell^2(\mathbb{C}^{m-k})} = 1$ such that

$$\mathsf{H}_{\boldsymbol{w}_k} \boldsymbol{a}_1^{(k)} = r_{k+1,k+1} \hat{\boldsymbol{e}}_1,$$

where $|r_{k+1,k+1}| = \left\|\boldsymbol{a}_1^{(k)}\right\|_{\ell^2(\mathbb{C}^{m-k})}$. Set

$$\mathsf{H}_{k+1} = \begin{bmatrix} \mathsf{I}_k & \mathsf{O} \\ \mathsf{O}^\intercal & \mathsf{H}_{\boldsymbol{w}_k} \end{bmatrix}.$$

If one or both of the conditions in (5.20) fails, set $\mathsf{H}_{k+1} = \mathsf{I}_m$. In either case,

$$\mathsf{H}_{k+1}\mathsf{H}_k \cdots \mathsf{H}_1 \mathsf{A} = \begin{bmatrix} \mathsf{R}^{(k+1)} & \mathsf{B}^{(k+1)} \\ \mathsf{O}^{(k+1)} & \mathsf{A}^{(k+1)} \end{bmatrix},$$

where $\mathsf{R}^{(k+1)} \in \mathbb{C}^{(k+1)\times(k+1)}$ is an upper triangular matrix with diagonal elements

$$r_{1,1}, \ldots, r_{k,k}, r_{k+1,k+1} \in \mathbb{C},$$

$\mathsf{B}^{(k+1)} \in \mathbb{C}^{(k+1)\times(m-k-1)}$, $\mathsf{O}^{(k+1)} \in \mathbb{C}^{(m-k-1)\times(k+1)}$ is a zero matrix, and

$$\mathsf{A}^{(k+1)} = \left[\boldsymbol{a}_1^{(k+1)}, \ldots, \boldsymbol{a}_{n-k-1}^{(k+1)}\right] \in \mathbb{C}^{(m-k-1)\times(n-k-1)}.$$

The following result follows directly from our last definition.

Theorem 5.52 (triangularization). *Let $\mathsf{A} \in \mathbb{C}^{m\times n}$ with $m \ge n$ be given. The Householder triangulation procedure always proceeds to completion. In other words, there exists a sequence of Householder reflectors $\mathsf{H}_1, \ldots, \mathsf{H}_n \in \mathbb{C}^{m\times m}$ such that*

$$\mathsf{H}_n \cdots \mathsf{H}_1 \mathsf{A} = \mathsf{R},$$

where $\mathsf{R} \in \mathbb{C}^{m\times n}$ *is upper triangular. Moreover, the product*

$$\mathsf{Q}^{\mathsf{H}} = \mathsf{H}_n \cdots \mathsf{H}_1 \in \mathbb{C}^{m\times m}$$

is a unitary matrix. Consequently, there is a unitary matrix

$$\mathsf{Q} = \mathsf{H}_1 \cdots \mathsf{H}_n$$

and an upper triangular matrix R *such that* $\mathsf{A} = \mathsf{QR}$.

A more practical version of the algorithm is presented in Listing 5.2. Notice that this procedure does not necessarily compute the matrix Q in the QR factorization. This is controlled by the parameter `computeQ`.

Let us now study the complexity of the Householder algorithm without computing the matrix Q.

Proposition 5.53 (complexity of Householder). *The Householder algorithm requires, approximately,* $2mn^2 - \frac{2}{3}n^3$ *operations.*

Proof. Let us look at the innermost step, i.e.,

$$\mathsf{A}(k\colon m, k\colon n) = \mathsf{A}(k\colon m, k\colon n) - 2\boldsymbol{w}_k\boldsymbol{w}_k^{\mathsf{H}}\mathsf{A}(k\colon m, k\colon n),$$

where the vector $\boldsymbol{w}_k$ is of length $t = m-k+1$ and the matrix $\mathsf{A}(k\colon m, k\colon n) \in \mathbb{C}^{t\times s}$ with $s = n-k+1$. Thus,

1. The product $\boldsymbol{w}_k^{\mathsf{H}}\mathsf{A}(k\colon m, k\colon n)$ requires about $2t\cdot s$ operations, thus yielding a row vector of length s.
2. Once this is obtained, the multiplication $\boldsymbol{w}_k(\boldsymbol{w}_k^{\mathsf{H}}\mathsf{A}(k\colon m, k\colon n))$ requires $t\cdot s$ operations and yields a matrix of size $t\times s$.
3. This is subtracted from $\mathsf{A}(k\colon m, k\colon n)$ at a cost of $t\cdot s$ operations.

This is done in the innermost part of the loop, which is done n times. Thus,

$$\begin{aligned}\sum_{k=1}^{n} 4t\cdot s &= 4\sum_{k=1}^{n}(m-k+1)(n-k+1)\\ &\approx 4\left(\sum_{k=1}^{n} k^2 - (n+m)\sum_{k=1}^{n} k + n^2 m\right)\\ &\approx 4\left(\frac{n^3}{3} - (n+m)\frac{n^2}{2} + n^2 m\right)\\ &= 2mn^2 - \frac{2}{3}n^3,\end{aligned}$$

as claimed. □

Problems

5.1 In this problem we will show the existence of a least squares solution using a method different from the one presented in Theorem 5.5. Let $\mathsf{A} \in \mathbb{R}^{m\times n}$ with $m \geq n$ and $\operatorname{rank}(\mathsf{A}) = n$ and let $\boldsymbol{f} \in \mathbb{R}^m$. Prove the following:

a) For any $\lambda > 0$, the set

$$K_\lambda = \left\{ x \in \mathbb{R}^n \mid \|f - \mathsf{A}x\|_{\ell^2(\mathbb{R}^m)} \le \lambda \right\}$$

is closed and bounded. In other words, the sub-level sets of $\|f - \mathsf{A}x\|_{\ell^2(\mathbb{R}^m)}$ are *compact*.

b) If $\lambda = 2\|f\|_{\ell^2(\mathbb{R}^m)}$, then

$$\inf\left\{ \|f - \mathsf{A}x\|_{\ell^2(\mathbb{R}^m)} \mid x \in \mathbb{R}^n \right\} = \inf\left\{ \|f - \mathsf{A}x\|_{\ell^2(\mathbb{R}^m)} \mid x \in K_\lambda \right\}.$$

c) The two items above allow us to conclude the result.

5.2 Complete the proof of Theorem 5.10.

5.3 Let $m = n = 2$ and consider the following rank-one system:

$$\mathsf{A}x = f, \quad \mathsf{A} = \begin{bmatrix} 0 & 0 \\ 1 & 0 \end{bmatrix}, \quad f = \begin{bmatrix} 1 \\ 1 \end{bmatrix}.$$

Consider also a perturbed version of this system that has rank-two, where the perturbation is assumed small $\varepsilon \ll 1$:

$$\mathsf{A}_\varepsilon x_\varepsilon = f, \quad \mathsf{A}_\varepsilon = \begin{bmatrix} 0 & \varepsilon \\ 1 & 0 \end{bmatrix}, \quad f = \begin{bmatrix} 1 \\ 1 \end{bmatrix}.$$

Find the least squares solution to both systems and characterize the sensitivity of the solution to the perturbation ε.

Hint: Consider

$$\mathsf{U} = \begin{bmatrix} 0 & 1 \\ 1 & 0 \end{bmatrix}, \qquad \mathsf{V} = \mathsf{I}_2$$

as factors in the SVD of A.

5.4 Give an example of two vectors $x, y \in \mathbb{C}^2$ such that $(x, y)_2 \neq 0$ and

$$\|x + y\|_2^2 = \|x\|_2^2 + \|y\|_2^2.$$

What if $x, y \in \mathbb{R}^2$?

5.5 Assume that $\mathsf{A} \in \mathbb{C}^{m\times n}$ with $m \ge n$ is of full rank. What is the analogue of the normal equation that arises from using $\Phi_{\mathsf{B}}(\cdot)$, defined in (5.7), where $\mathsf{B} \in \mathbb{C}^{n\times n}$ is HPD?

5.6 Prove Proposition 5.12.

5.7 Prove Proposition 5.17.

5.8 Prove Proposition 5.18.

5.9 Prove Theorem 5.23.

5.10 Show that if P is an orthogonal projector, then $\mathsf{I} - 2\mathsf{P}$ is unitary.

5.11 Determine the eigenvalues, determinant, and singular values of an orthogonal projector. Give algebraic proofs for your conclusions.

5.12 Suppose that $q \in \mathbb{C}^n$, $\|q\|_2 = 1$. Set $\mathsf{P} = \mathsf{I} - qq^{\mathsf{H}}$.

a) Find im(P).

b) Find ker(P).

c) Find the eigenvalues of P.

5.13 Suppose that $\mathsf{A} \in \mathbb{C}^{n\times n}$ is invertible. Let $\mathsf{A} = \mathsf{QR}$ and $\mathsf{A}^{\mathsf{H}}\mathsf{A} = \mathsf{U}^{\mathsf{H}}\mathsf{U}$ be the QR factorization of A and Cholesky factorization of $\mathsf{A}^{\mathsf{H}}\mathsf{A}$, respectively, with the normalizations $r_{j,j}, u_{j,j} > 0$. Prove that $\mathsf{R} = \mathsf{U}$.

5.14 Prove Theorem 5.29.

Hint: Consider the Cholesky factorization of $\mathsf{A}^{\mathsf{H}}\mathsf{A}$.

5.15 Let $\mathsf{A} \in \mathbb{C}^{m\times n}$ $(m \geq n)$. Let $\mathsf{A} = \hat{\mathsf{Q}}\hat{\mathsf{R}}$ be a reduced QR factorization.

a) Show that A has full rank if and only if all the diagonal entries of $\hat{\mathsf{R}}$ are nonzero.

b) Suppose that $\hat{\mathsf{R}}$ has k nonzero diagonal entries for some k with $0 \leq k < n$. What does this imply about the rank of A? Is it exactly k? At least k? At most k?

5.16 Let $\mathsf{A} \in \mathbb{C}^{n\times n}$ be a normal matrix with spectrum $\sigma(\mathsf{A}) = \{\lambda_1, \ldots, \lambda_n\}$. Let $\mathsf{A} = \mathsf{QR}$ be a QR factorization of A. Prove that

$$\min_{1\leq j\leq n} |\lambda_j| \leq |r_{i,i}| \leq \max_{1\leq j\leq n} |\lambda_j|$$

for all $i = 1, \ldots, n$.

5.17 Let $W \subseteq \mathbb{C}^n$ be a vector subspace. Prove that W has an orthonormal basis.

Hint: Use Gram–Schmidt.

5.18 Suppose that $\mathsf{A} \in \mathbb{C}^{n\times n}$ is HPD. Define the inner product $(\boldsymbol{x}, \boldsymbol{y})_{\mathsf{A}} = \boldsymbol{y}^{\mathsf{H}}\mathsf{A}\boldsymbol{x}$ for all $\boldsymbol{x}, \boldsymbol{y} \in \mathbb{C}^n$. We say that vectors $\boldsymbol{x}, \boldsymbol{y} \in \mathbb{C}^n$ are A-orthogonal if and only if $(\boldsymbol{x}, \boldsymbol{y})_{\mathsf{A}} = 0$. Let $W \subseteq \mathbb{C}^n$ be a subspace. Generalize the Gram–Schmidt process to show that there is an A-orthonormal basis for W.

5.19 Suppose that $\mathsf{A}, \mathsf{U} \in \mathbb{C}^{n\times n}$ and U is unitary. Prove that

$$\|\mathsf{UA}\|_2 = \|\mathsf{AU}\|_2 = \|\mathsf{A}\|_2 .$$

5.20 Let $\mathsf{A} \in \mathbb{R}^{m\times m}$ be SPD. Suppose that $\mathsf{P} \in \mathbb{R}^{m\times n}$ with $m \geq n$ is full rank.

a) Show that $\mathsf{A}_C = \mathsf{P}^{\mathsf{T}}\mathsf{AP}$ is invertible.

b) Define $\mathsf{Q}_{\mathsf{A}} = \mathsf{P}\mathsf{A}_C^{-1}\mathsf{P}^{\mathsf{T}}\mathsf{A}$. Show that $\mathsf{Q}_{\mathsf{A}}\boldsymbol{u} \in \operatorname{im}(\mathsf{P})$ is the best approximation of $\boldsymbol{u} \in \mathbb{R}^m$ with respect to the A-norm, $\|\cdot\|_{\mathsf{A}}$, which is defined as follows:

$$(\boldsymbol{v}, \boldsymbol{w})_{\mathsf{A}} = \boldsymbol{v}^{\mathsf{T}}\mathsf{A}\boldsymbol{w}, \ \forall \boldsymbol{v}, \boldsymbol{w} \in \mathbb{R}^m, \quad \|\boldsymbol{w}\|_{\mathsf{A}} = \sqrt{(\boldsymbol{w}, \boldsymbol{w})_{\mathsf{A}}}, \ \forall \boldsymbol{w} \in \mathbb{R}^m.$$

In other words, prove that

$$\mathsf{Q}_{\mathsf{A}}\boldsymbol{u} = \operatorname{argmin}\left\{ \|\boldsymbol{z} - \boldsymbol{u}\|_{\mathsf{A}}^2 \;\middle|\; \boldsymbol{z} \in \operatorname{im}(\mathsf{P}) \right\}.$$

5.21 Prove Theorem 5.32.

5.22 Let $\mathsf{A} \in \mathbb{C}^{m\times n}$ with $m \geq n$. Suppose that A has full rank and $\mathsf{A} = \hat{\mathsf{Q}}\hat{\mathsf{R}}$ is a reduced QR factorization of A. Prove that $\mathsf{P} = \hat{\mathsf{Q}}\hat{\mathsf{Q}}^{\mathsf{H}}$ is an orthogonal projection onto $\operatorname{im}(\mathsf{A})$.

5.23 Prove Theorem 5.34.

5.24 Prove Theorem 5.36.

5.25 Prove Proposition 5.42.

5.26 Determine the eigenvalues, determinant, and singular values of a Householder reflector. Give algebraic proofs for your conclusions.

5.27 Suppose that $\{q_1, \ldots, q_{k-1}\}$ is an orthonormal set. Let P_k be defined as in (5.16). Show that this matrix is an orthogonal projection. Verify, in addition, that if we define $\hat{\mathsf{Q}}_k = [q_1, \ldots, q_{k-1}]$, then

$$\mathsf{P}_k = \mathsf{I} - \hat{\mathsf{Q}}_k \hat{\mathsf{Q}}_k^{\mathsf{H}}.$$

Finally, show that, for $k > 1$,

$$\mathsf{P}_k = \mathsf{P}_{q_{k-1}^{\perp}} \cdots \mathsf{P}_{q_1^{\perp}},$$

where $P_{a^{\perp}}$ denotes the orthogonal projection onto $\operatorname{span}\{a\}^{\perp}$.

5.28 Given $x, y \in \mathbb{C}_\star^n = \mathbb{C}^n \backslash \{\mathbf{0}\}$, determine a vector $v \in \mathbb{C}_\star^n$ such that the associated Householder reflector H satisfies $\mathsf{H}x = \gamma y$ for some γ.

5.29 Let $u, v \in \mathbb{R}^m$ and $\sigma \in \mathbb{R}$. Define $\mathsf{H}(u, v, \sigma) := \mathsf{I}_m - \sigma u v^{\mathsf{T}}$, where I_m is the $m \times m$ identity matrix.

a) Find all nonzero values of σ for which $\mathsf{H}(u, u, \sigma)$ is orthogonal. For such σ, determine all the eigenvalues and the corresponding eigenvectors of $\mathsf{H}(u, u, \sigma)$.

b) Let $x \in \mathbb{R}^m$ and $x \neq \mathbf{0}$. Describe how to choose a vector $u \in \mathbb{R}^m$ such that $\mathsf{H} = \mathsf{H}(u, u, \sigma)$ has the property that $\mathsf{H}x$ is a multiple of $\hat{e}_1 = (1, 0, 0, \ldots, 0)^{\mathsf{T}}$, where σ is as defined in the previous item.

5.30 Prove Lemma 5.50.

5.31 Let $\mathsf{Q} \in \mathbb{R}^{n \times n}$ be orthogonal, i.e., $\mathsf{Q}^{\mathsf{T}} = \mathsf{Q}^{-1}$. Recall that a Householder reflector is a matrix of the form

$$\mathsf{H} = \mathsf{I}_n - 2\frac{v v^{\mathsf{H}}}{v^{\mathsf{H}} v},$$

where $v \in \mathbb{R}_\star^n$. Prove that Q can be written as the product of n standard Householder reflectors or less.

5.32 The conclusion of the previous Problem does not necessarily hold in the complex case. In other words, if $\mathsf{U} \in \mathbb{U}^{n \times n}$ is unitary, i.e., $\mathsf{U}^{\mathsf{H}} = \mathsf{U}^{-1}$, then it may not be possible to represent it as the product of Householder reflectors. Instead, show that

$$\mathsf{U} = \mathsf{H}_1 \cdots \mathsf{H}_k \mathsf{D}$$

for some $k \in \{1, \ldots, n-1\}$, where H_i is a Householder matrix for each $i \in \{1, \ldots, k\}$ and D is a diagonal matrix whose diagonal entries each have modulus one.

5.33 Let $v \in \mathbb{C}^m$ be a unit vector. Define $\mathsf{A}_\alpha = \mathsf{I} + \alpha v v^{\mathsf{H}}$ for $\alpha \in \mathbb{C}_\star$.

a) Prove that the Householder matrix $\mathsf{H} = \mathsf{A}_{-2}$ is the unique unitary matrix in the set $\{\mathsf{A}_\alpha \mid \alpha \in \mathbb{R}_\star\}$.

b) Is the statement still true for the set $\{\mathsf{A}_\alpha \mid \alpha \in \mathbb{C}_\star\}$? If so, prove it. If not, give a counterexample.

5.34 Prove Theorem 5.52.

5.35 Consider the 2×2 orthogonal matrices

$$\mathsf{F} = \begin{bmatrix} -c & s \\ s & c \end{bmatrix}, \qquad \mathsf{J} = \begin{bmatrix} c & s \\ -s & c \end{bmatrix},$$

where $s = \sin\theta$ and $c = \cos\theta$ for some θ. The first matrix has $\det(F) = -1$ and is a reflector. The second has $\det(J) = 1$ and effects a rotation instead of a reflection. Such a matrix is called a *Givens rotation*.[4]

a) Describe exactly what geometric effect left-multiplications by F and J have on points in the plane.
b) Describe an algorithm for QR factorization based on Givens rotations.

Listings

```
function [Q, R, err] = ModifiedGramSchmidt( A )
% The modified Gram-Schmidt orthogonalization process.
%
% Input
%   A(1:m,1:n) : a matrix representing a collection of n column
%                vectors of dimension m
%
% Output
%   Q(1:m,1:n) : a collection of k orthonormal vectors of
%                dimension n
%   R(1:n,1:n) : an upper triangular square matrix such that
%                A = QR
%   err : = 0, if the columns of A are linearly independent
%         = 1, if an error has occurred
  m = size(A)(1);
  n = size(A)(2);
  Q = zeros(m,n);
  R = zeros(n,n);
  err = 0;
  if n>m
    err = 1;
    return;
  end
  V = A;
  for i=1:n
    R(i,i) = norm( V(:,i) );
    if R(i,i) > eps( R(i,i) );
      Q(:,i) = (1./R(i,i)) * V(:,i);
    else
      err = 1;
      return;
    end
    for j = i+1:n
      R(i,j) = V(:,j)'*Q(:,i);
      V(:,j) = V(:,j) - R(i,j)*Q(:,i);
    end
  end
end
```

Listing 5.1 The modified Gram–Schmidt process.

[4] Named in honor of the American mathematician and computer scientist James Wallace Givens (1910–1993). Givens was, for a time, a professor at the University of Tennessee and he worked at Oak Ridge National Laboratory in Oak Ridge, Tennessee.

```
function [Q, R, err] = QRFact( A, computeQ )
% The QR factorization using Householder reflectors.
%
% Input
%   A(1:m,1:n) : a rectangular matrix representing a collection
%                of n column vectors of dimension m
%   computeQ : = 1, the matrix Q is computed
%              = 0, the matrix Q is not computed. The algorithm
%                returns the identity
%
% Output
%   Q(1:m,1:m) : a unitary matrix of dimension n
%   R(1:m,1:n) : an rectangular upper triangular square matrix
%                such that A = QR
%   err : = 0, if the columns of A are linearly independent
%         = 1, if an error has occurred
  m = size(A)(1);
  n = size(A)(2);
  Q = eye(m,m);
  R = A;
  err = 0;
  if n>m
    err = 1;
    return;
  end
  for k=1:n
    nn = norm( R(k:m,k) );
    e = zeros(m-k+1,1);
    e(1) = 1;
    v = nn*e + R(k:m,k);
    norm_v = norm( v );
    if norm_v > eps( norm_v )
      w = (1.0/norm_v)*v;
    else
      err = 1;
      return;
    end
    R(k:m,k:n) = R(k:m,k:n) - 2.0*(w*w')*R(k:m,k:n);
    if computeQ > 0
      for t=1:m
        Q(k:m,1:m) = Q(k:m,1:m) - 2.0*(w*w')*Q(k:m,1:m);
      end
    end
  end
  if computeQ > 0
    Q = Q';
  end
end
```

Listing 5.2 QR factorization using reflectors.

6 Linear Iterative Methods

In Chapter 3 we learned that, using direct methods such as Gaussian elimination, one could obtain an exact solution to the linear system of equations $\mathsf{A}\boldsymbol{x} = \boldsymbol{f}$ with $\mathsf{A} \in \mathbb{C}^{n\times n}$ and $\boldsymbol{f} \in \mathbb{C}^n$. (Of course, we are ignoring the effects of roundoff.) Unfortunately, these algorithms require $\mathcal{O}(n^3)$ operations, which are frequently too expensive in practice. The high cost begs the following questions: *Are there lower cost options? Is an approximation of $\boldsymbol{x}$ good enough? How would such an approximation be generated?* As we will see, oftentimes we can find methods that have a much lower cost of computing a good approximate solution to $\boldsymbol{x}$.

As an alternate to the direct methods that we studied in the previous chapters, in the present chapter we will describe the so-called *linear* iteration methods for constructing sequences, $\{\boldsymbol{x}_k\}_{k=1}^{\infty} \subset \mathbb{C}^n$, with the desire that $\boldsymbol{x}_k \to \boldsymbol{x} = \mathsf{A}^{-1}\boldsymbol{f}$, as $k \to \infty$. The idea is that, given some $\varepsilon > 0$, we look for a $k \in \mathbb{N}$ such that

$$\|\boldsymbol{x} - \boldsymbol{x}_k\| \leq \varepsilon$$

with respect to some norm. In this context, ε is called the *stopping tolerance*.

In other words, we want to make certain the error is small in norm. But a word of caution. Usually, we do not have a direct way of approximating the error. The residual is more readily available. Suppose that $\boldsymbol{x}_k$ is an approximation of $\boldsymbol{x} = \mathsf{A}^{-1}\boldsymbol{f}$. The error is $\boldsymbol{e}_k = \boldsymbol{x} - \boldsymbol{x}_k$ and the residual is $\boldsymbol{r}_k = \boldsymbol{f} - \mathsf{A}\boldsymbol{x}_k = \mathsf{A}\boldsymbol{e}_k$. Recall that

$$\frac{\|\boldsymbol{e}_k\|}{\|\boldsymbol{x}\|} \leq \kappa(\mathsf{A})\frac{\|\boldsymbol{r}_k\|}{\|\boldsymbol{f}\|}.$$

Thus, when $\kappa(\mathsf{A})$ is large, $\frac{\|\boldsymbol{r}_k\|}{\|\boldsymbol{f}\|}$, which is easily computable, may not be a good indicator of the size of the relative error $\frac{\|\boldsymbol{e}_k\|}{\|\boldsymbol{x}\|}$, which is not directly computable. One must be careful when measuring the error.

The material of this section — containing topics such as the Gauss–Seidel method and the (successive) over-relaxation (SOR) method — does not, for the most part, represent the leading edge of research in iterative solvers. We call such methods *classical*, though not in the sense of a pejorative. Indeed, while workers are not typically applying Gauss–Seidel methods to solve industrial strength problems, understanding such methods is vital to our investigation of more modern methods, like multigrid and conjugate gradient methods and also effective preconditioning strategies. Excellent references for the classical material of this section may be found in the books by Hageman and Young [36] and Young [103]. Another good reference is [81].

6.1 Linear Iterative Methods

Definition 6.1 (iterative method). Let $\mathsf{A} \in \mathbb{C}^{n\times n}$ with $\det(\mathsf{A}) \neq 0$ and $\boldsymbol{f} \in \mathbb{C}^n$. An **iterative method** to find an approximate solution to $\mathsf{A}\boldsymbol{x} = \boldsymbol{f}$ is a process to generate a sequence of approximations $\{\boldsymbol{x}_k\}_{k=1}^{\infty}$ via an iteration of the form

$$\boldsymbol{x}_k = \varphi(\mathsf{A}, \boldsymbol{f}, \boldsymbol{x}_{k-1}, \dots, \boldsymbol{x}_{k-r}),$$

given the starting values $\boldsymbol{x}_0, \dots, \boldsymbol{x}_{r-1} \in \mathbb{C}^n$. Here,

$$\varphi(\cdot, \cdot, \dots, \cdot) \colon \mathbb{C}^{n\times n} \times \mathbb{C}^n \times \cdots \times \mathbb{C}^n \to \mathbb{C}^n$$

is called the **iteration function**. If $r = 1$, we say that the process is a **two-layer** method; otherwise, we say it is a multilayer method.

Definition 6.2 (linear iterative method). Let $\mathsf{A} \in \mathbb{C}^{n\times n}$ with $\det(\mathsf{A}) \neq 0$ and $\boldsymbol{f} \in \mathbb{C}^n$. Set $\boldsymbol{x} = \mathsf{A}^{-1}\boldsymbol{f}$. The two-layer iterative method

$$\boldsymbol{x}_k = \varphi(\mathsf{A}, \boldsymbol{f}, \boldsymbol{x}_{k-1})$$

is said to be **consistent** if and only if $\boldsymbol{x} = \varphi(\mathsf{A}, \boldsymbol{f}, \boldsymbol{x})$, i.e., $\boldsymbol{x} = \mathsf{A}^{-1}\boldsymbol{f}$ is a fixed point of $\varphi(\mathsf{A}, \boldsymbol{f}, \cdot)$. The method is **linear** if and only if

$$\varphi(\mathsf{A}, \alpha\boldsymbol{f}_1 + \beta\boldsymbol{f}_2, \alpha\boldsymbol{x}_1 + \beta\boldsymbol{x}_2) = \alpha\varphi(\mathsf{A}, \boldsymbol{f}_1, \boldsymbol{x}_1) + \beta\varphi(\mathsf{A}, \boldsymbol{f}_2, \boldsymbol{x}_2)$$

for all $\alpha, \beta \in \mathbb{C}$ and $\boldsymbol{f}_1, \boldsymbol{f}_2, \boldsymbol{x}_1, \boldsymbol{x}_2 \in \mathbb{C}^n$.

Proposition 6.3 (general form). *Let $\mathsf{A} \in \mathbb{C}^{n\times n}$ with $\det(\mathsf{A}) \neq 0$ and $\boldsymbol{f} \in \mathbb{C}^n$. Any two-layer, linear, and consistent method can be written in the form*

$$\boldsymbol{x}_{k+1} = \boldsymbol{x}_k + \mathsf{C}\boldsymbol{r}(\boldsymbol{x}_k) = \boldsymbol{x}_k + \mathsf{C}(\boldsymbol{f} - \mathsf{A}\boldsymbol{x}_k) \tag{6.1}$$

for some matrix $\mathsf{C} \in \mathbb{C}^{n\times n}$, where $\boldsymbol{r}(\boldsymbol{z}) = \boldsymbol{f} - \mathsf{A}\boldsymbol{z}$ is the residual vector.

Proof. A two-layer method is defined by an iteration function

$$\varphi(\cdot, \cdot, \cdot) \colon \mathbb{C}^{n\times n} \times \mathbb{C}^n \times \mathbb{C}^n \to \mathbb{C}^n.$$

Given φ, define the operator

$$\mathsf{C}\boldsymbol{z} = \varphi(\mathsf{A}, \boldsymbol{z}, \boldsymbol{0}).$$

This is a linear operator owing to the assumed linearity of the iteration function. Consequently, C can be identified as a square matrix. It follows from this definition, using the consistency and linearity of φ, that

$$(\mathsf{I}_n - \mathsf{C}\mathsf{A})\boldsymbol{w} = \boldsymbol{w} - \varphi(\mathsf{A}, \mathsf{A}\boldsymbol{w}, \boldsymbol{0}) = \varphi(\mathsf{A}, \mathsf{A}\boldsymbol{w}, \boldsymbol{w}) - \varphi(\mathsf{A}, \mathsf{A}\boldsymbol{w}, \boldsymbol{0}) = \varphi(\mathsf{A}, \boldsymbol{0}, \boldsymbol{w}).$$

Furthermore, by linearity, we can write

$$\begin{aligned}\boldsymbol{x}_{k+1} &= \varphi(\mathsf{A}, \boldsymbol{f} + \boldsymbol{0}, \boldsymbol{0} + \boldsymbol{x}_k) \\ &= \varphi(\mathsf{A}, \boldsymbol{f}, \boldsymbol{0}) + \varphi(\mathsf{A}, \boldsymbol{0}, \boldsymbol{x}_k) \\ &= \mathsf{C}\boldsymbol{f} + (\mathsf{I}_n - \mathsf{C}\mathsf{A})\boldsymbol{x}_k \\ &= \boldsymbol{x}_k + \mathsf{C}(\boldsymbol{f} - \mathsf{A}\boldsymbol{x}_k),\end{aligned}$$

as we intended to show. □

If C is invertible, we can, if we like, write

$$\mathsf{C}^{-1}(\boldsymbol{x}_{k+1}-\boldsymbol{x}_k)+\mathsf{A}\boldsymbol{x}_k=\boldsymbol{f}.$$

Motivated by this form, we have the following generalization.

Definition 6.4 (two-layer methods). Let $\mathsf{A} \in \mathbb{C}^{n\times n}$ with $\det(\mathsf{A}) \neq 0$ and $\boldsymbol{f} \in \mathbb{C}^n$. A method of the form

$$\mathsf{B}_{k+1}(\boldsymbol{x}_{k+1}-\boldsymbol{x}_k)+\mathsf{A}\boldsymbol{x}_k=\boldsymbol{f},$$

where $\mathsf{B}_{k+1} \in \mathbb{C}^{n\times n}$ is invertible, is called an **adaptive two-layer method**. If $\mathsf{B}_{k+1} = \mathsf{B}$, where B is invertible and independent of k, then the method is called a **stationary two-layer method**. If $\mathsf{B}_{k+1} = \frac{1}{\alpha_{k+1}}\mathsf{I}_n$, where $\alpha_{k+1} \in \mathbb{C}_\star$, then we say that the adaptive two-layer method is **explicit**.

Remark 6.5 (iterator). Consider a stationary two-layer method and assume that B is invertible, then

$$\boldsymbol{x}_{k+1}=\boldsymbol{x}_k+\mathsf{B}^{-1}(\boldsymbol{f}-\mathsf{A}\boldsymbol{x}_k); \tag{6.2}$$

from this it follows that if $\{\boldsymbol{x}_k\}_{k\geq 0}$ converges, then it must converge to $\boldsymbol{x} = \mathsf{A}^{-1}\boldsymbol{f}$. Of course, this form is equivalent to (6.1) with $\mathsf{C} = \mathsf{B}^{-1}$. The matrix B in the stationary two-layer method is called the *iterator*.

Remark 6.6 (choice of B). Let us now consider an extreme case, namely $\mathsf{B} = \mathsf{A}$. In this case, we obtain

$$\boldsymbol{x}_{k+1}=\boldsymbol{x}_k+\mathsf{B}^{-1}(\boldsymbol{f}-\mathsf{A}\boldsymbol{x}_k)=\boldsymbol{x}_k+\mathsf{A}^{-1}(\boldsymbol{f}-\mathsf{A}\boldsymbol{x}_k)=\mathsf{A}^{-1}\boldsymbol{f},$$

i.e., we get the *exact* solution after one step.

The previous observation shows that the choice of an iterator comes with two conflicting requirements:

1. The iterator B should be easy/cheap to invert.
2. The iterator B should "approximate" the matrix A well.

In essence, the art of iterative methods is concerned with finding good iterators.

Definition 6.7 (error transfer). Let $\mathsf{A} \in \mathbb{C}^{n\times n}$ be invertible and $\boldsymbol{f} \in \mathbb{C}^n$. Suppose that $\boldsymbol{x} = \mathsf{A}^{-1}\boldsymbol{f}$ and consider the stationary two-layer method (6.2) defined by the invertible matrix $\mathsf{B} \in \mathbb{C}^{n\times n}$. The matrix $\mathsf{T} = \mathsf{I}_n - \mathsf{B}^{-1}\mathsf{A}$ is called the **error transfer matrix** and satisfies

$$\boldsymbol{e}_{k+1}=\mathsf{T}\boldsymbol{e}_k,$$

where $\boldsymbol{e}_k = \boldsymbol{x} - \boldsymbol{x}_k$ is the **error** at step k.

We need to find conditions on T to guarantee that $\{\boldsymbol{e}_k\}_{k=0}^{\infty}$ converges to zero.

6.2 Spectral Convergence Theory

The theory in this section is based on properties of the spectrum $\sigma(\mathsf{T})$ of the error transfer matrix. In particular, with the tools developed in Section 4.1, we can provide conditions for a linear method to be convergent.

Theorem 6.8 (convergence of linear methods)**.** *Suppose that* $\mathsf{A}, \mathsf{B} \in \mathbb{C}^{n\times n}$ *are invertible,* $\boldsymbol{f}, \boldsymbol{x}_0 \in \mathbb{C}^n$ *are given, and* $\boldsymbol{x} = \mathsf{A}^{-1}\boldsymbol{f}$.

1. *The sequence* $\{\boldsymbol{x}_k\}_{k=1}^{\infty}$ *defined by the linear two-layer stationary iterative method (6.2) converges to* $\boldsymbol{x}$ *for any starting point* $\boldsymbol{x}_0$ *if and only if* $\rho(\mathsf{T}) < 1$, *where* T *is the error transfer matrix* $\mathsf{T} = \mathsf{I}_n - \mathsf{B}^{-1}\mathsf{A}$.
2. *A sufficient condition for the convergence of* $\{\boldsymbol{x}_k\}_{k=1}^{\infty}$ *for any starting point* $\boldsymbol{x}_0$ *is the condition that* $\|\mathsf{T}\| < 1$ *for some induced matrix norm.*

Proof. Before we begin the proof, observe that

$$\boldsymbol{e}_k = \mathsf{T}\boldsymbol{e}_{k-1} = \mathsf{T}^2\boldsymbol{e}_{k-2} = \cdots = \mathsf{T}^k\boldsymbol{e}_0.$$

Also, observe that $\boldsymbol{x}_k \to \boldsymbol{x} = \mathsf{A}^{-1}\boldsymbol{f}$, as $k \to \infty$, if and only if $\boldsymbol{e}_k \to \boldsymbol{0}$, as $k \to \infty$.

Suppose that $\boldsymbol{x}_k \to \boldsymbol{x} = \mathsf{A}^{-1}\boldsymbol{f}$, as $k \to \infty$, for any $\boldsymbol{x}_0$. Then $\boldsymbol{e}_k \to \boldsymbol{0}$, as $k \to \infty$, for any $\boldsymbol{e}_0$. Set $\boldsymbol{e}_0 = \boldsymbol{w}$, where $(\lambda, \boldsymbol{w})$ is any eigenpair of T, with $\|\boldsymbol{w}\|_\infty = 1$. Then

$$\boldsymbol{e}_k = \lambda^k\boldsymbol{e}_0$$

and

$$|\lambda|^k = |\lambda|^k\|\boldsymbol{w}\|_\infty = \|\boldsymbol{e}_k\|_\infty \to 0.$$

It follows that $|\lambda| < 1$. Since λ was arbitrary, $\rho(\mathsf{T}) < 1$.

If $\rho(\mathsf{T}) < 1$, appealing to Theorem 4.8,

$$\lim_{k\to\infty} \boldsymbol{e}_k = \lim_{k\to\infty} \mathsf{T}^k\boldsymbol{e}_0 = \boldsymbol{0}$$

for any $\boldsymbol{e}_0$. Hence, $\boldsymbol{x}_k \to \boldsymbol{x} = \mathsf{A}^{-1}\boldsymbol{f}$, as $k \to \infty$, for any $\boldsymbol{x}_0$.

Suppose now that $\|\mathsf{T}\| < 1$ for some induced matrix norm. Since, for any induced matrix norm,

$$\rho(\mathsf{T}) \le \|\mathsf{T}\|,$$

it follows that $\rho(\mathsf{T}) < 1$. Again, by Theorem 4.8,

$$\lim_{k\to\infty} \boldsymbol{e}_k = \lim_{k\to\infty} \mathsf{T}^k\boldsymbol{e}_0 = \boldsymbol{0}$$

for any $\boldsymbol{e}_0$. □

We can also provide an error estimate.

Theorem 6.9 (error estimate)**.** *Let* $\mathsf{A}, \mathsf{B} \in \mathbb{C}^{n\times n}$ *be invertible,* $\boldsymbol{x}_0, \boldsymbol{f} \in \mathbb{C}^n$ *are given, and* $\boldsymbol{x} = \mathsf{A}^{-1}\boldsymbol{f}$. *Let* $\{\boldsymbol{x}_k\}_{k=1}^{\infty}$ *be the sequence generated by the linear two-*

layer stationary method (6.2). *Assume that, for some induced norm,* $\|\mathsf{T}\| < 1$. *Then the following estimates hold:*

$$\|\boldsymbol{x} - \boldsymbol{x}_k\| \le \|\mathsf{T}\|^k \|\boldsymbol{x} - \boldsymbol{x}_0\|,$$

$$\|\boldsymbol{x} - \boldsymbol{x}_k\| \le \frac{\|\mathsf{T}\|^k}{1 - \|\mathsf{T}\|} \|\boldsymbol{x}_1 - \boldsymbol{x}_0\|.$$

Proof. It follows that $\boldsymbol{e}_k = \mathsf{T}^k \boldsymbol{e}_0$. By using the consistency and sub-multiplicativity of the induced matrix norm, we find

$$\|\boldsymbol{e}_k\| \le \left\|\mathsf{T}^k\right\| \|\boldsymbol{e}_0\| \le \|\mathsf{T}\|^k \|\boldsymbol{e}_0\|,$$

which proves the first estimate. To see the second, observe that $\boldsymbol{e}_k = \mathsf{T}^{k-1}\boldsymbol{e}_1$, and thus $\mathsf{T}\boldsymbol{e}_k = \mathsf{T}^k \boldsymbol{e}_1$. Subtracting the last expression from $\boldsymbol{e}_k = \mathsf{T}^k \boldsymbol{e}_0$, we find $(\mathsf{I}_n - \mathsf{T})\, \boldsymbol{e}_k = \mathsf{T}^k (\boldsymbol{x}_1 - \boldsymbol{x}_0)$. Since $\|\mathsf{T}\| < 1$, Theorem 4.19 guarantees that $\mathsf{I}_n - \mathsf{T}$ is invertible, and

$$\left\|(\mathsf{I}_n - \mathsf{T})^{-1}\right\| \le \frac{1}{1 - \|\mathsf{T}\|}.$$

Hence,

$$\boldsymbol{e}_k = (\mathsf{I}_n - \mathsf{T})^{-1} \mathsf{T}^k (\boldsymbol{x}_1 - \boldsymbol{x}_0)$$

and, using the consistency and sub-multiplicativity of the norm, we get

$$\|\boldsymbol{e}_k\| \le \left\|(\mathsf{I}_n - \mathsf{T})^{-1}\right\| \|\mathsf{T}\|^k \|\boldsymbol{x}_1 - \boldsymbol{x}_0\| \le \frac{1}{1 - \|\mathsf{T}\|} \|\mathsf{T}\|^k \|\boldsymbol{x}_1 - \boldsymbol{x}_0\|.$$

The result is proven. □

6.3 Matrix Splitting Methods

Here, we present some methods that are based on the idea of *matrix splitting*. Namely, we assume that we can *split* the coefficient matrix A as

$$\mathsf{A} = \mathsf{M} + \mathsf{N},$$

where M is invertible and, hopefully, easy to invert. Since $\mathsf{A}\boldsymbol{x} = \boldsymbol{f}$ if and only if $\mathsf{M}\boldsymbol{x} + \mathsf{N}\boldsymbol{x} = \boldsymbol{f}$, the strategy that we follow is then to construct a method of the form

$$\mathsf{M}\boldsymbol{x}_{k+1} + \mathsf{N}\boldsymbol{x}_k = \boldsymbol{f}.$$

6.3.1 Jacobi Method

Let $\mathsf{A} = [a_{i,j}] \in \mathbb{C}^{n\times n}$ have nonzero diagonal elements and consider the following splitting of A:

$$\mathsf{A} = \mathsf{L} + \mathsf{D} + \mathsf{U},$$

where $D = \operatorname{diag}(a_{1,1}, \dots, a_{n,n})$ is the diagonal part of A, L is the strictly lower triangular part of A, and U is its strictly upper triangular part. Choosing the iterator to be

$$B = B_J = D$$

yields the so-called *Jacobi method.*[1] In this case, the error transfer matrix is

$$T = T_J = I_n - D^{-1}A, \tag{6.3}$$

so that

$$T_J = -\begin{bmatrix} 0 & \frac{a_{1,2}}{a_{1,1}} & \cdots & \frac{a_{1,n}}{a_{1,1}} \\ \frac{a_{2,1}}{a_{2,2}} & 0 & \ddots & \vdots \\ \vdots & \ddots & \ddots & \frac{a_{n-1,n}}{a_{n-1,n-1}} \\ \frac{a_{n,1}}{a_{n,n}} & \cdots & \frac{a_{n,n-1}}{a_{n,n}} & 0 \end{bmatrix}.$$

Alternately, we may write the Jacobi method in component form via

$$x_{i,k+1} = [\mathbf{x}_{k+1}]_i = [T_J\mathbf{x}_k]_i + \left[D^{-1}\mathbf{f}\right]_i = -\frac{1}{a_{i,i}}\sum_{\substack{j=1\\ j\neq i}}^{n} a_{i,j}x_{j,k} + \frac{1}{a_{i,i}}f_i,$$

where $x_{i,k} = [\mathbf{x}_k]_i$. Equivalently,

$$a_{i,i}x_{i,k+1} = -\sum_{\substack{j=1\\ j\neq i}}^{n} a_{i,j}x_{j,k} + f_i.$$

Theorem 6.10 (convergence). *Let* $A = [a_{i,j}] \in \mathbb{C}^{n\times n}$ *be strictly diagonally dominant (SDD) of magnitude* $\delta > 0$*, and* $\boldsymbol{f} \in \mathbb{C}^n$*. Then the Jacobi iteration method for approximating the solution to* $A\mathbf{x} = \boldsymbol{f}$ *is convergent.*

Proof. Since A is SDD of magnitude $\delta > 0$, it follows that D is invertible, and T_J, given in (6.3), is well defined. Then

[1] Named in honor of the German mathematician Carl Gustav Jacob Jacobi (1804–1851).

$$\begin{aligned}
\|\mathsf{T}_\mathsf{J}\|_\infty &= \max_{1\le i\le n} \sum_{j=1}^{n} |[\mathsf{T}_\mathsf{J}]_{i,j}| \\
&= \max_{1\le i\le n} \sum_{j=1}^{n} \left|\delta_{i,j} - \frac{1}{a_{i,i}} a_{i,j}\right| \\
&= \max_{1\le i\le n} \sum_{\substack{j=1\\ j\neq i}}^{n} \left|\frac{a_{i,j}}{a_{i,i}}\right| \\
&= \max_{1\le i\le n} \frac{1}{|a_{i,i}|} \sum_{\substack{j=1\\ j\neq i}}^{n} |a_{i,j}| \\
&\le \max_{1\le i\le n} \frac{1}{|a_{i,i}|} (|a_{i,i}| - \delta) \\
&= 1 - \delta \min_{1\le i\le n} \frac{1}{|a_{i,i}|}.
\end{aligned}$$

Hence, $\|\mathsf{T}_\mathsf{J}\|_\infty < 1$. By Theorem 6.8 the method converges. □

Theorem 6.11 (convergence)**.** *Suppose that* $\mathsf{A} \in \mathbb{C}^{n\times n}$ *is column-wise SDD of magnitude* $\delta > 0$*, i.e.,*

$$|a_{j,j}| - \delta \ge \sum_{\substack{i=1\\ i\neq j}}^{n} |a_{i,j}|, \quad j = 1, \dots, n,$$

and $\boldsymbol{f} \in \mathbb{C}^n$ *is given. Then the sequence* $\{\boldsymbol{x}_k\}_{k=1}^{\infty}$ *generated by the Jacobi iteration method converges, for any starting value* $\boldsymbol{x}_0$*, to the vector* $\boldsymbol{x} = \mathsf{A}^{-1}\boldsymbol{f}$*.*

Proof. It will suffice to prove that

$$\rho(\mathsf{T}_\mathsf{J}) < 1,$$

where T_J is given by (6.3). Since $\mathsf{A} \in \mathbb{C}^{n\times n}$ is column-wise SDD of magnitude $\delta > 0$, A^H is row-wise SDD of magnitude $\delta > 0$. Therefore, by Theorem 6.10,

$$\rho\left(\mathsf{I}_n - \mathsf{D}^{-\mathsf{H}}\mathsf{A}^\mathsf{H}\right) \le \left\|\mathsf{I}_n - \mathsf{D}^{-\mathsf{H}}\mathsf{A}^\mathsf{H}\right\|_\infty < 1.$$

Define $\tilde{\mathsf{T}} = \mathsf{I}_n - \mathsf{D}^{-\mathsf{H}}\mathsf{A}^\mathsf{H}$. Then

$$\tilde{\mathsf{T}}^\mathsf{H} = \mathsf{I}_n - \mathsf{A}\mathsf{D}^{-1}$$

and

$$\mathsf{D}^{-1}\tilde{\mathsf{T}}^\mathsf{H}\mathsf{D} = \mathsf{I}_n - \mathsf{D}^{-1}\mathsf{A} = \mathsf{T}_\mathsf{J}.$$

Therefore,

$$\sigma(\mathsf{T}_\mathsf{J}) = \sigma(\tilde{\mathsf{T}}^\mathsf{H}) = \overline{\sigma(\tilde{\mathsf{T}})}$$

and

$$\rho(\mathsf{T}_\mathsf{J}) = \rho(\tilde{\mathsf{T}}) < 1. \qquad \square$$

6.3.2 Gauss–Seidel Method

Recall that the Jacobi method can be written in the form

$$\sum_{j=1}^{i-1} a_{i,j}x_{j,k} + a_{i,i}x_{i,k+1} + \sum_{j=i+1}^{n} a_{i,j}x_{j,k} = f_i.$$

However, at this stage, we have already computed new approximations for components x_m, $m = 1, \dots, i-1$, which we are not using. The Gauss–Seidel method uses these newly computed approximations to obtain the method

$$\sum_{j=1}^{i} a_{i,j}x_{j,k+1} + \sum_{j=i+1}^{n} a_{i,j}x_{j,k} = f_i.$$

As before, we require $a_{i,i} \neq 0$ for all $i = 1, \dots, n$, so that the method is well defined.

Recall the splitting $\mathsf{A} = \mathsf{L} + \mathsf{D} + \mathsf{U}$. Choosing the iterator matrix as

$$\mathsf{B} = \mathsf{B}_{\mathrm{GS}} = \mathsf{L} + \mathsf{D}$$

results in the so-called *Gauss–Seidel method.*[2] The linear iteration process for the Gauss–Seidel method may be expressed as

$$(\mathsf{L} + \mathsf{D})\boldsymbol{x}_{k+1} + \mathsf{U}\boldsymbol{x}_k = \boldsymbol{f}.$$

Therefore, assuming that D is invertible,

$$\boldsymbol{x}_{k+1} = -(\mathsf{L} + \mathsf{D})^{-1}\mathsf{U}\boldsymbol{x}_k + (\mathsf{L} + \mathsf{D})^{-1}\boldsymbol{f}.$$

The error transfer matrix may be expressed as

$$\mathsf{T}_{\mathrm{GS}} = \mathsf{I}_n - \mathsf{B}_{GS}^{-1}\mathsf{A} = -(\mathsf{L} + \mathsf{D})^{-1}\mathsf{U} = -(\mathsf{A} - \mathsf{U})^{-1}\mathsf{U}. \tag{6.4}$$

We have the following convergence criterion.

Theorem 6.12 (convergence). *Suppose that $\mathsf{A} = [a_{i,j}] \in \mathbb{C}^{n \times n}$ is SDD and $\boldsymbol{f} \in \mathbb{C}^n$. Then, for any starting value $\boldsymbol{x}_0$, the sequence generated by the Gauss–Seidel method converges to $\boldsymbol{x} = \mathsf{A}^{-1}\boldsymbol{f}$.*

Proof. Let us define

$$\gamma = \max_{i=1}^{n} \left\{ \frac{\sum_{j=i+1}^{n} |a_{i,j}|}{|a_{i,i}| - \sum_{j=1}^{i-1} |a_{i,j}|} \right\}.$$

[2] Named in honor of the German mathematician and physicist Johann Carl Friedrich Gauss (1777–1855) and the German mathematician Philipp Ludwig von Seidel (1821–1896).

Owing to the fact that A is SDD,

$$|a_{i,i}| > \sum_{\substack{j=1\\ j\neq i}}^{n} |a_{i,j}| = \sum_{j=1}^{i-1} |a_{i,j}| + \sum_{j=i+1}^{n} |a_{i,j}|,$$

which implies that

$$|a_{i,i}| - \sum_{j=1}^{i-1} |a_{i,j}| > \sum_{j=i+1}^{n} |a_{i,j}|,$$

and thus $\gamma \in [0, 1)$. We will show convergence of the Gauss–Seidel method by proving that $\|\mathsf{T}_{\mathrm{GS}}\|_\infty \le \gamma$.

Let $\mathsf{A} = \mathsf{L} + \mathsf{D} + \mathsf{U}$ be the usual decomposition into lower triangular, diagonal, and upper triangular parts, respectively. Set $\boldsymbol{y} = \mathsf{T}_{\mathrm{GS}}\boldsymbol{x}$, i.e., $(\mathsf{L} + \mathsf{D})\boldsymbol{y} = -\mathsf{U}\boldsymbol{x}$. We have

$$\sum_{j=1}^{i-1} a_{i,j} y_j + a_{i,i} y_i = - \sum_{j=i+1}^{n} a_{i,j} x_j, \quad 1 \le i \le n.$$

Let i be such that $|y_i| = \|\boldsymbol{y}\|_\infty$. By the triangle inequality,

$$\left|\sum_{j=1}^{i-1} a_{i,j} y_j + a_{i,i} y_i\right| \ge |a_{i,i}||y_i| - \sum_{j=1}^{i-1} |a_{i,j}||y_j| \ge \left(|a_{i,i}| - \sum_{j=1}^{i-1} |a_{i,j}|\right) \|\boldsymbol{y}\|_\infty.$$

Also, we have

$$\left|\sum_{j+1}^{n} a_{i,j} x_j\right| \le \sum_{j+1}^{n} |a_{i,j}| \|\boldsymbol{x}\|_\infty.$$

Consequently,

$$\|\mathsf{T}_{\mathrm{GS}}\boldsymbol{x}\|_\infty = \|\boldsymbol{y}\|_\infty \le \frac{\sum_{j=i+1}^{n} |a_{i,j}|}{|a_{i,i}| - \sum_{j=1}^{i-1} |a_{i,j}|} \|\boldsymbol{x}\|_\infty \le \gamma \|\boldsymbol{x}\|_\infty.$$

This implies that, for all $\boldsymbol{x} \in \mathbb{C}^n_\star$,

$$\frac{\|\mathsf{T}_{\mathrm{GS}}\boldsymbol{x}\|_\infty}{\|\boldsymbol{x}\|_\infty} \le \gamma,$$

which shows that

$$\|T_{\mathrm{GS}}\|_\infty \le \gamma < 1. \qquad \square$$

In fact, we can prove more than what the last result suggests.

Theorem 6.13 (convergence). *Suppose that* $\mathsf{A} = [a_{i,j}] \in \mathbb{C}^{n\times n}$ *is SDD of magnitude* $\delta > 0$ *and* $\boldsymbol{f} \in \mathbb{C}^n$. *Then*

$$\|\mathsf{T}_{\mathrm{GS}}\|_\infty \le \|\mathsf{T}_{\mathrm{J}}\|_\infty < 1,$$

where T_{GS} *and* T_{J} *denote, respectively, the error transfer matrices of the Gauss–Seidel and Jacobi methods. In particular, for any starting value* $\boldsymbol{x}_0$, *the sequences generated by the Jacobi and the Gauss–Seidel methods both converge to* $\boldsymbol{x} = \mathsf{A}^{-1}\boldsymbol{f}$.

Proof. Set

$$\eta_{\mathrm{GS}} = \|\mathsf{T}_{\mathrm{GS}}\|_\infty, \quad \eta_{\mathrm{J}} = \|\mathsf{T}_{\mathrm{J}}\|_\infty,$$

and $\mathbf{1} = [1, \ldots, 1]^\intercal$.

Our argument will simplify significantly with some new notation, which may not be widely used outside of this proof. Define, for any matrix $\mathsf{C} = [c_{i,j}] \in \mathbb{C}^{n \times n}$, its absolute value

$$|\mathsf{C}| = [|c_{i,j}|]_{i,j=1}^n \in \mathbb{C}^{n\times n}.$$

It follows that

$$|\mathsf{C}|\,\mathbf{1} = \begin{bmatrix} \sum_{j=1}^n |c_{1,j}| \\ \vdots \\ \sum_{j=1}^n |c_{n,j}| \end{bmatrix} \qquad \|\,|\mathsf{C}|\mathbf{1}\,\|_\infty = \|\,|\mathsf{C}|\,\|_\infty.$$

Similarly, for any vector $\boldsymbol{c} = [c_i] \in \mathbb{C}^n$, $|\boldsymbol{c}| = [|c_i|]_{i=1}^n \in \mathbb{C}^n$. For $\boldsymbol{a} = [a_i], \boldsymbol{b} = [b_i] \in \mathbb{C}^n$, we write $\boldsymbol{a} \preceq \boldsymbol{b}$ if and only if $a_i \leq b_i$ for all $i = 1, \ldots, n$. Similarly, for matrices, $\mathsf{A} = [a_{i,j}], \mathsf{B} = [b_{i,j}] \in \mathbb{C}^{n\times n}$, we write $\mathsf{A} \preceq \mathsf{B}$ if and only if $a_{i,j} \leq b_{i,j}$ for all $i, j = 1, \ldots, n$. As usual, $\boldsymbol{a} \prec \boldsymbol{b}$ means $\boldsymbol{a} \preceq \boldsymbol{b}$ and $\boldsymbol{a} \neq \boldsymbol{b}$.

Under the assumption that $\eta_{\mathrm{J}} = \|\mathsf{T}_{\mathrm{J}}\|_\infty < 1$, it follows that

$$|\mathsf{T}_{\mathrm{J}}|\,\mathbf{1} \preceq \eta_{\mathrm{J}}\mathbf{1} \prec \mathbf{1}.$$

Let $\mathsf{A} = \mathsf{L} + \mathsf{D} + \mathsf{U}$ be the usual decomposition into lower triangular, diagonal, and upper triangular parts, respectively. Then

$$\mathsf{T}_{\mathrm{J}} = \mathsf{I}_n - \mathsf{D}^{-1}\mathsf{A} = -\mathsf{D}^{-1}(\mathsf{L} + \mathsf{U}) = -\tilde{\mathsf{L}} - \tilde{\mathsf{U}},$$

where $\tilde{\mathsf{L}} = \mathsf{D}^{-1}\mathsf{L}$ is strictly lower triangular and $\tilde{\mathsf{U}} = \mathsf{D}^{-1}\mathsf{U}$ is strictly upper triangular. Hence,

$$|\mathsf{T}_{\mathrm{J}}| = |\tilde{\mathsf{L}}| + |\tilde{\mathsf{U}}|.$$

From this it follows that

$$|\tilde{\mathsf{U}}|\,\mathbf{1} = |\mathsf{T}_{\mathrm{J}}|\mathbf{1} - |\tilde{\mathsf{L}}|\,\mathbf{1} \preceq \eta_{\mathrm{J}}\mathbf{1} - |\tilde{\mathsf{L}}|\,\mathbf{1} = \left(\eta_{\mathrm{J}}\mathsf{I}_n - |\tilde{\mathsf{L}}|\right)\mathbf{1}.$$

Since $|\tilde{\mathsf{L}}|$ and $\tilde{\mathsf{L}}$ are strictly lower triangular, it follows that

$$\tilde{\mathsf{L}}^n = |\tilde{\mathsf{L}}|^n = \mathsf{O} \in \mathbb{C}^{n\times n}.$$

Furthermore, $\mathsf{I}_n + \tilde{\mathsf{L}}$ is invertible and

$$\left(\mathsf{I}_n + \tilde{\mathsf{L}}\right)^{-1} = \mathsf{I}_n - \tilde{\mathsf{L}} + \tilde{\mathsf{L}}^2 - \tilde{\mathsf{L}}^3 + \cdots + \left(-\tilde{\mathsf{L}}\right)^{n-1}.$$

Then it is left to the reader as an exercise (Problem 6.1) to show that

$$\begin{aligned} \mathsf{O} &\preceq \left|(\mathsf{I}_n + \tilde{\mathsf{L}})^{-1}\right| \\ &= \left|\mathsf{I}_n - \tilde{\mathsf{L}} + \tilde{\mathsf{L}}^2 - \tilde{\mathsf{L}}^3 + \cdots + \left(-\tilde{\mathsf{L}}\right)^{n-1}\right| \\ &\preceq \mathsf{I}_n + |\tilde{\mathsf{L}}| + |\tilde{\mathsf{L}}|^2 + \cdots + |\tilde{\mathsf{L}}|^{n-1} \\ &= \left(\mathsf{I}_n - |\tilde{\mathsf{L}}|\right)^{-1}. \end{aligned}$$

Now recall and observe that

$$\begin{aligned}\mathsf{T}_{\mathrm{GS}} &= \mathsf{I}_n - (\mathsf{L}+\mathsf{D})^{-1}\mathsf{A}\\ &= -(\mathsf{L}+\mathsf{D})^{-1}\mathsf{U}\\ &= -(\mathsf{L}+\mathsf{D})^{-1}\mathsf{D}\mathsf{D}^{-1}\mathsf{U}\\ &= -\left(\mathsf{I}_n+\tilde{\mathsf{L}}\right)^{-1}\tilde{\mathsf{U}}.\end{aligned}$$

Thus,

$$\left|\tilde{\mathsf{U}}\right|\mathbf{1} \preceq \left(\eta_{\mathrm{J}}\mathsf{I}_n - \left|\tilde{\mathsf{L}}\right|\right)\mathbf{1}$$

and

$$\left(\mathsf{I}_n - \left|\tilde{\mathsf{L}}\right|\right)^{-1}\left|\tilde{\mathsf{U}}\right|\mathbf{1} \preceq \left(\mathsf{I}_n - \left|\tilde{\mathsf{L}}\right|\right)^{-1}\left(\eta_{\mathrm{J}}\mathsf{I}_n - \left|\tilde{\mathsf{L}}\right|\right)\mathbf{1},$$

since

$$\mathsf{O} \preceq \left(\mathsf{I}_n - \left|\tilde{\mathsf{L}}\right|\right)^{-1}.$$

Finally,

$$\begin{aligned}\left|\mathsf{T}_{\mathrm{GS}}\right|\mathbf{1} &\preceq \left|\left(\mathsf{I}_n+\tilde{\mathsf{L}}\right)^{-1}\right|\left|\tilde{\mathsf{U}}\right|\mathbf{1}\\ &\preceq \left(\mathsf{I}_n - \left|\tilde{\mathsf{L}}\right|\right)^{-1}\left|\tilde{\mathsf{U}}\right|\mathbf{1}\\ &\preceq \left(\mathsf{I}_n - \left|\tilde{\mathsf{L}}\right|\right)^{-1}\left(\eta_{\mathrm{J}}\mathsf{I}_n - \left|\tilde{\mathsf{L}}\right|\right)\mathbf{1}\\ &= \left(\mathsf{I}_n - \left|\tilde{\mathsf{L}}\right|\right)^{-1}\left(\mathsf{I}_n - \left|\tilde{\mathsf{L}}\right| + (\eta_{\mathrm{J}}-1)\mathsf{I}_n\right)\mathbf{1}\\ &= \left\{\mathsf{I}_n + (\eta_{\mathrm{J}}-1)\left(\mathsf{I}_n - \left|\tilde{\mathsf{L}}\right|\right)^{-1}\right\}\mathbf{1}.\end{aligned}$$

Now

$$\mathsf{I}_n \preceq \left(\mathsf{I}_n - \left|\tilde{\mathsf{L}}\right|\right)^{-1} = \mathsf{I}_n + \left|\tilde{\mathsf{L}}\right| + \left|\tilde{\mathsf{L}}\right|^2 + \cdots + \left|\tilde{\mathsf{L}}\right|^{n-1}$$

and $\eta_{\mathrm{J}} < 1$ from Theorem 6.10, so that

$$\left|\mathsf{T}_{\mathrm{GS}}\right|\mathbf{1} \preceq \eta_{\mathrm{J}}\mathbf{1},$$

which implies that

$$\eta_{\mathrm{GS}} = \left\|\mathsf{T}_{\mathrm{GS}}\right\|_\infty \leq \eta_{\mathrm{J}}. \qquad \square$$

Let us conclude this discussion by, for a special class of matrices, linking the convergence of the Jacobi method to that of the Gauss–Seidel method.

Theorem 6.14 (tridiagonal matrices). *Let $\mathsf{A} \in \mathbb{C}^{n\times n}$ be tridiagonal with nonzero diagonal elements. Denote by T_{J} and T_{GS} the error transfer matrices of the Jacobi and Gauss–Seidel methods, respectively. Then we have*

$$\rho(\mathsf{T}_{\mathrm{GS}}) = \rho(\mathsf{T}_{\mathrm{J}})^2.$$

In particular, one method converges if and only if the other method converges.

Proof. We begin with a preliminary statement about tridiagonal matrices. Suppose that

$$\mathsf{A} = \begin{bmatrix} b_1 & c_2 & 0 & \cdots & 0 \\ a_2 & b_2 & \ddots & \ddots & \vdots \\ 0 & \ddots & \ddots & c_{n-1} & 0 \\ \vdots & \ddots & a_{n-1} & b_{n-1} & c_n \\ 0 & \cdots & 0 & a_n & b_n \end{bmatrix},$$

where $b_i \neq 0$ for all $i = 1, \ldots, n$. Let $0 \neq \mu \in \mathbb{C}$. Define

$$\mathsf{M}(\mu) = \mathsf{SAS}^{-1},$$

where $\mathsf{S} = \operatorname{diag}(\mu, \mu^2, \ldots, \mu^n)$. Then

$$\mathsf{M}(\mu) = \begin{bmatrix} b_1 & \mu^{-1}c_2 & 0 & \cdots & 0 \\ \mu a_2 & b_2 & \ddots & \ddots & \vdots \\ 0 & \ddots & \ddots & \mu^{-1}c_{n-1} & 0 \\ \vdots & \ddots & \mu a_{n-1} & b_{n-1} & \mu^{-1}c_n \\ 0 & \cdots & 0 & \mu a_n & b_n \end{bmatrix}.$$

It is left to the reader as an exercise to show that

$$\det(\mathsf{M}(\mu)) = \det(\mathsf{M}(1)) = \det(\mathsf{A}).$$

Now let $\mathsf{A} = \mathsf{L} + \mathsf{D} + \mathsf{U}$, where, as usual, D is diagonal and L and U are, respectively, strictly lower and strictly upper triangular. From (6.3) we have that $\mathsf{T}_\mathsf{J} = \mathsf{I}_n - \mathsf{D}^{-1}\mathsf{A}$; therefore, the eigenvalues of T_J are the zeros of the characteristic polynomial

$$\begin{aligned} \chi_\mathsf{J}(\lambda) &= \det(\mathsf{T}_\mathsf{J} - \lambda \mathsf{I}_n) \\ &= \det\left(-\mathsf{D}^{-1}(\mathsf{L} + \mathsf{U}) - \lambda \mathsf{I}_n\right) \\ &= \det\left[-\mathsf{D}^{-1}(\mathsf{L} + \mathsf{U} + \lambda \mathsf{D})\right] \\ &= \det(-\mathsf{D}^{-1}) q_\mathsf{J}(\lambda), \end{aligned}$$

where we defined the polynomial

$$q_\mathsf{J}(\lambda) = \det(\mathsf{L} + \lambda \mathsf{D} + \mathsf{U}).$$

On the other hand, from (6.4) we have that $\mathsf{T}_{\mathsf{GS}} = -(\mathsf{L} + \mathsf{D})^{-1}\mathsf{U}$, so its eigenvalues are the zeros of

$$\chi_{\mathsf{GS}}(\lambda) = \det(\mathsf{T}_{\mathsf{GS}} - \lambda \mathsf{I}_n) = \det\left(-(\mathsf{L} + \mathsf{D})^{-1}\right) q_{\mathsf{GS}}(\lambda),$$

where

$$q_{\mathsf{GS}}(\lambda) = \det(\lambda \mathsf{L} + \lambda \mathsf{D} + \mathsf{U}).$$

Notice that both matrices involved in the definitions of q_J and q_{GS}, respectively, are tridiagonal. By the previous statement about determinants of tridiagonal matrices we have that, if $\lambda \neq 0$,

$$\begin{aligned} q_{GS}(\lambda^2) &= \det(\lambda^2 L + \lambda^2 D + U) \\ &= \lambda^n \det(\lambda L + \lambda D + \lambda^{-1} U) \\ &= \lambda^n \det(L + \lambda D + U) \\ &= \lambda^n q_J(\lambda), \end{aligned}$$

and so, by continuity, this holds for all $\lambda \in \mathbb{C}$.

The previous relation shows that

$$\lambda \in \sigma(T_{GS}) \quad \Longrightarrow \quad \lambda^{1/2}, -\lambda^{1/2} \in \sigma(T_J)$$

and

$$(\lambda \in \sigma(T_J) \iff -\lambda \in \sigma(T_J)) \quad \Longrightarrow \lambda^2 \in \sigma(T_{GS}),$$

as we needed to show. □

6.4 Richardson's Method

Let $A \in \mathbb{C}^{n\times n}$ be invertible and $\boldsymbol{f} \in \mathbb{C}^n$ be given. Let us consider now what is known as *Richardson's method*:[3]

$$\boldsymbol{x}_{k+1} = \boldsymbol{x}_k + \alpha(\boldsymbol{f} - A\boldsymbol{x}_k).$$

Clearly, this is a stationary two-layer method that results from choosing

$$B = B_R = \frac{1}{\alpha} I_n,$$

where $\alpha \in \mathbb{C}_\star$. In this case, $T_R = I_n - \alpha A$.

Theorem 6.15 (convergence). *Let $A \in \mathbb{C}^{n\times n}$ be Hermitian positive definite (HPD), $\sigma(A) = \{\lambda_1, \ldots, \lambda_n\}$ with $0 < \lambda_1 \leq \lambda_2 \leq \cdots \leq \lambda_n$. Assume that $\alpha \in \mathbb{R}_\star$. Then Richardson's method converges if and only if $\alpha \in (0, 2/\lambda_n)$. In this case, we have the estimate*

$$\|\boldsymbol{e}_k\|_2 \leq \rho^k \|\boldsymbol{e}_0\|_2, \quad \rho = \rho(\alpha) = \max\{|1 - \alpha\lambda_n|, |1 - \alpha\lambda_1|\}.$$

From this, it follows that, by setting

$$\alpha = \alpha_{\text{opt}} = \frac{2}{\lambda_1 + \lambda_n},$$

one obtains the smallest possible value of ρ and

$$\rho_{\text{opt}} = \rho(\alpha_{\text{opt}}) = \frac{\lambda_n - \lambda_1}{\lambda_n + \lambda_1} = \frac{\kappa_2(A) - 1}{\kappa_2(A) + 1}.$$

[3] Named in honor of the British mathematician, physicist, and meteorologist Lewis Fry Richardson (1881–1953).

Proof. Since A is HPD, we know that the eigenvalues of A are positive real numbers and

$$\lambda_n = \|A\|_2.$$

Notice also that $T_R = I_n - \alpha A = T_R^H$, which implies that the eigenvalues of T_R are real. Observe that $(\lambda_i, \boldsymbol{w}_i)$ is an eigenpair of A if and only if $(\nu_i = 1 - \alpha\lambda_i, \boldsymbol{w}_i)$ is an eigenpair of T_R. Assume that $0 < \alpha < 2/\lambda_n$. Then

$$0 < \lambda_i \alpha < 2\frac{\lambda_i}{\lambda_n}, \quad i = 1, \ldots, n,$$

which implies that

$$1 > 1 - \lambda_i \alpha > 1 - 2\frac{\lambda_i}{\lambda_n} \geq -1, \quad i = 1, \ldots, n.$$

It follows that

$$1 > \nu_1 \geq \cdots \geq \nu_n > -1, \qquad \nu_i = 1 - \alpha\lambda_i.$$

This guarantees that $\|T_R\|_2 = \rho(T_R) < 1$, which implies convergence. Conversely, if $\alpha \notin (0, 2/\lambda_n)$, then $\rho(T_R) \geq 1$ and the method does not converge.

By consistency,

$$\|\boldsymbol{e}_{k+1}\|_2 = \|T_R^k \boldsymbol{e}_0\|_2 \leq \rho^k \|\boldsymbol{e}_0\|_2.$$

Of course, it is easy to see that

$$\rho = \rho(T_R) = \max\{|\nu_1|, |\nu_n|\} = \max\{|1 - \alpha\lambda_n|, |1 - \alpha\lambda_1|\}.$$

Finally, showing optimality amounts to minimizing ρ; see Figure 6.1. From this we see that the minimum of ρ is attained when

$$|1 - \alpha\lambda_1| = |1 - \alpha\lambda_n|$$

or

$$1 - \alpha\lambda_n = \alpha\lambda_1 - 1,$$

which implies that

$$\alpha_{\text{opt}} = \frac{2}{\lambda_1 + \lambda_n}.$$

Therefore,

$$\rho_{\text{opt}} = 1 - \alpha_{\text{opt}}\lambda_1 = \frac{\lambda_1 + \lambda_n - 2\lambda_1}{\lambda_1 + \lambda_n} = \frac{\lambda_n - \lambda_1}{\lambda_1 + \lambda_n} = \frac{\kappa_2(A) - 1}{\kappa_2(A) + 1}.$$ □

Remark 6.16 (optimal α)**.** Notice that the rate of convergence for the optimal choice α_{opt} degrades as $\kappa_2(A)$ gets larger, since $\rho_{\text{opt}} \to 1$. We will see that this is a major problem for classical iterative methods.

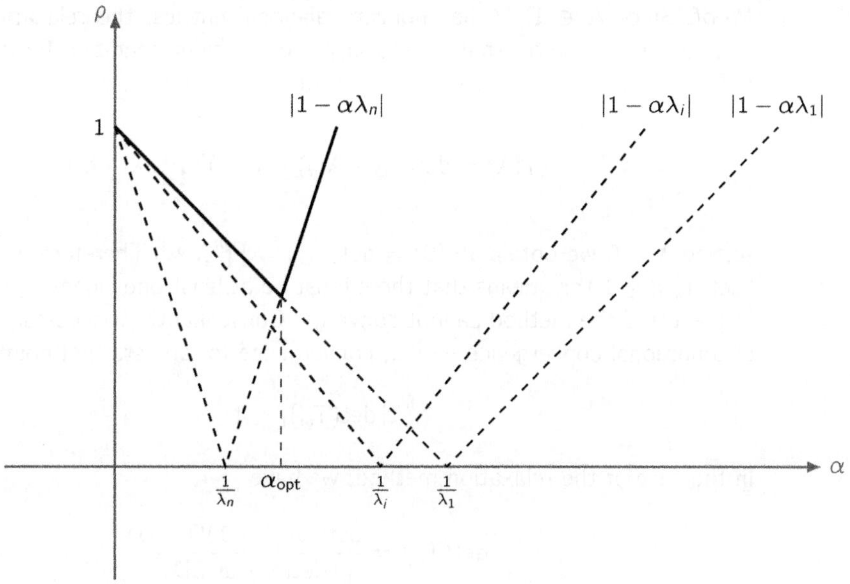

Figure 6.1 The curve $\rho(\mathsf{T_R})$ (solid line) as a function of α.

6.5 Relaxation Methods

Let us consider one last classical method, namely the relaxation method. Recall that Gauss–Seidel reads

$$\mathsf{L}\boldsymbol{x}_{k+1} + \mathsf{D}\boldsymbol{x}_{k+1} + \mathsf{U}\boldsymbol{x}_k = \boldsymbol{f},$$

where $\mathsf{A} = \mathsf{L} + \mathsf{D} + \mathsf{U}$ with the usual assumptions and notation. We will weight the contribution of the diagonal D by introducing the parameter $\omega > 0$:

$$\mathsf{L}\boldsymbol{x}_{k+1} + \omega^{-1}\mathsf{D}\boldsymbol{x}_{k+1} + (1 - \omega^{-1})\mathsf{D}\boldsymbol{x}_k + \mathsf{U}\boldsymbol{x}_k = \boldsymbol{f}.$$

In doing that, we obtain the method

$$(\mathsf{L} + \omega^{-1}\mathsf{D})\boldsymbol{x}_{k+1} = \left((\omega^{-1} - 1)\mathsf{D} - \mathsf{U}\right)\boldsymbol{x}_k + \boldsymbol{f}.$$

If we choose $\omega > 1$, the method is termed a (successive) *over-relaxation* (SOR) method; if $0 < \omega < 1$, the method is called an *under-relaxation* method. From the previous identity, we also find that the error transfer matrix is

$$\mathsf{T}_\omega = (\mathsf{L} + \omega^{-1}\mathsf{D})^{-1}\left((\omega^{-1} - 1)\mathsf{D} - \mathsf{U}\right)$$

and the iterator is

$$\mathsf{B}_\omega = \mathsf{L} + \omega^{-1}\mathsf{D}.$$

Theorem 6.17 (convergence). *Let $\mathsf{A} \in \mathbb{C}^{n\times n}$ have nonzero diagonal entries. A necessary condition for convergence of the relaxation method is that $\omega \in (0, 2)$.*

Proof. Since $A \in \mathbb{C}^{n\times n}$ has nonzero diagonal entries, the relaxation method is well defined. We know that a necessary and sufficient condition for convergence is $\rho(T_\omega) < 1$. Since the eigenvalues are roots of the characteristic polynomial,

$$\chi_T(\lambda) = \det(T_\omega - \lambda I_n) = (-1)^n \prod_{i=1}^{n} (\lambda - \lambda_i),$$

setting $\lambda = 0$ we obtain $\chi_T(0) = \det(T_\omega) = \prod_{i=1}^n \lambda_i$. Therefore, if we have that $|\det(T_\omega)| \geq 1$ this means that there must be at least one eigenvalue that satisfies $|\lambda_i| \geq 1$ and the method cannot converge. Consequently, a necessary condition for unconditional convergence — i.e., convergence for any starting point — is that

$$|\det(T_\omega)| < 1.$$

In the case of the relaxation method, we have

$$\begin{aligned}
\det(T_\omega) &= \frac{\det((\omega^{-1} - 1)D - U)}{\det(L + \omega^{-1}D)} \\
&= \frac{\prod_{i=1}^n (\omega^{-1} - 1) d_{i,i}}{\prod_{i=1}^n \omega^{-1} d_{i,i}} \\
&= \frac{(\omega^{-1} - 1)^n}{\omega^{-n}} \\
&= \omega^n (\omega^{-1} - 1)^n \\
&= (1 - \omega)^n.
\end{aligned}$$

If $\omega \notin (0, 2)$, then $|\det(T_\omega)| \geq 1$. In other words, if $\omega \notin (0, 2)$ the method cannot converge unconditionally, meaning that $\omega \in (0, 2)$ is a necessary condition for unconditional convergence. □

Let us show that, in a particular case, this condition is also sufficient.

Theorem 6.18 (convergence)**.** *Let* $A \in \mathbb{C}^{n\times n}$ *be HPD and* $\omega \in (0, 2)$. *Then the relaxation method converges. In particular, the Gauss–Seidel method converges, since* $T_{\omega=1} = T_{GS}$.

Proof. Recall that

$$T_\omega = B_\omega^{-1} \left((\omega^{-1} - 1)D - U \right)$$

and

$$I_n - T_\omega = I_n - B_\omega^{-1} \left((\omega^{-1} - 1)D - U \right) = B_\omega^{-1} \left(B_\omega - (\omega^{-1} - 1)D + U \right) = B_\omega^{-1} A.$$

Now suppose that $(\lambda, \boldsymbol{w})$ is an eigenpair of the matrix T_ω. Set

$$\boldsymbol{y} = (I_n - T_\omega)\boldsymbol{w} = (1 - \lambda)\boldsymbol{w}.$$

The previous computation implies that $B_\omega \boldsymbol{y} = A\boldsymbol{w}$. For this reason,

$$\begin{aligned}(B_\omega - A)\boldsymbol{y} &= (B_\omega - A)B_\omega^{-1}A\boldsymbol{w} \\ &= (B_\omega B_\omega^{-1}A - AB_\omega^{-1}A)\boldsymbol{w} \\ &= (A - AB_\omega^{-1}A)\boldsymbol{w} \\ &= A(I_n - B_\omega^{-1}A)\boldsymbol{w} \\ &= AT_\omega \boldsymbol{w} \\ &= \lambda A\boldsymbol{w}.\end{aligned}$$

Taking the inner product with $\boldsymbol{y}$, and using the fact that A is Hermitian, we find

$$(B_\omega \boldsymbol{y}, \boldsymbol{y})_2 = (A\boldsymbol{w}, \boldsymbol{y})_2 = (1 - \bar{\lambda})(\boldsymbol{w}, A\boldsymbol{w})_2,$$

which implies, using the explicit form of B_ω, that

$$(L\boldsymbol{y}, \boldsymbol{y})_2 + \omega^{-1}(D\boldsymbol{y}, \boldsymbol{y})_2 = (1 - \bar{\lambda})(\boldsymbol{w}, A\boldsymbol{w})_2. \tag{6.5}$$

Similarly,

$$(\boldsymbol{y}, (B_\omega - A)\boldsymbol{y})_2 = \bar{\lambda}(\boldsymbol{y}, A\boldsymbol{w})_2 = \bar{\lambda}(1 - \lambda)(\boldsymbol{w}, A\boldsymbol{w})_2,$$

which implies that

$$(\omega^{-1} - 1)(D\boldsymbol{y}, \boldsymbol{y})_2 - (\boldsymbol{y}, U\boldsymbol{y})_2 = \bar{\lambda}(1 - \lambda)(\boldsymbol{w}, A\boldsymbol{w})_2. \tag{6.6}$$

Adding (6.5) and (6.6) — and observing that, since A is Hermitian, $(L\boldsymbol{y}, \boldsymbol{y})_2 = (\boldsymbol{y}, U\boldsymbol{y})_2$ — we obtain

$$(2\omega^{-1} - 1)(D\boldsymbol{y}, \boldsymbol{y})_2 = (1 - |\lambda|^2)(\boldsymbol{w}, A\boldsymbol{w})_2.$$

Recall that, since A is HPD, so is its diagonal D. The expression on the left is positive, provided that $\omega \in (0, 2)$. This means that

$$1 - |\lambda|^2 = \frac{(2\omega^{-1} - 1)(D\boldsymbol{y}, \boldsymbol{y})_2}{(\boldsymbol{w}, A\boldsymbol{w})_2} > 0,$$

which implies that $|\lambda| < 1$. It follows that $\rho(T_\omega) < 1$. □

6.6 The Householder–John Criterion

This next result can be used to prove the convergence of a class of methods based on a verifiable criterion.

Theorem 6.19 (Householder–John criterion[4]). *Suppose that* $A \in \mathbb{C}^{n\times n}$ *is nonsingular and Hermitian and* $B \in \mathbb{C}^{n\times n}$ *is nonsingular. Assume that*

$$Q = B + B^H - A$$

is HPD. Then the two-layer stationary linear iteration method with error transfer matrix $T = I_n - B^{-1}A$ *converges unconditionally if and only if* A *is HPD.*

[4] Named in honor of the American mathematician Alston Scott Householder (1904–1993) and the German–American mathematician Fritz John (1910–1994).

Proof. For this proof, we use the standard spectral theory. Another method of proof is demonstrated in the next section.

Suppose that A is HPD. Let $(\lambda, \boldsymbol{w})$ be an arbitrary eigenpair of T. We want to show that $|\lambda| < 1$. Using the definition of T, we observe that

$$(1 - \lambda)\mathsf{B}\boldsymbol{w} = \mathsf{A}\boldsymbol{w}.$$

It follows that $\lambda \neq 1$. Otherwise, A would be singular. It follows that

$$\boldsymbol{w}^{\mathsf{H}}\mathsf{B}\boldsymbol{w} = \frac{1}{1-\lambda}\boldsymbol{w}^{\mathsf{H}}\mathsf{A}\boldsymbol{w}.$$

Taking the conjugate transpose of this equation, we have

$$\boldsymbol{w}^{\mathsf{H}}\mathsf{B}^{\mathsf{H}}\boldsymbol{w} = \frac{1}{1-\bar{\lambda}}\boldsymbol{w}^{\mathsf{H}}\mathsf{A}\boldsymbol{w},$$

using the fact that A is Hermitian. Combining the last two equations,

$$\boldsymbol{w}^{\mathsf{H}}\left(\mathsf{B}^{\mathsf{H}} + \mathsf{B} - \mathsf{A}\right)\boldsymbol{w} = \left(\frac{1}{1-\lambda} + \frac{1}{1-\bar{\lambda}} - 1\right)\boldsymbol{w}^{\mathsf{H}}\mathsf{A}\boldsymbol{w} = \frac{1-|\lambda|^2}{|1-\lambda|^2}\boldsymbol{w}^{\mathsf{H}}\mathsf{A}\boldsymbol{w}.$$

Since

$$\boldsymbol{w}^{\mathsf{H}}\mathsf{A}\boldsymbol{w} > 0, \qquad \boldsymbol{w}^{\mathsf{H}}\left(\mathsf{B}^{\mathsf{H}} + \mathsf{B} - \mathsf{A}\right)\boldsymbol{w} > 0,$$

it must be that

$$\frac{1-|\lambda|^2}{|1-\lambda|^2} > 0,$$

which implies that

$$|\lambda| < 1.$$

Using our previous calculations, but only assuming that A is nonsingular and Hermitian, if

$$\frac{1-|\lambda|^2}{|1-\lambda|^2} > 0$$

and

$$\boldsymbol{w}^{\mathsf{H}}\left(\mathsf{B}^{\mathsf{H}} + \mathsf{B} - \mathsf{A}\right)\boldsymbol{w} > 0,$$

it is easy to see that

$$\boldsymbol{w}^{\mathsf{H}}\mathsf{A}\boldsymbol{w} > 0$$

for every eigenvector $\boldsymbol{w} \in \mathbb{C}^n_\star$. This proves that A must be HPD. □

6.7 Symmetrization and Symmetric Relaxation

When the coefficient matrix, A, is symmetric, it is often desirable that the iterator matrix, B, is as well. For the standard Gauss–Seidel method, in particular, this is not the case. However, there is a simple way to symmetrize the iterator.

Definition 6.20 (symmetrized method)**.** Let $\mathsf{A} \in \mathbb{C}^{n\times n}$ be invertible and $\boldsymbol{f} \in \mathbb{C}^n$. Suppose that $\boldsymbol{x} = \mathsf{A}^{-1}\boldsymbol{f}$ and consider the stationary two-layer method (6.2) defined by the invertible iterator matrix $\mathsf{B} \in \mathbb{C}^{n\times n}$. The **symmetrized stationary two-layer method** is defined as follows:

$$\boldsymbol{x}_{k+\frac{1}{2}} = \boldsymbol{x}_k + \mathsf{B}^{-1}(\boldsymbol{f} - \mathsf{A}\boldsymbol{x}_k), \quad \boldsymbol{x}_{k+1} = \boldsymbol{x}_{k+\frac{1}{2}} + \mathsf{B}^{-\mathsf{H}}\left(\boldsymbol{f} - \mathsf{A}\boldsymbol{x}_{k+\frac{1}{2}}\right). \tag{6.7}$$

Lemma 6.21 (standard form)**.** *Let $\mathsf{A} \in \mathbb{C}^{n\times n}$ be invertible, $\boldsymbol{f} \in \mathbb{C}^n$, and $\boldsymbol{x} = \mathsf{A}^{-1}\boldsymbol{f}$. Suppose that $\mathsf{B} \in \mathbb{C}^{n\times n}$ is invertible and consider the symmetrized stationary two-layer method* (6.7)*. Then*

$$\boldsymbol{x}_{k+1} = \boldsymbol{x}_k + \mathsf{C}_\mathsf{S}(\boldsymbol{f} - \mathsf{A}\boldsymbol{x}_k),$$

where

$$\mathsf{C}_\mathsf{S} = \mathsf{B}^{-\mathsf{H}}\left(\mathsf{B} + \mathsf{B}^\mathsf{H} - \mathsf{A}\right)\mathsf{B}^{-1}.$$

If A is Hermitian, then C_S is as well.

Proof. See Problem 6.4. □

Definition 6.22 (symmetric relaxation)**.** Let $\mathsf{A} \in \mathbb{C}^{n\times n}$ be invertible with nonzero diagonal entries and consider the relaxation method with the iterator

$$\mathsf{B}_\omega = \mathsf{L} + \omega^{-1}\mathsf{D}, \quad \omega > 0,$$

where $\mathsf{A} = \mathsf{L} + \mathsf{D} + \mathsf{U}$ is the standard splitting of A into lower triangular, diagonal, and upper triangular parts, respectively. Note that B_ω is invertible. The **symmetric relaxation method** is the symmetrized stationary two-layer method with respect to B_ω, i.e.,

$$\mathsf{C}_{\omega,\mathsf{S}} = \mathsf{B}_\omega^{-\mathsf{H}}\left(\mathsf{B}_\omega + \mathsf{B}_\omega^\mathsf{H} - \mathsf{A}\right)\mathsf{B}_\omega^{-1}.$$

When $\omega = 1$, the method is called the **symmetric Gauss–Seidel method**.

Notice that if A is Hermitian, then $\mathsf{C}_{\omega,\mathsf{S}}$ is as well, according to Lemma 6.21; in particular,

$$\mathsf{C}_{\omega,\mathsf{S}} = \mathsf{B}_\omega^{-\mathsf{H}}\left(\mathsf{L} + \omega^{-1}\mathsf{D} + \mathsf{U} + \omega^{-1}\mathsf{D} - \mathsf{A}\right)\mathsf{B}_\omega^{-1} = \mathsf{B}_\omega^{-\mathsf{H}}\left((2\omega^{-1} - 1)\mathsf{D}\right)\mathsf{B}_\omega^{-1}.$$

Theorem 6.23 (convergence)**.** *Let $\mathsf{A} \in \mathbb{C}^{n\times n}$ be HPD and $\omega \in (0,2)$. Then the symmetric relaxation method converges. In particular, the symmetric Gauss–Seidel method, obtained by setting $\omega = 1$, converges.*

Proof. Since A is HPD, it has positive diagonal entries. Using this fact and the fact that $0 < \omega < 2$, it follows that $\mathsf{C}_{\omega,\mathsf{S}}$ is invertible. Then

$$\mathsf{C}_{\omega,\mathsf{S}}^{-1}(\boldsymbol{x}_{k+1} - \boldsymbol{x}_k) = \boldsymbol{f} - \mathsf{A}\boldsymbol{x}_k.$$

In other words, the iterator matrix for the symmetric relaxation method is precisely

$$\mathsf{C}_{\omega,\mathsf{S}}^{-1} = \left(\frac{2}{\omega} - 1\right)^{-1}\mathsf{B}_\omega\mathsf{D}^{-1}\mathsf{B}_\omega^\mathsf{H}.$$

One can show that

$$Q = C_{\omega,S}^{-1} + C_{\omega,S}^{-H} - A$$

is HPD, since A is HPD. Applying the Householder–John criterion, Theorem 6.19, we see that the method converges. The details are left to the reader as an exercise; see Problem 6.5. □

6.8 Convergence in the Energy Norm

In this section, we provide a powerful alternate method for proving convergence of some methods. This technique is called the energy method. Recall that, if A is HPD, we can define the so-called *energy norm* of the matrix by

$$\|\boldsymbol{x}\|_A^2 = (\boldsymbol{x}, \boldsymbol{x})_A = (A\boldsymbol{x}, \boldsymbol{x})_2.$$

We will give sufficient conditions for convergence in this norm. Before we do that, though, we need a technical result.

Lemma 6.24 (positive definite). *Suppose that* $Q \in \mathbb{C}^{n\times n}$ *is positive definite in the sense that*

$$\Re\left((Q\boldsymbol{y}, \boldsymbol{y})_2\right) > 0, \quad \forall \boldsymbol{y} \in \mathbb{C}_\star^n,$$

but Q *is not necessarily Hermitian. Then*

$$\|\boldsymbol{w}\|_Q = \sqrt{\Re\left((Q\boldsymbol{w}, \boldsymbol{w})_2\right)}, \quad \forall \boldsymbol{w} \in \mathbb{C}^n$$

defines a norm.

Proof. Suppose that Q is not Hermitian to avoid the simple case. Then

$$Q = Q_H + Q_A,$$

where

$$Q_H = \frac{1}{2}\left(Q + Q^H\right), \qquad Q_A = \frac{1}{2}\left(Q - Q^H\right)$$

are the Hermitian and anti-Hermitian parts, respectively. Observe that $Q_H^H = Q_H$ and $Q_A^H = -Q_A$. It follows that $(Q_A\boldsymbol{y}, \boldsymbol{y})_2$ is purely imaginary for any $\boldsymbol{y} \in \mathbb{C}^n$, because

$$\overline{(Q_A\boldsymbol{y}, \boldsymbol{y})_2} = \overline{\boldsymbol{y}^H Q_A \boldsymbol{y}} = \boldsymbol{y}^H Q_A^H \boldsymbol{y} = -\boldsymbol{y}^H Q_A \boldsymbol{y} = -(Q_A\boldsymbol{y}, \boldsymbol{y})_2.$$

Therefore, for all $\boldsymbol{y} \in \mathbb{C}_\star^n$,

$$0 < \Re\left((Q\boldsymbol{y}, \boldsymbol{y})_2\right) = \Re\left((Q_H\boldsymbol{y}, \boldsymbol{y})_2\right) + \Re\left((Q_A\boldsymbol{y}, \boldsymbol{y})_2\right) = (Q_H\boldsymbol{y}, \boldsymbol{y})_2,$$

since $(Q_H\boldsymbol{y}, \boldsymbol{y})_2$ is real. Therefore, Q_H is HPD. Since

$$\|\boldsymbol{w}\|_{Q_H} = \sqrt{(Q_H\boldsymbol{w}, \boldsymbol{w})_2}, \quad \forall \boldsymbol{w} \in \mathbb{C}^n$$

defines a norm, the result follows. □

Theorem 6.25 (convergence in energy). *Let* A *be HPD. Suppose that* $\mathsf{B} \in \mathbb{C}^{n\times n}$ *is the invertible iterator describing a two-layer stationary linear iteration method. If* $\mathsf{Q} = \mathsf{B} - \frac{1}{2}\mathsf{A}$ *is positive definite in the sense that*

$$\Re\left((\mathsf{Q}\boldsymbol{y}, \boldsymbol{y})_2\right) > 0, \quad \forall \boldsymbol{y} \in \mathbb{C}^n_\star,$$

but is not necessarily Hermitian, then method (6.2) *converges.*

Proof. Define the error $\boldsymbol{e}_k = \boldsymbol{x} - \boldsymbol{x}_k$, as usual, and notice that for any stationary two-layer linear iteration method we have

$$\mathsf{B}\left(\boldsymbol{e}_{k+1} - \boldsymbol{e}_k\right) + \mathsf{A}\boldsymbol{e}_k = \boldsymbol{0},$$

which is equivalent to

$$\mathsf{B}\boldsymbol{q}_{k+1} + \mathsf{A}\boldsymbol{e}_k = \boldsymbol{0},$$

where $\boldsymbol{q}_{k+1} = \boldsymbol{e}_{k+1} - \boldsymbol{e}_k$. Taking the inner product of this identity with $\boldsymbol{q}_{k+1}$ and using the fact that

$$\boldsymbol{e}_k = \tfrac{1}{2}(\boldsymbol{e}_{k+1} + \boldsymbol{e}_k) - \tfrac{1}{2}(\boldsymbol{e}_{k+1} - \boldsymbol{e}_k),$$

we obtain

$$\begin{aligned} 0 &= (\mathsf{B}\boldsymbol{q}_{k+1}, \boldsymbol{q}_{k+1})_2 + (\mathsf{A}\boldsymbol{e}_k, \boldsymbol{q}_{k+1})_2 \\ &= \left(\left(\mathsf{B} - \frac{1}{2}\mathsf{A}\right)\boldsymbol{q}_{k+1}, \boldsymbol{q}_{k+1}\right)_2 + \frac{1}{2}(\mathsf{A}\boldsymbol{e}_{k+1}, \boldsymbol{e}_{k+1})_2 - \frac{1}{2}(\mathsf{A}\boldsymbol{e}_k, \boldsymbol{e}_k)_2 \\ &= (\mathsf{Q}\boldsymbol{q}_{k+1}, \boldsymbol{q}_{k+1})_2 + \frac{1}{2}\|\boldsymbol{e}_{k+1}\|_{\mathsf{A}}^2 - \frac{1}{2}\|\boldsymbol{e}_k\|_{\mathsf{A}}^2. \end{aligned}$$

Consequently,

$$\Re\left((\mathsf{Q}\boldsymbol{q}_{k+1}, \boldsymbol{q}_{k+1})_2\right) + \frac{1}{2}\|\boldsymbol{e}_{k+1}\|_{\mathsf{A}}^2 = \frac{1}{2}\|\boldsymbol{e}_k\|_{\mathsf{A}}^2;$$

in other words,

$$\frac{1}{2}\|\boldsymbol{e}_{k+1}\|_{\mathsf{A}}^2 + \|\boldsymbol{q}_{k+1}\|_{\mathsf{Q}}^2 = \frac{1}{2}\|\boldsymbol{e}_k\|_{\mathsf{A}}^2.$$

This allows us to conclude that the sequence $\|\boldsymbol{e}_k\|_{\mathsf{A}}$ is nonincreasing. Since it is bounded below, by the monotone convergence theorem, it must have a limit, $\lim_{k\to\infty} \|\boldsymbol{e}_k\|_{\mathsf{A}} = \alpha$, say. Passing to the limit in this identity then tells us that

$$\|\boldsymbol{q}_{k+1}\|_{\mathsf{Q}} \to 0, \quad \text{as } k \to \infty.$$

But, since A is invertible,

$$\boldsymbol{e}_k = -\mathsf{A}^{-1}\mathsf{B}\boldsymbol{q}_{k+1}$$

and this implies that $\boldsymbol{e}_k \to \boldsymbol{0}$. Therefore, the method converges. □

Let us apply this result to obtain, by different means, convergence of two methods that we have previously studied.

Example 6.1 Convergence of Richardson's method. For Richardson's method we have $Q = \frac{1}{\alpha}I_n - \frac{1}{2}A$, $\alpha > 0$, and

$$(Qx, x)_2 = \frac{1}{\alpha}\|x\|_2^2 - \frac{1}{2}(Ax, x)_2 \geq \|x\|_2^2\left(\frac{1}{\alpha} - \frac{\|A\|_2}{2}\right).$$

A sufficient condition for Q to be positive definite, and, therefore, for the method to converge, is that

$$0 < \alpha < \frac{2}{\|A\|_2}.$$

But, since A is HPD, $\|A\|_2 = \lambda_n$, the largest eigenvalue. If $0 < \alpha\lambda_n < 2$, Richardson's method converges by Theorem 6.25.

Example 6.2 Convergence of the relaxation method. In the relaxation method, the iterator is

$$B_\omega = L + \frac{1}{\omega}D,$$

where $A = L + D + U$ is the standard matrix splitting and $\omega > 0$. Therefore,

$$Q_\omega = B_\omega - \frac{1}{2}A = \frac{1}{\omega}D + L - \frac{1}{2}(L + D + L^H) = \left(\frac{1}{\omega} - \frac{1}{2}\right)D + \frac{1}{2}\left(L - L^H\right)$$

and

$$\begin{aligned}\Re\left((Q_\omega x, x)_2\right) &= \left(\frac{1}{\omega} - \frac{1}{2}\right)(Dx, x)_2 + \Re\left(\frac{1}{2}((L - L^H)x, x)_2\right)\\ &= \left(\frac{1}{\omega} - \frac{1}{2}\right)(Dx, x)_2,\end{aligned}$$

where we used the fact that $\frac{1}{2}(L - L^H)$ is anti-Hermitian. If $\frac{1}{\omega} - \frac{1}{2} > 0$ or, equivalently, $\omega < 2$, Q_ω is positive definite — though not Hermitian — and the relaxation method converges.

We now state another theorem — which we have seen before and which is similar to the previous one — that can be proven using energy methods.

Theorem 6.26 (Householder–John). *Suppose that* $A \in \mathbb{C}^{n\times n}$ *is nonsingular and Hermitian and* $B \in \mathbb{C}^{n\times n}$ *is nonsingular. Assume that*

$$Q = B + B^H - A$$

is HPD. Then the two-layer stationary linear iteration method with error transfer matrix $T = I_n - B^{-1}A$ *converges unconditionally if and only if* A *is HPD.*

Proof. Let us prove one direction and save the other for an exercise; see Problem 6.6.

Suppose that A is HPD. Recall that $\|\cdot\|_{\mathsf{A}}$ defines a norm on $\mathbb{C}^n$. The error equation is precisely

$$e_{k+1} = \left(\mathsf{I} - \mathsf{B}^{-1}\mathsf{A}\right) e_k.$$

Then

$$\begin{aligned}
\|e_{k+1}\|_{\mathsf{A}}^2 &= \left(\left(\mathsf{I} - \mathsf{B}^{-1}\mathsf{A}\right) e_k\right)^{\mathsf{H}} \mathsf{A}\left(\left(\mathsf{I} - \mathsf{B}^{-1}\mathsf{A}\right) e_k\right) \\
&= \left(e_k{}^{\mathsf{H}}\left(\mathsf{I} - \mathsf{A}\mathsf{B}^{-\mathsf{H}}\right)\right) \mathsf{A}\left(\left(\mathsf{I} - \mathsf{B}^{-1}\mathsf{A}\right) e_k\right) \\
&= \left(e_k{}^{\mathsf{H}} - e_k{}^{\mathsf{H}}\mathsf{A}\mathsf{B}^{-\mathsf{H}}\right)\left(\mathsf{A}e_k - \mathsf{A}\mathsf{B}^{-1}\mathsf{A}e_k\right) \\
&= e_k{}^{\mathsf{H}}\mathsf{A}e_k - e_k{}^{\mathsf{H}}\mathsf{A}\mathsf{B}^{-\mathsf{H}}\mathsf{A}e_k - e_k{}^{\mathsf{H}}\mathsf{A}\mathsf{B}^{-1}\mathsf{A}e_k + e_k{}^{\mathsf{H}}\mathsf{A}\mathsf{B}^{-\mathsf{H}}\mathsf{A}\mathsf{B}^{-1}\mathsf{A}e_k \\
&= \|e_k\|_{\mathsf{A}}^2 - e_k{}^{\mathsf{H}}\mathsf{A}\left(\mathsf{B}^{-\mathsf{H}} + \mathsf{B}^{-1} - \mathsf{B}^{-\mathsf{H}}\mathsf{A}\mathsf{B}^{-1}\right)\mathsf{A}e_k \\
&= \|e_k\|_{\mathsf{A}}^2 - e_k{}^{\mathsf{H}}\mathsf{A}\mathsf{B}^{-\mathsf{H}}\left(\mathsf{B} + \mathsf{B}^{\mathsf{H}} - \mathsf{A}\right)\mathsf{B}^{-1}\mathsf{A}e_k \\
&= \|e_k\|_{\mathsf{A}}^2 - \left(\mathsf{B}^{-1}\mathsf{A}e_k\right)^{\mathsf{H}}\left(\mathsf{B} + \mathsf{B}^{\mathsf{H}} - \mathsf{A}\right)\mathsf{B}^{-1}\mathsf{A}e_k.
\end{aligned}$$

Since $\mathsf{B} + \mathsf{B}^{\mathsf{H}} - \mathsf{A}$ is HPD and $\mathsf{B}^{-1}\mathsf{A}e_k \neq \mathbf{0}$, in general, it follows that

$$\|e_{k+1}\|_{\mathsf{A}}^2 + \left\|\mathsf{B}^{-1}\mathsf{A}e_k\right\|_{\mathsf{Q}}^2 = \|e_k\|_{\mathsf{A}}^2$$

and $\|e_k\|_{\mathsf{A}}$ is a decreasing sequence. Therefore, by the monotone convergence theorem, $\|e_k\|_{\mathsf{A}}$ converges, i.e., there is some $\alpha \in [0, \infty)$ such that

$$\lim_{k\to\infty} \|e_k\|_{\mathsf{A}} = \alpha = \lim_{k\to\infty} \|e_{k+1}\|_{\mathsf{A}}.$$

This implies that

$$\lim_{k\to\infty} \left\|\mathsf{B}^{-1}\mathsf{A}e_k\right\|_{\mathsf{Q}} = 0,$$

which, in turn, implies that $e_k \to \mathbf{0}$, as $k \to \infty$. □

Example 6.3 Let us provide yet another proof of convergence of the relaxation method. Suppose that $\mathsf{A} \in \mathbb{C}^{n\times n}$ is HPD. We once again recall that the iterator matrix for the relaxation method is

$$\mathsf{B}_\omega = \mathsf{L} + \frac{1}{\omega}\mathsf{D},$$

where $\mathsf{A} = \mathsf{L} + \mathsf{D} + \mathsf{U}$ is the standard matrix splitting and $\omega > 0$. Therefore,

$$\mathsf{Q}_\omega = \mathsf{B}_\omega + \mathsf{B}_\omega^{\mathsf{H}} - \mathsf{A} = \frac{2}{\omega}\mathsf{D} + \mathsf{L} + \mathsf{L}^{\mathsf{H}} - \left(\mathsf{L} + \mathsf{D} + \mathsf{L}^{\mathsf{H}}\right) = \left(\frac{2}{\omega} - 1\right)\mathsf{D}.$$

If $0 < \omega < 2$, Q_ω is HPD and the method converges.

6.9 A Special Matrix

We end the discussion of classical, linear, stationary, iterative methods by studying their effect on the TST matrix defined in Theorem 3.38.

Theorem 6.27 (convergence). *Let* A *be the TST matrix defined in Theorem 3.38. Denote by* $\mathsf{T}_\mathsf{J}, \mathsf{T}_{\mathsf{GS}} \in \mathbb{R}^{(n-1)\times(n-1)}$ *the error transfer matrices of the Jacobi and Gauss–Seidel methods, respectively. Then* $\rho(\mathsf{T}_\mathsf{J}) < 1$ *and* $\rho(\mathsf{T}_{\mathsf{GS}}) < 1$.

Proof. From the proof of Theorem 3.38, we observe that the distinct eigenvalues of A are $\lambda_k = 2 - 2\cos(k\pi h)$, where $h = 1/n$, for $k = 1, \ldots, n-1$. This implies that

$$0 < \lambda_1 < \cdots < \lambda_{n-1} < 4.$$

In addition, we recall that the kth eigenvector of A, associated with λ_k, can be defined as

$$[\boldsymbol{w}_k]_i = \sin(k\pi i h).$$

The error transfer matrix of the Jacobi method is

$$\mathsf{T}_\mathsf{J} = \mathsf{D}^{-1}(\mathsf{D} - \mathsf{A}) = \frac{1}{2}(2\mathsf{I}_{n-1} - \mathsf{A}) = \frac{1}{2}\begin{bmatrix} 0 & 1 & 0 & \cdots & 0 \\ 1 & 0 & \ddots & \ddots & \vdots \\ 0 & \ddots & \ddots & 1 & 0 \\ \vdots & \ddots & 1 & 0 & 1 \\ 0 & \cdots & 0 & 1 & 0 \end{bmatrix}.$$

Then

$$\mathsf{T}_\mathsf{J}\boldsymbol{w}_k = \left(1 - \frac{1}{2}\lambda_k\right)\boldsymbol{w}_k.$$

In other words, the eigenvalues of T_J are $\mu_k = \cos(k\pi h)$, $k = 1, \ldots, n-1$. Hence,

$$\rho(\mathsf{T}_\mathsf{J}) = \mu_1 = -\mu_{n-1} = \cos(\pi h) < 1.$$

By Theorem 6.14, since A is symmetric positive definite (SPD) tridiagonal,

$$\rho(\mathsf{T}_{\mathsf{GS}}) = \rho^2(\mathsf{T}_\mathsf{J}) = \cos^2(\pi h).$$ □

Theorem 6.28 (tridiagonal matrices). *Suppose that* $\mathsf{A} \in \mathbb{C}^{n\times n}$ *is HPD and tridiagonal. Then*

$$\rho(\mathsf{T}_{\mathsf{GS}}) = \rho^2(\mathsf{T}_\mathsf{J}) < 1$$

and the optimal choice for ω *in the relaxation method is*

$$\omega_{\text{opt}} = \frac{2}{1 + \sqrt{1 - \rho(\mathsf{T}_{\mathsf{GS}})}}.$$

With this choice,

$$\omega_{\text{opt}} - 1 = \rho(\mathsf{T}_{\omega_{\text{opt}}}) = \min_{\omega\in(0,2)} \rho(\mathsf{T}_\omega).$$

Proof. See Problem 6.7. □

Example 6.4 Theorem 6.28 shows that, using the optimal choice for the relaxation method, this method can converge much faster than the Gauss–Seidel method. Suppose that A is the TST matrix defined in Theorem 3.38. Then we proved that

$$\rho(\mathsf{T}_{\mathrm{GS}}) = \rho^2(\mathsf{T}_{\mathrm{J}}) = \cos^2(\pi h).$$

Suppose that $n = 32$, so that $h = 1/32$. Then

$$\rho(\mathsf{T}_{\mathrm{GS}}) = \cos^2(\pi/32) \approx (0.995\,184\,72)^2 \approx 0.990\,392\,64.$$

But

$$\omega_{\mathrm{opt}} = \frac{2}{1 + \sqrt{1 - \cos^2(\pi/32)}} \approx 1.821\,465\,19.$$

Therefore,

$$\rho(\mathsf{T}_{\omega_{\mathrm{opt}}}) = \omega_{\mathrm{opt}} - 1 \approx 0.821\,465\,19.$$

This might not seem like a big deal, but it is. The convergence rate of the relaxation method with the optimal parameter ω_{opt} is much better than that for Gauss–Seidel.

6.10 Nonstationary Two-Layer Methods

We now consider a class of nonstationary methods that are usually obtained from a minimization condition.

6.10.1 Chebyshev's Method

Let us consider the explicit method:

$$\frac{\boldsymbol{x}_{k+1} - \boldsymbol{x}_k}{\alpha_{k+1}} + \mathsf{A}\boldsymbol{x}_k = \boldsymbol{f},$$

where $\alpha_k > 0$. This is a nonstationary method defined by setting

$$\mathsf{B}_k = \mathsf{B}_{C,k} = \frac{1}{\alpha_k}\mathsf{I}_n$$

and is known as *Chebyshev's method.*[5] It is like Richardson's method, except that we will choose the parameter α_k adaptively, in such a way that, after m steps, the quantity $\|\boldsymbol{e}_m\|_2 = \|\boldsymbol{x} - \boldsymbol{x}_m\|_2$ is as small as possible.

Theorem 6.29 (convergence of Chebyshev's method)**.** *Let* $\mathsf{A} \in \mathbb{C}^{n\times n}$ *be HPD with spectrum* $\sigma(\mathsf{A}) = \{\lambda_i\}_{i=1}^n$, $0 < \lambda_1 \le \lambda_2 \le \cdots \le \lambda_n$. *For a given* $m \in \mathbb{N}$, *the quantity* $\|\boldsymbol{e}_m\|_2$ *is minimized if*

$$\alpha_k = \frac{\alpha_0}{1 + \rho_0 t_k}, \qquad k = 1, \dots, m,$$

where

$$\alpha_0 = \frac{2}{\lambda_1 + \lambda_n}, \qquad \rho_0 = \frac{\kappa_2(\mathsf{A}) - 1}{\kappa_2(\mathsf{A}) + 1}, \qquad t_k = \cos\left[\frac{(2k-1)\pi}{2m}\right].$$

[5] Named in honor of the Russian mathematician Pafnuty Lvovich Chebyshev (1821–1894).

In this case, we have

$$\|\boldsymbol{e}_m\|_2 \le \frac{2\rho_1^m}{1+\rho_1^{2m}}\|\boldsymbol{e}_0\|_2, \quad \rho_1 = \frac{\sqrt{\kappa_2(\mathsf{A})}-1}{\sqrt{\kappa_2(\mathsf{A})}+1}.$$

Proof. Let us merely sketch the proof. The error is governed by

$$\boldsymbol{e}_{k+1} - \boldsymbol{e}_k + \alpha_{k+1}\mathsf{A}\boldsymbol{e}_k = \mathbf{0},$$

so that

$$\boldsymbol{e}_{k+1} = (\mathsf{I}_n - \alpha_{k+1}\mathsf{A})\boldsymbol{e}_k = \cdots = (\mathsf{I}_n - \alpha_{k+1}\mathsf{A})\cdots(\mathsf{I}_n - \alpha_1\mathsf{A})\boldsymbol{e}_0.$$

This implies that

$$\boldsymbol{e}_m = \mathsf{T}_m\boldsymbol{e}_0, \qquad \mathsf{T}_m = (\mathsf{I}_n - \alpha_m\mathsf{A})(\mathsf{I}_n - \alpha_{m-1}\mathsf{A})\cdots(\mathsf{I}_n - \alpha_1\mathsf{A}).$$

Since A is Hermitian, so is T_m; therefore, $\|\mathsf{T}_m\|_2 = \rho(\mathsf{T}_m)$. It follows that

$$\|\boldsymbol{e}_m\|_2 \le \rho(\mathsf{T}_m)\|\boldsymbol{e}_0\|_2$$

and it suffices to minimize $\rho(\mathsf{T}_m)$.

Notice that $\nu \in \sigma(\mathsf{T}_m)$ if and only if $\nu = \prod_{i=1}^m (1-\alpha_i\lambda)$ for some $\lambda \in \sigma(\mathsf{A})$. Now, since A is HPD,

$$\rho(\mathsf{T}_m) = \max_{i=1}^{n}\left|\prod_{j=1}^{m}(1-\alpha_j\lambda_i)\right| \le \max_{\zeta\in[\lambda_1,\lambda_n]} |p(\zeta)|,$$

where

$$p(\zeta) = \prod_{j=1}^{m}(1-\alpha_j\zeta).$$

Solving this problem is beyond the scope of our present discussion. For now, let us just say that this is possible to do. The solution is given by shifted and rescaled roots of the Chebyshev polynomials T_m, which will imply all the formulas given above; see Section 10.3. □

Chebyshev's method shows that, if we have bounds for the spectrum of A, it is possible to make the error as small as possible after a fixed number of iterations. What can we do if such bounds are not known? This gives rise to the following methods.

6.10.2 The Method of Minimal Residuals

Let $\mathsf{A} \in \mathbb{C}^{n\times n}$ be HPD. Given an arbitrary guess $\boldsymbol{x}_k$, the residual is defined by $\boldsymbol{r}_k = \boldsymbol{f} - \mathsf{A}\boldsymbol{x}_k$. Notice that

$$\mathsf{A}\boldsymbol{e}_k = \mathsf{A}(\boldsymbol{x} - \boldsymbol{x}_k) = \boldsymbol{f} - \mathsf{A}\boldsymbol{x}_k = \boldsymbol{r}_k.$$

Let us again consider the nonstationary method

$$\boldsymbol{x}_{k+1} = \boldsymbol{x}_k + \alpha_{k+1}(\boldsymbol{f} - \mathsf{A}\boldsymbol{x}_k) = \boldsymbol{x}_k + \alpha_{k+1}\boldsymbol{r}_k,$$

where we will choose the iteration parameter to minimize $\|r_{k+1}\|_2$. Let us apply A to the method to get

$$Ax_{k+1} = Ax_k + \alpha_{k+1}Ar_k,$$

which is equivalent to

$$r_{k+1} = f - Ax_{k+1} = f - Ax_k - \alpha_{k+1}Ar_k.$$

To sum up,

$$r_{k+1} = T_{k+1}r_k, \qquad T_{k+1} = I_n - \alpha_{k+1}A.$$

Notice that T_{k+1} has a familiar form. Computing the 2-norm of the residual, we find

$$\begin{aligned}\|r_{k+1}\|_2^2 &= (r_k - \alpha_{k+1}Ar_k, r_k - \alpha_{k+1}Ar_k)_2 \\ &= \|r_k\|_2^2 + \alpha_{k+1}^2\|Ar_k\|_2^2 - 2\alpha_{k+1}(Ar_k, r_k)_2,\end{aligned}$$

which, being a positive quadratic in α_{k+1}, is clearly minimized by setting

$$\alpha_{k+1} = \frac{(Ar_k, r_k)_2}{\|Ar_k\|_2^2}.$$

Thus, we arrive at the so-called *method of minimal residuals*. This method is presented in Listing 6.1. It will converge faster than Richardson's method, as we show in the following result.

Theorem 6.30 (convergence). *Let* A *be HPD with spectrum* $\sigma(A) = \{\lambda_i\}_{i=1}^n$, $0 < \lambda_1 \le \lambda_2 \le \cdots \le \lambda_n$. *Then*

$$\|Ae_k\|_2 \le \rho_\star^k \|Ae_0\|_2, \quad \rho_\star = \frac{\kappa_2(A) - 1}{\kappa_2(A) + 1}.$$

Proof. Since $r_{k+1} = T_{k+1}r_k$ in the 2-norm is minimized by the given choice of α_{k+1}, we must have that

$$\|r_{k+1}\|_2 = \|T_{k+1}r_k\|_2 \le \|Tr_k\|_2,$$

where $T = I_n - \alpha A$ and any choice of $\alpha \in \mathbb{C}$. In particular, we can set $\alpha = \alpha_\star = \frac{2}{\lambda_1+\lambda_n}$ to get the error transfer matrix of Richardson's method with its optimal choice of parameter. In this case, $\|T\|_2 = \rho_\star < 1$. Since $Ae_k = r_k$, the result follows. □

6.10.3 The Method of Minimal Corrections

Let us consider now a nonstationary two-layer method of the form

$$\frac{1}{\alpha_{k+1}}S(x_{k+1} - x_k) + Ax_k = f,$$

where $\alpha_{k+1} > 0$ and $S \in \mathbb{C}^{n\times n}$ is invertible. This conforms to our standard nonstationary iteration framework by setting $B_{k+1} = \frac{1}{\alpha_{k+1}}S$. We define the *correction* to be $w_k = S^{-1}r_k$ and notice that

$$x_{k+1} = x_k + \alpha_{k+1}S^{-1}(f - Ax_k) = x_k + \alpha_{k+1}S^{-1}r_k = x_k + \alpha_{k+1}w_k.$$

Let us now assume that S is HPD, so that it has an associated energy norm. The method of minimal corrections then chooses α_{k+1} so as to minimize $\|\boldsymbol{w}_{k+1}\|_{\mathsf{S}}^2$.

Theorem 6.31 (convergence). *Let* $\mathsf{A}, \mathsf{S} \in \mathbb{C}^{n\times n}$ *be HPD. Then*

$$\sigma(\mathsf{S}^{-1}\mathsf{A}) = \{\mu_i\}_{i=1}^n \subset (0, \infty).$$

Suppose that

$$0 < \mu_1 \leq \cdots \leq \mu_n \quad \text{and} \quad \kappa = \frac{\mu_n}{\mu_1}.$$

Then

$$\|\mathsf{A}\boldsymbol{e}_k\|_{\mathsf{S}^{-1}} \leq \rho_0^k \|\mathsf{A}\boldsymbol{e}_0\|_{\mathsf{S}^{-1}}, \qquad \rho_0 = \frac{\kappa - 1}{\kappa + 1}.$$

Proof. Since S is HPD, the object $(\boldsymbol{x}, \boldsymbol{y})_{\mathsf{S}} = (\mathsf{S}\boldsymbol{x}, \boldsymbol{y})_2$ defines an inner product and $\|\boldsymbol{x}\|_{\mathsf{S}} = \sqrt{(\boldsymbol{x}, \boldsymbol{x})_{\mathsf{S}}}$ defines a norm. Since the matrix $\mathsf{S}^{-1}\mathsf{A}$ is self-adjoint and positive definite with respect to the inner product $(\cdot, \cdot)_{\mathsf{S}}$, we leave it to the reader as an exercise to show that the eigenvalues of $\mathsf{S}^{-1}\mathsf{A}$ are all real and positive.

Furthermore, we can define the square root $\mathsf{S}^{1/2}$. In particular, since S is HPD, there exist a unitary matrix $\mathsf{U} \in \mathbb{C}^{n\times n}$ and a diagonal matrix $\mathsf{D} = \operatorname{diag}(\nu_1, \ldots, \nu_n)$, with positive real diagonal entries, such that

$$\mathsf{S} = \mathsf{U}\mathsf{D}\mathsf{U}^{\mathsf{H}}.$$

Consequently, we define $\mathsf{D}^{1/2} = \operatorname{diag}\left(\sqrt{\nu_1}, \ldots, \sqrt{\nu_n}\right)$. Set

$$\mathsf{S}^{1/2} = \mathsf{U}\mathsf{D}^{1/2}\mathsf{U}^{\mathsf{H}}.$$

Clearly, $\mathsf{S}^{1/2}$ is HPD and $\mathsf{S}^{1/2}\mathsf{S}^{1/2} = \mathsf{S}$. Let us introduce then the change of variables $\boldsymbol{v}_k = \mathsf{S}^{1/2}\boldsymbol{w}_k$ and notice that, from Problem 6.8, it follows that

$$\frac{1}{\alpha_{k+1}}(\boldsymbol{v}_{k+1} - \boldsymbol{v}_k) + \mathsf{C}\boldsymbol{v}_k = \boldsymbol{0} \tag{6.8}$$

with $\mathsf{C} = \mathsf{S}^{-1/2}\mathsf{A}\mathsf{S}^{-1/2}$. Finally, note that we must choose α_{k+1} in order to minimize

$$\|\boldsymbol{w}_{k+1}\|_{\mathsf{S}}^2 = (\mathsf{S}\boldsymbol{w}_{k+1}, \boldsymbol{w}_{k+1})_2 = (\mathsf{S}^{1/2}\boldsymbol{w}_{k+1}, \mathsf{S}^{1/2}\boldsymbol{w}_{k+1})_2 = \|\boldsymbol{v}_{k+1}\|_2^2.$$

We can repeat some steps from the proof of Theorem 6.30 to obtain

$$\alpha_{k+1} = \frac{(\mathsf{C}\boldsymbol{v}_k, \boldsymbol{v}_k)_2}{\|\mathsf{C}\boldsymbol{v}_k\|_2^2}.$$

Next, we observe the following identities:

$$\begin{aligned}
(\mathsf{C}\boldsymbol{v}_k, \boldsymbol{v}_k)_2 &= (\mathsf{S}^{-1/2}\mathsf{A}\mathsf{S}^{-1/2}\boldsymbol{w}_k, \mathsf{S}^{-1/2}\boldsymbol{w}_k)_2 = (\mathsf{A}\boldsymbol{w}_k, \boldsymbol{w}_k)_2, \\
\|\mathsf{C}\boldsymbol{v}_k\|_2^2 &= (\mathsf{S}^{-1/2}\mathsf{A}\boldsymbol{w}_k, \mathsf{S}^{-1/2}\mathsf{A}\boldsymbol{w}_k)_2 = \|\mathsf{A}\boldsymbol{w}_k\|_{\mathsf{S}^{-1}}^2, \\
\|\boldsymbol{v}_{k+1}\|_2 &= (\mathsf{S}\boldsymbol{w}_{k+1}, \boldsymbol{w}_{k+1})_2 = \|\mathsf{A}\boldsymbol{e}_{k+1}\|_{\mathsf{S}^{-1}}^2.
\end{aligned}$$

The desired error estimate follows by using these. □

Problems

6.1 Complete the proof of Theorem 6.13.

6.2 Let $\mathsf{A}, \mathsf{B} \in \mathbb{C}^{n\times n}$ with A and $\mathsf{A} - \mathsf{B}\mathsf{A}^{-1}\mathsf{B}$ being invertible. Consider solving the linear system

$$\mathsf{A}\boldsymbol{x} + \mathsf{B}\boldsymbol{y} = \boldsymbol{f}_1, \qquad \mathsf{B}\boldsymbol{x} + \mathsf{A}\boldsymbol{y} = \boldsymbol{f}_2$$

for the unknowns $\boldsymbol{x}$ and $\boldsymbol{y}$.

a) Use the Schur complement to show that this system has a unique solution.

b) Show that $\rho(\mathsf{A}^{-1}\mathsf{B}) < 1$ is a necessary and sufficient condition for convergence of the iteration method:

$$\mathsf{A}\boldsymbol{x}_{k+1} = \boldsymbol{f}_1 - \mathsf{B}\boldsymbol{y}_k, \qquad \mathsf{A}\boldsymbol{y}_{k+1} = \boldsymbol{f}_2 - \mathsf{B}\boldsymbol{x}_k$$

with an arbitrary initial guess.

c) Consider the following slightly modified iteration method:

$$\mathsf{A}\boldsymbol{x}_{k+1} = \boldsymbol{f}_1 - \mathsf{B}\boldsymbol{y}_k, \qquad \mathsf{A}\boldsymbol{y}_{k+1} = \boldsymbol{f}_2 - \mathsf{B}\boldsymbol{x}_{k+1}.$$

Does the conclusion of the previous part still hold? Why or why not?

6.3 Use the Householder–John criterion to prove that the Gauss–Seidel method converges for any initial guess if A is HPD. Can the criterion be used to establish convergence for the relaxation method?

6.4 Prove Lemma 6.21.

6.5 Prove Theorem 6.23.

6.6 Complete the proof of Theorem 6.26.

6.7 Prove Theorem 6.28.

6.8 Consider a nonstationary two-layer method of the form

$$\frac{1}{\alpha_{k+1}}\mathsf{S}\left(\boldsymbol{x}_{k+1} - \boldsymbol{x}_k\right) + \mathsf{A}\boldsymbol{x}_k = \boldsymbol{f},$$

where $\alpha_{k+1} > 0$ and $\mathsf{S} \in \mathbb{C}^{n\times n}$ is invertible. Show that $\boldsymbol{w}_k = \mathsf{S}^{-1}\boldsymbol{r}_k$ satisfies

$$\frac{1}{\alpha_{k+1}}\mathsf{S}\left(\boldsymbol{w}_{k+1} - \boldsymbol{w}_k\right) + \mathsf{A}\boldsymbol{w}_k = \boldsymbol{0}.$$

6.9 Define $\mathsf{A} \in \mathbb{R}^{n\times n}$ via

$$\mathsf{A} = \begin{bmatrix} 2 & -1 & 0 & \cdots & 0 \\ -1 & 2 & \ddots & & \vdots \\ 0 & \ddots & \ddots & -1 & 0 \\ \vdots & & -1 & 2 & -1 \\ 0 & \cdots & 0 & -1 & 2 \end{bmatrix}.$$

a) Find the eigenvalues and eigenvectors of A.

b) Suppose that $\mathsf{A} = \mathsf{D} + \mathsf{U} + \mathsf{U}^\intercal$, where D is the diagonal part of A and U is the strict upper triangular part. Find the eigenvalues and eigenvectors of $\mathsf{T}_\mathsf{J} = \mathsf{I}_n - \mathsf{D}^{-1}\mathsf{A}$, the error transfer matrix for the Jacobi method.

c) The damped Jacobi method has the error transfer matrix

$$\mathsf{T}(\omega) = -\omega \mathsf{D}^{-1}(\mathsf{U} + \mathsf{U}^{\intercal}) + (1-\omega)\mathsf{I}_n, \quad 0 < \omega < 2,$$

so that $\mathsf{T}(1) = \mathsf{T}_{\mathsf{J}}$. Write out the corresponding linear stationary two-layer method.

d) Find the eigenvalues and eigenvectors of $\mathsf{T}(\omega) \in \mathbb{C}^{n\times n}$.

e) Let $\mu_k(\omega)$ be the kth eigenvalue of $\mathsf{T}(\omega)$ with the ordering

$$\mu_1 \le \mu_2 \le \cdots \le \mu_n.$$

(You will observe in this problem that we may extend the definition of the eigenvalues in a natural way, so that $\mu_0 = 1$ and $\mu_{n+1} = 1 - 2\omega$.) Assume that $n+1$ is even. Prove that the quantity

$$S(\omega) = \max_{\frac{n+1}{2} \le k \le n+1} |\mu_k(\omega)|$$

is minimized by $\omega = \omega_0 = \frac{2}{3}$ and that

$$|\mu_k(\omega_0)| \le \frac{1}{3}$$

for all $\frac{n+1}{2} \le k \le n+1$.

6.10 Let $\mathsf{A} \in \mathbb{C}^{n\times n}$ be HPD and $\boldsymbol{f} \in \mathbb{C}^n$. Consider solving $\mathsf{A}\boldsymbol{x} = \boldsymbol{f}$ using the linear iterative method (6.2), where $\mathsf{B} \in \mathbb{C}^{n\times n}$ is invertible. Suppose that $\mathsf{Q} = \mathsf{B} + \mathsf{B}^{\mathsf{H}} - \mathsf{A}$ is positive definite. Let $\boldsymbol{e}_k = \boldsymbol{x} - \boldsymbol{x}_k$ be the error of the kth iteration. Show that each step of this method reduces the A-norm of $\boldsymbol{e}_k$, whenever $\boldsymbol{e}_k \neq \boldsymbol{0}$. Prove that this linear iteration method converges.

6.11 Let $\mathsf{A} \in \mathbb{C}^{n\times n}$ be Hermitian and $\boldsymbol{f} \in \mathbb{C}^n$. Consider solving $\mathsf{A}\boldsymbol{x} = \boldsymbol{f}$ using the linear iterative method (6.2), where B is invertible and $\boldsymbol{x}_0 \in \mathbb{C}^n$ is arbitrary.

a) If A and $\mathsf{B} + \mathsf{B}^{\mathsf{H}} - \mathsf{A}$ are HPD, prove that the method is convergent.

b) Conversely, suppose that $\mathsf{B} + \mathsf{B}^{\mathsf{H}} - \mathsf{A}$ is HPD and the linear iterative method above is convergent. Prove that A is HPD.

c) Prove that the Gauss–Seidel method converges when A is HPD.

d) Prove that the Jacobi method converges when A and $2\mathsf{D} - \mathsf{A}$ are HPD, where D is the diagonal part of A.

6.12 Suppose that $\mathsf{A} \in \mathbb{C}^{n\times n}$ is an upper triangular nonsingular matrix and $\boldsymbol{f} \in \mathbb{C}^n$. Show that both the Jacobi and Gauss–Seidel iteration methods always converge when used to solve $\mathsf{A}\boldsymbol{x} = \boldsymbol{f}$, and, moreover, that they will converge in finitely many steps.

6.13 Let $\mathsf{A} \in \mathbb{C}^{n\times n}$ have the property that there is $q \in (0,1)$ for which

$$q|a_{i,i}| > \sum_{\substack{j=1\\ j\neq i}}^{n} |a_{i,j}|, \quad 1 \le i \le n.$$

Show that the Gauss–Seidel method converges and that we have the error estimate

$$\|\boldsymbol{e}_k\|_\infty \le q^k \|\boldsymbol{e}_0\|_\infty.$$

6.14 Given the matrix $\mathsf{A} \in \mathbb{R}^{100\times 100}$ assume that its eigenvalues are known via the formula

$$\lambda_k = k^2, \quad k = 1, \ldots, 100.$$

The system $\mathsf{A}\boldsymbol{x} = \boldsymbol{f}$ is to be solved by the following *nonstationary* Richardson method:

$$\boldsymbol{x}_{k+1} = (\mathsf{I}_{100} - \tau_k \mathsf{A})\boldsymbol{x}_k + \tau_k \boldsymbol{f}$$

for some positive parameters τ_k. Find a particular set of iteration parameters $\{\tau_0, \ldots, \tau_{99}\}$ such that $\boldsymbol{x}_{100} = \boldsymbol{x}$, the exact solution of $\mathsf{A}\boldsymbol{x} = \boldsymbol{f}$.

Hint: Verify that $\boldsymbol{e}_{k+1} = (\mathsf{I}_n - \tau_k \mathsf{A})\boldsymbol{e}_k$, then expand the initial error

$$\boldsymbol{e}_0 = \alpha_1 \boldsymbol{u}_1 + \cdots + \alpha_{100}\boldsymbol{u}_{100},$$

where the $\boldsymbol{u}_i$ are the eigenvectors of A. Choose the iteration parameters to eliminate exactly one term of the expansion of $\boldsymbol{e}_0$.

6.15 Show that if $\mathsf{A} \in \mathbb{C}^{n\times n}$ is HPD, then the relaxation method converges for $0 < \omega < 2$.

6.16 Let X be a finite-dimensional vector space, $f \in X$, and $A \colon X \to X$ a linear operator. A general two-layer iterative process for the solution of $Ax = f$ can be understood as a map $\mathcal{J} \colon X \times X \to X$ as follows: given $x_0 \in X$ the sequence of iterates is defined by

$$x_{k+1} = \mathcal{J}(x_k, f).$$

a) The map $\mathcal{J}$ is said to be *consistent* (with $Ax = f$) if and only if the solution x is a fixed point, i.e., $x = \mathcal{J}(x, f)$.
b) The map $\mathcal{J}$ is linear if and only if

$$\mathcal{J}(\alpha x + \beta y, \alpha f + \beta g) = \alpha \mathcal{J}(x, f) + \beta \mathcal{J}(y, g).$$

Show that any linear and consistent iterative method can be written in the form

$$x_{k+1} = x_k + B(f - Ax_k),$$

where $B \colon X \to X$ is a linear operator.

Hint: Following Proposition 6.3, define $Bf = \mathcal{J}(0, f)$ and $(I - BA)x = \mathcal{J}(x, 0)$.

6.17 Suppose that $\mathsf{A} \in \mathbb{C}^{n\times n}$ is HPD and $\boldsymbol{f} \in \mathbb{C}^n$. Consider Richardson's method with an optimal choice of iterative parameter. How many iterations are needed to reduce the error by a predetermined factor? In other words, given $\sigma > 0$, how many iterations are needed to guarantee that

$$\|\boldsymbol{e}_k\|_2 \le \sigma \|\boldsymbol{e}_0\|_2.$$

6.18 Suppose that

$$\mathsf{A}_1 = \begin{bmatrix} 1 & -\frac{1}{2} \\ -\frac{1}{2} & 1 \end{bmatrix}, \quad \mathsf{A}_2 = \begin{bmatrix} 1 & -\frac{3}{4} \\ -\frac{1}{12} & 1 \end{bmatrix}.$$

Let T_1 and T_2 be the associated error transfer matrices for Jacobi's method for the respective coefficient matrices. Show that $\rho(\mathsf{T}_1) > \rho(\mathsf{T}_2)$, thereby refuting the claim that greater diagonal dominance implies faster convergence.

6.19 This problem deals with an acceleration procedure called *extrapolation*. Consider a linear stationary method

$$\mathbf{x}_{k+1} = \mathsf{T}\mathbf{x}_k + \mathbf{c}, \tag{6.9}$$

which can be embedded in a one-parameter family of methods

$$\mathbf{x}_{k+1} = \gamma(\mathsf{T}\mathbf{x}_k + \mathbf{c}) + (1 - \gamma)\mathbf{x}_k = \mathsf{T}_\gamma \mathbf{x}_k + \gamma \mathbf{c}$$

with $\mathsf{T}_\gamma = \gamma\mathsf{T} + (1 - \gamma)\mathsf{I}$ and $\gamma \neq 0$.

a) Show that any fixed point of T_γ is a fixed point of T.

b) Assume that we know that $\sigma(\mathsf{T}) \subset [a, b]$. Show that $\sigma(\mathsf{T}_\gamma) \subset [\gamma a+1-\gamma, \gamma b+1-\gamma]$.

c) Show that, from the previous item, it follows that

$$\rho(\mathsf{T}_\gamma) \leq \max_{a \leq \lambda \leq b} |\gamma\lambda + 1 - \gamma|.$$

d) Show that if $1 \notin [a, b]$, then the choice $\gamma^\star = 2/(2 - a - b)$ minimizes $\rho(\mathsf{T}_\gamma)$ and that, in this case,

$$\rho(\mathsf{T}_{\gamma^\star}) = 1 - |\gamma^\star| d,$$

where d is the distance between 1 and $[a, b]$. This shows that, using extrapolation, we can turn a nonconvergent method into a convergent one.

e) Suppose that A is HPD. Consider Richardson's method with $\alpha = 1$. If $\sigma(\mathsf{A}) \subset [m, M] \subset (0, \infty)$, show that choosing $\gamma = 2/(m + M)$ we obtain

$$\rho(\mathsf{T}_\gamma) = \frac{M - m}{M + m};$$

thus, Richardson's method with extrapolation converges.

6.20 This problem deals with *Chebyshev* acceleration. Consider, again, the linear stationary method (6.9) and assume that we have computed $\mathbf{x}_0, \ldots, \mathbf{x}_k$. We will compute a linear combination of them,

$$\mathbf{u}_k = \sum_{i=0}^{k} a_{k,i}\mathbf{x}_i,$$

in the hope that $\mathbf{u}_k$ is a better approximation to $\mathbf{x}$.

a) Show that if $\mathbf{x}_0 = \mathbf{x}$, then $\mathbf{x}_k = \mathbf{x}$. This tells us that we want $\sum_{i=0}^{k} a_{k,i} = 1$. Why?

b) Show that

$$\mathbf{u}_k - \mathbf{x} = p(\mathsf{T})\mathbf{e}_0,$$

where $p(z) = \sum_{i=0}^{k} a_{k,i}z^i$. Consequently, $\|\mathbf{u}_k - \mathbf{x}\| \leq \|p(\mathsf{T})\|\|\mathbf{e}_0\|$.

c) Show that, if $\sigma(\mathsf{T}) \subset [a, b]$,

$$\rho(p(\mathsf{T})) \leq \max_{z \in [a,b]} |p(z)|.$$

d) Thus, we need to find a polynomial of degree k such that $p(1) = 1$ (Why?) and minimizes

$$\max_{z\in[a,b]} |p(z)|.$$

It can be shown that shifted Chebyshev polynomials solve this problem and that, moreover,

$$u_1 = \gamma[\mathsf{T}u_0 + c] + (1-\gamma)u_0, \quad \gamma = \frac{2}{2-b-a},$$

and

$$u_k = \rho_k[\gamma(\mathsf{T}u_{k-1} + c) + (1-\gamma)u_{k-1}] + (1-\rho_k)u_{k-2},$$

with $\rho_1 = 2$, $\rho_k = (1 - \alpha\rho_{k-1})^{-1}$, and $\alpha = (2(2-b-a)/(b-a))^{-2}$.

6.21 Show that, for the method of minimal corrections,

$$\alpha_{k+1} = \frac{(\mathsf{A}w_k, w_k)_2}{(\mathsf{S}^{-1}\mathsf{A}w_k, w_k)_2}.$$

6.22 Suppose that $\mathsf{A} \in \mathbb{R}^{n\times n}$ is invertible and $f \in \mathbb{R}^n$. Consider the following iterative method to solve $\mathsf{A}x = f$:

$$x_{k+1} = x_k + \alpha r_k,$$

where $r_k = f - \mathsf{A}x_k$ is the residual, $x_0 \neq \mathsf{A}^{-1}f$ is arbitrary, and α is a scalar parameter to be determined.

a) Show that if all the eigenvalues of A have positive real part, then there will be some real α such that the method converges for any starting vector x_0.

b) Show how to choose α optimally in the case that A is SPD, and estimate the rate of convergence in the 2-norm.

c) Show that if some eigenvalues of A have negative real part and some have positive real part, then there is no real α for which the iterations converge.

6.23 Let $\mathsf{A} = \mathsf{B}+\mathsf{P} \in \mathbb{C}^{n\times n}$ be invertible. To approximate the solution of $\mathsf{A}x = f$, starting from an arbitrary $x_0 \in \mathbb{C}^n$, we apply the following iterative method (where we are implicitly assuming that B is invertible):

$$\mathsf{B}x_{k+1} + \mathsf{P}x_k = f.$$

Show that this iterative method is convergent if and only if all the roots of

$$\det(\kappa\mathsf{B} + \mathsf{P}) = 0, \quad \kappa \in \mathbb{C},$$

are of modulus strictly less than one. Use this to show that if A is HPD, then the Gauss–Seidel method converges.

Hint: You may use without proof that if M is skew-Hermitian, then for every $w \in \mathbb{C}^n$ we have

$$\Re\left(w^{\mathsf{H}}\mathsf{M}w\right) = 0.$$

6.24 Let $A \in \mathbb{R}^{n\times n}$ be SPD with eigenvalues $0 < \lambda_1 \le \lambda_2 \le \cdots \le \lambda_n$. To approximate the solution $\boldsymbol{x}$ of $A\boldsymbol{x} = \boldsymbol{f}$, we consider the following iterative method:

$$\boldsymbol{x}_{k+1} = \boldsymbol{x}_k + \alpha_k B(\boldsymbol{f} - A\boldsymbol{x}_k),$$

where $\alpha_k > 0$ and B is some nonsingular matrix.

a) For what choices of α_k and B does the above method reduce to:
   i) The Jacobi method?
   ii) The Gauss–Seidel method?

b) Let $B = I_n$ in the following and $\boldsymbol{e}_k = \boldsymbol{x} - \boldsymbol{x}_k$. Show that there is a polynomial p_k such that

$$\boldsymbol{e}_k = p_k(A)\boldsymbol{e}_0.$$

   What are the zeros of the polynomial p_k?

c) Prove the inequality

$$\|p_k(A)\|_2 \le \max_{\lambda_1 \le t \le \lambda_n} |p_k(t)|.$$

d) For the case where $\alpha_k = \alpha > 0$ independent of k, what conditions does α have to satisfy in order to ensure convergence?

6.25 Let $A \in \mathbb{C}^{n\times n}$ be HPD. Define $A_1 = \frac{1}{2}D + L$ and $A_2 = \frac{1}{2}D + U$, where $A = D+L+U$ is the usual decomposition of A into diagonal, strictly lower triangular, and strictly upper triangular parts, respectively. For the solution of $A\boldsymbol{x} = \boldsymbol{f}$, consider a linear stationary method with iterator

$$B = (I_n + \alpha A_1)(I_n + \alpha A_2), \qquad \alpha > 0.$$

a) Show that B is HPD.

b) Show that, for all $\alpha \ge \frac{1}{2}$, this method converges.

c) Why is this method efficient?

Listings

```
function [x, its, err] = MinRes( A, x0, f, maxit, tol )
% The method of minimal residuals to approximate the solution to
%
%    Ax = f
%
% with A HPD.
%
% Input
%    A(1:n,1:n) : the system matrix,
%    x0(1:n) : the initial guess
%    f(1:n) : the right hand side vector
%    maxit : the maximal number of iterations
%    tol : the tolerance
%
% Output
```

```
%   x(1:n) : the approximate solution to the linear system of
%            equations
%   its : the number of iterations
%   err : = 0, if the tolerance is reached in less than maxit
%           iterations
%         = 1, if the tolerance is not reached
  err = 0;
  x = x0;

  for its=1:maxit
     r = f-A*x;
     p = A*r;
     p_norm = norm( p );
     if p_norm < tol
        return;
     end
     alpha = r'*p/(p_norm*p_norm);
     x = x + alpha*r;
  end
  err = 1;
end
```

Listing 6.1 The method of minimal residuals.

7 Variational and Krylov Subspace Methods

In this chapter, we introduce gradient-type methods for solving the system of equations $\mathsf{A}\boldsymbol{x} = \boldsymbol{f}$, where $\mathsf{A} \in \mathbb{C}^{n\times n}$ is Hermitian positive definite (HPD). These are iterative methods that include the steepest descent and conjugate gradient (CG) methods. We will take advantage of the fact that solving this equation is equivalent to minimizing the quadratic function

$$E_{\mathsf{A}}(\boldsymbol{z}) = \frac{1}{2}\boldsymbol{z}^{\mathsf{H}}\mathsf{A}\boldsymbol{z} - \Re\left(\boldsymbol{z}^{\mathsf{H}}\boldsymbol{f}\right)$$

over $\mathbb{C}^n$, and will utilize some simple ideas from the theory of convex optimization, which we will study in more detail in Chapter 16.

7.1 Basic Facts about HPD Matrices

Let us collect some basic facts about HPD matrices and some connections between these and inner products. Many of these properties have been covered previously and will be familiar.

Theorem 7.1 (properties of HPD matrices). *Suppose that $\mathsf{A} \in \mathbb{C}^{n\times n}$ is HPD. Then the following are true.*

1. *The expression*

$$(\boldsymbol{x}, \boldsymbol{y})_{\mathsf{A}} = (\mathsf{A}\boldsymbol{x}, \boldsymbol{y})_2 = \boldsymbol{y}^{\mathsf{H}}\mathsf{A}\boldsymbol{x}, \quad \forall \boldsymbol{x}, \boldsymbol{y} \in \mathbb{C}^n$$

   defines an inner product on $\mathbb{C}^n$.
2. *The object $\|\boldsymbol{x}\|_{\mathsf{A}} = \sqrt{\boldsymbol{x}^{\mathsf{H}}\mathsf{A}\boldsymbol{x}}$, where $\boldsymbol{x} \in \mathbb{C}^n$ defines a norm on $\mathbb{C}^n$.*
3. *Let the eigenvalues of A be ordered, so that $0 < \lambda_1 \le \lambda_2 \le \cdots \le \lambda_n$. Then*

$$\sqrt{\lambda_1}\,\|\boldsymbol{x}\|_2 \le \|\boldsymbol{x}\|_{\mathsf{A}} \le \sqrt{\lambda_n}\,\|\boldsymbol{x}\|_2$$

   for any $\boldsymbol{x} \in \mathbb{C}^n$.
4. *Let $\boldsymbol{f} \in \mathbb{C}^n$ be given. Then $\boldsymbol{x} = \mathsf{A}^{-1}\boldsymbol{f}$ if and only if $\boldsymbol{x}$ minimizes the quadratic function $E_{\mathsf{A}} \colon \mathbb{C}^n \to \mathbb{R}$ defined by*

$$E_{\mathsf{A}}(\boldsymbol{z}) = \frac{1}{2}\boldsymbol{z}^{\mathsf{H}}\mathsf{A}\boldsymbol{z} - \Re\left(\boldsymbol{z}^{\mathsf{H}}\boldsymbol{f}\right).$$

Proof. The first and second parts are left to the reader as an exercise.

(3) Suppose that $\{\boldsymbol{w}_1, \ldots, \boldsymbol{w}_n\}$ is an orthonormal basis of $\mathbb{C}^n$ consisting of eigenvectors of A with the ordering $\mathsf{A}\boldsymbol{w}_i = \lambda_i \boldsymbol{w}_i$. Let $\boldsymbol{x} \in \mathbb{C}^n$ be arbitrary. Then there exist unique constants $c_1, \ldots, c_n \in \mathbb{C}$ such that $\boldsymbol{x} = \sum_{i=1}^n c_i \boldsymbol{w}_i$. Hence,

$$\begin{aligned}
\|\boldsymbol{x}\|_{\mathsf{A}}^2 &= \boldsymbol{x}^{\mathsf{H}}\mathsf{A}\boldsymbol{x} \\
&= \boldsymbol{x}^{\mathsf{H}} \sum_{i=1}^n c_i \lambda_i \boldsymbol{w}_i \\
&= \sum_{i=1}^n c_i \lambda_i \left(\boldsymbol{x}^{\mathsf{H}} \boldsymbol{w}_i\right) \\
&= \sum_{i=1}^n c_i \lambda_i \left(\sum_{j=1}^n \bar{c}_j \boldsymbol{w}_j^{\mathsf{H}} \boldsymbol{w}_i \right) \\
&= \sum_{i=1}^n c_i \lambda_i \left(\sum_{j=1}^n \bar{c}_j \delta_{ij} \right) \\
&= \sum_{i=1}^n |c_i|^2 \lambda_i \\
&\le \lambda_n \sum_{i=1}^m |c_i|^2 \\
&= \lambda_n \|\boldsymbol{x}\|_2^2 .
\end{aligned}$$

The inequality $\lambda_1 \|\boldsymbol{x}\|_2^2 \le \|\boldsymbol{x}\|_{\mathsf{A}}^2$ is obtained similarly and the result follows upon taking square roots.

(4: $\Longrightarrow$) Suppose that $\boldsymbol{x} = \mathsf{A}^{-1}\boldsymbol{f}$. Let $\boldsymbol{y} \in \mathbb{C}^n$ be arbitrary and consider

$$\begin{aligned}
E_{\mathsf{A}}(\boldsymbol{x}+\boldsymbol{y}) &= \frac{1}{2}(\boldsymbol{x}+\boldsymbol{y})^{\mathsf{H}} \mathsf{A} (\boldsymbol{x}+\boldsymbol{y}) - \Re\left(\boldsymbol{x}^{\mathsf{H}}\boldsymbol{f}\right) - \Re\left(\boldsymbol{y}^{\mathsf{H}}\boldsymbol{f}\right) \\
&= \frac{1}{2}\boldsymbol{x}^{\mathsf{H}}\mathsf{A}\boldsymbol{x} + \frac{1}{2}\boldsymbol{y}^{\mathsf{H}}\mathsf{A}\boldsymbol{y} + \Re\left(\boldsymbol{y}^{\mathsf{H}}(\mathsf{A}\boldsymbol{x} - \boldsymbol{f})\right) - \Re\left(\boldsymbol{x}^{\mathsf{H}}\boldsymbol{f}\right) \\
&= E_{\mathsf{A}}(\boldsymbol{x}) + \frac{1}{2}\boldsymbol{y}^{\mathsf{H}}\mathsf{A}\boldsymbol{y} \\
&\ge E_{\mathsf{A}}(\boldsymbol{x}),
\end{aligned}$$

where, in the last step, we used that A is HPD. Notice also that we have equality if and only if $\boldsymbol{y} = \boldsymbol{0}$. Hence, $\boldsymbol{x}$ minimizes E_{A}.

An alternate approach, one that needs some inspiration perhaps, is to establish the following equality (see Problem 7.7):

$$E_{\mathsf{A}}(\boldsymbol{z}) = \frac{1}{2}\left\|\boldsymbol{z} - \mathsf{A}^{-1}\boldsymbol{f}\right\|_{\mathsf{A}}^2 - \frac{1}{2}\boldsymbol{f}^{\mathsf{H}}\mathsf{A}^{-1}\boldsymbol{f} \qquad (7.1)$$

for arbitrary $\boldsymbol{z} \in \mathbb{C}^n$. The right-hand side is clearly strictly convex (as a function of $\boldsymbol{z}$) and has the minimizer $\boldsymbol{z} = \boldsymbol{x} = \mathsf{A}^{-1}\boldsymbol{f}$. Since the right- and left-hand sides must have the same minimizer, we are done.

(4: $\Longleftarrow$) Now suppose that $\boldsymbol{x}$ minimizes the function E_{A}. Let $\boldsymbol{u} \in \mathbb{C}^n$ be an arbitrary unit vector. Now define $g(s,t) = E_{\mathrm{A}}(\boldsymbol{x}+\alpha\boldsymbol{u})$, where $\alpha = s + \mathrm{i}t$, $s, t \in \mathbb{R}$. Then

$$\begin{aligned} g(s,t) &= \frac{1}{2}(\boldsymbol{x}+\alpha\boldsymbol{u})^{\mathrm{H}}\mathrm{A}(\boldsymbol{x}+\alpha\boldsymbol{u}) - \Re\left((\boldsymbol{x}+\alpha\boldsymbol{u})^{\mathrm{H}}\boldsymbol{f}\right) \\ &= \frac{1}{2}\boldsymbol{x}^{\mathrm{H}}\mathrm{A}\boldsymbol{x} + \Re\left(\bar{\alpha}\boldsymbol{u}^{\mathrm{H}}\mathrm{A}\boldsymbol{x}\right) + \frac{|\alpha|^2}{2}\boldsymbol{u}^{\mathrm{H}}\mathrm{A}\boldsymbol{u} - \Re\left(\boldsymbol{x}^{\mathrm{H}}\boldsymbol{f}\right) - \Re\left(\bar{\alpha}\boldsymbol{u}^{\mathrm{H}}\boldsymbol{f}\right) \\ &= E_{\mathrm{A}}(\boldsymbol{x}) + \Re\left(\bar{\alpha}\boldsymbol{u}^{\mathrm{H}}(\mathrm{A}\boldsymbol{x}-\boldsymbol{f})\right) + \frac{|\alpha|^2}{2}\boldsymbol{u}^{\mathrm{H}}\mathrm{A}\boldsymbol{u} \\ &= E_{\mathrm{A}}(\boldsymbol{x}) + \Re(\bar{\alpha})\Re\left(\boldsymbol{u}^{\mathrm{H}}(\mathrm{A}\boldsymbol{x}-\boldsymbol{f})\right) - \Im(\bar{\alpha})\Im\left(\boldsymbol{u}^{\mathrm{H}}(\mathrm{A}\boldsymbol{x}-\boldsymbol{f})\right) \\ &\quad + \frac{s^2+t^2}{2}\boldsymbol{u}^{\mathrm{H}}\mathrm{A}\boldsymbol{u} \\ &= E_{\mathrm{A}}(\boldsymbol{x}) + s\Re\left(\boldsymbol{u}^{\mathrm{H}}(\mathrm{A}\boldsymbol{x}-\boldsymbol{f})\right) + t\Im\left(\boldsymbol{u}^{\mathrm{H}}(\mathrm{A}\boldsymbol{x}-\boldsymbol{f})\right) + \frac{s^2+t^2}{2}\boldsymbol{u}^{\mathrm{H}}\mathrm{A}\boldsymbol{u}. \end{aligned}$$

Clearly, g is a strictly convex, quadratic function on $\mathbb{R}^2$. Moreover, g is minimized at $(s,t) = (0,0)$. Hence,

$$0 = \frac{\partial g}{\partial s}(0,0) = \Re\left(\boldsymbol{u}^{\mathrm{H}}(\mathrm{A}\boldsymbol{x}-\boldsymbol{f})\right)$$

and

$$0 = \frac{\partial g}{\partial t}(0,0) = \Im\left(\boldsymbol{u}^{\mathrm{H}}(\mathrm{A}\boldsymbol{x}-\boldsymbol{f})\right)$$

hold for any vector $\boldsymbol{u}$. It follows then that $\mathrm{A}\boldsymbol{x} = \boldsymbol{f}$. □

In the previous result we saw that an HPD matrix defines an inner product. It turns out that the converse is also true.

Proposition 7.2 (inner products and HPD matrices)**.** *Suppose that* $(\cdot,\cdot) : \mathbb{C}^n \times \mathbb{C}^n \to \mathbb{C}$ *is an inner product. There exists a unique HPD matrix* $\mathrm{A} \in \mathbb{C}^{n\times n}$ *such that*

$$(\boldsymbol{x},\boldsymbol{y}) = (\mathrm{A}\boldsymbol{x},\boldsymbol{y})_2 = (\boldsymbol{x},\boldsymbol{y})_{\mathrm{A}}, \quad \forall \boldsymbol{x},\boldsymbol{y} \in \mathbb{C}^n.$$

Proof. Let, for $j = 1,\ldots,n$, $\boldsymbol{e}_j$ denote the canonical unit basis vectors of $\mathbb{C}^n$. Define the matrix $\mathrm{A} = [a_{i,j}] \in \mathbb{C}^{n\times n}$ via

$$a_{i,j} = (\boldsymbol{e}_j, \boldsymbol{e}_i).$$

The reader must show that A has the desired properties; see Problem 7.2. □

Definition 7.3 (A-conjugate)**.** Suppose that $(\cdot,\cdot) : \mathbb{C}^n \times \mathbb{C}^n \to \mathbb{C}$ is an inner product and $\mathrm{A} \in \mathbb{C}^{n\times n}$ is its associated HPD matrix. We say that $\mathrm{B} \in \mathbb{C}^{n\times n}$ is **self-adjoint** with respect to this inner product if and only if

$$(\boldsymbol{x},\mathrm{B}\boldsymbol{y}) = (\boldsymbol{x},\mathrm{B}\boldsymbol{y})_{\mathrm{A}} = (\mathrm{B}\boldsymbol{x},\boldsymbol{y})_{\mathrm{A}} = (\mathrm{B}\boldsymbol{x},\boldsymbol{y}), \quad \forall \boldsymbol{x},\boldsymbol{y} \in \mathbb{C}^n.$$

We say that B is self-adjoint positive definite with respect to the inner product if and only if B is self-adjoint and satisfies

$$(\boldsymbol{x},\mathrm{B}\boldsymbol{x}) = (\boldsymbol{x},\mathrm{B}\boldsymbol{x})_{\mathrm{A}} > 0, \quad \forall \boldsymbol{x} \in \mathbb{C}^n_\star.$$

We say that two vectors $\boldsymbol{x}, \boldsymbol{y} \in \mathbb{C}^n$ are A-**orthogonal** (or A-**conjugate**) if and only if

$$(\boldsymbol{x}, \boldsymbol{y}) = (\boldsymbol{x}, \boldsymbol{y})_{\mathsf{A}} = 0.$$

We say that a set $S \subset \mathbb{C}^n$ of nonzero vectors is called A-**orthogonal** (or A-**conjugate**) if and only if whenever $\boldsymbol{x}, \boldsymbol{y} \in S$ and $\boldsymbol{x} \neq \boldsymbol{y}$, then

$$(\boldsymbol{x}, \boldsymbol{y}) = (\boldsymbol{x}, \boldsymbol{y})_{\mathsf{A}} = 0.$$

We say that $S \subset \mathbb{C}^n$ is A-**orthonormal** if and only if S is A-orthogonal and

$$\|\boldsymbol{x}\|_{\mathsf{A}} = 1, \ \forall \boldsymbol{x} \in S.$$

During the course of the proof of Theorem 6.31, we encountered the square root of an HPD matrix. Let us review this idea.

Proposition 7.4 (square root). *Suppose that* $\mathsf{A} \in \mathbb{C}^{n\times n}$ *is HPD. Then* A *is invertible and* A^{-1} *is HPD. Furthermore, there exists a unique HPD matrix* $\mathsf{B} \in \mathbb{C}^{n\times n}$ *with the property that* $\mathsf{BB} = \mathsf{A}$.

Proof. Since A is HPD, it is unitarily diagonalizable, i.e., there are a unitary matrix $\mathsf{U} \in \mathbb{C}^{n\times n}$ and a diagonal matrix with positive diagonal entries $\mathsf{D} = \operatorname{diag}(\lambda_1, \ldots, \lambda_n)$ such that

$$\mathsf{A} = \mathsf{U}\mathsf{D}\mathsf{U}^{\mathsf{H}}.$$

Set

$$\mathsf{B} = \mathsf{U}\mathsf{D}^{1/2}\mathsf{U}^{\mathsf{H}} \quad \text{and} \quad \mathsf{D}^{1/2} = \operatorname{diag}\left(\sqrt{\lambda_1}, \ldots, \sqrt{\lambda_n}\right).$$

Then B has the desired properties. Furthermore, it is easy to see that

$$\mathsf{A}^{-1} = \mathsf{U}\mathsf{D}^{-1}\mathsf{U}^{\mathsf{H}}, \quad \mathsf{D}^{-1} = \operatorname{diag}(\lambda_1^{-1}, \ldots, \lambda_n^{-1}).$$

We leave it to the reader to check the details. □

Proposition 7.5 (product of HPD matrices). *Suppose that* $\mathsf{A}, \mathsf{B} \in \mathbb{C}^{n\times n}$ *are HPD matrices. Then the product* BA *is self-adjoint and positive definite with respect to* $(\cdot, \cdot)_{\mathsf{A}}$, *i.e.,*

$$(\boldsymbol{x}, \mathsf{BA}\boldsymbol{y})_{\mathsf{A}} = (\mathsf{BA}\boldsymbol{x}, \boldsymbol{y})_{\mathsf{A}}, \quad \forall \boldsymbol{x}, \boldsymbol{y} \in \mathbb{C}^n,$$

and

$$(\mathsf{BA}\boldsymbol{x}, \boldsymbol{x})_{\mathsf{A}} > 0, \quad \forall \boldsymbol{x} \in \mathbb{C}^n_\star.$$

Proof. See Problem 7.3. □

Proposition 7.6 (similarity). *Suppose that* $\mathsf{A} \in \mathbb{C}^{n\times n}$ *is HPD and* $\mathsf{B} \in \mathbb{C}^{n\times n}$ *is self-adjoint with respect to* $(\cdot, \cdot)_{\mathsf{A}}$, *i.e.,*

$$(\boldsymbol{x}, \mathsf{B}\boldsymbol{y})_{\mathsf{A}} = (\mathsf{B}\boldsymbol{x}, \boldsymbol{y})_{\mathsf{A}}, \quad \forall \boldsymbol{x}, \boldsymbol{y} \in \mathbb{C}^n.$$

Then B *is similar to a matrix that is Hermitian, i.e., a matrix that is self-adjoint with respect to the Euclidean inner product* $(\cdot, \cdot)_2$.

Proof. Since $\mathsf{B} \in \mathbb{C}^{n\times n}$ is self-adjoint with respect to $(\cdot,\cdot)_{\mathsf{A}}$, then the reader should confirm that

$$\mathsf{B}^{\mathsf{H}}\mathsf{A} = \mathsf{A}\mathsf{B}.$$

Since A is HPD, there is an invertible matrix $\mathsf{L} \in \mathbb{C}^{n\times n}$ such that $\mathsf{A} = \mathsf{L}\mathsf{L}^{\mathsf{H}}$. Define

$$\mathsf{C} = \mathsf{L}^{\mathsf{H}}\mathsf{B}\mathsf{L}^{-\mathsf{H}}.$$

Then

$$\begin{aligned}
\mathsf{C}^{\mathsf{H}} &= \mathsf{L}^{-1}\mathsf{B}^{\mathsf{H}}\mathsf{L} \\
&= \mathsf{L}^{-1}\mathsf{B}^{\mathsf{H}}\mathsf{L}\mathsf{L}^{\mathsf{H}}\mathsf{L}^{-\mathsf{H}} \\
&= \mathsf{L}^{-1}\mathsf{B}^{\mathsf{H}}\mathsf{A}\mathsf{L}^{-\mathsf{H}} \\
&= \mathsf{L}^{-1}\mathsf{A}\mathsf{B}\mathsf{L}^{-\mathsf{H}} \\
&= \mathsf{L}^{-1}\mathsf{L}\mathsf{L}^{\mathsf{H}}\mathsf{B}\mathsf{L}^{-\mathsf{H}} \\
&= \mathsf{L}^{\mathsf{H}}\mathsf{B}\mathsf{L}^{-\mathsf{H}} \\
&= \mathsf{C}.
\end{aligned}$$

Since C is similar to B, the result follows. □

Knowing that the spectral decomposition, Theorem 1.47, is valid in the Euclidean case, the last result can be used to prove the validity of the more general version, Theorem 1.49.

Theorem 7.7 (spectral decomposition)**.** *Suppose that* $\mathsf{A} \in \mathbb{C}^{n\times n}$ *is HPD and* $\mathsf{B} \in \mathbb{C}^{n\times n}$ *is self-adjoint with respect to* $(\cdot,\cdot)_{\mathsf{A}}$. *Then all of the eigenvalues of* B *are real and there is an* A*-orthonormal basis of* $\mathbb{C}^n$ *consisting of eigenvectors of* B.

Proof. Applying the Euclidean spectral decomposition, Theorem 1.47, to the Hermitian matrix

$$\mathsf{C} = \mathsf{L}^{\mathsf{H}}\mathsf{B}\mathsf{L}^{-\mathsf{H}},$$

where the invertible matrix L is taken from the Cholesky-type decomposition $\mathsf{A} = \mathsf{L}\mathsf{L}^{\mathsf{H}}$ of A, there are a unitary matrix $\mathsf{U} \in \mathbb{C}^{n\times n}$ and a diagonal matrix with real entries $\mathsf{D} = \operatorname{diag}(\lambda_1, \ldots, \lambda_n)$ such that

$$\mathsf{L}^{\mathsf{H}}\mathsf{B}\mathsf{L}^{-\mathsf{H}} = \mathsf{C} = \mathsf{U}\mathsf{D}\mathsf{U}^{-1}.$$

Hence, B is similar to a diagonal matrix with real entries:

$$\mathsf{B} = \left(\mathsf{L}^{-\mathsf{H}}\mathsf{U}\right)\mathsf{D}\left(\mathsf{L}^{-\mathsf{H}}\mathsf{U}\right)^{-1}.$$

Moreover, setting $\mathsf{M} = \mathsf{L}^{-\mathsf{H}}\mathsf{U}$, we see that

$$\mathsf{B}\mathsf{M} = \mathsf{M}\mathsf{D},$$

which implies that the columns of the invertible matrix M are eigenvectors of B. It only remains to check that the columns of M form an A-orthonormal set. This is left to the reader as an exercise; see Problem 7.5. □

Corollary 7.8 (eigenvalues). *Suppose that* $\mathsf{A} \in \mathbb{C}^{n\times n}$ *is HPD and* $\mathsf{B} \in \mathbb{C}^{n\times n}$ *is self-adjoint and positive definite with respect to* $(\cdot\,,\,\cdot)_{\mathsf{A}}$. *Then the eigenvalues of* B *are all real and positive.*

Proof. See Problem 7.6. □

7.2 Gradient Descent Methods

Gradient descent methods are related to optimization methods. The basic idea is to solve, successively, several one-dimensional optimization problems called line searches.

Definition 7.9 (gradient descent). Suppose that $\mathsf{A} \in \mathbb{C}^{n\times n}$ is HPD and $\boldsymbol{f} \in \mathbb{C}^n$. Define the quadratic function $E_{\mathsf{A}}\colon \mathbb{C}^n \to \mathbb{R}$ via

$$E_{\mathsf{A}}(z) = \frac{1}{2}z^{\mathsf{H}}\mathsf{A}z - \Re\left(z^{\mathsf{H}}\boldsymbol{f}\right).$$

A **gradient descent method** is a two-layer iterative scheme to approximate $\boldsymbol{x} = \mathsf{A}^{-1}\boldsymbol{f}$. Starting from an arbitrary initial guess $\boldsymbol{x}_0$, the iterations proceed as

$$\boldsymbol{x}_k = \boldsymbol{x}_{k-1} + \alpha_k \boldsymbol{d}_{k-1}, \qquad k = 1, 2, 3, \ldots,$$

where $\boldsymbol{d}_{k-1} \in \mathbb{C}^n$ is the $(k-1)$st **search direction**, supplied by the algorithm, and $\alpha_k \in \mathbb{C}$ is the **step size** given by the condition

$$\alpha_k = \operatorname*{argmin}_{\alpha\in\mathbb{C}} E_{\mathsf{A}}\left(\boldsymbol{x}_{k-1} + \alpha\boldsymbol{d}_{k-1}\right),$$

which is called a **line search**.

Before attempting the more general case, let us first compute a formula for α_k in the case where real numbers are used exclusively.

Theorem 7.10 (gradient descent, real case). *Suppose that* $\mathsf{A} \in \mathbb{R}^{n\times n}$ *is symmetric positive definite (SPD) and* $\boldsymbol{f} \in \mathbb{R}^n$. *Define the quadratic function* $E_{\mathsf{A}}\colon \mathbb{R}^n \to \mathbb{R}$

$$E_{\mathsf{A}}(z) = \frac{1}{2}z^{\mathsf{T}}\mathsf{A}z - z^{\mathsf{T}}\boldsymbol{f}.$$

Suppose that the search direction $\boldsymbol{d}_{k-1} \in \mathbb{R}^n_\star$ *and previous iterate* $\boldsymbol{x}_{k-1} \in \mathbb{R}^n$ *in a gradient descent method are given. Define* $\boldsymbol{r}_{k-1} = \boldsymbol{f} - \mathsf{A}\boldsymbol{x}_{k-1}$. *Then the step size can be computed exactly via the formula*

$$\alpha_k = \operatorname*{argmin}_{\alpha\in\mathbb{R}} E_{\mathsf{A}}\left(\boldsymbol{x}_{k-1} + \alpha\boldsymbol{d}_{k-1}\right) = \frac{\boldsymbol{d}_{k-1}^{\mathsf{T}}\boldsymbol{r}_{k-1}}{\boldsymbol{d}_{k-1}^{\mathsf{T}}\mathsf{A}\boldsymbol{d}_{k-1}}.$$

In other words, a gradient descent method is well defined once a nontrivial search direction is specified.

Proof. Consider the quadratic

$$g(\alpha) = E_{\mathsf{A}}\left(\boldsymbol{x}_{k-1} + \alpha\boldsymbol{d}_{k-1}\right) = E_{\mathsf{A}}\left(\boldsymbol{x}_{k-1}\right) - \alpha\boldsymbol{d}_{k-1}^{\mathsf{T}}\boldsymbol{r}_{k-1} + \alpha^2\frac{1}{2}\boldsymbol{d}_{k-1}^{\mathsf{T}}\mathsf{A}\boldsymbol{d}_{k-1}.$$

A calculation with the first derivative locates the extremum:

$$0 = g'(\alpha_k) = -\boldsymbol{d}_{k-1}^{\mathsf{T}}\boldsymbol{r}_{k-1} + \alpha_k \boldsymbol{d}_{k-1}^{\mathsf{T}}\mathsf{A}\boldsymbol{d}_{k-1},$$

which implies that

$$\alpha_k = \frac{\boldsymbol{d}_{k-1}^{\mathsf{T}}\boldsymbol{r}_{k-1}}{\boldsymbol{d}_{k-1}^{\mathsf{T}}\mathsf{A}\boldsymbol{d}_{k-1}}.$$

The second derivative indicates that this is a minimum:

$$g''(\alpha) = \boldsymbol{d}_{k-1}^{\mathsf{T}}\mathsf{A}\boldsymbol{d}_{k-1} > 0,$$

provided that $\boldsymbol{d}_{k-1} \neq \boldsymbol{0}$. □

The more general case in the complex setting requires a bit more care.

Theorem 7.11 (gradient descent, complex case)**.** *Suppose that* $\mathsf{A} \in \mathbb{C}^{n\times n}$ *is HPD and* $\boldsymbol{f} \in \mathbb{C}^n$*. Define the quadratic function* $E_{\mathsf{A}} \colon \mathbb{C}^n \to \mathbb{R}$

$$E_{\mathsf{A}}(z) = \frac{1}{2}z^{\mathsf{H}}\mathsf{A}z - \Re\left(z^{\mathsf{H}}\boldsymbol{f}\right).$$

Suppose that the search direction $\boldsymbol{d}_{k-1} \in \mathbb{C}_\star^n$ *and previous iterate* $\boldsymbol{x}_{k-1} \in \mathbb{C}^n$ *in a gradient descent method are given. Define* $\boldsymbol{r}_{k-1} = \boldsymbol{f} - \mathsf{A}\boldsymbol{x}_{k-1}$*. Then the step size can be computed exactly via the formula*

$$\alpha_k = \underset{\alpha\in\mathbb{C}}{\operatorname{argmin}}\, E_{\mathsf{A}}\left(\boldsymbol{x}_{k-1} + \alpha\boldsymbol{d}_{k-1}\right) = \frac{\boldsymbol{d}_{k-1}^{\mathsf{H}}\boldsymbol{r}_{k-1}}{\boldsymbol{d}_{k-1}^{\mathsf{H}}\mathsf{A}\boldsymbol{d}_{k-1}}.$$

Proof. Let $\alpha = s + \mathrm{i}t$, where $s, t \in \mathbb{R}$. Define the function

$$\begin{aligned} g(s,t) &= E_{\mathsf{A}}\left(\boldsymbol{x}_{k-1} + \alpha\boldsymbol{d}_{k-1}\right) \\ &= E_{\mathsf{A}}\left(\boldsymbol{x}_{k-1}\right) - \Re\left(\bar{\alpha}\boldsymbol{d}_{k-1}^{\mathsf{H}}\boldsymbol{r}_{k-1}\right) + \frac{|\alpha|^2}{2}\boldsymbol{d}_{k-1}^{\mathsf{H}}\mathsf{A}\boldsymbol{d}_{k-1} \\ &= E_{\mathsf{A}}\left(\boldsymbol{x}_{k-1}\right) - s\Re\left(\boldsymbol{d}_{k-1}^{\mathsf{H}}\boldsymbol{r}_{k-1}\right) - t\Im\left(\boldsymbol{d}_{k-1}^{\mathsf{H}}\boldsymbol{r}_{k-1}\right) + \frac{s^2+t^2}{2}\boldsymbol{d}_{k-1}^{\mathsf{H}}\mathsf{A}\boldsymbol{d}_{k-1}. \end{aligned}$$

This is a strictly convex quadratic function of two variables. Setting the first derivatives equal to zero, we find

$$\begin{aligned} 0 &= \frac{\partial g}{\partial s}(s_k, t_k) = -\Re\left(\boldsymbol{d}_{k-1}^{\mathsf{H}}\boldsymbol{r}_{k-1}\right) + s_k\boldsymbol{d}_{k-1}^{\mathsf{H}}\mathsf{A}\boldsymbol{d}_{k-1}, \\ 0 &= \frac{\partial g}{\partial t}(s_k, t_k) = -\Im\left(\boldsymbol{d}_{k-1}^{\mathsf{H}}\boldsymbol{r}_{k-1}\right) + t_k\boldsymbol{d}_{k-1}^{\mathsf{H}}\mathsf{A}\boldsymbol{d}_{k-1}, \end{aligned}$$

which implies that

$$\alpha_k = s_k + \mathrm{i}t_k = \frac{\boldsymbol{d}_{k-1}^{\mathsf{H}}\boldsymbol{r}_{k-1}}{\boldsymbol{d}_{k-1}^{\mathsf{H}}\mathsf{A}\boldsymbol{d}_{k-1}}.$$

□

7.3 The Steepest Descent Method

Definition 7.12 (steepest descent)**.** Suppose that $A \in \mathbb{C}^{n\times n}$ is HPD and $\boldsymbol{f} \in \mathbb{C}^n$. The **steepest descent method** is a gradient descent method for which the search direction $\boldsymbol{d}_{k-1}$ is defined to be the residual, i.e.,

$$\boldsymbol{d}_{k-1} = \boldsymbol{r}_{k-1} = \boldsymbol{f} - A\boldsymbol{x}_{k-1},$$

so that the step size is precisely

$$\alpha_k = \frac{\boldsymbol{r}_{k-1}^{\mathsf{H}}\boldsymbol{r}_{k-1}}{\boldsymbol{r}_{k-1}^{\mathsf{H}}A\boldsymbol{r}_{k-1}}.$$

If $B \in \mathbb{C}^{n\times n}$ is an HPD matrix, the B-**preconditioned steepest descent** method is a gradient descent method with search direction

$$\boldsymbol{d}_{k-1} = B^{-1}\boldsymbol{r}_{k-1}, \tag{7.2}$$

so that the step size is precisely

$$\alpha_k = \frac{\boldsymbol{r}_{k-1}^{\mathsf{H}}B^{-1}\boldsymbol{r}_{k-1}}{\boldsymbol{r}_{k-1}^{\mathsf{H}}B^{-1}AB^{-1}\boldsymbol{r}_{k-1}}.$$

Remark 7.13 (preconditioning)**.** The idea with preconditioning is that the *preconditioner*, B, should be like A, but easier to invert. In fact, if we choose $B = A$, which is not practical, we would converge in a single iteration, because (7.2) would yield the error vector as the search direction. One way of realizing the preconditioned steepest descent, theoretically, is to observe that it is just the normal steepest descent applied to solve the equation

$$B^{-1}A\boldsymbol{x} = B^{-1}\boldsymbol{f}. \tag{7.3}$$

Proposition 7.14 (orthogonality)**.** *Suppose that $A \in \mathbb{C}^{n\times n}$ is HPD and $\boldsymbol{f} \in \mathbb{C}^n$. Suppose that $\{\boldsymbol{x}_k\}_{k=1}^{\infty}$ is computed using the steepest descent method with the starting vector $\boldsymbol{x}_0$. Then the sequence of residual vectors $\{\boldsymbol{r}_k\}_{k=1}^{\infty}$, $\boldsymbol{r}_k = \boldsymbol{f} - A\boldsymbol{x}_k$, has the property that*

$$(\boldsymbol{r}_k, \boldsymbol{r}_{k+1})_2 = \boldsymbol{r}_{k+1}^{\mathsf{H}}\boldsymbol{r}_k = 0$$

for $k = 0, 1, 2, \ldots$.

Proof. See Problem 7.9. □

Remark 7.15 (orthogonality)**.** Proposition 7.14 shows that the steepest descent method has the property that its next search direction is always orthogonal to the last. We will see that this can lead to some bad outcomes.

Theorem 7.16 (error equation)**.** *Suppose that $A \in \mathbb{C}^{n\times n}$ is HPD, $\boldsymbol{f} \in \mathbb{C}^n$, and $\boldsymbol{x} = A^{-1}\boldsymbol{f}$. Suppose that $\{\boldsymbol{x}_k\}_{k=1}^{\infty}$ is computed using the steepest descent method with the starting value $\boldsymbol{x}_0 \in \mathbb{C}^n$. Then the error $\boldsymbol{e}_k = \boldsymbol{x} - \boldsymbol{x}_k$ satisfies*

$$\|\boldsymbol{e}_{k+1}\|_A^2 = \gamma_k \|\boldsymbol{e}_k\|_A^2,$$

where

$$\gamma_k = 1 - \frac{\left(\boldsymbol{r}_k^{\mathsf{H}}\boldsymbol{r}_k\right)^2}{\left(\boldsymbol{r}_k^{\mathsf{H}}A\boldsymbol{r}_k\right)\left(\boldsymbol{r}_k^{\mathsf{H}}A^{-1}\boldsymbol{r}_k\right)}.$$

Proof. Suppose that $\boldsymbol{x} = \mathsf{A}^{-1}\boldsymbol{f}$. Then, for any $\boldsymbol{z} \in \mathbb{C}^n$,

$$E_{\mathsf{A}}(\boldsymbol{z}) = E_{\mathsf{A}}(\boldsymbol{x}) + \frac{1}{2}\left\|\boldsymbol{z} - \boldsymbol{x}\right\|_{\mathsf{A}}^2, \tag{7.4}$$

where, as usual,

$$E_{\mathsf{A}}(\boldsymbol{z}) = \frac{1}{2}\boldsymbol{z}^{\mathsf{H}}\mathsf{A}\boldsymbol{z} - \Re\left(\boldsymbol{z}^{\mathsf{H}}\boldsymbol{f}\right).$$

Since $\boldsymbol{r}_k = \boldsymbol{f} - \mathsf{A}\boldsymbol{x}_k$ and, for the standard steepest descent method,

$$\alpha_{k+1} = \frac{\boldsymbol{r}_k^{\mathsf{H}}\boldsymbol{r}_k}{\boldsymbol{r}_k^{\mathsf{H}}\mathsf{A}\boldsymbol{r}_k},$$

a brief calculation shows that

$$E_{\mathsf{A}}(\boldsymbol{x}_{k+1}) = E_{\mathsf{A}}(\boldsymbol{x}_k + \alpha_{k+1}\boldsymbol{r}_k) = E_{\mathsf{A}}(\boldsymbol{x}_k) - \frac{1}{2}\frac{\left(\boldsymbol{r}_k^{\mathsf{H}}\boldsymbol{r}_k\right)^2}{\boldsymbol{r}_k^{\mathsf{H}}\mathsf{A}\boldsymbol{r}_k}. \tag{7.5}$$

Combining (7.4) and (7.5), we get

$$\left\|\boldsymbol{e}_{k+1}\right\|_{\mathsf{A}}^2 = \left\|\boldsymbol{e}_k\right\|_{\mathsf{A}}^2 - \frac{\left(\boldsymbol{r}_k^{\mathsf{H}}\boldsymbol{r}_k\right)^2}{\boldsymbol{r}_k^{\mathsf{H}}\mathsf{A}\boldsymbol{r}_k}. \tag{7.6}$$

Since $\boldsymbol{r}_k = \mathsf{A}\boldsymbol{e}_k$, we have

$$\left\|\boldsymbol{e}_k\right\|_{\mathsf{A}}^2 = \boldsymbol{r}_k^{\mathsf{H}}\mathsf{A}^{-1}\boldsymbol{r}_k. \tag{7.7}$$

Combining (7.6) and(7.7), we get the desired result. □

The following technical lemma allows us to make some sense of the error equation derived in Theorem 7.16.

Lemma 7.17 (Kantorovich inequality[1]). *Let the matrix* $\mathsf{A} \in \mathbb{C}^{n\times n}$ *be HPD with spectrum* $\sigma(\mathsf{A}) = \{\lambda_i\}_{i=1}^n$, *with* $0 < \lambda_1 \leq \lambda_2 \leq \cdots \leq \lambda_n$, *and spectral condition number*

$$\kappa = \kappa_2(\mathsf{A}) = \frac{\lambda_n}{\lambda_1}.$$

Then, for any $\boldsymbol{x} \in \mathbb{C}_\star^n$,

$$\frac{\left(\boldsymbol{x}^{\mathsf{H}}\mathsf{A}\boldsymbol{x}\right)\left(\boldsymbol{x}^{\mathsf{H}}\mathsf{A}^{-1}\boldsymbol{x}\right)}{(\boldsymbol{x}^{\mathsf{H}}\boldsymbol{x})^2} \leq \frac{1}{4}\left(\sqrt{\kappa} + \sqrt{\kappa^{-1}}\right)^2.$$

Proof. Define $\mu = \sqrt{\lambda_1\lambda_n}$. Then

$$\kappa^{-1/2} \leq \frac{\lambda_i}{\mu} \leq \kappa^{1/2} \tag{7.8}$$

and

$$\kappa^{-1/2} \leq \frac{\mu}{\lambda_i} \leq \kappa^{1/2}.$$

Therefore, for all $i = 1, \ldots, n$,

$$2\kappa^{-1/2} \leq \frac{\lambda_i}{\mu} + \frac{\mu}{\lambda_i} \leq 2\kappa^{1/2}.$$

[1] Named in honor of the Russian mathematician Leonid Vitalyevich Kantorovich (1912–1986).

Next, observe that the function

$$f(x) = x + \frac{1}{x}$$

is strictly decreasing on $(0, 1)$ and strictly increasing on $(1, \infty)$. Set $x = \frac{\lambda_i}{\mu}$. Using (7.8), if

$$1 \leq x = \frac{\lambda_i}{\mu} \leq \kappa^{1/2},$$

then

$$2 = f(1) \leq f\left(\frac{\lambda_i}{\mu}\right) \leq f(\kappa^{1/2}).$$

This implies that

$$2 \leq \frac{\lambda_i}{\mu} + \frac{\mu}{\lambda_i} \leq \kappa^{1/2} + \kappa^{-1/2}.$$

On the other hand, if

$$\kappa^{-1/2} \leq x = \frac{\lambda_i}{\mu} \leq 1,$$

then

$$f(\kappa^{-1/2}) \geq f\left(\frac{\lambda_i}{\mu}\right) \geq f(1) = 2,$$

which implies that

$$\kappa^{-1/2} + \kappa^{1/2} \geq \frac{\lambda_i}{\mu} + \frac{\mu}{\lambda_i} \geq 2.$$

Therefore, it is always true that

$$2 \leq \frac{\lambda_i}{\mu} + \frac{\mu}{\lambda_i} \leq \kappa^{1/2} + \kappa^{-1/2}. \tag{7.9}$$

Now suppose that $(\lambda_i, \boldsymbol{w}_i)$ is an eigenpair of A, where $\{\boldsymbol{w}\}_{i=1}^n$ is an orthonormal basis for $\mathbb{C}^n$. Then

$$\left(\mu^{-1}\mathsf{A} + \mu\mathsf{A}^{-1}\right)\boldsymbol{w}_i = \left(\frac{\lambda_i}{\mu} + \frac{\mu}{\lambda_i}\right)\boldsymbol{w}_i.$$

Let $\boldsymbol{x} \in \mathbb{C}^n_\star$ be arbitrary. There exist unique constants $c_i \in \mathbb{C}$ such that

$$\boldsymbol{x} = \sum_{i=1}^n c_i \boldsymbol{w}_i.$$

Then

$$\frac{1}{\mu}\boldsymbol{x}^{\mathsf{H}}\mathsf{A}\boldsymbol{x} + \mu\boldsymbol{x}^{\mathsf{H}}\mathsf{A}^{-1}\boldsymbol{x} = \sum_{i=1}^n |c_i|^2\left(\frac{\lambda_i}{\mu} + \frac{\mu}{\lambda_i}\right) \leq \left(\kappa^{1/2} + \kappa^{-1/2}\right)\|\boldsymbol{x}\|_2^2.$$

We recall the standard inequality

$$|ab| \leq \frac{1}{2}|a|^2 + \frac{1}{2}|b|^2$$

for any $a, b \in \mathbb{R}$. From this, it follows that

$$ab \leq |ab| \leq \frac{1}{4}(|a| + |b|)^2.$$

Using the last inequality with

$$a = \frac{1}{\mu}\mathbf{x}^{\mathsf{H}}\mathsf{A}\mathbf{x}, \quad b = \mu\mathbf{x}^{\mathsf{H}}\mathsf{A}^{-1}\mathbf{x},$$

we obtain

$$\left(\mathbf{x}^{\mathsf{H}}\mathsf{A}\mathbf{x}\right)\left(\mathbf{x}^{\mathsf{H}}\mathsf{A}^{-1}\mathbf{x}\right) \leq \frac{1}{4}\left(\frac{1}{\mu}\mathbf{x}^{\mathsf{H}}\mathsf{A}\mathbf{x} + \mu\mathbf{x}^{\mathsf{H}}\mathsf{A}^{-1}\mathbf{x}\right)^2 \leq \frac{1}{4}\left(\kappa^{1/2} + \kappa^{-1/2}\right)^2\left(\mathbf{x}^{\mathsf{H}}\mathbf{x}\right)^2$$

and the proof is complete. □

Theorem 7.18 (convergence)**.** *Suppose that* $\mathsf{A} \in \mathbb{C}^{n\times n}$ *is HPD,* $\boldsymbol{f} \in \mathbb{C}^n$*, and* $\boldsymbol{x} = \mathsf{A}^{-1}\boldsymbol{f}$*. Suppose that* $\{\boldsymbol{x}_k\}_{k=1}^{\infty}$ *is computed using the steepest descent method with the starting value* $\boldsymbol{x}_0 \in \mathbb{C}^n$*. Then the following estimate holds:*

$$\|\boldsymbol{e}_k\|_{\mathsf{A}} \leq \left(\frac{\kappa - 1}{\kappa + 1}\right)^k \|\boldsymbol{e}_0\|_{\mathsf{A}},$$

where $\kappa = \kappa_2(\mathsf{A})$*.*

Proof. Using the Kantorovich inequality from Lemma 7.17 and the result of Theorem 7.16,

$$\gamma_k \leq 1 - \frac{4}{\left(\kappa^{-1/2} + \kappa^{1/2}\right)^2} = \left(\frac{\kappa - 1}{\kappa + 1}\right)^2,$$

which implies the result. □

Example 7.1 Let us show, by means of an example, that the convergence rate for the steepest descent method, presented in the previous result, is sharp. Let $\mathsf{A} = \mathrm{diag}(a, b) \in \mathbb{R}^{2\times2}$, $a > b > 0$, $\boldsymbol{f} = \mathbf{0}$, and $\boldsymbol{x} = \mathbf{0}$. We will prove the following: if

$$\boldsymbol{x}_i = c_i\,[b, sa]^{\mathsf{T}} \in V_1 = \langle [b, sa]^{\mathsf{T}} \rangle$$

for some $c_i \in \mathbb{R}_\star$, where $s = \pm 1$, then

$$\boldsymbol{x}_{i+1} = c_{i+1}\,[b, -sa]^{\mathsf{T}} \in V_2 = \langle [b, -sa]^{\mathsf{T}} \rangle$$

for some coefficient $c_{i+1} \in \mathbb{R}_\star$. In general, $\boldsymbol{x}_k \in V_1 \cup V_2$. First $\boldsymbol{r}_i = -c_i\,[ab, sab]^{\mathsf{T}}$ and $\mathsf{A}\boldsymbol{r}_i = -c_i\left[a^2b, sab^2\right]^{\mathsf{T}}$. Then

$$\alpha_{i+1} = \frac{\boldsymbol{r}_i^{\mathsf{T}}\boldsymbol{r}_i}{\boldsymbol{r}_i^{\mathsf{T}}\mathsf{A}\boldsymbol{r}_i} = \frac{2c_i^2a^2b^2}{c_i^2\left(a^3b^2 + a^2b^3\right)} = \frac{2}{a+b}.$$

Finally,

$$\begin{aligned}
\boldsymbol{x}_{i+1} &= \boldsymbol{x}_i + \alpha_i \boldsymbol{r}_i \\
&= \begin{bmatrix} c_i b - \frac{2c_i}{a+b} ab \\ c_i sa - \frac{2c_i}{a+b} sab \end{bmatrix} \\
&= \frac{c_i}{a+b} \begin{bmatrix} b(a+b) - 2ab \\ sa(a+b) - 2sab \end{bmatrix} \\
&= \frac{c_i}{a+b} \begin{bmatrix} b^2 - ab \\ sa^2 - sab \end{bmatrix} \\
&= \frac{c_i(b-a)}{a+b} \begin{bmatrix} b \\ -sa \end{bmatrix}.
\end{aligned}$$

Hence,

$$c_{i+1} = \frac{c_i(b-a)}{a+b} = -c_i \frac{\kappa - 1}{\kappa + 1},$$

where $\kappa = \frac{a}{b}$ is precisely the 2-norm condition number of the matrix A. From this, one can see that the convergence rate is the worst possible, according to our convergence theory, because

$$c_k = (-1)^k c_0 \left(\frac{\kappa - 1}{\kappa + 1} \right)^k.$$

Remark 7.19 (ill-conditioning). For the steepest descent method, with large spectral condition number κ, we observe that

$$\frac{\kappa - 1}{\kappa + 1} \approx 1 - \frac{2}{\kappa}.$$

In other words, the convergence rate deteriorates as $\kappa \to \infty$, i.e., as the system matrix becomes increasingly ill-conditioned. We will see, when we study numerical methods for certain types of differential equations, how a family of matrices can become more and more ill-conditioned as the size of the matrices increases.

Let us now study the convergence of the B-preconditioned steepest descent method defined in (7.2).

Theorem 7.20 (error equation). *Suppose that* $\mathsf{A}, \mathsf{B} \in \mathbb{C}^{n \times n}$ *are HPD,* $\boldsymbol{f} \in \mathbb{C}^n$, *and* $\boldsymbol{x} = \mathsf{A}^{-1}\boldsymbol{f}$. *Suppose that* $\{\boldsymbol{x}_k\}_{k=1}^{\infty}$ *is computed using the* B*-preconditioned steepest descent method with the starting value* $\boldsymbol{x}_0 \in \mathbb{C}^n$. *Then the error* $\boldsymbol{e}_k = \boldsymbol{x} - \boldsymbol{x}_k$ *satisfies*

$$\|\boldsymbol{e}_{k+1}\|_{\mathsf{A}}^2 = \beta_k \|\boldsymbol{e}_k\|_{\mathsf{A}}^2,$$

where

$$\beta_k = 1 - \frac{\left(\boldsymbol{d}_k^{\mathsf{H}} \boldsymbol{r}_k\right)^2}{\left(\boldsymbol{d}_k^{\mathsf{H}} \mathsf{A} \boldsymbol{d}_k\right)\left(\boldsymbol{r}_k^{\mathsf{H}} \mathsf{A}^{-1} \boldsymbol{r}_k\right)}$$

and

$$\boldsymbol{r}_k = \mathsf{B}\boldsymbol{d}_k.$$

Setting $\mathsf{B} = \mathsf{L}^{\mathsf{H}}\mathsf{L}$, *for some invertible matrix* $\mathsf{L} \in \mathbb{C}^{n\times n}$,

$$\boldsymbol{g}_k = \mathsf{L}\boldsymbol{d}_k, \qquad \mathsf{C} = \mathsf{L}^{-\mathsf{H}}\mathsf{A}\mathsf{L}^{-1},$$

the error equation may be expressed using

$$\beta_k = 1 - \frac{\left(\boldsymbol{g}_k^{\mathsf{H}}\boldsymbol{g}_k\right)^2}{\left(\boldsymbol{g}_k^{\mathsf{H}}\mathsf{C}\boldsymbol{g}_k\right)\left(\boldsymbol{g}_k^{\mathsf{H}}\mathsf{C}^{-1}\boldsymbol{g}_k\right)}.$$

Proof. See Problem 7.10. □

Using the Kantorovich inequality, we get the the following error estimate for the preconditioned steepest descent method.

Theorem 7.21 (convergence). *Suppose that* $\mathsf{A}, \mathsf{B} \in \mathbb{C}^{n\times n}$ *are HPD,* $\boldsymbol{f} \in \mathbb{C}^n$, *and* $\boldsymbol{x} = \mathsf{A}^{-1}\boldsymbol{f}$. *Suppose that* $\{\boldsymbol{x}_k\}_{k=1}^{\infty}$ *is computed using the* B*-preconditioned steepest descent method with the starting value* $\boldsymbol{x}_0 \in \mathbb{C}^n$. *Suppose that* B *has the Cholesky-type factorization* $\mathsf{B} = \mathsf{L}^{\mathsf{H}}\mathsf{L}$, *where* $\mathsf{L} \in \mathbb{C}^{n\times n}$ *is invertible. Define* $\mathsf{C} = \mathsf{L}^{-\mathsf{H}}\mathsf{A}\mathsf{L}^{-1}$. *Then the following error estimate holds:*

$$\|\boldsymbol{e}_k\|_{\mathsf{A}} \le \left(\frac{\kappa_{\mathsf{C}} - 1}{\kappa_{\mathsf{C}} + 1}\right)^k \|\boldsymbol{e}_0\|_{\mathsf{A}},$$

where

$$\kappa_{\mathsf{C}} = \kappa_2(\mathsf{C}) = \frac{\mu_n}{\mu_1}, \quad \sigma(\mathsf{C}) = \{\mu_i\}_{i=1}^n, \quad 0 < \mu_1 \le \cdots \le \mu_n.$$

Proof. Observe that $\mathsf{C} = \mathsf{L}^{-\mathsf{H}}\mathsf{A}\mathsf{L}^{-1}$ is HPD. The result now follows by applying the Kantorovich inequality to estimate the size β_k, the contraction factor from Theorem 7.20. □

Remark 7.22 (generalized condition number)**.** Consider the preconditioned system (7.3) and observe that $\mathsf{B}^{-1}\mathsf{A}$ is self-adjoint and positive definite with respect to $(\cdot,\cdot)_{\mathsf{A}}$, since B^{-1} is HPD. It, therefore, has positive real eigenvalues. Furthermore, one will find that $\mathsf{B}^{-1}\mathsf{A}$ is similar to $\mathsf{C} = \mathsf{L}^{-\mathsf{H}}\mathsf{A}\mathsf{L}^{-1}$. In particular,

$$\mathsf{L}\left(\mathsf{B}^{-1}\mathsf{A}\right)\mathsf{L}^{-1} = \mathsf{L}^{-\mathsf{H}}\mathsf{A}\mathsf{L}^{-1} = \mathsf{C}.$$

Therefore, the eigenvalues of C and the preconditioned coefficient matrix $\mathsf{B}^{-1}\mathsf{A}$ are the same. Therefore, we will often write the result of the last theorem in the following way:

$$\|\boldsymbol{e}_k\|_{\mathsf{A}} \le \left(\frac{\kappa - 1}{\kappa + 1}\right)^k \|\boldsymbol{e}_0\|_{\mathsf{A}},$$

where

$$\kappa = \kappa_{\mathsf{B}^{-1}\mathsf{A}} = \frac{\mu_n}{\mu_1}, \quad \sigma(\mathsf{B}^{-1}\mathsf{A}) = \{\mu_i\}_{i=1}^n, \quad 0 < \mu_1 \le \cdots \le \mu_n.$$

The number $\kappa_{\mathsf{B}^{-1}\mathsf{A}} \ge 1$ is called the *condition number* of the preconditioned coefficient matrix $\mathsf{B}^{-1}\mathsf{A}$.

Remark 7.23 (effect of preconditioning). The result of this last theorem is quite important. It shows that, if we select the preconditioner B in a careful way, it is possible to dramatically improve the convergence rate. In particular, it is possible, theoretically, to find B such that κ_{C} is nearly one. The closer κ_{C} is to one, the faster is the convergence rate.

7.4 The Conjugate Gradient Method

We have arrived at the so-called conjugate gradient (CG) method. Let $\mathsf{A} \in \mathbb{C}^{n\times n}$ be HPD and $\boldsymbol{f} \in \mathbb{C}^n$. Recall that we are interested in finding a solution to

$$\mathsf{A}\boldsymbol{x} = \boldsymbol{f},$$

or, equivalently,

$$\boldsymbol{x} = \underset{\boldsymbol{z}\in\mathbb{C}^n}{\operatorname{argmin}}\, E_{\mathsf{A}}(\boldsymbol{z}), \qquad E_{\mathsf{A}}(\boldsymbol{z}) = \frac{1}{2}\boldsymbol{z}^{\mathsf{H}}\mathsf{A}\boldsymbol{z} - \Re\left(\boldsymbol{z}^{\mathsf{H}}\boldsymbol{f}\right).$$

To solve $\mathsf{A}\boldsymbol{x} = \boldsymbol{f}$, we will, instead, try to solve the minimization problem in a clever way; namely, by doing it over an increasing sequence of subspaces of $\mathbb{C}^n$.

Definition 7.24 (Krylov subspace[2]). Given $\mathsf{A} \in \mathbb{C}^{n\times n}$ and $\boldsymbol{0} \neq \boldsymbol{q} \in \mathbb{C}^n$, the **Krylov subspace** of degree m is

$$\mathcal{K}_m(\mathsf{A}, \boldsymbol{q}) = \operatorname{span}\left\{\mathsf{A}^k\boldsymbol{q} \;\middle|\; k = 0, \dots, m-1\right\}.$$

Many students, when they learn the CG method for the first time, come away with the notion that the method is quite complicated. On the contrary, the idea is quite simple and it is easy to remember. Here, it is in its most natural form.

Definition 7.25 (CG). Suppose that $\mathsf{A} \in \mathbb{C}^{n\times n}$ is HPD, $\boldsymbol{f} \in \mathbb{C}^n_\star$, and $\boldsymbol{x} = \mathsf{A}^{-1}\boldsymbol{f}$. The **zero-start conjugate gradient method** is an iterative scheme for producing a sequence of approximations $\{\boldsymbol{x}_k\}_{k=1}^{\infty}$ from the starting point $\boldsymbol{x}_0 = \boldsymbol{0} \in \mathbb{C}^n$ according to the following prescription: setting $\mathcal{K}_k = \mathcal{K}_k(\mathsf{A}, \boldsymbol{f})$, the kth iterate is obtained by

$$\boldsymbol{x}_k = \underset{\boldsymbol{z}\in\mathcal{K}_k}{\operatorname{argmin}}\, E_{\mathsf{A}}(\boldsymbol{z}). \tag{7.10}$$

Notice that, by construction, $\mathcal{K}_m(\mathsf{A}, \boldsymbol{q}) \subseteq \mathcal{K}_{m+1}(\mathsf{A}, \boldsymbol{q})$. Thus, we are minimizing over a nondecreasing family of nested subspaces of $\mathbb{C}^n$.

Remark 7.26 (nonzero starting vector). Later on, we will mention the important case of starting the CG method with a nonzero starting vector $\boldsymbol{x}_0$. This is a very important issue, since, in many cases, the user may have some insight on how to start the CG method so as to obtain faster convergence.

[2] Named in honor of the Russian mathematician Aleksei Nikolaevich Krylov (1863–1945).

Definition 7.27 (Galerkin approximation[3]). Suppose that $\mathsf{A} \in \mathbb{C}^{n\times n}$ is HPD, $\boldsymbol{f} \in \mathbb{C}^n$, $\boldsymbol{x} = \mathsf{A}^{-1}\boldsymbol{f}$, and W is a subspace of $\mathbb{C}^n$. The vector $\boldsymbol{x}_W \in W$ is called the **Galerkin approximation** of $\boldsymbol{x}$ in W if and only if

$$(\mathsf{A}\boldsymbol{x}_W, \boldsymbol{w})_2 = (\boldsymbol{f}, \boldsymbol{w})_2, \quad \forall \boldsymbol{w} \in W. \tag{7.11}$$

Theorem 7.28 (existence and uniqueness). *Suppose that* $\mathsf{A} \in \mathbb{C}^{n\times n}$ *is HPD,* $\boldsymbol{f} \in \mathbb{C}^n$, $\boldsymbol{x} = \mathsf{A}^{-1}\boldsymbol{f}$, *and* W *is a subspace of* $\mathbb{C}^n$. *The Galerkin approximation* $\boldsymbol{x}_W \in W$ *exists and is unique.*

Proof. Let $B = \{\boldsymbol{w}_1, \dots, \boldsymbol{w}_k\}$ be an A-orthonormal basis for W, i.e.,

$$(\boldsymbol{w}_i, \boldsymbol{w}_j)_\mathsf{A} = (\mathsf{A}\boldsymbol{w}_i, \boldsymbol{w}_j)_2 = \delta_{i,j}$$

for all $1 \le i, j \le k \le n$. Then (7.11) holds if and only if

$$(\mathsf{A}\boldsymbol{x}_W, \boldsymbol{w}_i)_2 = (\boldsymbol{f}, \boldsymbol{w}_i)_2, \quad i = 1, \dots, k. \tag{7.12}$$

Since B is a basis, there are unique constants $c_1, \dots, c_k \in \mathbb{C}$ such that $\boldsymbol{x}_W = \sum_{j=1}^k c_j \boldsymbol{w}_j$. Plugging this into (7.12) we get the trivial diagonal system

$$c_i = \sum_{j=1}^{k} c_j (\boldsymbol{w}_j, \boldsymbol{w}_i)_\mathsf{A} = (\boldsymbol{f}, \boldsymbol{w}_i)_2, \quad i = 1, \dots, k. \tag{7.13}$$

This proves existence and uniqueness. □

Galerkin approximations have important properties that we will find useful.

Proposition 7.29 (properties of Galerkin approximations). *Suppose that* $\mathsf{A} \in \mathbb{C}^{n\times n}$ *is HPD,* $\boldsymbol{f} \in \mathbb{C}^n$, $\boldsymbol{x} = \mathsf{A}^{-1}\boldsymbol{f}$, W *is a subspace of* $\mathbb{C}^n$, *and* $\boldsymbol{x}_W$ *is the Galerkin approximation to* $\boldsymbol{x}$ *in* W.

1. *The residual is orthogonal to* W. *That is, if* $\boldsymbol{r} = \boldsymbol{f} - \mathsf{A}\boldsymbol{x}_W$, *we have*

$$(\boldsymbol{r}, \boldsymbol{w})_2 = 0, \quad \forall \boldsymbol{w} \in W.$$

2. *Galerkin orthogonality: Define the error* $\boldsymbol{e} = \boldsymbol{x} - \boldsymbol{x}_W$. *Then we have*

$$(\mathsf{A}\boldsymbol{e}, \boldsymbol{w})_2 = 0, \quad \forall \boldsymbol{w} \in W.$$

3. *Optimality:*

$$(\mathsf{A}\boldsymbol{e}, \boldsymbol{e})_2 \le (\mathsf{A}(\boldsymbol{x} - \boldsymbol{w}), \boldsymbol{x} - \boldsymbol{w})_2, \quad \forall \boldsymbol{w} \in W.$$

Proof. We only prove optimality, leaving the first two for the reader; see Problem 7.11. Let $\boldsymbol{w} \in W$ be arbitrary. Using Galerkin orthogonality, and the Cauchy–Schwarz inequality for the A-norm,

[3] Named in honor of the Russian mathematician Boris Grigorievich Galerkin (1871–1945).

$$\begin{aligned}
\|e\|_A^2 &= (Ae, e)_2 \\
&= (Ae, x - x_W)_2 \\
&= (Ae, x - x_W)_2 + (Ae, x_W - w)_2 \\
&= (Ae, x - x_W + x_W - w)_2 \\
&= (Ae, x - w)_2 \\
&\le \|e\|_A \|x - w\|_A .
\end{aligned}$$

If $\|e\|_A = 0$, the result is trivial. If $\|e\|_A > 0$,

$$\|e\|_A \le \|x - w\|_A ,$$

as we claimed. □

Theorem 7.30 (characterization of Galerkin approximations). *Suppose that $A \in \mathbb{C}^{n\times n}$ is HPD, $f \in \mathbb{C}^n$, and W is a subspace of $\mathbb{C}^n$. Then the following are equivalent.*

1. *The vector $x_W \in W$ is a minimizer of E_A over W:*

$$x_W = \underset{z\in W}{\operatorname{argmin}}\, E_A(z).$$

2. *The vector $x_W \in W$ is a Galerkin approximation of $x = A^{-1}f$:*

$$(Ax_W, w)_2 = (f, w)_2, \quad \forall w \in W.$$

Proof. See Problem 7.12. □

With these definitions at hand we can actually show that the CG method will converge in a finite number of steps.

Theorem 7.31 (convergence). *Let A be HPD, $f \in \mathbb{C}^n_\star$, and $x = A^{-1}f$. We have that $\dim \mathcal{K}_m(A, f) = m$ and, as a consequence, the sequence $\{x_k\}_{k=1}^\infty$, generated by the zero-start CG method, is such that there is an integer $m_\star \in \{1, \dots, n\}$ for which*

$$x_k \ne x, \quad k = 1, \dots, m_\star - 1, \qquad x_k = x, \quad k \ge m_\star.$$

Proof. Let $\mathcal{K}_m = \mathcal{K}_m(A, f)$. Notice that $\dim \mathcal{K}_m \le m$, so that, if we show that equality actually holds, all the statements will follow.

We will proceed by induction. Set $m = 1$ and notice that, since $f \ne 0$,

$$\mathcal{K}_1 = \operatorname{span}\{f\} \implies \dim \mathcal{K}_1 = 1.$$

Assume now that, for all $m = 1, \dots, k$ with $k < n - 1$, we have $\dim \mathcal{K}_k = k$ and $x_k \ne x$. Therefore, the residual $r_k = f - Ax_k \ne \mathbf{0}$ and $r_k \in \mathcal{K}_{k+1}$.

Notice that, using the characterization of Galerkin approximations given in Theorem 7.30, we have that $x_k \in \mathcal{K}_k$ is the Galerkin approximation of x in $\mathcal{K}_k$. Thus, the residual r_k must be orthogonal to $\mathcal{K}_k$, i.e.,

$$(r_k, w)_2 = 0, \quad \forall w \in \mathcal{K}_k.$$

In other words, we have shown that $\mathbf{0} \neq \boldsymbol{r}_k \in \mathcal{K}_{k+1} \setminus \mathcal{K}_k$. This is only possible if $\dim \mathcal{K}_{k+1} > \dim \mathcal{K}_k$. In other words, $\dim \mathcal{K}_{k+1} = k+1$ and the result follows. □

While the previous result shows that CG obtains the exact solution in at most n steps, we will consider this as an *iterative scheme* and study its properties as such. The reason for this is twofold. First, we want to consider the case when n is very large; therefore, we may wish to stop the iterations before the exact solution is found. Second, while in theory the exact solution can be obtained with CG, experience shows that rounding errors make this not possible.

We will begin then by rephrasing the CG method in an equivalent form, which is more convenient for practical computations. Indeed, as stated in Definition 7.25, it seems that at step k we need to store k vectors — a basis of $\mathcal{K}_k(\mathsf{A}, \boldsymbol{f})$ — as we need to minimize over it. The following equivalent formulation shows that this is not necessary.

Theorem 7.32 (equivalence). *Suppose that $\mathsf{A} \in \mathbb{C}^{n\times n}$ is HPD, $\boldsymbol{f} \in \mathbb{C}^n_\star$, and $\boldsymbol{x} = \mathsf{A}^{-1}\boldsymbol{f}$. The sequence generated by the zero-start CG method, $\{\boldsymbol{x}_k\}_{k=1}^{m_\star}$, is the same sequence as that generated by the following recursive algorithm: given $\boldsymbol{x}_0 = \mathbf{0}$, define $\boldsymbol{r}_0 = \boldsymbol{f} - \mathsf{A}\boldsymbol{x}_0 = \boldsymbol{f}$ and $\boldsymbol{p}_0 = \boldsymbol{r}_0 = \boldsymbol{f}$. For $k = 0, \ldots, m_\star - 1$, compute:*

1. *Update the iterate:*

$$\boldsymbol{x}_{k+1} = \boldsymbol{x}_k + \lambda_{k+1}\boldsymbol{p}_k, \quad \lambda_{k+1} = \frac{(\boldsymbol{r}_k, \boldsymbol{p}_k)_2}{(\mathsf{A}\boldsymbol{p}_k, \boldsymbol{p}_k)_2}.$$

2. *Update the residual:*

$$\boldsymbol{r}_{k+1} = \boldsymbol{r}_k - \lambda_{k+1}\mathsf{A}\boldsymbol{p}_k.$$

3. *Update the search direction:*

$$\boldsymbol{p}_{k+1} = \boldsymbol{r}_{k+1} - \mu_{k+1}\boldsymbol{p}_k, \quad \mu_{k+1} = \frac{(\mathsf{A}\boldsymbol{r}_{k+1}, \boldsymbol{p}_k)_2}{(\mathsf{A}\boldsymbol{p}_k, \boldsymbol{p}_k)_2}.$$

4. *If $k = m_\star - 1$, stop. Otherwise, index k and go to step 1.*

Clearly,

$$\boldsymbol{r}_{m_\star} = \mathbf{0} = \boldsymbol{p}_{m_\star},$$

but it is guaranteed that

$$\mathbf{0} \neq \boldsymbol{r}_k \in \mathcal{K}_{k+1} \setminus \mathcal{K}_k, \quad \mathbf{0} \neq \boldsymbol{p}_k \in \mathcal{K}_{k+1} \setminus \mathcal{K}_k, \quad k = 0, \ldots, m_\star - 1,$$

and the following orthogonalities hold:

$$(\boldsymbol{r}_j, \boldsymbol{r}_i)_2 = (\boldsymbol{p}_j, \boldsymbol{p}_i)_\mathsf{A} = 0$$

for all $0 \leq j < i \leq m_\star - 1$.

Proof. The proof is by induction on k, terminating at step $m_\star - 1$.

We leave the reader to check that the base cases, $k \leq 2$, are true; see Problem 7.13.

For the induction hypothesis we assume that the formulas and properties above are true for $0 \leq k \leq m-1$.

Now let $k \leq m \leq m_\star - 1$. Let us assume that $\{\boldsymbol{x}_k\}_{k=1}^{m+1}$ is generated by the zero-start CG algorithm. We know that $\boldsymbol{x}_m \neq \boldsymbol{x}$ and $\boldsymbol{r}_m = \boldsymbol{f} - \mathsf{A}\boldsymbol{x}_m \neq \boldsymbol{0}$. The vector $\boldsymbol{x}_{m+1} \in \mathcal{K}_{m+1}$ is obtained by solving the optimization problem (7.10) over the Krylov space $\mathcal{K}_{m+1}$. Notice that

$$\boldsymbol{r}_m = \boldsymbol{f} - \mathsf{A}\boldsymbol{x}_m \in \mathcal{K}_{m+1} = \mathcal{K}_{m+1}(\mathsf{A}, \boldsymbol{f}) = \operatorname{span}\left\{\mathsf{A}^k \boldsymbol{f} \;\middle|\; k = 0, \dots, m\right\},$$

because, by assumption, $\boldsymbol{x}_m \in \mathcal{K}_m$. Since $\boldsymbol{x}_m$ is a Galerkin approximation to $\boldsymbol{x}$, by Galerkin orthogonality, we observe that for all $\boldsymbol{y} \in \mathcal{K}_m$,

$$0 = (\mathsf{A}\boldsymbol{e}_m, \boldsymbol{y})_2 = (\boldsymbol{r}_m, \boldsymbol{y})_2.$$

In other words,

$$\boldsymbol{r}_m \in \mathcal{K}_{m+1} \backslash \mathcal{K}_m.$$

The induction hypothesis guarantees that $\{\boldsymbol{r}_0, \dots, \boldsymbol{r}_{m-1}\}$ is an orthogonal set and a basis for $\mathcal{K}_m$. Galerkin orthogonality implies that the vector $\boldsymbol{r}_m$ is orthogonal to $\{\boldsymbol{r}_0, \dots, \boldsymbol{r}_{m-1}\}$; therefore, $\{\boldsymbol{r}_0, \dots, \boldsymbol{r}_{m-1}, \boldsymbol{r}_m\}$ is an orthogonal basis for $\mathcal{K}_{m+1}$.

Let $\mathcal{K}_m^{\perp_\mathsf{A}}$ denote the orthogonal complement of $\mathcal{K}_m$ in $\mathcal{K}_{m+1}$ in the A-inner product, i.e.,

$$\mathcal{K}_m^{\perp_\mathsf{A}} = \{\boldsymbol{w} \in \mathcal{K}_{m+1} \mid (\mathsf{A}\boldsymbol{w}, \boldsymbol{y})_2 = 0, \ \forall \boldsymbol{y} \in \mathcal{K}_m\}.$$

It follows that $\mathcal{K}_m^{\perp_\mathsf{A}} \neq \{\boldsymbol{0}\}$; moreover, it is not difficult to see that $\mathcal{K}_m^{\perp_\mathsf{A}}$ is one dimensional. Since

$$\mathcal{K}_{m+1} = \mathcal{K}_m \overset{\perp}{\oplus} \mathcal{K}_m^{\perp_\mathsf{A}},$$

any element $\boldsymbol{\xi}_{m+1} \in \mathcal{K}_{m+1}$ can be written as

$$\boldsymbol{\xi}_{m+1} = \boldsymbol{\xi}_m + \mu \boldsymbol{p}_m$$

for some $\mu \in \mathbb{C}$, $\boldsymbol{p}_m \in \mathcal{K}_m^{\perp_\mathsf{A}} \cap \mathbb{C}_\star^n$, and $\boldsymbol{\xi}_m \in \mathcal{K}_m$. Now consider the element

$$\boldsymbol{w} = \boldsymbol{x}_m + \lambda_{m+1}\boldsymbol{p}_m, \quad \lambda_{m+1} = \frac{(\boldsymbol{r}_m, \boldsymbol{p}_m)_2}{(\mathsf{A}\boldsymbol{p}_m, \boldsymbol{p}_m)_2}.$$

Then

$$\begin{aligned}
(\mathsf{A}\boldsymbol{w} - \boldsymbol{f}, \boldsymbol{\xi}_{m+1})_2 &= (\mathsf{A}(\boldsymbol{x}_m + \lambda_{m+1}\boldsymbol{p}_m) - \boldsymbol{f}, \boldsymbol{\xi}_m + \mu\boldsymbol{p}_m)_2 \\
&= (\mathsf{A}\boldsymbol{x}_m - \boldsymbol{f}, \boldsymbol{\xi}_m)_2 + \bar{\mu}\,(\mathsf{A}\boldsymbol{x}_m - \boldsymbol{f}, \boldsymbol{p}_m)_2 + \lambda_{m+1}(\mathsf{A}\boldsymbol{p}_m, \boldsymbol{\xi}_m)_2 \\
&\quad + \bar{\mu}\lambda_{m+1}(\mathsf{A}\boldsymbol{p}_m, \boldsymbol{p}_m)_2 \\
&= (\mathsf{A}\boldsymbol{x}_m - \boldsymbol{f}, \boldsymbol{\xi}_m)_2 - \bar{\mu}\,[(\boldsymbol{r}_m, \boldsymbol{p}_m)_2 - \lambda_{m+1}(\mathsf{A}\boldsymbol{p}_m, \boldsymbol{p}_m)_2] \\
&= -\bar{\mu}\,[(\boldsymbol{r}_m, \boldsymbol{p}_m)_2 - \lambda_{m+1}(\mathsf{A}\boldsymbol{p}_m, \boldsymbol{p}_m)_2] \\
&= 0,
\end{aligned}$$

where we used that $(\mathsf{A}\boldsymbol{p}_m, \boldsymbol{\xi}_m)_2 = 0$ for every $\boldsymbol{\xi}_m \in \mathcal{K}_m$, the fact that $\boldsymbol{x}_m$ minimizes over $\mathcal{K}_m$ and so $(\mathsf{A}\boldsymbol{x}_m - \boldsymbol{f}, \boldsymbol{\xi}_m)_2 = 0$ for every $\boldsymbol{\xi}_m \in \mathcal{K}_m$, and, finally, the definition of λ_{m+1}. Therefore,

$$(\mathsf{A}\boldsymbol{w} - \boldsymbol{f}, \boldsymbol{\xi}_{m+1})_2 = 0, \quad \forall \boldsymbol{\xi}_{m+1} \in \mathcal{K}_{m+1}.$$

But $\boldsymbol{x}_{m+1} \in \mathcal{K}_{m+1}$, the Galerkin approximation, is the unique element that has the property exhibited in the last equation. Therefore, $\boldsymbol{w} = \boldsymbol{x}_{m+1}$. In other words, $\boldsymbol{x}_{m+1} \in \mathcal{K}_{m+1}$ is the Galerkin approximation defining the $(m+1)$st iterate in the zero-start CG algorithm if and only if

$$\boldsymbol{x}_{m+1} = \boldsymbol{x}_m + \lambda_{m+1}\boldsymbol{p}_m, \quad \lambda_{m+1} = \frac{(\boldsymbol{r}_m, \boldsymbol{p}_m)_2}{(\mathsf{A}\boldsymbol{p}_m, \boldsymbol{p}_m)_2}.$$

Furthermore, a simple computation involving the last equation then yields the $(m+1)$st residual:

$$\boldsymbol{r}_{m+1} = \boldsymbol{r}_m - \lambda_{m+1}\mathsf{A}\boldsymbol{p}_m.$$

By the induction hypothesis, $\{\boldsymbol{p}_0, \ldots, \boldsymbol{p}_{m-1}\}$ is an A-orthogonal set; hence, it forms a basis for $\mathcal{K}_m$. Since $\boldsymbol{r}_m$ has a component in $\mathcal{K}_m^{\perp_\mathsf{A}}$, its A-orthogonal projection, $\boldsymbol{q}$, into $\mathcal{K}_m^{\perp_\mathsf{A}}$ is nonzero:

$$\boldsymbol{q} = \boldsymbol{r}_m - \sum_{i=0}^{m-1} \frac{(\mathsf{A}\boldsymbol{r}_m, \boldsymbol{p}_i)_2}{(\mathsf{A}\boldsymbol{p}_i, \boldsymbol{p}_i)_2}\boldsymbol{p}_i = \boldsymbol{r}_m - \sum_{i=0}^{m-1} \frac{(\boldsymbol{r}_m, \boldsymbol{p}_i)_\mathsf{A}}{(\boldsymbol{p}_i, \boldsymbol{p}_i)_\mathsf{A}}\boldsymbol{p}_i.$$

But if $0 \le i \le m-2$, then $\mathsf{A}\boldsymbol{p}_i \in \mathcal{K}_m$ and so $(\mathsf{A}\boldsymbol{p}_i, \boldsymbol{r}_m)_2 = 0$. We are thus left with

$$\boldsymbol{q} = \boldsymbol{r}_m - \frac{(\mathsf{A}\boldsymbol{r}_m, \boldsymbol{p}_{m-1})_2}{(\mathsf{A}\boldsymbol{p}_{m-1}, \boldsymbol{p}_{m-1})_2}\boldsymbol{p}_{m-1} \in \mathcal{K}_m^{\perp_\mathsf{A}}.$$

Finally, $\boldsymbol{p}_m \in \mathcal{K}_m^{\perp_\mathsf{A}}$ is not yet completely determined. We are free to pick it to suit our purposes. We take

$$\boldsymbol{p}_m = \boldsymbol{q} = \boldsymbol{r}_m - \frac{(\mathsf{A}\boldsymbol{r}_m, \boldsymbol{p}_{m-1})_2}{(\mathsf{A}\boldsymbol{p}_{m-1}, \boldsymbol{p}_{m-1})_2}\boldsymbol{p}_{m-1} \in \mathcal{K}_m^{\perp_\mathsf{A}}.$$

It follows that $\{\boldsymbol{p}_0, \ldots, \boldsymbol{p}_{m-1}, \boldsymbol{p}_m\}$ is an A-orthogonal set. □

Corollary 7.33 (iterates). *Suppose that* $\mathsf{A} \in \mathbb{C}^{n\times n}$ *is HPD,* $\boldsymbol{f} \in \mathbb{C}^n_\star$*, and* $\boldsymbol{x} = \mathsf{A}^{-1}\boldsymbol{f}$*. The sequence generated by the zero-start CG method,* $\{\boldsymbol{x}_k\}_{k=1}^{m_\star}$*, has the following property: for all* $k \in \{1, \ldots, m_\star\}$*,*

$$\boldsymbol{x}_k \in \mathcal{K}_k \backslash \mathcal{K}_{k-1},$$

which implies that

$$\langle \boldsymbol{x}_1, \ldots, \boldsymbol{x}_k \rangle = \mathcal{K}_k.$$

Proof. See Problem 7.14. □

Corollary 7.34 (spanning properties). *Suppose that* $\mathsf{A} \in \mathbb{C}^{n\times n}$ *is HPD,* $\boldsymbol{f} \in \mathbb{C}^n_\star$*, and* $\boldsymbol{x} = \mathsf{A}^{-1}\boldsymbol{f}$*. If the zero-start CG algorithm is employed to produce the approximation sequence* $\{\boldsymbol{x}_j\}_{j=1}^{m_\star}$*, then, for all* $1 \le i \le m_\star$*,*

$$\mathcal{K}_i(\mathsf{A}, \boldsymbol{f}) = \langle \boldsymbol{f}, \mathsf{A}\boldsymbol{f}, \ldots, \mathsf{A}^{i-1}\boldsymbol{f} \rangle = \langle \boldsymbol{x}_1, \ldots, \boldsymbol{x}_i \rangle = \langle \boldsymbol{p}_0, \ldots, \boldsymbol{p}_{i-1} \rangle = \langle \boldsymbol{r}_0, \ldots, \boldsymbol{r}_{i-1} \rangle.$$

Proof. See Problem 7.15. □

A slightly more computationally efficient, but entirely equivalent, version of the CG algorithm is possible.

Corollary 7.35 (equivalent formulation). *Suppose that* $\mathsf{A} \in \mathbb{C}^{n\times n}$ *is HPD,* $\boldsymbol{f} \in \mathbb{C}^n_\star$, *and* $\boldsymbol{x} = \mathsf{A}^{-1}\boldsymbol{f}$. *The sequence generated by the zero-start CG method,* $\{\boldsymbol{x}_k\}_{k=1}^{m_\star}$, *is the same sequence as that generated by the following recursive algorithm: given* $\boldsymbol{x}_0 = \boldsymbol{0}$, *define* $\boldsymbol{r}_0 = \boldsymbol{f} - \mathsf{A}\boldsymbol{x}_0 = \boldsymbol{f}$ *and* $\boldsymbol{p}_0 = \boldsymbol{r}_0 = \boldsymbol{f}$. *For* $0 \le k \le m_\star - 1$, *compute:*

1. *Update the iterate:*
$$\boldsymbol{x}_{k+1} = \boldsymbol{x}_k + \lambda_{k+1}\boldsymbol{p}_k, \quad \lambda_{k+1} = \frac{(\boldsymbol{r}_k, \boldsymbol{r}_k)_2}{(\mathsf{A}\boldsymbol{p}_k, \boldsymbol{p}_k)_2}.$$
2. *Update the residual:*
$$\boldsymbol{r}_{k+1} = \boldsymbol{r}_k - \lambda_{k+1}\mathsf{A}\boldsymbol{p}_k.$$
3. *Update the search direction:*
$$\boldsymbol{p}_{k+1} = \boldsymbol{r}_{k+1} + \beta_{k+1}\boldsymbol{p}_k, \quad \beta_{k+1} = \frac{(\boldsymbol{r}_{k+1}, \boldsymbol{r}_{k+1})_2}{(\boldsymbol{r}_k, \boldsymbol{r}_k)_2}.$$
4. *If* $k = m_\star - 1$, *stop. Otherwise, index* k *and go to step 1.*

It follows that
$$\boldsymbol{r}_{m_\star} = \boldsymbol{0} = \boldsymbol{p}_{m_\star},$$
but it is guaranteed that
$$\boldsymbol{0} \neq \boldsymbol{r}_k \in \mathcal{K}_{k+1}\backslash\mathcal{K}_k, \quad \boldsymbol{0} \neq \boldsymbol{p}_k \in \mathcal{K}_{k+1}\backslash\mathcal{K}_k, \quad k = 0, \dots, m_\star - 1.$$
The following orthogonalities hold:
$$(\boldsymbol{r}_j, \boldsymbol{r}_i)_2 = (\boldsymbol{p}_j, \boldsymbol{p}_i)_\mathsf{A} = 0$$
for all $0 \le j < i \le m_\star - 1$.

Proof. See Problem 7.16. □

We have defined the CG method via minimization over certain subspaces of increasing size. But we could alternately define the algorithm via one of the recursive algorithms above. In this case, we could derive the minimization property as a consequence. In other words, there are a few equivalent ways of defining the method of CGs.

Theorem 7.36 (minimization). *Suppose that* $\mathsf{A} \in \mathbb{C}^{n\times n}$ *is HPD,* $\boldsymbol{f} \in \mathbb{C}^n_\star$ *is given, and* $\boldsymbol{x} = \mathsf{A}^{-1}\boldsymbol{f}$. *Let, for some* $m \in \{1, \dots, n\}$, $\{\boldsymbol{x}_i\}_{i=0}^m$ *denote any sequence of vectors with* $\boldsymbol{x}_0 = \boldsymbol{0}$ *— with associated residual vectors* $\boldsymbol{r}_j = \boldsymbol{f} - \mathsf{A}\boldsymbol{x}_j$ *— that satisfies*
$$\mathcal{K}_j = \mathcal{K}_j(\mathsf{A}, \boldsymbol{f}) = \langle \boldsymbol{f}, \mathsf{A}\boldsymbol{f}, \dots, \mathsf{A}^{j-1}\boldsymbol{f}\rangle = \langle \boldsymbol{x}_1, \dots, \boldsymbol{x}_j\rangle = \langle \boldsymbol{r}_0, \dots, \boldsymbol{r}_{j-1}\rangle, \quad \boldsymbol{r}_{j-1} \neq \boldsymbol{0}$$
for all $j = 1, \dots, n$, *with the orthogonality relations*
$$\boldsymbol{r}_k^\mathsf{H}\boldsymbol{r}_i = 0$$

for all $0 \le k < i \le m$. *Then the* j*th iterate* $\boldsymbol{x}_j$ *is the unique vector in* $\mathcal{K}_j$ *that minimizes the error function* $\phi(\boldsymbol{y}) = \|\boldsymbol{x} - \boldsymbol{y}\|_{\mathsf{A}}$. *Furthermore,* ϕ *is monotonically decreasing:*

$$\|\boldsymbol{e}_j\|_{\mathsf{A}} = \|\boldsymbol{x} - \boldsymbol{x}_j\|_{\mathsf{A}} = \phi(\boldsymbol{x}_j) \le \phi(\boldsymbol{x}_{j-1}) = \|\boldsymbol{x} - \boldsymbol{x}_{j-1}\|_{\mathsf{A}} = \|\boldsymbol{e}_{j-1}\|_{\mathsf{A}}.$$

Proof. Let $\boldsymbol{z} \in \mathcal{K}_j$ be arbitrary. Define $\boldsymbol{w} = \boldsymbol{x}_j - \boldsymbol{z} \in \mathcal{K}_j$. Then

$$\begin{aligned}
\phi^2(\boldsymbol{z}) &= \|\boldsymbol{x} - \boldsymbol{z}\|_{\mathsf{A}}^2 \\
&= \|\boldsymbol{x} - \boldsymbol{x}_j + \boldsymbol{x}_j - \boldsymbol{z}\|_{\mathsf{A}}^2 \\
&= \|\boldsymbol{e}_j + \boldsymbol{w}\|_{\mathsf{A}}^2 \\
&= (\boldsymbol{e}_j + \boldsymbol{w})^{\mathsf{H}}\mathsf{A}(\boldsymbol{e}_j + \boldsymbol{w}) \\
&= \|\boldsymbol{e}_j\|_{\mathsf{A}}^2 + 2\boldsymbol{w}^{\mathsf{H}}\mathsf{A}\boldsymbol{e}_j + \|\boldsymbol{w}\|_{\mathsf{A}}^2 \\
&= \|\boldsymbol{e}_j\|_{\mathsf{A}}^2 + 2\boldsymbol{w}^{\mathsf{H}}\boldsymbol{r}_j + \|\boldsymbol{w}\|_{\mathsf{A}}^2.
\end{aligned}$$

Since $\boldsymbol{w} \in \mathcal{K}_j = \langle \boldsymbol{r}_0, \ldots, \boldsymbol{r}_{j-1}\rangle$, there exist unique $\alpha_0, \ldots, \alpha_{j-1} \in \mathbb{C}$ such that

$$\boldsymbol{w} = \sum_{i=0}^{j-1} = \alpha_i \boldsymbol{r}_i.$$

Using the orthogonality $\boldsymbol{r}_i{}^{\mathsf{H}}\boldsymbol{r}_j = 0$ for all $0 \le i < j$, it is clear that $\boldsymbol{w}^{\mathsf{H}}\boldsymbol{r}_j = 0$. Hence,

$$\phi^2(\boldsymbol{z}) = \|\boldsymbol{e}_j\|_{\mathsf{A}}^2 + \|\boldsymbol{w}\|_{\mathsf{A}}^2 \ge \|\boldsymbol{e}_j\|_{\mathsf{A}}^2,$$

with equality in the last relation if and only if $\boldsymbol{w} = \boldsymbol{0}$, or, equivalently, if and only if $\boldsymbol{z} = \boldsymbol{x}_j$. Hence, $\boldsymbol{x}_j$ is the unique minimizer of ϕ over $\mathcal{K}_j$.

Since we have the space containment $\mathcal{K}_{j-1}(\mathsf{A}, \boldsymbol{f}) \subseteq \mathcal{K}_j(\mathsf{A}, \boldsymbol{f})$, we must have

$$\begin{aligned}
\|\boldsymbol{e}_j\|_{\mathsf{A}} &= \phi(\boldsymbol{x}_j) \\
&= \inf \{\phi(\boldsymbol{z}) \mid \boldsymbol{z} \in \mathcal{K}_j(\mathsf{A}, \boldsymbol{f})\} \\
&\le \inf \{\phi(\boldsymbol{z}) \mid \boldsymbol{z} \in \mathcal{K}_{j-1}(\mathsf{A}, \boldsymbol{f})\} \\
&= \phi(\boldsymbol{x}_{j-1}) \\
&= \|\boldsymbol{e}_{j-1}\|_{\mathsf{A}}.
\end{aligned}$$

□

The convergence of the CG method is given in the following few results.

Theorem 7.37 (polynomial bound). *Suppose that the zero-start CG algorithm is applied to solve* $\mathsf{A}\boldsymbol{x} = \boldsymbol{f}$, *where* $\mathsf{A} \in \mathbb{C}^{n\times n}$ *is HPD and* $\boldsymbol{f} \in \mathbb{C}_\star^n$. *Then, if the iteration has not already converged* ($\boldsymbol{r}_{i-1} \neq \boldsymbol{0}$), *there is a unique polynomial*

$$p_i \in \mathbb{P}_{i,\star} = \{p \in \mathbb{P}_i \mid p(0) = 1\}$$

that minimizes $\|p(\mathsf{A})\boldsymbol{e}_0\|_{\mathsf{A}}$ *over all* $p \in \mathbb{P}_{i,\star}$. *The iterate* $\boldsymbol{x}_i$ *has the error* $\boldsymbol{e}_i = p_i(\mathsf{A})\boldsymbol{e}_0$ *and, consequently,*

$$\frac{\|\boldsymbol{e}_i\|_{\mathsf{A}}}{\|\boldsymbol{e}_0\|_{\mathsf{A}}} \le \inf_{p\in\mathbb{P}_{i,\star}} \max_{\lambda\in\sigma(\mathsf{A})} |p(\lambda)|.$$

Proof. Recall that

$$E_{\mathsf{A}}(\boldsymbol{z}) = \frac{1}{2}\left\|\boldsymbol{z}-\boldsymbol{x}\right\|_{\mathsf{A}}^2 - \frac{1}{2}\boldsymbol{f}^{\mathsf{H}}\mathsf{A}^{-1}\boldsymbol{f},$$

where $\boldsymbol{x} = \mathsf{A}^{-1}\boldsymbol{f}$ is the exact solution. Therefore, by definition of the zero-start CG algorithm, we have

$$\boldsymbol{x}_i = \operatorname*{argmin}_{\boldsymbol{z}\in\mathcal{K}_i} \left\|\boldsymbol{x}-\boldsymbol{z}\right\|_{\mathsf{A}}, \quad \min_{\boldsymbol{z}\in\mathcal{K}_i} \left\|\boldsymbol{x}-\boldsymbol{z}\right\|_{\mathsf{A}} = \left\|\boldsymbol{e}_i\right\|_{\mathsf{A}}$$

and $\boldsymbol{x}_i \in \mathcal{K}_i$ is uniquely determined. Now observe that $\boldsymbol{e}_0 = \boldsymbol{x}$, since $\boldsymbol{x}_0 = \boldsymbol{0}$, and, consequently, $\boldsymbol{r}_0 = \boldsymbol{f}$. Thus, for any $\boldsymbol{z} \in \mathcal{K}_i$, there are constants $c_j \in \mathbb{C}$, $1 \le j \le i$, such that $\boldsymbol{z} = \sum_{j=1}^{i}(-c_j)\mathsf{A}^{j-1}\boldsymbol{f}$. Consequently,

$$\boldsymbol{x}-\boldsymbol{z} = \boldsymbol{x} + \sum_{j=1}^{i} c_j\mathsf{A}^{j-1}\boldsymbol{f} = \boldsymbol{e}_0 + \sum_{j=1}^{i} c_j\mathsf{A}^{j-1}\boldsymbol{r}_0 = \boldsymbol{e}_0 + \sum_{j=1}^{i} c_j\mathsf{A}^{j}\boldsymbol{e}_0 = p(\mathsf{A})\boldsymbol{e}_0,$$

where $p(t) = 1 + \sum_{j=1}^{i} c_j t^j \in \mathbb{P}_{i,\star}$. It follows, then, that the minimization problem above is equivalent to

$$p_i = \operatorname*{argmin}_{p\in\mathbb{P}_{i,\star}} \left\|p(\mathsf{A})\boldsymbol{e}_0\right\|_{\mathsf{A}}, \quad \min_{p\in\mathbb{P}_{i,\star}} \left\|p(\mathsf{A})\boldsymbol{e}_0\right\|_{\mathsf{A}} = \left\|\boldsymbol{e}_i\right\|_{\mathsf{A}}$$

and $p_i \in \mathbb{P}_{i,\star}$ is, of course, uniquely determined. It, therefore, follows that

$$\left\|\boldsymbol{e}_i\right\|_{\mathsf{A}} = \inf_{p\in\mathbb{P}_{i,\star}} \left\|p(\mathsf{A})\boldsymbol{e}_0\right\|_{\mathsf{A}} \le \inf_{p\in\mathbb{P}_{i,\star}} \left\|p(\mathsf{A})\right\|_{\mathsf{A}}\left\|\boldsymbol{e}_0\right\|_{\mathsf{A}}$$

and, consequently,

$$\frac{\left\|\boldsymbol{e}_i\right\|_{\mathsf{A}}}{\left\|\boldsymbol{e}_0\right\|_{\mathsf{A}}} \le \inf_{p\in\mathbb{P}_{i,\star}} \left\|p(\mathsf{A})\right\|_{\mathsf{A}}. \tag{7.14}$$

To finish up, suppose that $\boldsymbol{z} \in \mathbb{C}^n_\star$. Let $\{\boldsymbol{w}_1, \ldots, \boldsymbol{w}_n\}$ be an orthonormal basis of eigenvectors of A. Set $\sigma(A) = \{\lambda_1, \ldots, \lambda_n\} \subset (0, \infty)$, with $\mathsf{A}\boldsymbol{w}_j = \lambda_j\boldsymbol{w}_j$ for $j = 1, \ldots, n$. There exist constants $\alpha_j \in \mathbb{C}$, $j = 1, \ldots, n$ such that $\boldsymbol{z} = \sum_{j=1}^{n} \alpha_j\boldsymbol{w}_j$ and

$$\left\|\boldsymbol{z}\right\|_{\mathsf{A}}^2 = \boldsymbol{z}^{\mathsf{H}}\mathsf{A}\boldsymbol{z} = \sum_{j=1}^{n} |\alpha_j|^2\lambda_j.$$

Furthermore,

$$\left\|p(\mathsf{A})\boldsymbol{z}\right\|_{\mathsf{A}}^2 = \boldsymbol{z}^{\mathsf{H}}p(\mathsf{A})^{\mathsf{H}}\mathsf{A}p(\mathsf{A})\boldsymbol{z} = \sum_{j=1}^{n} |\alpha_j|^2\lambda_j|p(\lambda_j)|^2.$$

As a consequence,

$$\frac{\left\|p(\mathsf{A})\boldsymbol{z}\right\|_{\mathsf{A}}^2}{\left\|\boldsymbol{z}\right\|_{\mathsf{A}}^2} = \frac{\sum_{j=1}^{n} |\alpha_j|^2\lambda_j|p(\lambda_j)|^2}{\sum_{j=1}^{n} |\alpha_j|^2\lambda_j} \le \max_{\lambda\in\sigma(\mathsf{A})} |p(\lambda)|^2,$$

which implies that

$$\left\|p(\mathsf{A})\right\|_{\mathsf{A}} \le \max_{\lambda\in\sigma(\mathsf{A})} |p(\lambda)|.$$

Putting this estimate together with that in (7.14), we have

$$\frac{\|e_i\|_A}{\|e_0\|_A} \leq \inf_{p \in \mathbb{P}_{i,\star}} \max_{\lambda \in \sigma(A)} |p(\lambda)|,$$

which is the desired result. □

Theorem 7.38 (convergence)**.** *Suppose that the zero-start CG algorithm is applied to solve* $Ax = f$*, where* $A \in \mathbb{C}^{n \times n}$ *is HPD and* $f \in \mathbb{C}^n_\star$*. If* A *has only* k *distinct eigenvalues,* $k < n$*, then the algorithm converges in at most* k *steps.*

Proof. Let $\sigma(A) = \{\lambda_j\}_{j=1}^k$ denote the set of k distinct eigenvalues of A. From the last theorem, for $i = 1, \ldots, k$,

$$\frac{\|e_i\|_A}{\|e_0\|_A} \leq \max_{\lambda \in \sigma(A)} |q_i(\lambda)|$$

for any polynomial q_i of degree at most i with the property that $q_i(0) = 1$. Let us define, for any $i = 1, \ldots, k$,

$$q_i(x) = \prod_{j=1}^{i} \left(1 - \frac{x}{\lambda_j}\right).$$

Clearly, $q_i(0) = 1$ and for all $j = 1, \ldots, i$,

$$q_i(\lambda_j) = 0.$$

But, on the other hand, if $j = i + 1, \ldots, k$,

$$q_i(\lambda_j) \neq 0.$$

Now, once $i = k$,

$$q_i(\lambda) = 0, \quad \forall \lambda \in \sigma(A).$$

Thus, $\|e_k\|_A = 0$. Of course, convergence could happen at an earlier stage if, by chance, $x \in \mathcal{K}_i$ for some $i < k$. □

Theorem 7.39 (convergence of CG)**.** *Let* $A \in \mathbb{C}^{n \times n}$ *be HPD and* $f \in \mathbb{C}^n_\star$*. The error for the zero-start CG method satisfies*

$$\|e_k\|_A \leq 2 \left(\frac{\sqrt{\kappa_2(A)} - 1}{\sqrt{\kappa_2(A)} + 1} \right)^k \|e_0\|_A.$$

Proof. Suppose that $\sigma(A) = \{\lambda_1, \ldots, \lambda_n\}$, $0 < \lambda_1 \leq \cdots \leq \lambda_n$. From Theorem 7.37, it follows that

$$\|e_k\|_A \leq \max_{\lambda \in [\lambda_1, \lambda_n]} |q_k(\lambda)| \|e_0\|_A$$

for any polynomial q_k of degree at most k such that $q_k(0) = 1$. Since this polynomial is arbitrary, we may choose it to minimize the right-hand side of this expression. It turns out, as in the proof of Theorem 6.29, that shifted and rescaled versions of the classical Chebyshev polynomials minimize this choice; see Section 10.3. Namely, we set

$$q_k(t) = \frac{1}{T_k(1/\rho)} T_k\left(\frac{1}{\rho}\left(1 - \frac{2}{\lambda_1 + \lambda_n} t\right)\right), \quad \rho = \frac{\lambda_n - \lambda_1}{\lambda_n + \lambda_1},$$

and

$$T_k(t) = \begin{cases} \cos(k \arccos(t)), & |t| \le 1, \\ \cosh(k \cosh^{-1}(t)), & |t| > 1. \end{cases}$$

With this choice, we obtain the bound

$$\max_{\lambda \in [\lambda_1, \lambda_n]} |q_k(\lambda)| \le \frac{1}{T_k(1/\rho)}.$$

Now, since $\rho < 1$, we set $\sigma = \cosh^{-1}(1/\rho)$ to see that

$$T_k(1/\rho) = \frac{1}{2}(e^{k\sigma} + e^{-k\sigma}) \ge \frac{1}{2} e^{k\sigma}.$$

But, using that $\sigma = \cosh^{-1}(1/\rho) = \ln\left(1/\rho + \sqrt{1/\rho^2 - 1}\right)$, we get

$$e^{k\sigma} = \left[\frac{1}{\rho}\left(1 + \sqrt{1 - \rho^2}\right)\right]^k.$$

Using that $\rho = \frac{\kappa_2(\mathsf{A}) - 1}{\kappa_2(\mathsf{A}) + 1}$, we then obtain

$$\frac{1}{T_k(1/\rho)} \le 2e^{-k\sigma} = 2\left(\frac{\sqrt{\kappa_2(\mathsf{A})} - 1}{\sqrt{\kappa_2(\mathsf{A})} + 1}\right)^k$$

and the result follows. □

7.4.1 Nonzero Starting Vectors

In the absence of any knowledge about the solution to a linear system, it makes no difference what is chosen as the starting vector for an iterative scheme. Thus, $\boldsymbol{x}_0 = \boldsymbol{0}$ seems like a perfect candidate. There are situations, however, when something is known about the solution; therefore, we wish to use a starting vector other than the trivial one. In this case, we use the following algorithm.

Definition 7.40 (standard CG)**.** Suppose that $\mathsf{A} \in \mathbb{C}^{n \times n}$ is HPD, $\boldsymbol{f} \in \mathbb{C}^n_\star$, and $\boldsymbol{x} = \mathsf{A}^{-1}\boldsymbol{f}$. The **(standard) conjugate gradient method** is an iterative scheme for producing a sequence of approximations $\{\boldsymbol{x}_k\}_{k=1}^\infty$ from the starting point $\boldsymbol{x}_0 \in \mathbb{C}^n$ according to the following recursive formula: setting $\mathcal{K}_k = \mathcal{K}_k(\mathsf{A}, \boldsymbol{r}_0)$, where $\boldsymbol{r}_0 = \boldsymbol{f} - \mathsf{A}\boldsymbol{x}_0$, the kth iterate is obtained by

$$\boldsymbol{x}_k = \boldsymbol{x}'_k + \boldsymbol{x}_0, \quad \boldsymbol{x}'_k = \operatorname*{argmin}_{\boldsymbol{z} \in \mathcal{K}_k} E_{\mathsf{A}}(\boldsymbol{z} + \boldsymbol{x}_0). \tag{7.15}$$

The following result shows that the standard CG method essentially reduced to the zero-start one for a modified equation.

Proposition 7.41 (equivalence). *Suppose that* $\mathsf{A} \in \mathbb{C}^{n\times n}$ *is HPD,* $\boldsymbol{f} \in \mathbb{C}^n_\star$, $\boldsymbol{x} = \mathsf{A}^{-1}\boldsymbol{f}$, *and* $\boldsymbol{x}_0 \in \mathbb{C}^n$. *Set*

$$\boldsymbol{x}' = \boldsymbol{x} - \boldsymbol{x}_0, \qquad \boldsymbol{r}_0 = \boldsymbol{f} - \mathsf{A}\boldsymbol{x}_0.$$

The sequence $\{\boldsymbol{x}'_k\}_{k=1}^\infty$ *generated by the zero-start CG algorithm to approximate the solution to* $\mathsf{A}\boldsymbol{x}' = \boldsymbol{r}_0$ *is equivalent to the sequence* $\{\boldsymbol{x}_k\}_{k=1}^\infty$ *generated by the (standard) CG algorithm with the starting vector* $\boldsymbol{x}_0$ *to approximate the solution to* $\mathsf{A}\boldsymbol{x} = \boldsymbol{f}$ *in the sense that*

$$\boldsymbol{x}_k = \boldsymbol{x}_0 + \boldsymbol{x}'_k, \qquad \boldsymbol{x} - \boldsymbol{x}_k = \boldsymbol{x}' - \boldsymbol{x}'_k.$$

Furthermore, as long as $\boldsymbol{x}_0 \neq \boldsymbol{x}$, *there is an integer* $m_\star \in \{1, \ldots, n\}$ *such that*

$$\boldsymbol{x}_k \neq \boldsymbol{x}, \quad k = 1, \ldots, m_\star - 1, \qquad \boldsymbol{x}_k = \boldsymbol{x}, \quad k \geq m_\star.$$

Proof. Recall that

$$E_\mathsf{A}(\boldsymbol{z}) = \frac{1}{2}(\mathsf{A}\boldsymbol{z}, \boldsymbol{z})_2 - \Re\left((\boldsymbol{f}, \boldsymbol{z})_2\right), \quad \forall \boldsymbol{z} \in \mathbb{C}^n.$$

Define

$$\tilde{E}_\mathsf{A}(\boldsymbol{z}) = \frac{1}{2}(\mathsf{A}\boldsymbol{z}, \boldsymbol{z})_2 - \Re\left((\boldsymbol{r}_0, \boldsymbol{z})_2\right), \quad \forall \boldsymbol{z} \in \mathbb{C}^n,$$

and observe that

$$E_\mathsf{A}(\boldsymbol{z} + \boldsymbol{x}_0) = \tilde{E}_\mathsf{A}(\boldsymbol{z}) + E_\mathsf{A}(\boldsymbol{x}_0).$$

It follows that

$$\boldsymbol{x}'_k = \operatorname*{argmin}_{\boldsymbol{z} \in \mathcal{K}_k} E_\mathsf{A}(\boldsymbol{z} + \boldsymbol{x}_0) = \operatorname*{argmin}_{\boldsymbol{z} \in \mathcal{K}_k} \tilde{E}_\mathsf{A}(\boldsymbol{z}).$$

The rest of the details are left to the reader as an exercise; see Problem 7.21. □

Remark 7.42 (equivalence). This result states that approximating $\mathsf{A}\boldsymbol{x} = \boldsymbol{f}$ using the (standard) CG algorithm is exactly equivalent to approximating $\mathsf{A}\boldsymbol{x}' = \boldsymbol{r}_0$ using the zero-start CG algorithm. In short, their convergence properties are the same. Of course, often it is advantageous to use a nonzero starting vector, especially in the case where one already has a good approximation to the exact solution of an equation of interest.

7.4.2 Preconditioned Conjugate Gradient Method

As was the case for the steepest descent method, we can discuss a preconditioning strategy for the CG method, though the details are somewhat more complicated. We begin by collecting some simple facts.

Proposition 7.43 (some useful facts). *Suppose that* $\mathsf{A}, \mathsf{B} \in \mathbb{C}^{n\times n}$ *are HPD,* $\boldsymbol{f} \in \mathbb{C}^n_\star$, *and* $\boldsymbol{x} = \mathsf{A}^{-1}\boldsymbol{f} \in \mathbb{C}^n_\star$. *Let* $\mathsf{B} = \mathsf{L}^\mathsf{H}\mathsf{L}$ *be a Cholesky-type factorization for* B, *where* $\mathsf{L} \in \mathbb{C}^{n\times n}$ *is invertible. Define*

$$\mathsf{C} = \mathsf{L}^{-\mathsf{H}}\mathsf{A}\mathsf{L}^{-1}.$$

Then C *is HPD and* $B^{-1}A$ *is similar to* C*. Consequently,*

$$\sigma(B^{-1}A) = \sigma(C) \subset (0, \infty).$$

Furthermore, the following problems are equivalent.

1. *Find* $\boldsymbol{x} \in \mathbb{C}^n$ *such that*

$$A\boldsymbol{x} = \boldsymbol{f}.$$

2. *Find* $\boldsymbol{x} \in \mathbb{C}^n$ *such that*

$$B^{-1}A\boldsymbol{x} = B^{-1}\boldsymbol{f}.$$

3. *Find* $\boldsymbol{x} \in \mathbb{C}^n$ *such that*

$$L\boldsymbol{x} = \boldsymbol{y}, \quad C\boldsymbol{y} = \boldsymbol{q},$$

   where $\boldsymbol{q} = L^{-H}\boldsymbol{f}$*.*

The equation $C\boldsymbol{y} = \boldsymbol{q}$ *is called the* preconditioned system.

Proof. See Problem 7.22. □

The preconditioned conjugate gradient (PCG) method is a method for solving $A\boldsymbol{x} = \boldsymbol{f}$ by utilizing the preconditioned system $C\boldsymbol{y} = \boldsymbol{q}$.

Definition 7.44 (PCG)**.** Suppose that $A, B \in \mathbb{C}^{n\times n}$ are HPD, $\boldsymbol{f} \in \mathbb{C}^n_\star$, and $\boldsymbol{x} = A^{-1}\boldsymbol{f}$. Assume that C, L, $\boldsymbol{q}$, and $\boldsymbol{y}$ are as defined in Proposition 7.43. The **zero-start** B-**preconditioned conjugate gradient (PCG) method** is an iterative scheme for producing the sequence $\{\boldsymbol{y}_k\}_{k=1}^\infty$ by applying the standard zero-start CG method to approximate the solution to the preconditioned system, $C\boldsymbol{y} = \boldsymbol{q}$. The sequence of approximations for the solution of interest, $\boldsymbol{x}$, denoted $\{\boldsymbol{x}_k\}_{k=1}^\infty$, is defined by $\boldsymbol{x}_k = L^{-1}\boldsymbol{y}_k$.

In exact arithmetic, the B-PCG algorithm terminates in a finite number of steps. We will address this point momentarily. Using Corollary 7.35 and Theorem 7.39, we immediately get an equivalent formulation of the PCG method.

Proposition 7.45 (equivalence I)**.** *Suppose that* $A, B \in \mathbb{C}^{n\times n}$ *are HPD,* $\boldsymbol{f} \in \mathbb{C}^n_\star$*, and* $\boldsymbol{x} = A^{-1}\boldsymbol{f}$*. Assume that* C, L, $\boldsymbol{q}$*, and* $\boldsymbol{y}$ *are as defined in Proposition 7.43. The sequence* $\{\boldsymbol{y}_k\}_{k=1}^\infty$*, generated by the zero-start* B*-PCG method, is the same sequence as that generated by the following recursive algorithm: given* $\boldsymbol{y}_0 = \boldsymbol{0}$*, define* $\boldsymbol{s}_0 = \boldsymbol{q} - C\boldsymbol{y}_0 = \boldsymbol{q}$ *and* $\boldsymbol{d}_0 = \boldsymbol{s}_0 = \boldsymbol{q}$*. For* $k \geq 0$*, compute:*

1. *Update the iterate:*

$$\boldsymbol{y}_{k+1} = \boldsymbol{y}_k + \theta_{k+1}\boldsymbol{d}_k, \quad \theta_{k+1} = \frac{(\boldsymbol{s}_k, \boldsymbol{s}_k)_2}{(C\boldsymbol{d}_k, \boldsymbol{d}_k)_2}.$$

2. *Update the residual:*

$$\boldsymbol{s}_{k+1} = \boldsymbol{s}_k - \theta_{k+1}C\boldsymbol{d}_k.$$

3. *Update the search direction:*

$$\boldsymbol{d}_{k+1} = \boldsymbol{s}_{k+1} + \nu_{k+1}\boldsymbol{d}_k, \quad \nu_{k+1} = \frac{(\boldsymbol{s}_{k+1}, \boldsymbol{s}_{k+1})_2}{(\boldsymbol{s}_k, \boldsymbol{s}_k)_2}.$$

4. *If $\boldsymbol{s}_{k+1} = \boldsymbol{0}$, stop. Otherwise, index k and go to step 1.*

Define $\mathcal{M}_k = \mathcal{K}_k(\mathsf{C}, \boldsymbol{q})$. Then there is an integer $m_\star^{\mathsf{C}} \in \{1, \ldots, n\}$ such that

$$\boldsymbol{s}_{m_\star^{\mathsf{C}}} = \boldsymbol{0} = \boldsymbol{d}_{m_\star^{\mathsf{C}}}$$

and

$$\boldsymbol{0} \neq \boldsymbol{s}_k \in \mathcal{M}_{k+1} \backslash \mathcal{M}_k, \quad \boldsymbol{0} \neq \boldsymbol{d}_k \in \mathcal{M}_{k+1} \backslash \mathcal{M}_k, \quad k = 0, \ldots, m_\star^{\mathsf{C}} - 1.$$

Furthermore, the following orthogonalities hold:

$$(\boldsymbol{s}_j, \boldsymbol{s}_i)_2 = (\boldsymbol{d}_j, \boldsymbol{d}_i)_{\mathsf{C}} = 0$$

for all $0 \leq j < i \leq m_\star^{\mathsf{C}} - 1$. Finally, the following convergence estimate holds:

$$\|\boldsymbol{y} - \boldsymbol{y}_k\|_{\mathsf{C}} \leq 2 \left(\frac{\sqrt{\kappa_2(\mathsf{C})} - 1}{\sqrt{\kappa_2(\mathsf{C})} + 1} \right)^k \|\boldsymbol{y} - \boldsymbol{y}_0\|_{\mathsf{C}}.$$

Proof. See Problem 7.23. □

Corollary 7.46 (equivalence II)**.** *With the same assumptions and notation as in Proposition 7.45, define, for $0 \leq k \leq m_\star^{\mathsf{C}}$,*

$$\boldsymbol{x}_k = \mathsf{L}^{-1} \boldsymbol{y}_k,$$
$$\boldsymbol{r}_k = \mathsf{L}^{\mathsf{H}} \boldsymbol{s}_k,$$
$$\boldsymbol{p}_k = \mathsf{L}^{-1} \boldsymbol{d}_k.$$

These vectors may be generated directly by the following recursive algorithm: $\boldsymbol{x}_0 = \boldsymbol{0}$, $\boldsymbol{r}_0 = \boldsymbol{f}$, and $\boldsymbol{p}_0 = \mathsf{B}^{-1}\boldsymbol{f}$. For $0 \leq k \leq m_\star^{\mathsf{C}} - 1$,

1. *Update the iterate:*

$$\boldsymbol{x}_{k+1} = \boldsymbol{x}_k + \theta_{k+1} \boldsymbol{p}_k, \quad \theta_{k+1} = \frac{(\mathsf{B}^{-1} \boldsymbol{r}_k, \boldsymbol{r}_k)_2}{(\mathsf{A} \boldsymbol{p}_k, \boldsymbol{p}_k)_2}.$$

2. *Update the residual:*

$$\boldsymbol{r}_{k+1} = \boldsymbol{r}_k - \theta_{k+1} \mathsf{A} \boldsymbol{p}_k.$$

3. *Update the search direction:*

$$\boldsymbol{p}_{k+1} = \mathsf{B}^{-1} \boldsymbol{r}_{k+1} + \nu_k \boldsymbol{p}_k, \quad \nu_{k+1} = \frac{(\mathsf{B}^{-1} \boldsymbol{r}_{k+1}, \boldsymbol{r}_{k+1})_2}{(\mathsf{B}^{-1} \boldsymbol{r}_k, \boldsymbol{r}_k)_2}.$$

4. *If $k = m_\star^{\mathsf{C}} - 1$, stop. Otherwise, index k and go to step 1.*

The following error estimate is valid:

$$\|\boldsymbol{x} - \boldsymbol{x}_k\|_{\mathsf{A}} \leq 2 \left(\frac{\sqrt{\kappa_2(\mathsf{C})} - 1}{\sqrt{\kappa_2(\mathsf{C})} + 1} \right)^k \|\boldsymbol{x} - \boldsymbol{x}_0\|_{\mathsf{A}}.$$

Proof. See Problem 7.24. □

Remark 7.47 (equivalences). To summarize, the practical version of the B-PCG algorithm is defined by the recursive algorithm given in Corollary 7.46. It is equivalent, as we have seen, to applying the zero-start CG algorithm to the preconditioned system $\mathsf{C}\boldsymbol{y} = \boldsymbol{q}$. As with the preconditioned steepest decent method, the rate of convergence can increase if we can find a preconditioner matrix $\mathsf{B} \approx \mathsf{A}$ such that the condition number of $\mathsf{B}^{-1}\mathsf{A}$ or, equivalently, the condition number of C is sufficiently small. An implementation of the algorithm of Corollary 7.46 is given in Listing 7.1.

7.5 The Conjugate Gradient Method as a Three-Layer Scheme

To conclude the discussion of the CG method we begin with a very important observation. Since $\boldsymbol{x}_k - \boldsymbol{x}_0$, according to Proposition 7.41, is the Galerkin approximation to $\boldsymbol{x} - \boldsymbol{x}_0$ over $\mathcal{K}_k$, to compute $\boldsymbol{x}_k$ we would need to have at hand a basis for $\mathcal{K}_k$, i.e., we need to store k vectors; see Theorem 7.31. It is remarkable, however, that owing to the equivalences described in Corollary 7.35 one only needs to remember the previous two residuals and search directions. This means that the CG method can be seen as a three-layer scheme. We will explore this in more detail. The presentation in this section closely follows [80] and, for simplicity, is done under the assumption that we are operating over the reals.

Consider

$$\mathsf{B}\frac{(\boldsymbol{x}_{k+1} - \boldsymbol{x}_k) + (1 - \alpha_{k+1})(\boldsymbol{x}_k - \boldsymbol{x}_{k-1})}{\tau_{k+1}\alpha_{k+1}} + \mathsf{A}\boldsymbol{x}_k = \boldsymbol{f},$$

where we will choose the iterative parameters α_{k+1} and τ_{k+1}. To start this procedure we need $\boldsymbol{x}_0$, which can be chosen arbitrarily, and $\boldsymbol{x}_1$, which we will compute by

$$\mathsf{B}\frac{\boldsymbol{x}_1 - \boldsymbol{x}_0}{\tau_1} + \mathsf{A}\boldsymbol{x}_0 = \boldsymbol{f}.$$

From these formulas we can see that the error $\boldsymbol{e}_k = \boldsymbol{x} - \boldsymbol{x}_k$ satisfies

$$\begin{aligned} \boldsymbol{e}_{k+1} &= \alpha_{k+1}(\mathsf{I} - \tau_{k+1}\mathsf{B}^{-1}\mathsf{A})\boldsymbol{e}_k + (1 - \alpha_{k+1})\boldsymbol{e}_{k-1}, \\ \boldsymbol{e}_1 &= (\mathsf{I} - \tau_1\mathsf{B}^{-1}\mathsf{A})\boldsymbol{e}_0. \end{aligned}$$

Let us now introduce the change of variables $\boldsymbol{v}_k = \mathsf{A}^{1/2}\boldsymbol{e}_k$, so that $\|\boldsymbol{v}_k\|_2 = \|\boldsymbol{e}_k\|_{\mathsf{A}}$. Then, upon defining $\mathsf{C} = \mathsf{A}^{1/2}\mathsf{B}^{-1}\mathsf{A}^{1/2}$, we see that

$$\begin{aligned} \boldsymbol{v}_{k+1} &= \alpha_{k+1}(\mathsf{I} - \tau_{k+1}\mathsf{C})\boldsymbol{v}_k + (1 - \alpha_{k+1})\boldsymbol{v}_{k-1}, \\ \boldsymbol{v}_1 &= (\mathsf{I} - \tau_1\mathsf{C})\boldsymbol{v}_0. \end{aligned}$$

Using these formulas recursively, we conclude that the error satisfies

$$\boldsymbol{v}_k = p_k(\mathsf{C})\boldsymbol{v}_0$$

for some polynomial p_k of degree k with $p_k(0) = 1$.

We will now choose the iterative parameters α_{k+1} and τ_{k+1} in order to minimize $\|\boldsymbol{v}_k\|_2 = \|\boldsymbol{e}_k\|_{\mathsf{A}}$. Notice that this is different than the Chebyshev iterations described in Section 6.10.1, in that there the error is minimized *only* after m iterations. By contrast, here, we are minimizing it in *every* iteration. From this condition, it immediately follows that we must choose τ_1 so as to minimize $\|\boldsymbol{v}_1\|_2$; thus,

$$\tau_1 = \frac{(\mathsf{C}\boldsymbol{v}_0, \boldsymbol{v}_0)_2}{\|\mathsf{C}\boldsymbol{v}_0\|_2^2} = \frac{(\mathsf{B}^{-1}\boldsymbol{r}_0, \boldsymbol{r}_0)_2}{\|\mathsf{B}^{-1}\boldsymbol{r}_0\|_{\mathsf{A}}^2}.$$

In doing so, we obtain

$$(\mathsf{C}\boldsymbol{v}_1, \boldsymbol{v}_0)_2 = (\mathsf{C}(\mathsf{I} - \tau_1\mathsf{C})\boldsymbol{v}_0, \boldsymbol{v}_0)_2 = (\mathsf{C}\boldsymbol{v}_0, \boldsymbol{v}_0)_2 - \frac{(\mathsf{C}\boldsymbol{v}_0, \boldsymbol{v}_0)_2}{\|\mathsf{C}\boldsymbol{v}_0\|_2^2}(\mathsf{C}^2\boldsymbol{v}_0, \boldsymbol{v}_0)_2 = 0,$$

or that $\boldsymbol{v}_1$ and $\boldsymbol{v}_0$ are orthogonal in the C-inner product.

Now, for $k > 1$, we will write

$$p_k(\mathsf{C}) = \mathsf{I}_n + \sum_{i=1}^{k} \hat{a}_{k,i}\mathsf{C}^i,$$

where the coefficients $\hat{a}_{k,i}$ are defined by the parameters α_i and τ_i for $i = 1, \dots, k$. Therefore,

$$\boldsymbol{v}_k = \boldsymbol{v}_0 + \sum_{i=1}^{k} \hat{a}_{k,i}\mathsf{C}^i\boldsymbol{v}_0,$$

which implies that

$$\|\boldsymbol{v}_k\|_2^2 = \|\boldsymbol{v}_0\|_2^2 + 2\sum_{j=1}^{k} \hat{a}_{k,j}(\mathsf{C}^j\boldsymbol{v}_0, \boldsymbol{v}_0)_2 + \sum_{i,j=1}^{k} \hat{a}_{k,i}\hat{a}_{k,j}(\mathsf{C}^i\boldsymbol{v}_0, \mathsf{C}^j\boldsymbol{v}_0)_2.$$

To minimize the error, we choose the coefficients of p_k from the conditions

$$\frac{1}{2}\frac{\partial\|\boldsymbol{v}_k\|_2^2}{\partial\hat{a}_{k,j}} = (\mathsf{C}^j\boldsymbol{v}_0, \boldsymbol{v}_0)_2 + \sum_{i=1}^{k} \hat{a}_{k,i}(\mathsf{C}^i\boldsymbol{v}_0, \mathsf{C}^j\boldsymbol{v}_0)_2 = 0, \quad j = 1, \dots, k. \tag{7.16}$$

Thus, from now on, we will study the solvability of the system (7.16). We begin by observing that this equation is equivalent to the orthogonality condition

$$(\mathsf{C}^m\boldsymbol{v}_0, \boldsymbol{v}_k)_2 = 0, \quad m = 1, \dots, k. \tag{7.17}$$

Lemma 7.48 (orthogonality). *Condition* (7.17) *is equivalent to the condition*

$$(\mathsf{C}\boldsymbol{v}_m, \boldsymbol{v}_k)_2 = 0, \quad m = 0, \dots, k-1. \tag{7.18}$$

Proof. Since $\boldsymbol{v}_m = p_m(\mathsf{C})\boldsymbol{v}_0$, we can write

$$\mathsf{C}\boldsymbol{v}_m = \mathsf{C}\boldsymbol{v}_0 + \sum_{i=1}^{m} \hat{a}_{m,i}\mathsf{C}^{i+1}\boldsymbol{v}_0$$

and then

$$
\begin{aligned}
(C\boldsymbol{v}_m, \boldsymbol{v}_k)_2 &= (C\boldsymbol{v}_0, \boldsymbol{v}_k)_2 + \sum_{i=1}^{m} \hat{a}_{m,i}(C^{i+1}\boldsymbol{v}_0, \boldsymbol{v}_k)_2 \\
&= (C\boldsymbol{v}_0, \boldsymbol{v}_k)_2 + \sum_{i=2}^{m+1} \hat{a}_{m,i-1}(C^{i}\boldsymbol{v}_0, \boldsymbol{v}_k)_2.
\end{aligned}
$$

From this, it follows that if (7.17) holds, then (7.18) must follow.

To show the reverse implication, we proceed by induction on m. For $m = 1$, condition (7.18) coincides with condition (7.17), so that there is nothing to prove. Now, if (7.17) holds for all $j \leq m$, let us show that this implies that $(C^{m+1}\boldsymbol{v}_0, \boldsymbol{v}_k)_2 = 0$. Since, by (7.18),

$$
\begin{aligned}
0 &= (C\boldsymbol{v}_m, \boldsymbol{v}_k)_2 \\
&= (Cp_m(C)\boldsymbol{v}_0, \boldsymbol{v}_k)_2 \\
&= \left(C\boldsymbol{v}_0 + \sum_{i=1}^{m} \hat{a}_{m,i}C^{i+1}\boldsymbol{v}_0, \boldsymbol{v}_k\right)_2 \\
&= (C\boldsymbol{v}_0, \boldsymbol{v}_k)_2 + \sum_{i=2}^{m} \hat{a}_{m,i-1}(C^{i}\boldsymbol{v}_0, \boldsymbol{v}_k)_2 + \hat{a}_{m,m}(C^{m+1}\boldsymbol{v}_0, \boldsymbol{v}_k)_2 \\
&= \hat{a}_{m,m}(C^{m+1}\boldsymbol{v}_0, \boldsymbol{v}_k)_2,
\end{aligned}
$$

where in the last step we used (7.17). But $\hat{a}_{m,m} \neq 0$, since $\deg p_m = m$. □

We must remark that the number k in Lemma 7.48 is fixed, while we need $\|\boldsymbol{v}_k\|_2$ to be minimized for *every* k. In light of the equivalence given in Lemma 7.48, we will choose the iteration parameters using the condition

$$
(C\boldsymbol{v}_m, \boldsymbol{v}_k)_2 = 0, \quad k = 1, 2, \ldots, \quad m = 0, 1, \ldots, k-1.
$$

However, this shows that if we are able to find parameters then we are constructing a system of vectors that is orthogonal in the C-inner product. For this reason, we *necessarily* have that, after *at most* n steps, $\boldsymbol{v}_n = \boldsymbol{0}$ and so we have found the exact solution to the system.

After this observation, we can construct the iteration parameters. As mentioned before,

$$
\alpha_1 = 1, \qquad \tau_1 = \frac{(C\boldsymbol{v}_0, \boldsymbol{v}_0)_2}{\|C\boldsymbol{v}_0\|_2^2}.
$$

If we have already found $\alpha_1, \ldots, \alpha_k$ and $\tau_1, \ldots, \tau_k$, then from the orthogonality condition and the error representation we must have, for $m \leq k-2$,

$$
\begin{aligned}
(\boldsymbol{v}_{k+1}, C\boldsymbol{v}_m)_2 &= \alpha_{k+1}(\boldsymbol{v}_k, C\boldsymbol{v}_m)_2 - \alpha_{k+1}\tau_{k+1}(C\boldsymbol{v}_k, C\boldsymbol{v}_m)_2 \\
&\quad + (1-\alpha_{k+1})(\boldsymbol{v}_{k-1}, C\boldsymbol{v}_m)_2 \\
&= -\alpha_{k+1}\tau_{k+1}(C\boldsymbol{v}_k, C\boldsymbol{v}_m)_2.
\end{aligned}
$$

But, since

$$\mathsf{C}\boldsymbol{v}_m = \frac{1}{\tau_{m+1}}\boldsymbol{v}_m - \frac{1}{\alpha_{m+1}\tau_{m+1}}\left[\boldsymbol{v}_{m+1} - \boldsymbol{v}_{m-1}\right],$$

it follows that

$$(\mathsf{C}\boldsymbol{v}_k, \mathsf{C}\boldsymbol{v}_m)_2 = 0.$$

It remains then to verify the orthogonality condition for $m = k-1, k$. For $m = k-1$, we obtain

$$\begin{aligned} 0 &= (\boldsymbol{v}_{k+1}, \mathsf{C}\boldsymbol{v}_{k-1})_2 \\ &= \alpha_{k+1}(\boldsymbol{v}_k, \mathsf{C}\boldsymbol{v}_{k-1})_2 - \alpha_{k+1}\tau_{k+1}(\mathsf{C}\boldsymbol{v}_k, \mathsf{C}\boldsymbol{v}_{k-1})_2 + (1-\alpha_{k+1})(\boldsymbol{v}_{k-1}, \mathsf{C}\boldsymbol{v}_{k-1})_2 \\ &= -\alpha_{k+1}\tau_{k+1}(\mathsf{C}\boldsymbol{v}_k, \mathsf{C}\boldsymbol{v}_{k-1})_2 + (1-\alpha_{k+1})(\boldsymbol{v}_{k-1}, \mathsf{C}\boldsymbol{v}_{k-1})_2 \end{aligned}$$

and, for $m = k$, we obtain

$$\begin{aligned} 0 &= (\boldsymbol{v}_{k+1}, \mathsf{C}\boldsymbol{v}_k)_2 \\ &= \alpha_{k+1}(\boldsymbol{v}_k, \mathsf{C}\boldsymbol{v}_k)_2 - \alpha_{k+1}\tau_{k+1}(\mathsf{C}\boldsymbol{v}_k, \mathsf{C}\boldsymbol{v}_k)_2 + (1-\alpha_{k+1})(\boldsymbol{v}_{k-1}, \mathsf{C}\boldsymbol{v}_k)_2 \\ &= \alpha_{k+1}(\boldsymbol{v}_k, \mathsf{C}\boldsymbol{v}_k)_2 - \alpha_{k+1}\tau_{k+1}(\mathsf{C}\boldsymbol{v}_k, \mathsf{C}\boldsymbol{v}_k)_2. \end{aligned}$$

Solving this system yields

$$\tau_{k+1} = \frac{(\mathsf{C}\boldsymbol{v}_k, \boldsymbol{v}_k)_2}{\|\mathsf{C}\boldsymbol{v}_k\|_2^2} = \frac{(\mathsf{B}^{-1}\boldsymbol{r}_k, \boldsymbol{r}_k)_2}{\|\mathsf{B}^{-1}\boldsymbol{r}_k\|_{\mathsf{A}}^2} \tag{7.19}$$

and

$$\begin{aligned} \alpha_{k+1} &= \left[1 - \frac{\tau_{k+1}}{\tau_k}\frac{1}{\alpha_k}\frac{(\mathsf{C}\boldsymbol{v}_k, \boldsymbol{v}_k)_2}{(\mathsf{C}\boldsymbol{v}_{k-1}, \boldsymbol{v}_{k-1})_2}\right]^{-1} \\ &= \left[1 - \frac{\tau_{k+1}}{\tau_k}\frac{1}{\alpha_k}\frac{(\mathsf{B}^{-1}\boldsymbol{r}_k, \boldsymbol{r}_k)_2}{(\mathsf{B}^{-1}\boldsymbol{r}_{k-1}, \boldsymbol{r}_{k-1})_2}\right]^{-1}. \end{aligned} \tag{7.20}$$

7.6 Krylov Subspace Methods for Non-HPD Problems

As a final section in our discussion of variational and Krylov subspace methods we consider the linear system $\mathsf{A}\boldsymbol{x} = \boldsymbol{f}$ in the case that the system matrix A is not necessarily HPD. In this case, the solution of the linear system may not minimize E_{A}. In fact, this quadratic function may not have a minimum! We must, then, consider different properties of the solution.

7.6.1 Methods Based on the Normal Equation

A first attempt at solving a system with a non-HPD matrix is to transform it into a problem with a matrix that is HPD and apply to it the CG method that we already discussed. We begin by recalling that, since A is assumed to be nonsingular, the matrix $\mathsf{A}^{\mathsf{H}}\mathsf{A}$ is HPD and $\boldsymbol{x}$ solves $\mathsf{A}\boldsymbol{x} = \boldsymbol{f}$ if and only if it solves

$$\mathsf{A}^{\mathsf{H}}\mathsf{A}\boldsymbol{x} = \mathsf{A}^{\mathsf{H}}\boldsymbol{f}. \tag{7.21}$$

This gives rise to the following method.

Definition 7.49 (CGNR). Let $\mathsf{A} \in \mathbb{C}^{n\times n}$ be nonsingular, $\boldsymbol{f} \in \mathbb{C}^n_\star$, and $\boldsymbol{x}_0 \in \mathbb{C}^n$ be arbitrary. The sequence $\{\boldsymbol{x}_k\}_{k=1}^\infty$ obtained by applying the CG method of Definition 7.40 to the system (7.21) is called the **conjugate gradient normal equation residual (CGNR) method**.

The following is an immediate consequence of Theorem 7.39.

Corollary 7.50 (convergence of CGNR). *Let $\mathsf{A} \in \mathbb{C}^{n\times n}$ be nonsingular, $\boldsymbol{f} \in \mathbb{C}^n_\star$, and $\boldsymbol{x}_0 \in \mathbb{C}^n$ be arbitrary. The sequence $\{\boldsymbol{x}_k\}_{k=1}^\infty$ obtained by the CGNR method is such that the error $\boldsymbol{e}_k = \boldsymbol{x} - \boldsymbol{x}_k$ satisfies*

$$\|\boldsymbol{e}_k\|_2 \leq 2\kappa_2(\mathsf{A}) \left(\frac{\kappa_2(\mathsf{A}) - 1}{\kappa_2(\mathsf{A}) + 1}\right)^k \|\boldsymbol{e}_k\|_2,$$

where $\kappa_2(\mathsf{A})$ is the spectral condition number of A.

Proof. See Problem 7.31. □

A motivation for the naming of this method is explored in Problem 7.32.

We can consider another system with an HPD coefficient matrix. Since A is nonsingular, the vector $\boldsymbol{y} \in \mathbb{C}^n$ solves

$$\mathsf{A}\mathsf{A}^\mathsf{H}\boldsymbol{y} = \boldsymbol{f} \tag{7.22}$$

if and only if $\boldsymbol{x} = \mathsf{A}^\mathsf{H}\boldsymbol{y}$ solves $\mathsf{A}\boldsymbol{x} = \boldsymbol{f}$. The matrix of the previous system is HPD, and so we can apply to it the CG method, giving rise to the so-called CGNE scheme.

Definition 7.51 (CGNE). Let $\mathsf{A} \in \mathbb{C}^{n\times n}$ be nonsingular, $\boldsymbol{f} \in \mathbb{C}^n_\star$, and $\boldsymbol{y}_0 \in \mathbb{C}^n$ be arbitrary. Let $\{\boldsymbol{y}_k\}_{k=1}^\infty$ be the sequence obtained by applying the CG method of Definition 7.40 to system (7.22). Set $\boldsymbol{x}_k = \mathsf{A}^\mathsf{H}\boldsymbol{y}_k$. This gives rise to the **conjugate gradient normal equation error (CGNE) method**.

The reason for the naming of this method is explored in Problem 7.33. The following convergence result is, again, an immediate consequence of Theorem 7.39.

Corollary 7.52 (convergence of CGNE). *Let $\mathsf{A} \in \mathbb{C}^{n\times n}$ be nonsingular, $\boldsymbol{f} \in \mathbb{C}^n_\star$, and $\boldsymbol{y}_0 \in \mathbb{C}^n$ be arbitrary. The sequence $\{\boldsymbol{x}_k\}_{k=1}^\infty$ obtained by the CGNE method is such that the error $\boldsymbol{e}_k = \boldsymbol{x} - \boldsymbol{x}_k$ satisfies*

$$\|\boldsymbol{e}_k\|_2 \leq 2 \left(\frac{\kappa_2(\mathsf{A}) - 1}{\kappa_2(\mathsf{A}) + 1}\right)^k \|\boldsymbol{e}_k\|_2,$$

where $\kappa_2(\mathsf{A})$ is the spectral condition number of A.

Proof. See Problem 7.34. □

7.6.2 The Generalized Minimization of the Residual (GMRES) Method

As we observe from Corollaries 7.50 and 7.52, although methods based on the normal equation are rather convenient, they suffer from a slower convergence rate. The condition number of the system is effectively squared. Here, we explore the so-called generalized minimization of the residual (GMRES) method, which does not suffer from this issue.

Definition 7.53 (GMRES). Let $\mathsf{A} \in \mathbb{C}^{n\times n}$ be nonsingular, $\boldsymbol{f} \in \mathbb{C}^n_\star$, and $\boldsymbol{x}_0 \in \mathbb{C}^n$ be arbitrary. The sequence $\{\boldsymbol{x}_k\}_{k=1}^\infty$, obtained by minimizing

$$\|\boldsymbol{r}(\boldsymbol{z})\|_2 = \|\boldsymbol{f} - \mathsf{A}\boldsymbol{z}\|_2$$

over

$$\boldsymbol{x}_0 + \mathcal{K}_k(\mathsf{A}, \boldsymbol{f} - \mathsf{A}\boldsymbol{x}_0),$$

gives rise to the **generalized minimization of the residual (GMRES) method**.

Before we embark on the study of the properties of this method, we must make some observations regarding its practical implementation. Recall that CG also minimizes over a translation of the Krylov subspace $\mathcal{K}_k$; see Proposition 7.41. Thus, a priori it seemed necessary to construct and store a basis for this space. However, owing to the fact that the system matrix was HPD, this method can be reduced to a three-layer scheme, as shown in Proposition 7.45. For GMRES, however, this property no longer holds, and so an orthonormal basis of $\mathcal{K}_k$ must be computed and stored.

This is a major bottleneck of this method when n is large. For this reason, it is common practice to consider the GMRES method *with restarts*. This means that a number $m \ll n$ is chosen, and the GMRES algorithm with arbitrary initial guess $\boldsymbol{x}_0$ is run for at most m iterations. If the solution, or a suitable approximation, has not been found, the Krylov subspace and its basis are discarded and the GMRES algorithm is run again, but this time the initial guess is $\boldsymbol{x}_m$, the last approximate solution. This procedure is repeated until a suitable tolerance is reached.

Let us now discuss how an orthonormal basis $\langle \boldsymbol{q}_1, \dots, \boldsymbol{q}_m\rangle = \mathcal{K}_m(\mathsf{A}, \boldsymbol{r}_0)$, with $\boldsymbol{r}_0 = \boldsymbol{f} - \mathsf{A}\boldsymbol{x}_0$, can be efficiently constructed and how $\boldsymbol{x}_k$ can be efficiently found. A clever application of the modified Gram–Schmidt algorithm, presented in Section 5.6, can be used to compute an orthonormal basis of $\mathcal{K}_m$. This is known as the *Arnoldi algorithm*[4] and is presented in Listing 7.2. We observe that the algorithm may break down if, at some point, $[\mathsf{H}]_{j+1,j} = 0$. This means that the Krylov subspace is *maximal*, i.e., invariant under A. For this reason, the exact solution satisfies $\boldsymbol{x} - \boldsymbol{x}_0 \in \mathcal{K}_k$. In the literature, this unlikely occurrence is known as a *lucky breakdown*.

Having found a way to compute an orthonormal basis of the Krylov subspaces, we observe that, setting $\mathsf{Q}_k = [\boldsymbol{q}_1, \dots, \boldsymbol{q}_k]$, by construction we have

$$\mathsf{A}\mathsf{Q}_k = \mathsf{Q}_{k+1}\mathsf{H}_k, \tag{7.23}$$

[4] Named in honor of the American engineer Walter Edwin Arnoldi (1917–1995).

where $\mathsf{H}_k \in \mathbb{C}^{k+1,k}$ is the matrix with entries corresponding to the inner products in the algorithm. Consider now how $\boldsymbol{x}_k$ is defined. According to Definition 7.53, we have that

$$\boldsymbol{x}_k \in \operatorname*{argmin}_{z\in\mathcal{K}_k} \|\boldsymbol{f} - \mathsf{A}(\boldsymbol{x}_0 + \boldsymbol{z})\|_2 = \operatorname*{argmin}_{z\in\mathcal{K}_k} \|\boldsymbol{r}_0 - \mathsf{A}\boldsymbol{z}\|_2.$$

Now, since $\boldsymbol{z} \in \mathcal{K}_k$, we can write it as $\boldsymbol{z} = \sum_{i=1}^{k} y_i \boldsymbol{q}_i = \mathsf{Q}_k \boldsymbol{y}$ for some $\boldsymbol{y} \in \mathbb{C}^k$. Substituting, using (7.23) and recalling that $\boldsymbol{r}_0 = \|\boldsymbol{r}_0\|_2 \boldsymbol{q}_1$, we obtain that

$$\begin{aligned}
\|\boldsymbol{r}_0 - \mathsf{A}\boldsymbol{z}\|_2 &= \|\|\boldsymbol{r}_0\|_2\boldsymbol{q}_1 - \mathsf{A}\mathsf{Q}_k\boldsymbol{y}\|_2 \\
&= \|\|\boldsymbol{r}_0\|_2\boldsymbol{q}_1 - \mathsf{Q}_{k+1}\mathsf{H}_k\boldsymbol{y}\|_2 \\
&= \|\mathsf{Q}_{k+1}\left(\|\boldsymbol{r}_0\|_2\boldsymbol{e}_1 - \mathsf{H}_k\boldsymbol{y}\right)\|_2 \\
&= \|\|\boldsymbol{r}_0\|_2\boldsymbol{e}_1 - \mathsf{H}_k\boldsymbol{y}\|_{\ell^2(\mathbb{C}^{k+1})},
\end{aligned}$$

where $\boldsymbol{e}_1$ denotes the first canonical basis vector in $\mathbb{C}^{k+1}$ and, in the last step, we used that Q_{k+1} has orthonormal columns. In conclusion, finding $\boldsymbol{x}_k \in \mathcal{K}_k$ is equivalent to solving the minimization problem

$$\boldsymbol{y} \in \operatorname*{argmin}_{\tilde{\boldsymbol{y}}\in\mathbb{C}^k} \|\|\boldsymbol{r}_0\|_2\boldsymbol{e}_1 - \mathsf{H}_k\tilde{\boldsymbol{y}}\|_{\ell^2(\mathbb{C}^{k+1})} \tag{7.24}$$

and setting $\boldsymbol{x}_k = \boldsymbol{x}_0 + \mathsf{Q}_k\boldsymbol{y}$.

Equation (7.24) defines a linear least squares problem, which we studied in Chapter 5. As Problem 7.36 shows, the matrix H_k is full rank and, consequently, there is a unique $\boldsymbol{y} \in \mathbb{C}^k$ that solves (7.24). This can be found by computing the QR factorization of the matrix H_k using either Householder reflectors (Section 5.7) or Givens rotations (Problem 5.35).

Having briefly discussed the implementation of the GMRES method, we now turn to its convergence properties. As we are minimizing over subspaces of nondecreasing dimension, we must converge in a finite number of iterations.

Proposition 7.54 (minimization). *Let $\mathsf{A} \in \mathbb{C}^{n\times n}$ be nonsingular and $\boldsymbol{f} \in \mathbb{C}^n_\star$ be given. Set $\boldsymbol{x} = \mathsf{A}^{-1}\boldsymbol{f}$. For any $\boldsymbol{x}_0 \in \mathbb{C}^n$, there is an integer $m_\star \in \{1, \dots, n\}$ such that the sequence $\{\boldsymbol{x}_k\}_{k=1}^\infty$ generated by the GMRES method of Definition 7.53 satisfies*

$$\boldsymbol{x}_k \neq \boldsymbol{x}, \quad k = 1, \dots, m_\star - 1, \qquad \boldsymbol{x}_k = \boldsymbol{x}, \quad k \geq m_\star.$$

Proof. Clearly, $\dim \mathcal{K}_k \leq k$. Moreover, if the Arnoldi process does not break down after k steps, then we must have equality. Thus, if the Arnoldi process can reach n steps, $\mathcal{K}_n = \mathbb{C}^n$. On the other hand, if the Arnoldi process breaks down for some $m_\star < n$, then $\mathsf{A}\mathcal{K}_{m_\star} = \mathcal{K}_{m_\star}$. Thus, since

$$\mathsf{A}(\boldsymbol{x} - \boldsymbol{x}_0) = \boldsymbol{r}_0 \in \mathcal{K}_{m_\star},$$

it follows that $\boldsymbol{x} - \boldsymbol{x}_0 \in \mathcal{K}_{m_\star}$ and the minimization procedure yields the exact solution. □

As was the case with CG, we consider GMRES as an iterative scheme. The following result shows its convergence properties.

Theorem 7.55 (convergence of GMRES). *Let* $\mathsf{A} \in \mathbb{C}^{n\times n}$ *be nonsingular and* $\boldsymbol{f} \in \mathbb{C}^n_\star$ *be given. Set* $\boldsymbol{x} = \mathsf{A}^{-1}\boldsymbol{f}$*. Let* $\boldsymbol{x}_0$ *be arbitrary and let the sequence* $\{\boldsymbol{x}_k\}_{k=1}^\infty$ *be generated by the GMRES method of Definition 7.53. The residual* $\boldsymbol{r}_k = \boldsymbol{f} - \mathsf{A}\boldsymbol{x}_k$ *satisfies*

$$\|\boldsymbol{r}_k\|_2 \leq \inf_{p\in\mathbb{P}_{k,\star}} \|p(\mathsf{A})\|_2\|\boldsymbol{r}_0\|_2,$$

where $\mathbb{P}_{k,\star}$ *was defined in Theorem 7.37. In addition, if* A *is diagonalizable* $\mathsf{A} = \mathsf{V}\Lambda\mathsf{V}^{-1}$*, then*

$$\|\boldsymbol{r}_k\|_2 \leq \kappa_2(\mathsf{V}) \inf_{p\in\mathbb{P}_{k,\star}} \max_{\lambda\in\sigma(\mathsf{A})} |p(\lambda)|\|\boldsymbol{r}_0\|_2.$$

Proof. Notice that $\boldsymbol{x}_k - \boldsymbol{x}_0 \in \mathcal{K}_k$, so, as in Theorem 7.37, there is a polynomial $q_{k-1} \in \mathbb{P}_{k-1}$ for which $\boldsymbol{x}_k - \boldsymbol{x}_0 = q_{k-1}(\mathsf{A})\boldsymbol{r}_0$. Therefore,

$$\boldsymbol{r}_k = \boldsymbol{f} - \mathsf{A}\boldsymbol{x}_k = \boldsymbol{r}_0 - \mathsf{A}(\boldsymbol{x}_k - \boldsymbol{x}_0) = (\mathsf{I} - \mathsf{A}q_{k-1}(\mathsf{A}))\,\boldsymbol{r}_0 = p_k(\mathsf{A})\boldsymbol{r}_0,$$

where $p_k \in \mathbb{P}_{k,\star}$. The minimization property of the residual then is equivalent to minimizing over $p \in \mathbb{P}_{k,\star}$, i.e.,

$$\|\boldsymbol{r}_k\|_2 = \inf_{p\in\mathbb{P}_{k,\star}} \|p(\mathsf{A})\boldsymbol{r}_0\|_2,$$

and the first assertion follows.

To show the second estimate, we begin by observing that if $\mathsf{A} = \mathsf{V}\Lambda\mathsf{V}^{-1}$ and p is a polynomial, then

$$p(\mathsf{A}) = \mathsf{V}p(\Lambda)\mathsf{V}^{-1}.$$

Since $\mathsf{A} \asymp \Lambda$ and Λ is diagonal, the second estimate follows from the first one. □

If knowledge about the location of the spectrum of A is available, then the estimate on the convergence rate can be refined as we now discuss. Since A is not assumed to be Hermitian, $\sigma(\mathsf{A}) \subset \mathbb{C}$ in general. Thus, to quantify the location of the eigenvalues, we define, for $c \in \mathbb{R}$, the set

$$\mathcal{E}(c, d, a) \subset \mathbb{C}$$

of points contained in the ellipse with center c, major semiaxis a, and focal distance d. Moreover, we assume that the major semiaxis is parallel to one of the coordinate axes.

The proof of the following result is beyond our scope here, as it requires properties of Chebyshev polynomials which we will only consider in Part II of our text. We refer the reader to [78, Corollary 6.1] for a proof.

Corollary 7.56 (convergence rate). *Let* $\mathsf{A} \in \mathbb{C}^{n\times n}$ *be nonsingular and diagonalizable* $\mathsf{A} = \mathsf{V}\Lambda\mathsf{V}^{-1}$ *and* $\boldsymbol{f} \in \mathbb{C}^n_\star$ *be given. Set* $\boldsymbol{x} = \mathsf{A}^{-1}\boldsymbol{f}$*. Let* $\boldsymbol{x}_0$ *be arbitrary and let the sequence* $\{\boldsymbol{x}_k\}_{k=1}^\infty$ *be generated by the GMRES method of Definition 7.53. Assume that* $\sigma(\mathsf{A}) \subseteq \mathcal{E}(c, d, a)$ *and* $0 \notin \mathcal{E}(c, d, a)$*. Then the residual* $\boldsymbol{r}_k = \boldsymbol{f} - \mathsf{A}\boldsymbol{x}_k$ *satisfies*

$$\|\boldsymbol{r}_k\|_2 \leq \kappa_2(\mathsf{V}) \frac{C_k\left(\frac{a}{d}\right)}{\left|C_k\left(\frac{c}{d}\right)\right|}\|\boldsymbol{r}_0\|_2,$$

where

$$C_k(z) = \left(z + \sqrt{z^2 - 1}\right)^k + \left(z + \sqrt{z^2 - 1}\right)^{-k}.$$

Problems

7.1 Complete the proof of Theorem 7.1.

7.2 Complete the proof of Proposition 7.2.

7.3 Prove Proposition 7.5.

7.4 Let $A \in \mathbb{C}^{n\times n}$ be HPD. Define $\|x\|_A : \mathbb{C}^n \to \mathbb{R}$ via $\|x\|_A = \sqrt{x^H A x}$. Prove that this is a norm and satisfies the estimates

$$\sqrt{\lambda_1}\,\|x\|_2 \le \|x\|_A \le \sqrt{\lambda_n}\,\|x\|_2,$$

where $0 < \lambda_1 \le \cdots \le \lambda_n$ are the eigenvalues of A, with both inequalities attainable (though perhaps not simultaneously) for suitable choices of x.

7.5 Complete the proof of Theorem 7.7.

7.6 Prove Corollary 7.8.

7.7 Prove formula (7.1).

7.8 Suppose that $A \in \mathbb{C}^{n\times n}$ is HPD, $f \in \mathbb{C}^n$ is given, and $x = A^{-1}f$. Consider the quadratic function $E_A\colon \mathbb{C}^n \to \mathbb{R}$ defined by

$$E_A(z) = \frac{1}{2} z^H A z - \Re\left(z^H f\right).$$

Given $x_k \in \mathbb{C}^n$ and a search direction $d_k \in \mathbb{C}^n$, define

$$\alpha_{k+1} = \operatorname*{argmin}_{\alpha\in\mathbb{R}} E_A(x_k + \alpha d_k), \quad x_{k+1} = x_k + \alpha_{k+1} d_k.$$

Consider a method in which the first n search directions $d_0, \ldots, d_{n-1}$ are taken to be the standard unit vectors $e_1, \ldots, e_n$, the next n search directions $d_n, \ldots, d_{2n-1}$ are again taken to be $e_1, \ldots, e_n$, and so on.

a) Show that each group of n steps constitutes one complete iteration of the Gauss–Seidel method.

b) Show that it can happen that $E_A(x_{k+1}) = E_A(x_k)$ even if $x_k \neq x$.

c) Prove that the situation described in the previous part cannot persist for n consecutive steps. In other words, show that, for some k with $x_k \neq x$, $E_A(x_{k+1}) < E_A(x_k)$. Thus, each complete Gauss–Seidel iteration reduces the function E_A. This idea leads to yet another proof that the Gauss–Seidel method applied to a HPD system matrix always converges.

7.9 Prove Proposition 7.14.

7.10 Prove Theorem 7.20.

7.11 Complete the proof of Proposition 7.29.

7.12 Prove Theorem 7.30.

Hint: Follow the ideas in the proof of Theorem 7.1.

7.13 Complete the proof of Theorem 7.32.

7.14 Prove Corollary 7.33.

7.15 Prove Corollary 7.34.

7.16 Prove Corollary 7.35.

7.17 Show that, for the CG method, the following error estimate holds:

$$\|x - x_k\|_2 \leq 2\kappa(A) \left(\frac{\sqrt{\kappa(A)} - 1}{\sqrt{\kappa(A)} + 1} \right)^k \|x - x_0\|_2.$$

7.18 Show that if $\lambda \in \sigma(\mathsf{A})$ and $p \in \mathbb{P}_n$, then $p(\lambda) \in \sigma(p(\mathsf{A}))$.

7.19 Suppose that $\mathsf{A} \in \mathbb{C}^{n\times n}$ is HPD and $\boldsymbol{f} \in \mathbb{C}^n$. For any z, define the *residual* as $\boldsymbol{r}(z) = \boldsymbol{f} - \mathsf{A}z$ and the *error* as $\boldsymbol{e}(z) = \mathsf{A}^{-1}\boldsymbol{r}(z)$. Show that $(\boldsymbol{e}(z), \boldsymbol{r}(z))_2 > 0$ unless $z = \mathsf{A}^{-1}\boldsymbol{f}$.

7.20 Suppose that $\mathsf{A} = \mathsf{I} - \mathsf{B} \in \mathbb{R}^{n\times n}$ is SPD and $\operatorname{rank}(\mathsf{B}) = r < n$. Show that when the CG algorithm is applied to solve $\mathsf{A}\boldsymbol{x} = \boldsymbol{f}$, where $\boldsymbol{f} \in \mathbb{R}^n$, the method converges in at most $r+1$ steps.

7.21 Complete the proof of Proposition 7.41.

7.22 Prove Proposition 7.43.

7.23 Prove Proposition 7.45.

7.24 Prove Corollary 7.46.

7.25 This problem is about preconditioning for the CG method. Let

$$\mathsf{A} = \begin{bmatrix} \mathsf{A}_1 & \mathsf{A}_2 \\ \mathsf{A}_2^\intercal & \mathsf{A}_3 \end{bmatrix}, \quad \mathsf{S} = \begin{bmatrix} \mathsf{A}_1 & \mathsf{O} \\ \mathsf{O} & \mathsf{A}_3 \end{bmatrix},$$

where $\mathsf{A}_1, \mathsf{A}_2, \mathsf{A}_3, \mathsf{O} \in \mathbb{R}^{n\times n}$ and O is the zero matrix. Suppose that A is SPD.

a) Prove that A_1, A_3, and S are also SPD.

b) Let $\mathsf{L}_1\mathsf{L}_1^\intercal = \mathsf{A}_1$, $\mathsf{L}_3\mathsf{L}_3^\intercal = \mathsf{A}_3$, and $\mathsf{B}\mathsf{B}^\intercal = \mathsf{S}$ be the respective Cholesky factorizations. Writing B in terms of L_1 and L_3, prove that

$$\mathsf{C} = \mathsf{B}^{-1}\mathsf{A}\mathsf{B}^{-\intercal} = \begin{bmatrix} \mathsf{I}_n & \mathsf{F} \\ \mathsf{F}^\intercal & \mathsf{I}_n \end{bmatrix},$$

where $\mathsf{F} = \mathsf{L}_1^{-1}\mathsf{A}_2\mathsf{L}_3^{-\intercal}$.
Hint: Start by writing

$$\mathsf{A} = \mathsf{B}\mathsf{B}^\intercal + \begin{bmatrix} \mathsf{O} & \mathsf{A}_2 \\ \mathsf{A}_2^\intercal & \mathsf{O} \end{bmatrix}.$$

c) Let λ_i, $i = 1, \ldots, 2n$, and μ_j, $j = 1, \ldots, n$, be the eigenvalues of C and $\mathsf{F}\mathsf{F}^\intercal$, respectively. Prove that, with the appropriate numbering,

$$\lambda_k = 1 - \sqrt{\mu_k}, \quad \lambda_{k+d} = 1 + \sqrt{\mu_k}, \quad k = 1, \ldots, n.$$

Hint: Calculate $(\mathsf{C} - \mathsf{I})^2$ and find the eigenvalues of this matrix.

d) Let $\operatorname{rank}(\mathsf{A}_2) = r < n$. Prove that $\operatorname{rank}(\mathsf{F}\mathsf{F}^\intercal) = r$.

e) Deduce that C has at most $2r+1$ distinct eigenvalues. In what number of iterations is the CG algorithm (with exact arithmetic) guaranteed to converge if applied to solve $\mathsf{C}\boldsymbol{x} = \boldsymbol{f}$?

7.26 Suppose that $\mathsf{A}, \mathsf{B} \in \mathbb{C}^{n\times n}$ are HPD. Assume that there are positive constants $\gamma_1, \gamma_2 > 0$ such that

$$\gamma_1(\mathsf{B}\boldsymbol{x}, \boldsymbol{x})_2 \leq (\mathsf{A}\boldsymbol{x}, \boldsymbol{x})_2 \leq \gamma_2(\mathsf{B}\boldsymbol{x}, \boldsymbol{x})_2$$

for all $\boldsymbol{x} \in \mathbb{C}^n$. Prove the following.

a) The matrix $B^{-1}A$ is self-adjoint and positive definite with respect to the B inner product, which is defined via

$$(\boldsymbol{x}, \boldsymbol{y})_B = (B\boldsymbol{x}, \boldsymbol{y})_2, \quad \forall \boldsymbol{x}, \boldsymbol{y} \in \mathbb{C}^n.$$

b) The matrix $B^{-1}A$ has positive real eigenvalues:

$$\sigma(B^{-1}A) = \{\mu_1, \ldots, \mu_n\}, \quad 0 < \mu_1 \le \cdots \le \mu_n.$$

c) The condition number of the preconditioned matrix $B^{-1}A$ with respect to the B-norm satisfies

$$\kappa_B(B^{-1}A) = \left\|B^{-1}A\right\|_B \left\|\left(B^{-1}A\right)^{-1}\right\|_B = \frac{\mu_n}{\mu_1}.$$

d) Finally, the following condition number estimate holds:

$$\kappa_B(B^{-1}A) \le \frac{\gamma_2}{\gamma_1}.$$

7.27 Prove that if $A \in \mathbb{C}^{n\times n}$ is HPD with eigenvalues $\lambda_1, \ldots, \lambda_n$, and p is a polynomial, then

$$\|p(A)\|_A = \max_{1\le j\le n} |p(\lambda_j)|.$$

7.28 Suppose that $A \in \mathbb{C}^{n\times n}$ is HPD and $\boldsymbol{f} \in \mathbb{C}^n_\star$ is given. To solve the system $A\boldsymbol{x} = \boldsymbol{f}$, we employ the CG method, starting from the initial guess $\boldsymbol{x}_0 = \boldsymbol{0}$.

a) Suppose that A has only two distinct eigenvalues, $0 < \lambda_1 < \lambda_2$. Prove that, for every $\lambda \in (\lambda_1, \lambda_2)$,

$$\|\boldsymbol{e}_1\|_A \le q(\lambda)\|\boldsymbol{e}_0\|_A, \quad q(\lambda) = \max_{i=1}^{2} \left|1 - \frac{\lambda_i}{\lambda}\right|.$$

b) Prove that

$$\frac{1}{2}(\lambda_1 + \lambda_2) = \operatorname*{argmin}_{\lambda\in(\lambda_1,\lambda_2)} q(\lambda).$$

Hint: $\max\{|a|, |b|\} = \frac{1}{2}|a + b| + \frac{1}{2}|a - b|$.

c) Use the last fact to show that

$$\|\boldsymbol{e}_1\|_A \le \frac{\kappa_2 - 1}{\kappa_2 + 1}\|\boldsymbol{e}_0\|_A,$$

where κ_2 is the spectral condition number of A.

7.29 Let $A \in \mathbb{R}^{n\times n}$ be SPD and $\sigma(A) = \{\lambda_i\}_{i=1}^n$. Assume that $0 < \lambda_1 \le \lambda_2 \le \cdots \le \lambda_n$, so that $\kappa_2(A) = \lambda_n/\lambda_1$. As in the proof of Theorem 7.37, denote by $\mathbb{P}_{k,\star}$ the set of polynomials of degree at most k that have the value one at zero and set

$$\tilde{\mathbb{P}}_{k,\star} = \{p \in \mathbb{P}_{k,\star} | p(\lambda_n) = 0\}.$$

Show that:

a) The error $\{\boldsymbol{e}_k\}_{k\ge0}$ in CG satisfies

$$\|\boldsymbol{e}_k\|_A \le \|\boldsymbol{e}_0\|_A \inf_{p\in\tilde{\mathbb{P}}_{k,\star}} \max_{\lambda\in[\lambda_1,\lambda_{n-1}]} |p(\lambda)|.$$

b) Use the previous estimate to show that the error in CG also satisfies

$$\|e_k\|_A \le \|e_0\|_A \left(\frac{\lambda_n - \lambda_1}{\lambda_n}\right) \inf_{p \in \mathbb{P}_{k-1,\star}} \max_{\lambda \in [\lambda_1, \lambda_{n-1}]} |p(\lambda)|.$$

c) Define $\tilde{\kappa}_2(A) = \lambda_{n-1}/\lambda_1$. From the previous estimate, deduce that

$$\|e_k\|_A \le 2\|e_0\|_A \left(\frac{\sqrt{\tilde{\kappa}_2(A)} - 1}{\sqrt{\tilde{\kappa}_2(A)} + 1}\right)^{k-1}.$$

7.30 When deriving the CG method as a three-layer iterative scheme, we obtained the following formula:

$$x_{k+1} = \alpha_{k+1} x_k + (1 - \alpha_{k+1}) x_{k-1} + \alpha_{k+1} \tau_{k+1} w_k,$$

where $w_k = B^{-1} r_k$ and $r_k = f - Ax_k$. By contrast, when deriving it as the Galerkin solution to $Ax = f$ over Krylov subspaces, we arrived at

$$x_{k+1} = x_k + \theta_{k+1} p_k,$$
$$p_k = w_k + \nu_k p_{k-1}, \quad k > 0, \qquad p_0 = w_0.$$

Show that, indeed, these two methods are equivalent.

7.31 Prove Corollary 7.50.

7.32 Show that the sequence $\{x_k\}_{k \ge 0}$ generated by the CGNR method of Definition 7.49 minimizes

$$\phi_1(z) = \frac{1}{2} z^H A^H A z - \Re\left(z^H A^H f\right)$$

over $x_0 + \mathcal{K}_k(A^H A, f - A^H A x_0)$. Moreover,

$$\phi_1(z) + \frac{1}{2}\Re\left(f^H f\right) = \frac{1}{2}\|f - Az\|_2^2.$$

7.33 Show that the sequence $\{y_k\}_{k \ge 0}$ generated by the CGNE method of Definition 7.51 minimizes E_{AA^H} over $y_0 + \mathcal{K}_k(AA^H, f - AA^H y_0)$.

7.34 Prove Corollary 7.52.

7.35 Let $\{y_k\}_{k \ge 0}$ be generated by the CGNE method of Definition 7.51. Define $x_k = A^H y_k$. Show that the sequence $\{x_k\}_{k \ge 0}$ minimizes

$$\frac{1}{2}\|z - A^{-1} f\|_2^2 + \frac{1}{2}\|A^{-1} f\|_2^2$$

over $x_0 + \mathcal{K}_k(A^H A, A^H(f - Ax_0))$.

7.36 Show that the matrix H_k, defined in (7.23), is full rank.

Listings

```
function [x, its, err] = PCG( A, x0, f, Binv, maxit, tol )
% The preconditioned conjugate gradient method to approximate
% the solution to
%
%   Ax = f
%
% with A HPD.
%
% Input
%   A(1:n,1:n) : the system matrix
%   x0(1:n) : the initial guess
%   f(1:n) : the right hand side vector
%   Binv(1:n,1:n) : the inverse of the preconditioner
%   maxit : the maximal number of iterations
%   tol : the tolerance
%
% Output
%   x(1:n) : the approximate solution to the linear system of
%            equations
%   its : the number of iterations
%   err : = 0, if the tolerance is reached in less than maxit
%         iterations
%         = 1, if the tolerance is not reached
  err = 0;
  x = x0;
  r = f - A*x;
  p = Binv*r;
  z = p;
  initerror = sqrt( r'*p );
  for its=1:maxit
     Ap = A*p;
     denom = p'*Ap;
     if denom < tol
       err = 0;
       return;
     end
     theta = (1./denom)*(z'*p);
     x = x + theta*p;
     r = r - theta*Ap;
     z = Binv*r;
     nu = (1./denom)*(z'*p);
     p = z + nu*p;
     if sqrt( r'*z )/initerror < tol
       err = 0;
       return;
     end
  end
  err = 1;
end
```

Listing 7.1 The preconditioned conjugate gradient method.

```
function [Q, H, err] = ArnoldiGMRES( A, r, m )
% The Arnoldi algorithm to compute an orthonormal basis of
%
%   K_k(A, r) = span{ r, Ar, ..., A^{m-1} r }
%
% the Krylov subspace of order m. This will be eventually used
% in GMRES
%
% Input:
%   A(1:n,1:n) : the system matrix
%   r(1:n) : the initial residual
%   m : the size of the Krylov subspace
%
% Output:
%  Q(1:n, 1:m) : the columns of this matrix are the orthonormal
%                basis
%  H(1:m, 1:m-1) : the upper Hessenberg matrix that is defined
%                  by
%
%                    AQ_k = Q_{k+1} H
%
%                  with Q_i being the first columns of Q
%  err : = 0, if the algorithm proceeded to completion
%        = 1, if H(j+1,j) = 0 at some point
  err = 0;
  n = size(A,1);
  Q = zeros(n,m);
  H = zeros(m,m-1);
  norm_r = norm( r );
  if norm_r < eps( norm_r )
    err = 1;
    return;
  end
  Q(:,1) = r/norm_r;
  for j=1:m-1
    Q(:,j+1) = A*Q(:,j);
    for i=1:j
      H(i,j) = Q(:,i)'*Q(:,j+1);
      Q(:,j+1) = Q(:,j+1) - H(i,j)*Q(:,i);
    end
    norm_r = norm( Q(:,j+1) );
    if norm_r < eps( norm_r )
      err = 1;
      return;
    end
    H(j+1,j) = norm_r;
    Q(:,j+1) = Q(:,j+1)/norm_r;
  end
end
```

Listing 7.2 The Arnoldi algorithm.

8 Eigenvalue Problems

The focus of this chapter will be the eigenvalue problem: given $A \in \mathbb{C}^{n\times n}$, we will be interested in finding pairs $(\lambda, \boldsymbol{x}) \in \mathbb{C} \times \mathbb{C}^n_\star$ such that

$$A\boldsymbol{x} = \lambda \boldsymbol{x}.$$

Standard references for this topic are the classic texts [70, 102]. The possible applications of this problem are so vast that any attempt at listing them here will force us to misrepresent them.

One might argue that this chapter is in the *wrong* part of the book. Indeed, finding eigenvalues requires *nonlinear* methods, and it is a bit misleading to place it in the numerical *linear* algebra part of our discussion. However, this is done for historical reasons.

Why *nonlinear* rather than *linear*? Since, for a $\boldsymbol{x} \neq \boldsymbol{0}$, we have that $(A-\lambda I)\boldsymbol{x} = \boldsymbol{0}$, we necessarily have that $\det(A - \lambda I) = 0$. Recall that the characteristic polynomial of the matrix A is defined by

$$\chi_A(t) = \det(A - tI).$$

This shows that $\lambda \in \sigma(A)$ if and only if $\chi_A(\lambda) = 0$. This suggests a naive approach to the problem at hand: to find the eigenvalues of a matrix, it is enough to find the roots of the characteristic polynomial, a nonlinear process. We now immediately see two issues. To find eigenvalues, following this approach we must:

1. Compute the characteristic polynomial χ_A.
2. Find its roots.

Even if we forget about the complexity, which in general is $\mathcal{O}(n^3)$, of computing the determinant of a square matrix of size n, the first step in this approach is plagued with numerical difficulties, as the following examples illustrate.

Example 8.1 Instability of the determinant: Suppose that

$$A = \begin{bmatrix} 1 & 20 & & & \\ & 2 & 20 & & \\ & & \ddots & \ddots & \\ & & & 9 & 20 \\ & & & & 10 \end{bmatrix}.$$

There are zeros everywhere that there is not a whole number above. It follows that $\sigma(\mathsf{A}) = \{1, \dots, 10\}$. Now, for $0 < \varepsilon \ll 1$, consider also

$$\mathsf{A}_\varepsilon = \begin{bmatrix} 1 & 20 & & & & \\ & 2 & 20 & & & \\ & & \ddots & \ddots & & \\ & & & 9 & 20 \\ \varepsilon & & & & 10 \end{bmatrix}.$$

In other words, we have replaced one zero, in the bottom left corner, with a very small number. In any reasonable norm, $\|\mathsf{A}_\varepsilon\|$ will be close to $\|\mathsf{A}\|$. However, we can choose ε, so that $0 \in \sigma(\mathsf{A}_\varepsilon)$. To see this, consider

$$\begin{aligned} \det(\mathsf{A}_\varepsilon) &= \det \begin{bmatrix} 2 & 20 & & \\ & \ddots & \ddots & \\ & & 9 & 20 \\ & & & 10 \end{bmatrix} - \varepsilon \det \begin{bmatrix} 20 & & & & \\ 2 & 20 & & & \\ & \ddots & \ddots & & \\ & & 9 & 20 & \\ & & & 10 & 20 \end{bmatrix} \\ &= 10! - 20^9 \varepsilon. \end{aligned}$$

Therefore, if

$$\varepsilon = \frac{10!}{20^9} \approx \frac{3 \times 10^6}{5 \times 10^{11}} \approx 7 \times 10^{-6},$$

the matrix A_ε has zero as an eigenvalue.

Example 8.2 The famous *Wilkinson polynomial*[1] is defined as

$$p_W(t) = \prod_{i=1}^{20} (t - i).$$

The roots of this polynomial are, clearly, $t_i = i$, for $i = 1, \dots, 20$. However, if there is a nondiagonal matrix W such that $\chi_{\mathsf{W}} = p_W$, a numerical procedure to compute χ_{W} will never produce it in the factored form presented above. In addition, the computations will be affected by roundoff. Let us illustrate how catastrophic this can be. Figure 8.1 compares p_W with a perturbation,

$$\tilde{p}_W(t) = p_W(t) - 10^{-9} t^{19}.$$

We clearly see that, although the size of the perturbation in front of the coefficient t^{19} is rather small, namely 10^{-9}, several of the roots of this polynomial have become complex.

Even if we choose to ignore the inherent complexity and stability issues in computing a determinant that we have just described, we are still faced with an insurmountable difficulty, which we now detail.

[1] Named in honor of the British mathematician James Hardy Wilkinson (1919–1986).

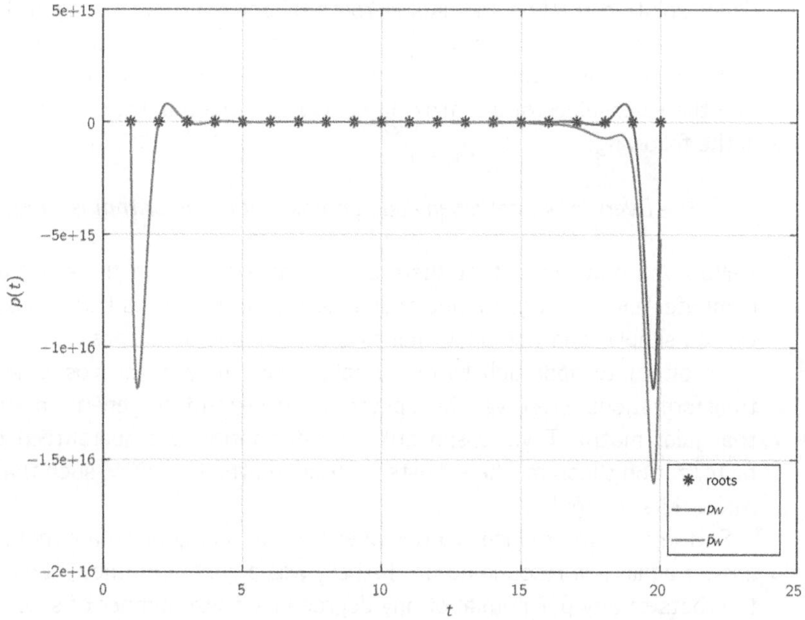

Figure 8.1 The Wilkinson polynomial p_W and its perturbation $\tilde{p}_W(x) = p_W(x) - 10^{-9}x^{19}$. Notice that the perturbation of one of the coefficients is rather small; the plot shows that we are missing several of the roots.

For a linear system $\mathsf{A}\boldsymbol{x} = \boldsymbol{f}$, or an even least squares problem, we showed that there are always direct methods for finding the solution vector $\boldsymbol{x}$. These may be quite expensive, requiring, say, $\mathcal{O}(n^3)$ operations; but, nonetheless, direct methods were available. However, direct methods are not typically available for finding eigenpairs. This fundamental difficulty can be seen from the following example, the details of which are left for Problem 8.1. Consider the matrix

$$\mathsf{F} = \begin{bmatrix} a_1 & a_2 & \cdots & a_{n-1} & a_n \\ 1 & 0 & \cdots & 0 & 0 \\ 0 & \ddots & \ddots & \vdots & \vdots \\ \vdots & \ddots & 1 & 0 & 0 \\ 0 & \cdots & 0 & 1 & 0 \end{bmatrix}.$$

F has the characteristic polynomial

$$\chi_{\mathsf{F}}(t) = (-1)^n\left(t^n - a_1 t^{n-1} - \cdots - a_n\right).$$

In other words, to any polynomial we can associate an eigenvalue problem, and vice versa. One of the most celebrated results of Galois theory[2] is the so-called

[2] Named in honor of the French mathematician Évariste Galois (1811–1832).

Abel's impossibility theorem,[3] which, in particular, states that there is no general direct solution method for finding roots of polynomials (of degree larger than or equal to 5); see, for example, [45, Proposition V.9.8]. There cannot be a direct method — an algorithmic method that completes in a finite number of steps — to find the eigenvalues of a matrix. In conclusion, the mantra of eigenvalue problems is the following:

Every practical eigenvalue approximation algorithm is iterative!

Now you may argue that there are matrices for which the spectrum is easy to compute, namely diagonal and triangular matrices. This is true: for such matrices, we can simply read the eigenvalues from the diagonal entries.

An alternate approach to our problem then may be proposed: since similarity transformations preserve the spectrum, transform a generic matrix A into a triangular matrix T via a similarity transformation. As guaranteed by the Schur factorization theorem, there exists a unitary matrix $\mathsf{U} \in \mathbb{C}^{n\times n}$ such that $\mathsf{A} = \mathsf{U}\mathsf{T}\mathsf{U}^{\mathsf{H}}$, where T is triangular.

So the question reduces to this one: *Can we find* U *using a direct method?* The answer again is a resounding no. If there was a such a method, then we could find the roots to any polynomial of any degree in a finite number of steps. Since this is impossible, in general, we get a contradiction.

Before we describe our iterative methods for finding eigenpairs, let us introduce some coarse approximation techniques.

8.1 Estimating Eigenvalues Using Gershgorin Disks

If all that is wanted are coarse approximations of the eigenvalues, this can be obtained easily using so-called Gershgorin disks.

Definition 8.1 (Gershgorin disks[4]). Let $n \geq 2$ and $\mathsf{A} \in \mathbb{C}^{n\times n}$. The **Gershgorin disks** D_i of A are

$$D_i = \{z \in \mathbb{C} \mid |z - a_{i,i}| \leq R_i\}, \quad R_i = \sum_{\substack{j=1 \\ j\neq i}}^{n} |a_{i,j}|, \quad i = 1, \ldots, n.$$

Theorem 8.2 (Gershgorin Circle Theorem). *Let $n \geq 2$ and $\mathsf{A} = [a_{i,j}] \in \mathbb{C}^{n\times n}$. Then*

$$\sigma(\mathsf{A}) \subset \bigcup_{i=1}^{n} D_i.$$

[3] Named in honor of the Norwegian mathematician Niels Henrik Abel (1802–1829).

[4] Named in honor of the Belarussian mathematician Semyon Aranovich Gershgorin (1901–1933).

Proof. Suppose that $(\lambda, \boldsymbol{w})$ is an eigenpair of A. Then

$$\sum_{j=1}^{n} a_{i,j} w_j = \lambda w_i, \quad i = 1, 2, \ldots, n.$$

Suppose that

$$|w_k| = \|\boldsymbol{w}\|_\infty = \max_{i=1}^{n} |w_i|.$$

Observe that $w_k \neq 0$, since $\boldsymbol{w} \neq \boldsymbol{0}$. Then

$$\begin{aligned} |\lambda - a_{k,k}| \cdot |w_k| &= |\lambda w_k - a_{k,k} w_k| \\ &= \left| \sum_{j=1}^{n} a_{k,j} w_j - a_{k,k} w_k \right| \\ &= \left| \sum_{\substack{j=1 \\ j \neq k}}^{n} a_{k,j} w_j \right| \\ &\leq \sum_{\substack{j=1 \\ j \neq k}}^{n} |a_{k,j} w_j| \\ &\leq \sum_{\substack{j=1 \\ j \neq k}}^{n} |a_{k,j}| \cdot |w_k| \\ &= R_k |w_k|. \end{aligned}$$

Thus, $\lambda \in D_k$. □

Theorem 8.3 (Gershgorin Second Theorem). *Let $n \geq 2$ and $\mathsf{A} \in \mathbb{C}^{n \times n}$. Suppose that $1 \leq p \leq n-1$ and that the Gershgorin disks of the matrix A can be divided into disjoint subsets $D^{(p)}$ and $D^{(q)}$ containing p and $q = n - p$ disks, respectively. Then the union of the disks in $D^{(p)}$ contains p eigenvalues, and the union of the disks in $D^{(q)}$ contains q eigenvalues, counting multiplicities. In particular, if one disk is disjoint from all the others, it contains exactly one eigenvalue. And, if all of the disks are disjoint, then each contains exactly one eigenvalue.*

Proof. We will proceed with a technique that is commonly known as a *homotopy argument.* In particular, we construct a family of matrices $\mathsf{B}(\varepsilon)$, parameterized by $\varepsilon \in [0, 1]$, such that we know the spectrum at $\varepsilon = 0$ completely. As ε increases from 0 to 1, we can follow the eigenvalue trajectories as they continuously deform. From this we will try to extract information for $\varepsilon = 1$. If we properly construct this family, then $\mathsf{B}(1) = \mathsf{A}$ and we will have proved the result.

Define the matrix $\mathsf{B}(\varepsilon) = [b_{i,j}(\varepsilon)]_{i,j=1}^{n}$ as follows:

$$b_{i,j}(\varepsilon) = \begin{cases} a_{i,i}, & i = j, \\ \varepsilon a_{i,j}, & i \neq j. \end{cases}$$

Then $\mathsf{B}(1) = \mathsf{A}$ and $\mathsf{B}(0) = \operatorname{diag}(a_{1,1}, \ldots, a_{n,n})$. Each eigenvalue of $\mathsf{B}(0)$ is the center of one of the Gershgorin disks of A. Thus, exactly p of the eigenvalues of $\mathsf{B}(0)$ lie in the union of the disks in $D^{(p)}$.

The eigenvalues of the matrix $\mathsf{B}(\varepsilon)$ are the zeros of the characteristic polynomial of $\mathsf{B}(\varepsilon)$. The coefficients of this characteristic polynomial are continuous functions of the parameter ε, and, in turn, the zeros of the polynomial are continuous functions of ε. We will accept this rather deep fact without proof; see, for example, [38].

As ε increases from 0 to 1, the eigenvalues of $\mathsf{B}(\varepsilon)$ move in continuous paths in the complex plane. Since the degree of the characteristic polynomial of $\mathsf{B}(\varepsilon)$ is always exactly n, none of the zeros of the characteristic polynomial diverge to infinity. At the same time, the radii of the Gershgorin disks increase from 0 to R_i, respectively. In particular, it is easy to see that they increase as

$$R_{\varepsilon,i} = \varepsilon R_i, \quad 0 \le \varepsilon \le 1, \quad i = 1, \ldots, n.$$

Since p of the eigenvalues of $\mathsf{B}(\varepsilon)$ lie in the union of the disks in $D^{(p)}$ when $\varepsilon = 0$, and these disks are disjoint from those in $D^{(q)}$, these p eigenvalues will stay within the union of the disks in $D^{(p)}$ for all values of ε, $0 \le \varepsilon \le 1$. □

Theorem 8.4 (almost diagonal). *Let $n \ge 2$. Suppose that in the matrix $\mathsf{A} \in \mathbb{C}^{n\times n}$ all off-diagonal elements are smaller in modulus than $\varepsilon > 0$, i.e.,*

$$|a_{i,j}| < \varepsilon, \quad 1 \le i, j \le n, \quad i \ne j.$$

Suppose that there is $\delta > 0$ such that, for some $r \in \{1, \ldots, n\}$, the diagonal element $a_{r,r}$ satisfies

$$|a_{r,r} - a_{i,i}| > \delta, \quad \forall i \ne r.$$

Then, provided that

$$\varepsilon < \frac{\delta}{2(n-1)},$$

there is an eigenvalue $\lambda \in \sigma(\mathsf{A})$ such that

$$|\lambda - a_{r,r}| < \frac{2(n-1)}{\delta}\varepsilon^2 < \frac{\delta}{2(n-1)}.$$

Proof. Let $\kappa > 0$. Define $\mathsf{K} = [k_{i,j}] \in \mathbb{C}^{n\times n}$ via

$$k_{i,j} = \begin{cases} \kappa, & i = j = r, \\ \delta_{i,j}, & \text{otherwise.} \end{cases}$$

Define the similar matrix $\mathsf{A}_\kappa = \mathsf{K}\mathsf{A}\mathsf{K}^{-1}$. For example, if $r = 3$, $n = 4$,

$$\mathsf{A} = \begin{bmatrix} a_{1,1} & a_{1,2} & a_{1,3} & a_{1,4} \\ a_{2,1} & a_{2,2} & a_{2,3} & a_{2,4} \\ a_{3,1} & a_{3,2} & a_{3,3} & a_{3,4} \\ a_{4,1} & a_{4,2} & a_{4,3} & a_{4,4} \end{bmatrix} \in \mathbb{C}^{4\times 4},$$

then

$$\mathsf{A}_\kappa = \begin{bmatrix} a_{1,1} & a_{1,2} & \kappa^{-1}a_{1,3} & a_{1,4} \\ a_{2,1} & a_{2,2} & \kappa^{-1}a_{2,3} & a_{2,4} \\ \kappa a_{3,1} & \kappa a_{3,2} & a_{3,3} & \kappa a_{3,4} \\ a_{4,1} & a_{4,2} & \kappa^{-1}a_{4,3} & a_{4,4} \end{bmatrix}.$$

The Gershgorin disk of A_κ, with respect to row r, has its center at $a_{r,r}$, and its radius is

$$R_{\kappa,r} = \sum_{\substack{j=1\\ j\neq r}}^{n} |\kappa a_{i,j}| = \kappa \sum_{\substack{j=1\\ j\neq r}}^{n} |a_{i,j}| < \kappa \sum_{\substack{j=1\\ j\neq r}}^{n} \varepsilon = \kappa(n-1)\varepsilon.$$

The disk of A_κ corresponding to row $i \neq r$ has $a_{i,i}$ as its center, and its radius is

$$R_{\kappa,i} = \sum_{\substack{j=1\\ j\neq i\\ j\neq r}}^{n} |a_{i,j}| + \kappa^{-1}|a_{i,r}| < (n-2)\varepsilon + \frac{\varepsilon}{\kappa}.$$

Now pick $\kappa = \frac{2\varepsilon}{\delta}$. Then

$$R_{\kappa,r} < \frac{2\varepsilon^2(n-1)}{\delta}, \qquad R_{\kappa,i} < \frac{\delta}{2} + (n-2)\varepsilon, \quad i \neq r.$$

Therefore, for $i \neq r$,

$$R_{\kappa,r} + R_{\kappa,i} < \frac{2\varepsilon^2(n-1)}{\delta} + \frac{\delta}{2} + (n-2)\varepsilon < \varepsilon + \frac{\delta}{2} + (n-2)\varepsilon < \delta$$

on the assumption that $\varepsilon < \frac{\delta}{2(n-1)}$. Finally, we have

$$\delta < |a_{r,r} - a_{i,i}|,$$

but $R_{\kappa,r} + R_{\kappa,i} < \delta$. Therefore, the two disks $D_{\kappa,r}$ and $D_{\kappa,i}$ must be disjoint. Hence, there is an eigenvalue $\lambda \in \sigma(\mathsf{A}_\kappa) = \sigma(\mathsf{A})$ with $\lambda \in D_{\kappa,r}$. In other words,

$$|\lambda - a_{r,r}| \leq R_{\kappa,r} < \frac{2\varepsilon^2(n-1)}{\delta} < \frac{\delta}{2(n-1)}.$$ □

8.2 Stability

In this section, we give a couple of results that are concerned with the stability of eigenvalue computation.

Theorem 8.5 (Bauer–Fike[5]). *Suppose that $\mathsf{A} \in \mathbb{C}^{n\times n}$ is diagonalizable, i.e., there is an invertible matrix $\mathsf{P} \in \mathbb{C}^{n\times n}$ and a diagonal matrix $\mathsf{D} = \text{diag}(\lambda_1, \dots, \lambda_n)$ such that $\mathsf{A} = \mathsf{PDP}^{-1}$. Suppose that $\lambda \notin \sigma(\mathsf{A}) = \{\lambda_i\}_{i=1}^n$ is an eigenvalue of the perturbed matrix $\mathsf{A} + \delta\mathsf{A}$ with $\delta\mathsf{A} \in \mathbb{C}^{n\times n}$. Then, for any $p \in [1, \infty]$, we have*

$$\min_{i=1}^{n} |\lambda - \lambda_i| \leq \|\mathsf{P}\|_p \|\mathsf{P}^{-1}\|_p \|\delta\mathsf{A}\|_p = \kappa_p(\mathsf{P}) \|\delta\mathsf{A}\|_p.$$

[5] Named in honor of the German mathematician Friedrich Ludwig Bauer (1924–2015) and the American mathematician and computer scientist Charles Theodore Fike (1933–2021).

Proof. There is some $\boldsymbol{x} \in \mathbb{C}^n_\star$ such that

$$(\mathsf{A} + \delta\mathsf{A})\boldsymbol{x} = \lambda\boldsymbol{x}.$$

It follows that

$$(\lambda\mathsf{I}_n - \mathsf{A})\boldsymbol{x} = \delta\mathsf{A}\boldsymbol{x}$$

and

$$\mathsf{P}(\lambda\mathsf{I}_n - \mathsf{D})\left(\mathsf{P}^{-1}\boldsymbol{x}\right) = \delta\mathsf{A}\mathsf{P}(\mathsf{P}^{-1}\boldsymbol{x}).$$

Note that $(\lambda\mathsf{I}_n - \mathsf{D})$ is invertible, since $\lambda \notin \sigma(\mathsf{A})$. Hence,

$$\left\|\mathsf{P}^{-1}\boldsymbol{x}\right\|_p \le \left\|(\lambda\mathsf{I}_n - \mathsf{D})^{-1}\right\|_p \left\|\mathsf{P}^{-1}\delta\mathsf{A}\mathsf{P}\right\|_p \left\|\mathsf{P}^{-1}\boldsymbol{x}\right\|_p.$$

Observe that, for any $p \in [1, \infty]$,

$$\left\|(\lambda\mathsf{I}_n - \mathsf{D})^{-1}\right\|_p = \max_{i=1}^{n} \frac{1}{|\lambda - \lambda_i|} = \frac{1}{\min_{i=1}^n |\lambda - \lambda_i|}.$$

Therefore,

$$\min_{i=1}^{n} |\lambda - \lambda_i| \le \|\mathsf{P}\|_p \left\|\mathsf{P}^{-1}\right\|_p \|\delta\mathsf{A}\|_p,$$

as desired. □

Corollary 8.6 (Hermitian matrix I). *If* $\mathsf{A} \in \mathbb{C}^{n\times n}$ *is Hermitian, then*

$$\min_{i=1}^{n} |\lambda - \lambda_i| \le \|\delta\mathsf{A}\|_2.$$

Proof. If A is Hermitian, it is unitarily diagonalizable. Since the 2-norm of a unitary matrix equals one, the result follows. □

Corollary 8.7 (Hermitian matrix II). *Suppose that* $\mathsf{A} \in \mathbb{C}^{n\times n}$ *is Hermitian,* $\lambda \in \mathbb{R}$, *and* λ *is closest to the eigenvalue* λ_r, *i.e.,*

$$r = \operatorname*{argmin}_{1\le i\le n} |\lambda_i - \lambda|.$$

Then, setting $\lambda = \lambda_r + \delta\lambda_r$,

$$|\delta\lambda_r| \le \|\delta\mathsf{A}\|_2.$$

Proof. See Problem 8.6 □

Remark 8.8 (stability). In other words, for Hermitian matrices, the eigenvalue problem is stable to perturbations in the coefficient matrix A. In the general case, the eigenvalue problem is stable, but a type of condition number, $\|\mathsf{P}\|_p \left\|\mathsf{P}^{-1}\right\|_p$, appears on the left-hand side.

Theorem 8.9 (eigenvector perturbation). *Suppose that* $\mathsf{A} \in \mathbb{C}^{n\times n}$ *is Hermitian and* $\sigma(\mathsf{A}) = \{\lambda_1, \ldots, \lambda_n\} \subset \mathbb{R}$. *Suppose that* $\boldsymbol{x} \in \mathbb{C}^n_\star$ *and* $\lambda \in \mathbb{C}$ *are given. Define*

$$\boldsymbol{w} = \mathsf{A}\boldsymbol{x} - \lambda\boldsymbol{x}.$$

Then

$$\min_{1\le i\le n} |\lambda_i - \lambda| \le \frac{\|\boldsymbol{w}\|_2}{\|\boldsymbol{x}\|_2}.$$

Proof. See Problem 8.7. □

Remark 8.10 (stability). The last result can be interpreted as follows: eigenvalue calculations are stable with respect to perturbations in the associated eigenvector.

8.3 The Rayleigh Quotient for Hermitian Matrices

In light of the stability results presented above, in this section, let us assume that A is Hermitian.

Definition 8.11 (Rayleigh quotient[6]). Suppose that $\mathsf{A} \in \mathbb{C}^{n\times n}$ is Hermitian. The **Rayleigh quotient** of $\boldsymbol{x} \in \mathbb{C}^n_\star$ is

$$R(\boldsymbol{x}) = \frac{(\mathsf{A}\boldsymbol{x}, \boldsymbol{x})_2}{(\boldsymbol{x}, \boldsymbol{x})_2}.$$

Proposition 8.12 (properties of the Rayleigh quotient). *Suppose that $\mathsf{A} \in \mathbb{C}^{n\times n}$ is Hermitian with spectrum $\sigma(\mathsf{A}) = \{\lambda_i\}_{i=1}^n \subset \mathbb{R}$, where the following ordering is imposed:*

$$\lambda_1 \le \lambda_2 \le \cdots \le \lambda_{n-1} \le \lambda_n.$$

Then

1. *$R(\boldsymbol{x}) \in \mathbb{R}$ for all $\boldsymbol{x} \in \mathbb{C}^n_\star$.*
2. *For all $\boldsymbol{x} \in \mathbb{C}^n_\star$,*

$$\min_{j=1}^{n} \lambda_j \le R(\boldsymbol{x}) \le \max_{j=1}^{n} \lambda_j.$$

3. *$R(\boldsymbol{x}) = \lambda_k$ if and only if $\boldsymbol{x}$ is an eigenvector associated with λ_k.*
4. *For a fixed $\boldsymbol{x} \in \mathbb{C}^n_\star$, the function*

$$Q(\alpha) = \|\mathsf{A}\boldsymbol{x} - \alpha\boldsymbol{x}\|_2^2, \quad \forall \alpha \in \mathbb{C}$$

has a unique global minimum; in fact, the Rayleigh quotient is the unique minimizer

$$R(\boldsymbol{x}) = \operatorname*{argmin}_{\alpha\in\mathbb{C}} \|\mathsf{A}\boldsymbol{x} - \alpha\boldsymbol{x}\|_2^2.$$

Proof. See Problem 8.8. □

To gain insight into what the previous result is saying, let us suppose that $\mathsf{A} \in \mathbb{R}^{n\times n}$ is symmetric. In light of the minimality property described in the last

[6] Named in honor of the British physicist John William Strutt (Lord Rayleigh) (1842–1919).

result, let us consider how the Rayleigh quotient behaves under perturbations of the vector $\boldsymbol{x}$. It is not difficult to show that

$$\nabla R(\boldsymbol{x}) = \frac{2}{(\boldsymbol{x}, \boldsymbol{x})_2}(\mathsf{A}\boldsymbol{x} - R(\boldsymbol{x})\boldsymbol{x}).$$

Thus, if $(\lambda, \boldsymbol{w})$ is an eigenpair of A, we have

$$\nabla R(\boldsymbol{w}) = \frac{2}{(\boldsymbol{w}, \boldsymbol{w})_2}(\mathsf{A}\boldsymbol{w} - R(\boldsymbol{w})\boldsymbol{w})_2 = \boldsymbol{0},$$

so that eigenvectors are stationary points of the Rayleigh quotient. In light of this, using Taylor's Theorem (B.54), we observe that

$$R(\boldsymbol{x}) - R(\boldsymbol{w}) = (\nabla R(\boldsymbol{w}), \boldsymbol{x} - \boldsymbol{w})_2 + \mathcal{O}(\|\boldsymbol{x} - \boldsymbol{w}\|_2^2) = \mathcal{O}(\|\boldsymbol{x} - \boldsymbol{w}\|_2^2), \quad \boldsymbol{x} \to \boldsymbol{w},$$

so that the Rayleigh quotient is a *quadratically accurate* approximation of λ.

Let us make the last statement rigorous before moving on.

Theorem 8.13 (eigenvalue estimate)**.** *Suppose that $\mathsf{A} \in \mathbb{C}^{n\times n}$ is Hermitian with spectrum $\sigma(\mathsf{A}) = \{\lambda_i\}_{i=1}^n \subset \mathbb{R}$ and an associated orthonormal basis of eigenvectors $\{\boldsymbol{w}_i\}_{i=1}^n$. Suppose that $\boldsymbol{x} \in \mathbb{C}_\star^n$ is a vector with the property that*

$$\|\boldsymbol{x} - \boldsymbol{w}_k\|_2 < \varepsilon, \quad \|\boldsymbol{x}\|_2 = 1$$

for some positive number ε. Then

$$|R(\boldsymbol{x}) - \lambda_k| < 2\rho(\mathsf{A})\varepsilon^2.$$

Proof. Suppose that $\boldsymbol{x} \neq \boldsymbol{w}_k$. There are unique constants $\alpha_i \in \mathbb{C}$ such that

$$\boldsymbol{x} = \sum_{j=1}^n \alpha_j \boldsymbol{w}_j, \qquad \|\boldsymbol{x}\|_2^2 = \sum_{j=1}^n |\alpha_j|^2 = 1.$$

It follows from orthonormality that

$$0 < \|\boldsymbol{x} - \boldsymbol{w}_k\|_2^2 = \sum_{\substack{j=1\\ j\neq k}}^n |\alpha_j|^2 + |\alpha_k - 1|^2.$$

Therefore, using our assumptions,

$$0 < \sum_{\substack{j=1\\ j\neq k}}^n |\alpha_j|^2 + |\alpha_k - 1|^2 = \|\boldsymbol{x} - \boldsymbol{w}_k\|_2^2 < \varepsilon^2. \tag{8.1}$$

Since $\|\boldsymbol{x}\|_2^2 = 1$, it follows from (8.1) that

$$0 \le \sum_{\substack{j=1\\ j\neq k}}^n |\alpha_j|^2 = 1 - |\alpha_k|^2 < \varepsilon^2. \tag{8.2}$$

Finally,

$$R(\boldsymbol{x}) = \frac{\sum_{j=1}^n \lambda_j |\alpha_j|^2}{\sum_{j=1}^n |\alpha_j|^2} = \lambda_k |\alpha_k|^2 + \sum_{\substack{j=1\\ j\neq k}}^n \lambda_j |\alpha_j|^2,$$

so that

$$\begin{aligned}
|R(\boldsymbol{x}) - \lambda_k| &= \left| \lambda_k \left(|\alpha_k|^2 - 1 \right) + \sum_{\substack{j=1 \\ j \neq k}}^{n} \lambda_j |\alpha_j|^2 \right| \\
&\leq |\lambda_k| \left(1 - |\alpha_k|^2 \right) + \sum_{\substack{j=1 \\ j \neq k}}^{n} |\lambda_j| \cdot |\alpha_j|^2 \\
&\leq \max_{j=1}^{n} |\lambda_j| \left(1 - |\alpha_k|^2 \right) + \max_{j=1}^{n} |\lambda_j| \sum_{\substack{j=1 \\ j \neq k}}^{n} |\alpha_j|^2 \\
&\leq \rho(\mathsf{A})\varepsilon^2 + \rho(\mathsf{A})\varepsilon^2 \\
&= 2\rho(\mathsf{A})\varepsilon^2,
\end{aligned}$$

where we used (8.2). □

This last property motivates the so-called *power iterations* to find the largest eigenvalue, which will be our next topic of discussion.

8.4 Power Iteration Methods

Definition 8.14 (power iteration). Suppose that $\mathsf{A} \in \mathbb{C}^{n \times n}$ and $\boldsymbol{q} \in \mathbb{C}_\star^n$ is given. The **power method** is an algorithm that generates a sequence of vectors $\{\boldsymbol{v}_k\}_{k=0}^{\infty} \subset \mathbb{C}_\star^n$ according to the following recursive formula:

$$\boldsymbol{v}_0 = \frac{\boldsymbol{q}}{\|\boldsymbol{q}\|_2}.$$

For $k \geq 1$,

$$\boldsymbol{q}_k = \mathsf{A}\boldsymbol{v}_{k-1},$$

and, provided that $\boldsymbol{q}_k \neq \boldsymbol{0}$,

$$\boldsymbol{v}_k = \frac{\boldsymbol{q}_k}{\|\boldsymbol{q}_k\|_2}.$$

If $\boldsymbol{q}_k = \boldsymbol{0}$, the algorithm terminates at step k.

Remark 8.15 (terminology). The name of the method becomes clear once we realize that

$$\boldsymbol{v}_k = \frac{1}{\|\boldsymbol{q}_k\|_2}\boldsymbol{q}_k = \frac{1}{\|\boldsymbol{q}_k\|_2}\mathsf{A}\boldsymbol{v}_{k-1} = \cdots = c_k \mathsf{A}^k \boldsymbol{v}_0.$$

The convergence of the method is as follows.

Theorem 8.16 (convergence of the power iteration). *Let $\mathsf{A} \in \mathbb{C}^{n \times n}$ be Hermitian and $\boldsymbol{q} \in \mathbb{C}_\star^n$ be given. Suppose that the spectrum $\sigma(\mathsf{A}) = \{\lambda_i\}_{i=1}^n$ has the ordering*

$$0 \leq |\lambda_1| \leq |\lambda_2| \leq \cdots \leq |\lambda_{n-1}| < |\lambda_n|.$$

Assume that $\{\boldsymbol{w}_i\}_{i=1}^n$ is the associated basis of orthonormal eigenvectors of A*. If $(\boldsymbol{q}, \boldsymbol{w}_n)_2 \neq 0$, where $\boldsymbol{w}_n$ is the eigenvector corresponding to the dominant eigenvalue λ_n, then, for some constant $C > 0$,*

$$\|\boldsymbol{v}_k - s_k \boldsymbol{w}_n\|_2 \leq C \left| \frac{\lambda_{n-1}}{\lambda_n} \right|^k, \qquad |R(\boldsymbol{v}_k) - \lambda_n| \leq 2|\lambda_n| C^2 \left| \frac{\lambda_{n-1}}{\lambda_n} \right|^{2k},$$

when k is sufficiently large. Here, s_k is a sequence of modulus one complex numbers.

Proof. The vector $\boldsymbol{v}_0$ is obtained by normalizing $\boldsymbol{q}$. Let us then expand $\boldsymbol{v}_0$ in the basis of eigenvectors: there exist unique constants $\alpha_j \in \mathbb{C}$ such that

$$\boldsymbol{v}_0 = \sum_{j=1}^n \alpha_j \boldsymbol{w}_j.$$

From our assumptions, $\alpha_n \neq 0$ and $\sum_{j=1}^n |\alpha_j|^2 = 1$. Now

$$\boldsymbol{q}_1 = \mathsf{A}\boldsymbol{v}_0 = \sum_{j=1}^n \lambda_j \alpha_j \boldsymbol{w}_j.$$

From this, it follows that

$$\boldsymbol{v}_1 = \frac{1}{\sqrt{\sum_{j=1}^n |\alpha_j|^2 \lambda_j^2}} \sum_{j=1}^n \lambda_j \alpha_j \boldsymbol{w}_j.$$

Proceeding this way, we observe that

$$\begin{aligned}
\boldsymbol{v}_k &= \frac{1}{\sqrt{\sum_{j=1}^n |\alpha_j|^2 \lambda_j^{2k}}} \sum_{j=1}^n \lambda_j^k \alpha_j \boldsymbol{w}_j \\
&= \frac{1}{\sqrt{\sum_{j=1}^n |\alpha_j|^2 \lambda_j^{2k}}} \left(\lambda_n^k \alpha_n \boldsymbol{w}_n + \sum_{j=1}^{n-1} \lambda_j^k \alpha_j \boldsymbol{w}_j \right) \\
&= \frac{\alpha_n \lambda_n^k}{|\alpha_n \lambda_n^k|} \frac{1}{\sqrt{1 + \sum_{j=1}^{n-1} \left| \frac{\alpha_j}{\alpha_n} \right|^2 \left(\frac{\lambda_j}{\lambda_n} \right)^{2k}}} \left(\boldsymbol{w}_n + \sum_{j=1}^{n-1} \frac{\alpha_j}{\alpha_n} \left(\frac{\lambda_j}{\lambda_n} \right)^k \boldsymbol{w}_j \right).
\end{aligned}$$

Define $s_k = \frac{\alpha_n \lambda_n^k}{|\alpha_n \lambda_n^k|}$. Given the assumptions on $\sigma(\mathsf{A})$, we observe, for $j = 1, \ldots, n-1$, that

$$\left(\frac{\lambda_j}{\lambda_n} \right)^k \to 0, \qquad k \to \infty.$$

It is clear that

$$\frac{1}{s_k} \boldsymbol{v}_k - \boldsymbol{w}_n \to \boldsymbol{0}, \qquad k \to \infty.$$

We leave it to the reader as an exercise to prove the estimate

$$\|\boldsymbol{v}_k - s_k \boldsymbol{w}_n\|_2 \leq C \left| \frac{\lambda_{n-1}}{\lambda_n} \right|^k$$

for some $C > 0$, provided that k is sufficiently large. If this holds, Theorem 8.13 yields the second-order convergence rate of the eigenvalue approximations. □

8.4.1 Inverse Iteration

We just showed how, using the power iteration method, we can approximate the dominant eigenvalue, provided that it is isolated. A small modification of this idea allows us to approximate any other isolated eigenvalue. Let us begin with an auxiliary result, whose proof is straightforward.

Proposition 8.17 (shifts). *Suppose that* $\mathsf{A} \in \mathbb{C}^{n\times n}$ *and* $\mu \in \mathbb{C}$ *has the the property that* $\mu \notin \sigma(\mathsf{A})$. *Then* $(\mathsf{A} - \mu\mathsf{I})^{-1}$ *exists, and* $(\lambda, \boldsymbol{w})$ *is an eigenpair of* A *if and only if* $\left((\lambda - \mu)^{-1}, \boldsymbol{w}\right)$ *is an eigenpair of* $(\mathsf{A} - \mu\mathsf{I})^{-1}$.

Proof. See Problem 8.11. □

This motivates the following observation: if μ is close to $\lambda_r \in \sigma(\mathsf{A})$, then $(\lambda_r - \mu)^{-1}$ is much larger in modulus than $(\lambda_j - \mu)^{-1}$, for any other eigenvalue. Thus, to approximate λ_r, we can apply power iterations to the matrix $(\mathsf{A} - \mu\mathsf{I})^{-1}$. This is the idea of the inverse iteration method.

Definition 8.18 (inverse iteration method). Suppose that $\mathsf{A} \in \mathbb{C}^{n\times n}$, $\boldsymbol{q} \in \mathbb{C}^n_\star$, and $\mu \in \mathbb{C}$ are given. Assume that $\mathsf{A} - \mu\mathsf{I}$ is invertible. The *inverse iteration method* is an algorithm that generates a sequence of vectors $\{\boldsymbol{v}_k\}_{k=0}^\infty \in \mathbb{C}^n_\star$ according to the following recursive formula:

$$\boldsymbol{v}_0 = \frac{\boldsymbol{q}}{\|\boldsymbol{q}\|_2}.$$

For $k \geq 1$,

$$\boldsymbol{q}_k = (\mathsf{A} - \mu\mathsf{I})^{-1}\, \boldsymbol{v}_{k-1},$$

and, provided that $\boldsymbol{q}_k \neq \boldsymbol{0}$,

$$\boldsymbol{v}_k = \frac{\boldsymbol{q}_k}{\|\boldsymbol{q}_k\|_2}.$$

If $\boldsymbol{q}_k = \boldsymbol{0}$, the algorithm terminates at step k.

The convergence properties of this algorithm are as follows.

Theorem 8.19 (convergence of the inverse iteration). *Suppose that* $\mathsf{A} \in \mathbb{C}^{n\times n}$ *is Hermitian with spectrum* $\sigma(\mathsf{A}) = \{\lambda_i\}_{i=1}^n \subset \mathbb{R}$. *Let* $\{\boldsymbol{w}_i\}_{i=1}^n$ *be an associated orthonormal basis of eigenvectors of* A. *Suppose that* $\mu \in \mathbb{C}$ *is given with the property that* $\mu \notin \sigma(\mathsf{A})$. *Define*

$$\lambda_r = \operatorname*{argmin}_{j=1}^{n} |\lambda_j - \mu|, \quad \lambda_s = \operatorname*{argmin}_{\substack{j=1\\ j\neq r}}^{n} |\lambda_j - \mu|.$$

Assume that

$$|\lambda_r - \mu| < |\lambda_s - \mu| \leq |\lambda_j - \mu|$$

for all $j \in \{1, \dots, n\} \setminus \{r, s\}$. If $\boldsymbol{q} \in \mathbb{C}^n_\star$ is given with the property that $(\boldsymbol{q}, \boldsymbol{w}_r)_2 \neq 0$, then the inverse iteration converges. Moreover, we have the following convergence estimates: there is some $C > 0$ such that

$$\|\boldsymbol{v}_k - s_k \boldsymbol{w}_r\|_2 \le C \left| \frac{\mu - \lambda_r}{\mu - \lambda_s} \right|^k, \qquad |R(\boldsymbol{v}_k) - \lambda_r| \le 2\rho(\mathsf{A}) C^2 \left| \frac{\mu - \lambda_r}{\mu - \lambda_s} \right|^{2k},$$

provided that k is sufficiently large. As before, s_k is a sequence of modulus one complex numbers.

Proof. See Problem 8.12. □

8.4.2 Deflation

The next result addresses the issue of computing a next-to-dominant eigenvalue.

Theorem 8.20 (deflation). *Suppose that $\mathsf{A} \in \mathbb{C}^{n \times n}$ is Hermitian and its spectrum is denoted $\sigma(\mathsf{A}) = \{\lambda_1, \dots, \lambda_n\} \subset \mathbb{R}$. Let $\{\boldsymbol{w}_1, \dots, \boldsymbol{w}_n\}$ be an orthonormal basis of eigenvectors of A with $\mathsf{A}\boldsymbol{w}_k = \lambda_k \boldsymbol{w}_k$, for $k = 1, \dots, n$. Assume that the eigenvalues of A are ordered as follows:*

$$0 \le |\lambda_1| \le |\lambda_2| \le \cdots \le |\lambda_{n-2}| < |\lambda_{n-1}| < |\lambda_n|.$$

Assume that λ_n is known and that $\boldsymbol{v}_0 \in \mathbb{C}^n_\star$ is such that $\boldsymbol{w}_{n-1}^{\mathsf{H}} \boldsymbol{v}_0 \neq 0$. Then the sequence

$$\boldsymbol{v}_{k+1} = \mathsf{A}\boldsymbol{v}_k, \qquad \boldsymbol{y}_{k+1} = \boldsymbol{v}_{k+1} - \lambda_n \boldsymbol{v}_k$$

satisfies

$$\frac{\boldsymbol{y}_{k+1}^{\mathsf{H}} \boldsymbol{y}_k}{\boldsymbol{y}_k^{\mathsf{H}} \boldsymbol{y}_k} \to \lambda_{n-1}, \qquad k \to \infty.$$

Proof. There exist unique constants $\alpha_j \in \mathbb{C}$, $j = 1, 2, \dots, n$ such that $\boldsymbol{v}_0 = \sum_{j=1}^n \alpha_j \boldsymbol{w}_j$. By assumption, $\alpha_{n-1} \neq 0$. We observe that $\boldsymbol{v}_k = \mathsf{A}^k \boldsymbol{v}_0$ and, consequently,

$$\boldsymbol{y}_k = \mathsf{A}^k \boldsymbol{v}_0 - \lambda_n \mathsf{A}^{k-1} \boldsymbol{v}_0 = \mathsf{A}^{k-1}(\mathsf{A} - \lambda_n \mathsf{I}_n)\, \boldsymbol{v}_0 = \sum_{j=1}^{n-1} \alpha_j \lambda_j^{k-1} (\lambda_j - \lambda_n)\, \boldsymbol{w}_j.$$

Therefore,

$$\begin{aligned}
\frac{\boldsymbol{y}_{k+1}^{\mathsf{H}} \boldsymbol{y}_k}{\boldsymbol{y}_k^{\mathsf{H}} \boldsymbol{y}_k} &= \frac{\sum_{j=1}^{n-1} |\alpha_j|^2 \lambda_j^{2k-1} (\lambda_j - \lambda_n)^2}{\sum_{j=1}^{n-1} |\alpha_j|^2 \lambda_j^{2(k-1)} (\lambda_j - \lambda_n)^2} \\
&= \frac{\sum_{j=1}^{n-1} \left| \frac{\alpha_j}{\alpha_{n-1}} \right|^2 \lambda_j \left(\frac{\lambda_j}{\lambda_{n-1}} \right)^{2(k-1)} \left(\frac{\lambda_j - \lambda_n}{\lambda_{n-1} - \lambda_n} \right)^2}{\sum_{j=1}^{n-1} \left| \frac{\alpha_j}{\alpha_{n-1}} \right|^2 \left(\frac{\lambda_j}{\lambda_{n-1}} \right)^{2(k-1)} \left(\frac{\lambda_j - \lambda_n}{\lambda_{n-1} - \lambda_n} \right)^2} \\
&= \frac{\lambda_{n-1} + \sum_{j=1}^{n-2} \left| \frac{\alpha_j}{\alpha_{n-1}} \right|^2 \lambda_j \left(\frac{\lambda_j}{\lambda_{n-1}} \right)^{2(k-1)} \left(\frac{\lambda_j - \lambda_n}{\lambda_{n-1} - \lambda_n} \right)^2}{1 + \sum_{j=1}^{n-2} \left| \frac{\alpha_j}{\alpha_{n-1}} \right|^2 \left(\frac{\lambda_j}{\lambda_{n-1}} \right)^{2(k-1)} \left(\frac{\lambda_j - \lambda_n}{\lambda_{n-1} - \lambda_n} \right)^2} \\
&\to \lambda_{n-1},
\end{aligned}$$

as $k \to \infty$, since $\left(\frac{\lambda_j}{\lambda_{n-1}} \right)^2 < 1$ for each $j = 1, 2, \dots, n-2$. □

8.5 Reduction to Hessenberg Form

We have seen that it is impossible, in general, to reduce a square matrix to a triangular or diagonal matrix through similarity transformations in a finite number of steps (via a direct method). We will now see that it is possible to reduce a matrix to a so-called *Hessenberg matrix*, one that has zeros below the first sub-diagonal.

Definition 8.21 (Hessenberg matrix[7]). The square matrix $\mathsf{A} = [a_{i,j}] \in \mathbb{C}^{n\times n}$ is called a **lower Hessenberg matrix** or is said to have **lower Hessenberg form** if and only if $a_{i,j} = 0$ for all $1 \le i,j \le n$ satisfying $i \ge j+2$. A is said to be **upper Hessenberg** if and only if $\mathsf{A}^\intercal$ is lower Hessenberg.

We want to come up with a similarity transformation that effects the following change, where by $\times$ we denote nonzero entries of the result:

$$\mathsf{A} \to \mathsf{Q}^{\mathsf{H}}\mathsf{A}\mathsf{Q} = \begin{bmatrix} \times & \times & \times & \cdots & \times \\ \times & \times & \times & \ddots & \vdots \\ 0 & \times & \times & \ddots & \times \\ \vdots & \ddots & \ddots & \ddots & \times \\ 0 & \cdots & 0 & \times & \times \end{bmatrix}.$$

We now explain how this is achieved.

Suppose that H_{n-1} is the Householder matrix that leaves the first row of A unchanged and introduces zeros below the second row of the first column:

$$\mathsf{A} \to \mathsf{H}_{n-1}^{\mathsf{H}}\mathsf{A} = \begin{bmatrix} \times & \times & \times & \cdots & \times \\ \otimes & \otimes & \otimes & \cdots & \otimes \\ 0 & \otimes & \otimes & \ddots & \vdots \\ \vdots & \vdots & \vdots & \ddots & \otimes \\ 0 & \otimes & \otimes & \cdots & \otimes \end{bmatrix},$$

where, as before, by $\times$ we denote entries of the matrix A that did not change and by $\otimes$ those that did change. When we calculate $\mathsf{H}_{n-1}^{\mathsf{H}}\mathsf{A}\mathsf{H}_{n-1}$ it leaves the first column unchanged, so that we obtain

$$\mathsf{A} \to \mathsf{H}_{n-1}^{\mathsf{H}}\mathsf{A} = \begin{bmatrix} \times & \times & \times & \cdots & \times \\ \otimes & \otimes & \otimes & \cdots & \otimes \\ 0 & \otimes & \otimes & \ddots & \vdots \\ \vdots & \vdots & \vdots & \ddots & \otimes \\ 0 & \otimes & \otimes & \cdots & \otimes \end{bmatrix} \to \mathsf{H}_{n-1}^{\mathsf{H}}\mathsf{A}\mathsf{H}_{n-1} \begin{bmatrix} \times & \otimes & \otimes & \cdots & \otimes \\ \otimes & \oplus & \oplus & \cdots & \oplus \\ 0 & \oplus & \oplus & \ddots & \vdots \\ \vdots & \vdots & \vdots & \ddots & \oplus \\ 0 & \oplus & \oplus & \cdots & \oplus \end{bmatrix}.$$

[7] Named in honor of the German mathematician and engineer Karl Adolf Hessenberg (1904–1959).

The symbol $\oplus$ indicates that the entry has been changed twice. Repeating this idea we can reduce any matrix to Hessenberg form. The algorithm is presented in Listing 8.1.

If $A \in \mathbb{C}^{n\times n}$ is Hermitian, then the Hessenberg transformation results in a tridiagonal Hermitian matrix similar to A. We present a result on the existence of this transformation for the Hermitian case. We leave it to the reader to prove the existence in the more general setting.

Theorem 8.22 (existence). *Suppose that $A \in \mathbb{C}^{n\times n}$ is Hermitian. There exists a unitary matrix $Q \in \mathbb{C}^{n\times n}$, the product of $n-2$ Householder matrices,*

$$Q = H_{n-1} \cdots H_2,$$

where $H_k \in \mathbb{C}^{n\times n}$ is a Householder matrix such that

$$Q^H A Q = T,$$

where $T \in \mathbb{C}^{n\times n}$ is a Hermitian tridiagonal matrix.

Proof. Let us express A as

$$A = \begin{bmatrix} \alpha & \boldsymbol{b}^H \\ \boldsymbol{b} & C \end{bmatrix},$$

where $\alpha \in \mathbb{C}$, $\boldsymbol{b} \in \mathbb{C}^{n-1}$, and $C \in \mathbb{C}^{(n-1)\times(n-1)}$ is Hermitian. To simplify notation, we define

$$\mathcal{E}_1^k = \operatorname{span}\{\boldsymbol{e}_1\} \subset \mathbb{C}^k.$$

If $\boldsymbol{b} \in \mathbb{C}_\star^{n-1}$, there is a Householder matrix $H_{n-1} \in \mathbb{C}^{(n-1)\times(n-1)}$ such that

$$H_{n-1}\boldsymbol{b} \in \mathcal{E}_1^{n-1}.$$

If $\boldsymbol{b} = \boldsymbol{0}$, then $H_{n-1}\boldsymbol{b} = \boldsymbol{0} \in \mathcal{E}_1^{n-1}$, regardless of our choice for H_{n-1}, and so for the sake of definiteness we take $H_{n-1} = I_{n-1}$.

Define

$$H_{n,n-1} = \begin{bmatrix} 1 & \boldsymbol{0}^H \\ \boldsymbol{0} & H_{n-1} \end{bmatrix} \in \mathbb{C}^{n\times n}. \tag{8.3}$$

This is also a Householder matrix, as we have seen before. It follows that

$$H_{n,n-1}^H A H_{n,n-1} = \begin{bmatrix} 1 & \boldsymbol{0}^H \\ \boldsymbol{0} & H_{n-1}^H \end{bmatrix} \begin{bmatrix} \alpha & \boldsymbol{b}^H \\ \boldsymbol{b} & C \end{bmatrix} \begin{bmatrix} 1 & \boldsymbol{0}^H \\ \boldsymbol{0} & H_{n-1} \end{bmatrix} = \begin{bmatrix} \alpha & \boldsymbol{d}^H \\ \boldsymbol{d} & D \end{bmatrix},$$

where $\boldsymbol{d} = H_{n-1}^H \boldsymbol{b} = H_{n-1}\boldsymbol{b} \in \mathcal{E}_1^{n-1}$ and

$$D = H_{n-1}^H C H_{n-1} \in \mathbb{C}^{(n-1)\times(n-1)},$$

which is clearly a Hermitian matrix. The proof from this point is by induction.

($k = 3$) Suppose that $A \in \mathbb{C}^{3\times 3}$. The matrix

$$H_{3,2}^H A H_{3,2} \in \mathbb{C}^{3\times 3}$$

is already tridiagonal, since

$$H_{3,2}\boldsymbol{b} = \boldsymbol{d} \in \mathcal{E}_1^2.$$

This case follows upon taking $Q_3 = H_{3,2}$. Observe that, for any $\boldsymbol{f} \in \mathcal{E}_1^3$, i.e., $\boldsymbol{f} = [\lambda, 0, 0]^\intercal$, we have

$$Q_3^H \boldsymbol{f} = H_{3,2}^H \boldsymbol{f} = H_{3,2} \boldsymbol{f} = \boldsymbol{f}.$$

($k = m-1$) For the induction hypothesis, let us assume that, for any $D \in \mathbb{C}^{(m-1)\times(m-1)}$ that is Hermitian, there is a unitary matrix $Q_{m-1} \in \mathbb{C}^{(m-1)\times(m-1)}$, the product of Householder matrices

$$Q_{m-1} = H_{m-1,m-2} \cdots H_{m-1,2},$$

such that the product

$$Q_{m-1}^H D Q_{m-1} = T_{m-1} \in \mathbb{C}^{(m-1)\times(m-1)}$$

is Hermitian and tridiagonal. Furthermore, let us assume that

$$Q_{m-1}^H \boldsymbol{f} \in \mathcal{E}_1^{m-1}, \quad \forall \boldsymbol{f} \in \mathcal{E}_1^{m-1}.$$

($k = m$) Suppose that $A \in \mathbb{C}^{m\times m}$ is Hermitian. Let us inflate each $H_{m-1,k} \in \mathbb{C}^{(m-1)\times(m-1)}$ from the induction hypothesis step by

$$H_{m,k} = \begin{bmatrix} 1 & \boldsymbol{0}^H \\ \boldsymbol{0} & H_{m-1,k} \end{bmatrix} \in \mathbb{C}^{m\times m}$$

for $k = 2, \ldots, m-2$. For $k = m-1$, define $H_{m,m-1}$ as in (8.3). Next, set

$$Q_m = H_{m,m-1} H_{m,m-2} \cdots H_{m,2}.$$

It follows that

$$\begin{aligned} Q_m^H A Q_m &= H_{m,2} \cdots H_{m,m-2} H_{m,m-1} A H_{m,m-1} H_{m,m-2} \cdots H_{m,2} \\ &= \begin{bmatrix} 1 & \boldsymbol{0}^H \\ \boldsymbol{0} & Q_{m-1}^H \end{bmatrix} \begin{bmatrix} \alpha & \boldsymbol{d}^H \\ \boldsymbol{d} & D \end{bmatrix} \begin{bmatrix} 1 & \boldsymbol{0}^H \\ \boldsymbol{0} & Q_{m-1} \end{bmatrix} \\ &= \begin{bmatrix} \alpha & \boldsymbol{d}^H Q_{m-1} \\ Q_{m-1}^H \boldsymbol{d} & T_{m-1} \end{bmatrix} \\ &= T_m. \end{aligned}$$

Since $Q_{m-1}^H \boldsymbol{d} \in \mathcal{E}_1^{m-1}$, T_m must be tridiagonal.

Finally, for any $\boldsymbol{f} \in \mathcal{E}_1^m$, we have $Q_m^H \boldsymbol{f} \in \mathcal{E}_1^m$, since the $(1,1)$ entry of Q_m is 1 and the remaining entries in the first column of Q_n are zeros. The proof by induction is complete. □

Next, we have a simple, and quite amazing, criterion for a Hermitian tridiagonal matrix to have distinct eigenvalues.

Theorem 8.23 (distinct eigenvalues)**.** *Suppose that $A \in \mathbb{C}^{n\times n}$ is Hermitian and tridiagonal. Assume that its super- and sub-diagonal entries are all nonzero. Then the eigenvalues of A must be distinct.*

Proof. See Problem 8.13. □

Of course, it is easy to show that the converse is not true. Can you think of a quick example that shows the converse is false?

8.6 The QR Method

Definition 8.24 (QR iteration method)**.** Let $A \in \mathbb{C}^{n\times n}$. The **QR iteration method** is a recursive algorithm for computing the sequence $\{A_k\}_{k=0}^{\infty} \subset \mathbb{C}^{n\times n}$ according to the following rules.

1. Set $A_0 = A$.
2. For $k = 0, 1, 2, \ldots$, given A_k,
   a. Compute the factorization
   $$Q_{k+1}R_{k+1} = A_k,$$
   where $Q_{k+1} \in \mathbb{C}^{n\times n}$ is unitary and $R_{k+1} \in \mathbb{C}^{n\times n}$ is upper triangular.
   b. Define the next iterate as
   $$A_{k+1} = R_{k+1}Q_{k+1}.$$

Listing 8.2 presents an implementation of the QR iteration method.

Recall that every square matrix has a (not necessarily unique) *Schur decomposition*

$$A = Q^{H}UQ,$$

where Q is unitary and U is upper triangular. We will show that, under suitable assumptions, the QR iteration method computes the factor U in a Schur decomposition of A as the limit of the sequence $\{A_k\}_{k=0}^{\infty}$. To minimize the cost of the QR iteration, first one converts the matrix A to Hessenberg form. This preserves the spectrum, and the QR iteration will preserve this form, among other properties.

Proposition 8.25 (invariance)**.** *Let $A \in \mathbb{C}^{n\times n}$. Suppose that $\{A_k\}_{k=0}^{\infty}$ is computed according to the QR iteration method. Then we have:*

1. *The matrix A_k is similar to A_{k-1}.*
2. *If A_{k-1} is Hermitian, so is A_k.*
3. *If A_{k-1} is Hessenberg, so is A_k.*
4. *If A_{k-1} is tridiagonal, so is A_k.*

Proof. Since $R_k = Q_k^{H}A_{k-1}$ and $A_k = R_kQ_k$, it follows that

$$A_k = R_kQ_k = Q_k^{H}A_{k-1}Q_k.$$

Since Q_k is unitary, this shows that A_k is similar to A_{k-1}, and that A_k is Hermitian if A_{k-1} is Hermitian as well.

The last two statements are left for the reader as an exercise; see Problem 8.14. □

Example 8.3 Consider the symmetric, tridiagonal matrix

$$A = \begin{bmatrix} 9 & 17 & 0 & 0 & 0 \\ 17 & 3 & 18 & 0 & 0 \\ 0 & 18 & 20 & 2 & 0 \\ 0 & 0 & 2 & 1 & 8 \\ 0 & 0 & 0 & 8 & 16 \end{bmatrix}.$$

Observe that the QR factorization, $QR = A$, has factors

$$Q = \begin{bmatrix} -0.467\,888 & 0.533\,293 & 0.700\,794 & -0.008\,036 & -0.074\,181 \\ -0.883\,788 & -0.282\,332 & -0.371\,009 & 0.004\,254 & 0.039\,272 \\ 0 & 0.797\,425 & -0.600\,026 & 0.006\,880 & 0.063\,515 \\ 0 & 0 & -0.105\,874 & -0.107\,091 & -0.988\,596 \\ 0 & 0 & 0 & -0.994\,184 & 0.107\,696 \end{bmatrix}$$

and

$$R = \begin{bmatrix} -19.235\,384 & -10.605\,455 & -15.908\,182 & 0 & 0 \\ 0 & 22.572\,645 & 10.866\,537 & 1.594\,851 & 0 \\ 0 & 0 & -18.890\,423 & -1.305\,926 & -0.846\,990 \\ 0 & 0 & 0 & -8.046\,802 & -16.763\,666 \\ 0 & 0 & 0 & 0 & -6.185\,635 \end{bmatrix}$$

to six decimal digits of precision. Notice that we have not demanded that R has positive diagonal components in this factorization, though that can be easily remedied. In any case, note the placement of the zeros in the factor matrices. Can you show that these entries are generically zero, based on the structure of A?

Recall that, in the QR algorithm, $A_1 = RQ$, and we obtain

$$A_1 = \begin{bmatrix} 18.372\,973 & -19.949\,431 & 0 & 0 & 0 \\ -19.949\,431 & 2.292\,279 & -15.063\,703 & 0 & 0 \\ 0 & -15.063\,703 & 11.473\,011 & 0.851\,945 & 0 \\ 0 & 0 & 0.851\,945 & 17.527\,905 & 6.149\,659 \\ 0 & 0 & 0 & 6.149\,659 & -0.666\,167 \end{bmatrix},$$

again showing six decimal digits of precision. The structure of A is preserved in this first step of the QR algorithm; namely, A_1 is symmetric and tridiagonal, as predicted by Proposition 8.25.

This leads us to the main result of convergence. Under suitable assumptions, for a generic matrix A, the sequence $\{A_k\}_{k=1}^{\infty}$ generated by the QR iteration converges to an upper triangular matrix that is similar to A.

Theorem 8.26 (convergence I). *Suppose that $A \in \mathbb{C}^{n\times n}$ is invertible and all its eigenvalues are distinct in modulus, i.e.,*

$$|\lambda_1| > |\lambda_2| > \cdots > |\lambda_n| > 0.$$

Let $\mathsf{P} \in \mathbb{C}^{n\times n}$ *be an invertible matrix such that*

$$\mathsf{A} = \mathsf{P}\mathsf{D}\mathsf{P}^{-1},$$

where $\mathsf{D} = \operatorname{diag}(\lambda_1, \lambda_2, \ldots, \lambda_n)$*. If* P^{-1} *has an LU factorization — i.e., there exists a unit lower triangular matrix* $\mathsf{L} \in \mathbb{C}^{n\times n}$ *and an upper triangular matrix* $\mathsf{U} \in \mathbb{C}^{n\times n}$ *such that* $\mathsf{P}^{-1} = \mathsf{L}\mathsf{U}$ *(Theorem 3.7) — then the sequence of matrices* $\{\mathsf{A}_k\}_{k=1}^{\infty}$ *produced by the QR iteration method is such that*

$$\lim_{k\to\infty} [\mathsf{A}_k]_{i,i} = \lambda_i, \quad 1 \le i \le n,$$

and

$$\lim_{k\to\infty} [\mathsf{A}_k]_{i,j} = 0, \quad 1 \le j < i \le n.$$

Proof. Let us define, for $k \ge 1$,

$$\mathcal{Q}_k = \mathsf{Q}_1 \cdots \mathsf{Q}_k, \quad \mathcal{R}_k = \mathsf{R}_k \cdots \mathsf{R}_1.$$

With this notation we observe that

$$\begin{aligned}
\mathsf{A}_{k+1} &= \mathsf{Q}_k^{\mathsf{H}} \mathsf{A}_k \mathsf{Q}_k \\
&= \mathsf{Q}_k^{\mathsf{H}} \mathsf{Q}_{k-1}^{\mathsf{H}} \mathsf{A}_{k-1} \mathsf{Q}_{k-1} \mathsf{Q}_k \\
&\;\;\vdots \\
&= \mathsf{Q}_k^{\mathsf{H}} \cdots \mathsf{Q}_1^{\mathsf{H}} \mathsf{A}_0 \mathsf{Q}_1 \cdots \mathsf{Q}_k \\
&= \mathcal{Q}_k^{\mathsf{H}} \mathsf{A} \mathcal{Q}_k.
\end{aligned}$$

In addition,

$$\mathsf{A} = \mathsf{A}_0 = \mathsf{Q}_1\mathsf{R}_1 = \mathcal{Q}_1\mathcal{R}_1,$$

$$\mathsf{A}^2 = \mathsf{Q}_1\mathsf{R}_1\mathsf{Q}_1\mathsf{R}_1 = \mathsf{Q}_1(\mathsf{R}_1\mathsf{Q}_1)\mathsf{R}_1 = \mathsf{Q}_1\mathsf{A}_1\mathsf{R}_1 = \mathsf{Q}_1(\mathsf{Q}_2\mathsf{R}_2)\mathsf{R}_1 = \mathcal{Q}_2\mathcal{R}_2,$$

and

$$\begin{aligned}
\mathsf{A}^3 &= \mathsf{Q}_1(\mathsf{R}_1\mathsf{Q}_1)\mathsf{R}_1\mathsf{Q}_1\mathsf{R}_1 \\
&= \mathsf{Q}_1\mathsf{A}_1\mathsf{R}_1\mathsf{Q}_1\mathsf{R}_1 \\
&= \mathsf{Q}_1\mathsf{Q}_2\mathsf{R}_2(\mathsf{R}_1\mathsf{Q}_1)\mathsf{R}_1 \\
&= \mathsf{Q}_1\mathsf{Q}_2\mathsf{R}_2\mathsf{A}_1\mathsf{R}_1 \\
&= \mathsf{Q}_1\mathsf{Q}_2\mathsf{R}_2(\mathsf{R}_1\mathsf{Q}_1)\mathsf{R}_1 \\
&= \mathsf{Q}_1\mathsf{Q}_2(\mathsf{R}_2\mathsf{Q}_2)\mathsf{R}_2\mathsf{R}_1 \\
&= \mathsf{Q}_1\mathsf{Q}_2\mathsf{A}_2\mathsf{R}_2\mathsf{R}_1 \\
&= \mathsf{Q}_1\mathsf{Q}_2\mathsf{Q}_3\mathsf{R}_3\mathsf{R}_2\mathsf{R}_1 \\
&= \mathcal{Q}_3\mathcal{R}_3.
\end{aligned}$$

Continuing in this fashion, we have

$$\mathsf{A}^k = \mathsf{Q}_1\mathsf{R}_1 \cdots \mathsf{Q}_1\mathsf{R}_1 = \cdots = \mathsf{Q}_1 \cdots \mathsf{Q}_k\mathsf{R}_k \cdots \mathsf{R}_1 = \mathcal{Q}_k\mathcal{R}_k.$$

Now, since A has distinct eigenvalues, it is diagonalizable. Therefore, there is an invertible matrix P such that $\mathsf{A} = \mathsf{P}\mathsf{D}\mathsf{P}^{-1}$. Assume that P^{-1} has an LU factorization:

$$\mathsf{P}^{-1} = \mathsf{L}\mathsf{U}.$$

According to Theorem 3.14, this factorization is unique, if it exists. Every matrix has a unique QR factorization of the following form:

$$\mathsf{P} = \mathsf{Q}\mathsf{R},$$

where $\mathsf{Q} \in \mathbb{C}^{n\times n}$ is unitary and $\mathsf{R} \in \mathbb{C}^{n\times n}$ is upper triangular with positive diagonal entries. Thus,

$$\mathsf{A}^k = \mathsf{P}\mathsf{D}^k\mathsf{P}^{-1} = \mathsf{Q}\mathsf{R}(\mathsf{D}^k\mathsf{L}\mathsf{D}^{-k})\mathsf{D}^k\mathsf{U}.$$

Observe that, since $\mathsf{L} = [\ell_{i,j}]$ is unit lower triangular,

$$\left[\mathsf{D}^k\mathsf{L}\mathsf{D}^{-k}\right]_{i,j} = \begin{cases} 0, & i < j, \\ 1, & i = j, \\ \left(\frac{\lambda_i}{\lambda_j}\right)^k \ell_{i,j}, & i > j. \end{cases}$$

Since $\lim_{k\to\infty}\left(\frac{\lambda_i}{\lambda_j}\right)^k = 0$, it follows that

$$\lim_{k\to\infty}\left(\mathsf{D}^k\mathsf{L}\mathsf{D}^{-k}\right) = \mathsf{I}.$$

Now let us set

$$\mathsf{D}^k\mathsf{L}\mathsf{D}^{-k} = \mathsf{I} + \mathsf{F}_k,$$

where

$$\lim_{k\to\infty} \mathsf{F}_k = \mathsf{O}.$$

It follows that

$$\mathsf{R}(\mathsf{D}^k\mathsf{L}\mathsf{D}^{-k}) = (\mathsf{I} + \mathsf{R}\mathsf{F}_k\mathsf{R}^{-1})\mathsf{R}.$$

Since $\mathsf{F}_k \to \mathsf{O}$, there is some $K \in \mathbb{N}$ such that, for every $k \geq K$, $\left\|\mathsf{R}\mathsf{F}_k\mathsf{R}^{-1}\right\|_\infty < 1$. Thus, owing to Theorem 4.19, if $k \geq K$, $\mathsf{I} + \mathsf{R}\mathsf{F}_k\mathsf{R}^{-1}$ is invertible, and therefore admits a unique QR factorization:

$$\mathsf{I} + \mathsf{R}\mathsf{F}_k\mathsf{R}^{-1} = \tilde{\mathsf{Q}}_k\tilde{\mathsf{R}}_k,$$

where $\tilde{\mathsf{Q}}_k \in \mathbb{C}^{n\times n}$ is unitary and $\tilde{\mathsf{R}}_k \in \mathbb{C}^{n\times n}$ is upper triangular with positive diagonal entries. Since the sequence $\left\{\tilde{\mathsf{Q}}_k\right\}_{k=1}^\infty$ is bounded — in particular, $\left\|\tilde{\mathsf{Q}}_k\right\|_2 = 1$ — owing to Theorem B.9, there is a convergent subsequence $\left\{\tilde{\mathsf{Q}}_{k_m}\right\}_{m=1}^\infty \subseteq \left\{\tilde{\mathsf{Q}}_k\right\}_{k=1}^\infty$. Set

$$\tilde{\mathsf{Q}} = \lim_{m\to\infty} \tilde{\mathsf{Q}}_{k_m}.$$

It is not difficult to see that $\tilde{\mathsf{Q}}$ must be unitary. Next, we observe that

$$\tilde{\mathsf{R}}_{k_m} = \tilde{\mathsf{Q}}_{k_m}^{\mathsf{H}}(\mathsf{I} + \mathsf{R}\mathsf{F}_{k_m}\mathsf{R}^{-1}) \to \tilde{\mathsf{R}},$$

where $\tilde{\mathsf{R}}$ must be upper triangular. Its diagonal elements must be nonnegative.

Taking the limit of the subsequence that we constructed, we find

$$\mathsf{I} = \tilde{\mathsf{Q}}\tilde{\mathsf{R}},$$

which implies that the diagonal elements of $\tilde{\mathsf{R}}$ must be positive. By uniqueness of the QR factorization, $\mathsf{I} = \tilde{\mathsf{Q}} = \tilde{\mathsf{R}}$. We leave it to the reader to prove that, in fact, the original sequences $\{\tilde{\mathsf{Q}}_k\}_{k\geq 1}$ and $\{\tilde{\mathsf{R}}_k\}_{k\geq 1}$ actually converge and

$$\tilde{\mathsf{Q}}_k, \to \mathsf{I}, \qquad \tilde{\mathsf{R}}_k \to \mathsf{I}, \qquad k \to \infty.$$

Taking into account all the different factorizations we have performed we observe that we have

$$\mathsf{A}^k = \mathcal{Q}_k\mathcal{R}_k = \left(\mathsf{Q}\tilde{\mathsf{Q}}_k\right)\left(\tilde{\mathsf{R}}_k\mathsf{R}\mathsf{D}^k\mathsf{U}\right).$$

Note that $\mathsf{Q}\tilde{\mathsf{Q}}_k$ is unitary and $\tilde{\mathsf{R}}_k\mathsf{R}\mathsf{D}^k\mathsf{U}$ is upper triangular. Furthermore, $\mathcal{Q}_k$ is unitary and $\mathcal{R}_k$ is upper triangular. We do not assert here that the upper triangular matrices have positive diagonal entries. Therefore, there is a diagonal matrix $\mathsf{S}_k = \operatorname{diag}\left(s_1^{(k)}, \ldots, s_n^{(k)}\right)$ with $|s_i^{(k)}| = 1$, for $i = 1, \ldots, n$, such that

$$\mathcal{Q}_k = \mathsf{Q}\tilde{\mathsf{Q}}_k\mathsf{S}_k.$$

This matrix, effectively, accounts for the loss of uniqueness.

Finally, some simple manipulations reveal that

$$\begin{aligned}
\mathsf{A}_{k+1} &= \mathcal{Q}_k^{\mathsf{H}}\mathsf{A}\mathcal{Q}_k \\
&= \mathcal{Q}_k^{\mathsf{H}}\mathsf{P}\mathsf{D}\mathsf{P}^{-1}\mathcal{Q}_k \\
&= \mathcal{Q}_k^{\mathsf{H}}\mathsf{Q}\mathsf{R}\mathsf{D}\mathsf{R}^{-1}\mathsf{Q}^{\mathsf{H}}\mathcal{Q}_k \\
&= \mathsf{S}_k^{\mathsf{H}}\tilde{\mathsf{Q}}_k^{\mathsf{H}}\mathsf{Q}^{\mathsf{H}}\mathsf{Q}\mathsf{R}\mathsf{D}\mathsf{R}^{-1}\mathsf{Q}^{\mathsf{H}}\mathsf{Q}\tilde{\mathsf{Q}}_k\mathsf{S}_k \\
&= \mathsf{S}_k^{\mathsf{H}}\tilde{\mathsf{Q}}_k^{\mathsf{H}}\mathsf{R}\mathsf{D}\mathsf{R}^{-1}\tilde{\mathsf{Q}}_k\mathsf{S}_k.
\end{aligned}$$

Since $\lim_{k\to\infty}\tilde{\mathsf{Q}}_k = \mathsf{I}$,

$$\lim_{k\to\infty}\left(\tilde{\mathsf{Q}}_k^{\mathsf{H}}\mathsf{R}\mathsf{D}\mathsf{R}^{-1}\tilde{\mathsf{Q}}_k\right) = \mathsf{R}\mathsf{D}\mathsf{R}^{-1} = \begin{bmatrix} \lambda_1 & \times & \cdots & \times \\ 0 & \lambda_2 & \ddots & \vdots \\ \vdots & \ddots & \ddots & \times \\ 0 & \cdots & 0 & \lambda_n \end{bmatrix}.$$

Define

$$\mathcal{D}_k = \tilde{\mathsf{Q}}_k^{\mathsf{H}}\mathsf{R}\mathsf{D}\mathsf{R}^{-1}\tilde{\mathsf{Q}}_k.$$

Since the matrix S_k is diagonal,

$$[\mathsf{A}_{k+1}]_{i,j} = \overline{s_i^{(k)}}s_j^{(k)}[\mathcal{D}_k]_{i,j}.$$

Consequently,

$$[\mathsf{A}_{k+1}]_{i,i} = [\mathcal{D}_k]_{i,i} \to \lambda_i, \qquad k \to \infty.$$

The proof is thus complete. □

Remark 8.27 (convergence). Observe that the previous result says nothing about the convergence of $[A_k]_{i,j}$ for $j > i$. Notice also that the order of the eigenvalues is preserved.

When the matrix is Hermitian, the previous result can be refined.

Theorem 8.28 (convergence II). *Suppose that the QR iteration method is applied to a Hermitian matrix $A \in \mathbb{C}^{n\times n}$, whose eigenvalues satisfy*

$$|\lambda_1| > |\lambda_2| > \cdots > |\lambda_n|$$

and whose corresponding unitary eigenvector matrix Q has all nonsingular leading principal sub-matrices. Then, as $k \to \infty$, the matrices A_k converge linearly, with constant

$$\max_j \left|\frac{\lambda_{j+1}}{\lambda_j}\right|,$$

to $\Lambda = \operatorname{diag}(\lambda_1, \lambda_2, \dots, \lambda_n)$ and Q_k converges, with the same rate, to Q.

Proof. Extend the previous proof by using the fact that the matrix A is unitarily diagonalizable, i.e., $A = Q\Lambda Q^H$, and Q has an LU factorization. The details are left for Problem 8.15. □

We can give a slightly different proof in the case that A is Hermitian positive definite (HPD), by taking advantage of the Cholesky factorization.

Theorem 8.29 (convergence III). *Suppose that $A \in \mathbb{C}^{n\times n}$ is HPD. Then the sequence of matrices $\{A_k\}_{k=1}^{\infty}$ produced by the QR iteration converges to a diagonal matrix $D = \operatorname{diag}(\lambda_n, \dots, \lambda_1)$, where*

$$0 < \lambda_1 \leq \cdots \leq \lambda_n$$

are the eigenvalues of A.

Proof. Suppose that $B = A^2$, so that B is HPD. Consider the following sequence: set $B_0 = B$ and, for all $k \geq 0$, if $B_k = U_{k+1}^H U_{k+1}$ is the unique Cholesky factorization of B_k, then

$$B_{k+1} = U_{k+1} U_{k+1}^H.$$

It is left to the reader as a simple exercise to show that if B_k is HPD, then B_{k+1} is always HPD; therefore, it will possess a unique Cholesky factorization. Here, $U_k \in \mathbb{C}^{n\times n}$ is an upper triangular matrix with strictly positive real diagonal entries.

For the QR iteration method, recall that $A_0 = A$ and, for $k \geq 1$, if $A_{k-1} = Q_k R_k$, then

$$A_k = R_k Q_k.$$

Each A_k is unitarily equivalent to A and is HPD. Furthermore,

$$A_k^2 = A_k^H A_k = R_{k+1}^H Q_{k+1}^H Q_{k+1} R_{k+1} = R_{k+1}^H R_{k+1}$$

is a Cholesky decomposition of A_k^2 and

$$A_{k+1}^2 = A_{k+1} A_{k+1}^H = R_{k+1} Q_{k+1} Q_{k+1}^H R_{k+1}^H = R_{k+1} R_{k+1}^H.$$

It follows by uniqueness that

$$\mathsf{B}_k = \mathsf{A}_k^2, \quad \mathsf{U}_k = \mathsf{R}_k, \quad \forall k = 1, 2, \ldots .$$

Also, it must be that each B_k is unitarily equivalent to $\mathsf{B} = \mathsf{A}^2$. In particular, since $\mathsf{R}_{k+1} = \mathsf{Q}_{k+1}^{\mathsf{H}} \mathsf{R}_k \mathsf{Q}_k$,

$$\mathsf{B}_{k+1} = \mathsf{Q}_{k+1}^{\mathsf{H}} \mathsf{R}_k \mathsf{Q}_k \mathsf{Q}_k^{\mathsf{H}} \mathsf{R}_k^{\mathsf{H}} \mathsf{Q}_{k+1} = \mathsf{Q}_{k+1}^{\mathsf{H}} \mathsf{B}_k \mathsf{Q}_{k+1}, \quad k = 0, 1, \ldots .$$

Now, since each R_k is upper triangular,

$$[\mathsf{B}_k]_{i,j} = \sum_{m=1}^{i} \overline{[\mathsf{R}_{k+1}]_{m,i}} \, [\mathsf{R}_{k+1}]_{m,j} \, .$$

Similarly,

$$[\mathsf{B}_{k+1}]_{i,j} = \sum_{m=i}^{n} [\mathsf{R}_{k+1}]_{i,m} \, \overline{[\mathsf{R}_{k+1}]_{j,m}} .$$

With these formulas we observe that, for $1 \leq m \leq n$,

$$\sum_{p=1}^{m} [\mathsf{B}_k]_{p,p} = \sum_{i=1}^{m} \sum_{j=1}^{m} \left| [\mathsf{R}_{k+1}]_{i,j} \right|^2$$

and

$$\sum_{p=1}^{m} [\mathsf{B}_{k+1}]_{p,p} = \sum_{i=1}^{m} \sum_{j=1}^{n} \left| [\mathsf{R}_{k+1}]_{i,j} \right|^2 .$$

Thus, for $1 \leq m < n$,

$$\sum_{p=1}^{m} [\mathsf{B}_{k+1}]_{p,p} - \sum_{p=1}^{m} [\mathsf{B}_k]_{p,p} = \sum_{i=1}^{m} \sum_{j=m+1}^{n} \left| [\mathsf{R}_{k+1}]_{i,j} \right|^2 , \tag{8.4}$$

and, for $m = n$,

$$\sum_{p=1}^{n} [\mathsf{B}_{k+1}]_{p,p} - \sum_{p=1}^{n} [\mathsf{B}_k]_{p,p} = 0,$$

since the traces of the matrices B_k and B_{k+1} are equal.

Thus, we deduce that the sequences

$$\left\{ \sum_{p=1}^{m} [\mathsf{B}_k]_{p,p} \right\}_{k=1}^{\infty} \tag{8.5}$$

are increasing for each $m = 1, \ldots, n$. Since each B_k is unitarily equivalent to $\mathsf{B} = \mathsf{A}^2$,

$$\|\mathsf{B}_k\|_2 = \left\| \mathsf{A}^2 \right\|_2 ;$$

consequently, the elements of B_k must be bounded. Therefore, the sequences in (8.5) must remain bounded. By the monotone convergence theorem (see

Theorem B.7), the sequence of (8.5) converges for each $m = 1, \dots, n$. It must be that the sequences

$$\left\{ [\mathsf{B}_k]_{p,p} \right\}_{k=1}^{\infty} \tag{8.6}$$

converge for each $p = 1, \dots, n$. We denote their limits as the diagonal elements of the matrix B_∞, i.e.,

$$\lim_{k\to\infty} [\mathsf{B}_k]_{p,p} = [\mathsf{B}_\infty]_{p,p}.$$

Since this is the case, (8.4) implies that all of the off-diagonal elements of R_k must tend to zero. Thus, this is also the case for the off-diagonal elements of B_k. This implies that $\mathsf{B}_\infty = \mathsf{D}$ is a diagonal matrix. The theorem is proven. □

8.6.1 QR with Shifts

We conclude by commenting that there is a variant of the QR iteration method, presented in Listing 8.3, that is known as *QR with shifts*, which, in practice, converges faster than the standard QR iteration. This topic is beyond the scope of our text, and we refer the reader to [26, 34, 70, 96, 102] for more information.

8.7 Computation of the SVD

In this final section concerning the computation of eigenvalues and eigenvectors, we will study the computation of the singular value decomposition (SVD), as the techniques are closely related to the QR iteration method described in the previous section. The approaches rely upon certain properties of the SVD, which we will recall as needed.

As a first approach, we see that if $\mathsf{A} \in \mathbb{C}^{m\times n}$ with $m \geq n$ and $\mathsf{A} = \mathsf{U}\Sigma\mathsf{V}^{\mathsf{H}}$ is an SVD of A, then

$$\mathsf{A}^{\mathsf{H}}\mathsf{A} = \mathsf{V}\Sigma^{\intercal}\Sigma\mathsf{V}^{\mathsf{H}}$$

is a Schur (eigenvalue) decomposition of the matrix $\mathsf{A}^{\mathsf{H}}\mathsf{A}$; see Theorem 2.3. This suggests the following approach.

1. Form the Hermitian matrix $\mathsf{B} = \mathsf{A}^{\mathsf{H}}\mathsf{A}$.
2. Use the QR iteration method to find a decomposition of the form $\mathsf{V}^{\mathsf{H}}\mathsf{B}\mathsf{V} = \mathsf{D} = \operatorname{diag}(\sigma_1^2, \dots, \sigma_n^2)$, where $\{\sigma_i\}_{i=1}^n$ are the singular values of A and $\mathsf{V} \in \mathbb{C}^{n\times n}$ is unitary.
3. Define $\Sigma = \operatorname{diag}(\sigma_1, \dots, \sigma_n) \in \mathbb{R}^{m\times n}$.
4. Find a QR decomposition of $\mathsf{A}\mathsf{V} \in \mathbb{C}^{m\times n}$ to obtain $\mathsf{U}\mathsf{R} = \mathsf{A}\mathsf{V}$ with $\mathsf{U} \in \mathbb{C}^{m\times m}$ unitary and $\mathsf{R} \in \mathbb{C}^{m\times n}$ upper triangular.
5. From the computations above, we see that $\mathsf{A} = \mathsf{U}\Sigma\mathsf{V}^{\mathsf{H}}$ is the desired SVD.

This approach is not satisfactory in practice, as it is rather sensitive to perturbations. A more computationally robust approach is motivated by Problem 2.4. Let $\mathsf{A} \in \mathbb{C}^{m\times n}$ with $m \geq n$ and $\mathsf{A} = \hat{\mathsf{U}}\hat{\Sigma}\mathsf{V}^{\mathsf{H}}$ be a reduced SVD of A. We recall that the matrix $V \in \mathbb{C}^{n\times n}$ is unitary, $\hat{\Sigma} = \operatorname{diag}(\sigma_1, \dots, \sigma_n) \in \mathbb{R}^{n\times n}$, and $\hat{U} \in \mathbb{C}^{m\times n}$ has orthonormal columns. Let $\tilde{U} \in \mathbb{C}^{m\times(m-n)}$ be such that

$$U = [\hat{U}\ \tilde{U}] \in \mathbb{C}^{m\times m}$$

is unitary. Define

$$C = \begin{bmatrix} O_{n\times n} & A^H \\ A & O_{m\times m} \end{bmatrix}, \quad Q = \frac{1}{\sqrt{2}} \begin{bmatrix} V & V & O_{n\times(m-n)} \\ \hat{U} & -\hat{U} & \sqrt{2}\tilde{U} \end{bmatrix} \in \mathbb{C}^{(m+n)\times(m+n)},$$

where $O_{k\times p} \in \mathbb{R}^{k\times p}$ is the zero matrix of size $k \times p$. An easy and direct computation shows that Q is unitary and that

$$Q^H C Q = \operatorname{diag}\left(\hat{\Sigma}, -\hat{\Sigma}, O_{(m-n)\times(m-n)}\right).$$

Thus, the computation of the SVD of A can be obtained from the eigenvalue decomposition of the Hermitian matrix C.

The last method is stable to perturbations, but dealing with the large matrix C is problematical. To circumvent the need for computing with C, we use an algorithm that is now known as the *Golub–Kahan method.*[8]

The Golub–Kahan algorithm proceeds in two stages.

1. Bidiagonalize A: Put A into upper bidiagonal form. This is obtained by applying different unitary transformations from the left and the right to A, so that

$$U_1^H A V_1 = T = \begin{bmatrix} \tilde{T} \\ O_{(m-n)\times n} \end{bmatrix}, \qquad \tilde{T} = \begin{bmatrix} \times & \times & & & \\ & \times & \times & & \\ & & \times & \ddots & \\ & & & \ddots & \times \\ & & & & \times \end{bmatrix} \in \mathbb{C}^{n\times n}. \quad (8.7)$$

   Notice that, unlike in a reduction to Hessenberg form, we are not applying a similarity transformation to this matrix. This allows more flexibility and the possibility to reduce the matrix to upper bidiagonal form. The matrices $U_1 \in \mathbb{C}^{m\times m}$ and $V_1 \in \mathbb{C}^{n\times n}$ are unitary and can be computed, for instance, as products of Householder reflectors.
2. Compute the SVD of T: Find the SVD of the upper bidiagonal matrix $T = U_2 \Sigma V_2^H$. From the previous computations,

$$A = U_1 U_2 \Sigma V_2^H V_1^H = U \Sigma V^H,$$

   with $U = U_1 U_2$, and $V = V_1 V_2$ is an SVD for A.

Evidently, to realize the Golub–Kahan method, one must be able to (i) bidiagonalize A and (ii) compute the SVD of the bidiagonal matrix T. The next result guarantees that any matrix A can be put into upper bidiagonal form using unitary matrices.

[8] Named in honor of American mathematician Gene Howard Golub (1932–2007) and the Canadian mathematician and computer scientist William Morton Kahan (1933–).

Lemma 8.30 (upper bidiagonal form). *Suppose that* $\mathsf{A} \in \mathbb{C}^{m\times n}$ *with* $m \geq n$. *There exist unitary matrices* $\mathsf{U}_1 \in \mathbb{C}^{m\times m}$ *and* $\mathsf{V}_1 \in \mathbb{C}^{n\times n}$ *such that* $\mathsf{U}_1^{\mathsf{H}}\mathsf{A}\mathsf{V}_1$ *is upper bidiagonal, as in* (8.7). *In particular,* U_1 *and* V_1 *are products of Householder reflectors.*

Proof. The idea is to mimic the similarity transformation of a square matrix to upper Hessenberg form. However, we have the added flexibility that we are allowed to use different alternating Householder matrices on the right and left; see Problem 8.16. □

The following result gives a way to reduce the computation of the SVD of an upper bidiagonal matrix to the computation of the eigenpairs of a Hermitian tridiagonal matrix.

Lemma 8.31 (reduction to eigenvalues). *Let* $\mathsf{T} \in \mathbb{C}^{m\times n}$ *with* $m \geq n$ *be bidiagonal. The matrix* $\mathsf{Z} = \mathsf{T}^{\mathsf{H}}\mathsf{T}$ *is tridiagonal and Hermitian. Moreover, the singular values of* T *are the square roots of the eigenvalues of* Z. *The right singular vectors of* T *are the eigenvectors of* Z.

Proof. See Problem 8.17. □

The previous result implies that, by applying the QR iteration method to $\mathsf{Z} = \mathsf{T}^{\mathsf{H}}\mathsf{T}$, one can compute the factors Σ and V_2 in step 2 above. The left singular vectors can then be computed by a QR factorization of $\mathsf{T}\mathsf{V}_2$. There are variants of this idea that do not require to form the matrix Z.

We refer the reader to [26, 34] for many more practical details, variants, and convergence properties on the Golub–Kahan method.

Problems

8.1 A Frobenius matrix is a matrix that has the form

$$\mathsf{F} = \begin{bmatrix} a_1 & a_2 & \cdots & a_{n-1} & a_n \\ 1 & 0 & \cdots & 0 & 0 \\ 0 & 1 & \ddots & \vdots & \vdots \\ \vdots & \ddots & \ddots & 0 & 0 \\ 0 & \cdots & 0 & 1 & 0 \end{bmatrix}.$$

Find χ_{F} for a Frobenius matrix.

8.2 Let $\mathsf{A} \in \mathbb{C}^{n\times n}$ have eigenvalues $\sigma(\mathsf{A}) = \{\lambda_1, \ldots, \lambda_n\}$. Define

$$R_i = \sum_{\substack{j=1\\ j\neq i}}^{n} |a_{i,j}|, \quad C_j = \sum_{\substack{i=1\\ i\neq j}}^{n} |a_{i,j}|.$$

a) Prove that

$$\sigma(\mathsf{A}) \subseteq \bigcup_{i=1}^{n} \{z \in \mathbb{C} \mid |z - a_{i,i}| \leq R_i\}.$$

b) Prove that

$$\sigma(\mathsf{A}) \subseteq \bigcup_{j=1}^{n} \{z \in \mathbb{C} \mid |z - a_{j,j}| \le C_j\}.$$

8.3 Let $\mathsf{A} = [a_{i,j}] \in \mathbb{C}^{n\times n}$. Let

$$R_i = \sum_{\substack{j=1 \\ j\neq i}}^{n} |a_{i,j}|.$$

Define the *Cassini ovals*[9] as

$$C_{i,j} = \{z \in \mathbb{C} \mid |z - a_{i,i}||z - a_{j,j}| < R_i R_j\}.$$

a) Show that

$$\bigcup_{i,j=1}^{n} C_{i,j} \subseteq \bigcup_{i=1}^{n} D_i,$$

where $D_1, \ldots, D_n$ denotes the Gershgorin disks.

b) Show that

$$\sigma(\mathsf{A}) \subset \bigcup_{i,j=1}^{n} C_{i,j}.$$

c) Let $0 < \varepsilon \ll 1$ and consider the 2×2 matrix

$$\mathsf{A} = \begin{bmatrix} 1 & -\varepsilon \\ \varepsilon & -1 \end{bmatrix}.$$

Find the eigenvalues of A and compare the predictions given by the Gershgorin disks and Cassini ovals. Which one seems more accurate?

8.4 Let $\mathsf{A} = [a_{i,j}] \in \mathbb{C}^{n\times n}$. Let

$$R_i = \sum_{\substack{j=1 \\ j\neq i}}^{n} |a_{i,j}|.$$

Show that if

$$|a_{i,i}a_{j,j}| > R_i R_j,$$

then the matrix A is nonsingular.

8.5 Let $\mathsf{A} = [a_{i,j}] \in \mathbb{C}^{n\times n}$ have the property that, for some k and all i,

$$|a_{k,k} - a_{i,i}| > \sum_{j\neq k} |a_{k,j}| + \sum_{j\neq i} |a_{i,j}|.$$

Show that the region

$$\mathcal{D} = \left\{ \beta \in \mathbb{C} \;\middle|\; |\beta - a_{k,k}| \le \sum_{j\neq k} |a_{k,j}| \right\}$$

[9] Named in honor of the Italian (naturalized French) mathematician, astronomer, and engineer Giovanni Domenico Cassini (1625–1712).

contains one, and only one, eigenvalue.

8.6 Prove Corollary 8.7.

8.7 Prove Theorem 8.9.

8.8 Prove Proposition 8.12.

8.9 Complete the proof of Theorem 8.16.

8.10 Suppose that $A \in \mathbb{C}^{n \times n}$ is diagonalizable (not necessarily Hermitian) and $\boldsymbol{x}_0 \in \mathbb{C}^n$. Define the sequence of vectors $\{\boldsymbol{x}_k\}_{k=0}^{\infty} \subset \mathbb{C}^n$ via

$$\boldsymbol{x}_k = \frac{A^k \boldsymbol{x}_0}{\|A^k \boldsymbol{x}_0\|_\infty}, \quad k = 1, 2, 3, \dots.$$

Suppose that the spectrum $\sigma(A) = \{\lambda_i\}_{i=1}^n \subset \mathbb{C}$ has the ordering

$$0 \le |\lambda_1| \le |\lambda_2| \le \cdots \le |\lambda_{n-1}| < |\lambda_n|.$$

Suppose that $\boldsymbol{v}_n$ is the eigenvector associated with the dominant eigenvector λ_n. Assuming that $\boldsymbol{x}_0$ has a nonzero component in the direction of $\boldsymbol{v}_n$, prove that $\boldsymbol{x}_k$ converges to a multiple of $\boldsymbol{v}_n$.

8.11 Prove Proposition 8.17.

8.12 Prove Theorem 8.19.

8.13 Prove Theorem 8.23.

8.14 Complete the proof of Proposition 8.25.

8.15 Complete the proof of Theorem 8.28.

8.16 Prove Lemma 8.30.

8.17 Prove Lemma 8.31.

Listings

```
function [M, Q, err] = Hessenberg(A)
% The following algorithm reduces the matrix A to Hessenberg
% form
%
% M = Q'AQ
%
% Input
%   A(1:n,1:n) : a square matrix
%
% Output
%  M(1:n,1:n) : a lower Hessenberg matrix that is similar to A
%  Q(1:n,1:n) : the unitary matrix that realizes the
%               transformation
%  err : = 0, if no error was encountered
%        = 1, if there was an error
  [m,n] = size(A);
  M = A;
  err = 0;
  if m ~= n
    err = 1;
    return;
```

```
    end
    Q = eye(n);
    for k=1:n-2
      nn = norm( M(k+1:n,k) );
      e = zeros(n-k,1);
      e(1) = 1;
      v = nn*e + M(k+1:n,k);
      norm_v = norm(v);
      if norm_v < eps( norm_v )
        err = 1;
        return;
      end
      v = (1.0/norm_v)*v;
      M(k+1:n,k:n) = M(k+1:n,k:n) - 2.0*(v*v')*M(k+1:n,k:n);
      M(1:n,k+1:n) = M(1:n,k+1:n) - 2.0*M(1:n,k+1:n)*(v*v');
      Q(1:n,k+1:n) = Q(1:n,k+1:n) - 2.0*Q(1:n,k+1:n)*(v*v');
    end
end
```

Listing 8.1 Reduction of a matrix to lower Hessenberg form via similarity transformations.

```
function [U, EigVecs, err] = QRIter( A, maxits )
% The QR iteration method to find eigenvalues.
%
% Input
%   A(1:n,1:n) : a square matrix
%   maxits : the maximal number of iterations
%
% Output
%  U(1:n,1:n) : an approximation to an upper triangular matrix
%               that is similar to A
%  EigVecs(1:n,1:n) : A unitary matrix whose columns are
%                     eigenvectors of A
%  err : = 0, if no error was encountered
%        = 1, if there was an error
  err = 0;
  [U, EigVecs, err] = Hessenberg(A);
  if err == 1
    return;
  end
  for i=1:maxits
    [Q, R, err] = QRFact( U, 1 );
    if err == 1
      return;
    end
    U = R*Q;
    EigVecs = EigVecs*Q;
  end
end
```

Listing 8.2 The QR iteration to find eigenvalues.

```
function [U, EigVecs, err] = QRIterShifts( A, maxits )
% The QR iteration method with shifts to find eigenvalues.
%
% Input
%   A(1:n,1:n) : a square matrix
%   maxits : the maximal number of iterations
%
% Output
%  U(1:n,1:n) : an approximation to an upper triangular matrix
%               that is similar to A
%  err : = 0, if no error was encountered
%        = 1, if there was an error
  err = 0;
  [U, EigVecs, err] = Hessenberg(A);
  if err == 1
    return;
  end
  n = size(A,1);
  id = eye(n,n);
  for i=1:maxits
    mu = U(n,n);
    [Q, R, err] = QRFact( U - mu*id, 1 );
    if err == 1
      return;
    end
    U = R*Q + mu*id;

    EigVecs = EigVecs*Q;
  end
end
```

Listing 8.3 The QR iteration with shifts to find eigenvalues.

Part II

Constructive Approximation Theory

9 Polynomial Interpolation

In this chapter, we begin the study of *constructive approximation theory*, which, as its name suggests, is concerned with methods to approximate a function (which may only be known approximately) by a simpler one. One of the recurring features in our discussion will be the interplay between the smoothness of a function, measured in an appropriate sense, and the quality of approximation that we are able to produce. Applications of approximation theory are plentiful, and we will see several of these throughout this book. For instance, in Chapter 14, we will see how this is used to approximate the value of integrals. It will also play a central role in Part V, where the performance of a numerical scheme for the approximation of the solution to a boundary value problem depends in a fundamental way not only on the method of choice but also on the *smoothness of the solution*.

We begin approximation theory with the topic of polynomial approximation. Given a, usually continuous, function we wish to construct a polynomial satisfying certain properties that, in a very definite sense, approximates the given function. In fact, most of this part of the book is about generating approximations with polynomials, of both the ordinary and the trigonometric kind.

Let us immediately remark that, since our discussion is now concerned with functions and their properties, the reader should be familiar with some basic facts in real analysis: continuity, compactness, etc. We refer to Appendix B for a review and guide to notation. Some facts about spaces of smooth functions will also be necessary, and Appendix D provides an overview. With this in mind, our discussion is started by first presenting a cornerstone approximation result, which the reader may have seen before. For the moment, this is stated without proof.

Theorem (Weierstrass Approximation Theorem[1]). *Let $[a, b] \subset \mathbb{R}$ be a compact interval and $f \in C([a, b])$. For every $\varepsilon > 0$, there exists an $n \in \mathbb{N}$ and a polynomial $p_n \in \mathbb{P}_n$ such that*

$$\|f - p_n\|_{L^\infty(a,b)} \leq \varepsilon.$$

In other words, there is a sequence of polynomials, $\{p_n\}_{n=0}^{\infty}$ with $p_n \in \mathbb{P}_n$, that converges uniformly to f as $n \to \infty$.

This theorem tells us that a continuous function on a compact interval can be well approximated by polynomials, but it does not tell us how to construct such a polynomial and it does not tell us the convergence rate.

[1] Named in honor of the German mathematician Karl Theodor Wilhelm Weierstrass (1815–1897).

On the other hand, Taylor's Theorem B.31 gives a precise way to construct a polynomial approximation, but requires the function f to be n-times differentiable to build an approximating polynomial of degree n. Moreover, we have to know the values of all the derivatives at a certain point, which is not always practical.

Suppose, for example, that one only knows function values at a finite collection of points in the domain. What can we do? The celebrated theorems mentioned above do not give any indication. For constructive approximation, another place to start is interpolation. Interpolation works by demanding that a simple function — usually, but not always, a polynomial — exactly matches the values of a function of interest at a given number of points. Typically, we only require that the function of interest is continuous, or piecewise continuous, in its domain of definition. Interpolation is simple and often works well. But sometimes it can go badly wrong. Paradoxically, this can happen when we try to interpolate a large number of points. One must be careful, as we will see.

9.1 The Vandermonde Matrix and the Vandermonde Construction

To understand interpolation, we need some basic definitions.

Definition 9.1 (nodal set)**.** Let $[a, b] \subset \mathbb{R}$ be a compact interval. X is called a **nodal set** of size $n+1 \in \mathbb{N}$ in $[a, b]$ if and only if $X = \{x_i\}_{i=0}^n \subset [a, b]$ is a set of distinct elements. The elements of X, x_i are called **nodes**.

Observe that we usually enumerate the nodes in a very particular way, starting with 0 and ending with n. Also note that the points of any nodal set are always, by design, distinct. Not all finite sets that we shall introduce will have this property. Typically the nodes are numbered in increasing order, for convenience; but this is not always the case.

Definition 9.2 (interpolating polynomial)**.** Suppose that $X = \{x_i\}_{i=0}^n$ is a nodal set of size $n+1 \in \mathbb{N}$ contained in the compact interval $[a, b] \subset \mathbb{R}$ and $f: [a, b] \to \mathbb{R}$ is a function. The function $I: [a, b] \to \mathbb{R}$ is called an **interpolant of** f subordinate to X if and only if $I(x_i) = f(x_i)$, $i = 0, \dots, n$. In this case, we write $I(X) = f(X)$, for short. Suppose that $Y = \{y_i\}_{i=0}^n \subset \mathbb{R}$ is a set of not necessarily distinct points. Define the set of ordered pairs

$$O = \{(x_i, y_i) \mid x_i \in X,\ y_i \in Y,\ i = 0, \dots, n\}.$$

We say that I is an **interpolant of** O if and only if $I(x_i) = y_i$, $i = 0, \dots, n$. Often, we write $I(X) = Y$ as a shorthand. If the interpolant I is a polynomial, it is called an **interpolating polynomial**.

In short, an interpolant is a function that agrees with another, given, function at a finite number of points in its domain.

Definition 9.3 (Vandermonde matrix[2])**.** Suppose that $X = \{x_i\}_{i=0}^n$ is a nodal set

[2] Named in honor of the French mathematician, musician and chemist Alexandre-Théophile Vandermonde (1735–1796).

of size $n+1 \in \mathbb{N}$ in the compact interval $[a, b] \subset \mathbb{R}$. The **Vandermonde matrix** subordinate to the nodal set X, denoted $\mathsf{V} = [v_{i,j}] \in \mathbb{R}^{(n+1)\times(n+1)}$, is the matrix with entries

$$v_{i,j} = x_{i-1}^{j-1}, \quad i, j = 1, \dots, n+1,$$

where the superscript represents exponentiation. In other words,

$$\mathsf{V} = \begin{bmatrix} 1 & x_0 & x_0^2 & \cdots & x_0^n \\ 1 & x_1 & x_1^2 & \cdots & x_1^n \\ \vdots & \vdots & \vdots & & \vdots \\ 1 & x_n & x_n^2 & \cdots & x_n^n \end{bmatrix}.$$

Theorem 9.4 (Vandermonde). *Suppose that $X = \{x_i\}_{i=0}^n$ is a nodal set in the compact interval $[a, b] \subset \mathbb{R}$ and $\mathsf{V} = [v_{i,j}] \in \mathbb{R}^{(n+1)\times(n+1)}$ is the Vandermonde matrix subordinate to X. Then V is invertible and, in particular,*

$$\det(\mathsf{V}) = \prod_{0 \le i < j \le n} (x_j - x_i) \neq 0.$$

Proof. See Problem 9.1. Of course, if one can establish the determinant formula, the invertibility of V follows because nodes are distinct. □

With this at hand we can establish the existence and uniqueness of an interpolating polynomial.

Proposition 9.5 (existence and uniqueness). *Suppose that $X = \{x_i\}_{i=0}^n$ is a nodal set in the compact interval $[a, b] \subset \mathbb{R}$ and $Y = \{y_i\}_{i=0}^n \subset \mathbb{R}$. There is a unique polynomial $p \in \mathbb{P}_n$ with the property that $p(X) = Y$.*

Proof. We can set this problem up in the following way. Express p as $p(x) = \sum_{j=0}^n c_j x^j$, where the coefficients $c_j \in \mathbb{R}$ must be determined. Observe that p satisfies $p(X) = Y$ if and only if, for all $i = 0, \dots, n$,

$$p(x_i) = \sum_{j=0}^n c_j x_i^j = y_i.$$

In matrix form, this can be rewritten as

$$\begin{bmatrix} 1 & x_0 & x_0^2 & \cdots & x_0^n \\ 1 & x_1 & x_1^2 & \cdots & x_1^n \\ \vdots & \vdots & \vdots & & \vdots \\ 1 & x_n & x_n^2 & \cdots & x_n^n \end{bmatrix} \begin{bmatrix} c_0 \\ c_1 \\ \vdots \\ c_n \end{bmatrix} = \begin{bmatrix} y_0 \\ y_1 \\ \vdots \\ y_n \end{bmatrix}, \tag{9.1}$$

where the vector $\boldsymbol{c} = [c_0, \dots, c_n]^\intercal \in \mathbb{R}^{n+1}$ is unknown. Since the coefficient matrix is the Vandermonde matrix, which is invertible, the result follows. □

The interpolating polynomial construction represented by the linear system (9.1) is called the *Vandermonde construction*.

Example 9.1 While, in theory, all that is needed to construct an interpolating polynomial is to solve system (9.1), it turns out that the Vandermonde matrix tends to be severely ill-conditioned. For example, define $X_n = \{i/n\}_{i=0}^n$, a set of uniform nodes in $[0, 1]$. The following table shows how the spectral condition number of V grows with n:

n	$\kappa_2(\mathsf{V})$
4	6.86×10^{02}
8	2.01×10^{06}
16	2.42×10^{13}

Clearly, one should avoid using the Vandermonde matrix in practical applications when n is large. We will look for other more practical ways of constructing interpolants.

Interpolation, from the theoretical perspective, can be thought of as a linear projection operator.

Definition 9.6 (interpolation operator). Suppose that $X = \{x_i\}_{i=0}^n$ is a nodal set in the compact interval $[a, b] \subset \mathbb{R}$. The **interpolation operator** subordinate to X, denoted

$$\mathcal{I}_X : C([a, b]) \to \mathbb{P}_n,$$

is defined as follows: for $f \in C([a, b])$, $\mathcal{I}_X[f] \in \mathbb{P}_n$ is the unique interpolating polynomial satisfying $\mathcal{I}_X[f](X) = f(X)$.

The reader can easily prove the following result.

Proposition 9.7 (projection). *Suppose that $X = \{x_i\}_{i=0}^n$ is a nodal set in the compact interval $[a, b] \subset \mathbb{R}$ and $\mathcal{I}_X : C([a, b]) \to \mathbb{P}_n$ is the interpolation operator subordinate to X. $\mathcal{I}_X$ is a linear projection operator, meaning that*

$$\mathcal{I}_X[\alpha f + \beta g] = \alpha \mathcal{I}_X[f] + \beta \mathcal{I}_X[g], \quad \forall f, g \in C([a, b]), \quad \forall \alpha, \beta \in \mathbb{R}$$

and

$$\mathcal{I}_X[p] = p, \quad \forall p \in \mathbb{P}_n.$$

Proof. See Problem 9.2. □

The norm of the interpolation operator has a special name.

Definition 9.8 (Lebesgue constant[3]). Suppose that $X = \{x_i\}_{i=0}^n$ is a nodal set in the compact interval $[a, b] \subset \mathbb{R}$ and $\mathcal{I}_X : C([a, b]) \to \mathbb{P}_n$ is the interpolation

[3] Named in honor of the French mathematician Henri Léon Lebesgue (1875–1941).

operator subordinate to X. The **Lebesgue constant subordinate to** X, denoted $\Lambda(X)$, is the operator norm of $\mathcal{I}_X$, i.e., $\Lambda(X) = \|\mathcal{I}_X\|_\infty$, where

$$\|\mathcal{I}_X\|_\infty = \sup_{0\neq f\in C([a,b])} \frac{\|\mathcal{I}_X[f]\|_{L^\infty(a,b)}}{\|f\|_{L^\infty(a,b)}} = \sup_{\substack{f\in C([a,b])\\ \|f\|_{L^\infty(a,b)}=1}} \|\mathcal{I}_X[f]\|_{L^\infty(a,b)}. \tag{9.2}$$

In the next section, we will show how one might compute the Lebesgue constant. At this point, just note that the constant depends upon our choice of the nodal set.

9.2 Lagrange Interpolation and the Lagrange Nodal Basis

It is sometimes more convenient, from the viewpoint of representation and construction, to use the so-called Lagrange basis to build our interpolating polynomials.

Definition 9.9 (Lagrange nodal basis[4]). Suppose that $X = \{x_i\}_{i=0}^n$ is a nodal set of size $n+1 \in \mathbb{N}$ in the compact interval $[a,b] \subset \mathbb{R}$. The **Lagrange nodal basis** subordinate to X is the set of polynomials $\mathcal{L}_X = \{L_\ell\}_{\ell=0}^n \subset \mathbb{P}_n$ defined via

$$L_\ell(x) = \prod_{\substack{i=0\\ i\neq \ell}}^n \frac{x - x_i}{x_\ell - x_i}. \tag{9.3}$$

Proposition 9.10 (properties of $\mathcal{L}_X$). *Suppose that $X = \{x_i\}_{i=0}^n$ is a nodal set of size $n+1 \in \mathbb{N}$ in the compact interval $[a,b] \subset \mathbb{R}$ and $\mathcal{L}_X = \{L_i\}_{i=0}^n \subset \mathbb{P}_n$ is the Lagrange basis subordinate to X. Then $\mathcal{L}_X$ is a basis of $\mathbb{P}_n$ with the properties*

$$L_i(x_j) = \delta_{i,j}, \quad i,j = 0,\ldots,n \tag{9.4}$$

and

$$\sum_{i=0}^n L_i(x) = 1, \quad \forall x \in \mathbb{R}. \tag{9.5}$$

Proof. See Problem 9.3 □

Example 9.2 Figure 9.1 shows the Lagrange nodal basis subordinate to the nodal set $X = \{0, 0.30, 0.42, 0.71, 1.00\} \subset [0,1]$. The figure also shows the sum of the basis functions, $\sum_{i=0}^4 L_i(x)$, which confirms (9.5) for this case.

Theorem 9.11 (interpolation polynomial). *Suppose that $X = \{x_i\}_{i=0}^n$ is a nodal set in the compact interval $[a,b] \subset \mathbb{R}$, and $Y = \{y_i\}_{i=0}^n \subset \mathbb{R}$. Let $\mathcal{L}_X = \{L_i\}_{i=0}^n \subset \mathbb{P}_n$*

[4] Named in honor of the Italian, later naturalized French, mathematician and astronomer Joseph-Louis Lagrange (1736–1813).

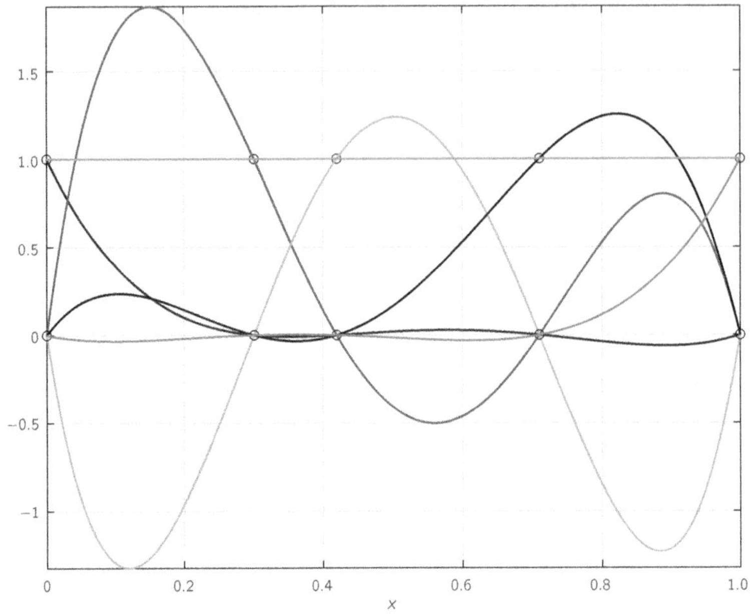

Figure 9.1 The Lagrange nodal basis of degree $n = 4$, subordinate to the nodal set $X = \{0, 0.30, 0.42, 0.71, 1.00\} \subset [0, 1]$. We also plot the sum of the basis functions, $\sum_{i=0}^{4} L_i(x)$, which confirms (9.5) for this case.

be the Lagrange nodal basis subordinate to X. The unique polynomial $p \in \mathbb{P}_n$, with the property that $p(X) = Y$, has the form

$$p(x) = \sum_{i=0}^{n} y_i L_i(x) \in \mathbb{P}_n. \tag{9.6}$$

Proof. From definition (9.6), we have, using property (9.4), that $p(x_i) = y_i$. □

Definition 9.12 (Lagrange interpolating polynomial). Suppose that $X = \{x_i\}_{i=0}^n$ is a nodal set in the compact interval $[a, b] \subset \mathbb{R}$, $\mathcal{L}_X = \{L_i\}_{i=0}^n \subset \mathbb{P}_n$ is the Lagrange nodal basis subordinate to X, and $f\colon [a, b] \to \mathbb{R}$. The **Lagrange interpolating polynomial** of the function f, subordinate to the nodal set X, is the polynomial

$$p(x) = \sum_{i=0}^{n} f(x_i) L_i(x) \in \mathbb{P}_n. \tag{9.7}$$

Observe that, by uniqueness, the Lagrange interpolating polynomial coincides with the interpolant we obtained in Proposition 9.5 via the Vandermonde construction. For historical reasons, the interpolating polynomial constructed by matching point values of the given function is called the Lagrange interpolating polynomial. We will give one more construction in Section 9.6 based on Newton's basis.

Definition 9.13 (nodal polynomial). Suppose that $X = \{x_i\}_{i=0}^n$ is a nodal set in the compact interval $[a, b] \subset \mathbb{R}$. The polynomial $\omega_{n+1} \in \mathbb{P}_{n+1}$, defined by

$$\omega_{n+1}(x) = \prod_{i=0}^{n}(x - x_i), \tag{9.8}$$

is called the **nodal polynomial subordinate to** X.

The following alternate formula for the elements of the Lagrange nodal basis is often useful.

Proposition 9.14 (Lagrange nodal basis). *Suppose that $X = \{x_i\}_{i=0}^n$ is a nodal set in the compact interval $[a, b] \subset \mathbb{R}$, $\mathcal{L}_X = \{L_i\}_{i=0}^n \subset \mathbb{P}_n$ is the Lagrange nodal basis subordinate to X, ω_{n+1} is the nodal polynomial subordinate to X, and $f\colon [a, b] \to \mathbb{R}$. Then, for all $i = 0, \ldots, n$, we have*

$$L_i(x) = \frac{\omega_{n+1}(x)}{(x - x_i)\omega'_{n+1}(x_i)}. \tag{9.9}$$

Consequently, the Lagrange interpolating polynomial, $p \in \mathbb{P}_n$, of the function f subordinate to the nodal set X is

$$p(x) = \sum_{i=0}^{n} f(x_i)\frac{\omega_{n+1}(x)}{(x - x_i)\omega'_{n+1}(x_i)}. \tag{9.10}$$

Proof. One can show that

$$\omega'_{n+1}(x_i) = \prod_{\substack{j=0 \\ j\neq i}}^{n}(x_i - x_j).$$

This and further details are left to the reader as an exercise; see Problem 9.5. □

Before we move on to the error analysis for Lagrange interpolation, let us compute the Lebesgue constant.

Theorem 9.15 ($\Lambda(X)$). *Suppose that $X = \{x_i\}_{i=0}^n$ is a nodal set in the compact interval $[a, b] \subset \mathbb{R}$. The Lebesgue constant satisfies*

$$\Lambda(X) = \max_{a\leq x\leq b} \lambda_X(x), \quad \lambda_X(x) = \sum_{i=0}^{n} |L_i(x)| \geq 1,$$

where $\mathcal{L}_X = \{L_i\}_{i=0}^n \subset \mathbb{P}_n$ is the Lagrange nodal basis subordinate to X. The function λ_X is called the Lebesgue function subordinate to X.

Proof. Suppose that $f \in C([a, b])$ is arbitrary. Then

$$|\mathcal{I}_X[f](x)| = \left|\sum_{i=0}^{n} f(x_i)L_i(x)\right| \leq \max_{0\leq i\leq n} |f(x_i)| \sum_{i=0}^{n} |L_i(x)| \leq \|f\|_{L^\infty(a,b)} \sum_{i=0}^{n} |L_i(x)| .$$

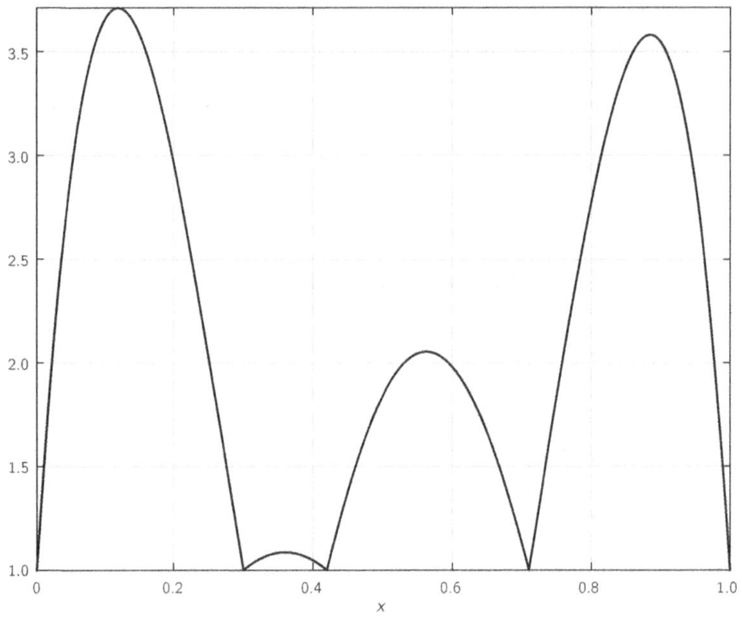

Figure 9.2 The Lebesgue function, $\lambda_X(x) = \sum_{i=0}^{4} |L_i(x)|$, subordinate to the nodal set $X = \{0, 0.30, 0.42, 0.71, 1.00\} \subset [0, 1]$. The basis functions, L_i, $i = 0, \ldots, 4$, are plotted in Figure 9.1, for comparison.

Thus,

$$\|\mathcal{I}_X\|_\infty \le \max_{a \le x \le b} \sum_{i=0}^{n} |L_i(x)| .$$

Now, on the other hand, there is a function $f \in C([a, b])$ with $\|f\|_{L^\infty(a,b)} = 1$ satisfying

$$\max_{a \le x \le b} |\mathcal{I}_X[f](x)| = \max_{a \le x \le b} \left| \sum_{i=0}^{n} f(x_i) L_i(x) \right| = \max_{a \le x \le b} \sum_{i=0}^{n} |L_i(x)| ,$$

which proves that

$$\|\mathcal{I}_X\|_\infty = \max_{a \le x \le b} \sum_{i=0}^{n} |L_i(x)| .$$

We leave it to the reader as an exercise to find the appropriate function f; see Problem 9.6. □

Example 9.3 The Lebesgue function, $\lambda_X(x) = \sum_{i=0}^{4} |L_i(x)|$, subordinate to the nodal set $X = \{0, 0.30, 0.42, 0.71, 1.00\} \subset [0, 1]$ is shown in Figure 9.2. The basis functions, L_i, $i = 0, \ldots, 4$, are plotted in Figure 9.1, for comparison.

In Chapter 10, we will give an error estimate for Lagrange interpolation that involves the Lebesgue constant and an object known as the minimax polynomial. For now, we give a more standard error formula.

Theorem 9.16 (Lagrange interpolation error). *Suppose that $X = \{x_i\}_{i=0}^n$ is a nodal set in the compact interval $[a, b] \subset \mathbb{R}$, $f \in C^{n+1}([a, b])$, and $p \in \mathbb{P}_n$ is the Lagrange interpolating polynomial of f subordinate to X. Then, for every $x \in [a, b] \setminus X$, there is a point $\xi = \xi(x) \in (a, b)$ with*

$$\min\{x_0, \dots, x_n, x\} < \xi < \max\{x_0, \dots, x_n, x\}$$

such that

$$f(x) - p(x) = \frac{f^{(n+1)}(\xi)}{(n+1)!}\omega_{n+1}(x), \tag{9.11}$$

where ω_{n+1} is the nodal polynomial introduced in Definition 9.13.

Proof. Fix $x \in [a, b] \setminus X$. Let us consider cases.

($n = 0$) In this case, $p(x) = f(x_0)$ for all $x \in [a, b]$. By the Mean Value Theorem B.30, there is a point ξ between x and x_0 such that

$$f(x) - p(x) = f(x) - f(x_0) = f'(\xi)(x - x_0),$$

which is the same as (9.11), since, in this case, $(n + 1)! = 1$ and $\omega_1(x) = x - x_0$.

($n \geq 1$) Define, for any $s \in [a, b]$,

$$e(s) = f(s) - p(s) - \frac{f(x) - p(x)}{\omega_{n+1}(x)}\omega_{n+1}(s).$$

Observe that, for all $i = 0, \dots, n$,

$$e(x_i) = 0 - \frac{f(x) - p(x)}{\omega_{n+1}(x)} \cdot 0 = 0.$$

Furthermore, $e(x) = 0$. Thus, $e(s)$ is zero at at least $n+2$ distinct points in $[a, b]$. By Rolle's Theorem B.29, there are $n+1$ distinct points in (a, b) where e' vanishes. Applying Rolle's Theorem B.29 again, there are n distinct points in (a, b) where e'' vanishes. Continuing in this fashion, there is one point (at least) ξ in (a, b) for which $e^{(n+1)}(\xi) = 0$. Now observe that

$$0 = e^{(n+1)}(\xi) = f^{(n+1)}(\xi) - \frac{f(x) - p(x)}{\omega_{n+1}(x)}(n + 1)!.$$

Rearranging terms we get (9.11). □

The interpolating polynomial, p, can become a surrogate for the original function f. If what is wanted is a derivative of f, we can approximate that with the derivative of p. If what is wanted is the integral of f, we can integrate p instead. This will be much further explored in Chapter 14, where we discuss numerical integration.

For the present, with a technique similar to that used in the last proof, we can also establish an error formula for derivatives.

Theorem 9.17 (error formula for derivatives). *Suppose that $X = \{x_i\}_{i=0}^n$ is a nodal set in the compact interval $[a, b] \subset \mathbb{R}$, $f \in C^{n+1}([a, b])$, and $p \in \mathbb{P}_n$ is the Lagrange interpolating polynomial of f subordinate to X. Then there are distinct points $\zeta_i \in (a, b)$, $i = 1, \dots, n$, whose values depend on f and X, such that, for every $x \in [a, b]\backslash X$, there is a point $\xi = \xi(x) \in (a, b)$ with*

$$\min\{x_0, \dots, x_n, x\} < \xi < \max\{x_0, \dots, x_n, x\}$$

for which

$$f'(x) - p'(x) = \frac{f^{(n+1)}(\xi)}{n!}\psi_n(x), \tag{9.12}$$

where

$$\psi_n(x) = \prod_{i=1}^{n}(x - \zeta_i). \tag{9.13}$$

Proof. By Rolle's Theorem B.29, there are points $\zeta_i \in (x_{i-1}, x_i)$, $i = 1, \dots, n$ such that $f'(\zeta_i) - p'(\zeta_i) = 0$. Clearly, all these ζ_i are distinct and only depend on f and the distribution of the nodes, x_i.

Suppose that $x \in [a, b]\backslash\{\zeta_i\}_{i=1}^n$. Define, for all $s \in [a, b]$,

$$e(s) = f'(s) - p'(s) - \frac{f'(x) - p'(x)}{\psi_n(x)}\psi_n(s),$$

where $\psi_n(x) = \prod_{i=1}^n(x - \zeta_i)$. Clearly, e vanishes at $n + 1$ distinct points: ζ_i, $i = 1, \dots, n$, and x. Repeated application of Rolle's Theorem B.29 gives the result, as before. □

9.3 The Runge Phenomenon

It is natural to think that, by increasing the size of the nodal set and, therefore, the degree of the interpolating polynomial, we should be able to get better and better approximations of the function of interest. But this is not always the case.

Proposition 9.18 (conditional convergence). *Suppose that $n \in \mathbb{N}$, $[a, b]$ is a compact interval, $h = \frac{b-a}{n}$, and $X = \{x_i\}_{i=0}^n$ is the uniformly spaced nodal set*

$$x_i = a + ih, \quad i = 0, 1, \dots, n.$$

Let $f \in C^\infty([a, b])$ and $p_n \in \mathbb{P}_n$ be the Lagrange interpolating polynomial of f subordinate to X. Then

$$\|f - p_n\|_{L^\infty(a,b)} = \max_{a \le x \le b}|f(x) - p_n(x)| \to 0, \qquad n \to \infty,$$

provided that

$$\lim_{n\to\infty}\frac{F_{n+1}\Omega_{n+1}}{(n+1)!} = 0,$$

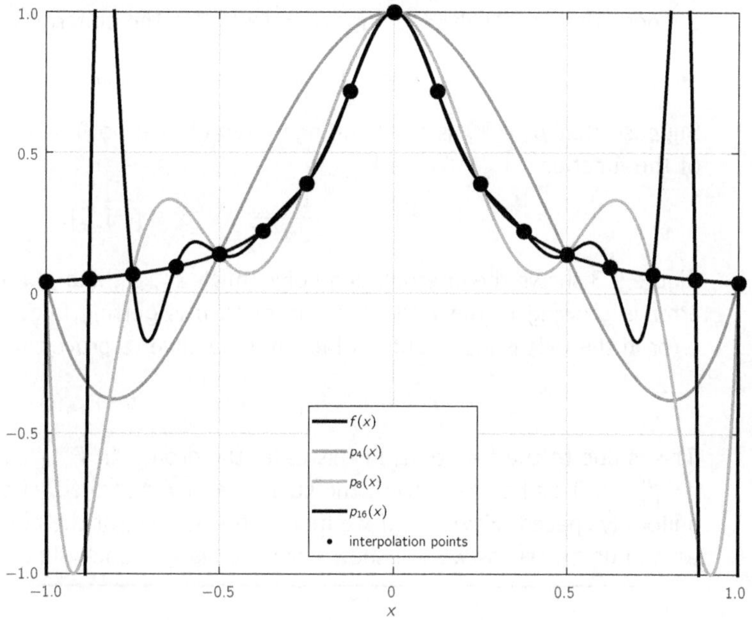

Figure 9.3 The Runge phenomenon: with uniformly spaced interpolation nodes, interpolation polynomials can oscillate wildly with increasing n, yielding inaccurate approximations.

where

$$F_{n+1} = \left\| f^{(n+1)} \right\|_{L^\infty(a,b)}, \qquad \Omega_{n+1} = \frac{n!}{4} h^{n+1}.$$

In other words, the sequence $\{p_n\}_{n\in\mathbb{N}}$ *converges uniformly to* f, *provided that*

$$\lim_{n\to\infty} \frac{F_{n+1} h^{n+1}}{4(n+1)} = 0.$$

Proof. It suffices to use (9.11) and prove that

$$\max_{a\le x\le b} |\omega_{n+1}(x)| \le \Omega_{n+1}.$$

We leave the remaining steps to the reader as an exercise; see Problem 9.8. □

The convergence condition of the previous result is not a trivial one, as the following example shows.

Example 9.4 The following is commonly referred to as the *Runge phenomenon*.[5] Suppose that $n \in \mathbb{N}$, $h = \frac{2}{n}$, and $X_n = \{x_i\}_{i=0}^n$ is the uniformly spaced nodal set

$$x_i = -1 + ih, \quad i = 0, 1, \dots, n.$$

Suppose that $p_n \in \mathbb{P}_n$ is the Lagrange interpolating polynomial subordinate to X_n of the function

$$f(x) = \frac{1}{1 + 25x^2}, \qquad x \in [-1, 1].$$

Figure 9.3 shows the interpolating polynomials p_4, p_8, and p_{16}. It appears that the error is growing in the $L^\infty(-1, 1)$ norm as n is getting larger. In particular, the error in the tails is increasing. In fact, it is possible to prove that, for this problem,

$$\|f - p_n\|_{L^\infty(-1,1)} \to \infty, \qquad n \to \infty.$$

This is due to the fact that, in this case, the growth in F_{n+1} cannot be controlled by $\frac{h^{n+1}}{4(n+1)}$. This has to do with the structure of f and the fact that our points are uniformly spaced. However, if we modify the spacing of the points in a smart way, we can do better, as we will show when we discuss Chebyshev interpolation.

9.4 Hermite Interpolation

One way to generalize Lagrange interpolation is to match not only the values of a given function at a nodal set but also the values of a certain number of derivatives. Matching point values and first derivatives of a function leads to Hermite interpolation.

Definition 9.19 (Hermite interpolating polynomial[6]). Suppose that $X = \{x_i\}_{i=0}^n$ is a nodal set of size $n+1 \in \mathbb{N}$ in the compact interval $[a, b] \subset \mathbb{R}$ and $f \in C^1([a, b])$. The polynomial $p \in \mathbb{P}_{2n+1}$ is called a **Hermite interpolating polynomial** of f if and only if

$$p(x_i) = f(x_i), \qquad p'(x_i) = f'(x_i), \qquad i = 0, \dots, n.$$

Theorem 9.20 (existence and uniqueness). *Suppose that $X = \{x_i\}_{i=0}^n$ is a nodal set of size $n+1 \in \mathbb{N}$ in the compact interval $[a, b] \subset \mathbb{R}$ and $f \in C^1([a, b])$. Then the Hermite interpolating polynomial is well defined, i.e., it exists and is unique.*

Proof. Since the case $n = 0$ is trivial, let us suppose that $n \geq 1$. Define, for $0 \leq \ell \leq n$,

$$F_{0,\ell}(x) = (L_\ell(x))^2(1 - 2L_\ell'(x_\ell)(x - x_\ell)), \qquad F_{1,\ell} = (L_\ell(x))^2(x - x_\ell), \quad (9.14)$$

[5] Named in honor of the German mathematician, physicist, and spectroscopist Carl David Tolmé Runge (1856–1927).

[6] Named in honor of the French mathematician Charles Hermite (1822–1901).

where L_ℓ is as in (9.3), and

$$p(x) = \sum_{\ell=0}^{n} \left[F_{0,\ell}(x)f(x_\ell) + F_{1,\ell}(x)f'(x_\ell)\right]. \tag{9.15}$$

We leave it to the reader as an exercise to prove that p has the desired properties and is unique; see Problem 9.9. □

Theorem 9.21 (Hermite interpolation error). *Suppose that $X = \{x_i\}_{i=0}^n$ is a nodal set of size $n+1 \in \mathbb{N}$ in the compact interval $[a,b] \subset \mathbb{R}$, $f \in C^{(2n+2)}([a,b])$, and $p \in \mathbb{P}_{2n+1}$ is the unique Hermite interpolating polynomial of f. Then, for every $x \in [a,b]\backslash X$, there is a point $\xi = \xi(x) \in (a,b)$, with*

$$\min\{x_0, \dots, x_n, x\} < \xi < \max\{x_0, \dots, x_n, x\}$$

such that

$$f(x) - p(x) = \frac{f^{(2n+2)}(\xi)}{(2n+2)!}(\omega_{n+1}(x))^2, \tag{9.16}$$

where ω_{n+1} is given in 9.8.

Proof. Fix $x \in [a,b]\backslash X$. Define, for all $s \in [a,b]$,

$$e(s) = f(s) - p(s) - \frac{f(x)-p(x)}{(\omega_{n+1}(x))^2}(\omega_{n+1}(s))^2.$$

As before, e vanishes at $n+2$ distinct points in $[a,b]$: the $n+1$ nodes x_i and the point x. By Rolle's Theorem B.29, e' vanishes at $n+1$ distinct points, which are distinct from the $n+2$ points in $X \cup \{x\}$. Furthermore, by construction, e' vanishes at the $n+1$ points of X. Thus, e' vanishes at a total of (at least) $2n+2$ distinct points in $[a,b]$. This is the key fact we need.

By repeated application of Rolle's Theorem B.29, the function $e^{(2n+2)}$, which exists and is continuous by assumption, vanishes at (at least) one point $\xi \in (a,b)$. Of course, since $p \in \mathbb{P}_{2n+1}$, $p^{(2n+2)} \equiv 0$ and, therefore,

$$0 = e^{(2n+2)}(\xi) = f^{(2n+2)}(\xi) - \frac{f(x)-p(x)}{(\omega_{n+1}(x))^2}(2n+2)!,$$

which proves the result. □

9.5 Complex Polynomial Interpolation

We now discuss the interpolation of functions that are holomorphic in a region of the complex plane. In doing so, we gain some further insight into the curious example provided by Runge; see Example 9.4.

9.5.1 Some Facts of Complex Analysis

In what follows, we will make use of some tools from complex analysis to make the discussion rigorous, but we supply these as needed. For further results, see, for example, [3, 24, 35, 60, 77].

Definition 9.22 (holomorphic function). Let $D \subset \mathbb{C}$ be a simply connected, open region and $g\colon D \to \mathbb{C}$. We say that g is **complex differentiable** at $a \in D$ if the limit

$$\lim_{z \to a} \frac{g(z) - g(a)}{z - a}$$

exists. In this case, we denote by $g'(a)$ the value of this limit and call it the **derivative** of g at a. If g is complex differentiable at every point of D, then we say that g is **holomorphic** in D.

As the reader may be aware, being complex differentiable is a very strong condition. It implies, in particular, that if g is complex differentiable in a neighborhood of a point $a \in \mathbb{C}$, then it is infinitely differentiable (again in the complex sense) at this point. Moreover, this also implies that this function is *analytic*, i.e., its power series representation

$$\sum_{k=0}^{\infty} g^{(k)}(a) \frac{(z-a)^k}{k!}$$

converges uniformly in a neighborhood of a. This can be proved with the help of Cauchy's integral theorem, which we discuss next. We will first need to define path integrals, and for that we need several definitions.

Definition 9.23 (path). Let $[a, b] \subset \mathbb{R}$ be a compact interval. A function $\gamma \in C([a, b]; \mathbb{C})$ is called a **path** or **curve**. We say that a path is **simple** or a **Jordan path**[7] if

$$\gamma(t_1) \neq \gamma(t_2), \quad \forall t_1, t_2 \in (a, b).$$

We say that a path is **closed**, or a **contour**, if $\gamma(a) = \gamma(b)$.

A powerful result known as the *Jordan Curve Theorem* (see [39, Section 2.B]), states that a simple contour divides the complex plane $\mathbb{C}$ into two parts: an "interior" region bounded by the path and an "exterior," so that every other path that goes from a point of the interior to the exterior must intersect the given contour. With the help of this, we can talk about points *inside* or *outside* a simple closed path.

Definition 9.24 (smooth path). Let $\gamma \in C([a, b]; \mathbb{C})$ be a path. We say that this path is **piecewise smooth** if there is $S \subset [a, b]$ of finite cardinality such that $\gamma \in C^1([a, b]\backslash S; \mathbb{C})$.

[7] Named in honor of the French mathematician Marie Ennemond Camille Jordan (1838–1922).

For piecewise smooth paths, the derivative exists at all, but possibly a finite number of, points. Thus, we can define the orientation of a path. This will be mostly used for contours, so we only define them this way.

Definition 9.25 (orientation). Let $\gamma \in C([a,b];\mathbb{C})$ be a simple, piecewise smooth contour and let $\Gamma = \gamma([a,b])$. Denote by $D \subset \mathbb{C}$ the collection of points inside γ, i.e., $\partial D = \Gamma$. Let $\boldsymbol{n}_D(t) = [n_1(t), n_2(t)]^\intercal$, $t \in [a,b]$ be the unit exterior normal to D at the point $\gamma(t)$. We say that γ is **traversed counterclockwise** if

$$\det \begin{bmatrix} n_1(t) & n_2(t) \\ \Re\gamma'(t) & \Im\gamma'(t) \end{bmatrix} > 0$$

for all $t \in [a,b]$.

Finally, we can define integrals over a path.

Definition 9.26 (path integral). Let $\gamma \in C([a,b];\mathbb{C})$ be a simple, piecewise smooth path; and let $\Gamma = \gamma([a,b])$. For $g\colon \Gamma \to \mathbb{C}$, the **path integral** of g over γ is defined as

$$\int_\gamma g(z)\mathrm{d}z = \int_a^b g(\gamma(t))\frac{\mathrm{d}\gamma(t)}{\mathrm{d}t}\,\mathrm{d}t.$$

It turns out that values of a holomorphic function can be obtained using path integrals.

Theorem 9.27 (Cauchy Integral Theorem[8]). *Suppose that $D \subset \mathbb{C}$ is a simply connected, open region and $g\colon D \to \mathbb{C}$ is holomorphic. Suppose that $\gamma\colon [0,2\pi] \to D$ is a simple closed contour traversed counterclockwise and the point $z_0 \in D$ is inside γ. Then*

$$g(z_0) = \frac{1}{2\pi\mathrm{i}}\int_\gamma \frac{g(z)}{z - z_0}\mathrm{d}z.$$

Let us now briefly discuss what can be said about functions that are holomorphic except at an isolated point.

Definition 9.28 (Laurent expansion[9]). Suppose that $D \subset \mathbb{C}$ is a simply connected, open region and $z_0 \in D$. Assume that $g\colon D\backslash\{z_0\} \to \mathbb{C}$ is holomorphic, i.e., g is holomorphic on D, except for the isolated singularity at z_0. In this setting, g admits the **Laurent expansion**

$$\begin{aligned} g(z) &= \sum_{k=1}^{\infty} \frac{b_k}{(z-z_0)^k} + \sum_{k=0}^{\infty} a_k(z-z_0)^k \\ &= \cdots + \frac{b_2}{(z-z_0)^2} + \frac{b_1}{(z-z_0)} + a_0 + a_1(z-z_0) + a_2(z-z_0)^2 + \cdots, \end{aligned}$$

where $b_j \in \mathbb{C}$, $j = 1, 2, \ldots$, (not all of which are zero) and $a_j \in \mathbb{C}$, $j = 0, 1, \ldots$.

[8] Named in honor of the French mathematician Augustin-Louis Cauchy (1789–1857).
[9] Named in honor of the French mathematician and engineer Pierre Alphonse Laurent (1813–1854).

Points like z_0 in the previous definition are called isolated singularities of a function.

Definition 9.29 (isolated singularity). Suppose that $D \subset \mathbb{C}$ is a simply connected, open region and $z_0 \in D$. Assume that $g\colon D\backslash\{z_0\} \to \mathbb{C}$ is holomorphic. Then we call z_0 an **isolated singularity** of g. Moreover, we say that z_0 is a **removable singularity** if the limit

$$\lim_{z \to z_0} g(z) \tag{9.17}$$

exists and it is finite. If the limit (9.17) exists but equals ∞, then we say that the point z_0 is a **pole**. Finally, if (9.17) does not exist, we call z_0 an **essential singularity**.

It turns out that poles and Laurent expansions are closely related.

Definition 9.30 (poles). Suppose that $D \subset \mathbb{C}$ is a simply connected, open region and $z_0 \in D$. Assume that $g\colon D\backslash\{z_0\} \to \mathbb{C}$ is holomorphic and that z_0 is a pole. Let g admit a Laurent expansion. If $b_1 \neq 0$, and $b_j = 0$ for $j > 1$, then we say that g has a **simple pole** at the singularity $z = z_0$. Similarly, if there is $k \in \mathbb{N}$ such that $b_k \neq 0$, but $b_j = 0$ for $j > k$, then we say that g has a **pole of degree** k at the singularity $z = z_0$. Finally, the **residue of** g **at** z_0, denoted $\operatorname{Res}(g, z_0)$, is the coefficient b_1, i.e.,

$$\operatorname{Res}(g, z_0) = b_1.$$

We comment that a function that is holomorphic in an open region except at a finite number of poles is called a *meromorphic* in this region.

Theorem 9.31 (Residue Theorem). *Let $D \subset \mathbb{C}$ be a simply connected, open, bounded region and $Z = \{z_i\}_{i=0}^n \subset D$ be a set of $n+1$ distinct points. Assume that $g\colon D\backslash Z \to \mathbb{C}$ is holomorphic, i.e., g is holomorphic on D, except for isolated singularities at $z_0, \ldots, z_n$. Suppose that $\gamma\colon [0, 2\pi] \to D$ is a simple closed contour traversed counterclockwise and the points of Z are inside γ. Then*

$$\int_\gamma g(z)\,\mathrm{d}z = 2\pi\mathrm{i}\sum_{i=0}^{n} \operatorname{Res}(g, z_i),$$

where $\operatorname{Res}(g, z_i)$ is the residue of g at z_i.

The following result is useful for calculating residues.

Proposition 9.32 (computation of residues). *Suppose that $D \subset \mathbb{C}$ is a simply connected, open region and $z_0 \in D$. Assume that $g, h\colon D \to \mathbb{C}$ are holomorphic, $g(z_0) \neq 0$, $h(z_0) = 0$, and $h'(z_0) \neq 0$. Then the function $f(z) = \frac{g(z)}{h(z)}$ is holomorphic in $D\backslash\{z_0\}$ and has a simple pole at z_0. Furthermore,*

$$\operatorname{Res}(f, z_0) = \frac{g(z_0)}{h'(z_0)}.$$

We conclude our slight detour into complex analysis by stating a result that will be used in Chapter 21 and is commonly referred to as the *Maximum Modulus Principle.*

Theorem 9.33 (Maximum Modulus Principle). *Let $D \subset \mathbb{C}$ be an open and connected region and $g\colon D \to \mathbb{C}$ be holomorphic. If there is a point $z_0 \in D$ such that $z \mapsto |g(z)|$ attains a (local) maximum at z_0, then g is constant.*

In other words, if $g \in C(\bar{D};\mathbb{C})$ is holomorphic in D, then it attains its maximum at some point of the boundary ∂D.

9.5.2 Lagrange Interpolation

We are now ready to study the problem of Lagrange interpolation in the complex case. The setting is similar to the real case: we let $D \subset \mathbb{C}$ be a simply connected, open region; $n \in \mathbb{N}$; and the nodal set is $Z = \{z_i\}_{i=0}^n \subset D$, where all the nodes are distinct. Given $f\colon D \to \mathbb{C}$, we need to find a (complex-valued) polynomial $p \in \mathbb{P}_n(\mathbb{C})$ such that

$$p(z_i) = f(z_i), \quad i = 0, \dots, n. \tag{9.18}$$

The following result is obtained by small variations of the real case.

Theorem 9.34 (Lagrange interpolation error). *Suppose that $D \subset \mathbb{C}$ is a simply connected, open region, $f\colon D \to \mathbb{C}$, and $Z = \{z_i\}_{i=0}^n \subset D$ is a set of $n+1 \in \mathbb{N}$ distinct points. There exists a unique polynomial $p \in \mathbb{P}_n(\mathbb{C})$ that satisfies* (9.18). *Suppose, in addition, that f is holomorphic, $\gamma\colon [0, 2\pi] \to D$ is a simple closed contour traversed counterclockwise, and the points of Z are inside γ. Then the error formula may be expressed as*

$$f(\zeta) - p(\zeta) = \frac{1}{2\pi \mathrm{i}} \int_\gamma \frac{f(z)}{z - \zeta} \prod_{i=0}^{n} \frac{\zeta - z_i}{z - z_i} \,\mathrm{d}z \tag{9.19}$$

for all ζ inside γ, but $\zeta \notin Z$.

Proof. The existence and uniqueness of the interpolating polynomial $p \in \mathbb{P}_n(\mathbb{C})$ follows from similar arguments to those used for the real case. One can either appeal to a Vandermonde-type construction, in which case the Vandermonde matrix is invertible; or one can appeal to a Lagrange-type construction. In fact, the complex analogue of the Lagrange interpolation formula (9.10) is valid, i.e., the interpolating polynomial is

$$p(z) = \sum_{i=0}^{n} f(z_i) \frac{\omega_{n+1}(z)}{(z - z_i)\omega'_{n+1}(z_i)}, \tag{9.20}$$

where

$$\omega_{n+1}(z) = \prod_{i=0}^{n} (z - z_i).$$

To get the error formula, consider now the function

$$g(z) = \frac{f(z)}{z - \zeta} \prod_{i=0}^{n} \frac{\zeta - z_i}{z - z_i} = \frac{f(z)}{z - \zeta} \cdot \frac{\omega_{n+1}(\zeta)}{\omega_{n+1}(z)},$$

which is the integrand in (9.19). Observe that the function g is meromorphic on D, i.e., it is holomorpic on D except at the isolated singularities $Z \cup \{\zeta\}$. Thus, by the residue Theorem 9.31, the integral

$$\frac{1}{2\pi \mathrm{i}} \int_\gamma g(z) \mathrm{d}z = \frac{1}{2\pi \mathrm{i}} \int_\gamma \frac{f(z)}{z-\zeta} \prod_{i=0}^{n} \frac{\zeta - z_i}{z - z_i} \mathrm{d}z$$

is the sum of the residues of g contained inside of γ. Using Proposition 9.32, we find

$$\operatorname{Res}(g, z_i) = \frac{f(z_i)}{z_i - \zeta} \cdot \frac{\omega_{n+1}(\zeta)}{\omega'_{n+1}(z_i)}, \qquad i = 0, \ldots, n,$$

and

$$\operatorname{Res}(g, \zeta) = f(\zeta) \cdot \frac{\omega_{n+1}(\zeta)}{\omega_{n+1}(\zeta)} = f(\zeta).$$

Thus,

$$\frac{1}{2\pi \mathrm{i}} \int_\gamma \frac{f(z)}{z-\zeta} \prod_{i=0}^{n} \frac{\zeta - z_i}{z - z_i} \mathrm{d}z = f(\zeta) - \sum_{i=0}^{n} \frac{f(z_i)}{\zeta - z_i} \cdot \frac{\omega_{n+1}(\zeta)}{\omega'_{n+1}(z_i)} = f(\zeta) - p(\zeta),$$

where in the last step we used (9.20). □

This last result provides a useful means to estimate errors.

Theorem 9.35 (error estimate). *Let $[a, b] \subset \mathbb{R}$ be a compact interval and $\varepsilon > 0$ be fixed. Define*

$$r = b - a + \varepsilon, \qquad C_\varepsilon = \{z \in \mathbb{C} \mid \operatorname{dist}(z, [a, b]) = r\}.$$

Suppose that $\gamma\colon [0, 2\pi] \to C_\varepsilon$ is the simple closed contour traversing the set C_ε in a counterclockwise fashion. Define D_ε as the open, simply connected set of all points inside γ. Suppose that $f\colon \overline{D_\varepsilon} \to \mathbb{C}$ is holomorphic on D_ε and there is $M > 0$ such that

$$|f(z)| \le M, \quad \forall x \in C_\varepsilon = \partial D_\varepsilon.$$

Let $n \in \mathbb{N}_0$, $X_n = \{x_i\}_{i=0}^n$ be a nodal set in the interval $[a, b]$, and $p_n \in \mathbb{P}_n(\mathbb{C})$ be the interpolating polynomial of f subordinate to the nodal set X_n. Then, for all $x \in [a, b] \backslash X_n$,

$$|f(x) - p_n(x)| \le \frac{(b - a + \pi r)M}{\pi} \left(\frac{b-a}{r} \right)^{n+1}.$$

Consequently, $p_n \to f$ uniformly on $[a, b]$, as $n \to \infty$, no matter how the nodes in X_n are chosen.

Proof. Observe that the length of γ is precisely $2(b - a) + 2\pi r$ and apply the last theorem. The details are left to the reader as an exercise; see Problem 9.10. □

Example 9.5 Let us revisit the Runge phenomenon. The function

$$f(x) = \frac{1}{1+25x^2}, \qquad x \in [-1,1],$$

which we have previously considered, and the uniformly spaced nodal set

$$X_n = \left\{-1 + \frac{2i}{n}\right\}_{i=0}^{n} \subset [-1,1], \qquad n \in \mathbb{N}.$$

As we mentioned in Example 9.4, if $p_n \in \mathbb{P}_n$ is the Lagrange interpolating polynomial of f subordinate to X_n, then one can show that $\|f - p_n\|_{L^\infty(-1,1)} \to \infty$, as $n \to \infty$. In light of Theorem 9.35, why do we fail to obtain uniform convergence? Clearly, one or more of the hypotheses of Theorem 9.35 must fail. In fact, as a complex function, f is not holomorphic, for any $\varepsilon > 0$, in the region D_ε defined in the hypotheses of Theorem 9.35. In fact, writing

$$f(z) = \frac{1}{1+25z^2} = \frac{1}{(1+5\mathrm{i}z)(1-5\mathrm{i}z)},$$

we observe that f has isolated singularities and simple poles at $z = \pm\frac{\mathrm{i}}{5}$. This explains is why Theorem 9.35 cannot be applied to guarantee uniform convergence in $[-1,1]$ with the uniformly spaced nodes X_n.

9.6 Divided Differences and the Newton Construction

In this section, we give another method for computing interpolating polynomials, using the so-called *Newton construction*. This method is based on an alternate basis for $\mathbb{P}_n$, namely the Newton basis.

Definition 9.36 (Newton basis[10]). Let $n \in \mathbb{N}_0$ and $X = \{x_i\}_{i=0}^n$ be a nodal set in the compact interval $[a,b] \subset \mathbb{R}$. The polynomial set $B_n = \{\omega_j\}_{j=0}^n$, defined as $\omega_0 \equiv 1$ and, if $n \geq 1$,

$$\omega_j(x) = \prod_{k=0}^{j-1}(x - x_k), \qquad j = 1, \ldots, n$$

is called the **Newton basis**. The polynomial $\omega_j \in \mathbb{P}_j$ is called the **nodal polynomial of order** j with respect to X.

The following result shows that B_n is indeed a basis.

Proposition 9.37 (basis). *For any compact interval $[a,b] \subset \mathbb{R}$, any $n \in \mathbb{N}_0$, and all nodal sets X in $[a,b]$, the Newton basis B_n is a basis of $\mathbb{P}_n$.*

Proof. See Problem 9.16. □

[10] Named in honor of the British mathematician, physicist, astronomer, theologian, and natural philosopher Sir Isaac Newton (1642–1726/27).

Since, as the previous result shows, B_n is a basis we can expand any interpolating polynomial with respect to it. It turns out that the coefficients in this basis expansion carry very useful information.

Definition 9.38 (divided differences)**.** Suppose that $n \in \mathbb{N}_0$ and $X = \{x_i\}_{i=0}^n$ is a nodal set in the compact interval $[a, b]$. Let $f \in C([a, b])$ and $p_n \in \mathbb{P}_n$ be the (unique) interpolating polynomial of f subordinate to X. Set

$$p_n(x) = \sum_{k=0}^{n} a_k \omega_k(x). \tag{9.21}$$

The coefficient,

$$a_k = f[x_0, x_1, \ldots, x_k], \quad k = 0, \ldots, n,$$

in the expansion with respect to the Newton basis $B_n = \{\omega_j\}_{j=0}^n$ is called the ***k*th order divided difference**.

The choice of the name, divided difference, will be clear only later when we discuss the properties of these objects.

Example 9.6 The Newton form of the interpolating polynomial lends itself to a very efficient algorithm for computation. Suppose, for example, that $n = 4$ and we wish to evaluate p_4 in (9.21) at x. Then, once the coefficients $\{a_k\}_{k=0}^4$ are known, we can express $p_4(x)$ as

$$p_4(x) = a_0 + (x - x_0)\{a_1 + (x - x_1)[a_2 + (x - x_2)\{a_3 + (x - x_3)[a_4]\}]\}.$$

This method of evaluation is known as *Horner's method.*[11]

While this representation is, in theory, completely equivalent to (9.21), its evaluation is much cheaper. In fact, it can be shown that evaluating a polynomial of degree n using Horner's method only requires $\mathcal{O}(n)$ arithmetic operations, which is optimal.

Proposition 9.39 (recursion)**.** *Suppose that $n \in \mathbb{N}_0$ and $X_n = \{x_i\}_{i=0}^n$ is a nodal set in the compact interval $[a, b] \subset \mathbb{R}$. Let $f \in C([a, b])$ and, for $k \in \{0, \ldots, n\}$, $p_k \in \mathbb{P}_k$ be the (unique) interpolating polynomial of f subordinate to $X_k = \{x_i\}_{i=0}^k$. Then*

$$p_k(x) = p_{k-1}(x) + b_k \omega_k, \quad k = 1, \ldots, n, \tag{9.22}$$

holds if and only if

$$b_k = \frac{f(x_k) - p_{k-1}(x_k)}{\omega_k(x_k)}, \quad k = 1, \ldots, n. \tag{9.23}$$

[11] Named in honor of the Britsh mathematician William George Horner (1786–1837).

As a consequence, it follows that, using the Newton basis representation (9.21), *we have* $f[x_0] = f(x_0)$ *and*

$$f[x_0, x_1, \ldots, x_k] = a_k = b_k = \frac{f(x_k) - p_{k-1}(x_k)}{\omega_k(x_k)}, \quad k = 1, \ldots, n.$$

Proof. We will sketch the proof and leave the details to the reader as an exercise; see Problem 9.17. Suppose that $1 \leq k \leq n$. Then

$$p_{k-1}(x_j) = f(x_j), \quad j = 0, \ldots, k-1,$$

and

$$p_k(x_j) = f(x_j), \quad j = 0, \ldots, k.$$

In general, however, $p_k(x_k) \neq p_{k-1}(x_k)$.

($\Longrightarrow$) Suppose that (9.22) holds. Then

$$f(x_j) = p_k(x_j) = p_{k-1}(x_j) + b_k\omega_k(x_j), \quad j = 0, \ldots, k.$$

This last equation holds identically for $j = 0, \ldots, k-1$, since $\omega_k(x_j) = 0$, for $j = 0, \ldots, k-1$. We get new information only from $j = k$. Thus,

$$f(x_k) = p_k(x_k) = p_{k-1}(x_k) + b_k\omega_k(x_k),$$

which implies, since $\omega_k(x_k) \neq 0$, that

$$b_k = \frac{f(x_k) - p_{k-1}(x_k)}{\omega_k(x_k)}.$$

($\Longleftarrow$) Suppose that (9.23) holds. Then, it is easy to see that (9.22) holds. □

This last result shows that the construction of the interpolating polynomial $p_n \in \mathbb{P}_n$ can be done in a particular order. First, set $p_0 \equiv f(x_0)$. Then p_1, the interpolant using the points x_0 and x_1, is constructed from p_0 by computing b_1. p_2, the interpolant using the points x_0, x_1, and x_2, is constructed from p_1 by computing b_2, and so on. The construction proceeds according to our labeling of the nodes: $x_0, x_1, \ldots, x_n$. But, as we have noted, there is no special numbering assigned to X_n. Thus, we could choose another order for our construction; for example, reverse order construction: $x_n, x_{n-1}, \ldots, x_0$.

Proposition 9.40 (divided differences). *Suppose that* $n \in \mathbb{N}_0$, $X_n = \{x_i\}_{i=0}^n$ *is a nodal set in the compact interval* $[a, b] \subset \mathbb{R}$, *and* $f\colon [a, b] \to \mathbb{R}$. *Then*

$$a_n = f[x_0, x_1, \ldots, x_n] = \sum_{j=0}^{n} \frac{f(x_j)}{\prod_{\substack{k=0 \\ k\neq j}}^{n}(x_j - x_k)}. \tag{9.24}$$

Proof. This follows by equating the Lagrange, Newton, and canonical forms of the interpolating polynomial. Suppose that $p_n \in \mathbb{P}_n$ is the unique interpolating polynomial of f subordinate to $X_n = \{x_i\}_{i=0}^n$. Then

$$p_n(x) = \sum_{j=0}^{n} L_j(x)f(x_j) = \sum_{k=0}^{n} a_k\omega_k(x) = \sum_{j=0}^{n} c_jx^j,$$

where $L_j \in \mathbb{P}_n$ is the jth Lagrange basis element; see (9.3). Let us figure out what are the coefficients of x^n when the Lagrange and Newton forms are expanded to canonical form. For the Newton form, clearly the coefficient of x^n is

$$c_n = a_n = f[x_0, x_1, \ldots, x_n].$$

On the other hand, for the Lagrange form,

$$c_n = \sum_{j=0}^{n} \frac{f(x_j)}{\prod_{\substack{k=0 \\ k \neq j}}^{n} (x_j - x_k)}. \qquad \square$$

Corollary 9.41 (invariance). *Suppose that $n \in \mathbb{N}_0$, $X_n = \{x_i\}_{i=0}^n$ is a nodal set in the compact interval $[a, b] \subset \mathbb{R}$, and $f\colon [a, b] \to \mathbb{R}$. Assume that $(i_0, i_1, \ldots, i_n)$ is a permutation of $(0, 1, \ldots, n)$. Define $\tilde{\omega}_0 \equiv 1$ and, if $n \geq 1$,*

$$\tilde{\omega}_k(x) = \prod_{j=0}^{k-1} (x - x_{i_j}), \quad k = 1, \ldots, n.$$

Suppose that $p_n \in \mathbb{P}_n$ is the interpolating polynomial of f subordinate to X_n, so that

$$p_n = \sum_{j=0}^{n} a_j \omega_j(x) = \sum_{j=0}^{n} \tilde{a}_j \tilde{\omega}_j(x). \tag{9.25}$$

Then

$$a_n = \tilde{a}_n$$

and

$$f[x_0, x_1, \ldots, x_n] = f\left[x_{i_0}, x_{i_1}, \ldots, x_{i_n}\right].$$

Proof. This follows directly from the right-hand side of (9.24), since the sum and product can rearranged arbitrarily. Alternately, one can simply expand the representations in (9.25) and consider the coefficient of x^n. $\square$

Remark 9.42 (invariance). Let us consider what the expansions in (9.25) represent. $\sum_{j=0}^n a_j \omega_j$ represents the construction of p_n in the order $x_0, x_1, \ldots, x_n$, whereas $\sum_{j=0}^n \tilde{a}_j \tilde{\omega}_j$ represents the construction of p_n in the permuted order $x_{i_0}, x_{i_1}, \ldots, x_{i_n}$. Though the order of the constructions may differ, the resulting interpolating polynomial must be the same if all of the same interpolation points are ultimately used.

Theorem 9.43 (divided difference formula). *Suppose that $n \in \mathbb{N}$, $X_n = \{x_i\}_{i=0}^n$ is a nodal set in the compact interval $[a, b] \subset \mathbb{R}$, and $f\colon [a, b] \to \mathbb{R}$. Assume that $p_n \in \mathbb{P}_n$ is the interpolating polynomial of f subordinate to X_n. When constructed in the order $x_0, x_1, \ldots, x_n$, let us write, as in (9.21),*

$$p_n(x) = \sum_{j=0}^{n} a_j \omega_j(x).$$

When p_n is constructed in the reverse order $x_n, x_{n-1}, \ldots, x_0$, let us write

$$p_n(x) = \sum_{j=0}^{n} \hat{a}_j \hat{\omega}_j(x),$$

where $\hat{\omega}_0 \equiv 1$ and, if $n \geq 1$,

$$\hat{\omega}_j(x) = \prod_{k=0}^{j-1}(x - x_{n-k}), \quad j = 1, \ldots, n,$$

and where, by definition,

$$\hat{a}_j = f[x_n, x_{n-1}, \ldots, x_{n-j}], \quad j = 0, \ldots, n.$$

Then it follows that

$$a_n = \hat{a}_n = \frac{a_{n-1} - \hat{a}_{n-1}}{x_0 - x_n},$$

which, in turn, implies that

$$f[x_0, x_1, \ldots, x_n] = \frac{f[x_0, x_1, \ldots, x_{n-1}] - f[x_1, x_2, \ldots, x_n]}{x_0 - x_n}, \quad \forall n \in \mathbb{N}. \tag{9.26}$$

Proof. Clearly, for all $x \in \mathbb{R}$,

$$\Delta(x) = \sum_{j=0}^{n} a_j \omega_j(x) - \sum_{j=0}^{n} \hat{a}_j \hat{\omega}_j(x) = 0$$

and, by Corollary 9.41, $a_n = \hat{a}_n$. Using $a_n = \hat{a}_n$, the function Δ may be re-expressed as

$$\begin{aligned} 0 = \Delta(x) = a_n[(x - x_0) - (x - x_n)](x - x_1)\cdots(x - x_{n-1}) + (a_{n-1} - \hat{a}_{n-1})x^{n-1} \\ + p_{n-2}(x), \end{aligned}$$

where $p_{n-2} \in \mathbb{P}_{n-2}$. The coefficient of x^{n-1} in the polynomial Δ is precisely

$$(x_n - x_0)a_n + (a_{n-1} - \hat{a}_{n-1}),$$

which must be zero since $\Delta \equiv 0$:

$$(x_n - x_0)a_n + (a_{n-1} - \hat{a}_{n-1}) = 0.$$

Equation 9.26 then follows immediately from the last equation and another application of Corollary 9.41, which guarantees that

$$\hat{a}_{n-1} = f[x_n, x_{n-1}, \ldots, x_1] = f[x_1, x_2, \ldots, x_n]. \qquad \square$$

We have the following, as a simple consequence of the last result.

Corollary 9.44 (divided difference formula)**.** *Let $n \in \mathbb{N}$. Suppose $X = \{x_j, \ldots, x_{j+n}\}$ is a nodal set in the compact interval $[a, b] \subset \mathbb{R}$. If $f \colon [a, b] \to \mathbb{R}$, then*

$$f[x_j, \ldots, x_{j+n}] = \frac{f[x_{j+1}, \ldots, x_{j+n}] - f[x_j, \ldots, x_{j+n-1}]}{x_{j+n} - x_j}. \tag{9.27}$$

Proof. See Problem 9.18. □

x_i	$f[x_i]$	$f[x_i, x_{i+1}]$	$f[x_i, x_{i+1}, x_{i+2}]$	$f[x_i, \ldots, x_{i+3}]$	$f[x_i, \ldots, x_{i+4}]$
x_0	$f[x_0]$				
x_1	$f[x_1]$	$f[x_0, x_1]$			
x_2	$f[x_2]$	$f[x_1, x_2]$	$f[x_0, x_1, x_2]$		
x_3	$f[x_3]$	$f[x_2, x_3]$	$f[x_1, x_2, x_3]$	$f[x_0, x_1, x_2, x_3]$	
x_4	$f[x_4]$	$f[x_3, x_4]$	$f[x_2, x_3, x_4]$	$f[x_1, x_2, x_3, x_4]$	$f[x_0, x_1, x_2, x_3, x_4]$

Table 9.1 A table of divided differences for computing up to quartic (fourth degree) interpolating polynomials.

Example 9.7 Suppose that $[a, b] \subset \mathbb{R}$ is a compact interval, $f\colon [a, b] \to \mathbb{R}$, and we wish to compute a quartic interpolating polynomial for f with the nodes $X_4 = \{x_0, \ldots, x_4\} \subset [a, b]$. Table 9.1 depicts the construction of the divided differences, where the order is assumed to proceed according to the numbering of the nodes: $0, 1, \ldots, 4$. The construction can be accomplished by recursively computing all of the divided differences in the table. For example, using 9.27, we compute

$$f[x_1, x_2, x_3] = \frac{f[x_2, x_3] - f[x_1, x_2]}{x_3 - x_1}.$$

Note that all of the values in the table are needed to compute the last entry, namely

$$f[x_0, x_1, x_2, x_3, x_4].$$

When all of the values are computed, the interpolating polynomial is

$$\begin{aligned} p_4(x) = f[x_0] & \\ &+ (x - x_0)f[x_0, x_1] \\ &+ (x - x_0)(x - x_1)f[x_0, x_1, x_2] \\ &+ (x - x_0)(x - x_1)(x - x_2)f[x_0, x_1, x_2, x_3] \\ &+ (x - x_0)(x - x_1)(x - x_2)(x - x_3)f[x_0, x_1, x_2, x_3, x_4]. \end{aligned}$$

If, on the other hand, one wished to stop with the cubic interpolating polynomial, for example, the last row of Table 9.1 need not be calculated. One will obtain

$$\begin{aligned} p_3(x) = f[x_0] & \\ &+ (x - x_0)f[x_0, x_1] \\ &+ (x - x_0)(x - x_1)f[x_0, x_1, x_2] \\ &+ (x - x_0)(x - x_1)(x - x_2)f[x_0, x_1, x_2, x_3]. \end{aligned}$$

Example 9.8 As a concrete example, consider the data in Table 9.2. There are five interpolation nodes, which allows us to construct an interpolating polynomial of degree up to four (4) for the function of interest, f. The interpolating polynomials of degrees three (3) and four (4) are plotted in Figure 9.4. Specifically,

$$\begin{aligned} p_3(x) = \mathcal{I}_{X_3}[f](x) = 5 & \\ & + (x-0.1)(-30) \\ & + (x-0.1)(x-0.3)\left(\frac{400}{3}\right) \\ & + (x-0.1)(x-0.3)(x-0.4)\left(-\frac{2125}{9}\right) \end{aligned}$$

and

$$p_4(x) = \mathcal{I}_{X_4}[f](x) = p_3(x) + (x-0.1)(x-0.3)(x-0.4)(x-0.7)\left(\frac{5125}{18}\right).$$

---

x_i	$f[x_i]$	$f[x_i, x_{i+1}]$	$f[x_i, x_{i+1}, x_{i+2}]$	$f[x_i, \dots, x_{i+3}]$	$f[x_i, \dots, x_{i+4}]$
0.1	5				
0.3	−1	−30			
0.4	0	10	$\frac{400}{3}$		
0.7	2	$\frac{20}{3}$	$-\frac{25}{3}$	$-\frac{2125}{9}$	
0.9	2	0	$-\frac{40}{3}$	$-\frac{25}{3}$	$\frac{5125}{18}$

Table 9.2 A table of divided differences for computing up to quartic (fourth degree) interpolating polynomials with real data. See Figure 9.4 for the interpolating polynomials of degrees three (3) and four (4) constructed from these data.

It turns out that divided differences are not only useful in practical computation, They can also be used to obtain an alternate error formula for interpolation.

Theorem 9.45 (Newton–Lagrange interpolation error). *Suppose that $n \in \mathbb{N}$, $X = \{x_i\}_{i=0}^n$ is a nodal set in the compact interval $[a, b] \subset \mathbb{R}$, and $f\colon [a, b] \to \mathbb{R}$. Let $p_n \in \mathbb{P}_n$ be the interpolating polynomial of f subordinate to X. Then, for any $x \in [a, b] \backslash X$,*

$$f(x) - p_n(x) = \omega_{n+1}(x) f[x_0, x_1, \dots, x_n, x]. \tag{9.28}$$

Proof. Since $x \notin X$, then $X' = X \cup \{x\}$ is a nodal set in $[a, b]$. The unique interpolating polynomial of f subordinate to X', which we label $p_{n+1} \in \mathbb{P}_{n+1}$, is

$$p_{n+1}(t) = p_n(t) + \omega_{n+1}(t) f[x_0, x_1, \dots, x_n, x], \quad \forall t \in [a, b],$$

according to Proposition 9.39. Since $x \in [a, b]$ is an interpolation node,

$$f(x) = p_{n+1}(x)$$

and the result follows. □

Comparing the Lagrange and Newton error formulas, we get the following result.

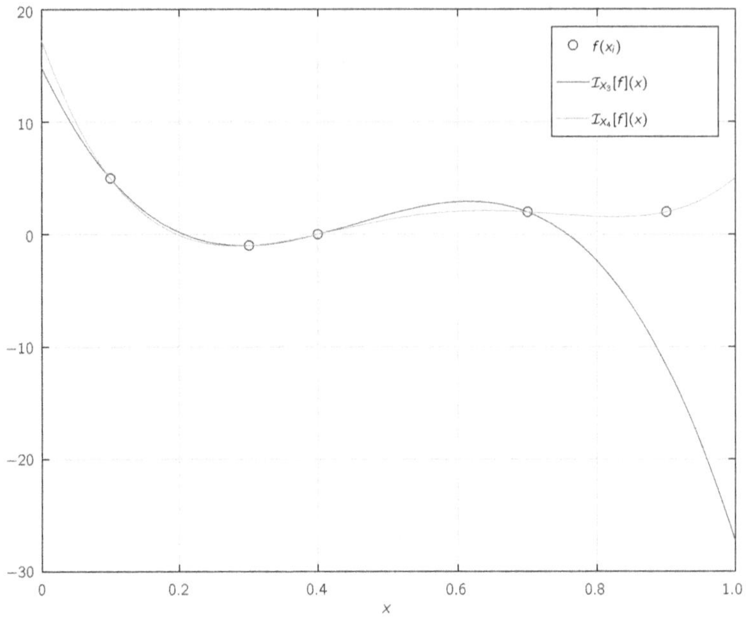

Figure 9.4 The third- and fourth-order interpolating polynomials constructed using Newton's divided differences and the data from Table 9.2.

Corollary 9.46 (divided differences)**.** *Suppose that* $n \in \mathbb{N}$, $X = \{x_i\}_{i=0}^{n}$ *is a nodal set in the compact interval* $[a, b] \subset \mathbb{R}$, *and* $f \in C^{n+1}([a, b])$. *Let* $x \in [a, b] \backslash X$ *be arbitrary. Then there is a point* $\xi = \xi(x_0, \dots, x_n, x) \in (a, b)$ *with*

$$\min\{x_0, x_1, \dots, x_n, x\} < \xi < \max\{x_0, x_1, \dots, x_n, x\}$$

such that

$$f[x_0, x_1, \dots, x_n, x] = \frac{f^{(n+1)}(\xi)}{(n+1)!}.$$

Proof. Compare the error representations in 9.28 from Theorem 9.45 and 9.11 from Theorem 9.16. □

The following is just a cosmetic, but quite useful, reformulation of the last result.

Corollary 9.47 (divided differences)**.** *Suppose that* $n \in \mathbb{N}$ *and* $X = \{x_i\}_{i=0}^{n}$ *is a nodal set in the compact interval* $[a, b] \subset \mathbb{R}$. *Let* $f \in C^n([a, b])$. *Then there is a point* $\xi = \xi(x_0, \dots, x_n) \in (a, b)$ *with*

$$\min\{x_0, x_1, \dots, x_n\} < \xi < \max\{x_0, x_1, \dots, x_n\}$$

such that

$$f[x_0, x_1, \ldots, x_n] = \frac{f^{(n)}(\xi)}{n!}.$$

Proof. See Problem 9.19. □

9.7 Extended Divided Differences

In this section, we give an alternate characterization of the divided difference function that will help us to extend its definition to the case where the points are not distinct, as well as to yield some other useful properties. First, we need a definition.

Definition 9.48 (n-simplex)**.** Suppose that $n \in \mathbb{N}$. The **canonical** n**-simplex** is the set

$$T_n = \{\boldsymbol{\tau} \in \mathbb{R}^n \mid \boldsymbol{\tau} \cdot \mathbf{1} \le 1,\ \boldsymbol{\tau} \cdot \boldsymbol{e}_i \ge 0,\ i = 1, \ldots, n\}, \tag{9.29}$$

where we recall that $\{\boldsymbol{e}_j\}_{j=1}^n$ is the canonical basis of $\mathbb{R}^n$ and $\mathbf{1} = [1, \ldots, 1]^\intercal \in \mathbb{R}^n$.

Lemma 9.49 (volume)**.** *Suppose that $n \in \mathbb{N}$. The n-dimensional volume of the canonical n-simplex is*

$$\operatorname{vol}(T_n) = \int_{T_n} \mathrm{d}^n\boldsymbol{\tau} = \frac{1}{n!}, \tag{9.30}$$

where $\mathrm{d}^n\boldsymbol{\tau} = \mathrm{d}\tau_1 \mathrm{d}\tau_2 \cdots \mathrm{d}\tau_n$.

Proof. The proof proceeds by induction. The crucial step is to show that, for $2 \le k \le n$,

$$\int_{T_k} \mathrm{d}^k\boldsymbol{\tau} = \int_{T_{k-1}} \int_{\tau_k=0}^{\tau_k = 1 - \sum_{j=1}^{k-1} \tau_j} \mathrm{d}\tau_k \mathrm{d}^{k-1}\boldsymbol{\tau}.$$

The details are left for the reader as an exercise; see Problem 9.22. □

Theorem 9.50 (Hermite–Genocchi Theorem[12])**.** *Suppose that $n \in \mathbb{N}$, $X = \{x_i\}_{i=0}^n$ is a nodal set in the compact interval $[a, b] \subset \mathbb{R}$, and $f \in C^n([a, b])$. We have*

$$f[x_0, x_1, \ldots, x_n] = \int_{T_n} f^{(n)}(\tau_0 x_0 + \tau_1 x_1 + \cdots + \tau_n x_n) \mathrm{d}^n\boldsymbol{\tau}, \tag{9.31}$$

where $\boldsymbol{\tau} = [\tau_1, \ldots, \tau_n]^\intercal \in T_n$, $\tau_0 = 1 - \boldsymbol{\tau} \cdot \mathbf{1}$, *and* $\mathrm{d}^n\boldsymbol{\tau} = \mathrm{d}\tau_1 \mathrm{d}\tau_2 \cdots \mathrm{d}\tau_n$.

Proof. The proof is by induction on n.

[12] Named in honor of the French mathematician Charles Hermite (1822–1901) and the Italian mathematician Angelo Genocchi (1817–1889).

($n = 1$) Suppose that $f \in C^1([a, b])$. Clearly, $T_1 = [0, 1]$ and

$$\begin{aligned}\int_{T_1} f'(\tau_0 x_0 + \tau_1 x_1) d\tau_1 &= \int_0^1 f'(x_0 + \tau_1(x_1 - x_0)) d\tau_1 \\ &= \frac{1}{x_1 - x_0} f(x_0 + \tau_1(x_1 - x_0))|_{\tau_1=0}^{\tau_1=1} \\ &= \frac{f(x_1) - f(x_0)}{x_1 - x_0} \\ &= f[x_0, x_1].\end{aligned}$$

($n = 2$) Suppose that $f \in C^2([a, b])$. The integration region T_2 is the triangle in $(\tau_1, \tau_2) \in \mathbb{R}^2$ with the vertices $(0, 0)$, $(1, 0)$, and $(0, 1)$. Then

$$\begin{aligned}&\int_{T_2} f''(\tau_0 x_0 + \tau_1 x_1 + \tau_2 x_2) d^2\boldsymbol{\tau} \\ &= \int_0^1 \int_0^{1-\tau_1} f''(x_0 + \tau_1(x_1 - x_0) + \tau_2(x_2 - x_0)) d\tau_2 d\tau_1 \\ &= \frac{1}{x_2 - x_0} \int_0^1 [f'(x_0 + \tau_1(x_1 - x_0) + \tau_2(x_2 - x_0))]|_{\tau_2=0}^{\tau_2=1-\tau_1} d\tau_1 \\ &= \frac{1}{x_2 - x_0} \left[\int_0^1 f'(x_2 + \tau_1(x_1 - x_2)) d\tau_1 - \int_0^1 f'(x_0 + \tau_1(x_1 - x_0)) d\tau_1\right] \\ &= \frac{1}{x_2 - x_0} [f[x_2, x_1] - f[x_0, x_1]] \\ &= \frac{1}{x_2 - x_0} [f[x_1, x_2] - f[x_0, x_1]] \\ &= f[x_0, x_1, x_2].\end{aligned}$$

($n = k - 1$) Let $k \in \mathbb{N}$, $k \geq 2$ be arbitrary. Suppose that $f \in C^{k-1}([a, b])$. Assume that $\{z_0, z_1, \dots, z_{k-1}\}$ is an arbitrary set of k distinct points in the compact interval $[a, b]$. For the induction hypothesis, let us assume that the formula holds for $n = k - 1$:

$$f[z_0, z_1, \dots, z_{k-1}] = \int_{T_{k-1}} f^{(k-1)}(\tau_0 z_0 + \tau_1 z_1 + \dots + \tau_{k-1} z_{k-1}) d^{k-1}\boldsymbol{\tau}.$$

($n = k$) Suppose that $f \in C^k([a, b])$. Using the induction hypothesis, Corollary 9.41, and 9.26 from Theorem 9.43, we have

$$
\begin{aligned}
&\int_{T_k} f^{(k)}(\tau_0 x_0 + \tau_1 x_1 + \cdots + \tau_k x_k) \mathrm{d}^k \boldsymbol{\tau} \\
&= \int_{T_{k-1}} \int_{\tau_k=0}^{\tau_k = 1 - \sum_{j=1}^{k-1} \tau_j} f^{(k)} \left(x_0 + \sum_{j=1}^{k} \tau_j (x_j - x_0) \right) \mathrm{d}\tau_k \mathrm{d}^{k-1} \boldsymbol{\tau} \\
&= \frac{1}{x_k - x_0} \int_{T_{k-1}} \left[f^{(k-1)} \left(x_0 + \sum_{j=1}^{k} \tau_j (x_j - x_0) \right) \right] \Bigg|_{\tau_n = 0}^{\tau_n = 1 - \sum_{j=1}^{k-1} \tau_j} \mathrm{d}^{k-1} \boldsymbol{\tau} \\
&= \frac{1}{x_k - x_0} \left[\int_{T_{k-1}} f^{(k-1)} \left(x_k + \sum_{j=1}^{k-1} \tau_j (x_j - x_k) \right) \mathrm{d}^{k-1} \boldsymbol{\tau} \right. \\
&\qquad \left. - \int_{T_{k-1}} f^{(k-1)} \left(x_0 + \sum_{j=1}^{k-1} \tau_j (x_j - x_0) \right) \mathrm{d}^{k-1} \boldsymbol{\tau} \right] \\
&= \frac{1}{x_k - x_0} \left[f[x_k, x_1, x_2, \ldots, x_{k-1}] - f[x_0, x_1, \ldots, x_{k-1}] \right] \\
&= \frac{1}{x_k - x_0} \left[f[x_1, x_2, \ldots, x_k] - f[x_0, x_1, \ldots, x_{k-1}] \right] \\
&= f[x_0, x_1, \ldots, x_k].
\end{aligned}
$$

The proof is complete. □

Let us now extend the definition of the divided difference function to the case where the nodes are not necessarily distinct.

Definition 9.51 (extended divided differences)**.** Suppose that $n \in \mathbb{N}$ and $[a, b] \subset \mathbb{R}$ is a compact interval. Set

$$
[a, b]^{n+1} = \left\{ [z_0, z_1, \ldots, z_n]^\intercal \in \mathbb{R}^{n+1} \;\middle|\; z_j \in [a, b],\ j = 0, \ldots, n \right\}.
$$

Let $f \in C^n([a, b])$. For every point $\boldsymbol{z} \in [a, b]^{n+1}$, we define

$$
f[\![\boldsymbol{z}]\!] = f[\![z_0, z_1, \ldots, z_n]\!] = \int_{T_n} f^{(n)}(\tau_0 z_0 + \tau_1 z_1 + \cdots + \tau_n z_n) \mathrm{d}^n \boldsymbol{\tau}, \tag{9.32}
$$

where T_n is the canonical n-simplex defined in (9.29), $\boldsymbol{\tau} = [\tau_1, \ldots, \tau_n]^\intercal \in T_n$, $\tau_0 = 1 - \boldsymbol{\tau} \cdot \mathbf{1}$, and $\mathrm{d}^n \boldsymbol{\tau} = \mathrm{d}\tau_1 \mathrm{d}\tau_2 \cdots \mathrm{d}\tau_n$. The function $f[\![\cdot]\!]\colon [a, b]^{n+1} \to \mathbb{R}$ is called the **extended divided difference of order** n.

Theorem 9.52 (continuity)**.** *Suppose that $n \in \mathbb{N}$, $[a, b]$ is a compact interval, and $f \in C^n([a, b])$. The extended finite difference function defined by* (9.32) *satisfies $f[\![\cdot]\!] \in C([a, b]^{n+1})$.*

Proof. Let $\boldsymbol{\tau} \in T_n$ and $\tau_0 = 1 - \boldsymbol{\tau} \cdot \mathbf{1}$. Define $\vec{\boldsymbol{\tau}} = [\tau_0, \boldsymbol{\tau}]^\intercal \in T_{n+1}$.

Notice now that $f^{(n)} \in C([a, b])$ and, consequently, it is uniformly continuous on $[a, b]$. In other words, given $\varepsilon > 0$, there is a $\delta > 0$, such that, if x and y are any points in $[a, b]$ satisfying $|x - y| \leq \delta$, then it follows that

$$\left| f^{(n)}(x) - f^{(n)}(y) \right| \leq \varepsilon n!.$$

Now suppose that $z_1, z_2 \in [a, b]^{n+1}$ satisfy $\|z_1 - z_2\|_\infty \leq \delta$. Then

$$|\vec{\boldsymbol{\tau}} \cdot z_1 - \vec{\boldsymbol{\tau}} \cdot z_2| = |\vec{\boldsymbol{\tau}} \cdot (z_1 - z_2)| \leq \|\vec{\boldsymbol{\tau}}\|_1 \|z_1 - z_2\|_\infty \leq \delta.$$

Therefore, uniformly with respect to $\vec{\boldsymbol{\tau}}$, we have

$$\left| f^{(n)}(\vec{\boldsymbol{\tau}} \cdot z_1) - f^{(n)}(\vec{\boldsymbol{\tau}} \cdot z_2) \right| \leq \varepsilon n!,$$

as long as $z_1, z_2 \in [a, b]^{n+1}$ satisfy $\|z_1 - z_2\|_\infty \leq \delta$. Then,

$$\begin{aligned} |f[\![z_1]\!] - f[\![z_2]\!]| &= \left| \int_{T_n} f^{(n)}(\vec{\boldsymbol{\tau}} \cdot z_1) \mathrm{d}^n \boldsymbol{\tau} - \int_{T_n} f^{(n)}(\vec{\boldsymbol{\tau}} \cdot z_2) \mathrm{d}^n \boldsymbol{\tau} \right| \\ &\leq \int_{T_n} \left| f^{(n)}(\vec{\boldsymbol{\tau}} \cdot z_1) - f^{(n)}(\vec{\boldsymbol{\tau}} \cdot z_2) \right| \mathrm{d}^n \boldsymbol{\tau} \\ &\leq \int_{T_n} \varepsilon n! \mathrm{d}^n \boldsymbol{\tau} \\ &= \varepsilon. \end{aligned}$$

This proves the result. □

We now prove a result for this extended definition, which is analogous to Corollary 9.47, but does not require the nodes to be distinct.

Proposition 9.53 (extended finite differences). *Let $n \in \mathbb{N}$, $[a, b] \subset \mathbb{R}$ be a compact interval, and $f \in C^n([a, b])$. Suppose that*

$$\{x_j\}_{j=0}^n \subset [a, b].$$

Then there is a point $\xi = \xi(x_0, \ldots, x_n) \in (a, b)$ with

$$\min\{x_0, x_1, \ldots, x_n\} < \xi < \max\{x_0, x_1, \ldots, x_n\}$$

such that

$$f[\![x_0, x_1, \ldots, x_n]\!] = \frac{f^{(n)}(\xi)}{n!}.$$

Proof. Set

$$\hat{a} = \min\{x_0, x_1, \ldots, x_n\}, \qquad \hat{b} = \max\{x_0, x_1, \ldots, x_n\},$$

and

$$m = \min_{\hat{a} \leq x \leq \hat{b}} f^{(n)}(x), \qquad M = \max_{\hat{a} \leq x \leq \hat{b}} f^{(n)}(x).$$

Then, using (9.30), we deduce

$$\frac{m}{n!} \leq f[\![x_0, x_1, \ldots, x_n]\!] \leq \frac{M}{n!},$$

where we have again used $\operatorname{vol}(T_n) = \int_{T_n} \mathrm{d}^n \boldsymbol{\tau} = \frac{1}{n!}$. In other words,

$$m \le f[\![x_0, x_1, \ldots, x_n]\!] n! \le M.$$

By the Intermediate Value Theorem B.27, there is a point $\xi \in [\hat{a}, \hat{b}]$ such that

$$f^{(n)}(\xi) = f[\![x_0, x_1, \ldots, x_n]\!] n!. \qquad \square$$

The following is a simple consequence.

Corollary 9.54 (extended divided differences). *Suppose that $[a, b]$ is a compact interval and $f \in C^n([a, b]; \mathbb{R})$. Suppose that $x \in [a, b]$. Then*

$$f[\![\underbrace{x, x, \ldots, x}_{n+1}]\!] = \frac{f^{(n)}(x)}{n!}.$$

Proof. See Problem 9.23. □

We showed that divided differences are invariant to permutations. Let us show that this is the case for extended finite differences as well. First, we need an approximation result.

Lemma 9.55 (density). *Let $n \in \mathbb{N}$. Suppose that $\boldsymbol{z} \in [a, b]^{n+1}$. There exists a sequence $\{\boldsymbol{z}_\ell\}_{\ell=1}^\infty \subset [a, b]^{n+1}$ with the following properties.*

1. *For each $\ell \in \mathbb{N}$ the coordinates of $\boldsymbol{z}_\ell$ are all distinct.*
2. *The sequence converges to $\boldsymbol{z}$ as $\ell \to \infty$.*

Proof. See Problem 9.24. □

Proposition 9.56 (invariance). *Suppose that $n \in \mathbb{N}$, $[a, b] \subset \mathbb{R}$ is a compact interval, and $f \in C^n([a, b])$. Let*

$$\boldsymbol{z} = [z_0, z_1, \ldots, z_n]^\intercal \in [a, b]^{n+1}$$

be arbitrary, and $(i_0, i_1, \ldots, i_n)$ be a permutation of $(0, 1, \ldots, n)$. Then

$$f[\![z_{i_0}, z_{i_1}, \ldots, z_{i_n}]\!] = f[\![z_0, z_1, \ldots, z_n]\!].$$

Proof. If the coordinates of $\boldsymbol{z}$, i.e., $z_0, z_1, \ldots, z_n$, are all distinct, there is nothing to prove.

Assume, then, that the coordinates of $\boldsymbol{z}$ are not distinct. Let $\{z_j\}_{j=1}^\infty$ be the sequence of Lemma 9.55. Since the divided difference function and the extended divided difference function agree for distinct points, by Corollary 9.41,

$$f[\![z_{\ell,i_0}, z_{\ell,i_1}, \ldots, z_{\ell,i_n}]\!] = f[\![z_{\ell,0}, z_{\ell,1}, \ldots, z_{\ell,n}]\!], \quad \forall \ell \in \mathbb{N},$$

where $\boldsymbol{z}_\ell = [z_{\ell,0}, z_{\ell,1}, \ldots, z_{\ell,n}]^\intercal$. Now Theorem 9.52 showed that $f[\![\cdot]\!]$ is continuous, i.e.,

$$\begin{aligned} f[\![z_{i_0}, z_{i_1}, \ldots, z_{i_n}]\!] &= \lim_{\ell\to\infty} f[\![z_{i_0,\ell}, z_{i_1,\ell}, \ldots, z_{i_n,\ell}]\!] \\ &= \lim_{\ell\to\infty} f[\![z_{0,\ell}, z_{1,\ell}, \ldots, z_{n,\ell}]\!] \\ &= f[\![z_0, z_1, \ldots, z_n]\!], \end{aligned}$$

and the claimed invariance follows. □

The following result will be needed for our analysis of numerical integration methods in Chapter 14.

Theorem 9.57 (differentiation). *Suppose that $n \in \mathbb{N}$, $[a, b] \subset \mathbb{R}$ is a compact interval, and $f \in C^{n+1}([a, b])$. Let $X = \{x_i\}_{i=0}^n \subset [a, b]$ be a nodal set in $[a, b]$. Then $f[\![x_0, x_1, \dots, x_n]\!]$ is continuously differentiable with respect to x_n and*

$$\frac{\partial}{\partial x_n} f[\![x_0, x_1, \dots, x_n]\!] = f[\![x_0, x_1, \dots, x_n, x_n]\!]. \tag{9.33}$$

Proof. Using (9.27),

$$\begin{aligned}
\frac{\partial}{\partial x_n} f[\![x_0, x_1, \dots, x_n]\!] &= \lim_{h \to 0} \frac{f[\![x_0, x_1, \dots, x_n + h]\!] - f[\![x_0, x_1, \dots, x_n]\!]}{h} \\
&= \lim_{h \to 0} \frac{f[\![x_0, x_1, \dots, x_n + h]\!] - f[\![x_n, x_0, x_1, \dots, x_{n-1}]\!]}{h} \\
&= \lim_{h \to 0} f[\![x_n, x_0, x_1, \dots, x_n + h]\!] \\
&= f[\![x_n, x_0, x_1, \dots, x_n]\!] \\
&= f[\![x_0, x_1, \dots, x_n, x_n]\!],
\end{aligned}$$

where the last equality came from Proposition 9.56. Now observe that

$$f[\![x_0, x_1, \dots, x_n, x_n]\!] = \int_{T_{n+1}} f^{(n+1)}(\tau_0 z_0 + \tau_1 z_1 + \dots + \tau_n z_n + \tau_{n+1} z_n) \mathrm{d}^{n+1} \boldsymbol{\tau}.$$

As long as $f \in C^{n+1}([a, b])$, $f[\![x_0, x_1, \dots, x_n, x_n]\!]$ is a continuous function of its arguments. □

We conclude this chapter with alternate formulas for the Hermite interpolating polynomial and its error in terms of divided differences.

Theorem 9.58 (Newton–Hermite interpolation error). *Suppose that $n \in \mathbb{N}$, $X = \{x_j\}_{j=0}^n$ is a nodal set in the compact interval $[a, b] \subset \mathbb{R}$, and $f \in C^{2n+2}([a, b])$. Then $p \in \mathbb{P}_{2n+1}$, defined by*

$$\begin{aligned}
p(x) = f(x_0) &+ (x - x_0) f[\![x_0, x_0]\!] + (x - x_0)^2 f[\![x_0, x_0, x_1]\!] \\
&+ (x - x_0)^2 (x - x_1) f[\![x_0, x_0, x_1, x_1]\!] \\
&+ (x - x_0)^2 (x - x_1)^2 f[\![x_0, x_0, x_1, x_1, x_2]\!] \\
&+ \dots + (x - x_0)^2 \cdots (x - x_{n-1})^2 (x - x_n) f[\![x_0, x_0, x_1, x_1, \dots, x_n, x_n]\!],
\end{aligned}$$

is the unique Hermite interpolating polynomial satisfying

$$p(x_j) = f(x_j), \qquad p'(x_j) = f'(x_j), \qquad j = 0, \dots, n.$$

For any $x \in [a, b]$, the interpolation error has the representation

$$f(x) - p(x) = \omega_{n+1}^2(x) f[\![x_0, x_0, x_1, x_1, \dots, x_n, x_n, x]\!]. \tag{9.34}$$

Proof. Suppose that $Z = \{z_0, \ldots, z_{2n+1}\}$ is a nodal set in $[a, b]$. Then $p \in \mathbb{P}_{2n+1}$, defined by

$$\begin{aligned} p(z) &= f(z_0) + (z - z_0) f[\![z_0, z_1]\!] + (z - z_0)(z - z_1) f[\![z_0, z_1, z_2]\!] \\ &\quad + (z - z_0)(z - z_1)(z - z_2) f[\![z_0, z_1, z_2, z_3]\!] \\ &\quad + \cdots + (z - z_0) \cdots (z - z_{2n}) f[\![z_0, z_1, \ldots, z_{2n+1}]\!], \end{aligned}$$

is the unique Lagrange interpolating polynomial satisfying

$$p(z_j) = f(z_j), \quad j = 0, \ldots, 2n+1.$$

Owing to Theorem 9.45, if $f \in C^{2n+2}([a, b])$, the error may be expressed as follows: for any $x \in [a, b]$,

$$f(x) - p(x) = (x - z_0) \cdots (x - z_{2n+1}) f[\![z_0, z_1, \ldots, z_{2n+1}, x]\!].$$

Use now the continuity of the extended divided differences to take the limits

$$z_0, z_1 \to x_0, \qquad z_2, z_3 \to x_1, \qquad \ldots, \qquad z_{2n}, z_{2n+1} \to x_n$$

and show that the desired properties hold. The details are left to the reader as an exercise; see Problem 9.25. □

Problems

9.1 Prove Theorem 9.4. In fact, prove the following more general result. Suppose that $Z = \{z_i\}_{i=0}^n \subset \mathbb{C}$ is a set of $n + 1 \in \mathbb{N}$ distinct points. Then the matrix

$$\mathsf{V} = \begin{bmatrix} 1 & z_0 & z_0^2 & \cdots & z_0^n \\ 1 & z_1 & z_1^2 & \cdots & z_1^n \\ \vdots & \vdots & \vdots & & \vdots \\ 1 & z_n & z_n^2 & \cdots & z_n^n \end{bmatrix} \in \mathbb{C}^{(n+1)\times(n+1)}$$

is invertible and, in particular,

$$\det(\mathsf{V}) = \prod_{0 \le i < j \le n} (z_j - z_i) \neq 0.$$

9.2 Prove Proposition 9.7.

9.3 Prove Proposition 9.10.

9.4 Complete the proof of Theorem 9.11.

9.5 Prove Proposition 9.14.

9.6 Complete the proof of Theorem 9.15.

9.7 Suppose that $X = \{x_i\}_{i=0}^n$ is a nodal set of size $n + 1 \in \mathbb{N}$ in the compact interval $[a, b] \subset \mathbb{R}$ and $\mathcal{L}_X = \{L_i\}_{i=0}^n \subset \mathbb{P}_n$ is the Lagrange basis subordinate to X. Prove that if $0 \le m \le n$, then

$$\sum_{i=0}^n x_i^m L_i(x) = x^m, \quad \forall x \in \mathbb{R}.$$

9.8 Prove Proposition 9.18 by showing the following. Let $n \in \mathbb{N}$. Define

$$h = \frac{b-a}{n},$$

and, for $i = 0, \dots, n$, set $x_i = a + ih$. Show that the nodal polynomial ω_{n+1}, introduced in Definition 9.13, satisfies

$$\max_{a \le x \le b} |\omega_{n+1}(x)| \le \Omega_{n+1},$$

where

$$\Omega_{n+1} = \frac{n!}{4} h^{n+1}.$$

9.9 Prove Theorem 9.20 by the following steps. Let $n \in \mathbb{N}$. Suppose that $X = \{x_i\}_{i=0}^n$ is a nodal set in the compact interval $[a, b] \subset \mathbb{R}$ and $f \in C^1([a, b])$. Define, for $0 \le \ell \le n$,

$$F_{0,\ell}(x) = (L_\ell(x))^2(1 - 2L'_\ell(x_\ell)(x - x_\ell)), \quad F_{1,\ell} = (L_\ell(x))^2(x - x_\ell),$$

where L_ℓ is the Lagrange basis polynomial of order ℓ, defined in (9.3), and

$$p(x) = \sum_{\ell=0}^{n} \left[F_{0,\ell}(x) f(x_\ell) + F_{1,\ell}(x) f'(x_\ell) \right].$$

a) Prove that

$$F_{0,\ell}(x_k) = \delta_{k,\ell}, \quad F'_{0,\ell}(x_k) = 0, \quad F_{1,\ell}(x_k) = 0, \quad F'_{1,\ell}(x_k) = \delta_{k,\ell},$$

and, therefore, that, for all $i = 0, \dots, n$, we have

$$f(x_i) = p(x_i), \qquad f'(x_i) = p'(x_i).$$

b) Prove that p is the unique polynomial in $\mathbb{P}_{2n+1}$ with the property above.

9.10 Prove Theorem 9.35.

9.11 Let $f \in C([-1, 1])$. Construct the Lagrange interpolation polynomial $p_1 \in \mathbb{P}_1$ for f using the interpolation nodes $x_0 = -1$ and $x_1 = 1$. Show that if $f \in C^2([-1, 1])$, then

$$|f(x) - p_1(x)| \le \frac{F_2}{2}\left(1 - x^2\right) \le \frac{F_2}{2}, \quad \forall x \in [-1, 1],$$

where $F_2 = \|f''\|_{L^\infty(-1,1)}$. Find a function f and a point x for which equality is achieved in the estimates above.

9.12 This problem is concerned with the estimate provided in Theorem 9.16.

a) Compute the Lagrange interpolating polynomial of degree one for the function $f(x) = x^3$ using the interpolation nodes $x_0 = 0$ and $x_1 = a$. Verify Theorem 9.16 by a direct calculation, showing that, in this case, ξ has the unique value $\xi = (x + a)/3$.

b) Repeat the calculation for the function $f(x) = (2x - a)^4$. Show that, in this case, there are two possible values for ξ and give their values.

9.13 Let $n \in \mathbb{N}_0$. Given the nodal set $X = \{x_i\}_{i=0}^{n+1}$ and $Y = \{y_i\}_{i=0}^{n+1} \subset \mathbb{R}$, let $q_n, r_n \in \mathbb{P}_n$ be the Lagrange interpolating polynomials for the coordinate sets

$$Q = \{(x_i, y_i) \mid i = 0, 1, \dots, n\}, \qquad R = \{(x_i, y_i) \mid i = 1, 2, \dots, n+1\},$$

respectively. Define

$$p_{n+1}(x) = \frac{(x - x_0)r_n(x) - (x - x_{n+1})q_n(x)}{x_{n+1} - x_0}.$$

Show that p_{n+1} is the Lagrange interpolating polynomial of degree $n+1$ for the coordinate set $P = \{(x_i, y_i) \mid i = 0, 1, \dots, n+1\}$.

9.14 Construct the Hermite interpolating polynomial of degree three (3) for the function $f(x) = x^5$, using the points $x_0 = 0$ and $x_1 = a$, and show that it has the form $p_3(x) = 3a^2x^3 - 2a^3x^2$. Verify Theorem 9.21 by direct calculation, showing that, in this case, ξ has the unique value $\xi = (x + 2a)/5$.

9.15 Let $n \in \mathbb{N}_0$ and $X = \{x_i\}_{i=0}^{n}$ be a nodal set in the compact interval $[a, b] \subset \mathbb{R}$. Let $f \in C^1([a, b])$ and $p \in \mathbb{P}_{2n+1}$ be the Hermite interpolating polynomial of f subject to X. State and prove an error estimate for $f'(x) - p'(x)$ with $x \in [a, b] \setminus X$.

9.16 Prove Proposition 9.37.

9.17 Complete the proof of Proposition 9.39.

9.18 Prove Corollary 9.44.

9.19 Prove Corollary 9.47.

9.20 Let $[a, b] \subset \mathbb{R}$ be a compact interval, $n \in \mathbb{N}_0$, and $f \in C([a, b])$. Show that $f \in \mathbb{P}_n$ if and only if for every nodal set $X = \{x_j\}_{j=0}^{n} \subset [a, b]$ we have

$$f[x_0, \dots, x_n] = 0.$$

9.21 Let $[a, b] \subset \mathbb{R}$ be a compact interval, $n \in \mathbb{N}_0$, and $g, h \in C([a, b])$. Define $f(x) = g(x)h(x)$. Show that, for every nodal set $X = \{x_j\}_{j=0}^{n} \subset [a, b]$, we have

$$f[x_0, \dots, x_n] = \sum_{j=0}^{n} g[x_0, \dots, x_j]h[x_{j+1}, \dots, x_n].$$

9.22 Prove Lemma 9.49.

9.23 Prove Corollary 9.54.

9.24 Prove Lemma 9.55.

9.25 Prove Theorem 9.58.

10 Minimax Polynomial Approximation

In this chapter — which is influenced by the presentations in the excellent books by Davis [24], Isaacson and Keller [46], Süli and Mayers [89], Trefethen [95], and Powell [71] — we look for best approximations of a given function of interest by polynomials in the max norm. This is called the *minimax problem.* For further insight, we also refer the reader to [1, 73].

While, the max norm is a simple-to-understand object, it turns out that finding minimax approximations is a highly nontrivial task, except for very simple cases. There is a way to characterize minimax approximations, based on a type of oscillation property, but this requires some complicated (and beautiful) theory. We will find an interesting connection between this characterization and what are called Chebyshev orthogonal polynomials and Chebyshev interpolation. This connection will give us some further insight into the Runge[1] phenomenon that we encountered in the last chapter.

To motivate this topic, let us consider a simple function of interest,

$$f(x) = x^2, \qquad 0 \leq x \leq 1.$$

Now let us search for the linear ($n = 1$) polynomial $p_1(x) = a_1 x + a_0$ that yields the best approximation of f, where the error is measured as

$$\|f - p_1\|_{L^\infty(0,1)} = \max_{0 \leq x \leq 1} |f(x) - p_1(x)|.$$

The best approximation will minimize the max norm of the error, hence the term *minimax.* The minimizer, as we see from Problem 10.11, is $p_1(x) = x - \frac{1}{8}$, and

$$\min_{p_1 \in \mathbb{P}_1} \|f - p_1\|_{L^\infty(0,1)} = \frac{1}{8}.$$

Interestingly, the error function, $e(x) = f(x) - p_1(x) = x^2 - x + \frac{1}{8}$ equi-oscillates, namely,

$$e(0) = \frac{1}{8}, \quad e\left(\frac{1}{2}\right) = -\frac{1}{8}, \quad e(1) = \frac{1}{8}.$$

In other words, with precisely alternating signs, the error function takes its largest values at exactly three ($n + 2$) points. We will see that this curious property is always present in the minimax approximations.

Before we get into the theory, we refer the reader to Appendix D for important notation and all of the properties of spaces of functions that we shall use in the the chapter.

[1] Named in honor of the German mathematician Carl David Tolmé Runge (1856–1927).

10.1 Minimax: Best Approximation in the ∞-Norm

Let us define our problem of interest and give a detailed study of it.

Definition 10.1 (minimax). Let $f \in C([a,b])$ and $n \in \mathbb{N}_0$. We say that the polynomial $p \in \mathbb{P}_n$ is a **best polynomial approximation of f in the L^∞-norm** or a **minimax polynomial** if and only if

$$\|f - p\|_{L^\infty(a,b)} = \inf_{q \in \mathbb{P}_n} \|f - q\|_{L^\infty(a,b)} .$$

If such a polynomial exists, we write

$$p \in \operatorname*{argmin}_{q \in \mathbb{P}_n} \|f - q\|_{L^\infty(a,b)} .$$

If, additionally, p is unique, we write

$$p = \operatorname*{argmin}_{q \in \mathbb{P}_n} \|f - q\|_{L^\infty(a,b)} .$$

Let us immediately show that a minimax polynomial always exists.

Theorem 10.2 (existence). *Let $f \in C([a,b])$ and $n \in \mathbb{N}_0$. There exists at least one minimax polynomial approximation to f, i.e., there is a polynomial, $p \in \mathbb{P}_n$, with*

$$p \in \operatorname*{argmin}_{q \in \mathbb{P}_n} \|f - q\|_{L^\infty(a,b)} .$$

Proof. We follow the proofs in [24] and [89]; see also Problem 5.1. Define the function $E \colon \mathbb{R}^{n+1} \to [0, \infty)$ via

$$E(c_0, \ldots, c_n) = \|f - q\|_{L^\infty(a,b)} , \quad q(x) = \sum_{i=0}^{n} c_i x^i .$$

We leave it to the reader to prove that $E \in C(\mathbb{R}^{n+1}; [0,\infty))$, i.e., it is continuous; see Problem 10.1. Now define

$$S = \left\{ [c_0, \ldots, c_n]^\intercal \in \mathbb{R}^{n+1} \,\middle|\, E(c_0, \ldots, c_n) \le \|f\|_{L^\infty(a,b)} + 1 \right\} .$$

The set S is nonempty: $\mathbf{0} \in S$, since $E(0, \ldots, 0) = \|f\|_{L^\infty(a,b)}$. In addition, the set S is closed and bounded in $\mathbb{R}^{n+1}$. In other words, S is a compact subset of $\mathbb{R}^{n+1}$. Therefore, Theorem B.47 guarantees that E attains its minimum on S. Let us label a point of attainment $\boldsymbol{c}^\star = [c_0^\star, \ldots, c_n^\star]^\intercal$, and define

$$p^\star(x) = \sum_{i=0}^{n} c_i^\star x^i \in \mathbb{P}_n .$$

Then

$$m = E(c_0^\star, \ldots, c_n^\star) = \|f - p^\star\|_{L^\infty(a,b)} .$$

It turns out that this is the minimum of E over all of $\mathbb{R}^{n+1}$, not just S. To see this, observe that

$$m = E(c_0^\star, \ldots, c_n^\star) \le E(0, \ldots, 0) = \|f\|_{L^\infty(a,b)} .$$

Suppose now that $[c_0, \dots, c_n]^\intercal \in \mathbb{R}^{n+1} \setminus S$. By definition,

$$E(c_0, \dots, c_n) > \|f\|_{L^\infty(a,b)} + 1 > \|f\|_{L^\infty(a,b)} \geq m.$$

This shows that

$$m \leq E(c_0, \dots, c_n), \quad \forall [c_0, \dots, c_n]^\intercal \in \mathbb{R}^{n+1},$$

and, in fact,

$$m = \|f - p^\star\|_{L^\infty(a,b)} = \inf_{[c_0,\dots,c_n]^\intercal \in \mathbb{R}^{n+1}} E(c_0, \dots, c_n) = \inf_{q \in \mathbb{P}_n} \|f - q\|_{L^\infty(a,b)} .$$

In other words,

$$p^\star \in \operatorname*{argmin}_{q \in \mathbb{P}_n} \|f - q\|_{L^\infty(a,b)} . \qquad \square$$

We shall only go into depth regarding best approximations in the L^∞-norm, but the following is of theoretical interest.

Definition 10.3 (best approximation)**.** Suppose that w is a weight function on $[a, b]$ and $1 \leq p < \infty$. Let $f \in C([a, b])$ and $n \in \mathbb{N}_0$. We say that the polynomial $q \in \mathbb{P}_n$ is a **best polynomial approximation of** f **in the** $L^p_w(a, b)$**-norm** if and only if

$$\|f - q\|_{L^p_w(a,b)} = \inf_{r \in \mathbb{P}_n} \|f - r\|_{L^p_w(a,b)} .$$

If such a polynomial exists, we write

$$q \in \operatorname*{argmin}_{r \in \mathbb{P}_n} \|f - r\|_{L^p_w(a,b)} ,$$

and if q is unique, we write

$$q = \operatorname*{argmin}_{r \in \mathbb{P}_n} \|f - r\|_{L^p_w(a,b)} .$$

A proof of the following can be obtained by adapting the proof of Theorem 10.2.

Theorem 10.4 (existence)**.** *Suppose that w is a weight function on $[a, b]$ and $1 \leq p < \infty$. Let $f \in C([a, b])$ and $n \in \mathbb{N}_0$ be fixed. There exists at least one polynomial, $q \in \mathbb{P}_n$, satisfying*

$$q \in \operatorname*{argmin}_{r \in \mathbb{P}_n} \|f - r\|_{L^p_w(a,b)} .$$

Proof. See Problem 10.2. $\square$

We note that the previous two results mention the existence of a best polynomial approximation, but say nothing about their uniqueness. It turns out that this is a delicate issue which heavily depends on the norm in question, as the following simple finite-dimensional examples show.

Example 10.1 The essence of best polynomial approximation is that we are trying to find the best (in some norm) approximation to an element of a vector space by elements of a subspace. If the ambient space is $\mathbb{R}^n$, for some $n \in \mathbb{N}$, with the

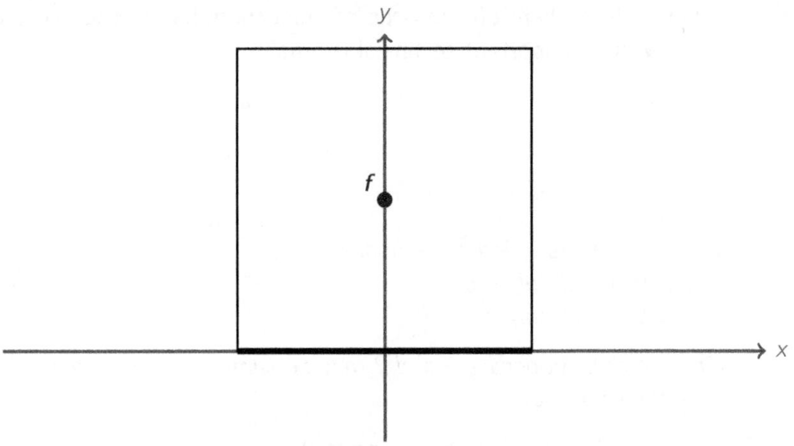

Figure 10.1 The best approximation in the ∞-norm of the vector $\boldsymbol{f} = [0, 1]^\intercal$ by vectors of the form $[x, 0]^\intercal$ is not unique.

2-norm then the results of Section 5.2 show that the best approximation is unique and realized by a projection matrix.

Example 10.2 Consider now $\mathbb{R}^2$ under the ∞-norm. Assume that we wish to approximate the vector $\boldsymbol{f} = [0, 1]^\intercal$ by elements of the subspace

$$S = \left\{ [x, y]^\intercal \in \mathbb{R}^2 \mid y = 0 \right\}.$$

In other words, we are trying to find

$$\inf_{\boldsymbol{v} \in S} \|\boldsymbol{f} - \boldsymbol{v}\|_\infty .$$

Figure 10.1 shows that all vectors of the form $[x, 0]^\intercal$ with $|x| \leq 1$ realize this infimum. On the other hand, we leave it to the reader, see Problem 10.3, to show that if the subspace is

$$W = \left\{ [x, y]^\intercal \in \mathbb{R}^2 \mid x = y \right\},$$

then the infimum is realized by a unique element.

The previous two examples show that the issue of uniqueness of best approximations is related to *convexity* of norms. It turns out that the 2-norm is *strictly convex*, whereas the ∞-norm is not. We will deal more in depth with the minimization of convex and strictly convex functions in Chapter 16.

Before we move on, we point out that, for all $p \in [1, \infty)$, the best polynomial approximation to a continuous function in the L^p_w norm — whose existence is guaranteed by Theorem 10.4 — is, in fact, unique. For a proof, we refer the reader to [24].

Let us focus here on the issue of uniqueness for the case of the L^∞-norm. We begin with an important technical lemma.

Lemma 10.5 (comparison). *Suppose that $a, b, c \in \mathbb{R}$ satisfy*

$$|a| > |b| \geq 0, \qquad c = a - b.$$

Then

1. *The numbers a and c are nonzero.*
2. *$a > 0$ if and only if $c > 0$.*
3. *$a < 0$ if and only if $c < 0$.*

Observe that properties 2 and 3 may be summarized by saying that a and c must have the same sign.

Proof. We begin by proving property 2.

($\Longleftarrow$) Suppose that $c > 0$. Then $a > b$. Since $|a| > |b|$,

$$a > b > -|a|,$$

which implies that

$$0 < a + |a| = \begin{cases} 0, & a < 0, \\ 2a, & a > 0. \end{cases}$$

The only possibility is that $a > 0$.

($\Longrightarrow$) Suppose that $a > 0$. Then $c > -b$. If $b < 0$, we get the conclusion we want, namely $c > 0$. On the other hand, if $b \geq 0$, then, by assumption,

$$a = |a| > |b| = b \geq 0.$$

So $a - b > 0$. But $c = a - b > 0$.

Next, we prove property 3.

($\Longleftarrow$) Suppose that $c < 0$. Then $b > a$. Since $|a| > |b|$,

$$a < b < |a|,$$

which implies that

$$0 < |a| - a = \begin{cases} 0, & a > 0, \\ -2a, & a < 0. \end{cases}$$

The only possibility is that $a < 0$.

($\Longrightarrow$) Suppose that $a < 0$. Then $c < -b$. If $b > 0$, we get the result we want, namely $c < 0$. If $b \leq 0$, then, by assumption,

$$-a = |a| > |b| = -b \geq 0.$$

So $a - b < 0$. But $c = a - b < 0$.

Finally, to prove property 1, observe that $|a| > |b| \geq 0$ implies that a cannot be zero. The equivalences above imply that c cannot be zero either. □

Theorem 10.6 (de la Vallée Poussin[2]). *Suppose that $f \in C([a, b])$, $n \in \mathbb{N}_0$, and $q \in \mathbb{P}_n$. Assume that there are $n+2$ distinct points in $[a, b]$, denoted*

$$a \le x_0 < x_1 < \cdots < x_n < x_{n+1} \le b$$

such that $\{f(x_i) - q(x_i)\}_{i=0}^{n+1}$ is a strictly alternating sequence, none of the differences vanishing. Let $p \in \mathbb{P}_n$ be a minimax polynomial. Then

$$\|f - p\|_{L^\infty(a,b)} \ge \min_{0 \le i \le n+1} |f(x_i) - q(x_i)| = \chi > 0. \tag{10.1}$$

Proof. To get a contradiction suppose that, in (10.1), $\chi > \|f - p\|_{L^\infty(a,b)}$. This implies that, for all $i = 0, \ldots, n+1$,

$$|f(x_i) - q(x_i)| > |f(x_i) - p(x_i)| \ge 0, \quad i = 0, 1, \ldots, n+1, \tag{10.2}$$

where $p \in \mathbb{P}_n$ is the minimax polynomial. Now write, for $i = 0, 1, \ldots, n+1$,

$$q(x_i) - p(x_i) = [q(x_i) - f(x_i)] - [p(x_i) - f(x_i)]. \tag{10.3}$$

Observe now that, for all $i = 0, \ldots, n+1$, $q(x_i) - p(x_i)$ and $q(x_i) - f(x_i)$ have the same signs, and neither are zero. This follows from the technical Lemma 10.5. This implies that $q - p \in \mathbb{P}_n$ changes sign $n+1$ times; in other words, $q - p$ has $n+1$ distinct zeros in $[a, b]$. This, in turn, implies that the difference is the zero polynomial, i.e., $q \equiv p$. But, if that is the case, (10.2) cannot be true. We have a contradiction. Thus, (10.1) must be true. □

We have then arrived at the celebrated *Chebyshev Oscillation Theorem*, which is a key to proving the uniqueness of the minimax polynomial [89, 95]. We remark that Davis [24] uses a convexity argument to prove uniqueness, a different tack from that used here.

Definition 10.7 (equi-oscillation). Suppose that $f \in C([a, b])$ and $n \in \mathbb{N}_0$. We say that a polynomial $q \in \mathbb{P}_n$ has the **equi-oscillation property** with respect to f if and only if there is a sequence of (at least) $n+2$ distinct points in $[a, b]$, called **critical points** and labeled

$$a \le x_0 < x_1 < \cdots < x_n < x_{n+1} \le b$$

such that

$$|f(x_i) - q(x_i)| = \|f - q\|_{L^\infty(a,b)} = R(f, q), \quad i = 0, \ldots, n+1, \tag{10.4}$$

and

$$f(x_i) - q(x_i) = \pm(-1)^i R(f, q), \quad i = 0, \ldots, n+1. \tag{10.5}$$

Let $k \in \mathbb{N}_0$ with $k \le n$. We say that $q \in \mathbb{P}_n$ has a **defective equi-oscillation property of degree** k if and only if there are at most k critical points

$$a \le x_0 < x_1 < \cdots < x_{k-1} < x_k \le b$$

[2] Named in honor of the Belgian mathematician Charles Jean de la Vallée Poussin (1866–1962).

such that

$$|f(x_i) - q(x_i)| = R(f, q), \quad f(x_i) - q(x_i) = \pm(-1)^i R(f, q), \quad i = 0, \dots, k.$$

Remark 10.8 (defective equi-oscillation). The defective equi-oscillation property is essentially the negation of the equi-oscillation property. Generically, since $f - p$ is continuous, there will always be at least one point where $\pm R(f, q)$ is attained. In other words, there will always be one critical point. The polynomial is defective if and only if there are not enough critical points where the polynomial equi-oscillates.

The following is the Chevyshev Oscillation Theorem.[3]

Theorem 10.9 (Chebyshev Oscillation Theorem). *Suppose that $f \in C([a, b])$ and $n \in \mathbb{N}_0$. $p \in \mathbb{P}_n$ is a minimax polynomial approximation of f if and only if p has the equi-oscillation property with respect to f.*

Proof. To avoid a trivial case, we assume that $f \notin \mathbb{P}_n$.

$(\Longleftarrow)$ Suppose that $p \in \mathbb{P}_n$ has the equi-oscillation property, i.e., that there is a sequence of $n + 2$ distinct points, denoted

$$a \leq x_0 < x_1 < \cdots < x_n < x_{n+1} \leq b,$$

such that (10.4) and (10.5) hold. Define

$$E(f) = \inf_{q \in \mathbb{P}_n} \|f - q\|_{L^\infty(a,b)} = \min_{q \in \mathbb{P}_n} \|f - q\|_{L^\infty(a,b)} . \tag{10.6}$$

This infimum is achieved as we have seen in Theorem 10.2, which justifies the use of the minimum. By Theorem 10.6, it must be that $E(f) \geq R(f, p)$ with $R(f, p)$ as defined in (10.4). On the other hand, since $E(f)$ is defined as a minimum, $E(f) \leq R(f, p)$. Hence, $E(f) = R(f, p)$ and p is a minimax polynomial for f.

$(\Longrightarrow)$ Suppose that $p \in \mathbb{P}_n$ is a minimax polynomial with respect to f. We want to prove that the equi-oscillation property holds.

$(n = 0)$ First, let us take care of the $n = 0$ case, for which we need two critical points. For this case, we can directly construct these points. Note that $|f(x) - p(x)|$ is continuous on $[a, b]$ and so it attains its maximum value, which is, of course, $R(f, p)$. This shows then that

$$S_0 = \{x \in [a, b] \mid |f(x) - p(x)| = R(f, p)\} \neq \emptyset.$$

Therefore, for the first critical point, we choose

$$x_0 = \inf S_0.$$

Observe now that, since $S_0 \subseteq [a, b]$ and $[a, b]$ is closed, we have that $x_0 \in [a, b]$. However, $x_0 \neq b$, as the contrary will yield a contradiction. To see this, let us suppose that $x_0 = b$. Observe that this implies that there can be no point $x \in [a, b)$ such that $|f(x) - p(x)| = R(f, p) > 0$. There are then two sub-cases.

[3] Named in honor of the Russian mathematician Pafnuty Lvovich Chebyshev (1821–1894).

1. If $f(x_0) - p(x_0) = R(f, p) > 0$, then

$$-R(f, p) < f(x) - p(x) \leq R(f, p), \quad \forall x \in [a, b]. \tag{10.7}$$

2. If $f(x_0) - p(x_0) = -R(f, p) < 0$, then

$$-R(f, p) \leq f(x) - p(x) < R(f, p), \quad \forall x \in [a, b].$$

Let us work with the first sub-case, as the second one can be treated in a similar fashion. If (10.7) holds, by continuity, there is $\delta \in (0, R(f, p))$ such that

$$-R(f, p) + \delta \leq f(x) - p(x) \leq R(f, p), \quad \forall x \in [a, b].$$

Subtracting $\delta/2$,

$$-(R(f, p) - \delta/2) \leq f(x) - (p(x) + \delta/2) \leq R(f, p) - \delta/2, \quad \forall x \in [a, b].$$

But this proves that, for $p + \delta/2 \in \mathbb{P}_n$, we have $R(f, p + \delta/2) < R(f, p)$. In other words, $p \in \mathbb{P}_n$ is not the minimax polynomial. This is the sought-after contradiction and, consequently, it must be that $a \leq x_0 < b$.

Without loss of generality, let us assume that $f(x_0) - p(x_0) = R(f, p) > 0$. Next, we need to show that there is a point $x_1 \in (x_0, b]$ such that $f(x_1) - p(x_1) = -R(f, p)$. To get a contradiction, suppose that no such point exists. If there is no such point, then, since $f(x_0) - p(x_0) = R(f, p)$,

$$-R(f, p) < f(x) - p(x) \leq R(f, p), \quad \forall x \in [a, b].$$

Since $f - p$ is continuous, there is $\delta \in (0, R(f, p))$ such that

$$-R(f, p) + \delta \leq f(x) - p(x) \leq R(f, p), \quad \forall x \in [a, b].$$

Subtracting $\delta/2$,

$$-(R(f, p) - \delta/2) \leq f(x) - (p(x) + \delta/2) \leq R(f, p) - \delta/2, \quad \forall x \in [a, b].$$

This, once again, shows that, for $p+\delta/2 \in \mathbb{P}_n$, we have $R(f, p+\delta/2) < R(f, p)$. In other words, $p \in \mathbb{P}_n$ is not the minimax polynomial, a contradiction. So $x_1 \in (x_0, b]$ must exist. In fact, we can define

$$x_1 = \inf S_1, \quad S_1 = \{x \in (x_0, b] \mid f(x) - p(x) = -R(f, p)\},$$

so that $f(x_1) - p(x_1) = -R(f, p)$, $x_1 \in (x_0, b]$. The proof is finished for the case $n = 0$.

($n \geq 1$) We define the critical points recursively as follows. Assume that $x_0 \in [a, b)$ has been found, as above, such that $|f(x_0) - p(x_0)| = R(f, p)$. Without loss of generality, let us assume that $f(x_0) - p(x_0) = -R(f, p)$. Then define

$$x_i = \inf S_i, \quad S_i = \left\{x \in (x_{i-1}, b] \;\middle|\; f(x) - p(x) = -(-1)^i R(f, p)\right\}$$

for $i = 1, \ldots, k$, where k is maximal. If $k \geq n + 1$, the proof is complete. If $1 \leq k \leq n$, then p has a defective equi-oscillation property. To get a contradiction, let us assume that $1 \leq k \leq n$, i.e., p is defective.

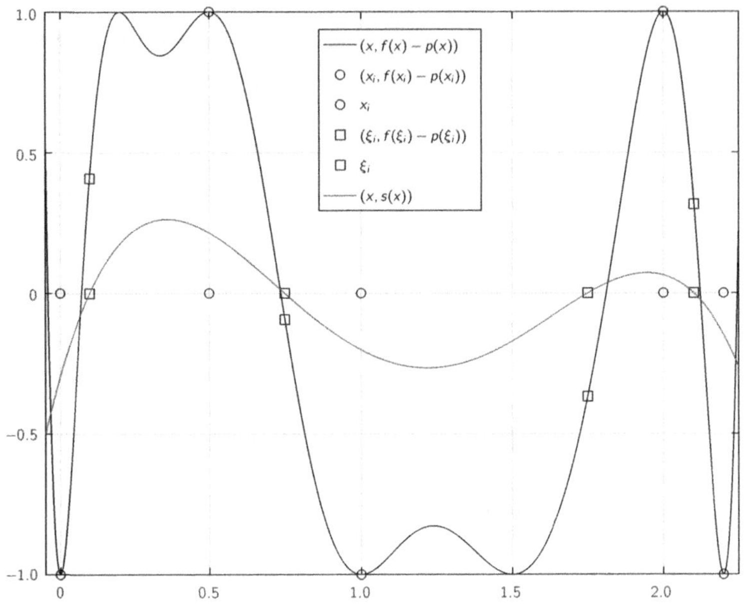

Figure 10.2 The equi–oscillation property. The notation used in the figure is defined in the proof of Theorem 10.9.

We have

$$f(x_i) - p(x_i) = -(-1)^i R(f, p), \quad i = 0, \dots, k \leq n;$$

see Figure 10.2. Let us define a sequence of points, $\{\xi_i\}_{i=0}^{k-1}$, as follows:

$$\xi_0 = \frac{1}{2}(\eta_0 + \zeta_0),$$

where

$$\eta_0 = \inf\{\eta \in [a, x_1] \mid f(\eta) - p(\eta) = R(f, p)\}$$

and

$$\zeta_0 = \sup\{\zeta \in [a, x_1] \mid f(\zeta) - p(\zeta) = -R(f, p)\}.$$

We claim that $x_0 \leq \zeta_0 < \eta_0 \leq x_1$. If this were not true, then $f - p$ would equi-oscillate at another pair of points previously unaccounted for. Next, define

$$\xi_1 = \frac{1}{2}(\eta_1 + \zeta_1),$$

where

$$\eta_1 = \inf\{\eta \in [\zeta_0, x_2] \mid f(\eta) - p(\eta) = -R(f, p)\}$$

and

$$\zeta_1 = \sup\{\zeta \in [\zeta_0, x_2] \mid f(\zeta) - p(\zeta) = R(f, p)\}.$$

We claim that $x_1 \le \zeta_1 < \eta_1 \le x_2$. The argument is the same as before. Likewise, define

$$\xi_2 = \frac{1}{2}(\eta_2 + \zeta_2),$$

where

$$\eta_2 = \inf\{\eta \in [\zeta_1, x_3] \mid f(\eta) - p(\eta) = R(f, p)\}$$

and

$$\zeta_2 = \sup\{\zeta \in [\zeta_1, x_3] \mid f(\zeta) - p(\zeta) = -R(f, p)\}.$$

We claim that $x_2 \le \zeta_2 < \eta_2 \le x_3$. The rest of the points can be defined similarly, so that the sequence $\{\xi_i\}_{i=0}^{k-1}$ *separates* $\{x_i\}_{i=0}^{k}$:

$$x_0 < \xi_0 < x_1 < \xi_1 < \cdots < x_{k-1} < \xi_{k-1} < x_k.$$

Again, see Figure 10.2.

Define the sets

$$S_- = \begin{cases} (a, \xi_0) \cup (\xi_1, \xi_2) \cup (\xi_3, \xi_4) \cup \cdots \cup (\xi_{k-1}, b), & k = 2m, \ m \in \mathbb{N}, \\ (a, \xi_0) \cup (\xi_1, \xi_2) \cup (\xi_3, \xi_4) \cup \cdots \cup (\xi_{k-2}, \xi_{k-1}), & k = 2m - 1, \ m \in \mathbb{N}, \end{cases}$$

and

$$S_+ = \begin{cases} (\xi_0, \xi_1) \cup (\xi_2, \xi_3) \cup (\xi_4, \xi_5) \cup \cdots \cup (\xi_{k-2}, \xi_{k-1}), & k = 2m, \ m \in \mathbb{N}, \\ (\xi_0, \xi_1) \cup (\xi_2, \xi_3) \cup (\xi_4, \xi_5) \cup \cdots \cup (\xi_{k-1}, b), & k = 2m - 1, \ m \in \mathbb{N}. \end{cases}$$

Because of the separation property, for each $0 \le i \le k - 1$,

$$|f(\xi_i) - p(\xi_i)| < R(f, p)$$

and there is a number $\varepsilon > 0$ such that

$$-R(f, p) \le f(x) - p(x) \le R(f, p) - \varepsilon, \quad x \in \overline{S_-} \tag{10.8}$$

and

$$-R(f, p) + \varepsilon \le f(x) - p(x) \le R(f, p), \quad x \in \overline{S_+}. \tag{10.9}$$

Thus, for each of the $k + 1$ intervals

$$[a, \xi_0], [\xi_0, \xi_1], \dots, [\xi_{k-2}, \xi_{k-1}], [\xi_{k-1}, b],$$

the error function $f(x) - p(x)$ takes on one and only one of the extreme values $+R(f, p)$ or $-R(f, p)$ and is bounded well away from the other extreme value. Again, take a careful look at Figure 10.2.

Now define

$$r(x) = (x - \xi_0)(x - \xi_1) \cdots (x - \xi_{k-1}) \in \mathbb{P}_k \subseteq \mathbb{P}_n.$$

This polynomial has one sign in each of the $k + 1$ separating intervals constructed above. In fact, the sign will clearly oscillate. Define

$$M = \max_{a \le x \le b} |r(x)|$$

and

$$q(x) = p(x) + \varepsilon s(x), \quad s(x) = (-1)^{k+1} \frac{r(x)}{2M}, \quad x \in \mathbb{R}$$

for $\varepsilon > 0$ sufficiently small. Clearly, $q \in \mathbb{P}_n$, as long as $1 \le k \le n$.

Notice that, by construction,

$$-\frac{1}{2} \le s(x) < 0, \quad \forall x \in S_-, \qquad 0 < s(x) \le \frac{1}{2}, \quad \forall x \in S_+. \tag{10.10}$$

Using inequalities (10.8) and (10.9) and the definition of q, we find

$$-R(f,p) - \varepsilon s(x) \le f(x) - q(x) \le R(f,p) - \varepsilon[1 + s(x)], \quad \forall x \in \overline{S_-}$$

and

$$-R(f,p) + \varepsilon[1 - s(x)] \le f(x) - q(x) \le R(f,p) - \varepsilon s(x), \quad \forall x \in \overline{S_+}.$$

Thus, applying (10.10) in the last two inequalities, we observe that

$$-R(f,p) < f(x) - q(x) < R(f,p), \quad \forall x \in [a,b].$$

It is straightforward now to see that $R(f,q) < R(f,p)$. Since $q \in \mathbb{P}_n$, this is a contradiction to the fact that p is a minimax polynomial. It must be that $k \ge n+1$ and, therefore, p has the equi-oscillation property. □

As a corollary of Chebyshev Oscillation Theorem, we can obtain uniqueness of a minimax polynomial.

Corollary 10.10 (uniqueness). *For every $n \in \mathbb{N}_0$ the minimax polynomial $p \in \mathbb{P}_n$ of $f \in C([a,b])$ is unique.*

Proof. Suppose that there are two minimax polynomials of degree n, labeled $p_1, p_2 \in \mathbb{P}_n$. Consistent with previous notation, define

$$E(f) = \inf_{q \in \mathbb{P}_n} \|f - q\|_{L^\infty(a,b)}, \quad \|f - p_i\|_{L^\infty(a,b)} = R(f, p_i), \quad i = 1, 2.$$

Then

$$E(f) = R(f, p_1) = R(f, p_2).$$

Set $q = \frac{1}{2}(p_1 + p_2)$. Then, for all $x \in [a,b]$,

$$|f(x) - q(x)| \le \frac{1}{2}|f(x) - p_1(x)| + \frac{1}{2}|f(x) - p_2(x)| \le E(f). \tag{10.11}$$

Therefore, q is yet another minimax polynomial of degree n. By the last theorem, q has the equi-oscillation property, i.e., there are $n+2$ distinct critical points

$$a \le x_0 < x_1 < \cdots < x_n < x_{n+1} \le b$$

such that (10.4) and (10.5) hold for q. From (10.11), it follows that, at the critical points of q,

$$\begin{aligned} E(f) &= |f(x_i) - q(x_i)| \\ &\leq \left| \frac{1}{2} f(x_i) + \frac{1}{2} p_1(x_i) + \frac{1}{2} f(x_i) + \frac{1}{2} p_2(x_i) \right| \\ &\leq \frac{1}{2} |f(x_i) - p_1(x_i)| + \frac{1}{2} |f(x_i) - p_2(x_i)| \\ &\leq E(f). \end{aligned}$$

The only possibility is that

$$|f(x_i) - p_1(x_i)| = E(f) = |f(x_i) - p_2(x_i)|, \quad i = 0, \ldots, n+1,$$

or, equivalently,

$$f(x_i) - p_1(x_i) = \pm E(f) = f(x_i) - p_2(x_i), \quad i = 0, \ldots, n+1,$$

such that the alternation pattern is in sync. Therefore,

$$p_1(x_i) - p_2(x_i) = f(x_i) - p_2(x_i) - [f(x_i) - p_1(x_i)] = 0, \quad i = 0, \ldots, n+1.$$

It follows that $p_1 \equiv p_2$ in $\mathbb{P}_n$. □

With the help of the equi-oscillation property, at least for smooth functions, we can relate the minimax and interpolating polynomials.

Corollary 10.11 (error). *Suppose that $n \in \mathbb{N}_0$ and $f \in C^{n+1}([a, b])$. Let $p \in \mathbb{P}_n$ be the unique minimax polynomial of f. There exist $n+1$ distinct points $x_0, \ldots, x_n$ in the interval $[a, b]$ such that $f(x_i) = p(x_i)$, $i = 0, \ldots, n$. Furthermore, for every $x \in [a, b] \backslash \{x_0, \ldots, x_n\}$, there is a point $\xi = \xi(x) \in (a, b)$ such that*

$$f(x) - p(x) = \frac{(x - x_0) \cdots (x - x_n)}{(n+1)!} f^{(n+1)}(\xi).$$

Proof. Because of equi-oscillation, there are $(n+1)$ distinct points $X = \{x_i\}_{i=0}^{n} \subset [a, b]$, such that $f(x_i) = p(x_i)$. (Note that these points are not the same as the critical points of equi-oscillation.) Thus, p is the unique interpolating polynomial for f that agrees with f at these $n+1$ distinct points. Now, we apply Theorem 9.16. □

10.2 Interpolation Error and the Lebesgue Constant

In this section, we state and prove a fundamental error estimate for interpolation that involves the Lebesgue constant and the minimax polynomial approximation.

Theorem 10.12 (interpolation versus minimax). *Suppose that $n \in \mathbb{N}_0$, $X = \{x_i\}_{i=0}^{n}$ is a nodal set in the compact interval $[a, b] \subset \mathbb{R}$, $f \in C([a, b])$, $p \in \mathbb{P}_n$ is the Lagrange interpolating polynomial for f subordinate to X, and $p^\star \in \mathbb{P}_n$ is the minimax polynomial approximation of f. Then*

$$\|f - p\|_{L^\infty(a,b)} \leq (1 + \Lambda(X)) \|f - p^\star\|_{L^\infty(a,b)},$$

where $\Lambda(X)$ is the Lebesgue constant subordinate to X, defined in (9.2).

Proof. We begin by writing $f - p = (f - p^\star) - (p - p^\star)$. Observe that $p - p^\star$ is the Lagrange interpolating polynomial for $f - p^\star$ subordinate to X. Thus,

$$\mathcal{I}_X[f - p^\star] = p - p^\star.$$

Thus,

$$\begin{aligned}
\|f - p\|_{L^\infty(a,b)} &= \|(f - p^\star) - (p - p^\star)\|_{L^\infty(a,b)} \\
&= \|(f - p^\star) - \mathcal{I}_X[f - p^\star]\|_{L^\infty(a,b)} \\
&\le \|(f - p^\star)\|_{L^\infty(a,b)} + \|\mathcal{I}_X[f - p^\star]\|_{L^\infty(a,b)} \\
&\le \|(f - p^\star)\|_{L^\infty(a,b)} + \|\mathcal{I}_X\|_\infty \|f - p^\star\|_{L^\infty(a,b)} \\
&= (1 + \Lambda(X)) \|f - p^\star\|_{L^\infty(a,b)}.
\end{aligned}$$

□

This theorem guarantees that if the Lebesgue constant is not large, then the interpolant subordinate to X is almost as good as the best approximation. Interestingly, for uniformly spaced nodes, the Lebesgue constant can be quite large, for large n. We will say more about this after we have described the so-called Chebyshev–Lagrange interpolation.

10.3 Chebyshev Polynomials

In light of the uniqueness result of Corollary 10.10, we can properly talk about *the* best polynomial approximation to a (continuous) function in the ∞-norm. For a few very particular cases, we can construct the minimax by using the equi-oscillation property guaranteed by Theorem 10.9.

Example 10.3 Let $n \in \mathbb{N}_0$, $f(x) = x^{n+1}$, and $p \in \mathbb{P}_n$ be its minimax approximation on $[-1, 1]$. Then there must be $X = \{x_j\}_{j=0}^{n+1} \subset [-1, 1]$ distinct critical points where

$$|f(x_j) - p(x_j)| = R(f, p) = E(f) > 0, \quad j = 0, \dots, n+1,$$

where $R(f, p)$ is defined in (10.4) and $E(f)$ is defined in (10.6). As usual, we assume that the critical points are ordered in an increasing fashion. Notice now that $e = f - p \in \mathbb{P}_{n+1}$ and, setting $E = E(f)$,

$$E^2 - e^2 \in \mathbb{P}_{2n+2}, \quad E^2 - e^2(x_j) = 0, \quad j = 0, \dots, n+1.$$

Of course, $e^2(x) \le E^2$ for all $x \in [-1, 1]$. This shows that the polynomial e^2 has a relative maximum at all the critical points; consequently, if $x_j \in (-1, 1)$,

$$\frac{de^2}{dx}(x_j) = 0,$$

which implies that

$$\frac{d(E^2 - e^2)}{dx}(x_j) = 0, \quad \forall x_j \in (-1, 1).$$

Observe, in addition, that each $x_j \in (-1, 1)$ is a zero of multiplicity (at least) two of the polynomial $E^2 - e^2 \in \mathbb{P}_{2n+2}$. By simple counting, there can only be n such interior double roots. But this implies that $x_0 = -1$ and $x_{n+1} = 1$ are the remaining critical points, and these are simple zeros. (You may have to think about that for a moment.)

On the other hand, the polynomial

$$(1 - x^2)[e'(x)]^2 \in \mathbb{P}_{2n+2}$$

has simple zeros at $x_0 = -1$ and $x_{n+1} = 1$ and zeros of multiplicity two for all x_j with $j = 1, \ldots, n$ (interior critical points). Thus, necessarily, we must have that

$$(1 - x^2)[e'(x)]^2 = c(E^2 - e^2)$$

for some constant $c \in \mathbb{R}$. The value of the constant can be easily identified by looking at the leading-order term. Thus, we have

$$(1 - x^2)[e'(x)]^2 = (n + 1)^2(E^2 - e(x)^2).$$

The previous differential equation can be solved on each interval for which e' does not change sign. If we assume that $e'(x) \geq 0$ for $x \in [-1, x_1]$, then

$$\frac{e'(x)}{\sqrt{E^2 - e^2(x)}} = \frac{n+1}{\sqrt{1 - x^2}}.$$

This is a separable equation with solution

$$\arccos\left(\frac{e}{E}\right) = (n + 1)\arccos(x) + c,$$

which implies that

$$e(x) = E\cos[(n + 1)\theta + c],$$

where $x = \cos\theta$ and c is an arbitrary constant. Now, since we assumed that $e'(-1) \geq 0$, we must have $e(-1) = -E$ and, consequently, $c = m\pi$ for $m \in \mathbb{Z}$. Therefore,

$$e(x) = \pm E\cos[(n + 1)\theta].$$

Example 10.3 motivates the following definition.

Definition 10.13 (Chebyshev polynomial[4]). Let $x \in [-1, 1]$ and $x = \cos\theta$. The polynomial

$$T_k(x) = \cos(k\theta), \quad k \in \mathbb{N}_0 \tag{10.12}$$

is called the **Chebyshev polynomial** of degree k.

From Definition 10.13, it is not immediately clear that the functions T_k are indeed polynomials. This can be proved by showing that the Chebyshev polynomials satisfy the following three-term recurrence relation.

[4] Named in honor of the Russian mathematician Pafnuty Lvovich Chebyshev (1821–1894).

n	$T_n(x)$
0	1
1	x
2	$2x^2 - 1$
3	$4x^3 - 3x$
4	$8x^4 - 8x^2 + 1$
5	$16x^5 - 20x^3 + 5x$
6	$32x^6 - 48x^4 + 18x^2 - 1$
7	$64x^7 - 112x^5 + 56x^3 - 7x$
8	$128x^8 - 256x^6 + 160x^4 - 32x^2 + 1$
9	$256x^9 - 576x^7 + 432x^5 - 120x^3 + 9x$
10	$512x^{10} - 1\,280x^8 + 1\,120x^6 - 400x^4 + 50x^2 - 1$

Table 10.1 The first 11 Chebyshev polynomials.

Proposition 10.14 (recurrence). *The functions T_k, $k \in \mathbb{N}_0$, defined as in* (10.12), *satisfy the three-term recurrence relation*

$$T_{n+1}(x) = 2xT_n(x) - T_{n-1}(x), \quad n = 1, 2, \ldots,$$

with $T_0(x) = 1$ and $T_1(x) = x$ for all $x \in [-1, 1]$. Thus, $T_k \in \mathbb{P}_k$, $k \in \mathbb{N}_0$. Furthermore, $T_k(1) = 1$ and $T_k(-1) = (-1)^k$.

Proof. Consider the trigonometric formula

$$\cos\left((n \pm 1)\theta\right) = \cos(n\theta)\cos(\theta) \mp \sin(n\theta)\sin(\theta).$$

Then

$$\begin{aligned} T_{n+1}(x) + T_{n-1}(x) &= \cos((n+1)\theta) + \cos((n-1)\theta) \\ &= 2\cos(n\theta)\cos(\theta) \\ &= 2T_n(x)x, \end{aligned}$$

which proves the first result. The rest of the details are left to the reader as an exercise; see Problem 10.4. □

The first 11 Chebyshev polynomials are shown in Table 10.1. Let us now explore some further properties of these polynomials.

Proposition 10.15 (orthogonality). *Let $w(x) = \frac{1}{\sqrt{1-x^2}}$ for $x \in (-1, 1)$. The Chebyshev polynomials $T_k \in \mathbb{P}_k$, $k \in \mathbb{N}_0$, defined in* (10.12) *are L^2_w-orthogonal, i.e.,*

$$(T_k, T_m)_{L^2_w(-1,1)} = \int_{-1}^{1} T_k(x)T_m(x)\frac{\mathrm{d}x}{\sqrt{1-x^2}} = \begin{cases} 0, & k \neq m, \\ \pi, & k = m = 0, \\ \dfrac{\pi}{2}, & k = m > 0. \end{cases}$$

Proof. Let us make the change of variables $\theta = \cos(x)$. Then

$$\int_{-1}^{1} T_k(x)T_m(x)\frac{\mathrm{d}x}{\sqrt{1-x^2}} = \int_0^{\pi} \cos(k\theta)\cos(m\theta)\mathrm{d}\theta.$$

The rest of the details are left to the reader as an exercise; see Problem 10.5. □

We will discuss orthogonal polynomials more in the next chapter. In the meantime, we can also prove the following.

Proposition 10.16 (differential equation). *The Chebyshev polynomials $T_k \in \mathbb{P}_k$, $k \in \mathbb{N}_0$, defined in (10.12), are solutions of the differential equation*

$$(1-x^2)y'' - xy' + k^2 y = 0.$$

Proof. See Problem 10.6. □

Theorem 10.17 (further properties). *The Chebyshev polynomials have the following further properties.*

1. *For $n \in \mathbb{N}$, $\deg(T_n) = n$ with leading coefficient 2^{n-1}.*
2. *T_n is even when $n \in \mathbb{N}_0$ is even, and odd otherwise.*
3. *For $n \in \mathbb{N}$, the zeros of the polynomial T_n are simple and given by the formula*

$$x_j = \cos\left(\frac{(2j-1)\pi}{2n}\right), \quad j = 1, \dots, n.$$

   In particular, the zeros are real, distinct, and lie in the interval $(-1, 1)$.
4. *$|T_n(x)| \le 1$, for all $x \in [-1, 1]$, and $n \in \mathbb{N}_0$.*
5. *For $n \in \mathbb{N}$, the extreme values of T_n satisfy*

$$T_n(x_k) = (-1)^{n+k}, \quad x_k = \cos\left(\frac{k\pi}{n}\right), \quad k = 0, \dots, n.$$

Proof. See Problem 10.7. □

Let us now give a full proof of the fact that our motivating calculations of Example 10.3 were pointing us toward.

Theorem 10.18 (minimax). *Suppose that $n \in \mathbb{N}_0$ and $f(x) = x^{n+1}$ on the interval $[-1, 1]$. Then the polynomial*

$$p(x) = x^{n+1} - 2^{-n}T_{n+1}(x) \in \mathbb{P}_n$$

is the minimax polynomial for f on $[-1, 1]$.

Proof. By Theorem 10.17, item 1, we are justified in our claim that $p \in \mathbb{P}_n$. Since

$$f(x) - p(x) = 2^{-n}T_{n+1}(x),$$

by Theorem 10.17, items 4 and 5, $f - p$ achieves its maximum value, 2^{-n}, in $[-1, 1]$, with alternating signs at the $n+2$ critical points $x_k = \cos(\frac{k\pi}{n+1})$, $k = 0, 1, \dots, n+1$. By the Chebyshev Oscillation Theorem 10.9, $p \in \mathbb{P}_n$ is the unique minimax polynomial for $f(x) = x^{n+1}$ in the interval $[-1, 1]$. □

Theorem 10.19 (least deviation). *Suppose that $n \in \mathbb{N}_0$. Among all polynomials in $\mathbb{P}_{n+1}$ whose leading coefficient is equal to one, the polynomial $p(x) = 2^{-n}T_{n+1}$ is the one with the smallest $L^\infty(-1,1)$ norm, i.e., it is the one that has the least deviation from zero.*

Proof. We can write any polynomial $e \in \mathbb{P}_{n+1}$, whose leading coefficient is equal to one, as the difference between $f(x) = x^{n+1}$ and a polynomial $q \in \mathbb{P}_n$. By Theorem 10.18,

$$\min_{q \in \mathbb{P}_n} \|f - q\|_{L^\infty(-1,1)} = \left\|2^{-n}T_{n+1}\right\|_{L^\infty(-1,1)}.$$

Therefore, among all polynomials in $\mathbb{P}_{n+1}$ whose leading coefficient equals one, the polynomial $p(x) = 2^{-n}T_{n+1}$ has the smallest deviation from zero, and the deviation is precisely 2^{-n}. □

10.4 Interpolation at Chebyshev Nodes

As motivation for the construction of this section, let us recall the result of Theorem 9.16. If $n \in \mathbb{N}_0$, $X = \{x_i\}_{i=0}^n$ is a nodal set in the compact interval $[a,b] \subset \mathbb{R}$, $f \in C^{n+1}([a,b])$, and $p \in \mathbb{P}_n$ is the Lagrange interpolating polynomial of f subordinate to X. then, for every $x \in [a,b] \backslash X$, there is a point $\xi \in (a,b)$ such that

$$f(x) - p(x) = \frac{f^{(n+1)}(\xi)}{(n+1)!}\omega_{n+1}(x),$$

where

$$\omega_{n+1}(x) = \prod_{i=0}^{n}(x - x_i).$$

Clearly, $\omega_{n+1} \in \mathbb{P}_{n+1}$ and has leading coefficient equal to one. We saw in Example 9.4 that a major problem with uniform interpolation is that $\|\omega_{n+1}\|_{L^\infty(a,b)}$ may get too large. One way to control this is to adjust the values of the nodes using facts we have just learned about Chebyshev polynomials.

Definition 10.20 (Chebyshev nodes). Let $n \in \mathbb{N}_0$ be given. Suppose that $[a,b] \subset \mathbb{R}$ is a compact interval. We call the nodal set $X = \{\xi_j\}_{j=0}^n$ the **Chebyshev nodal points** in $[a,b]$ of order n if and only if

$$\xi_j = \frac{1}{2}(b-a)\cos\left(\frac{(j+\frac{1}{2})\pi}{n+1}\right) + \frac{1}{2}(b+a), \quad j = 0, 1, 2, \ldots, n. \tag{10.13}$$

When $a = -1$, $b = 1$, these are just the $n+1$ distinct zeros of the Chebyshev polynomial T_{n+1}. The interpolating polynomial that results from using the Chebyshev nodal points is called the **Chebyshev–Lagrange interpolant** or just the **Chebyshev interpolant**.

This next result should be compared with Proposition 9.18, where uniformly spaced interpolation nodes are utilized. Note that a much better convergence rate is achieved.

Theorem 10.21 (interpolation error). *Let $n \in \mathbb{N}_0$ and $f \in C^{n+1}([a,b])$. Suppose that $p \in \mathbb{P}_n$ is the unique Chebyshev–Lagrange interpolating polynomial of f. In other words, p is the Lagrange interpolant subject to the Chebyshev nodal points $X = \{\xi_j\}_{j=0}^n$, which were defined in (10.13). Then*

$$\|f - p\|_{L^\infty(a,b)} \le \frac{(b-a)^{n+1}}{2^{2n+1}(n+1)!} \left\| f^{(n+1)} \right\|_{L^\infty(a,b)}.$$

Proof. By

$$\zeta_j = \cos\left(\frac{(j+\frac{1}{2})\pi}{n+1}\right), \quad j = 0, \dots, n,$$

we denote the $n+1$ distinct roots of T_{n+1} in the interval $(-1,1)$. Thus,

$$\prod_{j=0}^{n} (t - \zeta_j) = 2^{-n} T_{n+1}(t).$$

Define the affine function

$$x(t) = \frac{1}{2}(b-a)t + \frac{1}{2}(b+a).$$

Then $x([-1,1]) = [a,b]$ with $x(-1) = a$ and $x(0) = \frac{1}{2}(b+a)$, and $x(1) = b$. Furthermore,

$$x(\zeta_j) = \xi_j, \quad j = 0, \dots, n.$$

As with any affine map, x is invertible, and the inverse is precisely

$$t(x) = \frac{2x - a - b}{b - a}.$$

Therefore,

$$\prod_{j=0}^{n} (x - \xi_j) = \left(\frac{b-a}{2}\right)^{n+1} 2^{-n} T_{n+1}(t(x)).$$

Since $|T_{n+1}(t(x))| \le 1$ for all $x \in [a,b]$, the proof now follows from Theorem 9.16 and the fact that

$$|\omega_{n+1}(x)| = \left| \prod_{j=0}^{n} (x - \xi_j) \right| \le \frac{(b-a)^{n+1}}{2^{2n+1}}. \qquad \square$$

Example 10.4 (Runge phenomenon re-revisited) Let us now revisit (again) the Runge's function from Examples 9.4 and 9.5. See Figure 10.3, where, instead of uniformly spaced nodes, we use Chebyshev interpolation points. Note that the

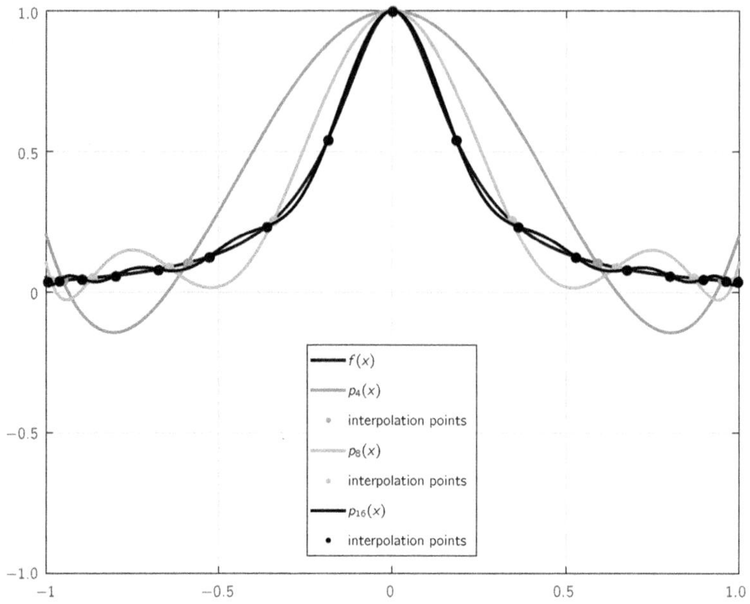

Figure 10.3 The Runge Phenomenon revisited. Using Chebyshev interpolation nodes, interpolation polynomials can behave better, i.e., oscillate less, than when uniformly spaced nodes are used. See Figure 9.3 for comparison.

interpolation polynomials oscillate less wildly at the tails, as n gets larger. The reason for this is explained in the result of Theorem 10.21 and in Figure 10.4.

Example 10.5 In Figure 10.5, we plot the Lagrange nodal bases for both uniformly spaced (a) and Chebyshev (b) nodes, for degrees $n = 4, 8, 16$. Observe that, for the uniformly spaced nodes, the absolute maxima are exploding in value, as n gets large, especially at the tails. By contrast, the basis functions for the Chebyshev nodes remain relatively well behaved, even as n becomes large. The Lebesgue functions, $\sum_{i=0}^{n} |L_i(x)|$, for uniformly spaced and Chebyshev nodes, for $n = 4, 8, 16$, are plotted in Figure 10.6. Not surprisingly, for the uniformly spaced nodes, these functions are becoming very large at the boundaries, as n increases.

Example 10.6 Recall that the Lebesgue constant is computed via

$$\Lambda(X) = \max_{a \le x \le b} \sum_{i=0}^{n} |L_i(x)| .$$

One can prove that the Lebesgue constant for the Chebyshev nodes has the asymptotic behavior

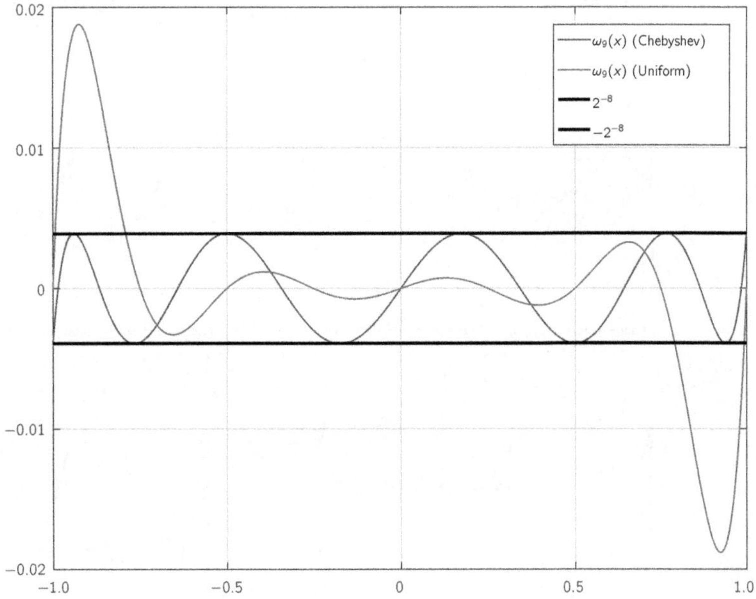

Figure 10.4 Plots of ω_{n+1}, for $n = 8$, using Chebyshev and uniformly spaced interpolation nodes. Observe that, as expected, the deviation from zero in $\omega_9(x)$ over $[-1, 1]$, using Chebyshev points, remains bounded in absolute value by 2^{-8}. This deviation grows for ω_9 when uniformly spaced nodes are used.

$$\Lambda(X_{n,\text{Cheb}}) \sim \frac{2}{\pi} \ln n, \qquad n \to \infty.$$

The asymptotic behavior for uniformly spaced nodes is

$$\Lambda(X_{n,\text{Unif}}) \sim \frac{2^{n+1}}{\exp(1) n \ln n}, \qquad n \to \infty.$$

These rates seem to be consistent with what we have observed in Figure 10.6.

While it is beyond the scope of our text to prove these relations [83, 95], it is clear that, as $n \to \infty$, the slow logarithmic growth of the Lebesgue constant for Chebyshev nodes, $X_{n,\text{Cheb}}$, is dominated by the exponential growth of the constant for the uniform nodal set, $X_{n,\text{Unif}}$. This gives another explanation for why interpolation with Chebyshev nodes is generally much better than that using uniformly spaced nodes.

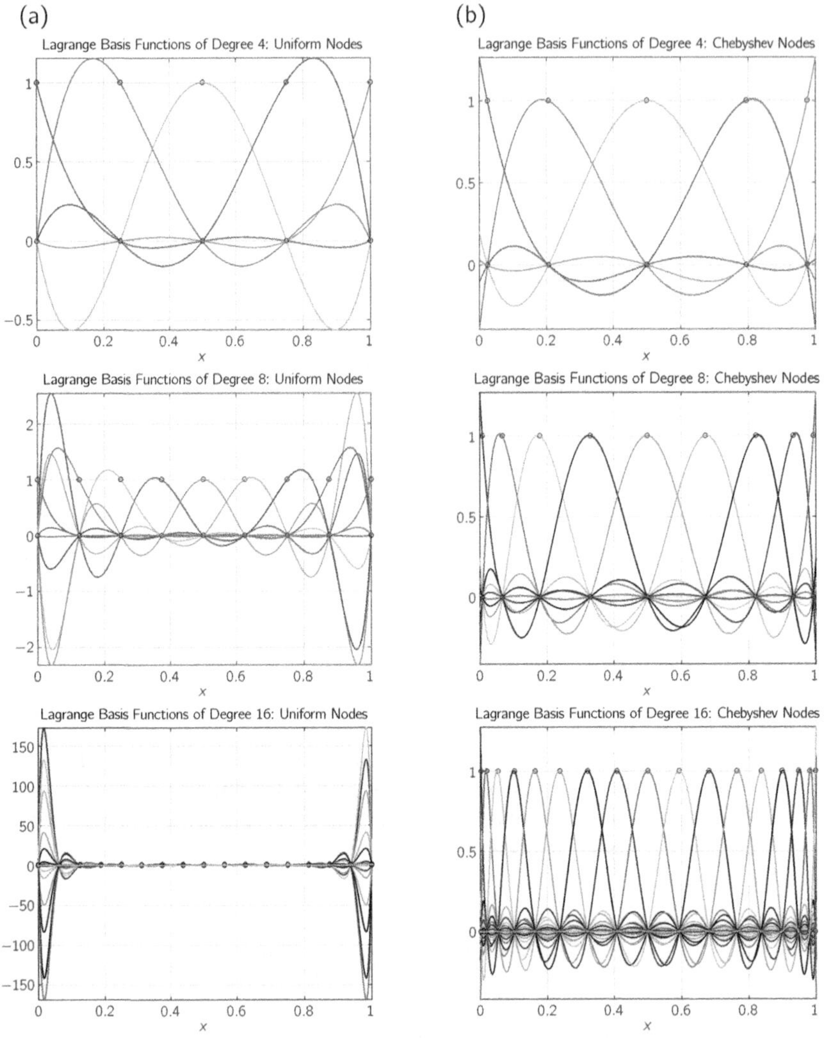

Figure 10.5 Lagrange nodal basis functions for uniformly spaced (a) and Chebyshev (b) nodes in $[0, 1]$. Observe that, for the uniformly spaced nodes, the absolute maxima are exploding in value, as n gets large, especially at the tails. By contrast, the basis functions for the Chebyshev nodes remain relatively well behaved, even as n becomes large.

10.5 Bernstein Polynomials and the Weierstrass Approximation Theorem

In this section, we give a constructive proof of the Weierstrass Approximation Theorem using so-called Bernstein polynomials. In the section after this one, we will use the Weierstrass Approximation Theorem to give a proof of convergence of the least squares approximation to a function in the weighted quadratic mean.

Before we introduce Bernstein polynomials, we need some definitions.

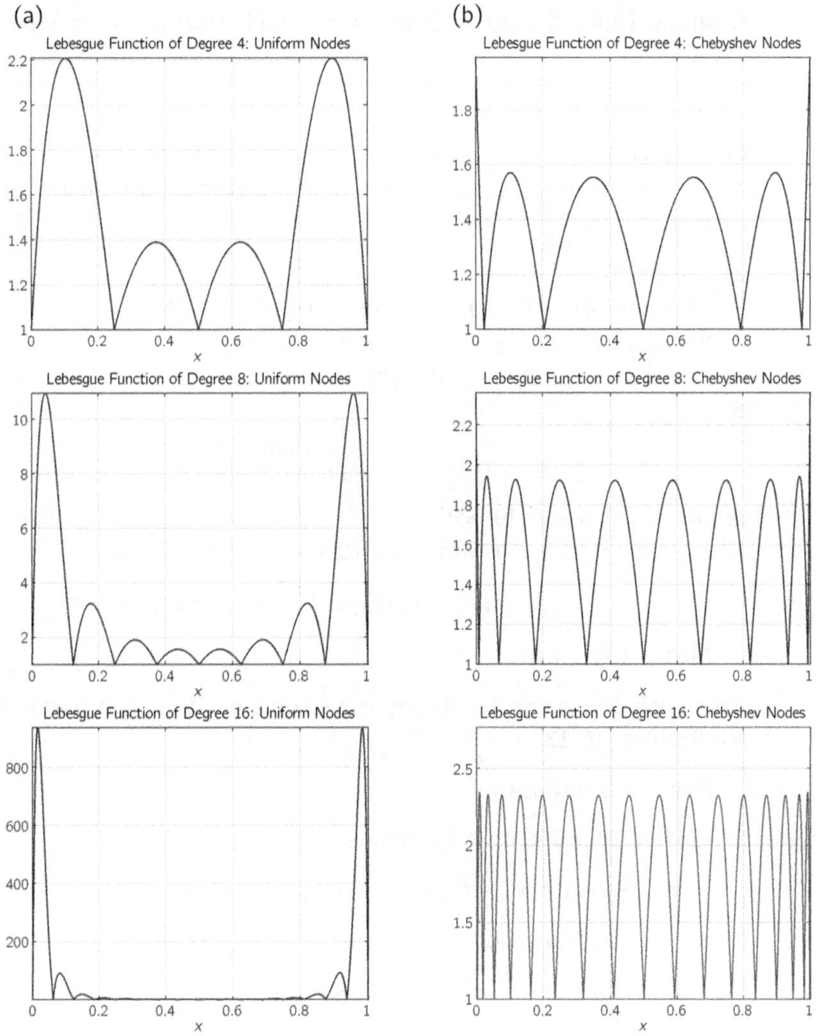

Figure 10.6 The Lebesgue function, $\lambda_X(x)$, for uniformly spaced (a) and Chebyshev (b) nodes in $[0, 1]$.

Definition 10.22 (modulus of continuity). Suppose that $I \subseteq \mathbb{R}$ is an interval and $f \in C(I; \mathbb{C})$. The (uniform) **modulus of continuity of** f is defined, for each $h \geq 0$, as

$$\omega_f(h) = \sup_{\substack{x,y \in I \\ |x-y| \leq h}} |f(x) - f(y)|.$$

Example 10.7 Suppose that $f(x) = x^2$ on the interval $I = (0, 1)$. Then $\omega_f(h) = 2h - h^2$. See Problem 10.14.

Example 10.8 If $f(x) = \frac{1}{x}$ and $I = (0,1)$, then $\omega_f(h) = +\infty$. See Problem 10.15.

Lemma 10.23 (properties of ω_f). *Suppose that $[a,b] \subset \mathbb{R}$ is a compact interval and $f \in C([a,b];\mathbb{C})$ is continuous. The modulus of continuity has the following properties.*

1. $\omega_f(0) = 0$.
2. *Monotonicity: If* $0 < h_1 < h_2$, *then* $\omega_f(h_1) \le \omega_f(h_2)$.
3. *Subadditivity:* $\omega_f(h_1 + h_2) \le \omega_f(h_1) + \omega_f(h_2)$.
4. *For all* $n \in \mathbb{N}$, $\omega_f(nh) \le n\omega_f(h)$.
5. $$\lim_{h\downarrow 0} \omega_f(h) = 0.$$
6. $\omega_f \in C([0, b-a]; [0,\infty))$.
7. *If* $f \in C^{0,1}([a,b];\mathbb{C})$ *with Lipschitz constant* $L > 0$, *i.e.,*
$$|f(x) - f(y)| \le L|x-y|, \quad \forall x, y \in [a,b],$$
then $\omega_f(h) \le Lh$.

Proof. We give some details and leave some points to the reader as an exercise; see Problem 10.13.

1: This one is obvious.

2: Suppose that $0 < h_1 < h_2$. Define

$$S_i = \{|f(x) - f(y)| \mid x, y \in [a,b],\ |x-y| \le h_i\}, \quad i = 1, 2.$$

If $|x-y| \le h_1$, then also $|x-y| \le h_2$. This implies that

$$S_1 \subseteq S_2,$$

which implies that

$$\omega_f(h_1) = \sup S_1 \le \sup S_2 = \omega_f(h_2).$$

3: Let $h_1 \ge 0$ and $h_2 \ge 0$ be given. Suppose that $0 \le y - x \le h_1$. Then,

$$|f(x) - f(y)| \le \omega_f(h_1) \le \omega_f(h_1) + \omega_f(h_2). \tag{10.14}$$

Now, suppose that $h_1 < y - x \le h_1 + h_2$. Then

$$x + h_1 < y, \qquad 0 < y - (x + h_1) < h_2.$$

Consequently,

$$\begin{aligned} |f(x) - f(y)| &\le |f(x) - f(x+h_1)| + |f(x+h_1) - f(y)| \\ &\le \omega_f(h_1) + \omega_f(y - (x+h_1)) \qquad (10.15) \\ &\le \omega_f(h_1) + \omega_f(h_2), \end{aligned}$$

where we have used the result of property 2 in the last estimate. If we swap the roles of x and y, then inequalities (10.14) and (10.15) still hold. Thus, if $|x-y| \le h_1+h_2$, we have

$$|f(x) - f(y)| \le \omega_f(h_1) + \omega_f(h_2).$$

Taking the supremum over all $x, y \in I$ such that $|x - y| \le h_1 + h_2$, we get

$$\omega_f(h_1 + h_2) = \sup_{\substack{x,y\in I \\ |x-y|\le h_1+h_2}} |f(x) - f(y)| \le \omega_f(h_1) + \omega_f(h_2).$$

4: From property 3,

$$\omega_f(2h) = \omega_f(h + h) \le \omega_f(h) + \omega_f(h) = 2\omega_f(h).$$

Property 4 then follows by a simple induction argument.

5: Since f is continuous on the compact interval $[a, b]$, f is uniformly continuous on $[a, b]$; see]Theorem B.20. Thus, given any $\varepsilon > 0$, there is a $\delta = \delta(\varepsilon) > 0$ such that if x and y are any points in $[a, b]$ satisfying $|x - y| < \delta$, then it follows that

$$|f(x) - f(y)| < \varepsilon.$$

Hence,

$$\omega_f(\delta) \le \varepsilon. \tag{10.16}$$

We know from 2 that $\omega_f(h)$ is an increasing function of h. Suppose that $h_n \downarrow 0$, as $n \to \infty$. Then

$$0 \le \omega_f(h_{n+1}) \le \omega_f(h_n) \le \omega_f(h_{n-1}) \le \cdots \le \omega_f(h_2) \le \omega_f(h_1).$$

By the Monotone Convergence Theorem B.7, the sequence $\omega_f(h_n)$ has a limit, say, α and

$$\omega_f(h_n) \downarrow \alpha \ge 0.$$

Suppose that $\alpha > 0$. Then by the argument above, namely (10.16), there is a $\delta > 0$ such that

$$\omega_f(\delta) \le \frac{\alpha}{2}.$$

Since $h_n \downarrow 0$, there is some $N \in \mathbb{N}$ such that $h_N < \delta$. Hence,

$$\omega_f(h_N) \le \omega(\delta) \le \frac{\alpha}{2} < \alpha.$$

This is a contradiction.

6: From property 2, for any $h \ge 0$ and $h_n > 0$,

$$\omega_f(h) \le \omega_f(h + h_n).$$

Thus,

$$0 \le \omega_f(h + h_n) - \omega_f(h) \le \omega_f(h) + \omega_f(h_n) - \omega_f(h) = \omega_f(h_n),$$

using property 3 in the last estimate. Now suppose that $h_n \downarrow 0$. Then $\omega_f(h_n) \downarrow 0$, from property 5. Hence, by the last estimate,

$$\omega_f(h + h_n) \downarrow \omega_f(h), \quad \text{as } n \to \infty.$$

This holds for any $h \geq 0$.

Next, suppose that $h > 0$. Assume that $h \geq h_n \downarrow 0$, as $n \to \infty$. By similar arguments,

$$\omega_f(h - h_f) \uparrow \omega_f(h),$$

as $n \to \infty$.

7: This is left to the reader as an exercise. □

The definition of the Bernstein polynomials is simplified if we restrict our focus to the interval $[0, 1]$. The general definition can be obtained by a translation and dilation of the interval.

Definition 10.24 (Bernstein polynomial[5]). Suppose that $n \in \mathbb{N}_0$. The **Bernstein basis polynomials**, with respect to the interval $[0, 1]$, are the $n + 1$ polynomials

$$\beta_{n,j}(x) = \binom{n}{j} x^j (1 - x)^{n-j} \in \mathbb{P}_n, \quad 0 \leq j \leq n, \tag{10.17}$$

where $\binom{n}{j}$ is the binomial coefficient

$$\binom{n}{j} = \frac{n!}{j!(n-j)!}.$$

We have the following simple properties of these polynomials.

Proposition 10.25 (properties of $\beta_{n,j}$). *The Bernstein basis polynomials form a basis for $\mathbb{P}_n$ and a partition of unity over $\mathbb{R}$, i.e.,*

$$\sum_{j=0}^{n} \beta_{n,j}(x) = 1, \quad \forall x \in \mathbb{R}. \tag{10.18}$$

Each basis polynomial is nonnegative on the interval $[0, 1]$. *Furthermore,*

$$\sum_{j=0}^{n} \frac{j}{n} \beta_{n,j}(x) = x, \quad \forall x \in \mathbb{R}, \tag{10.19}$$

and

$$\sum_{j=0}^{n} \frac{j^2}{n^2} \beta_{n,j}(x) = \left(1 - \frac{1}{n}\right) x^2 + \frac{1}{n} x, \quad \forall x \in \mathbb{R}. \tag{10.20}$$

[5] Named in honor of the Russian and Soviet mathematician Sergei Natanovich Bernstein (1880–1968).

Proof. The nonnegativity on $[0,1]$ is immediate from the definition. Recall the binomial theorem

$$(a+b)^n = \sum_{j=0}^{n} \binom{n}{j} a^j b^{n-j}.$$

Setting $a = x$ and $b = 1-x$, we have the partition of unity (10.18). The proofs of the other properties are left to the reader as an exercise; see Problem 10.20. □

Definition 10.26 (Bernstein approximation). Suppose that $f \in C([0,1])$. The **Bernstein polynomial approximation of** f is defined by

$$\mathcal{B}_n[f](x) = \sum_{j=0}^{n} f\left(\frac{j}{n}\right) \beta_{n,j}(x). \tag{10.21}$$

Definition 10.27 (monotonicity). A linear operator $\mathcal{A}\colon C([a,b]) \to C([a,b])$ is called **monotone** if and only if for any $f, g \in C([a,b])$, if $f \le g$, then $\mathcal{A}[f] \le \mathcal{A}[g]$.

Proposition 10.28 (properties of $\mathcal{B}_n$). *$\mathcal{B}_n\colon C([0,1]) \to \mathbb{P}_n$ is a linear, monotone operator. Unlike the interpolation operator, it is not a projection operator. In other words, there exists a polynomial $p \in \mathbb{P}_n$ such that $\mathcal{B}_n[p] \neq p$.*

Proof. The linearity is straightforward: if $f, g \in C([0,1])$, then

$$\mathcal{B}_n[\alpha f + \beta g] = \alpha \mathcal{B}_n[f] + \beta \mathcal{B}_n[g].$$

The reader can check the details. To show that $\mathcal{B}_n$ is monotone, since $\mathcal{B}_n$ is linear, it suffices to show that if $f \ge 0$ on $[0,1]$, then $\mathcal{B}_n[f] \ge 0$ on $[0,1]$. But this is clear from the fact that each $\beta_{n,j}$ is nonnegative on $[0,1]$. That $\mathcal{B}_n$ is not a projection follows from Proposition 10.25. □

Cheney [15] and others use the theory of monotone operators to prove the Weierstrass Approximation Theorem using Bernstein operators. We will take a more direct approach. This next theorem, due to Bernstein, will yield the Weierstrass Approximation Theorem as a corollary.

Theorem 10.29 (Bernstein). *Suppose that $n \in \mathbb{N}_0$ and $f \in C([0,1])$ with modulus of continuity ω_f. Then*

$$\|f - \mathcal{B}_n[f]\|_{L^\infty(0,1)} \le \frac{9}{4}\omega_f\left(\frac{1}{\sqrt{n}}\right).$$

Proof. Using (10.18), the error function can be written as

$$e_n[f](x) = f(x) - \mathcal{B}_n[f](x) = \sum_{j=0}^{n} \beta_{n,j}(x)\left(f(x) - f\left(\frac{j}{n}\right)\right).$$

Using the nonnegativity of the Bernstein basis polynomials and the triangle inequality, we get

$$|e_n[f](x)| \le \sum_{j=0}^{n} \beta_{n,j}(x)\left|f(x) - f\left(\frac{j}{n}\right)\right|.$$

Fix $x \in [0,1]$ and let $\delta > 0$ be arbitrary. If $|x - j/n| \leq \delta$, then observe that

$$\left| f(x) - f\left(\frac{j}{n}\right) \right| \leq \omega_f(\delta).$$

Define the sets

$$J_{1,n}(x,\delta) = \left\{ j \in \{0,\ldots,n\} \;\middle|\; \left| x - \frac{j}{n} \right| \leq \delta \right\},$$
$$J_{2,n}(x,\delta) = \{0,\ldots,n\} \backslash J_{1,n}(x;\delta) = \left\{ j \in \{0,\ldots,n\} \;\middle|\; \left| x - \frac{j}{n} \right| > \delta \right\}.$$

Then, since $\beta_{n,j} \geq 0$ on $[0,1]$ and $\sum_{j=0}^{n} \beta_{n,j} = 1$,

$$\begin{aligned}
|e_n[f](x)| &\leq \sum_{j \in J_{1,n}} \beta_{n,j}(x) \left| f(x) - f\left(\frac{j}{n}\right) \right| + \sum_{j \in J_{2,n}} \beta_{n,j}(x) \left| f(x) - f\left(\frac{j}{n}\right) \right| \\
&\leq \sum_{j \in J_{1,n}} \beta_{n,j}(x)\omega_f(\delta) + \sum_{j \in J_{2,n}} \beta_{n,j}(x) \left| f(x) - f\left(\frac{j}{n}\right) \right| \\
&\leq \omega_f(\delta) + \sum_{j \in J_{2,n}} \beta_{n,j}(x) \left| f(x) - f\left(\frac{j}{n}\right) \right|.
\end{aligned}$$

Now, for each $j \in J_{2,n}$, let $p \in \mathbb{N}$ be the unique number with the property that

$$p < \frac{1}{\delta} \left| x - \frac{j}{n} \right| \leq p + 1. \tag{10.22}$$

Define the sequence $\{y_k\}_{k=0}^{p+1} \subset [0,1]$ via

$$y_k = x + \frac{k}{p+1} \left(\frac{j}{n} - x \right), \quad k = 0, \ldots, p+1.$$

Notice that

$$y_0 = x, \quad y_{p+1} = \frac{j}{n},$$

and y_k is strictly increasing from x to $\frac{j}{n}$, if $x < \frac{j}{n}$, and strictly decreasing from x to $\frac{j}{n}$, if $x > \frac{j}{n}$. Furthermore, by construction,

$$|y_{k+1} - y_k| = \frac{1}{p+1} \left| x - \frac{j}{n} \right| \leq \delta.$$

Then, by the triangle inequality,

$$\begin{aligned}
\left| f(x) - f\left(\frac{j}{n}\right) \right| &\leq |f(x) - f(y_1)| + \cdots + \left| f(y_p) - f\left(\frac{j}{n}\right) \right| \\
&\leq (p+1)\omega_f(\delta).
\end{aligned}$$

For each $j \in J_{2,n}$, using (10.22), we have

$$\begin{aligned}\left|f(x) - f\left(\frac{j}{n}\right)\right| &< \omega_f(\delta)\left(1 + \frac{1}{\delta}\left|x - \frac{j}{n}\right|\right) \\ &< \omega_f(\delta)\left(1 + \frac{1}{\delta^2}\left(x - \frac{j}{n}\right)^2\right),\end{aligned}$$

the second inequality resulting from the fact that $\left|x - \frac{j}{n}\right| > \delta$. Therefore, using (10.18)–(10.20), we have

$$\begin{aligned}\sum_{j\in J_{2,n}} \beta_{n,j}(x)\left|f(x) - f\left(\frac{j}{n}\right)\right| &\leq \sum_{j\in J_{2,n}} \beta_{n,j}(x)\omega_f(\delta)\left(1 + \frac{1}{\delta^2}\left(x - \frac{j}{n}\right)^2\right) \\ &= \omega_f(\delta)\sum_{j\in J_{2,n}} \beta_{n,j}(x)\left(1 + \frac{1}{\delta^2}\left(x - \frac{j}{n}\right)^2\right) \\ &= \omega_f(\delta)\left(1 + \frac{1}{\delta^2}\frac{x(1-x)}{n}\right) \\ &\leq \omega_f(\delta)\left(1 + \frac{1}{4n\delta^2}\right).\end{aligned}$$

We now have

$$\begin{aligned}|e_n[f](x)| &\leq \omega_f(\delta) + \sum_{j\in J_{2,n}} \beta_{n,j}(x)\left|f(x) - f\left(\frac{j}{n}\right)\right| \\ &\leq \omega_f(\delta)\left(2 + \frac{1}{4n\delta^2}\right)\end{aligned} \tag{10.23}$$

for any $\delta > 0$ that we may choose. It follows, upon setting $\delta = n^{-1/2}$, that

$$|e_n[f](x)| \leq \frac{9}{4}\omega_f\left(n^{-1/2}\right).$$

Since $x \in [0, 1]$ was chosen arbitrarily in $[0, 1]$, we have the desired result. □

Our Bernstein construction was for real-valued functions. But it is not hard to see that we can generalize to complex-valued functions by treating the real and imaginary parts separately. Thus, we have the following corollary.

Corollary 10.30 (Bernstein). *Suppose that $n \in \mathbb{N}_0$ and $f \in C([0,1];\mathbb{C})$. Set $g = \Re(f)$ and $h = \Im(f)$. Define the Bernstein approximation of f via*

$$\mathcal{B}_n[f] = \mathcal{B}_n[g] + \mathrm{i}\mathcal{B}_n[h] \in \mathbb{P}_n(\mathbb{C}). \tag{10.24}$$

Then

$$\|f - \mathcal{B}_n[f]\|_{L^\infty(0,1;\mathbb{C})} \leq \frac{9}{4}\left(\omega_g\left(\frac{1}{\sqrt{n}}\right) + \omega_h\left(\frac{1}{\sqrt{n}}\right)\right).$$

Proof. See Problem 10.21. □

Corollary 10.31 (Lipschitz function). *Suppose that $f \in C^{0,1}([0,1];\mathbb{C})$ with Lipschitz constant $L > 0$, i.e.,*

$$|f(x) - f(y)| \le L|x - y|, \quad \forall x, y \in [0,1].$$

Then, for any $n \in \mathbb{N}_0$,

$$\|f - \mathcal{B}_n[f]\|_{L^\infty(0,1;\mathbb{C})} \le 2L\sqrt{\frac{2}{n}},$$

where $\mathcal{B}_n[f]$ is as defined in (10.24).

Proof. Set $g = \Re(f)$ and $h = \Im(f)$. Suppose that ω_f and ω_g are the moduli of continuity for the real and imaginary parts of f, respectively. Since f is Lipschitz, it follows that g and h are each Lipschitz continuous with the same Lipschitz constants, L. From Lemma 10.23.7 it follows that

$$\omega_g(\delta) \le L\delta, \qquad \omega_h(\delta) \le L\delta,$$

and from (10.23) we can obtain the estimate

$$|f(x) - \mathcal{B}_n[f](x)| \le 2L\delta\left(2 + \frac{1}{4n\delta^2}\right)$$

for any $x \in [0,1]$ and all $\delta > 0$. The choice $\delta = \frac{1}{2\sqrt{2n}}$ yields the desired result. □

The Weierstrass Approximation Theorem[6] follows easily from Theorem 10.29; it is essentially a corollary.

Theorem 10.32 (Weierstrass Approximation Theorem). *Let $[a,b] \subset \mathbb{R}$ be a compact interval and $f \in C([a,b];\mathbb{C})$. For every $\varepsilon > 0$, there exists an $n \in \mathbb{N}$ and a polynomial $p_n \in \mathbb{P}_n(\mathbb{C})$ such that*

$$\|f - p_n\|_{L^\infty(a,b;\mathbb{C})} \le \varepsilon.$$

In other words, there is a sequence of polynomial functions $\{p_n\}_{n=0}^\infty$, with $p_n \in \mathbb{P}_n(\mathbb{C})$, that converge uniformly to f as $n \to \infty$.

Proof. First, we need to make an affine change of coordinates from $[a,b]$ to $[0,1]$. In particular, consider

$$G(x) = \frac{x - a}{b - a}, \quad x \in [a,b],$$

whose inverse is, of course,

$$F(t) = (1 - t)a + tb, \quad t \in [0,1].$$

Use the affine change of coordinates to map f to the interval $[0,1]$; approximate this composite function by the appropriate Bernstein polynomial; then map the result back to the interval $[a,b]$. Finally, use the fact that $\omega_q(h) \to 0$ as $h \to 0$, from Lemma 10.23, for any continuous function q defined on a compact interval. The details are left to the reader as an exercise; see Problem 10.22. □

[6] Named in honor of the German mathematician Karl Theodor Wilhelm Weierstrass (1815–1897).

One may wonder how sharp the conclusion of Theorem 10.29 is. The following example, taken from [73], shows that using Bernstein polynomials, this is asymptotically sharp.

Example 10.9 Let $f(x) = |x - \frac{1}{2}|$. It is not difficult to see that $f \in C^{0,1}([0,1])$ with Lipschitz constant $L = 1$. Thus, owing to Corollary 10.31, we have

$$\|f - \mathcal{B}_n[f]\|_{L^\infty(0,1)} \leq \frac{C}{\sqrt{n}}, \qquad \forall n \in \mathbb{N}$$

for some constant $C > 0$ that is independent of n.

Notice, however, that at the point $x = \frac{1}{2}$ we have

$$\mathcal{B}_n[f](x) - f(x) = \mathcal{B}_n[f](x) = \left(\frac{1}{2}\right)^n \sum_{j=0}^{n} \left|\frac{j}{n} - \frac{1}{2}\right| \binom{n}{j}.$$

Then, if n is even, we have

$$\sum_{j=0}^{n} \left|\frac{j}{n} - \frac{1}{2}\right| \binom{n}{j} = \sum_{j=0}^{n/2} \left(\frac{1}{2} - \frac{j}{n}\right) \binom{n}{j} = \frac{1}{2}\binom{n}{n/2}.$$

Consequently,

$$\left|\mathcal{B}_n[f]\left(\frac{1}{2}\right) - f\left(\frac{1}{2}\right)\right| = \frac{1}{2^{n+1}}\binom{n}{n/2} > \frac{1}{2}n^{-1/2},$$

where the final estimate is a consequence of the well-known Stirling Formula[7]

$$\sqrt{2\pi m}m^m e^{-m} < m! < \sqrt{2\pi m}m^m e^{-m}\left(1 + \frac{1}{4m}\right), \qquad \forall m \in \mathbb{N}.$$

In conclusion, we have shown that

$$\|f - \mathcal{B}_n[f]\|_{L^\infty(0,1)} > \frac{1}{2}n^{-1/2};$$

therefore, we cannot expect to do better than as Theorem 10.29 states, at least using Bernstein polynomials.

Example 10.10 Although we do not discuss it here, *Jackson's Theorem*[8] shows that, for every compact interval $[a,b] \subset \mathbb{R}$, any $f \in C([a,b])$, and all $n \in \mathbb{N}_0$,

$$\inf_{p \in \mathbb{P}_n} \|f - p\|_{L^\infty(a,b)} \leq 6\omega_f\left(\frac{b-a}{2n}\right).$$

Interestingly, the so-called *Bernstein Theorem* shows that this is asymptotically sharp. The reader is referred to [73, 15] for details.

---

[7] Named in honor of the British mathematician James Stirling (1692–1770).
[8] Named in honor of the American mathematician Dunham Jackson (1888–1946).

10.5.1 Moduli of Smoothness

We conclude our discussion by an observation motivated by Problem 10.18. While the modulus of continuity proved rather useful, it is of no help in characterizing the smoothness of a function beyond Lipschitz continuity. To aid with this, *moduli of smoothness* are introduced.

Definition 10.33 (moduli of smoothness)**.** Let $[a, b] \subset \mathbb{R}$ be a compact interval and $f \in C([a, b]; \mathbb{C})$. For $k \in \mathbb{N}$ and $h \in [0, \frac{b-a}{k}]$, we define the **modulus of smoothness** of order k to be

$$\omega_f^{(k)}(h) = \sup\left\{\left|\Delta_t^k f(x)\right| \;\middle|\; |t| \leq h,\ x, x + kt \in [a, b]\right\},$$

where

$$\Delta_t^k f(x) = \sum_{i=0}^{k} (-1)^{k+i} \binom{k}{i} f(x + it).$$

Some of the basic properties of the modulus of smoothness are as follows.

Proposition 10.34 (properties of $\omega_f^{(k)}$)**.** *Let $[a, b] \subset \mathbb{R}$ be a compact interval and $f \in C([a, b]; \mathbb{C})$. For $k \in \mathbb{N}$, the modulus of smoothness has the following properties.*

1. *$\omega_f^{(1)} = \omega_f$, i.e., the modulus of smoothness of order one coincides with the modulus of continuity.*
2. *Monotonicity: If $0 < h_1 < h_2$, then $\omega_f^{(k)}(h_1) \leq \omega_f^{(k)}(h_2)$.*
3. *$\omega_f^{(k)}(h) \leq 2\omega_f^{(k-1)}(h)$.*
4. *For all $n \in \mathbb{N}$, $\omega_f^{(k)}(nh) \leq n^k \omega_f^{(k)}(h)$.*
5. *If $f \in C^k([a, b]; \mathbb{C})$, then*

$$\omega_f^{(k)}(h) \leq h^k \left\| f^{(k)} \right\|_{L^\infty(a,b;\mathbb{C})}.$$

6. *$\omega_f^{(k)}(h) \equiv 0$ if and only if $f \in \mathbb{P}_{k-1}$.*
7. *If f is differentiable on $[a, b]$ and f' is bounded, then $\omega_f^{(k)}(h) \leq h\omega_{f'}^{(k-1)}(h)$.*

Proof. Most of the proofs are similar to those for the modulus of continuity, and so we leave them to the reader as an exercise; see Problem 10.19. We will only prove the last one. For $|t| \leq h$, we have

$$
\begin{aligned}
|\Delta_t^k f(x)| &= \left|\Delta_t^{k-1}[f(x+t) - f(x)]\right| \\
&= \left|\Delta_t^{k-1} \int_x^{x+t} f'(y)\mathrm{d}y\right| \\
&= \left|\Delta_t^{k-1} \int_0^{t} f'(x+y)\mathrm{d}y\right| \\
&= \left|\int_0^{t} \Delta_t^{k-1} f'(x+y)\mathrm{d}y\right| \\
&\le \int_{\min\{0,t\}}^{\max\{0,t\}} |\Delta_t^{k-1} f'(x+y)|\mathrm{d}y \\
&\le \int_{\min\{0,t\}}^{\max\{0,t\}} \omega_{f'}^{(k-1)}(h)\mathrm{d}y \\
&\le h\omega_{f'}^{(k-1)}(h).
\end{aligned}
$$

Taking supremum over $|t| \le h$ on this inequality, the claim follows. □

With the help of moduli of smoothness of order larger than one, we can, for instance, provide a finer characterization of the error in Lagrange interpolation. The proof of the following result is beyond the scope of our text, but for context it should be compared with Theorems 9.16 and 9.45. The interested reader is referred to, for instance, [84].

Theorem 10.35 (Sendov[9]). *Let $f \in C([0,1])$. For every $n \in \mathbb{N}$, define the nodal set $X = \{\frac{i}{n+1}\}_{i=1}^n$. Let $p \in \mathbb{P}_{n-1}$ be the Lagrange interpolating polynomial subject to X. Then we have*

$$
\|f - p\|_{L^\infty(0,1)} \le 6\omega_f^{(n)}\left(\frac{1}{n+1}\right).
$$

Problems

10.1 Complete the proof of Theorem 10.2.

10.2 Prove Theorem 10.4.

10.3 Provide all the details for Example 10.2.

10.4 Complete the proof of Proposition 10.14.

10.5 Complete the proof of Proposition 10.15.

10.6 Prove Proposition 10.16.

10.7 Prove Theorem 10.17.

10.8 Construct the minimax polynomial $p \in \mathbb{P}_1$ on the interval $[-2,1]$ for the function $f(x) = |x|$.

10.9 Find the minimax polynomial $p \in \mathbb{P}_n$ on the interval $[-1,1]$ for the function $f(x) = \sum_{j=0}^{n+1} a_j x^j$, where $a_j \in \mathbb{R}$ and $a_{n+1} \neq 0$.

[9] Named after the Bulgarian mathematician, diplomat, and politician Blagovest Hristov Sendov (1932–2020).

10.10 Let $n \in \mathbb{N}$ be given. Show that, for any set of coefficients $a_j \in \mathbb{R}$, $j = 0, \dots, n-1$,

$$\max_{-1 \le x \le 1} \left| x^n + a_{n-1}x^{n-1} + \cdots + a_1 x + a_0 \right| \ge 2^{1-n}.$$

10.11 Suppose that $f \in C^2([a,b])$ and $f''(x) > 0$ for all $x \in [a,b]$. Suppose that $p \in \mathbb{P}_1$ is the minimax polynomial approximation of f and $p(x) = a_1 x + a_0$. Prove that

$$a_1 = \frac{f(b) - f(a)}{b-a} \quad \text{and} \quad a_0 = \frac{1}{2}(f(a) + f(c)) - \frac{f(b)-f(a)}{b-a} \cdot \frac{a+c}{2},$$

where $c \in (a,b)$ is the unique solution to

$$f'(c) = \frac{f(b)-f(a)}{b-a}.$$

10.12 Suppose that, for $a > 0$, $f \in C([-a,a])$ is even. Suppose that $p \in \mathbb{P}_n$ is the minimax polynomial approximation of f on $[-a,a]$. What can you say about the parity of p? Is p an even function? What can one say about the critical points in $[-a,a]$? How are they distributed? Answer the analogous questions in the case that f is odd. Prove your assertions.

10.13 Fill in any missing details in the proof of Lemma 10.23.

10.14 Suppose that $f(x) = x^2$ on the interval $I = (0,1)$. Show that $\omega_f(h) = 2h - h^2$.

10.15 If $f(x) = \frac{1}{x}$ and $I = (0,1)$, show that $\omega_f(h) = +\infty$.

10.16 Show that if $\alpha \in (0,1]$, and $f \in C^{0,\alpha}([0,1])$, then there is a constant $C > 0$, for which

$$\omega_f(h) \le C h^\alpha.$$

10.17 We say that a function $f \in LogLip([0,1])$ if there is a constant $M > 0$ such that

$$\omega_f(h) \le M h |\log(h)|.$$

a) Show that, for every $\alpha < 1$, $f \in LogLip([0,1])$ implies that $f \in C^{0,\alpha}([0,1])$.
b) Show that $LogLip([0,1]) \not\subset C^{0,1}([0,1])$.
Hint: Consider $f(x) = x \log x$.

10.18 Show that if $h^{-1}\omega_f(h) \to 0$ as $h \downarrow 0$, then f is constant.

10.19 Complete the proof of Proposition 10.34.

10.20 Recall that the Bernstein basis polynomials, with respect to the interval $[0,1]$, are the $n+1$ polynomials

$$\beta_{n,j}(x) = \binom{n}{j} x^j (1-x)^{n-j} \in \mathbb{P}_n, \quad 0 \le j \le n.$$

Prove the following facts; in doing so, prove Proposition 10.25.

a) $\beta_{n,j}(0) = \delta_{0,j}$ and $\beta_{n,j}(1) = \delta_{n,j}$.
b) $\beta_{n,j}(1-x) = \beta_{n,n-j}(x)$.
c) $\beta'_{n,j}(x) = n\left(\beta_{n-1,j-1}(x) - \beta_{n-1,j}(x)\right)$.
d) For $0 < j \le n$, $\beta_{n,j}$ has a root at $x = 0$ of multiplicity j.

e) For $0 \le j < n$, $\beta_{n,j}$ has a root at $x = 1$ of multiplicity $n - j$.
f) $\int_0^1 \beta_{n,j}(x)dx = \frac{1}{n+1}$ for all $0 \le j \le n$.
g) The set of Bernstein basis polynomials forms a basis for $\mathbb{P}_n$.
h) The Bernstein basis polynomials form a partition of unity over $\mathbb{R}$, i.e.,

$$\sum_{j=0}^{n} \beta_{n,j}(x) = 1, \quad \forall x \in \mathbb{R}.$$

i) Each basis polynomial is nonnegative on the interval $[0, 1]$.
j) The polynomials satisfy the identities

$$\sum_{j=0}^{n} \frac{j}{n} \beta_{n,j}(x) = x, \quad \forall x \in \mathbb{R},$$

$$\sum_{j=0}^{n} \frac{j-1}{n-1} \cdot \frac{j}{n} \beta_{n,j}(x) = x^2, \quad \forall x \in \mathbb{R},$$

$$\sum_{j=0}^{n} \frac{j^2}{n^2} \beta_{n,j}(x) = \left(1 - \frac{1}{n}\right) x^2 + \frac{1}{n} x, \quad \forall x \in \mathbb{R}.$$

Hint: Take the first and second derivatives of $(x + y)^n$, with respect to x, assuming y is constant, and then set $y = 1 - x$.

10.21 Prove Corollary 10.30.

10.22 Complete the proof of Theorem 10.32.

11 Polynomial Least Squares Approximation

In this chapter, we again want to approximate a function of interest with a polynomial. This time we will seek a polynomial that minimizes the error in the L^2-norm. This problem, known as the the polynomial least squares problem, is closely related to classical Fourier expansions. We saw in Chapter 10 that the minimax approximations were rather complicated objects. Polynomial least squares approximations, by contrast, are relatively easy to construct.

To motivate our work in this chapter, let us do some formal calculations. Suppose that J is a set of indices, of finite or infinite cardinality, and that $\Psi = \{\psi_j\}_{j\in J}$ is an orthonormal set of functions, i.e., $(\psi_i, \psi_j)_\star = \delta_{i,j}$, where $(\cdot\,,\cdot)_\star$ is an inner product on an appropriate function space V. Assume that $f \in V$ can be expanded in the functions of Ψ:

$$f = \sum_{j\in J} c_j \psi_j,$$

where $\{c_j\}_{j\in J}$ is a sequence of scalars that we wish to determine. Using orthonormality, we find

$$(f, \psi_i)_\star = \left(\sum_{j\in J} c_j \psi_j, \psi_i\right)_\star = \sum_{j\in J} c_j (\psi_j, \psi_i)_\star = \sum_{j\in J} c_j \delta_{i,j} = c_i.$$

In other words,

$$f = \sum_{j\in J} (f, \psi_j)_\star \psi_j.$$

Now, if the set J is finite, the series above are finite, and there would be no question about bringing the summation outside of the inner product. In fact, all of our calculations would be perfectly valid. However, if J is infinite, we need to rigorously justify our work.

Do our formal calculations make any sense if J is infinite? In what sense do the series above converge? If J is infinite, is $\Psi = \{\psi_j\}_{j\in J}$ some kind of basis for our space of functions, V? What properties should f possess for the calculations to work? Can all functions in V be represented in our "basis" Ψ? And, in what way are the previous calculations related to least squares approximation? In this and the next chapter, we seek to answer these questions.

11.1 Least Squares Polynomial Approximations

Let us begin by defining what we mean by least squares polynomial approximation.

Definition 11.1 (least squares polynomial). Let $n \in \mathbb{N}_0$, $[a, b] \subset \mathbb{R}$ be a compact interval, and w be a weight function on $[a, b]$. Given $f \in L^2_w(a, b; \mathbb{C})$, we say that the polynomial $p \in \mathbb{P}_n(\mathbb{C})$ is a **least squares polynomial approximation** of f of order n if and only if

$$\|f - p\|_{L^2_w(a,b;\mathbb{C})} = \inf_{q \in \mathbb{P}_n(\mathbb{C})} \|f - q\|_{L^2_w(a,b;\mathbb{C})}.$$

As usual, if such a $p \in \mathbb{P}_n(\mathbb{C})$ exists, we write

$$p \in \operatorname*{argmin}_{q \in \mathbb{P}_n} \|f - q\|_{L^2_w(a,b;\mathbb{C})};$$

if it is unique, we write

$$p = \operatorname*{argmin}_{q \in \mathbb{P}_n(\mathbb{C})} \|f - q\|_{L^2_w(a,b;\mathbb{C})}.$$

11.2 Orthogonal Polynomials

An essential concept related to least squares polynomial approximation is that of an orthogonal polynomial.

Definition 11.2 (orthogonal polynomial). Let $[a, b] \subset \mathbb{R}$ be a compact interval. Suppose that w is a weight function on $[a, b]$. The set $\Psi = \{\psi_j\}_{j=0}^{\infty}$ is called an **orthogonal polynomial system** on $[a, b]$ with respect to the weight w if and only if, for each $j \in \mathbb{N}$, we have $\psi_j \in \mathbb{P}_j(\mathbb{C})$ with $\deg \psi_j = j$ and

$$(\psi_k, \psi_\ell)_{L^2_w(a,b;\mathbb{C})} = \delta_{k,\ell} \|\psi_\ell\|^2_{L^2_w(a,b;\mathbb{C})}, \quad \forall k, \ell \in \mathbb{N}_0.$$

The set is called an **orthonormal polynomial system** if and only if in addition $\|\psi_\ell\|_{L^2_w(a,b;\mathbb{C})} = 1$ for every $\ell \in \mathbb{N}_0$.

The following result shows that for every weight function there is an orthonormal polynomial system.

Theorem 11.3 (existence). *Let $[a, b] \subset \mathbb{R}$ be a compact interval and w be a weight function on $[a, b]$. There exists an orthonormal polynomial system on $[a, b]$ with respect to the weight w. Moreover, the polynomials in the system can be taken to be real valued.*

Proof. We know that $\{x^n\}_{n=0}^{\infty}$ is a linearly independent set of continuous functions on $[a, b]$ such that, for all $n \in \mathbb{N}_0$, $x^n \in \mathbb{P}_n(\mathbb{C})$ and $\deg x^n = n$. The existence of the orthonormal system now follows from the Gram–Schmidt process, since $(\cdot, \cdot)_{L^2_w(a,b;\mathbb{C})}$ is an inner product on $C([a, b]; \mathbb{C})$. That the resulting polynomials are real valued is due to starting with $\{x^n\}_{n=0}^{\infty}$ and the details of the Gram–Schmidt process. □

Example 11.1 The Chebyshev polynomial system defined in Definition 10.13 (see also Proposition 10.15 and Table 10.1) is an orthogonal polynomial system with respect to the weight function $w(x) = \frac{1}{\sqrt{1-x^2}}$ on the interval $[-1, 1]$. Note that this polynomial system, as defined, is orthogonal but not orthonormal.

11.3 Existence and Uniqueness of the Least Squares Approximation

With the help of an orthonormal polynomial system, we can establish the existence and uniqueness of a least squares polynomial approximation.

Theorem 11.4 (existence and uniqueness). *Let $[a, b] \subset \mathbb{R}$ be a compact interval and $n \in \mathbb{N}_0$. Suppose that w is a weight function on $[a, b]$ and $f \in L^2_w(a, b; \mathbb{C})$. There exists a unique least squares polynomial approximation of order n of f on $[a, b]$. In particular, if $\Psi = \{\psi_j\}_{j=0}^{\infty}$ is an orthonormal polynomial system with respect to the weight function w on $[a, b]$, then the unique least squares polynomial is given by*

$$p(x) = \sum_{j=0}^{n} (f, \psi_j)_{L^2_w(a,b;\mathbb{C})} \psi_j(x) \in \mathbb{P}_n(\mathbb{C}).$$

Proof. By construction, the system $\Psi_n = \{\psi_j\}_{j=0}^{n}$ is a basis of $\mathbb{P}_n(\mathbb{C})$; consequently, each $q \in \mathbb{P}_n(\mathbb{C})$ can be expressed on this basis: there are unique constants $c_0, \dots, c_n \in \mathbb{C}$ such that

$$q(x) = \sum_{j=0}^{n} c_j \psi_j(x).$$

This allows us to define a function $E \colon \mathbb{C}^{n+1} \to [0, \infty)$ via

$$E(c_0, \dots, c_n) = \|f - q\|^2_{L^2_w(a,b;\mathbb{C})} \, .$$

In particular,

$$\begin{aligned} E(c_0, \dots, c_n) &= (f - q, f - q)_{L^2_w(a,b;\mathbb{C})} \\ &= \|f\|^2_{L^2_w(a,b;\mathbb{C})} - (f, q)_{L^2_w(a,b;\mathbb{C})} - (q, f)_{L^2_w(a,b;\mathbb{C})} + (q, q)_{L^2_w(a,b;\mathbb{C})} \\ &= \|f\|^2_{L^2_w(a,b;\mathbb{C})} - 2\Re\left((f, q)_{L^2_w(a,b;\mathbb{C})}\right) + \sum_{j=0}^{n} |c_j|^2 \\ &= \sum_{j=0}^{n} \left|c_j - (f, \psi_j)_{L^2_w(a,b;\mathbb{C})}\right|^2 + \|f\|^2_{L^2_w(a,b;\mathbb{C})} - \sum_{j=0}^{n} \left|(f, \psi_j)_{L^2_w(a,b;\mathbb{C})}\right|^2 . \end{aligned}$$

Clearly, E is minimized if and only if

$$c_j = c_j^\star = (f, \psi_j)_{L^2_w(a,b;\mathbb{C})}, \quad j = 0, \dots, n.$$

Thus,

$$p(x) = \sum_{j=0}^{n} c_j^\star \psi_j(x) \in \mathbb{P}_n(\mathbb{C})$$

is the unique minimizer, i.e., the unique least squares polynomial approximation to f. □

As a consequence of the previous result, we obtain the following corollary, which is commonly known as *Bessel's inequality*.

Corollary 11.5 (Bessel's inequality[1]). *Let $[a, b] \subset \mathbb{R}$ be a compact interval, w be a weight function on $[a, b]$, and $f \in L^2_w(a, b; \mathbb{C})$. If $\Psi = \{\psi_j\}_{j=0}^{\infty}$ is an orthonormal polynomial system with respect to the weight function w on $[a, b]$, then* Bessel's *inequality,*

$$\sum_{j=0}^{n} \left|(f, \psi_j)_{L^2_w(a,b;\mathbb{C})}\right|^2 \leq \|f\|^2_{L^2_w(a,b;\mathbb{C})}, \tag{11.1}$$

holds for any $n \in \mathbb{N}_0$. As a consequence, since $\|f\|^2_{L^2_w(a,b;\mathbb{C})} < \infty$, the series

$$\sum_{j=0}^{\infty} \left|(f, \psi_j)_{L^2_w(a,b;\mathbb{C})}\right|^2$$

converges absolutely.

Proof. In the proof of Theorem 11.4 we learned that if p is the least squares polynomial of order $n \in \mathbb{N}_0$, then

$$0 \leq \|f - p\|^2_{L^2_w(a,b;\mathbb{C})} = \|f\|^2_{L^2_w(a,b;\mathbb{C})} - \sum_{j=0}^{n} \left|(f, \psi_j)_{L^2_w(a,b;\mathbb{C})}\right|^2. \tag{11.2}$$

The result easily follows from this equation. □

Least squares polynomial approximations can be characterized by a fundamental orthogonality relation.

Theorem 11.6 (characterization). *Let $[a, b] \subset \mathbb{R}$ be a compact interval, w be a weight function on $[a, b]$, and $f \in L^2_w(a, b; \mathbb{C})$. The polynomial $p \in \mathbb{P}_n(\mathbb{C})$ is the least squares approximation to f of order $n \in \mathbb{N}_0$ if and only if*

$$(f - p, q)_{L^2_w(a,b;\mathbb{C})} = 0, \quad \forall q \in \mathbb{P}_n(\mathbb{C}). \tag{11.3}$$

[1] Named in honor of the German astronomer, mathematician, physicist, and geodesist Friedrich Wilhelm Bessel (1784–1846).

Proof. ($\Longleftarrow$) Suppose that (11.3) holds for some $p \in \mathbb{P}_n(\mathbb{C})$. We want to show that this p is the least squares approximation to f. To this end, since $p-q \in \mathbb{P}_n(\mathbb{C})$, for all $q \in \mathbb{P}_n(\mathbb{C})$, we find

$$\begin{aligned}\|f-p\|^2_{L^2_w(a,b;\mathbb{C})} &= (f-p, f-p)_{L^2_w(a,b;\mathbb{C})} \\ &= (f-p, f-p)_{L^2_w(a,b;\mathbb{C})} + (f-p, p-q)_{L^2_w(a,b;\mathbb{C})} \\ &= (f-p, f-q)_{L^2_w(a,b;\mathbb{C})} \\ &\le \|f-p\|_{L^2_w(a,b;\mathbb{C})} \|f-q\|_{L^2_w(a,b;\mathbb{C})},\end{aligned}$$

where the Cauchy–Schwarz inequality was used in the last step. Thus,

$$\|f-p\|_{L^2_w(a,b;\mathbb{C})} \le \|f-q\|_{L^2_w(a,b;\mathbb{C})}, \quad \forall q \in \mathbb{P}_n(\mathbb{C}),$$

which proves that p is the least squares approximation of f.

($\Longrightarrow$) Now suppose that $p \in \mathbb{P}_n(\mathbb{C})$ is the least squares approximation of f. Then

$$p(x) = \sum_{j=0}^{n} c_j^\star \psi_j(x) \in \mathbb{P}_n,$$

where $\Psi = \{\psi_j\}_{j=0}^\infty$ is an orthonormal polynomial system with respect to the weight function w on $[a, b]$ and

$$c_j^\star = (f, \psi_j)_{L^2_w(a,b;\mathbb{C})}, \quad j = 0, \dots, n.$$

This implies, using orthonormality, that if $0 \le j \le n$, then

$$(f-p, \psi_j)_{L^2_w(a,b;\mathbb{C})} = (f, \psi_j)_{L^2_w(a,b;\mathbb{C})} - (p, \psi_j)_{L^2_w(a,b;\mathbb{C})} = c_j^\star - c_j^\star = 0.$$

Finally, if $q \in \mathbb{P}_n(\mathbb{C})$, there is the unique representation

$$q = \sum_{j=0}^{n} a_j \psi_j \in \mathbb{P}_n(\mathbb{C})$$

and

$$(f-p, q)_{L^2_w(a,b;\mathbb{C})} = \sum_{j=0}^{n} \overline{a_j} (f-p, \psi_j)_{L^2_w(a,b;\mathbb{C})} = 0.$$

Since $q \in \mathbb{P}_n(\mathbb{C})$ is arbitrary, the result is proven. $\square$

Definition 11.7 (least squares projection)**.** Let $[a, b] \subset \mathbb{R}$ be a compact interval, w be a weight function on $[a, b]$, and $\Psi = \{\psi_j\}_{j=0}^\infty$ be an orthonormal polynomial system with respect to w on $[a, b]$. For every $n \in \mathbb{N}_0$, the mapping

$$\mathcal{P}_{w,n} \colon L^2_w(a, b; \mathbb{C}) \to \mathbb{P}_n(\mathbb{C}),$$

$$f \mapsto \mathcal{P}_{w,n}[f] = \sum_{j=0}^{n} (f, \psi_j)_{L^2_w(a,b;\mathbb{C})} \psi_j$$

is called the **polynomial least squares projection** or, sometimes, the **L^2_w-projection** of degree n.

The following result shows that this mapping is indeed a projection.

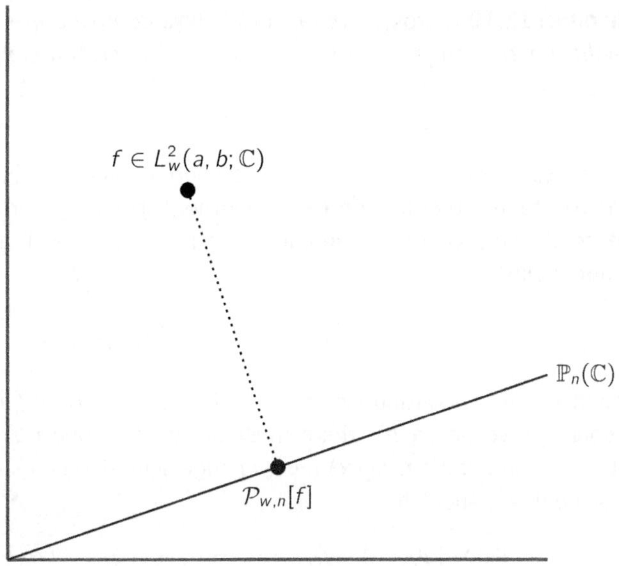

Figure 11.1 The fundamental orthogonality of the least squares projection.

Proposition 11.8 (projection). *For every compact interval $[a,b] \subset \mathbb{R}$, weight function w on $[a,b]$, and $n \in \mathbb{N}_0$, the polynomial least squares projection is a linear projection operator from $L^2_w(a,b;\mathbb{C})$ onto $\mathbb{P}_n(\mathbb{C})$, i.e.,*

$$\mathcal{P}_{w,n}[\alpha f + \beta g] = \alpha \mathcal{P}_{w,n}[f] + \beta \mathcal{P}_{w,n}[g], \quad \forall f,g \in L^2_w(a,b;\mathbb{C}), \quad \forall \alpha, \beta \in \mathbb{C},$$

and

$$\mathcal{P}_{w,n}[p] = p, \quad \forall p \in \mathbb{P}_n(\mathbb{C}).$$

Proof. See Problem 11.1. □

Remark 11.9 (orthogonality). Theorem 11.6 guarantees that the error between f and its projection $\mathcal{P}_{w,n}[f]$ is orthogonal to the subspace $\mathbb{P}_n(\mathbb{C})$, i.e.,

$$(f - \mathcal{P}_{w,n}[f], q)_{L^2_w(a,b;\mathbb{C})} = 0, \quad \forall q \in \mathbb{P}_n(\mathbb{C}). \tag{11.4}$$

Property (11.4) is sometimes referred to as the *fundamental orthogonality* of the projection. For a visual representation, see Figure 11.1, which can be compared with Figure 5.1. We will revisit this idea when we talk about Galerkin projections in the chapter on finite element methods, Chapter 25.

11.4 Properties of Orthogonal Polynomials

In this section we give some general properties of orthogonal polynomials with respect to a compact interval $[a,b] \subset \mathbb{R}$. The next result is really quite amazing. Take a moment to ponder it.

Theorem 11.10 (zeros). *Let $[a, b] \subset \mathbb{R}$ be a compact interval. Suppose that w is a weight function on $[a, b]$ and $\Psi = \{\psi_j\}_{j=0}^{\infty}$ is an orthogonal polynomial system with respect to the weight function w on $[a, b]$. For all $j \geq 1$, the polynomial $\psi_j \in \mathbb{P}_j$ has j real simple roots, all of which lie in the finite interval (a, b).*

Proof. Recall that, by definition, w is continuous on (a, b); w is nonnegative; and w is Riemann integrable on (a, b). Without loss of generality, let us assume that the coefficients of the polynomials are real. Fix $j \geq 1$ and observe that, using orthogonality,

$$(\psi_j, 1)_{L^2_w(a,b)} = \int_a^b w(x)\psi_j(x)\mathrm{d}x = 0.$$

The integrand is continuous and not identically zero on (a, b). Moreover, since w is nonnegative on (a, b), there must be at least one point $\xi \in (a, b)$ such that $\psi_j(\xi) = 0$. In addition, $\psi_j(x)$ must change sign at this point. In other words, there must be $\delta > 0$ such that

$$\psi_j(x_L)\psi_j(x_R) < 0, \quad \forall x_L \in (\xi - \delta, \xi), \quad \forall x_R \in (\xi, \xi + \delta).$$

Let us now suppose that there are exactly $k \in \mathbb{N}$ such points where ψ_j changes sign and denote them by $\{\xi_j\}_{j=1}^{k}$. Define

$$s_k(x) = (x - \xi_1) \cdots (x - \xi_k) \in \mathbb{P}_k.$$

The polynomial $\psi_j s_k$ does not change sign in the interval (a, b). Thus,

$$(\psi_j, s_k)_{L^2_w(a,b)} = \int_a^b w(x)\psi_j(x)s_k(x)\mathrm{d}x \neq 0.$$

Now, if $k < j$, this is a contradiction because ψ_j is orthogonal to every polynomial of strictly lesser degree. Thus, it must be that $k \geq j$. But k cannot be strictly greater than j either, since ψ_j cannot change signs more than j times. Thus, $k = j$ is the only option. Thus, all of the zeros lie in (a, b) and they are distinct. □

A similar argument yields an equally fascinating fact.

Theorem 11.11 (zeros of least squares). *Let $[a, b] \subset \mathbb{R}$ be a compact interval and $n \in \mathbb{N}_0$. Suppose that w is a weight function on $[a, b]$ and $f \in L^2_w(a, b)\backslash\mathbb{P}_n$. If the polynomial $p \in \mathbb{P}_n$ is the least squares approximation to f, then the function $f - p$ changes sign at no less than $n + 1$ distinct points in (a, b).*

Proof. Using the fundamental orthogonality property (for the error $f - p$), we have

$$0 = (f - p, 1)_{L^2_w(a,b)} = \int_a^b w(x)(f(x) - p(x))\mathrm{d}x = 0.$$

Once again, the integrand is not identically zero on (a, b); therefore, there are distinct points $\xi_i \in (a, b)$, $i = 1, \ldots, k$, such that $f - p$ changes sign at these points. Suppose that there are exactly k such points. Define, as before,

$$s_k(x) = (x - \xi_1) \cdots (x - \xi_k) \in \mathbb{P}_k.$$

We want to show that $k \geq n+1$. To see this, note that the function $(f-p)s_k w$ does not change sign in (a, b); consequently,

$$(f - p, s_k)_{L^2_w(a,b)} = \int_a^b w(x)(f(x) - p(x))s_k(s)\mathrm{d}x \neq 0.$$

It is not possible that $0 \leq k \leq n$, because of the fundamental orthogonality property and the fact that $s_k \in \mathbb{P}_k$. The only possibility is that $k > n$. □

As a consquence of these results, it turns out that orthonormal polynomial systems must satisfy certain recurrence relations.

Theorem 11.12 (recurrence). *Let $[a, b] \subset \mathbb{R}$ be a compact interval. Suppose that w is a weight function on $[a, b]$ and $\Psi = \{\psi_j\}_{j=0}^{\infty}$ is an orthonormal polynomial system with respect to the weight function w on $[a, b]$. Then the polynomials satisfy the following three-term recurrence relation:*

$$\psi_{n+1} = (A_n x + B_n)\psi_n - C_n \psi_{n-1}, \quad n = 1, 2, \dots,$$

for some sequences of constants $\{A_n\}_{n\in\mathbb{N}}$, $\{B_n\}_{n\in\mathbb{N}}$, and $\{C_n\}_{n\in\mathbb{N}}$.

Proof. See Problem 11.2. □

11.5 Convergence of Least Squares Approximations

In this section, we prove the convergence, in the weighted quadratic mean, of the least squares polynomial to the continuous function, f, that it approximates. We will make use of the Bernstein polynomial approximation of f established in Chapter 10.

Theorem 11.13 (convergence). *Suppose that $\Psi = \{\psi_j\}_{j=0}^{\infty}$ is an orthonormal polynomial system with respect to the weight function w on the compact interval $[a, b] \subset \mathbb{R}$. Assume that $f \in C([a, b]; \mathbb{C})$ and, for $n \in \mathbb{N}_0$, $p_n \in \mathbb{P}_n(\mathbb{C})$ is the least squares polynomial approximation of f. Then*

$$\lim_{n\to\infty} \|f - p_n\|_{L^2_w(a,b;\mathbb{C})} = 0.$$

Furthermore, Parseval's relation,[2]

$$\sum_{j=0}^{\infty} \left|(f, \psi_j)_{L^2_w(a,b;\mathbb{C})}\right|^2 = \int_a^b |f(x)|^2 w(x)\mathrm{d}x,$$

holds and

$$\lim_{j\to 0} (f, \psi_j)_{L^2_w(a,b;\mathbb{C})} = 0.$$

[2] Named in honor of the French mathematician Marc-Antoine Parseval des Chênes (1755–1836).

Proof. By definition, for any $r_n \in \mathbb{P}_n(\mathbb{C})$,

$$\|f - p_n\|^2_{L^2_w(a,b;\mathbb{C})} \leq \|f - r_n\|^2_{L^2_w(a,b;\mathbb{C})}.$$

By the Weierstrass Approximation Theorem 10.32, there is a sequence $\{q_n\}_{n=1}^{\infty}$, $q_n \in \mathbb{P}_n(\mathbb{C})$ such that

$$\lim_{n\to\infty} \|f - q_n\|_{L^\infty(a,b;\mathbb{C})} = 0.$$

Using (D.1) with $p = 2$, we have

$$\|f - p_n\|_{L^2_w(a,b;\mathbb{C})} \leq \|f - q_n\|_{L^2_w(a,b;\mathbb{C})} \leq \|f - q_n\|_{L^\infty(a,b)} \sqrt{\|w\|_{L^1(a,b)}} \to 0.$$

Thus, $\|f - p_n\|_{L^2_w(a,b;\mathbb{C})} \to 0$, as $n \to \infty$. The first part is proven.

Using (11.2), we observe that

$$\|f - p_n\|^2_{L^2_w(a,b;\mathbb{C})} = \|f\|^2_{L^2_w(a,b;\mathbb{C})} - \sum_{j=0}^{n} \left|(f, \psi_j)_{L^2_w(a,b;\mathbb{C})}\right|^2$$

or, equivalently,

$$\sum_{j=0}^{n} \left|(f, \psi_j)_{L^2_w(a,b;\mathbb{C})}\right|^2 = \|f\|^2_{L^2_w(a,b;\mathbb{C})} - \|f - p_n\|^2_{L^2_w(a,b;\mathbb{C})} \geq 0.$$

By Bessel's inequality (11.1),

$$0 \leq \|f\|^2_{L^2_w(a,b;\mathbb{C})} - \|f - p_n\|^2_{L^2_w(a,b;\mathbb{C})} = \sum_{j=0}^{n} \left|(f, \psi_j)_{L^2_w(a,b;\mathbb{C})}\right|^2 \leq \|f\|^2_{L^2_w(a,b;\mathbb{C})}.$$

By the Squeeze Theorem B.6, since $\|f - p_n\|_{L^2_w(a,b;\mathbb{C})} \to 0$,

$$\sum_{j=0}^{n} \left|(f, \psi_j)_{L^2_w(a,b;\mathbb{C})}\right|^2 \to \|f\|^2_{L^2_w(a,b;\mathbb{C})}$$

and Parseval's relation follows. □

Theorem 11.13 can be generalized for functions $f \in L^2_w(a, b; \mathbb{C})$. We first need a well-known result that can be found in a good book on integration theory; for example, [7] or [77].

Theorem 11.14 (density). *Let w be a weight function on the compact interval $[a, b] \subset \mathbb{R}$ and $f \in L^2_w(a, b; \mathbb{C})$. For any $\varepsilon > 0$, there is a $g \in C([a, b]; \mathbb{C})$ such that*

$$\|f - g\|_{L^2_w(a,b;\mathbb{C})} < \varepsilon.$$

In other words, $C([a, b]; \mathbb{C})$ is dense in $L^2_w(a, b; \mathbb{C})$ with respect to the norm $\|\cdot\|_{L^2_w(a,b;\mathbb{C})}$.

Then we can prove the following result, without any difficulty.

Theorem 11.15 (convergence). *Suppose that $\Psi = \{\psi_j\}_{j=0}^{\infty}$ is an orthonormal polynomial system with respect to the weight function w on the compact interval $[a, b] \subset \mathbb{R}$. Assume that $f \in L^2_w(a, b; \mathbb{C})$ and, for $n \in \mathbb{N}_0$, $p_n \in \mathbb{P}_n(\mathbb{C})$ is the least squares polynomial approximation of f. Then*

$$\lim_{n\to\infty} \|f - p_n\|_{L^2_w(a,b;\mathbb{C})} = 0.$$

Furthermore, Parseval's relation,

$$\sum_{j=0}^{\infty} \left|(f, \psi_j)_{L^2_w(a,b;\mathbb{C})}\right|^2 = \int_a^b |f(x)|^2 \, w(x)\mathrm{d}x,$$

holds.

Proof. Let $\varepsilon > 0$ be arbitrary. By Theorem 11.14, there is $g \in C([a, b]; \mathbb{C})$ such that

$$\|f - g\|_{L^2_w(a,b)} < \frac{\varepsilon}{2}.$$

Since $p_n \in \mathbb{P}_n(\mathbb{C})$ is the best approximation of f, for any $r_n \in \mathbb{P}_n(\mathbb{C})$,

$$\begin{aligned}
\|f - p_n\|_{L^2_w(a,b;\mathbb{C})} &\le \|f - r_n\|_{L^2_w(a,b;\mathbb{C})} \\
&= \|f - g + g - r_n\|_{L^2_w(a,b;\mathbb{C})} \\
&\le \|f - g\|_{L^2_w(a,b;\mathbb{C})} + \|g - r_n\|_{L^2_w(a,b;\mathbb{C})} \\
&< \frac{\varepsilon}{2} + \|g - r_n\|_{L^2_w(a,b;\mathbb{C})} .
\end{aligned}$$

Now use the Weierstrass Approximation Theorem 10.32. The remaining details are left to the reader as an exercise; see Problem 11.3. □

It is important to note what this result is saying, as well as what it is not. We have, as $n \to \infty$,

$$p_n = \sum_{j=0}^{n} (f, \psi_j)_{L^2_w(a,b;\mathbb{C})} \psi_j \longrightarrow f,$$

in the sense of $L^2_w(a, b; \mathbb{C})$, i.e., convergence in the weighted quadratic mean. This is a strange sort of convergence. With $L^2_w(a, b; \mathbb{C})$ convergence, it is possible that there are points $x \in [a, b]$ such that $p_n(x)$ does not converge to $f(x)$, as $n \to \infty$. It does not imply convergence in the sense of $L^\infty(a, b; \mathbb{C})$; in other words, uniform convergence. That topic is addressed in Section 11.6.

For those who have studied Fourier series, this topic should be familiar. In fact, we make the following definition.

Definition 11.16 (Fourier coefficients). Suppose that $\Psi = \{\psi_j\}_{j=0}^{\infty}$ is an orthonormal polynomial system with respect to the weight function w on the compact interval $[a, b] \subset \mathbb{R}$ and $f \in L^2_w(a, b; \mathbb{C})$. The numbers

$$c_j = (f, \psi_j)_{L^2_w(a,b;\mathbb{C})} , \quad j \in \mathbb{N}_0$$

n	$P_n(x)$
0	1
1	x
2	$\frac{1}{2}\left(3x^2-1\right)$
3	$\frac{1}{2}\left(5x^3-3x\right)$
4	$\frac{1}{8}\left(35x^4-30x^2+3\right)$
5	$\frac{1}{8}\left(63x^5-70x^3+15x\right)$
6	$\frac{1}{16}\left(231x^6-315x^4+105x^2-5\right)$
7	$\frac{1}{16}\left(429x^7-693x^5+315x^3-35x\right)$
8	$\frac{1}{128}\left(6\,435x^8-12\,012x^6+6\,930x^4-1\,260x^2+35\right)$
9	$\frac{1}{128}\left(12\,155x^9-25\,740x^7+18\,018x^5-4\,620x^3+315x\right)$
10	$\frac{1}{256}\left(46\,189x^{10}-109\,395x^8+90\,090x^6-30\,030x^4+3\,465x^2-63\right)$

Table 11.1 The first 11 Legendre polynomials.

are called the **generalized Fourier coefficients of** f.[3] The series

$$\sum_{j=0}^{\infty} c_j \psi_j$$

is called the **generalized Fourier series**. We say that this series **converges in** $L^2_w(a, b; \mathbb{C})$ if and only if there is a function $g \in L^2_w(a, b; \mathbb{C})$ such that the **sequence of partial sums**,

$$p_n = \sum_{j=0}^{n} c_j \psi_j \in \mathbb{P}_n,$$

converges to g in the $L^2_w(a, b; \mathbb{C})$ norm, i.e.,

$$\|p_n - g\|_{L^2_w(a,b;\mathbb{C})} \to 0.$$

In Chapter 12 we will discuss the classical Fourier series involving expansions of trigonometric functions. Here, we will focus our attention on another polynomial system.

Definition 11.17 (Legendre polynomials)**.** Suppose that $w \equiv 1$. The orthogonal system of polynomials on the interval $[-1, 1]$, denoted $\{P_n\}_{n=0}^{\infty}$ and satisfying the normalization

$$\int_{-1}^{1} P_k(x) P_\ell(x) \mathrm{d}x = \frac{2}{2k+1}\delta_{k,\ell}, \tag{11.5}$$

is called the **Legendre polynomial system**.[4]

The first few members of this system are given in Table 11.1. Let us now present some of the properties of these polynomials.

[3] Named in honor of the French mathematician and physicist Jean-Baptiste Joseph Fourier (1768–1830).

[4] Named in honor of the French mathematician Adrien-Marie Legendre (1752–1833).

Theorem 11.18 (properties of $\{P_n\}$). *Suppose that $\{P_n\}_{n=0}^{\infty}$ is the Legendre polynomial system given in Definition 11.17. Then the following properties hold.*

1. *Rodrigues formula:*[5] *For $n \in \mathbb{N}_0$,*

$$P_n(x) = \frac{1}{2^n n!} \frac{\mathrm{d}^n}{\mathrm{d}x^n} \left(x^2 - 1\right)^n.$$

2. *Recurrence relation: For $n \in \mathbb{N}$,*

$$(n+1)P_{n+1}(x) = (2n+1)xP_n(x) - nP_{n-1}(x).$$

3. *Recurrence relation for derivatives: For $n \in \mathbb{N}$,*

$$(2n+1)P_n(x) = \frac{\mathrm{d}}{\mathrm{d}x}\left(P_{n+1}(x) - P_{n-1}(x)\right).$$

4. *$P_n(1) = 1$ and $P_n(-1) = (-1)^n$ for $n \in \mathbb{N}_0$.*
5. *P_n satisfies the differential equation*

$$\frac{\mathrm{d}}{\mathrm{d}x}\left[(1-x^2)\frac{\mathrm{d}P_n}{\mathrm{d}x}(x)\right] + n(n+1)P_n(x) = 0.$$

6. *$|P_n(x)| \le 1$ for all $x \in [-1, 1]$ and $n \in \mathbb{N}_0$.*

Proof. See Problem 11.4. □

Example 11.2 Suppose that

$$f(x) = \begin{cases} 0, & -1 \le x < 0, \\ 1, & 0 \le x \le 1. \end{cases}$$

In this example, we will use the weight function $w \equiv 1$, so that the appropriate polynomial system is the Legendre polynomial system. The function f is not continuous, but $f \in L^2_w(-1,1)$. The least squares polynomial approximations of f in this setting are called *Fourier–Legendre expansions*. It is easy to see that the generalized Fourier coefficients (i.e., the Fourier–Legendre coefficients) satisfy

$$\begin{aligned} c_0 &= \frac{1}{2}, & \\ c_{2k} &= 0, & k \in \mathbb{N}, \\ c_{2k-1} &\neq 0, & k \in \mathbb{N}. \end{aligned}$$

We plot the approximations of degrees $n = 21, 31$, and 41 in Figure 11.2. One can prove that $p_n(0)$ does not converge to $f(0) = 1$, as $n \to \infty$. In other words, p_n does not converge uniformly to f, even though it must converge in the quadratic mean, according to Theorem 11.15. The overshoot phenomenon near the discontinuity in

[5] Named in honor of the French banker, mathematician, and social reformer Benjamin Olinde Rodrigues (1795–1851).

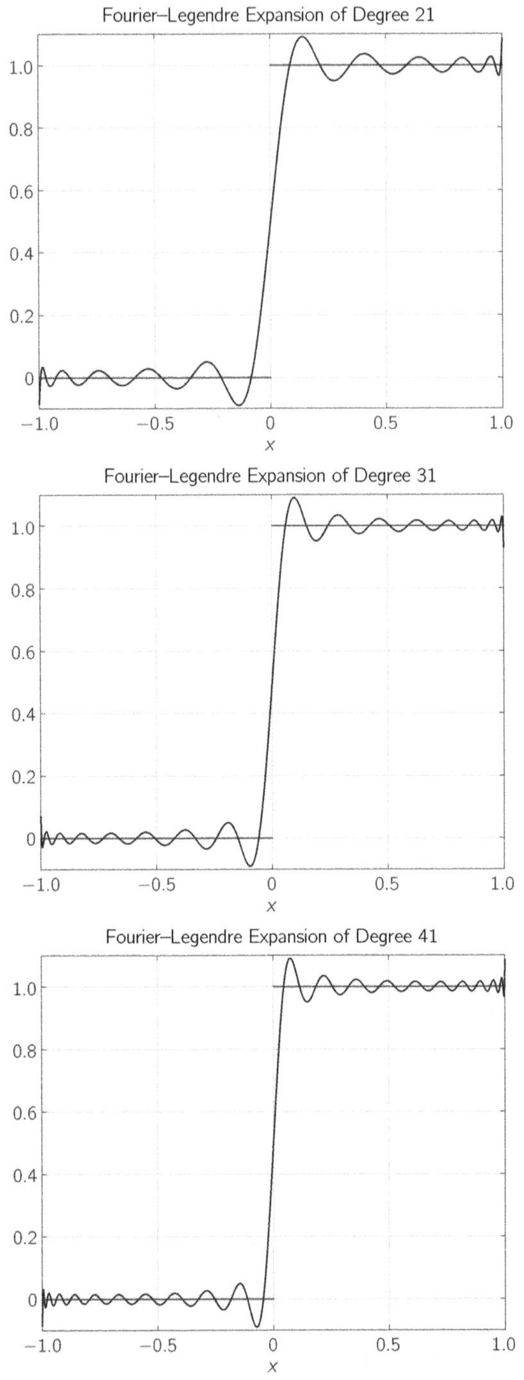

Figure 11.2 Fourier–Legendre expansions of the Heaviside function; see Example 11.2.

f that prevents uniform convergence is known as the *Gibbs phenomenon*.[6] We refer the interested reader to [42, 93, 104] for an in-depth analysis of this phenomenon.

11.6 Uniform Convergence of Least Squares Approximations

In this section, we will prove that, for suitable functions f, if p_n is its polynomial least squares approximation, then we have

$$p_n = \sum_{j=0}^{n} (f, \psi_j)_{L^2_w(a,b)} \psi_j \longrightarrow f, \qquad n \to \infty,$$

in the sense of $L^\infty(a, b)$, i.e., uniform convergence, for two common weight functions on the interval $[-1, 1]$. This is usually accomplished by adding some further conditions, such as $f \in C^2([a, b]; \mathbb{C})$, to the function of interest.

Let us begin with the simplest case, $w \equiv 1$, so that the orthonormal polynomial system is related to the Legendre polynomial system.

We first state and prove a useful, but simple, technical lemma that will be used several times in this chapter and Chapter 12.

Lemma 11.19 (convergent series)**.** *Suppose that $p \in (1, \infty)$. Then there is a constant $C > 0$, only depending on p, such that*

$$\sum_{j=n+1}^{\infty} \frac{1}{j^p} \leq \frac{C}{n^{p-1}}.$$

In particular, $C = \frac{1}{p-1}$.

Proof. First we note that, since $p > 1$, the series

$$\sum_{j=1}^{\infty} \frac{1}{j^p}$$

converges; therefore, the series in question converge by comparison. We further observe that

$$\sum_{j=n+1}^{\infty} \frac{1}{j^p} \leq \int_n^\infty \frac{1}{x^p}\,\mathrm{d}x = \left. \frac{1}{1-p} x^{1-p} \right|_n^\infty = \frac{C}{n^{p-1}}. \qquad \square$$

For Fourier–Legendre expansions, we have the following result.

[6] Although originally discovered by the British mathematician Henry Wilbraham (1825–1883), this is named in honor of the American scientist who made significant theoretical contributions to physics, chemistry, and mathematics, Josiah Willard Gibbs (1839–1903). For a fascinating historical account behind the controversy surrounding this phenomenon, we refer the interested reader to [42].

Theorem 11.20 (uniform convergence). *Suppose that $\Psi = \{\psi_j\}_{j=0}^{\infty}$ is an orthonormal polynomial system with respect to the weight function $w \equiv 1$ on the compact interval $[-1,1] \subset \mathbb{R}$. Assume that $f \in C^2([-1,1])$ with $f(-1) = f(1) = 0$ and $f'(-1) = f'(1) = 0$. Suppose that, for every $n \in \mathbb{N}_0$, $p_n \in \mathbb{P}_n$ is the least squares polynomial approximation of f on $[-1,1]$. Then*

$$\lim_{n\to\infty} \|f - p_n\|_{L^\infty(-1,1)} = 0.$$

In other words, p_n converges uniformly to f on $[-1,1]$. More precisely, for $n \geq 1$,

$$\|f - p_n\|_{L^\infty(-1,1)} \leq \frac{\sqrt{6}C_1}{\sqrt{n}}, \qquad |c_n| \leq \frac{C_1}{n^2},$$

where $c_n = (f, \psi_n)_{L^2(-1,1)}$ is the nth generalized Fourier coefficient of f for some constant $C_1 > 0$ that is independent of n.

Proof. In consideration of (11.5), the appropriate orthonormal system $\Psi = \{\psi_j\}_{j=0}^{\infty}$ is obtained by setting

$$\psi_j = \sqrt{\frac{2j+1}{2}} P_j,$$

in which case we are guaranteed that $(\psi_i, \psi_j)_{L^2(-1,1)} = \delta_{i,j}$. The least squares polynomial approximation of degree n is

$$p_n(x) = \sum_{j=0}^{n} c_j \psi_j(x),$$

where

$$c_j = (f, \psi_j)_{L^2(-1,1)}.$$

Now we define, for $j \in \mathbb{N}_0$,

$$c_j^{(1)} = (f', \psi_j)_{L^2(-1,1)}, \qquad c_j^{(2)} = (f'', \psi_j)_{L^2(-1,1)}.$$

Using integration by parts, and the fact that $f(-1) = f(1) = 0$, we find

$$c_{j\pm1}^{(1)} = -(f, \psi_{j\pm1}')_{L^2(-1,1)}.$$

Using the derivative recurrence relation obtained in Theorem 11.18, it follows that, for $j \in \mathbb{N}$,

$$\frac{1}{\sqrt{2j+3}}\psi_{j+1}'(x) - \frac{1}{\sqrt{2j-1}}\psi_{j-1}'(x) = \frac{2j+1}{\sqrt{2j+1}}\psi_j(x).$$

From this, we deduce that

$$-\frac{1}{\sqrt{2j+3}}c_{j+1}^{(1)} + \frac{1}{\sqrt{2j-1}}c_{j-1}^{(1)} = \frac{2j+1}{\sqrt{2j+1}}c_j$$

for $j \in \mathbb{N}$. Rearranging terms, we get

$$jc_j = -A_j c_{j+1}^{(1)} + B_j c_{j-1}^{(1)}, \tag{11.6}$$

where, for $j \in \mathbb{N}$,

$$A_j = \frac{j}{2j+1}\sqrt{\frac{2j+1}{2j+3}}, \qquad B_j = \frac{j}{2j+1}\sqrt{\frac{2j+1}{2j-1}}.$$

Now applying Theorem 11.13 for f', which is a continuous function on $[-1, 1]$, it follows that $c_j^{(1)} \to 0$, as $j \to \infty$. Since $A_j, B_j \to \frac{1}{2}$, it must be that

$$\lim_{j\to\infty} jc_j = 0.$$

We can apply the same arguments above to the derivative function, f', concluding that

$$jc_j^{(1)} = -A_j c_{j+1}^{(2)} + B_j c_{j-1}^{(2)} \tag{11.7}$$

and, since $c_j^{(2)} \to 0$, as $j \to \infty$, that

$$\lim_{j\to\infty} jc_j^{(1)} = 0.$$

But, appealing to (11.6), we see that

$$\lim_{j\to\infty} j^2 c_j = 0.$$

Since $\{j^2 c_j\}_{j=0}^{\infty}$ converges to 0, it must be bounded. Thus, there is a constant $C_1 > 0$ independent of n such that, for all $j \in \mathbb{N}$,

$$|c_j| \le \frac{C_1}{j^2}.$$

Next, we observe that, if $1 \le m < n$,

$$\begin{aligned}
\|p_m - p_n\|_{L^\infty(-1,1)} &= \left\| \sum_{j=m+1}^{n} c_j \psi_j \right\|_{L^\infty(-1,1)} \\
&\le C_1 \left\| \sum_{j=m+1}^{n} \frac{\psi_j}{j^2} \right\|_{L^\infty(-1,1)} \\
&\le C_1 \sum_{j=m+1}^{n} \frac{1}{j^2} \|\psi_j\|_{L^\infty(-1,1)} \\
&= C_1 \sum_{j=m+1}^{n} \frac{1}{j^2} \sqrt{\frac{2j+1}{2}} \|P_j\|_{L^\infty(-1,1)} \\
&\le C_1 \sum_{j=m+1}^{n} \frac{1}{j^2} \sqrt{\frac{2j+1}{2}} \\
&\le C_1 \sum_{j=m+1}^{\infty} \frac{1}{j^2} \sqrt{\frac{2j+1}{2}} \\
&\le \sqrt{\frac{3}{2}} C_1 \sum_{j=m+1}^{\infty} \frac{1}{j^{3/2}} \\
&\le 2\sqrt{\frac{3}{2}} \frac{C_1}{\sqrt{m}},
\end{aligned}$$

where we have used the fact that $\|P_j\|_{L^\infty(-1,1)} \le 1$, from Theorem 11.18, and Lemma 11.19 with $p = \frac{3}{2}$. Thus, if $1 \le m < n$,

$$\|p_m - p_n\|_{L^\infty(-1,1)} \le \frac{\sqrt{6}C_1}{\sqrt{m}}.$$

From this, we can conclude that $\{p_n\}_{n=0}^{\infty}$ is uniformly Cauchy. From Theorems B.61 and B.62, there is a continuous function $g \in C([-1,1])$ such that $p_n \to g$ uniformly on $[-1,1]$. By Theorems B.65 and 11.13,

$$0 = \lim_{n\to\infty} \int_{-1}^{1} |f(x) - p_n(x)|^2 \, dx = \int_{-1}^{1} |f(x) - g(x)|^2 \, dx.$$

This implies that $f \equiv g$. In other words, given any $\varepsilon > 0$, there is a number $N \in \mathbb{N}$ such that if $n > N$, then

$$\|f - p_n\|_{L^\infty(-1,1)} < \varepsilon.$$

Finally, if $1 \le m < n$ and $n > N$, then

$$\|f - p_m\|_{L^\infty(-1,1)} \le \|f - p_n\|_{L^\infty(-1,1)} + \|p_n - p_m\|_{L^\infty(-1,1)} < \varepsilon + \frac{\sqrt{6}C_1}{\sqrt{m}}.$$

Since ε can be made arbitrarily small,

$$\|f - p_m\|_{L^\infty(-1,1)} \le \frac{\sqrt{6}C_1}{\sqrt{m}},$$

provided that $m \ge 1$. □

Next, we remove the restrictions at the endpoints on f and its derivatives.

Corollary 11.21 (uniform convergence). *Suppose that $\Psi = \{\psi_j\}_{j=0}^\infty$ is an orthonormal polynomial system with respect to the weight function $w \equiv 1$ on the compact interval $[-1,1] \subset \mathbb{R}$. Assume that $f \in C^2([-1,1])$ and $p_n \in \mathbb{P}_n$, $n \in \mathbb{N}_0$, is the least squares polynomial approximation of f on $[-1,1]$. Then*

$$\lim_{n\to\infty} \|f - p_n\|_{L_\infty(-1,1)} = 0.$$

More precisely, if $n \ge 3$, then

$$\|f - p_n\|_{L^\infty(-1,1)} \le \frac{\sqrt{6}C_1}{\sqrt{n}}, \qquad |c_n| \le \frac{C_1}{n^2},$$

where $c_n = (f, \psi_n)_{L^2(-1,1)}$ is the nth generalized Fourier coefficient of f for some constant $C_1 > 0$ that is independent of n.

Proof. Let $g \in \mathbb{P}_3$ be the Hermite interpolating polynomial satisfying $g(\pm1) = f(\pm1)$, $g'(\pm1) = f'(\pm1)$. Then $f - g$ satisfies the hypotheses of Theorem 11.20. Thus, if $n \ge 3$, then

$$\|f - g - \mathcal{P}_{w,n}[f-g]\|_{L^\infty(-1,1)} \le \frac{\sqrt{6}C_1}{\sqrt{n}}.$$

But, since $\mathcal{P}_{w,n}$ is a linear projection and $n \ge 3$,

$$\mathcal{P}_{w,n}[f-g] = \mathcal{P}_{w,n}[f] - \mathcal{P}_{w,n}[g] = \mathcal{P}_{w,n}[f] - g.$$

Thus,

$$\|f - \mathcal{P}_{w,n}[f]\|_{L^\infty(-1,1)} = \|f - g - \mathcal{P}_{w,n}[f-g]\|_{L^\infty(-1,1)} \le \frac{\sqrt{6}C_1}{\sqrt{n}}.$$

Regarding the coefficients, if $n \ge 3$,

$$(f - g, \psi_n)_{L^2(-1,1)} = (f, \psi_n)_{L^2(-1,1)}$$

and the result follows. □

A similar result can be proven for Fourier–Chebyshev expansions, i.e., least squares approximations involving Chebyshev polynomials. Interestingly, we are able to prove a better rate of convergence for Fourier–Chebyshev expansions.

Theorem 11.22 (uniform convergence). *Suppose that $\Psi = \{\psi_j\}_{j=0}^\infty$ is an orthonormal polynomial system with respect to the weight function $w(x) = \frac{1}{\sqrt{1-x^2}}$ on the interval $[-1,1]$. Assume that $f \in C^2([-1,1])$ and $p_n \in \mathbb{P}_n$, $n \in \mathbb{N}_0$, is the least squares polynomial approximation of f on $[-1,1]$. Then*

$$\lim_{n\to\infty} \|f - p_n\|_{L^\infty(-1,1)} = 0.$$

More precisely, if $n > 1$, then

$$\|f - p_n\|_{L^\infty(-1,1)} \le \frac{C_2}{n}, \qquad |c_n| \le \frac{C_2}{n^2},$$

for some constant $C_2 > 0$ that is independent of n.

Proof. Recall that the Chebyshev polynomials $\{T_j\}_{j=0}^\infty$ were introduced in Definition 10.13. They have the normalization

$$(T_k, T_m)_{L^2_w(-1,1)} = \int_{-1}^1 T_k(x)T_m(x)\frac{dx}{\sqrt{1-x^2}} = \begin{cases} 0, & k \neq m, \\ \pi, & k = m = 0, \\ \frac{\pi}{2}, & k = m > 0, \end{cases}$$

where $w(x) = \frac{1}{\sqrt{1-x^2}}$, $x \in (-1,1)$. Thus, we set $\psi_0 = \frac{1}{\sqrt{\pi}}T_0$ and $\psi_j = \sqrt{\frac{2}{\pi}}T_j$, $j = 1, 2, \dots$ to obtain the appropriate orthonormal system. Therefore, $p_n = \sum_{j=0}^n c_j\psi_j$, where, for $j \in \mathbb{N}$,

$$c_j = \int_{-1}^1 f(x)\psi_j(x)\frac{dx}{\sqrt{1-x^2}} = \sqrt{\frac{2}{\pi}}\int_{-1}^1 f(x)T_j(x)\frac{dx}{\sqrt{1-x^2}}.$$

Making the change of variable, $x = \cos(\theta)$, for $0 \le \theta \le \pi$, we find

$$c_j = \sqrt{\frac{2}{\pi}}\int_0^\pi g(\theta)\cos(j\theta)d\theta,$$

where $g(\theta) = f(\cos(\theta))$. Now, integrating by parts, we obtain, for $j \in \mathbb{N}$,

$$\begin{aligned}
\sqrt{\frac{\pi}{2}}c_j &= g(\theta)\frac{1}{j}\sin(j\theta)\bigg|_0^\pi - \int_0^\pi g'(\theta)\frac{1}{j}\sin(j\theta)d\theta \\
&= -\int_0^\pi g'(\theta)\frac{1}{j}\sin(j\theta)d\theta \\
&= -\,g'(\theta)\frac{1}{j^2}\cos(j\theta)\bigg|_0^\pi + \int_0^\pi g''(\theta)\frac{1}{j^2}\cos(j\theta)d\theta \\
&= -\,f'(\cos(\theta))\sin(\theta)\frac{1}{j^2}\cos(j\theta)\bigg|_0^\pi + \int_0^\pi g''(\theta)\frac{1}{j^2}\cos(j\theta)d\theta \\
&= \int_0^\pi g''(\theta)\frac{1}{j^2}\cos(j\theta)d\theta.
\end{aligned}$$

Since $f \in C^2([-1,1])$, it follows that

$$|c_j| \le \sqrt{\frac{\pi}{2}}\frac{1}{j^2}\int_0^\pi |g''(\theta)|\,|\cos(j\theta)|\,d\theta \le \sqrt{\frac{\pi}{2}}\frac{1}{j^2}\int_0^\pi M d\theta = \sqrt{\frac{\pi}{2}}\frac{M\pi}{j^2},$$

where

$$M = \max_{0 \le \theta \le \pi}|g''(\theta)|.$$

Recall that, by Theorem 10.17, $\|T_j\|_{L^\infty(-1,1)} \le 1$. By the Weierstrass M-Test, Theorem B.63, since

$$|c_j\psi_j| \le \sqrt{\frac{\pi}{2}}\frac{M\pi}{j^2}|\psi_j| = \sqrt{\frac{\pi}{2}}\frac{M\pi}{j^2}\sqrt{\frac{2}{\pi}}|T_j| \le \frac{M\pi}{j^2}$$

and $\sum_{j=0}^{\infty}\frac{1}{j^2}$ converges absolutely, it follows that the series

$$\sum_{j=1}^{\infty} c_j\psi_j$$

converges uniformly and absolutely to $g \in C([-1,1])$. In other words, $p_n \to g$ uniformly on $[-1,1]$. To see that $f \equiv g$ on $[-1,1]$, observe that, by Theorems B.65 and 11.13,

$$0 = \lim_{n\to\infty}\int_{-1}^{1}|f(x) - p_n(x)|^2\frac{dx}{\sqrt{1-x^2}} = \int_{-1}^{1}|f(x) - g(x)|^2\frac{dx}{\sqrt{1-x^2}}.$$

This implies that $\|f - g\|_{L^2_w(-1,1)} = 0$, which implies the first part of the result. We leave the details of the second part to the reader as an exercise; see Problem 11.5. □

Problems

11.1 Prove Proposition 11.8.

11.2 Prove Theorem 11.12.

11.3 Prove Theorem 11.15.

11.4 Prove Theorem 11.18.

11.5 Complete the proof of Theorem 11.22.

11.6 Suppose that $c_j \in \mathbb{C}$, $j = 0, \ldots, n$ are given. Find a polynomial $q \in \mathbb{P}_n(\mathbb{C})$ such that

$$\int_a^b x^j q(x) = c_j, \quad j = 0, \ldots, n.$$

11.7 Suppose that $\Psi = \{\psi_j\}_{j=0}^m$ is an orthonormal polynomial system with respect to the weight function w on the compact interval $[a,b] \subset \mathbb{R}$ and $q \in \mathbb{P}_m$ satisfies

$$(q, \psi_j)_{L^2_w(a,b)} = 0, \quad j = 0, \ldots, m.$$

Prove that $q \equiv 0$.

12 Fourier Series

This chapter is a close companion to Chapter 11. Here, we will discuss least squares best approximation by trigonometric polynomials. Many topics from our discussion in the last chapter will reappear, but in a slightly different form. The big changes are that functions of interest in this chapter will be assumed periodic, and the orthonormal systems will be trigonometric functions. The topics covered here come under the classical subject of Fourier Analysis, which many students have encountered in their PDE or Physics courses when studying separation of variables. Now we want to make some details of that discussion rigorous. However, as numerical analysts, we have a slightly different perspective than, say, a physicist or a PDE analyst. In particular, we want to rigorously quantify the error in approximation of a function of interest by trigonometric functions, and this will be our ultimate goal in the chapter.

The subject was founded by Jean-Baptiste Joseph Fourier, who discovered what we would recognize as the basics of Fourier Analysis in his studies of heat flow in the 1820s. At its heart, the theory indicates that almost every signal is a linear combination of Fourier modes, i.e., the sum of trigonometric functions. Take, for example, the one-periodic square wave, f, as shown in Figure 12.1. We show approximations of the form

$$\mathcal{S}_n[f](x) = \sum_{j=1}^{n} c_j \sin(2\pi x j)$$

using $n = 15$, 17, and 19, where the coefficients, c_j, are chosen according to a rule that we will shortly describe. Notice that the square wave is qualitatively well approximated by our sums of sine waves. But there is a glaring problem. Near the discontinuities of the square wave, the approximations overshoot the function by a considerable amount. Will this overshoot diminish as n becomes larger? Of course, one would hope that this is the case. How are these Fourier approximations computed theoretically? How are they computed practically? When can we expect that Fourier approximations converge uniformly, as $n \to \infty$? What are the rates of convergence? Many of these and other questions will be answered in this chapter.

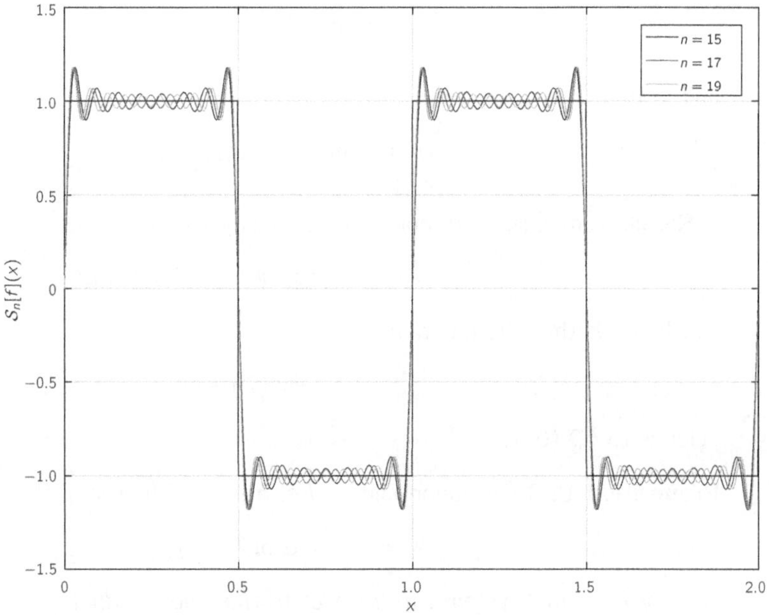

Figure 12.1 Fourier approximations of a square wave.

12.1 Least Squares Trigonometric Approximations

First, we want to precisely define the objects we will use to approximate periodic functions. The reader is referred to Appendix D for an overview of periodic functions and spaces of periodic functions. To simplify the description of general one-periodic signals, we use the family of complex exponential functions.

Definition 12.1 (trigonometric polynomial). Let $n \in \mathbb{N}_0$. The function

$$p_n(x) = \sum_{j=-n}^{n} c_j \exp(2\pi \mathrm{i} j x), \tag{12.1}$$

where $c_j \in \mathbb{C}$, $j = -n, \dots, n$ is called a **trigonometric polynomial** of degree at most n. The set of all such polynomials is labeled $\mathbb{T}_n(0,1)$, or just $\mathbb{T}_n$ when the period is understood. Under the usual function operations, this is a vector space over the field of complex numbers.

Definition 12.2 (least squares approximation). Let $f \in L^2_{\mathrm{p}}(0,1;\mathbb{C})$ and $n \in \mathbb{N}_0$. The trigonometric polynomial $p \in \mathbb{T}_n$, if it exists, is called a **least squares trigonometric polynomial approximation** of f of order n if and only if

$$\|f - p\|_{L^2_{\mathrm{p}}(0,1;\mathbb{C})} = \inf_{q \in \mathbb{T}_n} \|f - q\|_{L^2_{\mathrm{p}}(0,1;\mathbb{C})} .$$

As usual, if such a $p \in \mathbb{T}_n$ exists, we write

$$p \in \underset{q \in \mathbb{T}_n}{\operatorname{argmin}} \|f - q\|_{L^2_{\mathrm{p}}(0,1;\mathbb{C})},$$

and, if it is unique, we write

$$p = \underset{q \in \mathbb{T}_n}{\operatorname{argmin}} \|f - q\|_{L^2_{\mathrm{p}}(0,1;\mathbb{C})}.$$

Recall, from Euler's formula,[1] that, for any $x \in \mathbb{R}$,

$$\exp(2\pi i k x) = \cos(2\pi k x) + i \sin(2\pi k x),$$

which shows that the function

$$x \mapsto \exp(2\pi i k x)$$

belongs to $C^\infty_{\mathrm{p}}(0, 1; \mathbb{C})$.

Proposition 12.3 (orthonormality). *Let $n \in \mathbb{N}_0$. The set*

$$\Psi = \{\psi_j = \exp(2\pi i j \cdot)\}_{j=-n}^{n}$$

is an orthonormal system with respect to the inner product $(\cdot, \cdot)_{L^2(0,1;\mathbb{C})}$, i.e.,

$$(\psi_k, \psi_\ell)_{L^2(0,1;\mathbb{C})} = \delta_{k,\ell}, \quad -n \leq k, \ell \leq n.$$

Proof. Evaluate

$$(\psi_k, \psi_\ell)_{L^2(0,1;\mathbb{C})} = \int_0^1 \exp(2\pi i k x)\overline{\exp(2\pi i \ell x)} \mathrm{d}x.$$

The simple details are left to the reader as an exercise; see Problem 12.1 □

Using this orthonormal system, which we will call *standard*, it is rather easy to establish the existence and uniqueness of least squares trigonometric polynomial approximations.

Theorem 12.4 (existence and uniqueness). *Let $n \in \mathbb{N}_0$ and $f \in L^2_{\mathrm{p}}(0, 1; \mathbb{C})$. There exists a unique least squares trigonometric polynomial approximation of f. If $\Psi = \{\psi_j = \exp(2\pi i j \cdot)\}_{j=-n}^{n}$ is the standard orthonormal trigonometric polynomial system on $[0, 1]$, then the least squares approximation is given by*

$$p = \sum_{j=-n}^{n} \hat{f}_j \psi_j,$$

where $\hat{f}_j$ is the jth Fourier coefficient *of f, defined as*

$$\hat{f}_j = (f, \psi_j)_{L^2(0,1;\mathbb{C})} = \int_0^1 f(x) \exp(-2\pi i j x) \mathrm{d}x. \tag{12.2}$$

[1] Named in honor of the Swiss mathematician, physicist, astronomer, geographer, logician, and engineer Leonhard Euler (1707–1783).

Furthermore, Bessel's inequality,[2]

$$\sum_{j=-n}^{n} \left|\hat{f}_j\right|^2 \leq \|f\|^2_{L^2(0,1;\mathbb{C})}, \tag{12.3}$$

holds for any $n \in \mathbb{N}_0$. Since $\|f\|^2_{L^2(0,1;\mathbb{C})} < \infty$, the series

$$\sum_{j=-\infty}^{\infty} \left|\hat{f}_j\right|^2 = \lim_{n\to\infty} \sum_{j=-n}^{n} \left|\hat{f}_j\right|^2$$

converges absolutely.

Proof. The existence and uniqueness proof is identical to the proof of Theorem 11.4. In going through the details, we learn that

$$0 \leq \|f - p\|^2_{L^2(0,1;\mathbb{C})} = \|f\|^2_{L^2(0,1;\mathbb{C})} - \sum_{j=-n}^{n} \left|\hat{f}_j\right|^2, \tag{12.4}$$

where p is the least squares approximation. Bessel's inequality follows easily from this. We leave the rest of the details to the reader as an exercise; see Problem 12.2. □

By following the techniques in Chapter 11, the following results can be easily obtained.

Theorem 12.5 (characterization)**.** *Let $f \in L^2_{\mathrm{p}}(0,1;\mathbb{C})$. The trigonometric polynomial $p \in \mathbb{T}_n$ is the least squares approximation to f if and only if*

$$(f - p, q)_{L^2(0,1;\mathbb{C})} = 0, \quad \forall q \in \mathbb{T}_n. \tag{12.5}$$

Proof. See Problem 12.3. □

Definition 12.6 (Fourier projection)**.** Let $n \in \mathbb{N}_0$ and $\Psi = \{\psi_j = \exp(2\pi \mathrm{i} j\,\cdot)\}_{j=-n}^{n}$ be the standard orthonormal trigonometric polynomial system on $[0,1]$. The mapping

$$\mathcal{S}_n \colon L^2_{\mathrm{p}}(0,1;\mathbb{C}) \to \mathbb{T}_n,$$
$$f \mapsto \mathcal{S}_n[f] = \sum_{j=-n}^{n} \hat{f}_j \psi_j, \tag{12.6}$$

where $\hat{f}_j = (f, \psi_j)_{L^2(0,1;\mathbb{C})}$ is the jth Fourier coefficient of f, is called the **Fourier projection** of degree n.

Proposition 12.7 (projection)**.** *For every $n \in \mathbb{N}_0$, the Fourier projection is a linear projection operator from $L^2_{\mathrm{p}}(0,1;\mathbb{C})$ onto $\mathbb{T}_n$, i.e.,*

$$\mathcal{S}_n[\alpha f + \beta g] = \alpha \mathcal{S}_n[f] + \beta \mathcal{S}_n[g], \quad \forall f, g \in L^2_{\mathrm{p}}(0,1;\mathbb{C}), \quad \forall \alpha, \beta \in \mathbb{C},$$

and

$$\mathcal{S}_n[p] = p, \quad \forall p \in \mathbb{T}_n.$$

[2] Named in honor of the German astronomer, mathematician, physicist, and geodesist Friedrich Wilhelm Bessel (1784–1846).

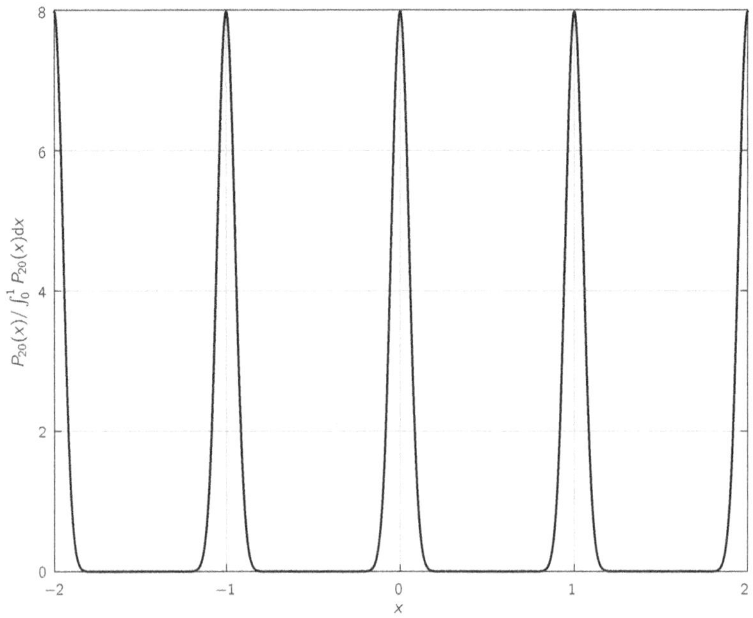

Figure 12.2 An approximation of the periodic Dirac delta function, the so-called *Dirac comb*, via (12.8) with $n = 20$.

Proof. See Problem 12.4. □

Remark 12.8 (orthogonality). Theorem 12.5 guarantees that the error between f and its Fourier projection, $\mathcal{S}_n[f]$, is orthogonal to the subspace $\mathbb{T}_n$, i.e.,

$$(f - \mathcal{S}_n[f], q)_{L^2(0,1;\mathbb{C})} = 0, \quad \forall q \in \mathbb{T}_n. \tag{12.7}$$

12.2 Density of Trigonometric Polynomials in the Space $C_{\mathrm{p}}(0, 1; \mathbb{C})$

Next, we want to prove the density of the trigonometric polynomials in the space $C_{\mathrm{p}}(0, 1; \mathbb{C})$ with respect to the norm $\|\cdot\|_{L^\infty(0,1;\mathbb{C})}$. This result is a sort of periodic analogue to the Weierstrass Approximation Theorem 10.32. To prove our result, we need a trigonometric approximation of the Dirac delta function; in fact, a periodic Dirac delta function, which is sometimes called a *Dirac comb*.

Lemma 12.9 (Dirac comb[3]). *For each $n \in \mathbb{N}_0$ and $x \in \mathbb{R}$, define*

$$Q_n(x) = \frac{P_n(x)}{\int_0^1 P_n(x)\mathrm{d}x}, \qquad P_n(x) = (1 + \cos(2\pi x))^n. \tag{12.8}$$

[3] Named in honor of the British theoretical physicist Paul Adrien Maurice Dirac (1902–1984).

Then $Q_n \in \mathbb{T}_n$*;* $Q_n(x) \geq 0$ *for all* $x \in \mathbb{R}$*;* $\int_0^1 Q_n(x)\mathrm{d}x = 1$*; and*

$$\lim_{n\to\infty} \int_\delta^{1-\delta} Q_n(x)\mathrm{d}x = 0 \tag{12.9}$$

for every $\delta \in (0, \frac{1}{2})$.

Proof. The proof is left to the reader as an exercise; see Problem 12.5. It may help to plot Q_n for a few values of n to get an idea of what one is trying to prove; see, for example, Figure 12.2. □

Definition 12.10 (periodic convolution)**.** Let $f, g \in C_p(0,1;\mathbb{C})$. The **periodic convolution** $[f \star g]_p$ is defined as the function

$$[f \star g]_p(x) = \int_0^1 f(x-y)g(y)\mathrm{d}y, \quad \forall x \in \mathbb{R}.$$

Lemma 12.11 (properties of convolution)**.** *Suppose that* $f, g \in C_p(0,1;\mathbb{C})$*. Then the periodic convolution* $[f \star g]_p$ *is well defined, linear in both entries, commutative, i.e.,*

$$[f \star g]_p(x) = [g \star f]_p(x), \quad \forall x \in \mathbb{R},$$

and an element of $C_p(0,1;\mathbb{C})$*. Furthermore, if* $g \in \mathbb{T}_n$ *for some* $n \in \mathbb{N}_0$*, then* $[f \star g]_p \in \mathbb{T}_n$*.*

Proof. Fix $x \in \mathbb{R}$. The function $C_x(y) = f(x-y)g(y)$ is continuous and one-periodic. Indeed,

$$C_x(y+m) = f(x-y-m)g(y+m) = f(x-y)g(y) = C_x(y), \quad \forall y \in \mathbb{R}, \quad \forall m \in \mathbb{Z}.$$

Therefore,

$$I(x) = \int_0^1 C_x(y)\mathrm{d}y = \int_a^{a+1} C_x(y)\mathrm{d}y, \quad \forall a \in \mathbb{R}.$$

The function I is certainly well defined. With the change of variable $t = x - y$ for integration, we have

$$\begin{aligned}
[f \star g]_p(x) = I(x) &= \int_0^1 f(x-y)g(y)\mathrm{d}y \\
&= -\int_x^{x-1} f(t)g(x-t)\mathrm{d}t \\
&= \int_{x-1}^{x} f(t)g(x-t)\mathrm{d}t \\
&= \int_0^1 f(t)g(x-t)\mathrm{d}t \\
&= [g \star f]_p(x).
\end{aligned}$$

Thus, we have shown that periodic convolution is commutative. Linearity is a straightforward property to show, and we skip the proof.

Next, we want to show that I is periodic. Let $\ell \in \mathbb{Z}$ be arbitrary. Then

$$\begin{aligned} I(x+\ell) &= \int_0^1 C_{x+\ell}(y)\mathrm{d}y \\ &= \int_0^1 f(x+\ell-y)g(y)\mathrm{d}y \\ &= \int_0^1 f(x-y)g(y)\mathrm{d}y \\ &= \int_0^1 C_x(y)\mathrm{d}y \\ &= I(x). \end{aligned}$$

To see that I is continuous, consider, for any $x_1, x_2 \in \mathbb{R}$,

$$\begin{aligned} |I(x_1) - I(x_2)| &= \left| \int_0^1 [f(x_1 - y) - f(x_2 - y)]g(y)\mathrm{d}y \right| \\ &\le \int_0^1 |f(x_1 - y) - f(x_2 - y)| \cdot |g(y)|\,\mathrm{d}y \\ &\le \max_{0 \le y \le 1} |f(x_1 - y) - f(x_2 - y)| \int_0^1 |g(y)|\,\mathrm{d}y. \end{aligned}$$

Let $\varepsilon > 0$ be given. Then, since f is continuous and one-periodic, it is uniformly continuous on any compact subset of $\mathbb{R}$. Therefore, there is a $\delta > 0$ such that if

$$|x_1 - x_2| = |(x_1 - y) - (x_2 - y)| < \delta,$$

then, uniformly with respect to y,

$$|f(x_1 - y) - f(x_2 - y)| < \varepsilon.$$

Hence,

$$|I(x_1) - I(x_2)| \le \varepsilon \int_0^1 |g(y)|\,\mathrm{d}y,$$

which proves that I is uniformly continuous. Thus, $I = [f \star g]_\mathrm{p} \in C_\mathrm{p}(0, 1; \mathbb{C})$.

Finally, suppose that $g \in \mathbb{T}_n$. By linearity, it suffices to prove our result with $g(x) = \exp(2\pi \mathrm{i} kx)$ for some $k \in \mathbb{Z}$, $|k| \le n$. Thus,

$$\begin{aligned} [f \star g]_\mathrm{p}(x) = [g \star f]_\mathrm{p}(x) &= \int_0^1 g(x-y)f(y)\mathrm{d}y \\ &= \exp(2\pi \mathrm{i} kx) \int_0^1 \exp(-2\pi \mathrm{i} ky)f(y)\mathrm{d}y \\ &= \exp(2\pi \mathrm{i} kx)\hat{f}_k, \end{aligned}$$

which proves that $[f \star g]_\mathrm{p} \in \mathbb{T}_n$, whenever $g \in \mathbb{T}_n$. □

We now come to the main result of the section, an analogue of the Weierstrass Approximation Theorem 10.32.

Theorem 12.12 (density). *The trigonometric polynomials of period one are dense in $C_p(0,1;\mathbb{C})$ in the following sense: given any $f \in C_p(0,1;\mathbb{C})$ and any $\varepsilon > 0$, there is an $n \in \mathbb{N}_0$ and a trigonometric polynomial $f_n \in \mathbb{T}_n$ such that*

$$\|f - f_n\|_{L^\infty(0,1;\mathbb{C})} \le \varepsilon.$$

Proof. For $n \in \mathbb{N}_0$, define $f_n = [f \star Q_n]_p$, where Q_n is defined in (12.8). From Lemma 12.11, we know that $f_n \in \mathbb{T}_n$. Fix $x \in \mathbb{R}$. Computing the difference $f - f_n$, we find

$$\begin{aligned}
|f(x) - f_n(x)| &= \left| \int_{-\frac{1}{2}}^{\frac{1}{2}} (f(x) - f(x-y))\, Q_n(y) dy \right| \\
&\le \int_{-\frac{1}{2}}^{\frac{1}{2}} |f(x) - f(x-y)|\, Q_n(y) dy \\
&\le \int_{-\frac{1}{2}}^{\frac{1}{2}} |f(x) - f(x-y)|\, Q_n(y) dy \\
&= I_1(x) + I_2(x),
\end{aligned}$$

where

$$\begin{aligned}
I_1(x) &= \int_{-\delta}^{\delta} |f(x) - f(x-y)|\, Q_n(y) dy, \\
I_2(x) &= \int_{\delta}^{1-\delta} |f(x) - f(x-y)|\, Q_n(y) dy.
\end{aligned}$$

Let $\varepsilon > 0$ be given. Since f is uniformly continuous, we can choose $\delta > 0$ small enough, so that if $-\delta \le y \le \delta$, then

$$|f(x) - f(x-y)| \le \frac{\varepsilon}{2}$$

uniformly with respect to x. Thus,

$$I_1(x) = \int_{-\delta}^{\delta} |f(x) - f(x-y)|\, Q_n(y) dy \le \int_{-\delta}^{\delta} \frac{\varepsilon}{2} Q_n(y) dy \le \frac{\varepsilon}{2}.$$

We now consider $\delta > 0$ to be fixed. Turning to I_2, let us assume, using (12.9), that n is sufficiently large, so that

$$\int_{\delta}^{1-\delta} Q_n(y) dy \le \frac{\varepsilon}{4 \|f\|_{L^\infty(0,1;\mathbb{C})}}.$$

If this is the case,

$$\begin{aligned} I_2(x) &= \int_\delta^{1-\delta} |f(x) - f(x-y)|\, Q_n(y)\mathrm{d}y \\ &\le 2\,\|f\|_{L^\infty(0,1;\mathbb{C})} \int_\delta^{1-\delta} Q_n(y)\mathrm{d}y \\ &\le 2\,\|f\|_{L^\infty(0,1;\mathbb{C})} \frac{\varepsilon}{4\,\|f\|_{L^\infty(0,1;\mathbb{C})}} \\ &= \frac{\varepsilon}{2}. \end{aligned}$$

Putting both of our estimates together, we find, for every $x \in [0,1]$,

$$|f(x) - f_n(x)| \le I_1(x) + I_2(x) \le \varepsilon,$$

provided that $n \in \mathbb{N}_0$ is sufficiently large. Since $\varepsilon > 0$ is arbitrary and independent of x, we have the desired result. □

12.3 Convergence of Fourier Series in the Quadratic Mean

We now focus on the trigonometric analogue of Section 11.5. In other words, we will show convergence of the least squares trigonometric polynomial.

Definition 12.13 (Fourier series). Suppose that $f \in L^2_{\mathrm{p}}(0,1;\mathbb{C})$. Let

$$\Psi = \{\psi_j = \exp(2\pi \mathrm{i} j\,\cdot)\}_{j=-\infty}^{\infty}$$

be the standard orthonormal trigonometric polynomial system on $[0,1]$ and $\hat{f}_j$ denote the jth Fourier coefficient of f, i.e., $\hat{f}_j = (f, \psi_j)_{L^2(0,1;\mathbb{C})}$. The **Fourier series** of f is defined as

$$\sum_{j=-\infty}^{\infty} \hat{f}_j \psi_j = \lim_{n\to\infty} \mathcal{S}_n[f],$$

where $\mathcal{S}_n$ is the Fourier projection defined in (12.6). We say that $\mathcal{S}_n[f]$ **converges to f in the quadratic mean** if and only if

$$\lim_{n\to\infty} \|f - \mathcal{S}_n[f]\|_{L^2(0,1;\mathbb{C})} = 0$$

and we write $\mathcal{S}_n[f] \to f$ in $L^2_{\mathrm{p}}(0,1;\mathbb{C})$.

Remark 12.14 (convergence). We are justified in writing

$$f = \sum_{j=-\infty}^{\infty} \hat{f}_j \psi_j$$

whenever the series on the right-hand side converges to f in the quadratic mean. However, as we saw in Chapter 11, this may not mean the series converges to f point-wise on $[0,1]$.

Theorem 12.15 (convergence). *Suppose that $f \in C_{\rm p}(0,1;\mathbb{C})$. Then $\mathcal{S}_n[f]$ converges to f in the quadratic mean, i.e.,*

$$\lim_{n\to\infty} \|f - \mathcal{S}_n[f]\|_{L^2(0,1;\mathbb{C})} = 0.$$

Furthermore, Parseval's relation,[4]

$$\int_0^1 |f(x)|^2 \,dx = \sum_{j=-\infty}^{\infty} |\hat{f}_j|^2 = \lim_{n\to\infty} \sum_{j=-n}^{n} |\hat{f}_j|^2,$$

holds.

Proof. For any $n \in \mathbb{N}_0$, set $f_n = [f \star Q_n]_{\rm p} \in \mathbb{T}_n$. We know that $\mathcal{S}_n[f]$ is the least squares trigonometric polynomial approximation for f. Therefore,

$$\|f - \mathcal{S}_n[f]\|^2_{L^2(0,1;\mathbb{C})} \le \|f - f_n\|^2_{L^2(0,1;\mathbb{C})}.$$

Since $f_n \to f$ uniformly,

$$\lim_{n\to\infty} \|f - f_n\|_{L^\infty(0,1;\mathbb{C})} = 0.$$

But recall, from inequality (D.1) (with $[a,b] = [0,1]$, $w \equiv 1$, and $p = 2$), that

$$\|f - f_n\|_{L^2(0,1;\mathbb{C})} \le \|f - f_n\|_{L^\infty(0,1;\mathbb{C})}.$$

This implies the convergence result. For Parseval's relation, we recall (12.4):

$$0 \le \|f - \mathcal{S}_n[f]\|^2_{L^2(0,1;\mathbb{C})} = \|f\|^2_{L^2(0,1;\mathbb{C})} - \sum_{j=-n}^{n} \left|\hat{f}_j\right|^2.$$

A little manipulation gives the result. □

The following result is a simple consequence of Theorem 11.14.

Theorem 12.16 (density). *Suppose that $f \in L^2_{\rm p}(0,1;\mathbb{C})$. For every $\varepsilon > 0$, there is a function $g \in C_{\rm p}(0,1;\mathbb{C})$ such that*

$$\|f - g\|_{L^2(0,1;\mathbb{C})} \le \varepsilon.$$

In other words, continuous one-periodic functions are dense in $L^2_{\rm p}(0,1;\mathbb{C})$.

From this, we easily get the following classical result by following the proof of Theorem 11.15.

Theorem 12.17 (Riesz–Fischer Theorem[5]). *Suppose that $f \in L^2_{\rm p}(0,1;\mathbb{C})$. Then $\mathcal{S}_n[f]$ converges to f in the quadratic mean, i.e.,*

$$\lim_{n\to\infty} \|f - \mathcal{S}_n[f]\|_{L^2(0,1;\mathbb{C})} = 0.$$

Conversely, let $\{a_j\}_{j\in\mathbb{Z}} \subset \mathbb{C}$. If the series $\sum_{j=-\infty}^{\infty} |a_j|^2$ converges, there is a function $f \in L^2_{\rm p}(0,1;\mathbb{C})$ such that $a_j = \hat{f}_j$ for all $j \in \mathbb{Z}$.

[4] Named in honor of the French mathematician Marc-Antoine Parseval des Chênes (1755–1836).

[5] Named in honor of the Hungarian mathematician Frigyes Riesz (1880–1956) and the Austrian mathematician Ernst Sigismund Fischer (1875–1954).

If $f \in L^2_{\mathrm{p}}(0,1;\mathbb{C})$, *Parseval's relation,*

$$\int_0^1 |f(x)|^2 \, dx = \sum_{j=-\infty}^{\infty} |\hat{f}_j|^2 = \lim_{n\to\infty} \sum_{j=-n}^{n} |\hat{f}_j|^2,$$

holds.

Proof. See Problem 12.6. □

The previous result established a unique correspondence between one-periodic square integrable functions and square summable sequences. Let us explore this in more detail. For that, we introduce the following classical sequence space.

Definition 12.18 ($\ell^2(\mathbb{Z};\mathbb{C})$)**.** We define

$$\ell^2(\mathbb{Z};\mathbb{C}) = \left\{ a \colon \mathbb{Z} \to \mathbb{C} \;\middle|\; \sum_{j=-\infty}^{\infty} |a_j|^2 < \infty \right\}.$$

This is a vector space. Moreover, this is a complex Hilbert space with inner product

$$(a, b)_{\ell^2(\mathbb{Z};\mathbb{C})} = \sum_{j=-\infty}^{\infty} a_j \overline{b_j}$$

and norm

$$\|a\|_{\ell^2(\mathbb{Z};\mathbb{C})} = \sqrt{(a,a)_{\ell^2(\mathbb{Z};\mathbb{C})}}.$$

Definition 12.19 (Fourier transform)**.** For $v \in L^2_{\mathrm{p}}(0,1;\mathbb{C})$ its **finite Fourier transform** is defined as

$$\mathcal{F}_{\mathrm{p}}[v] \in \ell^2(\mathbb{Z};\mathbb{C})$$

with

$$\mathcal{F}_{\mathrm{p}}[v]_j = \hat{v}_j = \int_0^1 f(x) \exp(-2\pi \mathrm{i} j x) dx.$$

Thus, $\mathcal{F}_{\mathrm{p}} \colon L^2_{\mathrm{p}}(0,1;\mathbb{C}) \to \ell^2(\mathbb{Z};\mathbb{C})$. The **inverse finite Fourier transform**, $\mathcal{F}_{\mathrm{p}}^{-1} \colon \ell^2(\mathbb{Z};\mathbb{C}) \to L^2_{\mathrm{p}}(0,1;\mathbb{C})$, is defined via

$$\mathcal{F}_{\mathrm{p}}^{-1}[c](x) = \sum_{j=-\infty}^{\infty} c_j \exp(2\pi \mathrm{i} j x).$$

Proposition 12.20 (isomorphism)**.** *The finite Fourier transform is an isometric isomorphism between* $L^2_{\mathrm{p}}(0,1;\mathbb{C})$ *and* $\ell^2(\mathbb{Z};\mathbb{C})$.

Proof. This immediately follows from Theorem 12.17. □

The finite Fourier transform is one of many transforms associated with the surname Fourier. For comparisons, see Definition 13.8, Proposition 23.16, and Definition 28.28.

12.4 Uniform Convergence of Fourier Series

When does the Fourier series converge uniformly to the function that it is designed to approximate? Let us answer this question for a rather simple case. Pay attention to the techniques of the following proof as they will be used several times.

Theorem 12.21 (uniform convergence). *Suppose that $f \in C^2_{\mathrm{p}}(0,1;\mathbb{C})$. Then*

$$\lim_{n\to\infty} \|f - \mathcal{S}_n[f]\|_{L^\infty(0,1;\mathbb{C})} = 0.$$

In other words, $\mathcal{S}_n[f] \to f$, uniformly on $[0,1]$. In particular,

$$|\hat{f}_j| \le \frac{C_1}{j^2}, \quad j \in \mathbb{Z}_\star,$$

and

$$\|f - \mathcal{S}_n[f]\|_{L^\infty(0,1;\mathbb{C})} \le \frac{2C_1}{n}, \quad \forall n \in \mathbb{N}$$

for some constant $C_1 > 0$ that is independent of n.

Proof. By $\hat{f}^{(2)}_j \in \mathbb{C}$, we denote the Fourier coefficients of f'':

$$\hat{f}^{(2)}_j = (f'', \psi_j)_{L^2(0,1;\mathbb{C})} = \int_0^1 f''(x)\exp(-2\pi \mathrm{i} jx)\mathrm{d}x, \quad j \in \mathbb{Z}.$$

Using integration by parts and the periodicity of the functions, we find

$$\begin{aligned}\hat{f}^{(2)}_j &= \int_0^1 f''(x)\exp(-2\pi \mathrm{i} jx)\mathrm{d}x \\ &= -(-2\pi \mathrm{i} j)\int_0^1 f'(x)\exp(-2\pi \mathrm{i} jx)\mathrm{d}x \\ &= (-2\pi \mathrm{i} j)^2 \int_0^1 f(x)\exp(-2\pi \mathrm{i} jx)\mathrm{d}x.\end{aligned}$$

Thus,

$$\hat{f}_j = -\frac{\hat{f}^{(2)}_j}{4\pi^2 j^2}, \quad j \in \mathbb{Z}_\star.$$

We know that $|\hat{f}^{(2)}_j|, |\hat{f}^{(2)}_{-j}| \to 0$, as $j \to \infty$, by Parseval's relation for f''. Since every convergent sequence is also bounded, there is a constant $C_1 > 0$ such that

$$|\hat{f}_j| = \frac{\left|\hat{f}^{(2)}_j\right|}{4\pi^2 j^2} \le \frac{C_1}{j^2}, \quad j \in \mathbb{Z}_\star.$$

By the Weierstrass M-Test, Theorem B.63, since

$$\left|\hat{f}_j \exp(2\pi \mathrm{i} jx)\right| \le \frac{C_1}{j^2}, \quad j \in \mathbb{Z}_\star,$$

and

$$\sum_{j\in\mathbb{Z}_\star} \frac{1}{j^2} = \sum_{\substack{j=-\infty \\ j\neq 0}}^{\infty} \frac{1}{j^2} = \frac{\pi^2}{3}$$

is convergent, $\mathcal{S}_n[f]$ converges uniformly to a continuous one-periodic function. Let us label the uniform limit $g \in C_{\mathrm{p}}(0,1;\mathbb{C})$.

Now we have $\mathcal{S}_n[f] \to g$ uniformly on $\mathbb{R}$ and $\mathcal{S}_n[f] \to f$ in $L^2(0,1;\mathbb{C})$. We want to show that $f \equiv g$. By Theorem B.65,

$$\begin{aligned} 0 &= \lim_{n\to\infty} \int_0^1 |f(x) - \mathcal{S}_n[f](x)|^2 \, \mathrm{d}x \\ &= \int_0^1 \lim_{n\to\infty} |f(x) - \mathcal{S}_n[f](x)|^2 \, \mathrm{d}x \\ &= \int_0^1 |f(x) - g(x)|^2 \, \mathrm{d}x \\ &= \|f - g\|^2_{L^2(0,1;\mathbb{C})} \,. \end{aligned}$$

Hence, $f - g \equiv 0$ since $\|\cdot\|_{L^2(0,1;\mathbb{C})}$ is a norm on $C_{\mathrm{p}}(0,1;\mathbb{C})$.

Finally, we want to prove the convergence rate estimate. Suppose that $n \in \mathbb{N}$ is fixed and $m > n$. Then, for any $x \in [0,1]$,

$$\begin{aligned} |\mathcal{S}_m[f](x) - \mathcal{S}_n[f](x)| &\le \sum_{j=n+1}^{m} \left(|\hat{f}_j| + |\hat{f}_{-j}|\right) \\ &\le 2C_1 \sum_{j=n+1}^{m} \frac{1}{j^2} \\ &\le 2C_1 \sum_{j=n+1}^{\infty} \frac{1}{j^2} \\ &\le \frac{2C_1}{n}, \end{aligned}$$

using Lemma 11.19 with $p = 2$. Now since $\mathcal{S}_m[f] \to f$ uniformly on $\mathbb{R}$, as $m \to \infty$, given any $\varepsilon > 0$, there is an $N \in \mathbb{N}$ such that if $m \ge N$, then

$$|f(x) - \mathcal{S}_m[f](x)| < \varepsilon$$

for all $x \in \mathbb{R}$. Therefore, if $m \ge N$ and $m > n$, then, for all $x \in \mathbb{R}$,

$$\begin{aligned} |f(x) - \mathcal{S}_n[f](x)| &= |f(x) - \mathcal{S}_m[f](x) + \mathcal{S}_m[f](x) - \mathcal{S}_n[f](x)| \\ &\le |f(x) - \mathcal{S}_m[f](x)| + |\mathcal{S}_m[f](x) - \mathcal{S}_n[f](x)| \\ &< \varepsilon + \frac{2C_1}{n}. \end{aligned}$$

Now since we can make $\varepsilon > 0$ as small as we like, we get

$$|f(x) - \mathcal{S}_n[f](x)| \le \frac{2C_1}{n}$$

for all $x \in \mathbb{R}$ and for all $n \in \mathbb{N}$. Since x is arbitrary, we get our result:

$$\|f - \mathcal{S}_n[f]\|_{L^\infty(0,1;\mathbb{C})} \le \frac{2C_1}{n}. \qquad \square$$

With similar techniques we can establish an improved rate of convergence for smoother functions.

Corollary 12.22 (uniform convergence). *Suppose that $r \geq 2$ and $f \in C_p^r(0,1;\mathbb{C})$. Then*

$$|\hat{f}_j| \leq \frac{C_2}{|j|^r}, \quad j \in \mathbb{Z}_\star,$$

and

$$\|f - \mathcal{S}_n[f]\|_{L^\infty(0,1;\mathbb{C})} \leq \frac{2C_2}{(r-1)n^{r-1}}, \quad n \in \mathbb{N},$$

for some constant $C_2 > 0$ that is independent of n.

Proof. See Problem 12.8. □

What if the function of interest, f, is not quite smooth enough so that Theorem 12.21 applies? Does its Fourier series converge uniformly? Let us answer a simpler question for the moment. The proof of the following results should be obvious after reading the proof of Theorem 12.21.

Theorem 12.23 (uniform limit). *Suppose that $f \in C_p(0,1;\mathbb{C})$ and $\mathcal{S}_n[f]$ converges uniformly in $\mathbb{R}$ to some function $g \colon \mathbb{R} \to \mathbb{C}$. Then it must be that $g \in C_p(0,1;\mathbb{C})$ and, in fact, $g \equiv f$.*

Proof. See Problem 12.9. □

Corollary 12.24 (uniform convergence). *Suppose that $f \in C_p(0,1;\mathbb{C})$ is such that $\sum_{j=-\infty}^{\infty} \left|\hat{f}_j\right| < \infty$. Then $\mathcal{S}_n[f]$ converges uniformly and absolutely in $\mathbb{R}$ to f.*

Proof. Use the Weierstrass M-Test, Theorem B.63. The details are left to the reader as an exercise; see Problem 12.10. □

To make further progress, we need a weakened definition of the derivative.

Definition 12.25 (piecewise continuity). Let $I \subseteq \mathbb{R}$ be an interval, finite or infinite in length. Suppose that $g \colon I \to \mathbb{C}$. We say that g is **piecewise continuous on** I if and only if:

1. There is $X \subset I$ of at most countable cardinality such that $g \in C(I \backslash X;\mathbb{C})$.
2. For any compact interval $[c,d] \subseteq I$, $X \cap [c,d]$ has finite cardinality.
3. For every $x \in X$, the right and left limits exist and are finite, i.e., there exist numbers $A_R, A_L \in \mathbb{C}$ such that

$$\lim_{y \downarrow x} g(y) = A_R, \qquad \lim_{y \uparrow x} g(y) = A_L.$$

Example 12.1 Consider the functions $f, g \colon \mathbb{R} \to \mathbb{R}$ defined by

$$f(x) = \begin{cases} \dfrac{1}{x}, & x \in (-\infty, 0) \cup (0, \infty), \\ 0, & x = 0, \end{cases}$$

and

$$g(x) = \begin{cases} x^3 + 1, & x \in (-\infty, 0), \\ x^2, & x \in [0, \infty). \end{cases}$$

The function f is not piecewise continuous, but the function g is piecewise continuous.

Definition 12.26 (piecewise differentiability). Let $I \subseteq \mathbb{R}$ be an interval, finite or infinite in length, and $f\colon I \to \mathbb{C}$. We say that f is **piecewise continuously differentiable on** I if and only if:

1. $f \in C(I; \mathbb{C})$.
2. There is a set $X \subset I$, of at most countable cardinality, such that f is classically differentiable in $I \backslash X$ and fails to be classically differentiable at the points of X.
3. For any compact interval $[c, d] \subseteq I$, $X \cap [c, d]$ has finite cardinality.
4. There is a piecewise continuous function $g\colon I \to \mathbb{C}$, whose points of discontinuity are precisely the points of X, and $g(x) = f'(x)$ at all points $x \in I \backslash X$.

In this case, we write $f'(x) = g(x)$ for all $x \in I$, and we call f' a **piecewise derivative of** f.

Example 12.2 We observe that, if f is piecewise continuously differentiable, the derivative is not unique. For example, suppose that $f\colon \mathbb{R} \to [0, \infty)$ is defined by $f(x) = |x|$. Of course, f is not differentiable at $x = 0$ in the classical sense. However, both

$$g_1(x) = \begin{cases} -1, & -\infty < x \le 0, \\ 1, & 0 < x < \infty \end{cases}$$

and

$$g_2(x) = \begin{cases} -1, & -\infty < x < 0, \\ 1, & 0 \le x < \infty \end{cases}$$

are candidates for our piecewise derivative. Which one should we choose? In fact, it does not matter for our purposes, as they differ only on a set of measure zero. Further, observe that if $[c, d]$ is any finite interval with $c < 0 < d$, then $g_1, g_2 \in L^2(c, d)$ and $g_1 \equiv g_2$ in the L^2-sense. In particular, for any $q \in C([c, d]; \mathbb{C})$,

$$(g_1, q)_{L^2(c,d;\mathbb{C})} = \int_c^d g_1(x)\overline{q(x)}\,dx = \int_c^d g_2(x)\overline{q(x)}\,dx = (g_2, q)_{L^2(c,d;\mathbb{C})}.$$

In fact, the last equality holds for any $q \in L^2(c, d; \mathbb{C})$.

We will make use of the following important integration by parts property for piecewise continuously differentiable functions.

Proposition 12.27 (integration by parts). *Let $I \subseteq \mathbb{R}$ be an open interval, finite or infinite in length. Suppose that $f \in C(I;\mathbb{C})$ is piecewise continuously differentiable in I. If $[c,d]$ is any compact subset of I, and $g \in C^1([c,d];\mathbb{C})$, then*

$$\int_c^d f(x)g'(x)\mathrm{d}x = f(x)g(x)|_c^d - \int_c^d f'(x)g(x)\mathrm{d}x.$$

Proof. See Problem 12.11. □

With these notions at hand we can provide a — more general — sufficient condition for uniform convergence of Fourier series.

Theorem 12.28 (uniform convergence). *Suppose that $f \in C_{\mathrm{p}}(0,1;\mathbb{C})$ is piecewise continuously differentiable in $\mathbb{R}$ and its piecewise derivative $f' \colon \mathbb{R} \to \mathbb{C}$ is one-periodic. Then $S_n[f] \to f$ uniformly and absolutely in $\mathbb{R}$.*

Proof. We first observe that $f' \in L^2_{\mathrm{p}}(0,1;\mathbb{C})$. Therefore, Parseval's relation holds for f':

$$\sum_{j=-\infty}^{\infty} \left|\widehat{f'}_j\right|^2 = \|f'\|^2_{L^2(0,1;\mathbb{C})} < \infty.$$

Using integration by parts,

$$\begin{aligned}
\hat{f}_j &= \int_0^1 f(x)\exp(-2\pi \mathrm{i} jx)\mathrm{d}x \\
&= -\frac{1}{2\pi \mathrm{i} j}\int_0^1 f'(x)\exp(-2\pi \mathrm{i} jx)\mathrm{d}x \\
&= -\frac{\widehat{f'}_j}{2\pi \mathrm{i} j}, \quad j \in \mathbb{Z}_\star.
\end{aligned}$$

Now for all $n \in \mathbb{N}$, using the Cauchy–Schwarz inequality for finite sums,

$$\begin{aligned}
\sum_{j=-n}^{n} |\hat{f}_j| &= |\hat{f}_0| + \frac{1}{2\pi}\sum_{\substack{j=-n\\ j\neq 0}}^{n} \frac{1}{j}\left|\widehat{f'}_j\right| \\
&\le |\hat{f}_0| + \frac{1}{2\pi}\left(\sum_{\substack{j=-n\\ j\neq 0}}^{n} \frac{1}{j^2}\right)^{1/2}\left(\sum_{\substack{j=-n\\ j\neq 0}}^{n} \left|\widehat{f'}_j\right|^2\right)^{1/2} \\
&\le |\hat{f}_0| + \frac{1}{2\pi}\left(\sum_{\substack{j=-\infty\\ j\neq 0}}^{\infty} \frac{1}{j^2}\right)^{1/2}\left(\sum_{j=-\infty}^{\infty} \left|\widehat{f'}_j\right|^2\right)^{1/2} \\
&= |\hat{f}_0| + \frac{1}{2\pi}\left(\frac{\pi^2}{3}\right)^{1/2} \|f'\|_{L^2(0,1;\mathbb{C})}\,.
\end{aligned}$$

Since the right-hand side does not depend on n,

$$\sum_{j=-\infty}^{\infty} |\hat{f}_j| \le |\hat{f}_0| + \frac{1}{2\sqrt{3}} \|f'\|_{L^2(0,1;\mathbb{C})} \, .$$

By Corollary 12.24, $\mathcal{S}_n[f] \to f$ uniformly and absolutely on $\mathbb{R}$. □

Unfortunately, our (now standard) trick for extracting a rate of uniform convergence using Lemma 11.19 will not work in the proof of the last theorem. This leaves us to wonder if a rate of convergence can be extracted for functions that are merely piecewise continuously differentiable. In numerical analysis, we always want, if possible, to get to the best, or optimal, rate of convergence. But this can be challenging in the present setting, and so we refer the reader to the classic work [48] to get a more complete picture. In the next example, we will see that a rate of convergence can be extracted for a particular piecewise continuously differentiable function, the triangle wave.

Example 12.3 Let us construct a one-periodic triangle wave. Suppose that $f_o \colon [0,1) \to \mathbb{R}$ is defined via

$$f_o(x) = \begin{cases} x, & 0 \le x \le \dfrac{1}{2}, \\ 1-x, & \dfrac{1}{2} < x < 1. \end{cases}$$

Now let us extend f_o to a one-periodic function that we label f:

$$f(x) = f_o(x - \lfloor x \rfloor), \quad x \in \mathbb{R},$$

where $\lfloor \cdot \rfloor$ is the floor function; see Figure 12.3. Then $f \in C_{\mathrm{p}}(0,1)$ and f is piecewise continuously differentiable. However, $f \notin C^1_{\mathrm{p}}(0,1)$. The Fourier coefficients of f are

$$\hat{f}_0 = \frac{1}{4}, \quad \hat{f}_j = -\frac{(\exp(\pi \mathrm{i} j) - 1)^2}{4\pi^2 j^2}, \quad j \in \mathbb{Z}_\star.$$

It is easy to see that the even coefficients vanish:

$$\hat{f}_j = 0, \quad j = \pm 2, \pm 4, \pm 6, \ldots .$$

The Fourier projections $\mathcal{S}_n[f]$, for $n = 3$, 5, and 9, are plotted in Figure 12.3. Observe that $\mathcal{S}_n[f]$ is a real-valued function. Uniform convergence is guaranteed by Theorem 12.28, but with no rate of convergence. On the other hand, since we can find (directly)

$$|\hat{f}_j| \le \frac{1}{\pi^2 j^2}, \quad j \in \mathbb{Z}_\star,$$

we can prove, using our usual technique, that

$$\|f - \mathcal{S}_n[f]\|_{L^\infty(0,1)} \le \frac{2}{\pi^2 n}, \quad n \in \mathbb{N}.$$

For the present example, we can show a rate of uniform convergence, even though

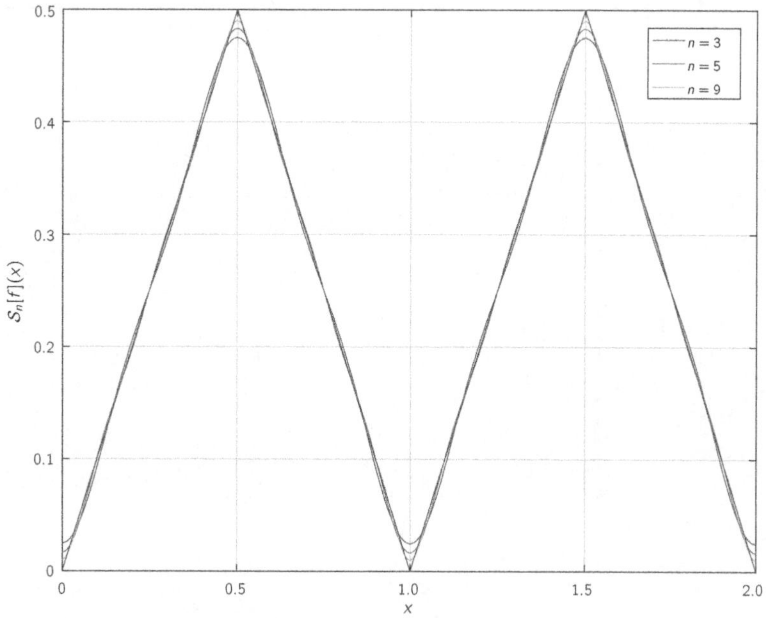

Figure 12.3 Fourier projections of a triangle wave. Uniform convergence is guaranteed by Theorem 12.28, but with no rate of convergence. However, in this case, we can show that $\|f - \mathcal{S}_n[f]\|_{L^\infty(0,1)} \to 0$ with rate $\mathcal{O}(n^{-1})$, as $n \to \infty$; see Example 12.3.

it is not covered by our limited theory. The errors $f - \mathcal{S}_n[f]$ are plotted in Figure 12.4, for $n = 3,\ 5$, and 9.

There is still a bit more that we can do using our simple set of tools. We can slightly weaken the assumption that $f \in C_{\mathrm{p}}^2(0,1)$ and still get a rate of convergence.

Theorem 12.29 (Hölder continuous derivative). *Let $\alpha \in (0,1]$. Suppose that $f \in C_{\mathrm{p}}^1(0,1;\mathbb{C})$. Assume, additionally, that $f' \in C^\alpha(\mathbb{R};\mathbb{C})$, i.e., there is a constant $C > 0$ such that*

$$|f'(x) - f'(y)| \le C|x - y|^\alpha$$

for all $x, y \in \mathbb{R}$. Then $\mathcal{S}_n[f] \to f$ uniformly and absolutely in $\mathbb{R}$. Furthermore,

$$|\hat{f}_j| \le \frac{C_3}{|j|^{1+\alpha}}, \quad j \in \mathbb{Z}_\star,$$

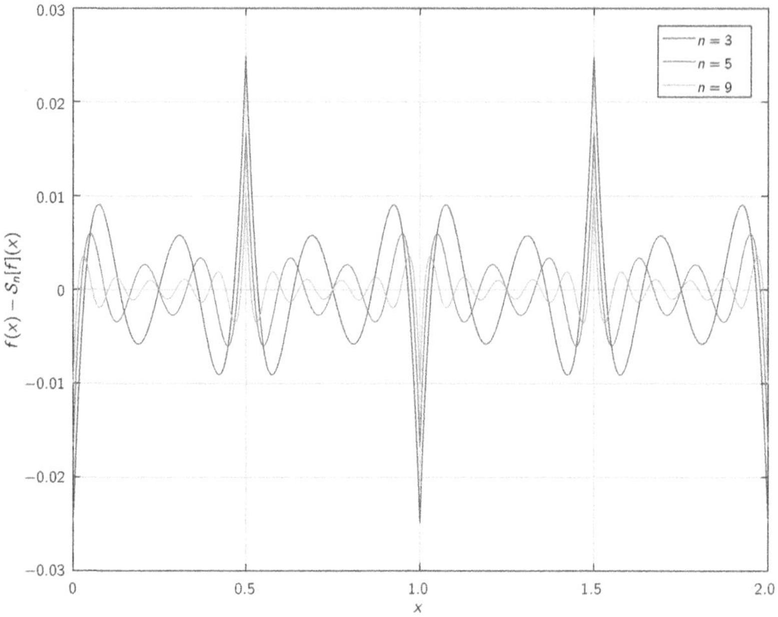

Figure 12.4 Errors in the Fourier projections of a triangle wave. We can show that $\|f - \mathcal{S}_n[f]\|_{L^\infty(0,1)} \to 0$ with rate $\mathcal{O}(n^{-1})$, as $n \to \infty$; see Example 12.3. Note that the error function oscillates about zero, though it does not "equi-oscillate."

and

$$\|f - \mathcal{S}_n[f]\|_{L^\infty(0,1;\mathbb{C})} \leq \frac{2C_3}{\alpha n^\alpha}, \qquad n \in \mathbb{N},$$

for some constant $C_3 > 0$ that is independent of n.

Proof. Using integration by parts and periodicity of the complex exponential, for $j \in \mathbb{Z}_\star$, we have

$$\begin{aligned}
\hat{f}_j &= \int_0^1 f(x)\exp(-2\pi \mathrm{i} jx)\mathrm{d}x \\
&= -\frac{1}{2\pi \mathrm{i} j}\int_0^1 f'(x)\exp(-2\pi \mathrm{i} jx)\mathrm{d}x \\
&= -\frac{1}{2\pi \mathrm{i} j}\int_0^1 f'\left(x + \frac{1}{2j}\right)\exp\left(-2\pi \mathrm{i} j\left(x + \frac{1}{2j}\right)\right)\mathrm{d}x \\
&= \frac{1}{2\pi \mathrm{i} j}\int_0^1 f'\left(x + \frac{1}{2j}\right)\exp(-2\pi \mathrm{i} jx)\mathrm{d}x.
\end{aligned}$$

Therefore, for $j \in \mathbb{Z}_\star$,

$$\begin{aligned}
|\hat{f}_j| &= \frac{1}{4\pi|j|}\left|\int_0^1 \left(f'\left(x+\frac{1}{2j}\right) - f'(x)\right)\exp(-2\pi i j x)\mathrm{d}x\right| \\
&\le \frac{1}{4\pi|j|}\int_0^1 \left|f'\left(x+\frac{1}{2j}\right) - f'(x)\right|\mathrm{d}x \\
&\le \frac{C}{4\pi|j|}\int_0^1 \left|\frac{1}{2j}\right|^\alpha \mathrm{d}x \\
&= \frac{C}{4\pi 2^\alpha |j|^{1+\alpha}} \\
&= \frac{C_3}{|j|^{1+\alpha}},
\end{aligned}$$

where $C_3 = \frac{C}{4\pi 2^\alpha}$.

From here, the details should be quite familiar to the reader. By the Weierstrass M-Test, Theorem B.63, since

$$\sum_{j\in\mathbb{Z}_\star} \frac{1}{|j|^{1+\alpha}}$$

is convergent, $\mathcal{S}_n[f]$ converges uniformly to a continuous one-periodic function. Let us label the uniform limit $g \in C_{\mathrm{p}}(0,1;\mathbb{C})$. But, since $\mathcal{S}_n[f] \to g$ uniformly on $\mathbb{R}$ and $\mathcal{S}_n[f] \to f$ in $L^2(0,1;\mathbb{C})$, it follows that $f \equiv g$.

Finally, we need to prove the convergence rate estimate. Suppose that $n \in \mathbb{N}$ is fixed and $m > n$. Then, for any $x \in [0,1]$,

$$\begin{aligned}
|\mathcal{S}_n[f](x) - \mathcal{S}_m[f](x)| &\le \sum_{j=n+1}^{m} \left(|\hat{f}_j| + |\hat{f}_{-j}|\right) \\
&\le 2C_3 \sum_{j=n+1}^{m} \frac{1}{j^{1+\alpha}} \\
&\le 2C_3 \sum_{j=n+1}^{\infty} \frac{1}{j^{1+\alpha}} \\
&\le \frac{2C_3}{\alpha n^\alpha},
\end{aligned}$$

using Lemma 11.19 with $p = 1+\alpha$. Since the right-hand side is independent of m and x, we have

$$\|f - \mathcal{S}_n[f]\|_{L^\infty(0,1;\mathbb{C})} \le \frac{2C_3}{\alpha n^\alpha}. \qquad \square$$

Remark 12.30 (Lipschitz continuity)**.** This last result shows that we do not actually need $f \in C^2_{\mathrm{p}}(0,1)$ in order to get a rate of convergence of order $\mathcal{O}(\frac{1}{n})$. In Theorem 12.29, if $\alpha = 1$, in which case f' is Lipschitz continuous, we still get a rate of convergence of order $\mathcal{O}(\frac{1}{n})$. Compare Theorem 12.29 with Theorem 12.21.

12.5 Convergence of Fourier Series in Sobolev Spaces

Next, let us discuss the convergence of Fourier series in spaces of so-called weakly differentiable functions, also known as *Sobolev spaces*; see Appendix D for an overview. These spaces are important in the study of partial differential equations.

Our main result regarding approximation in periodic Sobolev spaces reads as follows.

Theorem 12.31 (approximation). *Let $r \in \mathbb{N}$ and $f \in H_{\mathrm{p}}^r(0,1;\mathbb{C})$. For all $n \in \mathbb{N}$ and any $s \in \mathbb{N}_0$, $s < r$, we have*

$$\|f - \mathcal{S}_n[f]\|_{H^s(0,1;\mathbb{C})} \leq C \frac{1}{n^{r-s}} |f|_{H^r(0,1;\mathbb{C})}$$

for some constant $C > 0$ that is independent of f and n. If $f \in C_{\mathrm{p}}^\infty(0,1;\mathbb{C})$, then, for all $n, m \in \mathbb{N}$ and for any $s \in \mathbb{N}_0$,

$$\|f - \mathcal{S}_n[f]\|_{H^s(0,1;\mathbb{C})} \leq C \frac{1}{n^m} \tag{12.10}$$

for some constant $C > 0$ that is independent of n.

Proof. Suppose that $f \in L_{\mathrm{p}}^2(0,1;\mathbb{C})$. Let

$$\Psi = \{\psi_j = \exp(2\pi \mathrm{i} j \cdot)\}_{j=-\infty}^{\infty}$$

be the standard orthonormal trigonometric polynomial system on $[0,1]$ and $\hat{f}_j$ denote the jth Fourier coefficient of f. Then, from the Riesz–Fischer Theorem 12.17,

$$f = \lim_{n \to \infty} \sum_{j=-n}^{n} \hat{f}_j \psi_j,$$

in the sense of $L_{\mathrm{p}}^2(0,1;\mathbb{C})$. We will prove the first part for the special case $s = 0$ and leave the general case to the reader. For any $r \in \mathbb{N}$, since the numbers γ_k and γ_{-k}, $k \in \mathbb{N}$, defined in (D.3), are always strictly positive, and since $f - \mathcal{S}_n[f] \in L_{\mathrm{p}}^2(0,1;\mathbb{C})$,

$$\begin{aligned} f - \mathcal{S}_n[f] &= \sum_{k=n+1}^{\infty} \left(\hat{f}_{-k}\psi_{-k} + \hat{f}_k \psi_k \right) \\ &= \sum_{k=n+1}^{\infty} \left(\gamma_{-k}^{-r/2} \left(\gamma_{-k}^{r/2} \hat{f}_{-k} \right) \psi_{-k} + \gamma_k^{-r/2} \left(\gamma_k^{r/2} \hat{f}_k \right) \psi_k \right). \end{aligned}$$

Since, for $k \geq n+1$,

$$0 \leq \gamma_{-k}^{-r} = \gamma_k^{-r} = \frac{1}{(4\pi^2 k^2)^r} \leq \frac{1}{(4\pi^2 (n+1)^2)^r} = \gamma_{n+1}^{-r},$$

using Parseval's relation, we have

$$\begin{aligned}\|f - \mathcal{S}_n[f]\|_{L^2(0,1;\mathbb{C})}^2 &= \sum_{k=n+1}^{\infty} \left|\gamma_{-k}^{-r/2}\left(\gamma_{-k}^{r/2}\hat{f}_{-k}\right)\right|^2 + \sum_{k=n+1}^{\infty} \left|\gamma_k^{-r/2}\left(\gamma_k^{r/2}\hat{f}_k\right)\right|^2 \\ &\le \gamma_{n+1}^{-r}\left(\sum_{k=n+1}^{\infty} \gamma_{-k}^r \left|\hat{f}_{-k}\right|^2 + \sum_{k=n+1}^{\infty} \gamma_k^r \left|\hat{f}_k\right|^2\right) \\ &\le \gamma_{n+1}^{-r} |f|_{H^r(0,1;\mathbb{C})}^2,\end{aligned}$$

where, in the last step, we have used the fact that

$$|f|_{H^r(0,1;\mathbb{C})}^2 = \sum_{k=-\infty}^{\infty} \gamma_k^r \left|\hat{f}_k\right|^2 \ge \sum_{k=n+1}^{\infty} \gamma_{-k}^r \left|\hat{f}_{-k}\right|^2 + \sum_{k=n+1}^{\infty} \gamma_k^r \left|\hat{f}_k\right|^2,$$

which follows from (D.4) Theorem D.36. We have proven that

$$\|f - \mathcal{S}_n[f]\|_{L^2(0,1;\mathbb{C})} \le \gamma_{n+1}^{-r/2} |f|_{H^r(0,1;\mathbb{C})} = \frac{1}{2^r \pi^r (n+1)^r} |f|_{H^r(0,1;\mathbb{C})}.$$

The other cases can be proven in a similar fashion and are left to the reader as an exercise; see Problem 12.15. □

Remark 12.32 (spectral convergence)**.** The convergence property expressed in estimate (12.10) of Theorem 12.31 is known as *spectral convergence.*

In the periodic setting, we can, using Fourier series, provide a more fine-grained characterization of smoothness.

Definition 12.33 (fractional Sobolev space)**.** Suppose that $f \in L_{\mathrm{p}}^2(0,1;\mathbb{C})$ and $\alpha \in (0,\infty)$. Let

$$\Psi = \{\psi_j\}_{j=-\infty}^{\infty}, \quad \psi_j = \exp(2\pi i j \cdot)$$

be the standard orthonormal trigonometric polynomial system on $[0,1]$ and $\hat{f}_j$ denote the jth Fourier coefficient of f. We say that f is in the **periodic Sobolev**[6] **space of fractional order** α, denoted by $H_{\mathrm{p}}^{\alpha}(0,1;\mathbb{C})$, if and only if the series

$$\sum_{j=-\infty}^{\infty} \gamma_j^{\alpha} |\hat{f}_j|^2 = \lim_{n\to\infty} \sum_{j=-n}^{n} \gamma_j^{\alpha} |\hat{f}_j|^2$$

converges, where

$$\gamma_j = 4\pi^2 j^2, \quad j \in \mathbb{Z}.$$

If $f \in H_{\mathrm{p}}^{\alpha}(0,1;\mathbb{C})$, then, in the quadratic mean, as $n \to \infty$,

$$\sum_{j=-n}^{n} \hat{f}_j (2\pi i j)^{\alpha} \exp(2\pi i j \cdot) \longrightarrow g_{\alpha}$$

for some function $g_{\alpha} \in L_{\mathrm{p}}^2(0,1;\mathbb{C})$, and we are justified in writing

$$g_{\alpha} = \frac{\mathrm{d}^{\alpha} f}{\mathrm{d}x^{\alpha}} = \sum_{j=-\infty}^{\infty} \hat{f}_j (2\pi i j)^{\alpha} \exp(2\pi i j \cdot).$$

[6] Named in honor of the Soviet mathematician Sergei Lvovich Sobolev (1908–1989).

The function $\frac{d^\alpha f}{dx^\alpha}$ is called the **weak derivative of f of fractional order α**. The **Sobolev seminorm of f of fractional order α** is defined via

$$|f|^2_{H^\alpha(0,1;\mathbb{C})} = \left\| \frac{d^\alpha f}{dx^\alpha} \right\|^2_{L^2(0,1;\mathbb{C})} = \sum_{j=-\infty}^{\infty} \gamma_j^\alpha |\hat{f}_j|^2.$$

The following fundamental result connects spaces of continuous functions to fractional Sobolev spaces. This is a particular case of so-called *Sobolev embedding* theorems.

Theorem 12.34 (embedding). *Let $m \in \mathbb{N}_0$. Suppose that $\alpha > m + \frac{1}{2}$. Then $H^\alpha_p(0,1;\mathbb{C}) \subset C^m_p(0,1;\mathbb{C})$. Furthermore, there is a constant $C > 0$, which depends upon α, such that, for all $f \in H^\alpha_p(0,1;\mathbb{C})$, we have*

$$\left\| f^{(m)} \right\|_{L^\infty(0,1;\mathbb{C})} \le C \left(\|f\|_{L^2(0,1;\mathbb{C})} + |f|_{H^\alpha(0,1;\mathbb{C})} \right).$$

Finally, the Fourier series of $f \in H^\alpha_p(0,1;\mathbb{C})$ converges uniformly and absolutely to f, provided that $\alpha > \frac{1}{2}$.

Proof. Suppose that $f \in H^\alpha_p(0,1;\mathbb{C})$. Let

$$\Psi = \{\psi_j = \exp(2\pi i j \cdot)\}_{j=-\infty}^{\infty}$$

be the standard orthonormal trigonometric polynomial system on $[0,1]$ and $\hat{f}_j$ denote the jth Fourier coefficient of f.

We will prove the case $m = 0$ and leave the general case to the reader as an exercise; see Problem 12.16. Our goal is to show that the Fourier series of f converges uniformly and absolutely on $\mathbb{R}$. Using the Cauchy–Schwarz inequality, we find

$$\begin{aligned}
\sum_{j=-n}^{n} |\hat{f}_j| &= |\hat{f}_0| + \sum_{\substack{j=-n \\ j\neq 0}}^{n} |\hat{f}_j| \\
&= |\hat{f}_0| + \sum_{\substack{j=-n \\ j\neq 0}}^{n} \gamma_j^{-\alpha/2} \gamma_j^{\alpha/2} |\hat{f}_j| \\
&\le |\hat{f}_0| + \sqrt{\sum_{\substack{j=-n \\ j\neq 0}}^{n} \frac{1}{\gamma_j^\alpha}} \sqrt{\sum_{\substack{j=-n \\ j\neq 0}}^{n} \gamma_j^\alpha |\hat{f}_j|^2} \\
&\le |\hat{f}_0| + \sqrt{\sum_{\substack{j=-\infty \\ j\neq 0}}^{\infty} \frac{1}{\gamma_j^\alpha}} \sqrt{\sum_{j=-\infty}^{\infty} \gamma_j^\alpha |\hat{f}_j|^2} \\
&\le |\hat{f}_0| + \sqrt{\frac{2}{(4\pi^2)^\alpha} \sum_{j=1}^{\infty} \frac{1}{j^{2\alpha}}} \, |f|_{H^\alpha(0,1;\mathbb{C})} \\
&\le |\hat{f}_0| + \sqrt{\frac{2}{(4\pi^2)^\alpha} \zeta(2\alpha)} \, |f|_{H^\alpha(0,1;\mathbb{C})},
\end{aligned}$$

where

$$\zeta(s) = \sum_{j=1}^{\infty} \frac{1}{j^s}, \quad s > 1$$

is the Riemann zeta function.[7] Since the right-hand side is independent of n, it follows that

$$\sum_{j=-\infty}^{\infty} |\hat{f}_j| \le |\hat{f}_0| + \sqrt{\frac{2}{(4\pi^2)^\alpha}\zeta(2\alpha)}\ |f|_{H^\alpha(0,1;\mathbb{C})} < \infty.$$

By Corollary 12.24, the Fourier series of f converges uniformly and absolutely to f and $f \in C_{\mathrm{p}}(0,1;\mathbb{C})$. Finally, it follows that, for all $x \in \mathbb{R}$,

$$\begin{aligned}
|f(x)| &\le \sum_{j=-\infty}^{\infty} |\hat{f}_j| \\
&\le |\hat{f}_0| + \sqrt{\frac{2}{(4\pi^2)^\alpha}\zeta(2\alpha)}\ |f|_{H^\alpha(0,1;\mathbb{C})} \\
&\le \|f\|_{L^2(0,1)} + \sqrt{\frac{2}{(4\pi^2)^\alpha}\zeta(2\alpha)}\ |f|_{H^\alpha(0,1;\mathbb{C})} \\
&\le \left(1 + \sqrt{\frac{2}{(4\pi^2)^\alpha}\zeta(2\alpha)}\right)\left(\|f\|_{L^2(0,1;\mathbb{C})} + |f|_{H^\alpha(0,1;\mathbb{C})}\right).
\end{aligned}$$

Since the right-hand side of the last estimate is independent of x, the result follows. □

The following fractional version of Theorem 12.31 is easily proven. In fact, the proof is basically the same as that of Theorem 12.31.

Theorem 12.35 (approximation). *Let $\alpha > 0$ and $f \in H^\alpha_{\mathrm{p}}(0,1;\mathbb{C})$. Then, for all $n \in \mathbb{N}$,*

$$\|f - \mathcal{S}_n[f]\|_{L^2(0,1;\mathbb{C})} \le C\frac{1}{n^\alpha}|f|_{H^\alpha(0,1;\mathbb{C})}$$

for some constant $C > 0$ that is independent of f and n.

Proof. See Problem 12.17. □

Problems

12.1 Prove Proposition 12.3.

12.2 Complete the proof of Theorem 12.4.

12.3 Prove Theorem 12.5.

12.4 Prove Proposition 12.7.

12.5 Prove Lemma 12.9.

12.6 Prove Theorem 12.17.

[7] Named in honor of the German mathematician Georg Friedrich Bernhard Riemann (1826–1866).

12.7 Prove that $\ell^2(\mathbb{Z};\mathbb{C})$ is a complex Hilbert space.
12.8 Prove Corollary 12.22.
12.9 Prove Theorem 12.23.
12.10 Prove Corollary 12.24.
12.11 Prove Proposition 12.27.
12.12 Suppose that $f_o\colon [0,1) \to \mathbb{R}$ is defined via

$$f_o(x) = \begin{cases} -1, & 0 \le x \le \dfrac{1}{2}, \\ 1, & \dfrac{1}{2} < x < 1. \end{cases}$$

Extend f_o to a one-periodic function f:

$$f(x) = f_o(x - \lfloor x \rfloor), \quad \forall x \in \mathbb{R},$$

where $\lfloor \cdot \rfloor$ is the floor function. Compute the Fourier series of f and determine its convergence properties. Does the series converge uniformly?
12.13 Provide all the details for Example 12.3.
12.14 Suppose that $a > 1$. Prove that the function $\phi\colon (-a,a) \to \mathbb{R}$, defined via

$$\phi(x) = \begin{cases} \exp\left(-\dfrac{1}{1-x^2}\right), & x \in (-1,1), \\ 0, & x \in (-a,-1] \cup [1,a), \end{cases}$$

is a test function and $\operatorname{supp}(\phi) = [-1,1]$.
12.15 Complete the proof of Theorem 12.31.
12.16 Complete the proof of Theorem 12.34.
12.17 Prove Theorem 12.35.

13 Trigonometric Interpolation and the Fast Fourier Transform

In this chapter, we discuss trigonometric interpolation. To keep the discussion simple, we will only treat the case of uniformly spaced points. Surprisingly, we will see an interesting and deep connection between trigonometric interpolation and something called the *Discrete Fourier Transform (DFT)*.

Let us think about how interpolation would work in the periodic setting with an example. Consider the one-periodic function f shown in Figure 13.1. This function might perhaps describe the waveform of a piece of (very simple) music that repeats in time every one second. In this case, x will represent time in seconds. We wish to interpolate this function — which is often referred to as the the *signal* — at equally spaced points in $[0, 1)$ with a linear combination of trigonometric functions of the form $\sin(2\pi kx)$ and $\cos(2\pi kx)$. How can we do this? Suppose that $n \in \mathbb{N}$ is odd and $n \geq 3$. We use the interpolation points $x_j = j/n$, $j = 0, \ldots, n-1$. We could equally well use the interpolation points $x_j = j/n$, $j = 1, \ldots, n$, but the former is the usual convention. Let us assume that the interpolant has the form

$$q(x) = a_0 + \sum_{k=1}^{K} (a_k \cos(2\pi kx) + b_k \sin(2\pi kx)).$$

Of course, there are $2K + 1$ unknown coefficients to determine, and we should probably set $K = (n-1)/2$. To determine the coefficients, we use the $n = 2K + 1$ interpolation conditions

$$q(x_j) = f(x_j), \quad j = 0, \ldots, n-1.$$

Clearly, this yields a square linear system of n equations in n unknowns:

$$a_0 + \sum_{k=1}^{K} (a_k \cos(2\pi kx_j) + b_k \sin(2\pi kx_j)) = f(x_j), \quad j = 0, \ldots, n-1.$$

But the three tools that we introduced in earlier chapters for solving linear systems resulting from interpolation problems — the Vandermonde matrix, the Lagrange basis, and the Newton basis — are not directly applicable in this case. Is the system always uniquely solvable? How easy is it to solve this system? While the present system appears to be quite complicated, it turns out that we can exploit some hidden symmetry to get a unique solution to this system very rapidly; in particular, using an algorithm called the *Fast Fourier Transform (FFT)*.

One other thing that we should mention before going on is the interesting situation associated with the case that n is even. Do you see the problem? Every

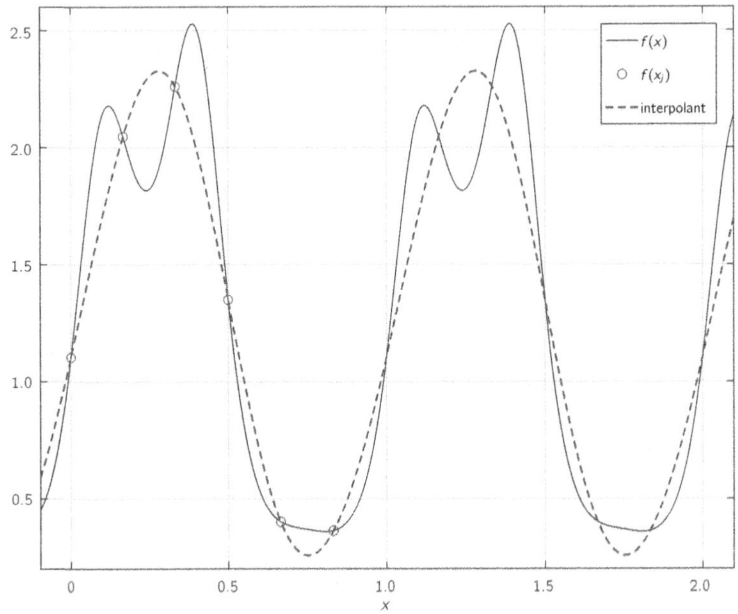

Figure 13.1 Trigonometric interpolation of a one-periodic function at six equally spaced points, $x_j = j/6$, $j = 0, \dots, 5$.

time we try to balance the number of sine and cosine terms in the interpolant, we end up with an odd number of them. (The sine term of index $k = 0$ is always identically zero.) So what should we do about this? The resolution to this issue is discussed in Example 13.1.

Before we get stuck worrying too much about practical matters associated with solving this system, let us address some other important issues. For example, how well does the interpolant approximate f, the one-periodic function of interest? In the present case, if $f \in H_p^m(0, 1)$, then the approximation is always reliably good. We indicated previously that one must be quite careful about using equally — spaced points. Remember? Again, if f is sufficiently nice, say $f \in H_p^m(0, 1)$, we need not worry about wild oscillations in the interpolant as n is increased. They do not appear. As in Chapter 12, we obtain approximations that are spectrally accurate for "nice" functions. For example, consider Figures 13.1–13.3, where we use $n = 6$ (Figure 13.1), $n = 8$ (Figure 13.2), and $n = 16$ (Figure 13.3). Notice that the interpolant is a better and better approximation to the one-periodic function f as n increases. For $n = 16$, the interpolant and the one-periodic function of interest f are essentially indistinguishable to the eye; see Example 13.1.

Finally, what if the signal f is corrupted by some noise? For example, suppose that the recorded waveform f was stored on an analogue cassette tape which has degraded over time. The signal that we now have can be written as

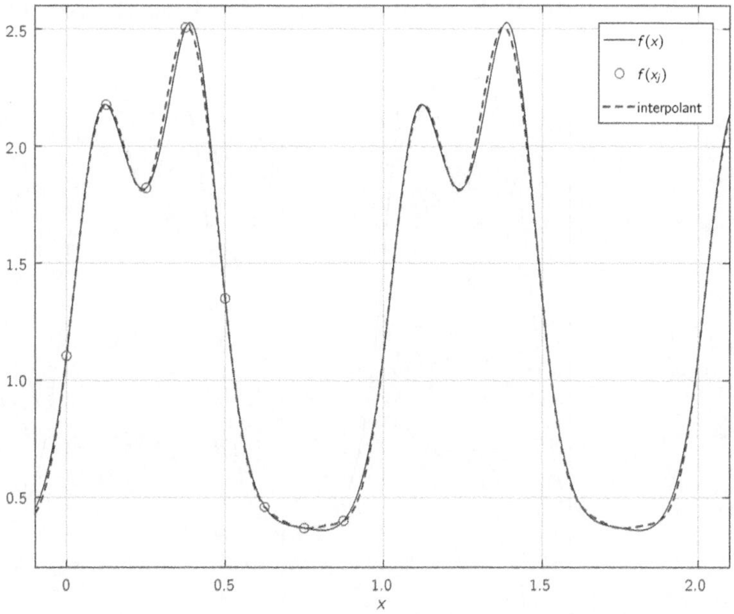

Figure 13.2 Trigonometric interpolation of a one-periodic function at eight equally spaced points, $x_j = j/8$, $j = 0, \dots, 7$.

$$g(x) = f(x) + \chi(x),$$

where χ represents random distortion of f. Using only the noisy data, how could we recover a waveform (a kind of pseudo-interpolant) that approximates the original signal f well? This is a complicated issue and a topic of the rich subject of signal processing and filtering, which we discuss very briefly in the final section of the chapter.

13.1 Periodic Interpolation and Periodic Grid Functions

To get started, let us be precise about what we want to obtain. We have the following definition.

Definition 13.1 (trigonometric interpolation). Let $f \in C_p(0, 1; \mathbb{C})$ and $n \in \mathbb{N}$. Suppose that $X = \{x_j\}_{j=0}^{n-1}$ is a uniformly spaced nodal set:

$$x_j = \frac{j}{n}, \quad j = 0, \dots, n-1.$$

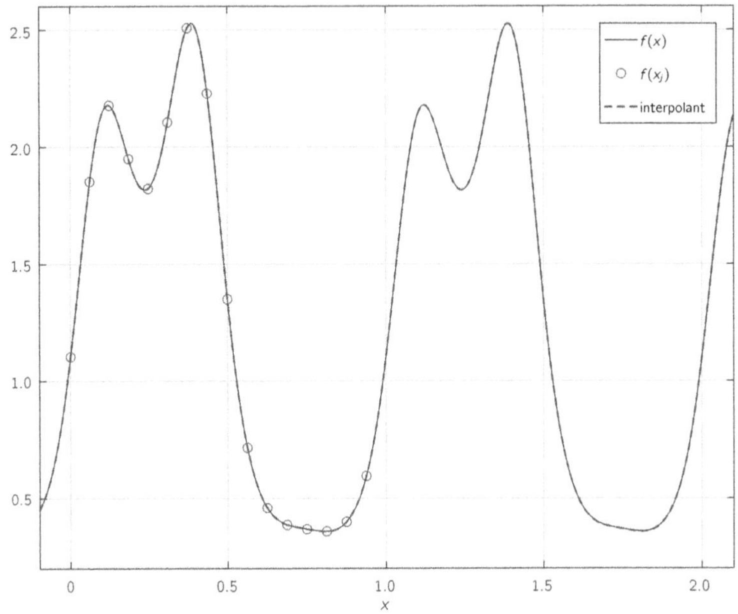

Figure 13.3 Trigonometric interpolation of a one-periodic function at 16 equally spaced points, $x_j = j/16$, $j = 0, \dots, 15$.

Let $K \in \mathbb{N}$ satisfy $n = 2K + 1$ if n is odd and $n = 2K$ if n is even. The function

$$q \in \mathbb{T}_K = \mathbb{T}_K(0,1) = \left\{ p(x) = \sum_{j=-K}^{K} c_j \exp(2\pi \mathrm{i} jx) \;\middle|\; \exists c_j \in \mathbb{C},\ j = -K, \dots, K \right\}$$

is called a **trigonometric interpolating polynomial of** f **subordinate to** X if and only if

$$q(x_j) = f(x_j), \quad j = 0, \dots, n-1.$$

Remark 13.2 (parity). As we mentioned above, in the odd case, there is exactly the same number of free coefficients in the trigonometric polynomial $(2K + 1)$ as there is interpolation nodes $(n = 2K + 1)$. For the even case, there are still $2K + 1$ coefficients, but only $n = 2K$ nodes. *How should we deal with this problem?*

Before we answer that last question and continue with the discussion of trigonometric interpolation, we need some machinery. Some of this involves grid functions and their properties, a topic that will be developed much further in Part V of this text. When we sample the one-periodic function f at the nodal points in X, we obtain a discrete function that has a kind of discrete periodicity.

Definition 13.3 (grid function). Suppose that $n \in \mathbb{N}$. Define

$$\mathcal{V}(\mathbb{C}) = \{ w \mid w \colon \mathbb{Z} \to \mathbb{C} \}.$$

Elements of this set are called **grid functions**. For grid functions, we use the notation $w_i = w(i)$, for $i \in \mathbb{Z}$. The subspace of n-**periodic grid functions** is defined as

$$\mathcal{V}_{n,\mathrm{p}}(\mathbb{C}) = \{w \in \mathcal{V}(\mathbb{C}) \mid w_{i+mn} = w_i, \ \forall i, m \in \mathbb{Z}\}.$$

One of the most important periodic functions is introduced in the following definition.

Definition 13.4 (periodic grid delta function)**.** Suppose that $n \in \mathbb{N}$. The **periodic grid delta function**, denoted $\delta^{n,\mathrm{p}} \in \mathcal{V}(\mathbb{C})$, is defined via

$$\delta_j^{n,\mathrm{p}} = \begin{cases} 1, & \exists m \in \mathbb{Z} \mid j = mn, \\ 0, & \forall m \in \mathbb{Z}\ j \neq mn \end{cases}$$

for all $j \in \mathbb{Z}$.

We have the following simple, but useful, property involving the periodic grid delta function.

Proposition 13.5 (convolution)**.** *Suppose that $n \in \mathbb{N}$. Then $\delta^{n,\mathrm{p}} \in \mathcal{V}_{n,\mathrm{p}}(\mathbb{C})$. Furthermore, for any $u \in \mathcal{V}_{n,\mathrm{p}}(\mathbb{C})$, it follows that*

$$\sum_{j=0}^{n-1} u_j \delta_{j-k}^{n,\mathrm{p}} = u_k \tag{13.1}$$

for all $k \in \mathbb{Z}$.

Proof. See Problem 13.1. □

Definition 13.6 (roots of unity)**.** Let $n \in \mathbb{N}$. Suppose that $X = \left\{x_j = \frac{j}{n}\right\}_{j=0}^{n-1}$ is the standard uniformly spaced nodal set. The grid function

$$\omega_{n,k} = \exp(2\pi\mathrm{i}x_k), \quad k \in \mathbb{Z}$$

is called the **nth root of unity**. To reference the nth root of unity grid function as a whole, independent of its individual components, we write $\omega_n \in \mathcal{V}(\mathbb{C})$.

The following elementary properties should be clear.

Proposition 13.7 (properties of ω_n)**.** *Let $n \in \mathbb{N}$. Then $\omega_n \in \mathcal{V}_{n,\mathrm{p}}(\mathbb{C})$ and*

$$\omega_{n,k}^n = 1, \quad \forall k \in \mathbb{Z},$$

where the superscript is an exponent. Furthermore, the functions $\omega_n^j \in \mathcal{V}(\mathbb{C})$, $j \in \mathbb{Z}$, that are defined (as one would expect) by

$$\omega_{n,k}^j = \exp(2\pi\mathrm{i}jx_k), \quad k \in \mathbb{Z},$$

are periodic, i.e.,

$$\omega_{n,k+mn}^j = \omega_{n,k}^j, \quad \forall j, k, m \in \mathbb{Z}$$

and

$$\left(\omega_{n,k}^j\right)^n = 1, \quad \forall j, k \in \mathbb{Z}.$$

Proof. See Problem 13.2. □

13.2 The Discrete Fourier Transform

Now we come to one of the most important tools in discrete Fourier analysis.

Definition 13.8 (DFT). Suppose that $n \in \mathbb{N}$. For all $u \in \mathcal{V}_{n,\mathrm{p}}(\mathbb{C})$ and all $k \in \mathbb{Z}$, define

$$\hat{u}_k = \frac{1}{n}\sum_{\ell=0}^{n-1} u_\ell e^{-2\pi \mathrm{i} k\ell/n}.$$

The grid function $\hat{u} \in \mathcal{V}(\mathbb{C})$ is called the **Discrete Fourier Transform** (DFT) of u, and we write $\hat{u} = \mathcal{F}_n[u]$.

Proposition 13.9 (periodicity). *Let $n \in \mathbb{N}$. For all $u \in \mathcal{V}_{n,\mathrm{p}}(\mathbb{C})$, we have $\hat{u} \in \mathcal{V}_{n,\mathrm{p}}(\mathbb{C})$, i.e., the DFT is a periodic grid function.*

Proof. See Problem 13.3. □

Proposition 13.10 (periodic grid delta). *Suppose that $n \in \mathbb{N}$. For any $k \in \mathbb{Z}$,*

$$\delta_k^{n,\mathrm{p}} = \frac{1}{n}\sum_{\ell=0}^{n-1} e^{2\pi \mathrm{i} k\ell/n}. \tag{13.2}$$

Proof. Suppose that $k \neq mn$ for any $m \in \mathbb{Z}$. Then

$$e^{2\pi \mathrm{i} k/n}\frac{1}{n}\sum_{\ell=0}^{n-1} e^{2\pi \mathrm{i} k\ell/n} = \frac{1}{n}\sum_{\ell=0}^{n-1} e^{2\pi \mathrm{i} k(\ell+1)/n} = \frac{1}{n}\sum_{\ell=1}^{n} e^{2\pi \mathrm{i} k\ell/n} = \frac{1}{n}\sum_{\ell=0}^{n-1} e^{2\pi \mathrm{i} k\ell/n},$$

where the last equality follows from the periodicity of $e^{2\pi \mathrm{i} k\ell/n}$ with respect to ℓ. Thus,

$$e^{2\pi \mathrm{i} k/n}\frac{1}{n}\sum_{\ell=0}^{n-1} e^{2\pi \mathrm{i} k\ell/n} = \frac{1}{n}\sum_{\ell=0}^{n-1} e^{2\pi \mathrm{i} k\ell/n} \iff \left(e^{2\pi \mathrm{i} k/n} - 1\right)\frac{1}{n}\sum_{\ell=0}^{n-1} e^{2\pi \mathrm{i} k\ell/n} = 0.$$

But, because it is assumed that $k \neq mn$,

$$e^{2\pi \mathrm{i} k/n} \neq 1.$$

Therefore, it follows that the sum must equal zero. On the other hand, when $k = mn$, for some $m \in \mathbb{Z}$,

$$e^{2\pi \mathrm{i} k\ell/n} = e^{2\pi \mathrm{i} mn\ell/n} = e^{2\pi m\ell} = 1.$$

Thus,

$$\frac{1}{n}\sum_{\ell=0}^{n-1} e^{2\pi \mathrm{i} k\ell/n} = \frac{1}{n}\sum_{\ell=0}^{n-1} e^{2\pi \mathrm{i} k\ell/n} = \frac{1}{n}\sum_{\ell=0}^{n-1} 1 = \frac{1}{n} n = 1.$$

The claim then follows. □

With this property at hand we can prove that the DFT is a bijection.

Theorem 13.11 (bijection). *Let $n \in \mathbb{N}$. The DFT mapping $\mathcal{F}_n[u] = \hat{u}$ is a linear, one-to-one, onto mapping from $\mathcal{V}_{n,\mathrm{p}}(\mathbb{C})$ to $\mathcal{V}_{n,\mathrm{p}}(\mathbb{C})$. Therefore, it has an inverse.*

Proof. Clearly, $\mathcal{F}_n$ is linear and maps into $\mathcal{V}_{n,\mathrm{p}}(\mathbb{C})$. To show that the mapping is one to one, suppose that $u, w \in \mathcal{V}_{n,\mathrm{p}}(\mathbb{C})$ have the same DFT, i.e., for every $k \in \mathbb{Z}$,

$$\frac{1}{n}\sum_{\ell=0}^{n-1} u_\ell e^{-2\pi \mathrm{i} k\ell/n} = \frac{1}{n}\sum_{\ell=0}^{n-1} w_\ell e^{-2\pi \mathrm{i} k\ell/n}.$$

Thus, for all $k \in \mathbb{Z}$,

$$\frac{1}{n}\sum_{\ell=0}^{n-1} (u_\ell - w_\ell) e^{-2\pi \mathrm{i} k\ell/n} = 0.$$

Then, for any $j \in \mathbb{Z}$,

$$\begin{aligned}
0 &= \frac{1}{n}\sum_{k=0}^{n-1} e^{2\pi \mathrm{i} jk/n} \sum_{\ell=0}^{n-1} (u_\ell - w_\ell) e^{-2\pi \mathrm{i} k\ell/n} \\
&= \frac{1}{n}\sum_{\ell=0}^{n-1} (u_\ell - w_\ell) \sum_{k=0}^{n-1} e^{2\pi \mathrm{i}(j-\ell)k/n} \\
&= \sum_{\ell=0}^{n-1} (u_\ell - w_\ell) \delta_{j-\ell}^{n,\mathrm{p}} \\
&= \sum_{\ell=0}^{n-1} (u_\ell - w_\ell) \delta_{\ell-j}^{n,\mathrm{p}} \\
&= u_j - w_j,
\end{aligned}$$

where we have used (13.1) and (13.2). Since $j \in \mathbb{Z}$ is arbitrary, $u = w$, which proves that $\mathcal{F}_n$ is one to one.

To show that $\mathcal{F}_n$ is onto, suppose that $v \in \mathcal{V}_{n,\mathrm{p}}(\mathbb{C})$ is arbitrary. We want to find some $w \in \mathcal{V}_{n,\mathrm{p}}(\mathbb{C})$ such that $\hat{w} = \mathcal{F}_n[w] = v$. Define

$$w_k = \sum_{\ell=0}^{n-1} v_\ell e^{2\pi \mathrm{i} k\ell/n}.$$

Then

$$\begin{aligned}
\hat{w}_j = \mathcal{F}_n[w]_j &= \frac{1}{n}\sum_{k=0}^{n-1} w_k e^{-2\pi \mathrm{i} jk/n} \\
&= \frac{1}{n}\sum_{k=0}^{n-1}\left(\sum_{\ell=0}^{n-1} v_\ell e^{2\pi \mathrm{i} k\ell/n}\right) e^{-2\pi \mathrm{i} jk/n} \\
&= \frac{1}{n}\sum_{k=0}^{n-1}\sum_{\ell=0}^{n-1} v_\ell e^{2\pi \mathrm{i}(\ell-j)k/n} \\
&= \sum_{\ell=0}^{n-1} v_\ell \frac{1}{n}\sum_{k=0}^{n-1} e^{2\pi \mathrm{i}(\ell-j)k/n} \\
&= \sum_{\ell=0}^{n-1} v_\ell \delta_{\ell-j}^{n,\mathrm{p}} \\
&= v_j.
\end{aligned}$$

We have again used the identities (13.1) and (13.2). The proof is complete. □

Definition 13.12 (IDFT). The **Inverse Discrete Fourier Transform** (IDFT) is defined via

$$\mathcal{F}_n^{-1}[w]_k = \sum_{\ell=0}^{n-1} w_\ell e^{2\pi \mathrm{i} k\ell/n}$$

for all $w \in \mathcal{V}_{n,\mathrm{p}}(\mathbb{C})$.

From Theorem 13.11, it immediately follows that the IDFT is a bijection as well.

Corollary 13.13 (bijection). *Let $n \in \mathbb{N}$. The IDFT mapping $\mathcal{F}_n^{-1}$ is a linear, one-to-one, onto mapping from $\mathcal{V}_{n,\mathrm{p}}(\mathbb{C})$ to $\mathcal{V}_{n,\mathrm{p}}(\mathbb{C})$, and it is the inverse of the DFT mapping $\mathcal{F}_n$:*

$$w = \mathcal{F}_n\left[\mathcal{F}_n^{-1}[w]\right] = \mathcal{F}_n^{-1}\left[\mathcal{F}_n[w]\right]$$

for all $w \in \mathcal{V}_{n,\mathrm{p}}(\mathbb{C})$.

Proof. It suffices to examine the proof of Theorem 13.11. □

Theorem 13.14 (discrete Parseval[1]). *Let $n \in \mathbb{N}$. Suppose that $w \in \mathcal{V}_{n,\mathrm{p}}(\mathbb{C})$ and $\hat{w} \in \mathcal{V}_{n,\mathrm{p}}(\mathbb{C})$ is its DFT. Then*

$$\frac{1}{n}\sum_{k=0}^{n-1} w_k \overline{w_k} = \sum_{k=0}^{n-1} \hat{w}_k \overline{\hat{w}_k}. \tag{13.3}$$

Proof. See Problem 13.4. □

[1] Named in honor of the French mathematician Marc-Antoine Parseval des Cheênes (1755–1836).

Definition 13.15 (discrete convolution)**.** Suppose that $n \in \mathbb{N}$ and $u, v \in \mathcal{V}_{n,\mathrm{p}}(\mathbb{C})$. Define, for all $k \in \mathbb{Z}$,

$$[u \star v]_k^{n,\mathrm{p}} = \frac{1}{n}\sum_{j=0}^{n-1} u_{k-j}v_j. \tag{13.4}$$

The operator $[\cdot \star \cdot]^{n,\mathrm{p}} : \mathcal{V}_{n,\mathrm{p}}(\mathbb{C}) \times \mathcal{V}_{n,\mathrm{p}}(\mathbb{C}) \to \mathcal{V}(\mathbb{C})$ is called the **discrete periodic convolution**.

Let us establish a couple of properties of the discrete convolution.

Proposition 13.16 (properties of discrete convolution)**.** *Suppose that $n \in \mathbb{N}$ and $u, v \in \mathcal{V}_{n,\mathrm{p}}(\mathbb{C})$ are arbitrary. Then $[u \star v]^{n,\mathrm{p}} \in \mathcal{V}_{n,\mathrm{p}}(\mathbb{C})$. Furthermore, the discrete convolution is commutative, i.e., $[u \star v]^{n,\mathrm{p}} = [v \star u]^{n,\mathrm{p}}$, and*

$$\mathcal{F}_n\left[[u \star v]^{n,\mathrm{p}}\right] = \mathcal{F}_n[u]\mathcal{F}_n[v].$$

Proof. See Problem 13.5. □

In addition to the standard grid delta function, we also need a singular version of this function.

Definition 13.17 (singular periodic grid delta)**.** Suppose that $n \in \mathbb{N}$. The **singular periodic grid delta function**, denoted $\tilde{\delta}^{n,\mathrm{p}} \in \mathcal{V}_{n,\mathrm{p}}(\mathbb{C})$, is defined via

$$\tilde{\delta}_j^{n,\mathrm{p}} = n\delta_j^{n,\mathrm{p}}.$$

The following is a simple consequence of (13.1).

Proposition 13.18 (convolution)**.** *Suppose that $n \in \mathbb{N}$ and $u \in \mathcal{V}_{n,\mathrm{p}}(\mathbb{C})$ is arbitrary. Then*

$$\left[u \star \tilde{\delta}^{n,\mathrm{p}}\right]^{n,\mathrm{p}} = u.$$

Proof. See Problem 13.6. □

Now let us relate periodic functions to periodic grid functions via a new type of projection.

Definition 13.19 (grid projection)**.** Suppose that $n \in \mathbb{N}$. The **periodic grid projection operator**, $\mathcal{G}_{n,\mathrm{p}}\colon C_\mathrm{p}(0,1;\mathbb{C}) \to \mathcal{V}_{n,\mathrm{p}}(\mathbb{C})$, is defined as follows: for any $f \in C_\mathrm{p}(0,1;\mathbb{C})$, $\mathcal{G}_{n,\mathrm{p}}[f] \in \mathcal{V}_{n,\mathrm{p}}(\mathbb{C})$ is the grid function

$$\mathcal{G}_{n,\mathrm{p}}[f]_j = f(j/n), \quad \forall j \in \mathbb{Z}.$$

Next, we can extend the DFT to continuous one-periodic functions in a natural way.

Definition 13.20 (DFT of continuous functions)**.** Let $n \in \mathbb{N}$ and $f \in C_\mathrm{p}(0,1;\mathbb{C})$. For $j \in \mathbb{Z}$, the numbers

$$\hat{f}_{n,j} = \frac{1}{n}\sum_{\ell=0}^{n-1} \mathcal{G}_{n,\mathrm{p}}[f]_\ell \exp(-2\pi\mathrm{i}j\ell/n) = \frac{1}{n}\sum_{\ell=0}^{n-1} f(\ell/n)\exp(-2\pi\mathrm{i}j\ell/n) \tag{13.5}$$

are called the **discrete Fourier coefficients of** f. The grid function

$$\mathcal{F}_n[\mathcal{G}_{n,\mathrm{p}}[f]] \in \mathcal{V}_{n,\mathrm{p}}(\mathbb{C})$$

is called the **Discrete Fourier Transform** of f.

To summarize, in order to compute the DFT of a continuous one-periodic function, we first project it into the space of periodic grid functions and then we apply the DFT. We can then take the IDFT of that object but, in general, we will not get back the original function, only its grid projection.

13.3 Existence and Uniqueness of the Interpolant

We have all of the necessary machinery in place. Let us get back to the subject of trigonometric interpolation.

Theorem 13.21 (existence and uniqueness). *Let $f \in C_{\mathrm{p}}(0, 1; \mathbb{C})$ and $n \in \mathbb{N}$. Suppose that*

$$X = \{x_j\}_{j=0}^{n-1}, \qquad x_j = \frac{j}{n}, \quad j = 0, \dots, n-1$$

is a uniformly spaced nodal set. Let $K \in \mathbb{N}$ satisfy $n = 2K + 1$ if n is odd and $n = 2K$ if n is even. If n is odd, there exists a unique trigonometric polynomial $q \in \mathbb{T}_K$ of the form

$$q(x) = \sum_{j=-K}^{K} c_j \exp(2\pi \mathrm{i} j x), \tag{13.6}$$

which interpolates f at the nodes X. If n is even, there exists a unique interpolating polynomial $q \in \mathbb{T}_K$, subordinate to X, of the form

$$\begin{aligned} q(x) &= \frac{c_K}{2}(\exp(2\pi \mathrm{i} K x) + \exp(-2\pi \mathrm{i} K x)) + \sum_{j=-K+1}^{K-1} c_j \exp(2\pi \mathrm{i} j x) \\ &= c_K \cos(2\pi K x) + \sum_{j=-K+1}^{K-1} c_j \exp(2\pi \mathrm{i} j x). \end{aligned} \tag{13.7}$$

Proof. Using the grid projection operator, set $F = \mathcal{G}_{n,p}[f] \in \mathcal{V}_{n,\mathrm{p}}(\mathbb{C})$. Then $F_\ell = f(x_\ell) = f(\ell/n)$ for $\ell = 0, \dots, n-1$. We require, in the odd case ($n = 2K + 1$), that

$$F_\ell = q(x_\ell) = \sum_{j=-K}^{K} c_j \exp(2\pi \mathrm{i} j x_\ell), \quad \ell = 0, \dots, n-1. \tag{13.8}$$

In the even case ($n = 2K$), we have, for $\ell = 0, \ldots, n-1$,

$$q(x_\ell) = \frac{c_K}{2}\left(\exp(2\pi i K x_\ell) + \exp(-2\pi i K x_\ell)\right) + \sum_{j=-K+1}^{K-1} c_j \exp(2\pi i j x_\ell)$$

$$= c_K \exp(2\pi i K x_\ell) + \sum_{j=-K+1}^{K-1} c_j \exp(2\pi i j x_\ell) \tag{13.9}$$

$$= \sum_{j=-K+1}^{K} c_j \exp(2\pi i j x_\ell).$$

Next, we can view the coefficients, c_j, as the discrete values of a grid function, which we label c. We can naturally extend c, so that it is also periodic, i.e., $c \in \mathcal{V}_{n,p}(\mathbb{C})$. These facts allow us to shift the summation indices in (13.8) and (13.9), so that we may write both expressions, regardless of parity, as

$$f(\ell/n) = F_\ell = \sum_{j=0}^{n-1} c_j \exp(2\pi i j x_\ell), \quad \ell = 0, \ldots, n-1.$$

Finally, using the DFT, the unique solution, regardless of parity, is given by

$$c_j = \hat{f}_{n,j} = \frac{1}{n}\sum_{\ell=0}^{n-1} f(\ell/n)\exp(-2\pi i j x_\ell), \quad j = 0, \ldots, n-1;$$

see Theorem 13.11. To get the coefficients with the desired indices, we can use the periodicity of the grid function c. □

Example 13.1 Consider the one-periodic function

$$f(x) = \exp(\sin(2\pi x) - 0.1\cos(2\pi x) + 0.2\cos(4\pi x) + 0.2\sin(6\pi x)),$$

which is shown in Figures 13.1, 13.2, and 13.3. We interpolate this function using $n = 6$ (Figure 13.1), $n = 8$ (Figure 13.2), and $n = 16$ (Figure 13.3) points, respectively. Note that it is straightforward to convert the coefficients from (13.7) to those for the expansion

$$q(x) = a_0 + \sum_{k=1}^{K-1}\left(a_k \cos(2\pi k x) + b_k \sin(2\pi k x)\right) + a_K \cos(2\pi K x), \quad n = 2K.$$

We leave it to the reader to derive the appropriate conversion. In any case, notice that the interpolant gets better as n gets larger. Of course, one might worry that adding more points might lead to some kind of undesired Runge phenomenon, where the interpolant oscillates more, as larger numbers of equally spaced points are sampled. In the periodic case, however, this does not happen for well-behaved functions f, as we will show in the next section.

We have demonstrated an interesting connection between trigonometric interpolation at evenly spaced points and the DFT, which we utilize later in Chapter 26 to solve certain differential equations. In the meantime, a number of standard questions arise: *How accurate is this interpolation? Does it converge as $n \to \infty$? In what norms can we expect convergence?* We will answer these questions in the next section.

13.4 Alias Error and Convergence of Trigonometric Interpolation

We can, as usual, think of the process of interpolation as an operation, mapping a function from $C_{\mathrm{p}}(0,1;\mathbb{C})$ to a function in $\mathbb{T}_K$. Let us make a useful definition.

Definition 13.22 (trigonometric interpolation)**.** Let $n \in \mathbb{N}$. Suppose that $K \in \mathbb{N}$ satisfies $n = 2K$ when n is even and $n = 2K+1$ when n is odd. The **trigonometric interpolation operator**, denoted $\mathcal{I}_{n,\mathrm{p}}$, is defined as follows:

$$\begin{aligned} \mathcal{I}_{n,\mathrm{p}} \colon C_{\mathrm{p}}(0,1;\mathbb{C}) &\to \mathbb{T}_K, \\ f &\mapsto \mathcal{I}_{n,\mathrm{p}}[f] = q_n, \end{aligned} \tag{13.10}$$

where $q_n \in \mathbb{T}_K$ is the unique trigonometric interpolating polynomial for f, subordinate to $X = \{j/n\}_{j=0}^{n-1}$, satisfying (13.7) when n is even and (13.6) when n is odd.

Proposition 13.23 (properties of $\mathcal{I}_{n,\mathrm{p}}$)**.** *Let $n \in \mathbb{N}$. Suppose that $K \in \mathbb{N}$ satisfies $n = 2K$ when n is even and $n = 2K+1$ when n is odd. $\mathcal{I}_{n,\mathrm{p}}$ is a linear projection operator, i.e.,*

$$\mathcal{I}_{n,\mathrm{p}}[f+g] = \mathcal{I}_{n,\mathrm{p}}[f] + \mathcal{I}_{n,\mathrm{p}}[g], \quad \forall f, g \in C_{\mathrm{p}}(0,1;\mathbb{C}),$$

and

$$\mathcal{I}_{n,\mathrm{p}}[q] = q, \quad \forall q \in \mathbb{T}_K.$$

Proof. See Problem 13.7. □

The goal of this section is to estimate the size of $\|f - \mathcal{I}_{n,\mathrm{p}}[f]\|_{L^2(0,1;\mathbb{C})}$.

Suppose that $n \in \mathbb{N}$ is fixed. Let $K \in \mathbb{N}$ satisfy $n = 2K+1$ if n is odd and $n = 2K$ if n is even. It follows from the proof of Theorem 13.21 and Definition 13.20 that, for $f \in C_{\mathrm{p}}(0,1;\mathbb{C})$, if n is odd,

$$\mathcal{I}_{n,\mathrm{p}}[f](x) = \sum_{j=-K}^{K} \hat{f}_{n,j} \exp(2\pi \mathrm{i} j x), \quad \forall x \in \mathbb{R}, \tag{13.11}$$

and, if n is even,

$$\mathcal{I}_{n,\mathrm{p}}[f](x) = \hat{f}_{n,K} \cos(2\pi K x) + \sum_{j=-(K-1)}^{K-1} \hat{f}_{n,j} \exp(2\pi \mathrm{i} j x), \quad \forall x \in \mathbb{R}. \tag{13.12}$$

Suppose that $n \in \mathbb{N}$ is fixed. To properly measure the accuracy of trigonometric interpolation, we need to compare $\hat{f}_{n,j}$ and $\hat{f}_j$. Recall that

$$\hat{f}_{n,j} = \frac{1}{n}\sum_{\ell=0}^{n-1} f(\ell/n)\exp(-2\pi \mathrm{i} j\ell/n), \qquad \hat{f}_j = \int_0^1 f(x)\exp(-2\pi \mathrm{i} jx)\mathrm{d}x.$$

The first object is just a particular Riemann sum approximation[2] — a trapezoidal rule approximation, in particular — of the second. Thus, we may be justified in writing $\hat{f}_{n,j} \approx \hat{f}_j$. *But how good an approximation is this?*

Before we answer that last question completely, let us compute a useful expression for the difference $\hat{f}_{n,j} - \hat{f}_j$, called the *alias error.*

Lemma 13.24 (alias error). *Let $n \in \mathbb{N}$. Suppose that $f \in H_{\mathrm{p}}^{\alpha}(0,1;\mathbb{C})$ with $\alpha > \frac{1}{2}$. Then*

$$\hat{f}_{n,j} = \hat{f}_j + \sum_{\ell \in \mathbb{Z}_\star} \hat{f}_{j+\ell}\delta_\ell^{n,\mathrm{p}},$$

or, equivalently,

$$\hat{f}_{n,j} - \hat{f}_j = \sum_{m \in \mathbb{Z}_\star} \hat{f}_{j+mn}.$$

Proof. By Theorem 12.34, the Fourier series of f converges uniformly and absolutely to f, and we may write

$$f(x) = \sum_{k \in \mathbb{Z}} \hat{f}_k \exp(2\pi \mathrm{i} kx), \quad \forall x \in \mathbb{R}.$$

By definition, i.e., 13.5, for all $j \in \mathbb{Z}$,

$$\begin{aligned}
\hat{f}_{n,j} &= \frac{1}{n}\sum_{\ell=0}^{n-1} f(\ell/n)\exp(-2\pi \mathrm{i} j\ell/n) \\
&= \frac{1}{n}\sum_{\ell=0}^{n-1}\sum_{k \in \mathbb{Z}} \hat{f}_k \exp(2\pi \mathrm{i} k\ell/n)\exp(-2\pi \mathrm{i} j\ell/n) \\
&= \sum_{k \in \mathbb{Z}} \hat{f}_k \frac{1}{n}\sum_{\ell=0}^{n-1} \exp(2\pi \mathrm{i}(k-j)\ell/n) \\
&= \sum_{k \in \mathbb{Z}} \hat{f}_k \delta_{k-j}^{n,\mathrm{p}},
\end{aligned}$$

using (13.2). Our manipulations above are justified because the Fourier series of f converges uniformly and absolutely. Shifting indices, we have

$$\hat{f}_{n,j} = \hat{f}_j + \sum_{\ell \in \mathbb{Z}_\star} \hat{f}_{j+\ell}\delta_\ell^{n,\mathrm{p}}.$$

Using the definition of the periodic grid delta function, we have, alternately,

$$\hat{f}_{n,j} = \hat{f}_j + \sum_{m \in \mathbb{Z}_\star} \hat{f}_{j+m \cdot n}.$$ □

[2] Named in honor of the German mathematician Georg Friedrich Bernhard Riemann (1826–1866).

Now we come to the main result of the present section.

Theorem 13.25 (convergence). *Let $n \in \mathbb{N}$. Suppose that $K \in \mathbb{N}$ satisfies $n = 2K$ when n is even and $n = 2K + 1$ when n is odd. Let $\mathcal{I}_{n,\mathrm{p}}$ denote the trigonometric interpolation operator defined in* (13.10)*. Assume that $\alpha > \frac{1}{2}$ and $f \in H^{\alpha}_{\mathrm{p}}(0,1;\mathbb{C})$. There is a constant $C > 0$ independent of f and n such that*

$$\|f - \mathcal{I}_{n,\mathrm{p}}[f]\|_{L^2(0,1;\mathbb{C})} \leq \frac{C}{n^{\alpha}} |f|_{H^{\alpha}(0,1;\mathbb{C})}.$$

Proof. Suppose that n is odd. Using Lemma 13.24 and the Cauchy–Schwarz inequality, we have

$$\begin{aligned}
\|\mathcal{S}_K[f] - \mathcal{I}_{n,\mathrm{p}}[f]\|^2_{L^2(0,1;\mathbb{C})} &= \sum_{j=-K}^{K} \left|\hat{f}_j - \hat{f}_{n,j}\right|^2 \\
&= \sum_{j=-K}^{K} \left| \sum_{m \in \mathbb{Z}_\star} \hat{f}_{j+mn} \right|^2 \\
&= \sum_{j=-K}^{K} \left| \sum_{m \in \mathbb{Z}_\star} \gamma_{j+mn}^{-\alpha/2} \gamma_{j+mn}^{\alpha/2} \hat{f}_{j+mn} \right|^2 \\
&\leq \sum_{j=-K}^{K} \left(\sqrt{\sum_{m \in \mathbb{Z}_\star} \gamma_{j+mn}^{-\alpha}} \sqrt{\sum_{m \in \mathbb{Z}_\star} \gamma_{j+mn}^{\alpha} \left|\hat{f}_{j+mn}\right|^2} \right)^2 \\
&= \sum_{j=-K}^{K} \left[\sum_{m \in \mathbb{Z}_\star} \gamma_{j+mn}^{-\alpha} \sum_{m \in \mathbb{Z}_\star} \gamma_{j+mn}^{\alpha} \left|\hat{f}_{j+mn}\right|^2 \right] \\
&\leq \max_{-K \leq j \leq K} \left\{ \sum_{m \in \mathbb{Z}_\star} \gamma_{j+mn}^{-\alpha} \right\} \sum_{j=-K}^{K} \sum_{m \in \mathbb{Z}_\star} \gamma_{j+mn}^{\alpha} \left|\hat{f}_{j+mn}\right|^2 .
\end{aligned}$$

Observe now that

$$\begin{aligned}
\max_{-K \leq j \leq K} \left\{ \sum_{m \in \mathbb{Z}_\star} \gamma_{j+mn}^{-\alpha} \right\} &= \frac{1}{(4\pi^2)^{\alpha}} \max_{-K \leq j \leq K} \sum_{m \in \mathbb{Z}_\star} \frac{1}{(j + mn)^{2\alpha}} \\
&= \frac{1}{(4\pi^2)^{\alpha}} \frac{1}{n^{2\alpha}} \max_{-K \leq j \leq K} \sum_{m \in \mathbb{Z}_\star} \frac{1}{(j/n + m)^{2\alpha}} \\
&\leq \frac{1}{(4\pi^2)^{\alpha}} \frac{1}{n^{2\alpha}} \sum_{m \in \mathbb{Z}_\star} \frac{1}{(|m| - 1/2)^{2\alpha}}
\end{aligned}$$

$$
\begin{aligned}
&\leq \frac{2}{(4\pi^2)^\alpha}\frac{1}{n^{2\alpha}}\sum_{m=1}^{\infty}\frac{1}{(m-1/2)^{2\alpha}} \\
&= \frac{2^{2\alpha+1}}{(4\pi^2)^\alpha}\frac{1}{n^{2\alpha}}\sum_{m=1}^{\infty}\frac{1}{(2m-1)^{2\alpha}} \\
&\leq \frac{2^{2\alpha+1}}{(4\pi^2)^\alpha}\frac{1}{n^{2\alpha}}\sum_{m=1}^{\infty}\frac{1}{m^{2\alpha}} \\
&= \frac{2^{2\alpha+1}\zeta(2\alpha)}{(4\pi^2)^\alpha}\frac{1}{n^{2\alpha}}.
\end{aligned}
$$

Next, we estimate the second term:

$$
\sum_{j=-K}^{K}\sum_{m\in\mathbb{Z}_\star}\gamma_{j+mn}^{\alpha}\left|\hat{f}_{j+mn}\right|^2 \leq \sum_{\ell\in\mathbb{Z}_\star}\gamma_\ell^\alpha\left|\hat{f}_\ell\right|^2 = |f|^2_{H^\alpha(0,1;\mathbb{C})}.
$$

Putting these last two estimates together, we have

$$
\|\mathcal{S}_K[f] - \mathcal{I}_{n,\mathrm{p}}[f]\|_{L^2(0,1;\mathbb{C})} \leq Cn^{-\alpha}|f|_{H^\alpha(0,1;\mathbb{C})} \tag{13.13}
$$

for some constant $C > 0$ that depends upon α but is independent of n.

Finally, using the triangle inequality and Theorem 12.35,

$$
\begin{aligned}
\|f - \mathcal{I}_{n,\mathrm{p}}[f]\|_{L^2(0,1;\mathbb{C})} &= \|f - \mathcal{S}_K[f] + \mathcal{S}_K[f] - \mathcal{I}_{n,\mathrm{p}}[f]\|_{L^2(0,1;\mathbb{C})} \\
&\leq \|f - \mathcal{S}_K[f]\|_{L^2(0,1;\mathbb{C})} + \|\mathcal{S}_K[f] - \mathcal{I}_{n,\mathrm{p}}[f]\|_{L^2(0,1;\mathbb{C})} \\
&\leq C\frac{1}{K^\alpha}|f|_{H^\alpha(0,1;\mathbb{C})} + C\frac{1}{n^\alpha}|f|_{H^\alpha(0,1;\mathbb{C})} \\
&= C\frac{1}{n^\alpha}|f|_{H^\alpha(0,1;\mathbb{C})}
\end{aligned}
$$

for some constant $C > 0$ that depends upon α but is independent of f and n.

The case for which n is even is only a little more tedious and is left to the reader as an exercise; see Problem 13.8. □

Before we end this section, let us give another estimate of the interpolation error. This estimate is not sharp, but it is very easy to produce.

Theorem 13.26 (uniform convergence). *Let $n \in \mathbb{N}$. Suppose that $K \in \mathbb{N}$ satisfies $n = 2K$ when n is even and $n = 2K + 1$ when n is odd. Let $\mathcal{I}_{n,\mathrm{p}}$ denote the trigonometric interpolation operator defined in* (13.10). *If $f \in C_{\mathrm{p}}^r(0,1;\mathbb{C})$ for some*

$r \geq 2$, then there is a constant $C > 0$ that depends on f and r but is independent of n such that

$$\|f - \mathcal{I}_{n,\mathrm{p}}[f]\|_{L^\infty(0,1;\mathbb{C})} \leq \frac{C}{n^{r-1}}.$$

Proof. Again, we only prove the estimate for the case that n is odd and leave the even case to the reader as an exercise; see Problem 13.9. Since $f \in C_{\mathrm{p}}^r(0,1;\mathbb{C}) \subset H_{\mathrm{p}}^r(0,1;\mathbb{C})$, $r \geq 2 > \frac{1}{2}$, Theorem 13.25 applies. More specifically, we use estimate (13.13) from Theorem 13.25 to get

$$\begin{aligned}
\|\mathcal{S}_K[f] - \mathcal{I}_{n,\mathrm{p}}[f]\|_{L^\infty(0,1;\mathbb{C})} &\leq \sum_{j=-K}^{K} |\hat{f}_j - \hat{f}_{n,j}| \\
&\leq \sqrt{\sum_{j=-K}^{K} 1^2}\sqrt{\sum_{j=-K}^{K} |\hat{f}_j - \hat{f}_{n,j}|^2} \\
&= \sqrt{n}\,\|\mathcal{S}_K[f] - \mathcal{I}_{n,\mathrm{p}}[f]\|_{L^2(0,1;\mathbb{C})} \\
&\leq Cn^{-r+1/2}|f|_{H^\alpha(0,1;\mathbb{C})} \\
&= Cn^{-r+1/2}.
\end{aligned}$$

Using the triangle inequality and Corollary 12.22,

$$\begin{aligned}
\|f - \mathcal{I}_{n,\mathrm{p}}[f]\|_{L^\infty(0,1;\mathbb{C})} &= \|f - \mathcal{S}_K[f] + \mathcal{S}_K[f] - \mathcal{I}_{n,\mathrm{p}}[f]\|_{L^\infty(0,1;\mathbb{C})} \\
&\leq \|f - \mathcal{S}_K[f]\|_{L^\infty(0,1;\mathbb{C})} + \|\mathcal{S}_K[f] - \mathcal{I}_{n,\mathrm{p}}[f]\|_{L^\infty(0,1;\mathbb{C})} \\
&\leq \frac{2C_2}{r-1}\frac{1}{K^{r-1}} + C\frac{1}{n^{r-1/2}} \\
&\leq C\frac{1}{n^{r-1}}
\end{aligned}$$

for some constant $C > 0$ that depends upon r and f but is independent of n. □

While the proof is too lengthy and complicated to be included here, the following, due to Jackson [48], gives the sharpest result known regarding uniform convergence of trigonometric interpolation.

Theorem 13.27 (Jackson[3]). *Let $n \in \mathbb{N}$. Suppose that $K \in \mathbb{N}$ satisfies $n = 2K$ when n is even and $n = 2K + 1$ when n is odd. Let $\mathcal{I}_{n,\mathrm{p}}$ denote the trigonometric interpolation operator defined in* (13.10). *If $f \in C_{\mathrm{p}}^r(0,1;\mathbb{C})$ for some $r \geq 1$, then there is a constant $C > 0$ that depends on r but is independent of f and n such that*

$$\|f - \mathcal{I}_{n,\mathrm{p}}[f]\|_{L^\infty(0,1;\mathbb{C})} \leq C\frac{\log(n)}{n^r}\left\|f^{(r)}\right\|_{L^\infty(0,1;\mathbb{C})}.$$

[3] Named in honor of the American mathematician Dunham Jackson (1888–1946).

13.5 Numerical Integration of Periodic Functions

Next, let us introduce a couple of results regarding the numerical integration, i.e., the numerical quadrature, of periodic functions. In particular, we show that a certain quadrature rule is spectrally accurate. This will nicely set the stage for the topic of the next chapter, which is numerical quadrature in general.

Theorem 13.28 (trapezoidal rule). *Let $n \in \mathbb{N}$. Suppose that, for some $r \geq 1$, $g \in C_{\mathrm{p}}^r(0,1;\mathbb{C})$. Let*

$$X = \{x_j\}_{j=0}^{n-1}, \quad x_j = \frac{j}{n}, \quad j = 0, \dots, n-1$$

be a uniformly spaced nodal set. Then

$$\left| \int_0^1 g(x)\mathrm{d}x - \frac{1}{n}\sum_{j=0}^{n-1} g(x_j) \right| \leq \frac{C}{n^r}$$

for some constant $C = C(g,r) > 0$ that is independent of n but is dependent upon g and r. If $g \in C_{\mathrm{p}}^\infty(0,1;\mathbb{C})$, then

$$\left| \int_0^1 g(x)\mathrm{d}x - \frac{1}{n}\sum_{j=0}^{n-1} g(x_j) \right| \leq \frac{1}{n^m} C(g,m)$$

for any $m \in \mathbb{N}$, $m \geq 2$.

Proof. To begin, let us compute the error for numerically integrating a trigonometric monomial: define, for any $k \in \mathbb{Z}$,

$$E_{n,k} = \int_0^1 \exp(2\pi\mathrm{i}kx)\mathrm{d}x - \frac{1}{n}\sum_{j=0}^{n-1} \exp(2\pi\mathrm{i}kx_j).$$

From (13.2), we have

$$\frac{1}{n}\sum_{j=0}^{n-1} \exp(2\pi\mathrm{i}kx_j) = \delta_k^{n,\mathrm{p}}.$$

On the other hand,

$$\int_0^1 \exp(2\pi\mathrm{i}kx)\mathrm{d}x = \delta_{0,k}, \quad k \in \mathbb{Z},$$

where $\delta_{i,j}$ is a standard Kronecker delta function[4]. Therefore,

$$E_{n,k} = \begin{cases} -1, & \exists m \in \mathbb{Z}_\star \mid k = mn, \\ 0, & \forall m \in \mathbb{Z}_\star \; k \neq mn. \end{cases}$$

Thus, if $q \in \mathbb{T}_K$, say

$$q(x) = \sum_{\ell=-K}^{K} c_\ell \exp(2\pi\mathrm{i}\ell x),$$

[4] Named in honor of the German mathematician Leopold Kronecker (1823–1891).

then

$$\int_0^1 q(x)\mathrm{d}x - \frac{1}{n}\sum_{j=0}^{n-1} q(x_j) = \sum_{\ell=-K}^{K} c_\ell E_{n,\ell} = -\sum_{\ell \in J_{K,n}} c_{\ell n},$$

where

$$J_{K,n} = \{\ell \in \mathbb{Z} \mid 0 < |\ell n| \le K\}.$$

Now if $g \in C_p^r(0,1;\mathbb{C})$ for $r \geq 2$, by Theorem 12.21, the Fourier partial sums $\mathcal{S}_k[f]$ converge uniformly to f, as $k \to \infty$. This allows us, by Theorem B.65, to interchange the limits of integration and summation below:

$$\begin{aligned}\int_0^1 g(x)\mathrm{d}x - \frac{1}{n}\sum_{j=0}^{n-1} g(x_j) &= \int_0^1 \sum_{k\in\mathbb{Z}} \hat{g}_k \exp(2\pi \mathrm{i} kx)\mathrm{d}x - \frac{1}{n}\sum_{j=0}^{n-1}\sum_{k\in\mathbb{Z}} \hat{g}_k \exp(2\pi \mathrm{i} kx_j)\\ &= \sum_{k\in\mathbb{Z}} \hat{g}_k E_{n,k}\\ &= -\sum_{k\in\mathbb{Z}_\star} \hat{g}_{kn}.\end{aligned}$$

By Corollary 12.22, since $g \in C_p^r(0,1;\mathbb{C})$,

$$\left| \int_0^1 g(x)\mathrm{d}x - \frac{1}{n}\sum_{j=0}^{n-1} g(x_j) \right| \le \sum_{k\in\mathbb{Z}_\star} |\hat{g}_{kn}| \le \frac{2C_2}{n^r}\sum_{k=1}^{\infty}\frac{1}{k^r} = \frac{2C_2}{n^r}\zeta(r). \qquad \square$$

To finish this section, we give a complementary result to the last, which requires a bit less regularity.

Theorem 13.29 (trapezoidal rule). *Let $n \in \mathbb{N}$. Suppose that*

$$X = \{x_j\}_{j=0}^{n-1}, \quad x_j = \frac{j}{n}, \quad j = 0, \dots, n-1$$

is a uniformly spaced nodal set. If $g \in H_p^\alpha(0,1;\mathbb{C})$ for some $\alpha > \frac{1}{2}$, then

$$\left| \int_0^1 g(x)\mathrm{d}x - \frac{1}{n}\sum_{j=0}^{n-1} g(x_j) \right| \le \frac{C}{n^\alpha}|g|_{H^\alpha(0,1;\mathbb{C})}$$

for some constant $C > 0$ that is independent of n and g but is dependent upon α.

Proof. By Theorem 12.34, since $\alpha > \frac{1}{2}$, the Fourier series of g converges uniformly and absolutely to g. Up to the last step, the computations in the proof of the last theorem hold. By the Cauchy–Schwarz inequality,

$$
\begin{aligned}
\left| \int_0^1 g(x)\mathrm{d}x - \frac{1}{n}\sum_{j=0}^{n-1} g(x_j) \right| &\le \sum_{k\in\mathbb{Z}_\star} |\hat{g}_{kn}| \\
&= \sum_{k\in\mathbb{Z}_\star} \gamma_{kn}^{-\alpha/2}\gamma_{kn}^{\alpha/2} |\hat{g}_{kn}| \\
&\le \sqrt{\sum_{k\in\mathbb{Z}_\star} \gamma_{kn}^{-\alpha}} \sqrt{\sum_{k\in\mathbb{Z}_\star} \gamma_{kn}^{\alpha} |\hat{g}_{kn}|^2} \\
&= \sqrt{\frac{1}{(4\pi^2)^\alpha n^{2\alpha}} \sum_{k\in\mathbb{Z}_\star} \frac{1}{k^{2\alpha}}} \sqrt{\sum_{k\in\mathbb{Z}_\star} \gamma_{kn}^{\alpha} |\hat{g}_{kn}|^2} \\
&= \frac{1}{(2\pi)^\alpha n^\alpha}\sqrt{2\zeta(2\alpha)} \sqrt{\sum_{k\in\mathbb{Z}_\star} \gamma_{kn}^{\alpha} |\hat{g}_{kn}|^2} \\
&= \frac{\sqrt{2\zeta(2\alpha)}}{(2\pi)^\alpha n^\alpha} |g|_{H^\alpha(0,1;\mathbb{C})}.
\end{aligned}
$$

The proof is complete. □

The previous two remarkable results show that the composite trapezoidal rule (14.16) is spectrally accurate; see Problem 14.17 for an alternate proof of Theorem 13.28 using the Euler–Maclaurin Theorem 14.40.

13.6 The Fast Fourier Transform (FFT)

Now let us discuss how the DFT can be computed efficiently — actually, very efficiently — in practice. In fact, what makes the DFT such a useful, ubiquitous tool in modern science, statistics, and mathematics is that it can be executed so efficiently in code. So it is hard to tell the story of the DFT without talking a bit about the fast algorithm used to compute it.

The DFT/IDFT is a very old object, known even before Fourier had his concept of trigonometric series. In fact, it was known to Gauss[5], who was interested in trigonometric interpolation applied to the prediction of the periodic motion of heavenly bodies [40]. The Fast Fourier Transform (FFT) is an algorithm for computing the DFT/IDFT using $\mathcal{O}(n\log_2(n))$ operations, rather than $\mathcal{O}(n^2)$ operations, the cost one would expect from straightforward application of Definition 13.8. The essence of the FFT algorithm was, it seems, invented by Gauss in 1805 in his work on trigonometric interpolation, though Gauss himself did not publish the discovery, it being to him such a trivial matter. It was rediscovered, in various forms, a couple of times over the years but more or less forgotten [40]. These days, the modern discovery of the FFT is usually attributed to James Cooley[6] and John Tukey[7] in a paper published in 1965 [20, 40].

[5] Johann Carl Friedrich Gauss (1777–1855) was a German mathematician.
[6] James William Cooley (1926–2016) was an American mathematician.
[7] John Wilder Tukey (1915–2000) was an American mathematician and statistician.

$n = 2^r$	r	Cost of FFT: (13.20) $T_n = 3rn + 2n$	Cost of standard DFT: (13.14) $2n^2$
64	6	1 280	8 192
128	7	2 944	32 768
256	8	6 656	131 072
512	9	14 848	524 288
1024	10	32 768	2 097 152
2048	11	71 680	8 388 608

Table 13.1 A comparison of the cost of the FFT algorithm and standard DFT for the case $n = 2^r$.

In this section, let us explain some of the details of the Cooley–Tukey FFT algorithm for the simplest case, i.e., when $n = 2^r$ for some $r \in \mathbb{N}$. This is sometimes called the *radix-2 case*. The algorithm is based on a divide and conquer strategy, and it involves rewriting the DFT as two separate DFTs of half the size. According to Definition 13.8, to compute the DFT, we must calculate the components

$$\hat{u}_k = \frac{1}{n}\sum_{\ell=0}^{n-1} u_\ell e^{-2\pi i k\ell/n}$$

for $k = 0, \dots, n-1 = 2^r - 1$. This can be construed as a matrix–vector multiplication. Specifically, define the matrix $\mathsf{W}_n = [w_{k,\ell}]_{k,\ell=0}^{n-1}$ via

$$w_{k,\ell} = \omega_{n,\ell}^{-k} = e^{-2\pi i k\ell/n}, \quad k, \ell = 0, \dots, n-1,$$

where $\omega_n \in \mathcal{V}_{n,\mathrm{p}}$ is the nth root of the unity grid function. We use the slightly unusual convention that column and row indexing start at 0 rather than 1. Then

$$\hat{u}_k = \frac{1}{n}\left[\mathsf{W}_n \boldsymbol{u}\right]_k,$$

where $\boldsymbol{u} = [u_0, u_1, \dots, u_{n-1}]^\intercal$; again, indexing starting with 0. The total cost of computing the DFT according to the straightforward formula above is, clearly,

$$\underbrace{n(n+1)}_{\text{multiplications}} + \underbrace{n(n-1)}_{\text{additions/subtractions}} = 2n^2 = \mathcal{O}(n^2). \tag{13.14}$$

This brute force approach, however, does not take any advantage of underlying symmetries. Exploiting the structure of the problem, we can lower the cost of computing the DFT dramatically, as is emphatically demonstrated in Table 13.1.

Using linearity, let us split this computation into two parts, one summing over the even indices and one over the odd indices:

$$\hat{u}_k = \frac{1}{n}\sum_{\ell=0}^{n/2-1} u_{2\ell} e^{-2\pi i k(2\ell)/n} + \frac{1}{n}\sum_{\ell=0}^{n/2-1} u_{2\ell+1} e^{-2\pi i k(2\ell+1)/n}.$$

We can extract a common factor out of the second summation, and, with just a little rewriting, we have

$$\hat{u}_k = \underbrace{\frac{1}{2}\frac{1}{n/2}\sum_{\ell=0}^{n/2-1} u_{2\ell} e^{-2\pi i \frac{k\ell}{n/2}}}_{\text{DFT even part of } u} + e^{-2\pi i k/n} \underbrace{\frac{1}{2}\frac{1}{n/2}\sum_{\ell=0}^{n/2-1} u_{2\ell+1} e^{-2\pi i \frac{k\ell}{n/2}}}_{\text{DFT odd part of } u}$$

for $k = 0, \ldots, n-1$. Now define

$$\hat{u}_k^{\mathrm{e}} = \frac{1}{n/2}\sum_{\ell=0}^{n/2-1} u_{2\ell} e^{-2\pi i \frac{k\ell}{n/2}}$$

and

$$\hat{u}_k^{\mathrm{o}} = \frac{1}{n/2}\sum_{\ell=0}^{n/2-1} u_{2\ell+1} e^{-2\pi i \frac{k\ell}{n/2}}$$

for $k = 0, \ldots, n-1$, so that

$$\hat{u}_k = \frac{1}{2}\hat{u}_k^{\mathrm{e}} + \frac{1}{2}e^{-2\pi i k/n}\hat{u}_k^{\mathrm{o}}, \quad k = 0, \ldots, n-1. \tag{13.15}$$

But observe that, for $k = 0, \ldots, n/2 - 1$,

$$\hat{u}_{k+n/2} = \frac{1}{2}\hat{u}_k^{\mathrm{e}} - \frac{1}{2}e^{-2\pi i k/n}\hat{u}_k^{\mathrm{o}}. \tag{13.16}$$

We leave it to the reader to prove the last equality using the periodicity of u and simple properties of the roots of unity; see Problem 13.11. Finally, we can conclude from (13.15) and (13.16) that

$$\begin{aligned} \hat{u}_k &= \frac{1}{2}\left(\hat{u}_k^{\mathrm{e}} + e^{-2\pi i k/n}\hat{u}_k^{\mathrm{o}}\right), \quad k = 0, \ldots, n/2-1, \\ \hat{u}_{k+n/2} &= \frac{1}{2}\left(\hat{u}_k^{\mathrm{e}} - e^{-2\pi i k/n}\hat{u}_k^{\mathrm{o}}\right), \quad k = 0, \ldots, n/2-1. \end{aligned} \tag{13.17}$$

In other words, the DFT of u can be constructed from the two DFTs of the even and odd parts of u, which are exactly half the size of the original. This is the basis of the FFT. It is computed recursively, using half-sized pieces at each level of recursion. In other words, we can compute $\hat{u}_k^{\mathrm{e}}$ and $\hat{u}_k^{\mathrm{o}}$ using quarter-sized pieces, and so on.

Using (13.17), let us compute the number of multiplications, M_n, required for the FFT algorithm at level n. Suppose that the number of multiplications required to compute the half-sized transforms, $\hat{u}_k^{\mathrm{e}}$ and $\hat{u}_k^{\mathrm{o}}$, via FFT is $M_{n/2}$. Then

$$M_n = 2M_{n/2} + 2n.$$

Of course, by recursion, we can assume that

$$M_{n/2} = 2M_{n/4} + 2n/2,$$

and so on, down to

$$M_2 = 2M_1 + 4, \quad M_1 = 2.$$

Naturally, we assume that factors such as $e^{-2\pi i k/n}$ are pre-computed and bring no extra cost to the algorithm. By induction, we can show that, if $n = 2^r$, $r \in \mathbb{N}$, the number of multiplications for the FFT at stage n is precisely

$$M_n = 2rn + 2n; \tag{13.18}$$

see Problem 13.12. Regarding the number of additions/subtractions, A_n, we clearly have

$$A_n = 2A_{n/2} + n.$$

By recursion, it follows that

$$A_{n/2} = 2A_{n/4} + n,$$

and so on, down to

$$A_2 = 2A_1 + 2, \quad A_1 = 0.$$

By induction, we can prove that, if $n = 2^r$, $r \in \mathbb{N}$,

$$A_n = rn; \tag{13.19}$$

see Problem 13.13. Thus, the total cost of the FFT is

$$T_n = M_n + A_n = 3n + 2n = \mathcal{O}(rn) = \mathcal{O}(n \log_2(n)). \tag{13.20}$$

A comparison of the cost of the FFT algorithm and standard DFT is presented in Table 13.1.

13.7 Fourier Matrices, Least Squares Approximation, and Basic Signal Processing

As we have indicated, the DFT can be computed via a matrix–vector multiplication procedure with the appropriate matrix. In this section, we want to explore the details of that matrix. Along the way, we will see that the FFT algorithm can be interpreted via an interesting matrix factorization. We will conclude with an interesting result about discrete least squares approximation using trigonometric polynomials and interpret the result in the context of signal processing.

Consider the following definition.

Definition 13.30 (Fourier matrices). Suppose that $n \in \mathbb{N}$. Define the matrices $\mathsf{F}_n = [f_{k\ell}]_{k,\ell=0}^{n-1}$ and $\mathsf{W}_n = [w_{k\ell}]_{k,\ell=0}^{n-1}$ via

$$f_{k,\ell} = \omega_{n,\ell}^{k} = e^{2\pi i k\ell/n}, \quad k, \ell = 0, \dots, n-1, \tag{13.21}$$

$$w_{k,\ell} = \omega_{n,\ell}^{-k} = e^{-2\pi i k\ell/n}, \quad k, \ell = 0, \dots, n-1, \tag{13.22}$$

respectively, where $\omega_n \in \mathcal{V}_{n,\mathrm{p}}$ is the nth root of the unity grid function and we use the convention that the row and column indices begin with 0 rather than 1. $\mathsf{F}_n \in \mathbb{C}^{n\times n}$ is called the **Fourier matrix**. $\mathsf{W}_n \in \mathbb{C}^{n\times n}$ is called the **DFT matrix**.

Of course, we met the DFT matrix in the last section. These two matrices are nearly unitary. We have the following.

Proposition 13.31 (properties of the Fourier matrices). *Suppose that $n \in \mathbb{N}$. Then*

1. $\mathsf{F}_n^{\mathsf{T}} = \mathsf{F}_n$ *and* $\mathsf{W}_n^{\mathsf{T}} = \mathsf{W}_n$.
2. $\overline{\mathsf{F}_n} = \mathsf{W}_n$ *and* $\overline{\mathsf{W}_n} = \mathsf{F}_n$, *where the overline indicates entry-wise complex conjugation.*
3. $\mathsf{F}_n^{\mathsf{H}} = \mathsf{W}_n$ *and* $\mathsf{W}_n^{\mathsf{H}} = \mathsf{F}_n$.
4. $\mathsf{F}_n\mathsf{W}_n = \mathsf{W}_n\mathsf{F}_n = n\mathsf{I}_n$.
5. $\frac{1}{\sqrt{n}}\mathsf{F}_n$ *and* $\frac{1}{\sqrt{n}}\mathsf{W}_n$ *are unitary matrices.*

Proof. These are all straightforward to show. We will prove property 4 and leave the others to the reader as an exercise; see Problem 13.14.

4: Using the identity (13.2), we find

$$
\begin{aligned}
[\mathsf{F}_n\mathsf{W}_n]_{k,\ell} &= \sum_{j=0}^{n-1} e^{2\pi \mathrm{i} kj/n} e^{-2\pi \mathrm{i} j\ell/n} \\
&= \sum_{j=0}^{n-1} e^{2\pi \mathrm{i} j(k-\ell)/n} \\
&= n\delta_{k-\ell}^{n,\mathrm{p}} \\
&= n\,[\mathsf{I}_n]_{k,\ell}\,.
\end{aligned}
$$

□

Example 13.2 Suppose that $n = 4$. Then, since $e^{2\pi \mathrm{i}/4} = \mathrm{i}$,

$$
\mathsf{F}_4 = \begin{bmatrix} 1 & 1 & 1 & 1 \\ 1 & \mathrm{i} & \mathrm{i}^2 & \mathrm{i}^3 \\ 1 & \mathrm{i}^2 & \mathrm{i}^4 & \mathrm{i}^6 \\ 1 & \mathrm{i}^3 & \mathrm{i}^6 & \mathrm{i}^9 \end{bmatrix} = \begin{bmatrix} 1 & 1 & 1 & 1 \\ 1 & \mathrm{i} & -1 & -\mathrm{i} \\ 1 & -1 & 1 & -1 \\ 1 & -\mathrm{i} & -1 & \mathrm{i} \end{bmatrix}
$$

and

$$
\mathsf{W}_4 = \begin{bmatrix} 1 & 1 & 1 & 1 \\ 1 & -\mathrm{i} & (-\mathrm{i})^2 & (-\mathrm{i})^3 \\ 1 & (-\mathrm{i})^2 & (-\mathrm{i})^4 & (-\mathrm{i})^6 \\ 1 & (-\mathrm{i})^3 & (-\mathrm{i})^6 & (-\mathrm{i})^9 \end{bmatrix} = \begin{bmatrix} 1 & 1 & 1 & 1 \\ 1 & -\mathrm{i} & -1 & \mathrm{i} \\ 1 & -1 & 1 & -1 \\ 1 & \mathrm{i} & -1 & -\mathrm{i} \end{bmatrix}.
$$

It is clear that these matrices have the properties outlined in Proposition 13.31.

Example 13.3 Observe that, in analogy with the work of the last section, we can write

$$
\mathsf{F}_4 = \begin{bmatrix} \mathsf{I}_2 & \mathsf{D}_2 \\ \mathsf{I}_2 & -\mathsf{D}_2 \end{bmatrix} \begin{bmatrix} \mathsf{F}_2 & \mathsf{O}_2 \\ \mathsf{O}_2 & \mathsf{F}_2 \end{bmatrix} \mathsf{P} \tag{13.23}
$$

and

$$
\mathsf{W}_4 = \begin{bmatrix} \mathsf{I}_2 & \tilde{\mathsf{D}}_2 \\ \mathsf{I}_2 & -\tilde{\mathsf{D}}_2 \end{bmatrix} \begin{bmatrix} \mathsf{W}_2 & \mathsf{O}_2 \\ \mathsf{O}_2 & \mathsf{W}_2 \end{bmatrix} \mathsf{P}, \tag{13.24}
$$

where I_2 is the 2×2 identity; O_2 is the 2×2 zero matrix; F_2 is the 2×2 Fourier matrix,

$$\mathsf{F}_2 = \begin{bmatrix} 1 & 1 \\ 1 & e^{\pi \mathrm{i}} \end{bmatrix} = \begin{bmatrix} 1 & 1 \\ 1 & -1 \end{bmatrix};$$

W_2 is the 2×2 DFT matrix,

$$\mathsf{W}_2 = \begin{bmatrix} 1 & 1 \\ 1 & e^{-\pi \mathrm{i}} \end{bmatrix} = \begin{bmatrix} 1 & 1 \\ 1 & -1 \end{bmatrix};$$

D_2 and $\tilde{\mathsf{D}}_2$ are diagonal matrices,

$$\mathsf{D}_2 = \operatorname{diag}\left(1, e^{2\pi \mathrm{i}/4}\right) = \operatorname{diag}(1, \mathrm{i}), \quad \tilde{\mathsf{D}}_2 = \operatorname{diag}\left(1, e^{-2\pi \mathrm{i}/4}\right) = \operatorname{diag}(1, -\mathrm{i});$$

and P is a permutation matrix,

$$\mathsf{P} = \begin{bmatrix} 1 & 0 & 0 & 0 \\ 0 & 0 & 1 & 0 \\ 0 & 1 & 0 & 0 \\ 0 & 0 & 0 & 1 \end{bmatrix}.$$

These factorizations are just another way of expressing the essence of the FFT algorithm. The reader is tasked with proving these results in Problem 13.15.

Example 13.4 The last example can be generalized. For any $n \in \mathbb{N}$ that is even, we have

$$\mathsf{F}_n = \begin{bmatrix} \mathsf{I}_{n/2} & \mathsf{D}_{n/2} \\ \mathsf{I}_{n/2} & -\mathsf{D}_{n/2} \end{bmatrix} \begin{bmatrix} \mathsf{F}_{n/2} & \mathsf{O}_{n/2} \\ \mathsf{O}_{n/2} & \mathsf{F}_{n/2} \end{bmatrix} \mathsf{P} \tag{13.25}$$

and

$$\mathsf{W}_n = \begin{bmatrix} \mathsf{I}_{n/2} & \tilde{\mathsf{D}}_{n/2} \\ \mathsf{I}_{n/2} & -\tilde{\mathsf{D}}_{n/2} \end{bmatrix} \begin{bmatrix} \mathsf{W}_{n/2} & \mathsf{O}_{n/2} \\ \mathsf{O}_{n/2} & \mathsf{W}_{n/2} \end{bmatrix} \mathsf{P}, \tag{13.26}$$

where $\mathsf{D}_{n/2} = \operatorname{diag}\left(1, \omega, \omega^2, \ldots, \omega^{n/2}\right)$, $\tilde{\mathsf{D}}_{n/2} = \operatorname{diag}\left(1, \omega^{-1}, \omega^{-2}, \ldots, \omega^{-n/2}\right)$, $\omega = e^{2\pi \mathrm{i}/n}$; and P is an $n \times n$ permutation matrix.

Now we come to a rather surprising result. We will deal with only the even case. The odd case is similar.

Theorem 13.32 (trigonometric least squares approximation). *Let $f \in C_{\mathrm{p}}(0, 1; \mathbb{C})$ and $n \in \mathbb{N}$. Suppose that*

$$X = \{x_j\}_{j=0}^{n-1}, \qquad x_j = \frac{j}{n}, \quad j = 0, \ldots, n-1$$

is a uniformly spaced nodal set. Assume that n is even. Let $K \in \mathbb{N}$ satisfy $n = 2K$. Suppose that $q_K \in \mathbb{T}_K$, expressed as

$$q_K(x) = c_K \cos(2\pi K x) + \sum_{j=-K+1}^{K-1} c_j \exp(2\pi \mathrm{i} j x), \tag{13.27}$$

is the unique interpolating polynomial of f, subordinate to X. Suppose that $M \in \mathbb{N}$, $M \leq K$, and $m = 2M$. Then, with the same coefficients as in (13.27), the polynomial $q_M \in \mathbb{T}_M$, expressed as

$$q_M(x) = c_M \cos(2\pi Mx) + \sum_{j=-M+1}^{M-1} c_j \exp(2\pi \mathrm{i} jx), \tag{13.28}$$

is the unique trigonometric polynomial that minimizes $\|\boldsymbol{r}\|_2^2$, where the vector $\boldsymbol{r} = [r_0, \dots, r_{n-1}]^\intercal$ is defined as

$$r_\ell = f(\ell/n) - c_M \cos(2\pi M x_\ell) - \sum_{j=-M+1}^{M-1} c_j \exp(2\pi \mathrm{i} j x_\ell), \quad \ell = 0, \dots, n-1. \tag{13.29}$$

Proof. Recall from the proof of Theorem 13.21, that we showed that, for the case of interpolation,

$$f(\ell/n) = \sum_{j=-K+1}^{K} c_j \exp(2\pi \mathrm{i} j x_\ell), \quad \ell = 0, \dots, n-1,$$

where c_j is viewed as an n-periodic grid function. Equivalently, in matrix–vector format,

$$\boldsymbol{f} = \tilde{\mathsf{F}}_n \boldsymbol{c},$$

where $\boldsymbol{c} = [c_j]_{j=-K+1}^{K}$, $\boldsymbol{f} = [f(\ell/n)]_{\ell=0}^{n-1} \in \mathbb{C}^n$. Here, $\tilde{\mathsf{F}}_n$ is merely a shifted version of F_n. In particular,

$$\left[\tilde{\mathsf{F}}_n\right]_{\ell,j} = \omega_{n,j}^{\ell} = e^{2\pi \mathrm{i} j\ell/n}, \quad \ell = 0, \dots, n-1, \quad j = -K+1, \dots, K.$$

Since the columns of the Fourier matrix are orthogonal with uniform normalization, the solution for the interpolation problem is

$$\boldsymbol{c} = \frac{1}{n} \tilde{\mathsf{F}}_n^{\mathsf{H}} \boldsymbol{f}. \tag{13.30}$$

For the least squares problem, the residual, defined in (13.29), can be expressed in matrix–vector form as

$$\boldsymbol{r} = \boldsymbol{f} - \hat{\mathsf{F}}_{n,m} \hat{\boldsymbol{c}},$$

where $\hat{\mathsf{F}}_{n,m} \in \mathbb{C}^{n \times m}$ is the matrix

$$\left[\hat{\mathsf{F}}_{n,m}\right]_{\ell,j} = \omega_{n,j}^{\ell} = e^{2\pi \mathrm{i} j\ell/n}, \quad \ell = 0, \dots, n-1, \quad j = -M+1, \dots, M,$$

and $\hat{\boldsymbol{c}} = [c_j]_{j=-M+1}^{M}$. The solution to the standard least squares problem, of course, satisfies the normal equation (Theorem 5.6):

$$\hat{\mathsf{F}}_{n,m}^{\mathsf{H}} \hat{\mathsf{F}}_{n,m} \hat{\boldsymbol{c}} = \hat{\mathsf{F}}_{n,m}^{\mathsf{H}} \boldsymbol{f}.$$

Since the columns of F_n (and $\hat{\mathsf{F}}_{n,m}$) are orthogonal with uniform normalization (Proposition 13.31), we have

$$\mathsf{I}_m \hat{\boldsymbol{c}} = \hat{\boldsymbol{c}} = \frac{1}{n} \hat{\mathsf{F}}_{n,m}^{\mathsf{H}} \boldsymbol{f}. \tag{13.31}$$

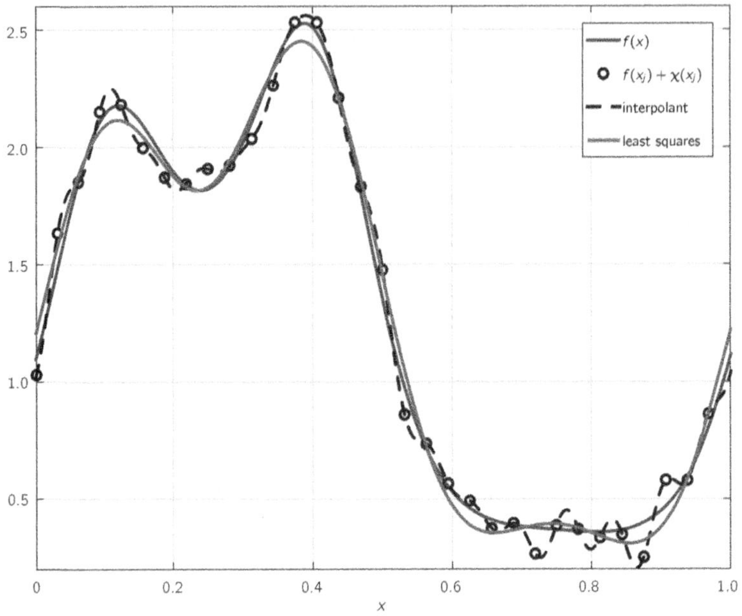

Figure 13.4 Trigonometric least squares approximation of corrupted data. f is the pure one-periodic uncorrupted signal and $f(x_j) + \chi(x_j)$ are the 32 sampled points of the noisy data. We use $m = 8$ to obtain a less oscillatory least squares approximation via Theorem 13.32; see also Figures 13.1, 13.2, and 13.3.

It is easy to see that entries $j = -M+1, \ldots, M$ of the solution $\boldsymbol{c}$ derived in (13.30) agree with the entries of the least squares solution $\hat{\boldsymbol{c}}$ derived in (13.31). □

This result is fascinating, because it indicates that a least squares approximation can be trivially obtained from an interpolant. Now one might argue that using the interpolant is always better than a lower–order least squares approximation. But this may not always be the case.

Example 13.5 (a simple filter) Consider the pure one-periodic signal

$$f(x) = \exp[\sin(2\pi x) - 0.1\cos(2\pi x) + 0.2\cos(4\pi x) + 0.2\sin(6\pi x)].$$

Suppose that the data f have been corrupted by some random noise, χ, that we want to filter out. In signal processing, we commonly have to work with the noisy signal $g(x) = f(x) + \chi(x)$, but we really want to know something about f. One simple tool for trying to remove the noise is to use Theorem 13.32. In other words, we interpolate the noisy signal using a large number of points, and then we truncate the interpolant, removing the highly oscillatory terms. What is left, the least squares approximation of lower order, is often a much better approximation of the original, uncorrupted signal, f, than the noisy interpolant; see Figure 13.4. The pure signal f is smooth, but the corrupted signal g, of which we have sampled $n = 32$ points, is rough. If we interpolate the sampled points, $f(x_j) + \chi(x_j)$, we get get an oscillatory interpolant. However, the least squares approximation ($m = 8$) is much less rough and arguably a better approximation of the uncorrupted signal, at least to the naked eye; see also Figures 13.1, 13.2, and 13.3 and Example 13.1.

Problems

13.1 Prove Proposition 13.5.
13.2 Prove Proposition 13.7.
13.3 Prove Proposition 13.9.
13.4 Prove Theorem 13.14.
13.5 Prove Proposition 13.16.
13.6 Prove Proposition 13.18.
13.7 Prove Proposition 13.23.
13.8 Complete the proof of Theorem 13.25.
13.9 Complete the proof of Theorem 13.26.
13.10 Suppose that the Fourier series for $f : \mathbb{R} \to \mathbb{R}$,

$$f(x) = \sum_{j=-\infty}^{\infty} \hat{f}_j e^{2\pi \mathrm{i} jx},$$

is absolutely convergent and

$$\psi(x) = \sum_{j=-n}^{n} c_j e^{2\pi \mathrm{i} jx} \in \mathbb{T}_n$$

satisfies

$$\psi(x_k) = f(x_k), \quad x_k = \frac{k}{2n+1}, \quad k = 0, \dots, 2n.$$

Prove that

$$c_k = \hat{f}_k + \sum_{j=-\infty}^{\infty} \left[\hat{f}_{j(2n+1)+k} + \hat{f}_{j(2n+1)-k}\right], \quad -n \le k \le n.$$

13.11 Prove that (13.15) and (13.16) hold.
13.12 Prove that, if $n = 2^r$, $r \in \mathbb{N}$, the number of multiplications for the FFT at stage n is precisely given by (13.18).
13.13 Prove that, if $n = 2^r$, $r \in \mathbb{N}$, the number of additions/subtractions for the FFT at stage n is precisely given by (13.19).
13.14 Prove Proposition 13.31.
13.15 Prove that the expressions in (13.23) and (13.24) are valid.

14 Numerical Quadrature

In the final chapter of this part, we want to approximate the value of a definite integral. Given, for example, $f \in C([a, b])$, we wish to compute an approximation of

$$I^{(a,b)}[f] = \int_a^b f(x)\mathrm{d}x.$$

If the antiderivative of f is not readily available, then an approximation of the integral may be a good alternate. Every calculus student already knows how to do this using Riemann sums. A small, and probably dwindling, number have learned Simpson's rule and the trapezoidal rule in calculus for approximating integrals. In this chapter, we want to estimate the sizes of the errors for such approximation schemes in a systematic way. The strategy we will adopt to accomplish this is as follows. Suppose that $g \in C([a, b])$, whose antiderivative is simply obtained, and $\|f - g\|_{L^\infty(a,b)} < \varepsilon$. Then

$$\left| \int_a^b f(x)\mathrm{d}x - \int_a^b g(x)\mathrm{d}x \right| \leq \varepsilon(b - a).$$

This estimate is the basis of most numerical integration methods.

In particular, suppose that $X = \{x_i\}_{i=0}^n \subset [a, b]$ is a nodal set and $p \in \mathbb{P}_n$ is the unique Lagrange interpolating polynomial of f subordinate to X. Then

$$f(x_i) = p(x_i), \quad i = 0, \dots, n,$$

and

$$f(x) = p(x) + E(x), \quad \forall x \in [a, b],$$

where E is an expression of the interpolation error. Then

$$\int_a^b f(x)\mathrm{d}x = \int_a^b p(x)\mathrm{d}x + \int_a^b E(x)\mathrm{d}x.$$

But

$$\int_a^b p(x)\mathrm{d}x = \int_a^b \sum_{i=0}^n L_i(x) f(x_i)\mathrm{d}x = \sum_{i=0}^n f(x_i) \int_a^b L_i(x)\mathrm{d}x = \sum_{i=0}^n f(x_i)\beta_i,$$

where $L_i \in \mathbb{P}_n$ is the ith Lagrange nodal basis element and β_i is its definite integral:

$$\beta_i = \int_a^b L_i(x)\mathrm{d}x.$$

The expression $\sum_{i=0}^{n} f(x_i)\beta_i$ is a typical numerical integration formula, requiring only certain point values of the integrand f. Regarding the error, we have

$$\left| \int_a^b f(x)\mathrm{d}x - \sum_{i=0}^{n} f(x_i)\beta_i \right| = \left| \int_a^b E(x)\mathrm{d}x \right| \le \int_a^b |E(x)|\, \mathrm{d}x.$$

In other words, we can generate an error formula or an error estimate for our *quadrature rule*, $\sum_{i=0}^{n} f(x_i)\beta_i$, by working with the term $\int_a^b E(x)\mathrm{d}x$, and much of this chapter will be devoted to it.

The *quadrature weights*, β_i, depend only on the positions of the nodes within $[a, b]$, as well as the interval $[a, b]$ itself. Suppose that $n = 1$, $x_0 = a$, and $x_1 = b$. Then

$$p(x) = f(a)\frac{x-b}{a-b} + f(b)\frac{x-a}{b-a}$$

and

$$\int_a^b p(x)\mathrm{d}x = \frac{f(a)}{a-b}\int_a^b (x-b)\mathrm{d}x + \frac{f(b)}{b-a}\int_a^b (x-a)\mathrm{d}x = \frac{b-a}{2}(f(a)+f(b)).$$

This is, essentially, the *trapezoidal rule* (or *trapezium rule*).

Now suppose that $n = 2$, $x_0 = a$, $x_1 = \frac{a+b}{2}$, and $x_2 = b$. It follows that

$$p(x) = f(a)\frac{\left(x - \frac{a+b}{2}\right)(x-b)}{\left(a - \frac{a+b}{2}\right)(a-b)} + f\left(\frac{a+b}{2}\right)\frac{(x-a)(x-b)}{\left(\frac{a+b}{2} - a\right)\left(\frac{a+b}{2} - b\right)} + f(b)\frac{(x-a)\left(x - \frac{a+b}{2}\right)}{(b-a)\left(b - \frac{a+b}{2}\right)}.$$

It follows that

$$\int_a^b p(x)\mathrm{d}x = \frac{b-a}{6}\left(f(a) + 4f\left(\frac{a+b}{2}\right) + f(b)\right),$$

which is *Simpson's rule*.[1]

14.1 Quadrature Rules for Weighted Integrals

To keep our discussion general, and because this also appears in applications, we will consider the approximation of a weighted integral

$$I_w^{(a,b)}[f] = \int_a^b f(x)w(x)\mathrm{d}x.$$

Here, w is a weight function on the compact interval $[a, b] \subset \mathbb{R}$. To avoid the trivial case, we always assume that $[a, b]$ has positive length.

[1] Named in honor of the British mathematician Thomas Simpson (1710–1761).

Definition 14.1 (quadrature rule). Suppose that $n, r \in \mathbb{N}_0$, w is a weight function on the compact interval $[a, b] \subset \mathbb{R}$, $h = b - a > 0$, and $f \in C^r([a, b])$. The expression

$$\begin{aligned} Q_{w,r}^{(a,b)}[f] &= \sum_{i=0}^{r}\sum_{j=0}^{n} \beta_{i,j} f^{(i)}(x_j) \\ &= \sum_{j=0}^{n}\left(\beta_{0,j}f(x_j) + \beta_{1,j}f'(x_j) + \cdots + \beta_{r,j}f^{(r)}(x_j)\right), \end{aligned} \tag{14.1}$$

where

$$\beta_{i,j} = h^{i+1}\hat{\beta}_{i,j}, \quad i = 0, \ldots, r, \quad j = 0, \ldots, n \tag{14.2}$$

and

$$x_j = a + h \cdot \hat{x}_j, \quad j = 0, \ldots, n, \tag{14.3}$$

is called a **quadrature rule of degree** r with **intrinsic nodes** $\hat{X} = \{\hat{x}_j\} \subset [0, 1]$ and **intrinsic weights** $\{\hat{\beta}_{i,j}\} \subset \mathbb{R}$. The sets $X = \{x_j\} \subset [a, b]$ and $\{\beta_{i,j}\} \subset \mathbb{R}$ are called the **effective nodes** and **effective weights**, respectively. A quadrature rule of degree $r = 0$ is called a **simple quadrature rule**, and we simplify the notation by writing $\beta_j = \beta_{0,j}$ and

$$Q_w^{(a,b)}[f] = \sum_{j=0}^{n} \beta_j f(x_j). \tag{14.4}$$

The **quadrature error** is defined as

$$E_Q[f] = I_w^{(a,b)}[f] - Q_{w,r}^{(a,b)}[f].$$

Definition 14.2 (consistency). The quadrature rule (14.1) is **consistent of order at least** $m \in \mathbb{N}_0$ if and only if $E_Q[q] = 0$ for all $q \in \mathbb{P}_m$. The quadrature rule (14.1) is **consistent of order exactly** m if and only if $E_Q[q] = 0$ for all $q \in \mathbb{P}_m$; however, for some $r \in \mathbb{P}_{m+1}$, $E_Q[r] \neq 0$.

Definition 14.3 (interpolatory quadrature rule). Assume that $n \in \mathbb{N}_0$, w is a weight function on the compact interval $[a, b] \subset \mathbb{R}$, and $f \in C([a, b])$. Suppose that $X = \{x_i\}_{i=0}^{n} \subset [a, b]$ is a nodal set and $p \in \mathbb{P}_n$ is the unique Lagrange interpolating polynomial of f subordinate to X, with

$$p(x) = \sum_{j=0}^{n} f(x_j)L_j(x),$$

where $L_j \in \mathbb{P}_n$ is the jth Lagrange nodal basis element defined in (9.3). The expression

$$Q_w^{(a,b)}[f] = \sum_{j=0}^{n} f(x_j)\beta_j, \tag{14.5}$$

where

$$\beta_j = \int_a^b L_j(x)w(x)dx,$$

is called an **interpolatory quadrature rule subordinate to** X **of Lagrange**[2] **type** for approximating $I_w^{(a,b)}[f]$. Suppose that $f \in C^1([a,b])$ and $p \in \mathbb{P}_{2n+1}$ is the unique Hermite interpolating polynomial of f subordinate to X, with

$$p(x) = \sum_{\ell=0}^{n} [F_{0,\ell}(x)f(x_\ell) + F_{1,\ell}(x)f'(x_\ell)],$$

where $F_{0,\ell}, F_{1,\ell}(x) \in \mathbb{P}_{2n+1}$ are defined in (9.14). The formula

$$Q_{w,1}^{(a,b)}[f] = \sum_{j=0}^{n} (f(x_j)\beta_{0,j} + f'(x_j)\beta_{1,j}), \tag{14.6}$$

where

$$\beta_{0,j} = \int_a^b F_{0,j}(x)w(x)dx, \quad \beta_{1,j} = \int_a^b F_{1,j}(x)w(x)dx,$$

is called an **interpolatory quadrature rule subordinate to** X **of Hermite**[3] **type**.

Clearly, we can use Lagrange and Hermite interpolation to construct quadrature rules of degree 0 and degree 1, respectively. Now suppose we look at the problem from a different perspective. Suppose that we want to construct a quadrature rule such that it is consistent to a certain order. What sort of quadrature rule will one obtain?

In the case of simple quadrature rules, we have the following result.

Proposition 14.4 (existence and uniqueness). *Suppose that* $X = \{x_j\}_{j=0}^n$ *is a nodal set in the compact interval* $[a,b] \subset \mathbb{R}$. *There exist unique weights* $\{\beta_j\}_{j=0}^n$ *such that*

$$\int_a^b q(x)w(x)dx = \sum_{j=0}^{n} \beta_j q(x_j), \quad \forall q \in \mathbb{P}_n, \tag{14.7}$$

or, equivalently,

$$E_Q[q] = 0, \quad \forall q \in \mathbb{P}_n.$$

Moreover, these weights are given by

$$\beta_j = \int_a^b L_j(x)w(x)dx, \quad j = 0, \ldots, n,$$

where L_j *is the* j*th Lagrange nodal basis polynomial subject to* X, *which is defined in* (9.3).

Proof. Recall that the Lagrange nodal basis polynomials satisfy

$$L_j(x_k) = \delta_{j,k}, \quad j,k \in \{0,1,\ldots,n\}.$$

[2] Named in honor of the Italian, later naturalized French, mathematician and astronomer Joseph-Louis Lagrange (1736–1813).
[3] Named in honor of the French mathematician Charles Hermite (1822–1901).

Furthermore, any polynomial $q \in \mathbb{P}_n$ can be written uniquely as

$$q(x) = \sum_{j=0}^{n} q(x_j) L_j(x).$$

Therefore,

$$\int_a^b q(x)w(x)\mathrm{d}x = \sum_{j=0}^{n} \left[\int_a^b L_j(x)w(x)\mathrm{d}x \right] q(x_j) = \sum_{j=0}^{n} \beta_j q(x_j)$$

if and only if

$$\beta_j = \int_a^b L_j(x)w(x)\mathrm{d}x, \quad j = 0, \ldots, n.$$

To prove that the weights are unique, suppose that

$$\int_a^b q(x)w(x)\mathrm{d}x = \sum_{j=0}^{n} \beta_j^{(1)} q(x_j) = \sum_{j=0}^{n} \beta_j^{(2)} q(x_j).$$

Then

$$\sum_{k=0}^{n} \left(\beta_k^{(1)} - \beta_k^{(2)} \right) q(x_k) = 0, \quad \forall q \in \mathbb{P}_n.$$

In particular, suppose that $q = L_j$. Then

$$0 = \sum_{k=0}^{n} \left(\beta_k^{(1)} - \beta_k^{(2)} \right) L_j(x_k) = \sum_{k=0}^{n} \left(\beta_k^{(1)} - \beta_k^{(2)} \right) \delta_{j,k} = \beta_j^{(1)} - \beta_j^{(2)}.$$

This proves uniqueness. □

Remark 14.5 (consistency)**.** The last result shows that the simple quadrature rule,

$$Q_w^{(a,b)}[f] = \sum_{i=0}^{n} \beta_i f(x_i),$$

is consistent of order at least n if and only if it is a quadrature rule of Lagrange type. An analogous result could be shown for a quadrature rule of Hermite type; see Problem 14.2.

It is important to note that, whether a quadrature rule of Lagrange or Hermite type, once the nodes are chosen the quadrature weights can be easily computed offline and tabulated. These form an important class of quadrature rules that we will investigate here.

14.2 Simple Estimates for Interpolatory Quadrature

In this short section, we give two easy estimates for the errors in interpolatory quadrature rules. It turns out that these estimates can be quite pessimistic in certain important cases, but they are easily derived.

Theorem 14.6 (error estimate). *Suppose that $n \in \mathbb{N}_0$, w is a weight function on the compact interval $[a, b] \subset \mathbb{R}$, $f \in C^{n+1}([a, b])$, and $X = \{x_i\}_{i=0}^n \subset [a, b]$ is a nodal set. Suppose that $Q_w^{(a,b)}[f]$ is the interpolatory quadrature rule subordinate to X of Lagrange type. Then*

$$|E_Q[f]| \le \frac{M_{n+1}}{(n+1)!} \int_a^b |\omega_{n+1}(x)| \, w(x) \mathrm{d}x, \tag{14.8}$$

where

$$\omega_{n+1}(x) = \prod_{j=0}^{n} (x - x_j)$$

and

$$M_{n+1} = \left\| f^{(n+1)} \right\|_{L^\infty(a,b)}.$$

Consequently, an interpolatory quadrature rule subordinate to X of Lagrange type is consistent of order at least n.

Proof. Suppose that $p \in \mathbb{P}_n$ is the unique interpolating polynomial of f subordinate to X. From Theorem 9.16, we have, for $x \in [a, b] \backslash X$,

$$f(x) - p(x) = \frac{f^{(n+1)}(\xi)}{(n+1)!} \omega_{n+1}(x)$$

for some $\xi = \xi(x) \in (a, b)$ satisfying

$$\min\{x_0, x_1, \dots, x_n, x\} < \xi < \max\{x_0, x_1, \dots, x_n, x\}.$$

Then

$$\begin{aligned} |E_Q[f]| &= \left| \int_a^b \frac{f^{(n+1)}(\xi)}{(n+1)!} \omega_{n+1}(x) w(x) \mathrm{d}x \right| \\ &\le \frac{1}{(n+1)!} \int_a^b \left| f^{(n+1)}(\xi) \omega_{n+1}(x) \right| w(x) \mathrm{d}x \\ &\le \frac{\left\| f^{(n+1)} \right\|_{L^\infty(a,b)}}{(n+1)!} \int_a^b |\omega_{n+1}(x)| \, w(x) \mathrm{d}x, \end{aligned}$$

using Theorem B.36 in the last step. □

The proof of the following result is similar, using Theorem 9.21.

Theorem 14.7 (error estimate). *Suppose that $n \in \mathbb{N}_0$, w is a weight function on the compact interval $[a, b] \subset \mathbb{R}$, $f \in C^{2n+2}([a, b])$, and $X = \{x_i\}_{i=0}^n \subset [a, b]$ is a nodal set. Suppose that $Q_{w,1}^{(a,b)}[f]$ is the interpolatory quadrature rule subordinate to X of Hermite type. Then*

$$|E_Q[f]| \le \frac{M_{2n+2}}{(2n+2)!} \int_a^b |\omega_{n+1}(x)|^2 \, w(x) \mathrm{d}x, \tag{14.9}$$

where

$$M_{2n+2} = \left\| f^{(2n+2)} \right\|_{L^\infty(a,b)}.$$

Consequently, an interpolatory quadrature rule subordinate to X of Hermite type is consistent of order at least $2n+1$.

Proof. See Problem 14.3. □

14.3 The Peano Kernel Theorem

The Peano Kernel Theorem is an important tool for analyzing quadrature rules. Before introducing it, let us remind the reader of a standard definition and state a needed technical lemma.

Definition 14.8 (characteristic function)**.** Suppose that $B \subseteq \mathbb{R}$. The **characteristic function of** B is the function

$$\chi_B(t) = \begin{cases} 1, & t \in B, \\ 0, & t \in \mathbb{R}\backslash B. \end{cases}$$

Lemma 14.9 (kernel)**.** *Suppose that $r \in \mathbb{N}_0$ and $m \in \mathbb{N}$, with $m > r$. Define the function $k_m \colon [a,b] \times [a,b] \to \mathbb{R}$ via*

$$k_m(x,y) = (x-y)^m \chi_{[a,x]}(y) = \begin{cases} (x-y)^m, & a \le y \le x \le b, \\ 0, & a \le x < y \le b. \end{cases} \tag{14.10}$$

Then, for each $i \in \{0,1,\ldots,r\}$,

$$\frac{\partial^i k_m}{\partial x^i} \in C([a,b] \times [a,b])$$

and

$$\frac{\partial^i k_m(x,y)}{\partial x^i} = \begin{cases} \prod_{k=0}^{i-1}(m-k)(x-y)^{m-i}, & a \le y \le x \le b, \\ 0, & a \le x < y \le b. \end{cases}$$

Proof. See Problem 14.4. □

Theorem 14.10 (Peano Kernel Theorem)**.** *Suppose that $r \in \mathbb{N}_0$, $m \in \mathbb{N}$, with $m > r$, w is a weight function on the compact interval $[a,b] \subset \mathbb{R}$, and $f \in C^{m+1}([a,b])$. Assume that $Q_{w,r}^{(a,b)}[f]$ is a quadrature rule of degree r, (14.1), that is consistent of order at least m. Let the function $k_m \colon [a,b] \times [a,b] \to \mathbb{R}$ be defined as in (14.10). Set*

$$K_m(y) = E_Q\left[k_m(\,\cdot\,,y)\right] = \int_a^b k_m(x,y)w(x)\mathrm{d}x - \sum_{j=0}^{n}\sum_{i=0}^{r}\beta_{i,j}\frac{\partial^i k_m(x_j,y)}{\partial x^i}. \tag{14.11}$$

Then the quadrature error satisfies

$$E_Q[f] = \frac{1}{m!}\int_a^b f^{(m+1)}(y)K_m(y)\mathrm{d}y.$$

The function $K_m(y)$ *is called the* Peano Kernel.[4]

Proof. Using Taylor's Theorem with an integral remainder, Theorem B.39,

$$f(x) = P(x) + R(x),$$

where

$$P(x) = \sum_{j=0}^{m} \frac{f^{(j)}(a)}{j!}(x-a)^j, \quad R(x) = \frac{1}{m!}\int_a^x f^{(m+1)}(y)(x-y)^m \, dy.$$

Observe that we can rewrite the remainder as

$$R(x) = \frac{1}{m!}\int_a^b f^{(m+1)}(y) k_m(x,y) dy.$$

By assumption, the quadrature rule is of order at least m, so that

$$E_Q[P] = 0.$$

As integration and quadrature are linear functionals,

$$E_Q[f] = E_Q[P] + E_Q[R] = E_Q[R].$$

Therefore, we need to evaluate the integral of R:

$$\begin{aligned}\int_a^b R(x)w(x)dx &= \int_a^b \left[\frac{1}{m!}\int_a^b f^{(m+1)}(y)k_m(x,y)dy\right] w(x)dx \\ &= \frac{1}{m!}\int_a^b \left[\int_a^b k_m(x,y)w(x)dx\right] f^{(m+1)}(y)dy.\end{aligned}$$

We also must apply the quadrature rule to R: using Lemma 14.9,

$$\begin{aligned}\sum_{j=0}^{n}\sum_{i=0}^{r} \beta_{i,j}\frac{d^i R(x_j)}{dx^i} &= \frac{1}{m!}\sum_{j=0}^{n}\sum_{i=0}^{r} \beta_{i,j}\frac{d^i}{dx^i}\left[\int_a^b f^{(m+1)}(y)k_m(x,y)dy\right]_{x=x_j} \\ &= \frac{1}{m!}\int_a^b \left[\sum_{j=0}^{n}\sum_{i=0}^{r} \beta_{i,j}\frac{\partial^i k_m(x_j,y)}{\partial x^i}\right] f^{(m+1)}(y)dy.\end{aligned}$$

The result now follows:

$$E_Q[f] = E_Q[R] = \frac{1}{m!}\int_a^b f^{(m+1)}(y)K_m(y)dy,$$

where K_m is given in (14.11). □

Remark 14.11 (Peano Kernel). Recall that, for $y \in [a,b]$, the Peano Kernel is defined by

$$K_m(y) = E_Q\left[k_m(\cdot\,,y)\right].$$

[4] Named in honor of the Italian mathematician and glottologist Giuseppe Peano (1858–1932).

In other words, the Peano Kernel itself is defined by a quadrature error. Furthermore, we have

$$E_Q[f] = \frac{1}{m!}\int_a^b f^{(m+1)}(y)E_Q\left[k_m(\,\cdot\,,y)\right]dy.$$

The following result is obvious.

Corollary 14.12 (quadrature error stability). *With the same hypotheses as for Theorem 14.10, we have*

$$|E_Q[f]| \le \frac{1}{m!}\left\|f^{(m+1)}\right\|_{L^\infty(a,b)}\|K_m\|_{L^1(a,b)}.$$

Since $\|K_m\|_{L^1(a,b)} < \infty$, there is a constant $C > 0$ that may depend on the size of the interval but is independent of f such that

$$|E_Q[f]| \le C\left\|f^{(m+1)}\right\|_{L^\infty(a,b)}. \tag{14.12}$$

Proof. See Problem 14.5. □

Corollary 14.13 (constant sign). *With the same hypotheses as for Theorem 14.10, if it is additionally known that K_m does not change sign on $[a, b]$, then*

$$E_Q[f] = \frac{f^{(m+1)}(\xi)}{m!}\int_a^b K_m(y)dy$$

for some $\xi \in (a, b)$. Furthermore, we have the simple representation for the error

$$E_Q[f] = \frac{E_Q\left[x^{m+1}\right]}{(m+1)!}f^{(m+1)}(\xi)$$

for some $\xi \in (a, b)$, where $E_Q\left[x^{m+1}\right]$ is the (computable) quadrature error for the function $x \mapsto x^{m+1}$.

Proof. Theorem 14.10 and the Integral Mean Value Theorem B.41 yield

$$E_Q[f] = \frac{1}{m!}\int_a^b f^{(m+1)}(y)K_m(y)dy = \frac{1}{m!}f^{(m+1)}(\xi)\int_a^b K_m(y)dy$$

for some $\xi \in (a, b)$.

Now notice that if K_m does not change sign on $[a, b]$, then we have a means for finding the quantity $\int_a^b K_m(y)dy$, which does not depend upon f. Namely, suppose that $f(x) = x^{m+1}$. Then, by the last result,

$$E_Q[f] = E_Q\left[x^{m+1}\right] = \frac{(m+1)!}{m!}\int_a^b K_m(y)dy.$$

Consequently,

$$\int_a^b K_m(y)dy = \frac{E_Q\left[x^{m+1}\right]}{m+1}.$$

Therefore, in general,

$$E_Q[f] = \frac{1}{m!}f^{(m+1)}(\xi)\frac{E_Q\left[x^{m+1}\right]}{m+1},$$

which gives the desired representation. □

14.3.1 Integral Representation of the Moduli of Smoothness

Let us, as a further application of the ideas presented above, derive an integral representation of the moduli of smoothness, introduced in Definition 10.33, which sometimes is important in applications and can be used to derive many deep results. We begin by defining the so-called Peano Kernel of Δ_t^k.

Definition 14.14 (Peano Kernel). Let $M_1 = \chi_{[0,1]}$. For $k \in \mathbb{N}$ with $k \geq 2$, set

$$M_k(x) = \int_{\mathbb{R}} M_{k-1}(x-y)M_1(y)\mathrm{d}y.$$

For $k \in \mathbb{N}$ and $h > 0$, we define

$$M_k(x,h) = \frac{1}{h}M_k\left(\frac{x}{h}\right).$$

The following properties of the kernel M_k are immediate.

Lemma 14.15 (properties of M_k). *Let $k \in \mathbb{N}$ and M_k be as in Definition 14.14. We have:*

1. *For all $h > 0$,*

$$\operatorname{supp} M_k(\cdot, h) = [0, kh].$$

2. *For all $x \in \mathbb{R}$ and $h > 0$, we have $0 \leq M_k(x,h) \leq 1$.*
3. *For all $h > 0$,*

$$\int_{\mathbb{R}} M_k(x,h)\mathrm{d}x = 1.$$

Proof. See Problem 14.6. □

We can now give an integral representation of the modulus of smoothness.

Proposition 14.16 (integral representation). *Let $k \in \mathbb{N}$. If $f \in C^k([a,b];\mathbb{C})$, then, for all $x[a,b]$ and $t \in \mathbb{R}$ such that $x + kt \in [a,b]$, we have*

$$\Delta_t^k f(x) = t^k \int_{\mathbb{R}} f^{(k)}(y)M_k(y-x,t)\mathrm{d}y.$$

Proof. We prove the result by induction. If $k = 1$, we have

$$\Delta_t^1 f(x) = f(x+t) - f(x) = \int_x^{x+t} f'(y)\mathrm{d}y = t\int_{\mathbb{R}} f'(y)M_1(y-x,t)\mathrm{d}y.$$

Assume now that the representation holds for $k-1$. Then

$$\begin{aligned}
t^{-k}\Delta_t^k f(x) &= t^{-1}\Delta_t^1\left[t^{-k+1}\Delta_t^{k-1}f(x)\right] \\
&= t^{-1}\Delta_t^1 \int_{\mathbb{R}} f^{(k-1)}(y)M_{k-1}(y-x,t)\mathrm{d}y \\
&= \int_{\mathbb{R}} \frac{\mathrm{d}}{\mathrm{d}z}\int_{\mathbb{R}} f^{(k-1)}(y)M_{k-1}(y-z,t)\mathrm{d}y M_1(z-x,t)\mathrm{d}z \\
&= \int_{\mathbb{R}} \frac{\mathrm{d}}{\mathrm{d}z}\int_{\mathbb{R}} f^{(k-1)}(\zeta+z)M_{k-1}(\zeta,t)\mathrm{d}\zeta M_1(z-x,t)\mathrm{d}z,
\end{aligned}$$

where, in the last step, we applied the change of variable $\zeta = y-z$. Consider now the term

$$\frac{\mathrm{d}}{\mathrm{d}z}\int_{\mathbb{R}} f^{(k-1)}(\zeta+z)M_{k-1}(\zeta,t)\mathrm{d}\zeta.$$

Since $f^{(k-1)}$ is differentiable, and M_{k-1} is bounded and of compact support, we can write

$$\begin{aligned}
\frac{\mathrm{d}}{\mathrm{d}z}\int_{\mathbb{R}} f^{(k-1)}(\zeta+z)M_{k-1}(\zeta,t)\mathrm{d}\zeta &= \int_{\mathbb{R}} f^{(k)}(\zeta+z)M_{k-1}(\zeta,t)\mathrm{d}\zeta \\
&= \int_{\mathbb{R}} f^{(k)}(y)M_{k-1}(y-z,t)\mathrm{d}y.
\end{aligned}$$

With this representation, we can continue our derivations and write

$$\begin{aligned}
t^{-k}\Delta_t^k f(x) &= \int_{\mathbb{R}}\int_{\mathbb{R}} f^{(k)}(y)M_{k-1}(y-z,t)\mathrm{d}y M_1(z-x,t)\mathrm{d}z \\
&= \int_{\mathbb{R}} f^{(k)}(y)\int_{\mathbb{R}} M_{k-1}(y-z,t)M_1(z-x,t)\mathrm{d}z\mathrm{d}y \\
&= \int_{\mathbb{R}} f^{(k)}(y)M_k(y-x,t)\mathrm{d}y,
\end{aligned}$$

where we exchanged the order of integration, carried out a change of variables, and used the definition of M_k. The continuity of $f^{(k)}$ and boundedness of the kernels justify these steps and imply the result. □

As a simple corollary of the last result, we provide a simple proof of two important properties of the modulus of smoothness.

Corollary 14.17 (properties of the modulus of smoothness)**.** *Let $[a,b] \subset \mathbb{R}$ be a compact interval, $k \in \mathbb{N}$, and $f \in C^k([a,b])$. Then, for all $h > 0$, we have*

$$\omega_f^{(k)}(h) \leq h^k \left\| f^{(k)} \right\|_{L^\infty(a,b)},$$

and, for all $m \in \mathbb{N}_0$,

$$\omega_f^{(k+m)}(h) \leq h^k \omega_{f^{(k)}}^{(m)}(h).$$

Proof. From the previous result we have, for any t,

$$\left|\Delta_t^k f(x)\right| \leq |t|^k \int_{\mathbb{R}} |f^{(k)}(y)| M_k(y-x,t)\mathrm{d}y \leq |t|^k \left\| f^{(k)} \right\|_{L^\infty(a,b)},$$

where we used the support and integration properties of M_k. Taking supremum over $|t| \leq h$, the first result follows.

For the second result, we can write

$$\Delta_t^{k+m} f(x) = t^k \Delta_t^m \int_{\mathbb{R}} f^{(k)}(y) M_k(y-x,t)\mathrm{d}y = t^k \Delta_t^m \int_{\mathbb{R}} f^{(k)}(x+y) M_k(y,t)\mathrm{d}y.$$

Therefore, if $|t| \leq h$,

$$\left|\Delta_t^{k+m} f(x)\right| \leq |t|^k \int_{\mathbb{R}} \left|\Delta_t^m f^{(k)}(x+y)\right| M_k(y,t)\mathrm{d}y \leq |t|^k \omega_{f^{(k)}}^{(m)}(h),$$

where we used the integration properties of M_k. Taking supremum over $|t| \leq h$, the second estimate follows. □

14.4 Proper Scaling and an Error Estimate Via a Scaling Argument

Proper scaling is related to the units of various quantities. Let us first consider what are the units of the integral of f,

$$\int_a^b f(x)\mathrm{d}x.$$

The type of units is not important; what is important is how units are transferred. For example, let us assume that the units of x are those of length and the units of f are those of a mass density, specifically mass per unit length. We express units in the following way:

$$[x] = \mathrm{m}, \quad [f] = \frac{\mathrm{kg}}{\mathrm{m}},$$

where m means meters and kg stands for kilograms. Therefore, the integral has units of mass:

$$\left[\int_a^b f(x)\mathrm{d}x\right] = \frac{\mathrm{kg}}{\mathrm{m}} \cdot \mathrm{m} = \mathrm{kg}.$$

Now consider a simple quadrature rule approximating the integral:

$$\int_a^b f(x)\mathrm{d}x \approx \sum_{j=0}^{n} \beta_j f(x_j).$$

For the approximation to make sense, it must be that the quadrature rule has the same units as the integral. This implies that

$$[\beta_j] = \mathrm{m}.$$

If one uses a quadrature rule of degree r to approximate the integral, we conclude that

$$\mathrm{kg} = \left[\sum_{j=0}^{n}\sum_{i=0}^{r} \beta_{i,j} f^{(i)}(x_j)\right] = [\beta_{i,j}] \cdot \left[f^{(i)}(x_j)\right] = [\beta_{i,j}] \cdot \frac{\mathrm{kg}}{\mathrm{m}^{i+1}},$$

so that

$$[\beta_{i,j}] = \mathrm{m}^{i+1}.$$

Now our intrinsic weights are defined so that they are *unitless*. This is usually expressed as

$$[\hat{\beta}_{i,j}] = 1.$$

And, naturally, $[h] = [b - a] = \mathrm{m}$. Thus, the expression (14.2) makes sense with respect to units:

$$\mathrm{m}^{i+1} = [\beta_{i,j}] = \left[h^{i+1}\hat{\beta}_{i,j}\right] = [h]^{i+1} \cdot \left[\hat{\beta}_{i,j}\right] = \mathrm{m}^{i+1}.$$

Furthermore, both numerical values of the effective quadrature weights (14.2) and effective quadrature nodes (14.3) agree with their intrinsic counterparts when $[a, b] = [0, 1]$.

With the stability result, Corollary 14.12, at hand, one can prove a type of convergence. This proof uses what is known as a scaling argument.

Theorem 14.18 (scaling argument)**.** *Suppose that w is a weight function on the compact interval $[a, b] \subset \mathbb{R}$, $h = b - a > 0$, $m \in \mathbb{N}_0$, $f \in C^{m+1}([a, b])$, and $Q_{w,r}^{(a,b)}[f]$ is a quadrature rule of degree $r < m$ (14.1) that is consistent of order at least m. Then*

$$|E_Q[f]| \leq C_0 h^{m+2} \left\| f^{(m+1)} \right\|_{L^\infty(a,b)}, \tag{14.13}$$

where $C_0 > 0$ is independent of h and f.

Proof. Let us use an affine transformation and calculate the integral on the reference interval $[0, 1]$. To this end, define

$$\hat{f}(\zeta) = f(a + \zeta h), \quad \hat{w}(\zeta) = w(a + \zeta h), \quad \zeta \in [0, 1].$$

The simple change of variables $x = a + h\zeta$ and the fact that $\hat{f}(\zeta) = f(x)$ reveals that

$$\frac{\mathrm{d}^i \hat{f}}{\mathrm{d}\zeta^i}(\hat{x}_j) = h^i \frac{\mathrm{d}^i f}{\mathrm{d}x^i}(x_j), \quad i = 0, \ldots, m + 1, \quad j = 0, \ldots, n.$$

Now on the reference interval $[0, 1]$, by the stability estimate (14.12) for $\hat{f}$, there is a constant $C_0 > 0$ such that

$$\left| \int_0^1 \hat{f}(\zeta)\hat{\omega}(\zeta)\,\mathrm{d}\zeta - \sum_{j=0}^{n} \sum_{i=1}^{r} \hat{\beta}_{i,j} \hat{f}^{(i)}(\hat{x}_j) \right| \leq C_0 \left\| \frac{\mathrm{d}^{m+1}\hat{f}}{\mathrm{d}\zeta^{m+1}} \right\|_{L^\infty(0,1)},$$

and C_0 must be independent of h, since we are working on the reference interval.

Using (14.2) and the stability estimate for $\hat{f}$, we have

$$\begin{aligned}\left|\int_a^b f(x)w(x)\mathrm{d}x - \sum_{j=0}^{n}\sum_{i=1}^{r}\beta_{i,j}f^{(i)}(x_j)\right| &= h\left|\int_0^1 \hat{f}(\zeta)\hat{\omega}(\zeta)\mathrm{d}\zeta - \sum_{j=0}^{n}\sum_{i=1}^{r}\hat{\beta}_{i,j}\hat{f}^{(i)}(\hat{x}_j)\right| \\ &\le hC_0\left\|\frac{\mathrm{d}^{m+1}\hat{f}}{\mathrm{d}\zeta^{m+1}}\right\|_{L^\infty(0,1)} \\ &\le C_0h^{m+2}\left\|\frac{\mathrm{d}^{m+1}f}{\mathrm{d}x^{m+1}}\right\|_{L^\infty(a,b)},\end{aligned}$$

where $C_0 > 0$ is independent of h and f. The desired result is proven. □

In other words, if the interval is small, i.e., h is small, we can realize a small error.

14.5 Newton–Cotes Formulas

A natural, although possibly not optimal, choice for the nodes is to set them equidistant. In the context of quadrature rules of Lagrange type, this gives rise to the so-called Newton–Cotes formulas; see, for example, [25, 46, 54].

Definition 14.19 (Newton–Cotes rules)**.** Suppose that w is a weight function on the compact interval $[a, b] \subset \mathbb{R}$ and $n \in \mathbb{N}$. Set $h = b - a > 0$ and $\hbar = \frac{h}{n}$. Suppose that, for the simple quadrature rule (14.4), the nodal set $X = \{x_j\}_{j=0}^n \subset [a, b]$ is defined by

$$x_j = a + j\hbar, \quad j = 0, 1, \dots, n, \tag{14.14}$$

and the weights are defined via

$$\beta_j = \int_a^b L_j(x)w(x)\mathrm{d}x, \quad L_j(x) = \prod_{\substack{k=0\\k\neq j}}^{n}\frac{x - x_k}{x_j - x_k}, \quad j = 0, \dots, n.$$

The resulting method, denoted $Q_n[f]$, is called a **closed Newton–Cotes quadrature rule of order** n.[5]

There are open Newton–Cotes rules as well, though we will not cover these in the text. For a description, see, for example, [46].

Example 14.1 In Table 14.1, we list closed Newton–Cotes quadrature rules of order $n = 1, 2, 3, 4$, assuming a weight function $w \equiv 1$ on the compact interval $[a, b] \subset \mathbb{R}$. Only the intrinsic nodes and weights are listed. To obtain the effective versions, use

$$x_j = a + h\hat{x}_j, \quad \beta_j = h\hat{\beta}_j, \quad h = b - a.$$

[5] Named in honor of the British mathematician, physicist, astronomer, theologian, and natural philosopher Sir Isaac Newton (1642–1726/27) and the British mathematician Roger Cotes (1682–1716).

n	rule	$\hat{x}_j$	$\hat{\beta}_j$	Error formula
1	Trapezoidal	0,1	$\frac{1}{2}, \frac{1}{2}$	$-\frac{1}{12}\hbar^3 f^{(2)}(\xi)$
2	Simpson's	$0, \frac{1}{2}, 1$	$\frac{1}{6}, \frac{4}{6}, \frac{1}{6}$	$-\frac{1}{90}\hbar^5 f^{(4)}(\xi)$
3	Simpson's $\frac{3}{8}$	$0, \frac{1}{3}, \frac{2}{3}, 1$	$\frac{1}{8}, \frac{3}{8}, \frac{3}{8}, \frac{1}{8}$	$-\frac{3}{80}\hbar^5 f^{(4)}(\xi)$
4	Boole's[a]	$0, \frac{1}{4}, \frac{1}{2}, \frac{3}{4}, 1$	$\frac{7}{90}, \frac{32}{90}, \frac{12}{90}, \frac{32}{90}, \frac{1}{90}$	$-\frac{8}{945}\hbar^7 f^{(6)}(\xi)$

Table 14.1 Common closed Newton–Cotes formulas with integration weight function $w \equiv 1$ on a compact interval $[a, b] \subset \mathbb{R}$ of length $h = b - a$. Recall, $\hbar = \frac{h}{n}$; see, for example, [46, 86]. Notice that, for $n = 2$ and 4, we observe the phenomenon of super-convergence, i.e., a higher than expected convergence considering, say, Theorem 14.20.

a Named in honor of the British mathematician, philosopher, and logician George Boole (1815–1864).

For example, consider the case $n = 3$, Simpson's $\frac{3}{8}$ rule, on the reference interval $[0, 1]$. The second Lagrange nodal basis element is

$$\hat{L}_1(x) = \frac{x\left(x - \frac{2}{3}\right)(x - 1)}{\frac{1}{3}\left(\frac{1}{3} - \frac{2}{3}\right)\left(\frac{1}{3} - 1\right)} = \frac{27}{2}\left(x^3 - \frac{5}{3}x^2 + \frac{2}{3}x\right).$$

Then

$$\hat{\beta}_1 = \int_0^1 \hat{L}_1(x) \mathrm{d}x = \frac{27}{2}\left(\frac{1}{4}x^4 - \frac{5}{9}x^3 + \frac{1}{3}x^2\right)\Bigg|_{x=0}^{x=1} = \frac{3}{8}.$$

Luckily, these weights need to be tabulated only once, as they clearly do not depend upon the function whose integral is being approximated.

Theorem 14.20 (error estimate). *Let $[a, b] \subset \mathbb{R}$ be a compact interval. Suppose that (14.4) is a closed Newton–Cotes quadrature rule of order $n \in \mathbb{N}$. Then the order of the quadrature rule is consistent of order at least n. Consequently, if $f \in C^{n+1}([a, b])$, then*

$$|E_{Q_n}[f]| \le C h^{n+2} \left\| f^{(n+1)} \right\|_{L^\infty(a,b)},$$

where $h = b - a$ and $C > 0$ is independent of h and f.

Proof. This follows from Theorem 14.18 and the fact that the quadrature rule was designed to be exact for all polynomials of degree at most n; see also Theorem 14.6. □

This last theorem only gives an error estimate. But, for even numbered rules, there is an interesting super-convergence that occurs. Specifically, the order of convergence is one higher than what one might expect from, say, Theorem 14.20; see Table 14.1.

Example 14.2 The reason for the super-convergence observed in Table 14.1 is that the even degree Newton–Cotes rules are exact for polynomials of degree $n+1$. For example, Simpson's rule is exact for cubics. To see this, it suffices to show that it is exact for $f(x) = x^3$ on the reference interval $[0, 1]$. Of course,

$$\int_0^1 x^3 \,\mathrm{d}x = \frac{1}{4}x^4\bigg|_{x=0}^{x=1} = \frac{1}{4}.$$

Approximating with Simpson's rule,

$$\frac{1}{6}f(0) + \frac{4}{6}f(1/2) + \frac{1}{6}f(1) = \frac{4}{6}(1/2)^3 + \frac{1}{6} = \frac{1}{4}.$$

This remarkable fact is due to the spacing of the nodes.

We wish to prove the error estimates appearing in Table 14.1, and even more general results, but we need some technical lemmas first.

Lemma 14.21 (symmetry). *Suppose that $n \in \mathbb{N}$ is even, $[a, b] \subset \mathbb{R}$ is a compact interval, and the nodal set $X = \{x_j\}_{j=0}^n \subset [a, b]$ is defined by (14.14). Set $\omega_{n+1}(x) = \prod_{j=0}^n (x - x_j)$. Then*

$$\omega_{n+1}(a + z) = -\omega_{n+1}(b - z), \quad \forall z \in \mathbb{R},$$

and

$$\omega_{n+1}(\mu + z) = -\omega_{n+1}(\mu - z), \quad \forall z \in \mathbb{R},$$

where $\mu = \frac{a+b}{2}$.

Proof. The proof is an exercise; see Figure 14.1 and Problem 14.8. □

Lemma 14.22 (positivity). *Suppose that $n \in \mathbb{N}$ is even, $[a, b] \subset \mathbb{R}$ is a compact interval, and the nodal set $X = \{x_j\}_{j=0}^n \subset [a, b]$ is defined by (14.14). Set $\omega_{n+1}(x) = \prod_{j=0}^n (x - x_j)$, as usual, and*

$$\Omega_{n+1}(x) = \int_a^x \omega_{n+1}(t)\mathrm{d}t. \tag{14.15}$$

Then $\Omega_{n+1}(a) = 0$, $\Omega_{n+1}(b) = 0$, and

$$\Omega_{n+1}(x) > 0, \quad \forall x \in (a, b).$$

Proof. The proof is an exercise; see Figures 14.1 and 14.2 and Problem 14.9. □

The following general and remarkable result can be found in [46] and in [54]. Our proof is inspired by those references.

Theorem 14.23 (error estimate for even order). *Let $[a, b] \subset \mathbb{R}$ be a compact interval. Suppose that (14.4) is a closed Newton–Cotes quadrature rule of order $n \in \mathbb{N}$ and $w \equiv 1$. If n is even and $f \in C^{n+2}([a, b])$, then for some $\xi \in (a, b)$,*

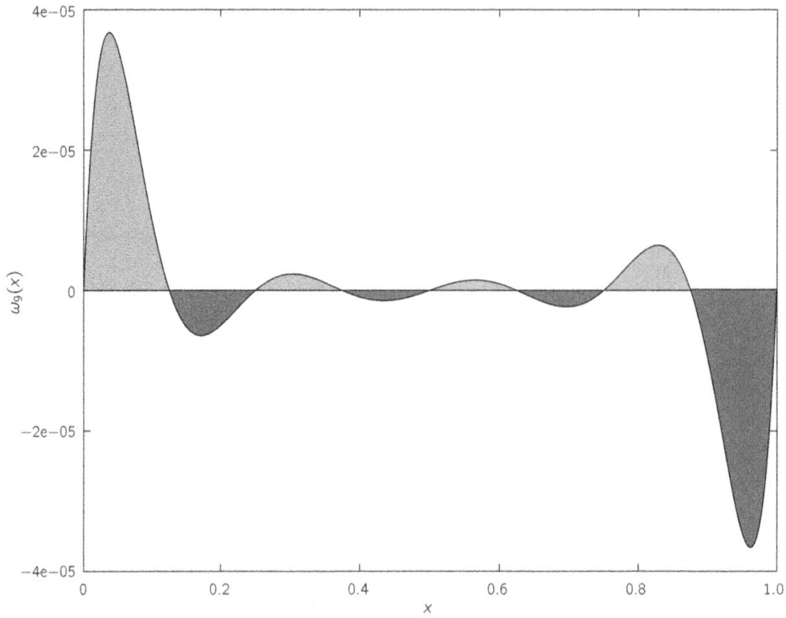

Figure 14.1 The nodal polynomial ω_9 for nine equally spaced nodes in $[0, 1]$, determined according to (14.14). The regions that are light shaded are positive-area regions, whereas the dark shaded regions are negative-area regions. From left to right, the first light shaded region is greater in area, in absolute value, than the first dark shaded region, which, in turn, is greater in area, in absolute value, than the second light shaded region, etc. This suggests that $\Omega_9(x)$, defined in (14.15), is positive for $x \in (a, b)$; see Figure 14.2.

$$E_{Q_n}[f] = C_n \frac{f^{(n+2)}(\xi)}{(n+2)!},$$

where

$$C_n = \int_a^b x\omega_{n+1}(x)\mathrm{d}x < 0.$$

Proof. By Theorem 9.45,

$$E_{Q_n}[f] = \int_a^b \omega_{n+1}(x) f[\![x_0, x_1, \ldots, x_n, x]\!]\mathrm{d}x.$$

Since, from Lemma 14.22, Ω_{n+1} does not change sign on $[a, b]$, using integration by parts and the Integral Mean Value Theorem B.41, we have

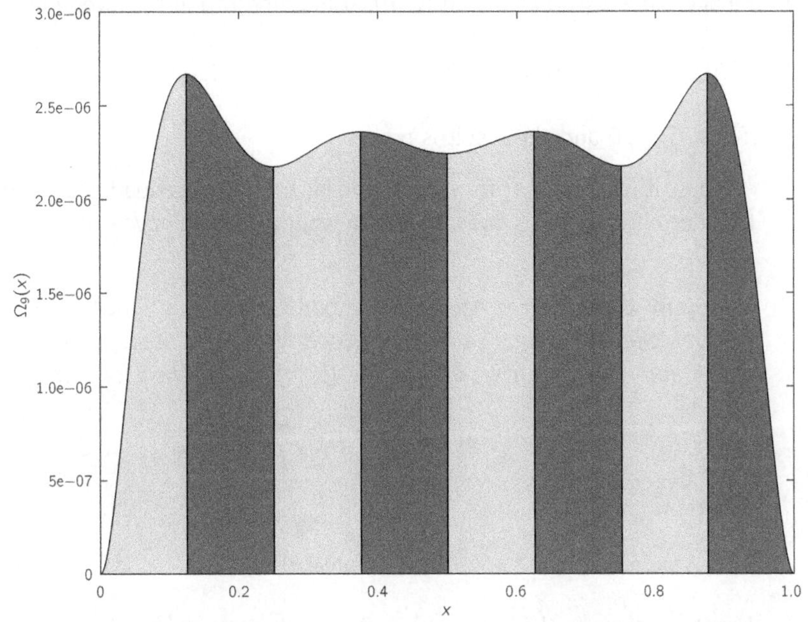

Figure 14.2 The indefinite integral $\Omega_9(x) = \int_0^x \omega_9(t)\mathrm{d}t$, where ω_9 is as shown in Figure 14.1; see (14.15). The light and dark shaded regions, respectively, correspond to those in Figure 14.1, where positive- or negative-area contributions are being added to the integral and the value of Ω_9 is increasing or decreasing.

$$\begin{aligned} E_{Q_n}[f] &= \Omega_{n+1}(x) f[\![x_0, x_1, \ldots, x_n, x]\!]\Big|_{x=a}^{x=b} - \int_a^b \frac{\partial}{\partial x} f[\![x_0, x_1, \ldots, x_n, x]\!]\Omega_{n+1}(x)\mathrm{d}x \\ &= -\int_a^b f[\![x_0, x_1, \ldots, x_n, x, x]\!]\Omega_{n+1}(x)\mathrm{d}x \\ &= -f[\![x_0, x_1, \ldots, x_n, \eta, \eta]\!] \int_a^b \Omega_{n+1}(x)\mathrm{d}x, \end{aligned}$$

where η is some number in (a, b). Using Proposition 9.53,

$$f[\![x_0, x_1, \ldots, x_n, \eta, \eta]\!] = \frac{f^{(n+2)}(\xi)}{(n+2)!}$$

for some $\xi \in (a, b)$ satisfying

$$\min\{x_0, x_1, \ldots, x_n, \eta\} < \xi < \max\{x_0, x_1, \ldots, x_n, \eta\}.$$

Finally, since $\Omega_{n+1}(x) > 0$ in (a, b), using integration by parts again, we find

$$0 < \int_a^b \Omega_{n+1}(x)\mathrm{d}x = x\Omega_{n+1}(x)|_{x=a}^{x=b} - \int_a^b x\omega_{n+1}(x)\mathrm{d}x = -\int_a^b x\omega_{n+1}(x)\mathrm{d}x.$$

Thus, $C_n < 0$ and the result is proven. □

Thus, it follows that, for n even, the Newton–Cotes quadrature rule is consistent to order $n+2$, which is one larger than what we would have expected from Theorem 14.20.

Theorem 14.24 (error estimate for odd order)**.** *Let $[a, b] \subset \mathbb{R}$ be a compact interval. Suppose that (14.4) is a closed Newton–Cotes quadrature rule of order $n \in \mathbb{N}$ and $w \equiv 1$. If n is odd and $f \in C^{n+1}([a, b])$, then for some $\xi \in (a, b)$,*

$$E_{Q_n}[f] = C_n \frac{f^{(n+1)}(\xi)}{(n+1)!},$$

where

$$C_n = \int_a^b \omega_{n+1}(x)\mathrm{d}x < 0.$$

Proof. First note that $\omega_{n+1}(x) < 0$ on the interval $(b - \hbar, b) = (x_{n-1}, x_n)$. This implies, by Theorem 9.45 and the Integral Mean Value Theorem B.41, that

$$\begin{aligned} E_{Q_n}[f] &= \int_a^b \omega_{n+1}(x) f[\![x_0, x_1, \ldots, x_n, x]\!]\mathrm{d}x \\ &= \int_a^{b-\hbar} \omega_{n+1}(x) f[\![x_0, x_1, \ldots, x_n, x]\!]\mathrm{d}x + \int_{b-\hbar}^b \omega_{n+1}(x) f[\![x_0, x_1, \ldots, x_n, x]\!]\mathrm{d}x \\ &= \int_a^{b-\hbar} \omega_{n+1}(x) f[\![x_0, x_1, \ldots, x_n, x]\!]\mathrm{d}x + f[\![x_0, x_1, \ldots, x_n, \eta]\!] \int_{b-\hbar}^b \omega_{n+1}(x)\mathrm{d}x \end{aligned}$$

for some $\eta \in (a, b)$. To simplify the first integral, let us write

$$\omega_{n+1}(x) = \omega_n(x)(x - x_n) = \omega_n(x)(x - b)$$

and define

$$\Omega_n(x) = \int_a^x \omega_n(x)\mathrm{d}x.$$

Using Corollary 9.44, we have

$$\begin{aligned} I_1 &= \int_a^{b-\hbar} \omega_{n+1}(x) f[\![x_0, x_1, \ldots, x_n, x]\!]\mathrm{d}x \\ &= \int_a^{b-\hbar} \omega_n(x)(x - x_n) \frac{f[\![x_0, x_1, \ldots, x_{n-1}, x]\!] - f[\![x_0, x_1, \ldots, x_n]\!]}{(x - x_n)} \mathrm{d}x \\ &= \int_a^{b-\hbar} \Omega_n'(x) \left(f[\![x_0, x_1, \ldots, x_{n-1}, x]\!] - f[\![x_0, x_1, \ldots, x_n]\!]\right)\mathrm{d}x \\ &= \int_a^{b-\hbar} \Omega_n'(x) f[\![x_0, x_1, \ldots, x_{n-1}, x]\!]\mathrm{d}x - f[\![x_0, x_1, \ldots, x_n]\!] \int_a^{b-\hbar} \Omega_n'(x)\mathrm{d}x. \end{aligned}$$

Now $n-1$ is even and, by Lemma 14.22, $\Omega_n(x) > 0$ for $x \in (a, b-\hbar)$, and $\Omega_n(a) = \Omega(b-\hbar) = 0$. Therefore,

$$\int_a^{b-\hbar} \Omega_n'(x)\mathrm{d}x = 0,$$

and, using integration by parts and the Integral Mean Value Theorem, again,

$$\begin{aligned} I_1 &= \int_a^{b-\hbar} \Omega_n'(x) f[\![x_0, x_1, \ldots, x_{n-1}, x]\!]\mathrm{d}x \\ &= \Omega_n(x) f[\![x_0, x_1, \ldots, x_{n-1}, x]\!]\big|_{x=a}^{x=b-\hbar} - \int_a^{b-\hbar} \Omega_n(x) f[\![x_0, x_1, \ldots, x_{n-1}, x, x]\!]\mathrm{d}x \\ &= -f[\![x_0, x_1, \ldots, x_{n-1}, \zeta, \zeta]\!] \int_a^{b-\hbar} \Omega_n(x)\mathrm{d}x \end{aligned}$$

for some $\zeta \in (a, b)$. Using Proposition 9.53, for some $\xi_1, \xi_2 \in (a, b)$,

$$\begin{aligned} E_{Q_n}[f] &= -f[\![x_0, x_1, \ldots, x_{n-1}, \zeta, \zeta]\!] \int_a^{b-\hbar} \Omega_n(x)\mathrm{d}x \\ &\quad + f[\![x_0, x_1, \ldots, x_n, \eta]\!] \int_{b-\hbar}^{b} \omega_{n+1}(x)\mathrm{d}x \\ &= -\frac{f^{(n+1)}(\xi_1)}{(n+1)!} \int_a^{b-\hbar} \Omega_n(x)\mathrm{d}x + \frac{f^{(n+1)}(\xi_2)}{(n+1)!} \int_{b-\hbar}^{b} \omega_{n+1}(x)\mathrm{d}x. \end{aligned}$$

Finally, by Lemma 14.22, since $n-1$ is even,

$$\begin{aligned} 0 &< \int_a^{b-\hbar} \Omega_n(x)\mathrm{d}x \\ &= (x-b)\Omega_n(x)\big|_{x=a}^{x=b-\hbar} - \int_a^{b-\hbar} (x-b)\omega_n(x)\mathrm{d}x \\ &= -\int_a^{b-\hbar} (x-b)\omega_n(x)\mathrm{d}x. \end{aligned}$$

Furthermore, since $\omega_{n+1} < 0$ on $(b-\hbar, b)$,

$$0 < -\int_{b-\hbar}^{b} \omega_{n+1}(x)\mathrm{d}x.$$

Using the Summation Mean Value Theorem B.40, there is a point $\xi \in (a, b)$ such that

$$\begin{aligned} E_{Q_n}[f] &= -\frac{f^{(n+1)}(\xi)}{(n+1)!} \left(-\int_a^{b-\hbar} (x-b)\omega_n(x)\mathrm{d}x - \int_{b-\hbar}^{b} \omega_{n+1}(x)\mathrm{d}x \right) \\ &= \frac{f^{(n+1)}(\xi)}{(n+1)!} \int_a^{b} \omega_{n+1}(x)\mathrm{d}x. \end{aligned}$$

The proof is complete. □

Example 14.3 In Table 14.1, we list error formulas for the closed Newton–Cotes quadrature rules of order $n = 1, 2, 3, 4$, assuming a weight function $w \equiv 1$ on the compact interval $[a, b] \subset \mathbb{R}$. Consider again Simpson's $\frac{3}{8}$ rule. Following Theorem 14.24, for some $\xi \in [a, b]$,

$$E_{Q_n}[f] = C_3 \frac{f^{(4)}(\xi)}{4!},$$

where

$$C_3 = \int_a^b \omega_4(x)dx$$

and

$$\omega_4(x) = (x-a)(x-a-\hbar)(x-a-2\hbar)(x-a-3\hbar), \quad \hbar = \frac{h}{3}, \quad h = b-a.$$

Make a change of variables

$$x = a + hz.$$

Thus,

$$\begin{aligned} C_3 &= h\int_0^1 \omega_4(a+hz)dz \\ &= h^5 \int_0^1 z\left(z-\frac{1}{3}\right)\left(z-\frac{2}{3}\right)(z-1)dz \\ &= -\frac{h^5}{270}. \end{aligned}$$

Therefore,

$$E_{Q_n}[f] = -\frac{h^5}{270}\frac{f^{(4)}(\xi)}{24} = -\hbar^5\frac{3}{80}f^{(4)}(\xi).$$

14.6 Peano Error Formulas for Trapezoidal, Midpoint, and Simpson's Rules

In this section, we show how to use the Peano Kernel Theorem 14.10, to compute errors for the trapezoidal rule and Simpson's rule, assuming, for simplicity, that the weight function is trivial, i.e., $w \equiv 1$ on the integration interval $[a, b]$. The error relations we derive are exactly the same as those found in Table 14.1; only the procedure for calculating the relations is different. Specifically, this procedure does not use any details of polynomial interpolation error or divided differences.

Recall that, if a method is exact for polynomials of degree m or less, the quadrature error satisfies

$$E_{Q_n}[f] = \frac{1}{m!}\int_a^b f^{(m+1)}(y)K_m(y)dy,$$

where K_m is the Peano Kernel, defined as

$$K_m(y) = E_{Q_n}[k_m(\cdot, y)] = \int_a^b k_m(x, y)dx - \sum_{j=0}^{n} \beta_j k_m(x_j, y).$$

We will show that, in both cases, K_m does not change sign on $[a, b]$, allowing us to use Corollary 14.13.

Theorem 14.25 (error estimates). *Suppose that $[a, b] \subset \mathbb{R}$ is a compact interval, the integration weight function is trivial, $w \equiv 1$, and $n = 1$ or 2. Set $h = [a, b]$ and $\hbar = \frac{h}{n}$. If $f \in C^2([a, b])$, then the error for the trapezoidal rule is*

$$E_{Q_1}[f] = -\frac{1}{12}\hbar^3 f^{(2)}(\xi)$$

for some $\xi \in [a, b]$. If $f \in C^4([a, b])$, then the error for Simpson's rule is

$$E_{Q_2}[f] = -\frac{1}{90}\hbar^5 f^{(4)}(\xi)$$

for some $\xi \in [a, b]$.

Proof. The trapezoidal rule is exact for polynomials of degree one. Invoking the Peano Kernel Theorem 14.10, we see that the error satisfies

$$E_{Q_1}[f] = \int_a^b f^{(2)}(y)K_1(y)dy,$$

where

$$K_1(y) = \int_a^b k_1(x, y)dx - \frac{h}{2}(k_1(a, y) + k_1(b, y))$$

and

$$k_1(x, y) = \begin{cases} x - y, & a \le y \le x \le b, \\ 0, & a \le x < y \le b. \end{cases}$$

It follows that

$$K_1(y) = \frac{1}{2}(b - y)^2 - \frac{1}{2}(b - y)(b - a),$$

which is nonpositive for $y \in [a, b]$. Integrating, we find

$$\begin{aligned} \int_a^b K_1(y)dy &= \left[\frac{1}{6}(y - b)^3 + \frac{1}{4}(y - b)^2(b - a)\right]_{y=a}^{y=b} \\ &= -\left[\frac{2}{12}(-h)^3 + \frac{3}{12}(-h)^2 h\right] \\ &= -\frac{h^3}{12} \\ &= -\frac{\hbar^3}{12}. \end{aligned}$$

Applying Corollary 14.13, we have, for some $\xi \in [a, b]$,

$$E_{Q_1}[f] = -\frac{1}{12}\hbar^3 f^{(2)}(\xi).$$

Simpson's rule, as we have seen in Example 14.2, is exact for polynomials of degree three. The Peano Kernel Theorem 14.10 guarantees that the error satisfies

$$E_{Q_2}[f] = \int_a^b f^{(4)}(y) K_3(y) dy,$$

where

$$K_3(y) = \int_a^b k_3(x, y) dx - \frac{h}{6}\left(k_3(a, y) + 4k_3(\mu, y) + k_3(b, y)\right),$$

with $\mu = \frac{a+b}{2}$, and

$$k_3(x, y) = \begin{cases} (x-y)^3, & a \leq y \leq x \leq b, \\ 0, & a \leq x < y \leq b. \end{cases}$$

It follows that

$$K_3(y) = \begin{cases} \frac{1}{72}(y-a)^3(3y-a-2b), & a \leq y \leq \mu, \\ \frac{1}{72}(b-y)^3(b+2a-3t), & \mu < y \leq b, \end{cases}$$

which, again, is nonpositive for $y \in [a, b]$. Integrating, we find

$$\int_a^b K_3(y) dy = -\frac{h^5}{5760} - \frac{h^5}{5760} = -\frac{h^5}{2880} = -\frac{\hbar^5}{90}.$$

Using Corollary 14.13, we see that, for some $\xi \in [a, b]$,

$$E_{Q_2}[f] = -\frac{1}{90}\hbar^5 f^{(4)}(\xi).$$

□

Now let us introduce a method that is not a closed Newton–Cotes rule.

Definition 14.26 (midpoint rule). Suppose that $[a, b] \subset \mathbb{R}$ is a compact interval of positive length. Set $\mu = \frac{a+b}{2}$ and $h = b - a$. Suppose that $f \in C([a, b])$. The **midpoint rule** is defined as

$$Q_\mu[f] = hf(\mu).$$

It is clear that the midpoint rule is consistent of order exactly one. Therefore, we have the following.

Theorem 14.27 (error estimate). *Suppose that $[a, b] \subset \mathbb{R}$ is a compact interval, and the integration weight function is trivial, $w \equiv 1$. Set $h = b - a$. If $f \in C^2([a, b])$, then the error for the midpoint rule is*

$$E_{Q_\mu}[f] = \frac{1}{24}h^3 f^{(2)}(\xi)$$

for some $\xi \in [a, b]$.

Proof. Since the midpoint rule is exact for polynomials of degree one, invoking the Peano Kernel Theorem 14.10, we have

$$E_{Q_\mu}[f] = \int_a^b f^{(2)}(y)K_1(y)\mathrm{d}y,$$

where

$$K_1(y) = \int_a^b k_1(x,y)\mathrm{d}x - hk_1(\mu, y),$$

with $\mu = \frac{a+b}{2}$, and

$$k_1(x,y) = \begin{cases} x - y, & a \le y \le x \le b, \\ 0, & a \le x < y \le b. \end{cases}$$

It follows that

$$K_1(y) = \begin{cases} \frac{1}{2}(b-y)^2 - (b-a)(\mu - y), & a \le y \le \mu < b, \\ \frac{1}{2}(b-y)^2, & a < \mu < y \le b, \end{cases}$$

which is nonnegative for $y \in [a,b]$. Integrating, we find

$$\begin{aligned} \int_a^b K_1(y)\mathrm{d}y &= 2\int_a^\mu K_1(y)\mathrm{d}y \\ &= 2\int_a^\mu \left(\frac{1}{2}(y-b)^2 + (b-a)(y-\mu)\right)\mathrm{d}y \\ &= \left[\frac{1}{3}(y-b)^3 + (b-a)(y-\mu)^2\right]_{y=a}^{y=\mu} \\ &= \frac{1}{24}h^3. \end{aligned}$$

Applying Corollary 14.13, we have, for some $\xi \in [a,b]$,

$$E_{Q_1}[f] = \frac{1}{24}h^3 f^{(2)}(\xi).$$

□

14.7 Composite Quadrature Rules

The typical way of using Newton–Cotes quadrature rules is to partition an interval into several smaller subintervals of equal size and apply low-order Newton–Cotes rules on each subinterval. In the case of the composite trapezoidal rule, this is equivalent to forming a piecewise linear interpolant of the function and integrating this interpolant. For Simpson's rule, the composite version of the rule may be obtained by interpolating the integrand by a piecewise quadratic function and integrating; see Figures 14.3 and 14.4.

We will assume everywhere in this section, for simplicity, that $w \equiv 1$.

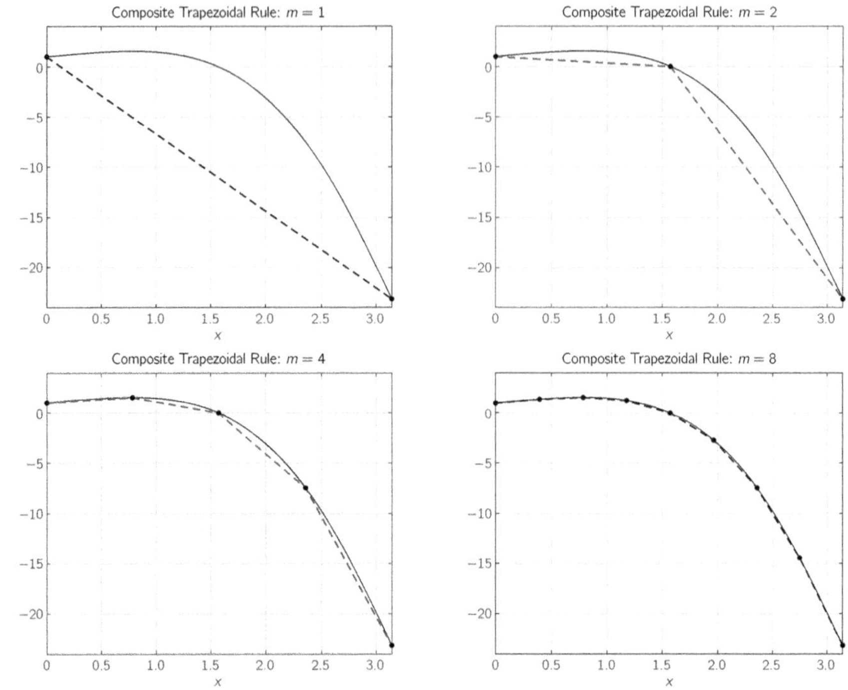

Figure 14.3 Composite trapezoidal rule approximation of $\int_0^\pi e^x \cos(x)\mathrm{d}x$. This rule is obtained by integrating the piecewise linear approximations of $e^x \cos(x)$, which are shown by dashed curves.

Definition 14.28 (composite Newton–Cotes rules). Let $[c, d] \subset \mathbb{R}$ be a compact interval, of positive length $L = d - c > 0$, and $m, n \in \mathbb{N}$. Suppose that $f \in C([c, d])$. Set $h = \frac{L}{m}$ and $\hbar = \frac{h}{n} = \frac{L}{mn}$. Define

$$x_{i,j} = c + (i - 1)h + j\hbar, \quad i = 1, \ldots, m, \quad j = 0, \ldots, n.$$

The **composite closed Newton–Cotes quadrature rule of order** n is the rule

$$Q_{n,m}^{(c,d)}[f] = \sum_{i=1}^{m} \sum_{j=1}^{n} \beta_j f(x_{i,j}),$$

where

$$\beta_j = \int_0^h L_j(x)\mathrm{d}x, \quad L_j(x) = \prod_{\substack{k=0 \\ k \neq j}}^{n} \frac{x - k\hbar}{(j - k)\hbar}, \quad j = 0, \ldots, n.$$

Remark 14.29 (nodes). Note that certain nodes repeat:

$$x_{i-1,n} = x_{i,0}, \quad i = 2, \ldots, m.$$

The following are the two most important composite rules.

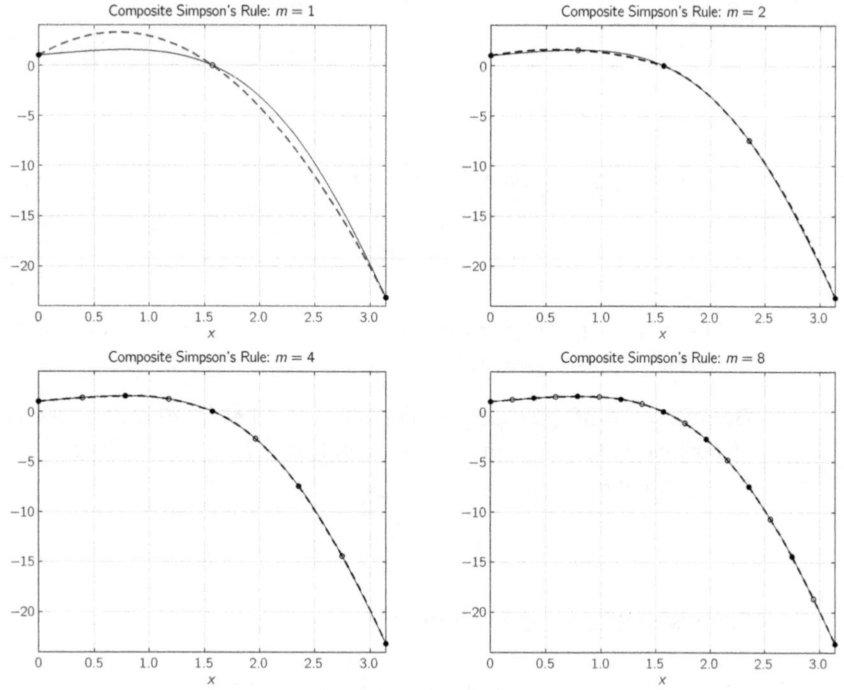

Figure 14.4 Composite Simpson's rule approximation of $\int_0^\pi e^x \cos(x)dx$. This rule is obtained by integrating a piecewise quadratic approximation of $e^x \cos(x)$.

Example 14.4 Suppose that $n = 1$. The composite closed Newton–Cotes quadrature rule of order $n = 1$ is the *composite trapezoidal rule*. The rule is

$$Q_{1,m}^{(c,d)}[f] = \sum_{i=1}^{m} \left(\frac{h}{2} f(x_{i,0}) + \frac{h}{2} f(x_{i,1}) \right).$$

Let us set

$$\xi_k = c + k\hbar, \quad f_k = f(\xi_k), \quad k = 0, \dots, m.$$

Then

$$Q_{1,m}^{(c,d)}[f] = \hbar \left(\frac{1}{2} f_0 + f_1 + \cdots + f_{m-1} + \frac{1}{2} f_m \right). \tag{14.16}$$

Example 14.5 Suppose that $n = 2$ and $w \equiv 1$. The composite closed Newton–Cotes quadrature rule of order $n = 2$ is the *composite Simpson's rule*. The rule is

$$Q_{2,m}^{(c,d)}[f] = \sum_{i=1}^{m} \left(\frac{h}{6} f(x_{i,0}) + \frac{4h}{6} f(x_{i,1}) + \frac{h}{6} f(x_{i,2}) \right).$$

Let us set

$$\xi_k = c + k\hbar, \quad f_k = f(\xi_k), \quad k = 0, \dots, 2m.$$

Then

$$Q_{2,m}^{(c,d)}[f] = \frac{\hbar}{3}\left(f_0 + 4f_1 + 2f_2 + \cdots + 2f_{2m-2} + 4f_{2m-1} + f_{2m}\right). \tag{14.17}$$

The composite midpoint rule cannot be realized as a composite closed Newton–Cotes rule but is closely related.

Definition 14.30 (composite midpoint rule). Let $[c, d] \subset \mathbb{R}$ be a compact interval, of positive length $L = d - c > 0$, and $m \in \mathbb{N}$. Suppose that $f \in C([c, d])$. Set $h = \frac{L}{m}$. Define

$$\mu_k = c + \left(k - \frac{1}{2}\right)h, \quad f_k = f(\mu_k), \quad k = 1, \dots, m.$$

The **composite midpoint rule** is

$$Q_{\mu,m}^{(c,d)}[f] = h\sum_{k=1}^{m} f_k = h\left(f_1 + \cdots + f_m\right). \tag{14.18}$$

The error formulas for the composite trapezoidal, Simpson's, and midpoint rules are easily obtained.

Theorem 14.31 (error formulas). *Suppose that $[c, d]$ is a compact interval of positive length $L = b - c > 0$, $m \in \mathbb{N}$, and $h = \frac{L}{m}$. If $f \in C^2([c, d])$, then*

$$E_{Q_{1,m}}[f] = \int_c^d f(x)\mathrm{d}x - Q_{1,m}^{(c,d)}[f] = -\frac{Lh^2}{12}f''(\eta_1)$$

for some $\eta_1 \in [c, d]$ and

$$E_{Q_{\mu,m}}[f] = \int_c^d f(x)\mathrm{d}x - Q_{\mu,m}^{(c,d)}[f] = \frac{Lh^2}{24}f''(\eta_\mu)$$

for some $\eta_\mu \in [c, d]$. If $f \in C^4([c, d])$ and $n = 2$, then

$$E_{Q_{2,m}}[f] = \int_c^d f(x)\mathrm{d}x - Q_{2,m}^{(c,d)}[f] = -\frac{L\hbar^4}{180}f^{(4)}(\eta_2)$$

for some $\eta_2 \in [c, d]$, where $\hbar = \frac{h}{2} = \frac{L}{2m}$.

Proof. For the composite trapezoidal rule, there are points $\eta_{i,1} \in [x_{i,0}, x_{i,1}]$, $i = 1, \dots, m$ such that

$$E_{Q_{1,m}}[f] = \sum_{i=1}^{m}\left(-\frac{1}{12}h^3 f''(\eta_{i,1})\right).$$

m	$Q_{1,m}$	$E_{Q_{1,m}}$	Rate
16	−12.148 004 099 896 829	7.765 778 350 719 543e−02	–
32	−12.089 742 117 014 199	1.939 580 062 456 514e−02	2.001 386
64	−12.075 194 099 202 140	4.847 782 812 506 196e−03	2.000 347
128	−12.071 558 189 102 351	1.211 872 712 717 721e−03	2.000 086
256	−12.070 649 280 005 421	3.029 636 157 876 325e−04	2.000 021
512	−12.070 422 057 008 418	7.574 061 878 479 199e−05	2.000 005

Table 14.2 Approximation of the integral (14.19) by the composite trapezoidal rule. The rate of convergence is computed according to (14.20).

m	$Q_{2,m}$	$E_{Q_{2,m}}$	Rate
16	−12.070 321 456 053 321	−2.486 033 631 221 574e−05	–
32	−12.070 344 759 931 452	−1.556 458 181 894 982e−06	3.997 507
64	−12.070 346 219 069 089	−9.732 054 451 205 840e−08	3.999 378
128	−12.070 346 310 306 443	−6.083 190 839 945 019e−09	3.999 844
256	−12.070 346 316 009 422	−3.802 114 179 052 296e−10	3.999 954
512	−12.070 346 316 365 873	−2.376 054 908 381 775e−11	4.000 161

Table 14.3 Approximation of the integral (14.19) by the composite Simpson's rule. The rate of convergence is computed according to (14.20).

Applying the Summation Mean Value Theorem B.40, there is a point $\eta_1 \in [c, d]$ such that

$$E_{Q_{1,m}}[f] = -f''(\eta)\sum_{i=1}^{m}\frac{1}{12}h^3 = -f''(\eta)\frac{1}{12}mh^3 = -f''(\eta_1)\frac{1}{12}Lh^2.$$

The two other results are derived similarly; see Problem 14.11. □

Example 14.6 In this example, let us apply the composite trapezoidal and Simpson's rule to approximate the value of the definite integral

$$\int_0^{\pi} e^x \cos(x)dx = -\frac{e^{\pi}+1}{2} \approx -12.070\,346\,316\,389\,633. \tag{14.19}$$

Approximations by the composite trapezoidal rule are shown in Table 14.2. The rate of convergence is computed by the formula

$$\log_2\left(\frac{|E_{Q_{n,m}}|}{|E_{Q_{n,2m}}|}\right). \tag{14.20}$$

Approximations by the composite Simpson's rule are shown in Table 14.3; see Figures 14.3 and 14.4.

14.8 Bernoulli Numbers and Euler–Maclaurin Error Formulas

The purpose of this section is to introduce error formulas for composite quadrature rules in terms of expansions in powers of h. To begin, we need to introduce special sequences of numbers and polynomials.

Theorem 14.32 (exponential generating function). *The function $g\colon \mathbb{C} \to \mathbb{C}$, defined by*

$$g(z) = \frac{z}{e^z - 1},$$

is holomorphic in the disk

$$D = \{z \in \mathbb{C} \mid |z| < 2\pi\},$$

and an expansion of the form

$$g(z) = \frac{z}{e^z - 1} = \sum_{j=0}^{\infty} \frac{B_n}{n!} z^n, \quad |z| < 2\pi \tag{14.21}$$

is, therefore, valid.

Proof. The point $z = 0$ is a removable singularity. However, the points $z = 2k\pi\mathrm{i}$, for $k \in \mathbb{Z}\backslash 0$ are essential singularities; see, for example, [60, 77]. □

The coefficients in the expansion of the function of the previous result bear an important name.

Definition 14.33 (Bernoulli numbers). The sequence $\{B_n\}_{n=0}^{\infty} \subset \mathbb{C}$, appearing in (14.21), is called the sequence of **Bernoulli numbers**.[6]

Proposition 14.34 (properties of Bernoulli numbers). *The Bernoulli numbers satisfy $B_0 = 1$ and*

$$\sum_{j=0}^{n} \frac{n!}{j!(n-j)!} B_j = B_n, \quad n \in \mathbb{N}. \tag{14.22}$$

Furthermore,

$$B_{2j+1} = 0, \quad j \in \mathbb{N}. \tag{14.23}$$

Proof. Since

$$z = g(z)\,(e^z - 1),$$

using the expansion of the exponential, it follows that

$$\left(\sum_{k=1}^{\infty} \frac{1}{k!} z^k\right) \left(\sum_{j=0}^{\infty} \frac{B_j}{j!} z^j\right) = z, \quad |z| < 2\pi.$$

[6] Named in honor of the Swiss mathematician Jacob Bernoulli (1654/5–1705).

n	$B_n(x)$	B_n
0	1	1
1	$x - \frac{1}{2}$	$-\frac{1}{2}$
2	$x^2 - x + \frac{1}{6}$	$\frac{1}{6}$
3	$x^3 - \frac{3}{2}x^2 + \frac{1}{2}x$	0
4	$x^4 - 2x^3 + x^2 - \frac{1}{30}$	$-\frac{1}{30}$
5	$x^5 - \frac{5}{2}x^4 + \frac{5}{3}x^3 - \frac{1}{6}x$	0
6	$x^6 - 3x^5 + \frac{5}{2}x^4 - \frac{1}{2}x^2 + \frac{1}{42}$	$\frac{1}{42}$
7	$x^7 - \frac{7}{2}x^6 + \frac{7}{2}x^5 - \frac{7}{6}x^3 + \frac{1}{6}x$	0
8	$x^8 - 4x^7 + \frac{14}{3}x^6 - \frac{7}{3}x^4 + \frac{2}{3}x^2 - \frac{1}{30}$	$-\frac{1}{30}$
9	$x^9 - \frac{9}{2}x^8 + 6x^7 - \frac{21}{5}x^5 + 2x^3 - \frac{3}{10}x$	0
10	$x^{10} - 5x^9 + \frac{15}{2}x^8 - 7x^6 + 5x^4 - \frac{3}{2}x^2 + \frac{5}{66}$	$\frac{5}{66}$

Table 14.4 The first 11 Bernoulli polynomials and Bernoulli numbers.

This can be used to prove (14.22). To prove (14.23), we use the fact that, for $|z| < 2\pi$,

$$g(-z) = \frac{-z}{e^{-z} - 1} = \sum_{j=0}^{\infty} \frac{(-1)^n B_n}{n!} z^n = z + \sum_{j=0}^{\infty} \frac{B_n}{n!} z^n.$$

The details are left to the reader as an exercise; see Problem 14.12. □

Theorem 14.35 (generating function)**.** *The function $G\colon \mathbb{R} \times \mathbb{C} \to \mathbb{C}$, defined by*

$$G(t, z) = e^{tz} g(z),$$

is, for each $t \in \mathbb{R}$, holomorphic in the disk $D = \{z \in \mathbb{C} \mid |z| < 2\pi\}$. Consequently, an expansion of the form

$$G(t, z) = e^{tz} \frac{z}{e^z - 1} = \sum_{j=0}^{\infty} \frac{B_n(t)}{n!} z^n, \quad |z| < 2\pi \tag{14.24}$$

is, therefore, valid.

Proof. This follows because e^{tz} is entire (holomporphic in $\mathbb{C}$), for every $t \in \mathbb{R}$, and $g(z)$ is holomorphic in D. □

Definition 14.36 (Bernoulli polynomials)**.** The coefficients $B_n(t)$ in (14.24) are called the **Bernoulli polynomials**.

Table 14.4 shows the first 11 Bernoulli numbers and polynomials. From the definition, however, it is not evident that these are indeed polynomials. The following result establishes this.

Proposition 14.37 (properties of Bernoulli polynomials)**.** *The Bernoulli polynomials are indeed real-valued polynomials of degree exactly n. They satisfy $B_0(t) = 1$,*

$$B_n(t) = \sum_{j=0}^{n} \frac{n!}{j!(n-j)!} B_{n-j} t^j, \quad n \in \mathbb{N},$$

and

$$B_n(t) = B_n + n\int_0^t B_{n-1}(t)\mathrm{d}t, \quad n \in \mathbb{N}.$$

Consequently, $B_n(0) = B_n$ *and* $B'(t) = nB_{n-1}(t)$.

Proof. See Problem 14.13. □

Some further properties of the Bernoulli polynomials are important for us.

Proposition 14.38 (further properties). *The Bernoulli polynomials* $\{B_n(t)\}_{n=0}^{\infty}$ *satisfy:*

1.
$$B_j(t+1) - B_j(t) = jt^{j-1}, \quad j \in \mathbb{N}.$$

2.
$$B_j(1) = B_j(0) = B_j, \quad j \in \mathbb{N}, \quad j > 1.$$

3.
$$\int_0^1 B_j(t)\,\mathrm{d}t = 0, \quad j \in \mathbb{N}.$$

4.
$$(-1)^j\left(B_{2j}(t) - B_{2j}\right) \geq 0, \quad \forall t \in [0,1], \quad j \in \mathbb{N}.$$

Proof. See Problem 14.14. □

With the help of Bernoulli numbers and polynomials, we can establish error representations for quadrature rules.

Theorem 14.39 (error representation). *Suppose that* $k \in \mathbb{N}$ *and* $f \in C^{2k+1}([0,1])$. *Then*

$$\int_0^1 f(x)\mathrm{d}x - \frac{1}{2}[f(0) + f(1)] = -\sum_{j=1}^{k}\frac{B_{2j}}{(2j)!}\left[f^{(2j-1)}(1) - f^{(2j-1)}(0)\right] + R_{2k}[f],$$

where

$$R_{2k}[f] = \frac{-1}{(2k+1)!}\int_0^1 f^{(2k+1)}(x)B_{2k+1}(x)\mathrm{d}x.$$

If $f \in C^{2k+2}([0,1])$, *then*

$$\begin{aligned} R_{2k}[f] &= \frac{1}{(2k+2)!}\int_0^1 f^{(2k+2)}(x)\left(B_{2k+2}(x) - B_{2k+2}\right)\mathrm{d}x \\ &= \frac{1}{(2k+2)!}f^{(2k+2)}(\eta)\int_0^1 \left(B_{2k+2}(x) - B_{2k+2}\right)\mathrm{d}x \\ &= -\frac{1}{(2k+2)!}f^{(2k+2)}(\eta)B_{2k+2} \end{aligned}$$

for some $\eta \in [0,1]$.

Proof. Using integration by parts and properties of the Bernoulli polynomials, we have

$$\begin{aligned}\int_0^1 f(x)\mathrm{d}x &= \int_0^1 f(x)B_1'(x)\mathrm{d}x \\ &= B_1(1)f(1) - B_1(0)f(0) - \int_0^1 f'(x)B_1(x)\mathrm{d}x \\ &= \frac{1}{2}[f(1)+f(0)] - \int_0^1 f'(x)\frac{1}{2}B_2'(x)\mathrm{d}x.\end{aligned}$$

To shorten notation, set

$$E = \int_0^1 f(x)\mathrm{d}x - \frac{1}{2}[f(1)+f(0)].$$

The previous computations have then shown that

$$E = -\int_0^1 f'(x)\frac{1}{2}B_2'(x)\mathrm{d}x.$$

Using integration by parts twice more, we obtain

$$\begin{aligned}E &= -\frac{1}{2}[B_2(1)f'(1) - B_2(0)f'(0)] + \int_0^1 f''(x)\frac{1}{2}B_2(x)\mathrm{d}x \\ &= -\frac{B_2}{2}[f'(1)-f'(0)] + \int_0^1 f''(x)\frac{1}{3!}B_3'(x)\mathrm{d}x \\ &= -\frac{B_2}{2}[f'(1)-f'(0)] - \int_0^1 f'''(x)\frac{1}{3!}B_3(x)\mathrm{d}x \\ &= -\frac{B_2}{2}[f'(1)-f'(0)] + \int_0^1 f'''(x)\frac{1}{4!}B_4'(x)\mathrm{d}x,\end{aligned}$$

where we have used the fact that

$$B_3(0) = B_3(1) = B_3 = 0.$$

Integrating by parts twice more gives

$$E = -\sum_{j=1}^{2}\frac{B_{2j}}{(2j)!}\left[f^{(2j-1)}(1) - f^{(2j-1)}(0)\right] - \frac{1}{5!}\int_0^1 f^{(5)}(x)B_5(x)\mathrm{d}x,$$

using

$$B_5(0) = B_5(1) = B_5 = 0.$$

Continuing in this fashion, we obtain the result with the remainder

$$R_{2k}[f] = \frac{-1}{(2k+1)!}\int_0^1 f^{(2k+1)}(x)B_{2k+1}(x)\mathrm{d}x.$$

Now if $f \in C^{2k+2}([0,1])$, using integration by parts once more, we have

$$\begin{aligned}
R_{2k}[f] &= \frac{-1}{(2k+2)!}\int_0^1 f^{(2k+1)}(x)\frac{\mathrm{d}}{\mathrm{d}x}\left(B_{2k+2}(x) - B_{2k+2}\right)\mathrm{d}x \\
&= \frac{1}{(2k+2)!}\int_0^1 f^{(2k+2)}(x)\left(B_{2k+2}(x) - B_{2k+2}\right)\mathrm{d}x \\
&\quad - \frac{1}{(2k+2)!}\left[f^{(2k+1)}(x)\left(B_{2k+2}(x) - B_{2k+2}\right)\right]_{x=0}^{x=1} \\
&= \frac{1}{(2k+2)!}\int_0^1 f^{(2k+2)}(x)\left(B_{2k+2}(x) - B_{2k+2}\right)\mathrm{d}x.
\end{aligned}$$

Since $B_{2k+2}(x) - B_{2k+2}$ does not change sign in $[0,1]$, using the integral mean value theorem, we have, for some $\eta \in [0,1]$,

$$\begin{aligned}
R_{2k}[f] &= \frac{1}{(2k+2)!}f^{(2k+2)}(\eta)\int_0^1 \left(B_{2k+2}(x) - B_{2k+2}\right)\mathrm{d}x \\
&= -\frac{1}{(2k+2)!}f^{(2k+2)}(\eta)B_{2k+2},
\end{aligned}$$

since $\int_0^1 B_{2k+2}(x)\mathrm{d}x = 0$. The proof is complete. □

The previous result was a special case of the so-called *Euler–Maclaurin Formula*.[7]

Theorem 14.40 (Euler–Maclaurin). *Suppose that $m, k \in \mathbb{N}$ are positive integers, $[c,d] \subset \mathbb{R}$ is a compact interval of positive length, and $f \in C^{2k+1}([c,d])$. Define $h = \frac{d-c}{m}$ and*

$$\overline{B}_n(t) = B_n(t - \lfloor t \rfloor).$$

Then

$$\begin{aligned}
E_{Q_{1,m}}[f] = &-\sum_{j=1}^{k}\frac{B_{2j}h^{2j}}{(2j)!}\left[f^{(2j-1)}(d) - f^{(2j-1)}(c)\right] \\
&- \frac{h^{2k+1}}{(2k+1)!}\int_c^d f^{(2k+1)}(x)\overline{B}_{2k+1}\left(\frac{x-c}{h}\right)\mathrm{d}x.
\end{aligned}$$

[7] Named in honor of the Swiss mathematician, physicist, astronomer, geographer, logician, and engineer Leonhard Euler (1707–1783) and the British mathematician Colin Maclaurin (1698–1746).

If $f \in C^{2k+2}([c,d])$, then

$$
\begin{aligned}
E_{Q_{1,m}}[f] &= -\sum_{j=1}^{k} \frac{B_{2j}h^{2j}}{(2j)!}\left[f^{(2j-1)}(d) - f^{(2j-1)}(c)\right] \\
&\quad + \frac{h^{2k+2}}{(2k+2)!}\int_c^d f^{(2k+2)}(x)\left(\overline{B}_{2k+2}\left(\frac{x-c}{h}\right) - B_{2k+2}\right)\mathrm{d}x \\
&= -\sum_{j=1}^{k} \frac{B_{2j}h^{2j}}{(2j)!}\left[f^{(2j-1)}(d) - f^{(2j-1)}(c)\right] \\
&\quad - \frac{h^{2k+2}B_{2k+2}}{(2k+2)!}(d-c)f^{(2k+2)}(\eta)
\end{aligned}
\tag{14.25}
$$

for some $\eta \in [c,d]$.

Proof. Define

$$\xi_i = c + ih, \quad i = 0, \ldots, m.$$

Now fix i and set

$$x = \xi_i + th, \quad 0 \le t \le 1,$$

and

$$g(t) = f(x) = f(\xi_i + th), \quad 0 \le t \le 1.$$

Then

$$\frac{\mathrm{d}^r g}{\mathrm{d}t^r}(t) = h^r \frac{\mathrm{d}^r f}{\mathrm{d}x^r}(x).$$

By Theorem 14.39,

$$
\begin{aligned}
\int_0^1 g(t)\mathrm{d}t - \frac{1}{2}[g(0) + g(1)] &= -\sum_{j=1}^{k} \frac{B_{2j}}{(2j)!}\left[g^{(2j-1)}(1) - g^{(2j-1)}(0)\right] \\
&\quad - \frac{1}{(2k+1)!}\int_0^1 g^{(2k+1)}(t)B_{2k+1}(t)\mathrm{d}t.
\end{aligned}
$$

Using our change of variable,

$$
\begin{aligned}
h^{-1}\int_{\xi_{i-1}}^{\xi_i} f(x)\mathrm{d}x - \frac{1}{2}[f(\xi_{i-1}) + f(\xi_i)] \\
= -\sum_{j=1}^{k} \frac{B_{2j}h^{2j-1}}{(2j)!}\left[f^{(2j-1)}(\xi_i) - f^{(2j-1)}(\xi_{i-1})\right] \\
- \frac{h^{2k}}{(2k+1)!}\int_{\xi_{i-1}}^{\xi_i} f^{(2k+1)}(x)B_{2k+1}\left(\frac{x-\xi_i}{h}\right)\mathrm{d}x.
\end{aligned}
$$

Thus, for each $i = 1, \dots, m$,

$$\int_{\xi_{i-1}}^{\xi_i} f(x)\mathrm{d}x - \frac{h}{2}\left[f(\xi_{i-1}) + f(\xi_i)\right]$$
$$= -\sum_{j=1}^{k} \frac{B_{2j}h^{2j}}{(2j)!}\left[f^{(2j-1)}(\xi_i) - f^{(2j-1)}(\xi_{i-1})\right]$$
$$- \frac{h^{2k+1}}{(2k+1)!}\int_{\xi_{i-1}}^{\xi_i} f^{(2k+1)}(x)B_{2k+1}\left(\frac{x-\xi_i}{h}\right)\mathrm{d}x.$$

Summing, from $i = 1$ to $i = m$, we have

$$\int_c^d f(x)\mathrm{d}x - \frac{h}{2}\left[f(c) + f(d)\right] - h\sum_{i=1}^{m-1} f(\xi_i)$$
$$= -\sum_{j=1}^{k} \frac{B_{2j}h^{2j}}{(2j)!}\left[f^{(2j-1)}(d) - f^{(2j-1)}(c)\right]$$
$$- \frac{h^{2k+1}}{(2k+1)!}\int_c^d f^{(2k+1)}(x)\overline{B}_{2k+1}\left(\frac{x-c}{h}\right)\mathrm{d}x.$$

The remaining details are left to the reader as an exercise; see Problem 14.15. □

Using simple algebraic manipulations, we can formulate Euler–Maclaurin–type expansions for other composite quadrature rules.

Lemma 14.41 (Simpson's rule)**.** *Suppose that $[c, d] \subset \mathbb{R}$ is a compact interval of positive length and $m \in \mathbb{N}$. Assume that $f \in C([c, d])$. Then*

$$-\frac{1}{3}E_{Q_{1,m}}[f] + \frac{4}{3}E_{Q_{1,2m}}[f] = E_{Q_{2,m}}[f].$$

Proof. Set $L = \frac{d-c}{m}$. Then

$$E_{Q_{1,m}}[f] = \int_c^d f(x)\mathrm{d}x - \frac{L}{m}\left(\frac{1}{2}f_0 + f_2 + f_4 + \cdots + \frac{1}{2}f_{2m}\right),$$
$$E_{Q_{1,2m}}[f] = \int_c^d f(x)\mathrm{d}x - \frac{L}{2m}\left(\frac{1}{2}f_0 + f_1 + f_2 + \cdots + \frac{1}{2}f_{2m}\right),$$

where

$$f_i = f(\xi_i), \quad \xi_i = c + \frac{L}{2m}i, \quad i = 1, \dots, 2m.$$

Therefore,

$$-\frac{1}{3}E_{Q_{1,m}}[f] = -\frac{1}{3}\int_c^d f(x)\mathrm{d}x + 2\frac{L}{3(2m)}\left(\frac{1}{2}f_0 + f_2 + f_4 + \cdots + \frac{1}{2}f_{2m}\right),$$
$$\frac{4}{3}E_{Q_{1,2m}}[f] = \frac{4}{3}\int_c^d f(x)\mathrm{d}x - 4\frac{L}{3(2m)}\left(\frac{1}{2}f_0 + f_1 + f_2 + \cdots + \frac{1}{2}f_{2m}\right).$$

Adding, we have

$$-\frac{1}{3}E_{Q_{1,m}}[f] + \frac{4}{3}E_{Q_{1,2m}}[f] = \int_c^d f(x)\mathrm{d}x - Q_{2,m}^{(c,d)}[f] = E_{Q_{2,m}}[f]. \qquad \square$$

Using the last results, we get the following Euler–Maclaurin–type formula for the composite Simpson's rule.

Theorem 14.42 (Euler–Maclaurin). *Suppose that $m, k \in \mathbb{N}$, $[c, d] \subset \mathbb{R}$ is a compact interval of positive length, and $f \in C^{2k+2}([c, d])$. Define, as usual,*

$$\hbar = \frac{d-c}{2m}.$$

Then

$$E_{Q_{2,m}}[f] = -\sum_{j=2}^{k} \frac{4-2^{2j}}{3}\frac{B_{2j}\hbar^{2j}}{(2j)!}\left[f^{(2j-1)}(d) - f^{(2j-1)}(c)\right] - \frac{4-2^{2k+2}}{3}\frac{\hbar^{2k+2}B_{2k+2}}{(2k+2)!}(d-c)f^{(2k+2)}(\eta)$$

for some $\eta \in [c, d]$.

Proof. See Problem 14.18. $\square$

Example 14.7 Using the Euler–Maclaurin Formula (14.25) in Theorem 14.40, we have

$$E_{Q_{1,m}}[f] = -\frac{h^2}{12}[f'(d) - f'(c)] + \mathcal{O}(h^4), \quad h = \frac{d-c}{m}.$$

This formula prompts the definition of the *corrected composite trapezoidal rule*

$$Q_{1,m}^{\text{corr}}[f] = Q_{1,m}^{(c,d)}[f] - \frac{h^2}{12}[f'(d) - f'(c)]. \tag{14.26}$$

Clearly,

$$\int_c^d f(x)\mathrm{d}x - Q_{1,m}^{\text{corr}}[f] = \mathcal{O}(h^4).$$

Similarly, from Theorem 14.42, we find

$$E_{Q_{2,m}}[f] = -\frac{\hbar^4}{180}\left[f^{(3)}(d) - f^{(3)}(c)\right] + \mathcal{O}(\hbar^6), \quad \hbar = \frac{d-c}{2m},$$

which motivates the definition of the *corrected composite Simpson's rule*

$$Q_{2,m}^{\text{corr}}[f] = Q_{2,m}^{(c,d)}[f] - \frac{\hbar^4}{180}[f'(d) - f'(c)]. \tag{14.27}$$

Clearly,

$$\int_c^d f(x)\mathrm{d}x - Q_{2,m}^{\text{corr}}[f] = \mathcal{O}(\hbar^6).$$

m	$Q^{\text{corr}}_{1,n}$	$E_{Q^{\text{corr}}_{1,m}}$	Rate
8	−12.071 929 244 529 246	1.582 928 139 612 250e−03	–
16	−12.070 445 803 590 246	9.948 720 061 281 335e−05	3.991 941
32	−12.070 352 542 937 552	6.226 547 919 041 536e−06	3.998 006
64	−12.070 346 705 682 978	3.892 933 442 273 261e−07	3.999 502
128	−12.070 346 340 722 560	2.433 292 678 460 930e−08	3.999 875
256	−12.070 346 317 910 474	1.520 840 342 550 400e−09	3.999 969

Table 14.5 Approximation of the integral (14.19) by the corrected composite trapezoidal rule (14.26). The rate of convergence is computed according to (14.20) and is, as predicted, 4.

m	$Q^{\text{corr}}_{2,m}$	$E_{Q^{\text{corr}}_{2,m}}$	Rate
8	−12.070 350 005 413 999	3.689 024 365 982 618e−06	–
16	−12.070 346 373 686 869	5.729 723 540 071 063e−08	6.008 629
32	−12.070 346 317 283 548	8.939 142 759 345 486e−10	6.002 185
64	−12.070 346 316 403 596	1.396 216 475 768 597e−11	6.000 541
128	−12.070 346 316 389 848	2.149 391 775 674 303e−13	6.021 450
256	−12.070 346 316 389 635	1.776 356 839 400 250e−15	6.918 863

Table 14.6 Approximation of the integral (14.19) by the corrected composite Simpson's rule (14.27). The rate of convergence is computed according to (14.20) and is, as predicted, 6.

We report the approximations and errors of computing the integral (14.19) by the corrected composite trapezoidal and Simpson's rules in Tables 14.5 and 14.6. Compare the results with those in Tables 14.2 and 14.3.

Example 14.8 Another application of the Euler–Maclaurin Formula (14.25) in Theorem 14.40 is an alternate proof of Theorem 13.28, i.e., showing that, in the numerical quadrature of periodic functions, the composite trapezoidal rule is spectrally accurate; see Problem 14.17.

14.9 Gaussian Quadrature Rules

Gaussian quadrature rules are based on certain polynomial orthogonality relations. We will use the following simple property as we develop these methods.

Proposition 14.43 (orthogonality). *Let $[a, b] \subset \mathbb{R}$ be a compact interval and suppose that w is a weight function on $[a, b]$. Suppose that $\{p_m\}_{m=0}^{\infty}$, where $p_m \in \mathbb{P}_m$ with $\deg p_m = m$, is a system of orthogonal polynomials on $[a, b]$ with respect to the weight function w. Then, for any $k \in \mathbb{N}$,*

$$\int_a^b p_k(x)q(x)w(x)\mathrm{d}x = 0, \quad \forall q \in \mathbb{P}_{k-1}.$$

Proof. See Problem 14.19. □

The existence of orthogonal polynomials allows us to construct so-called Gaussian quadrature rules.

Definition 14.44 (Gaussian rule). Let $n \in \mathbb{N}_0$ and $p_{n+1} \in \mathbb{P}_{n+1}$ be an orthogonal polynomial of order $n+1$ with respect to the weight function w on the compact interval $[a, b] \subset \mathbb{R}$. Then the simple quadrature rule (14.4) is called a **Gaussian quadrature rule**[8] **of degree** n if and only if:

1. The effective nodes $X = \{x_j\}_{j=0}^n$ are the zeros of p_{n+1} in (a, b). This choice is possible because of Theorem 11.10.
2. The effective weights $\{\beta_j\}_{j=0}^n$ are chosen to satisfy the exactness condition (14.7), i.e.,

$$\beta_j = \int_a^b L_j(x)w(x)\mathrm{d}x, \quad L_j(x) = \prod_{\substack{k=0 \\ k\neq j}}^n \frac{x - x_k}{x_j - x_k}, \quad j = 0, \ldots, n.$$

Gaussian quadrature rules possess a higher order of consistency, as the following result shows.

Theorem 14.45 (order of Gaussian rules). *A Gaussian quadrature rule of degree $n \in \mathbb{N}$ ($n+1$ nodes) is consistent of order at least $2n+1$.*

Proof. Let $p_{n+1} \in \mathbb{P}_{n+1}$ be an orthogonal polynomial of degree $n+1$ with respect to the weight function w on the compact interval $[a, b]$. Suppose that $\hat{p} \in \mathbb{P}_{2n+1}$ is arbitrary. Since $2n+1 \geq n+1$, there are polynomials $q, r \in \mathbb{P}_n$ such that

$$\hat{p} = p_{n+1}q + r.$$

If that is the case, then we have

$$\begin{aligned}\int_a^b \hat{p}(x)w(x)\mathrm{d}x &= \int_a^b p_{n+1}(x)q(x)w(x)\mathrm{d}x + \int_a^b r(x)w(x)\mathrm{d}x \\ &= \int_a^b r(x)w(x)\mathrm{d}x,\end{aligned}$$

where we used the orthogonality of p_{n+1}. In addition, by the choice of the nodes,

$$\sum_{j=0}^n \beta_j\hat{p}(x_j) = \sum_{j=0}^n \beta_j p_{n+1}(x_j)q(x_j) + \sum_{j=0}^n \beta_j r(x_j) = \sum_{j=0}^n \beta_j r(x_j).$$

[8] Named in honor of the German mathematician and physicist Johann Carl Friedrich Gauss (1777–1855).

Finally, since $r \in \mathbb{P}_n$, the choice of the weights guarantees that

$$\sum_{j=0}^{n} \beta_j r(x_j) = \int_a^b r(x)w(x)\mathrm{d}x.$$

Putting everything together, we have

$$\int_a^b \hat{p}(x)w(x)\mathrm{d}x = \sum_{j=0}^{n} \beta_j \hat{p}(x_j)$$

for any $\hat{p} \in \mathbb{P}_{2n+1}$, which shows that this formula is consistent of order $2n+1$. □

This is remarkable. Remember that we are using only $n+1$ nodes to construct the method. One might only expect that the method will be consistent of order n. But our nodes are special; this is the key point. Consequently, we obtain a method that is actually of order $2n+1$. It turns out that one cannot do better than this; see Theorem 14.46.

Theorem 14.46 (impossibility). *Let w be a weight function on the compact interval $[a, b] \subset \mathbb{R}$. There is no simple quadrature rule of Lagrange type constructed with $n+1$ nodes in $[a, b]$ that is consistent of order $2n+2$ or higher. Consequently, a Gaussian quadrature rule of degree n is consistent to exactly order $2n+1$.*

Proof. Suppose that there is a simple quadrature rule (14.4) constructed with $n+1$ nodes, $\{x_j\}_{j=0}^n \subset [a, b]$, that is consistent of order at least $2n+2$. This rule would be exact for any polynomial of degree not exceeding $2n+2$. Define

$$\hat{p}(x) = \prod_{j=0}^{n} (x - x_j)^2 \in \mathbb{P}_{2n+2}.$$

Then since $\hat{p}$ is strictly positive, except at the nodes where it is zero, we have

$$\int_a^b \hat{p}(x)w(x)\mathrm{d}x > 0.$$

But since $\hat{p}(x_j) = 0$, for all $j = 0, \ldots, n$,

$$\sum_{j=1}^{n} \beta_j \hat{p}(x_j) = 0.$$

Consequently,

$$0 = \sum_{j=1}^{n} \beta_j \hat{p}(x_j) \neq \int_a^b \hat{p}(x)w(x)\mathrm{d}x > 0,$$

which is a contradiction to our assumption that the rule is consistent of order at least $2n+2$. □

Modifying the previous arguments slightly, we can also prove the following.

Corollary 14.47 (improved order). *Suppose that w is a weight function on the compact interval $[a, b] \subset \mathbb{R}$. Let $r \in \mathbb{P}_{n+1}$ be such that*

1. *It has $n+1$ real, distinct roots in $[a,b]$.*
2. *There is an integer $m \in \{1,\dots,n,n+1\}$ for which*

$$(r,q)_{L^2_w(a,b)} = 0, \quad \forall q \in \mathbb{P}_{m-1},$$

but there is $\tilde{q} \in \mathbb{P}_m$ for which

$$(r,\tilde{q})_{L^2_w(a,b)} \neq 0.$$

Let the nodes $\{x_j\}_{j=0}^n \subset [a,b]$ be chosen as the roots of r and the weights $\{\beta_j\}_{j=0}^n$ be chosen to satisfy the exactness condition (14.7). Then the resulting simple quadrature rule (14.4) is consistent to exactly order $n+m$.

Proof. See Problem 14.20. □

The proof of Theorem 14.48 makes an interesting connection between Gaussian quadrature and Hermite interpolation.

Theorem 14.48 (error representation). *Suppose that w is a weight function on the compact interval $[a,b] \subset \mathbb{R}$ and (14.4) is a Gaussian quadrature of degree $n \in \mathbb{N}_0$, i.e., it has $n+1$ nodes. Assume that $f \in C^{2n+2}([a,b])$. Then there exists a point $\zeta \in (a,b)$ such that*

$$E_Q[f] = \frac{f^{(2n+2)}(\zeta)}{(2n+2)!}\int_a^b w(x)\left(\omega_{n+1}(x)\right)^2 \mathrm{d}x,$$

where, as usual,

$$\omega_{n+1}(x) = \prod_{i=0}^{n}(x-x_i)$$

and $\{x_i\}_{i=0}^n \subset [a,b]$ are the nodes for the quadrature rule.

Proof. Suppose that $p \in \mathbb{P}_{2n+1}$ is the Hermite interpolating polynomial with respect to the nodes $\{x_i\}_{i=0}^n \subset [a,b]$. This is the unique polynomial in $\mathbb{P}_{2n+1}$ with the property that

$$f(x_j) = p(x_j), \quad f'(x_j) = p'(x_j), \quad j = 0,\dots,n.$$

Recall that, to construct p, we define, for $0 \le \ell \le n$, the following polynomials from $\mathbb{P}_{2n+1}$:

$$F_{0,\ell}(x) = (L_\ell(x))^2\left(1 - 2L'_\ell(x_\ell)(x-x_\ell)\right), \quad F_{1,\ell} = (L_\ell(x))^2(x-x_\ell),$$

where L_ℓ is the Lagrange basis element (9.3). Then the Hermite interpolating polynomial is

$$p(x) = \sum_{\ell=0}^{n}\left[F_{0,\ell}(x)f(x_\ell) + F_{1,\ell}(x)f'(x_\ell)\right].$$

By consistency,

$$\begin{aligned}\int_a^b w(x)p(x)\mathrm{d}x &= \sum_{j=0}^n \beta_j p(x_j)\\ &= \sum_{j=0}^n \beta_j \left(\sum_{\ell=0}^n [F_{0,\ell}(x_j)f(x_\ell) + F_{1,\ell}(x_j)f'(x_\ell)]\right)\\ &= \sum_{\ell=0}^n A_\ell f(x_\ell) + \sum_{\ell=0}^n B_\ell f'(x_\ell),\end{aligned}$$

where

$$A_\ell = \sum_{j=0}^n \beta_j F_{0,\ell}(x_j), \quad B_\ell = \sum_{j=0}^n \beta_j F_{1,\ell}(x_j), \quad j = 0, \dots, n.$$

Now we prove the following remarkable facts:

$$A_\ell = \beta_\ell, \quad B_\ell = 0, \quad \ell = 0, \dots, n.$$

To see that $B_\ell = 0$, note that, again by consistency, and using the definition of L_ℓ,

$$\begin{aligned}B_\ell = \sum_{j=0}^n \beta_j F_{1,\ell}(x_j) &= \int_a^b w(x)F_{1,\ell}(x)\mathrm{d}x\\ &= \int_a^b w(x)L_\ell(x)(x - x_\ell)L_\ell(x)\mathrm{d}x\\ &= C_n \int_a^b w(x)\omega_{n+1}(x)L_\ell(x)\mathrm{d}x,\end{aligned}$$

where C_n is a constant and $\omega_{n+1} \in \mathbb{P}_{n+1}$ is an orthogonal polynomial of degree $n+1$, since the x_j were chosen as the roots of such an orthogonal polynomial. Since $L_\ell \in \mathbb{P}_n$,

$$\int_a^b w(x)\omega_{n+1}(x)L_\ell(x)\mathrm{d}x = 0, \quad \ell = 0, \dots, n.$$

Thus, $B_\ell = 0$ for $\ell = 0, \dots, n$. It is not hard to see that $A_\ell = \beta_\ell$ for $\ell = 0, \dots, n$, and we leave that to the reader as a short exercise; see Problem 14.21.

Regarding Hermite interpolation error, Theorem 9.21 guarantees that, for every $x \in [a, b]\backslash X$, there is a point $\xi = \xi(x) \in (a, b)$ such that

$$f(x) - p(x) = \frac{f^{(2n+2)}(\xi(x))}{(2n+2)!}(\omega_{n+1}(x))^2. \tag{14.28}$$

Alternately according to Theorem 9.58, for any $x \in [a, b]$,

$$f(x) - p(x) = (\omega_{n+1}(x))^2 f[\![x_0, x_0, x_1, x_1, \dots, x_n, x_n, x]\!]. \tag{14.29}$$

Of course, comparing (14.28) and (14.29), we see that

$$f[\![x_0, x_0, x_1, x_1, \dots, x_n, x_n, x]\!] = \frac{f^{(2n+2)}(\xi(x))}{(2n+2)!}.$$

The extended divided difference function $f[\![x_0, x_0, x_1, x_1, \ldots, x_n, x_n, x]\!]$ is a continuous function of its arguments, specifically of x. Using what we have just learned,

$$
\begin{aligned}
E_Q[f] &= \int_a^b w(x)f(x)\mathrm{d}x - \sum_{j=0}^{n} \beta_j f(x_j) \\
&= \int_a^b w(x)f(x)\mathrm{d}x - \sum_{j=0}^{n} \beta_j p(x_j) \\
&= \int_a^b w(x)\left(f(x) - p(x)\right)\mathrm{d}x \\
&= \int_a^b w(x) f[\![x_0, x_0, x_1, x_1, \ldots, x_n, x_n, x]\!](\omega_{n+1}(x))^2 \,\mathrm{d}x.
\end{aligned}
$$

Using the Integral Mean Value Theorem B.41, there is a point $\eta \in (a, b)$ such that

$$
\begin{aligned}
E_Q[f] &= f[\![x_0, x_0, x_1, x_1, \ldots, x_n, x_n, \eta]\!] \int_a^b w(x)\left(\omega_{n+1}(x)\right)^2 \mathrm{d}x \\
&= \frac{f^{(2n+2)}(\xi(\eta))}{(2n+2)!} \int_a^b w(x)\left(\omega_{n+1}(x)\right)^2 \mathrm{d}x \\
&= \frac{f^{(2n+2)}(\zeta)}{(2n+2)!} \int_a^b w(x)\left(\omega_{n+1}(x)\right)^2 \mathrm{d}x.
\end{aligned}
$$

□

To finish this section and chapter, let us state an interesting property of Gaussian quadratures.

Proposition 14.49 (positivity). *Suppose that w is a weight function on the compact interval $[a, b] \subset \mathbb{R}$ and (14.4) is the Gaussian quadrature rule of degree $n \in \mathbb{N}_0$ with respect to w. Then each of the weights is positive, i.e.,*

$$
\beta_j > 0, \quad j = 0, \ldots, n.
$$

Proof. Recall that a Gaussian quadrature of degree n has $n + 1$ nodes and is consistent of exactly order $2n + 1$. Therefore, the rule will be exact for the polynomials $q_j \in \mathbb{P}_{2n}$, defined by

$$
q_j(x) = \frac{\omega_{n+1}^2(x)}{(x - x_j)^2}, \quad j = 0, \ldots, n,
$$

where, as usual,

$$
\omega_{n+1}(x) = (x - x_0)(x - x_1)\cdots(x - x_n).
$$

Thus,

$$
\int_a^b q_j(x)w(x)\mathrm{d}x = \sum_{k=0}^{n} \beta_k q_j(x_k).
$$

But notice that

$$q_j(x_k) = \begin{cases} 0, & k \neq j, \\ \displaystyle\prod_{\substack{i=0 \\ i\neq j}}^{n} (x_j - x_i)^2 > 0, & k = j. \end{cases}$$

Therefore, for $j = 0, \dots, n$,

$$\begin{aligned} \beta_j &= \sum_{k=0}^{n} \beta_k \delta_{j,k} \\ &= \sum_{k=0}^{n} \beta_k \frac{q_j(x_k)}{q_j(x_j)} \\ &= \frac{1}{q_j(x_j)} \int_a^b q_j(x) w(x) \mathrm{d}x \\ &= \frac{1}{q_j(x_j)} \int_a^b \prod_{\substack{i=0 \\ i\neq j}}^{n} (x - x_i)^2 w(x) \mathrm{d}x > 0. \end{aligned}$$

□

Problems

14.1 Show that if a quadrature rule on $[a, b]$ for the weight $w \equiv 1$ is consistent of order at least one, then

$$\sum_{j=1}^{n} \beta_j = b - a.$$

What does one obtain for a general weight?

14.2 State and prove an analogue of Proposition 14.4 for quadrature rules of Hermite type.

14.3 Prove Theorem 14.7.

14.4 Prove Lemma 14.9.

14.5 Prove Corollary 14.12.

14.6 Prove Lemma 14.15.

14.7 Confirm the intrinsic weights for the Newton–Cotes rules listed in Table 14.1.

14.8 Prove Lemma 14.21.

14.9 Prove Lemma 14.22.

14.10 The closed Newton–Cotes quadrature rule of order n is designed with $n+1$ equally spaced nodes such that, when applied to any polynomial of degree at most n, there is no quadrature error. Show directly that, when n is even, there is no quadrature error for polynomials in $\mathbb{P}_{n+1}$.

14.11 Complete the proof of Theorem 14.31.

14.12 Complete the proof of Proposition 14.34.

14.13 Prove Proposition 14.37.

14.14 Prove Proposition 14.38.

14.15 Complete the proof of Theorem 14.40.

14.16 Derive an Euler–Maclaurin–type error formula for the composite midpoint rule (14.18).

14.17 Use Theorem 14.40 to give an alternate proof of Theorem 13.28.

14.18 Prove Theorem 14.42.

14.19 Prove Proposition 14.43.

14.20 Prove Corollary 14.47.

14.21 Complete the details of the proof of Theorem 14.48 by proving the following result. Let $[a, b] \subset \mathbb{R}$ be a compact interval, w be a weight on $[a, b]$, and $f \in C^1([a, b])$. Use the Hermite interpolating polynomial to design a quadrature method for approximating

$$\int_a^b f(x)w(x)dx$$

using $n+1$ nodes that is consistent of order $2n+1$. Show that one obtains precisely the Gaussian quadrature rule of order n.

14.22 Define the sequence of functions $\{q_m\}_{m=1}^{\infty}$ as follows.

a) $q_1(x) = -x$.
b) $q'_{m+1} = q_m$, for each $m \in \mathbb{N}$.
c) q_m is an odd function if m is odd.
d) q_m is an even function if m is even.
e) if $m > 1$ is odd, $q_m(-1) = 0$ and $q_m(1) = 0$.

Prove that the sequence $\{q_m\}_{m=1}^{\infty}$ is well defined and, in particular, that $q_m \in \mathbb{P}_m$, for each $m \in \mathbb{N}_0$. Show that the first five members are

$$\begin{aligned} q_1(x) &= -x; \\ q_2(x) &= -\frac{1}{2}x^2 + \frac{1}{6}; \\ q_3(x) &= -\frac{1}{6}x^3 + \frac{1}{6}x; \\ q_4(x) &= -\frac{1}{24}x^4 + \frac{1}{12}x^2 - \frac{7}{360}; \\ q_5(x) &= -\frac{1}{120}x^5 + \frac{1}{36}x^3 - \frac{7}{360}x. \end{aligned}$$

14.23 Suppose that, for $k \in \mathbb{N}$, $f \in C^{2k}([-1, 1])$. Denote

$$E = \int_{-1}^{1} f(x)dx - [f(-1) + f(1)].$$

Show that

$$E = \int_{-1}^{1} q_1(x)f(x)dx, \tag{14.30}$$

$$E = \sum_{m=1}^{k} q_{2m}(1)\left[f^{(2m-1)}(1) - f^{(2m-1)}(-1)\right] - \int_{-1}^{1} q_{2k}(x)f^{(2k)}(x)dx, \tag{14.31}$$

where $\{q_m\}_{m=1}^{\infty}$ is as in Problem 14.22.

14.24 Use the previous two problems to give an alternate proof of the Euler–Maclaurin Formula of Theorem 14.40.

14.25 Use Hermite interpolation to derive the corrected trapezoidal rule,

$$Q[f] = \frac{b-a}{2}(f(a)+f(b)) - \frac{(b-a)^2}{12}(f'(b)-f'(a)), \quad \forall f \in C^1([a,b]).$$

Derive the corrected composite trapezoidal rule using this result and show that the error for this rule (in approximating $\int_c^d f(x)dx$) is exactly as predicted by the Euler–Maclaurin Formula, provided that f is sufficiently smooth on $[c,d]$.

14.26 Derive a composite form of Boole's rule and obtain an error estimate analogous to those derived in Theorem 14.31; see Table 14.1.

14.27 Recall that

$$(f,g)_{L^2(-1,1;\mathbb{C})} = \int_{-1}^{1} f(x)\overline{g(x)}\,dx, \quad \forall f, g \in L^2(-1,1;\mathbb{C}).$$

Suppose that the elements of $X = \{x_i\}_{i=0}^n$ are chosen as the roots of the $(n+1)$st Legendre polynomial $P_{n+1} \in \mathbb{P}_{n+1}$; see Definition 11.17. Define

$$\beta_j = \int_{-1}^{1} L_j(x)dx,$$

where $L_j \in \mathbb{P}_n$ is the jth Lagrange nodal basis polynomial subordinate to X. Prove that

$$(p,q)_{L^2(-1,1;\mathbb{C})} = \sum_{j=0}^{n} \beta_j p(x_j)\overline{q(x_j)}, \quad \forall p, q \in \mathbb{P}_n(\mathbb{C}).$$

Part III

Nonlinear Equations and Optimization

15 Solution of Nonlinear Equations

In this chapter, we depart from *linear algebra* problems and concentrate on the study of *nonlinear* problems. We will focus on methods for solving a nonlinear system of equations. In other words, given $m, n \in \mathbb{N}$ and $\boldsymbol{f}\colon \mathbb{R}^n \to \mathbb{R}^m$, we wish to find a point $\boldsymbol{\xi} \in \mathbb{R}^n$ such that

$$\boldsymbol{f}(\boldsymbol{\xi}) = \mathbf{0}. \tag{15.1}$$

Such a point $\boldsymbol{\xi}$, if it exists, is called a *root of* $\boldsymbol{f}$. Of course, if $m = n$ and $\boldsymbol{f}$ so happens to be affine then this problem reduces to (3.1), and can be treated either by the direct methods of Chapter 3 or by the iterative ones of Chapters 6 and 7. If $m \neq n$, but the function $\boldsymbol{f}$ is still affine, then the least squares methods of Chapter 5 apply. The function $\boldsymbol{f}$ in this chapter, however, is not assumed to be affine. The importance of (15.1) cannot be overstated. Similar to (3.1), many problems can be reduced to this. We already saw an instance of this in Chapter 8: although this may not be the preferred choice, finding eigenvalues can be reduced to solving a problem like (15.1) with $m = n = 1$ and the function is a polynomial. Other examples will be seen in future chapters.

We must immediately remark that any method that attempts to find a solution to (15.1) must be iterative, unless a very special structure is assumed on the function $\boldsymbol{f}$. The general strategy that we will follow can be simply stated as:

- Show that the problem has at least one solution.
- Isolate a root, i.e., find an open region $D \subset \mathbb{R}^n$ for which there is $\boldsymbol{\xi} \in D$ that solves (15.1) and $\boldsymbol{f}(\boldsymbol{x}) \neq \mathbf{0}$ for all $\boldsymbol{x} \in D \backslash \{\boldsymbol{\xi}\}$.
- Iterate.

Unfortunately, there is no general strategy to treat the first two points, as these usually require analytical methods or additional knowledge about the problem at hand.

Before we begin, we must give a word of caution regarding iterations. Much as in Chapter 6, starting from some $\boldsymbol{x}_0 \in \mathbb{R}^n$, we will construct sequences $\{\boldsymbol{x}_k\}_{k=1}^{\infty} \subset \mathbb{R}^n$ which, hopefully, converge $\boldsymbol{x}_k \to \boldsymbol{\xi}$ as fast as possible. We will, as usual, stop the iteration when a prescribed tolerance $\varepsilon > 0$ is reached, i.e.,

$$\|\boldsymbol{x}_k - \boldsymbol{\xi}\| < \varepsilon.$$

How do we know when to stop the iterations? One might be tempted to say that, since $\boldsymbol{f}(\boldsymbol{\xi}) = \mathbf{0}$, we can stop them whenever

$$\|\boldsymbol{f}(\boldsymbol{x}_k)\| < C\varepsilon$$

for some suitable constant C. The following two examples, however, show that this is not always a viable strategy.

Example 15.1 The function $f(x) = e^x$ does not have a root in $\mathbb{R}$. However, for any $\varepsilon > 0$, there is x_ε such that

$$0 < f(x_\varepsilon) = e^{x_\varepsilon} < \varepsilon.$$

Example 15.2 The previous example may be misleading in the sense that the problem did not have a solution, and we were considering a function defined on an unbounded interval. This can be easily fixed. Let $\delta \in (0,1)$ and consider

$$f_\delta(x) = \begin{cases} \delta, & x \in [0, \frac{1}{2}), \\ 8(1-\delta)(x - \frac{1}{2}) + \delta, & x \in [\frac{1}{2}, \frac{5}{8}), \\ -\frac{16}{3}(x - \frac{5}{8}) + 1, & x \in [\frac{5}{8}, 1]. \end{cases}$$

This function is continuous and it has a unique root $\xi = \frac{13}{16} > \frac{1}{2}$. However, no matter what $\varepsilon > 0$ is, we can choose $\delta < \varepsilon$, so that any point $x \in [0, \frac{1}{2}]$ satisfies $0 < f_\delta(x) < \varepsilon$.

Let us now quickly and not very rigorously discuss the sensitivity of this problem to perturbations. To do so, we will assume that $m = n = 1$, and the function f is $k+1$ times continuously differentiable in a neighborhood of its unique root $\xi \in \mathbb{R}$. Moreover, we will assume that $f^{(p)}(\xi) = 0$ for $p < k$, but $f^{(k)}(\xi) \neq 0$. Then, given a perturbation δx, we let η be such that

$$f(\xi + \delta x) = \eta.$$

Taylor's Theorem then shows that

$$\begin{aligned} \eta &= f(\xi + \delta x) \\ &= f(\xi) + f'(\xi)\delta x + \frac{1}{2} f''(\xi)\delta x^2 + \cdots + \frac{1}{k!} f^{(k)}(\xi)\delta x^k + \mathcal{O}(|\delta x|^{k+1}) \\ &= \frac{1}{k!} f^{(k)}(\xi)\delta x^k + \mathcal{O}(|\delta x|^{k+1}). \end{aligned}$$

In other words, at least intuitively, the allowed relative size of the perturbation δx, to obtain an output of size η, is

$$\left| \frac{\delta x}{\eta} \right| \approx \left| \eta^{1-k} \frac{k!}{f^{(k)}(\xi)} \right|^{1/k}.$$

From this, we learn two things: the smaller the value of the first nonzero derivative, the larger δx can be; and, the higher the order of the first nonzero derivative, the larger δx can be. For this reason, of importance to us will be so-called *simple roots*.

Definition 15.1 (simple root). Let $f\colon \mathbb{R} \to \mathbb{R}$ have a root $\xi \in \mathbb{R}$, and assume that f is differentiable at ξ. We say that ξ is a **simple root** if $f'(\xi) \neq 0$. If this is not the case, we say that the root is **nonsimple**.

In this chapter, we will first present simple methods to tackle the case $m = n = 1$. As a rule, their convergence will be no better than linear (see Appendix B for definitions). We will then move on to Newton's method, for which we will show quadratic convergence. This method, and some of its variants, will be first presented in the one-dimensional case, and then for the multidimensional case, i.e., $m = n = d > 1$.

As always, we are barely scratching the surface of the subject. We refer the reader, for instance, to [28, 51, 68] for many more details.

15.1 Methods of Bisection and False Position

Here, we consider the case $m = n = 1$ and $f \in C([a, b])$ for some $-\infty < a < b < \infty$. The following result is a simple consequence of the Intermediate Value Theorem B.27.

Corollary 15.2 (existence). *Let $-\infty < a < b < \infty$ and $f \in C([a, b])$. If $f(a)f(b) < 0$, then f has a (not necessarily unique) root in $\xi \in (a, b)$.*

Proof. From the condition $f(a)f(b) < 0$ we infer that

$$\inf_{x\in[a,b]} f(x) \le \min\{f(a), f(b)\} < 0 \quad \text{and} \quad \sup_{x\in[a,b]} f(x) \ge \max\{f(a), f(b)\} > 0.$$

The Intermediate Value Theorem B.27 then implies the result. □

The idea of the *bisection method*, as its name suggests, is that we will successively subdivide the interval $[a, b]$ into two subintervals of equal length and check on which of the subintervals there must be a root.

Definition 15.3 (bisection method). Let $-\infty < a < b < \infty$ and $f \in C([a, b])$. Assume that $f(a)f(b) < 0$. The **bisection method** is an algorithm for generating an approximation sequence, $\{x_k\}_{k=0}^{\infty} \subset [a, b]$, via the following recursive procedure. Define $a_0 = a$, $b_0 = b$. For $k \ge 0$, set

$$x_k = \frac{1}{2}(a_k + b_k).$$

If $f(x_k) = 0$, the algorithm terminates. Otherwise,

$$(a_{k+1}, b_{k+1}) = \begin{cases} (x_k, b_k), & \text{if } f(x_k)f(b_k) < 0, \\ (a_k, x_k), & \text{if } f(x_k)f(b_k) > 0. \end{cases}$$

The convergence properties of this method are stated in the following result.

Theorem 15.4 (convergence). *Let $-\infty < a < b < \infty$ and $f \in C([a, b])$ be such that $f(a)f(b) < 0$. Then the sequence $\{x_k\}_{k=0}^{\infty}$ generated by the bisection method converges to a point $\xi \in [a, b]$ such that $f(\xi) = 0$. Moreover, this method converges linearly, with the following rate of convergence:*

$$|x_k - \xi| \le \frac{1}{2^{k+1}}(b - a).$$

Proof. The bisection method generates the sequences $\{a_k\}_{k=0}^{\infty}$, $\{b_k\}_{k=0}^{\infty}$, $\{x_k\}_{k=0}^{\infty}$, which satisfy

$$x_k \in (a_k, b_k), \quad [a_{k+1}, b_{k+1}] \subsetneq [a_k, b_k], \quad b_{k+1} - a_{k+1} = \frac{1}{2}(b_k - a_k).$$

Observe that $\{a_k\}_{k=0}^{\infty}$ is a bounded increasing sequence, and $\{b_k\}_{k=0}^{\infty}$ is a bounded decreasing sequence. By the Monotone Convergence Theorem (Theorem B.7), there exist limit points $\xi_a, \xi_b \in [a, b]$ such that

$$a_n \uparrow \xi_a \leq \xi_b \downarrow b_n.$$

But

$$b_n - a_n = \frac{1}{2^n}(b_0 - a_0) \downarrow 0,$$

which implies that $\xi_a = \xi_b$. Let us call the common point ξ. By the Squeeze Theorem B.6, since $x_k \in (a_k, b_k)$, we have also that $x_k \to \xi$. Moreover, since x_k is the midpoint of each interval, we observe that

$$|x_k - \xi| \leq \frac{1}{2}(b_k - a_k) = \cdots = \frac{1}{2^{k+1}}(b_0 - a_0) = \frac{1}{2^{k+1}}(b - a).$$

It remains then to show that $f(\xi) = 0$, and this follows from the continuity of f, since

$$0 \leq f(\xi)^2 = \lim_{k\to\infty} f(a_k) \lim_{k\to\infty} f(b_k) = \lim_{k\to\infty} f(a_k)f(b_k) \leq 0. \qquad \square$$

The bisection method requires very little from the function at hand, just continuity and that it takes values of different signs on the given interval. In this setting, it is always guaranteed to converge. The convergence, however, is only linear. Let us present a small modification, known as the *false position* method, which may improve the rate of convergence. The idea is simple. There is no reason why, in the bisection method, the approximation of the root must be the midpoint. Instead, suppose the update is chosen to be the zero of the line that connects $(a_k, f(a_k))$ and $(b_k, f(b_k))$.

Definition 15.5 (false position)**.** Let $-\infty < a < b < \infty$ and $f \in C([a, b])$. The sequence $\{x_k\}_{k=0}^{\infty} \subset [a, b]$ obtained by the following procedure defines the **false position method**. Define $a_0 = a$, $b_0 = b$. For $k \geq 0$, set

$$x_k = \frac{a_k f(b_k) - b_k f(a_k)}{f(b_k) - f(a_k)}.$$

If $f(x_k) = 0$, the algorithm terminates. Otherwise,

$$(a_{k+1}, b_{k+1}) = \begin{cases} (x_k, b_k), & \text{if } f(x_k)f(b_k) < 0, \\ (a_k, x_k), & \text{if } f(x_k)f(b_k) > 0. \end{cases}$$

The convergence of this method easily follows from that of the bisection method.

Theorem 15.6 (convergence)**.** *Let $-\infty < a < b < \infty$ and $f \in C([a, b])$ be such that $f(a)f(b) < 0$. Then the sequence $\{x_k\}_{k=0}^{\infty}$ generated by the false position method converges to a point $\xi \in [a, b]$ such that $f(\xi) = 0$.*

Proof. The details of the proof are left to the reader as an exercise; see Problem 15.2. Here, we merely sketch why $x_k \in [a_k, b_k]$. Consider,

$$x_k - a_k = -f(a_k)\frac{b_k - a_k}{f(b_k) - f(a_k)}.$$

Now if $f(a_k) > 0$, then the numerator of this expression is negative. If this is the case, by construction, $f(b_k) < 0$ and the denominator of the expression above is negative. Thus, $x_k - a_k > 0$. The remaining cases are treated similarly.

From this, it follows that we are constructing, again, a sequence $\{[a_k, b_k]\}_{k=0}^{\infty}$ of nested intervals whose lengths are strictly decreasing. □

We will not provide an analysis of the false position method. We will just mention that, in general, it is possible (Problem 15.3) for one of the endpoints of the interval to "*get stuck*," i.e., we can have, for all $k \geq 0$,

$$a_k = a_0 = a \qquad \text{or} \qquad b_k = b_0 = b.$$

Because of this, the rate of convergence of this method is no better than linear. In practice, however, this method seems to perform better than bisection. A partial explanation for this fact will be given in the analysis of the secant method provided in Section 15.4.4.

15.2 Fixed Points and Contraction Mappings

In this section, we will relate root finding to contraction mappings and fixed point iteration schemes. The reader should examine Appendix C and Theorem C.4 for the more general setting.

Definition 15.7 (fixed point iteration). Suppose that $-\infty < a < b < \infty$, $g \in C([a, b])$, and $g(x) \in [a, b]$ for all $x \in [a, b]$. In other words, $g([a, b]) \subseteq [a, b]$. Given $x_0 \in [a, b]$, the algorithm for constructing the recursive sequence $\{x_k\}_{k=0}^{\infty}$ via

$$x_{k+1} = g(x_k), \quad k \geq 0 \tag{15.2}$$

is called a **simple iteration scheme** or, sometimes, a **fixed point iteration scheme**.

The following fact follows easily from continuity.

Proposition 15.8 (fixed point). *Suppose that $-\infty < a < b < \infty$, $g \in C([a, b])$, and $g([a, b]) \subseteq [a, b]$. Assume that the sequence $\{x_k\}_{k=0}^{\infty}$ obtained by a simple iteration scheme converges to a limit $\xi \in [a, b]$. Then ξ is a* fixed point *of g, i.e., $g(\xi) = \xi$.*

Proof. Indeed, by continuity,

$$g(\xi) = g\left(\lim_{k\to\infty} x_k\right) = \lim_{k\to\infty} g(x_k) = \lim_{k\to\infty} x_{k+1} = \xi.$$ □

The following result provides sufficient conditions for the existence of a fixed point.

Theorem 15.9 (existence). *Suppose that $-\infty < a < b < \infty$, $g \in C([a, b])$, and $g([a, b]) \subseteq [a, b]$. Then there exists at least one fixed point $\xi \in [a, b]$ of g.*

Proof. Define

$$f(x) = x - g(x), \quad \forall x \in [a, b].$$

Then, since $g([a, b]) \subseteq [a, b]$,

$$f(b) = b - g(b) \geq 0 \quad \text{and} \quad f(a) = a - g(a) \leq 0.$$

By the Intermediate Value Theorem B.27, since $f(a) \leq 0 \leq f(b)$, there is a point $\xi \in [a, b]$ such that $f(\xi) = 0$. For this point, $\xi = g(\xi)$. □

In practice, it is very difficult to verify the condition that $g([a, b]) \subseteq [a, b]$. The following definition provides a sufficiently large class of functions for which this condition is almost automatically satisfied.

Definition 15.10 (contraction)**.** Suppose that $-\infty < a < b < \infty$ and $g \in C([a, b])$. We say that g is **Lipschitz continuous**[1] on $[a, b]$ if and only if there exists a constant $L > 0$ such that

$$|g(x) - g(y)| \leq L|x - y|, \quad \forall x, y \in [a, b],$$

and the associated L is called the **Lipschitz constant**. The function g is called a **contraction** on $[a, b]$ if and only if it is Lipschitz on $[a, b]$ and its associated Lipschitz constant, L, satisfies $L \in (0, 1)$.

Proposition 15.11 (translation)**.** *Let $-\infty < a < b < \infty$ and $g \in C([a, b])$ be a contraction on $[a, b]$. There is a constant $m \in \mathbb{R}$ such that the function $\tilde{g}\colon [a, b] \to \mathbb{R}$, where*

$$\tilde{g}(x) = g(x) + m, \quad \forall x \in [a, b]$$

is a contraction on $[a, b]$; moreover, $\tilde{g}([a, b]) \subseteq [a, b]$.

Proof. See Problem 15.5. □

We see also that, although Theorem 15.9 provides existence of fixed points, it says nothing about uniqueness. It turns out that, for contractions, fixed points must be unique.

Theorem 15.12 (uniqueness)**.** *Suppose that $-\infty < a < b < \infty$, $g \in C([a, b])$, and $g([a, b]) \subseteq [a, b]$. If g is a contraction on $[a, b]$, then g has a unique fixed point $\xi \in [a, b]$. Furthermore, the sequence $\{x_k\}_{k=0}^{\infty}$ generated by* (15.2) *converges to ξ for any starting value $x_0 \in [a, b]$.*

Proof. Theorem 15.9 guarantees the existence of at least one fixed point $\xi \in [a, b]$. Suppose that $\eta \in [a, b]$ is another fixed point. Since g is a contraction,

$$|\xi - \eta| = |g(\xi) - g(\eta)| \leq L|\xi - \eta|.$$

Therefore,

$$0 \leq (1 - L)|\xi - \eta| \leq 0,$$

which proves that $\xi = \eta$. Hence, the fixed point $\xi \in [a, b]$ is unique.

[1] Named in honor of the German mathematician Rudolf Otto Sigismund Lipschitz (1832–1903).

Now suppose that $\{x_k\}_{k=0}^{\infty}$ is generated by (15.2) for $x_0 \in [a, b]$. Then

$$|\xi - x_k| = |g(\xi) - g(x_{k-1})| \leq L|\xi - x_{k-1}|.$$

By induction, for any $k \in \mathbb{N}$,

$$|\xi - x_k| \leq L^k|\xi - x_0|.$$

By the Squeeze Theorem B.6, since $L \in (0, 1)$, we must have $x_k \to \xi$. □

Theorem 15.13 (local [at least linear] convergence). *Suppose that $-\infty < a < b < \infty$, $g \in C([a, b])$, and $g([a, b]) \subseteq [a, b]$. Let $\xi \in [a, b]$ be a fixed point of g, i.e., $\xi = g(\xi)$. Suppose that there is a constant $\delta > 0$ such that $I_\delta = (\xi - \delta, \xi + \delta) \subset [a, b]$, $g \in C^1(I_\delta)$, and $|g'(\xi)| < 1$. Then the sequence $\{x_k\}_{k=0}^{\infty}$ generated by (15.2) converges at least linearly to ξ, provided that x_0 is sufficiently close to ξ.*

Proof. Suppose that $\xi \in (a, b)$, i.e., ξ is in the interior of the set $[a, b]$. The reader can examine the other cases. Since $|g'(\xi)| < 1$, there is an $h \in (0, \delta)$ and an $L \in (0, 1)$ such that, for all $x \in I_h = (\xi - h, \xi + h)$,

$$|g'(x)| \leq L < 1.$$

The proof of this last fact is left to the reader as an exercise; see Problem 15.6.

Suppose that $x_k \in I_h$. Then, using the Mean Value Theorem B.30,

$$|\xi - x_{k+1}| = |g(\xi) - g(x_k)| = |g'(\eta_k)| \cdot |\xi - x_k| \leq L|\xi - x_k|$$

for some $\eta_k \in I_h$ between ξ and x_k. This proves that if $x_k \in I_h$, then $x_{k+1} \in I_h$. Using induction, if $x_0 \in I_h$,

$$|\xi - x_k| \leq L^k|\xi - x_0|.$$

By the Squeeze Theorem, since $L^k \to 0$ as $k \to \infty$, we have $x_k \to \xi$ as $k \to \infty$. Furthermore, the convergence is at least linear; see Definition B.10. □

We will now consider how simple fixed point iterations can be used to find roots. A general strategy is to find some function α that does not vanish and to consider a fixed point iteration scheme for

$$g(x) = x - \alpha(x)f(x).$$

The particular choice of α gives rise to the various methods we now consider.

15.2.1 Relaxation Method

Definition 15.14 (relaxation). Let $I \subseteq \mathbb{R}$ be an interval, $f \in C(I)$, and $x_0 \in I$ be given. The **relaxation method** is an algorithm for computing the terms of the sequence $\{x_k\}_{k=0}^{\infty}$ via the recursive formula

$$x_{k+1} = x_k - \lambda f(x_k), \tag{15.3}$$

where $\lambda \neq 0$. The method is **well defined** if and only if $x_k \in I$ for all $k = 1, 2, 3, \ldots$, and the relaxation method **converges** if and only if there is a $\xi \in I$, with $f(\xi) = 0$, such that $x_k \to \xi$ as $k \to \infty$.

Notice that the relaxation method is a simple fixed point iteration scheme with $g(x) = x - \lambda f(x)$. From this definition, it necessarily follows that a fixed point of g must be a root of f. Thus, the following result is just a translation of the theory of fixed point iterations.

Theorem 15.15 (convergence). *Let $I \subset \mathbb{R}$ be an interval. Suppose that $f\colon I \to \mathbb{R}$ and, for some $\xi \in I$, $f(\xi) = 0$, but $f'(\xi) \neq 0$. Assume that, for some $\delta > 0$, $f \in C^1(I_\delta)$, where $I_\delta = [\xi - \delta, \xi + \delta] \subseteq I$. Then there exists positive real numbers λ and $h \in (0, \delta)$ such that the sequence $\{x_k\}_{k=0}^{\infty}$ defined by the relaxation scheme (15.3) converges to ξ for any $x_0 \in I_h = [\xi - h, \xi + h]$.*

Proof. Suppose that $f'(\xi) = \alpha > 0$. The case $\alpha < 0$ is analogous and left to the reader. By continuity, we may assume that

$$0 < \frac{1}{2}\alpha \leq f'(x) \leq \frac{3}{2}\alpha, \quad \forall x \in I_\delta.$$

If this is not the case, we can just choose a smaller δ and redefine I_δ. Set

$$M = \max_{x \in I_\delta} f'(x).$$

Thus, $\frac{1}{2}\alpha \leq M \leq \frac{3}{2}\alpha$. For any $\lambda > 0$, it follows that

$$1 - \lambda M \leq 1 - \lambda f'(x) \leq 1 - \frac{1}{2}\lambda\alpha, \quad \forall x \in I_\delta.$$

We now choose, if possible, $\lambda > 0$ such that

$$1 - \lambda M = -\theta \quad \text{and} \quad 1 - \frac{1}{2}\lambda\alpha = \theta.$$

These equations are satisfied if and only if

$$\lambda M - 1 = 1 - \frac{1}{2}\lambda\alpha, \quad \theta = 1 - \frac{1}{2}\lambda\alpha$$

if and only if

$$\lambda = \frac{4}{2M + \alpha}, \quad \theta = \frac{2M - \alpha}{2M + \alpha}.$$

Now define the iteration function g via

$$g(x) = x - \lambda f(x) = x - \frac{4f(x)}{2M + \alpha}.$$

The rest of the details are left to the reader as an exercise; see Problem 15.7. Use Theorem 15.13 to conclude that $x_k \to \xi$, provided that x_0 is sufficiently close to ξ. □

15.2.2 Stationary Slope Approximation Methods

Let $I \subset \mathbb{R}$ be an interval and $f \in C^1(I)$. Assume that there is $\xi \in I$, which is a simple root of f. Then, by Taylor's Theorem B.31, we obtain that, for any $x \in I$,

$$0 = f(\xi) = f(x) + f'(\theta)(\xi - x)$$

for some θ between x and ξ. Thus, if $f'(\theta) \neq 0$, the root must satisfy

$$\xi = x - [f'(\theta)]^{-1} f(x).$$

This motivates the construction of a family of schemes. Let $s_0 \neq 0$ be a *slope approximation* and $x_0 \in I$ an initial guess. Then, for $k \geq 0$,

$$x_{k+1} = x_k - s_k^{-1} f(x_k), \tag{15.4}$$

with some rule to compute $s_{k+1} \neq 0$. Notice that if $s_k = s_0 = \frac{1}{\lambda}$, then this reduces to the relaxation method of Definition 15.14. Let us consider two particular examples.

Definition 15.16 (chord method). Let $I = [a, b]$ be an interval and $x_0 \in I$. The sequence $\{x_k\}_{k=0}^{\infty}$ is obtained by the **chord method** if is constructed via (15.4) with

$$s_k = \frac{f(b) - f(a)}{b - a}.$$

Theorem 15.17 (convergence). *Let $I = [a, b] \subset \mathbb{R}$ be an interval and $f \in C^1([a, b])$ is such that it has a unique simple root $\xi \in [a, b]$. If $b - a$ is sufficiently small, then the sequence $\{x_k\}_{k=1}^{\infty}$ obtained by the chord method of Definition 15.16 converges linearly to ξ as $k \to \infty$.*

Proof. See Problem 15.8. □

Definition 15.18 (simplified Newton). Let I be an interval and $x_0 \in I$ be such that $f'(x_0) \neq 0$. The sequence $\{x_k\}_{k=0}^{\infty}$ is obtained by the **simplified Newton method**[2] if it is constructed via (15.4) with

$$s_k = f'(x_0).$$

This method can be analyzed in two different ways. Here, we analyze it using the theory of fixed point iterations.

Proposition 15.19 (convergence). *Let $I \subset \mathbb{R}$ be an interval and $f \in C(I)$ be such that there is $\xi \in I$ for which $f(\xi) = 0$. Define, for $\delta > 0$, $I_\delta = [\xi - \delta, \xi + \delta] \subseteq I$. Assume that there is $\delta > 0$ such that $f \in C^1(I_\delta)$. Finally, assume that $f'(\xi) = \alpha > 0$. If x_0 is sufficiently close to ξ, then the simplified Newton method of Definition 15.18 converges linearly to ξ.*

Proof. See Problem 15.9. □

15.2.3 Fixed Point Iterations in Several Dimensions

Let us quickly comment that the idea of fixed point iterations can be generalized to several dimensions. For this, we consider a closed ball $B \subset \mathbb{R}^d$ and $\boldsymbol{g}: B \to \mathbb{R}^d$. The *fixed point iteration scheme*, in this setting, starts from $\boldsymbol{x}_0 \in B$ and proceeds, for $k \geq 0$, as

$$\boldsymbol{x}_{k+1} = \boldsymbol{g}(\boldsymbol{x}_k).$$

[2] Named in honor of the English mathematician, physicist, astronomer, theologian, and natural philosopher Sir Isaac Newton (1643–1726/27).

As in the one-dimensional case, we say that the fixed point iteration scheme is *well defined* if $\mathbf{x}_k \in B$ for all k. The convergence theory of this approach and its application to the solution of systems of nonlinear equations follow the same ideas we have presented here. We leave the details to the reader.

15.3 Newton's Method in One Space Dimension

We have now arrived at our preferred method of choice: Newton's method. Throughout our discussion, we will assume that the function f, for which we are trying to find a root, is at least twice continuously differentiable. One can prove these results using less regularity than this, but this assumption will greatly simplify our arguments.

Definition 15.20 (Newton's method[3]). Let $I \subseteq \mathbb{R}$ be an interval, $f \in C^1(I)$, and $x_0 \in I$, with $f'(x_0) \neq 0$, be given. **Newton's method** is an algorithm for computing the terms of the sequence $\{x_k\}_{k=0}^{\infty}$ via the recursive formula

$$x_{k+1} = x_k - \frac{f(x_k)}{f'(x_k)}. \tag{15.5}$$

We say that the method is **well defined** if and only if $x_k \in I$ and $f'(x_k) \neq 0$ for all $k = 1, 2, 3, \ldots$. We say that Newton's method **converges** if and only if there is a $\xi \in I$, with $f(\xi) = 0$, such that $x_k \to \xi$ as $k \to \infty$.

Newton's method can be studied with the theory of fixed point iterations presented in Section 15.2. Indeed, from the definition, it is clear that Newton's method is a fixed point iteration for the function

$$g(x) = x - \frac{f(x)}{f'(x)}. \tag{15.6}$$

Thus, for a simple root, one immediately obtains a linear convergence for Newton's method provided the initial guess, x_0, is sufficiently close to the root.

Proposition 15.21 (linear convergence). *Let $I \subseteq \mathbb{R}$ be an interval and $f \in C^2(I)$ with $|f'(x)| \geq \alpha > 0$ for all $x \in I$. Assume that there is $\xi \in I$ for which $f(\xi) = 0$. There is a constant $h > 0$ such that, if $x_0 \in I$ and $|x_0 - \xi| < h$, Newton's method converges at least linearly to ξ.*

Proof. For this proof, let us apply the theory of fixed points, as detailed in Section 15.2, to the function g defined in (15.6). Notice that since $f \in C^2(I)$, then

$$g'(x) = \frac{f(x)f''(x)}{[f'(x)]^2} \quad \Longrightarrow \quad g'(\xi) = 0.$$

Thus, by continuity, there is a constant $\delta > 0$ such that if $x \in I_\delta = [\xi - \delta, \xi + \delta]$, then

$$|g'(x)| < 1.$$

[3] Named in honor of the British mathematician, physicist, astronomer, theologian, and natural philosopher Sir Isaac Newton (1642–1726/27).

In addition, notice that, by Taylor's Theorem B.32, we have, for any $x \in I$,

$$0 = f(\xi) = f(x) + f'(x)(\xi - x) + \frac{1}{2}f''(\eta)(\xi - x)^2$$

for some η between x and ξ. Let us define

$$A = \frac{\max_{x\in I} |f''(x)|}{\alpha}, \quad h = \min\left\{\delta, \frac{1}{A}\right\}.$$

Then, for any $x \in I_h = [\xi - h, \xi + h]$, we have $|g'(x)| < 1$ and

$$\begin{aligned} |g(x) - \xi| &= \left| x - \xi - \frac{f(x)}{f'(x)} \right| \\ &= \frac{1}{2}\left| \frac{f''(\eta)}{f'(x)}(x - \xi)^2 \right| \\ &\le \frac{1}{2}Ah^2 \\ &\le \frac{1}{2}h, \end{aligned}$$

so that $g(x) \in I_h$. Theorem 15.13 then allows us to conclude the (at least linear) convergence of the fixed point iteration sequence defined in (15.2). □

It turns out that, under normal, reasonable circumstances, Newton's method converges quadratically.

Theorem 15.22 (quadratic convergence). *Let $I \subset \mathbb{R}$ be an interval. Suppose that $f\colon I \to \mathbb{R}$ and, for some $\xi \in I$, $f(\xi) = 0$, but $f'(\xi) \neq 0$ and $f''(\xi) \neq 0$. Assume that, for some $\delta > 0$, $f \in C^2(I_\delta)$, where $I_\delta = [\xi - \delta, \xi + \delta] \subseteq I$, and $0 < \alpha \le |f'(x)|$ for all $x \in I_\delta$. Set*

$$A = \frac{\max_{x\in I_\delta} |f''(x)|}{\alpha}, \quad h = \min\left\{\delta, \frac{1}{A}\right\}. \tag{15.7}$$

If $|\xi - x_0| \le h$, then the sequence $\{x_k\}_{k=0}^{\infty}$ defined by Newton's method (15.5) *converges quadratically, as $k \to \infty$, to the root ξ.*

Proof. We could use the result of Proposition 15.21 as our starting point for this proof. To give some variety, let us repeat the proof of linear convergence, but this time using a direct approach.

(Well-posedness) Suppose that $x_k \in I_\delta$. Then, by Taylor's Theorem B.32,

$$0 = f(\xi) = f(x_k) + (\xi - x_k)f'(x_k) + \frac{(\xi - x_k)^2}{2}f''(\eta_k)$$

for some η_k between x_k and ξ. Note that $f'(x_k) \neq 0$, and we have, using (15.5) (the definition of Newton's method),

$$x_{k+1} - \xi = \frac{(\xi - x_k)^2 f''(\eta_k)}{2f'(x_k)}. \tag{15.8}$$

Now, if $x_k \in I_h = [\xi - h, \xi + h]$,

$$|\xi - x_{k+1}| = \frac{1}{2}\frac{|f''(\eta_k)|}{|f'(x_k)|} \cdot |\xi - x_k| \cdot |\xi - x_k| \le \frac{A}{2} \cdot h \cdot |\xi - x_k| = \frac{1}{2}|\xi - x_k|,$$

and $x_{k+1} \in I_h$ as well. The algorithm is, therefore, well defined, since $x_{k+1} \in I_h$ and $f'(x_{k+1}) \neq 0$.

(Linear convergence) By induction, it is clear from the contraction estimate that

$$|\xi - x_k| \le \frac{h}{2^k},$$

which proves that the sequence converges to the root ξ as $k \to \infty$, at least linearly; see Definition B.10.

(Quadratic convergence) From (15.8) we obtain

$$\frac{|\xi - x_{k+1}|}{|\xi - x_k|^2} = \frac{1}{2}\frac{|f''(\eta_k)|}{|f'(x_k)|}.$$

Since $x_k \to \xi$, by the Squeeze Theorem B.6, $\eta_k \to \xi$ as $k \to \infty$ as well. Passing to limits, we have

$$\lim_{k\to\infty} \frac{|\xi - x_{k+1}|}{|\xi - x_k|^2} = \frac{1}{2}\frac{|f''(\xi)|}{|f'(\xi)|} = \sigma \in (0, \infty),$$

which establishes quadratic convergence; see Definition B.10. □

Remark 15.23 (faster convergence). We see immediately that it is possible for the convergence to be faster than quadratic if and only if $f''(\xi) = 0$.

One glaring fact about the last result for Newton's method is that it guarantees convergence only in a local region. To improve this to a more global result, we need additional assumptions.

Theorem 15.24 (global convergence). *Let $[a, b] \subset \mathbb{R}$ be an interval and $f \in C^2([a, b])$ be such that, for some $\xi \in [a, b]$, $f(\xi) = 0$. Assume further that f' and f'' are strictly positive on the interval $[a, b]$. For any starting value $x_0 \in (\xi, b]$, the sequence $\{x_k\}_{k=0}^{\infty}$ defined by Newton's method* (15.5) *converges quadratically to the root ξ as $k \to \infty$. Moreover, $x_k > \xi$ for all $k \in \mathbb{N}$.*

Proof. Since f is monotonically increasing on $[a, b]$, ξ is the only root in $[a, b]$. Otherwise, by Rolle's Theorem B.29, one could find a point where f' is zero, contradicting the assumptions. Also note that $f(x) > 0$ for all $x \in (\xi, b]$; likewise, $f(x) < 0$ for all $x \in [a, \xi)$.

Assume that $x_k > \xi$. Employing Newton's method (15.5) and using the positivity of $f(x_k)$ and $f'(x_k)$, we immediately obtain that

$$x_{k+1} = x_k - \frac{f(x_k)}{f'(x_k)} < x_k.$$

Furthermore, from the error equation (15.8), using the positivity of $f''(\eta_k)$ and $f'(x_k)$, we find out that $\xi - x_{k+1} < 0$. In other words, $\xi < x_{k+1} < x_k$. Thus, $\{x_k\}_{k=0}^{\infty}$ is a bounded, monotonically decreasing sequence in $[a, b]$. By Theorem B.7, it must, therefore, have a limit point in $[a, b]$, call it η, as $k \to \infty$. But this limit must be a fixed point of the function g, defined in (15.6). Therefore, $f(\eta) = 0$. But since ξ is the unique root of f in $[a, b]$, it must be that $\xi = \eta$. This shows that $x_k \to \xi$ as $k \to \infty$. Quadratic convergence follows as in the proof of Theorem 15.22. □

Remark 15.25 (other initial guesses). If in the setting of the last theorem one assumes that $x_0 < \xi$ to begin, one encounters some problems. One will conclude from (15.5) that $x_1 > x_0$. But from (15.8) one will conclude that $x_1 > \xi$. Thus, there is no "squeezing" action as before. However, one might recover if, by chance, $x_1 \leq b$. For then we could restart the argument in the last proof to guarantee convergence. Of course, if the interval in question is $(-\infty, +\infty)$ there is no problem at all.

Example 15.3 Let us use Newton's method to compute the square root of a positive real number. Suppose that we want to compute $\sqrt{5}$. Define $f(x) = x^2 - 5$. There are two solutions to $f(x) = 0$, namely $\xi_\pm = \pm\sqrt{5}$. Let us pick $x_0 = 5$. Theorem 15.24 guarantees that this is a suitable choice if we want to compute the zero $\xi_+ = \sqrt{5}$. The sequence of approximations for Newton's method is defined by

$$x_{k+1} = x_k - \frac{x_k^2 - 5}{2x_k}, \quad k = 0, 1, \ldots . \tag{15.9}$$

Below, we show the result of using the code presented in Listing 15.1. The correct digits are indicated using boldface.

k	x_k
0	5.000 000 000 000 000
1	3.000 000 000 000 000
2	**2**.333 333 333 333 333
3	**2.23**8 095 238 095 238
4	**2.236 06**8 895 643 363
5	**2.236 067 977 499** 978
6	**2.236 067 977 499 790**

This example illustrates an empirical fact that is a consequence of quadratic convergence: Newton's method doubles the number of correct digits with each iteration. A partial explanation for this fact is as follows: from the proof of Theorem 15.22, we see that

$$|\xi - x_{k+1}| \leq C|\xi - x_k|^2,$$

so that, by taking base-10 logarithms (which essentially counts the number of correct digits), we have

$$\log_{10}|\xi - x_{k+1}| \leq 2\log_{10}|\xi - x_k| + \log_{10} C.$$

Example 15.4 Define $f\colon [1,5] \to \mathbb{R}$ via $f(x) = (x-3)^3$. Observe that, for this simple example, $\xi = 3$ is a nonsimple root, i.e., $f(3) = f'(3) = 0$, which is something that the theory (up to this point) cannot handle. Nevertheless, Newton's method will still work. In particular, if Newton's method is employed with the starting point $x_0 = 4$ to approximate the root $\xi = 3$, then one can show directly that the convergence is exactly linear; see Problem 15.16.

15.3.1 Nonsimple roots

The next result describes what may happen when a certain number of derivatives vanish at the zero of interest, as in the Example 15.4.

Theorem 15.26 (nonsimple roots). *Let m be a positive integer and I be a closed and bounded interval. Suppose that $f \in C^m(I)$ is such that there is $\xi \in I$, for which $f(\xi) = f'(\xi) = \cdots = f^{(m-1)}(\xi) = 0$, but $f^{(m)}(\xi) \neq 0$. If $|\xi - x_0|$ is sufficiently small, the sequence $\{x_k\}_{k=0}^{\infty}$ defined by Newton's method* (15.5) *is well defined and converges to ξ exactly linearly with*

$$\lim_{k\to\infty} \frac{|x_{k+1}-\xi|}{|x_k-\xi|} = \frac{m-1}{m} = \sigma \in (0,1).$$

Proof. We give a sketch of the proof. The details of well-definedness and convergence, in particular, are left to the reader as an exercise; see Problems 15.17 and 15.18. By Taylor's Theorem,

$$\begin{aligned} f(x_k) &= f(\xi) + f'(\xi)(x_k-\xi) + \cdots + f^{(m-1)}(\xi)\frac{(x_k-\xi)^{m-1}}{(m-1)!} + f^{(m)}(\eta_k)\frac{(x_k-\xi)^m}{m!} \\ &= f^{(m)}(\eta_k)\frac{(x_k-\xi)^m}{m!} \end{aligned}$$

for some η_k between x_k and ξ. Another application of Taylor's Theorem gives

$$\begin{aligned} f'(x_k) &= f'(\xi) + f''(\xi)(x_k-\xi) + \cdots + f^{(m-1)}(\xi)\frac{(x_k-\xi)^{m-2}}{(m-2)!} \\ &\quad + f^{(m)}(\zeta_k)\frac{(x_k-\xi)^{m-1}}{(m-1)!} = f^{(m)}(\zeta_k)\frac{(x_k-\xi)^{m-1}}{(m-1)!} \end{aligned}$$

for some ζ_k between x_k and ξ. Thus, assuming $x_k \neq \xi$,

$$x_{k+1} - \xi = x_k - \xi - \frac{f(x_k)}{f'(x_k)} = x_k - \xi - \frac{f^{(m)}(\eta_k)\frac{(x_k-\xi)^m}{m!}}{f^{(m)}(\zeta_k)\frac{(x_k-\xi)^{m-1}}{(m-1)!}}$$

or

$$\frac{x_{k+1}-\xi}{x_k-\xi} = 1 - \frac{f^{(m)}(\eta_k)}{m\cdot f^{(m)}(\zeta_k)}.$$

Since $\eta_k, \zeta_k \to \xi$ as $k \to \infty$,

$$\lim_{k\to\infty} \frac{x_{k+1}-\xi}{x_k-\xi} = 1 - \frac{1}{m} = \frac{m-1}{m}.$$

□

The fact that, in Theorem 15.26, the constant σ depends only on the multiplicity m of the root hints at the fact that quadratic convergence for Newton's method can be recovered by a small modification. This is explored in Problem 15.19.

15.4 Quasi-Newton Methods

In the previous section, we developed the analysis of Newton's method of (15.5). We showed that, under suitable assumptions, this method converges quadratically. There are, however, two major drawbacks to Newton's method: it requires a sufficiently close initial approximation to the root, and it requires evaluating the derivative at every iteration. Requiring a good initial guess is mostly unavoidable, but such an approximation may be obtained by some other method. The evaluation of the derivative, on the other hand, may be an issue. In applications, this may be very costly, or not even at all possible. For this reason, here we propose several important variants.

The common feature of all these methods is that they take the form (15.4), where the slope approximation s_k will, in general, change on every iteration. Its construction will depend for instance, for $r \geq 0$, on $x_{k-r}, \ldots, x_k$, and the values of the function at these points, but not require evaluation of the derivative at these points. Another possibility is that we require fewer derivative evaluations than Newton does, i.e., not at every iteration.

15.4.1 Simplified Newton's Method

We already saw the simplified Newton method in Definition 15.18. Here, we provide an analysis for it. We recall that this method has the form (15.4) with $s_k = s = f'(x_0)$.

Theorem 15.27 (convergence). *Let $I \subseteq \mathbb{R}$ be an interval. Assume that $f \in C(I)$ is such that there is an $\xi \in I$ for which $f(\xi) = 0$, but $f'(\xi) \neq 0$ and $f''(\xi) \neq 0$. Set, for $\delta > 0$, $I_\delta = [\xi - \delta, \xi + \delta] \subseteq I$. Assume that, for some $\delta > 0$, $f \in C^2(I_\delta)$ and $0 < \alpha \leq |f'(x)|$ for all $x \in I_\delta$. Set*

$$A = \frac{\max_{x \in I_\delta} |f''(x)|}{\alpha}, \qquad h = \min\left\{\delta, \frac{1}{3A}\right\}. \tag{15.10}$$

If $|\xi - x_0| \leq h$, then the sequence $\{x_k\}_{k=0}^{\infty}$ defined by the simplified Newton method of Definition 15.18 converges linearly to the zero ξ of f as $k \to \infty$.

Proof. Suppose that $x_0, x_k \in [\xi - h, \xi + h]$. Then we have

$$\begin{aligned} x_{k+1} - \xi &= \frac{1}{f'(x_0)}\Big[f'(x_0)(x_k - \xi) - f(x_k)\Big] \\ &= \frac{1}{f'(x_0)}\Big[(f'(x_0) - f'(\xi))(x_k - \xi) - f(x_k) + f'(\xi)(x_k - \xi)\Big]. \end{aligned}$$

By the Mean Value Theorem B.30, there is $\beta \in [\xi - h, \xi + h]$ between x_0 and ξ, for which

$$f''(\beta)(x_0 - \xi) = f'(x_0) - f'(\xi).$$

Furthermore, for some $\eta_k \in [\xi - h, \xi + h]$ between ξ and x_k, we have

$$f(x_k) = f(\xi) + f'(\xi)(x_k - \xi) + \frac{f''(\eta_k)}{2}(x_k - \xi)^2$$

from Taylor's Theorem. Rearranging terms and using $f(\xi) = 0$ yields

$$-f(x_k) + f'(\xi)(x_k - \xi) = -\frac{f''(\eta_k)}{2}(x_k - \xi)^2.$$

Putting things together, we find

$$\begin{aligned} x_{k+1} - \xi &= \frac{1}{f'(x_0)}\left[f''(\beta)(x_0 - \xi)(x_k - \xi) - \frac{f''(\eta_k)}{2}(x_k - \xi)^2\right] \\ &= \frac{1}{f'(x_0)}\left[f''(\beta)(x_0 - \xi) - \frac{f''(\eta_k)}{2}(x_k - \xi)\right](x_k - \xi). \end{aligned}$$

Taking absolute values,

$$\begin{aligned} |x_{k+1} - \xi| &= \frac{1}{|f'(x_0)|}\left|f''(\beta)(x_0 - \xi) - \frac{f''(\eta_k)}{2}(x_k - \xi)\right| \cdot |x_k - \xi| \\ &\le \frac{1}{|f'(x_0)|}\left(|f''(\beta)| \cdot |x_0 - \xi| + \frac{1}{2}|f''(\eta_k)| \cdot |x_k - \xi|\right)|x_k - \xi| \\ &\le \left(A|x_0 - \xi| + \frac{1}{2}|x_k - \xi|A\right)|x_k - \xi| \\ &\le \left(A \cdot \frac{1}{3A} + \frac{1}{2}\frac{1}{3A}A\right)|x_k - \xi| = \frac{1}{2}|x_k - \xi|. \end{aligned}$$

Hence, $x_{k+1} \in [\xi - h, \xi + h]$ for any $k \in \mathbb{N}$, as long as $x_0, x_k \in [\xi - h, \xi + h]$. The simplified Newton algorithm is well defined. Furthermore, it is clear that

$$|x_k - \xi| \le \frac{h}{2^k},$$

which proves that $x_k \to \xi$ as $k \to \infty$.

The convergence is exactly linear, as can be seen from the error equation:

$$\lim_{k\to\infty} \frac{|x_{k+1} - \xi|}{|x_k - \xi|} = \frac{|f''(\beta)| \cdot |x_0 - \xi|}{|f'(x_0)|} = \mu.$$

By our assumptions, $0 < \mu \le 1/3 < 1$. □

15.4.2 Steffensen's Method

In the simplified Newton method, only one derivative evaluation is required. The trade-off is that the order of convergence is reduced. In the following, we can eliminate derivative evaluation altogether and still retain quadratic convergence.

Definition 15.28 (Steffensen's method[4]). Let $I \subseteq \mathbb{R}$ be an interval, $f \in C(I)$, and $x_0 \in I$. **Steffensen's method** is an algorithm for computing the terms of the sequence $\{x_k\}_{k=0}^{\infty}$ via (15.4) with

$$s_k = \frac{f(x_k + f(x_k)) - f(x_k)}{f(x_k)}.$$

We say that this method is **well defined** if and only if $x_0 \in I$ implies that $x_k \in I$ for all $k = 1, 2, \ldots$. We say that this method **converges** if and only if there is $\xi \in I$, with $f(\xi) = 0$, such that $x_k \to \xi$ as $k \to \infty$.

Before studying the convergence of this method, let us provide some intuition behind this slope approximation. If the method is to converge, then $h = f(x_k) \to f(\xi) = 0$, so that

$$s_k = \frac{f(x_k + h) - f(x_k)}{h}$$

is indeed a good approximation of the derivative $f'(x_k)$.

Let us now study the convergence of this method.

Theorem 15.29 (convergence). *Let $I \subseteq \mathbb{R}$ be an interval and $f \in C(I)$ be such that, for some $\xi \in I$, $f(\xi) = 0$. Define, for $\delta > 0$, $I_\delta = [\xi - \delta, \xi + \delta] \subseteq I$. Assume that there is $\delta > 0$ for which $f \in C^2(I_\delta)$, $f'(\xi) \neq 0$, and $f''(\xi) \neq 0$. If $|\xi - x_0|$ is sufficiently small, then the sequence $\{x_k\}_{k=0}^{\infty}$ defined by Steffensen's method of Definition 15.28 is well defined and converges quadratically to the zero ξ as $k \to \infty$.*

Proof. First observe that, using Taylor's Theorem B.32, there are points η_k and γ_k between x_k and $x_k + f(x_k)$ such that

$$s_k = \frac{f'(x_k)f(x_k) + \frac{1}{2}f''(\eta_k)f^2(x_k)}{f(x_k)} = f'(x_k) + \frac{1}{2}f''(\eta_k)f(x_k) = f'(\gamma_k). \quad (15.11)$$

From the definition of the scheme, we have

$$\begin{aligned} x_{k+1} - \xi &= x_k - \xi - \frac{f(x_k)}{s_k} \\ &= \frac{(x_k - \xi)\left(f'(x_k) + \frac{1}{2}f''(\eta_k)f(x_k)\right) - f(x_k)}{f'(x_k) + \frac{1}{2}f''(\eta_k)f(x_k)} \\ &= \frac{-\left[f(x_k) + f'(x_k)(\xi - x_k)\right] + \frac{1}{2}f''(\eta_k)f(x_k)(x_k - \xi)}{f'(x_k) + \frac{1}{2}f''(\eta_k)f(x_k)}. \end{aligned}$$

By Taylor's Theorem B.32, there is a point β_k between x_k and ξ such that

$$0 = f(\xi) = f(x_k) + f'(x_k)(\xi - x_k) + \frac{1}{2}f''(\beta_k)(\xi - x_k)^2,$$

so that

$$\frac{1}{2}f''(\beta_k)(\xi - x_k)^2 = -\left[f(x_k) + f'(x_k)(\xi - x_k)\right].$$

[4] Named in honor of the Danish mathematician and statistician Johan Frederik Steffensen (1873–1961).

Hence,

$$\begin{aligned}
x_{k+1} - \xi &= \frac{\frac{1}{2}f''(\beta_k)(\xi - x_k)^2 + \frac{1}{2}f''(\eta_k)f(x_k)(x_k - \xi)}{f'(x_k) + \frac{1}{2}f''(\eta_k)f(x_k)} \\
&= \frac{\frac{1}{2}f''(\beta_k)(\xi - x_k)^2 + \frac{1}{2}f''(\eta_k)(f(x_k) - f(\xi))(x_k - \xi)}{f'(x_k) + \frac{1}{2}f''(\eta_k)f(x_k)} \\
&= \frac{\frac{1}{2}f''(\beta_k)(\xi - x_k)^2 + \frac{1}{2}f''(\eta_k)f'(\alpha_k)(x_k - \xi)^2}{f'(x_k) + \frac{1}{2}f''(\eta_k)f(x_k)} \\
&= \left[\frac{\frac{1}{2}f''(\beta_k) + \frac{1}{2}f''(\eta_k)f'(\alpha_k)}{f'(x_k) + \frac{1}{2}f''(\eta_k)f(x_k)}\right](x_k - \xi)^2,
\end{aligned}$$

where α_k is some point between x_k and ξ.

Taking absolute values, using the triangle inequality, and using (15.11), we get

$$\begin{aligned}
|x_{k+1} - \xi| &= \left|\frac{\frac{1}{2}f''(\beta_k) + \frac{1}{2}f''(\eta_k)f'(\alpha_k)}{f'(\gamma_k)}\right| \cdot |x_k - \xi| \cdot |x_k - \xi| \\
&\le \frac{\frac{1}{2}|f''(\beta_k)| + \frac{1}{2}|f''(\eta_k)| \cdot |f'(\alpha_k)|}{|f'(\gamma_k)|} \cdot |x_k - \xi| \cdot |x_k - \xi|.
\end{aligned}$$

Notice now that the points η_k and γ_k are between x_k and $x_k + f(x_k)$. By continuity, there is an $h \in (0, \delta/2)$ such that

$$|f(x)| \le \delta/2, \quad \forall x \in I_h = [\xi - h, \xi + h].$$

Therefore, if $x_k \in I_h$, it easily follows that

$$|\xi - \eta_k| \le \delta, \qquad |\xi - \gamma_k| \le \delta.$$

We assume, as usual, that there are constants $m_2 \ge m_1 > 0$ such that

$$m_1 \le |f'(x)| \le m_2, \quad \forall x \in I_\delta,$$

and constants $m_4 \ge m_3 > 0$ such that

$$m_3 \le |f''(x)| \le m_4, \quad \forall x \in I_\delta.$$

Therefore, if $x_k \in I_s = [\xi - s, \xi + s]$, where

$$s < \min\left\{\frac{m_1}{m_4 + m_4 m_2}, h\right\},$$

$$\begin{aligned}
|x_{k+1} - \xi| &\le \frac{\frac{1}{2}|f''(\beta_k)| + \frac{1}{2}|f''(\eta_k)| \cdot |f'(\alpha_k)|}{|f'(\gamma_k)|} \cdot |x_k - \xi| \cdot |x_k - \xi| \\
&\le \frac{\frac{1}{2}m_4 + \frac{1}{2}m_4 m_2}{m_1} \cdot \frac{m_1}{m_4 + m_4 m_2} \cdot |x_k - \xi| \le \frac{1}{2}|x_k - \xi|.
\end{aligned}$$

Thus, the method is well defined and it converges at least linearly.

The proof of quadratic convergence follows from the fact that

$$\lim_{k\to\infty} \frac{|x_{k+1} - \xi|}{|x_k - \xi|^2} = \frac{1}{2}\left|\frac{f''(\xi) + f''(\xi)f'(\xi)}{f'(\xi)}\right| \ne 0,$$

where we used the fact that

$$\alpha_k, \beta_k, \gamma_k, \eta_k \to \xi, \qquad k \to \infty. \qquad \square$$

15.4.3 Two-Step Newton's Method

The next method is a variant of Newton's method which exhibits convergence that may be faster than quadratic.

Definition 15.30 (two-step Newton)**.** Let $I \subseteq \mathbb{R}$ be an interval and $f \in C^1(I)$. For $x_0 \in I$, with $f'(x_0) \neq 0$, the sequence $\{x_k\}_{k=0}^{\infty}$ defined by

$$y_k = x_k - \frac{f(x_k)}{f'(x_k)}, \qquad x_{k+1} = y_k - \frac{f(y_k)}{f'(x_k)} \tag{15.12}$$

is called the **two-step Newton** method. We say that the method is **well defined** if $x_k \in I$ and $f'(x_k) \neq 0$ for all $k \geq 0$. We say that the method **converges** if there is $\xi \in I$ such that $f(\xi) = 0$ and $x_k \to \xi$ as $k \to \infty$.

Theorem 15.31 (convergence)**.** *Let $I \subseteq \mathbb{R}$ be an interval, $f \in C(I)$ is such that there is $\xi \in I$ for which $f(\xi) = 0$, but $f'(\xi) \neq 0$, and $f''(\xi) \neq 0$. Set, for $\delta > 0$, $I_\delta = [\xi - \delta, \xi + \delta] \subseteq I$. Assume that is $\delta > 0$ for which $f \in C^2(I_\delta)$ and $0 < \alpha \leq |f'(x)|$ for all $x \in I_\delta$. Set*

$$A = \frac{\max_{x \in I_\delta} |f''(x)|}{\alpha}, \qquad h = \min\left\{\delta, \frac{1}{A}\right\}.$$

If $|\xi - x_0| \leq h$, then the sequence $\{x_k\}_{k=0}^{\infty}$ defined by the two-step Newton method converges exactly cubically to the zero ξ as $k \to \infty$.

Proof. Suppose that $x_k \in [\xi - h, \xi + h] \subseteq I_\delta$. Then, by Taylor's Theorem,

$$0 = f(\xi) = f(x_k) + f'(x_k)(\xi - x_k) + \frac{f''(\eta_k)}{2}(\xi - x_k)^2$$

for some η_k between x_k and ξ. Note that $f'(x_k) \neq 0$ and, using the first equation in (15.12), we have that

$$\xi - y_k = -\frac{(\xi - x_k)^2}{2} \frac{f''(\eta_k)}{f'(x_k)}. \tag{15.13}$$

Hence,

$$|\xi - y_k| = \frac{1}{2} \frac{|f''(\eta_k)|}{|f'(x_k)|} \cdot |\xi - x_k| \cdot |\xi - x_k| \leq \frac{A}{2} \cdot h \cdot |\xi - x_k| \leq \frac{1}{2}|\xi - x_k| \leq \frac{h}{2}.$$

We can conclude that if $x_k \in [\xi - h, \xi + h]$, then $y_k \in [\xi - h, \xi + h]$ as well.

Now, using the second equation in (15.12), we have

$$\begin{aligned} x_{k+1} - \xi &= \frac{1}{f'(x_k)} \Big[f'(x_k)\,(y_k - \xi) - f(y_k) \Big] \\ &= \frac{1}{f'(x_k)} \Big[\big(f'(x_k) - f'(\xi)\big)\,(y_k - \xi) - f(y_k) + f'(\xi)\,(y_k - \xi) \Big]. \end{aligned}$$

By the Mean Value Theorem B.30, there is $\beta_k \in [\xi - h, \xi + h]$ between x_k and ξ, for which

$$f''(\beta_k)(x_k - \xi) = f'(x_k) - f'(\xi).$$

Furthermore, for some $\gamma_k \in [\xi - h, \xi + h]$ between ξ and y_k, we have

$$f(y_k) = f(\xi) + f'(\xi)(y_k - \xi) + \frac{f''(\gamma_k)}{2}(y_k - \xi)^2$$

from Taylor's Theorem B.32. Rearranging terms and using $f(\xi) = 0$ yields

$$-f(y_k) + f'(\xi)(y_k - \xi) = -\frac{f''(\gamma_k)}{2}(y_k - \xi)^2.$$

Putting things together, we find

$$x_{k+1} - \xi = \frac{1}{f'(x_k)}\left[f''(\beta_k)(x_k - \xi)(y_k - \xi) - \frac{f''(\gamma_k)}{2}(y_k - \xi)^2\right]. \tag{15.14}$$

Taking absolute values,

$$\begin{aligned} |x_{k+1} - \xi| &= \frac{1}{|f'(x_k)|}\left|f''(\beta_k)(x_k - \xi)(y_k - \xi) - \frac{f''(\gamma_k)}{2}(y_k - \xi)^2\right| \\ &\le A|x_k - \xi| \cdot |y_k - \xi| + \frac{1}{2}|y_k - \xi| \cdot |y_k - \xi| A \\ &\le A|x_k - \xi|\frac{h}{2} + \frac{1}{2} \cdot \frac{h}{2} \cdot \frac{1}{2}|x_k - \xi| A \\ &\le \frac{1}{2}|x_k - \xi| + \frac{1}{8}|x_k - \xi| \\ &= \frac{5}{8}|x_k - \xi|. \end{aligned}$$

We can conclude that if $x_k \in [\xi - h, \xi + h]$, then $x_{k+1} \in [\xi - h, \xi + h]$ as well. More importantly, we see by induction that

$$|\xi - x_k| \le \left(\frac{5}{8}\right)^k h,$$

which proves that $x_k \to \xi$ as $k \to \infty$. Using this fact, it is easy to see that $y_k, \beta_k, \gamma_k, \eta_k \to \xi$ as $k \to \infty$ as well.

Now, using (15.14), we see that

$$\frac{x_{k+1} - \xi}{(x_k - \xi)(y_k - \xi)} = \frac{f''(\beta_k)}{f'(x_k)} - \frac{(y_k - \xi)}{2(x_k - \xi)}\frac{f''(\gamma_k)}{f'(x_k)}.$$

Making use of (15.13),

$$\frac{(x_{k+1} - \xi)}{(x_k - \xi)^3} = \frac{f''(\beta_k)f''(\eta_k)}{2(f'(x_k))^2} - \frac{1}{8}(x_k - \xi)\frac{(f''(\eta_k))^2}{(f'(x_k))^2}\frac{f''(\gamma_k)}{f'(x_k)}.$$

Taking limits, we have

$$\lim_{k\to\infty} \frac{|x_{k+1} - \xi|}{|x_k - \xi|^3} = \frac{1}{2}\left|\frac{f''(\xi)}{f'(\xi)}\right|^2 = \sigma \in (0, \infty).$$

This shows that the convergence is exactly cubic. □

15.4.4 The Secant Method

Definition 15.32 (secant method). Let $I \subseteq \mathbb{R}$ be an interval, $f \in C(I)$, and $x_0 \in I$. The **secant method** is an algorithm for computing the terms of the sequence $\{x_k\}_{k=0}^{\infty}$ via (15.4) with

$$s_k = \frac{f(x_k) - f(x_{k-1})}{x_k - x_{k-1}}, \quad k \geq 1. \tag{15.15}$$

We say that this method is **well defined** if and only if $x_0, x_1 \in I$ implies that $x_k \in I$ for all $k = 2, 3, \ldots$. We say that this method **converges** if and only if there is $\xi \in I$, with $f(\xi) = 0$, such that $x_k \to \xi$ as $k \to \infty$.

Theorem 15.33 (convergence). *Let $I \subseteq \mathbb{R}$ be an interval and $f \in C(I)$ be such that there is $\xi \in I$ for which $f(\xi) = 0$. Set, for $\delta > 0$, $I_\delta = [\xi - \delta, \xi + \delta] \subseteq I$. Assume that there is $\delta > 0$ for which $f \in C^1(I_\delta)$ and, for simplicity, $f'(\xi) > 0$. The sequence $\{x_k\}_{k=0}^{\infty}$ defined by the secant method converges (at least) linearly to the root ξ as $k \to \infty$, provided that x_0 and x_1 are sufficiently close to ξ.*

Proof. Set $f'(\xi) = \alpha > 0$. By continuity, there is no loss in generality in assuming that, for all $x \in I_\delta$,

$$0 < \frac{3\alpha}{4} \leq f'(x) \leq \frac{5\alpha}{4}.$$

Suppose now that $x_k, x_{k-1} \in I_\delta$. By the Mean Value Theorem B.30, there is η_k between x_k and x_{k-1} such that $s_k = f'(\eta_k)$. Then

$$x_{k+1} - \xi = x_k - \xi - \frac{f(x_k)}{f'(\eta_k)}.$$

By Taylor's Theorem, there is a γ_k between x_k and ξ such that

$$f(x_k) = f(\xi) + f'(\gamma_k)(x_k - \xi) = f'(\gamma_k)(x_k - \xi).$$

Thus,

$$x_{k+1} - \xi = x_k - \xi - \frac{f'(\gamma_k)(x_k - \xi)}{f'(\eta_k)} = (x_k - \xi)\left[1 - \frac{f'(\gamma_k)}{f'(\eta_k)}\right] \leq \frac{2}{5}(x_k - \xi).$$

If $|x_0 - \xi| \leq \delta$ and $|x_1 - \xi| \leq \delta$, then, by induction, we see that, for $k \geq 2$,

$$|x_k - \xi| \leq \left(\frac{2}{5}\right)^{k-1} \delta.$$

This proves that the method is well defined and that $x_k \to \xi$ at least linearly. □

It turns out that the convergence of the secant method is super-linear.

Theorem 15.34 (super-linear convergence). *Let $I \subseteq \mathbb{R}$ be an interval and $f \in C(I)$. In the setting of Theorem 15.33, assume, in addition, that $f \in C^2(I_\delta)$ and $f''(\xi) > 0$. Then the sequence $\{x_k\}_{k=0}^{\infty}$ generated by the secant method converges to ξ at the rate $q = \frac{1+\sqrt{5}}{2}$.*

Proof. See Problem 15.21. □

15.5 Newton's Method in Several Dimensions

In this section, we develop and analyze Newton's method for the solution of (15.1) in the case that $n = m = d > 1$. We will need several facts about basic calculus in several variables, and we refer the reader to Appendix B for a review.

Definition 15.35 (Newton's method)**.** Suppose that $d \in \mathbb{N}$, $\Omega \subseteq \mathbb{R}^d$ is an open, convex set, $\boldsymbol{x}_0 \in \Omega$ is given, and $\boldsymbol{f} \in C^1(\Omega;\mathbb{R}^d)$. **Newton's method in** d**-dimensions** is an algorithm for computing the terms of the sequence $\{\boldsymbol{x}_k\}_{k=0}^{\infty}$ via the recursive iteration

$$\mathsf{J}_{\boldsymbol{f}}(\boldsymbol{x}_k)\,(\boldsymbol{x}_{k+1} - \boldsymbol{x}_k) = -\boldsymbol{f}(\boldsymbol{x}_k), \tag{15.16}$$

where $\mathsf{J}_{\boldsymbol{f}}$ is the Jacobian matrix of $\boldsymbol{f}$. We say that the method is **well defined** if and only if $\boldsymbol{x}_k \in \Omega$ and $\mathsf{J}_{\boldsymbol{f}}(\boldsymbol{x}_k)$ is nonsingular, for all $k \in \mathbb{N}$. We say that Newton's method **converges** if and only if there is a $\boldsymbol{\xi} \in \Omega$, with $\boldsymbol{f}(\boldsymbol{\xi}) = \boldsymbol{0}$, such that $\boldsymbol{x}_k \to \boldsymbol{\xi}$ as $k \to \infty$.

As in the one-dimensional case, Newton's method converges quadratically to the root.

Theorem 15.36 (convergence)**.** *Let* $\boldsymbol{\xi} \in \mathbb{R}^d$ *and* $r > 0$ *be given. Suppose that*

$$\boldsymbol{f} \in C^2(\overline{B}(\boldsymbol{\xi}, r);\mathbb{R}^d),$$

$\boldsymbol{f}(\boldsymbol{\xi}) = \boldsymbol{0}$*, and for every* $\boldsymbol{x} \in \overline{B}(\boldsymbol{\xi}, r)$ *the Jacobian matrix* $\mathsf{J}_{\boldsymbol{f}}(\boldsymbol{x})$ *is invertible, with the estimate*

$$\left\| [\mathsf{J}(\boldsymbol{x})]^{-1} \right\|_2 \le \beta.$$

Then the sequence $\{\boldsymbol{x}_k\}_{k=0}^{\infty}$ *defined by Newton's method* (15.16) *converges (at least) quadratically to the root* $\boldsymbol{\xi}$ *as* $k \to \infty$*, provided that* $\boldsymbol{x}_0$ *is sufficiently close to* $\boldsymbol{\xi}$*.*

Proof. By Taylor's Theorem B.51, for each $i = 1, \ldots, d$, there is a point $\boldsymbol{\eta}_{k,i} \in B(\boldsymbol{\xi}, r)$ such that

$$0 = f_i(\boldsymbol{\xi}) = f_i(\boldsymbol{x}_k) + \nabla f_i(\boldsymbol{x}_k)^{\mathsf{T}}(\boldsymbol{\xi} - \boldsymbol{x}_k) + c_{k,i},$$

where $c_{k,i} = \frac{1}{2}(\boldsymbol{\xi} - \boldsymbol{x}_k)^{\mathsf{T}}\,\mathsf{H}_i(\boldsymbol{\eta}_{k,i})\,(\boldsymbol{\xi} - \boldsymbol{x}_k)$ and H_i is the Hessian matrix of f_i. To simplify notation in the proof, let us set $\boldsymbol{c}_k = [c_{k,1}, \ldots, c_{k,d}]^{\mathsf{T}}$, $\mathsf{J}_k = \mathsf{J}_{\boldsymbol{f}}(\boldsymbol{x}_k)$, and $\mathsf{H}^{(k,i)} = \mathsf{H}_i(\boldsymbol{\eta}_{k,i})$. Using the definition of Newton's method together with our Taylor expansion, we get

$$\nabla f_i(\boldsymbol{x}_k)^{\mathsf{T}}(\boldsymbol{x}_{k+1} - \boldsymbol{x}_k) = -f_i(\boldsymbol{x}_k) = \nabla f_i(\boldsymbol{x}_k)^{\mathsf{T}}(\boldsymbol{\xi} - \boldsymbol{x}_k) + c_{k,i},$$

which simplifies to

$$\boldsymbol{\xi} - \boldsymbol{x}_{k+1} = -\mathsf{J}_k^{-1}\boldsymbol{c}_k.$$

Using the Cauchy–Schwarz and other basic inequalities,

$$\begin{aligned}
\left\|\mathsf{J}_k^{-1}\mathbf{c}_k\right\|_2 &\le \left\|\mathsf{J}_k^{-1}\right\|_2 \|\mathbf{c}_k\|_2 \\
&\le \beta\sqrt{\sum_{i=1}^{d} c_{k,i}^2} \\
&= \frac{\beta}{2}\sqrt{\sum_{i=1}^{d}\left|(\boldsymbol{\xi}-\mathbf{x}_k)^\intercal \mathsf{H}^{(k,i)}(\boldsymbol{\xi}-\mathbf{x}_k)\right|^2} \\
&\le \frac{\beta}{2}\sqrt{\sum_{i=1}^{d}\|\boldsymbol{\xi}-\mathbf{x}_k\|_2^2\left\|\mathsf{H}^{(k,i)}(\boldsymbol{\xi}-\mathbf{x}_k)\right\|_2^2} \\
&= \frac{\beta}{2}\|\boldsymbol{\xi}-\mathbf{x}_k\|_2\sqrt{\sum_{i=1}^{d}\left\|\mathsf{H}^{(k,i)}(\boldsymbol{\xi}-\mathbf{x}_k)\right\|_2^2}.
\end{aligned}$$

Another application of Cauchy–Schwarz gives

$$\begin{aligned}
\left\|\mathsf{H}^{(k,i)}(\boldsymbol{\xi}-\mathbf{x}_k)\right\|_2^2 &= \sum_{j=1}^{d}\left|\sum_{m=1}^{d}\frac{\partial^2 \boldsymbol{f}_i}{\partial x_j\partial x_m}(\boldsymbol{\eta}_{k,i})(\xi_m - x_{k,m})\right|^2 \\
&\le \sum_{j=1}^{d}\left[\sum_{m=1}^{d}\left|\frac{\partial^2 \boldsymbol{f}_i}{\partial x_j\partial x_m}(\boldsymbol{\eta}_{k,i})\right|^2\|\boldsymbol{\xi}-\mathbf{x}_k\|_2^2\right] \\
&\le \|\boldsymbol{\xi}-\mathbf{x}_k\|_2^2\sum_{j=1}^{d}\left[\sum_{m=1}^{d}A^2\right] \\
&= \|\boldsymbol{\xi}-\mathbf{x}_k\|_2^2 A^2 d^2,
\end{aligned}$$

where A is an upper bound on the absolute values of the second derivatives of $\boldsymbol{f}$, which is available because $\boldsymbol{f} \in C^2\left(\overline{B}(\boldsymbol{\xi}, r); \mathbb{R}^d\right)$. We finally get the fundamental error estimate:

$$\|\boldsymbol{\xi}-\mathbf{x}_{k+1}\|_2 \le \frac{\beta A d^{3/2}}{2}\|\boldsymbol{\xi}-\mathbf{x}_k\|_2^2.$$

Therefore, if $\|\boldsymbol{\xi}-\mathbf{x}_0\|_2 \le \frac{1}{\beta A d^{3/2}} = h$, then $\|\boldsymbol{\xi}-\mathbf{x}_1\|_2 \le \frac{1}{2}\|\boldsymbol{\xi}-\mathbf{x}_0\|_2$. By induction, it follows that

$$\|\boldsymbol{\xi}-\mathbf{x}_k\|_2 \le h\left(\frac{1}{2}\right)^{2^k-1} = \varepsilon_k.$$

Thus, $\{\mathbf{x}_k\}_{k=0}^{\infty}$ is well defined, $\mathbf{x}_k \to \boldsymbol{\xi}$, and the order is at least quadratic. □

The next result, due to Kantorovich, is interesting in that it requires less regularity than what we have assumed in the previous results. Additionally, the existence of the zero point is not a required assumption, but is a consequence of the convergence. The following proof is similar to that in [86].

Theorem 15.37 (Kantorovich[5]). *Let $d \in \mathbb{N}$. Suppose that $\Omega \subset \mathbb{R}^d$ is an open, bounded, convex set, $\mathbf{x}_0 \in \Omega$, and $\boldsymbol{f} \in C^1\left(\overline{\Omega}; \mathbb{R}^d\right)$. Assume, additionally, with J_f denoting the Jacobian matrix of $\boldsymbol{f}$, that there is $\gamma > 0$ such that*

$$\|\mathsf{J}_f(\boldsymbol{x}) - \mathsf{J}_f(\boldsymbol{y})\|_2 \leq \gamma \|\boldsymbol{x} - \boldsymbol{y}\|_2, \quad \forall \boldsymbol{x}, \boldsymbol{y} \in \Omega.$$

Furthermore, let us assume the following.

a. *For all $\boldsymbol{x} \in \Omega$, the Jacobian matrix $\mathsf{J}_f(\boldsymbol{x})$ is invertible and there is $\beta > 0$ such that*

$$\left\|[\mathsf{J}_f(\boldsymbol{x})]^{-1}\right\|_2 \leq \beta, \quad \forall \boldsymbol{x} \in \Omega.$$

b. *The initial iterate, $\mathbf{x}_0 \in \Omega$, satisfies*

$$\left\|[\mathsf{J}_f(\mathbf{x}_0)]^{-1} \boldsymbol{f}(\mathbf{x}_0)\right\|_2 \leq \alpha.$$

c. *The parameters satisfy*

$$h = \frac{\alpha\beta\gamma}{2} < 1.$$

d. *The initial iterate is well inside Ω, in the sense that*

$$\overline{B}(\mathbf{x}_0, r) \subseteq \Omega,$$

where $r = \frac{\alpha}{1-h}$.

In this setting, the sequence $\mathbf{x}_k$ defined by Newton's method (15.16) *is well defined; in particular, $\mathbf{x}_k \in B(\mathbf{x}_0, r)$ for each $k \in \mathbb{N}$. Moreover, there exists a point $\boldsymbol{\xi} \in \overline{B}(\mathbf{x}_0, r)$ such that $\lim_{k\to\infty} \mathbf{x}_k = \boldsymbol{\xi}$, with the convergence estimate*

$$\|\mathbf{x}_k - \boldsymbol{\xi}\|_2 \leq \alpha \frac{h^{2^k - 1}}{1 - h^{2^k}}, \quad \forall k \in \mathbb{N}.$$

Since $0 < h < 1$, convergence is at least quadratic. Finally, the point $\boldsymbol{\xi}$ is a zero of the function $\boldsymbol{f}$, i.e., $\boldsymbol{f}(\boldsymbol{\xi}) = \mathbf{0}$.

Proof. We split the proof into several steps.

1. Since $[\mathsf{J}_f(\boldsymbol{x})]^{-1}$ exists for all $\boldsymbol{x} \in \Omega$, we will have that $\mathbf{x}_{k+1}$ is defined if $\mathbf{x}_k \in B(\mathbf{x}_0, r)$. Suppose that, for all $j = 0, 1, \ldots, k$, $\mathbf{x}_j \in B(\mathbf{x}_0, r)$. Then

$$\begin{aligned}
\|\mathbf{x}_{k+1} - \mathbf{x}_k\|_2 &= \left\|[\mathsf{J}_f(\mathbf{x}_k)]^{-1} \boldsymbol{f}(\mathbf{x}_k)\right\|_2 \\
&\leq \left\|[\mathsf{J}(\mathbf{x}_k)]^{-1}\right\|_2 \|\boldsymbol{f}(\mathbf{x}_k)\|_2 \\
&\leq \beta \|\boldsymbol{f}(\mathbf{x}_k)\|_2 \\
&= \beta \|\boldsymbol{f}(\mathbf{x}_k) - \boldsymbol{f}(\mathbf{x}_{k-1}) - \mathsf{J}(\mathbf{x}_{k-1})(\mathbf{x}_k - \mathbf{x}_{k-1})\|_2 \\
&\leq \frac{\beta\gamma}{2} \|\mathbf{x}_k - \mathbf{x}_{k-1}\|_2^2,
\end{aligned} \tag{15.17}$$

[5] Named in honor of the Soviet mathematician Leonid Vitalyevich Kantorovich (1912–1986).

using the result of Theorem B.56 in the last step. We claim that (15.17) implies that, for all $k \geq 0$,

$$\|\mathbf{x}_{k+1} - \mathbf{x}_k\|_2 \leq \alpha h^{2^k - 1}. \tag{15.18}$$

The proof is by induction. The case $k = 0$ holds because of assumption b:

$$\|\mathbf{x}_1 - \mathbf{x}_0\|_2 = \left\| [\mathsf{J}_f(\mathbf{x}_0)]^{-1} \boldsymbol{f}(\mathbf{x}_0) \right\|_2 \leq \alpha.$$

For the induction step, we suppose that (15.18) is valid for $k = j - 1$:

$$\|\mathbf{x}_j - \mathbf{x}_{j-1}\|_2 \leq \alpha h^{2^{j-1} - 1}.$$

Let $k = j$ now. Using (15.17) and the induction hypothesis,

$$\begin{aligned}\|\mathbf{x}_{j+1} - \mathbf{x}_j\|_2 &\leq \frac{\beta\gamma}{2} \|\mathbf{x}_j - \mathbf{x}_{j-1}\|_2^2 \leq \frac{\beta\gamma}{2} \alpha^2 \left(h^{2^{j-1}-1} \right)^2 = \frac{\beta\gamma}{2} \alpha^2 h^{2^j - 2} \\ &= \frac{\alpha\beta\gamma}{2} \alpha h^{2^j - 2} = \alpha h^{2^j - 1}.\end{aligned}$$

Hence, estimate (15.18) follows by induction.

Now, by the triangle inequality,

$$\begin{aligned}\|\mathbf{x}_{k+1} - \mathbf{x}_0\|_2 &\leq \|\mathbf{x}_{k+1} - \mathbf{x}_k\|_2 + \cdots + \|\mathbf{x}_1 - \mathbf{x}_0\|_2 \\ &\leq \alpha \left(1 + h + h^3 + h^7 + \cdots + h^{2^k - 1} \right) \\ &< \alpha \left(1 + h + h^2 + \cdots \right) \\ &= \frac{\alpha}{1 - h} \\ &= r.\end{aligned}$$

Thus, $\mathbf{x}_{k+1} \in B(\mathbf{x}_0, r)$. By induction, $\mathbf{x}_k \in B(\mathbf{x}_0, r)$ for all $k \in \mathbb{N}$.

2. Using (15.18), we can prove that $\{\mathbf{x}_k\}_{k=0}^{\infty}$ is a Cauchy sequence. Suppose that $m > n \geq 0$. Then

$$\begin{aligned}\|\mathbf{x}_m - \mathbf{x}_n\|_2 &\leq \|\mathbf{x}_m - \mathbf{x}_{m-1}\|_2 + \cdots + \|\mathbf{x}_{n+1} - \mathbf{x}_n\|_2 \\ &\leq \alpha h^{2^n - 1} \left(1 + h^{2^n} + h^{3 \cdot 2^n} + h^{5 \cdot 2^n} + \cdots \right) \\ &< \frac{\alpha h^{2^n - 1}}{1 - h^{2^n}} \\ &< \varepsilon,\end{aligned} \tag{15.19}$$

provided that n is sufficiently large. Since $\mathbf{x}_k$ is Cauchy, it converges to a unique limit point $\boldsymbol{\xi} \in \overline{B}(\mathbf{x}_0, r)$, appealing to Theorem B.8 and the fact that $\overline{B}(\mathbf{x}_0, r)$ is closed. It follows on taking $m \to \infty$ in (15.19) that

$$\|\boldsymbol{\xi} - \mathbf{x}_n\|_2 < \frac{\alpha h^{2^n - 1}}{1 - h^{2^n}}.$$

From this estimate, it follows that convergence is at least quadratic.

3. Finally, we prove that $\boldsymbol{f}(\boldsymbol{\xi}) = \mathbf{0}$. Since $\mathbf{x}_k \in B(\mathbf{x}_0, r)$,

$$\|\mathsf{J}_f(\mathbf{x}_k) - \mathsf{J}_f(\mathbf{x}_0)\|_2 \leq \gamma \|\mathbf{x}_0 - \mathbf{x}_k\|_2 \leq \gamma r.$$

Thus,

$$\|J_f(x_k)\|_2 = \|J_f(x_k) - J_f(x_0) + J_f(x_0)\|_2 \le \gamma r + \|J_f(x_0)\|_2 = R.$$

As a consequence,

$$\|f(x_k)\|_2 = \|-J_f(x_k)(x_{k+1} - x_k)\|_2 \le R\,\|x_{k+1} - x_k\|_2,$$

which implies that $\lim_{k\to\infty}\|f(x_k)\|_2 = 0$. It follows that $f(\xi) = \mathbf{0}$.

The proof is complete. □

Problems

15.1 Suppose that, for every $k \ge 0$,

$$|\xi - x_k| \le \alpha \frac{h^{2^k - 1}}{1 - h^{2^k}},$$

where $\alpha > 0$ and $0 < h < 1$. Show that the sequence $\{x_k\}_{k=0}^{\infty}$ converges to ξ at least quadratically.

15.2 Complete the proof of Theorem 15.6.

15.3 Let

$$f(x) = e^x - 2x - 1, \quad a = 1, \quad b = 2.$$

Show that the false position method in this setting will converge to a root of f. Show, in addition, that, for all k, we will have $b_k = 2$.

15.4 Can you generalize the example of the previous problem? In other words, let $-\infty < a < b < \infty$ and $f \in C^2([a,b])$ with $f(a)f(b) < 0$. Can you provide sufficient conditions on f, f', and f'', so that $b_k = b$ for all k?

15.5 Prove Proposition 15.11.

15.6 Complete the proof of Theorem 15.13.

15.7 Complete the proof of Theorem 15.15.

15.8 Prove Theorem 15.17.

15.9 Prove Proposition 15.19.

15.10 Let $\{x_k\}_{k=0}^{\infty}$ be the sequence generated, for some $g \in C([a,b])$, by the fixed point iteration scheme. In the setting of Theorem 15.12, assume, in addition, that $g'(x) < 0$ for all $x \in [a,b]$ and that $x_0 < \xi$, where ξ is the (unique) fixed point of g. Show that, for any $k \ge 0$,

$$x_{2k} < \xi < x_{2k+1}.$$

15.11 Let $a \in \mathbb{R}$ and, for some $r > 0$, $I = [a - r, a + r]$ be an interval. Let $g \in C(I)$ be such that there is $q \in (0,1)$ for which

$$|g'(x)| \le q, \quad \forall x \in I,$$

and

$$|g(a) - a| \le (1 - q)r.$$

Show that g has a unique fixed point $\xi \in I$ and that the fixed point iteration scheme converges for any starting value x_0. Moreover,

$$|x_k - \xi| \le q^k |x_0 - \xi|.$$

15.12 Consider the relaxation method (15.3). Show that if $f\colon \mathbb{R} \to \mathbb{R}$ is such that $f'(x) < 0$ and $|f'(x)| \in [m, M] \subset (0, \infty)$ for all $x \in \mathbb{R}$, then the choice

$$\lambda = \frac{2}{m+M}$$

is optimal for the relaxation parameter.

15.13 Let $\boldsymbol{f}\colon \mathbb{R}^d \to \mathbb{R}^d$ be given by

$$\boldsymbol{f}(\boldsymbol{x}) = \mathsf{A}\boldsymbol{x} + \boldsymbol{g}(\boldsymbol{x}),$$

where $\mathsf{A} \in \mathbb{R}^{d\times d}$ is invertible and $\boldsymbol{g} \in C^1(\mathbb{R}^d; \mathbb{R}^d)$. Assume that there is $\boldsymbol{\xi} \in \mathbb{R}^d$ for which $\boldsymbol{f}(\boldsymbol{\xi}) = \boldsymbol{0}$. To approximate it, consider the following *Picard-like* iteration method: given $\boldsymbol{x}_0 \in \mathbb{R}^d$, find $\boldsymbol{x}_{k+1}$, for $k \geq 0$, via

$$\mathsf{A}\boldsymbol{x}_{k+1} + \boldsymbol{g}(\boldsymbol{x}_k) = \boldsymbol{0}.$$

Provide sufficient conditions for the convergence of this approach.
Hint: Find an expression for the error $\boldsymbol{e}_k = \boldsymbol{\xi} - \boldsymbol{x}_k$. Then, use a version of the Mean Value Theorem in multiple dimensions.

15.14 Assume that $f \in C^2(\mathbb{R})$ with $f'(x) > 0$ and $f''(x) > 0$ for all $x \in \mathbb{R}$.

a) Exhibit a function that satisfies these assumptions, but has no root.
b) Show that, if a root $\xi \in \mathbb{R}$ exists, it is unique.
c) Prove that, for any starting guess $x_0 \in \mathbb{R}$, Newton's method converges and the convergence is quadratic.

15.15 Show that (15.26) will converge, for any $x_0 > 0$, to $\sqrt{5}$.

15.16 Define $f\colon [1, 5] \to \mathbb{R}$ via $f(x) = (x - 3)^3$. Use Newton's method with the starting point $x_0 = 4$ to approximate the root $\xi = 3$. Show directly that convergence is linear.

Hint: Show that, for $k \geq 0$,

$$x_k = 3 + \left(\frac{2}{3}\right)^k.$$

15.17 Suppose that $f \in C^2(I)$, where I is an interval. Let $\xi \in I$ be such that $f(\xi) = f'(\xi) = 0$, but $f''(\xi) \neq 0$.

a) Show that the sequence $\{x_k\}_{k=0}^{\infty}$ defined by Newton's method satisfies the relation

$$\xi - x_{k+1} = -\frac{1}{2}\frac{(\xi - x_k)^2 f''(\eta^k)}{f'(x_k)} = \frac{1}{2}(\xi - x_k)\frac{f''(\eta_k)}{f''(\chi_k)},$$

where η_k and χ_k lie between ξ and x_k.

b) Suppose that $0 < m \leq |f''(x)| \leq M$ for all $x \in [\xi - \delta, \xi + \delta] \subset I$ for some $\delta > 0$, where $0 < M < 2m$. Prove that if $x_0 \in [\xi - \delta, \xi + \delta]$, then $x_k \to \xi$.

15.18 Complete the proof of Theorem 15.26.

15.19 Suppose that $f \in C^2(\mathbb{R})$ with f'' Lipschitz continuous and $f(\zeta) = f'(\zeta) = 0$, but $f''(\zeta) \neq 0$.

a) Prove that the iterative method

$$x_{k+1} = x_k - 2\frac{f(x_k)}{f'(x_k)}$$

converges at least quadratically to ζ provided that x_0 is sufficiently near, but not equal, to ζ.

b) Can one extend the last result for the case

$$f(\zeta) = f'(\zeta) = f''(\zeta) = 0, \quad \text{but} \quad f'''(\zeta) \neq 0?$$

What method, if any, would still give quadratic convergence?

15.20 Use the secant method to show that the sequence, whose recursive definition is given below, converges to $\sqrt{Q}$, where $Q > 0$, given "good" starting values x_0 and x_1:

$$x_{k+1} = \frac{x_k x_{k-1} + Q}{x_k + x_{k-1}}.$$

Come up with a similar recursion for approximating $Q^{1/3}$ using the secant method.

15.21 Prove Theorem 15.34. To do so, proceed as follows.

a) Prove the iterations are well defined and converge.

b) Show that the secant method may be written in the equivalent form,

$$x_{k+1} = \frac{x_k f(x_{k-1}) - x_{k-1} f(x_k)}{f(x_{k-1}) - f(x_k)}, \quad k \geq 1.$$

c) Define

$$\phi(x_k, x_{k-1}) = \frac{x_{k+1} - \xi}{(x_k - \xi)(x_{k-1} - \xi)},$$

where x_{k+1} is expressed in terms of x_k and x_{k-1} through the recursive formula above. Find an expression for

$$\psi(x_{k-1}) = \lim_{x_k \to \xi} \phi(x_k, x_{k-1})$$

and then determine the value of

$$\lim_{x_{k-1} \to \xi} \psi(x_{k-1}).$$

d) Deduce that

$$\lim_{x_k, x_{k-1} \to \xi} \phi(x_k, x_{k-1}) = \frac{f''(\xi)}{2f'(\xi)}.$$

e) Next, suppose that

$$\lim_{k \to \infty} \frac{|x_{k+1} - \xi|}{|x_k - \xi|^q} = A > 0.$$

Prove that it must be that $q - 1 - 1/q = 0$ and, therefore, $q = (1 + \sqrt{5})/2$.

f) Finally, deduce that

$$\lim_{k \to \infty} \frac{|x_{k+1} - \xi|}{|x_k - \xi|^q} = \left| \frac{f''(\xi)}{2f'(\xi)} \right|^{q/(1+q)}.$$

15.22 Consider the nonlinear equation $\exp(x) = \sin(x)$.

a) With pencil and paper only, argue that there is one, and only one, solution $\xi \in \left(-\frac{3}{2}\pi, -\pi\right)$.

b) Show, in fact, that there is one, and only one, solution $\xi \in \left(-\frac{5}{4}\pi, -\pi\right)$.

c) Consider the following iterative methods:

$$x_{k+1} = \ln(\sin(x_k))$$

and

$$x_{k+1} = \sin^{-1}(\exp(x_k)),$$

where the inverse of the sine function is appropriately, and carefully, defined. What can you say about the local convergence of each of these methods to ξ and their convergence orders?

d) For estimating ξ, provide a method that is quadratically convergent. Will the method converge for any starting value in the interval $x_0 \in \left(-\frac{5}{4}\pi, -\pi\right)$? Why or why not?

15.23 In this method, we will explore a variant of the simplified Newton method. Let $f \in C(\mathbb{R})$. Let $\xi \in \mathbb{R}$ be such that $f(\xi) = 0$ and $f'(\xi) > 0$. Show that there is an $\varepsilon > 0$ such that if $|x_0 - \xi| < \varepsilon$, then the following iteration converges to ξ:

$$x_{k+1} = x_k - \frac{f(x_k)}{f'(x_m)},$$

where $m \leq k$ is chosen such that $|f'(x_m)| = \max_{j \leq k} |f'(x_j)|$.

15.24 In this problem, we will construct iterative schemes to find a solution of $f(\xi) = 0$ that converges with orders $q = 2$ and $q = 3$, respectively. Assume that $f(\xi) = 0$ can be rewritten as the fixed point of g, i.e., $\xi = g(\xi)$, so that we consider the iterative scheme

$$x_{k+1} = g(x_k).$$

a) Define the error $e_k = \xi - x_k$. Show that

$$|g(\xi - e_{k-1}) - g'(\xi)e_{k-1}| \leq C|e_{k-1}|^2$$

for some constant $C > 0$.

b) Show that if $g'(\xi) \neq 0$, then the method converges linearly.

c) Show that if $g'(\xi) = 0$, but $g''(\xi) \neq 0$, the method converges quadratically.

d) Consider, for a_1 and a_2, nonvanishing functions

$$g(x) = x + a_1(x)f(x) + a_2(x)[f(x)]^2. \tag{15.20}$$

Show that $x = g(x)$ if and only if $f(x) = 0$.

e) Let g be given by (15.20). Evaluate the first and second derivatives of g with respect to x. From them, show that if $q = 2$ (the method converges quadratically), then we obtain Newton's method.

f) Let g be given by (15.20). Show that if the method converges cubically, i.e., $q = 3$, then

$$a_1(x) = -\frac{1}{f'(x)}, \qquad a_2(x) = -\frac{f''(x)}{2[f'(x)]^3}.$$

Unfortunately, it turns out that this method is unstable. It is only used to accelerate the convergence once a good guess is already available.

15.25 Let $\boldsymbol{f}\colon \Omega \subset \mathbb{R}^n \to \mathbb{R}^n$ be twice continuously differentiable. Suppose that $\boldsymbol{\xi} \in \Omega$ is a solution of $\boldsymbol{f}(\boldsymbol{x}) = \boldsymbol{0}$ and the Jacobian matrix of $\boldsymbol{f}$, denoted $\mathsf{J}_{\boldsymbol{f}}$, is invertible at $\boldsymbol{\xi}$. Prove that if $\boldsymbol{x}_0 \in \Omega$ is sufficiently close to $\boldsymbol{\xi}$, then the following iteration converges to $\boldsymbol{\xi}$:

$$\boldsymbol{x}_{k+1} = \boldsymbol{x}_k - \mathsf{J}_{\boldsymbol{f}}\left(\boldsymbol{x}_0\right)^{-1}\boldsymbol{f}(\boldsymbol{x}_k).$$

15.26 Suppose that the function $\boldsymbol{f}\colon \mathbb{R}^2 \to \mathbb{R}^2$ is defined via

$$\boldsymbol{f}(x_1, x_2) = \begin{bmatrix} 16 - x_1^2 - x_2^2 \\ x_1^2 - 1 \end{bmatrix}.$$

How many real-valued (vector) solutions does the system $\boldsymbol{f}(\boldsymbol{x}) = \boldsymbol{0}$ have? Use Newton's method to obtain the approximations $\boldsymbol{x}_1$ and $\boldsymbol{x}_2$ when $\boldsymbol{x}_0 = [1, 1]^\intercal$.

15.27 Suppose that the function $\mathbf{f}\colon \mathbb{R}^2 \to \mathbb{R}^2$ is defined via

$$\boldsymbol{f}(x_1, x_2) = \begin{bmatrix} x_1^2 - 2x_1 + x_2^2 \\ x_1^2 + x_2^2 - 1 \end{bmatrix}.$$

How many real-valued (vector) solutions does the system $\boldsymbol{f}(\boldsymbol{x}) = \boldsymbol{0}$ have? Use Newton's method to obtain the approximations $\boldsymbol{x}_1$ and $\boldsymbol{x}_2$ when $\boldsymbol{x}_0 = [0, -1]^\intercal$.

15.28 By $B(\boldsymbol{x}, r) \subset \mathbb{R}^2$, denote the open ball of radius $r > 0$ centered at $\boldsymbol{x}$. Suppose that, for some $r > 0$, $f, g\colon B(\boldsymbol{\xi}, r) \to \mathbb{R}$ are nonlinear, twice continuously differentiable functions with

$$f(\boldsymbol{\xi}) = 0, \quad g(\boldsymbol{\xi}) = 0.$$

Consider the Gauss–Seidel-like iterative scheme: given

$$\boldsymbol{x}_k = [x_{1,k}, x_{2,k}]^\intercal \in B(\boldsymbol{\xi}, r),$$

find $\boldsymbol{x}_{k+1} = [x_{1,k+1}, x_{2,k+1}]^\intercal \in \mathbb{R}^2$ such that

$$f(x_{1,k+1}, x_{2,k}) = 0, \quad g(x_{1,k+1}, x_{2,k+1}) = 0.$$

a) Let $\boldsymbol{e}_k = \boldsymbol{\xi} - \boldsymbol{x}_k$ be the error.

b) Establish an iteration error equation of the form

$$\begin{bmatrix} \frac{\partial f}{\partial x_1}(\boldsymbol{\xi}) & \frac{\partial f}{\partial x_2}(\boldsymbol{\xi}) \\ \frac{\partial g}{\partial x_1}(\boldsymbol{\xi}) & \frac{\partial g}{\partial x_2}(\boldsymbol{\xi}) \end{bmatrix} \boldsymbol{e}_{k+1} - \begin{bmatrix} \frac{\partial f}{\partial x_1}(\boldsymbol{\xi})e_{1,k+1} + \frac{\partial f}{\partial x_2}(\boldsymbol{\xi})e_{2,k+1} \\ \frac{\partial g}{\partial x_1}(\boldsymbol{\xi})e_{1,k+1} + \frac{\partial g}{\partial x_2}(\boldsymbol{\xi})e_{2,k+1} \end{bmatrix} = \boldsymbol{r}_{k+1}.$$

Give a precise expression for the remainder term, $\boldsymbol{r}_{k+1}$.

c) Give sufficient conditions for the convergence of the scheme.

15.29 Let $\boldsymbol{f}: \mathbb{R}^2 \to \mathbb{R}^2$ be defined via

$$\boldsymbol{f}(x_1, x_2) = \begin{bmatrix} x_1^2 - 2x_1 + x_2 \\ 2x_1 - x_2^2 - 1 \end{bmatrix}.$$

Observe that $\boldsymbol{f}$ has the zero

$$\boldsymbol{\xi} = \begin{bmatrix} 1 \\ 1 \end{bmatrix}.$$

Consider the iteration

$$\boldsymbol{x}_{n+1} = \boldsymbol{x}_n - \mathsf{A}\boldsymbol{f}(\boldsymbol{x}_n), \qquad \mathsf{A} = \begin{bmatrix} 1 & 1/2 \\ 1 & 0 \end{bmatrix}. \tag{15.21}$$

a) Prove that $\boldsymbol{x}_n \to \boldsymbol{\xi}$, provided that $\boldsymbol{x}_0$ is sufficiently close to $\boldsymbol{\xi}$.
b) Show that the convergence is at least quadratic.
c) Is the iteration (15.21) equivalent to Newton's method?

Listings

```
function [root,count,err] = NewtonRoot(xin, p, q, tol, maxits)
%
% This function calculates the pth root of q ,
%
%   q^(1/p)
%
% using Newton's method. For simplicity, we assume that p is a
% positive integer and q is a positive real number
%
% Input:
%   xin : initial guess
%   p : the positive integer degree of the root
%   q : the positive number whose pth root is to estimated
%   tol : stopping tolerance
%   maxits : the maximal number of iterations
%
% Output:
%   root : approximation of the root
%   count : number of newton iterations required to compute the
%           root
%   err: = 0, if the algorithm proceeded to completion
%         = 1, if an error was encountered
%
  root = NaN;
  count = 0;
  err = 0;
  if int32(p) ~= p || p < 0
    disp('Error: p must be a positive integer');
    err = 1;
  return
  end
  if q < 0
```

```
    disp('Error: q must be positive');
    err = 1;
  return
  end
  diff = 1.0;
  x = xin;
  while diff > tol && count < maxits
    xo = x;
    x = xo - fn(xo,p,q)/dfn(xo,p,q);
    diff = abs(x-xo);
    count = count+1;
  end
  if count >= maxits
    err = 1;
  end
  root = x;
end

function y = fn(x,p,q)
  y = x^p-q;
end

function y = dfn(x,p,q)
  y = p*x^(p-1);
end
```

Listing 15.1 Newton's method for computing $\sqrt[p]{q}$.

16 Convex Optimization

Given an energy (or cost function, or objective function) E defined on a real Hilbert space $\mathcal{H}$, we consider the following *minimization (or optimization) problem*:

$$u = \operatorname*{argmin}_{v \in \mathcal{H}} E(v). \tag{16.1}$$

For mathematicians, the first thoughts must always be these: *Does this problem have a solution? Does the problem make sense as it is written?*

Such problems, called *unconstrained optimization problems*, are ubiquitous in science and engineering. For example, many physical problems have as their solution the minimization of some energy. Nature works by finding the minimum energy path through space and time. In other applications, E may be the cost to manufacture an item. To make the most profit, we will try to minimize the production cost.

In this chapter, we will work with the simplest version of this problem, by assuming that E is strongly convex, with some other nice properties as well. With our assumptions about E, we will be guaranteed that a unique minimizer always exists, *an important first step for a numerical analyst who wants to build an algorithm to approximate the minimizer.* In many real-world problems, much less may be known or assumed about E, and things can get messy, beyond the scope of our simple introduction.

A word of warning. This chapter may be challenging for readers not familiar with the beautiful theory of functional analysis and calculus on infinite-dimensional vector spaces. In fact, this chapter may serve as an introduction to the subject for many. Be prepared to read and think carefully. We use this theory not for the sake of pure abstraction, but because it is powerful, it will quickly lead us to deep insights about this problem, and it will allow us to cover many practical applications at once. For further reading about functional analysis and differential calculus in infinite-dimensional vector spaces, we refer the reader to [5, 18]. See [14] for a similar treatment of convex optimization.

16.1 Some Tools from Functional Analysis

Throughout this chapter, $\mathcal{H}$ will be a real Hilbert space, finite or infinite dimensional, which is nothing but an inner product space, in the sense of Section A.4, with an additional property that we now define.

Definition 16.1 (completeness). Let $\mathbb{V}$ be a normed vector space with norm $\|\cdot\|_{\mathbb{V}}$. We say that this space is **complete** if every Cauchy sequence converges. In other words, if $\{x_n\}_{n\in\mathbb{N}} \subset \mathbb{V}$ is a sequence that satisfies: for every $\varepsilon > 0$, there is $N \in \mathbb{N}$ such that for all $m, n \geq N$, we have

$$\|x_m - x_n\|_{\mathbb{V}} < \varepsilon,$$

then this sequence must **converge**, meaning that there is $x \in \mathbb{V}$ such that

$$\|x_n - x\|_{\mathbb{V}} \to 0$$

as $n \to \infty$.

Notice that, in this definition, we combine two, at first glance disparate, notions: *linear algebra* (through the notion of vector space) and *analysis* (through the notion of convergence). This is a recurring feature in functional analysis.

Now we are ready to define Hilbert spaces.

Definition 16.2 (Hilbert space[1]). A **Hilbert space** is an inner product space $\mathcal{H}$ with inner product $(\cdot,\cdot)_{\mathcal{H}}$ that is complete under the canonical norm $\|\cdot\|_{\mathcal{H}}$, i.e.,

$$\|v\|_{\mathcal{H}} = \sqrt{(v,v)_{\mathcal{H}}}, \quad \forall v \in \mathcal{H}.$$

We have already dealt with several examples of finite-dimensional Hilbert spaces.

Example 16.1 For $n \in \mathbb{N}$, $\mathbb{R}^n$ with the $(\cdot,\cdot)_2$ inner product is a Hilbert space.

Example 16.2 For $n \in \mathbb{N}$, $\mathbb{P}_n$ with the inner product

$$(p, q)_{L^2(-1,1)} = \int_0^1 p(x)q(x)\mathrm{d}x, \quad \forall p, q \in \mathbb{P}_n$$

is a Hilbert space.

Example 16.3 Let us now show an example of an inner product space that is not complete and, hence, is not a Hilbert space. Consider the vector space of continuous real-valued functions on $[-1,1]$ and define, on $C([-1,1])$, the inner product

$$(f, g)_{L^2(-1,1)} = \int_{-1}^1 f(x)g(x)\mathrm{d}x, \quad \forall f, g \in C([-1,1]).$$

It turns out that this is not a complete space. To see this, consider the following sequence:

$$f_n(x) = \begin{cases} 1, & x \leq 0, \\ 1 - nx, & x \in \left(0, \dfrac{1}{n}\right), \\ 0, & x \in \left[\dfrac{1}{n}, 1\right]. \end{cases}$$

[1] Named in honor of the German mathematician David Hilbert (1862–1943).

We now show that this sequence is Cauchy. Choose $\varepsilon > 0$. Notice that $0 \leq f_n(x) \leq 1$ for all $x \in [-1, 1]$. Thus, provided that $N > \frac{4}{\varepsilon^2}$, we have for any $m, n \geq N$,

$$\|f_n - f_m\|_{L^2(-1,1)} \leq \|f_n\|_{L^2(-1,1)} + \|f_m\|_{L^2(-1,1)} \leq 2\left(\int_0^{1/N} 1\mathrm{d}x\right)^{1/2} = \frac{2}{\sqrt{N}} < \varepsilon.$$

However, this sequence does not converge to a continuous function, as the only possible limit is

$$f_n(x) = \begin{cases} 1, & x \leq 0, \\ 0, & x \in [0, 1], \end{cases} \notin C([-1, 1]).$$

Notice that what made the difference in the previous two examples was the *dimension* of the space. While $\dim \mathbb{P}_n = n + 1$, we have that $C([-1, 1])$ is infinite dimensional. To see this, observe that $\mathbb{P} \leq C([-1, 1])$, where $\mathbb{P}$ is introduced in Example A.5.

Before moving forward, we quickly comment that complete normed spaces without an inner product can also be studied. These are called *Banach spaces.*[2] However, discussing them would stray too far from where we want to go. For this reason, for the remainder of this chapter, we assume that $\mathcal{H}$ is a real Hilbert space.

Because we are working in a fairly general setting — in particular, we are not assuming that $\mathcal{H}$ is finite dimensional — we will need some basic tools and facts from functional analysis.

The first one comes from realizing that the proof of Proposition A.31 never assumed that the inner product space was either finite dimensional or complete. Thus, the Cauchy–Schwarz inequality,

$$|(u, v)_{\mathcal{H}}| \leq \|u\|_{\mathcal{H}} \|v\|_{\mathcal{H}},$$

holds in a Hilbert space. Another identity, which shall prove useful in the sequel, is

$$2(u, v)_{\mathcal{H}} = \|u\|_{\mathcal{H}}^2 - \|v - u\|_{\mathcal{H}}^2 + \|v\|_{\mathcal{H}}^2, \quad \forall u, v \in \mathcal{H}; \tag{16.2}$$

see Problem 16.1.

With the aid of a norm, we can introduce the notions of convergence and continuity. In infinite-dimensional Hilbert spaces, there are multiple notions of convergence that are important.

Definition 16.3 (modes of convergence)**.** Let $\{u_n\}_{n=1}^\infty$ be a sequence in the Hilbert space $\mathcal{H}$. We say that $\{u_n\}_{n=1}^\infty$ is **bounded** if and only if there is a constant $C > 0$ such that

$$\|u_n\|_{\mathcal{H}} \leq C$$

for all $n \geq 1$. We say that $\{u_n\}_{n=1}^\infty$ **(strongly) converges** if and only if there is an element $u \in \mathcal{H}$ such that

$$\|u_n - u\|_{\mathcal{H}} \to 0, \qquad n \to \infty,$$

[2] Named in honor of the Polish mathematician Stefan Banach (1892–1945).

and we write $u_n \to u$. We say that $\{u_n\}_{n=1}^{\infty}$ **converges weakly** if and only if there is an element $u \in \mathcal{H}$ such that

$$(u_n, w)_{\mathcal{H}} \to (u, w)_{\mathcal{H}}, \quad \forall w \in \mathcal{H},$$

and we write $u_n \rightharpoonup u$.

The following result shows that weak convergence is indeed a weaker notion.

Proposition 16.4 (strong $\implies$ weak). *Suppose that $\{u_n\}_{n=1}^{\infty}$ is a sequence in the Hilbert space $\mathcal{H}$. Then $u_n \to u$ implies that $u_n \rightharpoonup u$. The converse is not necessarily true.*

Proof. See Problem 16.2. □

However, there are cases when from weak convergence, strong can be implied.

Theorem 16.5 (weak $\implies$ strong). *Suppose that $\{u_n\}_{n=1}^{\infty}$ is a sequence in the Hilbert space $\mathcal{H}$. If $u_n \rightharpoonup u$ then $\{u_n\}_{n=1}^{\infty}$ is bounded. Furthermore, if $u_n \rightharpoonup u$ and $\|u_n\|_{\mathcal{H}} \to \|u\|_{\mathcal{H}}$, then $u_n \to u$.*

Proof. The first result requires tools beyond the scope of our introduction. The second result follows from the following polarization identity:

$$\|u - u_n\|_{\mathcal{H}}^2 = 2(u - u_n, u)_{\mathcal{H}} + \|u_n\|_{\mathcal{H}}^2 - \|u\|_{\mathcal{H}}^2 .$$

□

In finite dimensions, the Heine–Borel Theorem, see Corollary B.15, ensures that every closed and bounded set is compact. But, in infinite dimensions, closed and bounded sets need not be compact. To further muddy the waters, as there are several notions of convergence, there are multiple notions on closedness.

Definition 16.6 (closed). Suppose that K is a subset of the Hilbert space $\mathcal{H}$. We say that K is **closed** if and only if whenever $\{u_n\}_{n=1}^{\infty} \subset K$ is such that $u_n \to u$, then we must have $u \in K$. We say that K is **weakly closed** if and only if whenever $\{u_n\}_{n=1}^{\infty} \subset K$ is such that $u_n \rightharpoonup u$, then this implies that $u \in K$.

Definition 16.7 (closure). Suppose that K is a subset of the Hilbert space $\mathcal{H}$. Define

$$L_s = \{u \in \mathcal{H} \mid \exists \{u_n\}_{n=1}^{\infty} \subset K \colon u_n \to u\}$$

and

$$L_w = \{u \in \mathcal{H} \mid \exists \{u_n\}_{n=1}^{\infty} \subset K \colon u_n \rightharpoonup u\} .$$

The set

$$\operatorname{clo}_s(K) = K \cup L_s$$

is called the **closure of** K and the set

$$\operatorname{clo}_w(K) = K \cup L_w$$

is called the **weak closure of** K.

These are the smallest closed and weakly closed sets, respectively, that contain K, as you would expect.

Next, let us remind the reader about subsequences, as these play an important role in the concept of compactness and the coming new concept of weak compactness.

Definition 16.8 (subsequence). Suppose that $\{n_k\}_{k=1}^{\infty} \subseteq \mathbb{N}$ is a strictly increasing sequence of numbers and $\{u_n\}_{n=1}^{\infty} \subset \mathcal{H}$. The sequence $\{u_{n_k}\}_{k=1}^{\infty}$ is called a **subsequence** of $\{u_n\}_{n=1}^{\infty}$ and we write $\{u_{n_k}\}_{k=1}^{\infty} \subseteq \{u_n\}_{n=1}^{\infty}$.

Definition 16.9 (sequential compactness). We say that a set $K \subset \mathcal{H}$ is **sequentially compact** if and only if every sequence $\{u_n\}_{n=1}^{\infty}$ has a subsequence $\{u_{n_k}\}_{k=1}^{\infty} \subseteq \{u_n\}_{n=1}^{\infty}$ that converges to a point $u \in K$. $K \subset \mathcal{H}$ is called **weakly sequentially compact** if and only if every sequence $\{u_n\}_{n=1}^{\infty}$ has a subsequence $\{u_{n_k}\}_{k=1}^{\infty} \subseteq \{u_n\}_{n=1}^{\infty}$ that converges weakly to a point $u \in K$.

Now we can state the so-called weak compactness property, a fundamental attribute of Hilbert spaces. The proof of this result is beyond the scope of this text.

Theorem 16.10 (weak compactness). *Suppose that $\mathcal{H}$ is a Hilbert space. Then every bounded sequence $\{u_n\}_{n=1}^{\infty} \subset \mathcal{H}$ has a weakly convergent subsequence, i.e., there is an element $u \in \mathcal{H}$ and a sequence of numbers $\{n_k\}_{k=1}^{\infty} \subset \mathbb{N}$ such that $u_{n_k} \rightharpoonup u$ as $k \to \infty$.*

16.1.1 The Dual Space

Functional analysis gets its name from the following definition.

Definition 16.11 (linear functional). A function $f\colon \mathcal{H} \to \mathbb{R}$ is called a **bounded linear functional** if and only if it is:

1. Linear, i.e.,

$$f(\alpha v + \beta w) = \alpha f(v) + \beta f(w), \quad \forall \alpha, \beta \in \mathbb{R}, \quad \forall v, w \in \mathcal{H}.$$

2. Bounded, i.e., there is some constant $C > 0$ such that for all $w \in \mathcal{H}$,

$$|f(w)| \leq C \|w\|_{\mathcal{H}}.$$

The set of all bounded linear functionals on $\mathcal{H}$ is called the **dual space** and is denoted $\mathcal{H}'$. Under point-wise addition and scalar multiplication this is a vector space.

The space of bounded linear functionals can be *normed*, as follows.

Definition 16.12 (operator norm). Suppose that $f \in \mathcal{H}'$. The **operator norm of** f is defined as

$$\|f\|_{\mathcal{H}'} = \sup_{\substack{v \in \mathcal{H} \\ \|v\|_{\mathcal{H}}=1}} |f(v)| = \sup_{v \in \mathcal{H}\setminus\{0\}} \frac{|f(v)|}{\|v\|_{\mathcal{H}}}. \tag{16.3}$$

Proposition 16.13 (completeness). *The function $\|\cdot\|_{\mathcal{H}'}\colon \mathcal{H}' \to \mathbb{R}$ defines a norm on $\mathcal{H}'$ and, with this norm, the dual space $\mathcal{H}'$ of a Hilbert space $\mathcal{H}$ is in fact a Banach space. Furthermore, if $f \in \mathcal{H}'$, then*

$$|f(v)| \le \|f\|_{\mathcal{H}'} \|v\|_{\mathcal{H}}, \quad \forall v \in \mathcal{H}.$$

Proof. The proof of the fact that $\|\cdot\|_{\mathcal{H}'}$ is indeed a norm and the estimate on $|f(v)|$ are left to the reader as an exercise; see Problem 16.3.

To show completeness, let $\{f_n\}_{n\in\mathbb{N}} \subset \mathcal{H}'$ be Cauchy, i.e., for every $\varepsilon > 0$, there is $N \in \mathbb{N}$ such that, whenever $m, n \ge N$, we have

$$\|f_n - f_m\|_{\mathcal{H}'} < \varepsilon.$$

By definition, this implies that, for any $v \in \mathcal{H}$ and all $m, n \ge N$, we have

$$|f_n(v) - f_m(v)| \le \|f_n - f_m\|_{\mathcal{H}'}\|v\|_{\mathcal{H}} < \varepsilon\|v\|_{\mathcal{H}}.$$

In other words, the sequence $\{f_n(v)\}_{n\in\mathbb{N}} \subset \mathbb{R}$ is Cauchy, and so it must converge. Denote the limit by $f(v)$.

By uniqueness of the limit, we have defined $f\colon \mathcal{H} \to \mathbb{R}$ via

$$f(v) = \lim_{n\to\infty} f_n(v).$$

This mapping is linear, since, whenever $\alpha, \beta \in \mathbb{R}$ and $v, w \in \mathcal{H}$, we have

$$f(\alpha v + \beta w) = \lim_{n\to\infty} f_n(\alpha v + \beta w) = \alpha \lim_{n\to\infty} f_n(v) + \beta \lim_{n\to\infty} f_n(w) = \alpha f(v) + \beta f(w).$$

We now show that $f \in \mathcal{H}'$, i.e., that it is bounded. First, it is left to the reader as an exercise to show that $\{\|f_n\|_{\mathcal{H}'}\}_{n\in\mathbb{N}}$ is bounded, i.e., there is $M > 0$ such that for all $n \in \mathbb{N}$,

$$\|f_n\|_{\mathcal{H}'} \le M.$$

With this at hand, let $v \in \mathcal{H}$ be arbitrary and observe that

$$|f(v)| = \left|\lim_{n\to\infty} f_n(v)\right| = \lim_{n\to\infty} |f_n(v)| \le \lim_{n\to\infty} \|f_n\|_{\mathcal{H}'}\|v\|_{\mathcal{H}} \le \lim_{n\to\infty} M\|v\|_{\mathcal{H}} \le M\|v\|_{\mathcal{H}}.$$

Finally, we show that $\|f_n - f\|_{\mathcal{H}'} \to 0$. For this, let $v \in \mathcal{H}$ be arbitrary and $m, n \ge N$, so that

$$|f_n(v) - f_m(v)| \le \|f_n - f_m\|_{\mathcal{H}'}\|v\|_{\mathcal{H}} < \frac{\varepsilon}{2}\|v\|_{\mathcal{H}}.$$

Passing to the limit $m \to \infty$, this implies that

$$\|f - f_n\|_{\mathcal{H}'} = \sup_{0\neq v\in\mathcal{H}} \frac{|f_n(v) - f(v)|}{\|v\|_{\mathcal{H}}} \le \frac{\varepsilon}{2} < \varepsilon,$$

and so $\mathcal{H}'$ is complete. □

The next result — commonly known as the Riesz Representation Theorem[3] — is classical, powerful, and famous. Its proof is beyond our scope here.

[3] Named in honor of the Hungarian mathematician Frigyes Riesz (1880–1956).

Theorem 16.14 (Riesz Representation Theorem). *Suppose that $f \in \mathcal{H}'$. There exists a unique element $\mathfrak{R}f \in \mathcal{H}$ such that*

$$f(v) = (\mathfrak{R}f, v)_{\mathcal{H}}, \quad \forall v \in \mathcal{H}.$$

Furthermore,

$$\|\mathfrak{R}f\|_{\mathcal{H}} = \|f\|_{\mathcal{H}'}.$$

The map $\mathfrak{R}\colon \mathcal{H}' \to \mathcal{H}$ is called the **canonical Riesz map**.

The previous result is important in two respects. First, it shows that $u_n \rightharpoonup u$ if and only if $f(u_n) \to f(u)$ for all $f \in \mathcal{H}'$. In addition, it gives a way to define an inner product on $\mathcal{H}'$ as well, so that $\mathcal{H}'$ can actually be viewed as a Hilbert space.

Definition 16.15 (canonical inner product). For all $f, g \in \mathcal{H}'$, define

$$(f, g)_{\mathcal{H}'} = (\mathfrak{R}f, \mathfrak{R}g)_{\mathcal{H}}.$$

The object $(\cdot,\cdot)_{\mathcal{H}'} : \mathcal{H}' \times \mathcal{H}' \to \mathbb{R}$ is called the **canonical inner product on** $\mathcal{H}'$.

Theorem 16.16 ($\mathcal{H}'$ is Hilbert). *Let $\mathcal{H}$ be a Hilbert space and $\mathcal{H}'$ be its dual. Then $(\cdot,\cdot)_{\mathcal{H}'} : \mathcal{H}' \times \mathcal{H}' \to \mathbb{R}$ is an inner product on $\mathcal{H}'$, and, with this inner product, $\mathcal{H}'$ is a Hilbert space. Moreover,*

$$\|f\|_{\mathcal{H}'} = \sqrt{(f, f)_{\mathcal{H}'}}, \quad \forall f \in \mathcal{H}',$$

where $\|\cdot\|_{\mathcal{H}'}$ is the operator norm defined in (16.3).

Proof. See Problem 16.5. □

Remark 16.17 (finite dimensions). Note that, as shown in Example 1.1, the content of the Riesz Representation Theorem 16.14 is essentially trivial in finite dimensions and that $(\mathbb{R}^n)'$ can be easily identified with $\mathbb{R}^n$ itself.

16.1.2 Bounded Linear Operators

We now introduce some notions regarding operators on a Hilbert space. We recall that for normed spaces $\mathbb{V}$ and $\mathbb{W}$, in Definition 1.1, we introduced $\mathfrak{L}(\mathbb{V}, \mathbb{W})$.

Definition 16.18 (bounded operator). Let $\mathbb{V}$ and $\mathbb{W}$ be normed spaces. We say that $T \in \mathfrak{L}(\mathbb{V}, \mathbb{W})$ is **bounded** if and only if there is a constant $C > 0$ such that, for every $v \in \mathbb{V}$,

$$\|Tv\|_{\mathbb{W}} \leq C\|v\|_{\mathbb{V}}.$$

We denote the set of all bounded linear operators $\mathbb{V} \to \mathbb{W}$ by $\mathfrak{B}(\mathbb{V}, \mathbb{W})$. If $\mathbb{V} = \mathbb{W}$, then we denote this by $\mathfrak{B}(\mathbb{V})$.

The point of introducing this new notion is that we are not assuming the normed spaces to be finite dimensional. Otherwise, we could have applied Proposition 1.2 and Definition 1.28 to conclude that $\mathfrak{B}(\mathbb{V}, \mathbb{W}) = \mathfrak{L}(\mathbb{V}, \mathbb{W})$ is a normed space with a very special norm. In infinite dimensions, however, not every linear operator is bounded.

Example 16.4 The space of (equivalence classes of) functions $f\colon (0,1) \to \mathbb{R}$, for which

$$\int_0^1 |f(x)|^2 \mathrm{d}x < \infty$$

is a vector space, which we denote $L^2(0,1)$. We can endow this space with the inner product

$$(f,g)_{L^2(0,1)} = \int_0^1 f(x)g(x)\mathrm{d}x, \quad \forall f,g \in L^2(0,1),$$

under which this is a Hilbert space. It is not difficult to show that $\mathbb{P} \subset L^2(0,1)$, so that $L^2(0,1)$ is infinite dimensional. Let us define, at least for $f \in \mathbb{P}$,

$$(Tf)(x) = f'(x).$$

This is a linear operator; however, it is not bounded. To see this, notice that we can set, for $n \in \mathbb{N}$, $f(x) = x^n$ to obtain

$$\|f\|_{L^2(0,1)}^2 = \int_0^1 |x|^{2n}\mathrm{d}x = \frac{1}{2n+1}, \quad \|f'\|_{L^2(0,1)}^2 = n^2 \int_0^1 |x|^{2(n-1)}\mathrm{d}x = \frac{n^2}{2n-1}.$$

Since n can be arbitrarily large, there is no constant $C > 0$ for which

$$\frac{n^2}{2n-1} \leq \frac{C^2}{2n+1}, \quad \forall n \in \mathbb{N}.$$

Even in infinite dimensions, for linear operators, boundedness and continuity are equivalent notions.

Proposition 16.19 (equivalence). *Let $\mathbb{V}$ and $\mathbb{W}$ be normed spaces and $T \in \mathfrak{L}(\mathbb{V},\mathbb{W})$. The following are equivalent.*

1. *The operator T is bounded.*
2. *The operator T is continuous, i.e., whenever $\{v_n\}_{n\in\mathbb{N}} \subset \mathbb{V}$ is such that $v_n \to v$, then*

$$\|Tv_n - Tv\|_{\mathbb{W}} \to 0.$$

3. *The operator T is continuous at zero, meaning that if $v_n \to 0$, then*

$$Tv_n \to 0.$$

Proof. Let us show that boundedness implies continuity. To do so, consider a sequence $v_n \to v$. By linearity and boundedness,

$$\|Tv_n - Tv\|_{\mathbb{W}} = \|T(v_n - v)\|_{\mathbb{W}} \leq C\|v_n - v\|_{\mathbb{V}} \to 0,$$

so that the operator is continuous.

Obviously, continuity implies that the operator T is continuous at a particular point, and by linearity $T0 = 0$.

To close the argument, assume that T is continuous at zero. This necessarily means that there is $\delta > 0$ such that, whenever $\|v\|_{\mathbb{V}} < \delta$, we must have $\|Tv\|_{\mathbb{W}} < 1$. Let now $w \in \mathbb{V}$ be nonzero, but otherwise arbitrary. Define

$$v = \frac{\delta}{2\|w\|_{\mathbb{V}}} w \in \mathbb{V} \quad \Longrightarrow \quad \|v\|_{\mathbb{V}} < \delta.$$

Thus, by assumption, $\|Tv\|_{\mathbb{W}} < 1$, which by linearity and properties of the norm implies that

$$\frac{\delta}{2\|w\|_{\mathbb{V}}} \|Tw\|_{\mathbb{W}} = \|Tv\|_{\mathbb{W}} < 1 \quad \Longleftrightarrow \quad \|Tw\|_{\mathbb{W}} < \frac{2}{\delta}\|w\|_{\mathbb{V}},$$

so that $C = 2/\delta$ is the needed constant. □

It turns out that $\mathfrak{B}(\mathbb{V}, \mathbb{W})$ is a vector space, that it can be normed, and that, if $\mathbb{W}$ is complete, this space is complete under this norm. Hence, it is a Banach space.

Definition 16.20 (operator norm)**.** Let $\mathbb{V}$, $\mathbb{W}$ be normed spaces and $T \in \mathfrak{B}(\mathbb{V}, \mathbb{W})$. We define the **operator norm**

$$\|T\|_{\mathfrak{B}(\mathbb{V},\mathbb{W})} = \sup_{0 \neq v \in \mathbb{V}} \frac{\|Tv\|_{\mathbb{W}}}{\|v\|_{\mathbb{V}}}.$$

Proposition 16.21 (completeness)**.** *Let $\mathbb{V}$, $\mathbb{W}$ be normed spaces. The set $\mathfrak{B}(\mathbb{V}, \mathbb{W})$ under the same rules of addition and scalar multiplication of $\mathfrak{L}(\mathbb{V}, \mathbb{W})$ is a vector space. Moreover, the quantity $\|\cdot\|_{\mathfrak{B}(\mathbb{V},\mathbb{W})}$ of Definition 16.20 is a norm on $\mathfrak{B}(\mathbb{V}, \mathbb{W})$. Finally, if $\mathbb{W}$ is complete, then so is $\mathfrak{B}(\mathbb{V}, \mathbb{W})$ under the operator norm.*

Proof. The fact that $\mathfrak{B}(\mathbb{V}, \mathbb{W})$ is a vector space and that the operator norm is indeed a norm are immediate and left to the reader. For completeness, one, essentially, needs to repeat the argument of completeness of $\mathcal{H}'$ presented in Proposition 16.13. The only difference worth noting is that, instead of the completeness of $\mathbb{R}$, we must invoke that of $\mathbb{W}$; see Problem 16.6. □

Even in finite dimensions, not every bounded operator is invertible. In infinite dimensions, one new issue arises; namely, even when an inverse exists and is linear, this may not be bounded. Here, we will only state a sufficient condition for an operator to have a bounded inverse. The proof of this fact is beyond the scope of our discussion here. Let us only mention that it follows from the celebrated *Banach Open Mapping Theorem.*

Proposition 16.22 (bounded inverse)**.** *Let $\mathcal{H}$ be a Hilbert space and $T \in \mathfrak{B}(\mathcal{H})$ be such that there is $m > 0$ such that*

$$\|Tv\|_{\mathcal{H}} \geq m\|v\|_{\mathcal{H}}, \quad \forall v \in \mathcal{H}.$$

Then the operator T has a bounded inverse, in the sense that there is a (necessarily) unique mapping $T^{-1} \in \mathfrak{B}(\mathcal{H})$ such that

$$T(T^{-1}v) = T^{-1}(Tv) = v, \quad \forall v \in \mathcal{H}.$$

Moreover,

$$\|T^{-1}\|_{\mathfrak{B}(\mathcal{H})} \leq \frac{1}{m}.$$

16.1.3 Convex Sets and Convex Functions

Definition 16.23 (convex set)**.** A subset $K \subseteq \mathcal{H}$ is called **convex** if and only if $u, v \in K$ implies that

$$tu + (1-t)v \in K, \quad \forall t \in [0,1].$$

Definition 16.24 (convex function)**.** Let K be a convex subset of the Hilbert space $\mathcal{H}$. We say that the function $E\colon K \to \mathbb{R}$ is **convex** if and only if

$$E(tu + (1-t)v) \leq tE(u) + (1-t)E(v), \quad \forall u, v \in K, \quad \forall t \in [0,1].$$

We say that E is **strictly convex** if and only if it is convex and for every $u, v \in K$, $u \neq v$,

$$E(tu + (1-t)v) < tE(u) + (1-t)E(v), \quad \forall t \in (0,1).$$

16.2 Existence and Uniqueness of a Minimizer

Let us now use some of the functional analysis tools we previously developed to discuss convex optimization.

Definition 16.25 (lower semi-continuity)**.** Suppose that K is a subset of the Hilbert space $\mathcal{H}$ and $\{u_n\}_{n=1}^{\infty} \subset K$ is a sequence. We say that $E\colon K \to \mathbb{R}$ is **lower semi-continuous** if and only if $u_n \to u \in K$ implies that

$$E(u) \leq \liminf_{n\to\infty} E(u_n). \tag{16.4}$$

We say that E is **weakly lower semi-continuous** if and only if $u_n \rightharpoonup u \in K$ implies (16.4).

The notions of lower semi-continuity are related as follows.

Proposition 16.26 (semi-continuity)**.** *Suppose that K is a subset of the Hilbert space $\mathcal{H}$ and $E\colon K \to \mathbb{R}$ is weakly lower semi-continuous. Then E is lower semi-continuous. If E is continuous, then it is lower semi-continuous.*

Proof. See Problem 16.7. □

Example 16.5 Let $\mathcal{H}$ be a Hilbert space. The norm $\|\cdot\|_{\mathcal{H}}$ is weakly lower semi-continuous. To see this, suppose that $u_n \rightharpoonup u \neq 0$ and observe that

$$\|u\|_{\mathcal{H}}^2 = (u,u)_{\mathcal{H}} = \lim_{n\to\infty}(u_n, u)_{\mathcal{H}} \leq \liminf_{n\to\infty} \|u_n\|_{\mathcal{H}} \|u\|_{\mathcal{H}}.$$

The following two, very general, results are commonly known as the *direct method* of calculus of variations.

Theorem 16.27 (existence I). *Suppose that $K \subset \mathcal{H}$ is bounded and weakly closed and $\mathcal{H}$ is a Hilbert space. Assume that $E\colon K \to \mathbb{R}$ is weakly lower semi-continuous. Then there is at least one element u such that*

$$E(u) = \inf_{v \in K} E(v). \tag{16.5}$$

Proof. Set $\alpha = \inf_{v \in K} E(v)$. Assume that $\alpha \in \mathbb{R}$. There exists a sequence $\{u_n\}_{n=1}^{\infty} \subset K$ such that

$$E(u_n) \to \alpha.$$

The sequence $\{u_n\}_{n=1}^{\infty} \subset K$ is called an infimizing or minimizing sequence. For completeness, let us prove its existence. Since α is the greatest lower bound, we know that, for each $n \in \mathbb{N}$, there must be an element $y_n \in E(K)$, where $E(K) \subseteq \mathbb{R}$ denotes the image of K under E such that

$$\alpha < y_n < \alpha + \frac{1}{n}.$$

Otherwise, α could not be the greatest lower bound. Now take $u_n \in K$, so that $y_n = E(u_n)$. This proves the existence of the minimizing sequence.

Having shown the existence of the infimizing sequence, we observe that, since K is bounded, $\{u_n\}_{n=1}^{\infty}$ is bounded. Therefore, by Theorem 16.10, there is a subsequence $\{u_{n_k}\}_{k=1}^{\infty} \subseteq \{u_n\}_{n=1}^{\infty}$ and a point $u \in \mathcal{H}$ such that $u_{n_k} \rightharpoonup u$ as $k \to \infty$. Since K is weakly closed, $u \in K$. Since E is weakly sequentially lower semi-continuous,

$$E(u) \leq \liminf_{k \to \infty} E(u_{n_k}).$$

The only possibility is that $\alpha = E(u)$ and the proof is complete.

Incidentally, we point out that the proof implies that $-\infty < \alpha$. □

Definition 16.28 (coercivity). Suppose that $E\colon \mathcal{H} \to \mathbb{R}$. E is called **coercive** if and only if

$$\lim_{n \to \infty} E(v_n) \to \infty$$

whenever $\|v_n\|_{\mathcal{H}} \to \infty$.

Theorem 16.29 (existence II). *Assume that $E\colon \mathcal{H} \to \mathbb{R}$ is weakly lower semi-continuous and coercive on the Hilbert space $\mathcal{H}$. Then there is at least one element u such that*

$$E(u) = \inf_{v \in \mathcal{H}} E(v). \tag{16.6}$$

Proof. Define

$$K_0 = \{v \in \mathcal{H} \mid E(v) \leq E(0)\}.$$

Since E is coercive, K_0 is bounded. Otherwise, there is a sequence $\{v_n\}_{n=1}^{\infty} \subset K$ such that $\|v_n\|_{\mathcal{H}} \to \infty$ as $n \to \infty$. But then $\lim_{n \to \infty} E(v_n) \to \infty$. So there is an $N \in \mathbb{N}$ such that, when $m \geq N$,

$$E(v_m) > E(0).$$

This is a contradiction and so K_0 must be bounded.

Now, since E is weakly lower semi-continuous, we claim that K_0 is weakly closed. To see this, suppose that $\{v_n\}_{n=1}^{\infty} \subset K_0$ is such that there is $v \in \mathcal{H}$ for which $v_n \rightharpoonup v$. We want to prove that $v \in K_0$. Since E is weakly lower semi-continuous,

$$E(v) \leq \liminf_{n\to\infty} E(v_n) \leq E(0).$$

Thus, $v \in K_0$.

Problem (16.6) then is equivalent to

$$E(u) = \inf_{v\in K_0} E(v).$$

By Theorem 16.27, there is at least one solution to (16.6). □

It is not easy to work directly with the concepts of weakly closed sets and weakly lower semi-continuous functions. Thankfully, we have the following result, whose proof is beyond our scope. It is commonly known as Mazur's Lemma.

Proposition 16.30 (Mazur's Lemma[4]). *Let $\mathcal{H}$ be a Hilbert space. Suppose that $K \subset \mathcal{H}$ is convex and closed. Then it is weakly closed. If $E\colon \mathcal{H} \to \mathbb{R}$ is convex and continuous (or lower semi-continuous), then it is weakly lower semi-continuous.*

Finally, we have this, the main result.

Theorem 16.31 (existence and uniqueness). *Assume that $E\colon \mathcal{H} \to \mathbb{R}$ is strictly convex, continuous, and coercive on the Hilbert space $\mathcal{H}$. Then there is a unique element $u \in \mathcal{H}$ such that*

$$E(u) = \inf_{v\in\mathcal{H}} E(v). \tag{16.7}$$

This allows us to write

$$u = \underset{v\in\mathcal{H}}{\operatorname{argmin}} E(v). \tag{16.8}$$

Proof. Define, as before,

$$K_0 = \{v \in \mathcal{H} \mid E(v) \leq E(0)\}.$$

Since E is continuous and convex, it is weakly lower semi-continuous on $\mathcal{H}$. The set K_0 is convex, since E is convex, and it is bounded, since E is coercive. Since E is continuous, if $v_n \to v$, with $\{v_n\}_{n=1}^{\infty} \subset K_0$, we have

$$E(v) = \lim_{n\to\infty} E(v_n) \leq E(0).$$

Thus, $v \in K_0$ and K_0 is closed. Since K_0 is closed and convex, it is weakly closed. The proof of existence now follows by the same argument as in the proof of Theorem 16.29.

[4] Named in honor of the Polish mathematician Stanisław Mieczysław Mazur (1905–1981).

Uniqueness follows from the strict convexity of E. Suppose that $u_1, u_2 \in \mathcal{H}$, $u_1 \neq u_2$ are minimizers. Define

$$\alpha = \min_{v \in \mathcal{H}} E(v).$$

Since E is strictly convex,

$$E(tu_1 + (1-t)u_2) < tE(u_1) + (1-t)E(u_2) = \alpha, \quad \forall t \in (0,1).$$

This implies that u_1 and u_2 are not minimizers, unless perhaps $u_1 = u_2$, the only possibility that does not lead to contradiction. □

16.3 The Euler Equation

Definition 16.32 (Fréchet derivative[5]). Suppose that $\mathcal{H}$ is a Hilbert space and $E\colon \mathcal{H} \to \mathbb{R}$. We say that E is **Fréchet differentiable at the point** $v \in \mathcal{H}$ if and only if there is a bounded linear functional $A \in \mathcal{H}'$ for which

$$\lim_{\|h\|_{\mathcal{H}} \to 0} \frac{|E(v+h) - E(v) - A(h)|}{\|h\|_{\mathcal{H}}} = 0.$$

A is called a **Fréchet derivative at** v. If E is Fréchet differentiable at every point v in a particular set $B \subseteq \mathcal{H}$, we say that E is **Fréchet differentiable on** B.

Proposition 16.33 (uniqueness). *Suppose that $E\colon \mathcal{H} \to \mathbb{R}$ is Fréchet differentiable at the point $v \in \mathcal{H}$. Then the derivative is unique. The (uniquely defined) derivative is denoted by $E'[v] \in \mathcal{H}'$.*

Proof. See Problem 16.8. □

Remark 16.34 (arguments). We observe that if E is Fréchet differentiable on a set B, then the bounded linear functional $E'[v]$ changes as $v \in B$ changes. Specifically, if v and w are two different points of B, then $E'[v]$ and $E'[w]$ are generally different linear functionals. Now, once $v \in B$ is fixed, we have $E'[v] \in \mathcal{H}'$, and the action of the functional on an element $w \in \mathcal{H}$ of the Hilbert space is denoted by $E'[v](w)$. Of course, $E'[v](\cdot)$ is linear, so that

$$E'[v]\left(\sum_{i=1}^{N} \alpha_i w_i\right) = \sum_{i=1}^{N} \alpha_i E'[v](w_i), \quad \forall \alpha_i \in \mathbb{R}, w_i \in \mathcal{H}, \ i = 1, \ldots, N.$$

Remark 16.35 (notation). We will use the duality pairing notation throughout this chapter; in particular, for any $f \in \mathcal{H}'$,

$$\langle f, w \rangle = f(w), \quad \forall w \in \mathcal{H}.$$

[5] Named in honor of the French mathematician Maurice René Fréchet (1878–1973).

The object on the left, $\langle \cdot, \cdot \rangle$, is not an inner product. It is called the **duality pairing between $\mathcal{H}'$ and $\mathcal{H}$**. The bounded linear functional always appears in the left-hand slot of $\langle \cdot, \cdot \rangle$. Thus, if E is Fréchet differentiable at a point $v \in \mathcal{H}$, we write

$$\langle E'[v], w \rangle = E'[v](w), \quad \forall w \in \mathcal{H}.$$

This usage will save some writing.

Now let us give a weaker version of the definition of differentiability.

Definition 16.36 (Gateaux derivative[6]). We say that $E \colon \mathcal{H} \to \mathbb{R}$ is **Gateaux differentiable at the point** $v \in \mathcal{H}$ if and only if there is a bounded linear functional $A \in \mathcal{H}'$ such that

$$\lim_{s \to 0} \frac{1}{s} \left(E(v + sw) - E(v) \right) = A(w)$$

for all $w \in \mathcal{H}$. A is called a **Gateaux derivative at** v. As before, if E is Gateaux differentiable at every point v of a set $B \subseteq \mathcal{H}$, then we say that E is **Gateaux differentiable on** B.

The reader should be able to prove the following without too much trouble.

Proposition 16.37 (uniqueness). *Suppose that $E \colon \mathcal{H} \to \mathbb{R}$ is Gateaux differentiable at the point $v \in \mathcal{H}$. Then the derivative is unique. The (uniquely defined) Gateaux derivative is denoted by $E'[v] \in \mathcal{H}'$, the same symbol as before.*

Proof. See Problem 16.9. □

Proposition 16.38 (Fréchet implies Gateaux). *If $E \colon \mathcal{H} \to \mathbb{R}$ is Fréchet differentiable at a point $v \in \mathcal{H}$, then it is Gateaux differentiable at v. The converse is not true, however.*

Proof. See Problem 16.10. □

For differentiable functions, we have various ways to express convexity.

Proposition 16.39 (convexity equivalences). *Let $K \subseteq \mathcal{H}$ be nonempty and convex. Assume that $E \colon \mathcal{H} \to \mathbb{R}$ is Fréchet or Gateaux differentiable on K. The following are equivalent.*

1. *E is convex on K.*
2. *E satisfies*

$$E(v) \geq E(u) + \langle E'[u], v - u \rangle, \quad \forall u, v \in K.$$

3. *E' satisfies the monotonicity condition*

$$\langle E'[v] - E'[u], v - u \rangle \geq 0, \quad \forall u, v \in K.$$

Proof. See Problem 16.11. □

There is a similar result for strict convexity.

[6] Named in honor of the French mathematician René Eugène Gateaux (1889–1914).

Proposition 16.40 (strict convexity equivalences). *Let $K \subseteq \mathcal{H}$ be nonempty and convex. Assume that $E\colon \mathcal{H} \to \mathbb{R}$ is Fréchet or Gateaux differentiable on K. The following are equivalent.*

1. *E is strictly convex on K.*
2. *E satisfies*

$$E(v) > E(u) + \langle E'[u], v - u \rangle, \quad \forall u, v \in K, \quad u \neq v.$$

3. *E' satisfies the strict monotonicity condition*

$$\langle E'[v] - E'[u], v - u \rangle > 0, \quad \forall u, v \in K, \quad u \neq v.$$

Proof. See Problem 16.12. □

Definition 16.41 (strong convexity). Suppose that $E\colon \mathcal{H} \to \mathbb{R}$ is Fréchet or Gateaux differentiable on the Hilbert space $\mathcal{H}$. We say that E is **strongly convex** if and only if there is a constant $\mu > 0$ such that

$$\mu \|w - v\|_{\mathcal{H}}^2 \leq \langle E'[w] - E'[v], w - v \rangle \tag{16.9}$$

for all $v, w \in \mathcal{H}$, where $\langle \cdot , \cdot \rangle$ is the dual pairing between $\mathcal{H}'$ and $\mathcal{H}$. The positive number μ is called the **convexity constant**. To emphasize the value of the convexity constant in connection with the strong convexity property, we say that E is μ-**strongly convex on** $\mathcal{H}$.

Remark 16.42 (terminology). Other authors, for example Ciarlet [16], use the term *elliptic* in place of strongly convex. The strong convexity property is also equivalent to the property that the derivative is *strongly monotone* [5], another common term.

Definition 16.43 (local Lipschitz smoothness). Suppose that $E\colon \mathcal{H} \to \mathbb{R}$ is Fréchet or Gateaux differentiable on the Hilbert space $\mathcal{H}$. We say that E is **locally — Lipschitz smooth on** $\mathcal{H}$ if and only if for every bounded convex set B, there is a constant $L_B > 0$ such that, for all $w, v \in B$,

$$\|E'[w] - E'[v]\|_{\mathcal{H}'} \leq L_B \|w - v\|_{\mathcal{H}}. \tag{16.10}$$

The positive number L_B is called the (local) **Lipschitz constant**. We say that E is L_B-**Lipschitz smooth on** B when we wish to emphasize the value of the Lipschitz constant on the convex set B.

For μ-strongly convex functions, much more can be said about the solution of a minimization problem.

Theorem 16.44 (Euler equation[7]). *If E is μ–strongly convex on $\mathcal{H}$, then, for all $w, v \in \mathcal{H}$,*

$$E(w) - E(v) \geq \langle E'[v], w - v \rangle + \frac{\mu}{2} \|w - v\|_{\mathcal{H}}^2. \tag{16.11}$$

[7] Named in honor of the Swiss mathematician, physicist, astronomer, geographer, logician, and engineer Leonhard Euler (1707–1783).

Consequently, E is strictly convex and coercive. Furthermore, there is a unique element $u \in \mathcal{H}$ with the property that

$$E(u) \le E(v), \quad \forall v \in \mathcal{H}, \qquad E(u) < E(v), \quad \forall v \neq u,$$

and this global minimizer *satisfies* Euler's equation

$$\langle E'[u], w \rangle = 0, \quad \forall w \in \mathcal{H}. \tag{16.12}$$

Proof. Using Taylor's Theorem with integral remainder (see Problem 16.13) guarantees that

$$\begin{aligned} E(w) - E(v) &= \int_0^1 \langle E'[v + t(w-v)], w - v \rangle dt \\ &= \langle E'[v], w - v \rangle + \int_0^1 \frac{\langle E'[v + t(w-v)] - E'[v], t(w-v) \rangle}{t} dt \\ &\ge \langle E'[v], w - v \rangle + \int_0^1 \frac{\mu \left\| t(w-v) \right\|_{\mathcal{H}}^2}{t} dt \\ &= \langle E'[v], w - v \rangle + \frac{\mu}{2} \left\| w - v \right\|_{\mathcal{H}}^2, \end{aligned}$$

which proves (16.11).

Inequality (16.11) implies that, for any $w, v \in \mathcal{H}$, $w \neq v$,

$$E(w) > E(v) + \langle E'[v], w - v \rangle,$$

which is an equivalent definition of strict convexity.

We say that E is *quadratically coercive* if and only if there are constants $C_1 \in \mathbb{R}$ and $C_2 > 0$ such that

$$E(w) \ge C_1 + C_2 \left\| w \right\|_{\mathcal{H}}^2, \quad \forall w \in \mathcal{H}.$$

Setting $v = 0$ in (16.11), we have

$$\begin{aligned} E(w) &\ge E(0) + \langle E'[0], w \rangle + \frac{\mu}{2} \left\| w \right\|_{\mathcal{H}}^2 \\ &\ge E(0) - \left\| E'[0] \right\|_{\mathcal{H}'} \left\| w \right\|_{\mathcal{H}} + \frac{\mu}{2} \left\| w \right\|_{\mathcal{H}}^2 \\ &\ge E(0) - \frac{1}{\mu} \left\| E'[0] \right\|_{\mathcal{H}'}^2 - \frac{\mu}{4} \left\| w \right\|_{\mathcal{H}}^2 + \frac{\mu}{2} \left\| w \right\|_{\mathcal{H}}^2 \\ &= C_1 + C_2 \left\| w \right\|_{\mathcal{H}}^2, \end{aligned}$$

where

$$C_1 = E(0) - \frac{1}{\mu} \left\| E'[0] \right\|_{\mathcal{H}'}^2, \qquad C_2 = \frac{\mu}{4}.$$

Of course, it is clear that quadratic coercivity implies coercivity.

The existence and uniqueness of a minimizer follows from Theorem 16.31.

Having shown the existence and uniqueness of a minimizer, let us now prove that this is equivalent to solving the Euler equation (16.12). First, suppose that $u \in \mathcal{H}$ satisfies Euler's equation (16.12). Strict convexity implies that

$$E(w) > E(u) + \langle E'[u], w - u \rangle = E(u)$$

for all $w \in \mathcal{H}$, $w \neq u$, and so it is a minimizer. On the other hand, suppose that $u \in \mathcal{H}$ is the unique global minimizer of E. Let $v \in \mathcal{H}$ be arbitrary. Then, for $s \neq 0$,

$$E(u + sv) - E(u) > 0.$$

Thus, the *directional derivatives* satisfy

$$\lim_{s \downarrow 0} \frac{1}{s}\left(E(u + sv) - E(u)\right) = \langle E'[u], v\rangle \geq 0$$

and

$$\lim_{s \uparrow 0} \frac{1}{s}\left(E(u + sv) - E(u)\right) = \langle E'[u], v\rangle \leq 0.$$

The only possibility is that $\langle E'[u], v\rangle = 0$. Since $v \in \mathcal{H}$ was arbitrary, u solves the Euler equation. □

Definition 16.45 (bounded energy set)**.** Fix $u_0 \in \mathcal{H}$. The set

$$\mathcal{E}_0 = \{v \in \mathcal{H} \mid E(v) \leq E(u_0)\} \tag{16.13}$$

is called the **bounded energy set**.

Proposition 16.46 (convexity)**.** *If E is convex on $\mathcal{H}$, the bounded energy set $\mathcal{E}_0$ is convex.*

Proof. Suppose that $v, w \in \mathcal{E}_0$. Then $E(w) \leq E(u_0)$ and $E(v) \leq E(u_0)$. Since E is convex, for any $t \in [0, 1]$,

$$E(u_0) \geq (1 - t)E(w) + tE(v) \geq E((1 - t)w + tv).$$

Thus, $(1 - t)w + tv \in \mathcal{E}_0$ for any $t \in [0, 1]$. □

If E is strongly convex and the Lipschitz smooth, then very useful estimates can be shown.

Lemma 16.47 (two-sided estimate)**.** *Suppose that E is μ-strongly convex on $\mathcal{H}$ and L-Lipschitz smooth on the bounded energy set $\mathcal{E}_0$. Then, for all $v, w \in \mathcal{E}_0$,*

$$\mu \left\|w - v\right\|_{\mathcal{H}}^2 \leq \langle E'[w] - E'[v], w - v\rangle \leq L \left\|w - v\right\|_{\mathcal{H}}^2.$$

Furthermore, the lower bound holds for all $v, w \in \mathcal{H}$.

Proof. The lower bound is just μ-strong convexity. To get the upper bound, observe that, for all $w, v \in \mathcal{E}_0$ and for any $z \in \mathcal{H}$,

$$\left|\langle E'[w] - E'[v], z\rangle\right| \leq \left\|E'[w] - E'[v]\right\|_{\mathcal{H}'} \left\|z\right\|_{\mathcal{H}} \leq L \left\|w - v\right\|_{\mathcal{H}} \left\|z\right\|_{\mathcal{H}}.$$

Setting $z = w - v$ gives the desired inequality. □

Now we consider the relation between the energy and the norm centered at the minimizer. The following estimates can be easily proved using Taylor's Theorem with integral remainder, as above; see, for example, [16].

Lemma 16.48 (quadratic energy trap)**.** *Suppose that E is μ-strongly convex on $\mathcal{H}$ and L-Lipschitz smooth on the bounded energy set $\mathcal{E}_0$. Then, for all $v, w \in \mathcal{E}_0$,*

$$\begin{aligned} \frac{\mu}{2} \|w - v\|_{\mathcal{H}}^2 + \langle E'[v], w - v \rangle &\le E(w) - E(v) \\ &\le \langle E'[v], w - v \rangle + \frac{L}{2} \|w - v\|_{\mathcal{H}}^2 . \end{aligned} \tag{16.14}$$

Furthermore, the lower bound holds for all $v, w \in \mathcal{H}$. In addition, suppose that $u \in \mathcal{E}_0$ is the minimizer of E. Then, for all $w \in \mathcal{E}_0$,

$$\frac{\mu}{2} \|w - u\|_{\mathcal{H}}^2 \le E(w) - E(u) \le \frac{L}{2} \|w - u\|_{\mathcal{H}}^2 . \tag{16.15}$$

Again, the lower bound holds for all $w \in \mathcal{H}$.

Proof. The lower bound for (16.14) has been established in Theorem 16.44. For the upper bound, we again apply Taylor's Theorem with integral remainder and use the fact that $\mathcal{E}_0$ is convex. The remaining details are left to the reader as an exercise; see Problem 16.14. □

If E is strongly convex, we can produce an upper bound for the energy that consists of the norm of the Fréchet (or Gateaux) derivative.

Lemma 16.49 (upper bound)**.** *Suppose that E is μ-strongly convex on $\mathcal{H}$ and $u \in \mathcal{H}$ is the minimizer of E. Then, for all $v \in \mathcal{H}$, we have*

$$0 \le E(v) - E(u) \le \frac{1}{2\mu} \|E'[v]\|_{\mathcal{H}'}^2 . \tag{16.16}$$

Proof. Fix the point $v \in \mathcal{H}$. Now, for any $w \in \mathcal{H}$, using the lower bound of (16.14), we have

$$E(w) \ge E(v) + \langle E'[v], w - v \rangle + \frac{\mu}{2} \|w - v\|_{\mathcal{H}}^2 = g(w).$$

For fixed $v \in \mathcal{H}$, the minimizer of g is $w^* = v - \frac{1}{\mu} \mathfrak{R} E'[v]$, where $\mathfrak{R} E'[v]$ is the Riesz representation in $\mathcal{H}$ of $E'[v]$. Therefore,

$$E(w) \ge g(w) \ge g(w^*) = E(v) - \frac{1}{2\mu} \|\mathfrak{R} E'[v]\|_{\mathcal{H}}^2 = E(v) - \frac{1}{2\mu} \|E'[v]\|_{\mathcal{H}'}^2 .$$

Then (16.16) is obtained by letting $w = u$ in the above inequality. □

We shall often use the following simple variant of Lemma 16.48.

Lemma 16.50 (convexity of energy sections)**.** *Suppose that E is μ-strongly convex on $\mathcal{H}$ and L-Lipschitz smooth on $\mathcal{E}_0$. Let $\xi \in \mathcal{E}_0$ be arbitrary. Assume that $\mathcal{W} \subseteq \mathcal{H}$ is a subspace. Define the function*

$$J(w) = E(\xi + w), \quad \forall w \in \mathcal{W},$$

which we refer to as an energy section (or energy slice) *of E. Then $J \colon \mathcal{W} \to \mathbb{R}$ is differentiable and strongly convex, and there exists a unique element $\eta \in \mathcal{W}$ such that $\xi + \eta \in \mathcal{E}_0$, η is the unique global minimizer of J, and*

$$\langle E'[\xi + \eta], w \rangle = \langle J'[\eta], w \rangle = 0, \quad \forall w \in \mathcal{W}.$$

Furthermore, for all $w \in \mathcal{W}$ with $w + \xi \in \mathcal{E}_0$,

$$\frac{\mu}{2} \|w - \eta\|_{\mathcal{H}}^2 \le J(w) - J(\eta) = E(\xi + w) - E(\xi + \eta) \le \frac{L}{2} \|w - \eta\|_{\mathcal{H}}^2 .$$

The lower bound holds for any $w \in \mathcal{W}$, without restriction.

Proof. See Problem 16.15. □

16.4 Preconditioners and Gradient Descent Methods

Next, we introduce an important operator, the *preconditioner*, and we use this to define a class of gradient descent methods.

Lemma 16.51 (preconditioner)**.** *Suppose that $\mathcal{L}\colon \mathcal{H} \to \mathcal{H}'$ is a linear and symmetric operator, in the sense that*

$$\langle \mathcal{L}[v], w\rangle = \mathcal{L}[v](w) = \mathcal{L}[w](v) = \langle \mathcal{L}[w], v\rangle, \quad \forall v, w \in \mathcal{H}.$$

Assume further that $\mathcal{L}$ is:

1. *Coercive, i.e., there exists a constant μ_1 such that, for all $v, w \in \mathcal{H}$,*

$$\langle \mathcal{L}[w], w\rangle \ge \mu_1 \|w\|_{\mathcal{H}}^2.$$

2. *Continuous, i.e., there exists a constant $L_1 > 0$ such that, for all $v, w \in \mathcal{H}$,*

$$\|\mathcal{L}[w]\|_{\mathcal{H}'} \le L_1 \|w\|_{\mathcal{H}}.$$

Then, for any $w \in \mathcal{H}$,

$$\mu_1 \|w\|_{\mathcal{H}}^2 \le \langle \mathcal{L}[w], w\rangle \le L_1 \|w\|_{\mathcal{H}}^2 ,$$

and, for all $v, w \in \mathcal{H}$,

$$\frac{\mu_1}{2} \|w - v\|_{\mathcal{H}}^2 + \langle \mathcal{L}[v], w - v\rangle \le E_1(w) - E_1(v) \le \langle \mathcal{L}[v], w - v\rangle + \frac{L_1}{2} \|w - v\|_{\mathcal{H}}^2 ,$$

where E_1 is the quadratic energy

$$E_1(v) = \frac{1}{2}\langle \mathcal{L}[v], v\rangle.$$

Proof. As before, the proof of the second set of estimates uses Taylor's Theorem with integral remainder; see Problem 16.16. □

Definition 16.52 (preconditioner)**.** A linear operator $\mathcal{L}\colon \mathcal{H} \to \mathcal{H}'$ that satisfies the hypotheses of the lemma is called a **preconditioner** or a **preconditioning operator**. When we wish to emphasize the values of μ_1 and L_1, we say that $\mathcal{L}$ is a (μ_1, L_1)-**preconditioner**.

Lemma 16.53 (properties of $\mathcal{L}$)**.** *Let $\mathcal{L}\colon \mathcal{H} \to \mathcal{H}'$ be a preconditioner. The object*

$$(u, v)_{\mathcal{L}} = \langle \mathcal{L}[u], v\rangle, \quad \forall u, v \in \mathcal{H}$$

is an inner product on the Hilbert space $\mathcal{H}$, and the norm that it induces is equivalent to the canonical norm $\|\cdot\|_{\mathcal{H}}$. Consequently, $\mathcal{L}$ is invertible.

Proof. The proof of the first fact is left to the reader; see Problem 16.17. Now, since $(\cdot,\cdot)_{\mathcal{L}}$ is an inner product on $\mathcal{H}$, by the Riesz Representation Theorem 16.14, there is a unique element $\mathfrak{R}_{\mathcal{L}} f \in \mathcal{H}$ such that

$$(\mathfrak{R}_{\mathcal{L}} f, w)_{\mathcal{L}} = \langle f, w \rangle, \quad \forall w \in \mathcal{H}.$$

The map $\mathfrak{R}_{\mathcal{L}} \colon \mathcal{H}' \to \mathcal{H}$ is called the *Riesz map subordinate to* $\mathcal{L}$, and we are justified in writing $\mathcal{L}^{-1} = \mathfrak{R}_{\mathcal{L}}$. □

Definition 16.54 ($\mathcal{L}^{-1}$-inner product)**.** Suppose that $\mathcal{L} \colon \mathcal{H} \to \mathcal{H}'$ is a preconditioner. The object

$$(f, g)_{\mathcal{L}^{-1}} = (\mathfrak{R}_{\mathcal{L}} f, \mathfrak{R}_{\mathcal{L}} g)_{\mathcal{L}}, \quad \forall f, g \in \mathcal{H}'$$

is called the **$\mathcal{L}^{-1}$-inner product**. The associated **$\mathcal{L}^{-1}$-norm** is defined as

$$\|f\|_{\mathcal{L}^{-1}} = \sqrt{(f, f)_{\mathcal{L}^{-1}}}, \quad \forall f \in \mathcal{H}'.$$

Proposition 16.55 ($\mathcal{L}^{-1}$-inner product)**.** *Suppose that $\mathcal{L} \colon \mathcal{H} \to \mathcal{H}'$ is a preconditioner. Then $(\cdot,\cdot)_{\mathcal{L}^{-1}}$ is an inner product on $\mathcal{H}'$.*

Proof. See Problem 16.18. □

Definition 16.56 (gradient descent)**.** Suppose that $E \colon \mathcal{H} \to \mathbb{R}$ is Fréchet differentiable on a convex open subset B of the Hilbert space $\mathcal{H}$. Let $u_0 \in B$ be given. A **gradient descent method** is an algorithm for the construction of the sequence $\{u_k\}_{k=0}^{\infty} \subset B$ according to the following rule: given $u_k \in B$, compute u_{k+1} via

$$u_{k+1} = u_k - \alpha_k \mathcal{L}_k^{-1} E'[u_k],$$

where $\alpha_k > 0$ is the **step size** and $\mathcal{L}_k \colon \mathcal{H} \to \mathcal{H}'$ is a family of symmetric, coercive, and continuous operators, i.e., a family of preconditioners. We also write

$$u_{k+1} = u_k + \varepsilon_k,$$

where

$$\varepsilon_k = \alpha_k s_k \in \mathcal{H}$$

is called the **correction**;

$$s_k = \mathcal{L}_k^{-1} r_k \in \mathcal{H}$$

is called the **search direction**; and

$$r_k = -E'[u_k] \in \mathcal{H}'$$

is called the **residual**. We say that the algorithm is **well defined** if and only if $u_{k+1} \in B$ whenever $u_k \in B$. The algorithm is **convergent** if and only if it is well defined, there is a point $u \in B$ such that $u_k \to u$, and u is at least a stationary point of E in B, i.e., $E'[u] = 0$.

Example 16.6 We can take $\mathcal{L}_k^{-1} = \mathfrak{R}\colon \mathcal{H}' \to \mathcal{H}$, the canonical Riesz map. This means that we precondition with whatever canonical metric we have chosen for the Hilbert space. The reader can check that $\mathcal{L}_k$ satisfies all of the requirements of a preconditioner. In this case, $\mu_1 = L_1 = 1$, as the reader can confirm.

16.5 The Golden Key

We will now study the convergence of the gradient descent method of Definition 16.56. To do so, we will utilize the following simple result in the proofs of convergence.

Theorem 16.57 (linear convergence). *Suppose that* $\{d_k\}_{k=0}^{\infty}$, $\{\delta_k\}_{k=0}^{\infty}$, $\{\beta_k\}_{k=0}^{\infty}$ *are sequences of nonnegative real numbers, the first two having the relationship*

$$\delta_k = d_k - d_{k+1}, \quad k = 0, 1, 2, \dots .$$

Assume that there are constants $C_L, C_U > 0$, *independent of* k, *such that*

$$C_L\beta_k \leq \delta_k \quad \text{and} \quad d_{k+1} \leq C_U\beta_k .$$

Then

$$d_{k+1} \leq \frac{C_U}{C_L + C_U} d_k, \quad k = 0, 1, 2, \dots . \tag{16.17}$$

Consequently, $\{d_k\}_{k=0}^{\infty}$ *converges monotonically and (at least) linearly to zero.*

Proof. Observe that

$$d_{k+1} \leq C_U\beta_k = \frac{C_U}{C_L} C_L\beta_k \leq \frac{C_U}{C_L}\delta_k = \frac{C_U}{C_L}(d_k - d_{k+1}),$$

which implies (16.17). Since $\rho = \frac{C_U}{C_L + C_U} \in (0, 1)$, the sequence $\{d_k\}_{k=0}^{\infty}$ is strictly decreasing. Moreover, it is bounded from below by zero. Thus, the Monotone Convergence Theorem B.7 implies that $d_k \to \gamma \geq 0$. Suppose that $\gamma > 0$, to get a contradiction. There is a positive integer K such that, if $k \geq K$,

$$\gamma \leq d_k < d_K \leq \gamma + \frac{(1-\rho)\gamma}{2\rho}.$$

Therefore,

$$d_{k+1} \leq \rho d_k \leq \rho\gamma + \frac{1}{2}(1-\rho)\gamma = \frac{1}{2}(\rho + 1)\gamma < \gamma.$$

This is a contradiction. The only possibility is that $\gamma = 0$. □

We will apply the last result with the following definitions:

$$d_k = E(u_k) - E(u), \qquad \delta_k = E(u_k) - E(u_{k+1}). \tag{16.18}$$

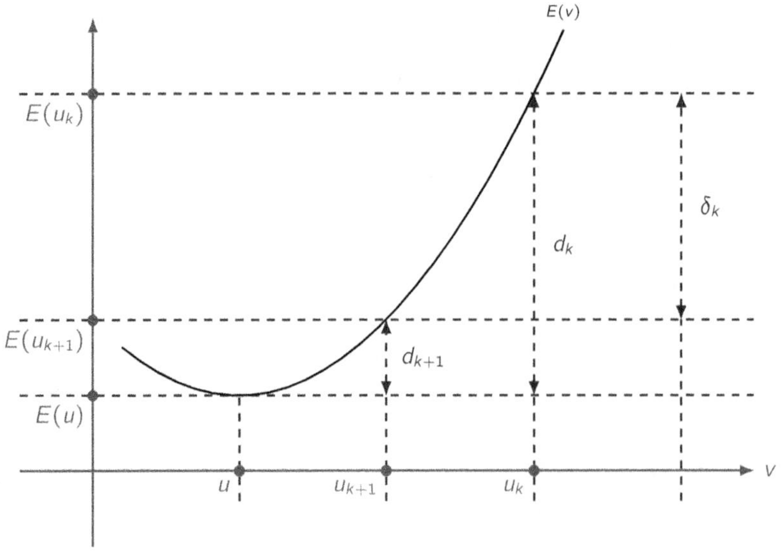

Figure 16.1 The sequences $\{d_k\}_{k=0}^{\infty}$ and $\{\delta_k\}_{k=0}^{\infty}$ used in the proof of the golden key; Corollary 16.58.

The quantity d_k is the difference between the current energy and the minimum energy, and δ_k is the energy decrease associated with the $(k+1)$st iteration. They are connected, as desired, by the trivial identity

$$\delta_k = d_k - d_{k+1}.$$

We define α_k in terms of the subspace corrections via

$$\beta_k = \|\varepsilon_k\|_{\mathcal{H}}^2,$$

and we assume the following upper and lower bounds.

- Lower bound on corrections. There exists a positive constant C_L such that, for any $k = 0, 1, 2, \ldots,$

$$E(u_k) - E(u_{k+1}) = \delta_k \geq C_L\beta_k = C_L\|\varepsilon_k\|_{\mathcal{H}}^2. \qquad (16.19)$$

- Upper bound on corrections. There exists a positive constant C_U such that, for any $k = 0, 1, 2, \ldots,$

$$E(u_{k+1}) - E(u) = d_{k+1} \leq C_U\beta_k = C_U\|\varepsilon_k\|_{\mathcal{H}}^2. \qquad (16.20)$$

These sequences, with the respective bounds, are depicted in Figure 16.1. If these bounds hold, then, as a corollary to Theorem 16.57, we have the following convergence result.

Corollary 16.58 (golden key). *Assume that the lower bound* (16.19) *and upper bound* (16.20) *hold with positive constants C_L and C_U, respectively. We then have*

$$E(u_{k+1}) - E(u) \leq \rho\left(E(u_k) - E(u)\right), \qquad \rho = \frac{C_U}{C_L + C_U},$$

and $E(u_k)$ converges monotonically and (at least) linearly to $E(u)$, at the linear rate ρ. Furthermore, $\{u_k\}_{k=0}^{\infty}$ converges at least linearly to u.

Proof. The linear convergence of $E(u_k)$ to $E(u)$ at the rate ρ is guaranteed by Theorem 16.57. Using (16.15), with $w = u_k$, we have

$$\frac{\mu}{2}\|u_k - u\|_{\mathcal{H}}^2 \le E(u_k) - E(u),$$

which guarantees the linear convergence of u_k to u. □

As we see, the essence of the analysis reduces to finding suitable bounds (16.19) and (16.20). Let us now obtain these for several particular instances.

16.6 Preconditioned Steepest Descent Method

The first gradient descent method that we examine is the preconditioned steepest descent (PSD) method, which is listed in Definition 16.59. There are two characteristics that make the PSD method a gradient descent method, as follows.

Definition 16.59 (PSD)**.** Suppose that B is an open and convex subset of the Hilbert space $\mathcal{H}$, $E\colon \mathcal{H} \to \mathbb{R}$ is Fréchet or Gateaux differentiable on B, and $\mathcal{L}\colon \mathcal{H} \to \mathcal{H}'$ is a fixed, invariant preconditioner. The **preconditioned steepest descent (PSD) algorithm** is a gradient descent method for which $\mathcal{L}_k = \mathcal{L}$, i.e., the search direction is computed as the solution to

$$\langle s_k, w\rangle_{\mathcal{L}} = \mathcal{L}[s_k](w) = \langle r_k, w\rangle, \quad \forall w \in \mathcal{H}, \tag{16.21}$$

and the step size is computed via line search along the search direction, i.e.,

$$\varepsilon_k = \alpha_k^* s_k, \tag{16.22}$$

where

$$\alpha_k^* = \operatorname*{argmin}_{\alpha\in\mathbb{R}} E(u_k + \alpha s_k) = \operatorname*{argzero}_{\alpha\in\mathbb{R}} \langle E'[u_k + \alpha s_k], s_k\rangle. \tag{16.23}$$

The correction ε_k is then applied via

$$u_{k+1} = u_k + \varepsilon_k. \tag{16.24}$$

Remark 16.60 (PSD)**.** We first observe that the PSD method of Definition 16.59 is a gradient descent algorithm where the preconditioner is invariant. The "steepest descent" part of the name refers to the orthogonalization step (16.23).

A first step in the analysis of PSD is to show that it is well defined. We have the following result.

Theorem 16.61 (well defined)**.** *Suppose that $E\colon \mathcal{H} \to \mathbb{R}$ is μ-strongly convex on the Hilbert space $\mathcal{H}$ and $\mathcal{L}\colon \mathcal{H} \to \mathcal{H}'$ is a fixed, invariant preconditioner. Let $u_0 \in \mathcal{H}$ be arbitrary. Suppose that $\mathcal{E}_0$ is the (convex) bounded energy set defined in* (16.13)*. Then the PSD algorithm in Definition 16.59 is well defined on $\mathcal{E}_0$.*

Proof. See Problem 16.19. □

To prove convergence of the PSD algorithm, we use the golden key of Corollary 16.58, which requires a certain lower bound and a certain upper bound. The lower bound is quite easy.

Theorem 16.62 (lower bound). *Suppose that E is μ-strongly convex on $\mathcal{H}$ and $\mathcal{L}$ is a (μ_1, L_1)-preconditioner. Let u_k be the kth iteration in the PSD algorithm of Definition 16.59. Then*

$$\delta_k = E(u_k) - E(u_{k+1}) \geq \frac{\mu}{2}\|\varepsilon_k\|_{\mathcal{H}}^2.$$

Proof. Owing to the line search (steepest descent step), we have the orthogonality property

$$\langle E'[u_{k+1}], w\rangle = 0, \quad w \in \operatorname{span}\{s_k\}; \tag{16.25}$$

see Problem 16.20. Applying Lemma 16.50, with the subspace $\mathcal{W} = \operatorname{span}\{s_k\}$, and noting that

$$u_{k+1} - u_k = \varepsilon_k = \alpha_k^* s_k \in \operatorname{span}\{s_k\},$$

we have

$$E(u_k) - E(u_{k+1}) \geq \frac{\mu}{2}\|u_k - u_{k+1}\|_{\mathcal{H}}^2 = \frac{\mu}{2}\|\varepsilon_k\|_{\mathcal{H}}^2.$$ □

The upper bound requires some preliminary estimates for α_k^*.

Lemma 16.63 (descent). *Let s_k be computed as in Definition 16.59. Suppose that $\mathcal{L}$ is a (μ_1, L_1)-preconditioner. Then s_k is a descent direction in the sense that*

$$\langle r_k, s_k\rangle = \langle -E'[u_k], s_k\rangle \geq \mu_1\|s_k\|_{\mathcal{H}}^2 > 0.$$

Proof. Recall the search direction problem: find $s_k \in \mathcal{H}$ such that

$$\langle \mathcal{L}[s_k], w\rangle = -\langle E'[u_k], w\rangle, \quad \forall w \in \mathcal{H}. \tag{16.26}$$

Choosing $w = s_k$, we obtain the inequality

$$\langle -E'[u_k], s_k\rangle = \langle \mathcal{L}[s_k], s_k\rangle \geq \mu_1\|s_k\|_{\mathcal{H}}^2 > 0.$$ □

In order to better understand the choice of the step size, we introduce the scalar function f_k; see Figure 16.2.

Proposition 16.64 (energy section). *Suppose that E is μ-strongly convex on $\mathcal{H}$ and $\mathcal{L}$ is a (μ_1, L_1)-preconditioner. Define the one-dimensional energy section*

$$f_k(\alpha) = E(u_k + \alpha s_k). \tag{16.27}$$

Then

$$f_k'(0) = \langle E'[u_k], s_k\rangle \leq -\mu_1 \|s_k\|_{\mathcal{H}}^2.$$

Furthermore, $\alpha_k^* > 0$ and, for all $\alpha \in (0, \alpha_k^*]$, $f_k(\alpha) < f_k(0)$.

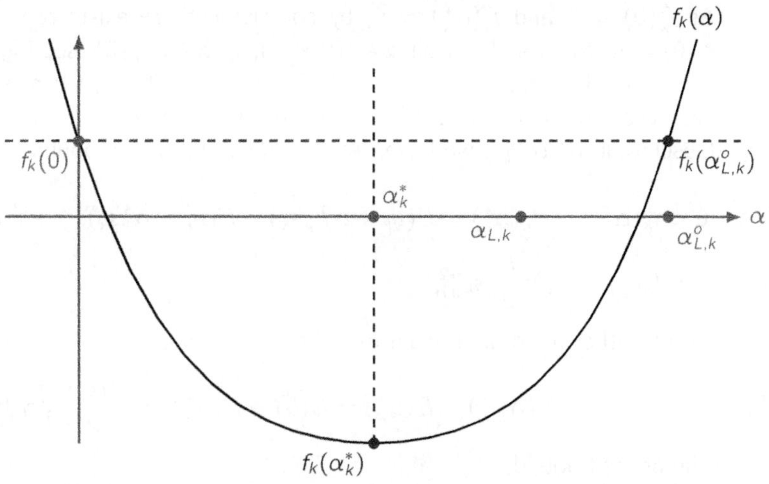

Figure 16.2 The function f_k is a one-dimensional energy section, a slice of the energy along the search direction s_k. It is straightforward to prove that its minimizer, α^*_k, is positive.

Proof. Lemma 16.63 implies that $f'_k(0) < 0$. As f'_k is continuous, we conclude that the minimizing point is positive, $\alpha^*_k > 0$, and, for all $\alpha \in (0, \alpha^*_k]$, $f_k(\alpha) < f_k(0) = E(u_k)$. □

It turns out that the energy sections f_k inherit many properties from the original energy E, as the following result shows.

Lemma 16.65 (properties of f_k). *Assume that E is μ-strongly convex on $\mathcal{H}$ and L-Lipschitz smooth on the bounded energy set $\mathcal{E}_0$. Then f_k, defined in (16.27), is differentiable and strongly convex in the following sense: for all $\alpha, \beta \in \mathbb{R}$,*

$$(f'_k(\alpha) - f'_k(\beta))(\alpha - \beta) \geq (\alpha - \beta)^2 \mu \|s_k\|^2_{\mathcal{H}}.$$

Furthermore, f'_k is Lipschitz in the following sense: for all $\alpha, \beta \in [0, \alpha_{L,k}]$,

$$|f'_k(\alpha) - f'_k(\beta)| \leq L\|s_k\|^2_{\mathcal{H}} |\alpha - \beta|,$$

where $\alpha_{L,k} = (1 + \sqrt{\mu/L})\alpha^*_k$.

Proof. The proof is based on the following identity:

$$f'_k(\alpha) - f'_k(\beta) = \langle E'[u_k + \alpha s_k] - E'[u_k + \beta s_k], s_k \rangle.$$

Then, by μ-strong convexity,

$$\begin{aligned}(f'_k(\alpha) - f'_k(\beta))(\alpha - \beta) &= \langle E'[u_k + \alpha s_k] - E'[u_k + \beta s_k], \alpha s_k - \beta s_k \rangle \\ &\geq \mu \|(\alpha - \beta) s_k\|^2_{\mathcal{H}}.\end{aligned}$$

To use the Lipschitz inequality involving E', we need to ensure that the points of evaluation are inside the set $\mathcal{E}_0$, which imposes an upper bound on α and β.

As $f_k'(0) < 0$ and $f_k'(\alpha_k^*) = 0$, by coercivity, there exists $\alpha_{L,k}^o > \alpha_k^*$ such that $f_k(0) = f_k(\alpha_{L,k}^o)$ and, for all $\alpha \in (0, \alpha_{L,k}^o)$, $f_k(\alpha) < f_k(0)$; see Figure 16.2.

This implies that $u_k + \alpha s_k \in \mathcal{E}_0$ for all $\alpha \in (0, \alpha_{L,k}^o)$, since the energy E is decreased for such values of α. We now estimate $\alpha_{L,k}^o$. As $f_k'(\alpha_k^*) = 0$ and f_k' is Lipschitz in $(0, \alpha_{L,k}^o)$, we have, from Lemma 16.50, the bound

$$0 < f_k(\alpha_{L,k}^o) - f_k(\alpha_i^*) = E(u_k + \alpha_{L,k}^o s_k) - E(u_k + \alpha_k^* s_k) \le \frac{L}{2} \left\|\alpha_{L,k}^o s_k - \alpha_k^* s_k\right\|_{\mathcal{H}}^2$$
$$= (\alpha_{L,k}^o - \alpha_k^*)^2 \frac{L}{2} \|s_k\|_{\mathcal{H}}^2.$$

On the other hand, and again from Lemma 16.50,

$$f_k(\alpha_{L,k}^o) - f_k(\alpha_k^*) = f_k(0) - f_k(\alpha_k^*) \ge \frac{\mu(\alpha_k^*)^2}{2} \|s_k\|_{\mathcal{H}}^2.$$

The desired bound,

$$\alpha_{L,k}^o \ge \alpha_{L,k} = \left(1 + \sqrt{\frac{\mu}{L}}\right) \alpha_k^* > \alpha_k^* > 0,$$

then follows. To finish up, since E' is Lipschitz, f_k' is Lipschitz with constant $L\|s_k\|_{\mathcal{H}}^2$ on the interval $[0, \alpha_{L,k}^o]$. Indeed, for all $\alpha, \beta \in (0, \alpha_{L,k}^o)$, $\alpha \neq \beta$,

$$\begin{aligned} |f_k'(\alpha) - f_k'(\beta)| &= |\langle E'[u_k + \alpha s_k] - E'[u_k + \beta s_k], s_k\rangle| \\ &= \frac{1}{|\alpha - \beta|} |\langle E'[u_k + \alpha s_k] - E'[u_k + \beta s_k], (\alpha - \beta) s_k\rangle| \\ &\le \frac{1}{|\alpha - \beta|} L\|(\alpha - \beta) s_k\|_{\mathcal{H}}^2 = L\|s_k\|_{\mathcal{H}}^2 |\alpha - \beta|. \end{aligned}$$

It is also Lipschitz with the same constant on the smaller interval $[0, \alpha_{L,k}] \subseteq [0, \alpha_{L,k}^o]$. The proof is complete. □

Remark 16.66 (computability)**.** The importance of using the right-hand endpoint

$$\alpha_{L,k} = \left(1 + \sqrt{\frac{\mu}{L}}\right) \alpha_k^*,$$

rather than $\alpha_{L,k}^o$, is that $\alpha_{L,k}$ is computable, provided that μ, L, and α_k^* are known.

We know that the optimal value α_k^* is positive. Let us provide a refined lower bound; see Figure 16.2.

Lemma 16.67 (refined lower bound)**.** *Assume that the energy E is μ-strongly convex on $\mathcal{H}$ and L-Lipschitz smooth on the bounded energy set $\mathcal{E}_0$ and $\mathcal{L}$ is a (μ_1, L_1)-preconditioner. Then we have the lower bound*

$$\frac{\mu_1}{L} \le \alpha_k^*.$$

Consequently,

$$\alpha_{L,k}^o \ge \alpha_{L,k} = \left(1 + \sqrt{\frac{\mu}{L}}\right) \alpha_k^* > \alpha_k^* \ge \frac{\mu_1}{L} > 0.$$

Proof. Recall that $\varepsilon_k = \alpha_k^* s_k \in \operatorname{span}\{s_k\}$ and, owing to the line search, we still have the orthogonality property (16.25). Thus, $E'[u_k + \varepsilon_k] = 0$ in the dual of $\operatorname{span}\{s_k\}$. The choice of search direction in Definition 16.59 implies that, in $\mathcal{H}'$, we have

$$-E'[u_k] = \mathcal{L}[s_k].$$

The lower bound is obtained by the coercivity of $\mathcal{L}$ and Lipschitz continuity of E':

$$\begin{aligned}
\alpha_k^* L \|s_k\|_{\mathcal{H}}^2 &= \frac{1}{\alpha_k^*} L \|\varepsilon_k\|_{\mathcal{H}}^2 \\
&\geq \frac{1}{\alpha_k^*} \langle E'[u_k + \varepsilon_k] - E'[u_k], \varepsilon_k \rangle \\
&= \langle E'[u_k + \varepsilon_k] - E'[u_k], s_k \rangle \\
&= -\langle E'[u_k], s_k \rangle \\
&= \langle \mathcal{L}[s_k], s_k \rangle \\
&\geq \mu_1 \|s_k\|_{\mathcal{H}}^2.
\end{aligned}$$

Note that $u_k + \varepsilon_k \in \mathcal{E}_0$ by Lemma 16.65, so that we can use Lipschitz continuity of E'. □

Next, we present a refined upper bound on the step size α_k^*.

Lemma 16.68 (refined upper bound). *Assume that the energy E is μ-strongly convex on $\mathcal{H}$ and L-Lipschitz smooth on $\mathcal{E}_0$ and $\mathcal{L}$ is a (μ_1, L_1)-preconditioner. Then we have the upper bound*

$$\alpha_k^* \leq \frac{L_1}{\mu}.$$

Proof. Since $E'[u_k + \varepsilon_k] = 0$ in the dual space of $\operatorname{span}\{s_k\}$,

$$\mu \|\varepsilon_k\|_{\mathcal{H}}^2 \leq \langle E'[u_k + \varepsilon_k] - E'[u_k], \varepsilon_k \rangle = \langle \mathcal{L}[s_k], \varepsilon_k \rangle \leq \frac{L_1}{\alpha_k^*} \|\varepsilon_k\|_{\mathcal{H}}^2. \qquad \square$$

With our estimates of α_k^* in place, we are now ready to establish the upper bound for the PSD algorithm of Definition 16.59 needed by the golden key of Corollary 16.58.

Theorem 16.69 (upper bound). *Suppose that the energy E is μ-strongly convex on $\mathcal{H}$ and L-Lipschitz smooth on the set $\mathcal{E}_0$ and $\mathcal{L}$ is a (μ_1, L_1)-preconditioner. Then we have the upper bound*

$$E(u_{k+1}) - E(u) \leq C_U \|\varepsilon_k\|_{\mathcal{H}}^2,$$

where

$$C_U = \frac{[L(1 + L_1/\mu_1)]^2}{2\mu}.$$

Proof. Note that, for any $w \in \mathcal{H}$,

$$\langle E'[u_{k+1}], w \rangle = \langle E'[u_{k+1}] - E'[u_k], w \rangle + \langle E'[u_k], w \rangle = \mathrm{I}_1 + \mathrm{I}_2,$$

where

$$\mathrm{I}_1 = \langle E'[u_{k+1}] - E'[u_k], w\rangle, \qquad \mathrm{I}_2 = \langle E'[u_k], w\rangle.$$

Since we know a priori that $u_k, u_{k+1} \in \mathcal{E}_0$, we can use the Lipschitz smoothness of E to get

$$\mathrm{I}_1 \le L \, \|\varepsilon_k\|_{\mathcal{H}} \| \, w\|_{\mathcal{H}}.$$

For I_2, we have

$$\begin{aligned}
\mathrm{I}_2 &= -\langle \mathcal{L}[s_k], w\rangle \\
&\le |\langle \mathcal{L}[s_k], w\rangle| \\
&\le \|\mathcal{L}[s_k]\|_{\mathcal{H}'} \, \|w\|_{\mathcal{H}} \\
&\le L_1 \, \|s_k\|_{\mathcal{H}} \, \|w\|_{\mathcal{H}} \\
&\le \frac{L_1}{\alpha_k^*} \, \|\varepsilon_k\|_{\mathcal{H}} \, \|w\|_{\mathcal{H}} \\
&\le \frac{L_1 L}{\mu_1} \, \|\varepsilon_k\|_{\mathcal{H}} \, \|w\|_{\mathcal{H}} \, .
\end{aligned}$$

In the last estimate, we used the relation $s_k = \alpha_k^{*\,-1} \varepsilon_k$ and the lower bound of α_k^* given in Lemma 16.67.

Putting the estimates together, we have, for any $w \in \mathcal{H}$,

$$\langle E'[u_{k+1}], w\rangle \le L \left(1 + \frac{L_1}{\mu_1}\right) \|\varepsilon_k\|_{\mathcal{H}} \|w\|_{\mathcal{H}},$$

which implies that

$$\|E'[u_{k+1}]\|_{\mathcal{H}'}^2 \le L^2 \left(1 + \frac{L_1}{\mu_1}\right)^2 \|\varepsilon_k\|_{\mathcal{H}}^2.$$

Using inequality (16.16) in Lemma 16.49 with $v = u_{k+1}$, the result follows. □

Using Theorems 16.62 and 16.69, and the golden key of Corollary 16.58, we obtain the following linear convergence result.

Corollary 16.70 (linear convergence)**.** *Let u_k be the kth iteration and u_{k+1} the next iteration in the PSD algorithm. Suppose that the energy E is μ-strongly convex on $\mathcal{H}$ and L-Lipschitz smooth on the set $\mathcal{E}_0$ and $\mathcal{L}$ is a (μ_1, L_1)-preconditioner. Then*

$$E(u_{k+1}) - E(u) \le \rho(E(u_k) - E(u)),$$

with

$$\rho = \frac{L^2 \, (1 + L_1/\mu_1)^2}{L^2 \, (1 + L_1/\mu_1)^2 + \mu^2}.$$

If we take $\mathcal{L} = \mathfrak{R}^{-1}$, then $\mu_1 = L_1 = 1$ and

$$\rho = \frac{4L^2}{\mu^2 + 4L^2}.$$

Proof. See Problem 16.21. □

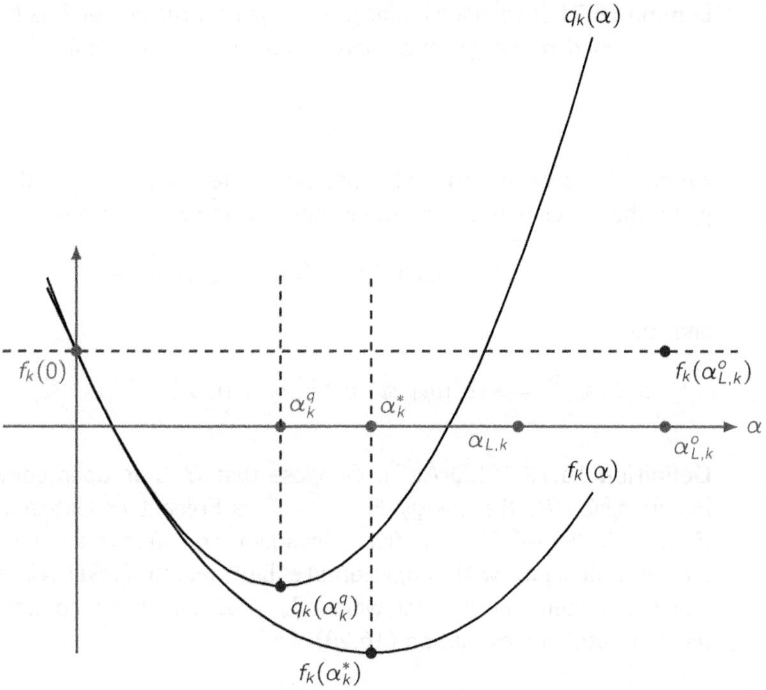

Figure 16.3 The function f_k, defined in (16.27), and its quadratic approximation q_k, which is defined in (16.28). The quadratic minimizer, α^q_k, is always to the left of α^*_k by construction.

16.7 PSD with Approximate Line Search

Next, we introduce another preconditioned gradient descent method. This one, which we will call the *preconditioned steepest descent method with approximate line search* (PSD-ALS) — a real mouthful — is pretty much as the name suggests. For this, we need a quadratic approximation of the function f_k, which was defined in (16.27). Before we go any further, we should point out that, in the optimization literature, this method is usually just called the *gradient descent method* or *preconditioned gradient descent method*; see, for example, [64].

From its definition, we see that $f_k(0) = E(u_k)$, $f_k'(0) = \langle E'[u_k], s_k\rangle < 0$. Using $f_k(0)$ and $f_k'(0)$, we define the quadratic function

$$q_k(\alpha) = f_k(0) + f_k'(0)\alpha + \frac{L\|s_k\|^2_{\mathcal{H}}}{2}\alpha^2; \tag{16.28}$$

see Figure 16.3. The optimal step size for PSD is, of course, $\alpha^*_k = \operatorname{argmin}_{\alpha\in\mathbb{R}} f_k(\alpha)$. Our choice for this new algorithm is

$$\alpha^q_k = \operatorname*{argmin}_{\alpha\in\mathbb{R}} q_k(\alpha) = -\frac{\langle E'[u_k], s_k\rangle}{L\|s_k\|^2_{\mathcal{H}}} = -\frac{f_k'(0)}{L\|s_k\|^2_{\mathcal{H}}}, \tag{16.29}$$

which satisfies the following estimate.

Lemma 16.71 (two-sided bound)**.** *Assume that the energy E is L-Lipschitz smooth on the bounded energy set $\mathcal{E}_0$ and $\mathcal{L}$ is a (μ_1, L_1)-preconditioner. Then*

$$\frac{\mu_1}{L} \le \alpha_k^q \le \alpha_k^*.$$

Proof. The lower bound is obtained by the definition of α_k^q and Lemma 16.63. To prove the upper bound, we notice that, owing to line search,

$$f_k'(\alpha_k^*) = \langle E'[u_k + \alpha_k^* s_k], s_k\rangle = 0$$

and, thus,

$$\alpha_k^q L\|s_k\|_{\mathcal{H}}^2 = -\langle E'[u_k], s_k\rangle = \langle E'[u_k + \alpha_k^* s_k] - E'[u_k], s_k\rangle \le \alpha_k^* L\|s_k\|_{\mathcal{H}}^2. \qquad \square$$

Definition 16.72 (PSD-ALS)**.** Suppose that B is an open convex subset of the Hilbert space $\mathcal{H}$, the energy $E\colon \mathcal{H} \to \mathbb{R}$ is Fréchet or Gateaux differentiable on B, and $\mathcal{L}\colon \mathcal{H} \to \mathcal{H}'$ is a fixed, invariant preconditioner. The **preconditioned steepest descent with approximate line search (PSD-ALS) algorithm** is a gradient descent method for which $\mathcal{L}_k = \mathcal{L}$, and the step size is computed via the quadratic approximation (16.29), i.e.,

$$\varepsilon_k = \alpha_k^q s_k, \qquad \alpha_k^q = \frac{\langle r_k, s_k\rangle}{L\|s_k\|_{\mathcal{H}}^2}. \tag{16.30}$$

Thus,

$$u_{k+1} = u_k + \varepsilon_k, \qquad \varepsilon_k = \alpha_k^q s_k.$$

Let us now elucidate the properties of the PSD-ALS algorithm. We will show that it is well defined and convergent.

Remark 16.73. In the literature, the PSD-ALS method is also called the preconditioned gradient descent method (PGD); see [64].

Theorem 16.74 (well defined)**.** *Suppose that $E\colon \mathcal{H} \to \mathbb{R}$ is μ-strongly convex on the Hilbert space $\mathcal{H}$ and $\mathcal{L}\colon \mathcal{H} \to \mathcal{H}'$ is a fixed, invariant preconditioner. Let $u_0 \in \mathcal{H}$ be arbitrary. Suppose that $\mathcal{E}_0$ is the (convex) bounded energy set defined in* (16.13)*. Then the PSD-ALS algorithm of Definition 16.72 is well defined on $\mathcal{E}_0$.*

Proof. See Problem 16.22. $\square$

Let us now show that, as Figure 16.3 suggests, the quadratic approximation q_k of f_k is larger than the energy section.

Lemma 16.75 (upper approximation)**.** *Suppose that the energy E is L-Lipschitz on the bounded energy set $\mathcal{E}_0$. Then*

$$f_k(\alpha) \le q_k(\alpha), \quad \forall \alpha \in [0, \alpha_{L,k}].$$

Proof. By Lemma 16.65, for $\alpha \in [0, \alpha_{L,k}]$, f_k' is Lipschitz continuous with constant $L\|s_k\|_{\mathcal{H}}^2$. We recall a classical result: since f_k' is Lipschitz continuous with constant $L\|s_k\|_{\mathcal{H}}^2$ on $\alpha \in [0, \alpha_{L,k}]$, then

$$|f_k(\alpha) - f_k(\beta) - (\alpha - \beta)f_k'(\beta)| \le \frac{L\|s_k\|_{\mathcal{H}}^2}{2}|\alpha - \beta|^2, \quad \forall \alpha, \beta \in [0, \alpha_{L,k}]; \quad (16.31)$$

see Problem 16.23. Setting $\beta = 0$ above, we get, for all $\alpha \in [0, \alpha_{L,k}]$,

$$\begin{aligned} f_k(\alpha) - q_k(\alpha) &= f_k(\alpha) - f_k(0) - \alpha f_k'(0) - \frac{L\|s_k\|_{\mathcal{H}}^2}{2}\alpha^2 \\ &\le |f_k(\alpha) - f_k(0) - \alpha f_k'(0)| - \frac{L\|s_k\|_{\mathcal{H}}^2}{2}\alpha^2 \\ &\le \frac{L\|s_k\|_{\mathcal{H}}^2}{2}\alpha^2 - \frac{L\|s_k\|_{\mathcal{H}}^2}{2}\alpha^2 \\ &= 0, \end{aligned}$$

which implies the result. □

Now, since the optimal linear search procedure is broken, the orthogonality condition (16.25) with respect to the corrections is broken. Thus, establishing the lower bound is a little more complicated.

Theorem 16.76 (lower bound). *Let $\{u_k\}_{k=0}^{\infty}$ be computed by the PSD-ALS algorithm given in Definition 16.72. Suppose that E is μ-strongly convex on $\mathcal{H}$ and L-Lipschitz smooth on $\mathcal{E}_0$ and $\mathcal{L}$ is a (μ_1, L_1)-preconditioner. Then we have*

$$E(u_k) - E(u_{k+1}) \ge C_L\|\alpha_k^q s_k\|_{\mathcal{H}}^2, \quad C_L = \frac{L}{2}.$$

Proof. It suffices to prove that

$$E(u_k) - E(u_{k+1}) = f_k(0) - f_k(\alpha_k^q) \ge \frac{L}{2}\|\alpha_k^q s_k\|^2.$$

By Lemma 16.75, $f_k(\alpha) \le q_k(\alpha)$ for all $\alpha \in [0, \alpha_{L,k}]$. As $\alpha_k^q = \operatorname{argmin}_{\alpha\in\mathbb{R}} q_k(\alpha)$ and $\alpha_k^q \le \alpha_k^*$, we get

$$f_k(\alpha_k^q) \le q_k(\alpha_k^q) = \min_{\alpha\in\mathbb{R}} q_k(\alpha) = f_k(0) - \frac{1}{2L\|s_k\|_{\mathcal{H}}^2}|f_k'(0)|^2 = f_k(0) - \frac{L}{2}\|\alpha_k^q s_k\|_{\mathcal{H}}^2.$$

In the last step, we have used the definition of α_k^q and this completes the proof. □

Since α_k^q has the same lower bound as α_k^*, we can derive the upper bound in exactly the same way as the proof of Theorem 16.69, only replacing $\varepsilon_k = \alpha_k^* s_k$ by $\alpha_k^q s_k$ and using the lower bound from Lemma 16.71.

Theorem 16.77 (upper bound). *Let $u_k, u_{k+1} \in \mathcal{H}$ be two consecutive iterations of the PSD-ALS algorithm of Definition 16.72. Suppose that E is μ-strongly convex on $\mathcal{H}$ and L-Lipschitz smooth on $\mathcal{E}_0$ and $\mathcal{L}$ is a (μ_1, L_1)-preconditioner. Then we have the upper bound*

$$E(u_{k+1}) - E(u) \le C_U\|\alpha_k^q s_k\|_{\mathcal{H}}^2, \quad C_U = \frac{L^2(1 + L_1/\mu_1)^2}{2\mu}.$$

Proof. See Problem 16.24. □

Finally, using the golden key of Corollary 16.58, we can easily compute the linear rate of convergence.

Corollary 16.78 (convergence). *Let $u_k, u_{k+1} \in \mathcal{H}$ be two consecutive iterations of the PSD-ALS algorithm of Definition 16.72. Suppose that E is μ-strongly convex on $\mathcal{H}$ and L-Lipschitz smooth on $\mathcal{E}_0$ and $\mathcal{L}$ is a (μ_1, L_1)-preconditioner. Then we have*

$$E(u_{k+1}) - E(u) \leq \rho(E(u_k) - E(u))$$

with

$$\rho = \frac{L^2 (1 + L_1/\mu_1)^2}{L^2 (1 + L_1/\mu_1)^2 + L\mu}.$$

Proof. See Problem 16.25. □

16.8 Newton's Method

In Section 15.5, we considered Newton's method with the purpose of solving a nonlinear system of $d \in \mathbb{N}$ equations in d unknowns. Given $\boldsymbol{f}\colon \mathbb{R}^d \to \mathbb{R}^d$, we wish to find a solution to

$$\boldsymbol{f}(\boldsymbol{x}) = \boldsymbol{0}.$$

Here, we will explore how we can use Newton's method to solve optimization problems as well. The idea is to think about solving the Euler equation,

$$\langle E'[v], w \rangle = 0, \quad \forall w \in \mathcal{H},$$

with the method. This might not be a finite-dimensional problem anymore, but this is not a problem. Newton's method continues to make sense for an infinite-dimensional Hilbert space $\mathcal{H}$. To see this, we observe that the Euler equation is actually an equation on $\mathcal{H}'$. Indeed, we define the function $F\colon \mathcal{H} \to \mathcal{H}'$ as $F(v) = E'[v]$, so that we are seeking

$$F(v) = 0 \in \mathcal{H}'.$$

Thus, Newton's method, at this formal stage, reads

$$F'[u_k](u_{k+1} - u_k) = -F(u_k).$$

However, here we see two issues. First, we must make sense of F', where F maps an infinite-dimensional normed space to another. Second, since the function F itself is the derivative of a functional, we must define higher order derivatives. Let us then present some facts about this theory.

First, we generalize the concept of Fréchet differentiability, introduced in Definition 16.32 for mappings between normed spaces.

Definition 16.79 (Fréchet derivative)**.** Let $\mathbb{V}, \mathbb{W}$ be normed spaces and $F\colon \mathbb{V} \to \mathbb{W}$. We say that F if **Fréchet differentiable at the point** $v \in \mathbb{V}$ if and only if there exists $T \in \mathfrak{B}(\mathbb{V}, \mathbb{W})$ such that

$$\lim_{\|h\|_{\mathbb{V}} \to 0} \frac{\|F(v+h) - F(v) - T(h)\|_{\mathbb{W}}}{\|h\|_{\mathbb{V}}} \to 0.$$

If F is Fréchet differentiable at every point of a set $B \subseteq \mathbb{V}$, then we say that F is **Fréchet differentiable on** B.

As in Proposition 16.33, we can show that, if it exists, the Fréchet derivative is unique, and so we denote this as $F'[v] \in \mathfrak{B}(\mathbb{V}, \mathbb{W})$. Observe now that, if $\mathcal{H}$ is a Hilbert space and $E\colon \mathcal{H} \to \mathbb{R}$ is a convex energy that, in addition, is Fréchet differentiable, we can then define $F\colon \mathcal{H} \to \mathcal{H}'$ via $F(v) = E'[v]$. With the definition given above, we can then speak about the differentiability of F, which in turn will give us the *second derivatives* of the energy E.

Definition 16.80 (second derivative)**.** Let $\mathcal{H}$ be a Hilbert space and $E\colon \mathcal{H} \to \mathbb{R}$ be an energy that is Fréchet differentiable on $\mathcal{H}$. If the mapping

$$E'\colon \mathcal{H} \to \mathcal{H}', \qquad v \in \mathcal{H} \mapsto E'[v] \in \mathcal{H}'$$

is Fréchet differentiable at $v \in \mathcal{H}$, then its derivative

$$E''[v] \in \mathfrak{B}(\mathcal{H}, \mathfrak{B}(\mathcal{H}, \mathbb{R})) = \mathfrak{B}(\mathcal{H}, \mathcal{H}')$$

is called the **second Fréchet derivative** of E at v, and E is said to be **twice Fréchet differentiable at the point** v. If the energy E is twice differentiable at every point of a set $B \subseteq \mathcal{H}$, then we say that E is **twice Fréchet differentiable on** B.

Remark 16.81 (notation)**.** The definition of the second derivative and what it represents requires some explanation. First, we noted that $\mathfrak{B}(\mathcal{H}, \mathbb{R}) = \mathcal{H}'$, a fact that easily follows from the definition. More importantly, we noted that, for $v \in \mathcal{H}$, we have

$$E''[v] \in \mathfrak{B}(\mathcal{H}, \mathcal{H}').$$

This means that, given $w \in \mathcal{H}$, we can define the continuous linear functional $E''[v](w) \in \mathcal{H}'$, so that

$$E''[v](w, x) = \langle E''[v](w), x \rangle, \quad \forall x \in \mathcal{H}$$

is meaningful and it satisfies

$$|E''[v](w, x)| \leq \|E''[v](w)\|_{\mathcal{H}'} \|x\|_{\mathcal{H}} \leq \|E''[v]\|_{\mathfrak{B}(\mathcal{H}, \mathcal{H}')} \|w\|_{\mathcal{H}} \|x\|_{\mathcal{H}}.$$

Finally, we note that, a priori, the order of the arguments in $E''[v](\cdot, \cdot)$ matters. We will see below, however, that this is not the case.

Before we define Newton's method, it is useful to have some assumptions on the second derivative.

Proposition 16.82 (second derivative). *Suppose that $E\colon \mathcal{H} \to \mathbb{R}$ is twice Fréchet differentiable on $\mathcal{H}$. Assume that:*

1. *E'' is μ-*coercive *on $\mathcal{H}$, i.e., for all $v \in \mathcal{H}$, there is a constant $\mu > 0$ such that*
$$\mu \|w\|_{\mathcal{H}}^2 \le E''[v](w, w), \quad \forall w \in \mathcal{H}.$$
2. *E''* is locally uniformly continuous *on $\mathcal{H}$, i.e., given any bounded convex subset $B \subset \mathcal{H}$, there is a constant $L_B > 0$ such that, for all $v \in B$,*
$$E''[v](w, x) \le L_B \|w\|_{\mathcal{H}} \|x\|_{\mathcal{H}}, \quad \forall w, x \in \mathcal{H}.$$

Then E'' is symmetric, in the sense that
$$E''[v](w, x) = E''[v](x, w), \quad \forall v, w, x \in \mathcal{H},$$
and, for a fixed $v \in \mathcal{H}$, the second derivative $E''[v](\cdot, \cdot)$ is bilinear. Furthermore, E is μ-strongly convex and locally Lipschitz smooth on $\mathcal{H}$.

Proof. The symmetry and bilinearity are standard properties of the second Fréchet derivative and can be found in the literature under the name *Schwarz Lemma*.[8] The last two properties follow from Problem 16.26. First, let us fix $w \in \mathcal{H}$. For every $v \in \mathcal{H}$, there is a point $x \in \mathcal{H}$ on the line segment between w and v such that
$$E(v) = E(w) + E'[w](v - w) + \frac{1}{2} E''[x](v - w, v - w). \tag{16.32}$$
Thus, for any $v, w \in \mathcal{H}$,
$$E(v) - E(w) - E'[w](v - w) = \frac{1}{2} E''[x](v - w, v - w) \ge \frac{\mu}{2} \|v - w\|_{\mathcal{H}}^2 .$$
Reversing the roles of v and w above, we have
$$E(w) - E(v) - E'[v](w - v) \ge \frac{\mu}{2} \|v - w\|_{\mathcal{H}}^2 .$$
Adding these inequalities gives
$$\langle E'[v] - E'[w], v - w \rangle \ge \mu \|v - w\|_{\mathcal{H}}^2 ,$$
which implies that E is μ-strongly convex.

For the upper bound, let us use the following version of Taylor's Theorem: fix $w \in B$. For every $v \in B$, there is a point $x \in B$ on the line segment between w and v such that
$$\langle E'[v], z \rangle = \langle E'[w], z \rangle + E''[x](w - v, z), \quad \forall z \in \mathcal{H}; \tag{16.33}$$
see Problem 16.27. Thus,
$$\langle E'[v] - E'[w], z \rangle = E''[x](w - v, z) \le L_B \|w - v\|_{\mathcal{H}} \|z\|_{\mathcal{H}} , \quad \forall z \in \mathcal{H}.$$

[8] Named in honor of the German mathematician Karl Hermann Amandus Schwarz (1843–1921).

Consequently,

$$\frac{\langle E'[v] - E'[w], z\rangle}{\|z\|_{\mathcal{H}}} \leq L_B \|w - v\|_{\mathcal{H}}, \quad \forall z \in \mathcal{H},$$

which implies that, for any $v, w \in B$,

$$\|E'[v] - E'[w]\|_{\mathcal{H}'} \leq L_B \|w - v\|_{\mathcal{H}}. \qquad \square$$

We are now ready to define Newton's method.

Definition 16.83 (Newton's method)**.** Suppose that B is an open convex subset of the Hilbert space $\mathcal{H}$ and $E\colon \mathcal{H} \to \mathbb{R}$ is twice Fréchet differentiable on B. **Newton's method** is a type of gradient descent method for which the preconditioner $\mathcal{L}_k$ changes at every iteration. It is given by

$$\mathcal{L}_k[v](\cdot) = E''[u_k](v, \cdot), \quad \forall v \in \mathcal{H},$$

i.e., the search direction is computed as the solution to

$$E''[u_k](s_k, w) = \langle r_k, w\rangle, \quad \forall w \in \mathcal{H}, \tag{16.34}$$

and the step size is $\alpha_k = 1$, which implies that the correction and the search direction are the same: $\varepsilon_k = s_k$. To sum up, Newton's method can be expressed as

$$E''[u_k](u_{k+1} - u_k, w) = -E'[u_k](w), \quad \forall w \in \mathcal{H}.$$

We immediately notice that, for Newton's method to be well defined, we require $E''[u_k](\cdot, \cdot)$ to be invertible. We will see that this is the case for a certain class of problems.

Lemma 16.84 (well defined)**.** *Suppose that $E\colon \mathcal{H} \to \mathbb{R}$ is twice Fréchet differentiable on $\mathcal{H}$, E'' is μ-coercive on $\mathcal{H}$, and E'' is locally uniformly continuous on $\mathcal{H}$. Let $u_0 \in \mathcal{H}$ be arbitrary. Suppose that $\mathcal{E}_0$ is the (convex) bounded energy set defined in* (16.13)*. Assume that uniform continuity constant for E'' on $\mathcal{E}_0$ is $L > 0$. Then, for any $u_k \in \mathcal{E}_0$,* (16.34) *always has a unique solution.*

Proof. Because of the assumptions on E'', we can use the Riesz Representation Theorem to prove the unique existence of the solution $s_k \in \mathcal{H}$ to the problem defined in (16.34). This is because $E''[u_k](\cdot, \cdot)$ defines an inner product on $\mathcal{H}$, equivalent to the canonical inner product. $\square$

The previous statement is far from giving a complete answer to the properties of Newton's method. It only says that we can advance *one iteration*. The complete analysis of Newton's method is more involved and will be given in the following sections.

16.8.1 Affine Invariance

Before we embark on the discussion of the convergence of Newton's method, let us show an important feature of Newton's method in this context. Namely, that of affine invariance. Simply put, this means that this method is independent of a linear (or affine) change of coordinates.

Let $T \in \mathfrak{B}(\mathcal{H})$ be such that there is $m_T > 0$ for which $\|Tv\|_{\mathcal{H}} \geq m_T \|v\|_{\mathcal{H}}$ for all $v \in \mathcal{H}$. Proposition 16.22 shows that this operator has a bounded inverse and provides a bound for the norm of the inverse. The action of T can be understood as a change of coordinates in T. Given an energy $E\colon \mathcal{H} \to \mathbb{R}$, we define

$$E_T(v) = E(Tv). \tag{16.35}$$

It turns out that many of the properties of E are inherited by E_T.

Lemma 16.85 (differentiability). *Let the energy $E_T\colon \mathcal{H} \to \mathbb{R}$ be defined as in* (16.35). *Then we have that E_T is twice Fréchet differentiable, with*

$$\langle E_T'[v], w\rangle = \langle E'[Tv], Tw\rangle, \qquad E_T''[v](w,x) = E''[Tv](Tw, Tx)$$

for all $v, w, x \in \mathcal{H}$. Moreover, if E satisfies all the hypotheses of Proposition 16.82, then E_T is μm_T^2-coercive on $\mathcal{H}$ and locally uniformly continuous with constant $L_{T(B)}\|T\|^2_{\mathfrak{B}(\mathcal{H})}$, where we denoted by $T(B)$ the image of B under the mapping T. Finally,

$$u = \operatorname*{argmin}_{v\in\mathcal{H}} E(v) \qquad \Longleftrightarrow \qquad T^{-1}u = \operatorname*{argmin}_{v\in\mathcal{H}} E_T(v).$$

Proof. See Problem 16.28. □

The affine invariance of Newton's method is the content of the next result.

Theorem 16.86 (affine invariance). *Let $\mathcal{H}$ be a Hilbert space and $E\colon \mathcal{H} \to \mathbb{R}$ satisfy all the hypotheses of Proposition 16.82. Let $E_T\colon \mathcal{H} \to \mathbb{R}$ be defined as in* (16.35). *Let $u_0 \in \mathcal{H}$. Denote by $\{u_k\}_{k\in\mathbb{N}}$ the result of applying Newton's method of Definition 16.83 to E_T. Starting from Tu_0, denote by $\{\tilde{u}_k\}_{k\in\mathbb{N}}$ the result of Newton's method, when applied to E. Then, for all $k \in \mathbb{N}_0$, we have*

$$\tilde{u}_k = Tu_k.$$

Proof. We will proceed by induction on k, with the obvious starting point $\tilde{u}_0 = Tu_0$. Assume then that, for all $k \leq n$, we have the thesis. If this is the case, then $u_{n+1} = u_n + s_n$ with

$$E_T''[u_n](s_n, w) = -E_T'[u_n](w), \quad \forall w \in \mathcal{H},$$

which by Lemma 16.85 is equivalent to, for all $w \in \mathcal{H}$,

$$E''[Tu_n](Ts_n, Tw) = -E'[Tu_n](Tw) \quad \Longleftrightarrow \quad E''[Tu_n](Ts_n, w) = -E'[Tu_n](w),$$

where we used that T is invertible. On the other hand, the Newton iterations for the energy E yield that $\tilde{u}_{n+1} = \tilde{u}_n + \tilde{s}_n$ with

$$E''[Tu_n](\tilde{s}_n, w) = E''[\tilde{u}_n](\tilde{s}_n, w) = -E'[\tilde{u}_n](w) = -E'[Tu_n](w), \quad \forall w \in \mathcal{H},$$

where we used the induction hypothesis. By uniqueness, we must then have that $\tilde{s}_n = Ts_n$, so that

$$\tilde{u}_{n+1} = \tilde{u}_n + \tilde{s}_n = Tu_n + Ts_n = T(u_n + s_n) = Tu_{n+1},$$

as we claimed. □

16.8.2 Local Convergence

Recall that, with Newton's method, we are, essentially, looking for the zero of a function. Thus, the analysis of local convergence of Newton's method presented in Section 15.5 can be generalized to this context.

Theorem 16.87 (local convergence). *Let $\mathcal{H}$ be a Hilbert space and $E\colon \mathcal{H} \to \mathbb{R}$ be a strongly convex and locally Lipschitz smooth energy. Assume that:*

1. *E is twice Fréchet differentiable on $\mathcal{H}$.*
2. *E'' is μ-coercive on $\mathcal{H}$.*
3. *E'' is locally uniformly continuous on $\mathcal{H}$.*
4. *E'' is locally Lipschitz continuous on $\mathcal{H}$; i.e., for every $B \subseteq \mathcal{H}$ that is bounded, there is γ_B such that*

$$\|E''[v] - E''[w]\|_{\mathfrak{B}(\mathcal{H},\mathcal{H}')} \leq \gamma_B \|v - w\|_{\mathcal{H}}.$$

Let $u \in \mathcal{H}$ be the (necessarily unique) minimum of E and, for $r > 0$, define

$$\overline{B}(u, r) = \{v \in \mathcal{H} \mid \|v - u\|_{\mathcal{H}} \leq r\}.$$

Denote by γ_r and L_r, respectively, the local Lipschitz and uniform continuity constants of E'' on $\overline{B}(u, r)$. Assume that:

$$r \leq \lambda \frac{L_r}{\gamma_r}, \qquad \lambda < \frac{1}{4}.$$

Then, for every $u_0 \in \overline{B}(u, r)$, Newton's method of Definition 16.83 is well defined, $\{u_k\}_{k=1}^{\infty} \subseteq \overline{B}(u, r)$, and, finally, Newton's method converges at least linearly.

Proof. The proof is nothing but a restatement of the results in Section 15.5 by carefully extending the results from $\mathbb{R}^d$ to this infinite-dimensional setting. We leave the results to the reader; see [16] or Problem 16.29. □

16.8.3 Damped Newton's Method. Forcing Global Convergence

We conclude our discussion of Newton's method by presenting a variant of this method that is *globally convergent*. We will only present the method and sketch the main results detailing its properties. For further details, we refer to [8, 27, 68].

Definition 16.88 (damped Newton)**.** Suppose that B is an open convex subset of the Hilbert space $\mathcal{H}$ and $E\colon \mathcal{H} \to \mathbb{R}$ is twice Fréchet differentiable on B. The **damped Newton's method** is a type of gradient descent method for which the search direction is computed as the solution to (16.34) and the step size is obtained by a **backtracking line search** algorithm, i.e., $\alpha_k \in (0, 1]$ is the largest step size for which

$$E(u_k + \alpha_k s_k) \leq E(u_k) + \frac{\alpha_k}{3}\langle E'[u_k], s_k\rangle.$$

Remark 16.89 (implementation). In the backtracking line search step the factor $\frac{1}{3}$ is somewhat arbitrary; any number in the interval $(0, \frac{1}{2})$ is suitable. In addition, the backtracking is usually implemented as follows. One chooses a parameter $\beta \in (0, 1)$ and sets $\alpha = 1$. While

$$E(u_k + \alpha s_k) > E(u_k) + \frac{\alpha}{3}\langle E'[u_k], s_k\rangle,$$

one replaces $\alpha \leftarrow \beta\alpha$.

The idea behind backtracking is that one performs an approximate line search in the direction s_k, and the step size is reduced until the exit condition is satisfied. Intuitively, the factor $\frac{1}{3}$ is by how much we allow the line search to be inexact.

Let us state and sketch the proof of the global convergence of the damped Newton's method.

Theorem 16.90 (global convergence). *Let $\mathcal{H}$ be a Hilbert space and $E\colon \mathcal{H} \to \mathbb{R}$ be a strongly convex and locally Lipschitz smooth energy. Assume that:*

1. *E is twice Fréchet differentiable on $\mathcal{H}$.*
2. *E'' is μ-coercive on $\mathcal{H}$.*
3. *E'' is uniformly continuous on $\mathcal{E}_0$ with constant M.*
4. *E'' is Lipschitz continuous on $\mathcal{E}_0$ with constant L.*

Then, for any u_0, the sequence $\{u_k\}_{k=1}^{\infty}$ obtained by the damped Newton's method of Definition 16.88 is well defined, i.e., $\{u_k\}_{k=1}^{\infty} \subset \mathcal{E}_0$. Moreover, we have that

$$u_k \to u, \qquad u = \operatorname*{argmin}_{v \in \mathcal{H}} E(v).$$

Finally, there is a $K \geq 0$ such that, for $k \geq K$, the convergence of the damped Newton method is quadratic.

Proof. (sketch) It is possible to show that there are numbers $\gamma > 0$ and $\eta \in (0, \mu^2/L)$ such that if

$$\|E'[u_k]\|_{\mathcal{H}'} \geq \eta, \tag{16.36}$$

then

$$E(u_{k+1}) - E(u_k) \leq -\gamma.$$

Otherwise, i.e., if $\|E'[u_k]\|_{\mathcal{H}'} < \eta$, the step size is chosen as $\alpha_k = 1$ and, moreover,

$$\frac{L}{2\mu^2}\|E'[u_{k+1}]\|_{\mathcal{H}'} \leq \left(\frac{L}{2\mu^2}\|E'[u_k]\|_{\mathcal{H}'}\right)^2.$$

Notice that if (16.36) is false, then we also have that $\|E'[u_{k+1}]\|_{\mathcal{H}'} < \eta$ and so, whenever $\ell \geq k$, we must have

$$\frac{L}{2\mu^2}\|E'[u_\ell]\|_{\mathcal{H}'} \leq \left(\frac{L}{2\mu^2}\|E'[u_k]\|_{\mathcal{H}'}\right)^{2^{\ell-k}} \leq \left(\frac{1}{2}\right)^{2^{\ell-k}}.$$

By Lemma 16.49, this implies that

$$E(u_\ell) - E(u) \le \frac{1}{2\mu}\|E'[u_\ell]\|_{\mathcal{H}'}^2 \le \frac{2\mu^2}{L^2}\left(\frac{1}{2}\right)^{2^{\ell-k+1}},$$

from which the quadratic convergence follows.

We finish by commenting that the backtracking guarantees the existence of $\gamma > 0$, so that, by continuity, since $E'[u] = 0$ we will eventually violate condition (16.36). □

Remark 16.91 (quasi-Newton methods)**.** The most difficult step in either Newton or damped Newton methods is finding the search direction via (16.34), since the preconditioner changes at every iteration. There are variants, known as *quasi-Newton methods*, in which this problem is replaced by simpler ones. We will not discuss these here.

16.9 Accelerated Gradient Descent Methods

We have so far presented two classes of methods: steepest descent, and its variants, and Newton. We also developed a theory regarding the convergence, both global and local, of each of these schemes. At this stage, the avid reader may have noticed a fundamental difference between these two methods, one that is of utmost relevance in practice. Namely, steepest descent is a *first-order method*, whereas Newton is a second-order one. By this, we mean that the order of derivatives of the objective that we assume known and computable for each one of these schemes is one and two, respectively.

In simple problems, computing derivatives may not be an issue. However, in practice, the second derivative of the objective function may not be available or it may be extremely difficult to compute. Even if the second derivative is available, as Remark 16.91 has pointed out, the computation of the search direction, essentially, entails the solution of a large linear system of equations. The reasons outlined above show that, for large-scale optimization problems, first-order methods may be preferable.

Having set our attention to first-order methods, we must now think about their performance. To make matters concrete, we will consider schemes without preconditioning and operate under the assumptions that our objective is μ-strongly convex and *globally* Lipschitz smooth with constant L. We must note, however, that, while strong convexity is a realistic assumption, and many objectives are *locally* Lipschitz smooth, making this a global assumption is rather unrealistic. In fact, most of the objectives that we encounter in practice will not satisfy this condition. Nevertheless, let us assume this for the sake of argument.

Motivated by Problem 16.30, to quantify the performance of our schemes we will introduce the *nonlinear* condition number

$$\hat{\kappa}(E) = \frac{L}{\mu} \ge 1.$$

With this quantity at hand, we observe that both Corollary 16.70 and Corollary 16.78 show that

$$E(u_{k+1}) - E(u) = \mathcal{O}\left(\left(1 - \frac{1}{\hat{\kappa}(E)}\right)^k\right).$$

Notice also that this is consistent with the convergence rates we obtained for gradient descent methods when we discussed the iterative solution of linear systems with a symmetric positive definite (SPD) matrix; see Theorems 7.18 and 7.21. We also refer to Remark 7.19 for a discussion on the meaning of this estimate, and its limitation, particularly in the case of a large $\hat{\kappa}(E)$.

This was one of the motivations, and the major advantage, of the conjugate gradient method; see Definitions 7.25 and 7.40 and Theorem 7.39. The convergence rate depends not on the condition number itself, but rather on its square root. This motivates the search for general first-order optimization schemes whose rate of convergence depends not on the nonlinear condition number, but on some other, smaller, quantity. In this regard, we mention that there are, for instance, nonlinear variants of the conjugate gradient method; see [65, Section 5.2] and [16, Section 8.5].

We begin with a sort of threshold on what can be expected from a first-order scheme. The following result can be found in [64, Theorem 2.1.13].

Theorem 16.92 (lower bound)**.** *Let $\mathcal{H}$ be a Hilbert space. For any $u_0 \in \mathcal{H}$ and constants $0 < \mu < L$, there is an objective $E\colon \mathcal{H} \to \mathbb{R}$ that is μ-strongly convex and globally Lipschitz smooth with constant L such that, for any first-order optimization scheme satisfying, for $k \geq 0$,*

$$u_{k+1} - u_0 \in \operatorname{span}\{\mathfrak{R}E'[u_0], \dots, \mathfrak{R}E'[u_k]\}, \tag{16.37}$$

we have

$$E(u_k) - E(u) \geq \frac{\mu}{2}\left(\frac{\sqrt{\hat{\kappa}(E)} - 1}{\sqrt{\hat{\kappa}(E)} + 1}\right)^{2k} \|u_0 - u\|_{\mathcal{H}}^2,$$

where $u \in \mathcal{H}$ is the unique minimizer of E.

Proof. (sketch) Consider the space

$$\ell^2 = \left\{ \boldsymbol{v} \in \ell^2(\mathbb{Z};\mathbb{C}) \middle| [\boldsymbol{v}]_i \in \mathbb{R},\ \forall i \in \mathbb{Z}, \quad [\boldsymbol{v}]_i = 0, \forall i \leq 0 \right\},$$

where $\ell^2(\mathbb{Z};\mathbb{C})$ was introduced in Definition 12.18, and we made an obvious extension of notation from the finite-dimensional case. Given positive constants $\mu < L$, define the objective $E_{\mu,L}\colon \ell^2 \to \mathbb{R}$ via

$$E_{\mu,L}(\boldsymbol{v}) = \frac{L-\mu}{8}\left[[\boldsymbol{v}]_1^2 + \sum_{i=1}^{\infty}([\boldsymbol{v}]_{i+1} - [\boldsymbol{v}]_i)^2 - 2[\boldsymbol{v}]_1 \right] + \frac{\mu}{2}\|\boldsymbol{v}\|_{\ell^2}^2.$$

Let us introduce the (infinite) matrix

$$\mathsf{A} = \begin{bmatrix} 2 & -1 & 0 & 0 & \cdots \\ -1 & 2 & -1 & 0 & \cdots \\ 0 & -1 & 2 & -1 & \cdots \\ 0 & 0 & \ddots & \ddots & \ddots \\ \ddots & \ddots & \ddots & \ddots & \ddots \end{bmatrix},$$

which is an infinite-dimensional analogue of the Toeplitz symmetric tridiagonal (TST) matrix introduced in Theorem 3.38. It is possible to show that

$$E'_{\mu,L}[\boldsymbol{v}](\boldsymbol{w}) = \boldsymbol{v}^\intercal\left(\frac{L-\mu}{4}\mathsf{A} + \mu\mathsf{I}\right)\boldsymbol{w} - \frac{L-\mu}{4}[\boldsymbol{w}]_1,$$

$$E''_{\mu,L}[\boldsymbol{x}](\boldsymbol{v},\boldsymbol{w}) = \frac{L-\mu}{4}\boldsymbol{v}^\intercal\mathsf{A}\boldsymbol{w} + \mu\boldsymbol{v}^\intercal\boldsymbol{w}.$$

Therefore, an extension of Theorem 3.38 reveals that the objective $E_{\mu,L}$ is μ-strongly convex and globally Lipschitz smooth with constant L.

Let $\hat{\kappa} = L/\mu$. The Euler equation for this objective then reads

$$\left(\mathsf{A} + \frac{4}{\hat{\kappa}-1}\mathsf{I}\right)\boldsymbol{u} = \boldsymbol{e}_1,$$

or, in coordinate form,

$$2\frac{\hat{\kappa}+1}{\hat{\kappa}-1}[\boldsymbol{u}]_1 - [\boldsymbol{u}]_2 = 1,$$

$$[\boldsymbol{u}]_{i+1} - 2\frac{\hat{\kappa}+1}{\hat{\kappa}-1}[\boldsymbol{u}]_i + [\boldsymbol{u}]_{i+1} = 0, \qquad i = 2, 3, \dots.$$

Setting $\lambda = \frac{\sqrt{\hat{\kappa}}-1}{\sqrt{\hat{\kappa}}+1} \in (0,1)$, we see that $[\boldsymbol{u}]_i = \lambda^i$ solves the system and, therefore, is the minimizer of $E_{\mu,L}$.

Let now $\{\boldsymbol{u}_k\}_{k=0}^\infty$ be a sequence generated by a first-order optimization scheme that satisfies (16.37). Without loss of generality, we can assume that $\boldsymbol{u}_0 = \boldsymbol{0}$, so that

$$\|\boldsymbol{u}_0 - \boldsymbol{u}\|_{\ell^2}^2 = \sum_{i=1}^\infty |\lambda^i|^2 = \sum_{i=1}^\infty \lambda^{2i} = \frac{\lambda^2}{1-\lambda^2}.$$

Notice also that

$$\boldsymbol{u}_1 \in \operatorname{span}\{\Re E'_{\mu,L}[\boldsymbol{u}_0]\} = \operatorname{span}\{\Re E'_{\mu,L}[\boldsymbol{0}]\} \qquad \Longrightarrow \qquad [\boldsymbol{u}_1]_i = 0,\ \forall i > 1,$$

and by induction,

$$[\boldsymbol{u}_k]_j = 0,\ \forall j > k.$$

Consequently,

$$\|\boldsymbol{u}_k - \boldsymbol{u}\|_{\ell^2}^2 \geq \sum_{i=k+1}^\infty |[\boldsymbol{u}]_i|^2 = \sum_{i=k+1}^\infty \lambda^{2i} = \frac{\lambda^{2(k+1)}}{1-\lambda^2} \geq C\lambda^{2k}.$$

The lower quadratic energy trap (16.14) then implies the result. □

Notice that (16.37) is satisfied by both PSD and PSD-ALS and that a variation of this result for their preconditioned versions is not difficult to show. Having obtained a lower bound on the rate of convergence for certain first-order schemes, the following question arises: *Are there any optimal methods?* If the objective is quadratic, then the conjugate gradient method gives exactly this result. In the general case, however, this is much more complicated.

We now present the so-called accelerated gradient descent scheme devised by Nesterov[9] [63]. To simplify the discussion, we present our results under the restrictive assumption that the objective is *globally* Lipschitz smooth with constant L. The extension to locally Lipschitz smooth objectives is presented in [69].

Definition 16.93 (PAGD)**.** Let $\mathcal{H}$ be a Hilbert space; $E\colon \mathcal{H} \to \mathbb{R}$ be a μ-strongly convex and globally L-smooth objective; and $\mathcal{L}\colon \mathcal{H} \to \mathcal{H}'$ be a fixed, invariant preconditioner. The **preconditioned accelerated gradient descent** (PAGD) or **Nesterov method** is an algorithm that constructs a sequence $\{u_k\}_{k=0}^{\infty} \subset \mathcal{H}$ via the following rules: u_0 is given and we set $\lambda > 1$ and $y_0 = u_0$. Then, for $k \geq 0$,

$$u_{k+1} = y_k - \alpha\mathcal{L}^{-1}E'[y_k], \qquad y_{k+1} = u_{k+1} + \frac{\lambda-1}{\lambda+1}\left(u_{k+1} - u_k\right),$$

where $\alpha > 0$ is the **step size**.

Notice that the first part of the scheme is rather similar to a gradient descent scheme. The difference lies in the second step, which we call an *extrapolation* step. Indeed, since $\frac{\lambda-1}{\lambda+1} > 0$, we have that y_{k+1} lies on the line that passes through u_k and u_{k+1}, *but it is not on the segment* with endpoints u_k and u_{k+1}. This hints at the fact that the decay of the energy may not be monotone, as we will later see. It also shows why the analysis of this method for a merely locally L-smooth objective is rather involved. Since we are extrapolating, we may leave any bounded energy set we prescribe beforehand, and thus we cannot assume that the constant L remains fixed.

We will now show that AGD, i.e., PAGD with $\mathcal{L}^{-1}$ being the Riesz map, is optimal under the assumption that the objective is globally L-smooth. Our presentation will closely follow [98]. The proofs in the preconditioned case are similar. We begin with a simple algebraic identity.

Lemma 16.94 (auxiliary variable)**.** *Let $x, y, u \in \mathcal{H}$ and $\lambda > 1$. Define*

$$\hat{z} = (\lambda+1)y - \lambda x - u, \qquad z = \lambda y - (\lambda-1)x - u.$$

Then we have

$$\left(1 - \frac{1}{\lambda^2}\right)\|z\|_{\mathcal{H}}^2 = \left(1 - \frac{1}{\lambda}\right)\|\hat{z}\|_{\mathcal{H}}^2 + \frac{\lambda-1}{\lambda^2}\|x-u\|_{\mathcal{H}}^2 - \left(\lambda - \frac{1}{\lambda}\right)\|y-x\|_{\mathcal{H}}^2.$$

Proof. See Problem 16.31. □

Next, we relate the value of the objective at the approximations $\{u_k\}_{k=0}^{\infty}$ to the value at any other point via the extrapolations $\{y_k\}_{k=0}^{\infty}$. Notice that, since we

[9] Yurii Evgenevich Nesterov is a Russian mathematician (1956–). He is currently a professor at the Université catholique de Louvain, Belgium.

assume that the objective E is globally L-smooth, the second condition in this lemma holds whenever $\tau \le 1/L$.

Lemma 16.95 (descent condition). *Let $E\colon \mathcal{H} \to \mathbb{R}$ be μ-strongly convex and globally L-smooth. For any $v, w \in \mathcal{H}$ and $\tau > 0$ that satisfy*

$$v = w - \tau \mathfrak{R} E'[w], \qquad E(v) \le E(w) + E'[w](v-w) + \frac{1}{2\tau}\|v-w\|_{\mathcal{H}}^2, \quad (16.38)$$

we have that

$$\frac{\tau}{2}\|E'[w]\|_{\mathcal{H}'}^2 \le E(w) - E(v)$$

and

$$E(v) + \frac{1}{2\tau}\|v-z\|_{\mathcal{H}}^2 \le E(z) + \frac{1-\mu\tau}{2\tau}\|w-z\|_{\mathcal{H}}^2, \quad \forall z \in \mathcal{H}.$$

Proof. By observing that $v - w = -\tau \mathfrak{R} E'[w]$, we obtain that

$$E(v) \le E(w) + (\mathfrak{R}E'[w], -\tau\mathfrak{R}E'[w])_{\mathcal{H}} + \frac{\tau}{2}\|\mathfrak{R}E'[w]\|_{\mathcal{H}}^2 = E(w) - \frac{\tau}{2}\|E'[w]\|_{\mathcal{H}'}^2,$$

thus proving the first claim.

To obtain the second estimate, we subtract an arbitrary $z \in \mathcal{H}$ from the first identity in (16.38) and then take the inner product of the result with $v - z$ to obtain

$$\|v-z\|_{\mathcal{H}}^2 = (w-z, v-z)_{\mathcal{H}} - \tau\langle E'[w], v-z\rangle.$$

Identity (16.2) then implies that

$$\|v-z\|_{\mathcal{H}}^2 + \|w-v\|_{\mathcal{H}}^2 = \|w-z\|_{\mathcal{H}}^2 + 2\tau E'[w](z-w) + 2\tau E'[w](w-v).$$

Recall now that, by μ-strong convexity, we have

$$E'[w](z-w) \le E(z) - E(w) - \frac{\mu}{2}\|z-w\|_{\mathcal{H}}^2,$$

so that

$$\begin{aligned}\|v-z\|_{\mathcal{H}}^2 + \|w-v\|_{\mathcal{H}}^2 &= (1-\mu\tau)\|w-z\|_{\mathcal{H}}^2 + 2\tau\left[E(z) - E(w) + E'[w](w-v)\right] \\ &\le (1-\mu\tau)\|w-z\|_{\mathcal{H}}^2 + 2\tau\left[E(z) - E(v)\right],\end{aligned}$$

where in the last step we used the second condition in (16.38). The claim follows. □

We are now ready to prove convergence of AGD.

Theorem 16.96 (convergence). *Let $\mathcal{H}$ be a Hilbert space and $E\colon \mathcal{H} \to \mathbb{R}$ be μ-strongly convex and globally L-smooth. Assume that $u \in \mathcal{H}$ is the unique minimizer of E and $\{u_k\}_{k=0}^\infty \subset \mathcal{H}$ is the sequence obtained by the method of Definition 16.93 with $\mathcal{L}^{-1} = \mathfrak{R}$, $\alpha \in [\frac{1}{2L}, \frac{1}{L}]$, and $\lambda \ge \sqrt{\frac{2L}{\mu}}$. Then we have*

$$\begin{aligned}&\frac{\lambda^2}{L}(E(u_k) - E(u)) + \|(\lambda+1)y_k - \lambda u_k - u\|_{\mathcal{H}}^2 \\ &\qquad \le \left(1 - \frac{1}{\lambda}\right)^k \left[2\alpha\lambda^2(E(u_0) - E(u)) + \|u_0 - u\|_{\mathcal{H}}^2\right],\end{aligned}$$

where the sequence $\{y_k\}_{k=0}^\infty \subset \mathcal{H}$ is introduced in Definition 16.93.

Proof. Notice, first of all, that, for every $k \geq 0$, the vectors $v = u_{k+1}$, $w = y_k$ satisfy the assumptions of Lemma 16.95 with $\tau = \alpha$. Thus, setting

$$z = \left(1 - \frac{1}{\lambda}\right) u_k + \frac{1}{\lambda} u,$$

we get

$$\begin{aligned} &E(u_{k+1}) + \frac{1}{2\alpha} \left\| u_{k+1} - \left(1 - \frac{1}{\lambda}\right) u_k - \frac{1}{\lambda} u \right\|_{\mathcal{H}}^2 \\ &\quad \leq E\left(\left(1 - \frac{1}{\lambda}\right) u_k + \frac{1}{\lambda} u\right) + \frac{1 - \alpha\mu}{2\alpha} \left\| y_k - \left(1 - \frac{1}{\lambda}\right) u_k - \frac{1}{\lambda} u \right\|_{\mathcal{H}}^2. \end{aligned} \tag{16.39}$$

Let us consider the second term in the previous estimate in more detail. Defining

$$\hat{z}_{k+1} = (\lambda + 1) y_{k+1} - \lambda u_{k+1} - u,$$

and using the definition of y_{k+1} given in Definition 16.93, we get that

$$\begin{aligned} u_{k+1} - \left(1 - \frac{1}{\lambda}\right) u_k - \frac{1}{\lambda} u &= \frac{1}{\lambda} \left[2\lambda u_{k+1} - (\lambda - 1) u_k - \lambda u_{k+1} - u\right] \\ &= \frac{1}{\lambda} \left[(\lambda + 1) y_{k+1} - \lambda u_{k+1} - u\right]. \end{aligned}$$

With this notation, (16.39) then becomes

$$\begin{aligned} &E(u_{k+1}) + \frac{1}{2\alpha\lambda^2} \|\hat{z}_{k+1}\|_{\mathcal{H}}^2 \\ &\quad \leq E\left(\left(1 - \frac{1}{\lambda}\right) u_k + \frac{1}{\lambda} u\right) + \frac{1 - \alpha\mu}{2\alpha} \left\| y_k - \left(1 - \frac{1}{\lambda}\right) u_k - \frac{1}{\lambda} u \right\|_{\mathcal{H}}^2. \end{aligned} \tag{16.40}$$

We now consider the last term in (16.40). Define

$$z_k = \lambda y_k - (\lambda - 1) u_k - u,$$

so that

$$y_k - \left(1 - \frac{1}{\lambda}\right) u_k - \frac{1}{\lambda} u = \frac{1}{\lambda} \left[\lambda y_k - (\lambda - 1) u_k - u\right] = \frac{1}{\lambda} z_k,$$

and (16.40) becomes

$$E(u_{k+1}) + \frac{1}{2\alpha\lambda^2} \|\hat{z}_{k+1}\|_{\mathcal{H}}^2 \leq E\left(\left(1 - \frac{1}{\lambda}\right) u_k + \frac{1}{\lambda} u\right) + \frac{1 - \alpha\mu}{2\alpha\lambda^2} \|z_k\|_{\mathcal{H}}^2. \tag{16.41}$$

Since $\lambda > 1$, by the μ-strong convexity of E and the fact that u is its minimizer, we have that

$$E\left(\left(1 - \frac{1}{\lambda}\right) u_k + \frac{1}{\lambda} u\right) \leq \left(1 - \frac{1}{\lambda}\right) E(u_k) + \frac{1}{\lambda} E(u) - \frac{\mu}{2} \left(1 - \frac{1}{\lambda}\right) \|u_k - u\|_{\mathcal{H}}^2.$$

This, combined with (16.41), yields

$$
\begin{aligned}
&E(u_{k+1}) - E(u) + \frac{1}{2\alpha\lambda^2}\|\hat{z}_{k+1}\|_{\mathcal{H}}^2 \\
&\quad\le \left(1-\frac{1}{\lambda}\right)(E(u_k)-E(u)) + \frac{1-\alpha\mu}{2\alpha\lambda^2}\|z_k\|_{\mathcal{H}}^2 - \frac{\mu}{2}\left(1-\frac{1}{\lambda}\right)\|u_k-u\|_{\mathcal{H}}^2.
\end{aligned}
\tag{16.42}
$$

We can now set $x = u_k$, $y = y_k$ in Lemma 16.94, and use the definition of y_{k+1} to see that

$$
\begin{aligned}
&E(u_{k+1}) - E(u) + \frac{1}{2\alpha\lambda^2}\left[\|\hat{z}_{k+1}\|_{\mathcal{H}}^2 + \left(\lambda - \frac{1}{\lambda}\right)\|y_k-u_k\|_{\mathcal{H}}^2\right] \\
&\quad\le \left(1-\frac{1}{\lambda}\right)\left(E(u_k)-E(u)+\frac{1}{2\alpha\lambda^2}\|\hat{z}_k\|_{\mathcal{H}}^2\right) \\
&\qquad + \frac{1}{2\alpha\lambda^2}\left(\frac{1}{\lambda^2}-\alpha\mu\right)(\lambda-1)\left(\|u_k-u\|_{\mathcal{H}}^2+\|z_k\|_{\mathcal{H}}^2\right).
\end{aligned}
\tag{16.43}
$$

Finally, we observe that the choice of parameters α and λ implies that

$$
\frac{1}{\lambda^2} \le \frac{\mu}{2L} \le \alpha\mu,
$$

so that, finally, we conclude

$$
\begin{aligned}
&E(u_{k+1}) - E(u) + \frac{1}{2\alpha\lambda^2}\left[\|\hat{z}_{k+1}\|_{\mathcal{H}}^2 + \left(\lambda - \frac{1}{\lambda}\right)\|y_k-u_k\|_{\mathcal{H}}^2\right] \\
&\quad\le \left(1-\frac{1}{\lambda}\right)\left(E(u_k)-E(u)+\frac{1}{2\alpha\lambda^2}\left(\|\hat{z}_k\|_{\mathcal{H}}^2+\|u_k-u\|_{\mathcal{H}}^2+\|z_k\|_{\mathcal{H}}^2\right)\right).
\end{aligned}
\tag{16.44}
$$

Iterating this estimate then yields the claimed convergence estimate. □

Remark 16.97 (rate of convergence)**.** Notice that setting, in Theorem 16.96, $\alpha = L^{-1}$ and $\lambda = \sqrt{\frac{2L}{\mu}}$, we get

$$
E(u_k) - E(u) = \mathcal{O}\left(\left(1 - \frac{1}{\sqrt{\hat{\kappa}(E)}}\right)^k\right).
$$

It is in this sense that this method is optimal; compare with Theorem 16.92.

Remark 16.98 (monotonicity)**.** Notice that, in Theorem 16.96, the rate of convergence is proved *not for the objective E itself*, but rather for the extended functional

$$
E(u_{k+1}) - E(u) + \frac{1}{2\alpha\lambda^2}\left[\|\hat{z}_{k+1}\|_{\mathcal{H}}^2 + \left(\lambda - \frac{1}{\lambda}\right)\|y_k-u_k\|_{\mathcal{H}}^2\right].
$$

Numerical illustrations, see Figure 16.4, confirm that the value of the objective $\{E(u_k)\}_{k=1}^{\infty}$ does not necessarily monotonically decay for PAGD.

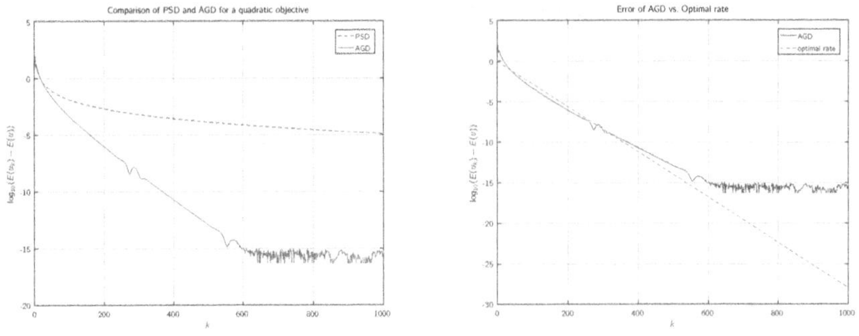

Figure 16.4 Comparison of the performance of PSD (dashed) and AGD (solid) for the quadratic objective functional defined in (16.45) with $\mu = 1$, $L = 10^3$, and $N = 100$. Observe that, for AGD, the objective does not decay monotonically.

16.10 Numerical Illustrations

In this section, we illustrate and compare the performance of the methods that we have developed. Listing 16.1 shows the implementation of each of these methods. Starting from an initial guess, each method is run for a fixed number of iterations. The choice of parameters is in accordance with the theory. Indeed, for steepest descent,

$$\alpha = \frac{1}{2L},$$

whereas for accelerated gradient descent,

$$\alpha = \frac{1}{2L}, \qquad \lambda = \sqrt{\frac{2L}{\mu}}.$$

We compare the performance of these methods in a series of different objectives.

16.10.1 Quadratic Objectives

In an attempt to replicate the construction obtained in the proof of Theorem 16.92, we consider, for different values of $N \in \mathbb{N}$, the objective $E\colon \mathbb{R}^N \to \mathbb{R}$ defined as

$$E(\boldsymbol{v}) = \frac{L-\mu}{8}(\boldsymbol{v} - \boldsymbol{e}_1)^\intercal \mathsf{A}(\boldsymbol{v} - \boldsymbol{e}_1) + \frac{\mu}{2}\|\boldsymbol{v}\|_2^2, \tag{16.45}$$

where A is the TST matrix defined in Theorem 3.38 for $n = N+1$. The description of $\sigma(\mathsf{A})$, provided in the proof of Theorem 3.38, reveals that this objective is μ-strongly convex and globally L-smooth.

Listing 16.2 implements this objective and compares the performance of each one of the methods for $\mu = 1$ and $L = 10^3$. In addition, it compares the rate of convergence of AGD with the lower bound provided in Theorem 16.92.

Figure 16.4 shows the results for $N = 100$. The figure clearly shows how AGD outperforms PSD. In addition, we observe that, for AGD, the decay of the objective

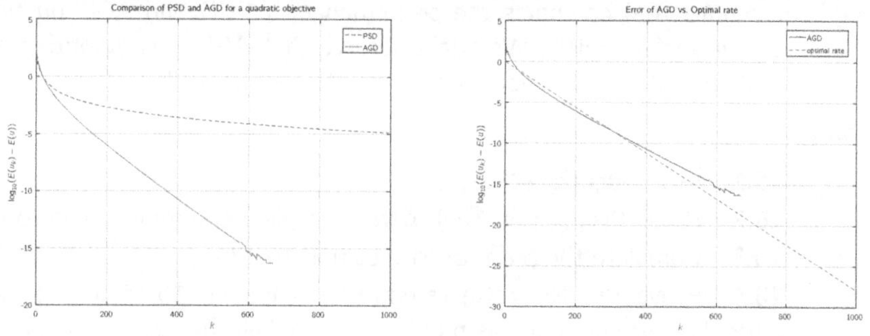

Figure 16.5 Comparison of the performance of PSD (dashed) and AGD (solid) for the quadratic objective functional defined in (16.45) with $\mu = 1$, $L = 10^3$, and $N = 10^6$. Observe that, for AGD, the objective does not decay monotonically.

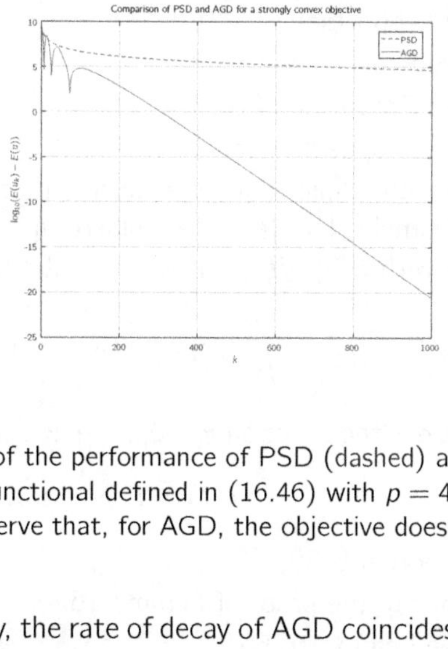

Figure 16.6 Comparison of the performance of PSD (dashed) and AGD (solid) for the nonquadratic objective functional defined in (16.46) with $p = 4$ and $N = 10^6$. We set $\mu = 1$ and $L = 400$. Observe that, for AGD, the objective does not decay monotonically.

is not monotone. Finally, the rate of decay of AGD coincides with the optimal decay established in Theorem 16.92.

In Figure 16.5, we present the results for $N = 10^6$. The same observations apply.

16.10.2 Nonquadratic Objectives

Let us now consider a nonquadratic, strongly convex objective that is locally, but not globally, Lipschitz smooth. Let $N \in \mathbb{N}$ and $p > 2$. We define the objective $E\colon \mathbb{R}^N \to \mathbb{R}$ as

$$E(\boldsymbol{v}) = \frac{1}{p}\|\boldsymbol{v}\|_p^p + \frac{1}{2}\|\boldsymbol{v}\|_2^2. \tag{16.46}$$

Clearly, the minimizer is $\boldsymbol{u} = 0$. By construction, this objective is μ-strongly convex with $\mu \geq 1$. As we do not know the value of any local L-smoothness constant, we set $L = 400$. This objective is implemented in Listing 16.3.

Figure 16.6 compares the performance of PSD and AGD on this problem for $p = 4$ and $N = 10^6$. We again observe that AGD greatly outperforms PSD and that the value of the objective does not monotonically decay for AGD.

Problems

16.1 Prove identity (16.2).

16.2 Prove Proposition 16.4. Give an explicit counterexample to the converse.

16.3 Complete the proof of Proposition 16.13.

16.4 Prove the Riesz Representation Theorem 16.14 in the case that $\mathcal{H}$ is finite dimensional. For this problem, it may help to recall some properties of SPD matrices.

16.5 Prove Theorem 16.16.

16.6 Prove Proposition 16.21.

16.7 Prove Proposition 16.26.

16.8 Prove Proposition 16.33.

16.9 Prove Proposition 16.37.

16.10 Prove Proposition 16.38.

16.11 Prove Proposition 16.39.

16.12 Prove Proposition 16.40.

16.13 Prove the following generalization of Taylor's Theorem with integral remainder. Namely, let $\mathcal{H}$ be a Hilbert space and $E \colon \mathcal{H} \to \mathbb{R}$ be Fréchet differentiable with $E'[\cdot] \colon \mathcal{H} \to \mathcal{H}'$ being continuous. Then

$$E[w] - E[v] = \int_0^1 \langle E'[v + t(w - v)], w - v \rangle \mathrm{d}t, \quad \forall v, w \in \mathcal{H}.$$

Hint: Consider the function $\varphi \colon [0,1] \to \mathbb{R}$ defined by

$$\varphi(t) = E\left(v + t(w - v)\right)$$

and apply Theorem B.39.

16.14 Complete the proof of Lemma 16.48.

16.15 Prove Lemma 16.50.

16.16 Complete the proof of Lemma 16.51.

16.17 Complete the proof of Lemma 16.53.

16.18 Prove Proposition 16.55.

16.19 Prove Theorem 16.61.

16.20 Prove the orthogonality (16.25).

16.21 Prove Corollary 16.70.

16.22 Prove Theorem 16.74. Note that this result is more subtle than the proof for PSD, which was the content of Problem 16.19.

16.23 Prove (16.31).

Hint: Use a variant of Taylor's Theorem.

16.24 Prove Theorem 16.77.

16.25 Prove Corollary 16.78.

16.26 Prove the following generalization of Theorem B.31. Let $\mathcal{H}$ be a Hilbert space and $E\colon \mathcal{H} \to \mathbb{R}$ be continuously Fréchet differentiable on $\mathcal{H}$ and, for $v, w \in \mathcal{H}$, twice Fréchet differentiable on the set $S = \{v + t(w-v) | t \in (0,1)\}$. Then there is $t \in (0,1)$ such that

$$E(w) = E(v) + \langle E'[v], w - v\rangle + \frac{1}{2}E''[v + t(w-v)](w - v, w - v).$$

Hint: Consider the function $\varphi\colon [0,1] \to \mathbb{R}$ defined by

$$\varphi(t) = E\left(v + t(w - v)\right)$$

and apply Theorem B.31.

16.27 Prove (16.33).

16.28 Prove Lemma 16.85.

16.29 Complete the proof of Theorem 16.87.

16.30 Let $N \in \mathbb{N}$ and consider the energy $E\colon \mathbb{R}^N \to \mathbb{R}$

$$E_{\mathsf{A}}(\boldsymbol{v}) = \frac{1}{2}\boldsymbol{v}^{\intercal}\mathsf{A}\boldsymbol{v} - \boldsymbol{f}^{\intercal}\boldsymbol{v},$$

where $\mathsf{A} \in \mathbb{R}^{N\times N}$ and $\boldsymbol{f} \in \mathbb{R}^N$ are given and we assume that A is SPD with $\sigma(\mathsf{A}) \subset [\lambda, \Lambda]$. Show that E_{A} is λ-coercive and globally Lipschitz smooth with constant Λ. Consequently,

$$\hat{\kappa}(E_{\mathsf{A}}) = \frac{\Lambda}{\lambda} \leq \kappa_2(\mathsf{A}).$$

16.31 Prove Lemma 16.94.

Listings

```
function [energyPSD, energyAGD, upsd, uagd] = RunMethods( ...
  E, Ep, mu, L, MaxIts, uinit )
% This function runs both the Steepest Descent with approximate
% line search (PSD)
%
% u_{k+1} = u_k - alpha*Ep( u_k )
%
% and the Accelerated Gradient Descent (AGD)
%
% u_{k+1} = y_k - alpha*Ep( y_k )
%
% y_{k+1} = u_{k+1} + ((lambda-1)/(lambda+1))*( u_{k+1} - u_k );
%
% methods for a stongly convex and (locally) Lipschitz smooth
% objective E with derivative Ep.
%
% Input
%   E : the objective function
%   Ep : the derivative of the objective function
%   mu : (an estimate of) the strong convexity constant for E
```

```
%   L : (an estimate of) the (local) Lipschitz smoothness
%       constant for E
%   MaxIts : the maximal number of iterations
%   uinit : an initial guess for the minimizer
%
% Output
%   energyPSD(1:MaxIts+1) : the value of the objective at every
%                           iteration of PSD
%   energyAGD(1:MaxIts+1) : the value of the objective at every
%                           iteration of AGD
%   upsd : the approximate minimizer after MaxIts iterations
%          of PSD
%   uagd : the approximate minimizer after MaxIts iterations
%          of AGD
%

% We first run PSD
  alphaPSD = 0.5/L;

  energyPSD = zeros(MaxIts+1,1);
  energyPSD(1) = E( uinit );

  upsd = uinit;
  for i=1:MaxIts
    upsd = PSD( alphaPSD, upsd, Ep );
    energyPSD(i+1) = E( upsd );
  end

% Next we run AGD
  alphaAGD = 0.5/L;
  lambdaAGD = sqrt( 2*L/mu );

  energyAGD = energyPSD;
  uagd = uinit;
  yagd = uinit;
  for i=1:MaxIts
    [uagd, yagd ] = AGD(alphaAGD, lambdaAGD, uagd, yagd, Ep );
    energyAGD(i+1) = E( uagd );
  end
end


function u = PSD( alpha, uold, Ep )
% This function does one iteration of the PSD scheme
%
% Input
%   alpha : the line search parameter
%   uold : the previous approximate minimizer
%   Ep : the derivative of the objective
%
% Output
%   u : the next approximate minimizer
%
  u = uold - alpha.*Ep( uold );
end

```

```
function [u, y] = AGD(alpha, lambda, uold, yold, Ep )
% This function does one iteration of the AGD scheme
%
% Input
%   alpha : the line search parameter
%   lambda : the extrapolation parameter
%   uold : the previous approximate minimizer
%   yold : the previous extrapolated value
%   Ep : the derivative of the objective
%
% Output
%   u : the next approximate minimizer
%   y : the next approximate extrapolation
%
  u = yold - alpha.*Ep( yold );
  y = u + ((lambda-1.)/(lambda+1.)).*( u - uold );
end
```

Listing 16.1 Implementation of gradient descent and accelerated gradient descent methods. Starting from an initial guess they are run for a fixed number of iterations.

```
% This code defines a strongly convex and globally Lipschitz
% smooth quadratic objective and uses it to compare the
% performance of PSD and AGD for its minimization:
clear all
clc

% We begin by defining the space dimension and the parameters
% that are used to define the objective function:
dim = 100000; % the dimension: the number of variables
mu = 1.; % the strong convexity constant
L = 1000.; % the global Lipschitz constant

% Since the objective is quadratic, part of it is defined as
%
% v'*AA*v
%
% where AA is a tridiagonal matrix that we now define.
mult = 0.25*( L - mu );
ii = zeros(dim + 2*( dim - 1 ), 1 );
jj = ii;
vv = ii;
vvEuler = ii;
for i = 1:dim
  ii(i) = i;
  jj(i) = i;
  vv(i) = 2;
  vvEuler(i) = mult*vv(i) + mu;
  if i<dim
    ii(i+dim) = i;
    jj(i+dim) = i+1;
    vv(i+dim) = -1;
    vvEuler(i+dim) = mult*vv(i+dim);
  end

```

```
    if i>1
      ii(i+2*(dim-1)) = i;
      jj(i+2*(dim-1)) = i-1;
      vv(i+2*(dim-1)) = -1;
      vvEuler(i+2*(dim-1)) = mult*vv(i+2*(dim-1));
    end
  end

AA = sparse(ii,jj, vv );
AAEuler = sparse( ii, jj, vvEuler );

e1 = zeros(dim,1);
e1(1,1) = 1;

% We define two anonymous function handles to describe the
% objective and its derivative:
E = @(u) Objective( u, mu, L, AA, e1 );
Ep = @(u) ObjectivePrime( u, mu, L, AA, e1 );

% The minimum of the objective can be found by solving the
% Euler equations. Since the objective is quadratic these turn
% out to be a linear system of equations:
uexact = AA*e1;
uexact = AAEuler\uexact;
uexact = 0.25*( L - mu )*uexact;
Eexact = E( uexact );

% We now can start with the optimization schemes. Maximum
% number of iterations and initial guess:
MaxIts = 1000; % Maximum number of iterations
uinit = zeros( dim, 1 ); % Initial guess

% We run the methods
[energyPSD, energyAGD, upsd, uagd] = RunMethods( E, Ep, mu, ...
  L, MaxIts, uinit );
energyPSD = energyPSD - Eexact;
energyAGD = energyAGD - Eexact;

% We plot to compare
its = 0:MaxIts;
hf = figure();
clf;
plot( its, log10( energyPSD ), its, log10( energyAGD ) );
grid on;
xlabel('k');
ylabel('log_{10}( E(u_k) - E(u) )');
title('Comparison of PSD and AGD for a quadratic objective');
legend('PSD', 'AGD');
exportgraphics( gca, 'ComparisonQuadratic.pdf');

% We now compare the rate of convergence of AGD with the
% theoretically optimal one:
hff = figure(2);
clf;
kappa = sqrt( L/mu );
rate =  2.*log10( 1 - 1/kappa )*its;
```

```
plot( its, log10( energyAGD ), its, rate );
grid on;
xlabel('k');
ylabel('log_{10}( E(u_k) - E(u) )');
legend('AGD','optimal rate')
title('Error of AGD vs. Optimal rate');
exportgraphics( gca,'ComparisonAGDandOptimal.pdf');

%%%%%%%%%%%%%%%%%%%%%%%

function E = Objective( u, mu, L, AA, e1 )
% This function defines a quadratic objective
%
% Input
%   u : the argument of the Objective function
%   mu : the strong convexity constant of the objective
%   L : the Lipschitz smoothness constnat of the objective
%   AA : an SPD matrix used to define the objective
%   e1 : A fixed vector to define the objective
%
% Output
%   E : the value of the objective function
%
  E = 0.125*( L - mu )*( ( u - e1 )'*AA*( u - e1 ) ) ...
    + 0.5*mu*( norm(u)^2 );
end

function Ep = ObjectivePrime( u, mu, L, AA, e1 )
% This function defines the derivative of a quadratic objective
%
% Input
%   u : the argument of the Objective function
%   mu : the strong convexity constant of the objective
%   L : the Lipschitz smoothness constnat of the objective
%   AA : an SPD matrix used to define the objective
%   e1 : A fixed vector to define the objective
%
% Output
%   Ep : the value of the derivative of the objective function
%
  Ep = 0.25*(L-mu)*AA*(u - e1) + mu*u;
end
```

Listing 16.2 Implementation of the quadratic objective defined in (16.45) and comparison of the performace of PSD and AGD on this objective.

```
% This code defines an objective that not quadratic, it is
% strongly convex, and locally but not globally Lipschitz
% smooth. This objective is used to compare the performance of
% PSD and AGD.
clear all
clc

% We begin by defining the space dimension and the parameters
```

```
% that are used to define the objective function:
dim = 10000; % the dimension: the number of variables
p = 4; % this is used to define the norm
mu = 1; % the strong convexity constant

% We do not know exactly the (local) Lipschitz smoothness
% constant. This is an estimate:
L = 400;

% We define two anonymous function handles to describe the
% objective and its derivative:
E = @(u) Objective( u, p );
Ep = @(u) ObjectivePrime( u, p, dim );

% We know the exact minimizer:
uexact = zeros( dim, 1 );
Eexact = E( uexact );

% We now can start with the optimization schemes. Maximum
% number of iterations and initial guess:
MaxIts = 500; % Maximum number of iterations
uinit = 10*ones(dim,1); % Initial guess

% We run the methods:
[errorPSD, errorAGD, upsd, uagd] = RunMethods( E, Ep, mu, ...
  L, MaxIts, uinit );
errorPSD = errorPSD - Eexact;
errorAGD = errorAGD - Eexact;

% We plot to compare
its = 0:MaxIts;
hf = figure();
clf;
plot( its, log10( errorPSD ), its, log10( errorAGD ) );
grid on;
xlabel('k');
ylabel('log_{10}( E(u_k) - E(u) )');
title( ...
  'Comparison of PSD and AGD for a strongly convex objective');
legend('PSD', 'AGD');
exportgraphics( gca, 'ComparisonStrongConvex.pdf');

%%%%%%%%%%%%%%%%%%%%%%

function E = Objective( u, p )
% This defines a non quadratic, strongly convex, and locally
% but not globally Lipschitz smooth objective.
%
% Input
%   u : the argument of the Objective function
%   p : the order of the norm used to make this non quadratic
%
% Output
%   E : the value of the objective function
%
  E = ( norm( u, p )^p )/p + 0.5*norm( u )^2;
```

```
end

function Ep = ObjectivePrime( u, p, dim )
% This defines the derivative of the nonquadratic objective
%
% Input
%   u(1,dim) : the argument of the Objective function
%   p : the order of the norm used to make this non quadratic
%   dim : the dimension of u
%
% Output
%   Ep : the value of the derivative of the objective function
%
  Ep = u + ( abs( u ).^(p-2) ).*u;
end
```

Listing 16.3 Implementation of the nonquadratic, stongly convex, locally but not globally Lipschitz smooth objective defined in (16.46) and comparison of the performace of PSD and AGD on this objective.

Part IV

Initial Value Problems for Ordinary Differential Equations

17 Initial Value Problems for Ordinary Differential Equations

In this chapter, we begin the study of methods for approximating the solution of an ordinary differential equation (ODE),

$$\boldsymbol{u}'(t) = \boldsymbol{f}(t, \boldsymbol{u}(t)), \quad t \in (0, T],$$

supplemented with a given initial condition, $\boldsymbol{u}(0) = \boldsymbol{u}_0$. This is usually known as a *Cauchy problem*[1] or an *initial value problem* (IVP) for an ODE.

Ordinary differential equations, and their Cauchy problems, arise in many problems related to the sciences and engineering. They essentially describe how the *change* of one quantity — which above we denoted by t, since it is usually thought of as time — affects another quantity, $\boldsymbol{u}$. For instance, everyone is familiar with Newton's second law of motion,[2] which in elementary textbooks is presented as

$$\boldsymbol{F} = m\boldsymbol{a},$$

where $\boldsymbol{F}$ is the net force exerted on a particle, m is its mass, and $\boldsymbol{a}$ is its acceleration. Let us assume that the force depends only upon time. We know that acceleration is the rate of change (with respect to time) of velocity $\boldsymbol{v}$, which in turn is the rate of change of position $\boldsymbol{x}$, i.e., $\boldsymbol{a}(t) = \boldsymbol{v}'(t) = \boldsymbol{x}''(t)$. Let us introduce the variable

$$\boldsymbol{u}(t) = \begin{bmatrix} \boldsymbol{x}(t) \\ \boldsymbol{v}(t) \end{bmatrix},$$

so that Newton's second law can then be written as

$$\boldsymbol{u}'(t) = \begin{bmatrix} \boldsymbol{v}(t) \\ \boldsymbol{a}(t) \end{bmatrix} = \begin{bmatrix} \boldsymbol{v}(t) \\ \frac{1}{m}\boldsymbol{F}(t) \end{bmatrix} = \begin{bmatrix} \mathsf{O}_3 & \mathsf{I}_3 \\ \mathsf{O}_3 & \mathsf{O}_3 \end{bmatrix} \begin{bmatrix} \boldsymbol{x}(t) \\ \boldsymbol{v}(t) \end{bmatrix} + \begin{bmatrix} \boldsymbol{0} \\ \frac{1}{m}\boldsymbol{F}(t) \end{bmatrix} = \boldsymbol{f}(t, \boldsymbol{u}(t)).$$

Of course, in this case, we have

$$\boldsymbol{f}(t, \boldsymbol{u}(t)) = \mathsf{A}\boldsymbol{u}(t) + \boldsymbol{b}(t),$$

where A is a matrix and $\boldsymbol{b}$ is a time-dependent vector. If, say, the force depends upon the position, $\boldsymbol{x}$, and velocity, $\boldsymbol{v}$, one would arrive at a more complicated form. In any case, assuming that the initial position $\boldsymbol{x}(0) = \boldsymbol{x}_0$ and velocity $\boldsymbol{v}(0) = \boldsymbol{v}_0$ are known, we set $\boldsymbol{u}(0) = \boldsymbol{u}_0 = [\boldsymbol{x}_0, \boldsymbol{v}_0]^\intercal$ and arrive at a Cauchy problem.

The previous elementary discussion illustrates the meaning of such a problem: it describes how the position of a system evolves over time, provided that we know its initial position, initial velocity, and the forces that are acting on it. In addition,

[1] Named in honor of the French mathematician Augustin Louis Cauchy (1789–1857).

[2] Named in honor of the British mathematician, physicist, astronomer, theologian, and natural philosopher Sir Isaac Newton (1642–1726/27).

it illustrates a very important point: the problem we will study is a first-order ODE, since we only see first derivatives of the *dependent variable*, $\boldsymbol{u}$, with respect to the *independent* one, t. However, there is no significant loss in generality in doing this, as *many* higher order ODEs (those that have derivatives of order larger than one) can be reduced to a (system of) first-order ODEs.

The IVP, as stated above, is rather vague, as we did not clearly state what it is we mean by *solving* this problem, nor did we provide conditions that guarantee that a solution exists. Because of this, before we start discussing numerical schemes for such problems, we discuss some of the theory about the solutions to IVPs. For further developments and complete proofs of some of the results we omit, we refer the reader to [19, 37].

17.1 Existence of Solutions

To start, let us define precisely what we mean by a solution to an IVP. In what follows, we will assume that the following are fixed: d is a positive integer; $\Omega \subseteq \mathbb{R}^d$ is an open set; $\boldsymbol{u}_0$ is a point in Ω; $I \subseteq \mathbb{R}$ is a closed interval; t_0 is a point in I; $S = I \times \overline{\Omega}$; and $\boldsymbol{f}\colon S \to \mathbb{R}^d$ is a given function, which we call the *slope function.* The IVP that we consider seeks a function $\boldsymbol{u}\colon I \to \Omega$ that, in some sense, satisfies the *initial condition* $\boldsymbol{u}(t_0) = \boldsymbol{u}_0$ and the equation

$$\boldsymbol{u}'(t) = \boldsymbol{f}(t, \boldsymbol{u}(t)). \tag{17.1}$$

Definition 17.1 (classical solution)**.** The function $\boldsymbol{u} \in C^1(I; \Omega)$ is called a **classical solution** on I to the IVP if and only if (17.1) holds point-wise for all $t \in I$ and $\lim_{t\to t_0} \boldsymbol{u}(t) = \boldsymbol{u}_0$.

Definition 17.2 (mild solution)**.** We say that $\boldsymbol{u} \in C(I; \Omega)$ is a **mild solution** on I to the IVP (17.1) if and only if, for all $t \in I$, we have

$$\boldsymbol{u}(t) = \boldsymbol{u}_0 + \int_{t_0}^{t} \boldsymbol{f}(s, \boldsymbol{u}(s))\mathrm{d}s. \tag{17.2}$$

Theorem 17.3 (equivalence)**.** *Assume that* $\boldsymbol{f} \in C\left(S; \mathbb{R}^d\right)$. *A function is a mild solution on I to problem* (17.1) *if and only if it is a classical solution to problem* (17.1).

Proof. See Problem 17.1. □

Definition 17.4 ($\boldsymbol{u}$-Lipschitz)**.** We say that the slope function $\boldsymbol{f}\colon S \to \mathbb{R}^d$ is **$\boldsymbol{u}$-Lipschitz** on S if and only if there is a constant $L > 0$ such that

$$\|\boldsymbol{f}(t, \boldsymbol{v}_1) - \boldsymbol{f}(t, \boldsymbol{v}_2)\|_2 \le L \|\boldsymbol{v}_1 - \boldsymbol{v}_2\|_2 \tag{17.3}$$

for all $t \in I$ and for all $\boldsymbol{v}_1, \boldsymbol{v}_2 \in \Omega$. If (17.3) holds with $\Omega = \mathbb{R}^d$, we say that $\boldsymbol{f}$ is **globally $\boldsymbol{u}$-Lipschitz.**[3]

[3] The term *Lipschitz* is used in honor of the German mathematician Rudolf Otto Sigismund Lipschitz (1832–1903).

We have the following local existence and uniqueness result whose brief but elegant proof is based on the Banach Fixed Point Theorem C.4; see Appendix C for some background.

Theorem 17.5 (Picard–Lindelöf Theorem[4]). *Suppose that there exist constants* $\beta, \delta_0 > 0$*, such that* $I_0 = [t_0 - \delta_0, t_0 + \delta_0] \cap I \neq \emptyset$*, and* $\overline{B}(\boldsymbol{u}_0, \beta) \subset \Omega$*. Define*

$$S_0 = I_0 \times \overline{B}(\boldsymbol{u}_0, \beta).$$

Assume that $\boldsymbol{f} \in C(S_0; \mathbb{R}^d)$*; there is a constant* $M > 0$ *such that, for all* $(t, \boldsymbol{v}) \in S_0$*,* $\|\boldsymbol{f}(t, \boldsymbol{v})\|_2 \leq M$*; and* $\boldsymbol{f}$ *is* $\boldsymbol{u}$*-Lipschitz on* S_0 *with constant* $L > 0$*. Let*

$$\delta_1 = \min\left\{\delta_0, \frac{\delta_0}{2L}, \frac{\beta}{M}\right\}, \qquad I_1 = [t_0 - \delta_1, t_0 + \delta_1].$$

Then there is a unique mild solution on I_1 *to* (17.1)*. Moreover,* $\boldsymbol{u} \in C(I_1; B(\boldsymbol{u}_0, \beta))$*.*

Proof. We will apply the the Banach Fixed Point Theorem C.4. Let $J \subseteq I_0$ be an interval with $t_0 \in J$. Define the mapping $\mathcal{A} \colon C(J; B(\boldsymbol{u}_0, \beta)) \to C(J; B(\boldsymbol{u}_0, \beta))$ by

$$\mathcal{A}[\boldsymbol{v}](t) = \boldsymbol{u}_0 + \int_{t_0}^{t} \boldsymbol{f}(s, \boldsymbol{v}(s)) ds.$$

Let us show first that if J is suitably chosen and $\boldsymbol{v} \in C(J; B(\boldsymbol{u}_0, \beta))$, then, indeed, $\mathcal{A}[\boldsymbol{v}] \in C(J; B(\boldsymbol{u}_0, \beta))$. The continuity of $\mathcal{A}[\boldsymbol{v}]$ is guaranteed by the Fundamental Theorem of Calculus (Theorem B.37). We need only show that this function takes values in $B(\boldsymbol{u}_0, \beta)$. Note that

$$\begin{aligned} \|\mathcal{A}[\boldsymbol{v}](t) - \boldsymbol{u}_0\|_2 &= \left\| \int_{t_0}^{t} \boldsymbol{f}(s, \boldsymbol{v}(s)) ds \right\|_2 \\ &\leq \int_{\min\{t_0, t\}}^{\max\{t_0, t\}} \|\boldsymbol{f}(s, \boldsymbol{v}(s))\|_2 \, ds \\ &\leq M|t - t_0|, \end{aligned}$$

so that if $J = [t_0 - \delta, t_0 + \delta]$ with $\delta = \min\{\delta_0, \beta/M\}$, then $\mathcal{A}[\boldsymbol{v}](t) \in B(\boldsymbol{u}_0, \beta)$ for all $t \in J$.

Now, given $\boldsymbol{v}_1, \boldsymbol{v}_2 \in C(J; B(\boldsymbol{u}_0, \beta))$, consider

$$\begin{aligned} \|\mathcal{A}[\boldsymbol{v}_1] - \mathcal{A}[\boldsymbol{v}_2]\|_{L^\infty(J)} &= \sup_{t \in J} \left\| \int_{t_0}^{t} (\boldsymbol{f}(s, \boldsymbol{v}_1(s)) - \boldsymbol{f}(s, \boldsymbol{v}_2(s))) ds \right\|_2 \\ &\leq \sup_{t \in J} \int_{\min\{t_0, t\}}^{\max\{t_0, t\}} \|\boldsymbol{f}(s, \boldsymbol{v}_1(s)) - \boldsymbol{f}(s, \boldsymbol{v}_2(s))\|_2 \, ds. \end{aligned}$$

[4] Named in honor of the French mathematician Charles Émile Picard (1856–1941) and the Finnish mathematician Ernst Leonard Lindelöf (1870–1946).

Since $\boldsymbol{f}$ is $\boldsymbol{u}$-Lipschitz on S, we then obtain

$$\begin{aligned}\|\mathcal{A}[\boldsymbol{v}_1] - \mathcal{A}[\boldsymbol{v}_2]\|_{L^\infty(J)} &\le \sup_{t\in J} \int_{\min\{t_0,t\}}^{\max\{t_0,t\}} L\,\|\boldsymbol{v}_1(s) - \boldsymbol{v}_2(s)\|_2\, ds \\ &\le L\|\boldsymbol{v}_1 - \boldsymbol{v}_2\|_{L^\infty(J)} \sup_{t\in J} |t - t_0|.\end{aligned}$$

Consequently, if we restrict our study to the interval

$$I_1 = \left[t_0 - \frac{\delta_0}{2L}, t_0 + \frac{\delta_0}{2L}\right] \cap J,$$

then we obtain

$$\|\mathcal{A}[\boldsymbol{v}_1] - \mathcal{A}[\boldsymbol{v}_2]\|_{L^\infty(I_1)} \le \frac{1}{2}\|\boldsymbol{v}_1 - \boldsymbol{v}_2\|_{L^\infty(I_1)}.$$

This proves that the mapping $\mathcal{A}\colon C(I_1; B(\boldsymbol{u}_0, \beta)) \to C(I_1; B(\boldsymbol{u}_0, \beta))$ is a contraction. Since

$$\left(C(I_1; B(\boldsymbol{u}_0, \beta)), \|\cdot\|_{L^\infty(I_1)}\right)$$

is a complete metric space, see Example C.6, the contraction mapping principle for metric spaces, as stated in Theorem C.11, yields the existence and uniqueness of a fixed point for $\mathcal{A}$. Call the fixed point $\boldsymbol{u} \in C(I_1; B(\boldsymbol{u}_0, \beta))$. Then

$$\boldsymbol{u}(t_0) = \mathcal{A}[\boldsymbol{u}](t_0) = \boldsymbol{u}_0 + \int_{t_0}^{t_0} \boldsymbol{f}(s, \boldsymbol{u}(s)) ds = \boldsymbol{u}_0;$$

otherwise, for all $t \in I_1$,

$$\boldsymbol{u}(t) = \mathcal{A}[\boldsymbol{u}](t) = \boldsymbol{u}_0 + \int_{t_0}^{t} \boldsymbol{f}(s, \boldsymbol{u}(s)) ds.$$

Clearly, $\boldsymbol{u}$ is a mild solution on I_1 to (17.1). □

Example 17.1 Suppose that $u_0 > 0$. Observe that $u(t) = \left(u_0^{-1} - t\right)^{-1}$ is a classical solution on the interval $[0, u_0)$ to the IVP

$$u'(t) = u^2(t), \quad u(0) = u_0.$$

The autonomous slope function $f(t, u(t)) = u^2(t)$ is not globally u-Lipschitz on $S = [0, T] \times \mathbb{R}$, regardless of the size of $T > 0$. Clearly, a global solution, i.e., a (classical or mild) solution on $\mathbb{R}$, cannot be guaranteed. In any case, Theorem 17.5 is applicable, and a unique solution, locally defined around $t = 0$, can be guaranteed.

Observe that Theorem 17.5 only guarantees the existence of a solution on a possibly small interval of time. For example, if M or L is very large, the interval of existence may be quite small. To guarantee existence on a larger interval of time, we need to strengthen the Lipschitz condition on the slope function $\boldsymbol{f}$.

To simplify notation, here and in what follows we will assume that $t_0 = 0$ and $I = [0, T]$ for some $T > 0$. A simple translation and possible rescaling can bring us to the general case. In addition, as we mentioned before, this justifies the name: *initial* value problem.

Theorem 17.6 (global existence). *Assume that the slope function $\boldsymbol{f} \in C(S; \mathbb{R}^d)$ is globally $\boldsymbol{u}$-Lipschitz with constant $L > 0$. Then there is at least one mild solution on $[0, T]$ to (17.1), which we denote by $\boldsymbol{u} \in C([0, T]; \mathbb{R}^d)$. Moreover, this solution satisfies the estimate*

$$\|\boldsymbol{u}(t) - \boldsymbol{u}_0\|_2 \leq \frac{M}{L}\left(e^{Lt} - 1\right), \quad \forall t \in [0, T],$$

where $M = \|\boldsymbol{f}(\cdot, \boldsymbol{u}_0)\|_{L^\infty(0,T)}$.

Proof. We could use an argument based on the Banach Fixed Point Theorem again. But, for variety, let us follow a slightly different path; see, for example, [83].

Again, define, for $\boldsymbol{v} \in C([0, T]; \mathbb{R}^d)$,

$$\mathcal{A}[\boldsymbol{v}](t) = \boldsymbol{u}_0 + \int_0^t \boldsymbol{f}(s, \boldsymbol{v}(s))\mathrm{d}s.$$

Since the slope function is globally $\boldsymbol{u}$-Lipschitz,

$$\|\mathcal{A}[\boldsymbol{v}_1](t) - \mathcal{A}[\boldsymbol{v}_2](t)\|_2 \leq L \int_0^t \|\boldsymbol{v}_1(s) - \boldsymbol{v}_2(s)\|_2 \,\mathrm{d}s. \tag{17.4}$$

Now recursively define a sequence of functions as $\boldsymbol{u}^{(0)}(t) = \boldsymbol{u}_0$ for all $t \in [0, T]$, and, for $k \geq 0$, $\boldsymbol{u}^{(k+1)}(t) = \mathcal{A}\left[\boldsymbol{u}^{(k)}\right](t)$ for all $t \in [0, T]$. Therefore,

$$\left\|\boldsymbol{u}^{(1)}(t) - \boldsymbol{u}^{(0)}(t)\right\|_2 = \left\|\int_0^t \boldsymbol{f}(s, \boldsymbol{u}_0)\mathrm{d}s\right\|_2 \leq \int_0^t \left\|\boldsymbol{f}(s, \boldsymbol{u}_0)\right\|_2 \mathrm{d}s \leq Mt.$$

More generally, using (17.4), for $k \geq 1$,

$$\begin{aligned}\left\|\boldsymbol{u}^{(k)}(t) - \boldsymbol{u}^{(k-1)}(t)\right\|_2 &= \left\|\int_0^t \left(\boldsymbol{f}(s, \boldsymbol{u}^{(k-1)}(s)) - \boldsymbol{f}(s, \boldsymbol{u}^{(k-2)}(s))\right)\mathrm{d}s\right\|_2 \\ &\leq L \int_0^t \left\|\boldsymbol{u}^{(k-1)}(s) - \boldsymbol{u}^{(k-2)}(s)\right\|_2 \mathrm{d}s.\end{aligned}$$

From these two estimates, we deduce, by induction, that, for all $t \in [0, T]$, we have

$$\left\|\boldsymbol{u}^{(k)}(t) - \boldsymbol{u}^{(k-1)}(t)\right\|_2 \leq \frac{ML^{k-1}t^k}{k!}.$$

Next, we show that the infinite series

$$\boldsymbol{s}(t) = \boldsymbol{u}(t) - \boldsymbol{u}_0 = \sum_{j=1}^{\infty}\left(\boldsymbol{u}^{(j)}(t) - \boldsymbol{u}^{(j-1)}(t)\right) \tag{17.5}$$

is absolutely and uniformly convergent on $[0, T]$. To see this, note that, for each $t \in [0, T]$, the kth partial sum is

$$\boldsymbol{s}_k(t) = \sum_{j=1}^{k}\left(\boldsymbol{u}^{(j)}(t) - \boldsymbol{u}^{(j-1)}(t)\right) = \boldsymbol{u}^{(k)}(t) - \boldsymbol{u}_0.$$

Thus, for $m > n$, using the triangle inequality,

$$\|\boldsymbol{s}_m(t) - \boldsymbol{s}_n(t)\|_2 \leq \sum_{j=n+1}^{m} \left\|\boldsymbol{u}^{(j)}(t) - \boldsymbol{u}^{(j-1)}(t)\right\|_2 \leq \frac{M}{L} \sum_{j=n+1}^{m} \frac{L^j t^j}{j!}. \tag{17.6}$$

Since the series

$$\sum_{j=1}^{\infty} \frac{L^j t^j}{j!} = e^{Lt} - 1$$

converges absolutely and uniformly on $[0, T]$, the sequence of partial sums $\{\boldsymbol{s}_k\}_{k=1}^{\infty}$ is uniformly Cauchy on $[0, T]$. This proves that the series (17.5) converges absolutely and uniformly on $[0, T]$. Hence, $\boldsymbol{s}$, and therefore $\boldsymbol{u}$, is continuous on $[0, T]$. Now $\boldsymbol{s}_k \to \boldsymbol{s} = \boldsymbol{u} - \boldsymbol{u}_0$, uniformly on $[0, T]$, and

$$\|\boldsymbol{s}_k(t)\|_2 \leq \frac{M}{L} \sum_{j=1}^{k} \frac{L^j t^j}{j!} \leq \frac{M}{L}\left(e^{Lt} - 1\right).$$

This implies that, for all $t \in [0, T]$,

$$\|\boldsymbol{u}(t) - \boldsymbol{u}_0\|_2 \leq \frac{M}{L}\left(e^{Lt} - 1\right).$$

To conclude the proof, we must show that the $\boldsymbol{u}$ we constructed in (17.5) is precisely the mild solution that we are seeking. To this end, taking $m \to \infty$ in (17.6), we have, for any $t \in [0, T]$,

$$\begin{aligned}
\left\|\boldsymbol{u}(t) - \boldsymbol{u}^{(n)}(t)\right\|_2 &= \|\boldsymbol{u}(t) - \boldsymbol{u}_0 - \boldsymbol{s}_n(t)\|_2 \\
&\leq \frac{M}{L} \sum_{j=n+1}^{\infty} \frac{L^j t^j}{j!} \\
&\leq \frac{M}{L} \frac{(Lt)^{n+1}}{(n+1)!} \sum_{j=0}^{\infty} \frac{L^j t^j}{j!} \\
&\leq \frac{M}{L} \kappa_{n+1} e^{Lt},
\end{aligned}$$

where $\kappa_n = \frac{(LT)^n}{n!}$. For every $\gamma > 0$, there is a constant $C = C(\gamma) > 0$ such that

$$\kappa_n \leq C(\gamma)\gamma^n.$$

The proof of this fact is left to the reader as an exercise; see Problem 17.2. Below, we will want to take $0 < \gamma < 1$.

Using the bounds we just established, the $\boldsymbol{u}$-Lipschitz property of $\boldsymbol{f}$, and the definition of the sequence $\{\boldsymbol{u}^{(k)}\}_{k=0}^{\infty}$, we have, for all $t \in [0, T]$,

$$\begin{aligned}
&\left\| \boldsymbol{u}(t) - \boldsymbol{u}_0 - \int_0^t \boldsymbol{f}(s, \boldsymbol{u}(s)) \mathrm{d}s \right\|_2 \\
&\quad = \left\| \boldsymbol{u}(t) - \boldsymbol{u}^{(k+1)}(t) + \int_0^t \left(\boldsymbol{f}(s, \boldsymbol{u}^{(k)}(s)) - \boldsymbol{f}(s, \boldsymbol{u}(s)) \right) \mathrm{d}s \right\|_2 \\
&\quad \le \left\| \boldsymbol{u}(t) - \boldsymbol{u}^{(k+1)}(t) \right\|_2 + \left\| \int_0^t \left(\boldsymbol{f}(s, \boldsymbol{u}^{(k)}(s)) - \boldsymbol{f}(s, \boldsymbol{u}(s)) \right) \mathrm{d}s \right\|_2 \\
&\quad \le \left\| \boldsymbol{u}(t) - \boldsymbol{u}^{(k+1)}(t) \right\|_2 + \int_0^t \left\| \boldsymbol{f}(s, \boldsymbol{u}^{(k)}(s)) - \boldsymbol{f}(s, \boldsymbol{u}(s)) \right\|_2 \mathrm{d}s \\
&\quad \le \left\| \boldsymbol{u}(t) - \boldsymbol{u}^{(k+1)}(t) \right\|_2 + L \int_0^t \left\| \boldsymbol{u}^{(k)}(s) - \boldsymbol{u}(s) \right\|_2 \mathrm{d}s \\
&\quad \le C(\gamma)\gamma^{k+2} \frac{M}{L} e^{Lt} + L \int_0^t \frac{M}{L} C(\gamma)\gamma^{k+1} e^{Ls} \mathrm{d}s \\
&\quad \le C(\gamma)\gamma^{k+2} \frac{M}{L} e^{Lt} + Lt \frac{M}{L} C(\gamma)\gamma^{k+1} e^{Lt} \\
&\quad \le C(\gamma) \frac{M}{L} e^{LT} (\gamma + LT) \gamma^{k+1}
\end{aligned}$$

for all $t \in [0, T]$. Picking $\gamma < 1$ and letting $k \to \infty$, we see that $\boldsymbol{u}$ is a mild solution. □

17.2 Uniqueness and Regularity of Solutions

We have just established the global existence of a mild solution, and therefore also a classical solution. To prove uniqueness, we can use an a posteriori argument. We will need the following estimates for the uniqueness proof.

Lemma 17.7 (Grönwall-type inequalities[5]). *Let $T > 0$, $K_1 \ge 0$, $K_2 \ge 0$, and $\Phi \in C^1([0, T])$. If $\Phi(0) = 0$, $\Phi(t) \ge 0$ for all $t \in [0, T]$, and*

$$\Phi'(t) \le K_1 \Phi(t) + K_2,$$

then

$$\Phi(t) \le \frac{K_2}{K_1} \left[e^{K_1 t} - 1 \right]$$

and

$$\Phi'(t) \le K_2 e^{K_1 t}.$$

Proof. The solution to the IVP

$$\Phi'(t) - K_1 \Phi(t) = \alpha(t), \quad t \in [0, T], \quad \Phi(0) = 0$$

[5] Named in honor of the Swedish–American mathematician Thomas Hakon Grönwall (1877–1932).

is

$$\Phi(t) = e^{K_1 t} \int_0^t \alpha(s) e^{-K_1 s} ds.$$

In the present case, $\alpha(t) \leq K_2$ for all $t \in [0, T]$. Hence,

$$\begin{aligned}
\Phi(t) &\leq e^{K_1 t} K_2 \int_0^t e^{-K_1 s} ds \\
&= -\frac{e^{K_1 t} K_2}{K_1} e^{-K_1 s} \Big|_{s=0}^{s=t} \\
&= e^{K_1 t} \frac{K_2}{K_1} \left[1 - e^{-K_1 t}\right] \\
&= \frac{K_2}{K_1} \left[e^{K_1 t} - 1\right].
\end{aligned}$$

Finally, using the last estimate,

$$\Phi'(t) = \alpha(t) + K_1 \Phi(t) \leq K_2 + K_1 \frac{K_2}{K_1} \left[e^{K_1 t} - 1\right] = K_2 e^{K_1 t},$$

and the proof is complete. □

The next ingredient that is needed is a continuous-dependence result with respect to initial data.

Theorem 17.8 (continuous dependence). *Let $\Omega_0 \subseteq \Omega$ (and one or both possibly equal to $\mathbb{R}^d$). Assume that $\boldsymbol{f} \in C(S; \mathbb{R}^d)$ is $\boldsymbol{u}$-Lipschitz on S with Lipschitz constant $L > 0$. Assume that, for each $\boldsymbol{q} \in \Omega_0$, there exists a classical solution, $\boldsymbol{u}(\cdot\,; \boldsymbol{q}) \in C^1([0, T]; \Omega)$, to the parameterized IVP*

$$\boldsymbol{u}'(t; \boldsymbol{q}) = \boldsymbol{f}(t, \boldsymbol{u}(t; \boldsymbol{q})), \quad \boldsymbol{u}(0; \boldsymbol{q}) = \boldsymbol{q}. \tag{17.7}$$

Then, for all $\boldsymbol{q}_1, \boldsymbol{q}_2 \in \Omega_0$ and $t \in [0, T]$, we have

$$\|\boldsymbol{u}(t; \boldsymbol{q}_1) - \boldsymbol{u}(t; \boldsymbol{q}_2)\|_2 \leq \exp(Lt) \|\boldsymbol{q}_1 - \boldsymbol{q}_2\|_2 . \tag{17.8}$$

Proof. Owing to Theorem 17.3, a classical solution is a mild solution. Thus, the corresponding parameterized mild solution satisfies, for all $t \in [0, T]$,

$$\boldsymbol{u}(t; \boldsymbol{q}) = \boldsymbol{q} + \int_0^t \boldsymbol{f}(s, \boldsymbol{u}(s; \boldsymbol{q})) ds.$$

Hence,

$$\boldsymbol{u}(t; \boldsymbol{q}_1) - \boldsymbol{u}(t; \boldsymbol{q}_2) = \boldsymbol{q}_1 - \boldsymbol{q}_2 + \int_0^t [\boldsymbol{f}(s, \boldsymbol{u}(s; \boldsymbol{q}_1)) - \boldsymbol{f}(s; \boldsymbol{u}(s; \boldsymbol{q}_2))] ds,$$

by the triangle inequality, and the fact that $\boldsymbol{f}$ is $\boldsymbol{u}$-Lipschitz,

$$\|\boldsymbol{u}(t; \boldsymbol{q}_1) - \boldsymbol{u}(t; \boldsymbol{q}_2)\|_2 \leq \|\boldsymbol{q}_1 - \boldsymbol{q}_2\|_2 + L \int_0^t \|\boldsymbol{u}(s; \boldsymbol{q}_1) - \boldsymbol{u}(s; \boldsymbol{q}_2)\|_2 \, ds. \tag{17.9}$$

Define

$$\Phi(t) = \int_0^t \|\boldsymbol{u}(s; \boldsymbol{q}_1) - \boldsymbol{u}(s; \boldsymbol{q}_2)\|_2 \, ds;$$

in which case,

$$\Phi'(t) = \|\boldsymbol{u}(t;\boldsymbol{q}_1) - \boldsymbol{u}(t;\boldsymbol{q}_2)\|_2 .$$

By estimate (17.9), for all $t \in [0, T]$, we have that

$$\Phi'(t) - L\Phi(t) \le \|\boldsymbol{q}_1 - \boldsymbol{q}_2\|_2 .$$

The final result now follows from the second Grönwall inequality in Lemma 17.7

$$\|\boldsymbol{u}(t;\boldsymbol{q}_1) - \boldsymbol{u}(t;\boldsymbol{q}_2)\|_2 = \Phi'(t) \le \|\boldsymbol{q}_1 - \boldsymbol{q}_2\|_2 \, e^{Lt},$$

as we intended to show. □

With these two results, we are ready to prove uniqueness.

Corollary 17.9 (uniqueness). *With the same hypotheses as in Theorem 17.6, the solution $\boldsymbol{u} \in C^1([0,T];\mathbb{R}^d)$ is unique.*

Proof. See Problem 17.4. □

Following classic references, such as [86], we define specialized classes of slope functions.

Definition 17.10 (classes of slope functions). Let $\boldsymbol{f} \in C\left(S;\mathbb{R}^d\right)$ be a slope function. We say that $\boldsymbol{f} \in F^1(S)$ if and only if $\boldsymbol{f} \in C^1\left(S;\mathbb{R}^d\right)$, and there is a real number $A > 0$ such that, for any $i,j = 1,\ldots,d$, and all $(t,\boldsymbol{v}) \in S$,

$$\left|\partial_{u_j} f_i(t,\boldsymbol{v})\right| \le A.$$

We also define, for $m \in \mathbb{N}$,

$$\mathcal{F}^m(S) = F^1(S) \cap C^m(S;\mathbb{R}^d).$$

Proposition 17.11 (F^1 implies Lipschitz). *Let $\boldsymbol{f}$ be a slope function in $F^1(S)$. Then, $\boldsymbol{f}$ is $\boldsymbol{u}$-Lipschitz on S.*

Proof. See Problem 17.6. □

Remark 17.12 (simplification). The assumption that $\boldsymbol{f} \in F^1(S)$ is not often verified in practice. For example, consider the autonomous differential equation

$$u'(t) = -u^3 + u, \quad t \in [0,\infty)$$

with $u(0) = u_0 \in \mathbb{R}$. In this case, $f(t,u) = -u^3 + u$. Clearly, the first derivative of the slope function f with respect to u is unbounded; consequently, $f \notin F^1(S)$. Yet, this autonomous ODE has a bounded classical solution on $[0,\infty)$. In fact, one can show that $\lim_{t\to\infty} u(t) = 1$, if $u_0 > 0$, and $\lim_{t\to\infty} u(t) = -1$, if $u_0 < 0$.

The introduction of the class $F^1(S)$ is merely for convenience, as the assumption $\boldsymbol{f} \in F^1(S)$ makes the analysis much simpler. We will not explain what is required to remove this assumption; rather, we will content ourselves with saying that this can often be accomplished, though usually with a little more sophisticated machinery and a lot more effort.

Theorem 17.13 (higher differentiability). *Let $m \in \mathbb{N}$. Suppose that the slope function satisfies $\boldsymbol{f} \in \mathcal{F}^m(S)$. Then the unique classical solution on I to (17.1), which we denote $\boldsymbol{u} \in C^1(I; \mathbb{R}^d)$, actually belongs to $C^{m+1}(I; \mathbb{R}^d)$.*

Proof. As $\boldsymbol{u}$ is a classical solution, for all $t \in I$, we have

$$\boldsymbol{u}'(t) = \boldsymbol{f}(t, \boldsymbol{u}(t)).$$

The right-hand side of this identity is differentiable on I; in particular,

$$\frac{\mathrm{d}}{\mathrm{d}t}\left[\boldsymbol{f}(t, \boldsymbol{u}(t))\right] = \partial_t \boldsymbol{f}(t, \boldsymbol{u}(t)) + D_{\boldsymbol{u}}\boldsymbol{f}(t, \boldsymbol{u}(t))\boldsymbol{f}(t, \boldsymbol{u}(t)),$$

where $D_{\boldsymbol{u}}\boldsymbol{f} = \left[\partial_{u_j} f_i\right]_{i,j=1}^d$ is the $d \times d$ matrix of partial derivatives of $\boldsymbol{f}$ with respect to $\boldsymbol{u}$. Consequently, $\boldsymbol{u}''(t)$ exists and is continuous on I. The higher order derivatives exist and are continuous on I, as may be seen via an induction argument; see Problem 17.7. $\square$

17.3 The Flow Map and the Alekseev–Gröbner Lemma

In Theorem 17.8, we studied the continuous dependence of the solution to an IVP with respect to initial data. Our purpose here will be to study how the solution depends on the changes in the slope function. We will only state the main results in this direction and refer the reader to [37, Chapter 1] for more details.

We begin with a definition that might seem a bit odd.

Definition 17.14 (flow map). Suppose that, for some $m \in \mathbb{N}$, the slope function satisfies $\boldsymbol{f} \in \mathcal{F}^m(S)$. The **flow map of $z'(t) = \boldsymbol{f}(t, z(t))$**, denoted

$$\boldsymbol{U} \colon I \times \Omega \times I \to \mathbb{R}^d,$$

is defined by

$$\boldsymbol{U}(s, \boldsymbol{v}, t) = \boldsymbol{u}_{s,\boldsymbol{v}}(t),$$

where $\boldsymbol{u}_{s,\boldsymbol{v}} \in C^1(I; \mathbb{R}^d)$ is the unique solution to the ODE problem

$$\frac{\mathrm{d}\boldsymbol{u}_{s,\boldsymbol{v}}}{\mathrm{d}t}(t) = \boldsymbol{f}(t, \boldsymbol{u}_{s,\boldsymbol{v}}(t)), \quad \boldsymbol{u}_{s,\boldsymbol{v}}(s) = \boldsymbol{v}. \tag{17.10}$$

This definition emphasizes that the solution of an IVP is a function not only of the final time but also of the initial time, and the value of the initial condition. If we change any of these, the solution may change. First, we have the following simple properties of the flow map.

Proposition 17.15 (properties of the flow map). *Suppose that, for some $m \in \mathbb{N}$, the slope function satisfies $\boldsymbol{f} \in \mathcal{F}^m(S)$. Denote by $\boldsymbol{U}$ the flow map of $z'(t) = \boldsymbol{f}(t, z(t))$. Then*

$$\boldsymbol{U}(s, \boldsymbol{v}, s) = \boldsymbol{v}, \quad \forall (s, \boldsymbol{v}) \in S.$$

For any $t_1, t_2 \in I$ *and all* $\mathbf{v} \in \Omega$, *we have*

$$\boldsymbol{U}(s, \mathbf{v}, t_2) = \boldsymbol{U}(t_1, \boldsymbol{U}(s, \mathbf{v}, t_1), t_2).$$

In addition, for any $s, t \in I$ *and all* $\mathbf{v} \in \Omega$,

$$\frac{\partial}{\partial t}\boldsymbol{U}(s, \mathbf{v}, t) = \boldsymbol{f}(t, \boldsymbol{U}(s, \mathbf{v}, t)).$$

Finally, $\boldsymbol{U}$ *is continuously differentiable with respect to its second variable, i.e., the initial condition. This derivative, denoted* $D_{\mathbf{v}}\boldsymbol{U}$, *is a* $d \times d$ *matrix at every point* $(s, \mathbf{v}, t) \in I \times \Omega \times I$, *and we have*

$$[D_{\mathbf{v}}\boldsymbol{U}(s, \mathbf{v}, t)]_{i,j} = \frac{\partial U_i}{\partial v_j}(s, \mathbf{v}, t).$$

Furthermore, $D_{\mathbf{v}}\boldsymbol{U}$ *is differentiable with respect to its third argument,* t, *and satisfies the differential equation*

$$\frac{\partial}{\partial t}D_{\mathbf{v}}\boldsymbol{U}(s, \mathbf{v}, t) = D_{\boldsymbol{u}}\boldsymbol{f}(t, \boldsymbol{U}(s, \mathbf{v}, t))D_{\mathbf{v}}\boldsymbol{U}(s, \mathbf{v}, t)$$

subject to the initial data

$$D_{\mathbf{v}}\boldsymbol{U}(s, \mathbf{v}, s) = \mathrm{I}_d,$$

where $D_{\boldsymbol{u}}\boldsymbol{f}$ *is the derivative of* $\boldsymbol{f}$ *with respect to its second argument.*

Theorem 17.16 (Alekseev–Gröbner Lemma[6])**.** *Let* $t_0 \in I$ *and, for some* $m \in \mathbb{N}$, $\boldsymbol{f}, \boldsymbol{g} \in \mathcal{F}^m(S)$. *Denote by* $\boldsymbol{U}$ *the flow map of* $\boldsymbol{z}'(t) = \boldsymbol{f}(t, \boldsymbol{u}(t))$. *Suppose that* $\boldsymbol{u}, \mathbf{v} \in C^1(I; \mathbb{R}^d)$ *are the unique classical solutions on* I *of*

$$\begin{aligned} \boldsymbol{u}'(t) &= \boldsymbol{f}(t, \boldsymbol{u}(t)), && t \in I, \quad \boldsymbol{u}(t_0) = \boldsymbol{u}_0, \\ \mathbf{v}'(t) &= \boldsymbol{f}(t, \mathbf{v}(t)) + \boldsymbol{g}(t, \mathbf{v}(t)), && t \in I, \quad \mathbf{v}(t_0) = \boldsymbol{u}_0. \end{aligned}$$

Then

$$\mathbf{v}(t) = \boldsymbol{u}(t) + \int_{t_0}^{t} D_{\mathbf{v}}\boldsymbol{U}(s, \mathbf{v}(s), t)\boldsymbol{g}(s, \mathbf{v}(s))\mathrm{d}s. \tag{17.11}$$

We can think of $\boldsymbol{g}$ as a perturbation function or a residual. This result then measures the difference between two solutions: one a solution to the original system and the second a solution to a perturbed system. Clearly, if $\boldsymbol{g} \equiv \mathbf{0}$, there is no difference in the solutions. Furthermore, we can imagine that $\boldsymbol{g}$ represents a small difference resulting from some numerical approximation, and $\mathbf{v}$ is an approximate solution. In this case, (17.11) measures the error in our approximation. We will use this latter characterization to quantify the error of certain numerical methods.

[6] Named in honor of the Soviet mathematician Vladimir Mikhailovich Alekseev (1932–1980) and the Austrian mathematician Wolfgang Gröbner (1899–1980).

17.4 Dissipative Equations

In this section, we briefly present some facts about the theory of so-called *dissipative equations*, which is a class of problems that often appears in practice. Let us, first of all, note that all the theory that we developed so far regarding existence, uniqueness, and regularity only relies on the fact that $\mathbb{R}^d$, with $d \in \mathbb{N}$, is a finite-dimensional normed space. No other structural properties were used on the range of the slope function or the solution. For this reason, and to allow for sufficient generality, in this section we let $d \in \mathbb{N}$ and consider IVPs posed on subsets of $\mathbb{C}^d$. Finally, in this section, we let $(\cdot, \cdot)$ be an inner product on $\mathbb{C}^d$ and $\|\cdot\|$ be the induced norm.

Definition 17.17 (monotonicity). Let $\boldsymbol{f}\colon [0,T] \times \mathbb{C}^d \to \mathbb{C}^d$. We say that $\boldsymbol{f}$ is **monotone** with respect to $(\,\cdot\,,\,\cdot\,)$ if and only if

$$\Re\left[(\boldsymbol{v}_1 - \boldsymbol{v}_2, \boldsymbol{f}(t,\boldsymbol{v}_1) - \boldsymbol{f}(t,\boldsymbol{v}_2))\right] \le 0$$

for all $t \in [0,T]$ and every $\boldsymbol{v}_1, \boldsymbol{v}_2 \in \mathbb{C}^d$.

Theorem 17.18 (dissipativity). *Assume that $\boldsymbol{f}\colon [0,T] \times \mathbb{C}^d \to \mathbb{C}^d$ is monotone with respect to $(\cdot,\cdot)$. Let $\boldsymbol{u}_0, \boldsymbol{v}_0 \in \mathbb{C}^d$. Assume that $\boldsymbol{f}$ is such that there are unique classical solutions on $[0,T]$ to the problems*

$$\boldsymbol{u}'(t) = \boldsymbol{f}(t,\boldsymbol{u}(t)), \quad \boldsymbol{u}(0) = \boldsymbol{u}_0, \qquad\qquad \boldsymbol{v}'(t) = \boldsymbol{f}(t,\boldsymbol{v}(t)), \quad \boldsymbol{v}(0) = \boldsymbol{v}_0.$$

In this setting, we have that, for all $t \in [0,T]$,

$$\frac{\mathrm{d}}{\mathrm{d}t}\|\boldsymbol{u}(t) - \boldsymbol{v}(t)\|^2 \le 0.$$

Furthermore, for all $0 \le t_1 \le t_2 \le T$,

$$\|\boldsymbol{u}(t_2) - \boldsymbol{v}(t_2)\| \le \|\boldsymbol{u}(t_1) - \boldsymbol{v}(t_1)\| \le \|\boldsymbol{u}_0 - \boldsymbol{v}_0\|.$$

Proof. Set $E(t) = \frac{1}{2}\|\boldsymbol{u}(t) - \boldsymbol{v}(t)\|^2$. Then

$$\begin{aligned}\frac{\mathrm{d}}{\mathrm{d}t}E(t) &= \Re\left[(\boldsymbol{u}(t) - \boldsymbol{v}(t), \boldsymbol{u}'(t) - \boldsymbol{v}'(t))\right]\\ &= \Re\left[(\boldsymbol{u}(t) - \boldsymbol{v}(t), \boldsymbol{f}(t,\boldsymbol{u}(t)) - \boldsymbol{f}(t,\boldsymbol{v}(t)))\right]\\ &\le 0.\end{aligned}$$

This proves that the function E is nonincreasing. Thus, the second inequality follows; see Problem 19.7. □

Remark 17.19 (dissipativity). An ODE problem with these properties is said to be *dissipative* with respect to $(\,\cdot\,,\,\cdot\,)$.

Example 17.2 Let $\mathsf{A} \in \mathbb{C}^{d\times d}$ be such that if $\lambda \in \sigma(A)$, then $\Re\lambda \le 0$. The slope function

$$\boldsymbol{f}(t,\boldsymbol{v}) = \mathsf{A}\boldsymbol{v}$$

is monotone.

17.5 Lyapunov Stability

In this section, we present some elements of the *qualitative* theory of IVP for ODEs. In other words, we will try to extract information about the behavior of the solution to an IVP, specially for large times, without necessarily knowing the exact solution to this problem. A particular instance of this is the theory of *Lyapunov stability*, which we will present here. As always, we are barely touching the subject of a deep and rich subject; namely, *stability of dynamical systems*. The interested reader is referred to [4, 57, 62, 74, 90] for much more insight and further developments.

We will be mostly interested in, for a fixed slope function $\boldsymbol{f}$, how the — mild or classical — solution to (17.1) depends on the initial condition $\boldsymbol{u}_0$, specially for large times. Thus, we will operate under the implicit assumption that the slope function is such that, for every initial condition, a — mild or classical — solution exists and is unique for all positive times, i.e., $T = \infty$.

Definition 17.20 (trajectory)**.** Let $\boldsymbol{f}$ be a slope function such that, for every initial condition, problem (17.1) has a unique — mild or classical — solution in $C([0, \infty); \mathbb{R}^d)$. The mapping

$$\mathbb{R}^d \to C([0, \infty); \mathbb{R}^d),$$
$$\boldsymbol{u}_0 \mapsto \boldsymbol{u}(\,\cdot\,; \boldsymbol{u}_0) = \boldsymbol{U}(0, \boldsymbol{u}_0, \cdot),$$

where $\boldsymbol{U}$ is the flow map, is called the **trajectory** associated with $\boldsymbol{u}_0$.

The beginning of our considerations is the content of Theorem 17.8. This result shows that, for a fixed time $t > 0$, the trajectory is a Lipschitz mapping. The Lipschitz constant, however, depends (exponentially) on t. While this is enough to guarantee uniqueness of solutions, it does not say anything about how close (for large t) two trajectories stay, provided they originate from points nearby. Having this property is related to *stability*, a rigorous definition of which we provide below.

Definition 17.21 (Lyapunov stability[7])**.** Let $\boldsymbol{u}_0 \in \mathbb{R}^d$. The trajectory $\boldsymbol{u}(\cdot; \boldsymbol{u}_0)$ associated with $\boldsymbol{u}_0$ is called **Lyapunov stable** if, for every $\varepsilon > 0$, there is $\delta > 0$ such that, whenever

$$\|\boldsymbol{u}_0 - \boldsymbol{q}\|_2 < \delta,$$

we have

$$\|\boldsymbol{u}(t; \boldsymbol{u}_0) - \boldsymbol{u}(t; \boldsymbol{q})\|_2 < \varepsilon, \quad \forall t > 0.$$

A trajectory that is not Lyapunov stable is called *unstable*.

There are other notions of stability. Some are more general, some more specialized, but we shall not concern ourselves with those here. We are now ready to state our goal of this section: given a trajectory, we wish to determine whether it is Lyapunov stable. For special cases, we can already provide a positive answer, as the following example shows.

[7] Named in honor of the Russian mathematician Aleksandr Mikhailovich Lyapunov (1857–1918).

Example 17.3 Let the slope function $\boldsymbol{f}$ be monotone in the sense of Definition 17.17. Theorem 17.18 shows that any trajectory is Lyapunov stable.

Given $\boldsymbol{u}_0 \in \mathbb{R}^d$, assume that $\boldsymbol{u}(\cdot\,; \boldsymbol{u}_0)$ is its associated trajectory. Suppose that we wish to study its Lyapunov stability. To do so, we now introduce some useful reductions. Let $\boldsymbol{q} \in \mathbb{R}^d$ be arbitrary and $\boldsymbol{u}(\cdot\,, \boldsymbol{q})$ be its associated trajectory. Let us introduce the change of variables

$$\boldsymbol{v}(t) = \boldsymbol{u}(t; \boldsymbol{q}) - \boldsymbol{u}(t; \boldsymbol{u}_0). \tag{17.12}$$

In this new coordinate system, we have, using the differential equation,

$$\boldsymbol{v}'(t) = \boldsymbol{u}'(t; \boldsymbol{q}) - \boldsymbol{u}'(t; \boldsymbol{u}_0) = \boldsymbol{f}(t, \boldsymbol{u}(t; \boldsymbol{q})) - \boldsymbol{f}(t, \boldsymbol{u}(t; \boldsymbol{u}_0)) = \mathsf{A}(t; \boldsymbol{q}, \boldsymbol{u}_0)\boldsymbol{v}(t),$$

where, in the last step, we used the Mean Value Theorem B.57 for vector-valued functions, so that

$$\mathsf{A}(t; \boldsymbol{q}, \boldsymbol{u}_0) = \int_0^1 D_u \boldsymbol{f}(t, s\boldsymbol{u}(t; \boldsymbol{q}) + (1 - s)\boldsymbol{u}(t; \boldsymbol{u}_0))ds. \tag{17.13}$$

In short, we have reduced the study of the Lyapunov stability of an arbitrary trajectory $\boldsymbol{u}(\cdot\,; \boldsymbol{u}_0)$ to the study of the Lyapunov stability of the *trivial stationary point* $\boldsymbol{v} \equiv \boldsymbol{0}$ for the *linear* ODE

$$\boldsymbol{v}'(t) = \mathsf{A}(t; \boldsymbol{q}, \boldsymbol{u}_0)\boldsymbol{v}(t), \tag{17.14}$$

where the matrix-valued function $\mathsf{A}(\cdot; \boldsymbol{q}, \boldsymbol{u}_0)$ is given by (17.13).

Remark 17.22 (stationary points)**.** Let us provide some intuition into the value of stationary points, and specially of the trivial one, i.e., $\boldsymbol{v} \equiv \boldsymbol{0}$. To do this, let us consider the *autonomous* problem

$$\boldsymbol{u}'(t) = \boldsymbol{f}(\boldsymbol{u}(t)), \qquad t > 0.$$

It is important to notice that most laws of physics are described by an autonomous system. This, for instance, is a consequence of *Galilean invariance*.[8] In layman's terms, this means that the laws of physics, unlike many unfortunate things in life, are independent of the day of the week. For the same reason, if we are dealing with a physical system, it is not unreasonable to assume that $\boldsymbol{f}(\boldsymbol{0}) = \boldsymbol{0}$ (zero input equals zero output), so that $\boldsymbol{u}(t; \boldsymbol{0}) \equiv \boldsymbol{0}$. In this case, all our derivations reduce to studying the Lyapunov stability of the origin with respect to the system

$$\boldsymbol{v}'(t) = \mathsf{A}(t)\boldsymbol{v}(t), \qquad \mathsf{A}(t) = \int_0^1 D_u \boldsymbol{f}(s\boldsymbol{u}(t; \boldsymbol{q}))ds. \tag{17.15}$$

One final reduction can get us to where we want to be, so that we can develop our desired stability theory. Let us assume that not only is the slope function

[8] Named in honor of the Italian astronomer, physicist, engineer, and Rennaissance man Galileo di Vincenzo Bonaiuti dé Galilei (1564–1642).

continuously differentiable with respect to its second argument but also that this derivative is constant on the trivial solution, i.e.,

$$\frac{\mathrm{d}}{\mathrm{d}t} D_u \boldsymbol{f}(t, \boldsymbol{0}) \equiv \mathrm{O}.$$

If this is the case, we can write

$$\boldsymbol{v}'(t) = \mathsf{A}(t; \boldsymbol{q}, \boldsymbol{u}_0)\boldsymbol{v}(t) = \mathsf{A}\boldsymbol{v}(t) + \boldsymbol{g}(t, \boldsymbol{v}),$$

where, for every t, $\frac{1}{\|\boldsymbol{v}\|_2}\boldsymbol{g}(t, \boldsymbol{v}) \to \boldsymbol{0}$ as $\boldsymbol{v} \to \boldsymbol{0}$. In conclusion, to study the stability of trajectories it is enough, at this level, to study the stability of the origin for the system

$$\boldsymbol{v}'(t) = \mathsf{A}\boldsymbol{v}(t), \tag{17.16}$$

where $\mathsf{A} \in \mathbb{R}^{d \times d}$. This is called the *first or linear approximation* to (17.1).

Theorem 17.23 (stability). *Let* $\mathsf{A} \in \mathbb{R}^{d \times d}$. *The origin is a stable solution of* (17.16) *if and only if*

$$\lambda \in \sigma(\mathsf{A}) \quad \Longrightarrow \quad \Re\lambda \leq 0;$$

if $\lambda \in \sigma(\mathsf{A})$ *is such that* $\Re\lambda = 0$, *then its algebraic and geometric multiplicities coincide.*

Proof. For a given initial condition $\boldsymbol{v}_0 \in \mathbb{R}^d$, we can explicitly write the solution to the IVP for (17.16). Namely,

$$\boldsymbol{v}(t) = \exp(t\mathsf{A})\boldsymbol{v}_0,$$

where $\exp(t\mathsf{A})$ denotes the matrix exponential. If the conditions on the spectrum of A are satisfied, then

$$\alpha = \sup_{t>0} \|\exp(t\mathsf{A})\|_2 < \infty.$$

Let $\varepsilon > 0$. Then, for any $\boldsymbol{v}_0$ with $\|\boldsymbol{v}_0\|_2 < \frac{\varepsilon}{\alpha}$,

$$\|\boldsymbol{v}(t)\|_2 = \|\exp(t\mathsf{A})\boldsymbol{v}_0\|_2 \leq \alpha \|\boldsymbol{v}_0\|_2 < \varepsilon,$$

and the origin is stable.

Assume now that there is $\lambda \in \sigma(\mathsf{A})$ with $\Re\lambda > 0$. Let $\boldsymbol{q}$ be a corresponding eigenvalue. Given $\varepsilon \in (0, 1)$, define $\boldsymbol{v}_0 = \varepsilon\boldsymbol{q}$ to see that the solution

$$\boldsymbol{v}(t) = \exp(t\mathsf{A})\boldsymbol{v}_0 = \varepsilon e^{\lambda t}\boldsymbol{q}$$

grows unboundedly large, regardless of the value of ε. This shows that, in this case, the origin is unstable.

If $\lambda \in \sigma(\mathsf{A})$ has geometric multiplicity strictly smaller than its algebraic, then we can find $\boldsymbol{w} \in C^\infty([0, \infty); \mathbb{R}^d)$ such that

$$\boldsymbol{v}(t) = \exp(t\mathsf{A})\boldsymbol{v}_0 + te^{\lambda t}\boldsymbol{w}(t).$$

Now, since $\Re\lambda = 0$, $|e^{\lambda t}| = 1$ for all t and, regardless of $\boldsymbol{w}$, the second term grows unboundedly large. Thus, in this case, the origin is unstable as well. □

Problems

17.1 Prove Theorem 17.3.

17.2 Let $L, T > 0$, $n \in \mathbb{N}$, and define $\kappa_n = (LT)^n/n!$. Show that, for every $\gamma > 0$, there is $C = C(\gamma) > 0$ such that

$$\kappa_n \leq C\gamma^n, \quad \forall n \in \mathbb{N}.$$

17.3 In the setting of Theorem 17.8, assume, in addition, that $d = 1$ and that the slope function satisfies $f \in C^1(S; \mathbb{R})$

$$\partial_u f(t, v) \leq 0, \quad \forall (t, v) \in S.$$

Show that, in this case, estimate (17.8) can be improved to

$$|u(t, q_1) - u(t, q_2)| \leq |q_1 - q_2|.$$

17.4 Prove Corollary 17.9.

17.5 Show that $u = -t^2/4$ and $u = 1 - t$ are solutions of the IVP

$$2u'(t) = \sqrt{t^2 + 4u} - t, \qquad u(2) = -1.$$

Why does this not contradict the uniqueness of Corollary 17.9?

17.6 Prove Proposition 17.11.

17.7 Complete the proof of Theorem 17.13.

17.8 Consider the scalar equation

$$u'(t) = au,$$

where $a \in \mathbb{R}_\star$. Show that if $a < 0$ any trajectory is Lyapunov stable, whereas if $a > 0$ all trajectories are unstable.

17.9 Show that the change of variables (17.12) is nonsingular.

17.10 Justify the derivations that lead to (17.15).

18 Single-Step Methods

In this chapter, we begin the study of approximation methods for initial value problems (IVPs). The approach we will follow is motivatived by (17.2), which is the defining relation that a mild solution must satisfy. Indeed, if $\boldsymbol{u}'(t) = \boldsymbol{f}(t, \boldsymbol{u}(t))$ and $\boldsymbol{u}(t_0) = \boldsymbol{u}_0$, we can approximate $\boldsymbol{u}(t)$ via

$$\boldsymbol{u}(t) = \boldsymbol{u}_0 + \int_{t_0}^{t} \boldsymbol{f}(s, \boldsymbol{u}(s))\mathrm{d}s \approx u_0 + Q_{1,0}^{(t_0,t)}[\boldsymbol{f}(\,\cdot\,, \boldsymbol{u}(\,\cdot\,))],$$

where $Q_{1,0}^{(t_0,t)}$ is a quadrature formula. Notice that, in doing so, we now only require knowledge of $\boldsymbol{f}(\,\cdot\,, \boldsymbol{u}(\,\cdot\,))$ at the quadrature nodes. Thus, for instance, if we use the simple left-hand Riemann sum approximation:

$$Q_{1,0}^{(t_0,t)}[\boldsymbol{f}(\cdot, \boldsymbol{u}(\cdot))] = (t - t_0)\boldsymbol{f}(t_0, \boldsymbol{u}(t_0)).$$

Then the approximation becomes

$$\boldsymbol{u}(t) \approx \boldsymbol{u}_0 + (t - t_0)\boldsymbol{f}(t_0, \boldsymbol{u}(t_0))$$

or

$$\frac{1}{t - t_0}(\boldsymbol{u}(t) - \boldsymbol{u}_0) \approx \boldsymbol{f}(t_0, \boldsymbol{u}(t_0)).$$

This is Euler's famous approximation method. In this chapter, we will examine the convergence of methods of this type.

To simplify the discussion, in this and upcoming chapters, we consider the IVP over the "time" interval $I = [0, T]$ for $T > 0$. A simple linear transformation can be used to reduce the general case to this one. In addition, we will suppose that $d \in \mathbb{N}$, $\Omega \subseteq \mathbb{R}^d$ is open, and $\boldsymbol{u}_0 \in \Omega$. We set $S = [0, T] \times \overline{\Omega}$ and assume that the slope function satisfies, at least, $\boldsymbol{f} \in C(S; \overline{\Omega})$. We denote by $\boldsymbol{u} \in C^1([0, T]; \Omega)$ a classical solution on $[0, T]$ to the IVP

$$\boldsymbol{u}'(t) = \boldsymbol{f}(t, \boldsymbol{u}(t)), \quad \boldsymbol{u}(0) = \boldsymbol{u}_0. \tag{18.1}$$

The methods we will present here do not give us a function that approximates $\boldsymbol{u}$ but rather a sequence of vectors that approximate this function at a particular collection of points in time. More precisely, we let $K \in \mathbb{N}$, set $\tau = \frac{T}{K}$, which we call the *time step size*, and $t_k = k\tau$. We will then produce a finite sequence $\{\boldsymbol{w}^k\}_{k=0}^K \subset \mathbb{R}^d$ such that $\boldsymbol{w}^k \approx \boldsymbol{u}(t_k)$.

18.1 Single-Step Approximation Methods

We begin with a definition.

Definition 18.1 (single-step method). The finite sequence $\left\{\boldsymbol{w}^k\right\}_{k=0}^K \subset \mathbb{R}^d$ is called a **single-step approximation** to $\boldsymbol{u}$ if and only if $\boldsymbol{w}^0 = \boldsymbol{u}_0$ and

$$\boldsymbol{w}^{k+1} = \boldsymbol{w}^k + \tau \boldsymbol{G}\left(t_k, \tau, \boldsymbol{w}^k, \boldsymbol{w}^{k+1}\right), \quad k = 0, \dots, K-1, \tag{18.2}$$

where $\boldsymbol{G}$, called the **slope approximation**, satisfies $\boldsymbol{G}(t, 0, \boldsymbol{v}, \boldsymbol{v}) = \boldsymbol{f}(t, \boldsymbol{v})$ and

$$\boldsymbol{G} \in C([0, T] \times [0, T] \times \mathbb{R}^d \times \mathbb{R}^d; \mathbb{R}^d).$$

The single-step approximation is called **explicit** if and only if $\boldsymbol{G}$ is independent of the last variable; otherwise, the approximation is called **implicit**. The **global error** of the single-step approximation is a finite sequence $\left\{\boldsymbol{e}^k\right\}_{k=0}^K$ defined via

$$\boldsymbol{e}^k = \boldsymbol{u}(t_k) - \boldsymbol{w}^k.$$

Let us present some examples of single-step approximation methods.

Example 18.1 The *forward (or explicit) Euler method:*[1]

$$\boldsymbol{G}(t, s, \boldsymbol{v}_1, \boldsymbol{v}_2) = \boldsymbol{G}_{FE}(t, s, \boldsymbol{v}_1) = \boldsymbol{f}(t, \boldsymbol{v}_1). \tag{18.3}$$

Example 18.2 The *backward (or implicit) Euler method*:

$$\boldsymbol{G}(t, s, \boldsymbol{v}_1, \boldsymbol{v}_2) = \boldsymbol{G}_{BE}(t, s, \boldsymbol{v}_2) = \boldsymbol{f}(t + s, \boldsymbol{v}_2). \tag{18.4}$$

Example 18.3 The *trapezoidal method*:

$$\boldsymbol{G}(t, s, \boldsymbol{v}_1, \boldsymbol{v}_2) = \boldsymbol{G}_{TR}(t, s, \boldsymbol{v}_1, \boldsymbol{v}_2) = \frac{1}{2}\boldsymbol{f}(t, \boldsymbol{v}_1) + \frac{1}{2}\boldsymbol{f}(t + s, \boldsymbol{v}_2). \tag{18.5}$$

Example 18.4 *Taylor's method:*[2]

$$\begin{aligned}\boldsymbol{G}(t, s, \boldsymbol{v}_1, \boldsymbol{v}_2) &= \boldsymbol{G}_{TM}(t, s, \boldsymbol{v}_1) \\ &= \boldsymbol{f}(t, \boldsymbol{v}_1) + \frac{s}{2}\left[\partial_t \boldsymbol{f}(t, \boldsymbol{v}_1) + D_{\boldsymbol{u}}\boldsymbol{f}(t, \boldsymbol{v}_1)\boldsymbol{f}(t, \boldsymbol{v}_1)\right],\end{aligned} \tag{18.6}$$

where $D_{\boldsymbol{u}}\boldsymbol{f} = \left[\partial_{u_j} f_i\right]_{i,j=1}^d$ is the $d \times d$ Jacobian matrix of partial derivatives of $\boldsymbol{f}$ with respect to $\boldsymbol{u}$.

Example 18.5 The *midpoint method*:

$$\boldsymbol{G}(t, s, \boldsymbol{v}_1, \boldsymbol{v}_2) = \boldsymbol{G}_{MR}(t, s, \boldsymbol{v}_1, \boldsymbol{v}_2) = \boldsymbol{f}\left(t + \frac{s}{2}, \frac{1}{2}\boldsymbol{v}_1 + \frac{1}{2}\boldsymbol{v}_2\right). \tag{18.7}$$

---

[1] Named in honor of the Swiss mathematician, physicist, astronomer, geographer, logician, and engineer Leonhard Euler (1707–1783).

[2] Named in honor of the British mathematician Brook Taylor (1685–1731).

Definition 18.2 (LTE and convergence). Let $\{\boldsymbol{w}^k\}_{k=0}^K$ be a single-step approximation to $\boldsymbol{u}$ generated by the slope approximation $\boldsymbol{G}$. The **local truncation error** (LTE) or **consistency error** of the single-step approximation is defined as

$$\boldsymbol{\mathcal{E}}[\boldsymbol{u}](t,s) = \frac{\boldsymbol{u}(t) - \boldsymbol{u}(t-s)}{s} - \boldsymbol{G}(t-s, s, \boldsymbol{u}(t-s), \boldsymbol{u}(t))$$

for any $t \in [s, T]$. We make frequent use of the notation $\boldsymbol{\mathcal{E}}^k[\boldsymbol{u}] = \boldsymbol{\mathcal{E}}[\boldsymbol{u}](t_k, \tau)$ for $k = 1, \dots, K$. We say that the approximation method is **consistent to at least order** $p \in \mathbb{N}$ if and only if, whenever

$$\boldsymbol{u} \in C^{p+1}([0,T]; \Omega),$$

there is a constant $\tau_0 \in (0, T]$ and a constant $C > 0$ independent of t and τ such that

$$\|\boldsymbol{\mathcal{E}}[\boldsymbol{u}](t,\tau)\|_2 \le C\tau^p \tag{18.8}$$

for all $\tau \in (0, \tau_0]$ and $t \in [\tau, T]$. We say that the single-step approximation is **consistent to exactly order** p if and only if p is the largest positive integer for which (18.8) holds regardless of how smooth the exact solution $\boldsymbol{u}$ is.

We say that the single-step approximation method **converges globally** if

$$\lim_{K \to \infty} \max_{k=0,\dots,K} \|\boldsymbol{e}^k\|_2 = 0.$$

In addition, we say that it converges globally, with at least order $p \in \mathbb{N}$, if and only if, when

$$\boldsymbol{u} \in C^{p+1}([0,T]; \Omega),$$

there is some $\tau_1 \in (0, T]$ and a constant $C > 0$ independent of k and τ such that

$$\left\|\boldsymbol{e}^k\right\|_2 \le C\tau^p$$

for all $k = 1, \dots, K$ and any $\tau \in (0, \tau_1]$.

18.2 Consistency and Convergence

Let us now study the consistency and convergence of some single-step approximation methods. We will present a few illustrative cases that highlight the type of techniques and ideas that are needed. The consistency and convergence of many other methods are found in the Problems at the end of the chapter.

A useful tool in this will be a discrete analogue of the Grönwall[3] inequality proved in Lemma 17.7.

Lemma 18.3 (discrete Grönwall). *Let $K \in \mathbb{N}$. Suppose that the finite sequence $\{a_k\}_{k=0}^K \subset \mathbb{R}_+ \cup \{0\}$ satisfies $a_0 = 0$ and, for some $b > 1$, $c \ge 0$,*

$$a_{k+1} \le b a_k + c, \quad k = 0, \dots, K-1.$$

[3] Named in honor of the Swedish–American mathematician Thomas Hakon Grönwall (1877–1932).

Then, for all $k = 0, \ldots, K$,

$$a_k \leq \frac{c}{b-1}\left[b^k - 1\right].$$

Proof. The proof is by induction. The base case, $k = 0$, is trivial since $a_0 = 0$.
For the induction hypothesis, we assume that

$$a_k \leq \frac{c}{b-1}\left[b^k - 1\right]$$

holds for every $k = 0, \ldots, n$.

For the induction step, we observe that, by assumption,

$$\begin{aligned}
a_{n+1} &\leq ba_n + c \\
&\leq b\frac{c}{b-1}\left[b^n - 1\right] + c \\
&= \frac{c}{b-1}\left[b^{n+1} - b\right] + c \\
&= \frac{c}{b-1}b^{n+1} - \frac{bc}{b-1} + \frac{(b-1)c}{b-1} \\
&= \frac{c}{b-1}b^{n+1} - \frac{c}{b-1} \\
&= \frac{c}{b-1}\left[b^{n+1} - 1\right],
\end{aligned}$$

which completes the induction argument. □

18.2.1 Forward Euler Method

Proposition 18.4 (consistency). *Suppose that* $d = 1$ *and* $f \in \mathcal{F}^1(S)$. *Then, for all* $s \in (0, T]$ *and* $t \in [s, T]$,

$$u(t) = u(t-s) + sf(t-s, u(t-s)) + s\mathcal{E}[u](t,s),$$

where $\mathcal{E}[u](t,s)$ *satisfies*

$$|\mathcal{E}[u](t,s)| \leq Cs$$

and $C > 0$ *is a constant that is independent of* t *and* s.

Proof. Since $f \in \mathcal{F}^1(S)$, $u \in C^2([0,T])$. Fix $s \in (0, T]$ and $t \in [s, T]$. By Taylor's Theorem B.31, for some $\eta \in (t-s, t)$,

$$u(t) = u(t-s) + u'(t-s)s + \frac{1}{2}u''(\eta)s^2.$$

Hence,

$$u(t) = u(t-s) + sf(t-s, u(t-s)) + s\mathcal{E}[u](t,s),$$

where

$$\mathcal{E}[u](t,s) = \frac{1}{2}u''(\eta)s.$$

The result is proved using that

$$|\mathcal{E}[u](t,s)| \leq \frac{s}{2}\max_{t-s\leq\eta\leq t}|u''(\eta)| \leq \frac{s}{2}\max_{0\leq\eta\leq T}|u''(\eta)|$$

and taking

$$C = \frac{1}{2} \max_{0 \le \eta \le T} |u''(\eta)|. \qquad \square$$

An analogous result, for $d > 1$, immediately implies consistency of the forward Euler method.

Corollary 18.5 (consistency). *The forward Euler method* (18.3) *is of exactly order* $p = 1$. *In other words, if* $\boldsymbol{f} \in \mathcal{F}^1(S)$ *and* $\boldsymbol{u} \in C^2([0, T]; \mathbb{R}^d)$ *is the unique solution to* (18.1), *then, for all* $\tau \in (0, T]$ *and* $t \in (\tau, T]$,

$$\|\boldsymbol{\mathcal{E}}[\boldsymbol{u}](t, \tau)\|_2 \le C_{FE}\tau \tag{18.9}$$

for some $C_{FE} > 0$ *that is independent of* t *and* τ.

Proof. See Problem 18.3. $\square$

Theorem 18.6 (convergence). *Suppose that* $\boldsymbol{f} \in \mathcal{F}^1(S)$. *Let* $L > 0$ *be its* $\boldsymbol{u}$*-Lipschitz constant on* S. *Suppose that the forward Euler method* (18.3) *is used to approximate* $\boldsymbol{u}$, *the unique solution to* (18.1). *Then, for all* $k = 0, \ldots, K$,

$$\|\boldsymbol{e}^k\|_2 \le \frac{C_{FE}}{L}\left[e^{TL} - 1\right]\tau,$$

where $C_{FE} > 0$ *is the LTE constant from* (18.9). *Consequently,*

$$\max_{k=1,\ldots,K} \|\boldsymbol{e}^k\|_2 \le \frac{C_{FE}}{L}\left[e^{TL} - 1\right]\tau.$$

Proof. Recall that $\boldsymbol{e}^k = \boldsymbol{u}(t_k) - \boldsymbol{w}^k$ and notice that we have the error equation

$$\boldsymbol{e}^{k+1} = \boldsymbol{e}^k + \tau\boldsymbol{f}(t_k, \boldsymbol{u}(t_k)) - \tau\boldsymbol{f}(t_k, \boldsymbol{w}^k) + \tau\boldsymbol{\mathcal{E}}^{k+1}[\boldsymbol{u}]$$

for $k = 0, \ldots, K-1$ with $\boldsymbol{e}^0 = \boldsymbol{0}$. Using the triangle inequality, the Lipschitz condition, and the LTE bound (18.9), we have, for $k = 0, \ldots, K-1$,

$$\left\|\boldsymbol{e}^{k+1}\right\|_2 \le \left\|\boldsymbol{e}^k\right\|_2 + \tau L\left\|\boldsymbol{e}^k\right\|_2 + \tau\left\|\boldsymbol{\mathcal{E}}^{k+1}[u]\right\|_2 \le (1 + \tau L)\left\|\boldsymbol{e}^k\right\|_2 + C_{FE}\tau^2.$$

Using Lemma 18.3, we find, for $k = 0, \ldots, K$,

$$\left\|\boldsymbol{e}^k\right\|_2 \le \frac{C_{FE}}{L}\left[(1 + \tau L)^k - 1\right]\tau.$$

Now, since $\tau L > 0$, $1 + \tau L < e^{\tau L}$. Hence,

$$\left\|\boldsymbol{e}^k\right\|_2 \le \frac{C_{FE}}{L}\left[e^{\tau k L} - 1\right]\tau \le \frac{C_{FE}}{L}\left[e^{TL} - 1\right]\tau$$

for all $k = 0, \ldots, K$. $\square$

18.2.2 Trapezoidal Method

Proposition 18.7 (consistency). *Suppose that* $d = 1$ *and* $f \in \mathcal{F}^2(S)$. *Then, for all* $s \in (0, T]$ *and* $t \in [s, T]$, *we have*

$$u(t) = u(t - s) + \frac{s}{2}\left[f(t - s, u(t - s)) + f(t, u(t))\right] + s\mathcal{E}[u](t, s),$$

where $\mathcal{E}[u](t,s)$ satisfies

$$|\mathcal{E}[u](t,s)| \leq Cs^2,$$

where $C > 0$ is a constant that is independent of s and t.

Proof. Since $f \in \mathcal{F}^2(S)$, we have that $u \in C^3([0,T])$. By Taylor's Theorem B.31, for some $\eta \in (t-s, t-s/2)$,

$$u(t-s/2) = u(t-s) + u'(t-s)\frac{s}{2} + \frac{1}{2}u''(\eta)\frac{s^2}{4}.$$

Likewise, for some $\zeta \in (t-s/2, t)$,

$$u(t-s/2) = u(t) + u'(t)\frac{(-s)}{2} + \frac{1}{2}u''(\zeta)\frac{(-s)^2}{4}.$$

Subtracting, we have

$$u(t) = u(t-s) + \frac{s}{2}\left(u'(t-s) + u'(t)\right) + \frac{s^2}{8}\left(u''(\eta) - u''(\zeta)\right).$$

Using the Mean Value Theorem B.30, for some $\chi \in (\eta, \zeta)$,

$$u''(\eta) - u''(\zeta) = u'''(\chi)(\eta - \zeta).$$

Hence,

$$u(t) = u(t-s) + \frac{s}{2}\left[f(t-s, u(t-s)) + f(t, u(t))\right] + s\mathcal{E}[u](t,s),$$

where

$$\mathcal{E}[u](t,s) = s\frac{\eta - \zeta}{8}u'''(\chi).$$

Since $|\eta - \zeta| \leq s$ and $u \in C^3([0,T])$, the result is proved via the following estimate:

$$|\mathcal{E}[u](t,s)| \leq \frac{s^2}{8}\max_{t-s\leq\chi\leq t}|u'''(\chi)| \leq \frac{s^2}{8}\max_{t\in[0,T]}|u'''(t)|. \qquad \square$$

An extension, to $d > 1$, of this result implies consistency of the trapezoidal method.

Corollary 18.8 (consistency). *The trapezoidal method* (18.5) *is of order exactly $p = 2$. Precisely, if $\boldsymbol{f} \in \mathcal{F}^2(S)$, then, for all $\tau \in (0,T]$ and $t \in [\tau, T]$,*

$$\|\mathcal{E}[\boldsymbol{u}](t,\tau)\|_2 \leq C_{TR}\tau^2 \tag{18.10}$$

for some $C_{TR} > 0$ that is independent of t and s.

Proof. See Problem 18.6. $\square$

We can now obtain global convergence of the trapezoidal method.

Theorem 18.9 (convergence). *Let $\boldsymbol{f} \in \mathcal{F}^2(S)$ and $L > 0$ be its $\boldsymbol{u}$-Lipschitz constant on S. Suppose that the trapezoidal method* (18.5) *is used to approximate the solution to* (18.1)*. Then, for all $k = 0, \ldots, K$, we have*

$$\left\|\boldsymbol{e}^k\right\|_2 \leq \frac{C_{TR}}{L}\left[\exp(2TL) - 1\right]\tau^2,$$

provided that $0 < \tau L < 1$*, where* $C_{TR} > 0$ *is the LTE constant from* (18.10).

Proof. As in the case of the forward Euler method, we begin by identifying an equation for the error $\boldsymbol{e}^k = \boldsymbol{u}(t_k) - \boldsymbol{w}^k$. In this case, we have

$$\boldsymbol{e}^{k+1} = \boldsymbol{e}^k + \frac{\tau}{2}\left[\boldsymbol{f}(t_k, \boldsymbol{u}(t_k)) - \boldsymbol{f}(t_k, \boldsymbol{w}^k)\right] + \frac{\tau}{2}\left[\boldsymbol{f}(t_{k+1}, \boldsymbol{u}(t_{k+1})) - \boldsymbol{f}(t_{k+1}, \boldsymbol{w}^{k+1})\right] + \tau\boldsymbol{\mathcal{E}}^{k+1}[\boldsymbol{u}].$$

We take norms and apply the triangle inequality, the $\boldsymbol{u}$-Lipschitz condition on $\boldsymbol{f}$ and (18.10), to obtain

$$\left(1 - \frac{\tau L}{2}\right)\|\boldsymbol{e}^{k+1}\|_2 \leq \left(1 + \frac{\tau L}{2}\right)\|\boldsymbol{e}^k\|_2 + C_{TR}\tau^3.$$

Since, by assumption, $\frac{1}{2} < 1 - \frac{\tau L}{2} < 1$, Lemma 18.3 then implies that

$$\begin{aligned}\|\boldsymbol{e}^{k+1}\|_2 &\leq \frac{C_{TR}\tau^3}{1 - \frac{\tau L}{2}}\left[\left(\frac{1 + \frac{\tau L}{2}}{1 - \frac{\tau L}{2}}\right) - 1\right]^{-1}\left[\left(\frac{1 + \frac{\tau L}{2}}{1 - \frac{\tau L}{2}}\right)^k - 1\right] \\ &= \frac{C_{TR}}{L}\left[\left(\frac{1 + \frac{\tau L}{2}}{1 - \frac{\tau L}{2}}\right)^k - 1\right]\tau^2.\end{aligned}$$

Finally, notice that

$$\frac{1 + \frac{\tau L}{2}}{1 - \frac{\tau L}{2}} = 1 + \frac{\tau L}{1 - \frac{\tau L}{2}} \leq 1 + 2\tau L \leq e^{2\tau L}$$

to, in conclusion, obtain that

$$\|\boldsymbol{e}^{k+1}\|_2 \leq \frac{C_{TR}}{L}[e^{2k\tau L} - 1]\tau^2 \leq \frac{C_{TR}}{L}[e^{2TL} - 1]\tau^2,$$

as claimed. □

18.2.3 Taylor's Method

As a last example, we consider the consistency and convergence of Taylor's method.

Theorem 18.10 (consistency). *Suppose that* $\boldsymbol{f} \in \mathcal{F}^2(S)$*. For any* $s \in (0, T]$ *and* $t \in (s, T]$*, we have that the LTE of Taylor's method satisfies*

$$\|\boldsymbol{\mathcal{E}}[\boldsymbol{u}](t, s)\|_2 \leq C_{TM}s^2 \tag{18.11}$$

for some $C_{TM} > 0$ *that is independent of t and s. In other words, Taylor's method is consistent to exactly order* $p = 2$.

Proof. For simplicity, we will only consider the case $d = 1$. Since $f \in \mathcal{F}^2(S)$, we know that $u \in C^3([0, T])$. Thus, using Taylor's Theorem, we get that

$$u(t) = u(t - s) + su'(t - s) + \frac{s^2}{2}u''(t - s) + \frac{s^3}{6}u'''(\xi)$$

for some $\xi \in (t-s, t)$. Now we note that $u'(t-s) = f(t-s, u(t-s))$ and also

$$u''(t-s) = \frac{\partial f}{\partial t}(t-s, u(t-s)) + \frac{\partial f}{\partial u}(t-s, u(t-s))f(t-s, u(t-s)).$$

Hence,

$$\begin{aligned} s\mathcal{E}[u](t,s) &= u(t) - u(t-s) - sf(t-s, u(t-s)) - \frac{s^2}{2}\frac{\partial f}{\partial t}(t-s, u(t-s)) \\ &\quad - \frac{s^2}{2}\frac{\partial f}{\partial u}(t-s, u(t-s))f(t-s, u(t-s)) \\ &= \frac{s^3}{6}u'''(\xi). \end{aligned}$$

The result follows. □

Theorem 18.11 (convergence). *Suppose that $\boldsymbol{f} \in \mathcal{F}^2(S)$ and there is some constant $B > 0$ such that*

$$|D^{\boldsymbol{\alpha}} f_i(t, \boldsymbol{v})| \leq B$$

for all multi-indices $\boldsymbol{\alpha} \in \mathbb{N}^{d+1}$ with $|\boldsymbol{\alpha}| = 2$, for all $i = 1, \ldots, d$, and for all $(t, \boldsymbol{v}) \in S$. (In other words, all second derivatives are bounded on S.) Then Taylor's method (18.6) is convergent and the global rate of convergence is $p = 2$. In particular,

$$\left\|\boldsymbol{e}^k\right\|_2 \leq \frac{C_{TM}}{L'}\left[e^{TL'} - 1\right] s^2,$$

where $C_{TM} > 0$ is the LTE constant from (18.11) and $L' > 0$ is a Lipschitz constant given below.

Proof. The key step in this proof is to establish a global $\boldsymbol{u}$-Lipschitz continuity for the slope approximation, i.e., an estimate of the form

$$\|\boldsymbol{G}_{TM}(t, s, \boldsymbol{v}_1) - \boldsymbol{G}_{TM}(t, s, \boldsymbol{v}_2)\|_2 \leq L' \|\boldsymbol{v}_1 - \boldsymbol{v}_2\|_2,$$

for any $s \in (0, T]$, for any $t \in [s, T]$, and for all $\boldsymbol{v}_1, \boldsymbol{v}_2 \in \mathbb{R}^d$. This requires the extra assumptions on the slope function that are included in the hypotheses. The details are left to the reader as an exercise; see Problem 18.9. □

18.3 Linear Slope Functions

In the case that the slope function is linear, we can provide more direct proofs of convergence. The following example will be of interest in Chapter 28, where we examine numerical methods for the heat equation.

Theorem 18.12 (convergence). *Let $\mathsf{A} \in \mathbb{R}^{d\times d}$ be symmetric. Suppose that $\boldsymbol{u}\colon [0, T] \to \mathbb{R}^d$ is the solution to*

$$\boldsymbol{u}'(t) = \mathsf{A}\boldsymbol{u}(t), \qquad \boldsymbol{u}(0) = \boldsymbol{u}_0 \in \mathbb{R}^d.$$

Let $K \in \mathbb{N}$. Suppose that the sequence $\{\boldsymbol{w}^k\}_{k=0}^K$ is generated using the forward Euler method (18.3). Then, for all $k = 0, \ldots, K$, we have

$$\left\|\boldsymbol{e}^k\right\|_2 \leq \|\boldsymbol{u}_0\|_2 \max_{\lambda \in \sigma(\mathsf{A})} \left|e^{\lambda k \tau} - (1+\lambda\tau)^k\right|. \tag{18.12}$$

Suppose that the maximum on the right-hand side of (18.12) is achieved at $\lambda_{\max} \in \sigma(\mathsf{A})$ and, furthermore, that $\lambda_{\max} < 0$. Then there is a constant $\tau_0 \in (0, T]$ such that, for all $\tau \in (0, \tau_0]$ and all $k = 1, 2, \ldots, K$,

$$\left\|\boldsymbol{e}^k\right\|_2 \leq \frac{T}{2}\lambda_{\max}^2 \|\boldsymbol{u}_0\|_2 \tau.$$

Proof. Since $\mathsf{A} \in \mathbb{R}^{d\times d}$ is symmetric, it is orthogonally diagonalizable. In other words, there exists a diagonal matrix D (whose diagonal entries $[\mathsf{D}]_{i,i} = \lambda_i$ are the eigenvalues of A) and an orthogonal matrix Q such that $\mathsf{A} = \mathsf{Q}\mathsf{D}\mathsf{Q}^\intercal$. The exact solution of the equation is then given by

$$\boldsymbol{u}(t) = \mathsf{Q}e^{t\mathsf{D}}\mathsf{Q}^\intercal \boldsymbol{u}_0,$$

where $e^{t\mathsf{D}}$ is a diagonal matrix whose diagonal entries are precisely $\left[e^{t\mathsf{D}}\right]_{i,i} = e^{t\lambda_i}$. Now using the forward Euler method, it is easy to see that

$$\boldsymbol{w}^k = (\mathsf{I} + \tau\mathsf{A})^k \boldsymbol{u}_0 = \mathsf{Q}(\mathsf{I} + \tau\mathsf{D})^k \mathsf{Q}^\intercal \boldsymbol{u}_0.$$

Thus,

$$\boldsymbol{e}^k = \mathsf{Q}e^{k\tau\mathsf{D}}\mathsf{Q}^\intercal\boldsymbol{u}_0 - \mathsf{Q}(\mathsf{I}+\tau\mathsf{D})^k\mathsf{Q}^\intercal\boldsymbol{u}_0 = \mathsf{Q}\left(e^{k\tau\mathsf{D}} - (\mathsf{I}+\tau\mathsf{D})^k\right)\mathsf{Q}^\intercal\boldsymbol{u}_0.$$

Taking norms, we get

$$\left\|\boldsymbol{e}^k\right\|_2 \leq \left\|\mathsf{Q}\left(e^{k\tau\mathsf{D}} - (\mathsf{I}+\tau\mathsf{D})^k\right)\mathsf{Q}^\intercal\right\|_2 \|\boldsymbol{u}_0\|_2 \leq \left\|e^{k\tau\mathsf{D}} - (\mathsf{I}+\tau\mathsf{D})^k\right\|_2 \|\boldsymbol{u}_0\|_2.$$

Notice that $(\mathsf{I} + \tau\mathsf{D})^k$ is a diagonal matrix with the entries $(1+\tau\lambda_i)^k$. To conclude, we use the fact that the 2-norm of a diagonal matrix is simply the largest diagonal element in absolute value. Hence,

$$\left\|\boldsymbol{e}^k\right\|_2 \leq \|\boldsymbol{u}_0\|_2 \max_{\lambda\in\sigma(\mathsf{A})} \left|e^{k\tau\lambda} - (1+\tau\lambda)^k\right|.$$

Now, using Taylor expansions, we see that, for any $x \leq 0$,

$$1 + x \leq e^x \leq 1 + x + \frac{1}{2}x^2.$$

Equivalently,

$$1 + x - \frac{1}{2}x^2 \leq e^x - \frac{1}{2}x^2 \leq 1 + x \leq e^x.$$

Using the binomial expansion, it follows that, if $n \geq 2$,

$$(1-\alpha)^n = 1 - n\alpha + \binom{n}{2}\alpha^2 + \sum_{j=3}^{n}\binom{n}{j}(-1)^j\alpha^j.$$

There is an $\alpha_0 \in (0,1)$ such that if $\alpha \in (0, \alpha_0)$, then

$$\binom{n}{2}\alpha^2 + \sum_{j=3}^{n} \binom{n}{j}(-1)^j \alpha^j \geq 0.$$

Essentially, the first term, which is positive, dominates the others. Thus, if $n \geq 2$ and $\alpha \in (0, \alpha_0)$,

$$(1-\alpha)^n \geq 1 - n\alpha.$$

We apply the last estimate with

$$\alpha = \frac{x^2}{2e^x}.$$

Thus,

$$1 - n\frac{x^2}{2e^x} \leq \left(1 - \frac{x^2}{2e^x}\right)^n.$$

Multiplying by e^{nx}, we get

$$e^{nx} - n\frac{x^2}{2}e^{(n-1)x} \leq \left(e^x - \frac{x^2}{2}\right)^n \leq (1+x)^n \leq e^{nx},$$

provided that $x \in (-1, 0]$ and x is sufficiently small in absolute value.

Now, if $\tau\lambda_{\max} \in [-1, 0)$ is sufficiently small in absolute value,

$$-k\frac{(\tau\lambda_{\max})^2}{2}e^{(k-1)\tau\lambda_{\max}} \leq (1+\tau\lambda_{\max})^k - e^{k\tau\lambda_{\max}} \leq 0,$$

or, equivalently,

$$0 \leq e^{k\tau\lambda_{\max}} - (1+\tau\lambda_{\max})^k \leq k\frac{(\tau\lambda_{\max})^2}{2}e^{(k-1)\tau\lambda_{\max}} \leq \frac{T}{2}\lambda_{\max}^2\tau.$$

The result follows. □

Problems

18.1 Alternate discrete Grönwall inequality: Let $K \in \mathbb{N}$. Suppose that the sequence $\{a_k\}_{k=0}^K \subset \mathbb{R}_+$ is such that there are $b > 0$ and $c \geq 0$ for which

$$a_{k+1} \leq b a_k + c, \quad k = 0, \ldots, K-1.$$

Prove that, for all $k = 0, \ldots, K$,

$$a_k \leq b^k a_0 + \left(\sum_{j=0}^{k-1} b^j\right) c.$$

For $b \neq 1$, show that

$$a_k \leq b^k a_0 + \frac{b^k - 1}{b - 1}c.$$

18.2 Yet another discrete Grönwall inequality: Let $\gamma, \beta \in \mathbb{R}$ with $\beta > 0$ and $\gamma > -1$. Let $\{a_n\}_{n\in\mathbb{N}}$ and $\{f_n\}_{n\in\mathbb{N}}$ be two sequences of nonnegative real numbers satisfying

$$(1+\gamma)a_{n+1} \leq a_n + \beta f_n.$$

Prove that

$$a_{n+1} \leq \frac{a_0}{(1+\gamma)^{n+1}} + \beta \sum_{k=0}^{n} \frac{f_k}{(1+\gamma)^{n-k+1}}.$$

18.3 Prove Corollary 18.5.

18.4 Suppose that $\boldsymbol{f} \in \mathcal{F}^1(S)$. Prove that the backward Euler method is:

a) Consistent to exactly order $p = 1$.

b) Globally convergent with order $p = 1$.

18.5 Show that a sharper LTE estimate than the Proposition 18.7 estimate can be obtained. Specifically, under the same assumptions, prove that

$$|\mathcal{E}[u](t, s)| \leq \frac{s^2}{12} \max_{t\in[0,T]} |u'''(t)|.$$

18.6 Prove Corollary 18.8.

18.7 Suppose that $\boldsymbol{f} \in \mathcal{F}^2(S)$. Prove that the midpoint method is:

a) Consistent to exactly order $p = 2$.

b) Globally convergent with order $p = 2$.

18.8 Let $\theta \in [0, 1]$. The θ-method is defined as

$$\boldsymbol{w}^{k+1} = \boldsymbol{w}^k + \tau \boldsymbol{f}(t_k + (1-\theta)\tau, \theta \boldsymbol{w}^k + (1-\theta)\boldsymbol{w}^{k+1}).$$

Assuming that $\boldsymbol{f} \in \mathcal{F}^2(S)$, find the consistency order of this method and show that it is convergent.

Hint: The order of consistency depends on the value of θ.

18.9 Complete the proof of Theorem 18.11.

18.10 Consider the IVP

$$u'(t) = u(t), \quad t \in (0, 1], \qquad u(0) = 1.$$

For $K \in \mathbb{N}$, set $\tau = \frac{1}{K}$. Apply the following methods to obtain approximations of $u(1) = e$.

a) Forward Euler method.

b) Taylor's method.

c) Heun's[4] method:

$$w^{k+1} = w^k + \frac{\tau}{2}\left[f(t_k, w^k) + f\left(t_{k+1}, w^k + \tau f(t_k, w^k)\right)\right].$$

d) Modified Euler's method:

$$w^{k+1} = w^k + \tau f\left[t_k + \frac{\tau}{2}, w^k + \frac{\tau}{2} f(t_k, w^k)\right].$$

For these approximations, show directly (without appealing to any convergence theorems) that $w^K \to u(1) = e$ as $K \to \infty$.

[4] Named in honor of the German mathematician Karl Heun (1859–1929).

19 Runge–Kutta Methods

In this chapter, we introduce Runge–Kutta (RK) approximation methods for initial value problems (IVPs). These are single-step methods that have multiple *stages*, and the form and function of these methods are rather distinct from those we have previously defined. For simplicity, we will largely neglect the convergence theory of RK methods, though it would not present any new technical difficulties, only tedious calculations.

We begin by recalling that we are trying to approximate the solution to (18.1). To achieve this, as before, we introduce a discretization of the time interval $[0, T]$ via the following procedure. Let $K \in \mathbb{N}$, $\tau = \frac{T}{K}$, and $t_k = \tau k$ for $k = 0, \dots, K$. As before, we will produce a sequence $\{\boldsymbol{w}^k\}_{k=0}^K$ such that $\boldsymbol{u}(t_k) \approx \boldsymbol{w}^k$.

The main motivation behind these methods can be explained by taking a second look at Taylor's method (18.6), where the slope approximation function is given by

$$\boldsymbol{G}_{TM}(t, s, \boldsymbol{v}_1) = \boldsymbol{f}(t, \boldsymbol{v}_1) + \frac{s}{2}\left[\partial_t \boldsymbol{f}(t, \boldsymbol{v}_1) + D_{\boldsymbol{u}}\boldsymbol{f}(t, \boldsymbol{v}_1)\boldsymbol{f}(t, \boldsymbol{v}_1)\right].$$

Clearly, this approximation comes from an approximation of $\boldsymbol{u}'(t+s)$ via a second-order Taylor expansion. As we saw in Theorem 18.11, this method is convergent with rate $p = 2$. While this is a perfectly acceptable method, it has one major drawback. Namely, it requires knowledge not only of the slope function $\boldsymbol{f}$ but also of its partial derivatives with respect to time, $\partial_t \boldsymbol{f}$, and $\boldsymbol{u}$, $D_{\boldsymbol{u}}\boldsymbol{f}$. In practice, these functions may not be available, or they may be very difficult to compute.

If we are to allow dealing with derivatives of the slope function, the next logical step may be trying to devise a method, via a Taylor expansion, that is consistent to order at least $p = 3$. The procedure is clear. We do a Taylor expansion of $\boldsymbol{u}(t+s)$ about the point t and use (18.1) to arrive at

$$\begin{aligned}\boldsymbol{u}(t+s) &= \boldsymbol{u}(t) + s\boldsymbol{u}'(t) + \frac{s^2}{2}\boldsymbol{u}''(t) + \frac{s^3}{6}\boldsymbol{u}''(t) + \mathcal{O}(|s|^4) \\ &\approx \boldsymbol{u}(t) + s\Big\{\boldsymbol{f} + \frac{s}{2}\left[\partial_t \boldsymbol{f} + D_{\boldsymbol{u}}\boldsymbol{f}\boldsymbol{f}\right] \\ &\quad + \frac{s^2}{6}\left[\partial_t^2 \boldsymbol{f} + D_{\boldsymbol{u}}\partial_t \boldsymbol{f}\cdot\boldsymbol{f} + D_{\boldsymbol{u}}\boldsymbol{f}\left(\partial_t \boldsymbol{f} + D_{\boldsymbol{u}}\boldsymbol{f}\boldsymbol{f}\right) + \left(\partial_t D_{\boldsymbol{u}}\boldsymbol{f} + D_{\boldsymbol{u}}^2\boldsymbol{f}\cdot\boldsymbol{f}\right)\boldsymbol{f}\right]\Big\},\end{aligned}$$

where, for simplicity, we have suppressed the arguments of the slope function and its derivatives. Hopefully, the reader appreciates the difficulties we have encountered. Not only will such a procedure require evaluating many derivatives of the slope function, but also the number of terms that is involved grows extremely large.

The idea behind RK methods is that, instead of differentiating the slope function, we evaluate it at a special collection of points in $[t_k, t_{k+1}] \times \bar{\Omega}$ called *stages*, so that, for instance, in the case of a two-stage RK method we have

$$\boldsymbol{\kappa}_1 = \boldsymbol{f}(t_k, \boldsymbol{w}^k), \qquad \boldsymbol{\kappa}_2 = \boldsymbol{f}(t_k + c\tau, \boldsymbol{w}^k + a\tau\boldsymbol{\kappa}_1);$$

then the approximation at t_{k+1} is given by a *linear combination* of the stages:

$$\boldsymbol{w}^{k+1} = \boldsymbol{w}^k + \tau\,(b_1\boldsymbol{\kappa}_1 + b_2\boldsymbol{\kappa}_2).$$

By properly choosing the coefficients a, c, b_1, and b_2, a method that is consistent to order at least $p = 2$ (like Taylor's method) can be obtained.

19.1 Simple Two-Stage Methods

Before we give a general definition of RK methods, let us elaborate on the previous idea for a simple, scalar, autonomous problem. We leave it to the reader to show that the following fits the general definition given later.

Theorem 19.1 (RK2). *Let $T > 0$ be given. Consider the general two-stage explicit RK method, defined by*

$$\xi^k = w^k + a\tau f(w^k), \quad w^{k+1} = w^k + \tau\big[b_1 f(w^k) + b_2 f(\xi^k)\big],$$

for approximating the solution to the scalar autonomous IVP

$$u'(t) = f(u(t)), \quad t \in [0, T], \quad u(0) = u_0.$$

Assume that $f \in \mathcal{F}^2(S)$ and, therefore, $u \in C^3([0, T])$. If the coefficients satisfy $b_1 + b_2 = 1$, $b_1, b_2 \geq 0$, and $ab_2 = \frac{1}{2}$, then the method is consistent of order $p = 2$ and the method is convergent to second order.

Proof. Let us first show consistency. By Taylor's Theorem, for some ζ between $u(t-\tau)$ and $u(t-\tau) + a\tau f(u(t-\tau))$,

$$\begin{aligned} f\big(u(t-\tau) + a\tau f(u(t-\tau))\big) &= f(u(t-\tau)) + a\tau f(u(t-\tau))f'(u(t-\tau)) \\ &\quad + \frac{1}{2}\big(a\tau f(u(t-\tau))\big)^2 f''(\zeta). \end{aligned}$$

Upon setting $b_1 + b_2 = 1$ and $ab_2 = \frac{1}{2}$, the LTE satisfies

$$\begin{aligned} \mathcal{E}[u](t,\tau) &= \frac{u(t) - u(t-\tau)}{\tau} \\ &\quad - \big[b_1 f(u(t-\tau)) - b_2 f\big(u(t-\tau) + a\tau f(u(t-\tau))\big)\big] \\ &= \frac{u(t) - u(t-\tau)}{\tau} - b_1 f(u(t-\tau)) - b_2 f(u(t-\tau)) \\ &\quad - \tau a b_2 f(u(t-\tau))f'(u(t-\tau)) - \tau^2\frac{a^2 b_2}{2} f^2(u(t-\tau))f''(\zeta) \\ &= \frac{u(t) - u(t-\tau)}{\tau} - f(u(t-\tau)) \\ &\quad - \frac{\tau}{2} f(u(t-\tau))\, f'(u(t-\tau)) - \tau^2\frac{a^2 b_2}{2} f^2(u(t-\tau))f''(\zeta). \end{aligned}$$

On the other hand, using Taylor's Theorem, the exact solution must satisfy

$$\begin{aligned}u(t) &= u(t-\tau) + \tau u'(t-\tau) + \frac{\tau^2}{2}u''(t-\tau) + \frac{\tau^3}{6}u'''(\sigma)\\ &= u(t-\tau) + \tau f(u(t-\tau)) + \frac{\tau^2}{2}f'(u(t-\tau))f(u(t-\tau)) + \frac{\tau^3}{6}u'''(\sigma)\end{aligned}$$

for some $\sigma \in (t-\tau, t)$. Comparing the expansions,

$$\mathcal{E}[u](t,\tau) = \frac{\tau^2}{6}u'''(\sigma) - \tau^2\frac{a^2 b_2}{2}f^2(u(t-\tau))f''(\zeta)$$

provided that $f \in C^2((-\infty,\infty))$, $u \in C^3([0,T])$, $b_1 + b_2 = 1$, and $ab_2 = \frac{1}{2}$. There is some $C > 0$ such that

$$|\mathcal{E}[u](t,\tau)| \le C\tau^2$$

for any $\tau \in (0,T]$ and $t \in [\tau, T]$. The proof for this is subtle and relies on the fact that u is bounded over $[0,T]$.

We will only show convergence in the case that $a = 1/2$, $b_1 = 0$, and $b_2 = 1$, leaving the general case to the reader as an exercise; see Problem 19.1. With this simplification, the method reads

$$w^{k+1} = w^k + \tau f\left(w^k + \frac{\tau}{2}f(w^k)\right).$$

The exact solution satisfies

$$u(t_{k+1}) = u(t_k) + \tau f\left(u(t_k) + \frac{\tau}{2}f(u(t_k))\right) + \tau\mathcal{E}^{k+1}[u].$$

Therefore,

$$e^{k+1} = e^k + \tau f\left(u(t_k) + \frac{\tau}{2}f(u(t_k))\right) - \tau f\left(w^k + \frac{\tau}{2}f(w^k)\right) + \tau\mathcal{E}^{k+1}[u].$$

Taking absolute values and using the triangle inequality and the Lipschitz continuity of the slope function f, we have

$$\begin{aligned}|e^{k+1}| &\le |e^k| + \tau L\left|e^k + \frac{\tau}{2}(f(u(t_k)) - f(w^k))\right| + \tau\left|\mathcal{E}^{k+1}[u]\right|\\ &\le (1+\tau L)|e^k| + \frac{\tau^2L^2}{2}|e^k| + C\tau^3\\ &= \left(1 + \tau L + \frac{\tau^2 L^2}{2}\right)|e^k| + C\tau^3.\end{aligned}$$

Using the discrete Grönwall inequality from Lemma 18.3,

$$|e^k| \le \frac{C\tau^2}{L + \frac{\tau L^2}{2}}\left[\left(1 + \tau L + \frac{\tau^2L^2}{2}\right)^k - 1\right].$$

Now, since $\tau L > 0$,

$$1 + \tau L + \frac{\tau^2 L^2}{2} < e^{\tau L};$$

therefore, for any $m = 1, \dots, K$,

$$(1+\tau L)^m < e^{m\tau L} \le e^{K\tau L} = e^{TL},$$

where we used that $K\tau = T$. It follows that, for all $k = 0, \dots, K$,

$$|e^k| \leq \frac{C}{L}\left[e^{TL} - 1\right]\tau^2. \qquad \square$$

19.2 General Definition and Basic Properties

We now embark upon the study of general RK methods and their properties. We begin with their definition.

Definition 19.2 (RK). Let $r \in \mathbb{N}$. A general r**-stage Runge–Kutta method** (RK method)[1] is a recursive algorithm for generating an approximation $\left\{\boldsymbol{w}^k\right\}_{k=0}^{K}$ to the solution of (18.1), via $\boldsymbol{w}^0 = \boldsymbol{u}_0$ and, for $k = 0, \dots, K-1$,

$$\boldsymbol{\xi}_i = \boldsymbol{w}^k + \tau \sum_{j=1}^{r} a_{i,j} \boldsymbol{f}(t_k + c_j\tau, \boldsymbol{\xi}_j), \quad i = 1, \dots, r, \tag{19.1}$$

$$\boldsymbol{w}^{k+1} = \boldsymbol{w}^k + \tau \sum_{j=1}^{r} b_j \boldsymbol{f}(t_k + c_j\tau, \boldsymbol{\xi}_j). \tag{19.2}$$

Here, $a_{i,j} \in \mathbb{R}$ and $b_j, c_j \in [0,1]$ for $i, j = 1, \dots, r$. An RK method is completely determined by its weights $\mathsf{A} = [a_{i,j}]_{i,j=1}^r \in \mathbb{R}^{r\times r}$, $\boldsymbol{b} = [b_i]_{i=1}^r \in \mathbb{R}^r$, and $\boldsymbol{c} = [c_i]_{i=1}^r \in \mathbb{R}^r$, which are often expressed in **tableau** form

$$\begin{array}{c|c} \boldsymbol{c} & \mathsf{A} \\ \hline & \boldsymbol{b}^\intercal \end{array},$$

which is commonly referred to as the **Butcher tableau**[2] of the method. The RK method is called **explicit** (ERK) if and only if $a_{i,j} = 0$ for all $i \leq j$, and is called **implicit** (IRK) otherwise. The RK method is called **diagonally implicit** (DIRK) if and only if $a_{i,j} = 0$ for all $i < j$.

Remark 19.3 (equivalent definition). As we mentioned above, there is an alternate, but equivalent, definition of the general r-stage RK method. Let us write out this equivalent form. First, define

$$\boldsymbol{\kappa}_j = \boldsymbol{f}(t_k + c_j\tau, \boldsymbol{\xi}_j), \quad j = 1, \dots, r.$$

Then, from (19.2),

$$\boldsymbol{w}^{k+1} = \boldsymbol{w}^k + \tau \sum_{j=1}^{r} b_j \boldsymbol{\kappa}_j, \tag{19.3}$$

where, from (19.1),

$$\boldsymbol{\kappa}_i = \boldsymbol{f}\left(t_k + c_i\tau, \boldsymbol{w}^k + \tau \sum_{j=1}^{r} a_{i,j}\boldsymbol{\kappa}_j\right), \quad i = 1, \dots, r. \tag{19.4}$$

[1] Named in honor of the German mathematicians Carl David Tolmé Runge (1856–1927) and Martin Wilhelm Kutta (1867–1944).

[2] Named in honor of the New Zealand mathematician John Charles Butcher (1933–).

The following theorem describes constraints on the weights of an RK method, so that the method satisfies certain consistency requirements.

Theorem 19.4 (properties of weights). *Assume that $\boldsymbol{f} \in \mathcal{F}^1(S)$. Consider the general r-stage RK method given by the weights $\mathsf{A} = [a_{i,j}]_{i,j=1}^r \in \mathbb{R}^{r\times r}$, $\boldsymbol{b} = [b_i]_{i=1}^r \in [0,1]^r$, and $\boldsymbol{c} = [c_i]_{i=1}^r \in [0,1]^r$. Let $\mathbf{1} = [1]_{i=1}^r \in \mathbb{R}^r$.*

1. *For the method to be at least first order, it is necessary that*
$$\boldsymbol{b}^\intercal \mathbf{1} = 1.$$
2. *For the jth RK stage $\boldsymbol{\xi}_j$ to be at least a first-order approximation of $\boldsymbol{u}(t_k + c_j\tau)$, it is necessary that*
$$\mathsf{A}\mathbf{1} = \boldsymbol{c}. \tag{19.5}$$
3. *Suppose that $\boldsymbol{f} \in \mathcal{F}^2(S)$ and (19.5) holds. For the method to be at least second order, it is necessary that*
$$\boldsymbol{b}^\intercal \boldsymbol{c} = \frac{1}{2}.$$
4. *Suppose that $\boldsymbol{f} \in \mathcal{F}^3(S)$ and (19.5) holds. For the method to be at least third order, it is necessary that*
$$\boldsymbol{b}^\intercal \mathsf{A}\boldsymbol{c} = \frac{1}{6}.$$

Proof. We sketch the proof and leave the details to the reader; see Problem 19.2. To prove the result, use the general r-step method to approximate the solution to the linear scalar problem $u'(t) = u(t)$, $u(0) = 1$, whose exact solution is $u(t) = e^t$. At time $t = \tau$, the solution may be expressed as
$$u(\tau) = 1 + \tau + \frac{\tau^2}{2} + \frac{\tau^3}{6} + \frac{\tau^4}{24}e^\eta$$
for some $\eta \in (0, \tau)$. For the RK stages, assume that the matrix $\mathsf{I} - \tau\mathsf{A}$ is invertible — this will always will be the case provided that τ is sufficiently small — and
$$(\mathsf{I} - \tau\mathsf{A})^{-1} = \mathsf{I} + \tau\mathsf{A} + \tau^2\mathsf{A}^2 + \tau^3\mathsf{A}^3 + \cdots.$$
It is possible to show that the vector of stages satisfies
$$\boldsymbol{\xi} = (\mathsf{I} - \tau\mathsf{A})^{-1}\mathbf{1}.$$
Solve explicitly for w^1 and compare the result with the expansion $u(\tau)$ above. □

The following expressions for the ERK and IRK approximations, respectively, of $u' = \lambda u$ will be needed in subsequent chapters.

Theorem 19.5 (amplification factor I). *Applying an r-stage explicit RK method to approximate the solution of the differential equation $u'(t) = \lambda u(t)$, $u(0) = u_0$, one obtains*
$$w^{k+1} = g(\lambda\tau)w^k, \quad g(z) = \sum_{j=0}^r \beta_j z^j, \quad w^0 = u_0,$$

where $\beta_j \in \mathbb{R}$, $j = 0, \ldots, r$. If the method is consistent to exactly order r, then

$$g(z) = \sum_{j=0}^{r} \frac{z^j}{j!},$$

i.e., $\beta_j = \frac{1}{j!}$ *for* $j = 0, \ldots, r$.

Proof. See Problem 19.3. □

Theorem 19.6 (amplification factor II). *Applying an r-stage implicit RK method to approximate the solution of the differential equation $u'(t) = \lambda u(t)$, $u(0) = u_0$, one obtains*

$$w^{k+1} = g(\lambda\tau)w^k, \quad g(z) = \frac{p_1(z)}{p_2(z)}, \quad w^0 = u_0,$$

where $p_1, p_2 \in \mathbb{P}_r$ and $p_2 \not\equiv 0$. In particular, g is the rational polynomial

$$g(z) = 1 + z\boldsymbol{b}^{\intercal}(\mathsf{I} - z\mathsf{A})^{-1}\mathbf{1} = \frac{\det(\mathsf{I} - z\mathsf{A} + z\mathbf{1}\,\boldsymbol{b}^{\intercal})}{\det(\mathsf{I} - z\mathsf{A})},$$

where $\mathbf{1} = [1]_{i=1}^{r} \in \mathbb{R}^r$.

Proof. See Problem 19.4. □

Remark 19.7 (amplification factor). The function g (a polynomial in the ERK case and a rational function in the IRK case) that appears in Theorems 19.5 and 19.6 is called the *linear amplification factor* or just the *amplification factor*.

Remark 19.8 (LTE). The *consistency error* for any explicit RK method is defined in a straightforward way, since the RK stages can be computed explicitly in terms of the approximations. Specifically,

$$\tau\mathcal{E}[\boldsymbol{u}](t,s) = \boldsymbol{u}(t) - \boldsymbol{u}(t-s) - s\sum_{i=1}^{r} b_i \boldsymbol{f}(t - \tau + c_i\tau, \boldsymbol{\xi}_{e,i}),$$

where $\boldsymbol{\xi}_{e,1} = \boldsymbol{u}(t-\tau)$ and, for $i = 2, \ldots, r$,

$$\boldsymbol{\xi}_{e,i} = \boldsymbol{u}(t-\tau) + \tau\sum_{j=1}^{i-1} a_{i,j}\boldsymbol{f}(t - \tau + c_j\tau, \boldsymbol{\xi}_{e,j}).$$

Notice that, in the end, the $\boldsymbol{\xi}_{e,i}$ can be completely eliminated, which is a key insight. For implicit RK methods, the situation is a bit more complicated.

We have already encountered the following, but in a slightly simpler form, in Theorem 19.1.

Theorem 19.9 (two-stage RK method). *Suppose that $\boldsymbol{f} \in \mathcal{F}^2(S)$, so that $\boldsymbol{u} \in C^3([0,T];\mathbb{R}^d)$ is a classical solution to the IVP* (18.1). *Consider an explicit two-stage RK method given by the tableau*

$$\begin{array}{c|cc} 0 & 0 & 0 \\ c_2 & a_{2,1} & 0 \\ \hline & b_1 & b_2 \end{array}.$$

The method is consistent to order $p = 2$ if and only if

$$b_1 + b_2 = 1, \quad a_{2,1} = c_2, \quad b_2 c_2 = \frac{1}{2}.$$

Proof. For simplicity of notation, we will give a proof in the scalar case ($d = 1$).

We begin with a two-dimensional Taylor expansion of $f \in C^2(S)$. Set $\boldsymbol{q} = [c_2, a_{2,1} f(t-\tau, u(t-\tau))]^\intercal$. Then

$$\begin{aligned} f\big(t - \tau + c_2\tau, u(t-\tau) + a_{2,1}\tau f(t-\tau, u(t-\tau))\big) \\ = f(t-\tau, u(t-\tau)) + \tau q_1 \partial_t f(t-\tau, u(t-\tau)) \\ + \tau q_2 \partial_u f(t-\tau, u(t-\tau)) + \frac{\tau^2}{2} \boldsymbol{q}^\intercal \mathsf{H}_f(\eta, \gamma) \boldsymbol{q}, \end{aligned} \tag{19.6}$$

where H_f is the 2×2 Hessian matrix of second derivatives of f, η is some number between $t - \tau$ and $t - \tau + c_2\tau$, and γ is some number between $u(t-\tau)$ and $u(t-\tau) + a_{2,1}\tau f(t-\tau, u(t-\tau))$. For $t \in [\tau, T]$, it follows that, since the solution u is twice continuously differentiable,

$$|\boldsymbol{q}^\intercal \mathsf{H}_f(\eta, \gamma)\boldsymbol{q}| \leq C,$$

where $C > 0$ is independent of t.

Using the expansion (19.6) above, the local truncation error, which is defined as usual, may be expressed as

$$\begin{aligned} \tau\mathcal{E}[u](t,\tau) &= u(t) - u(t-\tau) - \tau b_1 f(t-\tau, u(t-\tau)) \\ &\quad - \tau b_2 f\big(t - \tau + c_2\tau, u(t-\tau) + \tau a_{2,1} f(t-\tau, u(t-\tau))\big) \\ &= u(t) - u(t-\tau) - \tau(b_1 + b_2) f(t-\tau, u(t-\tau)) \\ &\quad - \tau^2 b_2 c_2 \partial_t f(t-\tau, u(t-\tau)) \\ &\quad - \tau^2 b_2 a_{2,1} \partial_u f(t-\tau, u(t-\tau)) f(t-\tau, u(t-\tau)) \\ &\quad - \tau^3 \frac{b_2}{2} \boldsymbol{q}^\intercal \mathsf{H}_f(\eta, \gamma)\boldsymbol{q}. \end{aligned} \tag{19.7}$$

On the other hand, applying Taylor's Theorem to the solution, we have, for some $\beta \in (t - \tau, t)$,

$$\begin{aligned} 0 &= u(t) - u(t-\tau) - \tau u'(t-\tau) - \frac{\tau^2}{2} u''(t-\tau) - \frac{\tau^3}{6} u'''(\beta) \\ &= u(t) - u(t-\tau) - \tau f(t-\tau, u(t-\tau)) - \frac{\tau^2}{2} \partial_t f(t-\tau, u(t-\tau)) \\ &\quad - \frac{\tau^2}{2} \partial_u f(t-\tau, u(t-\tau)) f(t-\tau, u(t-\tau)) - \frac{\tau^3}{6} u'''(\beta). \end{aligned} \tag{19.8}$$

($\Longrightarrow$) Suppose that $b_1 + b_2 = 1$, $a_{2,1} = c_2$, $b_2 c_2 = \frac{1}{2}$ in (19.7). Then, combining (19.7) and (19.8),

$$\tau\mathcal{E}[u](t,\tau) = \frac{\tau^3}{6} u'''(\beta) - \tau^3 \frac{b_2}{2} \boldsymbol{q}^\intercal \mathsf{H}_f(\eta, \gamma)\boldsymbol{q}$$

and the method is clearly consistent to order $p = 2$.

($\Longleftarrow$) On the other hand, according to Theorem 19.4, for the method to be consistent to exactly order $p = 2$, it must be that $b_1 + b_2 = 1$, $a_{2,1} = c_2$, $b_2 c_2 = \frac{1}{2}$. Otherwise, the method would be of order $p = 1$, or, perhaps, inconsistent. □

The following three explicit two-stage RK methods are consistent to order $p = 2$ and conform to Theorem 19.9.

Example 19.1 Midpoint method:

$$\begin{array}{c|cc} 0 & 0 & 0 \\ \frac{1}{2} & \frac{1}{2} & 0 \\ \hline & 0 & 1 \end{array}.$$

Example 19.2 Heun's method:[3]

$$\begin{array}{c|cc} 0 & 0 & 0 \\ 1 & 1 & 0 \\ \hline & \frac{1}{2} & \frac{1}{2} \end{array}.$$

Example 19.3 Ralston's method:[4]

$$\begin{array}{c|cc} 0 & 0 & 0 \\ \frac{2}{3} & \frac{2}{3} & 0 \\ \hline & \frac{1}{4} & \frac{3}{4} \end{array}.$$

Theorem 19.10 (three-stage ERK methods). *Suppose that $\boldsymbol{f} \in \mathcal{F}^3(S)$, so that $\boldsymbol{u} \in C^4\left([0, T]; \mathbb{R}^d\right)$ is a classical solution to the IVP (18.1). Consider an explicit three-stage RK method given by the tableau*

$$\begin{array}{c|ccc} 0 & 0 & 0 & 0 \\ c_2 & a_{2,1} & 0 & 0 \\ c_3 & a_{3,1} & a_{3,2} & 0 \\ \hline & b_1 & b_2 & b_3 \end{array}.$$

The method is consistent to order $p = 3$ if and only if

$$b_1 + b_2 + b_3 = 1, \quad b_2 c_2 + b_3 c_3 = \frac{1}{2}, \quad b_2 c_2^2 + b_3 c_3^2 = \frac{1}{3}, \quad b_3 a_{3,2} c_2 = \frac{1}{6}.$$

Proof. The proof can be found in [12]; see also [47]. □

[3] Named in honor of the German mathematician Karl Heun (1859–1929).
[4] Named in honor of the American mathematician Anthony Ralston (1930–).

Example 19.4 The following three-stage explicit RK method is consistent to exactly order $p = 3$:

$$
\begin{array}{c|ccc}
0 & 0 & 0 & 0 \\
\frac{1}{2} & \frac{1}{2} & 0 & 0 \\
1 & -1 & 2 & 0 \\
\hline
 & \frac{1}{6} & \frac{2}{3} & \frac{1}{6}
\end{array}
.
$$

This method is called the classical RK method.

Example 19.5 The following four-stage explicit RK method is consistent to exactly order $p = 4$:

$$
\begin{array}{c|cccc}
0 & 0 & 0 & 0 & 0 \\
\frac{1}{2} & \frac{1}{2} & 0 & 0 & 0 \\
\frac{1}{2} & 0 & \frac{1}{2} & 0 & 0 \\
1 & 0 & 0 & 1 & 0 \\
\hline
 & \frac{1}{6} & \frac{1}{3} & \frac{1}{3} & \frac{1}{6}
\end{array}
.
$$

For a proof of the consistency, see [12]. The usual approach, which can be quite tedious, is to use Taylor expansions to prove the result.

19.3 Collocation Methods

In this section we introduce collocation methods, which form a very general class of numerical methods for differential equations. We show that these are related to RK methods, in certain cases.

Definition 19.11 (RK collocation method)**.** Let $\boldsymbol{f} \in C\left(S; \mathbb{R}^d\right)$. Suppose that the so-called **collocation points** satisfy

$$0 \le c_1 < c_2 < \cdots < c_r \le 1.$$

Let $\boldsymbol{w}^k \in \mathbb{R}^d$ be given. Assume that $\boldsymbol{p}_k \in [\mathbb{P}_r]^d$ satisfies, if possible,

$$\boldsymbol{p}_k(t_k) = \boldsymbol{w}^k, \quad \boldsymbol{p}_k'(t_k + c_j\tau) = \boldsymbol{f}(t_k + c_j\tau, \boldsymbol{p}_k(t_k + c_j\tau)) \tag{19.9}$$

for $j = 1, \dots, r$. Define $\boldsymbol{w}^{k+1} = \boldsymbol{p}_k(t_{k+1})$, for $k = 0, \dots, K-1$, with $\boldsymbol{w}^0 = \boldsymbol{u}_0$. This algorithm for producing the approximation sequence $\left\{\boldsymbol{w}^k\right\}_{k=0}^K \subset \mathbb{R}^d$ is called a **Runge–Kutta collocation method**.

Remark 19.12 (existence)**.** The previous definition only makes sense if we can find a vector-valued polynomial

$$\boldsymbol{p}_k(t) = \sum_{j=0}^{r} \boldsymbol{a}_j t^j, \quad \boldsymbol{a}_j \in \mathbb{R}^d, \quad j = 0, \dots, r$$

that satisfies (19.9). If so, we say that the implicit r-stage RK collocation method is *well defined.* In fact, it may be the case that such a polynomial will not exist or will not be uniquely determined unless $\tau > 0$ is sufficiently small.

Theorem 19.13 (collocation). *Let $\{c_j\}_{j=1}^r \subset [0,1]$ be a set of distinct collocation points. Suppose that a unique polynomial $\boldsymbol{p}_k \in [\mathbb{P}_r]^d$ satisfying* (19.9) *exists. Define*

$$\boldsymbol{\xi}_i = \boldsymbol{p}_k(t_k + c_i\tau), \quad i = 1, \dots, r,$$

$$L_j(t) = \prod_{\substack{i=1\\ i\neq j}}^{r} \frac{(t - c_i)}{(c_j - c_i)}, \quad j = 1, \dots, r,$$

$$a_{i,j} = \int_0^{c_i} L_j(s)\mathrm{d}s, \quad b_j = \int_0^1 L_j(s)\mathrm{d}s, \quad i,j = 1, \dots, r.$$

Then the collocation method of Definition 19.11 is a standard implicit RK method, as in Definition 19.2, with the weights $\mathsf{A} = [a_{i,j}]$, $\boldsymbol{b} = [b_j]$, and $\boldsymbol{c} = [c_j]$, the last weights being precisely the collocation points.

Proof. Suppose that $\boldsymbol{p}_k \in [\mathbb{P}_r]^d$ satisfies (19.9). Consider the unique Lagrange interpolating polynomial of degree at most $r-1$, $\boldsymbol{\rho} \in [\mathbb{P}_r]^d$ such that

$$\boldsymbol{\rho}(t_k + c_j\tau) = \boldsymbol{p}_k'(t_k + c_j\tau) = \boldsymbol{f}(t_k + c_j\tau, \boldsymbol{p}_k(t_k + c_j\tau)) = \boldsymbol{\nu}_j, \quad j = 1, \dots, r.$$

Theorem 9.11 guarantees that

$$\boldsymbol{\rho}(t) = \sum_{j=1}^{r} L_j\left(\frac{t - t_k}{\tau}\right)\boldsymbol{\nu}_j.$$

Observe that $\boldsymbol{p}_k' \in [\mathbb{P}_{r-1}]^d$ and, in fact,

$$\boldsymbol{p}_k'(t_k + c_j\tau) = \boldsymbol{\rho}(t_k + c_j\tau), \quad j = 1, \dots, r.$$

Therefore, $\boldsymbol{p}_k' \equiv \boldsymbol{\rho}$, since these polynomials (of degree at most $r-1$) agree at r points. By (19.9),

$$\boldsymbol{p}_k'(t) = \sum_{j=1}^{r} L_j\left(\frac{t - t_k}{\tau}\right)\boldsymbol{f}(t_k + c_j\tau, \boldsymbol{p}_k(t_k + c_j\tau)).$$

Integrating the last expression and using the condition $\boldsymbol{p}_k(t_k) = \boldsymbol{w}^k$, we observe that

$$\begin{aligned} \boldsymbol{p}_k(t) &= \boldsymbol{w}^k + \int_{t_k}^{t}\sum_{j=1}^{r} L_j\left(\frac{s - t_k}{\tau}\right)\boldsymbol{f}(t_k + c_j\tau, \boldsymbol{p}_k(t_k + c_j\tau))\mathrm{d}s \\ &= \boldsymbol{w}^k + \tau\sum_{j=1}^{r}\boldsymbol{f}(t_k + c_j\tau, \boldsymbol{\xi}_j)\int_0^{\frac{t-t_k}{\tau}} L_j(s)\mathrm{d}s. \end{aligned} \tag{19.10}$$

Setting $t = t_k + c_i\tau$ in (19.10), we have

$$\boldsymbol{\xi}_i = \boldsymbol{w}^k + \tau\sum_{j=1}^{r} a_{i,j}\boldsymbol{f}(t_k + c_j\tau, \boldsymbol{\xi}_j).$$

Setting $t = t_{k+1}$ in (19.10), we find

$$\boldsymbol{w}^{k+1} = \boldsymbol{w}^k + \tau \sum_{j=1}^{r} b_j \boldsymbol{f}(t_k + c_j\tau, \boldsymbol{\xi}_j),$$

and the proof is finished. □

For collocation methods, everything is determined by picking the collocation points. These are usually chosen as the roots of certain orthogonal polynomials.

Theorem 19.14 (collocation order). *Suppose that $\{c_j\}_{j=1}^r \subset [0,1]$ is the set of r distinct collocation points that determine the RK collocation method of Definition 19.11. Define*

$$q(t) = \prod_{n=1}^{r} (t - c_n) \in \mathbb{P}_r.$$

If, for some $m \in \{1, \ldots, r\}$,

$$\int_0^1 q(s)p(s)\mathrm{d}s = 0, \quad \forall p \in \mathbb{P}_{m-1},$$

but there is some $\tilde{p} \in \mathbb{P}_m$ such that

$$\int_0^1 q(s)\tilde{p}(s)\mathrm{d}s \neq 0,$$

then the RK collocation method is consistent to exactly order $p = r + m$.

Proof. Consider the simple quadrature rule,

$$Q_1^{(0,1)}[f] = \sum_{j=1}^{r} b_j f(c_j),$$

with quadrature nodes $\{c_j\}_{j=1}^r$ and quadrature weights $\{b_j\}_{j=1}^r$. Suppose that the quadrature weights b_j are chosen to satisfy the exactness condition (14.7); namely,

$$b_j = \int_0^1 L_j(s)\mathrm{d}s, \quad L_j(s) = \prod_{\substack{k=1 \\ k\neq j}}^{r} \frac{s - c_k}{c_j - c_k}, \quad j = 1, \ldots, r.$$

Appealing to Corollary 14.47, the quadrature rule is consistent of order exactly $r + m - 1$. In other words,

$$\int_0^1 \psi(s)\mathrm{d}s = \sum_{j=1}^{r} b_j \psi(c_j), \quad \forall \psi \in \mathbb{P}_{r+m-1}.$$

By the Alekseev–Gröbner Lemma, stated in Theorem 17.16,

$$\boldsymbol{p}_k(t_{k+1}) - \boldsymbol{u}(t_{k+1}) = \int_{t_k}^{t_{k+1}} D_v \boldsymbol{U}(s, \boldsymbol{p}_k(s), t_{k+1}) \boldsymbol{g}(s, \boldsymbol{p}_k(s))\mathrm{d}s,$$

where $\boldsymbol{p}_k$ is the vector-valued polynomial in the definition of the collocation method, assuming that $\boldsymbol{p}_k(t_k) = \boldsymbol{u}(t_k)$ and that the deviation, $\boldsymbol{g}$, in the Alekseev–Gröbner Lemma satisfies

j	$\tilde{P}_j(t)$
0	1
1	$2t-1$
2	$6t^2-6t+1$
3	$20t^3-30t^2+12t-1$
4	$70t^4-140t^3+90t^2-20t+1$
5	$252t^5-630t^4+560t^3-210t^2+30t-1$

Table 19.1 The first six transformed Legendre polynomials.

$$\boldsymbol{g}(s,\boldsymbol{p}_k(s)) = \boldsymbol{p}_k'(s) - \boldsymbol{f}(s,\boldsymbol{p}_k(s)) = \tilde{\boldsymbol{g}}(s).$$

Observe that the collocation rule requires that

$$\boldsymbol{p}_k'(t_k + c_j\tau) = \boldsymbol{f}(t_k + c_j\tau, \boldsymbol{p}_k(t_k + c_j\tau)), \quad j = 1,\ldots,r,$$

which implies that $\tilde{\boldsymbol{g}}$ vanishes at the collocation points:

$$\tilde{\boldsymbol{g}}(t_k + c_j\tau) = \boldsymbol{0}, \quad j = 1,\ldots,r.$$

Applying the quadrature rule,

$$\begin{aligned}\boldsymbol{p}_k(t_{k+1}) - \boldsymbol{u}(t_{k+1}) &= \int_{t_k}^{t_{k+1}} D_{\boldsymbol{v}}\boldsymbol{U}(s,\boldsymbol{p}_k(s),t_{k+1})\boldsymbol{g}(s,\boldsymbol{p}_k(s))\mathrm{d}s \\ &= \sum_{j=1}^{r} b_j D_{\boldsymbol{v}}\boldsymbol{U}(t_k+\tau c_j, \boldsymbol{p}_k(t_k+\tau c_j), t_{k+1})\tilde{\boldsymbol{g}}(t_k+\tau c_j) + \boldsymbol{E}_Q \\ &= \boldsymbol{E}_Q,\end{aligned}$$

where $\boldsymbol{E}_Q$ denotes the quadrature error. If we assume that

$$\hat{\boldsymbol{g}}(\cdot) = D_{\boldsymbol{v}}\boldsymbol{U}(\cdot,p_k(\cdot),t_{k+1})\tilde{\boldsymbol{g}}(\cdot) \in C^{r+m}([t_k,t_{k+1}];\mathbb{R}^d),$$

then, by Theorem 14.18,

$$\|\boldsymbol{E}_Q\|_2 \le C\tau^{r+m+1}\left\|\hat{\boldsymbol{g}}^{(r+m)}\right\|_{L^\infty(t_k,t_{k+1};\mathbb{R}^d)}$$

for some constant $C > 0$. Hence,

$$\left\|\boldsymbol{w}^{k+1} - \boldsymbol{u}(t_{k+1})\right\|_2 = \|\boldsymbol{p}_k(t_{k+1}) - \boldsymbol{u}(t_{k+1})\|_2 \le \tilde{C}\tau^{r+m+1},$$

where we assume that $\boldsymbol{w}^k = \boldsymbol{u}(t_k)$. We leave it to the reader as an exercise to prove that the local truncation error must be of order $r+m$; see Problem 19.6. □

Definition 19.15 (transformed Legendre polynomials)**.** By $\{\tilde{P}_j\}_{j\in\mathbb{N}_0}$ we denote the set of **transformed Legendre polynomials,**[5] which have the property that

$$\int_0^1 \tilde{P}_i(s)\tilde{P}_j(s)\mathrm{d}s = \frac{1}{2j+1}\delta_{i,j}.$$

The first few transformed Legendre polynomials are given in Table 19.1.

[5] Named in honor of the French mathematician Adrien-Marie Legendre (1752–1833).

Corollary 19.16 (Gauss–Legendre–RK). *Let the collocation points $c_1, \ldots, c_r$ be precisely the zeros of the transformed Legendre polynomial $\tilde{P}_r \in \mathbb{P}_r$. According to Theorem 11.10, these lie in the open interval $(0,1)$. Then the corresponding collocation method of Definition 19.11 is consistent to order exactly $p = 2r$.*

Proof. In this case, $q \equiv C_r \tilde{P}_r$, where $0 \neq C_r \in \mathbb{R}$. Since the $\tilde{P}_i$ form an orthogonal basis for $\mathbb{P}_r$, for any $j \in \{0, 1, 2, \ldots, r\}$, we can express

$$t^j = \sum_{m=0}^{j} \beta_{j,m} \tilde{P}_m(t)$$

for some constants $\beta_{j,1}, \ldots, \beta_{j,j}$. Therefore,

$$\int_0^1 q(s) s^j \mathrm{d}s = C_r \sum_{m=0}^{j} \beta_{j,m} \int_0^1 \tilde{P}_r(s) \tilde{P}_m(s) \mathrm{d}s = 0,$$

provided that $j \leq r - 1$. By Theorem 19.14, the method is exactly of order $p = r + r = 2r$. □

Definition 19.17 (Gauss–Legendre–RK method). The implicit r-stage RK methods constructed as collocation methods whose collocation points are the zeros of the transformed Legendre polynomial $\tilde{P}_r$ are called **Gauss–Legendre–Runge–Kutta methods**.[6]

The following Gauss–Legendre–RK methods are collocation methods constructed using Corollary 19.16; see Table 19.1.

Example 19.6 *The midpoint rule*: Suppose that $r = 1$. The transformed Legendre polynomial of order one is

$$\tilde{P}_1(t) = 2t - 1 \implies c_1 = \frac{1}{2}.$$

The corresponding Gauss–Legendre IRK method is given by

$$\begin{array}{c|c} \frac{1}{2} & \frac{1}{2} \\ \hline & 1 \end{array}$$

and is of order $2r = 2$. For a scalar autonomous system, $u' = f(u)$, the method can be expressed as

$$w^{k+1} = w^k + \tau f\left(w^k + \frac{\tau}{2}\kappa_1\right), \quad \kappa_1 = f\left(w^k + \frac{\tau}{2}\kappa_1\right). \tag{19.11}$$

It is a simple exercise to show that this is equivalent to the midpoint rule,

$$w^{k+1} = w^k + \tau f\left(\frac{w^{k+1} + w^k}{2}\right). \tag{19.12}$$

[6] Named in honor of the German mathematician and physicist Johann Carl Friedrich Gauss (1777–1855) and the French mathematician Adrien-Marie Legendre (1752–1833).

But let us write this another way. Define

$$\tilde{w}^{k+\frac{1}{2}} = \frac{w^{k+1} + w^k}{2}.$$

Then we can express the midpoint rule as

$$\tilde{w}^{k+\frac{1}{2}} = w^k + \frac{\tau}{2} f\left(\tilde{w}^{k+\frac{1}{2}}\right), \quad w^{k+1} = 2\tilde{w}^{k+\frac{1}{2}} - w^k. \tag{19.13}$$

Still another, equivalent, way of writing this method is as follows:

$$\tilde{w}^{k+\frac{1}{2}} = w^k + \frac{\tau}{2} f\left(\tilde{w}^{k+\frac{1}{2}}\right), \quad w^{k+1} = \tilde{w}^{k+\frac{1}{2}} + \frac{\tau}{2} f\left(\tilde{w}^{k+\frac{1}{2}}\right). \tag{19.14}$$

Observe that method (19.13) shows that the midpoint rule is essentially a backward (implicit) Euler method with half the time step size followed by an extrapolation. Method (19.14) expresses the midpoint rule as a half-step-size backward Euler method followed by a half-step-size forward (explicit) Euler method.

In either case, the simple modification of a backward Euler method, which is only first-order accurate, leads to a second-order accurate method; for more insight on this, see [11].

Example 19.7 Suppose that $r = 2$. The transformed Legendre polynomial of order two is

$$\tilde{P}_2(t) = 6t^2 - 6t + 1 \implies c_1 = \frac{1}{2} - \frac{\sqrt{3}}{6}, \quad c_2 = \frac{1}{2} + \frac{\sqrt{3}}{6}.$$

The Gauss–Legendre IRK method is given by

$$\begin{array}{c|cc} \frac{1}{2} - \frac{\sqrt{3}}{6} & \frac{1}{4} & \frac{1}{4} - \frac{\sqrt{3}}{6} \\ \frac{1}{2} + \frac{\sqrt{3}}{6} & \frac{1}{4} + \frac{\sqrt{3}}{6} & \frac{1}{4} \\ \hline & \frac{1}{2} & \frac{1}{2} \end{array}$$

and is of order $2r = 4$.

Example 19.8 Suppose that $r = 3$. The transformed Legendre polynomial of order three is

$$\tilde{P}_3(t) = 20t^3 - 30t^2 + 12t - 1 \implies c_1 = \frac{1}{2} - \frac{\sqrt{15}}{10}, \quad c_2 = \frac{1}{2}, \quad c_3 = \frac{1}{2} + \frac{\sqrt{15}}{10}.$$

The Gauss–Legendre IRK method is given by

$$\begin{array}{c|ccc} \frac{1}{2} - \frac{\sqrt{15}}{10} & \frac{5}{36} & \frac{2}{9} - \frac{\sqrt{15}}{15} & \frac{5}{36} - \frac{\sqrt{15}}{30} \\ \frac{1}{2} & \frac{5}{36} + \frac{\sqrt{15}}{24} & \frac{2}{9} & \frac{5}{36} - \frac{\sqrt{15}}{24} \\ \frac{1}{2} + \frac{\sqrt{15}}{10} & \frac{5}{36} + \frac{\sqrt{15}}{30} & \frac{2}{9} + \frac{\sqrt{15}}{15} & \frac{5}{36} \\ \hline & \frac{5}{18} & \frac{4}{9} & \frac{5}{18} \end{array}$$

and is of order $2r = 6$.

Example 19.9 Not all IRK methods are of collocation type. Consider, for example, the methods given by the tables

$$
\begin{array}{c|cc}
0 & \frac{1}{4} & -\frac{1}{4} \\
\frac{2}{3} & \frac{1}{4} & \frac{5}{12} \\
\hline
 & \frac{1}{4} & \frac{3}{4}
\end{array}
\qquad\qquad
\begin{array}{c|cc}
\frac{1}{3} & \frac{5}{12} & -\frac{1}{12} \\
1 & \frac{3}{4} & \frac{1}{4} \\
\hline
 & \frac{3}{4} & \frac{1}{4}
\end{array}.
$$

For both methods, the necessary conditions of Theorem 19.4 are satisfied. The method on the left, which is consistent to exactly order $p = 3$, is not of collocation type. (How do we know this?) The method on the right is of collocation type. One can check that the collocation points $c_1 = \frac{1}{3}$ and $c_2 = 1$ completely determine the other weights. The method on the right is consistent to exactly order $p = 3$. To see this, observe that

$$\int_0^1 \left(s - \frac{1}{3}\right)(s-1)s^j ds = 0$$

only for $j = 0$. Thus, invoking Theorem 19.14, we have $m = 1$ and $p = r + m = 3$.

19.4 Dissipative Methods

In this section, we demonstrate how some IRK methods are particularly well suited to approximate dissipative equations, as defined in Section 17.4. We follow the notation introduced there and consider IVPs posed on $\mathbb{C}^d$, with $d \in \mathbb{N}$. The reader will easily verify that all the numerical methods, and theory for them, we have constructed so far extend to this case without difficulty.

One may wish to have a numerical method that preserves the dissipation property presented in Theorem 17.18. To achieve this, we begin with a definition.

Definition 19.18 (algebraic stability). Assume that $\boldsymbol{f}\colon [0, T] \times \mathbb{C}^d \to \mathbb{C}^d$ is monotone with respect to $(\cdot,\cdot)$. Let $\boldsymbol{u}_0, \boldsymbol{v}_0 \in \mathbb{C}^d$. Assume that $\boldsymbol{f}$ is such that there are unique classical solutions on $[0, T]$ to the problems

$$\boldsymbol{u}'(t) = \boldsymbol{f}(t, \boldsymbol{u}(t)), \quad \boldsymbol{u}(0) = \boldsymbol{u}_0, \qquad \boldsymbol{v}'(t) = \boldsymbol{f}(t, \boldsymbol{v}(t)), \quad \boldsymbol{v}(0) = \boldsymbol{v}_0.$$

Let $\left\{\boldsymbol{w}^k\right\}_{k=0}^K$ and $\left\{\boldsymbol{z}^k\right\}_{k=0}^K$ be approximations to $\boldsymbol{u}$ and $\boldsymbol{v}$, respectively, obtained by the same numerical method; for example, some RK method. Notice that we must have $\boldsymbol{w}^0 = \boldsymbol{u}_0$ and $\boldsymbol{z}^0 = \boldsymbol{v}_0$. We say that the method is **dissipative** or **algebraically stable** if and only if, for any K and for all starting values $\boldsymbol{u}_0, \boldsymbol{v}_0 \in \mathbb{C}^d$,

$$\left\|\boldsymbol{w}^k - \boldsymbol{z}^k\right\| \le \left\|\boldsymbol{w}^{k-1} - \boldsymbol{z}^{k-1}\right\| \le \|\boldsymbol{u}_0 - \boldsymbol{v}_0\|$$

for all $k = 1, \dots, K$.

In the same way that an RK method can be encoded in a Butcher tableau, its properties can be encoded in a particular matrix.

Definition 19.19 (M-matrix)**.** Suppose that the r-stage RK method is defined by the weights $\mathsf{A} = [a_{i,j}]_{i,j=1}^{r}$, $\boldsymbol{b} = [b_i]_{i=1}^{r}$, and $\boldsymbol{c} = [c_i]_{i=1}^{r}$. The **M-matrix** of the method is the matrix $\mathsf{M} = [m_{i,j}]_{i,j=1}^{r} \in \mathbb{R}^{r\times r}$ defined via

$$m_{i,j} = b_i a_{i,j} + b_j a_{j,i} - b_i b_j, \quad i,j = 1, \ldots, r.$$

The importance of the M-matrix of an RK method is in the dissipativity condition given in the following result.

Theorem 19.20 (dissipativity condition)**.** *Suppose that* $\boldsymbol{f}\colon [0,T] \times \mathbb{C}^d \to \mathbb{C}^d$ *is monotone with respect to* $(\cdot,\cdot)$*. Let* $\boldsymbol{u}_0, \boldsymbol{v}_0 \in \mathbb{C}^d$*. Assume that* $\boldsymbol{f}$ *is such that there are unique classical solutions on* $[0,T]$ *to the problems*

$$\boldsymbol{u}'(t) = \boldsymbol{f}(t, \boldsymbol{u}(t)), \quad \boldsymbol{u}(0) = \boldsymbol{u}_0, \qquad \boldsymbol{v}'(t) = \boldsymbol{f}(t, \boldsymbol{v}(t)), \quad \boldsymbol{v}(0) = \boldsymbol{v}_0.$$

Let the r*-stage RK method be defined by the weights* $\mathsf{A} = [a_{i,j}]_{i,j=1}^{r}$, $\boldsymbol{b} = [b_i]_{i=1}^{r}$, *and* $\boldsymbol{c} = [c_i]_{i=1}^{r}$*. If its M-matrix is positive semi-definite and* $b_j \geq 0$, $j = 1, \ldots, r$, *then the RK method is dissipative.*

Proof. Define, for $i,j = 1, \ldots, r$,

$$\boldsymbol{\rho}_j = \boldsymbol{f}(t_k + c_j\tau, \boldsymbol{\xi}_j), \quad \boldsymbol{\xi}_i = \boldsymbol{w}^k + \tau \sum_{j=1}^{r} a_{i,j}\boldsymbol{\rho}_j, \quad \boldsymbol{w}^{k+1} = \boldsymbol{w}^k + \tau \sum_{j=1}^{r} b_j \boldsymbol{\rho}_j,$$

and

$$\boldsymbol{\sigma}_j = \boldsymbol{f}(t_k + c_j\tau, \boldsymbol{\zeta}_j), \quad \boldsymbol{\zeta}_i = \boldsymbol{z}^k + \tau \sum_{j=1}^{r} a_{i,j}\boldsymbol{\sigma}_j, \quad \boldsymbol{z}^{k+1} = \boldsymbol{z}^k + \tau \sum_{j=1}^{r} b_j \boldsymbol{\sigma}_j.$$

Then observe that

$$\left\|\boldsymbol{w}^{k+1} - \boldsymbol{z}^{k+1}\right\|^2 = \left\|\boldsymbol{w}^k - \boldsymbol{z}^k\right\|^2 + 2\tau\Re\left[\left(\boldsymbol{w}^k - \boldsymbol{z}^k, \sum_{j=1}^{r} b_j \boldsymbol{d}_j\right)\right] + \tau^2 \left\|\sum_{j=1}^{r} b_j \boldsymbol{d}_j\right\|^2,$$

where, for $j = 1, \ldots, r$,

$$\boldsymbol{d}_j = \boldsymbol{\rho}_j - \boldsymbol{\sigma}_j.$$

Thus, we will have proven the result if we can show that

$$2\tau\Re\left[\left(\boldsymbol{w}^k - \boldsymbol{z}^k, \sum_{j=1}^{r} b_j \boldsymbol{d}_j\right)\right] + \tau^2 \left\|\sum_{j=1}^{r} b_j \boldsymbol{d}_j\right\|^2 \leq 0.$$

Since, for $j = 1, \ldots, r$,

$$\boldsymbol{w}^k = \boldsymbol{\xi}_j - \tau \sum_{i=1}^{r} a_{j,i}\boldsymbol{\rho}_i, \quad \boldsymbol{z}^k = \boldsymbol{\zeta}_j - \tau \sum_{i=1}^{r} a_{j,i}\boldsymbol{\sigma}_i,$$

we have

$$\Re\left[\left(\boldsymbol{w}^k - \boldsymbol{z}^k, \sum_{j=1}^{r} b_j \boldsymbol{d}_j\right)\right] = \sum_{j=1}^{r} b_j \Re\left[\left(\boldsymbol{\xi}_j - \boldsymbol{\zeta}_j - \tau \sum_{i=1}^{r} a_{j,i} \boldsymbol{d}_i, \boldsymbol{d}_j\right)\right]$$

$$= \sum_{j=1}^{r} b_j \Re\left[(\boldsymbol{\xi}_j - \boldsymbol{\zeta}_j, \boldsymbol{d}_j)\right] - \tau \sum_{j=1}^{r}\sum_{i=1}^{r} b_j a_{j,i} \Re\left[(\boldsymbol{d}_i, \boldsymbol{d}_j)\right].$$

Since $\boldsymbol{f}$ is monotone, for each $j = 1, \dots, r$,

$$\Re\left[(\boldsymbol{\xi}_j - \boldsymbol{\zeta}_j, \boldsymbol{d}_j)\right] \leq 0.$$

Since the weights b_j are all nonnegative,

$$\sum_{j=1}^{r} b_j \Re\left[(\boldsymbol{\xi}_j - \boldsymbol{\zeta}_j, \boldsymbol{d}_j)\right] \leq 0.$$

It then follows that

$$\Re\left[\left(\boldsymbol{w}^k - \boldsymbol{z}^k, \sum_{j=1}^{r} b_j \boldsymbol{d}_j\right)\right] \leq -\tau \sum_{j=1}^{r}\sum_{i=1}^{r} b_j a_{j,i} \Re\left[(\boldsymbol{d}_i, \boldsymbol{d}_j)\right],$$

or, equivalently, after swapping summation indices,

$$\Re\left[\left(\boldsymbol{w}^k - \boldsymbol{z}^k, \sum_{j=1}^{r} b_j \boldsymbol{d}_j\right)\right] \leq -\tau \sum_{i=1}^{r}\sum_{j=1}^{r} b_i a_{i,j} \Re\left[(\boldsymbol{d}_j, \boldsymbol{d}_i)\right].$$

Thus,

$$2\tau \Re\left[\left(\boldsymbol{w}^k - \boldsymbol{z}^k, \sum_{j=1}^{r} b_j \boldsymbol{d}_j\right)\right] + \tau^2 \left\| \sum_{j=1}^{r} b_j \boldsymbol{d}_j \right\|^2$$

$$\leq -\tau^2 \sum_{j=1}^{r}\sum_{i=1}^{r} b_j a_{j,i} \Re\left[(\boldsymbol{d}_i, \boldsymbol{d}_j)\right] - \tau^2 \sum_{i=1}^{r}\sum_{j=1}^{r} b_i a_{i,j} \Re\left[(\boldsymbol{d}_j, \boldsymbol{d}_i)\right] + \tau^2 \left\| \sum_{j=1}^{r} b_j \boldsymbol{d}_j \right\|^2$$

$$= -\tau^2 \sum_{i=1}^{r}\sum_{j=1}^{r} m_{i,j} \Re\left[(\boldsymbol{d}_i, \boldsymbol{d}_j)\right].$$

Since M is symmetric positive semi-definite, there is an orthogonal matrix Q and a diagonal matrix $\mathsf{D} = \operatorname{diag}[\lambda_1, \dots, \lambda_r]$, with nonnegative diagonal entries, $\lambda_j \geq 0$, such that

$$\mathsf{M} = \mathsf{Q}\mathsf{D}\mathsf{Q}^\intercal.$$

In terms of coordinates,

$$m_{i,j} = \sum_{k=1}^{r} \lambda_k q_{i,k} q_{j,k}.$$

Thus,

$$
\begin{aligned}
2\tau\Re&\left[\left(\boldsymbol{w}^k - \boldsymbol{z}^k, \sum_{j=1}^r b_j \boldsymbol{d}_j\right)\right] + \tau^2 \left\| \sum_{j=1}^r b_j \boldsymbol{d}_j \right\|^2 \\
&\le -\tau^2 \sum_{i=1}^r \sum_{j=1}^r \sum_{k=1}^r \lambda_k q_{i,k} q_{j,k} \Re\left[(\boldsymbol{d}_i, \boldsymbol{d}_j)\right] \\
&= -\tau^2 \sum_{k=1}^r \lambda_k \Re\left[\left(\sum_{i=1}^r q_{i,k}\boldsymbol{d}_i, \sum_{j=1}^r q_{j,k}\boldsymbol{d}_j\right)\right] \\
&= -\tau^2 \sum_{k=1}^r \lambda_k \left\| \sum_{j=1}^r w_{j,k}\boldsymbol{d}_j \right\|^2 \\
&\le 0.
\end{aligned}
$$

The result is proved:

$$
\left\|\boldsymbol{w}^{k+1} - \boldsymbol{z}^{k+1}\right\|^2 \le \left\|\boldsymbol{w}^k - \boldsymbol{z}^k\right\|^2 . \qquad \square
$$

We will conclude this section by proving that the Gauss–Legendre IRK methods are dissipative. To do this, we need some definitions.

Definition 19.21 (type)**.** Let $q \in \mathbb{N}$. An r-stage RK method defined by the weights $\mathsf{A} = [a_{i,j}]_{i,j=1}^r$, $\boldsymbol{b} = [b_i]_{i=1}^r$, and $\boldsymbol{c} = [c_i]_{i=1}^r$ is said to be of **type** $B(q)$ if and only if

$$
\sum_{i=1}^r b_i c_i^{k-1} = \frac{1}{k}, \quad k = 1, \dots, q.
$$

The method is said to be of **type** $C(q)$ if and only if

$$
\sum_{j=1}^r a_{i,j} c_j^{k-1} = \frac{c_i^k}{k}, \quad i = 1, \dots, r, \quad k = 1, \dots, q.
$$

Theorem 19.22 (dissipativity criterion)**.** *Consider an r-stage RK method defined by the weights $\mathsf{A} = [a_{i,j}]_{i,j=1}^r$, $\boldsymbol{b} = [b_i]_{i=1}^r$, and $\boldsymbol{c} = [c_i]_{i=1}^r$. Assume that the entries of $\boldsymbol{c} = [c_i]_{i=1}^r$ are distinct. If the method is of type $B(2r)$ and of type $C(r)$, then the M-matrix for this method is the zero matrix.*

Proof. Let $\mathsf{M} = [m_{i,j}] \in \mathbb{R}^{r\times r}$ denote the M-matrix for the method. Define the matrix $\mathsf{N} = [n_{i,j}] \in \mathbb{R}^{r\times r}$ via

$$
n_{k,m} = \sum_{i=1}^r \sum_{j=1}^r c_i^{k-1} m_{i,j} c_j^{m-1} = \sum_{i=1}^r \sum_{j=1}^r c_i^{k-1} \left(b_i a_{i,j} + b_j a_{j,i} - b_i b_j\right) c_j^{m-1} = 0,
$$

where $k, m = 1, \dots, r$ and we used the fact that the method is of type $B(2r)$ and type $C(r)$. The details are left to the reader as an exercise; see Problem 19.9. In conclusion, $\mathsf{N} = \mathsf{O}$, the zero matrix. But observe that

$$
\mathsf{N} = \mathsf{V}^\intercal \mathsf{M} \mathsf{V},
$$

where $\mathsf{V} = [c_i^{j-1}]$ is a variant of the Vandermonde matrix, which is nonsingular provided that the entries of $\boldsymbol{c} = [c_i]_{i=1}^r$ are distinct (Theorem 9.4). It follows that $\mathsf{M} = \mathsf{O}$. □

Theorem 19.23 (dissipativity). *All Gauss–Legendre RK methods are dissipative.*

Proof. The idea is to show that all Gauss–Legendre RK methods are of type $B(2r)$ and type $C(r)$, thus implying that their M-matrices are $r \times r$ zero matrices. Finally, one needs to show that the weights b_j are all nonnegative. The result then follows from Theorem 19.20. The details are left to the reader as an exercise; see Problem 19.10. □

Problems

19.1 Let $T > 0$ be given. Consider the general two-stage explicit RK method, defined by

$$\xi^k = w^k + a\tau f(w^k), \quad w^{k+1} = w^k + \tau(b_1 f(w^k) + b_2 f(\xi^k)),$$

for approximating the solutions to the autonomous IVP

$$u'(t) = f(u(t)), \quad t \in [0, T], \quad u(0) = u_0.$$

Assume that $f \in \mathcal{F}^2(S)$ and the coefficients satisfy $b_1 + b_2 = 1$ and $ab_2 = \frac{1}{2}$. Prove that the method is convergent to second order.

19.2 Provide all the details for the proof of Theorem 19.4.

19.3 Prove Theorem 19.5.

19.4 Prove Theorem 19.6.

19.5 Consider the implicit RK methods given by the tableaux

$$\begin{array}{c|cc} 0 & \frac{1}{4} & -\frac{1}{4} \\ \frac{2}{3} & \frac{1}{4} & \frac{5}{12} \\ \hline & \frac{1}{4} & \frac{3}{4} \end{array} \qquad \begin{array}{c|cc} \frac{1}{3} & \frac{5}{12} & -\frac{1}{12} \\ 1 & \frac{3}{4} & \frac{1}{4} \\ \hline & \frac{3}{4} & \frac{1}{4} \end{array}.$$

a) Show that the necessary conditions of Theorem 19.4 are all satisfied.
b) Show that one of the methods is a collocation method and that the other is not. For the one that is a collocation method, find its order of consistency.

19.6 Complete the proof of Theorem 19.14.

19.7 Complete the proof of Theorem 17.18.

19.8 Show that no explicit RK method can be dissipative.

19.9 Complete the proof of Theorem 19.22.

19.10 Complete the proof of Theorem 19.23.

20 Linear Multi-step Methods

The fundamental formula that defines mild solutions

$$u(t_2) = u(t_1) + \int_{t_1}^{t_2} f(s, u(s))ds$$

was used in the previous chapter with $t_1 = t_k$ and $t_2 = t_{k+1}$. To approximate the integral, we used information about the slope function in the time interval $[t_k, t_{k+1}]$ only. However, we could use more information to build an approximation of the integral. For instance, for some $q \in \mathbb{N}_0$, we set $X_q = \{t_{k-q}, \dots, t_k, t_{k+1}\}$ and replace the slope function by its interpolant on X_q

$$\mathcal{I}_{X_q}[f(\cdot, u(\cdot))](t) \approx f(t, u(t)).$$

Then

$$u(t_2) \approx u(t_1) + \int_{t_1}^{t_2} \mathcal{I}_{X_q}[f(\cdot, u(\cdot))](s)ds.$$

This is the basis of Adams-type methods and other multi-step methods, which we examine in this chapter.

Before we begin, we recall that we are trying to approximate the solution to (18.1) by choosing $K \in \mathbb{N}$ and letting $\tau = T/K$ and $t_k = k\tau$. We will produce $\{w^k\}_{k=0}^K \subset \mathbb{R}^d$ such that $w^k \approx u(t_k)$.

20.1 Consistency of Linear Multi-step Methods

In Chapters 18 and 19, we discussed single-step methods. With those, all that was needed to obtain the approximation at time level $k+1$ was the approximate solution at time level k. In this chapter, we will introduce multi-step methods, which use approximations at additional past time levels, $k-1$, $k-2$, etc.

Definition 20.1 (linear multi-step method)**.** Let $K, q \in \mathbb{N}$ with $q < K$. The finite sequence $\left\{w^k\right\}_{k=0}^K \subset \mathbb{R}^d$ is called a **linear** q**-step approximation** (or just a **linear multi-step approximation**) to u, solution of (18.1), with starting values $w^0, w^1, \dots, w^{q-1}$ if and only if, for $k = 0, \dots, K-q$,

$$\sum_{j=0}^{q} a_j w^{k+j} = \tau \sum_{j=0}^{q} b_j f(t_{k+j}, w^{k+j}), \tag{20.1}$$

where $\{a_j\}_{j=0}^q, \{b_j\}_{j=0}^q \subseteq \mathbb{R}$. The multi-step approximation is called **explicit** if $b_q = 0$; otherwise, it is called **implicit**. As before, the **global error** of the multi-step approximation is the finite sequence $\left\{e^k\right\}_{k=0}^K \subseteq \mathbb{R}^d$ defined via

$$e^k = u(t_k) - w^k.$$

Remark 20.2 (convention)**.** Notice that, for the multi-step method (20.1) to make sense, we must have $a_q \neq 0$. In addition, observe that the coefficients $\{a_j\}_{j=0}^q, \{b_j\}_{j=0}^q$ are defined up to multiplication by a common constant. Because of these two considerations, here and in what follows we will assume that

$$a_q = 1.$$

Definition 20.3 (LTE and consistency)**.** Let $u \in C^1([0,T];\Omega)$ be a classical solution on $[0,T]$ to (18.1). Let the sequence $\left\{w^k\right\}_{k=0}^K$ be obtained with the q-step method (20.1) with starting values $w^0, w^1, \ldots, w^{q-1}$. The **local truncation error** (LTE) or **consistency error** of the multi-step approximation is defined as

$$\begin{aligned}\mathcal{E}[u](t,\tau)&\\ &= \frac{1}{\tau}\sum_{j=0}^{q}\left[a_j u(t+(j-q)\tau) - \tau b_j f(t+(j-q)\tau, u(t+(j-q)\tau))\right] \quad (20.2)\end{aligned}$$

for any $t \in [t_q, T]$. We make frequent use of the notation $\mathcal{E}^k[u] = \mathcal{E}[u](t_k,\tau)$ for $k = q, \ldots, K$. We say that the linear multi-step approximation is **consistent to at least order** $p \in \mathbb{N}$ if and only if, when

$$u \in C^{p+1}([0,T];\mathbb{R}^d),$$

there is a constant $\tau_0 \in (0,T]$ and a constant $C > 0$ such that, for all $\tau \in (0,\tau_0]$ and all $t \in [t_q, T]$,

$$\|\mathcal{E}[u](t,\tau)\|_2 \leq C\tau^p. \quad (20.3)$$

We say that the linear q-step approximation is **consistent to exactly order** p if and only if p is the largest positive integer for which (20.3) holds.

We say that the multi-step approximation **converges globally**, with at least order $p \in \mathbb{N}$, if and only if, when

$$u \in C^{p+1}([0,T];\mathbb{R}^d),$$

there is some $\tau_1 \in (0,T]$ and a constant $C > 0$ such that

$$\left\|e^k\right\|_2 \leq C\tau^p$$

for all $\tau \in (0,\tau_1]$ and any $k = 0, \ldots, K$.

The following result can be used to determine the order of a linear multi-step method.

Theorem 20.4 (method of C's). *Let $\boldsymbol{f} \in \mathcal{F}^p(S)$ and $\boldsymbol{u} \in C^{p+1}([0,T];\mathbb{R}^d)$ be a classical solution on $[0,T]$ to (18.1). Suppose that $\boldsymbol{u}$ is approximated by the linear q-step method (20.1). Define*

$$C_m = \begin{cases} \sum_{j=0}^{q} a_j, & m = 0, \\ \sum_{j=0}^{q} \left(\frac{j^m}{m!} a_j - \frac{j^{m-1}}{(m-1)!} b_j \right), & m \in \{1,2,3,\dots\}, \end{cases} \tag{20.4}$$

with the convention that $0^0 = 1$. The method is consistent to exactly order p if and only if $C_0 = 0 = C_1 = \cdots = C_p$, but $C_{p+1} \neq 0$.

Proof. For simplicity of notation, let us suppose that $d = 1$. Consider $t \in [t_q, T]$ and we extend, for any $k \in \mathbb{Z}$, the definition $t_k = k\tau$. Using Taylor's Theorem, with the expansion point $t - t_q$, one finds, for each $j = 0, 1, \dots, q$,

$$u(t + t_{j-q}) = \sum_{m=0}^{p} u^{(m)}(t - t_q) \frac{(j\tau)^m}{m!} + u^{(p+1)}(\zeta_j) \frac{(j\tau)^{p+1}}{(p+1)!}$$

and

$$\begin{aligned} u'(t + t_{j-q}) &= \sum_{m=0}^{p-1} u^{(m+1)}(t - t_q) \frac{(j\tau)^m}{m!} + u^{(p+1)}(\xi_j) \frac{(j\tau)^p}{p!} \\ &= \sum_{m=1}^{p} u^{(m)}(t - t_q) \frac{(j\tau)^{m-1}}{(m-1)!} + u^{(p+1)}(\xi_j) \frac{(j\tau)^p}{p!}. \end{aligned}$$

Observe that the $j = 0$ case holds if we agree that $0^0 = 1$. Therefore,

$$\begin{aligned} \tau\mathcal{E}[u](t,\tau) &= \sum_{j=0}^{q} a_j u(t + t_{j-q}) - \tau \sum_{j=0}^{q} b_j u'(t + t_{j-q}) \\ &= \sum_{j=0}^{q} a_j \sum_{m=0}^{p} u^{(m)}(t - t_q) \frac{(j\tau)^m}{m!} - \tau \sum_{j=0}^{q} b_j \sum_{m=1}^{p} u^{(m)}(t - t_q) \frac{(j\tau)^{m-1}}{(m-1)!} \\ &\quad + \tau^{p+1} \sum_{j=0}^{q} \left[a_j u^{(p+1)}(\zeta_j) \frac{j^{p+1}}{(p+1)!} - b_j u^{(p+1)}(\xi_j) \frac{j^p}{p!} \right]. \end{aligned}$$

Interchanging the summations, so that we sum by powers of τ, we have

$$\begin{aligned}\tau\mathcal{E}[u](t,\tau) &= u(t-t_q)\sum_{j=0}^{q} a_j + \sum_{m=1}^{p}\tau^m u^{(m)}(t-t_q)\sum_{j=0}^{q}\left[a_j\frac{j^m}{m!} - b_j\frac{j^{m-1}}{(m-1)!}\right] \\ &\quad + \tau^{p+1}\sum_{j=0}^{q}\left[a_j u^{(p+1)}(\zeta_j)\frac{j^{p+1}}{(p+1)!} - b_j u^{(p+1)}(\xi_j)\frac{j^p}{p!}\right] \\ &= C_0 u(t-t_q) + \sum_{m=1}^{p} C_m\tau^m u^{(m)}(t-t_q) \\ &\quad + \tau^{p+1}\sum_{j=0}^{q}\left[a_j u^{(p+1)}(\zeta_j)\frac{j^{p+1}}{(p+1)!} - b_j u^{(p+1)}(\xi_j)\frac{j^p}{p!}\right]\end{aligned} \tag{20.5}$$

for some constants $\zeta_j, \xi_j \in [t-t_q, t]$, $j = 0, \ldots, q$.

($\Longrightarrow$) Suppose that the method is of exactly order p. Then, from (20.5), we must have $C_0 = C_1 = \cdots = C_p = 0$. If the true solution has higher regularity, say $u \in C^{p+2}([0,T])$, then we can extend the Taylor expansion by one term to obtain

$$\begin{aligned}\tau\mathcal{E}[u](t,\tau) &= C_{p+1}\tau^{p+1}u^{(p+1)}(t-t_q) \\ &\quad + \tau^{p+2}\sum_{j=0}^{q}\left[a_j u^{(p+2)}(\tilde{\zeta}_j)\frac{j^{p+2}}{(p+2)!} - b_j u^{(p+2)}(\tilde{\xi}_j)\frac{j^{p+1}}{(p+1)!}\right].\end{aligned}$$

Since the method does not exceed order p, it must be true that $C_{p+1} \neq 0$ generically, by the definition of the local truncation error.

($\Longleftarrow$) Suppose that $C_0 = C_1 = \cdots = C_p = 0$, but $C_{p+1} \neq 0$. Then

$$\mathcal{E}[u](t,\tau) = \tau^p\sum_{j=0}^{q}\left[a_j u^{(p+1)}(\zeta_j)\frac{j^{p+1}}{(p+1)!} - b_j u^{(p+1)}(\xi_j)\frac{j^p}{p!}\right]$$

and the method is consistent to at least order p. Since $C_{p+1} \neq 0$, the order of accuracy cannot exceed p, even if the true solution has higher regularity, say $u \in C^{p+2}([0,T])$. □

The consistency criterion given in Theorem 20.4 is usually known as the *method of C's*. The algorithmic description of the computation of each one of the involved C's is given in Listing 20.1.

Definition 20.5 (characteristic polynomials)**.** For the linear q-step method (20.1), we define the **first** and **second characteristic polynomials**, respectively, as

$$\psi(z) = \sum_{j=0}^{q} a_j z^j \in \mathbb{P}_q, \qquad \chi(z) = \sum_{j=0}^{q} b_j z^j \in \mathbb{P}_q.$$

Corollary 20.6 (first-order consistency)**.** *Let $\boldsymbol{f} \in \mathcal{F}^1(S)$. Assume that the function $\boldsymbol{u} \in C^2([0,T];\mathbb{R}^d)$ is a classical solution to the initial value problem (IVP)* (18.1).

Suppose that $\boldsymbol{u}$ is approximated by the linear q-step method (20.1). *The method is consistent to at least first order if and only if*

$$\psi(1) = 0, \qquad \psi'(1) - \chi(1) = 0.$$

Proof. This follows from Theorem 20.4; see Problem 20.1. □

The following result provides another way to verify the consistency of a linear multi-step method.

Theorem 20.7 (the log-method). *Let $\boldsymbol{f} \in \mathcal{F}^p(S)$. Assume that the function $\boldsymbol{u} \in C^{p+1}([0,T];\mathbb{R}^d)$ is a classical solution to the IVP* (18.1). *Suppose that $\boldsymbol{u}$ is approximated by the linear q-step method* (20.1). *The method is consistent to exactly order p if and only if the function*

$$\phi(\mu) = \frac{\psi(\mu)}{\ln(\mu)} - \chi(\mu),$$

which is complex analytic in a neighborhood of $\mu = 1$, has the property that $\mu = 1$ is a p-fold zero or, equivalently, that

$$\tilde{\phi}(\mu) = \psi(\mu) - \chi(\mu)\ln(\mu),$$

which is also complex analytic in a neighborhood of $\mu = 1$, has the property that $\mu = 1$ is a $p+1$-fold zero.

Proof. Once again, for simplicity of notation, we consider $d = 1$ and assume that u is real analytic. From Theorem 20.4, we observe that the method is consistent to order p if and only if $C_0 = C_1 = \cdots = C_p = 0$, but $C_{p+1} \neq 0$. Using the techniques developed in the proof of Theorem 20.4, we can expand to all orders to obtain

$$\tau\mathcal{E}[u](t,\tau) = \sum_{m=p+1}^{\infty} C_m u^{(m)}(t - t_q)\tau^m.$$

In particular, setting $u(t) = \exp(t)$, which is certainly real analytic, we find

$$\tau\mathcal{E}[\exp(\cdot)](t,\tau) = \exp(t - t_q)\sum_{m=p+1}^{\infty} C_m\tau^m.$$

On the other hand, by the definition of the local truncation error, we have

$$\begin{aligned}
\tau\mathcal{E}[\exp(\cdot)](t,\tau) &= \exp(t - t_q)\sum_{j=0}^{q}\{a_j\exp(t_j) - \tau b_j\exp(t_j)\} \\
&= \exp(t - t_q)\sum_{j=0}^{q}\{a_j\exp(j\tau) - \tau b_j\exp(j\tau)\} \\
&= \exp(t - t_q)\left[\psi(\exp(\tau)) - \tau\chi(\exp(\tau))\right].
\end{aligned}$$

Equating terms, we have

$$\tilde{\phi}(\exp(\tau)) = \tau\phi(\exp(\tau)) = \psi(\exp(\tau)) - \tau\chi(\exp(\tau)) = \sum_{m=p+1}^{\infty} C_m\tau^m.$$

Thus, $\tau = 0$ is a p-fold zero of the function $\phi(\exp(\tau))$. Here, in fact, we can assume that τ is any complex number. Setting $\mu = \exp(\tau)$ and using the fact that, in a neighborhood of $\mu = 1$,

$$\begin{aligned}\ln(\mu) &= \sum_{m=1}^{\infty} \frac{(-1)^{m+1}}{m}(\mu - 1)^m \\ &= (\mu - 1) - \frac{1}{2}(\mu - 1)^2 + \frac{1}{3}(\mu - 1)^3 - \frac{1}{4}(\mu - 1)^4 + \frac{1}{5}(\mu - 1)^5 + \cdots,\end{aligned}$$

it follows that this is equivalent to the condition that $\phi(\mu)$ has a p-fold zero at $\mu = 1$, which is equivalent to the condition that $\tilde{\phi}(\mu)$ has a $p+1$-fold zero at $\mu = 1$. □

The consistency criterion given in Theorem 20.7 is usually known as the *log-method.*

Example 20.1 Consider the linear two-step method

$$\boldsymbol{w}^{k+2} - \boldsymbol{w}^k = \frac{\tau}{3}\left[\boldsymbol{f}(t_{k+2}, \boldsymbol{w}^{k+2}) + 4\boldsymbol{f}(t_{k+1}, \boldsymbol{w}^{k+1}) + \boldsymbol{f}(t_k, \boldsymbol{w}^k)\right].$$

This method is implicit and consistent to exactly order $p = 4$. To prove this, assuming that the slope function is sufficiently regular, $\boldsymbol{f} \in \mathcal{F}^4(S)$, we need only to show that $C_0 = C_1 = \cdots = C_4 = 0$, but $C_5 \neq 0$, where these constants are defined in (20.4). Clearly, $C_0 = 0$. Now

$$C_1 = \sum_{j=0}^{2}(ja_j - b_j) = 2\cdot 1 + 1\cdot 0 + 0\cdot(-1) - \left(\frac{1}{3} + \frac{4}{3} + \frac{1}{3}\right) = 2 - 2 = 0,$$

$$C_2 = \sum_{j=0}^{2}\left(\frac{j^2}{2!}a_j - jb_j\right) = \frac{2^2}{2}\cdot 1 - \left(2\cdot\frac{1}{3} + 1\cdot\frac{4}{3}\right) = 2 - 2 = 0,$$

$$C_3 = \sum_{j=0}^{2}\left(\frac{j^3}{6}a_j - \frac{j^2}{2}b_j\right) = \frac{2^3}{6}\cdot 1 - \left(\frac{2^2}{2}\cdot\frac{1}{3} + \frac{1^2}{2}\cdot\frac{4}{3}\right) = \frac{8}{6} - \frac{8}{6} = 0,$$

$$C_4 = \sum_{j=0}^{2}\left(\frac{j^4}{24}a_j - \frac{j^3}{6}b_j\right) = \frac{2^4}{24}\cdot 1 - \left(\frac{2^3}{6}\cdot\frac{1}{3} + \frac{1^3}{6}\cdot\frac{4}{3}\right) = \frac{2}{3} - \frac{2}{3} = 0.$$

But

$$C_5 = \sum_{j=0}^{2}\left(\frac{j^5}{120}a_j - \frac{j^4}{24}b_j\right) = \frac{2^5}{120}\cdot 1 - \left(\frac{2^4}{24}\cdot\frac{1}{3} + \frac{1^4}{24}\cdot\frac{4}{3}\right) = \frac{4}{15} - \frac{5}{18} \neq 0.$$

Example 20.2 In this example, we use Theorem 20.7 to show that the two-step implicit method

$$\boldsymbol{w}^{k+2} - \boldsymbol{w}^{k+1} = \tau\left[\frac{5}{12}\boldsymbol{f}(t_{k+2}, \boldsymbol{w}^{k+2}) + \frac{8}{12}\boldsymbol{f}(t_{k+1}, \boldsymbol{w}^{k+1}) - \frac{1}{12}\boldsymbol{f}(t_k, \boldsymbol{w}^k)\right]$$

is consistent to exactly order $p = 3$. To do so, it is convenient to make the change of variables $z = \mu - 1$. Then

$$\begin{aligned}\psi(\mu) &= \mu^2 - \mu \\ &= (z+1)^2 - (z+1) \\ &= z^2 + z, \\ \chi(\mu) &= \frac{5}{12}\mu^2 + \frac{8}{12}\mu - \frac{1}{12} \\ &= \frac{5}{12}(z+1)^2 + \frac{8}{12}(z+1) - \frac{1}{12} \\ &= \frac{5}{12}z^2 + \frac{3}{2}z + 1,\end{aligned}$$

and

$$\begin{aligned}\ln(\mu) &= (\mu-1) - \frac{1}{2}(\mu-1)^2 + \frac{1}{3}(\mu-1)^3 - \frac{1}{4}(\mu-1)^4 + \frac{1}{5}(\mu-1)^5 + \cdots \\ &= z - \frac{1}{2}z^2 + \frac{1}{3}z^3 - \frac{1}{4}z^4 + \frac{1}{5}z^5 + \cdots .\end{aligned}$$

We need to consider the difference

$$\frac{\psi(\mu)}{\ln(\mu)} - \chi(\mu).$$

To do so, we first find an expansion, in terms of z, for $\frac{\psi(\mu)}{\ln(\mu)}$:

$$\frac{\psi(\mu)}{\ln(\mu)} = c_0 + c_1 z + c_2 z^2 + c_3 z^3 + c_4 z^4 + \cdots .$$

Then

$$\left(c_0 + c_1 z + c_2 z^2 + c_3 z^3 + \cdots\right)\left(z - \frac{1}{2}z^2 + \frac{1}{3}z^3 - \frac{1}{4}z^4 + \cdots\right) = z + z^2,$$

which implies that

$$\begin{aligned} c_0 &= 1, & & \\ c_1 - \frac{1}{2}c_0 &= 1 & &\implies c_1 = \frac{3}{2}, \\ c_2 - \frac{1}{2}c_1 + \frac{1}{3}c_0 &= 0 & &\implies c_2 = \frac{5}{12}, \\ c_3 - \frac{1}{2}c_2 + \frac{1}{3}c_1 - \frac{1}{4}c_0 &= 0 & &\implies c_3 = -\frac{1}{24}. \end{aligned}$$

Finally,

$$\begin{aligned}\frac{\psi(\mu)}{\ln(\mu)} - \chi(\mu) &= 1 + \frac{3}{2}z + \frac{5}{12}z^2 - \frac{1}{24}z^3 + c_4 z^4 + \cdots - \left(1 + \frac{3}{2}z + \frac{5}{12}z^2\right) \\ &= -\frac{1}{24}z^3 + c_4 z^4 + \cdots ,\end{aligned}$$

which proves that the method is of exactly third order.

20.2 Adams–Bashforth and Adams–Moulton Methods

In this section, we derive some examples of the so-called Adams–Moulton and Adams–Bashforth multi-step methods. We make extensive use of the Lagrange interpolation techniques of Chapter 9. Let $q < K$. Assume that we have computed $k+q-1 < K$ approximations $\{\boldsymbol{w}^j\}_{j=0}^{k+q-1}$. Now, from the definition of mild solution (17.2), we have that

$$\boldsymbol{u}(t_{k+q}) - \boldsymbol{u}(t_{q+k-1}) = \int_{t_{k+q-1}}^{t_{k+q}} \boldsymbol{f}(s, \boldsymbol{u}(s))\mathrm{d}s.$$

We could approximate this integral using the values $\boldsymbol{w}^k$ and $\boldsymbol{w}^{k+1}$, and this is the idea behind the single-step methods studied in Chapter 18. However, in doing so, we are not making use of all the information that we had computed before, i.e., $\{\boldsymbol{w}^j\}_{j=0}^{k+q-1}$. The idea of the *Adams methods* is to use a subset of $\{\boldsymbol{f}(t_j, \boldsymbol{w}^j)\}_{j=0}^{k+q-1}$ to construct an interpolating polynomial and use this polynomial to approximate the integral in the previous identity. Two important classes of methods here are:

1. *Adams–Bashforth* methods,[1] which use $\{\boldsymbol{f}(t_j, \boldsymbol{w}^j)\}_{j=k}^{k+q-1}$ and thus are *explicit* methods.
2. *Adams–Moulton* methods,[2] which use $\{\boldsymbol{f}(t_j, \boldsymbol{w}^j)\}_{j=k}^{k+q}$ and thus are *implicit*.

The general strategy is clear, so we confine ourselves to presenting a few examples.

Example 20.3 The *Adams–Bashforth four-step method* (AB4) is defined as follows: for $k = 0, \ldots, K-4$,

$$\boldsymbol{w}^{k+4} - \boldsymbol{w}^{k+3} = \tau \left[\frac{55}{24}\boldsymbol{f}^{k+3} - \frac{59}{24}\boldsymbol{f}^{k+2} + \frac{37}{24}\boldsymbol{f}^{k+1} - \frac{9}{24}\boldsymbol{f}^{k} \right], \tag{20.6}$$

where $\boldsymbol{f}^j = \boldsymbol{f}(t_j, \boldsymbol{w}^j)$. This requires the starting values $\boldsymbol{w}^0$, $\boldsymbol{w}^1$, $\boldsymbol{w}^2$, $\boldsymbol{w}^3$. The coefficients are $a_4 = 1$, $a_3 = -1$, $a_2 = a_1 = a_0 = 0$, and $b_4 = 0$, $b_3 = \frac{55}{24}$, $b_2 = -\frac{59}{24}$, $b_1 = \frac{37}{24}$, $b_0 = -\frac{9}{24}$.

Theorem 20.8 (LTE of AB4). *Suppose that $\boldsymbol{f} \in \mathcal{F}^4(S)$ and $\boldsymbol{u} \in C^5([0, T]; \mathbb{R}^d)$ is the classical solution to* (18.1). *Then the local truncation error for the AB4 method* (20.6) *may be expressed as*

$$\mathcal{E}^{k+4}[\boldsymbol{u}] \le \frac{251 d^{1/2}}{720} \max_{\eta \in [t_k, t_{k+4}]} \|\boldsymbol{u}^{(5)}(\eta)\|_2 \tau^4$$

for every $k = 0, 1, \ldots, K-4$.

[1] Named in honor of the British mathematicians John Couch Adams (1819–1892) and Francis Bashforth (1819–1912).

[2] Named in honor of the British mathematician John Couch Adams (1819–1892) and American astronomer Forest Ray Moulton (1872–1952).

Proof. Suppose that $0 \le k \le K-4$. Let $\boldsymbol{p}_k \in [\mathbb{P}_3]^d$ be the vector-valued Lagrange interpolating polynomial with respect to the four interpolation points

$$\{(t_{k+j}, \boldsymbol{f}(t_{k+j}, \boldsymbol{u}(t_{k+j})))\}_{j=0}^{3}.$$

Then, for all $t \in [t_k, t_{k+4}]$,

$$\boldsymbol{f}(t, \boldsymbol{u}(t)) = \boldsymbol{p}_k(t) + \boldsymbol{E}_k(t),$$

where $\boldsymbol{E}_k$ is an error function. Thus,

$$\boldsymbol{u}(t_{k+4}) - \boldsymbol{u}(t_{k+3}) = \int_{t_{k+3}}^{t_{k+4}} \boldsymbol{f}(t, \boldsymbol{u}(t))\mathrm{d}t = \int_{t_{k+3}}^{t_{k+4}} \boldsymbol{p}_k(t)\mathrm{d}t + \int_{t_{k+3}}^{t_{k+4}} \boldsymbol{E}_k(t)\mathrm{d}t.$$

According to the error theory for Lagrange interpolation (Theorem 9.16) we have, for all $i = 1, \dots, d$,

$$[\boldsymbol{E}_k]_i(t) = \frac{1}{4!}\frac{\mathrm{d}^4}{\mathrm{d}t^4}[\boldsymbol{f}(t, \boldsymbol{u}(t))]_i|_{t=\xi_i} \prod_{j=0}^{3}(t - t_{k+j}) = \frac{1}{24}[\boldsymbol{u}^{(5)}]_i(\xi_i(t)) \prod_{j=0}^{3}(t - t_{k+j})$$

for all $t \in [t_k, t_{k+4}]$ and some $\xi_i = \xi_i(t) \in (t_k, t_{k+4})$. We find then that, after the change of variables $t = r\tau + t_{k+3}$,

$$\int_{t_{k+3}}^{t_{k+4}} [\boldsymbol{E}_k]_i(t)\mathrm{d}t = \frac{\tau^5}{24}\int_0^1 r(r+1)(r+2)(r+3)[\boldsymbol{u}^{(5)}]_i(\xi_i(t))\mathrm{d}r.$$

Evidently, $\boldsymbol{E}_k$ is a continuous function on $[0, T]$ since it is the difference of continuous functions. Now set

$$g(r) = r(r+1)(r+2)(r+3)$$

and observe that $g \ge 0$ on $[0, 1]$. Then, taking norms,

$$\begin{aligned}
\left\| \int_{t_{k+3}}^{t_{k+4}} \boldsymbol{E}_k(t)\mathrm{d}t \right\|_2 &\le \frac{\tau^5}{24}\int_0^1 g(r)\left(\sum_{i=1}^{d}\left|[\boldsymbol{u}^{(5)}]_i(\xi_i(t))\right|^2\right)^{1/2}\mathrm{d}r \\
&\le d^{1/2}\max_{\xi\in[t_k, t_{k+4}]}\left\|\boldsymbol{u}^{(5)}(\xi)\right\|_2 \frac{\tau^5}{24}\int_0^1 g(r)\mathrm{d}r \\
&= \max_{\xi\in[t_k, t_{k+4}]}\left\|\boldsymbol{u}^{(5)}(\xi)\right\|_2 \frac{\tau^5}{24}\frac{251 d^{1/2}}{30} \\
&= \frac{251 d^{1/2}}{720}\tau^5 \max_{\xi\in[t_k, t_{k+4}]}\left\|\boldsymbol{u}^{(5)}(\xi)\right\|_2.
\end{aligned}$$

Let us use the notation $\boldsymbol{f}_e^{k+j} = \boldsymbol{f}(t_{k+j}, \boldsymbol{u}(t_{k+j}))$, $j = 0, 1, 2, 3$. Then

$$\begin{aligned}
\boldsymbol{p}_k(t) = \boldsymbol{f}_e^k & \frac{(t - t_{k+1})(t - t_{k+2})(t - t_{k+3})}{(t_k - t_{k+1})(t_k - t_{k+2})(t_k - t_{k+3})} \\
&+ \boldsymbol{f}_e^{k+1} \frac{(t - t_k)(t - t_{k+2})(t - t_{k+3})}{(t_{k+1} - t_k)(t_{k+1} - t_{k+2})(t_{k+1} - t_{k+3})} \\
&+ \boldsymbol{f}_e^{k+2} \frac{(t - t_k)(t - t_{k+1})(t - t_{k+3})}{(t_{k+2} - t_k)(t_{k+2} - t_{k+1})(t_{k+2} - t_{k+3})} \\
&+ \boldsymbol{f}_e^{k+3} \frac{(t - t_k)(t - t_{k+1})(t - t_{k+2})}{(t_{k+3} - t_k)(t_{k+3} - t_{k+1})(t_{k+3} - t_{k+2})}.
\end{aligned}$$

Observe that

$$\boldsymbol{p}_k(t_{k+j}) = \boldsymbol{f}_e^{k+j} = \boldsymbol{f}(t_{k+j}, \boldsymbol{u}(t_{k+j})), \quad j = 0, 1, 2, 3.$$

Integrating $\boldsymbol{p}_k$ on the interval $[t_{k+3}, t_{k+4}]$, the reader can confirm that

$$\int_{t_{k+3}}^{t_{k+4}} \boldsymbol{p}_k(t) \mathrm{d}t = \tau \left[\frac{55}{24} \boldsymbol{f}_e^{k+3} - \frac{59}{24} \boldsymbol{f}_e^{k+2} + \frac{37}{24} \boldsymbol{f}_e^{k+1} - \frac{9}{24} \boldsymbol{f}_e^{k} \right].$$

Using the definition of the local truncation error (20.2), the result is proven. □

Remark 20.9 (consistency). With the same hypotheses as in Theorem 20.8, we can use the result of Theorem 20.4 to come to the same conclusion as above. In particular, for the AB4 method, we find

$$C_0 = C_1 = C_2 = C_3 = C_4 = 0, \qquad C_5 = \frac{251}{720}.$$

Example 20.4 The *Adams–Moulton four-step method* (AM4) is defined as follows: for $k = 0, \ldots, K - 4$,

$$\begin{aligned}
\boldsymbol{w}^{k+4} - \boldsymbol{w}^{k+3} = \tau \Bigg[& \frac{251}{720} \boldsymbol{f}^{k+4} + \frac{646}{720} \boldsymbol{f}^{k+3} - \frac{264}{720} \boldsymbol{f}^{k+2} \\
& + \frac{106}{720} \boldsymbol{f}^{k+1} - \frac{19}{720} \boldsymbol{f}^{k} \Bigg],
\end{aligned} \tag{20.7}$$

where $\boldsymbol{f}^j = \boldsymbol{f}(t_j, \boldsymbol{w}^j)$. This requires the starting values $\boldsymbol{w}^0$, $\boldsymbol{w}^1$, $\boldsymbol{w}^2$, $\boldsymbol{w}^3$. The coefficients are $a_4 = 1$, $a_3 = -1$, $a_2 = a_1 = a_0 = 0$, and $b_4 = \frac{251}{720}$, $b_3 = \frac{646}{720}$, $b_2 = -\frac{264}{720}$, $b_1 = \frac{106}{720}$, $b_0 = -\frac{19}{720}$.

As we will see in the proof of the following result, the difference between the AB4 and AM4 methods is that the interpolating polynomial in the AM4 method uses the additional implicit time level point $\boldsymbol{f}(t_{k+4}, \boldsymbol{u}(t_{k+4}))$. This results in the AM4 method being more accurate, but comes at the cost of producing an implicit method. Implicit methods are usually always more complicated in practice than explicit methods.

Theorem 20.10 (LTE for AM4). *Assume that $\boldsymbol{f} \in \mathcal{F}^5(S)$. Let the function $\boldsymbol{u} \in C^6([0,T];\mathbb{R}^d)$ be the classical solution to (18.1). Then the local truncation error for the AM4 method (20.7) may be expressed as*

$$\|\boldsymbol{\mathcal{E}}^{k+4}[\boldsymbol{u}]\|_2 \le \frac{3d^{1/2}}{160} \max_{\eta\in[t_k,t_{k+4}]} \left\|\boldsymbol{u}^{(6)}(\eta)\right\|_2 \tau^5$$

for every $k = 0, 1, \dots, K-4$.

Proof. The proof is similar to that of Theorem 20.8. Suppose that $0 \le k \le K-4$. Let $\boldsymbol{p}_k \in [\mathbb{P}_4]^d$ be the vector-valued Lagrange interpolating polynomial uniquely determined by the five interpolation points

$$\{(t_{k+j}, \boldsymbol{f}(t_{k+j}, \boldsymbol{u}(t_{k+j})))\}_{j=0}^{4}.$$

Then, for all $t \in [t_k, t_{k+4}]$,

$$\boldsymbol{f}(t, \boldsymbol{u}(t)) = \boldsymbol{p}_k(t) + \boldsymbol{E}_k(t)$$

and, for all $i = 1, \dots, d$,

$$[\boldsymbol{E}_k]_i(t) = \frac{1}{5!}\frac{\mathrm{d}^5}{\mathrm{d}t^5}[\boldsymbol{f}(t,\boldsymbol{u}(t))]_{i|t=\xi}\prod_{j=0}^{4}(t-t_{k+j}) = \frac{1}{120}[\boldsymbol{u}^{(6)}]_i(\xi_i(t))\prod_{j=0}^{4}(t-t_{k+j})$$

for all $t \in [t_k, t_{k+4}]$ and some $\xi_i = \xi_i(t) \in (t_k, t_{k+4})$. After the change of variables $t = r\tau + t_{k+3}$,

$$\int_{t_{k+3}}^{t_{k+4}} [\boldsymbol{E}_k]_i(t)\mathrm{d}t = \frac{\tau^6}{120}\int_0^1 (r-1)r(r+1)(r+2)(r+3)[\boldsymbol{u}^{(6)}]_i(\xi_i(t))\mathrm{d}r.$$

Setting

$$g(r) = -(r-1)r(r+1)(r+2)(r+3),$$

we observe that $g \ge 0$ on $[0,1]$ and, as before,

$$\left\|\int_{t_{k+3}}^{t_{k+4}} \boldsymbol{E}_k(t)\mathrm{d}t\right\|_2 \le \frac{3d^{1/2}}{160}\tau^6 \max_{\eta\in[t_k,t_{k+4}]}\left\|\boldsymbol{u}^{(6)}(\eta)\right\|_2.$$

For the Lagrange interpolating polynomial, the reader can confirm that

$$\int_{t_{k+3}}^{t_{k+4}} \boldsymbol{p}_k(t)\mathrm{d}t = \tau\left[\frac{251}{720}\boldsymbol{f}_e^{k+4} + \frac{646}{720}\boldsymbol{f}_e^{k+3} - \frac{264}{720}\boldsymbol{f}_e^{k+2} + \frac{106}{720}\boldsymbol{f}_e^{k+1} - \frac{19}{720}\boldsymbol{f}_e^{k}\right],$$

where $\boldsymbol{f}_e^{k+j} = \boldsymbol{f}(t_{k+j}, \boldsymbol{u}(t_{k+j}))$, $j = 0, \dots, 4$. The result follows from the definition of the local truncation error (20.2). □

20.3 Backward Differentiation Formula Methods

Another important class of multi-step methods is the Backward Differentiation Formula (BDF) methods. These differ from the Adams–Bashforth (AB) and Adams–Moulton (AM) methods in a fundamental way. Whereas the AB and AM methods are derived via an integration procedure, the BDF methods are derived via differentiation. We will demonstrate this point shortly. But before that, we give a general definition for these methods and derive some of their properties.

Definition 20.11 (BDF). A linear q-step method (20.1) is called a **BDF method** (or a **BDFq method**) if and only if it is of order q, exactly, and

$$b_q \neq 0, \quad b_{q-1} = b_{q-2} = \cdots = b_1 = b_0 = 0.$$

The reader should recall that, herein, we always assume that $a_q = 1$.

Theorem 20.12 (construction of BDF). *Let $q \in \mathbb{N}$ and set $\beta = \left[\sum_{j=1}^{q} \frac{1}{j}\right]^{-1}$. Suppose that the linear q-step method* (20.1) *is a BDF method. Then $b_q = \beta$ and*

$$\psi(z) = \sum_{j=0}^{q} a_j z^j = \beta \sum_{j=1}^{q} \frac{1}{j} z^{q-j}(z-1)^j$$

or, equivalently,

$$a_q = 1, \qquad a_{q-m} = \beta \sum_{j=m}^{q} \frac{(-1)^m}{j} \binom{j}{m}, \qquad m = 1, \ldots, q.$$

Proof. Appealing to Theorem 20.7, we see that, in a neighborhood of $\mu = 1$,

$$\psi(\mu) - b_q \mu^q \ln(\mu) = \sum_{m=q+1}^{\infty} \tilde{C}_m (\mu - 1)^m$$

with $\tilde{C}_{q+1} \neq 0$. In other words, $\mu = 1$ is a $(q+1)$-fold zero. Now we make the substitution $\mu = \nu^{-1}$. In a neighborhood of $\nu = 1$,

$$\nu^q \psi(\nu^{-1}) + b_q \ln(\nu) = \sum_{m=q+1}^{\infty} \hat{C}_m (\nu - 1)^m$$

with $\hat{C}_{q+1} \neq 0$. Thus, in a neighborhood of $\nu = 1$,

$$\nu^q \psi(\nu^{-1}) = b_q \sum_{m=1}^{\infty} \frac{(-1)^m (\nu - 1)^m}{m} + \sum_{m=q+1}^{\infty} \hat{C}_m (\nu - 1)^m \in \mathbb{P}_q.$$

This implies that the tail of the series vanishes:

$$\nu^q \psi(\nu^{-1}) = b_q \sum_{m=1}^{q} \frac{(-1)^m (\nu - 1)^m}{m}.$$

Therefore,

$$\psi(\mu) = b_q \sum_{m=1}^{q} \frac{(-1)^m \mu^q (\mu^{-1} - 1)^m}{m} = b_q \sum_{m=1}^{q} \frac{\mu^{q-m} (\mu - 1)^m}{m} = \sum_{j=0}^{q} a_j \mu^j.$$

It is clear at this point that $a_q = b_q \beta^{-1}$. Thus, to achieve our standard normalization, we require $a_q = 1$, $b_q = \beta$.

Now we establish the equivalence. Using the binomial theorem,

$$
\begin{aligned}
\sum_{j=0}^{q} a_j \mu^j &= b_q \sum_{j=1}^{q} \frac{1}{j} \mu^{q-j} (\mu - 1)^j \\
&= b_q \sum_{j=1}^{q} \frac{1}{j} \mu^{q-j} \sum_{m=0}^{j} \binom{j}{m} \mu^{j-m} (-1)^m \\
&= b_q \sum_{j=1}^{q} \sum_{m=0}^{j} \frac{1}{j} \binom{j}{m} (-1)^m \mu^{q-m} \\
&= b_q \sum_{m=0}^{q} \sum_{j=\max\{1,m\}}^{q} \frac{(-1)^m}{j} \binom{j}{m} \mu^{q-m}.
\end{aligned}
$$

Thus, we have $a_q = 1$ and

$$
a_{q-m} = \beta \sum_{j=m}^{q} \frac{(-1)^m}{j} \binom{j}{m}, \qquad m = 1, \dots, q.
$$

□

Owing to the previous result, we have the following BDF coefficients.

Example 20.5 For $q = 1$, we find

$$
b_1 = 1, \quad a_1 = 1, \quad a_0 = -1.
$$

Of course, this corresponds to the single-step backward Euler method.

Example 20.6 For $q = 2$, we find

$$
b_2 = \frac{2}{3}, \quad a_2 = 1, \quad a_1 = -\frac{4}{3}, \quad a_0 = \frac{1}{3}.
$$

Example 20.7 For $q = 3$, we find

$$
b_3 = \frac{6}{11}, \quad a_3 = 1, \quad a_2 = -\frac{18}{11}, \quad a_1 = \frac{9}{11}, \quad a_0 = -\frac{2}{11}.
$$

We conclude by commenting that the traditional way to develop BDF methods is via differentiation of the Lagrange interpolating polynomials studied in Chapter 9. The following example illustrates how to obtain the BDFq method using this approach.

Example 20.8 Let $d = 1$ and $u \in C^1([0, T])$ be a classical solution to (18.1). For $0 \le k \le K-2$, and any $t \in [t_k, t_{k+2}]$,

$$u(t) = u(t_k)\frac{(t - t_{k+1})(t - t_{k+2})}{(t_k - t_{k+1})(t_k - t_{k+2})} + u(t_{k+1})\frac{(t - t_k)(t - t_{k+2})}{(t_{k+1} - t_k)(t_{k+1} - t_{k+2})}$$
$$+ u(t_{k+2})\frac{(t - t_k)(t - t_{k+1})}{(t_{k+2} - t_k)(t_{k+2} - t_{k+1})} + \mathcal{E}(t),$$

where $\mathcal{E}$ is an error term. Differentiating and evaluating at $t = t_{k+2}$, we get

$$\begin{aligned} u'(t_{k+2}) &= u(t_k)\frac{(t_{k+2} - t_{k+1}) + (t_{k+2} - t_{k+2})}{(t_k - t_{k+1})(t_k - t_{k+2})} \\ &\quad + u(t_{k+1})\frac{(t_{k+2} - t_k) + (t_{k+2} - t_{k+2})}{(t_{k+1} - t_k)(t_{k+1} - t_{k+2})} \\ &\quad + u(t_{k+2})\frac{(t_{k+2} - t_k) + (t_{k+2} - t_{k+1})}{(t_{k+2} - t_k)(t_{k+2} - t_{k+1})} + \mathcal{E}'(t_{k+2}) \\ &= \frac{1}{2\tau}u(t_k) - \frac{2}{\tau}u(t_{k+1}) + \frac{3}{2\tau}u(t_{k+2}) + \mathcal{E}'(t_{k+2}) \\ &= f(t_{k+2}, u(t_{k+2})). \end{aligned}$$

Equivalently,

$$u(t_{k+2}) - \frac{4}{3}u(t_{k+1}) + \frac{1}{3}u(t_k) + \frac{2\tau}{3}\mathcal{E}'(t_{k+2}) = \frac{2\tau}{3}f(t_{k+2}, u(t_{k+2})).$$

This yields the BDF2 method, as claimed.

20.4 Zero Stability

We now examine the issue of zero stability of multi-step methods. It turns out that not all consistent multi-step approximation methods are zero stable. This was not an issue for single-step methods; they are generally always stable. For linear multi-step methods, we need to take great care: if a consistent method is not also stable, it will not be a convergent method.

Definition 20.13 (zero stability)**.** Suppose that $\boldsymbol{f} \in \mathcal{F}^1(S)$ and $\boldsymbol{u} \in C^2([0, T]; \mathbb{R}^d)$ is a classical solution to (18.1). Let, for $i = 1, 2$, $\left\{\boldsymbol{w}_i^k\right\}_{k=0}^{K}$ be approximations generated by the linear q-step method (20.1) with the starting values $\left\{\boldsymbol{w}_i^k\right\}_{k=0}^{q-1}$, $i = 1, 2$, respectively. The method is called **zero stable** if and only if there is a $C > 0$ independent of $\tau > 0$ and the starting values such that, for any $k = q, \dots, K$,

$$\left\|\boldsymbol{w}_1^k - \boldsymbol{w}_2^k\right\|_2 \le C \max_{m=0,\dots,q-1} \left\|\boldsymbol{w}_1^m - \boldsymbol{w}_2^m\right\|_2.$$

Definition 20.14 (root condition)**.** The linear q-step method (20.1) satisfies the **root condition** if and only if:

1. All of the roots of the first characteristic polynomial $\psi(z) = \sum_{j=0}^{q} a_j z^j$ are inside the unit disk
$$\{z \in \mathbb{C} \mid |z| \leq 1\} \subset \mathbb{C}.$$
2. If $\psi(\xi) = 0$ and $|\xi| = 1$, then ξ is a simple root, i.e., its multiplicity is exactly one, i.e., $\psi'(\xi) \neq 0$.

Definition 20.15 (homogeneous zero stability). Suppose that $\boldsymbol{f} \equiv \boldsymbol{0}$ and $\boldsymbol{u}_0 = \boldsymbol{0}$, so that the unique solution to (18.1) is $\boldsymbol{u}(t) = \boldsymbol{0}$ for all $t \geq 0$. Let $\left\{\boldsymbol{w}^k\right\}_{k=0}^{K}$ be the approximation generated by the linear q-step method (20.1) with the starting values $\left\{\boldsymbol{w}^k\right\}_{k=0}^{q-1}$. The method is called **homogeneous zero stable** if and only if there is a $C > 0$ independent of $\tau > 0$ and the starting values such that, for any $k = q, \ldots, K$,
$$\left\|\boldsymbol{w}^k\right\|_2 \leq C \max_{m=0,\ldots,q-1} \|\boldsymbol{w}^m\|_2 .$$

Definition 20.16 (stable solutions). Suppose that $\{a_j\}_{j=0}^{q-1} \subset \mathbb{C}$ are given. An equation of the form
$$\zeta_{k+q} + \sum_{j=0}^{q-1} a_j \zeta_{k+j} = 0, \quad k = 0, 1, 2, \ldots \tag{20.8}$$
is called a **homogeneous difference equation**. We say that solutions to (20.8) are **stable** if and only if, given any starting values $\{\zeta_k\}_{k=0}^{q-1} \subset \mathbb{R}$, the sequence $\{\zeta_k\}_{k=0}^{\infty} \subset \mathbb{R}$ is bounded by a constant $C > 0$ that only depends upon the starting values.

We will see that the concepts of stability, homogeneous zero stability, and the root condition are all actually equivalent.

Example 20.9 In this example, we exhibit a method that does not satisfy the root condition and is *not* homogeneously zero stable. Consider the method $q = 2$, $a_2 = 1$, $a_1 = -3$, $a_0 = 2$ and $b_2 = 0$, $b_1 = 0$, $b_0 = -1$. In other words,
$$\boldsymbol{w}^{k+2} - 3\boldsymbol{w}^{k+1} + 2\boldsymbol{w}^k = -\tau \boldsymbol{f}(t_k, \boldsymbol{w}^k)$$
with the starting values $\boldsymbol{w}^0$, $\boldsymbol{w}^1$. The method is consistent. We find
$$C_0 = 0 = C_1, \quad C_2 = \frac{1}{2},$$
which implies that the method is consistent to order $p = 1$.

The first characteristic polynomial is $\psi(z) = z^2 - 3z + 2 = (z-1)(z-2)$. Thus, the method fails to satisfy the root condition. Since we are considering homogeneous zero stability, we take $\boldsymbol{f} \equiv \boldsymbol{0}$ and $\boldsymbol{u}_0 = \boldsymbol{0}$. The solution of the homogeneous linear constant coefficient difference equation,
$$\zeta_{k+2} - 3\zeta_{k+1} + 2\zeta_k = 0, \quad k = 0, 1, \ldots,$$

is precisely

$$\zeta_k = 2\zeta_0 - \zeta_1 + 2^k(\zeta_1 - \zeta_0).$$

This can be verified by a simple induction argument. For starting values, let us take $\zeta_0 = 0$, $\zeta_1 = \tau$. Then $\zeta_k = \tau(2^k - 1)$, $k = 0, 1, 2, \ldots$. Let us examine the approximation at time $T = 1$. In this case, $\tau = 1/K$ and we have, as $K \to \infty$,

$$w^K = \frac{2^K - 1}{K} \to \infty.$$

Thus, the method is not homogeneously zero stable.

We need the following technical lemma in order to construct general solutions of linear homogeneous difference equations. Here, we follow the exposition in the book by Kress [52].

Lemma 20.17 (operator Q). *Suppose that $q \in \mathbb{N}$ and $\nu(t) = \sum_{j=0}^{q} \beta_j t^j \in \mathbb{P}_q$ with coefficients $\beta_j \in \mathbb{C}$, $j = 0, \ldots, q$. Assume that $\beta_q \neq 0$, $\beta_0 \neq 0$. Define the operator $Q \colon \mathbb{P}_q \to \mathbb{P}_q$ via $Q[u](t) = tu'(t)$. The number $\lambda \in \mathbb{C}$ is a root of the polynomial ν of multiplicity m, $m = 1, \ldots, q$ if and only if, for every $p = 0, \ldots, m-1$,*

$$Q^p[\nu](\lambda) = 0, \tag{20.9}$$

where Q^0 is the identity operator but $Q^m[\nu](\lambda) \neq 0$.

Proof. We start with a couple of observations. First, since $\beta_0 \neq 0$, $\lambda = 0$ is not a root of ν. Next, as the case $m = 1$ is trivial, we will assume that $m > 1$. Finally, the repeated application of Q results in

$$Q^p[\nu](t) = \sum_{j=0}^{q} \beta_j j^p t^j \in \mathbb{P}_q,$$

which holds for all $p = 0, 1, 2, \ldots$, provided we interpret $0^0 = 1$.

($\Longrightarrow$) Suppose that $\lambda \neq 0$ is a root of ν of multiplicity $m > 1$. Then $\nu(t) = (t - \lambda)^m \phi_0(t)$, where $\phi_0 \in \mathbb{P}_{q-m}$ and $\phi_0(\lambda) \neq 0$. Then

$$Q[\nu](t) = t\phi_0'(t)(t-\lambda)^m + t\phi_0(t)m(t-\lambda)^{m-1} = (t-\lambda)^{m-1}\phi_1(t),$$

where $\phi_1 \in \mathbb{P}_{q-m+1}$ and $\phi_1(\lambda) \neq 0$. Clearly, $Q[\nu](\lambda) = 0$. Continuing recursively, we observe that

$$Q^p[\nu](t) = (t-\lambda)^{m-p}\phi_p(t),$$

where $\phi_p \in \mathbb{P}_{q-m+p}$ and $\phi_p(\lambda) \neq 0$, for all $p = 1, \ldots, m-1$, with $Q^p[\nu](\lambda) = 0$. However, it is clear that $Q^m[\nu](\lambda) \neq 0$.

($\Longleftarrow$) To obtain a contradiction, suppose that $\lambda \neq 0$ is *not* a root of multiplicity $m > 1$, but property (20.9) holds. Then we proceed in three steps.

1. If λ is not a root of ν at all, then we arrive at a contradiction, namely $Q^0[\nu](\lambda) = \nu(\lambda) \neq 0$.

2. If λ is a root of ν of multiplicity $n < m$, then we again get a contradiction, since $Q^n[\nu](\lambda) \neq 0$.
3. Lastly, if λ is a root of ν of multiplicity $n > m$, we get a contradiction, since $Q^m[\nu](\lambda) = 0$. It must be that property (20.9) implies that $\lambda \neq 0$ is a root of multiplicity exactly m.

This concludes the proof. □

Theorem 20.18 (solution of difference equations). *Suppose that $q \in \mathbb{N}$, $a_q = 1$, $\{a_j\}_{j=0}^{q-1} \subset \mathbb{R}$, with $a_0 \neq 0$, and $\{\zeta_j\}_{j=0}^{q-1} \subset \mathbb{R}$ are given. The linear homogeneous difference* equation (20.8) *has a unique solution $\{\zeta_k\}_{k=0}^{\infty}$. Assume that $\lambda_j \in \mathbb{C}$, $j = 1, \ldots, r$, where $r \leq q$, are the distinct roots of the characteristic polynomial ψ, with multiplicities m_j, respectively. Hence,*

$$\psi(z) = \prod_{j=1}^{r}(z - \lambda_j)^{m_j}, \quad \sum_{j=1}^{r} m_j = q.$$

Then the solution to (20.8) *is given by*

$$\zeta_k = \sum_{j=1}^{r} p_j(k)\lambda_j^k,$$

where $p_j \in \mathbb{P}_{m_j-1}$. In particular,

$$p_j(z) = \sum_{m=0}^{m_j-1} \alpha_{j,m} z^m$$

for some coefficients $\alpha_{j,m} \in \mathbb{C}$, $\#\{\alpha_{j,m}\} = q$, that are uniquely determined by the q starting values $\zeta_0, \ldots, \zeta_{q-1}$.

Proof. By linearity, uniqueness follows from the fact that the only possible solution to (20.8) with zero starting values: $\zeta_j = 0$ for $j = 0, \ldots, q-1$, is a sequence of zeros $\zeta_k = 0$, $k = 0, 1, \ldots$.

To show existence, suppose that λ is the root of ψ of multiplicity $m \geq 1$. Consider the sequence

$$\zeta_k = k^n \lambda^k \tag{20.10}$$

for some $n = 0, \ldots, m-1$. Then

$$\begin{aligned}
\zeta_{k+q} + \sum_{j=0}^{q-1} a_j \zeta_{k+j} &= \sum_{j=0}^{q} a_j (k+j)^n \lambda^{k+j} \\
&= \sum_{j=0}^{q} a_j \sum_{i=0}^{n} \binom{n}{i} k^i j^{n-i} \lambda^{k+j} \\
&= \lambda^k \sum_{i=0}^{n} \binom{n}{i} k^i \sum_{j=0}^{q} a_j j^{n-i} \lambda^j \\
&= \lambda^k \sum_{i=0}^{n} \binom{n}{i} k^i Q^{n-i}[\psi](\lambda),
\end{aligned}$$

where Q is the operator defined in Lemma 20.17. Observe that, since $\psi(z) = (z-\lambda)^m \phi(z)$, where $\phi \in \mathbb{P}_{q-m}$,

$$Q^{n-i}[\psi](\lambda) = 0$$

for all $n = 0, \dots, m-1$ and $i = 0, \dots, n$, appealing to Lemma 20.17. This proves that (20.10) is a solution to (20.8).

Next, we aim to prove that the general solution is just a linear combination of the solutions from above. To establish that the general solution has the form

$$\zeta_k = \sum_{j=1}^{r} \sum_{m=0}^{m_j-1} \alpha_{j,m} k^m \lambda_j^k, \quad k = 0, 1, \dots, \tag{20.11}$$

we need to determine the q free parameters $\alpha_{1,0}, \alpha_{1,1}, \dots, \alpha_{r,m_r-1} \in \mathbb{C}$ such that (20.11) holds for the q starting values, i.e., (20.11) holds for $k = 0, \dots, q-1$. This forms a square $q \times q$ system of linear equations that is uniquely solvable if and only if the corresponding homogeneous system has only the trivial solution. In other words, we want to show that

$$\sum_{j=1}^{r} \sum_{m=0}^{m_j-1} \alpha_{j,m} k^m \lambda_j^k = 0, \quad k = 0, 1, \dots, q-1 \tag{20.12}$$

implies that $\alpha_{j,m} = 0$ for $j = 1, \dots, r$ and $m = 0, \dots, m_j - 1$.

Note that we can represent the homogeneous system above as $\mathsf{B}\boldsymbol{\alpha} = \mathbf{0}$, where $\boldsymbol{\alpha}^\intercal = [\alpha_{1,0}, \alpha_{1,2}, \dots, \alpha_{r,m_r-1}]^\intercal$. We recall that B is nonsingular if and only if $\mathsf{B}^\intercal$ is nonsingular if and only if B^{H} is nonsingular. In order to prove that $\mathsf{B}^\intercal$ is nonsingular, we consider the homogeneous adjoint equation $\mathsf{B}^\intercal \boldsymbol{\beta} = \mathbf{0}$. In particular, suppose that $\beta_k \in \mathbb{C}$, $k = 0, \dots, q-1$ satisfy the homogeneous adjoint equations

$$\sum_{k=0}^{q-1} \beta_k k^m \lambda_j^k = 0, \quad j = 1, \dots, r, \quad m = 0, \dots, m_j - 1. \tag{20.13}$$

Now define

$$\eta(t) = \sum_{k=0}^{q-1} \beta_k t^k.$$

Then, for any $m \in \mathbb{N}$,

$$Q^m[\eta](t) = \sum_{k=0}^{q-1} \beta_k k^m t^k,$$

and it is clear from (20.13) that

$$Q^m[\eta](\lambda_j) = 0, \quad j = 1, \dots, r, \quad m = 0, \dots, m_j - 1.$$

Appealing to Lemma 20.17, this proves that λ_j is a root η of multiplicity m_j for $j = 1, \dots, r$. All told, $\eta \in \mathbb{P}_{q-1}$ has q roots, counting multiplicities. The only possibility, therefore, is that $\eta \equiv 0$. In other words, $\beta_k = 0$, $k = 0, \dots, q-1$. $\square$

Because of the form of the general solution, there is a clear connection between the root condition and the stability of solutions to the homogeneous difference equation (20.8).

Theorem 20.19 (root condition and stability). *Suppose that $q \in \mathbb{N}$. Consider a linear q-step method (20.1) with coefficients $a_j, b_j \in \mathbb{R}$, $j = 0, \ldots, q$, with $a_q = 1$ and $a_0 \neq 0$. The solutions to the corresponding homogeneous difference equation (20.8) are bounded, i.e., stable, if and only if the root condition is satisfied.*

Proof. Given the starting values ζ_j, $j = 0, \ldots, q-1$, the general solution to (20.8) is

$$\zeta_k = \sum_{j=1}^{r} \sum_{m=0}^{m_j - 1} \alpha_{j,m} k^m \lambda_j^k, \quad k = 0, 1, \ldots,$$

where the q coefficients $\alpha_{1,1}, \alpha_{1,2}, \ldots, \alpha_{r,m_r} \in \mathbb{C}$ are determined uniquely by the q starting values.

($\Longrightarrow$) From the form of the general solution, it is clear that, if the root condition is not satisfied, the approximations can grow unboundedly, as in Example 20.9.

($\Longleftarrow$) Conversely, if the approximations remain bounded, the only possibility is that the root condition is satisfied. The details are left to the reader as an exercise; see Problem 20.8. □

Theorem 20.20 (nonhomogeneous difference equations). *Let $q \in \mathbb{N}$ and, for $m = 0, \ldots, q-1$, the sequence $\left\{ g_k^{(m)} \right\}_{k=0}^{\infty}$ be the unique solution to the homogeneous linear difference equation (20.8) with $a_j \in \mathbb{R}$, $j = 0, \ldots, q$, $a_q = 1$, and $a_0 \neq 0$ and starting values*

$$g_k^{(m)} = \delta_{k,m}, \quad k, m = 0, 1, 2, \ldots, q-1. \tag{20.14}$$

Let $\{c_k\}_{k=q}^{\infty} \subset \mathbb{C}$ be a given sequence. Then there is a unique solution to the linear difference equation

$$\zeta_{k+q} + \sum_{j=0}^{q-1} a_j \zeta_{k+j} = c_{k+q}, \quad k = 0, 1, \ldots, \tag{20.15}$$

with starting values $\{\zeta_k\}_{k=0}^{q-1}$. This solution is given by

$$\zeta_{k+q} = \sum_{j=0}^{q-1} \zeta_j g_{k+q}^{(j)} + \sum_{j=0}^{k} c_{j+q} g_{k+q-j-1}^{(q-1)}, \quad k = 0, 1, 2, \ldots. \tag{20.16}$$

Proof. See Kress [52, Chapter 10] or Gautschi [32, Chapter 6]. □

Theorem 20.21 (zero stability). *A linear q-step method (20.1) is homogeneous zero stable if and only if solutions to the corresponding homogeneous equations (20.8) are stable.*

Proof. It suffices to prove the result for the scalar case, i.e., $d = 1$.

($\Longrightarrow$) Suppose that the linear q-step method (20.1) with the first characteristic polynomial $\psi(z) = \sum_{j=0}^{q} a_j z^j$ is homogeneous zero stable, i.e., if $\{w^k\}_{k=q}^{K}$ solves (20.1) with starting values $\{w^m\}_{m=0}^{q-1}$, then

$$|w^k| \le C \max_{m=0,\dots,q-1} |w^m|.$$

Let now the sequence $\{\zeta_k\}_{k=q}^\infty$ solve (20.8) with starting values $\zeta_k = w^k$ for $k = 0, \dots, q-1$. Notice that we must necessarily have $\zeta_k = w^k$ for $k = q, \dots, K$, where $\tau K = T$. In other words, the product τK is always the same fixed constant. The homogeneous zero stability of (20.1) then shows that there exists a constant $C > 0$ independent of $\tau > 0$ such that, for any $k = q, \dots, K$,

$$|\zeta_k| \le C \max_{m=0,\dots,q-1} |\zeta_m| . \tag{20.17}$$

Since C is independent of τ, it must also be independent of K. In other words, $K \in \mathbb{N}$ may be arbitrarily large. It follows that (20.17) holds for any $k \in \mathbb{N}$. It must be that $\{\zeta_k\}_{k=q}^\infty$ is bounded for any given set of starting values $\{\zeta_k\}_{k=0}^{q-1}$.

($\Longleftarrow$) Let, for $m = 0, \dots, q-1$, $\left\{g_k^{(m)}\right\}_{k=0}^\infty$ be the unique solution to the homogeneous linear difference equation (20.8) with the "impulse" starting values (20.14). Then, for all $m = 0, \dots, q-1$,

$$\left|g_k^{(m)}\right| \le C_m$$

for all $k \in \mathbb{N}$, where $C_m > 0$ is independent of k. We can then define

$$\widehat{C} = \max_{m=0,\dots,q-1} C_m$$

to obtain a constant that is independent of m. Suppose that, with the starting values $\left\{w^k\right\}_{k=0}^{q-1}$, the sequence $\left\{w^k\right\}_{k=q}^\infty$ satisfies (20.1) with $f \equiv 0$. Then, using (20.16), we have

$$w^{k+q} = \sum_{j=0}^{q-1} w^j g_{k+q}^{(j)}, \quad k = 0, 1, 2, \dots .$$

Taking absolute values and using the triangle inequality, we get

$$\left|w^{k+q}\right| \le \sum_{j=0}^{q-1} \left|w^j\right| \left|g_{k+q}^{(j)}\right| \le \widehat{C} \sum_{j=0}^{q-1} \left|w^j\right| \le q\widehat{C} \max_{m=0,\dots,q-1} |w^m| .$$

The proof is complete. □

Remark 20.22 ($a_0 = 0$)**.** We have only discussed the case for which $a_0 \neq 0$. What if $a_0 = 0$?

20.5 Convergence of Linear Multi-step Methods

Having discussed the notions of consistency and stability for multi-step methods, we are now ready to present the theory regarding their convergence. We begin with yet another discrete incarnation of Grönwall's[3] Lemma.

[3] Named in honor of the Swedish–American mathematician Thomas Hakon Grönwall (1877–1932).

Lemma 20.23 (discrete Grönwall). *Let $\{a_n\}_{n=0}^{\infty} \subset \mathbb{R}_+ \cup \{0\}$ be a sequence with the property that, for $n = 1, 2, \dots$,*

$$a_n \leq b \sum_{m=0}^{n-1} a_m + c$$

for some constants $b > 0$ and $c \geq 0$. Then, for all $n = 1, 2, \dots$,

$$a_n \leq (b a_0 + c)\, e^{(n-1)b}.$$

Proof. The result follows by an induction argument, which is left to the reader as an exercise; see Problem 20.11. □

Theorem 20.24 (convergence). *Let $p, K, q \in \mathbb{N}$ with $q < K$. Suppose that $\boldsymbol{f} \in \mathcal{F}^p(S)$, so that $\boldsymbol{u} \in C^{p+1}([0,T]; \mathbb{R}^d)$ satisfies the IVP (18.1). Let $\left\{\boldsymbol{w}^k\right\}_{k=0}^{K} \subset \mathbb{R}^d$ be an approximation generated by the linear q-step method (20.1) with the starting values $\left\{\boldsymbol{w}^k\right\}_{k=0}^{q-1} \subset \mathbb{R}^d$. Assume that the starting values are such that, for some constant $\tau_0 \in (0, T]$ and some $C_0 > 0$, we have*

$$\max_{k=0,\dots,q-1} \left\|\boldsymbol{e}^k\right\|_2 \leq C_0 \tau^p, \quad \forall \tau \in (0, \tau_0].$$

Assume, in addition, that the multi-step method is consistent to order p and satisfies the root condition. In this setting, there are constants $\tau_1 \in (0, T]$ and $C_1 > 0$ such that

$$\max_{k=0,\dots,K} \left\|\boldsymbol{e}^k\right\|_2 \leq C_1 \tau^p, \quad \forall \tau \in (0, \tau_1].$$

Proof. Let us assume that the method is implicit. The explicit case is simpler. Then we have the following error equation: for $k = 0, \dots, K - q$,

$$\begin{aligned}\boldsymbol{e}^{k+q} + \sum_{j=0}^{q-1} a_j \boldsymbol{e}^{k+j} &= \tau \sum_{j=0}^{q} b_j \left(\boldsymbol{f}(t_{k+j}, \boldsymbol{u}(t_{k+j})) - \boldsymbol{f}(t_{k+j}, \boldsymbol{w}^{k+j})\right) + \tau \boldsymbol{\mathcal{E}}^{k+q}[\boldsymbol{u}] \\ &= \tau \boldsymbol{c}^{k+q}.\end{aligned}$$

Since $\boldsymbol{f}$ satisfies a global $\boldsymbol{u}$-Lipschitz condition, we can estimate $\boldsymbol{c}^{k+q}$, as follows: by the triangle inequality and the Lipschitz continuity,

$$\left\|\boldsymbol{c}^{k+q}\right\|_2 \leq LB \sum_{j=0}^{q} \left\|\boldsymbol{e}^{k+j}\right\|_2 + \left\|\boldsymbol{\mathcal{E}}^{k+q}[\boldsymbol{u}]\right\|_2, \tag{20.18}$$

where $B = \max_{j=0,\dots,q} |b_j|$ and $L > 0$ is the standard Lipschitz constant.

By Theorem 20.20, the solution of the error equation can be represented as

$$\boldsymbol{e}^{k+q} = \sum_{j=0}^{q-1} g_{k+q}^{(j)} \boldsymbol{e}^j + \tau \sum_{j=0}^{k} g_{k+q-j-1}^{(q-1)} \boldsymbol{c}^{j+q}$$

for all $k = 0, \dots, K - q$. Because the q-step method satisfies the root condition and is, therefore, homogeneous zero stable, the solutions $g_k^{(j)}$ are bounded for every

$j = 0, \dots, q-1$ and every $k = 0, 1, 2, \dots$. In other words, there is a constant $C > 0$ such that, for all $j = 0, \dots, q-1$ and every $k = 0, 1, 2, \dots$,

$$\left|g_k^{(j)}\right| \le C.$$

So, we can estimate the error as

$$\left\|\boldsymbol{e}^{k+q}\right\|_2 \le \sum_{j=0}^{q-1} C\tau^p + \tau \sum_{j=0}^{k} C \left\|\boldsymbol{c}^{j+q}\right\|_2 \le C\left(\tau^p + \tau \sum_{j=0}^{k} \left\|\boldsymbol{c}^{j+q}\right\|_2\right),$$

provided that $0 < \tau \le \tau_0$. Here and in what follows, the constant C may change value from line to line. The important point is that it is a constant, and it is independent of all the involved quantities. Now we can use our estimate (20.18) above to obtain

$$\begin{aligned}
\left\|\boldsymbol{e}^{k+q}\right\|_2 &\le C\left\{\tau^p + \tau BL \sum_{j=0}^{k}\sum_{m=0}^{q} \left\|\boldsymbol{e}^{j+m}\right\|_2 + \tau \sum_{j=0}^{k} \left\|\boldsymbol{\mathcal{E}}^{j+q}[\boldsymbol{u}]\right\|_2\right\} \\
&\le C\left\{\tau^p + BL\tau \sum_{j=0}^{k}\sum_{m=0}^{q} \left\|\boldsymbol{e}^{j+m}\right\|_2 + \tau(k+1)C_1\tau^p\right\} \qquad (20.19)\\
&\le C\left\{\tau^p + BL\tau \sum_{j=0}^{k}\sum_{m=0}^{q} \left\|\boldsymbol{e}^{j+m}\right\|_2 + TC_1\tau^p\right\}
\end{aligned}$$

for all $k = 0, \dots, K-q$ and for all $0 < \tau \le \tau_1$. Now observe that

$$\begin{aligned}
\sum_{j=0}^{k}\sum_{m=0}^{q} \left\|\boldsymbol{e}^{j+m}\right\|_2 &= \sum_{m=0}^{q}\sum_{j=0}^{k} \left\|\boldsymbol{e}^{j+m}\right\|_2 \\
&\le (q+1)\sum_{j=0}^{k+q} \left\|\boldsymbol{e}^{j}\right\|_2 \\
&\le (q+1)\sum_{j=0}^{q-1} \left\|\boldsymbol{e}^{j}\right\|_2 + (q+1)\sum_{j=q}^{k+q} \left\|\boldsymbol{e}^{j}\right\|_2 \qquad (20.20)\\
&\le q(q+1)C\tau^p + (q+1)\sum_{j=q}^{k+q} \left\|\boldsymbol{e}^{j}\right\|_2 .
\end{aligned}$$

Combining estimates (20.19) and (20.20), we get

$$\begin{aligned}
\left\|\boldsymbol{e}^{k+q}\right\|_2 &\le C\left\{\tau^p + \bar{C}\tau^{p+1} + BL(q+1)\tau \sum_{j=q}^{k+q} \left\|\boldsymbol{e}^{j}\right\|_2 + T\bar{\bar{C}}\tau^p\right\} \\
&\le C\tau^p + \tau\tilde{C} \sum_{j=q}^{k+q} \left\|\boldsymbol{e}^{j}\right\|_2 .
\end{aligned}$$

Provided that τ is sufficiently small, in particular

$$0 < \tau\tilde{C} < 1,$$

we have, for all $k = 0, 1, \ldots, K - q$,

$$\left\|e^{k+q}\right\|_2 \leq \frac{C}{1-\tau\tilde{C}}\tau^p + \tau\frac{\tilde{C}}{1-\tau\tilde{C}}\sum_{j=q}^{k+q-1}\left\|e^j\right\|_2.$$

Notice that we have made the sum on the right-hand side explicit with respect to the left-hand side. Reindexing the summation, we have, for every $m = q, q+2, \ldots, K$,

$$\|e^m\|_2 \leq \frac{C}{1-\tau\tilde{C}}\tau^p + \tau\frac{\tilde{C}}{1-\tau\tilde{C}}\sum_{j=q}^{m-1}\left\|e^j\right\|_2.$$

If we further restrict the time step so that

$$0 < \tau\tilde{C} \leq \frac{1}{2},$$

then it follows that

$$\frac{1}{1-\tau\tilde{C}} \leq 2,$$

and, for every $m = q, q+2, \ldots, K$,

$$\|e^m\|_2 \leq 2\tilde{C}\tau^p + 2\tau\tilde{C}\sum_{j=q}^{m-1}\left\|e^j\right\|_2.$$

Applying Lemma 20.23, we have

$$\|e^m\|_2 \leq \left(2\tau\tilde{C}\left\|e^q\right\|_2 + 2C\tau^p\right)\exp\left[2\tau(m-q-1)\tilde{C}\right] \leq C\tau^p e^{2T\tilde{C}}.$$

The theorem is proven with $C_1 = Ce^{2T\tilde{C}}$. □

The reader will notice that, in the proof of Theorem 20.24, we only used the homogeneous zero stability of the method. In fact, we now have the tools to prove that, if $\boldsymbol{f}$ satisfies the usual global Lipschitz condition, the notions of zero stability and homogeneous zero stability are equivalent.

Corollary 20.25 (equivalence). *Let $K, q \in \mathbb{N}$ with $q < K$. Suppose that $\boldsymbol{f} \in \mathcal{F}^1(S)$, so that $\boldsymbol{u} \in C^2([0, T]; \mathbb{R}^d)$ satisfies the IVP (18.1). Let $\left\{\boldsymbol{w}^k\right\}_{k=0}^{K} \subset \mathbb{R}^d$ be an approximation generated by the linear q-step method (20.1) with the starting values $\left\{\boldsymbol{w}^k\right\}_{k=0}^{q-1} \subset \mathbb{R}^d$. The multi-step method is zero stable if and only if it is homogeneously zero stable.*

Proof. See Problem 20.12. □

20.6 Dahlquist Theorems

As a last topic in our discussion of the linear multi-step method, we present a series of results due to Dahlquist. The first one is known as the *Dahlquist Equivalence Theorem.* This essentially gives a converse to Theorem 20.24. The proof can be found in [32].

Theorem 20.26 (Dahlquist Equivalence Theorem [4]). *Let $p, K, q \in \mathbb{N}$ with $q < K$. Suppose that $\boldsymbol{f} \in \mathcal{F}^p(S)$, so that $\boldsymbol{u} \in C^{p+1}([0,T];\mathbb{R}^d)$ satisfies the IVP* (18.1). *Let $\{\boldsymbol{w}^k\}_{k=0}^K \subset \mathbb{R}^d$ be an approximation generated by the linear q-step method* (20.1) *with the starting values $\{\boldsymbol{w}^k\}_{k=0}^{q-1} \subset \mathbb{R}^d$. Suppose that the method satisfies the root condition. Then the multi-step method is consistent to order p if and only if it is globally convergent with order p.*

The next theorem is known as the *Dahlquist First Barrier Theorem*; see, for example, [12] or [32]. The result gives us a firm upper limit on the order of a *zero stable* multi-step method.

Theorem 20.27 (Dahlquist First Barrier Theorem). *The order of accuracy (consistency) of a zero stable linear q-step method* (20.1) *cannot exceed $q+1$ if q is odd or $q+2$ if q is even.*

Problems

20.1 Complete the proof of Corollary 20.6.

20.2 Show that the two-step (implicit) Adams–Moulton method,

$$\boldsymbol{w}^{k+2} - \boldsymbol{w}^{k+1} = \tau\left[\frac{5}{12}\boldsymbol{f}(t_{k+2}, \boldsymbol{w}^{k+2}) + \frac{8}{12}\boldsymbol{f}(t_{k+1}, \boldsymbol{w}^{k+1}) - \frac{1}{12}\boldsymbol{f}(t_k, \boldsymbol{w}^k)\right],$$

is at least order one using the conditions $\psi(1) = 0$ and $\psi'(1) - \chi(1) = 0$. Prove that, in fact, the method is exactly third order.

20.3 Show that the three-step (implicit) Adams–Moulton method,

$$\boldsymbol{w}^{k+3} - \boldsymbol{w}^{k+2} = \tau\left[\frac{9}{24}\boldsymbol{f}(t_{k+3}, \boldsymbol{w}^{k+3}) + \frac{19}{24}\boldsymbol{f}(t_{k+2}, \boldsymbol{w}^{k+2}) - \frac{5}{24}\boldsymbol{f}(t_{k+1}, \boldsymbol{w}^{k+1}) + \frac{1}{24}\boldsymbol{f}(t_k, \boldsymbol{w}^k)\right],$$

is at least order one using the conditions $\psi(1) = 0$ and $\psi'(1) - \chi(1) = 0$. Prove that, in fact, the method is exactly fourth order using both the method of C's and the log-method.

20.4 Find all of the values of α and β, so that the three-step method,

$$\boldsymbol{w}^{k+3} + \alpha(\boldsymbol{w}^{k+2} - \boldsymbol{w}^{k+1}) - \boldsymbol{w}^k = \tau\beta\left[\boldsymbol{f}(t_{k+2}, \boldsymbol{w}^{k+2}) + \boldsymbol{f}(t_{k+1}, \boldsymbol{w}^{k+1})\right],$$

is of order four.

20.5 Show that the BDF3 method,

$$\boldsymbol{w}^{k+3} - \frac{18}{11}\boldsymbol{w}^{k+2} + \frac{9}{11}\boldsymbol{w}^{k+1} - \frac{2}{11}\boldsymbol{w}^k = \frac{6}{11}\tau\boldsymbol{f}(t_{k+3}, \boldsymbol{w}^{k+3}),$$

is consistent to order $p = 3$ using both the method of C's and the log-method.

20.6 The Adams–Bashforth two-step method is given by

$$\boldsymbol{w}^{k+2} = \boldsymbol{w}^{k+1} + \tau\left[\frac{3}{2}\boldsymbol{f}(t_{k+1}, \boldsymbol{w}^{k+1}) - \frac{1}{2}\boldsymbol{f}(t_k, \boldsymbol{w}^k)\right].$$

[4] Named in honor of the Swedish mathematician Germund Dahlquist (1925–2005).

Derive this method by an integration procedure and give an exact expression for the local truncation error.

20.7 Derive the general form of a q-step BDF method using the method of C's.

20.8 Complete the proof of Theorem 20.19.

20.9 Show that the BDF3 method,

$$w^{k+3} - \frac{18}{11}w^{k+2} + \frac{9}{11}w^{k+1} - \frac{2}{11}w^k = \frac{6}{11}\tau f(t_{k+3}, w^{k+3}),$$

satisfies the root condition.

Hint: One of the roots is $w = 1$.

20.10 Consider the method

$$w^{k+2} - w^k = \frac{\tau}{3}\left[f(t_{k+2}, w^{k+2}) + 4f(t_{k+1}, w^{k+1}) + f(t_k, w^k)\right].$$

Show that it is of fourth order and it obeys the root condition.

20.11 Prove Lemma 20.23.

20.12 Prove Corollary 20.25.

20.13 Show that, for all the values of α and β that make the three-step method

$$w^{k+3} + \alpha(w^{k+2} - w^{k+1}) - w^k = \tau\beta\left[f(t_{k+2}, w^{k+2}) + f(t_{k+1}, w^{k+1})\right]$$

of order four, the resulting method does not satisfy the root condition and is, therefore, not convergent.

20.14 Show that a linear multi-step method is of order $p \geq 1$ if and only if it yields the exact solution to an ordinary differential equation problem whose solution is a polynomial of degree no greater than p.

20.15 Recall Simpson's quadrature rule:

$$\int_a^b f(x)\mathrm{d}x = \frac{b-a}{6}\left[f(a) + 4f\left(\frac{a+b}{2}\right) + f(b)\right] + E[f](a,b),$$

where $E[f](a,b)$ is an error term that satisfies

$$|E[f](a,b)| \leq C(b-a)^4$$

and $C > 0$ is a constant that depends on f. Starting from the identity

$$u(t_{k+1}) = u(t_{k-1}) + \int_{t_{k-1}}^{t_{k+1}} f(s, u(s))\mathrm{d}s,$$

use Simpson's rule to derive a two-step method. Determine its order and whether it is convergent.

20.16 Show that the method

$$w^{k+2} - 3w^{k+1} + 2w^k = \tau\left[\frac{13}{12}f(t_{k+2}, w^{k+2}) - \frac{5}{3}f(t_{k+1}, w^{k+1}) - \frac{5}{12}f(t_k, w^k)\right]$$

is of order two. However, this method does not converge. Why?

20.17 Show that the explicit multi-step method,

$$\begin{aligned} &w^{k+3} + \alpha_2 w^{k+2} + \alpha_1 w^{k+1} + \alpha_0 w^k \\ &\quad = \tau\left[\beta_2 f(t_{k+2}, w^{k+2}) + \beta_1 f(t_{k+1}, w^{k+1}) + \beta_0 f(t_k, w^k)\right], \end{aligned}$$

is fourth order only if $\alpha_0 + \alpha_2 = 8$ and $\alpha_1 = -9$. Prove that this method cannot be both fourth order and convergent.

20.18 Consider the method

$$w^{k+2} + w^{k+1} - 2w^k = \tau \left[f(t_{k+2}, w^{k+2}) + f(t_{k+1}, w^{k+1}) + f(t_k, w^k) \right].$$

What is the order of the method? Is it a convergent method?

20.19 Study the order and convergence of the method

$$w^{k+1} - w^k = \frac{\tau}{12} \left[5f(t_{k+1}, w^{k+1}) + 8f(t_k, w^k) - f(t_{k-1}, w^{k-1}) \right].$$

Listings

```
function res = MethodCs( a, b, m )
% The method of Cs to determine the order of consistency of a
% linear multistep method.
%
% Input
% a : The coefficients of the first characteristic polynomial
% b : The coefficients of the second characteristic polynomial
% m : The number of C that one wants to compute
%
% Output
% res: The number C_m. A method is consistent to order exactly
%        p if C_0 = ... = C_p = 0, but C_{p+1} != 0
  if m == 0
    res = sum( a );
  else
    res = 0.;
    q = length(a);
    factmminusone = factorial(m-1);
    factm = m*factmminusone;
    for j=1:q
      res = res + a(j)*(j-1)^m/factm - b(j)*(j-1)^(m-1) ...
        /factmminusone;
    end
  end
end
```

Listing 20.1 Algorithmic description of the method of C's.

21 Stiff Systems of Ordinary Differential Equations and Linear Stability

In Section 19.4, we studied the approximation of dissipative equations for which, as we know, all solutions are Lyapunov stable; see Example 17.3. In this chapter, we continue with the topic of stability of solutions, and try to understand which methods are suited to approximate Lyapunov stable solutions. As the discussion of Section 17.5 shows, it is sufficient to study linear equations and the stability of the origin. With this in mind, we begin by presenting an example from [47].

Example 21.1 Suppose that $\boldsymbol{u}'(t) = \mathsf{A}\boldsymbol{u}(t)$ for $t \in [0, T]$ with $\boldsymbol{u}(0) = \boldsymbol{u}_0$, where

$$\mathsf{A} = \begin{bmatrix} -100 & 1 \\ 0 & -\frac{1}{10} \end{bmatrix}.$$

As Theorem 17.23 shows, the origin is stable for this problem. Let us apply the forward Euler method to approximate the solutions. Then

$$\boldsymbol{w}^k = (\mathsf{I} + \tau\mathsf{A})^k \boldsymbol{u}_0, \quad k = 0, 1, \dots, K,$$

where $\tau = T/K$. Now observe that we have the diagonalization $\mathsf{A} = \mathsf{X}\mathsf{D}\mathsf{X}^{-1}$, where

$$\mathsf{X} = \begin{bmatrix} 1 & 1 \\ 0 & \frac{999}{10} \end{bmatrix}, \quad \mathsf{D} = \begin{bmatrix} -100 & 0 \\ 0 & -\frac{1}{10} \end{bmatrix}.$$

The solution to the initial value problem (IVP) is

$$\boldsymbol{u}(t) = \mathsf{X}e^{t\mathsf{D}}\mathsf{X}^{-1}\boldsymbol{u}_0,$$

where

$$e^{t\mathsf{D}} = \begin{bmatrix} e^{-100t} & 0 \\ 0 & e^{-\frac{1}{10}t} \end{bmatrix}.$$

Decomposing $\boldsymbol{u}_0$ in the basis of eigenvectors (columns of X), we have

$$\boldsymbol{u}_0 = \alpha_1\boldsymbol{x}_1 + \alpha_2\boldsymbol{x}_2, \quad \boldsymbol{x}_1 = \begin{bmatrix} 1 \\ 0 \end{bmatrix}, \quad \boldsymbol{x}_2 = \begin{bmatrix} 1 \\ \frac{999}{10} \end{bmatrix},$$

for some $\alpha_1, \alpha_2 \in \mathbb{R}$. Thus,

$$\boldsymbol{u}(t) = \alpha_1 e^{-100t}\boldsymbol{x}_1 + \alpha_2 e^{-\frac{1}{10}t}\boldsymbol{x}_2.$$

Meanwhile,

$$\boldsymbol{w}^k = \mathsf{X}(\mathsf{I} + \tau\mathsf{D})^k \mathsf{X}^{-1}\mathsf{X}\boldsymbol{\alpha} = \alpha_1 (1 - 100\tau)^k \boldsymbol{x}_1 + \alpha_2 \left(1 - \frac{1}{10}\tau\right)^k \boldsymbol{x}_2. \quad (21.1)$$

Suppose that $\tau > \frac{1}{50}$. Then

$$100\tau > 2 \iff 100\tau - 1 > 1 \implies |1 - 100\tau| > 1,$$

and the first term in the forward Euler approximation blows up rapidly.

Now, suppose that $\boldsymbol{u}_0$ is the eigenvector associated with the eigenvalue $\lambda_2 = -\frac{1}{10}$:

$$\boldsymbol{u}_0 = \begin{bmatrix} 1 \\ \frac{999}{10} \end{bmatrix} = \boldsymbol{x}_2.$$

In this case, the solution satisfies

$$\boldsymbol{u}(t) = e^{-\frac{1}{10}t}\boldsymbol{x}_2 \to \boldsymbol{0},$$

as $t \to \infty$. Likewise, as $k \to \infty$,

$$\boldsymbol{w}^k = \left(1 - \frac{1}{10}\tau\right)^k \boldsymbol{x}_2 \to \boldsymbol{0},$$

provided that

$$\left|1 - \frac{1}{10}\tau\right| < 1 \iff 0 < \tau < 20.$$

All seems well, but there is a problem. Suppose that

$$\frac{1}{50} < \tau < 20,$$

and in the general numerical approximation (21.1) $\alpha_2 = 1$, but $\alpha_1 = \varepsilon \in (0, 1)$, with $\varepsilon \ll 1$. This corresponds to the case that $\boldsymbol{u}_0 \approx \boldsymbol{x}_2$, where the ε represents a small error.

Roundoff errors will guarantee that $\varepsilon(1 - 100\tau)^k$ will eventually dominate, regardless of the smallness of ε. So, while we get the correct asymptotic behavior for the second term, the present method does not capture the correct behavior for the first, unless the time step τ is sufficiently small.

Example 21.2 Not all methods behave in this way. For example, applying the trapezoidal rule to approximate solutions to the problem in the previous example, one obtains

$$\boldsymbol{w}^k = \alpha_1 \left(\frac{1 - 50\tau}{1 + 50\tau}\right)^k \boldsymbol{x}_1 + \left(\frac{1 - \frac{1}{20}\tau}{1 + \frac{1}{20}\tau}\right)^k \boldsymbol{x}_2.$$

For every $\tau > 0$,

$$\max\left\{\left|\frac{1 - 50\tau}{1 + 50\tau}\right|, \left|\frac{1 - \frac{1}{20}\tau}{1 + \frac{1}{20}\tau}\right|\right\} < 1,$$

and the correct behavior is predicted; namely, the approximation decays to $\boldsymbol{0}$ as $k \to \infty$.

Definition 21.1 (stiffness)**.** Suppose that the matrix $\mathsf{A} \in \mathbb{R}^{d\times d}$ is diagonalizable, with spectrum $\sigma(\mathsf{A}) = \{\lambda_1, \dots, \lambda_d\} \subset \mathbb{C}$. Assume that

$$\Re\lambda_j < 0, \quad j = 1, \dots, d,$$

and, more specifically, that the eigenvalues are ordered such that

$$0 > \Re\lambda_1 \geq \Re\lambda_2 \geq \cdots \geq \Re\lambda_{d-1} \geq \Re\lambda_d.$$

Let $\boldsymbol{g} \in C\left(\mathbb{R}; \mathbb{R}^d\right)$. The system of ordinary differential equations (ODEs)

$$\boldsymbol{u}'(t) = \mathsf{A}\boldsymbol{u}(t) + \boldsymbol{g}(t),$$

which is called a **linearly dissipative system**, is said to have **stiffness ratio**

$$Q = \frac{-\Re\lambda_d}{-\Re\lambda_1} \geq 1.$$

Solutions of a linear dissipative system are easily expressed. Suppose that the linearly independent eigenpairs are denoted $(\lambda_i, \boldsymbol{x}_i)$ for $i = 1, \dots, d$. Then the general solution has the form

$$\boldsymbol{u}(t) = \sum_{i=1}^{d} \alpha_i e^{\lambda_i t} \boldsymbol{x}_i + \boldsymbol{q}(t),$$

where the α_i are constants and the function $\boldsymbol{q}$ is determined by an integral. If $\boldsymbol{g} \equiv \boldsymbol{0}$, then $\boldsymbol{q} \equiv \boldsymbol{0}$; otherwise, it is not important to know the precise form of $\boldsymbol{q}$. Now, since $\Re\lambda_j < 0$ for all $j = 1, \dots, d$, each term $e^{\lambda_i t}$ decays to zero. Thus, the solution tends asymptotically to $\boldsymbol{q}(t)$ as $t \to \infty$. We call the sum $\sum_{i=1}^{d} \alpha_i e^{\lambda_i t} \boldsymbol{x}_i$ the **dissipative part**, or the **transient part**, of the solution.

The idea is basically the same as before. If Q is large, then there is a good chance that the forward Euler method, say, will not accurately capture the dissipative nature of each exponentially decaying term. In other words, our numerical method will fail to predict that the solution should tend to $\boldsymbol{q}(t)$ as $t \to \infty$, which is a big problem.

Remark 21.2 (stiff systems)**.** Loosely, we say that $\boldsymbol{u}'(t) = \mathsf{A}\boldsymbol{u}(t) + \boldsymbol{g}(t)$ is **stiff** when $Q \gg 1$. The greater the value of Q, the greater the risk that we will fail to capture the correct asymptotic decay of the dissipative part of the true solution, especially the rapidly decaying terms, as $t \to \infty$.

Remark 21.3 (local stiffness)**.** For more general dissipative systems

$$\boldsymbol{u}'(t) = \boldsymbol{f}(t, \boldsymbol{u}(t)),$$

we can examine stiffness locally, if not globally, by considering the eigenvalues of the matrix $D_{\boldsymbol{u}}\boldsymbol{f} = \left[\partial_{u_j} f_i\right]$. Assuming that the eigenvalues all have negative real parts, etc., if the ratio Q is large at some point, then we say that the equation is *locally stiff* in a neighborhood of that point.

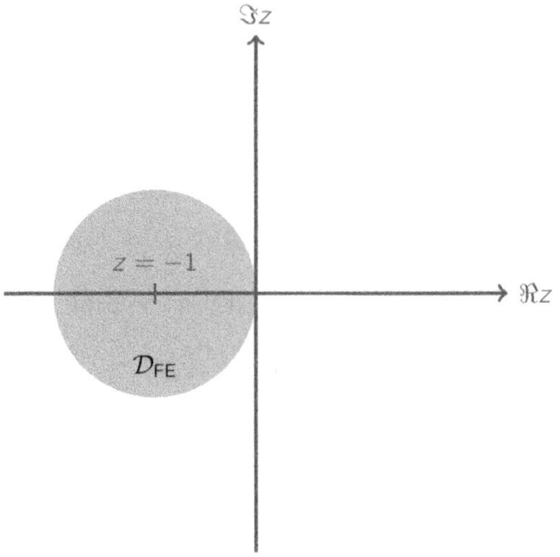

Figure 21.1 Linear stability domain $\mathcal{D}_{\mathrm{FE}}$ for the forward Euler method. This method is not A-stable.

21.1 The Linear Stability Domain and A-Stability

Definition 21.4 (linear stability domain)**.** Let $\lambda \in \mathbb{C}$. Consider the IVP

$$u'(t) = \lambda u(t), \qquad u(0) = 1. \tag{21.2}$$

Suppose that $\{w^k\}_{k=0}^{\infty}$ is a numerical approximation of u generated with the time step size $\tau > 0$ by a single-step or Runge–Kutta (RK) method, with $w^0 = 1$, or a linear q-step method, with consistent starting values, such as $w^j = 1$ for $j = 0, \ldots, q-1$. The **linear stability domain** of this method is the set

$$\mathcal{D} = \left\{ z = \lambda\tau \in \mathbb{C} \;\middle|\; w^k \to 0,\ k \to \infty \right\}.$$

The approximation method is called **A-stable** if and only if

$$\mathbb{C}^- = \{z \in \mathbb{C} \mid \Re(z) < 0\} \subseteq \mathcal{D}.$$

In words, a method is A-stable if and only if its linear stability domain contains the left-half complex plane.

Remark 21.5 (history and context)**.** The term A-stable is due to the Swedish mathematician Germund Dahlquist (1925–2005). Some authors use the term linearly stable. If a method is A-stable, then it may be a good choice for approximating the solutions of a simple linearly dissipative system with a large stiffness ratio. This is due to the fact that the method will accurately predict the asymptotic decay of each term of the dissipative part of the solution.

Example 21.3 The forward Euler method is not A-stable; therefore, it would not be an appropriate choice for a stiff system. To see this, we compute its linear stability domain. The forward Euler approximation applied to (21.2) yields

$$w^k = g^k(\lambda\tau), \quad g(z) = 1 + z.$$

Clearly, $w^k \to 0$ as $k \to \infty$ if and only if $|1 + \lambda\tau| < 1$. Thus,

$$\mathcal{D}_{\mathrm{FE}} = \{z \in \mathbb{C} \mid |1 + z| < 1\},$$

which is a unit disk centered at $z = -1$. The function g is called the *amplification factor*; see Figure 21.1.

Example 21.4 The backward Euler method is A-stable. Its linear stability domain is

$$\mathcal{D}_{\mathrm{BE}} = \{z \in \mathbb{C} \mid |1 - z| > 1\},$$

as the reader can show; see Problem 21.2. Thus, $\mathcal{D}_{\mathrm{BE}}$ contains every point $z \in \mathbb{C}$ that is outside of the unit disk centered at $z = 1$. In particular, $\mathcal{D}_{\mathrm{BE}}$ contains $\mathbb{C}^-$; therefore, the backward Euler method is A-stable.

Example 21.5 The trapezoidal rule is A-stable. Its linear stability domain is

$$\mathcal{D}_{\mathrm{TR}} = \left\{ z \in \mathbb{C} \;\middle|\; \frac{\left|1 + \frac{z}{2}\right|}{\left|1 - \frac{z}{2}\right|} < 1 \right\}.$$

The reader should prove that $\mathcal{D}_{\mathrm{TR}} = \mathbb{C}^-$, and, therefore, that the trapezoidal rule is A-stable.

21.2 A-Stability of Runge–Kutta Methods

Let us now study the A-stability of RK methods. We begin with an impossibility result.

Theorem 21.6 (impossibility). *No consistent explicit RK is A-stable.*

Proof. This follows from Theorem 19.5. In particular, applying an explicit r-stage RK method to approximate the solution to the IVP (21.2), one obtains

$$w^k = g^k(\lambda\tau),$$

where the amplification factor g is a polynomial of degree at most r. Clearly, there is a point $z \in \mathbb{C}^-$ such that $|g(z)| > 1$. At all such points, $|g^k(z)| = |g(z)|^k \to \infty$ as $k \to \infty$. □

The situation for implicit RK methods is a bit more complicated. Many implicit RK methods are A-stable, but some are not. The proof of the following should be clear.

Theorem 21.7 (linear stability criterion). *Suppose that an implicit r-stage RK method is applied to approximate the solutions of the IVP* (21.2). *Then the method is A-stable if and only if $|g(z)| < 1$ for all $z \in \mathbb{C}^-$, where the amplification factor g is the rational polynomial from Theorem 19.6.*

Proof. See Problem 21.4. □

Example 21.6 The implicit RK methods given by the tables

$$\begin{array}{c|cc} 0 & \frac{1}{4} & -\frac{1}{4} \\ \frac{2}{3} & \frac{1}{4} & \frac{5}{12} \\ \hline & \frac{1}{4} & \frac{3}{4} \end{array} \qquad\qquad \begin{array}{c|cc} \frac{1}{3} & \frac{5}{12} & -\frac{1}{12} \\ 1 & \frac{3}{4} & \frac{1}{4} \\ \hline & \frac{3}{4} & \frac{1}{4} \end{array}$$

are both A-stable and both order three. The reader can prove the A-stability by considering the amplification factors g. In this case, the amplification factors are the same rational polynomial:

$$g(z) = \frac{1 + \frac{1}{3}z}{1 - \frac{2}{3}z + \frac{1}{6}z^2}.$$

It can be a rather difficult task to prove (or disprove) that $|g(z)| < 1$ for all $z \in \mathbb{C}^-$. The next theorem can make the task considerably simpler.

Theorem 21.8 (A-stability criterion). *Let g be a rational polynomial that is not a constant. Then*

$$|g(z)| < 1, \quad \forall z \in \mathbb{C}^- \tag{21.3}$$

if and only if all of the poles of g have positive real parts and $|g(\mathrm{i}t)| \leq 1$ for all $t \in \mathbb{R}$.

Proof. ($\Longrightarrow$) By continuity, $|g(z)| \leq 1$ for all $z \in \overline{\mathbb{C}^-}$. Thus, there are no poles (singular points) in $\overline{\mathbb{C}^-}$ and $|g(\mathrm{i}t)| \leq 1$ for all $t \in \mathbb{R}$.

($\Longleftarrow$) If all of the poles of g are in the right-half complex plane,

$$\mathbb{C}^+ = \{z \in \mathbb{C} \mid \Re z > 0\},$$

then g is analytic in $\overline{\mathbb{C}^-}$. If it is nonconstant, then, by the Maximum Modulus Principle (see Theorem 9.33), its maximum in modulus in $\overline{\mathbb{C}^-}$ occurs on the boundary, namely the imaginary axis. Therefore, $|g(\mathrm{i}t)| \leq 1$ for all $t \in \mathbb{R}$ implies that $|g(z)| < 1$ for all $z \in \mathbb{C}^-$. □

Definition 21.9 (A-acceptable). A rational polynomial r is called **A-acceptable** if and only if it satisfies (21.3). Such rational polynomials are associated with A-stable implicit Runge–Kutta (IRK) methods. In particular, if the amplification factor g for an IRK method is A-acceptable, we call that method **A-acceptable**.

Example 21.7 Let us apply Theorem 21.8 to investigate the A-stability of the IRK methods from Example 21.6, whose amplification factors are

$$g(z) = \frac{1 + \frac{1}{3}z}{1 - \frac{2}{3}z + \frac{1}{6}z^2}.$$

We first find its poles, i.e., the roots of the denominator. They are

$$z_{1,2} = \frac{\frac{2}{3} \pm \sqrt{\frac{4}{9} - 4 \cdot 1 \cdot \frac{1}{6}}}{\frac{1}{3}} = 2 \pm \sqrt{2}\mathrm{i},$$

and we see that they have positive real part.

On the other hand, $|g(\mathrm{i}t)| \leq 1$ is equivalent to requiring that

$$|1 + \tfrac{1}{3}\mathrm{i}t|^2 \leq |1 - \tfrac{2}{3}\mathrm{i}t - \tfrac{1}{6}t^2|^2,$$

or that

$$1 + \frac{1}{9}t^2 \leq 1 + \frac{1}{9}t^2 + \frac{1}{36}t^4,$$

which is always true and we, again, conclude that the method is A-stable.

Example 21.8 For the two-step Gauss–Legendre–RK method, which is consistent to exactly order $p = 4$, the amplification factor is

$$g(z) = \frac{1 + \frac{1}{2}z + \frac{1}{12}z^2}{1 - \frac{1}{2}z + \frac{1}{12}z^2}.$$

Using Theorem 21.8, it is easy to see that the method is A-stable/A-acceptable. The details are left to the reader as an exercise; see Problem 21.7.

To find rational polynomials that are A-acceptable, we need the following result.

Lemma 21.10 (approximate exponential). *Let g be the amplification factor for an IRK method of order p, then there are constants $C_0, C_1 > 0$ such that, for all $z \in \mathbb{C}$, with $|z| \leq C_0$,*

$$|g(z) - e^z| \leq C_1|z^{p+1}|.$$

As a shorthand, we write $g(z) = e^z + \mathcal{O}(z^{p+1})$ as $z \to 0$.

Proof. This comes precisely from the definition of order. Since for the linear problem (21.2), the solution is $u(t_{k+1}) = e^{\lambda\tau}u(t_k)$ and the method gives $w^{k+1} = g(\lambda\tau)w^k$. □

We introduce one final definition.

Definition 21.11 (exponential order). Let g be a nonconstant rational polynomial. If $g(z) = e^z + \mathcal{O}(z^{p+1})$ as $z \to 0$, then we say that g is of **exponential order** p.

The following results essentially narrow down the possible A-acceptable, exponential order p rational polynomials. We will denote, for $m, n \in \mathbb{N}_0$, by $\mathbb{P}_{m/n}$ the set of rational polynomials $q(t) = p(t)/r(t)$ whose numerator $p \in \mathbb{P}_m$ and denominator $r \in \mathbb{P}_n$.

Theorem 21.12 (Padé approximation). *For every $\alpha, \beta \in \mathbb{N}_0$, there is a unique $g_{\alpha/\beta} \in \mathbb{P}_{\alpha/\beta}$ that is of exponential order $\alpha + \beta$. Moreover, $g_{\alpha/\beta} = p_{\alpha/\beta}/q_{\alpha/\beta}$, where*

$$p_{\alpha/\beta}(z) = \sum_{k=0}^{\alpha} \binom{\alpha}{k} \frac{(\alpha+\beta-k)!}{(\alpha+\beta)!} z^k, \qquad q_{\alpha/\beta}(z) = p_{\beta/\alpha}(-z).$$

Note that $g_{\alpha/\beta}(0) = 1$. This $g_{\alpha/\beta}$ is the only element in $\mathbb{P}_{\alpha/\beta}$ that is of exponential order $\alpha + \beta$ and no function in $\mathbb{P}_{\alpha/\beta}$ exceeds this order.

Definition 21.13 (Padé approximation[1]). The functions $g_{\alpha/\beta} \in \mathbb{P}_{\alpha/\beta}$ of the previous theorem are called **Padé approximations** of the exponential.

The following result elucidates when a Padé approximation is A-acceptable.

Theorem 21.14 (Wanner–Hairer–Nørsett Theorem[2]). *The Padé approximation of the exponential $p_{\alpha/\beta}$ is A-acceptable if and only if $\alpha \le \beta \le \alpha + 2$.*

With the help of this result, we can show that Gauss–Legendre methods are not only of maximal order, but also A-stable.

Corollary 21.15 (A-stability). *Every r-stage implicit Gauss–Legendre–RK method is A-stable.*

Proof. For a method with r stages, Theorem 19.6 shows that the associated amplification factor $g \in \mathbb{P}_{r/r}$. On the other hand, a Gauss–Legendre method has order $2r$. This implies that g is of order $2r$. The uniqueness then implies that $g = g_{r/r}$, the Padé approximation of the exponential. Finally, since $\alpha = \beta = r$, the Padé approximation is A-acceptable. □

Remark 21.16 (alternate proof). Theorem 19.23 furnishes another proof of the fact that every r-stage implicit Gauss–Legendre–RK method is A-stable. This follows because the linear problem (21.2) with $\Re\lambda < 0$ is a dissipative ODE system. In fact, every dissipative numerical method is A-stable.

[1] Named in honor of the French mathematician Henri Eugène Padé (1863–1953).

[2] Named in honor of the Austrian mathematician Gerhard Wanner (1942–), the Austrian mathematician Ernst Hairer (1949–), and the Norwegian mathematician Syvert Paul Nørsett (1944–).

21.3 A-Stability of Linear Multi-step Methods

We now study the stability of linear multi-step methods. The situation here is more delicate since, to start a q-step method, we need q initial values $\{w^k\}_{k=0}^{q-1}$. From the problem, we only have the initial condition u_0, so it makes sense to set $w^0 = u_0$. However, there is no natural way to set the remaining starting values. For this reason, for A-stability, we need to require that the method produces the correct asymptotic behavior *for all* starting values $\{w^k\}_{k=0}^{q-1}$.

Theorem 21.17 (linear stability criterion). *Consider the linear q-step method given by* (20.1) *with the associated first and second characteristic polynomials*

$$\psi(\mu) = \sum_{j=0}^{q} a_j \mu^j \in \mathbb{P}_q, \quad \chi(\mu) = \sum_{j=0}^{q} b_j \mu^j \in \mathbb{P}_q.$$

Define the parameterized stability polynomial

$$\rho(z, \mu) = \psi(\mu) - z\chi(\mu).$$

Suppose that, at $z \in \mathbb{C}$, $a_q \neq z b_q$ and

$$\rho(z, \mu) = c(z) \prod_{i=1}^{n(z)} (\mu - \omega_i(z))^{m_i(z)}, \quad \sum_{i=1}^{n(z)} m_i(z) = q.$$

Then $z \in \mathcal{D}$, i.e., z is a point of linear stability if and only if $|\omega_i(z)| < 1$ for all $i = 1, \ldots, n(z)$.

Proof. Applying the approximation method to (21.2) and setting $z = \lambda\tau$, we have

$$\sum_{j=0}^{q} (a_j - z b_j) w^{k+j} = 0, \quad k = 0, 1, \ldots. \tag{21.4}$$

We have already constructed solutions of this equation in Section 20.4. In particular, the solution of (21.4) is given by

$$w^k = \sum_{j=1}^{n(z)} p_j(k) \omega_j^k(z),$$

where $p_j \in \mathbb{P}_{m_j(z)-1}$. In particular,

$$p_j(\zeta) = \sum_{i=0}^{m_j(z)-1} \alpha_{j,i} \zeta^i$$

for some coefficients $\alpha_{j,i} \in \mathbb{C}$, q a number, that are uniquely determined by the q starting values $w^0, \ldots, w^{q-1}$. Clearly, $w^k \to 0$ if and only if $|\omega_i(z)| < 1$ for all $i = 1, \ldots, n(z)$. □

Remark 21.18 (degeneracies). We point out that it is possible for degeneracies to arise. This happens, in particular, when $a_q = z b_q$; consequently,

$$\sum_{i=1}^{n(z)} m_i(z) < q.$$

Special care will be required at such points of degeneracy.

Corollary 21.19 (A-stability). *The linear q-step method* (20.1) *is A-stable if and only if*

$$|\omega_i(z)| < 1$$

for all $i = 1, \dots, n(z)$ *and all* $z \in \mathbb{C}^-$.

Theorem 21.20 guarantees that no explicit linear multi-step method is A-stable and it simplifies the conditions on the roots of the parameterized polynomial ρ, but only slightly.

Theorem 21.20 (A-stability criterion). *A linear q-step method* (20.1) *is A-stable if and only if* $b_q > 0$ *and* $|\omega_j(\mathrm{i}t)| \leq 1$ *for all* $j = 1, \dots, n(z)$ *and all* $t \in \mathbb{R}$.

Proof. The proof is similar to that of Theorem 21.8; see the books by Butcher [12] or Gautschi [32] for the details. □

The following general and powerful theorem is called the *Dahlquist Second Barrier Theorem.*

Theorem 21.21 (Dahlquist Second Barrier Theorem[3]). *Regarding q–step methods, we have that:*

1. *No explicit method is A-stable.*
2. *No A-stable method can have order (of consistency) greater than* $p = 2$.
3. *The second-order A-stable method with the smallest local truncation error is the trapezoidal rule.*

Proof. See the books by Butcher [12] or Gautschi [32] for a proof. □

21.4 The Boundary Locus Method

The *boundary locus method* is a tool for determining the linear stability domain for linear multi-step methods. We present the boundary locus method without thorough theoretical justification, though the details can be made rigorous. The general idea is to approximate the boundary of the linear stability domain, $\partial\mathcal{D}$, rather than $\mathcal{D}$ itself.

Suppose that $z \in \partial\mathcal{D}$. Then, by continuity, it is reasonable to assume that one of the roots of the stability polynomial $\rho(z, \cdot)$ satisfies $\omega(z) = e^{\mathrm{i}\theta}$, for some $\theta \in (0, 2\pi]$. In other words,

$$\rho(z, e^{\mathrm{i}\theta}) = 0.$$

Equivalently, provided that $\chi(e^{\mathrm{i}\theta}) \neq 0$,

$$z = \frac{\psi(e^{\mathrm{i}\theta})}{\chi(e^{\mathrm{i}\theta})},$$

which gives an explicit parameterization of the boundary, with respect to θ.

[3] Named in honor of the Swedish mathematician Germund Dahlquist (1925–2005).

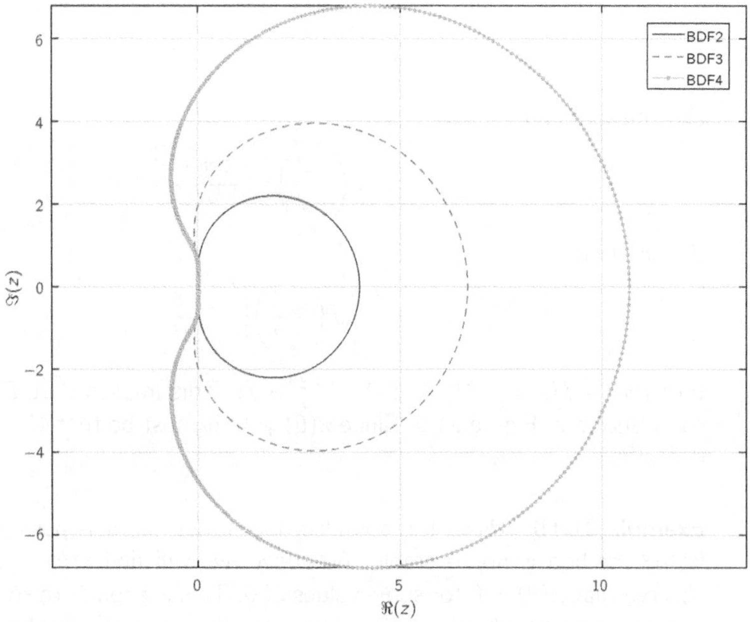

Figure 21.2 The boundary of the linear stability domain, $\partial\mathcal{D}_{\mathrm{BDF}q}$, for the BDF$q$ method, $q = 2, 3, 4$.

Example 21.9 Let us use the boundary locus method to find the linear stability domain of the two-step Backward Differentiation Formula (BDF2) method and, in doing so, show that the method is A-stable. For this method, we have the following first and second characteristic polynomials:

$$\psi(\mu) = \mu^2 - \frac{4}{3}\mu + \frac{1}{3}, \quad \chi(\mu) = \frac{2}{3}\mu^2.$$

Therefore, $z \in \partial\mathcal{D}$ if and only if

$$z = \frac{e^{i2\theta} - \frac{4}{3}e^{i\theta} + \frac{1}{3}}{\frac{2}{3}e^{i2\theta}} = \frac{3}{2} - 2e^{-i\theta} + \frac{1}{2}e^{-i2\theta}.$$

Now, setting $z(\theta) = x(\theta) + iy(\theta)$, we have

$$x(\theta) = \frac{3}{2} - 2\cos(\theta) + \frac{1}{2}\cos(2\theta), \quad y(\theta) = 2\sin(\theta) - \frac{1}{2}\sin(2\theta).$$

The set $\partial\mathcal{D}$ is plotted in Figure 21.2. This was obtained with the help of Listing 21.1. It is straightforward to show that $x(\theta) \geq 0$ for all $\theta \in (0, 2\pi]$. Furthermore, by checking a single point off of the boundary — as long as it is not a point of degeneracy (at which $a_q = zb_q$) — we can determine whether $\mathcal{D}$ is "inside" or "outside" of the curve. Checking the outside, regular point $z = -\frac{3}{2}$, we find

$$\rho\left(-\frac{3}{2}, \mu\right) = 2\mu^2 - \frac{4}{3}\mu + \frac{1}{3},$$

and the roots are

$$\omega\left(-\frac{3}{2}\right) = \frac{1 \pm i\frac{\sqrt{8}}{4}}{3}.$$

We see that

$$\left|\omega\left(-\frac{3}{2}\right)\right|^2 = \frac{3}{18};$$

therefore, $\left|\omega\left(z = -\frac{3}{2}\right)\right| < 1$. So $-\frac{3}{2} \in \mathcal{D}$. This implies that $\mathcal{D}$ is outside of the curve shown in Figure 21.2. Since $x(\theta) \geq 0$, it must be that $\mathbb{C}^- \subset \mathcal{D}$. The method is A-stable.

Example 21.10 Using the boundary locus method, it is possible to see that the BDF3 method is not A-stable. As above, we will find $z(\theta) = x(\theta) + iy(\theta)$, but observe that $x(\theta) < 0$ for some values of θ. This is enough to show that the BDF3 method cannot be A-stable. This fact is confirmed by the Dahlquist Second Barrier Theorem, since the BDF3 method is of order three.

Remark 21.22 (degeneracy)**.** Note that, for the BDF2 method, $z = \frac{3}{2}$ is a point of degeneracy:

$$\rho\left(\frac{3}{2}, \mu\right) = -\frac{4}{3}\left(\mu - \frac{1}{4}\right).$$

The point $z = \frac{3}{2}$ is inside the curve in Figure 21.2. It appears that the single root $\omega\left(z = \frac{3}{2}\right) = \frac{1}{4}$, which is clearly less than one in modulus, indicates a point of stability. In fact, the second root at this point is at ∞, and this is not a point of linear stability.

To see what is going on, suppose that we check the regular point $z = \frac{3}{2} - \frac{3}{2}\varepsilon$, where $0 < \varepsilon < 1$. Then

$$\rho(z, \mu) = \varepsilon\mu^2 - \frac{4}{3}\mu + \frac{1}{3}$$

and

$$\omega(z) = \frac{2 \pm \sqrt{4 - 3\varepsilon}}{3\varepsilon}.$$

Then note, as $\varepsilon \to 0$,

$$\frac{2 + \sqrt{4 - 3\varepsilon}}{3\varepsilon} \to \infty, \qquad \frac{2 - \sqrt{4 - 3\varepsilon}}{3\varepsilon} \to \frac{1}{4}.$$

Problems

21.1 Suppose that the matrix $\mathsf{A} \in \mathbb{R}^{d\times d}$ is diagonalizable, with spectrum $\sigma(\mathsf{A}) = \{\lambda_1, \ldots, \lambda_d\} \subset \mathbb{C}$. Assume that

$$\Re\lambda_j < 0, \quad j = 1, \ldots, d,$$

and, more specifically, that the eigenvalues are ordered such that

$$0 > \Re\lambda_1 \geq \Re\lambda_2 \geq \cdots \geq \Re\lambda_{d-1} \geq \Re\lambda_d.$$

Let $\boldsymbol{g} \in C(\mathbb{R}; \mathbb{R}^d)$ be a given function of time.

a) Prove that the general solution to the linear dissipative system of ordinary differential equations,

$$\boldsymbol{u}'(t) = \mathsf{A}\boldsymbol{u}(t) + \boldsymbol{g}(t),$$

is given by

$$\boldsymbol{u}(t) = \sum_{i=1}^{d} \alpha_i e^{\lambda_i t} \boldsymbol{x}_i + \boldsymbol{q}(t),$$

where $\{\boldsymbol{x}_1, \ldots, \boldsymbol{x}_d\}$ is the linearly independent set of corresponding eigenvectors, the α_i are constants, and $\boldsymbol{q}$ is some function of time. Find the form of the function $\boldsymbol{q}$.

b) Apply the forward Euler method to approximate the solutions of the linearly dissipative system. For what value of the time step size, $\tau > 0$, does the numerical approximation accurately capture the asymptotic behavior of the dissipative part of the solution?

c) Apply the backward Euler method to approximate the solutions of the linearly dissipative system. Show that, for any value of the step size, $\tau > 0$, the numerical approximation correctly captures the asymptotic behavior of the solution.

21.2 Compute the linear stability domains of the backward Euler method.

21.3 For the trapezoidal rule, show that the linear stability domain is precisely equal to the left-half complex plane

$$\mathbb{C}^- = \{z \in \mathbb{C} \mid \Re z < 0\}.$$

21.4 Prove Theorem 21.7.

21.5 Complete the details of Example 21.6.

21.6 Consider the implicit RK methods given by the tables

$$\begin{array}{c|cc} 0 & \frac{1}{4} & -\frac{1}{4} \\ \frac{2}{3} & \frac{1}{4} & \frac{5}{12} \\ \hline & \frac{1}{4} & \frac{3}{4} \end{array} \qquad \begin{array}{c|cc} \frac{1}{3} & \frac{5}{12} & -\frac{1}{12} \\ 1 & \frac{3}{4} & \frac{1}{4} \\ \hline & \frac{3}{4} & \frac{1}{4} \end{array}.$$

a) Show that the necessary conditions of Theorem 19.4 are all satisfied.

b) Show that these methods are both A-stable.

c) Show that one of the methods is a collocation method and that the other is not. For the one that is a collocation method, find its order of consistency.

21.7 Provide all the details of Example 21.8.

21.8 Use the boundary locus method to prove that BDF2 is A-stable.

21.9 If you have not already done so, prove that the BDF3 method,

$$\boldsymbol{w}^{k+3} - \frac{18}{11}\boldsymbol{w}^{k+2} + \frac{9}{11}\boldsymbol{w}^{k+1} - \frac{2}{11}\boldsymbol{w}^{k} = \frac{6}{11}\tau \boldsymbol{f}(t_{k+3}, \boldsymbol{w}^{k+3}),$$

satisfies the root condition and is of order three. Conclude, therefore, that it must be a convergent method. Use the boundary locus method to prove that BDF3 cannot be A-stable.

21.10 Using the boundary locus method, determine all the values of $\theta \in [0, 1]$ for which the θ-method,

$$\boldsymbol{w}^{k+1} = \boldsymbol{w}^{k} + \tau\left[\theta \boldsymbol{f}(t_k, \boldsymbol{w}^k) + (1-\theta)\boldsymbol{f}(t_{k+1}, \boldsymbol{w}^{k+1})\right],$$

is A-stable.

21.11 Consider the implicit method

$$\boldsymbol{w}^{k+2} - \boldsymbol{w}^{k} = \frac{2}{3}\tau\left[\boldsymbol{f}(t_{k+2}, \boldsymbol{w}^{k+2}) + \boldsymbol{f}(t_{k+1}, \boldsymbol{w}^{k+1}) + \boldsymbol{f}(t_k, \boldsymbol{w}^k)\right].$$

What is the order of the method? Is it a convergent method? (This method is A-stable, though it is a bit difficult to prove. Can you prove it?)

21.12 If you have not already done so, prove that the method

$$\boldsymbol{w}^{k+2} - \boldsymbol{w}^{k} = \frac{\tau}{3}\left[\boldsymbol{f}(t_{k+2}, \boldsymbol{w}^{k+2}) + 4\boldsymbol{f}(t_{k+1}, \boldsymbol{w}^{k+1}) + \boldsymbol{f}(t_k, \boldsymbol{w}^k)\right]$$

is of fourth order and that it obeys the root condition, which implies that it is convergent. Use a (high-powered) theorem to argue that the method is not A–stable. Prove directly, by calculating roots of the stability polynomial, that it is not A-stable.

Listings

```
% A procedure to plot the boundary of the linear stability
% domains for the BDF2, BDF3, and BDF4 methods:
t = 0:0.01:2*pi;
%
% BDF2
x2 = 1.5-2.0*cos(t)+0.5*cos(2.0*t);
y2 = 2.0*sin(t)-0.5*sin(2.0*t);
%
% BDF3
x3 = 11/6-3*cos(t)+3/2*cos(2*t)-1/3*cos(3*t);
y3 = 3*sin(t)-3/2*sin(2*t)+1/3*sin(3*t);
%
% BDF4
x4 = 25/12-4*cos(t)+3*cos(2*t)-4/3*cos(3*t)+1/4*cos(4*t);
y4 = 4*sin(t)-3*sin(2*t)+4/3*sin(3*t)-1/4*sin(4*t);

plot(x2,y2,x3,y3,x4,y4)
grid on
```

Listing 21.1 A simple procedure to plot the boundary of the linear stability domains, $\partial\mathcal{D}_{\text{BDF}q}$, for the BDF2, BDF3, and BDF4 methods.

22 Galerkin Methods for Initial Value Problems

So far, to approximate the solution to an initial value problem (IVP), we have produced a sequence of vectors $\{\boldsymbol{w}^k\}_{k=0}^N$, meant to approximate the solution at a finite collection of instances of time $\{t_k = k\tau\}_{k=0}^N$. In this chapter, we will see how to use Galerkin techniques to approximate the solution of IVPs. Galerkin methods are based upon a *weak formulation* of the problem and have a flavor different from the methods that we have introduced up to now. In a Galerkin technique, we approximate the spaces of possible solutions and test functions in the weak formulation of the problem. In particular, our approximate solutions will have values for all instances of time, not just at the *grid points* $\{t_k\}_{k=0}^N$. This chapter will serve as a starting point for the discussion in subsequent chapters, especially when we consider the finite element method in Chapter 25. The methods we present here can be generalized to approximate solutions to time-dependent partial differential equations (PDEs), such as those we will study in Chapter 28. To make the transition to PDEs simple, we keep the treatment general, working in a Hilbert $\mathcal{H}$ space, rather than in $\mathbb{R}^d$.

22.1 Assumptions and Basic Definitions

To keep matters sufficiently general, yet simple enough, we will assume here that $\mathcal{H}$ is a real Hilbert space with inner product $(\cdot,\cdot)_{\mathcal{H}}$ and associated norm $\|\cdot\|_{\mathcal{H}}$; see Definition 16.2.

Definition 22.1 (coercivity and monotonicity)**.** Let $A\colon \mathcal{H} \to \mathcal{H}$ be a (not necessarily linear) mapping. We say that A is **coercive** if and only if there is a constant $C_1 > 0$ such that

$$C_1\|v\|_{\mathcal{H}}^2 \le (Av, v)_{\mathcal{H}}, \quad \forall v \in \mathcal{H}.$$

We say that A is **monotone** if and only if there is a convex differentiable functional $E\colon \mathcal{H} \to \mathbb{R}$ such that

$$E'[v](\phi) = (Av, \phi)_{\mathcal{H}}, \quad \forall v, \phi \in \mathcal{H},$$

where $E'[v]$ is the Gateaux derivative of E at $v \in \mathcal{H}$, in the sense of Definition 16.36.

We recall that, owing to Proposition 16.39, a monotone operator satisfies

$$(Av_1 - Av_2, v_1 - v_2)_{\mathcal{H}} \ge 0, \quad \forall v_1, v_2 \in \mathcal{H}. \tag{22.1}$$

Remark 22.2 (monotonicity). We remark that our definition of a monotone operator does not coincide with the one usually adopted in the literature. Usually, (22.1) is taken as the definition of monotonicity, and then it is shown that derivatives of convex functionals are monotone. Since we will only deal with derivatives of convex functionals here, we will stick with this definition.

We will assume in this chapter that A is either coercive or monotone. Given this mapping A, we are interested in approximating the solution to the following IVP: given $u_0 \in \mathcal{H}$ and $f \in C([0,T];\mathcal{H})$, find $u\colon [0,T] \to \mathcal{H}$ satisfying

$$u'(t) + Au(t) = f(t), \qquad u(0) = u_0. \tag{22.2}$$

We will assume that this problem always has a unique solution with the regularity $u \in C^1([0,T];\mathcal{H})$. This can be proven easily for the case that $\mathcal{H}$ is finite dimensional; see Chapter 17. The infinite-dimensional case requires techniques that are beyond the scope of this text.

Notice that we can take the inner product of the equation with an arbitrary $v \in \mathcal{H}$ to obtain the following.

Definition 22.3 (weak form). The **weak form** of the IVP (22.2) is given as follows: given $u_0 \in \mathcal{H}$ and $f \in C([0,T];\mathcal{H})$, find $u\colon [0,T] \to \mathcal{H}$, satisfying $u(0) = u_0 \in \mathcal{H}$, and

$$(u', v)_{\mathcal{H}} + (Au, v)_{\mathcal{H}} = (f, v)_{\mathcal{H}}, \quad \forall v \in \mathcal{H}, \quad \forall t \in [0,T]. \tag{22.3}$$

The arbitrary function $v \in \mathcal{H}$ in the previous identity is called a **test function**. A solution $u \in C^1([0,T];\mathcal{H})$ to (22.3) is called a **weak solution**.

We will always assume that a weak solution exists and is unique. Now, owing to the structure of A, the following a priori estimates can be obtained.

Theorem 22.4 (a priori estimate, coercive case). *Suppose that $u \in C^1([0,T];\mathcal{H})$ is a solution to* (22.3) *with $u(0) = u_0 \in \mathcal{H}$. If A is coercive, then, for any $t \in [0,T]$,*

$$\|u(t)\|_{\mathcal{H}}^2 + C_1 \int_0^t \|u(s)\|_{\mathcal{H}}^2 \mathrm{d}s \le \|u_0\|_{\mathcal{H}}^2 + \frac{1}{C_1} \int_0^t \|f(s)\|_{\mathcal{H}}^2 \mathrm{d}s. \tag{22.4}$$

Proof. First, observe that, since $u \in C^1([0,T];\mathcal{H})$,

$$\frac{\mathrm{d}}{\mathrm{d}t}\|u\|_{\mathcal{H}}^2 = 2\,(u', u)_{\mathcal{H}}.$$

In the weak form, start by setting $v = u(t)$, for every $t \in [0,T]$. Using the definition of coercivity and the Cauchy–Schwarz and Young inequalities, we have, for every $t \in [0,T]$,

$$\frac{1}{2}\frac{\mathrm{d}}{\mathrm{d}t}\|u\|_{\mathcal{H}}^2 + C_1\|u\|_{\mathcal{H}}^2 \le (f, u)_{\mathcal{H}} \le \frac{1}{2C_1}\|f\|_{\mathcal{H}}^2 + \frac{C_1}{2}\|u\|_{\mathcal{H}}^2.$$

Integrating in time on the interval $[0, t]$ gives the result. □

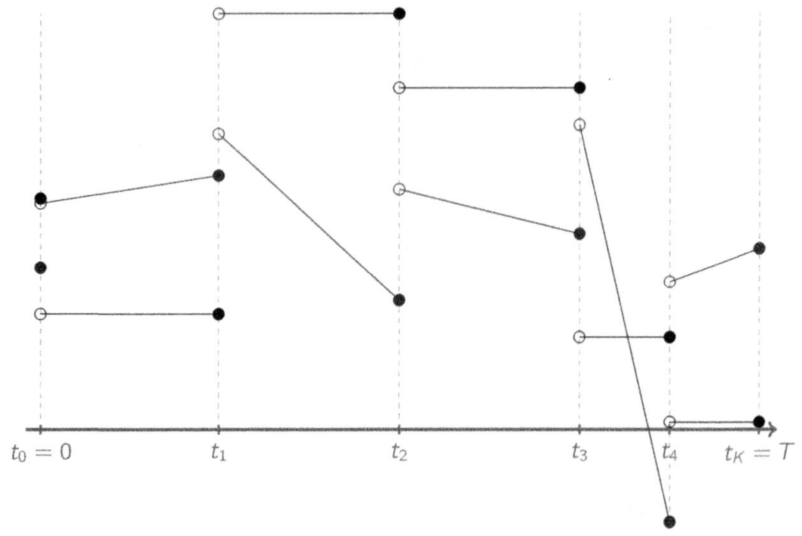

Figure 22.1 A typical function in the space $\mathscr{S}^{p,-1}(\boldsymbol{\tau})$ for $p = 0$ (thin lines) and $p = 1$ (thick lines).

Theorem 22.5 (a priori estimate, monotone case)**.** *Suppose that $u \in C^1([0, T]; \mathcal{H})$ is a solution to (22.3) with $u(0) = u_0 \in \mathcal{H}$. If A is monotone, then, for all $t \in [0, T]$,*

$$\frac{1}{2}\int_0^t \|u'(s)\|_{\mathcal{H}}^2 \,\mathrm{d}s + E(u(t)) \leq E(u_0) + \frac{1}{2}\int_0^t \|f(s)\|_{\mathcal{H}}^2 \,\mathrm{d}s. \tag{22.5}$$

Proof. This time, set $v = u'$ in the weak form and use the Cauchy–Schwarz and Young inequalities to get

$$\|u'\|_{\mathcal{H}}^2 + (Au, u')_{\mathcal{H}} = (f, u')_{\mathcal{H}} \leq \frac{1}{2}\|f\|_{\mathcal{H}}^2 + \frac{1}{2}\|u'\|_{\mathcal{H}}^2.$$

Next, notice that

$$\frac{\mathrm{d}}{\mathrm{d}t}E(u(t)) = (E'[u], u')_{\mathcal{H}} = (Au, u')_{\mathcal{H}}.$$

Using the monotonicity of the mapping and integrating over the interval $[0, t]$, we get the second result. □

Remark 22.6 (convexity)**.** Note that we have not yet used the fact that E is convex. This property will be needed later.

We briefly comment that estimates (22.4) and (22.5) can be used to prove the existence and uniqueness of solutions in their respective cases. This, however, is beyond the scope of our discussion here. Instead, we will focus on the development of methods that preserve or mimic these estimates. We will do so by using Galerkin techniques.

We begin the discretization procedure by introducing a partition of the interval $[0, T]$.

Definition 22.7 (partition). A **partition** of $[0, T]$ of size $K \in \mathbb{N}$ is a set hbox$\boldsymbol{\tau} = \{t_0, \dots, t_K\}$ with the property

$$0 = t_0 < t_1 < \cdots < t_K = T.$$

The sets $I_k = (t_{k-1}, t_k]$ for $k \in \{1, \dots, K\}$ are called the **intervals** of the partition.

On the basis of this partition, we introduce a space of discrete functions.

Definition 22.8 (space $\mathscr{S}^{p,-1}(\boldsymbol{\tau}; \mathcal{H})$). Let $p \in \mathbb{N}_0$. The space of **$\mathcal{H}$-valued piecewise polynomials** of **degree** p is defined via

$$\mathscr{S}^{p,-1}(\boldsymbol{\tau}; \mathcal{H}) = \left\{ v_{\boldsymbol{\tau}} \colon [0, T] \to \mathcal{H} \,\middle|\, v_{\boldsymbol{\tau}}|_{I_k}(t) = \sum_{j=0}^{p} v_{k,j} t^j, \ v_{k,j} \in \mathcal{H}, \ k = 1, \dots, K \right\}. \tag{22.6}$$

Note that the elements of $\mathscr{S}^{p,-1}(\boldsymbol{\tau}; \mathcal{H})$ are generically discontinuous at the partition points. That is the reason for the second super-index: we use it to indicate that functions in this space have *smoothness of order minus one.* Example functions from $\mathscr{S}^{p,-1}(\boldsymbol{\tau})$ are shown in Figure 22.1 for the piecewise constant ($p = 0$) and piecewise linear ($p = 1$) cases.

22.2 Coercive Operators: The Discontinuous Galerkin Method

We first deal with the case of a coercive operator. We begin by observing that no continuity requirements are imposed on $\mathscr{S}^{p,-1}(\boldsymbol{\tau}; \mathcal{H})$, but, since the intervals I_k are semi-open, functions on $\mathscr{S}^{p,-1}(\boldsymbol{\tau}; \mathcal{H})$ are only left continuous. Owing to this, we introduce the following notation.

Definition 22.9 (jump). Let $\mathscr{S}^{p,-1}(\boldsymbol{\tau}; \mathcal{H})$ be as in (22.6). For $v_{\boldsymbol{\tau}} \in \mathscr{S}^{p,-1}(\boldsymbol{\tau}; \mathcal{H})$, and $k = 0, \dots, K-1$, define the quantities

$$v^k = v_{\boldsymbol{\tau}}(t_k), \qquad v_+^k = \lim_{t \downarrow t_k} v_{\boldsymbol{\tau}}(t), \qquad [\![v^k]\!] = v_+^k - v^k.$$

The last quantity is known as the **jump**.

We note the following simple identity.

Proposition 22.10 (tensor product). *Suppose, $\mathscr{S}^{p,-1}(\boldsymbol{\tau}; \mathcal{H})$ is given by (22.6). Then we have the decomposition*

$$\mathscr{S}^{p,-1}(\boldsymbol{\tau}; \mathcal{H}) = \bigotimes_{k=1}^{K} \mathscr{S}^{p,\infty}(I_k; \mathcal{H}),$$

$$\mathscr{S}^{p,\infty}(I_k; \mathcal{H}) = \left\{ v_k \colon I_k \to \mathcal{H} \,\middle|\, v_k(t) = \sum_{j=0}^{p} \psi_{k,j} t^j, \ \psi_{k,j} \in \mathcal{H} \right\}.$$

Proof. See Problem 22.1. □

Now, in identity (22.3), set the test function $v = v(t) \in \mathcal{H}$. Integrating over the interval $[0, t_K]$, we obtain

$$\int_0^{t_K} (u', v)_{\mathcal{H}} \, \mathrm{d}t + \int_0^{t_K} (Au, v)_{\mathcal{H}} \, \mathrm{d}t = \int_0^{t_K} (f, v)_{\mathcal{H}} \, \mathrm{d}t.$$

Let us assume now that $v(t_K) = 0$. Integrating by parts, we conclude that

$$-\int_0^{t_K} (u, v')_{\mathcal{H}} \, \mathrm{d}t + \int_0^{t_K} (Au, v)_{\mathcal{H}} \, \mathrm{d}t = (u_0, v(0))_{\mathcal{H}} + \int_0^{t_K} (f, v)_{\mathcal{H}} \, \mathrm{d}t. \quad (22.7)$$

We will use identity (22.7) to derive our numerical method. If we replace the exact solution by a function $u_{\boldsymbol{\tau}} \in \mathscr{S}^{p,-1}(\boldsymbol{\tau}; \mathcal{H})$ while keeping v as it is, and integrate back by parts on each partition interval I_k, we get

$$-\int_{t_{k-1}}^{t_k} (u_{\boldsymbol{\tau}}, v')_{\mathcal{H}} \, \mathrm{d}t = \int_{I_k} (u_{\boldsymbol{\tau}}', v)_{\mathcal{H}} \, \mathrm{d}t - \left(u^k, v_+^k\right)_{\mathcal{H}} + \left(u_+^{k-1}, v_+^{k-1}\right)_{\mathcal{H}},$$

where we used that v is smooth, so $v^k = v_+^k$. Adding over k and recalling that $v(t_K) = 0$, we obtain

$$\begin{aligned} \int_0^{t_K} (u_{\boldsymbol{\tau}}', v)_{\mathcal{H}} \, \mathrm{d}t + \sum_{k=1}^{K-1} \left([\![u^k]\!], v_+^k\right)_{\mathcal{H}} + \left(u_+^0, v_+^0\right)_{\mathcal{H}} + \int_0^{t_K} (Au_{\boldsymbol{\tau}}, v)_{\mathcal{H}} \, \mathrm{d}t \\ = \left(u_0, v_+^0\right)_{\mathcal{H}} + \int_0^{t_K} (f, v)_{\mathcal{H}} \, \mathrm{d}t. \end{aligned}$$

Finally, we replace in the previous expression the test function by a discrete one, i.e., $v_{\boldsymbol{\tau}} \in \mathscr{S}^{p,-1}(\boldsymbol{\tau}; \mathcal{H})$, to obtain the final form of our method.

Definition 22.11 (DG approximation)**.** The **discontinuous Galerkin (DG) approximation**[1] to (22.3) is defined as follows: find $u_{\boldsymbol{\tau}} \in \mathscr{S}^{p,-1}(\boldsymbol{\tau}; \mathcal{H})$ such that, $u^0 = u_0$ and

$$\begin{aligned} \int_0^{t_K} (u_{\boldsymbol{\tau}}', v_{\boldsymbol{\tau}})_{\mathcal{H}} \, \mathrm{d}t + \sum_{k=1}^{K-1} \left([\![u^k]\!], v_+^k\right)_{\mathcal{H}} + \left(u_+^0, v_+^0\right)_{\mathcal{H}} + \int_0^{t_K} (Au_{\boldsymbol{\tau}}, v_{\boldsymbol{\tau}})_{\mathcal{H}} \, \mathrm{d}t \\ = \left(u_0, v_+^0\right)_{\mathcal{H}} + \int_0^{t_K} (f, v_{\boldsymbol{\tau}})_{\mathcal{H}} \, \mathrm{d}t, \qquad \forall v_{\boldsymbol{\tau}} \in \mathscr{S}^{p,-1}(\boldsymbol{\tau}; \mathcal{H}). \end{aligned} \quad (22.8)$$

Notice that we can set, in (22.8), $v_{\boldsymbol{\tau}} = v_k \chi_{I_k}$ with $v_k \in \mathscr{S}^{p,\infty}(I_k; \mathcal{H})$ and χ_{I_k} being the characteristic function of the interval I_k, $k = 1, \dots, K$, to obtain a local version of our method.

Definition 22.12 (local version)**.** The **local version of the DG approximation** to (22.3) is defined as follows: find $u_{\boldsymbol{\tau}} \in \mathscr{S}^{p,-1}(\boldsymbol{\tau}; \mathcal{H})$ such that $u^0 = u_0$ and, for every $k \in \{1, \dots, K\}$ and all $v_k \in \mathscr{S}^{p,\infty}(I_k; \mathcal{H})$, it holds that

$$\int_{I_k} \left[(u_{\boldsymbol{\tau}}', v_k)_{\mathcal{H}} + (Au_{\boldsymbol{\tau}}, v_k)_{\mathcal{H}}\right] \mathrm{d}t + \left([\![u^{k-1}]\!], v_{k,+}^{k-1}\right)_{\mathcal{H}} = \int_{I_k} (f, v_k)_{\mathcal{H}} \, \mathrm{d}t. \quad (22.9)$$

[1] Named in honor of the Russian mathematician and engineer Boris Grigorievich Galerkin (1871–1945).

Remark 22.13 (upwinding). Throughout the course of the derivation of the method, when dealing with the smooth test function v, it may have seemed arbitrary to use v_+^k, as it makes no difference at this stage. However, this is crucial when we replace this test function by a discrete one, and it is what will allow us to obtain stability of our method. This particular choice, against the direction of time, corresponds to a so-called *upwinding method*, as the value is taken against the direction of flow of information. We will learn more about upwinding in Section 24.4.2 and Chapter 29.

22.2.1 Stability

Let us now pause a moment to see what have we gained. First of all, we must notice that we are working with *functions* instead of point values, as we did in previous chapters. Thus, we can repeat many of the arguments for the actual continuous problem to analyze our method.

Theorem 22.14 (stability). *For any partition $\boldsymbol{\tau}$, the function $u_{\boldsymbol{\tau}} \in \mathscr{S}^{p,-1}(\boldsymbol{\tau};\mathcal{H})$ that solves* (22.8) *satisfies*

$$\|u^K\|_{\mathcal{H}}^2 + \sum_{k=0}^{K-1} \|[\![u^k]\!]\|_{\mathcal{H}}^2 + C_1 \int_0^T \|u_{\boldsymbol{\tau}}\|_{\mathcal{H}}^2 \,\mathrm{d}t \le \|u_0\|_{\mathcal{H}}^2 + \frac{1}{C_1}\int_0^T \|f\|_{\mathcal{H}}^2 \,\mathrm{d}t. \quad (22.10)$$

Proof. Set, in (22.9), $v_k = 2u_{\boldsymbol{\tau}}\chi_{I_k}$ to obtain

$$\|u^k\|_{\mathcal{H}}^2 - \|u_+^{k-1}\|_{\mathcal{H}}^2 + 2\int_{I_k} (Au_{\boldsymbol{\tau}}, u_{\boldsymbol{\tau}})_{\mathcal{H}} \,\mathrm{d}t + 2\left(u_+^{k-1} - u^{k-1}, u_+^{k-1}\right)_{\mathcal{H}} = 2\int_{I_k} (f, u_{\boldsymbol{\tau}})_{\mathcal{H}} \,\mathrm{d}t.$$

The elementary identity

$$2a(a-b) = a^2 - b^2 + (a-b)^2$$

then allows us to conclude that

$$2\left(u_+^{k-1} - u^{k-1}, u_+^{k-1}\right)_{\mathcal{H}} = \|u_+^{k-1}\|_{\mathcal{H}}^2 - \|u^{k-1}\|_{\mathcal{H}}^2 + \|[\![u^{k-1}]\!]\|_{\mathcal{H}}^2.$$

Consequently, using the coercivity property, we have

$$\|u^k\|_{\mathcal{H}}^2 - \|u^{k-1}\|_{\mathcal{H}}^2 + C_1 \int_{I_k} \|u_{\boldsymbol{\tau}}\|_{\mathcal{H}}^2 \,\mathrm{d}t + \|[\![u^{k-1}]\!]\|_{\mathcal{H}}^2 \le \frac{1}{C_1}\int_{I_k} \|f\|_{\mathcal{H}}^2 \,\mathrm{d}t;$$

adding over k, we thus obtain (22.10). □

This is the discrete version of the stability estimate (22.4). It is remarkable to notice that this estimate was also obtained by the same technique as (22.4), i.e., choosing the test function to be equal to the solution.

Remark 22.15 (stability). Notice that if we examine estimate (22.10) more closely, this only gives us control on $u_{\boldsymbol{\tau}}$ *at the partition points* t_k. How can we say then that this is a stability result? If $p = 0$, then $u_{\boldsymbol{\tau}}$ is constant on I_k and, thus, this gives control on the whole function. On the other hand, if $p = 1$, notice that we have

$$
\begin{aligned}
u_{\boldsymbol{\tau}|I_k}(t) &= u_+^{k-1}\frac{t_k - t}{t_k - t_{k-1}} + u^k\frac{t - t_{k-1}}{t_k - t_{k-1}} \\
&= u^{k-1}\frac{t_k - t}{t_k - t_{k-1}} + u^k\frac{t - t_{k-1}}{t_k - t_{k-1}} + [\![u^k]\!]\frac{t_k - t}{t_k - t_{k-1}},
\end{aligned}
$$

so that

$$
\|u_{\boldsymbol{\tau}|I_k}(t)\|_{\mathcal{H}} \le \|u^{k-1}\|_{\mathcal{H}}\frac{t_k - t}{t_k - t_{k-1}} + \|u^k\|_{\mathcal{H}}\frac{t - t_{k-1}}{t_k - t_{k-1}} + \|[\![u^k]\!]\|_{\mathcal{H}}\frac{t_k - t}{t_k - t_{k-1}},
$$

and it is not only the point value *but also the jump* that gives us control over the function over the whole interval I_k.

22.2.2 Convergence

We have obtained then the stability of the method, and Problem 22.2 shows that it is consistent. As we have seen before, and we will see time and time again, this implies convergence.

We begin the analysis with the construction of a *canonical* interpolant for the space $\mathscr{S}^{p,-1}(\boldsymbol{\tau};\mathcal{H})$.

Proposition 22.16 (interpolation)**.** *Let $p \in \mathbb{N}_0$ and $\boldsymbol{\tau}$ be a partition of $[0,T]$. For any $v \in C([0,T];\mathcal{H})$, there is a unique $\tilde{v}_{\boldsymbol{\tau}} \in \mathscr{S}^{p,-1}(\boldsymbol{\tau};\mathcal{H})$ that satisfies, for all $k = 0,\dots,K$,*

$$
\tilde{v}^k = v(t_k), \qquad \int_{I_k}(v - \tilde{v}_{\boldsymbol{\tau}})t^m \mathrm{d}t = 0, \quad \forall m < p.
$$

Moreover, if v is sufficiently smooth, this unique function $\tilde{v}_{\boldsymbol{\tau}}$ satisfies

$$
\|v(t) - \tilde{v}_{\boldsymbol{\tau}}(t)\|_{\mathcal{H}} \le C\tau^{p+1}, \quad \forall t \in I_k.
$$

Proof. See Problem 22.3. □

A useful, and important, property of this interpolant is the following.

Lemma 22.17 (orthogonality)**.** *Let $p \in \mathbb{N}_0$, $\boldsymbol{\tau}$ be a partition of $[0,T]$, and $v \in C([0,T];\mathcal{H})$. For any $k \in \{1,\dots,K\}$, the interpolant $\tilde{v}_{\boldsymbol{\tau}}$, defined in Proposition 22.16, satisfies*

$$
\int_{I_k}(v - \tilde{v}_{\boldsymbol{\tau}})\phi \mathrm{d}t = 0, \qquad \forall \phi \in \mathscr{S}^{p,\infty}(I_k;\mathcal{H}).
$$

Proof. See Problem 22.4. □

We are now in a position to show convergence of our method. Although it is not necessary, to simplify the presentation, we will assume that our partition $\boldsymbol{\tau}$ is *uniform*. In other words, all the points are equidistant, so that

$$
t_k = k\tau, \quad \tau = T/K.
$$

One final, simplifying, set of assumptions we will make are that the operator $A \in \mathfrak{B}(\mathcal{H})$, see Definition 16.18, and self-adjoint in the sense of Definition 1.17.

Theorem 22.18 (convergence I). *Assume that u, the solution of (22.2), is such that $u \in C^{p+1}([0,T];\mathcal{H})$. Let $p \in \mathbb{N}_0$, $K \in \mathbb{N}$, and $\boldsymbol{\tau}$ be a uniform partition of $[0,T]$ of size $\tau = T/K$. Let $u_{\boldsymbol{\tau}} \in \mathscr{S}^{p,-1}(\boldsymbol{\tau};\mathcal{H})$ be the solution to (22.8). There is a constant, C, that depends only on u such that*

$$\max_{t\in\boldsymbol{\tau}} \|u(t) - u_{\boldsymbol{\tau}}(t)\|_{\mathcal{H}} \le C\tau^{p+1}.$$

Proof. Let $\tilde{u}_{\boldsymbol{\tau}} \in \mathscr{S}^{p,-1}(\boldsymbol{\tau};\mathcal{H})$ be the interpolant of u, defined via Proposition 22.16. We now decompose the error $e = u - u_{\boldsymbol{\tau}}$ into $e = (u - \tilde{u}_{\boldsymbol{\tau}}) + (\tilde{u}_{\boldsymbol{\tau}} - u_{\boldsymbol{\tau}}) = \rho + \theta$, where $\theta \in \mathscr{S}^{p,-1}(\boldsymbol{\tau};\mathcal{H})$, and, by Lemma 22.17, we have that

$$\rho(t_k) = 0, \quad \forall k = 0,\dots,K, \qquad \|\rho(t)\|_{\mathcal{H}} \le C\tau^{p+1}.$$

It remains then to handle θ. As will happen many times in our discussion, the strategy is to find an equation for this function. Notice that we can integrate (22.3) over I_k and set $v \in \mathscr{S}^{p,\infty}(I_k;\mathcal{H})$ to obtain

$$\int_{I_k} [(u',v)_{\mathcal{H}} + (Au,v)_{\mathcal{H}}]\mathrm{d}t = \int_{I_k} (f,v)_{\mathcal{H}}\,\mathrm{d}t;$$

we then subtract from this the local version of our method (22.9) to obtain

$$\begin{aligned}\int_{I_k} [(\theta',v)_{\mathcal{H}} + (A\theta,v)_{\mathcal{H}}]\mathrm{d}t + \left([\![\theta^{k-1}]\!], v_+^{k-1}\right)_{\mathcal{H}} \\ = -\int_{I_k} [(\rho',v)_{\mathcal{H}} + (A\rho,v)_{\mathcal{H}}]\mathrm{d}t - \left([\![\rho^{k-1}]\!], v_+^{k-1}\right)_{\mathcal{H}}.\end{aligned}$$

Notice that this previous identity is exactly the method. The only difference lies in what the right-hand side is. For this reason, we can just apply the stability estimate (22.10) to obtain

$$\begin{aligned}\|\theta^k\|_{\mathcal{H}}^2 - \|\theta^{k-1}\|_{\mathcal{H}}^2 + \|[\![\theta^{k-1}]\!]\|_{\mathcal{H}}^2 + 2\int_{I_k} (A\theta,\theta)_{\mathcal{H}}\,\mathrm{d}t \\ = -2\int_{I_k} [(\rho',\theta)_{\mathcal{H}} + (A\rho,\theta)_{\mathcal{H}}]\mathrm{d}t - 2\left([\![\rho^{k-1}]\!], \theta_+^{k-1}\right)_{\mathcal{H}}.\end{aligned}$$

However, for every $v \in \mathscr{S}^{p,\infty}(I_k;\mathcal{H})$, we have

$$\begin{aligned}\int_{I_k} (\rho',v)_{\mathcal{H}}\,\mathrm{d}t + \left([\![\rho^{k-1}]\!], v_+^{k-1}\right)_{\mathcal{H}} \\ = -\int_{I_k} (\rho,v')_{\mathcal{H}}\,\mathrm{d}t + \left(\rho^k, v^k\right)_{\mathcal{H}} - \left(\rho_+^{k-1}, v_+^{k-1}\right)_{\mathcal{H}} \\ + \left(\rho_+^{k-1}, v_+^{k-1}\right)_{\mathcal{H}} - \left(\rho^{k-1}, v_+^{k-1}\right)_{\mathcal{H}} \\ = 0,\end{aligned}$$

where we used that $\rho^k = \rho(t_k) = 0$ and the fact that v' is a polynomial of degree at most $p-1$ with coefficients in $\mathcal{H}$. Notice also that the coercivity of A implies that

$$0 \le (A(\rho-\theta), \rho-\theta)_{\mathcal{H}};$$

combining this with the fact that A is self-adjoint, we get

$$2\int_{I_k} (A\rho, \theta)_{\mathcal{H}}\, \mathrm{d}t \leq \int_{I_k} (A\rho, \rho)_{\mathcal{H}}\, \mathrm{d}t + \int_{I_k} (A\theta, \theta)_{\mathcal{H}}\, \mathrm{d}t \leq C\tau^{2(p+1)+1} + \int_{I_k} (A\theta, \theta)_{\mathcal{H}}\, \mathrm{d}t,$$

where we used that the operator A is bounded and that $\|\rho(t)\|_{\mathcal{H}} \leq C\tau^{p+1}$ and is being integrated over an interval of length τ.

Gathering all the previous estimates, we get

$$\|\theta^k\|_{\mathcal{H}}^2 - \|\theta^{k-1}\|_{\mathcal{H}}^2 + \|[\![\theta^{k-1}]\!]\|_{\mathcal{H}}^2 + \int_{I_k} (A\theta, \theta)_{\mathcal{H}}\, \mathrm{d}t \leq C\tau^{2(p+1)+1};$$

summing over k, we then get

$$\|\theta^K\|_{\mathcal{H}}^2 + \sum_{k=0}^{K-1} \|[\![\theta^k]\!]\|_{\mathcal{H}}^2 + \int_0^T (A\theta, \theta)_{\mathcal{H}}\, \mathrm{d}t \leq C\tau^{2(p+1)}\tau \sum_{k=1}^{K} = CT\tau^{2(p+1)},$$

and this implies the result. □

The previous result only gives convergence at the partition points. The next result, however, shows that this is enough to obtain convergence at every point in $[0, T]$. Its proof relies on the fact that, much as in the real (or complex) case, for every $p \in \mathbb{N}$ and all I_k, we can define a least squares polynomial approximation $\mathcal{P}_{1,p}[u] \in \mathscr{S}^{p,\infty}(I_k; \mathcal{H})$. This defines the mapping

$$\Pi_{\boldsymbol{\tau}}[u] \in \mathscr{S}^{p,-1}(\boldsymbol{\tau}; \mathcal{H}), \qquad \Pi_{\boldsymbol{\tau}}[u]_{|I_k} = \mathcal{P}_{1,p}[u],$$

which satisfies

$$\max_{t\in[0,T]} \|u(t) - \Pi_{\boldsymbol{\tau}} u(t)\|_{\mathcal{H}} + \tau \max_{t\in[0,T]} \|(u - \Pi_{\boldsymbol{\tau}} u)'(t)\| \leq c\tau^{p+1} \max_{t\in[0,T]} \|u^{(p+1)}(t)\|_{\mathcal{H}},$$

whenever u is sufficiently smooth.

Theorem 22.19 (convergence II). *Assume that u, the solution of (22.2), is such that $u \in C^{p+1}([0, T]; \mathcal{H})$. Let $p \in \mathbb{N}_0$, $K \in \mathbb{N}$, and $\boldsymbol{\tau}$ be a uniform partition of $[0, T]$ of size $\tau = T/K$. Let $u_{\boldsymbol{\tau}} \in \mathscr{S}^{p,-1}(\boldsymbol{\tau}; \mathcal{H})$ be the solution to (22.8). There are constants C_1 and C_2 such that, for every $k = 1, \dots, K$, we have that*

$$\max_{t\in I_k} \|u(t) - u_{\boldsymbol{\tau}}(t)\|_{\mathcal{H}} \leq \|u^k - u(t_k)\|_{\mathcal{H}} + C_1\|u^{k-1} - u(t_{k-1})\|_{\mathcal{H}} + C_2\tau^{p+1}.$$

Example 22.1 Let us consider the simplest case, namely $p = 0$ and $\mathcal{H} = \mathbb{R}$. In this case, $u'_{\boldsymbol{\tau}|I_k} = 0$ and

$$u_{\boldsymbol{\tau}|I_k} = u^k = u_+^{k-1}.$$

In this case, by setting $v = 1$, the local version (22.9) of our method reduces to

$$u^k - u^{k-1} + \tau A u^k = \int_{I_k} f(t)\mathrm{d}t$$

or

$$u^k = u^{k-1} + \tau \left[\frac{1}{\tau} \int_{I_k} f(t)\mathrm{d}t - Au^k \right].$$

This is nothing but a variant of the implicit Euler method that we considered in Chapter 18, in which we replace $f(t_k)$ by its average over I_k.

Example 22.2 For $p \geq 1$, possibly after quadrature, we arrive at particular classes of RK methods; see Problem 22.5.

22.3 Monotone Operators: The Continuous Petrov–Galerkin Method

There is something troubling in the constructions of Section 22.2. Namely, if we assume that u is a mild solution to (22.2), then we must at least have that $u \in C([0,T];\mathcal{H})$. However, the approximate solution $u_{\boldsymbol{\tau}}$, defined via (22.8), is not continuous in time. To retain this continuity property, we must consider a *Petrov–Galerkin*[2] method, i.e., one where the test function does not necessarily lie in the space where we seek the solution. Essentially, we will consider an approximation of (22.3) but when the solution and test function lie in different spaces. To implement this idea, we keep the notation of the previous section and define another space of polynomials.

Definition 22.20 (space $\mathscr{S}^{p,0}(\boldsymbol{\tau};\mathcal{H})$). Let $p \in \mathbb{N}$. The space of **continuous $\mathcal{H}$-valued piecewise polynomials** of **degree** p is defined via

$$\mathscr{S}^{p,0}(\boldsymbol{\tau};\mathcal{H}) = \mathscr{S}^{p,-1}(\boldsymbol{\tau};\mathcal{H}) \cap C([0,T];\mathcal{H}). \tag{22.11}$$

Remark 22.21 (splines). As we mentioned above, the second super-index, in the definition of $\mathscr{S}^{p,0}(\boldsymbol{\tau};\mathcal{H})$, indicates the smoothness of functions in this space with the zero here indicating continuity. We can similarly define, for $r \in \mathbb{N}$,

$$\mathscr{S}^{p,r}(\boldsymbol{\tau};\mathcal{H}) = \mathscr{S}^{p,-1}(\boldsymbol{\tau};\mathcal{H}) \cap C^r([0,T];\mathcal{H}).$$

Spaces of functions like this one are usually called *splines* in the literature.

In the continuous Galerkin method, we will now seek $u_{\boldsymbol{\tau}} \in \mathscr{S}^{p,0}(\boldsymbol{\tau};\mathcal{H})$. Before even defining the method, notice that continuity implies that, for all $k = 0,\dots,K$, we have $u_{\boldsymbol{\tau}}(t_k) = u^k$ and so, once we have reached I_k, we only have p conditions left to fully define $u_{\boldsymbol{\tau}}$ on I_k. For this reason, we will test the equation with functions from $\mathscr{S}^{p-1,-1}(\boldsymbol{\tau};\mathcal{H})$. This means that we have the following method.

Definition 22.22 (CG approximation). The **continuous Galerkin (CG) approximation** to (22.3) is defined as follows: find $u_{\boldsymbol{\tau}} \in \mathscr{S}^{p,0}(\boldsymbol{\tau};\mathcal{H})$ such that $u^0 = u_0$ and, for all $v_{\boldsymbol{\tau}} \in \mathscr{S}^{p-1,-1}(\boldsymbol{\tau};\mathcal{H})$,

$$\int_0^T \left[(u_{\boldsymbol{\tau}}', v_{\boldsymbol{\tau}})_{\mathcal{H}} + (Au_{\boldsymbol{\tau}}, v_{\boldsymbol{\tau}})_{\mathcal{H}}\right] \mathrm{d}t = \int_0^T (f, v_{\boldsymbol{\tau}})_{\mathcal{H}} \, \mathrm{d}t. \tag{22.12}$$

[2] Named in honor of the Soviet engineer Georgiy Ivanovich Petrov (1912–1987) and the Russian mathematician and engineer Boris Grigorievich Galerkin (1871–1945).

22.3.1 Stability

We now study the stability of (22.12). The following result provides a discrete analogue of (22.5).

Theorem 22.23 (stability). *For any partition $\boldsymbol{\tau}$, the function $u_{\boldsymbol{\tau}} \in \mathscr{S}^{p,0}(\boldsymbol{\tau};\mathcal{H})$ that solves* (22.12) *satisfies*

$$\frac{1}{2}\int_{I_k} \|u_{\boldsymbol{\tau}}'\|_{\mathcal{H}}^2 \mathrm{d}t + E(u_{\boldsymbol{\tau}}(t_k)) \le \frac{1}{2}\int_{I_k} \|f\|_{\mathcal{H}}^2 \mathrm{d}t + E(u_{\boldsymbol{\tau}}(t_{k-1})).$$

Proof. As we did in the continuous case, we will proceed by suitably choosing a test function $v_{\boldsymbol{\tau}}$. Namely, we observe that $v_{\boldsymbol{\tau}} = u_{\boldsymbol{\tau}}'\chi_{I_k} \in \mathscr{S}^{p-1,-1}(\boldsymbol{\tau};\mathcal{H})$ and so it is an admissible function for the method (22.12). This will yield

$$\int_{I_k} \|u_{\boldsymbol{\tau}}'\|_{\mathcal{H}}^2 \mathrm{d}t + \int_{I_k} (Au_{\boldsymbol{\tau}}, u_{\boldsymbol{\tau}}')_{\mathcal{H}}\,\mathrm{d}t = \int_{I_k} (f, u_{\boldsymbol{\tau}}')_{\mathcal{H}}\,\mathrm{d}t \le \frac{1}{2}\int_{I_k} \|f\|_{\mathcal{H}}^2 \mathrm{d}t + \frac{1}{2}\int_{I_k} \|u_{\boldsymbol{\tau}}'\|_{\mathcal{H}}^2 \mathrm{d}t;$$

if A is monotone, this implies that

$$\frac{1}{2}\int_{I_k} \|u_{\boldsymbol{\tau}}'\|_{\mathcal{H}}^2 \mathrm{d}t + E(u_{\boldsymbol{\tau}}(t_k)) \le \frac{1}{2}\int_{I_k} \|f\|_{\mathcal{H}}^2 \mathrm{d}t + E(u_{\boldsymbol{\tau}}(t_{k-1})),$$

as we intended to show. □

We will not carry out a convergence analysis of this method. Instead, we will try to elucidate the form of this method in some simple cases.

Example 22.3 Let $\mathcal{H} = \mathbb{R}$. Notice that we must necessarily then have $0 < A \in \mathbb{R}$, so that the convex functional is

$$E(u) = \frac{1}{2}Au^2.$$

In addition, since we must require continuity, we must have $p \ge 1$. Let us set then $p = 1$ and $\tau = T/K$. Observe then that elements of $v_{\boldsymbol{\tau}} \in \mathscr{S}^{1,0}(\boldsymbol{\tau};\mathbb{R})$ are piecewise linear functions, and those of $v_{\boldsymbol{\tau}} \in \mathscr{S}^{0,-1}(\boldsymbol{\tau};\mathbb{R})$ piecewise constant ones. Consequently,

$$u_{\boldsymbol{\tau}}' = \frac{1}{\tau}(u^k - u^{k-1}).$$

Setting $v_{\boldsymbol{\tau}} = 1$ in (22.12), we then get

$$u^k - u^{k-1} + A\int_{I_k} u_{\boldsymbol{\tau}} \mathrm{d}t = \int_{I_k} f(t)\mathrm{d}t.$$

In addition, since $u_{\boldsymbol{\tau}}$ is linear, we can apply the midpoint rule to get

$$\int_{I_k} u_{\boldsymbol{\tau}}\,\mathrm{d}t = \frac{1}{2}(u^k + u^{k-1})\tau.$$

In summary, we obtained the method

$$u^k = u^{k-1} + \tau\left[\frac{1}{\tau}\int_{I_k} f(t)\mathrm{d}t - \frac{1}{2}A\left(u^k + u^{k-1}\right)\right].$$

This can be thought of as a variant of either the trapezoidal or midpoint methods, where the average of f can be approximated via

$$\frac{1}{\tau}\int_{I_k} f(t)\mathrm{d}t \approx \frac{1}{2}\left[f(t_k) + f(t_{k-1})\right]$$

or

$$\frac{1}{\tau}\int_{I_k} f(t)\mathrm{d}t \approx f\left(t_k + \frac{1}{2}\tau\right),$$

respectively.

Problems

22.1 Prove Proposition 22.10.
22.2 Replace, in (22.9), the Galerkin approximation u_τ by the exact solution u and obtain an identity that expresses the *consistency* of our method.
22.3 Prove Proposition 22.16.
22.4 Prove Lemma 22.17.
22.5 Show that, in the case that $\mathcal{H} = \mathbb{R}$, A is linear, $f \equiv 0$, and $p = 1$, the DG method (22.8) satisfies

$$u^k = g_{2/1}(\tau A)u^{k-1},$$

where $g_{2/1}$ is the Padé approximation of the exponential introduced in Definition 21.13.

Part V

Boundary and Initial Boundary Value Problems

23 Boundary and Initial Boundary Value Problems for Partial Differential Equations

In Chapter 17, we presented some facts about the theory of initial value problems (IVPs) for ordinary differential equations (ODEs). Many, but very far from all, ODE problems can fit into that framework. Consider the following two problems:

$$ay''(t) + by'(t) + cy(t) = 0, \ t \in (0, 1), \qquad y(0) = 0, \ y'(0) = 0,$$

and

$$ay''(t) + by'(t) + cy(t) = 0, \ t \in (0, 1), \qquad y(0) = 0, \ y(1) = 0,$$

with $0 \neq a, b, c \in \mathbb{R}$. The ODEs in these two problems are exactly the same; they are linear, constant-coefficient, second-order equations. The ODE can be transformed into a system of first-order equations for the dependent variable $\boldsymbol{u} = [y, y']^\intercal$. The first problem is an IVP, and it has the initial condition $\boldsymbol{u}(0) = [0, 0]^\intercal$. However, the second problem is fundamentally different. It is not an IVP because it has *boundary conditions*, i.e., conditions on the dependent variable at the distinct points $t = 0$ and $t = 1$. This fact prevents us from recasting this problem as an IVP.

To make matters worse, many equations derived from physics, chemistry, economics, etc. involve rates of change with respect to more than one independent variable. Such equations are called *partial differential equations* (PDEs). As one might expect, the theory can be much more complicated for PDEs than for ODEs. PDEs can have complicated boundary conditions, and some require initial conditions as well.

The theory of PDEs is a very deep and rich subject, one that we cannot fully cover here. Our purpose will be to present a few simple, yet prototypical, example problems, and some of the most relevant theory, especially that which will be important for numerical approximation. There are several extremely good references for these subjects, at various levels of depth. Good starting references are [31, 49, 72, 87, 99]. We must reiterate that, since this is an overview, we will often state our results under severely restrictive assumptions.

23.1 Heuristic Derivation of the Common Partial Differential Equations

Before we get into the theory, it is useful to see how some of the most common PDEs are derived.

23.1.1 Conservation Laws and the Conservation Equation

To get started, let us imagine a familiar phenomenon. Suppose a drop of dye falls upon the surface of a flowing stream. How would we model the motion of the dye particles? Of course, the drop is advected by the motion of the stream, but it simultaneously spreads out, until, eventually, the drop is so diffused that it is no longer discernible.

Suppose that $\Omega \subset \mathbb{R}^3$ is a bounded, open subset of three-dimensional space, and $T > 0$ is a final time. Let $\phi \colon \Omega \times [0, T] \to \mathbb{R}$ be a density of dye particles. We will assume that the particle density ϕ is at least continuous, which is called the *continuum assumption*. Of course, in reality, this assumption eventually breaks down if we zoom in around any point $\boldsymbol{x} \in \Omega$ with a sufficiently large magnification, because then we would see the individual particles. The units of ϕ are

$$[\phi] = \frac{\text{number}}{\text{volume}}.$$

Let us suppose that there is background flow of some fluid medium, by which the dye particles are advected, described by the time-independent velocity field $\boldsymbol{u} \colon \Omega \to \mathbb{R}^3$. The units of the velocity vector field are

$$[\boldsymbol{u}] = \frac{\text{length}}{\text{time}}.$$

Oftentimes, we can and will assume that this flow field is *solenoidal*, or *divergence free*, meaning that

$$\nabla \cdot \boldsymbol{u}(\boldsymbol{x}) = \sum_{i=1}^{3} \frac{\partial u_i}{\partial x_i}(\boldsymbol{x}) = 0, \quad \forall \boldsymbol{x} \in \Omega.$$

This is a good approximation when the fluid is nearly incompressible, as for the case of water. The differential operator just defined, $\nabla \cdot$, acts upon differentiable vector-valued functions and is called the *divergence operator*. There is a companion differential operator that we will use frequently. The *gradient operator* acts upon differentiable scalar functions and is defined as

$$\nabla \phi(\boldsymbol{x}) = \sum_{i=1}^{3} \frac{\partial \phi}{\partial x_i}(\boldsymbol{x}) \boldsymbol{e}_i,$$

where $\boldsymbol{e}_i$ is the canonical unit vector.

Let $V \subset \Omega$ be a stationary control volume in Ω. The total number of dye particles in V is given by the formula

$$N_V(t) = \int_V \phi(\boldsymbol{x}, t) \mathrm{d}\boldsymbol{x}.$$

Now we can ponder by what mechanisms the number of particles in V changes with time. When we consider all of the ways that particles are created, destroyed, transported in, and/or transported out of V, we arrive at

$$\frac{\mathrm{d}N_V}{\mathrm{d}t}(t) = \int_V \frac{\partial \phi}{\partial t}(\boldsymbol{x}, t) \mathrm{d}\boldsymbol{x} = \int_{\partial V} -\boldsymbol{K} \cdot \boldsymbol{n}_V \, \mathrm{d}\boldsymbol{S} + \int_V S(\phi(\boldsymbol{x}, t), \boldsymbol{x}, t) \mathrm{d}\boldsymbol{x},$$

where ∂V is the bounding surface of V (its boundary); $\boldsymbol{n}_V$ is its outward-pointing unit normal vector; $\boldsymbol{K}$ is a vector-valued *total flux*; and S is a volumetric source term of particles. The term $\int_{\partial V} -\boldsymbol{K} \cdot \boldsymbol{n}_V \mathrm{d}\boldsymbol{S}$ accounts for the time rate of change of particles via a flux of particles across the boundary of the control volume. The term $\int_V S(\phi(\boldsymbol{x}, t), \boldsymbol{x}, t)\mathrm{d}\boldsymbol{x}$ accounts for the time rate of change of particles via a volumetric source term inside of V, as from a chemical reaction, or a spontaneous decay. In the case of our thought experiment with the dye in a stream, S would likely be identically zero.

The *Divergence Theorem*, which follows from Theorem B.59, states that

$$\int_V \nabla \cdot \boldsymbol{F}(\boldsymbol{x})\mathrm{d}\boldsymbol{x} = \int_{\partial V} \boldsymbol{F}(\boldsymbol{x}) \cdot \boldsymbol{n}_V \mathrm{d}\boldsymbol{S} \tag{23.1}$$

for all functions $\boldsymbol{F} \in C^1(V; \mathbb{R}^3) \cap C(\overline{V}; \mathbb{R}^3)$, provided that the boundary of V is sufficiently regular. Using the Divergence Theorem, we find

$$\int_V \left(\frac{\partial \phi}{\partial t}(\boldsymbol{x}, t) + \nabla \cdot \boldsymbol{K} - S(\phi(\boldsymbol{x}, t), \boldsymbol{x}, t) \right) \mathrm{d}\boldsymbol{x} = 0.$$

Now we argue that, since this result holds for any arbitrary control volume V, it must be that

$$\frac{\partial \phi}{\partial t}(\boldsymbol{x}, t) + \nabla \cdot \boldsymbol{K} - S(\phi(\boldsymbol{x}, t), \boldsymbol{x}, t) = 0, \quad \forall \boldsymbol{x} \in \Omega, \quad \forall t \in [0, T].$$

Let us now consider the total flux, $\boldsymbol{K}$, in more detail. It is a vector-valued function, as we have indicated, and its units are

$$[K] = \frac{\text{number}}{\text{area} \cdot \text{time}}.$$

Since there is a background flow field $\boldsymbol{u}$, it is common to decompose $\boldsymbol{K}$ as follows:

$$\boldsymbol{K} = \phi \boldsymbol{u}(\boldsymbol{x}) + \boldsymbol{J}(\boldsymbol{x}, \phi, \nabla \phi),$$

where $\boldsymbol{J}$ is called the *diffusive flux*. The first term is called the *linear advection flux*; and observe that it has the correct units to be a flux:

$$[\phi \boldsymbol{u}] = \frac{\text{number}}{\text{area} \cdot \text{time}}.$$

In the case of our thought experiment with the dye in a stream, this term accounts for the motion of the drop as it is carried by the background flow. The second term in the decomposition, $\boldsymbol{J}(\boldsymbol{x}, \phi, \nabla \phi)$, accounts for the random diffusive motion of the particles, as, for example, via *Brownian motion*.[1] Thus, we arrive at the general *conservation equation*: for all $\boldsymbol{x} \in \Omega$ and all $t \in [0, T]$,

$$\frac{\partial \phi}{\partial t}(\boldsymbol{x}, t) + \nabla \cdot (\phi(\boldsymbol{x}, t)\boldsymbol{u}(\boldsymbol{x})) = -\nabla \cdot \boldsymbol{J}(\boldsymbol{x}, \phi, \nabla \phi) + S(\phi, \boldsymbol{x}, t).$$

In our present example, this equation followed from the universal law of mass conservation. But it applies to any conserved quantity, like energy or electric charge;

[1] Named in honor of the British botanist and paleobotanist Robert Brown (1773–1858).

see Table 23.1. To construct a usable equation, we need to make some constitutive assumptions about the form of the diffusion flux, $\boldsymbol{J}$, and the source term.

In general, the diffusion flux $\boldsymbol{J}$ and the conserved quantity ϕ are related by a constitutive law of the form

$$\boldsymbol{J}(\boldsymbol{x}) = -\mathsf{A}(\boldsymbol{x}, \boldsymbol{J}(\boldsymbol{x}))\Psi(\phi, \nabla\phi),$$

which encodes information about the object in question and the physics that are taking place in it. It is said that the process is *physically linear* if

$$\mathsf{A} = \mathsf{A}(\boldsymbol{x}),$$

and *geometrically linear* if $\Psi = \nabla\phi$. In summary, the constitutive law that describes the behavior of a linear process is

$$\boldsymbol{J} = -\mathsf{A}(\boldsymbol{x})\nabla\phi.$$

In this setting, A is usually called the *material modulus*; see Table 23.1 for its interpretation in several physical scenarios.

23.1.2 Advection–Reaction–Diffusion Equation

As mentioned above, we must pay particularly close attention to the physics of the problem. In the mass conservation case, motivated by our thought experiment with the dye particles, when diffusion is dominated by random, Brownian-like motion, the standard model is *Fick's Law*, which states that

$$\boldsymbol{J}(\boldsymbol{x}, \phi, \nabla\phi) = -\mathsf{D}(\boldsymbol{x}, \phi(\boldsymbol{x}, t))\nabla\phi(\boldsymbol{x}, t),$$

where $\mathsf{D}\colon \mathbb{R}^3 \times \mathbb{R} \to \mathbb{R}^{3\times 3}$ is a diffusion coefficient matrix. This law implies that particles move, on average, from regions where their density is high to regions where their density is low. Intuitively, this makes sense.

Now we will typically assume that D is symmetric positive definite (SPD) at every point at which it is defined. One very common and simple model is

$$\mathsf{D} = a(\boldsymbol{x})\mathsf{I},$$

where I is the 3×3 identity matrix and $a(\boldsymbol{x}) > 0$ for all $\boldsymbol{x} \in \overline{\Omega}$. Now, regarding the source term, it is common to choose the linear model

$$S = -c(\boldsymbol{x})\phi(\boldsymbol{x}, t) + f(\boldsymbol{x}, t),$$

where f is a density-independent source of particles. When $c(\boldsymbol{x}) > 0$, the term $-c(\boldsymbol{x})\phi(\boldsymbol{x}, t)$ describes the decay of the particles, via degradation or evaporation, say. When $c(\boldsymbol{x}) < 0$, the term $-c(\boldsymbol{x})\phi(\boldsymbol{x}, t)$ describes the growth of particles, due, for example, to reproduction or a chemical reaction. The units of the source term are

$$[S] = \frac{\text{number}}{\text{volume} \cdot \text{time}},$$

which implies that c must have units

$$[c] = \frac{1}{\text{time}}.$$

Problem	Conserved quantity	ϕ	J	a	S	Law
Deformation of an elastic rod	linear momentum	displacement	stress	Young's modulus	body forces	Hooke[a]
Flow of particles in a material	mass	number density	mass flux	mobility	mass sources	Fick[b]
Heat conduction in a rod	energy	temperature	heat flux	thermal conductivity	heat sources	Fourier[c]
Fluid flow through a channel	linear momentum	velocity	shear stress	viscosity	body forces	Stokes[d]
Electrostatics	electric flux	electric potential	electric flux	dielectric permittivity	charge	Coulomb[e]
Flow through porous medium	mass	hydraulic head	flow rate	permeability	fluid source	Darcy[f]

Table 23.1 Interpretation of the variables of (23.2) for various types of physical problems. The last column gives the name of the constitutive law in that particular context.

a Named in honor of the English scientist, architect, and polymath Robert Hooke (1635–1703).
b Named in honor of the German physician and physiologist Adolf Eugen Fick (1829–1901).
c Named in honor of the French mathematician and physicist Jean-Baptiste Joseph Fourier (1768–1830).
d Named in honor of the Anglo-Irish physicist and mathematician Sir George Gabriel Stokes, 1st Baronet (1819–1903).
e Named in honor of the French military engineer and physicist Charles-Augustin de Coulomb (1736–1806).
f Named in honor of the French engineer Henry Philibert Gaspard Darcy (1803–1858).

Gathering all of our assumptions, we have, for all $\boldsymbol{x} \in \Omega$ and $t \in [0, T]$,

$$\frac{\partial \phi}{\partial t}(\boldsymbol{x}, t) + \nabla \cdot (\phi(\boldsymbol{x}, t)\boldsymbol{u}(\boldsymbol{x})) = \nabla \cdot (a(\boldsymbol{x})\nabla \phi(\boldsymbol{x}, t)) - c(x)\phi(\boldsymbol{x}, t) + f(\boldsymbol{x}, t).$$

But observe that

$$\nabla \cdot (\phi \boldsymbol{u}) = \boldsymbol{u} \cdot \nabla \phi + \phi(\nabla \cdot \boldsymbol{u}).$$

Thus, if the flow field $\boldsymbol{u}$ is solenoidal,

$$\frac{\partial \phi}{\partial t}(\boldsymbol{x}, t) + \boldsymbol{u}(\boldsymbol{x}) \cdot \nabla \phi(\boldsymbol{x}, t) = \nabla \cdot (a(\boldsymbol{x})\nabla \phi(\boldsymbol{x}, t)) - c(x)\phi(\boldsymbol{x}, t) + f(\boldsymbol{x}, t). \quad (23.2)$$

This specific conservation equation is called the *linear advection–reaction–diffusion equation*. When $\boldsymbol{u} \equiv \boldsymbol{0}$, it is called the *linear reaction–diffusion equation*; and, when $c \equiv 0$, it is called the *linear advection–diffusion equation*. Perhaps the most well-known variant is the *diffusion equation*, which is obtained by setting $\boldsymbol{u} \equiv \boldsymbol{0}$, $c \equiv 0$, and $f \equiv 0$:

$$\frac{\partial \phi}{\partial t}(\boldsymbol{x}, t) = \nabla \cdot (a(\boldsymbol{x})\nabla \phi(\boldsymbol{x}, t)). \quad (23.3)$$

To form a well-defined problem, i.e., one that has a unique solution, we need to specify initial and boundary conditions. An initial condition is one of the form

$$\phi(\boldsymbol{x}, 0) = q(\boldsymbol{x}), \quad \forall \boldsymbol{x} \in \Omega.$$

Boundary conditions can come in a few different varieties. Suppose that $\boldsymbol{n}$ is the outward-pointing unit normal vector on the boundary of Ω, denoted by $\partial\Omega$. We will tacitly assume that $\partial\Omega$ is sufficiently smooth (nice), so that $\boldsymbol{n}$ is well defined and single valued, except possibly at a set of surface area zero. We define the so-called *boundary normal derivative* as

$$\frac{\partial \phi}{\partial n}(\boldsymbol{x}, t) = \boldsymbol{n}(\boldsymbol{x}) \cdot \nabla \phi(\boldsymbol{x}, t), \quad \forall \boldsymbol{x} \in \partial\Omega, \quad \forall t \in (0, T].$$

There are three common types of boundary conditions for an advection–reaction–diffusion problem. The first is called a *Neumann boundary condition*[2] (or a *natural boundary condition*), which is defined as

$$\frac{\partial \phi}{\partial n}(\boldsymbol{x}, t) = \alpha(\boldsymbol{x}, t), \quad \forall \boldsymbol{x} \in \partial\Omega, \quad \forall t \in (0, T],$$

where $\alpha\colon \partial\Omega \times (0, T] \to \mathbb{R}$ is a given function. The second is called a *Dirichlet boundary condition*[3] (or an *essential boundary condition*), which is defined as

$$\phi(\boldsymbol{x}, t) = \beta(\boldsymbol{x}), \quad \forall \boldsymbol{x} \in \partial\Omega, \quad \forall t \in (0, T],$$

where $\beta\colon \partial\Omega \times (0, T] \to \mathbb{R}$ is a given function. The third is called a *Robin boundary condition*,[4] which is a kind of combination of the previous two:

$$\frac{\partial \phi}{\partial n}(\boldsymbol{x}, t) + \gamma(\boldsymbol{x})\phi(\boldsymbol{x}, t) = \chi(\boldsymbol{x}, t), \quad \forall \boldsymbol{x} \in \partial\Omega, \quad \forall t \in (0, T],$$

[2] Named in honor of the German mathematician Carl Gottfried Neumann (1832–1925).

[3] Named in honor of the German mathematician Johann Peter Gustav Lejeune Dirichlet (1805–1859).

[4] Named in honor of the French mathematician Victor Gustave Robin (1855–1897).

where $\gamma, \chi \colon \partial\Omega \times (0,T] \to \mathbb{R}$ are given functions.

Boundary conditions can be mixed. For example, the boundary can be split into two or more pairwise disjoint regions, on which different boundary conditions are imposed. Which boundary condition is required where, and when, is a matter of the underlying physics (or characteristics) of the problem, as the following example suggests.

Example 23.1 Suppose that $\boldsymbol{u} \equiv \boldsymbol{0}$ and $S \equiv 0$. Then the motion of the dye particles in our thought experiment is only by Fickian diffusion, and the density of particles satisfies the diffusion equation,

$$\frac{\partial \phi}{\partial t}(\boldsymbol{x},t) = \nabla \cdot (a(\boldsymbol{x})\nabla\phi(\boldsymbol{x},t)), \quad \forall \boldsymbol{x} \in \Omega, \quad \forall t \in (0,T]. \tag{23.4}$$

Suppose that Ω describes the spatial location of a closed container of fluid at rest from the macroscopic point of view. The function $a \colon \overline{\Omega} \to \mathbb{R}$ measures some known heterogeneity of the fluid that affects the diffusion of particles. The initial condition

$$\phi(\boldsymbol{x},0) = q(\boldsymbol{x}), \quad \forall \boldsymbol{x} \in \Omega$$

describes the initial concentration of the dye particles. Perhaps the particles are highly concentrated in some location initially because we have carefully inserted the drop via a syringe. Of course, after the initial preparation of the dye particles, they diffuse throughout the background fluid. But, importantly, there can be no flux of particles through the boundary $\partial\Omega$, since the container is closed. These *no-flux boundary conditions* are described via

$$-a(\boldsymbol{x})\nabla\phi(\boldsymbol{x},t)\cdot\boldsymbol{n}(\boldsymbol{x}) = \boldsymbol{J}\cdot\boldsymbol{n}(\boldsymbol{x}) = 0, \quad \forall \boldsymbol{x} \in \partial\Omega, \quad \forall t \in (0,T].$$

Since $a(\boldsymbol{x}) > 0$, for all $\boldsymbol{x} \in \overline{\Omega}$, the boundary condition must be of Neumann type:

$$\frac{\partial \phi}{\partial n}(\boldsymbol{x},t) = 0, \quad \forall \boldsymbol{x} \in \partial\Omega, \quad \forall t \in (0,T].$$

We must reiterate that equation (23.2) was derived as a mass conservation equation, but it is much more general than that. It applies for any conserved quantity, mass, momentum, charge, energy, etc. Table 23.1 describes the interpretation of ϕ, $\boldsymbol{J}$, a, and S for various types of physical problems.

23.1.3 The Linear Transport Equation

Suppose again that $\boldsymbol{u}$ is solenoidal, i.e., $\nabla \cdot \boldsymbol{u} \equiv 0$. Let us assume that random diffusion may be neglected, i.e., $\boldsymbol{J} \equiv \boldsymbol{0}$. In this case, passive advection via the background flow is the only means by which the particles are transported from one spatial location to another. Finally, let us assume that $S \equiv 0$, which is appropriate when there are no reactions to produce or degrade the particles, no evaporation, and no creation or destruction of particles via any other processes. Under these assumptions, (23.2) becomes

$$\frac{\partial \phi}{\partial t}(\boldsymbol{x}, t) + \boldsymbol{u}(\boldsymbol{x}) \cdot \nabla \phi(\boldsymbol{x}, t) = 0, \tag{23.5}$$

which is known as the *linear transport equation.* Formally, we can write down a very simple solution to this equation; namely,

$$\phi(\boldsymbol{x}, t) = q(\boldsymbol{x} - \boldsymbol{u}(\boldsymbol{x})t),$$

where $\phi(\,\cdot\,, 0) = q$ is the initial condition. Observe that (23.5) is satisfied pointwise using our proposed solution, provided that q is differentiable. One can prove this using the Chain Rule in multiple dimensions.

Let us now consider what type of boundary conditions are meaningful in the present case. Unlike the case of the previous section, where diffusion could not be neglected, we do not need boundary conditions on all parts of the boundary. In fact, let us divide $\partial\Omega$ into two disjoint parts,

$$\Gamma_{\text{inflow}} = \{\boldsymbol{x} \in \partial\Omega \mid \boldsymbol{n}(\boldsymbol{x}) \cdot \boldsymbol{u}(\boldsymbol{x}) < 0\}, \quad \Gamma_{\text{outflow}} = \{\boldsymbol{x} \in \partial\Omega \mid \boldsymbol{n}(\boldsymbol{x}) \cdot \boldsymbol{u}(\boldsymbol{x}) \geq 0\},$$

called the *inflow* and *outflow* boundaries, respectively. To fully define the linear transport problem, we need an initial condition, i.e.,

$$\phi(\boldsymbol{x}, 0) = q(\boldsymbol{x}), \quad \forall \boldsymbol{x} \in \Omega,$$

and we need to specify values for ϕ only on the inflow boundary,

$$\phi(\boldsymbol{x}, t) = r(\boldsymbol{x}, t), \quad \forall \boldsymbol{x} \in \Gamma_{\text{inflow}}, \quad \forall t \in (0, T].$$

These are called *inflow boundary conditions.*

23.1.4 Stationary Conservation Equations and Boundary Value Problems

Let us assume that, in (23.2),

$$\lim_{t\to\infty} \frac{\partial f}{\partial t}(\,\cdot\,, t) \equiv 0.$$

In other words, eventually, the source term f becomes independent of time. In this case, it is possible — depending upon a few other factors, including what boundary conditions are imposed — that the density of particles reaches a *steady state* at very large times. This means, specifically, that

$$\lim_{t\to\infty} \frac{\partial \phi}{\partial t}(\,\cdot\,, t) \equiv 0.$$

In this case, the density of particles is essentially time invariant after enough time has elapsed. This does not mean that the density at steady state is constant; indeed, it can, and usually will, still have spatial variation. In any case, at steady state, we have the equation

$$-\nabla \cdot (a(\boldsymbol{x})\nabla \phi_\infty(\boldsymbol{x})) + \boldsymbol{u}(\boldsymbol{x}) \cdot \nabla \phi_\infty(\boldsymbol{x}) + c(x)\phi_\infty(\boldsymbol{x}) = f_\infty(\boldsymbol{x}), \tag{23.6}$$

where

$$\phi_\infty(\boldsymbol{x}) = \lim_{t\to\infty} \phi(\boldsymbol{x}, t), \quad f_\infty(\boldsymbol{x}) = \lim_{t\to\infty} f(\boldsymbol{x}, t), \quad \forall \boldsymbol{x} \in \Omega,$$

assuming that these limits make sense. Dropping the subscripts, we obtain

$$-\nabla \cdot (a(\boldsymbol{x})\nabla\phi(\boldsymbol{x})) + \boldsymbol{u}(\boldsymbol{x}) \cdot \nabla\phi(\boldsymbol{x}) + c(\boldsymbol{x})\phi(\boldsymbol{x}) = f(\boldsymbol{x}), \tag{23.7}$$

which is the *stationary advection–reaction–diffusion equation*. In the case that $\boldsymbol{u} \equiv \boldsymbol{0}$, we obtain what is called the *stationary reaction–diffusion equation*

$$-\nabla \cdot (a(\boldsymbol{x})\nabla\phi(\boldsymbol{x}, t)) + c(\boldsymbol{x})\phi(\boldsymbol{x}) = f(\boldsymbol{x}). \tag{23.8}$$

When, additionally, $c \equiv 0$, we obtain the *stationary diffusion equation*

$$-\nabla \cdot (a(\boldsymbol{x})\nabla\phi(\boldsymbol{x})) = f(\boldsymbol{x}). \tag{23.9}$$

The last equation, in the particular case that $a \equiv 1$, becomes the *Poisson equation*:[5]

$$-\Delta\phi(\boldsymbol{x}) = f(\boldsymbol{x}), \tag{23.10}$$

where we used the so-called Laplacian,[6] which, for a scalar-valued function w that is twice differentiable at the point $\boldsymbol{x} \in \mathbb{R}^d$, is defined as:

$$\Delta w(\boldsymbol{x}) = \nabla \cdot (\nabla w(\boldsymbol{x})) = \sum_{i=1}^{d} \frac{\partial^2 w}{\partial x_i^2}(\boldsymbol{x}).$$

To solve any of the stationary problems represented by (23.7)–(23.9), we do not need initial conditions. We do, however, need boundary conditions of one of the three types mentioned earlier: Neumann, Dirichlet, or Robin. The combination of (23.7) with appropriate boundary conditions is called a *boundary value problem* (BVP).

23.1.5 The Heat Equation

Let us derive another famous equation, the *heat equation*. We will see that it is related to an equation that we have already derived. Consider the internal energy of a solid material, like a block of copper, that occupies the region Ω:

$$E = \int_\Omega e(\theta(\boldsymbol{x}), \rho(\boldsymbol{x}))\mathrm{d}\boldsymbol{x},$$

where e is the internal energy density, $\theta\colon \Omega \to \mathbb{R}$ is the temperature field, and $\rho\colon \Omega \to \mathbb{R}$ is the mass density. The units are as follows:

$$[e] = \frac{\text{energy}}{\text{volume}}, \quad [\theta] = \text{degrees}, \quad [\rho] = \frac{\text{mass}}{\text{volume}}.$$

[5] Named in honor of the French mathematician, engineer, and physicist Baron Siméon Denis Poisson (1781–1840).

[6] Named in honor of the French scholar and polymath Pierre-Simon, Marquis de Laplace (1749–1827).

Incidentally, in SI units, energy is measured in joules, length is measured in meters, and temperature is measured in kelvin. One common, and quite simple, model for the internal energy is given by

$$e(\theta, \rho) = C\rho\theta,$$

where $C > 0$ is the specific heat capacity, which we assume is constant. It has units

$$[C] = \frac{\text{energy}}{\text{mass} \cdot \text{degrees}}.$$

Suppose that our material is insulated from the outside world. This means that no energy can pass to or from Ω. The internal energy must be conserved. Let $V \subset \Omega$ be an arbitrary control volume. Define

$$E_V(t) = \int_V e(\theta(\boldsymbol{x}, t), \rho(\boldsymbol{x}, t))\mathrm{d}\boldsymbol{x}.$$

Then

$$\begin{aligned}\frac{\mathrm{d}E_V}{\mathrm{d}t}(t) &= \int_V \frac{\partial}{\partial t} e(\theta(\boldsymbol{x}, t), \rho(\boldsymbol{x}, t))\mathrm{d}\boldsymbol{x} \\ &= \int_V \left(C\frac{\partial \rho}{\partial t}(\boldsymbol{x}, t)\theta(\boldsymbol{x}, t) + C\rho(\boldsymbol{x}, t)\frac{\partial \theta}{\partial t}(\boldsymbol{x}, t) \right) \mathrm{d}\boldsymbol{x}.\end{aligned}$$

To simplify the situation, let us assume that the change in the mass density with respect to time is negligible, i.e.,

$$\frac{\partial \rho}{\partial t}(\,\cdot\,, t) \equiv 0.$$

Similar to the mass conservation case, we assume that energy can move into the control volume V through a diffusive flux $\boldsymbol{J}$. However, to keep the discussion simple, we assume that there are no energy sources within V. And, of course, since the material is solid, there is no background flow. Thus,

$$\frac{\mathrm{d}E_V}{\mathrm{d}t}(t) = -\int_{\partial V} \boldsymbol{n}_V \cdot \boldsymbol{J}(\boldsymbol{x})\mathrm{d}\boldsymbol{S} = -\int_V \nabla \cdot \boldsymbol{J}\,\mathrm{d}\boldsymbol{x},$$

where we have used the Divergence Theorem (23.1). Consequently,

$$\int_V \left(C\rho(\boldsymbol{x})\frac{\partial \theta}{\partial t}(\boldsymbol{x}, t) + \nabla \cdot \boldsymbol{J} \right) \mathrm{d}\boldsymbol{x} = 0.$$

Since the control volume is arbitrary,

$$C\rho(\boldsymbol{x})\frac{\partial \theta}{\partial t}(\boldsymbol{x}, t) = -\nabla \cdot \boldsymbol{J}, \quad \forall \boldsymbol{x} \in \Omega, \quad \forall t \in (0, T],$$

where $T > 0$ is the final time (the end of our thought experiment). To finish the derivation of the heat equation, we invoke *Fourier's Law*, which states

$$\boldsymbol{J} = -a(\boldsymbol{x})\nabla\theta(\boldsymbol{x}, t).$$

Consistent with the *Second Law of Thermodynamics*, heat energy flows, or fluxes, from hot regions into cold regions, and is proportional to $\nabla\theta$. Recall, from your vector calculus class, that the gradient of θ points in the direction of the greatest increase of θ. The coefficient $a\colon \overline{\Omega} \to \mathbb{R}$ is positive and point-wise, and is called the diffusion coefficient. Thus, we have

$$C\rho(\boldsymbol{x})\frac{\partial \theta}{\partial t}(\boldsymbol{x}, t) = \nabla \cdot (a(\boldsymbol{x})\nabla\theta), \tag{23.11}$$

which is called the *heat equation*.

Let us simplify our model a bit further. Suppose that, to a good approximation, $\rho \equiv \rho_o$, where $\rho_o > 0$ is a constant, and $a \equiv a_o$, where $a_o > 0$ is a constant. Then

$$\frac{\partial \theta}{\partial t}(\boldsymbol{x}, t) = D\Delta\theta(\boldsymbol{x}, t), \tag{23.12}$$

where Δ is the Laplacian operator and $D = \frac{a_o}{C\rho_o} > 0$ is the *diffusion constant*, which has units

$$[D] = \frac{\text{area}}{\text{time}}.$$

Regarding boundary conditions, recall that we assumed that the material was thermally insulated. Mathematically, this means that there can be no flux of thermal energy across $\partial\Omega$:

$$-a(\boldsymbol{x})\frac{\partial \theta}{\partial n}(\boldsymbol{x}, t) = \boldsymbol{n} \cdot \boldsymbol{J} = 0, \quad \boldsymbol{x} \in \partial\Omega, \quad \forall t \in (0, T].$$

This implies that the appropriate boundary condition is one of Neumann type:

$$\frac{\partial \theta}{\partial n}(\boldsymbol{x}, t) = 0, \quad \forall \boldsymbol{x} \in \partial\Omega, \quad \forall t \in (0, T].$$

To close the system, we only need to add an initial condition; namely, the initial temperature profile of the material in Ω.

It is not difficult to show that energy is conserved in our system with insulating boundary conditions. Indeed, using the Divergence Theorem, given in (23.1), we obtain

$$\begin{aligned}
\frac{\mathrm{d}E_\Omega}{\mathrm{d}t} &= \frac{\mathrm{d}}{\mathrm{d}t}\int_\Omega e(\theta, \rho)\mathrm{d}\boldsymbol{x} \\
&= \int_\Omega \left(\frac{\partial e}{\partial \theta}\frac{\partial \theta}{\partial t}(\boldsymbol{x}, t) + \frac{\partial e}{\partial \rho}\frac{\partial \rho}{\partial t}(\boldsymbol{x}, t)\right)\mathrm{d}\boldsymbol{x} \\
&= \int_\Omega C\rho(\boldsymbol{x})\frac{\partial \theta}{\partial t}(\boldsymbol{x}, t)\mathrm{d}\boldsymbol{x} \\
&= \int_\Omega C\rho_o\frac{a_o}{C\rho_o}\Delta\theta(\boldsymbol{x}, t)\mathrm{d}\boldsymbol{x} \\
&= \int_{\partial\Omega} a_o\boldsymbol{n}(\boldsymbol{x}) \cdot \nabla\theta(\boldsymbol{x}, t)\mathrm{d}\boldsymbol{a} \\
&= 0,
\end{aligned}$$

where we have used the homogeneous Neumann boundary condition in the last step.

Interestingly, while the energy of the system is conserved, the entropy of the system, which measures the amount of disorder in the system, increases until the steady state, or equilibrium, is reached. This makes sense intuitively. Consider the following example.

Example 23.2 Consider a long, cylindrical rod of solid material of constant cross-sectional area A, with diffusion constant $D = 1$. (We will suppress units to make the calculations simpler.) Suppose that it is oriented along the x_1 axis. In this thought experiment, we heat up the end of the rod at $x_1 = 10$ to the temperature $\theta_R = 200$, and we cool the rod at the other end, $x_1 = -10$, to the temperature $\theta_L = 0$. Assume, for simplicity, that the initial temperature profile in the rod is

$$\theta(\boldsymbol{x}, 0) = 100 + 100 \sin\left(\frac{\pi x_1}{20}\right).$$

In other words, we assume that the initial profile depends only upon the long axis position, x_1. Now, after heating, let us quickly insulate the rod on all sides.

To further simplify the problem, let us assume that the variation of the temperature with respect to the coordinates x_2 and x_3 can always be neglected. Thus, the equation that we need to solve is

$$\frac{\partial \theta}{\partial t}(x_1, t) = \frac{\partial^2 \theta}{\partial x_1^2}(x_1, t),$$

subject to the boundary conditions

$$\frac{\partial \theta}{\partial x}(x_1, t) = 0, \quad x_1 = \pm 10.$$

This problem has an exact solution; namely,

$$\theta(x_1, t) = 100 + 100 \exp\left(-\frac{\pi^2 t}{400}\right) \sin\left(\frac{\pi x_1}{20}\right).$$

Thus, if we heat up one end of a cylindrical solid block of material, cool the other end, and then quickly insulate the entire system, heat energy, which is conserved in the system, flows from the hotter to the colder side, until a constant temperature is reached in the system. In fact, the reader can check that

$$\lim_{t \to \infty} \theta(x_1, t) = 100, \quad \forall x_1 \in [-10, 10].$$

If we wait long enough, we will, to a high degree of accuracy, observe this steady state, which can also be termed the equilibrium state, in this case. When equilibrium is reached, we cannot tell which end of the system was heated and which was cooled. Information has been forever lost, and disorder (entropy) has increased to its maximum possible value.

23.1.6 The Wave Equation

The last equation that we will consider is the *wave equation*. Consider an initially flat elastic sheet of material that occupies the domain $\Omega \subset \mathbb{R}^2$. Let $u \colon \Omega \times [0, T] \to$

$\mathbb{R}$ describe the displacement of the sheet in the vertical direction (the direction perpendicular to the initially flat sheet). The units are

$$[u] = \text{length}.$$

Suppose that the sheet is pinned along the boundary $\partial\Omega$. (Think about a drum head, for example.) In other words,

$$u(\mathbf{x}, t) = 0, \quad \forall \mathbf{x} \in \partial\Omega, \quad \forall t \in (0, T].$$

Now, the energy of the sheet — to a good approximation, provided the displacement is small, always and everywhere — can be written as

$$E(t) = \int_\Omega \left[\frac{\rho_o}{2} \left| \frac{\partial u}{\partial t}(\mathbf{x}, t) \right|^2 + \frac{\tau_o}{2} \|\nabla u(\mathbf{x}, t)\|_2^2 \right] d\mathbf{x},$$

where $\rho_o > 0$ is the constant density per unit area of the sheet, and $\tau_o > 0$ is the surface tension, We recall that

$$\nabla u(\mathbf{x}, t) = \left[\frac{\partial u}{\partial x_1}(\mathbf{x}, t) \; \frac{\partial u}{\partial x_2}(\mathbf{x}, t) \right]^\intercal;$$

consequently,

$$\|\nabla u(\mathbf{x}, t)\|_2^2 = \left(\frac{\partial u}{\partial x_1}(\mathbf{x}, t) \right)^2 + \left(\frac{\partial u}{\partial x_2}(\mathbf{x}, t) \right)^2.$$

The first term in the energy is the *kinetic energy*, whereas the second represents the *potential* or *restorative energy* of the sheet. The units are

$$[\rho_o] = \frac{\text{mass}}{\text{area}}, \quad [\tau_o] = \frac{\text{energy}}{\text{area}} = \frac{\text{force}}{\text{length}}.$$

Note that $\frac{\tau_o}{\rho_o}$ has units of velocity squared:

$$\left[\frac{\tau_o}{\rho_o} \right] = \frac{\text{length}^2}{\text{time}^2}.$$

The *wave equation*, which models the motion of our sheet, is the following:

$$\frac{\partial^2 u}{\partial t^2}(\mathbf{x}, t) = c^2 \Delta u(\mathbf{x}, t), \quad \forall \mathbf{x} \in \Omega, \quad \forall t \in (0, T], \tag{23.13}$$

where

$$c^2 = \frac{\tau_o}{\rho_o}$$

is the *wave speed* or *speed of propagation*. Several references derive this equation from Newton's second law, or via energy methods; see, for example, [49, 99]. We will not derive the equation here, but will describe some energy properties of the solutions. As we have indicated, the boundary conditions are of homogeneous Dirichlet type, since the deformation is zero at the boundary. The wave equation requires two initial conditions, which is not surprising, since the equation involves a second partial derivative with respect to time. The initial conditions are usually of the form

$$u(\boldsymbol{x},0) = \alpha(\boldsymbol{x}), \quad \frac{\partial u}{\partial t}(\boldsymbol{x},0) = \beta(\boldsymbol{x}), \quad \forall \boldsymbol{x} \in \Omega,$$

where $\alpha, \beta\colon \Omega \to \mathbb{R}$ are given functions.

Let us, at least formally, show that the wave equation is energy conservative, i.e., $\frac{\mathrm{d}}{\mathrm{d}t}E = 0$. This will be made rigorous in Theorem 23.39. We will need the following integration-by-parts formula, which is presented in Theorem B.59:

$$\int_\Omega \boldsymbol{v}(\boldsymbol{x}) \cdot \nabla u(\boldsymbol{x}) \mathrm{d}\boldsymbol{x} = \int_{\partial\Omega} \boldsymbol{n}(\boldsymbol{x}) \cdot \boldsymbol{v}(\boldsymbol{x}) u(\boldsymbol{x}) \mathrm{d}\boldsymbol{S} - \int_\Omega \nabla \cdot \boldsymbol{v}(\boldsymbol{x}) u(\boldsymbol{x}) \mathrm{d}\boldsymbol{x}$$

for all $\boldsymbol{v} \in C^1(\bar{\Omega}; \mathbb{R}^2)$ and $u \in C^1(\bar{\Omega}; \mathbb{R})$. Applying this last result with the vector $\boldsymbol{v} = \nabla q$, we have

$$\int_\Omega \nabla q(\boldsymbol{x}) \cdot \nabla u(\boldsymbol{x}) \mathrm{d}\boldsymbol{x} = \int_{\partial\Omega} \boldsymbol{n}(\boldsymbol{x}) \cdot \nabla q(\boldsymbol{x}) u(\boldsymbol{x}) \mathrm{d}\boldsymbol{S} - \int_\Omega \Delta q(\boldsymbol{x}) u(\boldsymbol{x}) \mathrm{d}\boldsymbol{x}.$$

Taking the time derivative of the energy and using the second integration-by-parts formula, we have

$$\begin{aligned}
\frac{\mathrm{d}E}{\mathrm{d}t}(t) &= \frac{\mathrm{d}}{\mathrm{d}t} \int_\Omega \left\{ \frac{\rho_o}{2} \left(\frac{\partial u}{\partial t}\right)^2 + \frac{\tau_o}{2} \left(\left(\frac{\partial u}{\partial x_1}\right)^2 + \left(\frac{\partial u}{\partial x_2}\right)^2 \right) \right\} \mathrm{d}\boldsymbol{x} \\
&= \int_\Omega \left\{ \rho_o \frac{\partial u}{\partial t} \frac{\partial^2 u}{\partial t^2} + \tau_o \nabla u \cdot \nabla \left(\frac{\partial u}{\partial t}\right) \right\} \mathrm{d}\boldsymbol{x} \\
&= \int_\Omega \left\{ \rho_o \frac{\partial u}{\partial t} \frac{\partial^2 u}{\partial t^2} - \tau_o \Delta u \frac{\partial u}{\partial t} \right\} \mathrm{d}\boldsymbol{x} + \tau_o \int_{\partial\Omega} \boldsymbol{n} \cdot \nabla u \frac{\partial u}{\partial t} \mathrm{d}\boldsymbol{x}.
\end{aligned}$$

Since u vanishes on the boundary, its time derivative is zero there as well. Hence, the boundary integral term is zero. With a little rewriting, we have

$$\frac{\mathrm{d}E}{\mathrm{d}t}(t) = \int_\Omega \left\{ \rho_o \frac{\partial u}{\partial t} \left(\frac{\partial^2 u}{\partial t^2} - c^2 \Delta u \right) \right\} \mathrm{d}\boldsymbol{x} = 0.$$

Solutions to the wave equation conserve energy. They do this by transmitting energy (without loss) at speed c to various parts of the domain; see, for example, the discussions in [49, 99]. Solutions to the wave equation also conserve entropy. There is no loss of information in the system as time progresses. This is related to the fact that solutions of the wave equation are time reversible.

Let us, from the wave equation, derive another important equation. To do so, we recall the physical origins of the wave equation; it is related to wave propagation. In other words, periodic, in time, solutions to this equation are of importance. For this reason, we propose the following *solution ansatz*:

$$u(\boldsymbol{x}, t) = e^{\mathrm{i}\omega t} \widehat{u}(\boldsymbol{x}),$$

where ω can be regarded as the wave frequency. Substituting into (23.13), we see that the function $\hat{u}$ must be a solution to the so-called *Helmholtz equation*:[7]

[7] Named in honor of the Prussian mathematician and physician Hermann Ludwig Ferdinand von Helmholtz (1821–1894).

$$\Delta u(\boldsymbol{x}) + k^2 u(\boldsymbol{x}) = 0, \tag{23.14}$$

where $k = \frac{\omega}{c}$. Notice that this equation bears resemblance to the reaction–diffusion equation (23.8), but we have a definite sign on the zero-order term, i.e., the term that does not involve partial derivatives of our unknown function.

A close relative of the wave equation is the *damped wave equation*:

$$\frac{\partial^2 u}{\partial t^2}(x, t) = c^2 \Delta u(x, t) - \frac{\beta}{\rho_o}\frac{\partial u}{\partial t}(x, t), \quad \forall x \in \Omega, \quad \forall t \in (0, T], \tag{23.15}$$

where $\beta \geq 0$ is a damping parameter with units

$$[\beta] = \frac{\text{mass}}{\text{time} \cdot \text{area}}.$$

The last term measures the effect of the frictional force per unit area and is proportional to the velocity of the displacement. In other words, this measures a drag effect. This type of term is responsible for the fact that a guitar string does not vibrate forever. Instead, its vibration is damped over time. It is not hard to show, see Problem 23.1, that energy is dissipated in the damped system at the rate

$$\frac{dE}{dt}(t) = \int_\Omega \left\{\rho_o \frac{\partial u}{\partial t}\left(\frac{\partial^2 u}{\partial t^2} - c^2 \Delta u\right)\right\} d\boldsymbol{x} = -\beta \int_\Omega \left(\frac{\partial u}{\partial t}\right)^2 d\boldsymbol{x}, \tag{23.16}$$

assuming, as before, pinned displacement (homogeneous Dirichlet) boundary conditions.

23.1.7 Classification of Linear Second-Order Equations

In all the examples above, owing to our assumptions, we obtained a linear, second-order PDE which is supplemented by boundary, and, possibly, initial, conditions. In the simplest case, this equation has constant coefficients, so that, for some $m \in \mathbb{N}$ and $\boldsymbol{x} \in \mathbb{R}^m$, it reads $\mathcal{D}u(\boldsymbol{x}) = f(\boldsymbol{x})$, where

$$\begin{aligned}\mathcal{D}v(\boldsymbol{x}) &= \sum_{i,j=1}^m a_{i,j}\frac{\partial^2 v(\boldsymbol{x})}{\partial x_i \partial x_j} + \sum_{i=1}^m b_i \frac{\partial v(\boldsymbol{x})}{\partial x_i} + cv(\boldsymbol{x}) \\ &= \mathsf{A} : D^2 v(\boldsymbol{x}) + \boldsymbol{b} \cdot \nabla v(\boldsymbol{x}) + cv(\boldsymbol{x}),\end{aligned} \tag{23.17}$$

where $\mathsf{A} = [a_{i,j}] \in \mathbb{R}^{m \times m}$; $\boldsymbol{b} = [b_i]^\intercal \in \mathbb{R}^m$; the symbol : is the so-called Frobenius[8] inner product on $\mathbb{R}^{m\times m}$, which we recall is defined as

$$\mathsf{A} : \mathsf{B} = \operatorname{tr}(\mathsf{A}\mathsf{B}^\intercal), \quad \forall \mathsf{A}, \mathsf{B} \in \mathbb{R}^{m\times m};$$

$D^2 v(\boldsymbol{x})$ denotes the Hessian of v at the point $\boldsymbol{x}$; and $\nabla v(\boldsymbol{x})$ denotes its gradient at the same point. We comment that, since for smooth functions the Hessian matrix is symmetric, there is no loss of generality in assuming that the matrix A is symmetric as well.

[8] Named in honor of the German mathematician Ferdinand Georg Frobenius (1849–1917).

The behavior of the solution to an equation like $\mathcal{D}u = f$ heavily depends on the spectrum of A. For this reason, we give the following definition.

Definition 23.1 (classification)**.** Consider the second-order, constant-coefficient partial differential operator $\mathcal{D}$. We say this operator is:

1. **Elliptic:** If $\sigma(\mathsf{A}) \subset \mathbb{R}_+$ or $\sigma(\mathsf{A}) \subset \mathbb{R}_-$.
2. **Parabolic:** If A contains exactly $m-1$ either positive or negative eigenvalues, and zero is an eigenvalue of multiplicity one.
3. **Hyperbolic:** If A has $m-1$ either positive or negative eigenvalues, and the remaining one is nonzero and of opposite sign.
4. **Ultra-parabolic:** If zero is a multiple eigenvalue, and all the remaining ones have the same sign.
5. **Ultra-hyperbolic:** If zero is not an eigenvalue, and there is more than one positive and more than one negative eigenvalue.

If the matrix A depends on the spatial variable $\boldsymbol{x} \in \mathbb{R}^d$, then the classification is done at each point.

Example 23.3 The conservation equation, presented in (23.7) for $d = 3$, and its higher-dimensional ($d > 3$) counterparts are elliptic, as the material modulus $a(\boldsymbol{x})$ is assumed to always be strictly positive. The case $a \equiv 1$ in (23.7) corresponds to $\mathsf{A} \equiv \mathsf{I}_d$ in (23.17). In this case, we obtain the so-called *minus Laplacian operator*[9] in the highest order term:

$$-\Delta v(\boldsymbol{x}) = -\mathsf{I}_d : D^2 v(\boldsymbol{x}) = -\sum_{i=1}^{d} \frac{\partial^2 v(\boldsymbol{x})}{\partial x_i^2}.$$

Example 23.4 Equation (23.2) ($d = 3$) and its higher-dimensional ($d > 3$) counterparts are parabolic. Notice that, in this case, $m = d+1$ and $(\boldsymbol{x}, t) \in \mathbb{R}^{d+1}$ are the independent variables. The case $a \equiv 1$, $\boldsymbol{u} \equiv \boldsymbol{0}$, and $c \equiv 0$ in (23.2) corresponds to the following *heat* or *diffusion* operator:

$$\mathcal{D}v(\boldsymbol{x}, t) = Hv(\boldsymbol{x}, t) = \frac{\partial v(\boldsymbol{x}, t)}{\partial t} + \begin{bmatrix} -\mathsf{I}_d & \boldsymbol{0} \\ \boldsymbol{0}^\intercal & 0 \end{bmatrix} : D^2 v(\boldsymbol{x}, t) = \frac{\partial v(\boldsymbol{x}, t)}{\partial t} - \Delta v(\boldsymbol{x}, t),$$

where the Laplacian, Δ, is taken over the spatial variables $\boldsymbol{x}$ only.

Example 23.5 Equation (23.13) ($d = 2$) and its higher-dimensional ($d > 3$) counterparts are hyperbolic. Again, we have $m = d + 1$, and $(\boldsymbol{x}, t) \in \mathbb{R}^{d+1}$ are the independent variables. The appropriate operator is

$$\mathcal{D}v(\boldsymbol{x}, t) = \Box v(\boldsymbol{x}, t) = \begin{bmatrix} -\mathsf{I}_d & \boldsymbol{0} \\ \boldsymbol{0}^\intercal & 1 \end{bmatrix} : D^2 v(\boldsymbol{x}, t) = \frac{\partial^2 v(\boldsymbol{x}, t)}{\partial t^2} - \Delta v(\boldsymbol{x}, t),$$

[9] Named in honor of the French scholar and polymath Pierre-Simon, Marquis de Laplace (1749–1827).

which is known as the *wave operator*. Again, the Laplacian, Δ, is taken over the spatial variables $\boldsymbol{x}$ only.

Example 23.6 Let us consider an equation that changes type. The so-called Euler–Tricomi equation[10] reads

$$\frac{\partial^2 u(x,y)}{\partial x^2} + x\frac{\partial^2 u(x,y)}{\partial y^2} = 0.$$

This equation is elliptic for $x > 0$, parabolic for $x = 0$, and hyperbolic when $x < 0$.

23.2 Elliptic Equations

In this section, we will focus on the theory of BVPs for elliptic equations. Let $d \in \mathbb{N}$ and $\Omega \subseteq \mathbb{R}^d$ be a bounded domain with boundary $\partial\Omega$. Given $\mathsf{A} \in C(\bar{\Omega}; \mathbb{R}^{d\times d})$, $\boldsymbol{b} \in C(\bar{\Omega}; \mathbb{R}^d)$, and $c \in C(\bar{\Omega})$, we consider the operator

$$Lv(\boldsymbol{x}) = -\mathsf{A}(\boldsymbol{x}) : D^2 v(\boldsymbol{x}) + \boldsymbol{b}(\boldsymbol{x}) \cdot \nabla v(\boldsymbol{x}) + c(\boldsymbol{x})v(\boldsymbol{x}). \tag{23.18}$$

In addition, we assume that, for all $\boldsymbol{x} \in \bar{\Omega}$, the matrix $\mathsf{A}(\boldsymbol{x})$ is symmetric, and there are constants $\lambda, \Lambda \in \mathbb{R}_+$ such that

$$\sigma(\mathsf{A}(\boldsymbol{x})) \subset [\lambda, \Lambda], \quad \forall \boldsymbol{x} \in \bar{\Omega},$$

so that our operator is elliptic at every point. The prototypical example of this is the Laplacian, which we introduced in Example 23.3.

As we have mentioned before, given $f \in C(\Omega)$, we supplement the equation

$$Lu = f, \qquad \text{in } \Omega,$$

with boundary conditions on $\partial\Omega$, which can be of either Dirichlet, Neumann, or Robin boundary conditions.

Before we embark on the study of BVPs, let us study some properties of solutions to the equation itself, which follow from ellipticity.

23.2.1 The Maximum Principle

Recall that, if a function $f\colon \mathbb{R} \to \mathbb{R}$ is convex and smooth, then $-f''(x) \leq 0$ and, for any interval $[a,b] \subset \mathbb{R}$, we have

$$f(x) \leq \max\{f(a), f(b)\}, \quad \forall x \in (a,b).$$

In other words, it cannot have a maximum in the interior of a domain. Similar reasoning can be made in several dimensions, but now the Hessian, $D^2 v(x)$, is

[10] Named in honor of the Swiss mathematician, physicist, astronomer, geographer, logician, and engineer Leonhard Euler (1707–1783) and the Italian mathematician Francesco Giacomo Tricomi (1897–1978).

assumed to be positive semi-definite at every point. Since, for an elliptic operator, the coefficient $-\mathsf{A}$ is negative definite, we expect that the product $-\mathsf{A} : D^2 v$ has a sign. These statements are the intuition behind the so-called maximum principle.

Theorem 23.2 (maximum principle). *Let $v \in C^2(\Omega) \cap C(\bar{\Omega})$ be such that, for all $\boldsymbol{x} \in \Omega$,*

$$Lv(\boldsymbol{x}) \leq 0.$$

1. *If $c \equiv 0$, then the function v attains its maximum at the boundary, i.e.,*

$$\max_{\boldsymbol{x}\in\bar{\Omega}} v(\boldsymbol{x}) \leq \max_{\boldsymbol{x}\in\partial\Omega} v(\boldsymbol{x}).$$

2. *If $c(\boldsymbol{x}) \geq 0$ for all $\boldsymbol{x} \in \Omega$, then*

$$\max_{\boldsymbol{x}\in\bar{\Omega}} v(\boldsymbol{x}) \leq \max\left\{0, \max_{\boldsymbol{x}\in\partial\Omega} v(\boldsymbol{x})\right\}.$$

Proof. Before we proceed with the proof, it is first instructive to sketch the idea of what happens when $c \equiv 0$. Let $\boldsymbol{x}_0 \in \Omega$ be a point where v attains its maximum. This means that, at this point, we must have

$$\nabla v(\boldsymbol{x}_0) = \boldsymbol{0}, \qquad \sigma(D^2 v(\boldsymbol{x}_0)) \subset (-\infty, 0].$$

This means that

$$Lv(\boldsymbol{x}_0) = -\mathsf{A}(\boldsymbol{x}_0) : D^2 v(\boldsymbol{x}_0) + \boldsymbol{b}(\boldsymbol{x}_0) \cdot \nabla v(\boldsymbol{x}_0) \geq 0,$$

which is *almost* a contradiction. The way to push this to an actual contradiction is by a so-called *comparison* function.

We now proceed with the proof. Let $\phi(\boldsymbol{x}) = e^{\alpha x_1}$, where $\alpha > 0$ and sufficiently large, so that

$$L\phi(\boldsymbol{x}) = \left(-a_{1,1}(\boldsymbol{x})\alpha^2 + b_1(\boldsymbol{x})\alpha\right)\phi(\boldsymbol{x}) < 0, \quad \forall \boldsymbol{x} \in \bar{\Omega}.$$

This will be the desired function. Now, since $\mathsf{A}(\boldsymbol{x})$ is SPD, and we have bounds on the spectrum,

$$a_{1,1}(\boldsymbol{x}) \geq \lambda, \quad \forall \boldsymbol{x} \in \bar{\Omega}.$$

Also, since $\boldsymbol{b} \in C(\bar{\Omega}; \mathbb{R}^d)$,

$$|b_1(\boldsymbol{x})| \leq \|\boldsymbol{b}\|_{L^\infty(\Omega;\mathbb{R}^d)}, \quad \forall \boldsymbol{x} \in \bar{\Omega}.$$

Therefore,

$$-a_{1,1}(\boldsymbol{x})\alpha^2 + b_1(\boldsymbol{x})\alpha \leq -\lambda\alpha^2 + \|\boldsymbol{b}\|_{L^\infty(\Omega;\mathbb{R}^d)}\alpha,$$

and the choice of α is now clear.

Assume now that $c \equiv 0$. Define the function $\tilde{v} = v + \varepsilon\phi$, where $\varepsilon > 0$ is to be chosen later. Notice now that

$$L\tilde{v}(\boldsymbol{x}) = Lv(\boldsymbol{x}) + \varepsilon L\phi(\boldsymbol{x}) < 0.$$

If v attains its maximum at some point $\boldsymbol{x}_0 \in \Omega$, then, for ε sufficiently small, the function $\tilde{v}$ also attains its maximum at some point $\boldsymbol{x}_1 \in \Omega$. At this point, we must then have

$$L\tilde{v}(\boldsymbol{x}_1) \geq 0,$$

which is the desired contradiction.

Let now $c(\boldsymbol{x}) \geq 0$ for all $\boldsymbol{x} \in \Omega$. Notice that if $v(\boldsymbol{x}) \leq 0$ for all $\boldsymbol{x} \in \bar{\Omega}$, then there is nothing to prove. Thus, assume that $\boldsymbol{x}_0 \in \Omega$ is such that

$$v(\boldsymbol{x}_0) = \max_{\boldsymbol{x} \in \bar{\Omega}} v(\boldsymbol{x}) > 0.$$

This, by continuity, implies that there is a neighborhood of the point $\boldsymbol{x}_0$ where $v > 0$. Denote by $\Omega(\boldsymbol{x}_0)$ the largest open and connected set where this holds. Then, for every $\boldsymbol{x} \in \Omega(\boldsymbol{x}_0)$, we notice that the operator

$$\tilde{L}v(\boldsymbol{x}) = -\mathsf{A}(\boldsymbol{x}) : D^2 v(\boldsymbol{x}) + \boldsymbol{b}(\boldsymbol{x}) \cdot \nabla v(\boldsymbol{x}) = Lv(\boldsymbol{x}) - c(\boldsymbol{x})v(\boldsymbol{x}) \leq 0.$$

Since this operator is elliptic and it has no zero-order coefficient, the previous case implies that

$$0 < v(\boldsymbol{x}_0) = \max_{\boldsymbol{x} \in \partial\Omega(\boldsymbol{x}_0)} v(\boldsymbol{x}).$$

Notice then that this implies that $\partial\Omega(\boldsymbol{x}_0)$ cannot be contained in Ω. Otherwise, there would be a point in $\partial\Omega(\boldsymbol{x}_0)$ where v is positive, and this contradicts that $\Omega(\boldsymbol{x}_0)$ is maximal. □

Remark 23.3 (minimum principle). If $Lv \geq 0$, Theorem 23.2 remains valid if we replace maxima by minima. This can be seen using linearity and the function $-v$.

From the maximum principle, we can easily obtain a stability result.

Corollary 23.4 (stability). *There is a constant $C > 0$ such that, for every $v \in C^2(\bar{\Omega})$, we have*

$$\|v\|_{L^\infty(\Omega)} \leq \|v\|_{L^\infty(\partial\Omega)} + C\|Lv\|_{L^\infty(\Omega)}.$$

Proof. As before, we let $\phi(x) = Ce^{\alpha x_1}$, where $C > 0$ and $\alpha > 0$ are chosen, so that $\phi(x) \geq 0$ and $L\phi(\boldsymbol{x}) \leq -1$ for all $\boldsymbol{x} \in \Omega$.

Define

$$\tilde{v}_\pm = \pm v + \|Lv\|_{L^\infty(\Omega)}\phi.$$

Notice that

$$L\tilde{v}_\pm(\boldsymbol{x}) = \pm Lv(\boldsymbol{x}) + \|Lv\|_{L^\infty(\Omega)} L\phi(\boldsymbol{x}) \leq \pm Lv(\boldsymbol{x}) - \|Lv\|_{L^\infty(\Omega)} \leq 0.$$

Consequently, by the maximum principle of Theorem 23.2,

$$\max_{\boldsymbol{x} \in \bar{\Omega}} \tilde{v}_\pm(\boldsymbol{x}) = \max_{\boldsymbol{x} \in \partial\Omega} \tilde{v}_\pm(\boldsymbol{x}),$$

so that, for every $x \in \Omega$,

$$\begin{aligned}\pm v(x) &\leq \tilde{v}_{\pm}(x) \\ &\leq \max_{x \in \partial\Omega} \tilde{v}_{\pm}(x) \\ &\leq \max_{x \in \partial\Omega} |v(x)| + \|Lv\|_{L^\infty(\Omega)} \|\phi\|_{L^\infty(\partial\Omega)} \\ &\leq \|v\|_{L^\infty(\partial\Omega)} + C\|Lv\|_{L^\infty(\Omega)},\end{aligned}$$

as we intended to show. □

As a final application, we present a sort of monotonicity result.

Corollary 23.5 (monotonicity). *Let $v_1, v_2 \in C^2(\Omega) \cap C(\bar{\Omega})$ be such that $v_1 \leq v_2$ on $\partial\Omega$ and $Lv_1 \leq Lv_2$ in Ω. Then we must have $v_1 \leq v_2$ in $\bar{\Omega}$.*

Proof. See Problem 23.5. □

23.2.2 The Dirichlet Problem: Classical Solutions

We now pose the first BVP for an elliptic operator. Given $f \in C(\Omega)$ and $g \in C(\partial\Omega)$, we will seek a solution to

$$\begin{cases} Lu(x) = f(x), & x \in \Omega, \\ u(x) = g(x), & x \in \partial\Omega. \end{cases} \tag{23.19}$$

Since the boundary conditions are of Dirichlet type, this is sometimes known as the *Dirichlet problem*. When $L = -\Delta$, this problem is usually referred to as the *Poisson problem*.

We must specify what we mean by a solution to this problem, and there are many ways of doing so. Our focus here is on classical solutions.

Definition 23.6 (classical solution). The function $u \in C^2(\Omega) \cap C(\bar{\Omega})$ is a **classical solution** of the Dirichlet problem (23.19) if and only if the equation and boundary condition are satisfied point-wise.

An immediate consequence of the maximum principle is the uniqueness of classical solutions.

Corollary 23.7 (uniqueness). *Problem* (23.19) *cannot have more than one classical solution.*

Proof. See Problem 23.6. □

Existence, unfortunately, is a topic that is beyond our discussion here. We refer the reader to the references given at the beginning of the chapter for details. Let us, nevertheless, state a simplified version.

Theorem 23.8 (existence). *Let $d = 2$, $\Omega = (0,1)^2$, and $L = -\Delta$. Assume that $f \in C(\bar{\Omega})$, $g \in C(\partial\Omega)$ are, in addition, sufficiently smooth, so that their Fourier series converge uniformly and absolutely. Then the Poisson problem* (23.19) *has a solution.*

Proof. Since we are on a square, the idea is to use separation of variables. The smoothness assumptions on the data allow us to justify the series representations; see [49, 99] for further details. □

To conclude the discussion, we comment on the existence of results concerning the further *regularity* of a classical solution. In essence, these results provide sufficient conditions on the domain Ω and the data f, g to guarantee that a classical solution satisfies $u \in C^k(\bar{\Omega})$, for $k > 2$.

23.2.3 The Dirichlet Problem: Weak Solutions

To motivate this treatment of a BVP, we observe that, in Section 23.1, when we derived PDEs, these were usually in *divergence form*, as this was the natural form in which they arose from a conservation. The divergence form of an equation was valid, even at points where the material modulus was not smooth.

For this reason, here, we will consider problems in divergence form. The operator we will consider here is then

$$Lv(\boldsymbol{x}) = -\nabla \cdot (\mathsf{A}(\boldsymbol{x})\nabla v(\boldsymbol{x})) + \boldsymbol{b}(\boldsymbol{x}) \cdot \nabla v(\boldsymbol{x}) + c(\boldsymbol{x})v(\boldsymbol{x}), \tag{23.20}$$

under the same assumptions on the coefficients as before. It is not difficult to see that this operator is elliptic, in the sense of Definition 23.1. We will once again consider the Dirichlet problem (23.19), but where the operator now is as in (23.20). The prototypical example here, again, is $L = -\Delta$.

Before we define our notion of solution, we will first motivate it. Let us multiply the equation in (23.19) by a function $v \in C^1(\bar{\Omega})$ that vanishes in a neighborhood of $\partial\Omega$ and integrate. An application of the Divergence Theorem reveals that

$$-\int_\Omega \nabla \cdot (\mathsf{A}\nabla u)\, v\, \mathrm{d}\boldsymbol{x} = \int_\Omega \nabla v^\intercal \mathsf{A}\nabla u\, \mathrm{d}\boldsymbol{x} - \int_{\partial\Omega} \boldsymbol{n}^\intercal \mathsf{A}\nabla u v\, \mathrm{d}\boldsymbol{S} = \int_\Omega \nabla v^\intercal \mathsf{A}\nabla u\, \mathrm{d}\boldsymbol{x},$$

where we used that v vanishes on $\partial\Omega$ and $\boldsymbol{n}$ denotes the outer unit normal to $\partial\Omega$. In short, if u is a classical solution to (23.19) with the operator L given by (23.20), then it must also satisfy

$$\int_\Omega [\nabla v^\intercal \mathsf{A}\nabla u + \boldsymbol{b} \cdot \nabla u v + c u v]\, \mathrm{d}\boldsymbol{x} = \int_\Omega f v\, \mathrm{d}\boldsymbol{x} \tag{23.21}$$

for every $v \in C^1(\bar{\Omega})$ that vanishes on $\partial\Omega$.

Notice that the integral identity (23.21) makes sense under much less restrictive assumptions than those we used to derive it. In fact, repeated applications of the Cauchy–Schwarz inequality yield the following:

$$\begin{aligned}
\left|\int_\Omega \nabla v^\intercal \mathsf{A} \nabla u \,\mathrm{d}\mathbf{x}\right| &\leq \|\mathsf{A}\|_{L^\infty(\Omega;\mathbb{R}^{d\times d})}\|\nabla u\|_{L^2(\Omega;\mathbb{R}^d)}\|\nabla v\|_{L^2(\Omega;\mathbb{R}^d)},\\
\left|\int_\Omega \boldsymbol{b}\cdot\nabla u v \,\mathrm{d}\mathbf{x}\right| &\leq \|\boldsymbol{b}\|_{L^\infty(\Omega;\mathbb{R}^d)}\|\nabla u\|_{L^2(\Omega;\mathbb{R}^d)}\|v\|_{L^2(\Omega)},\\
\left|\int_\Omega cuv \,\mathrm{d}\mathbf{x}\right| &\leq \|c\|_{L^\infty(\Omega)}\|u\|_{L^2(\Omega)}\|v\|_{L^2(\Omega)},\\
\left|\int_\Omega fv \,\mathrm{d}\mathbf{x}\right| &\leq \|f\|_{L^2(\Omega)}\|v\|_{L^2(\Omega)}.
\end{aligned} \tag{23.22}$$

These estimates show that all that is needed from the functions u and v is that they, together with all of their derivatives, belong to the space $L^2(\Omega)$. In fact, this can be made even weaker. All that is needed is that these functions can be approximated *in the* $L^2(\Omega)$ *sense* by smooth ones. In other words, we require that u and v belong to the Sobolev space $H^1(\Omega)$, which is discussed in Appendix D.

Having realized this, we immediately observe that (23.22) are the correct bounds to define a weaker notion of solution, one that is based on (23.21). Notice that we have already seen similar notions before. Namely, a weak solution to a rectangular system of equations was introduced in Chapter 5, and a weak formulation of a problem (for a differential equation) was given in Chapter 22.

Definition 23.9 (weak solution)**.** We say that the function $u \in H_0^1(\Omega)$ is a **weak solution** to the Dirichlet problem (23.19) with $g = 0$ and L given as in (23.20) if and only if, for every **test function** $v \in H_0^1(\Omega)$, we have

$$\int_\Omega [\nabla v^\intercal \mathsf{A}\nabla u + \boldsymbol{b}\cdot\nabla u v + cuv]\,\mathrm{d}\mathbf{x} = \int_\Omega fv\,\mathrm{d}\mathbf{x}.$$

Let us now present the following result concerning the existence and uniqueness of solutions. In the literature, this is commonly known as the *Lax–Milgram Theorem.*

Theorem 23.10 (Lax–Milgram Theorem[11])**.** *Let* $\mathcal{H}$ *be a Hilbert space and* $F \in \mathcal{H}'$. *Assume that we have a function* $\mathcal{A}\colon \mathcal{H}\times\mathcal{H}\to\mathbb{R}$ *that is*

1. *Bilinear: In other words, linear in each argument.*
2. *Bounded: There is a constant* $M > 0$ *such that, for every* $v_1, v_2 \in \mathcal{H}$, *we have*

$$|\mathcal{A}(v_1, v_2)| \leq M\|v_1\|_{\mathcal{H}}\|v_2\|_{\mathcal{H}}.$$

3. *Coercive: There is a constant* $\alpha > 0$ *such that, for every nonzero* $v \in \mathcal{H}$,

$$\alpha\|v\|_{\mathcal{H}}^2 \leq \mathcal{A}(v, v).$$

[11] Named in honor of the Hungarian-born American mathematician Peter David Lax (1926–) and the American mathematician Arthur Norton Milgram (1912–1961).

In this setting, the problem: Find $u \in \mathcal{H}$ such that

$$\mathcal{A}(u, v) = F(v), \quad \forall v \in \mathcal{H} \tag{23.23}$$

has a unique solution. This solution satisfies the estimate

$$\|u\|_{\mathcal{H}} \leq \frac{1}{\alpha}\|F\|_{\mathcal{H}'}.$$

If, in addition, the bilinear form is symmetric, i.e., $\mathcal{A}(v_1, v_2) = \mathcal{A}(v_2, v_1)$ for all $v_1, v_2 \in \mathcal{H}$, then the element $u \in \mathcal{H}$ minimizes the quadratic energy

$$E(v) = \frac{1}{2}\mathcal{A}(v, v) - F(v)$$

if and only if it solves problem (23.23).

Proof. Notice, first of all, that if $u \in \mathcal{H}$ is a solution to (23.23), then setting $v = u$ yields

$$\alpha\|u\|_{\mathcal{H}}^2 \leq \mathcal{A}(u, u) = F(u) \leq \|F\|_{\mathcal{H}'}\|u\|_{\mathcal{H}},$$

which is the claimed estimate.

Consider now the case where $\mathcal{A}$ is symmetric. It is not difficult to see that E is a strictly convex, continuous, and coercive energy on $\mathcal{H}$, and that (23.23) is nothing but its Euler equation. Theorem 16.31 gives then the existence and uniqueness of a minimizer. Finally, Theorem 16.44 shows that $u \in \mathcal{H}$ solves (23.23) if and only if it minimizes the energy E.

We now address the general case. We will proceed via a *homotopy technique*. Define the symmetric and skew-symmetric parts of $\mathcal{A}$ via

$$\mathcal{A}_{\mathrm{sym}}(v_1, v_2) = \frac{1}{2}\left[\mathcal{A}(v_1, v_2) + \mathcal{A}(v_2, v_1)\right], \quad \mathcal{A}_{\mathrm{sk}}(v, w) = \frac{1}{2}\left[\mathcal{A}(v_1, v_2) - \mathcal{A}(v_2, v_1)\right].$$

Let now $t \in [0, 1]$. Define

$$\mathcal{A}(v_1, v_2; t) = \mathcal{A}_{\mathrm{sym}}(v_1, v_2) + t\mathcal{A}_{\mathrm{sk}}(v_1, v_2).$$

Notice that this parametrized bilinear form satisfies

$$\mathcal{A}(v_1, v_2; 1) = \mathcal{A}(v_1, v_2),$$

$\mathcal{A}(\cdot, \cdot; t)$ is bounded with constant $2M$, and $\mathcal{A}(v, v; t) = \mathcal{A}(v, v)$, so it is also coercive with constant α.

Our theorem will be proved if we can prove the following claim: Let $t \in [0, 1]$ and consider, for $G \in \mathcal{H}'$, the problem: find $u_t \in \mathcal{H}$ such that

$$\mathcal{A}(u_t, v; t) = G(v), \quad \forall v \in \mathcal{H}.$$

Assume that, for $t_0 \in [0, 1]$ and every $G \in \mathcal{H}'$, this problem has a unique solution. Then this problem is uniquely solvable for $t \in [t_0, t_0 + \frac{\alpha}{2M}]$ and any $G \in \mathcal{H}'$.

If the previous statement holds, then we obtain the result. This is because $\mathcal{A}(\cdot, \cdot; 0) = \mathcal{A}_{\mathrm{sym}}(\cdot, \cdot)$ is symmetric, and the result has already been proved for this case. By the claim, the problem is uniquely solvable for $t \in [0, \frac{\alpha}{2M}]$. Set $t_0 = \frac{\alpha}{2M}$ and apply the claim again. In a finite number of steps, we will reach that the problem is uniquely solvable for $t = 1$, but $\mathcal{A}(v_1, v_2; 1) = \mathcal{A}(v_1, v_2)$, and so our original problem has a unique solution.

It remains then to prove the claim. To do so, we define the mapping $T\colon \mathcal{H} \to \mathcal{H}$ as $u_\phi = T\phi$, where u_ϕ solves

$$\mathcal{A}(u_\phi, v; t_0) = G(v; t), \quad \forall v \in \mathcal{H}, \qquad G(v; t) = F(v) + (t_0 - t)\mathcal{A}_{\mathrm{sk}}(\phi, v),$$

with $t \in [t_0, t_0 + \frac{\alpha}{2M}]$. Notice that $G(\cdot; t) \in \mathcal{H}'$, so by the assumption of the claim the mapping T is well defined. In addition, notice that if it has a fixed point, i.e., $u = Tu$, then

$$\mathcal{A}_{\mathrm{sym}}(u, v) + t_0\mathcal{A}_{\mathrm{sk}}(u, v) = F(v) + (t_0 - t)\mathcal{A}_{\mathrm{sk}}(u, v),$$

so that the parameterized problem for t is well posed, and this will imply the claim. Thus, we will show that the mapping T satisfies all the assumptions of the Banach Fixed Point Theorem C.4. Let $\phi_1, \phi_2 \in \mathcal{H}$ and set $u_i = T\phi_i$, for $i = 1, 2$. We can then subtract the corresponding equations and, by coercivity, obtain

$$\begin{aligned} \alpha\|u_1 - u_2\|_{\mathcal{H}}^2 &\le \mathcal{A}(u_1 - u_2, u_1 - u_2; t_0) \\ &\le |t_0 - t||\mathcal{A}_{\mathrm{sk}}(\phi_1 - \phi_2, u_1 - u_2)| \\ &\le \frac{\alpha}{2M} M\|\phi_1 - \phi_2\|_{\mathcal{H}}\|u_1 - u_2\|_{\mathcal{H}}, \end{aligned}$$

so that T is a contraction. □

Let us now apply the Lax–Milgram Theorem 23.10 to the existence and uniqueness of weak solutions.

Corollary 23.11 (existence and uniqueness). *Let $d \in \mathbb{N}$ and $\Omega \subset \mathbb{R}^d$ be a bounded domain with sufficiently nice boundary. Let $\mathsf{A} \in C(\bar{\Omega}; \mathbb{R}^{d\times d})$, $\boldsymbol{b} \in C^1(\bar{\Omega}; \mathbb{R}^d)$, and $c \in C(\bar{\Omega})$. Assume that there are constants $\lambda, \Lambda \in \mathbb{R}_+$ such that, for all points $\boldsymbol{x} \in \bar{\Omega}$, matrix $\mathsf{A}(\boldsymbol{x})$ is symmetric and satisfies*

$$\sigma(\mathsf{A}(\boldsymbol{x})) \subset [\lambda, \Lambda].$$

If $f \in L^2(\Omega)$, $g = 0$, and

$$c(\boldsymbol{x}) - \frac{1}{2}\nabla \cdot \boldsymbol{b}(\boldsymbol{x}) \ge 0, \quad \forall \boldsymbol{x} \in \Omega,$$

then the Dirichlet problem (23.19) with the operator L given by (23.20) has a unique weak solution which, moreover, satisfies

$$\|\nabla u\|_{L^2(\Omega;\mathbb{R}^d)} \le \frac{1}{\lambda}(C + \Lambda)C_P\|f\|_{L^2(\Omega)},$$

where the constant C depends on the coefficients $\boldsymbol{b}$ and c and C_P is the constant in the Poincaré inequality (D.2). Finally, if $\boldsymbol{b} \equiv \boldsymbol{0}$, this solution minimizes, over $H_0^1(\Omega)$, the quadratic energy

$$E(v) = \frac{1}{2}\int_\Omega \left[\nabla v^{\mathsf{T}}\mathsf{A}\nabla v + c|v|^2\right] \mathrm{d}\boldsymbol{x} - \int_\Omega f v \mathrm{d}\boldsymbol{x}.$$

Proof. We only need to verify the assumptions of Theorem 23.10. Over the Hilbert space $H_0^1(\Omega)$, we define

$$\mathcal{A}(v_1, v_2) = \int_\Omega [\nabla v_2^\intercal \mathsf{A} \nabla v_1 + \boldsymbol{b} \cdot \nabla v_1 v_2 + c v_1 v_2] \, \mathrm{d}\boldsymbol{x}, \quad F(v) = \int_\Omega f v \, \mathrm{d}\boldsymbol{x}.$$

Estimates (23.22) show that the bilinear form $\mathcal{A}$ is bounded and that if $\boldsymbol{b} \equiv \mathbf{0}$, then $\mathcal{A}$ is symmetric. Thus, it remains to prove its coercivity.

Using the properties of A, $\boldsymbol{b}$, and c and integrating back by parts, we see that, for every $v \in H_0^1(\Omega)$,

$$\begin{aligned}
\mathcal{A}(v, v) &= \int_\Omega \left[\nabla v^\intercal \mathsf{A} \nabla v + \boldsymbol{b} \cdot \nabla v v + c|v|^2\right] \mathrm{d}\boldsymbol{x} \\
&= \int_\Omega \left[\nabla v^\intercal \mathsf{A} \nabla v + \boldsymbol{b} \cdot \nabla \left(\frac{1}{2}|v|^2\right) + c|v|^2\right] \mathrm{d}\boldsymbol{x} \\
&\geq \lambda \int_\Omega |\nabla v|^2 \, \mathrm{d}\boldsymbol{x} + \int_\Omega [c - \frac{1}{2}\nabla \cdot \boldsymbol{b}]|v|^2 \mathrm{d}\boldsymbol{x} \\
&\geq \lambda \|\nabla v\|^2_{L^2(\Omega;\mathbb{R}^d)},
\end{aligned}$$

which is the coercivity we were looking for.

Finally, using Cauchy–Schwarz and Poincaré inequalities, we get

$$\left| \int_\Omega f v \, \mathrm{d}\boldsymbol{x} \right| \leq \|f\|_{L^2(\Omega)} \|v\|_{L^2(\Omega)} \leq C_P \|f\|_{L^2(\Omega)} \|\nabla v\|_{L^2(\Omega;\mathbb{R}^d)},$$

so that $\|F\|_{H_0^1(\Omega)'} \leq C_P \|f\|_{L^2(\Omega)}$. □

We conclude by discussing further regularity of weak solutions, at least in a particular case.

Theorem 23.12 (regularity). *Let $d \in \mathbb{N}$ and $\Omega \subset \mathbb{R}^d$ be a bounded domain which is either convex or has a smooth boundary. There is a constant $C > 0$ such that, whenever $f \in L^2(\Omega)$ and $g = 0$, the unique solution to (23.19) with $L = -\Delta$ satisfies*

$$|u|_{H^2(\Omega)} \leq C\|f\|_{L^2(\Omega)}, \quad i, j = 1, \ldots, d.$$

To conclude the discussion, since it will be needed to develop the theory of evolution problems, we present one result regarding the eigenvalue problem for elliptic operators. The proof of this result, however, is beyond the scope of our discussion. For a proof, we refer the reader to [31]. We begin with a definition.

Definition 23.13 (eigenvalue). Let L be given in (23.20). We say that $\lambda \in \mathbb{C}$ is an **eigenvalue** of the differential operator L provided that the problem

$$\begin{cases} L\varphi = \lambda\varphi, & \text{in } \Omega, \\ \varphi = 0, & \text{on } \partial\Omega \end{cases}$$

has a nontrivial weak solution $\varphi \in H_0^1(\Omega)$, which is known as an **eigenfunction**.

Theorem 23.14 (eigenvalue problem). *Let $d \in \mathbb{N}$ and $\Omega \subset \mathbb{R}^d$ be a bounded domain with sufficiently nice boundary. Let $\mathsf{A} \in C(\bar{\Omega}; \mathbb{R}^{d \times d})$, $\boldsymbol{b} \equiv \mathbf{0}$, and $c \in C(\bar{\Omega})$. Assume that there are constants $\lambda, \Lambda \in \mathbb{R}_+$ such that, for all points $\boldsymbol{x} \in \bar{\Omega}$, the matrix $\mathsf{A}(\boldsymbol{x})$ is symmetric and satisfies*

$$\sigma(\mathsf{A}(\boldsymbol{x})) \subset [\lambda, \Lambda].$$

Consider the differential operator L, given in (23.20), supplemented with homogeneous Dirichlet boundary conditions. In this setting:

1. *The set of all eigenvalues of L is at most countable.*
2. *All eigenvalues are real and positive.*
3. *Let $\{\lambda_k\}_{k=1}^{\infty}$ be the collection of all eigenvalues, repeated according to their (finite) multiplicity. Then*

$$0 < \lambda_1 < \lambda_2 < \cdots < \lambda_k < \cdots, \qquad \lambda_k \to \infty, \quad k \to \infty.$$

4. *There exists an orthonormal basis of $L^2(\Omega)$ consisting of eigenfunctions of L, i.e., there is $\{\varphi_k\}_{k=1}^{\infty} \subset H_0^1(\Omega)$ that form an orthonormal basis of $L^2(\Omega)$ and*

$$\begin{cases} L\varphi_k = \lambda_k \varphi_k, & \text{in } \Omega, \\ \varphi_k = 0, & \text{on } \partial\Omega \end{cases}$$

for all $k \in \mathbb{N}$.

Example 23.7 Let $\Omega = (0,1)^d$. The eigenpairs of the Dirichlet Laplacian, i.e., nontrivial solutions of

$$\begin{cases} -\Delta\varphi = \lambda\varphi, & \text{in } \Omega, \\ \varphi = 0, & \text{on } \partial\Omega, \end{cases}$$

are the pairs $\{(\lambda_{\boldsymbol{k}}, \varphi_{\boldsymbol{k}})\}_{\boldsymbol{k}\in\mathbb{N}^d} \subset \mathbb{R}_+ \times H_0^1(\Omega)$ given by

$$\lambda_{\boldsymbol{k}} = \pi^2 \|\boldsymbol{k}\|_2^2, \qquad \varphi_{\boldsymbol{k}} = 2^{d/2} \prod_{i=1}^{d} \sin\left(\pi[\boldsymbol{k}]_i [\boldsymbol{x}]_i\right).$$

23.3 Parabolic Equations

In this section, we will consider initial value problems and initial boundary value problems (IBVPs) for parabolic equations, the prototype of which is the heat equation introduced in Example 23.4. As always, we will just present the most relevant notions and estimates that will be used later in the numerical treatment of these problems.

23.3.1 The Initial Value Problem

Let us begin by considering the pure IVP for the heat equation in $\mathbb{R}^d$ with $d \in \mathbb{N}$. In this case, we let $T > 0$. Given a sufficiently smooth function $u_0 \colon \mathbb{R}^d \to \mathbb{R}$, we will seek a function $u \colon \mathbb{R}^d \times [0,T] \to \mathbb{R}$ that satisfies, in some sense,

$$\begin{cases} \dfrac{\partial u(\boldsymbol{x},t)}{\partial t} - \Delta u(\boldsymbol{x},t) = 0, & (\boldsymbol{x},t) \in \mathbb{R}^d \times (0,T], \\ u(\boldsymbol{x},0) = u_0(\boldsymbol{x}), & \boldsymbol{x} \in \mathbb{R}^d. \end{cases} \tag{23.24}$$

We begin by making precise our notion of solution.

Definition 23.15 (classical solution). A function $u\colon \mathbb{R}^d \times [0,T]$ is a **classical solution** of (23.24) if, for every $t > 0$, $u(\cdot, t) \in C^2(\mathbb{R}^d)$, for every $\boldsymbol{x} \in \mathbb{R}^d$ $u(\boldsymbol{x}, \cdot) \in C^1([0,T])$, the equation is satisfied point-wise, and as $t \downarrow 0$ we have $u(\boldsymbol{x}, t) \to u_0(\boldsymbol{x})$ for all $\boldsymbol{x} \in \mathbb{R}^d$.

We will seek a solution using the *Fourier transform*,[12] which, for a sufficiently nice function v, is defined via

$$\hat{v}(\boldsymbol{\xi}) = \mathcal{F}[v](\boldsymbol{\xi}) = \int_{\mathbb{R}^d} e^{-\mathrm{i}\boldsymbol{x}\cdot\boldsymbol{\xi}} v(\boldsymbol{x}) \mathrm{d}\boldsymbol{x}. \tag{23.25}$$

We refer the reader, for instance, to [77, 97] for a complete account of the properties of this mapping. Here, we simply state the following.

Proposition 23.16 (properties of $\mathcal{F}$). *Let the functions $v, v_1, v_2 \colon \mathbb{R}^d \to \mathbb{R}$ be sufficiently smooth and decay sufficiently fast at infinity. Then we have:*

1. *Differentiation: For any $j = 1, \ldots, d$,*
$$\mathcal{F}\left[\frac{\partial v}{\partial x_j}\right](\boldsymbol{\xi}) = \mathrm{i}\xi_j \mathcal{F}[v](\boldsymbol{\xi}).$$
2. *Convolution: The convolution of v_1 and v_2 is defined by*
$$(v_1 \star v_2)(\boldsymbol{x}) = \int_{\mathbb{R}^d} v_1(\boldsymbol{x} - \boldsymbol{y}) v_2(\boldsymbol{y}) \mathrm{d}\boldsymbol{y}.$$
The transform of the convolution satisfies
$$\mathcal{F}[v_1 \star v_2](\boldsymbol{\xi}) = \mathcal{F}[v_1](\boldsymbol{\xi})\mathcal{F}[v_2](\boldsymbol{\xi}).$$
3. *Gaussian:*[13] *Let $G(\boldsymbol{x}) = e^{-\|\boldsymbol{x}\|_2^2}$. Then we have*
$$\mathcal{F}[G](\boldsymbol{\xi}) = \pi^{d/2} e^{-\|\boldsymbol{\xi}\|_2^2/4}.$$
4. *Plancherel identity:*[14] *If $v \in L^2(\mathbb{R}^d)$,*
$$\|v\|_{L^2(\mathbb{R}^d)} = \|\hat{v}\|_{L^2(\mathbb{R}^d)}.$$
5. *Shift: For any $\boldsymbol{a} \in \mathbb{R}^d$,*
$$\mathcal{F}[v(\cdot - \boldsymbol{a})](\boldsymbol{\xi}) = e^{-\mathrm{i}\boldsymbol{a}\cdot\boldsymbol{\xi}} \mathcal{F}[v](\boldsymbol{\xi}).$$

Proof. We will only sketch the proof of the shift property, as later we will need a discrete analogue. By definition, we have

$$\mathcal{F}[v(\cdot - \boldsymbol{a})](\boldsymbol{\xi}) = \int_{\mathbb{R}^d} e^{-\mathrm{i}\boldsymbol{x}\cdot\boldsymbol{\xi}} v(\boldsymbol{x} - \boldsymbol{a}) \mathrm{d}\boldsymbol{x} = e^{-\mathrm{i}\boldsymbol{a}\cdot\boldsymbol{\xi}} \int_{\mathbb{R}^d} e^{-\mathrm{i}\boldsymbol{y}\cdot\boldsymbol{\xi}} v(\boldsymbol{y}) \mathrm{d}\boldsymbol{y} = e^{-\mathrm{i}\boldsymbol{a}\cdot\boldsymbol{\xi}} \mathcal{F}[v](\boldsymbol{\xi}),$$

where we applied a simple change of variables. □

[12] Named in honor of the French mathematician and physicist Jean-Baptiste Joseph Fourier (1768–1830).

[13] Named in honor of the German mathematician and physicist Johann Carl Friedrich Gauss (1777–1855).

[14] Named in honor of the Swiss mathematician Michel Plancherel (1885–1967).

We incorporate the Fourier transform into our discussion because, at least formally, this can help us construct a solution to (23.24). To see this, take the Fourier transform of this problem to obtain the parameterized linear ODE

$$\frac{d\hat{u}(\boldsymbol{\xi}, t)}{dt} = -\|\boldsymbol{\xi}\|_2^2 \hat{u}(\boldsymbol{\xi}, t), \quad \hat{u}(\boldsymbol{\xi}, 0) = \hat{u}_0(\boldsymbol{\xi}),$$

which can be solved exactly by

$$\hat{u}(\boldsymbol{\xi}, t) = \hat{u}_0(\boldsymbol{\xi}) e^{-t\|\boldsymbol{\xi}\|_2^2}.$$

Let us define the *fundamental solution* to the heat equation to be the function $\Phi\colon \mathbb{R}^d \times (0, T]$ that is defined via

$$\hat{\Phi}(\boldsymbol{\xi}, t) = e^{-t\|\boldsymbol{\xi}\|_2^2} \quad \Longrightarrow \quad \Phi(\boldsymbol{x}, t) = \frac{1}{(4\pi t)^{d/2}} e^{-\frac{\|\boldsymbol{x}\|_2^2}{4t}}. \tag{23.26}$$

The convolution property of the Fourier transform then shows that the solution to (23.24) is given by

$$u(\boldsymbol{x}, t) = \left(\Phi(\,\cdot\,, t) \star u_0(\cdot)\right)(\boldsymbol{x}) = \frac{1}{(4\pi t)^{d/2}} \int_{\mathbb{R}^d} e^{-\frac{\|\boldsymbol{x}-\boldsymbol{y}\|_2^2}{4t}} u_0(\boldsymbol{y}) d\boldsymbol{y}. \tag{23.27}$$

These arguments can be made rigorous to provide the *existence* of a solution.

Theorem 23.17 (existence). *Let $u_0 \in C_b(\mathbb{R}^d) \cap L^2(\mathbb{R}^d)$. Then the function u, defined by* (23.27)*, is a classical solution to* (23.24)*. Moreover, for any $\boldsymbol{x}_0 \in \mathbb{R}^d$, we have*

$$u(\boldsymbol{x}, t) \to u_0(\boldsymbol{x}_0), \qquad (\boldsymbol{x}, t) \to (\boldsymbol{x}_0, 0).$$

Proof. It suffices to observe that, for $t > 0$, it is legitimate to differentiate the integral representation (23.27) to obtain the requisite smoothness, and the fact that this function solves the equation.

To show the sense in which the initial condition is attained, recall that

$$\frac{1}{\pi^{d/2}} \int_{\mathbb{R}^d} e^{-\|\boldsymbol{x}\|_2^2} d\boldsymbol{x} = 1,$$

use this in the expression,

$$u(\boldsymbol{x}, t) - u_0(\boldsymbol{x}_0) = \frac{1}{(4\pi t)^{d/2}} \int_{\mathbb{R}^d} e^{-\frac{\|\boldsymbol{x}-\boldsymbol{y}\|_2^2}{4t}} u_0(\boldsymbol{y}) d\boldsymbol{y} - u_0(\boldsymbol{x}_0),$$

and carry out the change of variables $\boldsymbol{z} = \frac{1}{\sqrt{4t}}(\boldsymbol{y} - \boldsymbol{x})$. We leave the details to the reader as an exercise; see Problem 23.14. □

Example 23.8 Let us illustrate the solution of (23.24), described in (23.27), in the case $d = 1$ and

$$u_0(x) = e^{-x^2/2}, \quad x \in \mathbb{R}.$$

The derivations that led to (23.27) showed that

$$\hat{u}(\xi, t) = \hat{u}_0(\xi) e^{-t\xi^2}, \qquad \hat{u}_0(\xi) = \sqrt{2\pi} e^{-\frac{\xi^2}{2}},$$

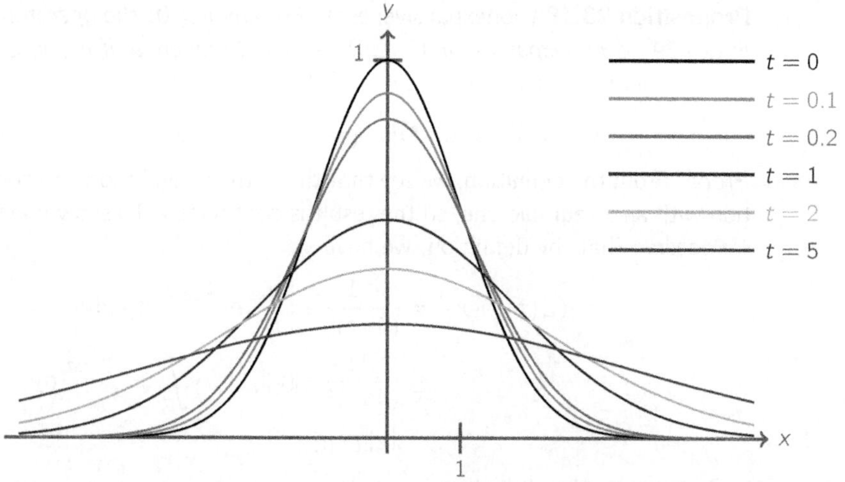

Figure 23.1 Solution to the IVP for the heat equation (23.24) with $d = 1$ and Gaussian initial data $u_0(x) = e^{-x^2/2}$.

where we used the Fourier transform of a Gaussian. Therefore,

$$\hat{u}(\xi, t) = \sqrt{2\pi} e^{-\xi^2\left(t+\frac{1}{2}\right)} \quad \Longrightarrow \quad u(x,t) = \frac{e^{-\frac{x^2}{4t+2}}}{\sqrt{1+2t}}. \tag{23.28}$$

This solution is illustrated in Figure 23.1.

Example 23.9 The previous example helps to explain why the backward diffusion equation,

$$\frac{\partial u(\boldsymbol{x}, t)}{\partial t} + \Delta u(\boldsymbol{x}, t) = 0,$$

is not well posed. Formally, (23.28) is a solution to the backward diffusion equation. But, as $t \to -\frac{1}{2}$, the solution approaches the Dirac delta function. In other words, solutions to the backward diffusion equation can blow up in finite time.

Let us now work on showing continuous dependence in the class of bounded continuous functions $C_b(\mathbb{R}^d \times [0, T])$. To achieve this, we begin by defining, for $t > 0$, the operator

$$\begin{aligned} &E(t) \colon C_b(\mathbb{R}^d) \to C_b(\mathbb{R}^d), \\ &(E(t)v)(\boldsymbol{x}) = \frac{1}{(4\pi t)^{d/2}} \int_{\mathbb{R}^d} e^{-\frac{\|\boldsymbol{x}-\boldsymbol{y}\|_2^2}{4t}} v(\boldsymbol{y}) \mathrm{d}\boldsymbol{y}. \end{aligned} \tag{23.29}$$

Proposition 23.18 (nonexpansiveness). *For any $t > 0$, the operator $E(t)$, defined in (23.29) is nonexpansive in $C_b(\mathbb{R}^d)$. As a consequence, if $u_0^1, u_0^2 \in C_b(\mathbb{R}^d)$, then the corresponding solutions, u_1, u_2, of (23.24) satisfy*

$$\|u_1(\cdot\,, t) - u_2(\cdot\,, t)\|_{L^\infty(\mathbb{R}^d)} \le \|u_0^1 - u_0^2\|_{L^\infty(\mathbb{R}^d)}, \quad \forall t > 0.$$

Proof. From the definition, we see that this is the convolution of a continuous function with an integrable one, so the result is continuous. To show nonexpansiveness, we observe that, by definition, we have

$$\begin{aligned} |(E(t)v)(\boldsymbol{x})| &= \frac{1}{(4\pi t)^{d/2}} \left| \int_{\mathbb{R}^d} e^{-\frac{\|x-y\|_2^2}{4t}} v(\boldsymbol{y}) \mathrm{d}\boldsymbol{y} \right| \\ &\le \frac{1}{(4\pi t)^{d/2}} \|v\|_{L^\infty(\mathbb{R}^d)} \int_{\mathbb{R}^d} e^{-\frac{\|x-y\|_2^2}{4t}} \mathrm{d}\boldsymbol{y} \\ &= \|v\|_{L^\infty(\mathbb{R}^d)}. \end{aligned}$$

By linearity, the stability result follows. □

From the representation formula for the solution, it is also possible to obtain smoothness of the solution.

Theorem 23.19 (regularity). *Let $u_0 \in C_b(\mathbb{R}^d)$. Then, for all $t > 0$, we have that $u(\cdot\,, t) \in C_b(\mathbb{R}^d)$, the solution of (23.24), is infinitely differentiable in space and time. In fact, for any $j \in \mathbb{N}_0$ and any multi-index $\alpha \in \mathbb{N}_0^d$ of length $k \ge 0$, there is a constant $C > 0$ such that we have*

$$\left\| \frac{\partial^j}{\partial t^j} D_x^\alpha u(\cdot\,, t) \right\|_{L^\infty(\mathbb{R}^d)} \le C t^{-j-k/2} \|u_0\|_{L^\infty(\mathbb{R}^d)}, \quad \forall t > 0.$$

Proof. Let $t > 0$. Since the integral from the representation $u(\boldsymbol{x}, t) = (E(t)u_0)(\boldsymbol{x})$ converges absolutely, we can write

$$\frac{\partial^j}{\partial t^j} D_x^\alpha u(\boldsymbol{x}, t) = \int_{\mathbb{R}^d} \frac{\partial^j}{\partial t^j} D_x^\alpha \Phi(\boldsymbol{x} - \boldsymbol{y}, t) u_0(\boldsymbol{y}) \mathrm{d}\boldsymbol{y}.$$

It is not difficult to see now that

$$\left| \frac{\partial^j}{\partial t^j} D_x^\alpha \Phi(\boldsymbol{x}, t) \right| \le t^{-j-k/2-d/2} P\left(\frac{\|\boldsymbol{x}\|_2}{\sqrt{4t}} \right) e^{-\frac{\|x\|_2^2}{4t}}, \tag{23.30}$$

where P is a polynomial. Thus, we have

$$\begin{aligned} \left| \frac{\partial^j}{\partial t^j} D_x^\alpha u(\boldsymbol{x}, t) \right| &\le \frac{t^{-j-k/2} \|u_0\|_{L^\infty(\mathbb{R}^d)}}{\sqrt{t^d}} \int_{\mathbb{R}^d} P\left(\frac{\|\boldsymbol{x}\|_2}{\sqrt{4t}} \right) e^{-\frac{\|x\|_2^2}{4t}} \mathrm{d}\boldsymbol{x} \\ &\le C t^{-j-k/2} \|u_0\|_{L^\infty(\mathbb{R}^d)}, \end{aligned}$$

as we intended to show. □

We conclude by establishing *Duhamel's formula*[15] for the solution of the inhomogeneous IVP: let $T > 0$, given sufficiently smooth $u_0 \colon \mathbb{R}^d \to \mathbb{R}$ and $f \colon \mathbb{R}^d \times (0, T] \to \mathbb{R}$, find $u \colon \mathbb{R}^d \times [0, T] \to \mathbb{R}$ such that

[15] Named in honor of the French mathematician and physicist Jean-Marie Constant Duhamel (1797–1872).

$$\begin{cases} \dfrac{\partial u(\boldsymbol{x},t)}{\partial t} - \Delta u(\boldsymbol{x},t) = f(\boldsymbol{x},t), & (\boldsymbol{x},t) \in \mathbb{R}^d \times \mathbb{R}_+, \\ u(\boldsymbol{x},0) = u_0(\boldsymbol{x}), & \boldsymbol{x} \in \mathbb{R}^d. \end{cases} \tag{23.31}$$

The notion of a classical solution is as before. At least formally, the solution of this problem is given by

$$\begin{aligned} u(\boldsymbol{x},t) &= \int_{\mathbb{R}^d} \Phi(\boldsymbol{x}-\boldsymbol{y},t)u_0(\boldsymbol{y})\mathrm{d}\boldsymbol{y} + \int_0^t \int_{\mathbb{R}^d} \Phi(\boldsymbol{x}-\boldsymbol{y},s)f(\boldsymbol{y},s)\mathrm{d}\boldsymbol{y}\mathrm{d}s \\ &= (E(t)u_0)(\boldsymbol{x}) + \int_0^t (E(t-s)f(\,\cdot\,,s))(\boldsymbol{x})\mathrm{d}s. \end{aligned}$$

This solution representation can be made rigorous provided that $u_0 \in C_b(\mathbb{R}^d)$ and $f \in C_b^2(\mathbb{R}^d \times (0,T])$.

23.3.2 The Initial Boundary Value Problem

We will now consider the IBVP for the heat equation. We assume that $d \in \mathbb{N}$, $\Omega \subset \mathbb{R}^d$ is a bounded domain with sufficiently smooth boundary. Given $T > 0$, $u_0 \colon \bar{\Omega} \to \mathbb{R}$, and $f \colon \Omega \times (0,T] \to \mathbb{R}$, we seek a function $u \colon \bar{\Omega} \times [0,T] \to \mathbb{R}$ that, in some sense, satisfies

$$\begin{cases} \dfrac{\partial u(\boldsymbol{x},t)}{\partial t} - \Delta u(\boldsymbol{x},t) = f(\boldsymbol{x},t), & (\boldsymbol{x},t) \in \Omega \times (0,T], \\ u(\boldsymbol{x},t) = 0, & (\boldsymbol{x},t) \in \partial\Omega \times (0,T], \\ u(\boldsymbol{x},0) = u_0(\boldsymbol{x}), & \boldsymbol{x} \in \bar{\Omega}. \end{cases} \tag{23.32}$$

The definition of the classical solution is rather standard.

Definition 23.20 (classical solution)**.** A function $u \colon \bar{\Omega} \times \mathbb{R}_+ \cup \{0\}$ is a **classical solution** of (23.32) if, for every $t > 0$, $u(\,\cdot\,,t) \in C^2(\Omega) \cap C(\bar{\Omega})$, for every $\boldsymbol{x} \in \Omega$ $u(\boldsymbol{x},\cdot) \in C^1((0,T])$, the equation is satisfied point-wise, and as $t \downarrow 0$ we have $u(\boldsymbol{x},t) \to u_0(\boldsymbol{x})$ for all $\boldsymbol{x} \in \bar{\Omega}$.

While it is possible to develop a theory regarding existence, uniqueness, and further regularity of such classical solutions, this imposes rather restrictive assumptions on the domain Ω and the data u_0 and f. Thus, we will seek an alternate notion of the solution.

To motivate it, we will find a solution representation. To achieve this, we recall that Theorem 23.14 showed that there is a family of pairs $\{(\lambda_k,\varphi_k)\}_{k\in\mathbb{N}} \subset \mathbb{R}_+ \times H_0^1(\Omega)$, called the *eigenpairs of the Laplacian*, where the family $\{\varphi_k\}_{k\in\mathbb{N}}$ forms an orthonormal basis of the space $L^2(\Omega)$.

We consider the following ansatz for the solution of (23.32):

$$u(\boldsymbol{x},t) = \sum_{k=1}^{\infty} u_k(t)\varphi_k(\boldsymbol{x}).$$

Let us assume that the data u_0 and f are sufficiently smooth that they admit the following representations:

$$u_0(\mathbf{x}) = \sum_{k=1}^{\infty} u_{0,k}\varphi_k(\mathbf{x}), \qquad f(\mathbf{x}, t) = \sum_{k=1}^{\infty} f_k(t)\varphi_k(\mathbf{x}).$$

Substituting these representations in (23.32), we obtain, for all $t > 0$,

$$\sum_{k=1}^{\infty} \left[u_k'(t) + \lambda_k u_k(t) - f_k(t)\right] \varphi_k(\mathbf{x}) = 0.$$

As we mentioned above, the functions φ_k are orthonormal, and hence linearly independent. Thus, we obtain a countable collection of linear ODEs,

$$u_k'(t) + \lambda_k u_k(t) = f_k(t), \quad t > 0, \qquad u_k(0) = u_{0,k}.$$

This will allow us to obtain a solution representation and several estimates on it.

Theorem 23.21 (existence). *Let $u_0 = \sum_{k=1}^{\infty} u_{0,k}\varphi_k \in L^2(\Omega)$. The function*

$$u(\mathbf{x}, t) = \sum_{k=1}^{\infty} u_{0,k} e^{-\lambda_k t}\varphi_k(\mathbf{x}) \tag{23.33}$$

satisfies (23.32) *with $f \equiv 0$. Moreover, for all $t > 0$, we have*

$$\|u(\,\cdot\,, t)\|_{L^2(\Omega)} \leq \|u_0\|_{L^2(\Omega)}, \tag{23.34}$$

$$\|\nabla u(\,\cdot\,, t)\|_{L^2(\Omega;\mathbb{R}^d)} \leq C t^{-1/2}\|u_0\|_{L^2(\Omega)}, \tag{23.35}$$

$$\|\Delta^m u(\,\cdot\,, t)\|_{L^2(\Omega)} \leq C_m t^{-m}\|u_0\|_{L^2(\Omega)}, \tag{23.36}$$

where the constants C and C_m are independent of t, but C_m may depend on $m \in \mathbb{N}$.

Proof. From the solution representation, with $f_k \equiv 0$ for all $k \in \mathbb{N}$, we obtain

$$u_k(t) = u_{0,k}e^{-\lambda_k t}.$$

Consequently, using that $0 < \lambda_1 \leq \lambda_k$, for all $k \in \mathbb{N}$,

$$\|u(\,\cdot\,, t)\|_{L^2(\Omega)}^2 = \sum_{k=1}^{\infty} |u_k(t)|^2 = \sum_{k=1}^{\infty} |u_{0,k}|^2 e^{-2\lambda_k t} \leq e^{-2\lambda_1 t}\|u_0\|_{L^2(\Omega)}^2,$$

which shows that $u(\,\cdot\,, t) \in L^2(\Omega)$ and proves (23.34). Notice now that, for $t > 0$, the function defined in (23.33) is smooth and the properties of φ_k show that this function satisfies the heat equation and the boundary condition.

Let us now show in which sense the initial condition is attained. Let $\varepsilon > 0$. Since $u_0 \in L^2(\Omega)$, there is $N \in \mathbb{N}$ such that

$$\sum_{k>N} |u_{0,k}|^2 < \frac{\varepsilon}{8}.$$

Choose now $t_0 > 0$, but small enough, so that, for all $t \in [0, t_0]$, we have

$$\max_{k=1,\dots,N} \left(e^{-\lambda_k t} - 1\right)^2 < \frac{\varepsilon}{2(\|u_0\|_{L^2(\Omega)}^2 + 1)}.$$

We now consider

$$\begin{aligned}
\|u(\cdot,t)-u_0\|_{L^2(\Omega)}^2 &= \sum_{k=1}^{\infty} |u_{0,k}|^2 |e^{-\lambda_k t}-1|^2 \\
&= \sum_{k=1}^{N} |u_{0,k}|^2 |e^{-\lambda_k t}-1|^2 + \sum_{k>N} |u_{0,k}|^2 |e^{-\lambda_k t}-1|^2 \\
&\le \max_{k=1,\dots,N} \left(e^{-\lambda_k t}-1\right)^2 \sum_{k=1}^{N} |u_{0,k}|^2 + 2(e^{-2\lambda_1 t}+1) \sum_{k>N} |u_{0,k}|^2 \\
&\le \max_{k=1,\dots,N} \left(e^{-\lambda_k t}-1\right)^2 \|u_0\|_{L^2(\Omega)}^2 + 4\sum_{k>N} |u_{0,k}|^2 \\
&< \varepsilon.
\end{aligned}$$

In other words, the initial condition is attained in the $L^2(\Omega)$-sense.

Let us now show that (23.35)

$$\begin{aligned}
\|\nabla u(\cdot,t)\|_{L^2(\Omega;\mathbb{R}^d)}^2 &= \int_\Omega \nabla u(\mathbf{x},t)\cdot\nabla u(\mathbf{x},t)\mathrm{d}\mathbf{x} \\
&= \sum_{k,m=1}^{2} u_k(t)u_m(t)\int_\Omega \nabla\varphi_k(\mathbf{x})\cdot\nabla\varphi_m(\mathbf{x})\mathrm{d}\mathbf{x} \\
&= \sum_{k=1}^{2} \lambda_k |u_k(t)|^2 \\
&= \sum_{k=1}^{\infty} \lambda_k e^{-2\lambda_k t}|u_{0,k}|^2 \\
&= t^{-1}\sum_{k=1}^{\infty} (\lambda_k t)e^{-2\lambda_k t}|u_{0,k}|^2 \\
&\le Ct^{-1}\|u_0\|_{L^2(\Omega)}^2.
\end{aligned}$$

The proof of (23.36) is left to the reader as an exercise; see Problem 23.17. □

The previous result shows that although, for any $t>0$, we have $u(\cdot,t)\in H_0^1(\Omega)$, this property, in general, does not survive as $t\downarrow 0$. The following result gives sufficient conditions for this to be valid for all times.

Proposition 23.22 (limit $t\downarrow 0$). *Assume that $u_0\in H_0^1(\Omega)$. Then we have*

$$\|\nabla u(\cdot,t)\|_{L^2(\Omega;\mathbb{R}^d)} \le \|\nabla u_0\|_{L^2(\Omega;\mathbb{R}^d)}, \quad \forall t\ge 0.$$

Proof. From the proof of (23.35), we have

$$\|\nabla u(\cdot,t)\|_{L^2(\Omega;\mathbb{R}^d)}^2 = \sum_{k=1}^{\infty} e^{-2\lambda_k t}\lambda_k|u_{0,k}|^2 \le e^{-2\lambda_1 t}\sum_{k=1}^{\infty}\lambda_k|u_{0,k}|^2 \le \|\nabla u_0\|_{L^2(\Omega;\mathbb{R}^d)}^2.$$

□

Let us finally consider the case $f \neq 0$ and the analogous Duhamel's formula. Let us, for $t > 0$, define the operator

$$E_\Omega(t) \colon L^2(\Omega) \to L^2(\Omega), \quad (E_\Omega(t)v)(\mathbf{x}) = \sum_{k=1}^{\infty} v_k e^{-\lambda_k t} \varphi_k(\mathbf{x}),$$

where

$$v(\mathbf{x}) = \sum_{k=1}^{\infty} v_k \varphi_k(\mathbf{x}).$$

Then the solution to (23.32) can be written as

$$u(\mathbf{x}, t) = (E_\Omega(t)u_0)(\mathbf{x}) + \int_0^t (E_\Omega(t-s)f(\cdot, s))(\mathbf{x}) \mathrm{d}s, \tag{23.37}$$

provided that $f \in L^2(\Omega \times \mathbb{R}_+)$. Notice that this representation immediately implies that

$$\|u(\cdot, t)\|_{L^2(\Omega)} \leq \|u_0\|_{L^2(\Omega)} + \int_0^t \|f(\cdot, s)\|_{L^2(\Omega)} \, \mathrm{d}s.$$

All these previous computations and estimates serve to motivate our definition of a weak solution. To properly define the notion of a weak solution, it will be necessary to define some classes of functions. The intuition behind this definition is that a function $v \colon \Omega \times \mathbb{R}_+ \to \mathbb{R}$ can be thought of either as a function of two variables or as a function of one variable (time), which at every instance $t \in \mathbb{R}_+$ produces a function over the spatial domain:

$$t \mapsto (v(\cdot, t) \colon \Omega \to \mathbb{R}).$$

Definition 23.23 (vector-valued L^p)**.** Let $T \in (0, \infty]$, $p \in [1, \infty]$, and $\mathbb{V}$ be a (separable) Banach space with norm $\|\cdot\|_{\mathbb{V}}$. We define

$$L^p(0, T; \mathbb{V})$$

as the vector space of functions $v \colon (0, T) \to \mathbb{V}$ such that the mapping $t \mapsto \|v(t)\|_{\mathbb{V}}$ is measurable and the quantity

$$\int_0^T \|v(t)\|_{\mathbb{V}}^p \, \mathrm{d}t < \infty.$$

This is a Banach space with norm

$$\|v\|_{L^p(0,T;\mathbb{V})} = \left(\int_0^T \|v(t)\|_{\mathbb{V}}^p \, \mathrm{d}t \right)^{1/p}, \quad p \in [1, \infty),$$

and

$$\|v\|_{L^\infty(0,T;\mathbb{V})} = \operatorname*{ess\,sup}_{t \in [0,T]} \|v(t)\|_{\mathbb{V}}.$$

Let u now be a classical solution to (23.32). Multiply the equation by $v \in C_0^1(\bar{\Omega})$ and integrate in Ω to obtain, after integration by parts,

$$\int_\Omega \left[\frac{\partial u(\mathbf{x}, t)}{\partial t} v(\mathbf{x}) + \nabla u(\mathbf{x}, t) \cdot \nabla v(\mathbf{x}) \right] \mathrm{d}\mathbf{x} = \int_\Omega f(\mathbf{x}, t) v(\mathbf{x}) \mathrm{d}\mathbf{x}.$$

Notice now that, as in the elliptic case, much less regularity is needed on the solution and the test function (in this case in space and time) for the integrals given above to make sense. This motivates the definition of a weak solution.

Definition 23.24 (weak solution)**.** Let $T \in (0, \infty]$. The function

$$u \in L^2(0, T; H_0^1(\Omega)), \qquad \frac{\partial u}{\partial t} \in L^2\left(0, T; H_0^1(\Omega)'\right)$$

is a **weak solution** of (23.32) if $u(\cdot, t) \to u_0$ in $L^2(\Omega)$ as $t \downarrow 0$ and, for almost every $t \in (0, T)$, we have

$$\left\langle \frac{\partial u(\cdot, t)}{\partial t}, v \right\rangle + \int_\Omega \nabla u(\boldsymbol{x}, t) \cdot \nabla v(\boldsymbol{x}) \mathrm{d}\boldsymbol{x} = \langle f(\cdot, t), v \rangle, \quad \forall v \in H_0^1(\Omega),$$

where $\langle \cdot, \cdot \rangle$ denotes the duality pairing of $H_0^1(\Omega)$.

The following result, concerning existence, essentially shows that the solution representation (23.33) is a weak solution.

Theorem 23.25 (well-posedness)**.** *Let $T \in (0, \infty)$, the initial condition satisfy $u_0 \in L^2(\Omega)$, and the right-hand side satisfy $f \in L^2(0, T; H_0^1(\Omega)')$. Then there is a unique weak solution to the IBVP* (23.32)*. This solution is given by Duhamel's formula* (23.37)*; moreover, it satisfies, for all $t > 0$,*

$$\|u(\cdot, t)\|_{L^2(\Omega)}^2 + \int_0^t \|\nabla u(\cdot, s)\|_{L^2(\Omega;\mathbb{R}^d)}^2 \mathrm{d}s \le \|u_0\|_{L^2(\Omega)}^2 + \int_0^t \|f(\cdot, s)\|_{H_0^1(\Omega)'}^2 \mathrm{d}s.$$

If, in addition, we have $u_0 \in H_0^1(\Omega)$ and $f \in L^2(0, T; L^2(\Omega))$, then, for all $t > 0$,

$$\|\nabla u(\cdot, t)\|_{L^2(\Omega;\mathbb{R}^d)}^2 + \int_0^t \left\| \frac{\partial u(\cdot, s)}{\partial t} \right\|_{L^2(\Omega)}^2 \mathrm{d}s$$
$$\le \|\nabla u_0\|_{L^2(\Omega;\mathbb{R}^d)}^2 + \int_0^t \|f(\cdot, s)\|_{L^2(\Omega)}^2 \mathrm{d}s.$$

Proof. By linearity, we see that uniqueness follows from the first estimate. Existence is provided by Duhamel's formula. Let us then show, at least formally, the energy estimates. To achieve this, we set $v = u(\cdot, t) \in H_0^1(\Omega)$ in the definition of a weak solution and integrate in time to obtain

$$\frac{1}{2} \int_0^t \frac{\mathrm{d}}{\mathrm{d}t} \|u(\cdot, s)\|_{L^2(\Omega)}^2 \mathrm{d}s + \int_0^t \|\nabla u(\cdot, s)\|_{L^2(\Omega;\mathbb{R}^d)}^2 \mathrm{d}s + \int_0^t \langle f(\cdot, s), u(\cdot, t) \rangle \mathrm{d}s.$$

Clearly,

$$\int_0^t \frac{\mathrm{d}}{\mathrm{d}t} \|u(\cdot, s)\|_{L^2(\Omega)}^2 \mathrm{d}s = \|u(\cdot, t)\|_{L^2(\Omega)}^2 - \|u_0\|_{L^2(\Omega)}^2.$$

By definition of the dual norm and Young's inequality,

$$\int_0^t \langle f(\cdot, s), u(\cdot, t) \rangle \mathrm{d}s \le \int \|f(\cdot, s)\|_{H_0^1(\Omega)'} \|\nabla u(\cdot, s)\|_{L^2(\Omega;\mathbb{R}^d)} \mathrm{d}s$$
$$\le \frac{1}{2} \int_0^t \|f(\cdot, s)\|_{H_0^1(\Omega)'}^2 \mathrm{d}s + \frac{1}{2} \int_0^t \|\nabla u(\cdot, s)\|_{L^2(\Omega;\mathbb{R}^d)}^2 \mathrm{d}s.$$

Combining these bounds, we get the first estimate.

The second estimate is obtained, formally, by setting $v = \frac{\partial u(\cdot,t)}{\partial t}$. The details are left to the reader as an exercise; see Problem 23.18. □

23.3.3 The Maximum Principle

As a final property of solutions to parabolic equations, we will show a maximum principle. We will consider a slighly more general version of (23.32). With the same notation as in the previous section, we seek u such that

$$\begin{cases} \dfrac{\partial u(\boldsymbol{x},t)}{\partial t} - \Delta u(\boldsymbol{x},t) = f(\boldsymbol{x},t), & (\boldsymbol{x},t) \in \Omega \times (0,T], \\ u(\boldsymbol{x},t) = g(\boldsymbol{x},t), & (\boldsymbol{x},t) \in \partial\Omega \times (0,T], \\ u(\boldsymbol{x},0) = u_0(\boldsymbol{x}), & \boldsymbol{x} \in \bar{\Omega}, \end{cases} \tag{23.38}$$

where the initial condition u_0, the right-hand side f, and the boundary condition g are given. We need to introduce some terminology.

Definition 23.26 (parabolic boundary)**.** The **space–time cylinder** for the IBVP (23.38) is

$$\mathcal{C} = \Omega \times (0,T).$$

The **parabolic boundary** of $\mathcal{C}$ is

$$\partial_p\mathcal{C} = (\partial\Omega \times [0,T]) \cup \left(\bar{\Omega} \times \{0\}\right).$$

The following result is known as the *maximum principle*. It provides rigor to our intuition behind diffusion: it is the tendency of a substance to evenly spread.

Theorem 23.27 (maximum principle)**.** *Let the function $v \colon \bar{\mathcal{C}} \to \mathbb{R}$ be sufficiently smooth and, for $(\boldsymbol{x},t) \in \Omega \times (0,T)$, satisfy*

$$\frac{\partial v(\boldsymbol{x},t)}{\partial t} - \Delta v(\boldsymbol{x},t) \leq 0.$$

Then the function u attains its maximum on the parabolic boundary $\partial_p\mathcal{C}$.

Proof. If this is not the case, the maximum is attained at some point $(\bar{\boldsymbol{x}},\bar{t}) \in \Omega \times (0,T]$, i.e.,

$$M = v(\bar{\boldsymbol{x}},\bar{t}) = \max_{\bar{\mathcal{C}}} v > \max_{\partial_p\mathcal{C}} v = m.$$

Define $\tilde{v}(\boldsymbol{x},t) = v(\boldsymbol{x},t) + \frac{\varepsilon}{2}\|\boldsymbol{x}\|_2^2$, where $\varepsilon > 0$ is sufficiently small, so that $\tilde{v}$ also attains its maximum on $\Omega \times (0,T]$. Indeed,

$$\max_{\partial_p\mathcal{C}} \tilde{v} \leq m + \frac{\varepsilon}{2}\max_{\partial_p\mathcal{C}} \|x\|_2^2 < M \leq \max_{\bar{\mathcal{C}}} v.$$

Since $\Delta(\|\boldsymbol{x}\|_2^2) = 2d$, we have

$$\frac{\partial\tilde{v}}{\partial t} - \Delta\tilde{v} = \frac{\partial v}{\partial t} - \Delta v - \varepsilon d < 0.$$

Let $(\boldsymbol{y}, s) \in \Omega \times (0, T]$ be the point where $\tilde{v}$ attains its maximum. We must have

$$-\Delta \tilde{v}(\boldsymbol{y}, s) \geq 0,$$

and, if $s < T$,

$$\frac{\partial \tilde{v}(\boldsymbol{y}, s)}{\partial t} = 0,$$

or, if $s = T$,

$$\frac{\partial \tilde{v}(\boldsymbol{y}, s)}{\partial t} \geq 0.$$

In conclusion,

$$\frac{\partial \tilde{v}(\boldsymbol{y}, s)}{\partial t} - \Delta \tilde{v}(\boldsymbol{y}, s) \geq 0,$$

which is a contradiction. □

The maximum principle gives us stability of continuous solutions.

Corollary 23.28 (stability). *Let u be a classical solution to* (23.38). *If $f \equiv 0$, then we have*

$$\|u\|_{L^\infty(C)} \leq \max\left\{\|g\|_{L^\infty(\partial\Omega \times (0,T))}, \|u_0\|_{L^\infty(\Omega)}\right\}.$$

If $f \neq 0$ and there is $R > 0$ for which $\Omega \subseteq B(0, R)$, then

$$\|u\|_{L^\infty(C)} \leq \max\left\{\|g\|_{L^\infty(\partial\Omega \times (0,T))}, \|u_0\|_{L^\infty(\Omega)}\right\} + \frac{R^2}{2d}\|f\|_{L^\infty(C)}.$$

Proof. See Problem 23.20. □

23.4 Hyperbolic Equations

In this section, we will consider hyperbolic problems, the prototype of which is the wave equation, which we presented in Example 23.5. We will consider, mostly, IVPs. For reasons that will become evident later, we must begin the discussion with the transport equation that we described in Section 23.1.3.

23.4.1 The Initial Value Problem for the Transport Equation

We begin our discussion with the Cauchy, or initial value, problem for the transport equation. In other words, given $T > 0$, $c \in \mathbb{R}\backslash\{0\}$, $f : \mathbb{R} \times (0, T] \to \mathbb{R}$, and $u_0 : \mathbb{R} \to \mathbb{R}$, we seek $u : \mathbb{R} \times [0, T] \to \mathbb{R}$ such that

$$\begin{cases} \dfrac{\partial u(x,t)}{\partial t} + c\dfrac{\partial u(x,t)}{\partial x} = f(x,t), & (x,t) \in \mathbb{R} \times (0, T], \\ u(x, 0) = u_0(x), & x \in \mathbb{R}. \end{cases} \tag{23.39}$$

Let us define what we mean by a classical solution to this problem.

Definition 23.29 (classical solution). The function $u \in C^1(\mathbb{R} \times [0, T])$ is called a **classical solution** to (23.39) if and only if the equation and initial condition hold point-wise.

Let us consider, first, the homogeneous case, i.e., $f \equiv 0$. The following result gives us the existence, uniqueness, and stability of classical solutions.

Theorem 23.30 (well-posedness). *Assume that $f \equiv 0$ and $u_0 \in C^1(\mathbb{R})$. Then there is a classical solution u to the Cauchy problem (23.39). This solution is given by*

$$u(x, t) = u_0(x - ct). \tag{23.40}$$

If, in addition, there is $p \in [1, \infty]$ such that $u_0 \in L^p(\mathbb{R})$, then, for any $t > 0$,

$$\|u(\cdot, t)\|_{L^p(\mathbb{R})} = \|u_0\|_{L^p(\mathbb{R})},$$

which, in particular, implies uniqueness.

Proof. Existence is clear from (23.40) and the fact that the assumed smoothness on the initial condition allows us to differentiate u to show that the equation is satisfied point-wise.

To show the norm invariance property, for $p < \infty$, we multiply the equation by $p|u|^{p-2}u$ to obtain

$$p|u(x,t)|^{p-2}u(x,t)\frac{\partial u(x,t)}{\partial t} = -cp|u(x,t)|^{p-2}u(x,t)\frac{\partial u(x,t)}{\partial x},$$

or, equivalently,

$$\frac{\partial |u(x,t)|^p}{\partial t} = -c\frac{\partial |u(x,t)|^p}{\partial x},$$

which, integrating with respect to the x-variable, yields

$$\begin{aligned}\frac{\mathrm{d}}{\mathrm{d}t}\left[\int_{-\infty}^{\infty} |u(x,t)|^p \,\mathrm{d}x\right] &= -c\int_{-\infty}^{\infty} \frac{\partial}{\partial x}|u(x,t)|^p \,\mathrm{d}x \\ &= -c\left[\lim_{x_R\to\infty}|u(x_R,t)|^p - \lim_{x_L\to-\infty}|u(x_L,t)|^p\right] \\ &= 0,\end{aligned}$$

where the last identity follows from the fact that, since $u_0 \in L^p(\mathbb{R})$, we have $u(x,t) = u_0(x - ct) \to 0$ as $x \to \pm\infty$. Taking the limit $p \to \infty$ of the norm invariance yields the case $p = \infty$.

Finally, suppose that there are two classical solutions, u_1 and u_2. Then, by linearity, $v = u_1 - u_2$ solves the same problem, but with the initial data $u_0 \equiv 0 \in L^2(\mathbb{R})$. Therefore, using the norm invariance, we have, for any $t \geq 0$,

$$\|v(\cdot, t)\|_{L^2(\mathbb{R})} = \|u_0\|_{L^2(\mathbb{R})} = 0.$$

Uniqueness follows. □

The solution of the homogeneous transport equation, given by (23.40), helps illustrate the so-called *method of characteristics*, which we will use to, formally, give a solution to (23.39) for an inhomogeneous right-hand side. We begin by observing that we can introduce the *characteristic equation*

$$X'(s) = c, \qquad X(t) = x \qquad \Longleftrightarrow \qquad X(s) = x + c(s - t),$$

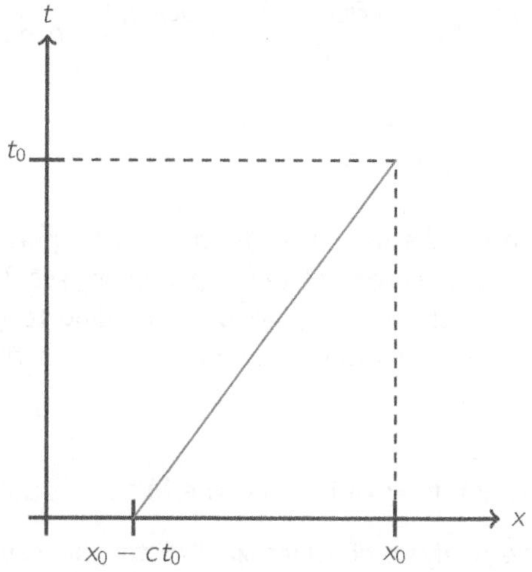

Figure 23.2 Domain of dependence for the transport equation. The solution to (23.39) with $c > 0$ at point (x_0, t_0) depends only on the information in the shaded region.

so that, by setting $x = X(s)$, the transport equation (23.39) can be written as

$$\frac{\mathrm{d}}{\mathrm{d}s}u(X(s), s+t) = \frac{\partial u(X(s), s+t)}{\partial t} + c\frac{\partial u(x, s+t)}{\partial x} = f(X(s), s+t)$$

with $u(X(0), 0) = u_0(X(0))$. The solution to this ODE is

$$\begin{aligned} u(x,t) &= u(X(t), t) \\ &= u(X(0), 0) + \int_0^t f(X(s), s)\mathrm{d}s \\ &= u_0(x - ct) + \int_0^t f(x + c(s-t), s)\mathrm{d}s. \end{aligned} \tag{23.41}$$

Remark 23.31 (domain of dependence)**.** Formulas (23.40) and (23.41) not only give us an explicit solution representation but also show two important things. First, (23.40) motivates the name of this equation, it *transports* or *advects* the initial condition with speed c. Second, these formulas show that, as opposed to elliptic or parabolic equations, there is a finite and well-defined *domain of dependence.* In other words, the value of the solution to (23.39) at a point $(x, t) \in \mathbb{R} \times (0, T)$ depends only on the values of the initial condition and the right-hand side of the cone depicted in Figure 23.2.

One important thing to note is that the solution representation (23.40) makes sense even if $u_0 \notin C^1(\mathbb{R})$. In this case, however, we cannot speak about a classical solution, because the equation cannot hold point-wise. For this reason, we introduce the notion of a *weak solution* of the transport equation.

Definition 23.32 (weak solution)**.** A function $u \colon \mathbb{R} \times (0, T) \to \mathbb{R}$ such that, for every $t > 0$ and every bounded interval, $I \subset \mathbb{R}$ $u(\cdot, t) \in L^1(I)$ is a **weak solution** to (23.39) with $f \equiv 0$ if and only if

$$\int_{\mathbb{R}} \int_0^T u(x,t) \left[\frac{\partial \varphi(x,t)}{\partial t} + c \frac{\partial \varphi(x,t)}{\partial x} \right] \mathrm{d}t\mathrm{d}x + \int_{\mathbb{R}} u_0(x)\varphi(x,0)\mathrm{d}x = 0$$

for every $\varphi \in C^\infty(\mathbb{R} \times [0,T])$ such that there is $M > 0$ for which $|x| + t > M$ implies that $\varphi(x,t) = 0$.

We finish the discussion of the transport equation by noticing that, as opposed to elliptic (see Theorem 23.12) or parabolic (see Theorem 23.19) equations, there is no regularity theory to speak of. According to (23.40) or (23.41), the solution to the transport equation is as good, or bad, as the inital condition is.

23.4.2 The Initial Value Problem for the Wave Equation

We are now ready to consider the IVP for the wave equation of Example 23.5. To convey the essential ideas, we will focus on the one-dimensional case. Thus, given $T > 0$, $a > 0$, and $u_0, v_0 \colon \mathbb{R} \to \mathbb{R}$, we seek a function $u \colon \mathbb{R} \times [0,T] \to \mathbb{R}$ such that

$$\begin{cases} \dfrac{\partial^2 u(x,t)}{\partial t^2} - a \dfrac{\partial^2 u(x,t)}{\partial x^2} = 0, & (x,t) \in \mathbb{R} \times (0,T], \\ u(x,0) = u_0(x), & x \in \mathbb{R}, \\ \dfrac{\partial u(x,0)}{\partial t} = v_0(x), & x \in \mathbb{R}. \end{cases} \tag{23.42}$$

A classical solution is defined as usual.

Definition 23.33 (classical solution)**.** The function $u \in C^2(\mathbb{R} \times [0,T])$ is called a **classical solution** to (23.42) if and only if the equation and initial condition hold point-wise.

Before we embark on the study of existence of solutions, we will present an energy-type estimate that shows that the solution to the wave equation has a very well-defined *domain of dependence*; see Figure 23.3.

Theorem 23.34 (energy estimate)**.** *Let u be a classical solution to* (23.42). *Given $(x_0, t_0) \in \mathbb{R} \times (0,T]$, define the cone*

$$K(x_0,t_0) = \left\{ (x,t) \in \mathbb{R} \times [0,T] \;\middle|\; t \leq t_0,\ |x - x_0| \leq \sqrt{a}(t_0 - t) \right\},$$

and, for $t \in [0, t_0]$, its t-section

$$B_t(x_0,t_0) = \{ x \in \mathbb{R} \mid (x,t) \in K(x_0,t_0) \}.$$

Then, for every $t \leq t_0$,

$$\int_{B_t(x_0,t_0)} \left(\left| \frac{\partial u(x,t)}{\partial t} \right|^2 + a \left| \frac{\partial u(x,t)}{\partial x} \right|^2 \right) \mathrm{d}x \leq \int_{x_0 - \sqrt{a}t_0}^{x_0 + \sqrt{a}t_0} \left(|v_0(x)|^2 + a|u_0'(x)|^2 \right) \mathrm{d}x.$$

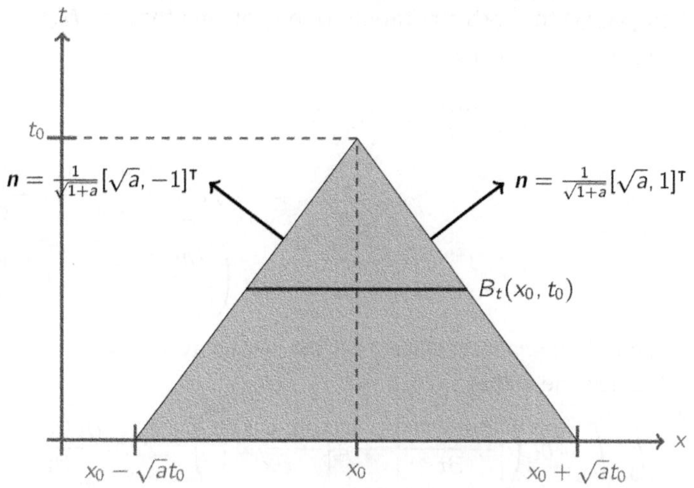

Figure 23.3 Domain of dependence for the wave equation. The solution to (23.42) at point (x_0, t_0) depends only on the information in the region $K(x_0, t_0)$ (shaded). A t-section of $K(x_0, t_0)$ is denoted by $B_t(x_0, t_0)$. The exterior unit normal to M_s is $\boldsymbol{n}$. Refer to Theorem 23.34 and its proof for notation.

Proof. Multiply the differential equation by $2\frac{\partial u(x,t)}{\partial t}$ to obtain

$$\begin{aligned} 0 &= 2\left(\frac{\partial^2 u(x,t)}{\partial t^2} - a\frac{\partial^2 u(x,t)}{\partial x^2}\right)\frac{\partial u(x,t)}{\partial t} \\ &= \frac{\partial}{\partial t}\left(\left|\frac{\partial u(x,t)}{\partial t}\right|^2 + a\left|\frac{\partial u(x,t)}{\partial x}\right|^2\right) - 2a\frac{\partial}{\partial x}\left(\frac{\partial u(x,t)}{\partial t}\frac{\partial u(x,t)}{\partial x}\right). \end{aligned}$$

Let $t \le t_0$. Integrating over $\{(x,s) \in K(x_0,t_0) \mid s \le t\}$ and using the Divergence Theorem, we obtain

$$\begin{aligned} 0 &= \int_0^t \int_{B_s}\left[\frac{\partial}{\partial t}\left(\left|\frac{\partial u(x,s)}{\partial t}\right|^2 + a\left|\frac{\partial u(x,s)}{\partial x}\right|^2\right)\right. \\ &\quad \left. - 2a\frac{\partial}{\partial x}\left(\frac{\partial u(x,s)}{\partial t}\frac{\partial u(x,s)}{\partial x}\right)\right]\mathrm{d}x\mathrm{d}s \\ &= \int_{B_t}\left(\left|\frac{\partial u(x,t)}{\partial t}\right|^2 + a\left|\frac{\partial u(x,t)}{\partial x}\right|^2\right)\mathrm{d}x - \int_{x_0-\sqrt{a}t_0}^{x_0+\sqrt{a}t_0}\left(|v_0(x)|^2 + a\,|u_0'(x)|^2\right)\mathrm{d}x \\ &\quad + \int_0^t \int_{M_s}\left[n_t\left(\left|\frac{\partial u(x,s)}{\partial t}\right|^2 + a\left|\frac{\partial u(x,s)}{\partial x}\right|^2\right) - 2an_x\frac{\partial u(x,s)}{\partial t}\frac{\partial u(x,s)}{\partial x}\right]\mathrm{d}x\mathrm{d}s, \end{aligned}$$

where $M_s(x_0,t_0) = \{(x,t) \in \mathbb{R} \times [0,T] \mid |x - x_0| = \sqrt{a}(t_0 - s)\}$, $\boldsymbol{n} = [n_x, n_t]^\intercal$ denotes the unit exterior normal to $K(x_0,t_0)$, and we suppressed the dependence on (x_0, t_0) to shorten notation. Upon noticing that, on $M_s(x_0, t_0)$, we have $n_t = \frac{1}{\sqrt{2}}$ and $n_t^2 = a n_x^2$. Thus,

$$-2an_x \frac{\partial u(x,s)}{\partial t}\frac{\partial u(x,s)}{\partial x} \geq -2a|n_x|\left|\frac{\partial u(x,s)}{\partial t}\right|\left|\frac{\partial u(x,s)}{\partial x}\right|$$
$$= -2\sqrt{a}n_t\left|\frac{\partial u(x,s)}{\partial t}\right|\left|\frac{\partial u(x,s)}{\partial x}\right|$$
$$\geq n_t\left(\left|\frac{\partial u(x,s)}{\partial t}\right|^2 + a\left|\frac{\partial u(x,s)}{\partial x}\right|^2\right),$$

where, in the last step, we used the inequality $2\xi\eta \leq \xi^2 + \eta^2$. Having this estimate, it is now clear that

$$\int_0^t \int_{M_s} \left[n_t\left(\left|\frac{\partial u(x,s)}{\partial t}\right|^2 + a\left|\frac{\partial u(x,s)}{\partial x}\right|^2\right) - 2an_x\frac{\partial u(x,s)}{\partial t}\frac{\partial u(x,s)}{\partial x}\right] dxds \geq 0$$

and the claimed energy estimate immediately follows. □

The previous result can be used to not only establish uniqueness but also give a very precise description of the domain of dependence. If $u_0 = v_0 \equiv 0$ on $(x_0 - \sqrt{a}t_0, x_0 + \sqrt{a}t_0)$, then the solution to (23.42) will be zero on the whole set $K(x_0, t_0)$, in particular at (x_0, t_0). In other words, the value of $u(x_0, t_0)$ is not affected by the values of the inital data outside of the interval $(x_0 - \sqrt{a}t_0, x_0 + \sqrt{a}t_0)$.

Let us now show the existence of solutions by means of the so-called *d'Alembert formula*.[16]

Theorem 23.35 (existence). *Assume that $u_0 \in C^2(\mathbb{R})$ and $v_0 \in C^1(\mathbb{R})$. Then problem* (23.42) *has a unique classical solution, which is given by*

$$u(x,t) = \frac{1}{2}\left[u_0(x + \sqrt{a}t) + u_0(x - \sqrt{a}t)\right] + \frac{1}{2\sqrt{a}}\int_{x-\sqrt{a}t}^{x+\sqrt{a}t} v_0(y)dy. \quad (23.43)$$

Proof. Uniqueness follows from Theorem 23.34. Given the assumed regularity of the initial data, verifying that (23.43) is indeed a classical solution is merely a calculation. Let us here, instead, provide a formal derivation of this formula, which will help elucidate the behavior of the solution.

Let us note that, at least formally, the PDE can be factored into

$$\left(\frac{\partial}{\partial t} + \sqrt{a}\frac{\partial}{\partial x}\right)\left(\frac{\partial}{\partial t} - \sqrt{a}\frac{\partial}{\partial x}\right) u(x,t) = 0.$$

Define

$$w(x,t) = \left(\frac{\partial}{\partial t} - \sqrt{a}\frac{\partial}{\partial x}\right) u(x,t)$$

to see that w must be a solution of

$$\frac{\partial w(x,t)}{\partial t} + \sqrt{a}\frac{\partial w(x,t)}{\partial x} = 0.$$

[16] Named in honor of the French mathematician Jean-Baptiste le Rond d'Alembert (1717–1783).

By (23.40),

$$w(x, t) = w_0(x - \sqrt{a}t),$$

where $w_0(x) = w(x, 0)$. With this at hand, we see that

$$\frac{\partial u(x, t)}{\partial t} - \sqrt{a}\frac{\partial u(x, t)}{\partial x} = w(x, t) = w_0(x - \sqrt{a}t),$$

which, by (23.41), gives us the solution to (23.42) in the form

$$\begin{aligned} u(x, t) &= u_0(x + \sqrt{a}t) + \int_0^t w_0(x - \sqrt{a}(s - t) - \sqrt{a}s)ds \\ &= u_0(x + \sqrt{a}t) + \int_0^t w_0(x + \sqrt{a}t - 2\sqrt{a}s)ds \\ &= u_0(x + \sqrt{a}t) + \frac{1}{2\sqrt{a}}\int_{x-\sqrt{a}t}^{x+\sqrt{a}t} w_0(y)dy, \end{aligned}$$

where, in the last step, we applied the change of variables $y = x + \sqrt{a}t - 2\sqrt{a}s$.

To conclude, we must find the function w_0. Notice that

$$w_0(x) = w(x, 0) = \frac{\partial u(x, 0)}{\partial t} - \sqrt{a}\frac{\partial u(x, 0)}{\partial x} = v_0(x) - \sqrt{a}u_0'(x).$$

Substituting,

$$\begin{aligned} u(x, t) &= u_0(x + \sqrt{a}t) + \frac{1}{2\sqrt{a}}\int_{x-\sqrt{a}t}^{x+\sqrt{a}t} \left[v_0(y) - \sqrt{a}u_0'(y)\right] dy \\ &= u_0(x + \sqrt{a}t) + \frac{1}{2\sqrt{a}}\int_{x-\sqrt{a}t}^{x+\sqrt{a}t} v_0(y)dy - \frac{1}{2}\left[u_0(x + \sqrt{a}t) - u_0(x - \sqrt{a}t)\right] \\ &= \frac{1}{2}\left[u_0(x + \sqrt{a}t) + u_0(x - \sqrt{a}t)\right] + \frac{1}{2\sqrt{a}}\int_{x-\sqrt{a}t}^{x+\sqrt{a}t} v_0(y)dy, \end{aligned}$$

as we claimed. □

We comment that similar representation formulas exist in more dimensions $d > 1$. For $d = 2$, this bears the name of the *Poisson formula*,[17] whereas, for $d = 3$, this is known as the *Kirchhoff formula*.[18]

Similar to the IVP for the heat equation, we have Duhamel's formula for the solution of the inhomogeneous IVP for the wave equation: given $T > 0, u_0, v_0 : \mathbb{R} \to \mathbb{R}$, $f : \mathbb{R} \times (0, T] \to \mathbb{R}$, find $u : \mathbb{R} \times [0, T] \to \mathbb{R}$ such that

$$\begin{cases} \dfrac{\partial^2 u(x, t)}{\partial t^2} - a\dfrac{\partial^2 u(x, t)}{\partial x^2} = f(x, t), & (x, t) \in \mathbb{R} \times (0, T], \\ u(x, 0) = u_0(x), & x \in \mathbb{R}, \\ \dfrac{\partial u(x, 0)}{\partial t} = v_0(x), & x \in \mathbb{R}. \end{cases}$$

[17] Named in honor of the French mathematician, engineer, and physicist Baron Siméon Denis Poisson (1781–1840).

[18] Named in honor of the German physicist Gustav Robert Kirchhoff (1824–1887).

The notion of a classical solution is as before. A formal representation formula is given by

$$u(x,t) = \frac{1}{2}\left[u_0(x+\sqrt{a}t) + u_0(x-\sqrt{a}t)\right] + \frac{1}{2\sqrt{a}}\int_{x-\sqrt{a}t}^{x+\sqrt{a}t} v_0(y)\mathrm{d}y \\ + \frac{1}{2\sqrt{a}}\int_0^t \int_{x-\sqrt{a}s}^{x+\sqrt{a}s} f(y,t-s)\mathrm{d}y\mathrm{d}s. \tag{23.44}$$

Remark 23.36 (domain of dependence)**.** The representation formulas (23.43) and (23.44) illustrate an important point about the solution of the IVP for the wave equation. The solution at a point (x_0, t_0) only depends on the initial data and right-hand side within the cone $K(x_0, t_0)$ defined in Theorem 23.34; see Figure 23.3.

Remark 23.37 (regularity)**.** Notice that, as in the case of the transport equation, there is no regularity gain from the regularity of the initial data or right-hand side. This, again, is in stark contrast to Theorems 23.12 and 23.19.

23.4.3 The Initial Boundary Value Problem for the Wave Equation

Let us now study the IBVP for the wave equation. We let $d \in \mathbb{N}$, $\Omega \subset \mathbb{R}^d$ be a bounded domain with sufficiently smooth boundary, $T > 0$, and $a > 0$. Given $u_0 \colon \bar{\Omega} \to \mathbb{R}$, $v_0 \colon \bar{\Omega} \to \mathbb{R}$, and $f \colon \Omega \times (0,T] \to \mathbb{R}$, we seek $u \colon \bar{\Omega} \times [0,T] \to \mathbb{R}$ such that

$$\begin{cases} \dfrac{\partial^2 u(\boldsymbol{x},t)}{\partial t^2} - a\Delta u(\boldsymbol{x},t) = f(\boldsymbol{x},t), & (\boldsymbol{x},t) \in \Omega \times (0,T], \\ u(\boldsymbol{x},t) = 0, & \boldsymbol{x} \in \partial\Omega \times (0,T], \\ u(\boldsymbol{x},0) = u_0(\boldsymbol{x}), & \boldsymbol{x} \in \Omega, \\ \dfrac{\partial u(\boldsymbol{x},0)}{\partial t} = v_0(\boldsymbol{x}), & \boldsymbol{x} \in \Omega. \end{cases} \tag{23.45}$$

As is the case with the heat equation, while it is possible to provide a notion of classical solutions and develop a theory for them, it is more instructive to introduce the notion of a weak solution.

Definition 23.38 (weak solution)**.** Let $T > 0$. The function

$$u \in L^2(0,T;H_0^1(\Omega)), \quad \frac{\partial u}{\partial t} \in L^2(0,T;L^2(\Omega)), \quad \frac{\partial^2 u}{\partial t^2} \in L^2(0,T;H_0^1(\Omega)')$$

is a **weak solution** to (23.45) if and only if, for almost every $t \in (0,T)$, we have

$$\left\langle \frac{\partial^2 u(\cdot,t)}{\partial t^2}, v \right\rangle + a\int_\Omega \nabla u(\boldsymbol{x},t)\cdot\nabla v(\boldsymbol{x})\mathrm{d}\boldsymbol{x} = \int_\Omega f(\boldsymbol{x},t)v(\boldsymbol{x})\mathrm{d}\boldsymbol{x}, \quad \forall v \in H_0^1(\Omega),$$

$u(\cdot,t) \to u_0$ in $L^2(\Omega)$, and $\frac{\partial u(\cdot,t)}{\partial t} \to v_0$ in $H_0^1(\Omega)'$.

The motivation for this definition is as usual; see Problem 23.25. Let us show the uniqueness of weak solutions via the so-called energy conservation property for (23.45).

Theorem 23.39 (conservation). *Let u be a weak solution to (23.45) with $f \equiv 0$. The quantity*

$$E(t) = \int_\Omega \left[\left| \frac{\partial u(\mathbf{x},t)}{\partial t} \right|^2 + a \|\nabla u(\mathbf{x},t)\|_2^2 \right] \mathrm{d}\mathbf{x}$$

is constant in time, i.e., $E(t) = E(0)$ for all $t \in (0,T]$. As a consequence, weak solutions are unique.

Proof. We leave the proof of uniqueness to the reader as an exercise; see Problem 23.27.

To prove the energy conservation property, we formally set, in the definition of a weak solution, $v = \frac{\partial u(\cdot,t)}{\partial t}$ to obtain

$$\frac{1}{2}\frac{\mathrm{d}}{\mathrm{d}t} \int_\Omega \left[\left| \frac{\partial u(\mathbf{x},t)}{\partial t} \right|^2 + a \|\nabla u(\mathbf{x},t)\|_2^2 \right] \mathrm{d}\mathbf{x} = 0$$

from which energy conservation follows. □

To prove existence, we follow a similar approach to that used for the heat equation, namely an eigenvalue expansion.

Theorem 23.40 (existence). *Let $a > 0$, $T > 0$, $u_0 \in H_0^1(\Omega)$, $v_0 \in L^2(\Omega)$, and $f \in L^2(0,T;L^2(\Omega))$. In this setting, problem (23.45) has a weak solution, which, in addition, for all $t \in [0,T]$, satisfies*

$$\begin{aligned} &\left\| \frac{\partial u(\cdot,t)}{\partial t} \right\|_{L^2(\Omega)}^2 + a \|\nabla u(\cdot,t)\|_{L^2(\Omega;\mathbb{R}^d)}^2 \\ &\quad \leq C \left(\|\nabla u_0\|_{L^2(\Omega;\mathbb{R}^d)}^2 + \|v_0\|_{L^2(\Omega)}^2 + \int_0^t \|f(\cdot,s)\|_{L^2(\Omega)}^2 \, \mathrm{d}s \right) \end{aligned}$$

for some constant that depends on T, but not the rest of the problem data.

Proof. We propose the following solution ansatz:

$$u(\mathbf{x},t) = \sum_{k=1}^{\infty} u_k(t)\varphi_k(\mathbf{x}),$$

where the functions $\{\varphi_k\}_{k=1}^{\infty}$ are the eigenfunctions of the Laplacian, which we can use owing to Theorem 23.14. Thus, by their orthonormality in $L^2(\Omega)$, we obtain, for each $k \geq 1$, the following ODE:

$$u_k''(t) + a\lambda_k u_k(t) = f_k(t), \quad u_k(0) = u_{0,k}, \quad u_k'(0) = v_{0,k},$$

where

$$u_0(\mathbf{x}) = \sum_{k=1}^{\infty} u_{0,k}\varphi(\mathbf{x}), \quad v_0(\mathbf{x}) = \sum_{k=1}^{\infty} v_{0,k}\varphi(\mathbf{x}), \quad f(\mathbf{x},t) = \sum_{k=1}^{\infty} f_k(t)\varphi_k(\mathbf{x}).$$

Thus,

$$\begin{aligned} u_k(t) &= u_{0,k}\cos(\sqrt{a\lambda_k}t) + \frac{v_{0,k}}{\sqrt{a\lambda_k}}\sin(\sqrt{a\lambda_k}t) \\ &\quad + \frac{1}{\sqrt{a\lambda_k}} \int_0^t f_k(s) \sin\left(\sqrt{a\lambda_k}(t-s)\right) \mathrm{d}s. \end{aligned}$$

The assumed regularity on the problem data guarantees that the series representations converges and, moreover, gives the claimed estimate. □

We conclude our discussion with a regularity result; see [31] for a proof.

Theorem 23.41 (regularity)**.** *Assume that the boundary of* Ω *is sufficiently smooth,* $u_0 \in H^2(\Omega)$, $v_0 \in H_0^1(\Omega)$, *and* $\frac{\partial f}{\partial t} \in L^2(0, T; L^2(\Omega))$. *Then the weak solution of* (23.45) *satisfies*

$$u \in L^\infty(0, T; H^2(\Omega)), \qquad \frac{\partial u}{\partial t} \in L^\infty(0, T; H_0^1(\Omega)),$$
$$\frac{\partial^2 u}{\partial t^2} \in L^\infty(0, T; L^2(\Omega)), \qquad \frac{\partial^3 u}{\partial t^3} \in L^2(0, T; H_0^1(\Omega)').$$

23.4.4 Hyperbolic Systems

As a last topic in this (not so) short review of PDE theory, we consider the Cauchy problem for so-called *symmetric hyperbolic* or *Friedrichs systems.*[19] This is a generalization of the Cauchy problem (23.39) for the transport equation. Let $d, m \in \mathbb{N}$. Here, d will be the spatial dimension, whereas m will denote the size of our system. We assume that we are given a final time $T > 0$, an initial condition $\boldsymbol{u}_0 \colon \mathbb{R}^d \to \mathbb{R}^m$, a forcing function $\boldsymbol{f} \colon \mathbb{R}^d \times (0, T] \to \mathbb{R}^m$, and a collection $\{\mathsf{A}_j\}_{j=1}^d \subset \mathbb{R}^{m\times m}_{\text{sym}}$. We seek $\boldsymbol{u} \colon \mathbb{R}^d \times [0, T] \to \mathbb{R}^m$ such that

$$\begin{cases} \dfrac{\partial \boldsymbol{u}(\boldsymbol{x}, t)}{\partial t} + \displaystyle\sum_{j=1}^d \mathsf{A}_j \frac{\partial \boldsymbol{u}(\boldsymbol{x}, t)}{\partial x_j} = \boldsymbol{f}(\boldsymbol{x}, t), & (\boldsymbol{x}, t) \in \mathbb{R}^d \times (0, T], \\ \boldsymbol{u}(\boldsymbol{x}, 0) = \boldsymbol{u}_0(\boldsymbol{x}), & \boldsymbol{x} \in \mathbb{R}^d. \end{cases} \tag{23.46}$$

Let us give a definition.

Definition 23.42 (hyperbolicity)**.** For $\boldsymbol{y} \in \mathbb{R}^d$, define

$$\mathsf{B}(\boldsymbol{y}) = \sum_{j=1}^d [\boldsymbol{y}]_j \mathsf{A}_j \in \mathbb{R}^{m\times m}.$$

We say that problem (23.46) is **symmetric hyperbolic** if, for every $\boldsymbol{y} \in \mathbb{R}^d$, the matrix $\mathsf{B}(\boldsymbol{y})$ is diagonalizable, i.e., it has m real eigenpairs $\{\lambda_j(\boldsymbol{y}), \boldsymbol{q}(\boldsymbol{y})\}_{j=1}^m \subset \mathbb{R} \times \mathbb{R}^m$,

$$\lambda_1(\boldsymbol{y}) \leq \cdots \leq \lambda_m(\boldsymbol{y}),$$

and the eigenvectors $\{\boldsymbol{q}(\boldsymbol{y})\}_{j=1}^m$ form a basis of $\mathbb{R}^m$. If, in addition, all the eigenvalues are distinct, i.e.,

$$\lambda_1(\boldsymbol{y}) < \cdots < \lambda_m(\boldsymbol{y}),$$

then we say that the problem is **strictly hyperbolic**.

[19] Named in honor of the German–American mathematician Kurt Otto Friedrichs (1901–1982).

Before we proceed with any theory, we present some examples of systems that fit this framework.

Example 23.10 The wave equation in one dimension is actually a symmetric strictly hyperbolic system. Indeed, let us recall that, for $a > 0$, the one-dimensional wave equation reads

$$\frac{\partial^2 u(x,t)}{\partial t^2} = a\frac{\partial^2 u(x,t)}{\partial x^2}.$$

Define the variable

$$\boldsymbol{u} = \begin{bmatrix} u_1 \\ u_2 \end{bmatrix} : \mathbb{R} \times [0,T] \to \mathbb{R}^2,$$

via

$$u_1(x,t) = \frac{\partial u(x,t)}{\partial t}, \qquad u_2(x,t) = \sqrt{a}\frac{\partial u(x,t)}{\partial x}.$$

Then, using the wave equation, it follows that

$$\begin{aligned}
\frac{\partial \boldsymbol{u}(x,t)}{\partial t} &= \begin{bmatrix} \frac{\partial^2 u(x,t)}{\partial t^2} \\ \sqrt{a}\frac{\partial^2 u(x,t)}{\partial t \partial x} \end{bmatrix} \\
&= \begin{bmatrix} a\frac{\partial^2 u(x,t)}{\partial x^2} \\ \sqrt{a}\frac{\partial^2 u(x,t)}{\partial x \partial t} \end{bmatrix} \\
&= \frac{\partial}{\partial x}\begin{bmatrix} 0 & \sqrt{a} \\ \sqrt{a} & 0 \end{bmatrix}\begin{bmatrix} \frac{\partial u(x,t)}{\partial t} \\ \sqrt{a}\frac{\partial u(x,t)}{\partial x} \end{bmatrix} \\
&= \begin{bmatrix} 0 & \sqrt{a} \\ \sqrt{a} & 0 \end{bmatrix}\frac{\partial \boldsymbol{u}(x,t)}{\partial x}.
\end{aligned}$$

This is a symmetric hyperbolic system with

$$\mathsf{A} = \begin{bmatrix} 0 & -\sqrt{a} \\ -\sqrt{a} & 0 \end{bmatrix}.$$

The initial condition is given by

$$\boldsymbol{u}_0(x) = \begin{bmatrix} v_0(x) \\ \sqrt{a}u_0'(x) \end{bmatrix}.$$

Example 23.11 Maxwell's equations[20] describe the evolution of an electromagnetic field in the vaccuum. Upon choosing suitable units, they read

[20] Named in honor of the British scientist James Clerk Maxwell (1831–1879).

$$\begin{cases} \dfrac{\partial \boldsymbol{B}(\boldsymbol{x},t)}{\partial t} + \nabla \times \boldsymbol{E}(\boldsymbol{x},t) = 0, \\ \dfrac{\partial \boldsymbol{E}(\boldsymbol{x},t)}{\partial t} - \nabla \times \boldsymbol{B}(\boldsymbol{x},t) + \boldsymbol{J}(\boldsymbol{x},t) = 0, \\ \nabla \cdot \boldsymbol{B}(\boldsymbol{x},t) = 0, \\ \nabla \cdot \boldsymbol{E}(\boldsymbol{x},t) = \rho(\boldsymbol{x},t), \end{cases}$$

and are supplemented with initial conditions for $\boldsymbol{E}$ and $\boldsymbol{B}$. The unknowns in these equations are the electric, $\boldsymbol{E}\colon \mathbb{R}^3 \times (0,\infty) \to \mathbb{R}^3$, and magnetic, $\boldsymbol{B}\colon \mathbb{R}^3 \times (0,\infty) \to \mathbb{R}^3$, fields. The quantities $\rho\colon \mathbb{R}^3 \times (0,\infty) \to \mathbb{R}$ and $\boldsymbol{J}\colon \mathbb{R}^3 \times (0,\infty) \to \mathbb{R}^3$ are the charge and current density, respectively, and they are related by

$$\frac{\partial \rho(\boldsymbol{x},t)}{\partial t} + \nabla \cdot \boldsymbol{J}(\boldsymbol{x},t) = 0.$$

In these equations, we introduced the *curl or rotor* of a vector field. If $\boldsymbol{v} \in C^1(\mathbb{R}^3;\mathbb{R}^3)$, then

$$\nabla \times \boldsymbol{v}(\boldsymbol{x}) = \det \begin{bmatrix} \boldsymbol{e}_1 & \boldsymbol{e}_2 & \boldsymbol{e}_3, \\ \frac{\partial}{\partial x_1} & \frac{\partial}{\partial x_2} & \frac{\partial}{\partial x_3} \\ v_1(\boldsymbol{x}) & v_2(\boldsymbol{x}) & v_3(\boldsymbol{x}) \end{bmatrix} = \begin{bmatrix} \frac{\partial v_3(\boldsymbol{x})}{\partial x_2} - \frac{\partial v_2(\boldsymbol{x})}{\partial x_3} \\ \frac{\partial v_1(\boldsymbol{x})}{\partial x_3} - \frac{\partial v_3(\boldsymbol{x})}{\partial x_1} \\ \frac{\partial v_2(\boldsymbol{x})}{\partial x_1} - \frac{\partial v_1(\boldsymbol{x})}{\partial x_2} \end{bmatrix}.$$

This is also a hyperbolic system. While this is true in general, we only show this in the so-called $2\frac{1}{2}$-*dimensional* case, i.e., when $\boldsymbol{E}$ and $\boldsymbol{B}$ are independent of x_3, so that $d = 2$ and $m = 3$. Upon introducing

$$\boldsymbol{U} = \begin{bmatrix} E_1 \\ E_2 \\ B_3 \end{bmatrix}, \quad \boldsymbol{V} = \begin{bmatrix} B_1 \\ B_2 \\ E_3 \end{bmatrix} : \mathbb{R}^2 \times (0,\infty) \to \mathbb{R}^3,$$

we obtain that (see Problem 23.28)

$$\frac{\partial \boldsymbol{U}(\boldsymbol{x},t)}{\partial t} + \mathsf{A}_1 \frac{\partial \boldsymbol{U}(\boldsymbol{x},t)}{\partial x_1} + \mathsf{A}_2 \frac{\partial \boldsymbol{U}(\boldsymbol{x},t)}{\partial x_2} = \boldsymbol{f}(\boldsymbol{x},t)$$

with

$$\mathsf{A}_1 = \begin{bmatrix} 0 & 0 & 0 \\ 0 & 0 & 1 \\ 0 & 1 & 0 \end{bmatrix}, \quad \mathsf{A}_2 = \begin{bmatrix} 0 & 0 & -1 \\ 0 & 0 & 0 \\ -1 & 0 & 0 \end{bmatrix}, \quad \boldsymbol{f}(\boldsymbol{x},t) = \begin{bmatrix} -J_1(\boldsymbol{x},t) \\ -J_2(\boldsymbol{x},t) \\ 0 \end{bmatrix}.$$

Similarly,

$$\frac{\partial \boldsymbol{V}(\boldsymbol{x},t)}{\partial t} + \mathsf{B}_1 \frac{\partial \boldsymbol{V}(\boldsymbol{x},t)}{\partial x_1} + \mathsf{B}_2 \frac{\partial \boldsymbol{V}(\boldsymbol{x},t)}{\partial x_2} = \boldsymbol{g}(\boldsymbol{x},t)$$

with

$$\mathsf{B}_1 = \begin{bmatrix} 0 & 0 & 0 \\ 0 & 0 & -1 \\ 0 & -1 & 0 \end{bmatrix}, \quad \mathsf{B}_2 = \begin{bmatrix} 0 & 0 & 1 \\ 0 & 0 & 0 \\ 1 & 0 & 0 \end{bmatrix}, \quad \boldsymbol{g}(\boldsymbol{x},t) = \begin{bmatrix} 0 \\ 0 \\ -J_3(\boldsymbol{x},t) \end{bmatrix}.$$

Example 23.12 Let us consider a general symmetric hyperbolic system in one dimension, $d = 1$, but we allow $m \in \mathbb{N}$

$$\begin{cases} \dfrac{\partial \boldsymbol{u}(x,t)}{\partial t} + \mathsf{A}\dfrac{\partial \boldsymbol{u}(x,t)}{\partial x} = \boldsymbol{f}(x,t), & (x,t) \in \mathbb{R} \times (0,T], \\ \boldsymbol{u}(x,0) = \boldsymbol{u}_0(x), & x \in \mathbb{R}. \end{cases}$$

Since A is diagonalizable, there is an invertible matrix $\mathsf{X} \in \mathbb{R}^{m \times m}$ and a diagonal matrix $\Lambda = \operatorname{diag}(\lambda_1, \ldots, \lambda_m) \in \mathbb{R}^{m \times m}$ such that

$$\mathsf{AX} = \mathsf{X}\Lambda.$$

Define

$$\boldsymbol{w} = \mathsf{X}^{-1}\boldsymbol{u} \colon \mathbb{R} \times [0,T] \to \mathbb{R}^m.$$

Then

$$\begin{cases} \dfrac{\partial \boldsymbol{w}(x,t)}{\partial t} + \Lambda\dfrac{\partial \boldsymbol{w}(x,t)}{\partial x} = \boldsymbol{g}(x,t), & (x,t) \in \mathbb{R} \times (0,T], \\ \boldsymbol{w}(x,0) = \boldsymbol{w}_0(x), & x \in \mathbb{R}, \end{cases}$$

with $\boldsymbol{g} = \mathsf{X}^{-1}\boldsymbol{f}$ and $\boldsymbol{w}_0 = \mathsf{X}^{-1}\boldsymbol{u}_0$. Notice that, since Λ is diagonal, this is nothing but a collection of m independent Cauchy problems for a linear transport equation.

We refer the reader to [23] and [29, Chapter 7] for several other examples of models that can be written as Friedrichs systems.

As Example 23.12 has shown, to obtain a theory of existence and uniqueness for (23.46) for $d = 1$, all that is needed is a change of coordinates. The case $d > 1$ can be treated with techniques that we have developed before.

For instance, we can obtain the existence of solutions via the Fourier transform. Take, component-wise, the Fourier transform of (23.46) to obtain, using Proposition 23.16, that

$$\begin{cases} \dfrac{\mathrm{d}\hat{\boldsymbol{u}}(\boldsymbol{\xi},t)}{\mathrm{d}t} + \mathrm{i}\mathsf{B}(\boldsymbol{\xi})\hat{\boldsymbol{u}}(\boldsymbol{\xi},t) = \hat{\boldsymbol{f}}(\boldsymbol{\xi},t), & t \in (0,T], \\ \hat{\boldsymbol{u}}(0;\boldsymbol{\xi}) = \hat{\boldsymbol{u}}_0(\boldsymbol{\xi}). \end{cases}$$

Thus, at least formally, the solution is given by the inverse Fourier transform of

$$\hat{\boldsymbol{u}}(\boldsymbol{\xi},t) = \exp(-\mathrm{i}t\mathsf{B}(\boldsymbol{\xi}))\hat{\boldsymbol{u}}_0(\boldsymbol{\xi}) + \int_0^t \exp(\mathrm{i}(s-t)\mathsf{B}(\xi))\hat{\boldsymbol{f}}(\boldsymbol{\xi},t)\mathrm{d}s.$$

In addition, we can use energy methods to provide suitable a priori estimates.

Proposition 23.43 (energy estimates). *Let $T > 0$ and $d, m \in \mathbb{N}$. Assume that $\boldsymbol{f} \in L^\infty(0,T;L^2(\mathbb{R}^d;\mathbb{R}^m))$ and $\boldsymbol{u}_0 \in L^2(\mathbb{R}^d;\mathbb{R}^m)$. Then any sufficiently smooth solution to* (23.46)*, which decays sufficiently fast at infinity, satisfies*

$$\|\boldsymbol{u}\|_{L^\infty(0,T;L^2(\mathbb{R}^d;\mathbb{R}^m))} \leq C\left(\|\boldsymbol{f}\|_{L^\infty(0,T;L^2(\mathbb{R}^d;\mathbb{R}^m))} + \|\boldsymbol{u}_0\|_{L^2(\mathbb{R}^d;\mathbb{R}^m)}\right),$$

where the constant C depends on T.

Proof. Take the $L^2(\mathbb{R}^d;\mathbb{R}^m)$ inner product of the equation with $\boldsymbol{u}(\boldsymbol{x},t)$ to obtain

$$\frac{1}{2}\frac{\mathrm{d}}{\mathrm{d}t}\|\boldsymbol{u}(\cdot,t)\|^2_{L^2(\mathbb{R}^d;\mathbb{R}^m)}+\sum_{j=1}^{d}\int_{\mathbb{R}^d}\left(\mathsf{A}_j\frac{\partial \boldsymbol{u}(\boldsymbol{x},t)}{\partial x_j}\right)\cdot\boldsymbol{u}(\boldsymbol{x},t)\mathrm{d}\boldsymbol{x}=\int_{\mathbb{R}^d}\boldsymbol{f}(\boldsymbol{x},t)\cdot\boldsymbol{u}(\boldsymbol{x},t)\mathrm{d}\boldsymbol{x}.$$

Notice now that, since A_j is symmetric and does not depend on $\boldsymbol{x}$,

$$\left(\mathsf{A}_j\frac{\partial \boldsymbol{u}(\boldsymbol{x},t)}{\partial x_j}\right)\cdot\boldsymbol{u}(\boldsymbol{x},t)=\frac{\partial \boldsymbol{u}(\boldsymbol{x},t)}{\partial x_j}\cdot\mathsf{A}_j\boldsymbol{u}(\boldsymbol{x},t).$$

Therefore, since we assumed that $\boldsymbol{u}$ decays sufficiently fast at infinity,

$$\int_{\mathbb{R}^d}\mathsf{A}_j\frac{\partial \boldsymbol{u}(\boldsymbol{x},t)}{\partial x_j}\cdot\boldsymbol{u}(\boldsymbol{x},t)\mathrm{d}\boldsymbol{x}=\frac{1}{2}\int_{\mathbb{R}^d}\frac{\partial}{\partial x_j}\left[(\mathsf{A}_j\boldsymbol{u}(\boldsymbol{x},t))\cdot\boldsymbol{u}(\boldsymbol{x},t)\right]\mathrm{d}\boldsymbol{x}=0.$$

Consequently, we have obtained that

$$\frac{1}{2}\frac{\mathrm{d}}{\mathrm{d}t}\|\boldsymbol{u}(\cdot,t)\|^2_{L^2(\mathbb{R}^d;\mathbb{R}^m)}\le\frac{1}{2}\|\boldsymbol{f}(\cdot,t)\|^2_{L^2(\mathbb{R}^d;\mathbb{R}^m)}+\frac{1}{2}\|\boldsymbol{u}(\cdot,t)\|^2_{L^2(\mathbb{R}^d;\mathbb{R}^m)}.$$

Define $K_1=1$,

$$\Phi(t)=\frac{1}{2}\left(\|\boldsymbol{u}(\cdot,t)\|^2_{L^2(\mathbb{R}^d;\mathbb{R}^m)}-\|\boldsymbol{u}_0\|^2_{L^2(\mathbb{R}^d;\mathbb{R}^m)}\right),$$

$$K_2=\frac{1}{2}\left(\|\boldsymbol{f}(\cdot,t)\|^2_{L^\infty(0,T;L^2(\mathbb{R}^d;\mathbb{R}^m))}+\|\boldsymbol{u}_0\|^2_{L^2(\mathbb{R}^d;\mathbb{R}^m)}\right).$$

The previous estimates then show that

$$\Phi'(t)\le K_1\Phi(t)+K_2,$$

which, by Grönwall's inequality of Lemma 17.7, implies the claimed estimate. The constant C depends (exponentially) on T. □

Problems

23.1 Show (23.16).

23.2 Show that the Frobenius inner product is indeed an inner product.

23.3 Let $m\in\mathbb{N}$ and consider $\mathbb{R}^{m\times m}$ with the Frobenius inner product. Define

$$\mathbb{S}=\left\{\mathsf{A}\in\mathbb{R}^{m\times m}\middle|\mathsf{A}=\mathsf{A}^\intercal\right\},$$
$$\mathbb{H}=\left\{\mathsf{A}\in\mathbb{R}^{m\times m}\middle|\mathsf{A}=-\mathsf{A}^\intercal\right\}.$$

a) Show that $\mathbb{S},\mathbb{H}\le\mathbb{R}^{m\times m}$.
b) Show that $\mathbb{S}$ and $\mathbb{H}$ are complementary subspaces.
c) Show that $\mathbb{S}$ and $\mathbb{H}$ are orthogonal subspaces.
d) Can you describe the projection $\mathbb{R}^{m\times m}\to\mathbb{S}$?

23.4 Let $\mathcal{D}$ be a linear, second-order partial differential operator on $\mathbb{R}^m$, for some $m\in\mathbb{N}$, that has variable, but continuous, coefficients.

a) Show that if the operator is elliptic (hyperbolic) at a point, then it is elliptic (hyperbolic) in a neighborhood of this point.
b) Does the previous statement hold for a parabolic operator?
c) Let $\boldsymbol{x},\boldsymbol{y}\in\mathbb{R}^m$ with $\boldsymbol{x}\ne\boldsymbol{y}$. Show that if $\mathcal{D}$ is elliptic at $\boldsymbol{x}$ and hyperbolic at $\boldsymbol{y}$, then there must be a third point where the equation is parabolic.

23.5 Prove Corollary 23.5.
23.6 Prove Corollary 23.7.
23.7 Consider the one-dimensional, nonlinear BVP

$$-u'' + u = e^u, \quad x \in (0,1), \qquad u(0) = u(1) = 0.$$

Show that any solution to this problem is nonnegative, i.e., $u(x) \geq 0$ for all $x \in [0,1]$.
23.8 Consider the divergence form operator (23.20), where $\mathsf{A} \in C^1(\bar{\Omega}; \mathbb{R}^{d\times d})$, $\boldsymbol{b} \in C(\bar{\Omega}; \mathbb{R}^d)$, and $c \in C(\bar{\Omega})$. Assume that there are constants $\lambda, \Lambda \in \mathbb{R}_+$ such that, for all points $\boldsymbol{x} \in \mathbb{R}^d$, the matrix $\mathsf{A}(\boldsymbol{x})$ is symmetric and

$$\sigma(\mathsf{A}(\boldsymbol{x})) \subset [\lambda, \Lambda].$$

Show that this operator is elliptic, in the sense of Definition 23.1.
23.9 Consider the Dirichlet problem (23.19) with $g = 0$ and $L = -\Delta$. Show that if $u \in C^2(\Omega) \cap C(\bar{\Omega})$ is a classical solution to this problem, then it is also a weak solution. Conversely, show that if $u \in H_0^1(\Omega)$ is a weak solution, which happens to be sufficiently smooth, then it is also a classical solution.
23.10 Consider two bounded domains Ω_1 and Ω_2 with a common boundary S. Let $\Gamma_i = \partial\Omega_i \backslash S$ for $i = 1, 2$. Give a variational formulation of the problem: for $i = 1, 2$, find $u_i \colon \Omega_i \to \mathbb{R}$ such that

$$\begin{aligned} -a_1 \Delta u_1 &= f_1, \quad \text{in } \Omega_1, & \qquad -a_2 \Delta u_2 &= f_2, \quad \text{in } \Omega_2, \\ u_1 &= 0, \quad \text{on } \Gamma_1, & u_2 &= 0, \quad \text{on } \Gamma_2, \end{aligned}$$

and, on S,

$$u_1 = u_2, \qquad [\![(a_1 \nabla u_1 - a_2 \nabla u_2)\cdot \boldsymbol{n}]\!] = 0.$$

Here, a_1, a_2 are positive constants, $f_i \in L^2(\Omega_i)$, and $\boldsymbol{n}$ is a unit normal to S.
23.11 Give a variational formulation of the problem

$$-\Delta u = f, \quad \text{in } \Omega, \qquad \nabla u \cdot \boldsymbol{n} + u = g, \quad \text{on } \partial\Omega.$$

Here, $f \in L^2(\Omega)$ and $g \in L^2(\partial\Omega)$. Use the Friedrichs inequality to show that this problem has a unique weak solution in $H^1(\Omega)$.
23.12 Let $i = 1, 2$ and $u_i \in H_0^1(\Omega)$ be weak solutions of

$$-\nabla \cdot (\mathsf{A}_i \nabla u_i) = f, \qquad \text{in } \Omega, \qquad u = 0, \quad \text{on } \partial\Omega,$$

where the matrices A_i are SPD and $\sigma(\mathsf{A}_i) \subset [\lambda, \Lambda]$. Show that

$$\|u_1 - u_2\|_{H_0^1(\Omega)} \leq \frac{C_P}{\lambda^2} \|\mathsf{A}_1 - \mathsf{A}_2\|_{L^\infty(\Omega;\mathbb{R}^{d\times d})} \|f\|_{L^2(\Omega)}.$$

23.13 Prove identity (23.26).
Hint: Use the Fourier transform of the Gaussian.
23.14 Complete the proof of Theorem 23.17.
23.15 Prove (23.30) for $d = 1$. Use this to deduce that there is a constant $C > 0$ such that

$$|P(y)e^{-y^2}| \leq Ce^{-y^2/2}, \quad \forall y > 0.$$

23.16 Let

$$u(x,t) = \begin{cases} xt^{-3/2}e^{-x^2/4t}, & t > 0, \\ 0, & t = 0. \end{cases}$$

a) Show that, for all $x \in \mathbb{R}$,

$$u(x,t) \to 0, \qquad t \downarrow 0.$$

b) Show that this function solves (23.24) with $d = 1$ and $f \equiv 0$.

c) Why is this not a counterexample to uniqueness?

Hint: Set $x = t$.

23.17 Prove (23.36).

23.18 Let u be a weak solution to (23.32) with $u_0 \in H_0^1(\Omega)$. Assume that $f \in L^2(0,T;L^2(\Omega))$. Show that

$$\|\nabla u(\cdot,t)\|_{L^2(\Omega;\mathbb{R}^d)}^2 + \int_0^t \left\| \frac{\partial u(\cdot,s)}{\partial t} \right\|_{L^2(\Omega)}^2 \mathrm{d}s \leq \|\nabla u_0\|_{L^2(\Omega;\mathbb{R}^d)}^2 + \int_0^t \|f(\cdot,s)\|_{L^2(\Omega)}^2 \mathrm{d}s.$$

23.19 Let $T \in (0,\infty)$, u be a weak solution to (23.32) with $u_0 \in H_0^1(\Omega)$, and $f \in L^2(0,T;L^2(\Omega))$. Show that there is a constant $C = C(T)$ such that, for every $t \in (0,T]$,

$$\int_0^t s \left\| \frac{\partial u(\cdot,s)}{\partial t} \right\|_{L^2(\Omega)}^2 \mathrm{d}s \leq C \left(\|v\|_{L^2(\Omega)}^2 + \int_0^t \|f(\cdot,s)\|_{L^2(\Omega)}^2 \mathrm{d}s \right),$$

$$\|\nabla u(\cdot,t)\|_{L^2(\Omega;\mathbb{R}^d)}^2 \leq C t^{-1} \left(\|v\|_{L^2(\Omega)}^2 + \int_0^t \|f(\cdot,s)\|_{L^2(\Omega)}^2 \mathrm{d}s \right).$$

23.20 Prove Corollary 23.28.

23.21 Consider the homogeneous IBVP for the heat equation with Neumann boundary conditions, i.e.,

$$\begin{cases} \dfrac{\partial u(\boldsymbol{x},t)}{\partial t} - \Delta u(\boldsymbol{x},t) = 0, & (\boldsymbol{x},t) \in \Omega \times (0,T], \\ \nabla u(\boldsymbol{x},t) \cdot \boldsymbol{n} = 0, & (\boldsymbol{x},t) \in \partial\Omega \times (0,T], \\ u(\boldsymbol{x},0) = u_0(\boldsymbol{x}), & \boldsymbol{x} \in \bar{\Omega}. \end{cases}$$

Show that, for all $t > 0$,

$$\frac{1}{|\Omega|} \int_\Omega u(\boldsymbol{x},t)\mathrm{d}\boldsymbol{x} = \frac{1}{|\Omega|} \int_\Omega u_0(\boldsymbol{x})\mathrm{d}\boldsymbol{x}.$$

23.22 Show that (23.41) indeed solves (23.39). Give sufficient conditions on u_0 and f, so that this is a classical solution.

23.23 Derive the notion of a weak solution, in the sense of Definition 23.32, for the transport equation. That is, multiply the equation by φ, integrate, and integrate by parts to arrive at the desired identity.

23.24 Show that if $u_0 \in L^1(\mathbb{R})$, then (23.40) is a weak solution, in the sense of Definition 23.32, of (23.39) with $f \equiv 0$.

23.25 Derive the notion of a weak solution, in the sense of Definition 23.38, for (23.45). That is, multiply the equation by a sufficiently smooth function φ that vanishes on the boundary, integrate, and integrate by parts to arrive at the desired identity.

23.26 Let u be a weak solution, in the sense of Definition 23.38, for (23.45) with $f \equiv 0$. Show that if $u_0 \in H^2(\Omega) \cap H_0^1(\Omega)$ and $v_0 \in H_0^1(\Omega)$, then, for $t > 0$, we have

$$
\begin{aligned}
\|u(t)\|_{L^2(\Omega)} &\leq C\big(\|u_0\|_{L^2(\Omega)} + \|v_0\|_{L^2(\Omega)}\big),\\
\|\nabla u(t)\|_{L^2(\Omega;\mathbb{R}^d)} &\leq C\big(\|\nabla u_0\|_{L^2(\Omega;\mathbb{R}^d)} + \|v_0\|_{L^2(\Omega)}\big),\\
\|\Delta u(t)\|_{L^2(\Omega)} &\leq C\big(\|\Delta u_0\|_{L^2(\Omega)} + \|\nabla v_0\|_{L^2(\Omega;\mathbb{R}^d)}\big),\\
\|u(t) - u_0\|_{L^2(\Omega)} &\leq Ct\big(\|\nabla u_0\|_{L^2(\Omega;\mathbb{R}^d)} + \|v_0\|_{L^2(\Omega)}\big),\\
\left\| \frac{\partial u(\cdot, t)}{\partial t} - v_0 \right\|_{L^2(\Omega)} &\leq Ct\big(\|\Delta u_0\|_{L^2(\Omega)} + \|\nabla v_0\|_{L^2(\Omega;\mathbb{R}^d)}\big).
\end{aligned}
$$

23.27 Complete the proof of Theorem 23.39.

23.28 Provide all the details for Example 23.11.

24 Finite Difference Methods for Elliptic Problems

In this chapter, we explore finite difference methods (FDMs) for solving elliptic boundary value problems (BVPs). To present the essential ideas, without obscuring the discussion with technical details, we will assume that our grids are uniform; the domain is simple, say $\Omega = (0,1)^d$, with $d \in \mathbb{N}$; and we will mostly discuss the one- ($d = 1$) and two- ($d = 2$) dimensional cases.

As the name suggests, the main idea behind FDMs is to replace the derivatives appearing in a BVP by differences of values at nearby points, thus reducing the BVP to a system of algebraic equations (for the point values). This seems very natural. After all, if a function $v\colon \mathbb{R}^d \to \mathbb{R}$ is sufficiently smooth, for any $i = 1, \dots, d$, we have

$$\frac{\partial v(\boldsymbol{x})}{\partial x_i} = \lim_{h \to 0} \frac{v(\boldsymbol{x} + h\boldsymbol{e}_i) - v(\boldsymbol{x})}{h},$$

where $\boldsymbol{e}_i$ is the ith canonical basis vector in $\mathbb{R}^d$, thus we can propose the following *finite difference approximations* of this partial derivative:

$$\frac{v(\boldsymbol{x} + h\boldsymbol{e}_i) - v(\boldsymbol{x})}{h}, \qquad \frac{v(\boldsymbol{x}) - v(\boldsymbol{x} - h\boldsymbol{e}_i)}{h}, \qquad h > 0.$$

Evidently, these are not the only two possibilities. For instance, the reader can easily verify that, for a sufficiently smooth function v, the quantities

$$\frac{v(\boldsymbol{x} + h\boldsymbol{e}_i) - v(\boldsymbol{x} - h\boldsymbol{e}_i)}{2h}, \qquad \frac{3v(\boldsymbol{x}) - 4v(\boldsymbol{x} - h\boldsymbol{e}_i) + v(\boldsymbol{x} - 2h\boldsymbol{e}_i)}{2h}$$

converge, as $h \downarrow 0$, to this partial derivative. Which finite difference approximation must be used then? The answer to this comes from trying to satisfy two, often contradictory, requirements, as follows.

- *Consistency:* We wish the finite difference approximation to be as accurate as possible. The way this is usually verified is by assuming that the function v is sufficiently smooth. If this is the case, a Taylor expansion can show that, for instance,

$$\frac{v(\boldsymbol{x} + h\boldsymbol{e}_i) - v(\boldsymbol{x})}{h} = \frac{\partial v(\boldsymbol{x})}{\partial x_i} + \mathcal{O}(h),$$
$$\frac{v(\boldsymbol{x} + h\boldsymbol{e}_i) - v(\boldsymbol{x} - h\boldsymbol{e}_i)}{2h} = \frac{\partial v(\boldsymbol{x})}{\partial x_i} + \mathcal{O}(h^2),$$

  as $h \downarrow 0$. For this reason, we say that the first finite difference approximation is *consistent to order one*, whereas the second one is *consistent to order two*.

From this consideration alone, it seems that we need to find a finite difference approximation of the derivatives that is consistent to as high order as possible.

- *Stability:* We are not trying to approximate the action of a derivative, but the solution to a BVP. For this reason, the solution of our FDM must be stable, not only with respect to perturbations of the data but, more importantly, in (discrete versions of) the same norms in which the solution to the continuous problem possesses its own stability properties. This will allow us to compare the continuous and discrete solutions (via consistency) and establish convergence.

It turns out that, at least for linear problems, the satisfaction of these two requirements is enough to obtain convergence. In the literature, this is commonly known as *Lax's Principle*:[1]

$$\text{consistency} + \text{stability} \implies \text{convergence}.$$

This is one of the most fundamental mantras in the numerical approximation of differential equations. For instance, the Dahlquist Equivalence Theorem 20.26 is stating precisely this fact. Of course, at this stage, for us this is more a guiding principle than a rigorous statement, as we have not properly defined any of these three notions. The development, and justification, of these ideas for elliptic problems will be the main focus in this chapter.

As a final comment, we mention that here we will present the easiest and most elementary incarnation of the theory of finite differences, where we assume that the problem data are sufficiently smooth, and we are concerned with the approximation of *classical solutions.* We refer the reader to [50], [56, Chapter VI], and [79] for the study of FDMs that are able to handle nonsmooth data and approximate weak solutions.

24.1 Grid Functions and Finite Difference Operators

Before we develop the approximation of BVPs via FDMs, we must introduce some notations and notions regarding grid functions in $\mathbb{R}^d$, $d \in \mathbb{N}$. Thus, we let $N \in \mathbb{N}$, set $h = 1/(N+1)$, called the *grid size*, and define the *uniform grid*

$$\mathbb{Z}_h^d = \left\{ hz \in \mathbb{R}^d \;\middle|\; z \in \mathbb{Z}^d \right\}.$$

Definition 24.1 (grid function)**.** A **grid domain** in $\mathbb{R}^d$ is any nonempty set $\mathcal{G}_h \subseteq \mathbb{Z}_h^d$. The points $\boldsymbol{x} \in \mathcal{G}_h$ are the **nodes** or **grid points**. A **grid function** on $\mathcal{G}_h$ is any function $v \colon \mathcal{G}_h \to \mathbb{R}$. For a grid function v and $h\boldsymbol{i} \in \mathcal{G}_h$, we set $v_{\boldsymbol{i}} = v(h\boldsymbol{i})$. The set of grid functions on $\mathcal{G}_h$ is denoted by

$$\mathcal{V}(\mathcal{G}_h) = \{ v \mid v \colon \mathcal{G}_h \to \mathbb{R} \}.$$

[1] Named in honor of the Hungarian-born American mathematician Peter David Lax (1926–).

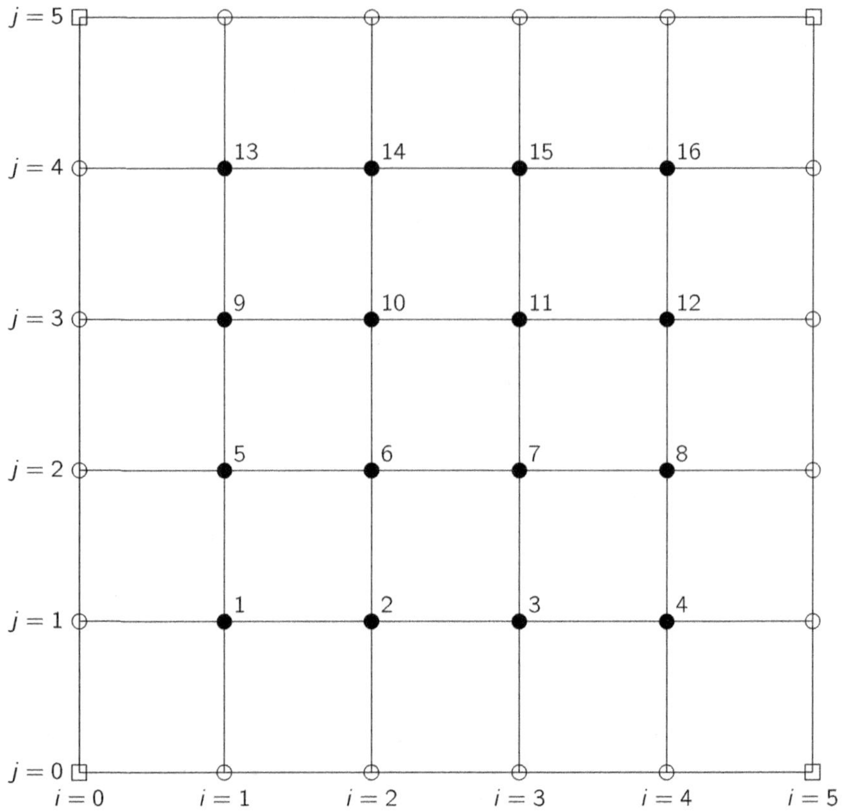

Figure 24.1 A uniform grid of size 5×5 and spacing $h = \frac{1}{5}$ covering $\Omega = (0, 1)^2$. The discrete interior, Ω_h, is the collection of filled circles, and the discrete boundary, $\partial\Omega_h$, consists of the remaining ones. The space of grid functions, $\mathcal{V}(\bar{\Omega}_h)$, is the space of functions from all points in this grid into $\mathbb{R}$. The numbers next to the filled circles denote their lexicographical ordering, which is used to provide the isomorphism $\mathcal{V}(\bar{\Omega}_h) \longleftrightarrow \mathbb{R}^{16}$. The filled circles constitute the domain Ω_h^I for the two-dimensional discrete Laplacian of Example 24.7, whereas the unfilled circles constitute Ω_h^B. Note that, although the unfilled squares on the four corner points belong to $\bar{\Omega}_h$, they are not used in a finite difference approximation of the Poisson problem with Dirichlet boundary conditions.

This is a vector space. If $\#\mathcal{G}_h < \infty$, this space is finite dimensional; otherwise, it is infinite dimensional. In a similar manner, if $\mathbb{V}$ is a, finite-dimensional, vector space, we set

$$\mathcal{V}(\mathcal{G}_h; \mathbb{V}) = \{v \mid v \colon \mathcal{G}_h \to \mathbb{V}\}.$$

Example 24.1 Let $d = 1$ and $\Omega = (0, 1)$ be the interval where we will usually consider our elliptic BVPs. For $N \in \mathbb{N}$ and $h = 1/(N + 1)$, we set

$$\bar{\Omega}_h = \bar{\Omega} \cap \mathbb{Z}_h = \{hi \mid i = 0, \dots, N + 1\}.$$

Notice that, in this case, $\mathcal{V}(\bar{\Omega}_h)$ is nothing but the collection of functions $v\colon \{0, \dots, N+1\} \to \mathbb{R}$.

Notice also that $\partial\Omega = \{0, 1\}$ and, similarly, we can define the *discrete boundary* and *discrete interior* of $\bar{\Omega}_h$, respectively, as

$$\partial\Omega_h = \partial\Omega \cap \mathbb{Z}_h, \qquad \Omega_h = \bar{\Omega}_h \backslash \partial\Omega_h.$$

Example 24.2 The previous example can be easily generalized to any $d \in \mathbb{N}$. For instance, if $\Omega = (0, 1)^2$, we can define

$$\bar{\Omega}_h = \bar{\Omega} \cap \mathbb{Z}_h^2 = \{(hi, hj) \mid i, j = 0, \dots, N+1\}.$$

The discrete interior is then

$$\Omega_h = \Omega \cap \mathbb{Z}_h^2 = \{(hi, hj) \mid i, j = 1, \dots, N\}.$$

The discrete boundary can also be defined accordingly; see Figure 24.1. The space $\mathcal{V}(\bar{\Omega}_h)$ is defined accordingly.

Since we will be mostly concerned with approximating BVPs in $\Omega = (0, 1)^d$ with $d \in \mathbb{N}$, we introduce the following notation.

Definition 24.2 (space $\mathcal{V}_0(\bar{\Omega}_h)$). Let $d \in \mathbb{N}$, $\Omega = (0, 1)^d$, $N \in \mathbb{N}$, $h = 1/(N+1)$, and $\bar{\Omega}_h = \bar{\Omega} \cap \mathbb{Z}_h^d$. To treat Dirichlet boundary conditions, we also introduce the space of grid functions that vanish on the boundary

$$\mathcal{V}_0(\bar{\Omega}_h) = \left\{ v \in \mathcal{V}(\bar{\Omega}_h) \;\middle|\; v_i = 0, \ \forall h\boldsymbol{i} \in \partial\Omega_h \right\}.$$

There is another natural way to understand functions in $\mathcal{V}(\bar{\Omega}_h)$, which, depending on the context, may also prove useful.

Proposition 24.3 (isomorphism). *Let $d, N \in \mathbb{N}$, $h = \frac{1}{N+1}$, $\Omega = (0, 1)^d$, and $\bar{\Omega}_h = \bar{\Omega} \cap \mathbb{Z}_h^d$. There is a one-to-one correspondence between the space of grid functions $\mathcal{V}(\Omega_h)$ and $\mathbb{R}^{N^d}$. Likewise, there is a one-to-one correspondence between the space of grid functions $\mathcal{V}_0(\bar{\Omega}_h)$ and $\mathbb{R}^{N^d}$. Therefore, $\dim(\mathcal{V}(\Omega_h)) = \dim(\mathcal{V}_0(\bar{\Omega}_h)) = N^d$.*

Proof. Let $d = 1$. We make the canonical correspondence. If $v \in \mathcal{V}(\Omega_h)$, then define $\boldsymbol{v} \in \mathbb{R}^N$ component-wise via

$$[\boldsymbol{v}]_i = v_i, \qquad ih \in \Omega_h.$$

Conversely, if $\boldsymbol{v} \in \mathbb{R}^N$ is given, define the grid function $v \in \mathcal{V}(\Omega_h)$ via

$$v_i = [\boldsymbol{v}]_i.$$

The grid functions with zero values on the boundary (those from $\mathcal{V}_0(\bar{\Omega}_h)$) are handled similarly.

To illustrate the general procedure, consider now $d = 2$. If $v \in \mathcal{V}(\Omega_h)$, then define $\boldsymbol{v} \in \mathbb{R}^{N^2}$ component-wise via

$$[\boldsymbol{v}]_{i+(j-1)N} = v_{i,j}, \qquad (ih, jh) \in \Omega_h.$$

Conversely, if $\boldsymbol{v} \in \mathbb{R}^{N^2}$ is given, define the grid function $v \in \mathcal{V}(\Omega_h)$ component-wise via

$$v_{i,j} = [\boldsymbol{v}]_{i+(j-1)N}, \quad (ih, jh) \in \Omega_h.$$

Figure 24.1 shows an illustration of this equivalence, which is commonly called *lexicographical ordering.* □

Remark 24.4 (convention)**.** Because of this canonical (and trivial) correspondence, we need not be too careful about whether we are talking about a grid function or its vector representation. We will often write

$$\boldsymbol{v} \in \mathbb{R}^{N^d} \longleftrightarrow v \in \mathcal{V}(\Omega_h), \qquad \boldsymbol{v} \in \mathbb{R}^{N^d} \longleftrightarrow v \in \mathcal{V}_0(\bar{\Omega}_h)$$

to express the fact that our object may be viewed in either setting. We use the convention of denoting a grid function by a lowercase Greek or Roman character and its corresponding canonical vector representative by the boldface of the same Greek or Roman character.

Having introduced spaces of grid functions, we can define basic operators on them.

Definition 24.5 (shift operator)**.** Let $d \in \mathbb{N}$ and $\boldsymbol{e} \in \mathbb{Z}^d$. The **shift operator** in the direction of $\boldsymbol{e}$,

$$\mathcal{S}_{\boldsymbol{e}} \colon \mathcal{V}(\mathbb{Z}_h^d) \to \mathcal{V}(\mathbb{Z}_h^d),$$

is given, for $\boldsymbol{x} = h\boldsymbol{z} \in \mathbb{Z}_h^d$, by

$$(\mathcal{S}_{\boldsymbol{e}} v)(\boldsymbol{x}) = v(\boldsymbol{x} + h\boldsymbol{e}), \quad (\mathcal{S}_{\boldsymbol{e}} v)_{\boldsymbol{z}} = v_{\boldsymbol{z}+\boldsymbol{e}}.$$

Proposition 24.6 (properties of shifts)**.** *Let $d \in \mathbb{N}$. For any $\boldsymbol{e} \in \mathbb{Z}^d$, the shift operator satisfies $\mathcal{S}_{\boldsymbol{e}} \in \mathfrak{L}(\mathcal{V}(\mathbb{Z}_h^d))$. Moreover, this operator is invertible and its inverse is given by*

$$\mathcal{S}_{\boldsymbol{e}}^{-1} = \mathcal{S}_{-\boldsymbol{e}}.$$

In particular, $\mathcal{S}_{\boldsymbol{0}}$ is the identity operator.

Proof. See Problem 24.1. □

Shifts will be the building blocks of what we will call *finite difference operators*, which, simply put, will be linear combinations of shifts.

Definition 24.7 (finite difference operator)**.** The mapping

$$\mathcal{F}_h \colon \mathcal{V}(\mathbb{Z}_h^d) \to \mathcal{V}(\mathbb{Z}_h^d)$$

is called a (linear) **finite difference (FD) operator** if and only if it has the form

$$(\mathcal{F}_h v)(\boldsymbol{x}) = \sum_{\boldsymbol{e} \in S} a_{\boldsymbol{e}}(\boldsymbol{x}, h)(\mathcal{S}_{\boldsymbol{e}} v)(\boldsymbol{x}), \quad \boldsymbol{x} \in \mathbb{Z}_h^d,$$

where the set $S \subset \mathbb{Z}^d$ is such that $\#S < \infty$, $\boldsymbol{0} \in S$, and it is called the **stencil** of the operator $\mathcal{F}_h$. The **stencil size** is

$$\max \left\{ \|\boldsymbol{j}\|_{\ell^\infty} \mid \boldsymbol{j} \in S \right\}.$$

Finally, for $\boldsymbol{e} \in S$, we have $a_{\boldsymbol{e}} \colon \mathbb{Z}_h^d \times \mathbb{R}_+ \to \mathbb{R}$.

Remark 24.8 (stencil). The previous definition is sufficiently general for our purposes. However, some sources, such as [79], allow the stencil to depend on $\boldsymbol{x}$, i.e.,

$$(\mathcal{F}_h v)(\boldsymbol{x}) = \sum_{e \in S(x)} a_e(\boldsymbol{x}, h)(\mathcal{S}_e v)(\boldsymbol{x}),$$

where, for each $\boldsymbol{x} \in \mathbb{Z}_h^d$, the set $S(\boldsymbol{x})$ is a stencil according to our definition. We will not consider such operators.

Let us now present, for $d = 1, 2$, several examples of finite difference operators.

Example 24.3 The *forward difference* operator is defined as

$$\delta_h v(x) = \frac{v(x+h) - v(x)}{h} = \frac{v_{i+1} - v_i}{h},$$

where we assumed that $x = ih$. Its stencil is $\{0, 1\}$.

Example 24.4 The *backward difference* operator is defined as

$$\bar{\delta}_h v(x) = \frac{v(x) - v(x-h)}{h} = \frac{v_i - v_{i-1}}{h},$$

where we assumed that $x = ih$. Its stencil is $\{-1, 0\}$.

Example 24.5 The *centered difference* operator is defined as

$$\mathring{\delta}_h v(x) = \frac{v(x+h) - v(x-h)}{2h} = \frac{v_{i+1} - v_{i-1}}{2h},$$

where we assumed that $x = ih$. Its stencil is $\{-1, 0, 1\}$.

Example 24.6 The one-dimensional *discrete Laplace* operator[2] is defined as

$$\Delta_h v(x) = \bar{\delta}_h \delta_h v(x) = \frac{v(x+h) - 2v(x) + v(x-h)}{h^2} = \frac{v_{i+1} - 2v_i + v_{i-1}}{h^2},$$

where we assumed that $x = ih$. Its stencil is $\{-1, 0, 1\}$.

Example 24.7 The two-dimensional discrete Laplace operator is defined, for $(i, j) \in \mathbb{Z}^2$, as

$$\Delta_h v_{i,j} = \frac{v_{i-1,j} + v_{i+1,j} + v_{i,j-1} + v_{i,j+1} - 4v_{i,j}}{h^2}.$$

Its stencil is the *five points* $\{(0, 0), (0, \pm 1), (\pm 1, 0)\}$.

Example 24.8 The discrete mixed derivative operator is defined, for $(i, j) \in \mathbb{Z}^2$, as

$$\delta_h^{\diamondsuit} v_{i,j} = \frac{v_{i-1,j-1} + v_{i+1,j+1} - v_{i+1,j-1} + v_{i-1,j-1}}{4h^2}.$$

Its stencil is the *five points* $\{(0, 0), (\pm 1, \pm 1)\}$.

[2] Named in honor of the French scholar and polymath Pierre-Simon, Marquis de Laplace (1749–1827).

Example 24.9 The two-dimensional skew Laplacian operator is defined, for $(i,j) \in \mathbb{Z}^2$, as

$$\Delta_h^{\square} v_{i,j} = \frac{v_{i+1,j+1} + v_{i+1,j-1} + v_{i-1,j-1} + v_{i-1,j+1} - 4v_{i,j}}{2h^2}.$$

Its stencil is the *five points* $\{(0,0), (\pm 1, \pm 1)\}$.

It turns out that the operators of the previous examples possess several interesting properties that resemble those of the derivatives that, at least intuitively at this stage, are meant to approximate.

Proposition 24.9 (properties of difference operators). *Let $d = 1$ and consider the difference operators of Examples 24.3—24.6. For any $v, v_1, v_2 \in \mathcal{V}(\mathbb{Z}_h)$, we have the following identities:*

1. *Product rule I:*

$$\delta_h(v_1 v_2)(x) = \delta_h v_1(x) v_2(x) + v_1(x+h)\bar{\delta}_h v_2(x+h).$$

2. *Product rule II:*

$$\bar{\delta}_h(v_1 v_2)(x) = \bar{\delta}_h v_1(x) v_2(x) + v_1(x-h)\delta_h v_2(x-h).$$

3. *Abel transformation:*[3]

$$h\sum_{k=0}^{N-1} \delta_h v_1(kh) v_2(kh) = v_1(Nh)v_2(Nh) - v_1(0)v_2(0) - h\sum_{k=1}^{N} v_1(kh)\bar{\delta}_h v_2(kh).$$

4. *Symmetry:*

$$\Delta_h v(x) = \delta_h \bar{\delta}_h v(x).$$

Proof. See Problem 24.2. □

Eventually, we will use finite difference operators to approximate the derivatives appearing in a BVP on a domain. Notice, however, that when we deal with grid domains certain shifts will not be admissible, as they will take us outside of the domain of a function. For this reason, we must make a distinction between two types of points, and this is dependent on the finite difference operator in question.

Definition 24.10 (interior and boundary grid points). Let $\mathcal{G}_h \subseteq \mathbb{Z}_h^d$ be a grid domain and $\mathcal{F}_h$ be a finite difference operator with stencil S. The set of points

$$\mathcal{G}_h^I = \{\boldsymbol{x} \in \mathcal{G}_h \mid \boldsymbol{x} + \boldsymbol{s} \in \mathcal{G}_h, \ \forall \boldsymbol{s} \in S\}$$

is called the set of **interior grid points** with respect to the operator $\mathcal{F}_h$. Similarly,

$$\mathcal{G}_h^B = \mathcal{G}_h \backslash \mathcal{G}_h^I$$

is the set of **boundary grid points** with respect to the operator $\mathcal{F}_h$.

[3] Named in honor of the Norwegian mathematician Niels Henrik Abel (1802–1829).

Example 24.10 The stencil of the forward difference operator of Example 24.3 is $\{0, 1\}$. Thus,

$$\bar{\Omega}_h^I = [0, 1) \cap \mathbb{Z}_h = \{ih \mid i = 0, \dots, N\}.$$

Example 24.11 The stencil of the backward difference operator of Example 24.4 is $\{-1, 0\}$. Thus,

$$\bar{\Omega}_h^I = (0, 1] \cap \mathbb{Z}_h = \{ih \mid i = 1, \dots, N+1\}.$$

Example 24.12 The stencil of the centered difference operator of Example 24.5 is $\{-1, 0, 1\}$. Thus,

$$\bar{\Omega}_h^I = \Omega_h.$$

Example 24.13 The stencil of the one-dimensional discrete Laplace operator of Example 24.6 is $\{-1, 0, 1\}$. Thus,

$$\bar{\Omega}_h^I = \Omega_h.$$

Example 24.14 The domains Ω_h^I and Ω_h^B for the two-dimensional discrete Laplace operator of Example 24.7 are illustrated in Figure 24.1.

We have then defined the objects we will use to approximate differential operators and their solutions. However, to properly define the crucial notions of *consistency* and *stability* we alluded to at the beginning of this chapter, we must provide a notion of convergence for functions in $\mathcal{V}(\mathcal{G}_h)$, where $\mathcal{G}_h$ is a grid domain. If $\mathcal{V}(\mathcal{G}_h)$ is finite dimensional, which will happen if $\#\mathcal{G}_h < \infty$, all norms on this space are equivalent (recall Theorem A.29). The issue here, however, is that we need to establish notions of stability and convergence that are *robust* with respect to $h > 0$. In addition, the norm we choose for this space must, in a sense, resemble the natural norm for the space we are trying to approximate. The following definition makes these considerations rigorous.

Definition 24.11 (approximation property). Let $d \in \mathbb{N}$ and $\Omega = (0, 1)^d$. For each $N \in \mathbb{N}$, define $h = 1/(N+1)$ and let $\bar{\Omega}_h = \bar{\Omega} \cap \mathbb{Z}_h^d$ be the corresponding grid domain. Let $(\mathbb{V}, \|\cdot\|_{\mathbb{V}})$ be a normed space of functions on Ω, i.e.,

$$\mathbb{V} = \left\{v \colon \bar{\Omega} \to \mathbb{R} \;\middle|\; \|v\|_{\mathbb{V}} < \infty\right\}.$$

Assume that, for each $N \in \mathbb{N}$, we endow the space $\mathcal{V}(\bar{\Omega}_h)$ with a norm $\|\cdot\|_h$. We say that the family $\{(\mathcal{V}(\bar{\Omega}_h), \|\cdot\|_h)\}_{h>0}$ possesses the **approximation property** if there is a (linear) operator, called the **grid projection operator**, such that

$$\pi_h \colon \mathbb{V} \to \mathcal{V}(\bar{\Omega}_h), \qquad \lim_{h \downarrow 0} \|\pi_h v\|_h = \|v\|_{\mathbb{V}}, \quad \forall v \in \mathbb{V}.$$

Let us give some examples of this in a somewhat simplified setting.

Example 24.15 Let $d = 1$, $\Omega = (0,1)$, and $\mathcal{V}(\bar{\Omega}_h)$ be defined as before. Let us define, for $p \in [1, \infty)$, the norms

$$\|v\|_{L^p_h} = \left[h \sum_{i=1}^{N} |v_i|^p + \frac{h}{2} \left(|v_0|^p + |v_{N+1}|^p \right) \right]^{1/p}, \quad v \in \mathcal{V}(\bar{\Omega}_h).$$

These have the approximation property with respect to the norms of $L^p(0,1)$, $p \in (1, \infty)$. Indeed, if $v \in L^p(0,1)$, we define

$$(\pi_h v)_i = \begin{cases} \dfrac{1}{h} \displaystyle\int_{(i-1/2)h}^{(i+1/2)h} v(x) \mathrm{d}x, & i \in \{1, \dots, N\}, \\ \dfrac{2}{h} \displaystyle\int_{0}^{h/2} v(x) \mathrm{d}x, & i = 0, \\ \dfrac{2}{h} \displaystyle\int_{1-h/2}^{1} v(x) \mathrm{d}x, & i = N+1, \end{cases}$$

which we call the *averaging operator*. Now

$$\|\pi_h v\|_{L^p_h}^p = h \sum_{i=1}^{N} |\pi_h v_i|^p + \frac{h}{2} \left(|\pi_h v_0|^p + |\pi_h v_{N+1}|^p \right),$$

and we consider, separately, the interior and boundary indices.

For the interior indices, using Young's inequality and setting $p' = p/(p-1)$, we have

$$\begin{aligned} h \sum_{i=1}^{N} |\pi_h v_i|^p &= h^{1-p} \sum_{i=1}^{N} \left| \int_{(i-1/2)h}^{(i+1/2)h} v(x) \mathrm{d}x \right|^p \\ &\leq h^{1-p-p/p'} \sum_{i=1}^{N} \int_{(i-1/2)h}^{(i+1/2)h} |v(x)|^p \,\mathrm{d}x \\ &= \sum_{i=1}^{N} \int_{(i-1/2)h}^{(i+1/2)h} |v(x)|^p \,\mathrm{d}x \\ &= \int_{h/2}^{1-h/2} |v(x)|^p \,\mathrm{d}x. \end{aligned}$$

Similarly,

$$\frac{h}{2} |\pi_h v_0|^p \leq \int_0^{h/2} |v(x)|^p \,\mathrm{d}x, \qquad \frac{h}{2} |\pi_h v_{N+1}|^p \leq \int_{1-h/2}^{1} |v(x)|^p \,\mathrm{d}x,$$

so that, in conclusion, we have, for any $v \in L^p(0,1)$,

$$\|\pi_h v\|_{L^p_h} \leq \|v\|_{L^p(0,1)} \quad \Longrightarrow \quad \limsup_{h \downarrow 0} \|\pi_h v\|_{L^p_h} \leq \|v\|_{L^p(0,1)}.$$

The reverse inequality is the content of the so-called *Lebesgue Differentiation Theorem* and so we have the approximation property.

Example 24.16 The previous example, with few modifications, shows that $\mathcal{V}(\bar{\Omega}_h)$ with $\|\cdot\|_{L^1_h}$ has the approximation property with respect to $L^1(0,1)$. We leave the details to the reader as an exercise; see Problem 24.4.

Example 24.17 The case $p = \infty$ requires a little modification. As usual, we define, for $v \in \mathcal{V}(\bar{\Omega}_h)$,

$$\|v\|_{L^\infty_h} = \max_{x \in \bar{\Omega}_h} |v(x)|$$

and

$$(\pi_h v)_i = v(ih), \quad i = 0, \ldots, N+1,$$

which we call the *sampling operator*. This has the approximation property on $C([0,1])$ with the norm $\|\cdot\|_{L^\infty(0,1)}$. We leave the details to the reader as an exercise; see Problem 24.5.

Remark 24.12 (scaling). One may wonder what, for $p \in [1, \infty)$, is the purpose of the factor h in the definitions of the L^p_h-norms. After all, after the identification of Remark 24.4, we could use, as norms of $\mathcal{V}(\bar{\Omega}_h)$, any of the norms $\|\cdot\|_{\ell^p(\mathbb{R}^N)}$, for $p \in [1, \infty]$. It is precisely to attain the approximation property. For instance, set $p = 1$ and consider the constant function $v \equiv \alpha > 0$ on $(0,1)$. Then the averaging operator yields $\pi_h v_i = \alpha$ for all i. Consequently,

$$\|\pi_h v\|_{L^1_h} = h \sum_{i=1}^{N} v_i + h\alpha = \frac{N}{N+1}\alpha + \frac{1}{N+1}\alpha = \|v\|_{L^1(0,1)}.$$

If we were to use the $\ell^1(\mathbb{R}^N)$-norm, we would not attain approximation

$$\|\boldsymbol{\pi}_h \boldsymbol{v}\|_1 = \sum_{i=0}^{N+1} v_i = (N+2)\alpha \to \infty, \qquad N \to \infty.$$

Notice that, as expected, the L^2_h-norm of the previous examples comes from an inner product. Thus, we define the following inner products.

Definition 24.13 (discrete inner products). Define, on $\mathcal{V}(\bar{\Omega}_h)$, the bilinear forms

$$[v^{(1)}, v^{(2)}]_{L^2_h} = \frac{h}{2}\left(v_0^{(1)} v_0^{(2)} + v_{N+1}^{(1)} v_{N+1}^{(2)}\right) + h \sum_{i=1}^{N} v_i^{(1)} v_i^{(2)},$$

$$(v^{(1)}, v^{(2)})_{L^2_h} = h \sum_{i=1}^{N} v_i^{(1)} v_i^{(2)},$$

where $v^{(1)}, v^{(2)} \in \mathcal{V}(\bar{\Omega}_h)$.

These expressions satisfy the expected properties.

Proposition 24.14 (inner products). *The expression $[\cdot,\cdot]_{L^2_h}$, introduced in Definition 24.13, is an inner product on $\mathcal{V}(\bar{\Omega}_h)$ and it induces the L^2_h-norm. Similarly, the* *expression* $(\cdot,\cdot)_{L^2_h}$, *introduced in Definition 24.13, is an inner product on* $\mathcal{V}_0(\bar{\Omega}_h)$ *and it induces the* L^2_h*-norm. In addition, we have*

$$(\delta_h v_1, v_2)_{L^2_h} = -\left(v_1, \bar{\delta}_h v_2\right)_{L^2_h}, \quad \forall\, v_1, v_2 \in \mathcal{V}_0(\bar{\Omega}_h); \tag{24.1}$$

as a consequence,

$$(-\Delta_h v_1, v_2)_{L^2_h} = \left(\bar{\delta}_h v_1, \bar{\delta}_h v_2\right)_{L^2_h} = (\delta_h v_1, \delta_h v_2)_{L^2_h}, \quad \forall v_1, v_2 \in \mathcal{V}_0(\bar{\Omega}_h).$$

Proof. See Problem 24.6. □

Remark 24.15 (summation by parts)**.** The identity (24.1) is a discrete analogue of the integration by parts formula of Theorem B.38. For this reason, it is commonly referred to as the *summation by parts formula.*

24.2 Consistency and Stability of Finite Difference Methods

We are finally in a position to state the important notions of consistency, stability, and convergence for a finite difference problem. Recall that our grand goal is to approximate the solution of a BVP. Thus, we assume that we have at hand the following problem. Let $\mathbb{V}$ be some normed space of functions defined on $\bar{\Omega} = [0,1]^d$, $d \in \mathbb{N}$, with norm $\|\cdot\|_{\mathbb{V}}$. We need to find $u \in \mathbb{V}$, such that

$$\begin{cases} Lu = f, & \text{in } \Omega, \\ \ell u = g, & \text{in } \partial\Omega, \end{cases} \tag{24.2}$$

where L and ℓ are some differential operators. The operator L encodes the differential equation, whereas ℓ represents the boundary conditions. Here, the functions $f\colon \Omega \to \mathbb{R}$ and $g\colon \partial\Omega \to \mathbb{R}$ are assumed to be given and belong to some normed spaces $f \in \mathbb{F}$ and $g \in \mathbb{G}$, with norms $\|\cdot\|_{\mathbb{F}}$ and $\|\cdot\|_{\mathbb{G}}$, respectively.

Problem (24.2) is replaced by the finite difference problem: Find $w \in \mathcal{V}(\bar{\Omega}_h)$ such that

$$\begin{cases} L_h w = f_h, & \text{in } \Omega_h^I, \\ \ell_h w = g_h, & \text{in } \Omega_h^B. \end{cases} \tag{24.3}$$

Here, L_h and ℓ_h are finite difference operators; Ω_h^I and Ω_h^B are the interior and boundary grid points, respectively, with respect to the operator L_h; and $f_h \in \mathcal{V}(\Omega_h^I)$ and $g_h \in \mathcal{V}(\Omega_h^B)$ are assumed to be given.

We can now state all our needed notions. We begin with stability.

Definition 24.16 (stability)**.** Let $\{(\mathcal{V}(\bar{\Omega}_h), \|\cdot\|_h)\}_{h>0}$ have the approximation property with respect to $(\mathbb{V}, \|\cdot\|_{\mathbb{V}})$. We say that the FDM (24.3) is **stable** if and only if there are constants $h_0 > 0$ and $C > 0$ such that, for all $h \in (0, h_0]$, problem (24.3) has a unique solution for any pair $(f_h, g_h) \in \mathcal{V}(\Omega_h^I) \times \mathcal{V}(\Omega_h^B)$; additionally, for $h \in (0, h_0]$, we have

$$\|w\|_h \leq C\left(\|f_h\|_{\mathcal{V}(\Omega_h^I)} + \|g_h\|_{\mathcal{V}(\Omega_h^B)}\right).$$

where $\|\cdot\|_{\mathcal{V}(\Omega_h^I)}$ and $\|\cdot\|_{\mathcal{V}(\Omega_h^B)}$ are norms that have the approximation property with respect to $(\mathbb{F}, \|\cdot\|_{\mathbb{F}})$ and $(\mathbb{G}, \|\cdot\|_{\mathbb{G}})$, respectively.

We can now state consistency.

Definition 24.17 (operator consistency)**.** Let $\{(\mathcal{V}(\mathbb{Z}_h^d), \|\cdot\|_h)\}_{h>0}$ have the approximation property with respect to $(\mathbb{V}, \|\cdot\|_{\mathbb{V}})$. Let $L\colon \mathbb{V} \to \mathbb{V}$ be a linear, not necessarily bounded, operator and $L_h\colon \mathcal{V}(\mathbb{Z}_h^d) \to \mathcal{V}(\mathbb{Z}_h^d)$ be a finite difference operator. We define the **consistency error** at $v \in \mathbb{V}$ to be

$$\mathcal{E}_h[L, v](x) = (L_h \pi_h v - \pi_h(Lv))\,(x), \quad x \in \mathbb{Z}_h^d.$$

We say that the finite difference operator L_h is **consistent to at least order** $p \in \mathbb{N}$ with L if and only if there is $k \in \mathbb{N}$, usually $k > p$, such that there is a constant $h_1 > 0$ such that, whenever $h \in (0, h_1]$ and $v \in C^k(\mathbb{R}^d)$, there is $C > 0$, possibly depending on v, for which

$$\|\mathcal{E}_h[L, v]\|_h \le Ch^p.$$

Finally, we say that the operator is **consistent with exactly order** p if there is at least one v for which the previous estimate cannot be improved upon, i.e., the value of p cannot be increased, by assuming that v is smoother.

Definition 24.18 (consistency)**.** Let $\{(\mathcal{V}(\bar{\Omega}_h), \|\cdot\|_h)\}_{h>0}$ have the approximation property with respect to $(\mathbb{V}, \|\cdot\|_{\mathbb{V}})$. We define the **consistency error** of (24.3) to be the function $\mathcal{E}_h \in \mathcal{V}(\bar{\Omega}_h)$, defined as

$$\mathcal{E}_h[u](x) = \begin{cases} (L_h \pi_h u - f_h)(x), & x \in \Omega_h^I, \\ (\ell_h \pi_h u - g_h)(x), & x \in \Omega_h^B, \end{cases}$$

where $u \in \mathbb{V}$ solves (24.2). We say that the FDM (24.3) is **consistent to at least order** $p \in \mathbb{N}$ with (24.2) if and only if there is $k \in \mathbb{N}$, usually $k > p$, such that if $u \in C^k(\bar{\Omega})$, then there are constants $h_1 > 0$ and $C > 0$ such that, whenever $h \in (0, h_1]$, we have

$$\|\mathcal{E}_h[u]\|_{\mathcal{V}(\Omega_h^I)} + \|\mathcal{E}_h[u]\|_{\mathcal{V}(\Omega_h^B)} \le Ch^p.$$

Finally, we say that the method is **consistent with exactly order** p if the previous estimate cannot be improved upon, i.e., the value of p cannot be increased, by assuming that u is smoother.

Remark 24.19 (relation between notions)**.** Notice that we are calling consistency error two seemingly unrelated concepts. To show how they are related, let us assume, for the sake of illustration, that $f_h = \pi_h f$. Then, for all points $x \in \Omega_h^I$, we have

$$\begin{aligned}(L_h \pi_h u - f_h)\,(x) &= (L_h \pi_h u - \pi_h(Lu))\,(x) + \pi_h(Lu)(x) - f_h(x) \\ &= \mathcal{E}_h[L, u](x) + \pi_h(Lu - f)(x) \\ &= \mathcal{E}_h[L, u](x).\end{aligned}$$

In other words, when $f_h = \pi_h f$, the consistency error of the method and the operator coincide (at interior points). In general, we see that the consistency error of a method has two components: the *operator consistency* error, $\mathcal{E}_h[L,u](\boldsymbol{x})$, and the grid projection error, $\pi_h(Lu)(\boldsymbol{x}) - f_h(\boldsymbol{x})$. A similar consideration can be made for the boundary points and the operator ℓ.

Before we proceed, let us illustrate the notion of consistency with a few examples.

Proposition 24.20 (consistency). *Let $d = 1$. We have:*

1. *The forward difference operator of Example 24.3 is consistent, on $C_b(\mathbb{R})$, to exactly order one with*
$$\frac{\mathrm{d}w(x)}{\mathrm{d}x}, \quad x \in \mathbb{R}.$$
2. *The backward difference operator of Example 24.4 is consistent, on $C_b(\mathbb{R})$, to exactly order one with*
$$\frac{\mathrm{d}w(x)}{\mathrm{d}x}, \quad x \in \mathbb{R}.$$
3. *The centered difference operator of Example 24.5 is consistent, on $C_b(\mathbb{R})$, to exactly order two with*
$$\frac{\mathrm{d}w(x)}{\mathrm{d}x}, \quad x \in \mathbb{R}.$$
4. *The one-dimensional discrete Laplacian operator of Example 24.3 is consistent, on $C_b(\mathbb{R})$, to exactly order two with*
$$\Delta w(x) = \frac{\mathrm{d}^2 w(x)}{\mathrm{d}x^2}, \quad x \in \mathbb{R}.$$
As a consequence, if $P \in \mathbb{P}_1$,
$$\Delta_h P(x) = 0, \quad \forall x \in \Omega_h.$$
5. *The operator*
$$(\mathrm{BDF}\, w)_i = \frac{1}{2h}(3w_i - 4w_{i-1} + w_{i-2})$$
is consistent, on $C_b(\mathbb{R})$, to exactly order two with
$$\frac{\mathrm{d}w(x)}{\mathrm{d}x}, \quad x \in \mathbb{R}.$$

Proof. See Problem 24.7. □

Let us return to our general notions now and define convergence.

Definition 24.21 (convergence). Let $\{(\mathcal{V}(\bar{\Omega}_h), \|\cdot\|_h)\}_{h>0}$ have the approximation property with respect to $(\mathbb{V}, \|\cdot\|_{\mathbb{V}})$. Let $u \in \mathbb{V}$ and $w \in \mathcal{V}(\bar{\Omega}_h)$ solve (24.2) and (24.3), respectively. The **error** of the method is the function $e \in \mathcal{V}(\bar{\Omega}_h)$, defined by
$$e = \pi_h u - w.$$
We say that the FDM (24.3) is **convergent** if and only if
$$\|e\|_h \to 0, \qquad h \downarrow 0.$$

We say that the method is **convergent with rate** $p \in \mathbb{N}$ if there is $C > 0$ such that, for sufficiently small $h > 0$, we have

$$\|e\|_h \leq Ch^p.$$

We can now state and prove Lax's Principle.

Theorem 24.22 (Lax). *Let problem* (24.3) *be stable and consistent. Then it is convergent. In addition, if* (24.3) *is consistent to order* $p \in \mathbb{N}$, *then the approximation is convergent with rate* p.

Proof. This is nothing but an exercise in notation. By definition, we have

$$\begin{cases} L_h \pi_h u(x) = f_h(x) + \mathcal{E}_h[u](x), & x \in \Omega_h^I, \\ \ell_h \pi_h u(x) = g_h(x) + \mathcal{E}_h[u](x), & x \in \Omega_h^B. \end{cases}$$

By linearity, we have

$$\begin{cases} L_h(\pi_h u - w)(x) = \mathcal{E}_h[u](x), & x \in \Omega_h^I, \\ \ell_h(\pi_h u - w)(x) = \mathcal{E}_h[u](x), & x \in \Omega_h^B, \end{cases}$$

and stability then implies that

$$\|\pi_h u - w\|_h \leq C \left(\|\mathcal{E}_h[u]\|_{\mathcal{V}(\Omega_h^I)} + \|\mathcal{E}_h[u]\|_{\mathcal{V}(\Omega_h^B)} \right).$$

Thus, convergence, even with the prescribed rate, follows from consistency. □

Example 24.18 Before we proceed any further, it is instructive to observe that we have already seen particular examples of all these constructions before. Indeed, the methods for initial value problems for ordinary differential equations we saw in Chapters 18 and 20 fit this framework.

In these methods, we subdivided the interval $[0, T] \subseteq \mathbb{R}$ into $K \in \mathbb{N}$ equal parts of length $\tau = T/K$. In doing so, we defined the grid domain

$$[0, T]_\tau = [0, T] \cap \mathbb{Z}_\tau = \{k\tau \mid k = 0, \dots, K\}.$$

Our approximation methods computed sequences $\{\boldsymbol{w}^k\}_{k=0}^K \subset \mathbb{R}^d$ which can be identified with elements of $\mathcal{V}([0, T]_\tau; \mathbb{R}^d)$ via $\boldsymbol{w}(t_k) = \boldsymbol{w}^k$. The approximation is obtained by a nonlinear version of an FDM

$$\mathcal{D}_\tau \boldsymbol{w}(x) = \boldsymbol{G}(x), \quad x \in [0, T]_\tau^I, \qquad \boldsymbol{w}(x) = \boldsymbol{g}(x), \quad x \in [0, T]_\tau^B,$$

where the interior and boundary domains $[0, T]_\tau^I$ and $[0, T]_\tau^B$, respectively, are relative to the operator $\mathcal{D}_\tau$.

For single-step methods, the operator $\mathcal{D}_\tau$ is given by

$$(\mathcal{D}_\tau \boldsymbol{w})(t_{k+1}) = \frac{1}{\tau} \left(\boldsymbol{w}^{k+1} - \boldsymbol{w}^k \right), \quad k \geq 0,$$

so that $[0, T]_\tau^I = (0, T] \cap \mathbb{Z}_\tau$, $[0, T]_\tau^B = \{0\}$, and $\boldsymbol{g} = \boldsymbol{u}_0$. The function $\boldsymbol{G}$ is given by the slope approximation; see Definition 18.1.

For q-step methods ($q \in \mathbb{N}$), we have

$$(\mathcal{D}_\tau \boldsymbol{w})(t_{k+q}) = \frac{1}{\tau}\sum_{j=0}^{q} a_j \boldsymbol{w}^{k+j}, \quad k \geq 0,$$

so that $[0,T]^I_\tau = [t_q, T] \cap \mathbb{Z}_\tau$, $[0,T]^B_\tau = \{t_0, t_1, \dots, t_{q-1}\}$, and $\boldsymbol{g}(t_j) = \boldsymbol{w}^j$ for $j = 0, \dots, q-1$, i.e., the starting values of the multi-step method. The function $\boldsymbol{G}$ is given by

$$\boldsymbol{G}(t_{k+q}) = \sum_{j=0}^{q} b_j \boldsymbol{f}(t_{k+j}, \boldsymbol{w}^{k+j});$$

see (20.1).

24.3 The Poisson Problem in One Dimension

Let us now apply all the general theory that we have developed to the approximation of the one-dimensional Poisson problem. Let $\Omega = (0,1)$. We wish to approximate the solution to

$$-\frac{\mathrm{d}^2 u(x)}{\mathrm{d}x^2} = f(x), \ x \in \Omega, \qquad u(0) = u_L, \quad u(1) = u_R, \tag{24.4}$$

where $u_L, u_R \in \mathbb{R}$ and $f \in C(\bar{\Omega})$.

We recall that Theorem 23.8 provided existence and uniqueness of classical solutions to this problem. We will operate under the assumptions that guarantee this and will approximate this solution via finite differences.

Definition 24.23 (finite differences)**.** Suppose that $u \in C^2(\Omega) \cap C(\bar{\Omega})$ is a classical solution to (24.4). Let $N \in \mathbb{N}$. We call $\tilde{w} \in \mathcal{V}(\bar{\Omega}_h)$ a **finite difference approximation** to (24.4) if and only if

$$-\Delta_h \tilde{w}(x) = f_h(x), \quad x \in \Omega_h, \qquad \tilde{w}(0) = u_L, \ \tilde{w}(1) = u_R, \tag{24.5}$$

where $f_h \in \mathcal{V}(\Omega_h)$ is defined as $f_h = \pi_h f$ and π_h is the sampling operator.

One of our first concerns should be to determine whether our finite difference approximation is well defined. We can get rid of the Dirichlet boundary conditions via a change of variables. Indeed, let $P \in \mathcal{V}(\bar{\Omega}_h)$ be given by

$$P_i = u_L + (u_R - u_L)ih \qquad \Longrightarrow \qquad P_0 = u_L, \ P_{N+1} = u_R.$$

Owing to Proposition 24.20, we have $\Delta_h P = 0$. Let us define $w = \tilde{w} - P$. Then $w \in \mathcal{V}_0(\bar{\Omega}_h)$ and it solves

$$-\Delta_h w(x) = f_h(x), \quad x \in \Omega_h, \qquad w(0) = 0, \ w(1) = 0, \tag{24.6}$$

if and only if $\tilde{w}$ solves (24.5). Now, to find w, we use the identification of Remark 24.4 to realize that problem (24.6) can be equivalently rewritten as

$$\mathsf{A}\boldsymbol{w} = h^2 \boldsymbol{f},$$

where $\boldsymbol{w} \in \mathbb{R}^N \longleftrightarrow w \in \mathcal{V}_0(\bar{\Omega}_h)$, $\boldsymbol{f} \in \mathbb{R}^N \longleftrightarrow f_h \in \mathcal{V}(\Omega_h)$, and the matrix $\mathsf{A} \in \mathbb{R}^{N \times N}$ is the so-called *stiffness matrix*, which is defined as

$$\mathsf{A} = \begin{bmatrix} 2 & -1 & 0 & \dots & 0 \\ -1 & 2 & \ddots & & \vdots \\ 0 & \ddots & \ddots & -1 & 0 \\ \vdots & & -1 & 2 & -1 \\ 0 & \dots & 0 & -1 & 2 \end{bmatrix}. \tag{24.7}$$

This observation is important from both practical and theoretical points of view. First, in practice, we have reduced the approximation of a BVP to the solution of a linear system of equations. Not only is such a problem more tractable, but also we see that the stiffness matrix A is tridiagonal. Thus, to compute $\boldsymbol{w}$, one only needs to apply the (linear complexity) methods of Section 3.1.3. Second, and this is what we are concerned with here, to analyze the FDM we only need to study the properties of a linear system of equations.

Let us then analyze this method. First, assuming that problem (24.6) has a solution, let us study its consistency. This is an easy consequence of what we have done so far.

Proposition 24.24 (consistency). *Let $u \in C^4(\Omega)$ be a classical solution of* (24.4) *with $u_L = u_R = 0$. Let $w \in \mathcal{V}_0(\bar{\Omega}_h)$ be its finite difference approximation defined by* (24.6). *This method is consistent, in $C(\bar{\Omega})$, to exactly order $p = 2$.*

Proof. Clearly, $\mathcal{E}_h[u]_0 = \mathcal{E}_h[u]_{N+1} = 0$, so we only need to study the consistency of approximating the differential equation. But Proposition 24.20 has shown that our approximation is of exactly order $p = 2$. □

We now need to show that the discrete problem (24.6) is well posed and stable, so that we can infer convergence. To do this, we observe that, in fact, we have already studied the matrix A. This is the Toeplitz symmetric tridiagonal matrix, with $n = N + 1$, of Theorem 3.38. The following is nothing but a restatement of that result with a few easy additional consequences.

Theorem 24.25 (spectrum of A). *Let $N \in \mathbb{N}$. Suppose that $\mathsf{A} \in \mathbb{R}^{N \times N}$ is the stiffness matrix defined in* (24.7). *Consider the set of grid functions $\{\varphi_k\}_{k=1}^N \subset \mathcal{V}_0(\bar{\Omega}_h)$, defined by*

$$[\boldsymbol{\varphi}_k]_i = \sin(k\pi i h), \quad i = 1, \dots, N.$$

1. *The set $\{\boldsymbol{\varphi}_k\}_{k=1}^N \subset \mathbb{R}^N$ with $\boldsymbol{\varphi}_k \longleftrightarrow \varphi_k$ is an orthogonal set of eigenvectors of the stiffness matrix* A.
2. *The eigenvalue λ_k corresponding to the eigenvector $\boldsymbol{\varphi}_k$ is given by*

$$\lambda_k = 2(1 - \cos(k\pi h)) = 4\sin^2\left(\frac{k\pi h}{2}\right).$$

Since $0 < \lambda_k < 4$, for all $k = 1, \dots, N$, the stiffness matrix A *is symmetric positive definite (SPD) and is, therefore, invertible.*

3. *There is a constant $C_1 > 0$, independent of h, such that, if $0 < h < \frac{1}{2}$,*

$$\left\|\mathsf{A}^{-1}\right\|_2 = \frac{1}{4\sin^2\left(\frac{h\pi}{2}\right)} \leq C_1 h^{-2}.$$

4. *The (spectral) condition number of* A *satisfies the estimate*

$$\kappa_2(\mathsf{A}) = \|\mathsf{A}\|_2 \left\|\mathsf{A}^{-1}\right\|_2 \leq 4C_1 h^{-2}.$$

Proof. The first two statements are the content of, or were shown during the course of the proof of, Theorem 3.38. Let us then focus on the norm and condition number estimates.

The largest eigenvalue of A is λ_N and the smallest is λ_1. Since the eigenvalues of A^{-1} are the reciprocals of the eigenvalues of A, it follows that

$$\left\|\mathsf{A}^{-1}\right\|_2 = \frac{1}{\lambda_1} = \frac{1}{4\sin^2\left(\frac{\pi h}{2}\right)} = \frac{1}{2\left(1 - \cos(\pi h)\right)}.$$

Now, by Taylor's Theorem B.31, for any $x \in (0, \pi/2)$, there is $\eta \in (0, x)$ such that

$$1 - \cos(x) = \frac{x^2}{2!} - \frac{x^4}{4!} + \frac{x^6}{6!}\cos(\eta) \geq \frac{x^2}{2} - \frac{x^4}{24}.$$

Consequently, if $0 < \pi h < \pi/2$, or, equivalently, $0 < h < 1/2$,

$$2h^{-2}(1 - \cos(\pi h)) \geq \pi^2 - \frac{\pi^4 h^2}{12}.$$

Since

$$\frac{\sqrt{6}}{\pi} > \frac{1}{2} > h > 0,$$

it follows that

$$2h^{-2}(1 - \cos(\pi h)) \geq \pi^2 - \frac{\pi^4 h^2}{12} \geq \frac{\pi^2}{2}.$$

Equivalently,

$$\frac{2}{\pi^2 h^2} \geq \frac{1}{2(1 - \cos(\pi h))} = \frac{1}{\lambda_1}$$

and the norm estimate proof is completed upon taking $C_1 = \frac{2}{\pi^2}$. The condition number estimate follows immediately from

$$\kappa_2(\mathsf{A}) = \frac{\lambda_N}{\lambda_1} \leq 4C_1 h^{-2}.$$ □

As an immediate consequence, problem (24.6) always has a unique solution.

Corollary 24.26 (stability). *For every $N \in \mathbb{N}$, the finite difference problem* (24.6) *has a unique solution $w \in \mathcal{V}_0(\bar{\Omega}_h) \longleftrightarrow \boldsymbol{w} \in \mathbb{R}^N$, which, moreover, satisfies the estimate*

$$\|w\|_{L_h^2} \leq C_1 \|f_h\|_{L_h^2}.$$

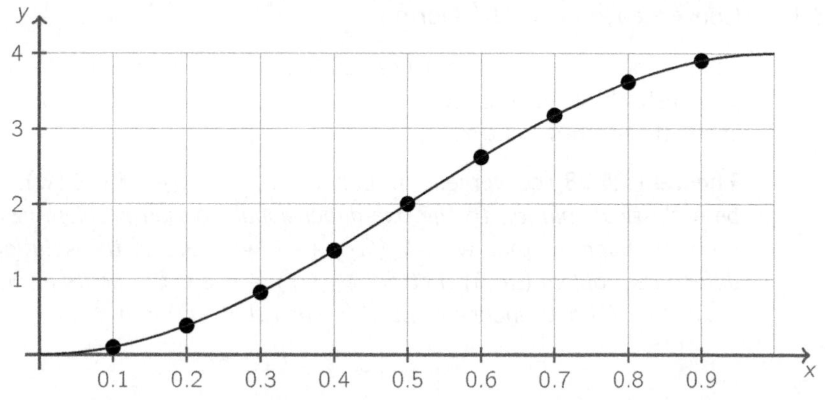

Figure 24.2 The eigenvalues of the stiffness matrix A are shown (filled circles) for $N = 9$. The horizontal axis is $x = kh$. The solid curve is the plot of $4\sin^2\left(\frac{\pi x}{2}\right)$. Observe that $0 < \lambda_k < 4$.

Proof. For every $N \in \mathbb{N}$, the matrix A is invertible; thus, there is a unique $\boldsymbol{w}$. Moreover, this satisfies

$$\|\boldsymbol{w}\|_2 \leq h^2 \left\|\mathsf{A}^{-1}\right\|_2 \|\boldsymbol{f}\|_2 \leq C_1 \|\boldsymbol{f}\|_2,$$

where we used the estimate on the norm of the inverse. Since $\|w\|_{L_h^2} = h^{1/2}\|\boldsymbol{w}\|_2$, the result follows. □

Remark 24.27 (consistency). In one dimension, $\{\tilde{\varphi}_k, \mu_k\}_{k\in\mathbb{N}}$ is the family of eigenpairs of the Laplacian:

$$-\tilde{\varphi}_k''(x) = \mu_k \varphi_k(x), \quad x \in (0,1), \qquad \tilde{\varphi}_k(0) = \tilde{\varphi}_k(1) = 0,$$

where

$$\tilde{\varphi}_k(x) = \sin(k\pi x), \quad \mu_k = k^2\pi^2.$$

Theorem 24.25 shows that the matrix $h^{-2}\mathsf{A}$ has eigenvectors

$$\varphi_k = \pi_h \tilde{\varphi}_k,$$

where π_h is the sampling operator, and eigenvalues (see Figure 24.2)

$$\frac{1}{h^2}\lambda_k^2 = 4(N+1)^2 \sin^2\left(\frac{k\pi}{2(N+1)}\right)^2.$$

For $N \uparrow \infty$ ($h \downarrow 0$), we have

$$\frac{1}{h^2}\lambda_k^2 = 4(N+1)^2\left(\frac{k^2\pi^2}{4(N+1)^2} + \mathcal{O}((N+1)^{-4})\right) = \mu_k + \mathcal{O}(h^2).$$

This shows that this method is not only point-wise consistent, but also, in a sense, *spectrally* consistent.

24.3.1 Convergence in the L^2_h-Norm

Let us now prove a convergence result in the L^2_h-norm. To do this, we use the previously obtained consistency error, and then we use some facts (already proven) about the stiffness matrix.

Theorem 24.28 (convergence). *Let $N \in \mathbb{N}$, $h = \frac{1}{N+1}$, $f \in C(\Omega)$, and $u \in C^4(\bar{\Omega})$ be a classical solution to the one-dimensional Poisson problem (24.4) with $u_L = u_R = 0$. Suppose that $w \in \mathcal{V}_0(\bar{\Omega}_h) \longleftrightarrow \boldsymbol{w} \in \mathbb{R}^N$ is the solution to the finite difference problem (24.6). Let $e \in \mathcal{V}_0(\bar{\Omega}_h) \longleftrightarrow \boldsymbol{e} \in \mathbb{R}^N$ be its error. Then there is a constant $C_2 > 0$, independent of h, such that, if $0 < h < \frac{1}{2}$,*

$$\|e\|_{L^2_h} \leq C_1 C_2 h^2,$$

where $C_1 > 0$ is the constant from Theorem 24.25.

Proof. From Proposition 24.20, we know that

$$\|\mathcal{E}_h[u]\|_{L^\infty_h} \leq C_2 h^2, \qquad \|\boldsymbol{\mathcal{E}}_h[u]\|_\infty \leq C_2 h^2.$$

We now proceed in the usual way. By definition of the consistency error, we have, on Ω_h,

$$-\Delta_h \pi_h u = f_h + \mathcal{E}_h[u],$$

so that, subtracting from the method (24.6), we obtain an equation that controls the error

$$-\Delta_h e = \mathcal{E}_h[u] \qquad \longleftrightarrow \qquad \mathsf{A}\boldsymbol{e} = h^2 \boldsymbol{\mathcal{E}}_h[u].$$

Thus, using Theorem 24.25 and proceeding as in Corollary 24.26,

$$\|e\|_{L^2_h} = h^{1/2}\|\boldsymbol{e}\|_2 \leq h^{5/2}\left\|\mathsf{A}^{-1}\right\|_2 \|\boldsymbol{\mathcal{E}}_h[u]\|_2 \leq C_1\sqrt{hN}\|\boldsymbol{\mathcal{E}}_h[u]\|_\infty \leq C_1 C_2 h^2. \quad \square$$

Remark 24.29 (suboptimal estimate). Using Problem 24.8, we can immediately get an error estimate in the L^∞_h-norm using our L^2_h-estimate:

$$\|e\|_{L^\infty_h} \leq \frac{1}{\sqrt{h}}\|e\|_{L^2_h} \leq C_1 C_2 h^{\frac{3}{2}}.$$

As we will see, this estimate is suboptimal.

24.3.2 The Discrete Maximum Principle and Convergence in the L^∞_h-Norm

We will now obtain optimal error estimates for our finite difference approximations in the L^∞_h-norm. For this purpose, we need a Discrete Maximum Principle, which serves as the discrete analogue of the maximum principle of Theorem 23.2.

Theorem 24.30 (Discrete Maximum Principle). *Suppose that $v \in \mathcal{V}(\bar{\Omega}_h)$ satisfies*

$$-\Delta_h v(x) \leq 0, \quad \forall x \in \Omega_h.$$

Then, for all $x \in \bar{\Omega}_h$,

$$v(x) \leq \max\{v(0), v(1)\}.$$

In other words, the maximum must occur on the boundary.

Proof. To obtain a contradiction, suppose that a strict maximum occurs in the interior. If this is true, there is some $x_k \in \Omega_h$ such that

$$v_k = \max_{x_j \in \Omega_h} v_j > \max\{v_0, v_{N+1}\}.$$

Let us suppose, for simplicity, that $2 \leq k \leq N-1$. Then

$$0 \geq -h^2 \Delta_h v_k = 2v_k - v_{k-1} - v_{k+1} \geq 0.$$

This implies that $-\Delta_h v_k = 0$ and, therefore,

$$v_k = \frac{1}{2}(v_{k-1} + v_{k+1}).$$

The only way to satisfy the last equation and the fact that $v_k \geq v_{k\pm1}$ is to have $v_k = v_{k\pm1}$.

We can now repeat our argument at all neighboring points; we conclude that

$$v_1 = \cdots = v_k = \cdots = v_N.$$

Next to the left boundary point, we have, for example,

$$0 = -h^2 \Delta v_1 = 2v_1 - v_0 - v_2 > 0,$$

since $v_1 > v_0$. This is a contradiction. □

From this result, a stability result analogous to Corollary 23.4 can be easily obtained.

Theorem 24.31 (stability). *Suppose that $w \in \mathcal{V}(\bar{\Omega}_h)$ and $f_h \in \mathcal{V}(\Omega_h)$ and $g_0, g_1 \in \mathbb{R}$ are such that*

$$-\Delta_h w = f_h, \quad \Omega_h, \qquad w_0 = g_0, \quad w_{N+1} = g_1.$$

There is some constant $C > 0$, independent of h and w, such that

$$\|w\|_{L_h^\infty} \leq \max_{j \in \{0,1\}} |g_j| + C\|f_h\|_{L_h^\infty}.$$

Proof. Define the comparison function $\Phi\colon [0,1] \to \mathbb{R}$ via

$$\Phi(x) = \left(x - \frac{1}{2}\right)^2 \geq 0.$$

Define the grid function $\phi = \pi_h \Phi$, where π_h is the sampling operator. Then, in Ω_h, we have

$$-\Phi'' \equiv -2 \equiv -\Delta_h \phi.$$

Now define the grid functions

$$\psi_\pm = \pm w + \frac{\|f_h\|_{L_h^\infty}}{2}\phi.$$

Observe that, in Ω_h, we have

$$-\Delta_h \psi_\pm = \pm f_h - \|f_h\|_{L_h^\infty} \leq 0.$$

By the Discrete Maximum Principle, in Ω_h,

$$\pm w \le \pm\psi \le \max\{\psi_\pm(0), \psi_\pm(1)\} \le \max_{j\in\{0,1\}} |g_j| + \frac{\|f_h\|_{L_h^\infty}}{8}.$$

Putting the two results together, we get

$$\|w\|_{L_h^\infty} \le \max_{j\in\{0,1\}} |g_j| + \frac{1}{8}\|f_h\|_{L_h^\infty}. \qquad \square$$

This serves as a stability result in the L_h^∞-norm. Thus, we can now obtain convergence.

Corollary 24.32 (convergence). *Let $f \in C(\Omega)$ and $u \in C^4(\bar{\Omega})$ be a classical solution to the one-dimensional Poisson problem (24.4) with $u_L = u_R = 0$. Let $N \in \mathbb{N}$. Suppose that $w \in \mathcal{V}_0(\bar{\Omega}_h) \longleftrightarrow \boldsymbol{w} \in \mathbb{R}^N$ is the solution to the finite difference problem (24.6). Then there is a constant $C_3 > 0$, independent of h, such that*

$$\|e\|_{L_h^\infty} \le C_2 C_3 h^2,$$

where $C_2 > 0$ is the consistency error constant.

Proof. In the course of the proof of Theorem 24.28, we found that the error $e \in \mathcal{V}_0(\bar{\Omega}_h)$ satisfies, in Ω_h,

$$-\Delta_h e = \mathcal{E}_h[u].$$

By the previous stability result, we have

$$\|e\|_{L_h^\infty} \le \frac{1}{8}\|\mathcal{E}_h[u]\|_{L_h^\infty} \le \frac{C_2}{8}h^2.$$

The result follows with $C_3 = \frac{1}{8}$. $\square$

24.4 Elliptic Problems in One Dimension

Let us consider here more general elliptic problems in one dimension and their finite difference approximation. We will focus on the Dirichlet problem

$$\begin{cases} Lu = f, & \text{in } \Omega = (0,1), \\ u(0) = u(1) = 0, \end{cases} \qquad (24.8)$$

where $f \in C([0,1])$ and the operator L is a second-order elliptic operator to be specified below. We will be interested in constructing finite difference operators L_h that are consistent and have stencil $\{-1, 0, 1\}$, so that the finite difference problem will read: Find $w \in \mathcal{V}_0(\bar{\Omega}_h)$ such that

$$L_h w = f_h, \quad \text{in } \Omega_h, \qquad L_h w_i = -A_i w_{i-1} + C_i w_i - B_i w_{i+1}; \qquad (24.9)$$

here, $f_h \in \mathcal{V}(\Omega_h)$ is, as usual, defined as $f_h(ih) = f(ih)$.

We will call methods like (24.9) *homogeneous* FDMs. Although the coefficients of the equation and the method are variable, the method has the same representation at every point x_i of the grid domain Ω_h.

To make matters simple, we will proceed under the assumption that the coefficients of L are sufficiently smooth. We comment, however, that this is not necessary and we refer the reader to [50], [56, Chapter VI], and [79] for detailed accounts of the general theory.

24.4.1 Divergence Form Operators

Here, we consider a differential operator in *divergence form*, i.e.,

$$Lu(x) = -\frac{\mathrm{d}}{\mathrm{d}x}\left(a(x)\frac{\mathrm{d}u(x)}{\mathrm{d}x}\right) + c(x)u(x), \tag{24.10}$$

where we assume that the coefficients satisfy $a \in C^1([0,1])$, $0 \le c \in C([0,1])$; in addition, there are constants $\lambda, \Lambda \in \mathbb{R}$ such that

$$0 < \lambda \le a(x) \le \Lambda, \quad \forall x \in [0,1].$$

In this setting, the operator is elliptic in the sense of Definition 23.1, and the theory of weak solutions presented in Section 23.2.3 provides the existence, uniqueness, and stability of the solution to (24.8). The stability, in particular, is expressed by Corollary 23.11.

We now wish to construct the difference method, i.e., find the coefficients A_i, B_i, C_i. We do so arguing from consistency considerations. Namely, we consider, for $v \in C^4([0,1])$ such that $v(0) = v(1) = 0$, the consistency error

$$\mathcal{E}_h[v] = L_h\pi_h v - \pi_h(Lv),$$

where π_h is the sampling operator, and require that it satisfies $\|\mathcal{E}_h[v]\|_{L_h^\infty} \le Ch^2$.

We begin by introducing the following change of notation. Setting $\alpha_i = h^2 A_i$, $\beta_i = h^2 B_i$, and $\gamma_i = h^2 C_i$, we get

$$L_h v_i = -\frac{1}{h}\left[\beta_i\frac{v_{i+1}-v_i}{h} - \alpha_i\frac{v_i - v_{i-1}}{h}\right] + \kappa_i v_i = -\frac{1}{h}\left[\beta_i\delta_h v_i - \alpha_i\bar{\delta}_h v_i\right] + \kappa_i v_i,$$

where $\kappa_i = h^{-2}(\gamma_i - \beta_i - \alpha_i)$. Notice that, at least symbolically, the finite difference operator begins to resemble the divergence form operator L. Now, to achieve consistency, we must have

$$\mathcal{E}_h[v]_i = -\frac{1}{h}\left[\beta_i\delta_h v(x_i) - \alpha_i\bar{\delta}_h v(x_i)\right] + \kappa_i v(x_i) + (a(x_i)v'(x_i))' - c(x_i)v(x_i) = \mathcal{O}(h^2).$$

From Taylor expansions, we know, see Proposition 24.20,

$$\delta_h v(x_i) = v'(x_i) + \frac{1}{2}v''(x_i)h + \frac{1}{6}v'''(x_i)h^2 + \mathcal{O}(h^3),$$
$$\bar{\delta}_h v(x_i) = v'(x_i) - \frac{1}{2}v''(x_i)h + \frac{1}{6}v'''(x_i)h^2 + \mathcal{O}(h^3),$$

so that, substituting in $\mathcal{E}_h[v]_i$, we get

$$\mathcal{E}_h[v]_i = \left(a'(x_i) - \frac{\beta_i - \alpha_i}{h}\right) v'(x_i) + \left(a(x_i) - \frac{\alpha_i + \beta_i}{2}\right) v''(x_i)$$
$$- \frac{\beta_i - \alpha_i}{6} h v'''(x_i) + (\kappa_i - c(x_i))v(x_i) + \mathcal{O}(h^2).$$

Thus, we require

$$\frac{\beta_i - \alpha_i}{h} = a'(x_i) + \mathcal{O}(h^2), \quad \frac{\alpha_i + \beta_i}{2} = a(x_i) + \mathcal{O}(h^2), \quad \kappa_i = c(x_i) + \mathcal{O}(h^2). \tag{24.11}$$

There are several ways this can be achieved. For instance,

$$\beta_i = a(x_i + h/2), \qquad \alpha_i = a(x_i - h/2), \qquad \kappa_i = c(x_i), \tag{24.12}$$
$$\beta_i = \frac{a(x_{i+1}) + a(x_i)}{2}, \qquad \alpha_i = \frac{a(x_i) + a(x_{i-1})}{2}, \qquad \kappa_i = c(x_i) \tag{24.13}$$

are possible choices; see Problem 24.9. Let us write the final operator with the first choice

$$L_h v_i = -\frac{1}{h}\left(a_{i+1/2}\delta_h v_i - a_{i-1/2}\bar{\delta}_h v_i\right) + c_i w_i = -\delta_h(\hat{a}_i \bar{\delta}_h v_i) + c_i w_i, \tag{24.14}$$

where $a_{i\pm1/2} = a(x_i \pm h/2)$, $c_i = c(x_i)$, and $\hat{a}_i = a_{i-1/2}$. Notice the resemblance to the divergence form differential operator L.

Let us now study the stability of this method. We will do so in a norm that is natural for this problem.

Definition 24.33 (H_h^1-seminorm)**.** The H_h^1**-seminorm**, on $\mathcal{V}(\bar{\Omega}_h)$, is defined as

$$\|v\|_{H_h^1}^2 = h \sum_{i=1}^{N+1} |\bar{\delta}_h v_i|^2.$$

Notice that, indeed, this is not a norm, but only a seminorm. A grid function that takes constant, nonzero values satisfies $\|v\|_{H_h^1} = 0$. However, it turns out that on $\mathcal{V}_0(\bar{\Omega}_h)$ this is a norm. Compare the following result with (D.2).

Theorem 24.34 (discrete Poincaré)**.**[4] *There is a constant, independent of $h > 0$, such that, for all $v \in \mathcal{V}_0(\bar{\Omega}_h)$, we have*

$$\|v\|_{L_h^2} \le C\|v\|_{H_h^1}.$$

Consequently, the quantity $\|\cdot\|_{H_h^1}$ is a norm on $\mathcal{V}_0(\bar{\Omega}_h)$.

Proof. See Problem 24.10. □

We can now state the stability of the method. It is important to compare the estimate in this problem with that of Corollary 23.11.

Theorem 24.35 (stability)**.** *There is a constant $C > 0$ that depends only on the coefficients a and c such that any solution to* (24.9) *with the operator defined as in* (24.14) *satisfies*

[4] Named in honor of the French mathematician, theoretical physicist, engineer, and philosopher of science Jules Henri Poincaré (1854–1912).

$$\|w\|_{H_h^1} \leq C\|f_h\|_{L_h^2}.$$

As a consequence, the solution to this problem is unique and convergent with order $p = 2$ in the H_h^1-norm.

Proof. Since the FDM is a square system of linear equations, the estimate implies uniqueness, and this in turn implies existence.

Let us now show the estimate. We can take the L_h^2-inner product of the method with w itself to obtain

$$-\left(\delta_h(\hat{a}\bar{\delta}_h w), w\right)_{L_h^2} + (cw, w)_{L_h^2} = (f_h, w)_{L_h^2} \leq \|f_h\|_{L_h^2}\|w\|_{L_h^2}.$$

Since, by assumption, $c \geq 0$, this inequality reduces to

$$-\left(\delta_h(\hat{a}\bar{\delta}_h w), w\right)_{L_h^2} \leq \|f_h\|_{L_h^2}\|w\|_{L_h^2}.$$

We now invoke the Abel transformation of Proposition 24.9 to obtain, since $w \in \mathcal{V}_0(\bar{\Omega}_h)$,

$$-\left(\delta_h(\hat{a}\bar{\delta}_h w), w\right)_{L_h^2} = h\sum_{i=1}^{N} \hat{a}_i|\bar{\delta}_h w_i|^2 \geq \lambda\|w\|_{H_h^1}^2,$$

where we used that $a_i = a(x_i - h/2) \geq \lambda$.

Finally, applying the discrete Poincaré inequality of Theorem 24.34 and Young's inequality, we conclude that

$$\lambda\|w\|_{H_h^1}^2 \leq C\|f_h\|_{L_h^2}\|w\|_{H_h^1} \leq \frac{C^2}{2\lambda}\|f_h\|_{L_h^2}^2 + \frac{\lambda}{2}\|w\|_{H_h^1}^2,$$

as we intended to show. □

24.4.2 Nondivergence Form Operators. Upwinding

Here, we will consider a *nondivergence form* operator

$$Lu(x) = -a(x)\frac{\mathrm{d}^2u(x)}{\mathrm{d}x^2} + b(x)\frac{\mathrm{d}u(x)}{\mathrm{d}x} - c(x)u(x),$$

where the coefficients satisfy $a, b, c \in C([0,1])$. In addition, we assume that $c(x) \geq 0$ for $x \in [0,1]$ and there are constants $\Lambda, \lambda > 0$ such that

$$\lambda \leq a(x) \leq \Lambda, \quad \forall x \in [0,1].$$

With these assumptions, this operator is elliptic and satisfies all the conditions of the maximum principle of Theorem 23.2. Our main objective here is to construct finite difference operators that are *monotone*, i.e., they satisfy an analogue of this result. For simplicity, we will assume also that $b(x) \geq 0$ for all $x \in [0,1]$. The general case can be treated accordingly.

The main tool we will use to show monotonicity is the following auxiliary result.

Lemma 24.36 (comparison). *Let $\{y_i\}_{i=0}^{N+1} \subset \mathbb{R}$. Assume that there are three families of real numbers $\{A_i\}_{i=1}^N, \{B_i\}_{i=1}^N, \{C_i\}_{i=1}^N \subset \mathbb{R}$ such that*

$$-A_i y_{i-1} + C_i y_i - B_i y_{i+1} \leq 0, \quad i = 1, \dots, N.$$

If

$$A_i > 0, \qquad B_i > 0, \qquad A_i + B_i \leq C_i, \qquad i = 1, \dots, N, \tag{24.15}$$

then $\{y_i\}_{i=0}^{N+1}$ *is either constant or it satisfies*

$$\max_{i=1,\dots,N} y_i \leq \max\{0, y_0, y_{N+1}\}.$$

Proof. See Problem 24.11. □

Notice that this comparison estimate immediately implies several stability estimates for the solution of (24.9) with a finite difference operator whose coefficients satisfy (24.15).

Corollary 24.37 (existence and uniqueness). *Assume that the method* (24.9) *is such that the coefficients of the operator satisfy* (24.15). *Then this method has a unique solution.*

Proof. To show existence, it is sufficient to show uniqueness of a solution. To do so, consider (24.9) with $f_h \equiv 0$. Appealing to Lemma 24.36, we obtain, for any $i = 1, \dots, N$,

$$0 \leq \min_{j=1,\dots,N} w_j \leq w_i \leq \max_{j=1,\dots,N} w_j \leq 0.$$

□

Let us construct a monotone FDM, i.e., one that satisfies (24.15). From the form of the operator, we see that, essentially, the only design choice that we can have is the discretization of the first derivative. We will choose an *upwind* discretization. In other words, since it is assumed that $b \geq 0$, we will choose a backward difference to approximate the first derivative, meaning that we take information from the direction that is opposite to b. We then obtain

$$\begin{aligned} L_h v_i &= -a_i \Delta_h v_i + b_i \bar{\delta}_h v_i + c_i v_i \\ &= -\left(\frac{a_i}{h^2} + \frac{b_i}{h}\right) v_{i-1} + \left(\frac{2a_i}{h^2} + \frac{b_i}{h} + c_i\right) v_i - \left(\frac{a_i}{h^2}\right) v_{i+1}, \end{aligned}$$

where $a_i = a(x_i)$, $b_i = b(x_i)$, and $c_i = c(x_i)$. We see that this method has the form (24.9) with

$$A_i = \frac{a_i}{h^2} + \frac{b_i}{h}, \qquad B_i = \frac{a_i}{h^2}, \qquad C_i = \frac{2a_i}{h^2} + \frac{b_i}{h} + c_i. \tag{24.16}$$

Clearly, this operator is consistent, on $C_b(\mathbb{R})$, with L to order exactly one. Let us now show that it is monotone.

Lemma 24.38 (monotonicity). *For every* $h > 0$, *the coefficients* (24.16) *satisfy* (24.15).

Proof. First of all, since we assumed $b \geq 0$, we clearly have

$$A_i > 0, \qquad B_i > 0.$$

Finally,

$$A_i + B_i = \frac{2a_i}{h^2} + \frac{b_i}{h} = C_i - c_i \leq C_i,$$

where we used that $c(x) \geq 0$. □

Remark 24.39 (downwind and centered discretizations). Let us define the so-called *mesh Péclet number*[5]

$$\mathrm{Pe} = \frac{h\|b\|_{L^\infty(0,1)}}{\lambda},$$

which measures how strong convection (characterized by b) is with respect to diffusion (characterized by λ). Notice that, if we choose a downwind or centered discretization, we obtain monotonicity only for Pe small; see Problems 24.12 and 24.13, respectively. In practice, this means that, although the exact solution may be monotone, the approximate solution will experience spurious oscillations for a grid size h that is not sufficiently small.

Having obtained monotonicity, stability follows ideas we have established before. To make the arguments more tractable, we separate the cases $c \equiv 0$ and $0 \not\equiv c(x) \geq 0$ for all $x \in [0,1]$.

Theorem 24.40 (stability for $c \equiv 0$). *Let the finite difference operator in* (24.9) *be given by* (24.16) *with* $c_i \equiv 0$. *Then, for all* $h > 0$, *this method has a unique solution. Moreover, there is a constant* $C > 0$, *independent of* $h > 0$ *and* f_h, *such that this solution satisfies the estimate*

$$\|w\|_{L_h^\infty} \leq C\|f_h\|_{L_h^\infty}.$$

Proof. Lemma 24.38 has shown that the coefficients (24.16) satisfy (24.15). From Corollary 24.37, existence and uniqueness follow.

It remains then to obtain the stability estimate. The approach is similar to that of Theorem 24.31, in that we need to construct a comparison function $\phi \in \mathcal{V}_0(\bar{\Omega}_h)$ that, for some positive constants $C_0, C_1 > 0$, satisfies

$$0 \leq \phi \leq C_0, \qquad L_h\phi \leq -C_1.$$

If we are able to construct this function, then we define

$$\psi_\pm = \pm w + \frac{\|f_h\|_{L_h^\infty}}{C_1}\phi$$

and observe that

$$L_h\psi_\pm = \pm f_h + \frac{\|f_h\|_{L_h^\infty}}{C_1} L_h\phi \leq \pm f_h - \|f_h\|_{L_h^\infty} \leq 0.$$

The comparison of Lemma 24.36 implies then that

$$\max_{i=1,\ldots,N} \pm w_i \leq \max_{i=1,\ldots,N} \psi_{\pm,i} \leq \frac{C_0}{C_1}\|f_h\|_{L_h^\infty},$$

which is precisely the claimed norm estimate.

[5] Named in honor of the French physicist Jean Claude Eugène Péclet (1793–1857).

We thus need to construct the comparison function. The candidate is, as always, a quadratic

$$\phi_i = (x_i - 1)^2 \geq 0, \qquad C_0 = 1.$$

Let us then apply the operator L_h to this function. For this, observe that

$$h^2 \Delta\phi_i = (x_{i+1} - 1)^2 - 2(x_i - 1)^2 + (x_{i-1} - 1)^2 = 2h^2$$

and that

$$h\bar{\delta}_h\phi_i = (x_i - 1)^2 - (x_{i-1} - 1)^2 = 2h\left(\frac{x_i + x_{i-1}}{2} - 1\right) < 0,$$

because $x_{i-1}, x_i \in [0, 1)$.

Therefore,

$$L_h\phi \leq -2\lambda,$$

and the method is monotone. □

The case $c \not\equiv 0$ is more elaborate and requires several steps.

Theorem 24.41 (stability for $c \geq 0$). *Let the finite difference operator in* (24.9) *be given by* (24.16) *such that there is at least one point $x_i \in \Omega_h$ for which $c_i > 0$. Then, for all $h > 0$, this method has a unique solution. Moreover, there is a constant $C > 0$, independent of $h > 0$ and f_h, such that this solution satisfies the estimate*

$$\|w\|_{L_h^\infty} \leq C\|f_h\|_{L_h^\infty}.$$

Proof. In a similar way to Theorem 24.40, we obtain existence and uniqueness, so that we only focus on the estimate.

We decompose $w = \mathring{w} + \bar{w}$, where $\mathring{w} \in \mathcal{V}_0(\bar{\Omega}_h)$ satisfies

$$-a_i\Delta\mathring{w}_i + b_i\bar{\delta}_h\mathring{w}_i = f_{h,i},$$

so that Theorem 24.40 shows that

$$\|\mathring{w}\|_{L_h^\infty} \leq C\|f_h\|_{L_h^\infty}.$$

Now, by linearity, the function $\bar{w} \in \mathcal{V}_0(\bar{\Omega}_h)$ must satisfy

$$L_h\bar{w}_i = -c_i\mathring{w}_i.$$

We will estimate $\bar{w}$ using the (discrete) comparison function $\phi \in \mathcal{V}_0(\bar{\Omega}_h)$, which solves

$$L_h\phi_i = c_i\|\mathring{w}_i\|_{L_h^\infty} \geq 0.$$

First, we note that, by Lemma 24.36, we have $\phi \geq 0$. In addition,

$$L_h(\pm\bar{w} + \phi)_i = c_i\left(\mp\mathring{w}_i + \|\mathring{w}\|_{L_h^\infty}\right) \geq 0.$$

Lemma 24.36 then implies that, for any i,

$$0 \leq \min\{\pm\bar{w} + \phi\} \leq \pm\bar{w}_i + \phi_i;$$

therefore,

$$\|\bar{w}\|_{L_h^\infty} \leq \|\phi\|_{L_h^\infty}.$$

It remains then to estimate the function ϕ.

Recall that $\phi \geq 0$. Let $x_m \in \Omega_h$ be the point where ϕ attains its, necessarily positive, maximum. At this point, we must have

$$\Delta_h \phi_m \leq 0, \qquad \bar{\delta}_h \phi_m \geq 0.$$

Consequently,

$$0 \leq c_m \phi_m \leq L_h \phi_m = -a_m \Delta_h \phi_m + b_m \bar{\delta}_h \phi_m + c_m \phi_m = c_m \|\mathring{w}\|_{L_h^\infty}.$$

If $c_m \neq 0$, then

$$\|\phi\|_{L_h^\infty} = \phi_m \leq \|\mathring{w}\|_{L_h^\infty},$$

which is the needed estimate.

If, on the other hand, $c_m = 0$, then we can rewrite the equation for ϕ as

$$\frac{a_m}{h^2}(\phi_m - \phi_{m+1}) + \left(\frac{a_m}{h^2} + \frac{b_m}{h}\right)(\phi_m - \phi_{m-1}) = 0,$$

which is only possible if $\phi_{m-1} = \phi_m = \phi_{m+1}$. We can now repeat the argument at the points $x_{m\pm1}$ to reach either the desired conclusion or that $\phi_{m-2} = \phi_{m-1} = \phi_m = \phi_{m+1} = \phi_{m+2}$. Since $c \not\equiv 0$, we will eventually conclude that either $\phi \equiv 0$ or reach an index, say j, for which $c_j > 0$, and obtain the needed estimate. □

24.5 The Poisson Problem in Two Dimensions

In this section, we introduce the two-dimensional Poisson problem on, for simplicity, a square domain $\Omega = (0, 1)^2$. Recall that, for $v \in C^2(\mathbb{R}^2)$,

$$\Delta v(x_1, x_2) = \frac{\partial^2 v(x_1, x_2)}{\partial x_1^2} + \frac{\partial^2 v(x_1, x_2)}{\partial x_2^2}.$$

Thus, we are trying to approximate the solution to

$$-\Delta u(x_1, x_2) = f(x_1, x_2), \quad (x_1, x_2) \in \Omega, \qquad u_{|\partial\Omega} = 0, \tag{24.17}$$

where $f \in C(\bar{\Omega})$ is given. We again recall that Theorem 23.8 provided sufficient conditions for the existence and uniqueness of a classical solution. This is the object that we will try to approximate via finite differences.

Let us then define our finite difference approximation.

Definition 24.42 (finite difference approximation). Let $d = 2$, $f \in C(\bar{\Omega})$, and $u \in C^2(\Omega) \cap C(\overline{\Omega})$ be a classical solution to the two-dimensional Poisson problem (24.17). Let $N \in \mathbb{N}$ and $h = \frac{1}{N+1}$. We call $w \in \mathcal{V}_0(\bar{\Omega}_h)$ a **finite difference approximation** to u if and only if

$$-\Delta_h w_{i,j} = f_{i,j}, \quad (ih, jh) \in \Omega_h, \tag{24.18}$$

where $f_{i,j} = f(ih, jh)$ and Δ_h denotes the two-dimensional discrete Laplace operator of Example 24.7.

We must begin by studying the consistency properties of the two-dimensional discrete Laplace operator of Example 24.7. To do so, we must define suitable norms, so that our setting has the approximation property. The following are analogues of the norms we introduced in Examples 24.15—24.17. The reason for the scaling can be inferred from similar arguments to the one-dimensional case.

Definition 24.43 (discrete L^p_h-norms)**.** Let $d = 2$ and $p \in [1, \infty)$. The **L^p_h-norm on $\mathcal{V}_0(\bar{\Omega}_h)$ or $\mathcal{V}(\Omega_h)$** is

$$\|v\|_{L^p_h} = \left(h^2 \sum_{i,j=1}^{N} |v_{i,j}|^p \right)^{1/p}.$$

For $p = 2$, this norm comes from the **L^2_h-inner product**

$$(v, \phi)_{L^2_h} = h^2 \sum_{i,j=1}^{N} v_{i,j}\phi_{i,j}.$$

The **L^∞_h-norm** on these spaces is

$$\|v\|_{L^\infty_h} = \max_{i,j=1,\dots,N} |v_{i,j}| .$$

The space $\mathcal{V}_0(\bar{\Omega}_h)$ with these norms has the approximation property in $L^p(\Omega)$, for $p < \infty$, and $C_0(\bar{\Omega})$, for $p = \infty$, respectively. For further properties of these norms, see Problem 24.8.

We can now study consistency.

Proposition 24.44 (consistency)**.** *The two-dimensional discrete Laplace operator of Example 24.7 is consistent, in $C_b(\mathbb{R}^2)$, to order exactly two with the Laplacian.*

Proof. See Problem 24.14. □

With consistency at hand, the program to analyze our method should by now be clear. We need to study existence, uniqueness, and stability of the finite difference approximation (24.18), and from these conclude convergence. We begin by showing that this problem has a unique solution.

Theorem 24.45 (stiffness matrix)**.** *Let $N \in \mathbb{N}$. Define $\mathsf{A}_N \in \mathbb{R}^{N \times N}$ via*

$$\mathsf{A}_N = \begin{bmatrix} 4 & -1 & 0 & \dots & 0 \\ -1 & 4 & \ddots & & \vdots \\ 0 & \ddots & \ddots & -1 & 0 \\ \vdots & & -1 & 4 & -1 \\ 0 & \dots & 0 & -1 & 4 \end{bmatrix}.$$

Let $\mathsf{O}_N, \mathsf{I}_N \in \mathbb{R}^{N\times N}$ *denote the zero and identity matrices, respectively. Define the matrix* $\mathsf{A} \in \mathbb{R}^{N^2\times N^2}$ *via*

$$\mathsf{A} = \begin{bmatrix} \mathsf{A}_N & -\mathsf{I}_N & \mathsf{O}_N & \dots & \mathsf{O}_N \\ -\mathsf{I}_N & \mathsf{A}_N & \ddots & & \vdots \\ \mathsf{O}_N & \ddots & \ddots & -\mathsf{I}_N & \mathsf{O}_N \\ \vdots & & -\mathsf{I}_N & \mathsf{A}_N & -\mathsf{I}_N \\ \mathsf{O}_N & \dots & \mathsf{O}_N & -\mathsf{I}_N & \mathsf{A}_N \end{bmatrix}. \tag{24.19}$$

The grid function $w \in \mathcal{V}_0(\bar{\Omega}_h) \longleftrightarrow \boldsymbol{w} \in \mathbb{R}^{N^2}$ *is a solution to the finite difference problem* (24.18) *if and only if it is a solution to the problem*

$$\mathsf{A}\boldsymbol{w} = h^2 \boldsymbol{f}, \tag{24.20}$$

with $f \in \mathcal{V}(\Omega_h) \longleftrightarrow \boldsymbol{f} \in \mathbb{R}^{N^2}$. *By linearity, the error* $e \in \mathcal{V}_0(\bar{\Omega}_h) \longleftrightarrow \boldsymbol{e} \in \mathbb{R}^{N^2}$ *and the consistency error* $\mathcal{E}_h[u] \in \mathcal{V}_0(\bar{\Omega}_h) \longleftrightarrow \boldsymbol{\mathcal{E}}_h[u] \in \mathbb{R}^{N^2}$ *are related by*

$$\mathsf{A}\boldsymbol{e} = h^2 \boldsymbol{\mathcal{E}}_h[u]. \tag{24.21}$$

Proof. See Problem 24.15. □

The matrix A constructed in the previous result is usually called the *two-dimensional stiffness matrix*. Let us now prove that this matrix has suitable properties.

Theorem 24.46 (spectrum of A). *Let* $N \in \mathbb{N}$. *Suppose that* $\mathsf{A} \in \mathbb{R}^{N^2\times N^2}$ *is the stiffness matrix defined in* (24.19). *Consider the set of vectors*

$$S = \left\{ \boldsymbol{\varphi}_{k+(n-1)N} \mid (kh, nh) \in \Omega_h \right\},$$

where the components of $\boldsymbol{\varphi}_{k+(n-1)N}$, *for* $(ih, jh) \in \Omega_h$, *are*

$$\left[\boldsymbol{\varphi}_{k+(n-1)N}\right]_{i+(j-1)N} = \varphi_{k+(n-1)N, i+(j-1)N} = \sin(k\pi i h)\sin(n\pi j h).$$

Then we have:

1. *S is an orthogonal set of eigenvectors of* A.
2. *The eigenvalue* $\lambda_{k+(n-1)N}$ *corresponding to the eigenvector* $\boldsymbol{\varphi}_{k+(n-1)N}$ *is given by*

$$\lambda_{k+(n-1)N} = 2\left(2 - \cos(k\pi h) - \cos(n\pi h)\right) = 4\sin^2\left(\frac{k\pi h}{2}\right) + 4\sin^2\left(\frac{n\pi h}{2}\right).$$

   Therefore, $0 < \lambda_{k+(n-1)N} < 8$, *for all* $(kh, nh) \in \Omega_h$; *consequently,* A *is an SPD matrix.*
3. *There is a constant* $C_1 > 0$, *independent of* h, *such that, if* $0 < h < \frac{1}{2}$,

$$\left\|\mathsf{A}^{-1}\right\|_2 = \frac{1}{8\sin^2\left(\frac{h\pi}{2}\right)} \le C_1 h^{-2}.$$

4. *The spectral condition number of* A *satisfies the estimate*

$$\kappa_2(\mathsf{A}) = \|\mathsf{A}\|_2 \left\|\mathsf{A}^{-1}\right\|_2 \le 8C_1 h^{-2}.$$

Proof. See Problem 24.16. □

From this, existence and uniqueness immediately follow.

Corollary 24.47 (well-posedness). *For every $N \in \mathbb{N}$, there is a unique solution $w \in \mathcal{V}_0(\bar{\Omega}_h) \longleftrightarrow \boldsymbol{w} \in \mathbb{R}^{N^2}$ to the finite difference problem (24.18).*

24.5.1 Convergence in the L^2_h-Norm

As in the one-dimensional case, we have almost everything that is needed to present a convergence result in the L^2_h-norm.

Theorem 24.48 (convergence). *Let $u \in C^4(\bar{\Omega})$ be a classical solution to the two-dimensional Poisson problem (24.17). Let $N \in \mathbb{N}$. Suppose that $w \in \mathcal{V}_0(\bar{\Omega}_h) \longleftrightarrow \boldsymbol{w} \in \mathbb{R}^{N^2}$ is a solution to the finite difference problem (24.18). Let $e \in \mathcal{V}_0(\bar{\Omega}_h) \longleftrightarrow \boldsymbol{e} \in \mathbb{R}^{N^2}$ be its error. Then there is a constant $C_2 > 0$, independent of h, such that*

$$\|\mathcal{E}_h[u]\|_{L^\infty_h} \le C_2 h^2.$$

Furthermore, if $0 < h < \frac{1}{2}$,

$$\|e\|_{L^2_h} \le C_1 C_2 h^2,$$

where $C_1 > 0$ is the constant from Theorem 24.46.

Proof. The consistency estimate follows from Proposition 24.44.

To obtain convergence, we recall that the consistency error $\mathcal{E}_h[u]$ and error e are related by (24.21). Therefore, $\boldsymbol{e} = h^2 \mathsf{A}^{-1} \boldsymbol{\mathcal{E}}_h[u]$ and

$$\|\boldsymbol{e}\|_2 \le h^2 \left\|\mathsf{A}^{-1}\right\|_2 \|\boldsymbol{\mathcal{E}}_h[u]\|_2 \le C_1 \|\boldsymbol{\mathcal{E}}_h[u]\|_2 \le C_1 h^{-1} \|\boldsymbol{\mathcal{E}}_h[u]\|_\infty \le C_1 C_2 h.$$

Using the fact that $\|e\|_{L^2_h} = h \|\boldsymbol{e}\|_2$, the result follows. □

Remark 24.49 (suboptimality). Once again, we get a suboptimal error estimate in the L^∞_h-norm using our L^2_h estimate:

$$\|e\|_{L^\infty_h} \le \frac{1}{h} \|e\|_{L^2_h} \le C_1 C_2 h.$$

We sharpen this estimate in the next section.

24.5.2 A Discrete Maximum Principle and Convergence in the L^∞_h-Norm

We now state and prove a Discrete Maximum Principle for two-dimensional grid functions.

Theorem 24.50 (Discrete Maximum Principle). *Let $d = 2$. Suppose that $v \in \mathcal{V}(\bar{\Omega}_h)$ is such that*

$$-\Delta_h v(\boldsymbol{x}) \le 0, \quad \forall \boldsymbol{x} \in \Omega_h.$$

Then

$$\max_{\boldsymbol{x} \in \bar{\Omega}_h} v(\boldsymbol{x}) \le \max_{\boldsymbol{x} \in \partial\Omega_h} v(\boldsymbol{x}).$$

In other words, the maximum must occur on the boundary.

Proof. To obtain a contradiction, suppose that a strict maximum occurs in the interior. If this is true, there is some $(kh, \ell h) \in \Omega_h$ such that

$$v(kj, \ell h) = \max_{(ih,jh)\in\Omega_h} v(ih, jh) > \max_{(ih,jh)\in\partial\Omega_h} v(ih, jh).$$

For simplicity, let us suppose that $2 \le k, \ell \le N - 1$. Then

$$0 \ge -h^2 \Delta_h v_{k,\ell} = -v_{k-1,\ell} - v_{k+1,\ell} - v_{k,\ell-1} - v_{k,\ell+1} + 4v_{k,\ell} \ge 0.$$

This implies that $\Delta_h w_{k,\ell} = 0$ and

$$v_{k,\ell} = \frac{1}{4}(v_{k-1,\ell} + v_{k+1,\ell} + v_{k,\ell-1} + v_{k,\ell+1}).$$

The only way to satisfy the last equation and the fact that $v_{k,\ell} \ge v_{k\pm1,\ell}, v_{k,\ell\pm1}$ is to have $v_{k,\ell} = v_{k\pm1,\ell} = v_{k,\ell\pm1}$.

We can now repeat our argument at neighboring points, and we conclude that

$$v_{k,\ell} = v_{i,j}, \quad \forall (ih, jh) \in \tilde{\Omega}_h = \Omega_h \backslash \{(h, h), (h, Nh), (Nh, h), (N, N)\}.$$

Next to the left boundary, we have, assuming that $(h, jh) \in \tilde{\Omega}_h$,

$$0 \ge -h^2 \Delta_h v_{1,j} = -v_{0,j} - v_{2,j} - v_{1,j-1} - v_{1,j+1} + 4v_{1,j} > 0,$$

because $v_{1,j} > v_{0,j}$. This is a contradiction. The other possible cases are treated similarly. □

With the help of the Discrete Maximum Principle, we can obtain stability.

Theorem 24.51 (stability). *Let $d = 2$. Given $f \in \mathcal{V}(\Omega_h)$, suppose that $w \in \mathcal{V}(\bar{\Omega}_h)$ satisfies*

$$-\Delta_h w(\boldsymbol{x}) = f(\boldsymbol{x}), \quad \forall \boldsymbol{x} \in \Omega_h.$$

Then there is some constant $C > 0$, independent of h and w, such that

$$\|w\|_{L_h^\infty} \le \max_{(ih,jh)\in\partial\Omega_h} |w_{i,j}| + C\|f\|_{L_h^\infty}.$$

Proof. The strategy, as in previous cases, is to construct a comparison function. This time, the function $\Phi\colon [0, 1]^2 \to \mathbb{R}$ is

$$\Phi(\boldsymbol{x}) = \left\| \boldsymbol{x} - \begin{bmatrix} \frac{1}{2} \\ \frac{1}{2} \end{bmatrix} \right\|_2^2 \ge 0.$$

Define the grid function $\phi_{i,j} = \Phi(ih, jh)$. Then, for all $(ih, jh) \in \Omega_h$,

$$-\Delta\Phi(ih, jh) \equiv -4 = -\Delta_h \Phi_{i,j}.$$

Define the grid function

$$\Psi_\pm = \pm w + \frac{\|f\|_{L_h^\infty}}{4}\Phi.$$

Notice that, in Ω_h, we have

$$-\Delta_h \Psi = \pm f - \|f\|_{L_h^\infty} \le 0.$$

By the Discrete Maximum Principle then, for all $(ih, jh) \in \Omega_h$,

$$\pm w_{i,j} \leq \Psi_{i,j} \leq \max_{\partial\Omega_h} \Psi \leq \max_{\partial\Omega_h} w + \frac{\|f\|_{L_h^\infty}}{8},$$

as we needed to show. □

Corollary 24.52 (stability). *Let $d = 2$. Suppose that $v \in \mathcal{V}(\bar{\Omega}_h)$ satisfies*

$$-\Delta_h v(\mathbf{x}) = 0, \quad \forall \mathbf{x} \in \Omega_h.$$

Then

$$\|v\|_{L_h^\infty} \leq \max_{\partial\Omega_h} |v|.$$

Proof. See Problem 24.17. □

With these tools, we can now conclude convergence.

Corollary 24.53 (convergence). *Suppose that $u \in C^4(\overline{\Omega})$ is a classical solution to the two-dimensional Poisson problem (24.17). Let $N \in \mathbb{N}$ and $w \in \mathcal{V}_0(\bar{\Omega}_h) \longleftrightarrow \mathbf{w} \in \mathbb{R}^{N^2}$ be a solution to the finite difference problem (24.18). Let $e \in \mathcal{V}_0(\bar{\Omega}_h) \longleftrightarrow \mathbf{e} \in \mathbb{R}^{N^2}$ be its error. Then there is a constant $C_3 > 0$, independent of h, such that*

$$\|e\|_{L_h^\infty} \leq C_2 C_3 h^2,$$

where $C_2 > 0$ is the local truncation error constant from Theorem 24.48.

Proof. Repeat the proof of Corollary 24.32. □

Problems

24.1 Prove Proposition 24.6.
24.2 Prove Proposition 24.9.
24.3 Complete the proof of Theorems 24.25 and 24.45.
24.4 Provide all the details for Example 24.16.
24.5 Provide all the details for Example 24.17.
24.6 Prove Proposition 24.14.
24.7 Prove Proposition 24.20.
24.8 This problem is about properties of the L_h^p-norms introduced in Example 24.15 for $d = 1$ and Definition 24.43 for $d = 2$. Let $p, q, r \in [1, \infty]$. Show that, for every $w \in \mathcal{V}_0(\bar{\Omega}_h)$, we have:

a) *Discrete embedding:* If $p < q$, then

$$\|w\|_{L_h^p} \leq \|w\|_{L_h^q}.$$

b) *Interpolation:* If $p \leq r \leq q$, then

$$\|w\|_{L_h^r} \leq \|w\|_{L_h^p}^{\theta} \|w\|_{L_h^q}^{1-\theta}, \quad \frac{1}{r} = \frac{\theta}{p} + \frac{1-\theta}{q}.$$

c) *Inverse inequality:* If $p < q$, then

$$\|w\|_{L_h^q} \leq h^{d/q - d/p} \|w\|_{L_h^p}.$$

24.9 Show that (24.12) and (24.13) satisfy (24.11).

24.10 Prove Theorem 24.34.

24.11 Prove Lemma 24.36.

24.12 Let the operator L be defined in (24.10) with $b(x) \geq 0$ and $a(x) \geq \lambda > 0$ for all $x \in [0, 1]$. Consider a *downwind* discretization, i.e., one where the first derivative is discretized using forward differences,

$$L_h v_i = -a_i \Delta v_i + b_i \delta_h v_i + c_i v_i.$$

Show that this method is monotone, in the sense that its coefficients satisfy the assumptions of Lemma 24.36, provided that

$$\frac{\|b\|_{L^\infty(0,1)} h}{\lambda} \leq 1.$$

24.13 Let the operator L be defined in (24.10) with $b(x) \geq 0$ and $a(x) \geq \lambda > 0$ for all $x \in [0, 1]$. Consider a *centered* discretization, i.e., one where the second derivative is discretized using centered differences,

$$L_h v_i = -a_i \Delta v_i + b_i \mathring{\delta}_h v_i + c_i v_i.$$

Show that this method is monotone, in the sense that its coefficients satisfy the assumptions of Lemma 24.36, provided that

$$\frac{\|b\|_{L^\infty(0,1)} h}{2\lambda} \leq 1.$$

24.14 Prove Proposition 24.44.

24.15 Prove Theorem 24.45.

Hint: Review the one-dimensional case.

24.16 Prove Theorem 24.46.

24.17 Prove Corollary 24.52.

24.18 The purpose of this problem is to introduce the *method of undetermined coefficients* to construct finite difference operators with a prescribed stencil that are consistent with some differential operator. Thus, assume that we wish to construct a finite difference operator $\mathcal{D}_h \colon \mathcal{V}_{\bar{N}}(\mathbb{Z}_h) \to \mathcal{V}_{\bar{N}}(\mathbb{Z}_h)$ that is consistent, on $C_b(\mathbb{R})$, with the second derivative and has stencil $\{-2, -1, 0, 1, 2\}$. The method of undetermined coefficients consists of expressing, for $v \in \mathcal{V}_{\bar{N}}(\mathbb{Z}_h)$, $\mathcal{D}_h v_i$ as

$$\mathcal{D}_h v_i = A v_{i-2} + B v_{i-1} + C v_i + D v_{i+1} + E v_{i+2},$$

where the coefficients $\{A, B, C, D, E\}$ are to be determined.

a) By using Taylor expansions, find the coefficients.

b) Show that there is $k \geq 2$ such that, if $u \in C_b^k(\mathbb{R})$, the finite difference operator $\mathcal{D}_h$ is consistent, on $C_b(\mathbb{R})$, with the second derivative to order four.

c) What is the minimal value of k for the previous item to hold?

24.19 Here we will construct higher order approximations to the solution to (24.4).

a) Armed with the solution to Problem 24.18, propose an FDM whose consistency error satisfies

$$\|\mathcal{E}_h[u]\|_{L_h^\infty} \le Ch^4$$

for some constant $C > 0$ independent of h. This method, however, has a major drawback compared with (24.6). What is this drawback?

b) Another approach to obtain a higher order discretization, while preserving the stencil, is to use the differential equation. We will sketch this procedure, and your job will be to fill in the blanks.

   i) Observe that, during the course of the solution to Problem 24.7, we actually obtained that

$$\Delta_h v = v'' + \frac{h^2}{12} v^{(4)} + \mathcal{O}(h^4).$$

   ii) If u and f are sufficiently smooth, using the differential equation, we see that

$$u^{(4)} = -f'';$$

   thus, the method: Find $w \in \mathcal{V}_0(\bar{\Omega}_h)$ such that

$$-\Delta_h w = f + \frac{h^2}{12} f'' \quad \text{in } \Omega_h$$

   will satisfy

$$\|\mathcal{E}_h[u]\|_{L_h^\infty} \le Ch^4.$$

24.20 The purpose of this problem is to treat other types of boundary conditions besides Dirichlet. Consider the Neumann problem

$$-u''(x) + u(x) = f(x), \quad x \in \Omega = (0,1), \qquad -u'(0) = g_1, \qquad u'(1) = 1.$$

A finite difference approximation will read: Find $w \in \mathcal{V}(\bar{\Omega}_h)$ such that

$$-\Delta_h w(x) + w(x) = f_h(x), \quad x \in \Omega_h,$$

where $f_h \in \mathcal{V}(\Omega_h)$ satisfies $f_h(x_i) = f(x_i)$. The issue at hand is to discretize the boundary conditions:

a) Since the boundary conditions are first-order derivatives, one can use left and right finite difference approximations, respectively. What will be the effect of this on the consistency error?

b) Another option is to use the method of undetermined coefficients, introduced in Problem 24.18, to construct, for instance,

$$Av_0 + Bv_1 + Cv_2 = -v'(0) + \mathcal{O}(h^2).$$

Find $\{A, B, C\}$ and a corresponding discretization for the boundary condition at $x = 1$, so that the consistency error has order two. This approach, however, destroys a fundamental structural property of the FDM. Which one?

c) As in Problem 24.19, one can use the differential equation to construct a method whose consistency order is higher order, but the ensuing matrix is still tridiagonal. Do this.

24.21 The estimates obtained in this chapter allow us to provide convergence in other L^p_h-norms. Consider problem (24.4) and its finite difference approximation given by (24.6).

a) Show that, for $p \in (1,2)$,

$$\|\pi_h u - w\|_{L^p_h} \leq Ch^2.$$

Hint: Recall the embedding properties given in Problem 24.8.

b) Show that, for $p \in (2,\infty)$,

$$\|\pi_h u - w\|_{L^p_h} \leq Ch^2.$$

Hint: Recall the interpolation inequalities given in Problem 24.8.

24.22 Prove the following discrete embedding: There is a constant $C > 0$, independent of $h > 0$, such that, for all $v \in \mathcal{V}_0(\bar{\Omega}_h)$,

$$\|v\|_{L^\infty_h} \leq C\|v\|_{H^1_h}.$$

24.23 Prove the following inverse inequality: There is a constant $C_I > 0$, independent of $h > 0$, such that, for all $v \in \mathcal{V}_0(\bar{\Omega}_h)$,

$$\|v\|_{H^1_h} \leq C_I h^{-1}\|v\|_{L^2_h}.$$

25 Finite Element Methods for Elliptic Problems

The purpose of this chapter is to present the most rudimentary facts of the theory of finite element methods for the approximation of partial differential equations. Finite element methods have established themselves as the de facto method to approximate the solution to most problems in the sciences and engineering. This is due to several natural advantages:

- Its formulation is based on integral — read weak — formulations of the problems to be discretized. As we have seen, weak formulations allow for more general problem data.
- The idea of decomposing the domain into a finite number of pieces (the elements) and treating the properties and behavior of each piece as known and fixed not only has roots in the physical origins of many problems but also allows us to easily treat general geometries, unstructured meshes, and local mesh refinements without great complication; see, for example, Figure 25.4.
- The approximate solution to our problem is a piecewise polynomial function, i.e., a polynomial on each of the elements. Having an actual function instead of, say, a grid one is sometimes desirable, as this function can be easily evaluated, differentiated, etc., without any special considerations.
- As opposed to a finite difference methodology, in finite elements one is not discretizing the partial differential operators, but rather searching for an approximation of the solution in a subspace of the solution space. For this reason, many of the properties of the discrete problem are automatically inherited from the continuous one: finite element methods are, in their simplest form, automatically consistent and stable. The only remaining question is then how well elements of the said subspace can approximate generic members of the solution space.

Several works can be credited with originally developing the idea of finite element methods, and we will not make any attempt here at a historical account. We refer the interested reader to, for instance, [67, 92] for a detailed history of the origins of the finite element method and its analysis.

The organization of this chapter will be as follows. First, we will present the general theory of Galerkin methods, which serves as a tool not only for the analysis of finite element methods but for other techniques as well, like the spectral methods of Chapter 26 and classes of collocation methods discussed in Chapter 27. Then we proceed with the construction and detailed analysis of a finite element method for a one-dimensional problem. The two-dimensional case is illustrated in the last section, but to avoid technicalities we present several results without proofs. The reader is referred to [9, 10, 17, 30] for a full account of the theory.

25.1 The Galerkin Method

As we mentioned above, the finite element method aims to approximate the weak solution to an elliptic boundary value problem. Let us recall that, in Section 23.2.3, the general theory of variational problems in Hilbert spaces was used to provide an analysis of weak solutions. With the help of the Lax–Milgram Lemma (Theorem 23.10), we obtained existence, uniqueness, and the stability of solutions. The Galerkin method, in its general form, aims to approximate the solution to a variational problem in a Hilbert space via approximations that lie in (finite-dimensional) subspaces. We have already seen incarnations of this principle when we studied the conjugate gradient method for the iterative solution of linear systems of algebraic equations, see Definition 7.27, and when we studied Galerkin methods for ordinary differential equations in Chapter 22. Here, we will present the general theory behind this idea.

Let us then operate in the setting of Section 23.2.3 and assume that $\mathcal{H}$ is a Hilbert space, $\mathcal{A}\colon \mathcal{H}\times\mathcal{H} \to \mathbb{R}$ is bounded and of coercive bilinear form, and $F \in \mathcal{H}'$. We seek to approximate the solution to: Find $u \in \mathcal{H}$ such that

$$\mathcal{A}(u, v) = F(v), \quad \forall v \in \mathcal{H}. \tag{25.1}$$

Let us now define Galerkin approximations in this general setting.

Definition 25.1 (Galerkin approximation[1]). Let $n \in \mathbb{N}$ and $\mathcal{H}_n \leq \mathcal{H}$, with $\dim \mathcal{H}_n = n$. We say that $u_n \in \mathcal{H}_n$ is a **Galerkin approximation** to u, solution of (25.1), if and only if

$$\mathcal{A}(u_n, v_n) = F(v_n), \quad \forall v_n \in \mathcal{H}_n.$$

We recall also that, if $\mathcal{A}$ is symmetric, we can define the energy

$$E(v) = \frac{1}{2}\mathcal{A}(v, v) - F(v), \tag{25.2}$$

and Theorem 23.10 showed that $u \in \mathcal{H}$ solves (25.1) if and only if it minimizes E over $\mathcal{H}$. The idea behind a Ritz approximation is to minimize this energy over a subspace.

Definition 25.2 (Ritz approximation[2]). Let $n \in \mathbb{N}$ and $\mathcal{H}_n \leq \mathcal{H}$, with $\dim \mathcal{H}_n = n$. We say that $u_n \in \mathcal{H}_n$ is a **Ritz approximation** to u, solution of (25.1), if and only if

$$u_n = \operatorname*{argmin}_{v_n \in \mathcal{H}_n} E(v_n).$$

Before we begin to provide an analysis of these methods, we must realize what we have gained with a Galerkin, or Ritz, approach. Let us introduce a basis of $\mathcal{H}_n$,

[1] Named in honor of the Russian mathematician and engineer Boris Grigorievich Galerkin (1871–1945).

[2] Named in honor of the Swiss theoretical physicist Walther Heinrich Wilhelm Ritz (1878–1909).

$$B_n = \{\phi_i\}_{i=1}^n, \qquad \mathcal{H}_n = \operatorname{span} B_n.$$

Then every $v_n \in \mathcal{H}_n$ has a unique representation of the form

$$v_n = \sum_{i=1}^n v_i \phi_i \quad \longleftrightarrow \quad \boldsymbol{v} = [v_1, \ldots, v_n]^\intercal \in \mathbb{R}^n.$$

Since this representation is dependent on the basis, we will often write

$$v_n \in \mathcal{H}_n \overset{B_n}{\longleftrightarrow} \boldsymbol{v} \in \mathbb{R}^n$$

to emphasize this connection. The object on the right is called the coordinate vector.

Now let us use such representation in Definition 25.1 and notice that, by linearity, it is sufficient to set $v_n = \phi_j \in B_n$. We obtain

$$\mathcal{A}(u_n, \phi_j) = \mathcal{A}\left(\sum_{i=1}^n u_i \phi_i, \phi_j\right) = \sum_{i=1}^n \mathcal{A}(\phi_i, \phi_j) u_i = F(\phi_j), \quad j = 1, \ldots, n.$$

We have obtained a linear system of equations for $\boldsymbol{u} \overset{B_n}{\longleftrightarrow} u_n$, namely

$$\mathsf{A}\boldsymbol{u} = \boldsymbol{f},$$

where $f_i = F(\phi_i)$, for $i = 1, \ldots, n$. All the properties of this system are contained in the so-called *stiffness matrix*.

Definition 25.3 (stiffness matrix)**.** Suppose that $\mathcal{H}_n$ is a finite-dimensional subspace of $\mathcal{H}$ with basis $B_n = \{\phi_1, \ldots, \phi_n\}$. The square matrix $\mathsf{A} = [a_{i,j}] \in \mathbb{R}^{n\times n}$ with entries

$$a_{i,j} = \mathcal{A}(\phi_j, \phi_i), \quad i, j \in \{1, \ldots, n\}$$

is called the **stiffness matrix** relative to the basis B_n. The vector $\boldsymbol{f} = [f_i] \in \mathbb{R}^n$ with entries

$$f_i = F(\phi_i), \qquad i = 1, \ldots, n$$

is called the **load vector** relative to the basis B_n.

It turns out that many of the properties of the bilinear form $\mathcal{A}$ imply useful properties of the stiffness matrix, as the following result shows.

Lemma 25.4 (properties of A)**.** *Let $\mathcal{H}_n$ be a finite-dimensional subspace of $\mathcal{H}$. The stiffness matrix $\mathsf{A} = [a_{i,j}] \in \mathbb{R}^{n\times n}$ relative to the basis $B_n = \{\phi_1, \ldots, \phi_n\}$ is positive definite. In addition, if $\mathcal{A}$ is symmetric, so is A.*

Proof. Let $v_n \in \mathcal{H}_n \overset{B_n}{\longleftrightarrow} \boldsymbol{v} \in \mathbb{R}^n$ be arbitrary. Then, using the coercivity of $\mathcal{A}(\cdot, \cdot)$, we get

$$\boldsymbol{v}^\intercal \mathsf{A} \boldsymbol{v} = \sum_{i=1}^n \sum_{j=1}^n \mathcal{A}(\phi_j, \phi_i) v_i v_j = \mathcal{A}(v, v) \geq \alpha \|v\|_{\mathcal{H}}^2.$$

The symmetry of A is obtained by observing that, since $\mathcal{A}$ is symmetric,

$$a_{i,j} = \mathcal{A}(\phi_j, \phi_i) = \mathcal{A}(\phi_i, \phi_j) = a_{j,i}.$$

□

The stiffness matrix is invertible; therefore, we obtain existence and uniqueness of Galerkin approximations. The following result should be compared with Theorem 7.28.

Theorem 25.5 (uniform well-posedness). *Suppose that $\mathcal{H}_n$ is a finite-dimensional subspace of $\mathcal{H}$, with basis $B_n = \{\phi_1, \ldots, \phi_n\}$. There is a unique $u_n \in \mathcal{H}_n \stackrel{B_n}{\longleftrightarrow} \boldsymbol{u} \in \mathbb{R}^n$ that is a Galerkin approximation of the solution to (25.1), which, moreover, satisfies*

$$\mathsf{A}\boldsymbol{u} = \boldsymbol{f}.$$

If, in addition, $\mathcal{A}$ is symmetric, then $u_n \in \mathcal{H}_n$ is the Galerkin approximation to u if and only if it is the Ritz approximation to u. Finally, we have the estimate

$$\|u_n\|_{\mathcal{H}} \leq \frac{1}{\alpha} \|F\|_{\mathcal{H}'}.$$

Proof. See Problem 25.1. □

25.1.1 Cèa's Lemma and Galerkin Orthogonality

Having obtained that, regardless of the dimension $n \in \mathbb{N}$ and the choice of subspace $\mathcal{H}_n$, a Galerkin approximation is always well posed, we can proceed to obtain an error analysis for it. The following results, which usually bear the names of *Céa's Lemma* and *Galerkin orthogonality*, express that a Galerkin approximation is, in a sense, the best possible. We have seen instances of these facts before; see Proposition 7.29.

Theorem 25.6 (Céa's Lemma[3]). *Suppose that $\mathcal{H}_n$ is a finite-dimensional subspace of $\mathcal{H}$, with basis $B_n = \{\phi_1, \ldots, \phi_n\}$. Suppose that $u \in \mathcal{H}$ is the unique solution to (25.1) and $u_n \in \mathcal{H}_n \stackrel{B_n}{\longleftrightarrow} \boldsymbol{u} \in \mathbb{R}^n$ is the unique Galerkin approximation in the sense of Definition 25.1. Then*

$$\mathcal{A}(u - u_n, v_n) = 0, \quad \forall v_n \in \mathcal{H}_n, \tag{25.3}$$

which is called Galerkin orthogonality. *Furthermore, we have the* quasi-best approximation *property*

$$\|u - u_h\|_{\mathcal{H}} \leq \frac{M}{\alpha} \min_{v \in \mathcal{H}_n} \|u - v\|_{\mathcal{H}},$$

where M and α denote the boundedness and coercivity constants, respectively, of the bilinear form $\mathcal{A}$.

Proof. Notice that, since $\mathcal{H}_n \subset \mathcal{H}$, we can set, in (25.1), the test function $v = v_n \in \mathcal{H}_n$. Subtracting this from the definition of Galerkin approximation, we then obtain the Galerkin orthogonality relation (25.3).

[3] Named in honor of the French mathematician Jean Céa (1932–).

Now, for $v_n \in \mathcal{H}_n$ arbitrary, using coercivity and Galerkin orthogonality, we have

$$\begin{aligned}\alpha\|u-u_n\|_{\mathcal{H}}^2 &\le \mathcal{A}(u-u_n, u-u_n)\\ &= \mathcal{A}(u-u_n, u-v_n) + \mathcal{A}(u-u_n, v_n-u_n)\\ &= \mathcal{A}(u-u_n, u-v_n)\\ &\le M\|u-u_n\|_{\mathcal{H}}\|u-v_n\|_{\mathcal{H}},\end{aligned}$$

where in the last step we used the boundedness of $\mathcal{A}$. Since $v_n \in \mathcal{H}_n$ is arbitrary,

$$\|u-u_n\|_{\mathcal{H}} \le \frac{M}{\alpha}\inf_{v_n\in\mathcal{H}_n}\|u-v_n\|_{\mathcal{H}}.$$

But, of course, the infimum is achieved at $v_n = u_n$,

$$\|u-u_n\|_{\mathcal{H}} = \frac{M}{\alpha}\min_{v_n\in\mathcal{H}_n}\|u-v_n\|_{\mathcal{H}},$$

which is the claimed quasi-best approximation property. □

Remark 25.7 (energy norm)**.** Notice that if $\mathcal{A}$ is symmetric, then it defines the so-called energy norm

$$\|v\|_E = \mathcal{A}(v,v)^{1/2}, \quad \forall v \in \mathcal{H}.$$

The proof of the previous result shows that, in this case, we actually have the *best approximation* property

$$\|u-u_h\|_E = \min_{v\in\mathcal{H}}\|u-v\|_E.$$

Remark 25.8 (best approximation)**.** Notice that the content of Céa's Lemma reduced the analysis of a Galerkin approximation to a question in approximation theory. In other words, after this result, it is no longer of relevance that we are trying to approximate the solution to (25.1). The only thing that matters is how well an object in $\mathcal{H}$ can be approximated by elements in the subspace $\mathcal{H}_n$.

25.2 The Finite Element Method in One Dimension

In this section, we will present the construction of a finite element method and provide all the details of its analysis. To keep the theoretical constructions, and technical details, to a minimum, we will do so in a simple one-dimensional problem. The two-dimensional case will be presented, mostly without proofs, in the following section. For a complete analysis of the finite element method, with various degrees of generality and abstraction, we refer the reader to [9, 10, 17, 30].

Let $d = 1$, $\Omega = (0,1)$, and consider the linear differential operator $Lv = -(av')'$, for some function $a \in C(\bar{\Omega})$, such that, for some $\alpha, M > 0$,

$$0 < \alpha \le a(x) \le M, \quad \forall x \in [0,1].$$

We will consider the Dirichlet problem for this operator, i.e., given $f \in L^2(\Omega)$, we seek a function $u \in H_0^1(\Omega)$ such that

$$\mathcal{A}(u, v) = \int_0^1 a u' v' \, dx = \int_0^1 f v \, dx, \quad \forall v \in H_0^1(\Omega). \tag{25.4}$$

Notice that, owing to Corollary 23.11, this problem is well posed. We can thus study approximations to its solution.

The finite element method is nothing but a Galerkin approximation with a particular choice of subspace and basis. The construction of a finite element space, in this setting, is detailed below.

Definition 25.9 (mesh). Let $\Omega = (0, 1)$, a **triangulation** or **mesh** of Ω is a partition of Ω into subintervals I_i, which we call **elements**:

$$\mathscr{T}_h = \{I_i\}_{i=0}^N, \qquad I_i = (x_i, x_{i+1}), \qquad h_i = x_{i+1} - x_i,$$

where the **nodes** are given by

$$0 = x_0 < x_1 < \cdots < x_{N+1} = 1.$$

Notice that, in the previous definition, we are not assuming that the nodes are equally spaced, i.e., h_i is not constant. We set

$$h = \max_{i=0,\ldots,N} h_i,$$

which we call the mesh size. Subordinate to the mesh $\mathscr{T}_h$ we define a finite element space.

Definition 25.10 (finite element spaces). Given a mesh $\mathscr{T}_h$, we define the finite element spaces of **continuous piecewise linear** functions

$$\mathscr{S}^{1,0}(\mathscr{T}_h) = \left\{ v_h \in C([0,1]) \mid v_{h|I_j} \in \mathbb{P}_1, \ 0 \le j \le N \right\},$$
$$\mathscr{S}_0^{1,0}(\mathscr{T}_h) = \mathscr{S}^{1,0}(\mathscr{T}_h) \cap H_0^1(\Omega).$$

The **Lagrange nodal basis**[4] of $\mathscr{S}^{1,0}(\mathscr{T}_h)$ is given by

$$\{\phi_i\}_{i=0}^{N+1} \subset \mathscr{S}^{1,0}(\mathscr{T}_h), \qquad \phi_i(x_j) = \delta_{i,j},$$

and the **Lagrange nodal basis** of $\mathscr{S}_0^{1,0}(\mathscr{T}_h)$ is $\{\phi_i\}_{i=1}^N$.

Remark 25.11 (hat function). A typical depiction of a function ϕ_i is given in Figure 25.1. This motivates us to call them *hat functions*.

Remark 25.12 (implementation). Recall that the *support* of a function is defined as

$$\operatorname{supp} v = \overline{\{x \mid v(x) \neq 0\}}.$$

Notice that

$$\operatorname{supp} \phi_i \cap \operatorname{supp} \phi_j \neq \emptyset \quad \Longleftrightarrow \quad |i - j| \le 1.$$

This means that the stiffness matrix will be very sparse; in fact, tridiagonal in the present case. In addition, because of the way the basis functions are defined, when

[4] Named in honor of the Italian, later naturalized French, mathematician and astronomer Joseph-Louis Lagrange (1736–1813).

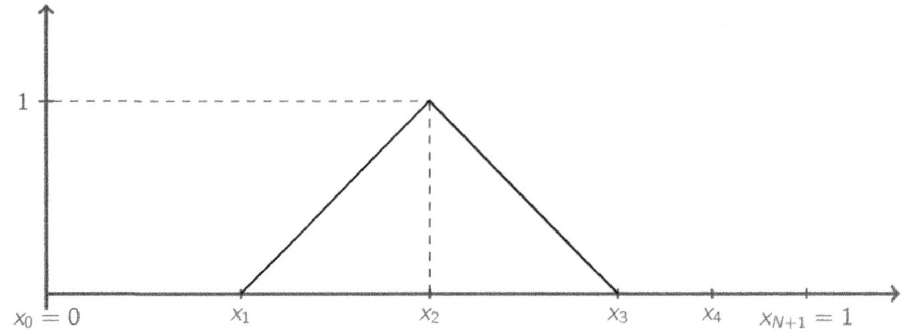

Figure 25.1 A one-dimensional hat function, i.e., a member of the Lagrange nodal basis for $\mathscr{S}^{1,0}(\mathscr{T}_h)$.

computing the entries of the stiffness matrix or load vector, we can subdivide the integral into elements and operate locally, where the basis functions are linear. For instance,

$$[A]_{i,j} = a_{i,j} = \int_0^1 a(x)\phi_j'\phi_i' \, dx = \sum_{k=0}^{N} \int_{I_k} a(x)\phi_j'\phi_i' \, dx.$$

Now, because of how the hat functions are defined, we observe that

$$\phi_i'(x) = \begin{cases} \frac{1}{h_{i-1}}, & x_{i-1} < x < x_i, \\ -\frac{1}{h_i}, & x_i < x < x_{i+1}, \\ 0, & x \notin [x_{i-1}, x_{i+1}]. \end{cases}$$

This can be used to efficiently implement the finite element method.

Let us now proceed to analyze the finite element method in one dimension. Recall that, from Céa's Lemma (Theorem 25.6), we immediately obtain that

$$\|u - u_h\|_{H_0^1(0,1)} \le C \inf_{v_h \in \mathscr{S}_0^{1,0}(\mathscr{T}_h)} \|u - v_h\|_{H_0^1(0,1)}.$$

It is now necessary to find the *best approximation error*, i.e., the right-hand side of the previous inequality. To do so, we will construct an *interpolation operator*

$$\Pi_h \colon H_0^1(0,1) \to \mathscr{S}_0^{1,0}(\mathscr{T}_h),$$

and we will show that it has suitable approximation properties. That the interpolation operator is defined for $H_0^1(0,1)$ functions is a consequence of the following observation.

Remark 25.13 (one-dimensional embedding). In one space dimension, and in one dimension only, we have

$$H^1(\Omega) \hookrightarrow C(\overline{\Omega});$$

see Problem D.6. This means that $H^1(\Omega) \subset C(\overline{\Omega})$, and there exists a constant $C > 0$, independent of u, such that

$$\|u\|_{L^\infty(0,1)} \le C\,\|u\|_{H^1(0,1)}\,.$$

The one-dimensional interpolation operator Π_h is defined as follows.

Definition 25.14 (Lagrange interpolant). The **Lagrange nodal interpolation** operator

$$\Pi_h \colon C([0,1]) \to \mathscr{S}^{1,0}(\mathscr{T}_h)$$

is (uniquely) defined by $\Pi_h v(x_i) = v(x_i)$ for all $i = 0, \dots, N+1$.

Let us now prove error estimates for the Lagrange nodal interpolant. Many of the ideas behind the proof of the following result bear a lot of resemblance to what was done in Part II, in particular in Chapter 9. The main difference here is that we are measuring smoothness in Sobolev spaces, instead of spaces of continuously differentiable functions.

Theorem 25.15 (properties of Π_h). *Let Π_h be as in Definition 25.14. There is a constant $C > 0$, independent of the mesh spacings, such that, for all $j = 0, \dots, N$, and all $v \in H^1(0,1)$ such that $v'' \in L^2(0,1)$,*

$$\|v - \Pi_h v\|_{L^2(I_j)} \le C h_j^2 \|v''\|_{L^2(I_j)},$$
$$\|v' - \Pi_h v'\|_{L^2(I_j)} \le C h_j \|v''\|_{L^2(I_j)}.$$

Proof. To alleviate the notation, let us denote by $I = (x_l, x_r)$ a generic element of the mesh and $h = |I| = x_r - x_l$.

Consider now $v \in H^1(0,1)$ such that $v'' \in L^2(0,1)$. An argument similar to that of Problem D.6 shows that $v \in C^1([0,1])$. Define the first-order Taylor polynomial of v about x_l

$$Q_1 v(x) = v(x_l) + v'(x_l)(x - x_l) \in \mathbb{P}_1.$$

Notice that:

1. Π_h is *polynomial space preserving* in the following sense: $\Pi_h Q_1 v$, when restricted to I, equals $Q_1 v$.
2. The operator Π_h is max-norm *stable*: i.e.,

$$\|\Pi_h v\|_{L^\infty(I)} = \max\{|v(x_l)|, |v(x_r)|\} \le \|v\|_{L^\infty(I)}.$$

Now, to bound the interpolation error $v - \Pi_h v$, we proceed as follows:

$$\|v - \Pi_h v\|_{L^\infty(I)} \le \|v - Q_1 v\|_{L^\infty(I)} + \|Q_1 v - \Pi_h v\|_{L^\infty(I)} \le 2\|v - Q_1 v\|_{L^\infty(I)}.$$

By Taylor's Theorem with integral remainder, see Theorem B.39,

$$(v - Q_1 v)(x) = \int_{x_l}^{x} (x - y) v''(y) \mathrm{d}y,$$

which implies that

$$\|v - Q_1 v\|_{L^\infty(I)} \le h \int_I |v''| \,\mathrm{d}y;$$

as a consequence,

$$\begin{aligned}\|v - Q_1 v\|_{L^2(I)}^2 &= \int_I |v - Q_1 v|^2 \,\mathrm{d}x \le h\|v - Q_1 v\|_{L^\infty(I)}^2 \\ &\le h^3 \left(\int_I |v''| \,\mathrm{d}y \right)^2 \\ &\le h^4 \int_I |v''|^2 \,\mathrm{d}y,\end{aligned}$$

where, in the last step, we applied the Cauchy–Schwarz inequality. Taking square roots proves the first result.

To bound the derivatives, we must note that

$$(\Pi_h v)'(x) = \frac{1}{h} \int_I v'(y) \mathrm{d}y,$$

so that

$$(\Pi_h v - v)'(x) = \frac{1}{h} \int_I (v'(y) - v'(x)) \mathrm{d}y,$$

which implies that

$$\begin{aligned}\|(\Pi_h v - v)'\|_{L^2(I)}^2 &= \frac{1}{h^2} \int_I \left(\int_I (v'(y) - v'(x)) \mathrm{d}y \right)^2 \mathrm{d}x \\ &= \frac{1}{h^2} \int_I \left(\int_I \int_x^y v''(s) \mathrm{d}s \mathrm{d}y \right)^2 \mathrm{d}x \\ &\le \frac{1}{h^2} \int_I \left(\int_I |x - y|^{1/2} \left(\int_{\min\{x,y\}}^{\max\{x,y\}} |v''(s)|^2 \,\mathrm{d}s \right)^{1/2} \mathrm{d}y \right)^2 \mathrm{d}x.\end{aligned}$$

Now, since $x, y \in I$, we can bound $|x - y| \le h$ and

$$\int_{\min\{x,y\}}^{\max\{x,y\}} |v''(s)|^2 \,\mathrm{d}s \le \int_I |v''(s)|^2 \,\mathrm{d}s.$$

Using these upper bounds, we obtain

$$\|(\Pi_h v - v)'\|_{L^2(I)}^2 \le \frac{1}{h} \int_I |v''(s)|^2 \,\mathrm{d}s \int_I \left(\int_I \mathrm{d}y \right)^2 \mathrm{d}x \le h^2 \int_I |v''(s)|^2 \,\mathrm{d}s,$$

as we needed to show. □

Remark 25.16 (stable and space-preserving operator)**.** Notice that, in the course of the proof of Theorem 25.15, the particular form of the Lagrange interpolation operator was, ultimately, inconsequential. All that was needed was that the operator was stable and that it preserved the polynomial space. In consequence, Theorem 25.15 also holds for any other operator that satisfies these two properties.

This approximation result immediately implies a convergence estimate for finite element methods.

Corollary 25.17 (convergence of finite element method). *Let $u \in H_0^1(0,1)$ solve (25.4) and $u_h \in \mathscr{S}_0^{1,0}(\mathscr{T}_h)$ be its finite element approximation. If u is such that $u'' \in L^2(0,1)$, then we have*

$$\|u - u_h\|_{H_0^1(0,1)} \leq Ch\|u''\|_{L^2(0,1)},$$

where the constant $C > 0$ is independent of $h > 0$, u, and u_h.

Proof. By Céa's Lemma, we have that

$$\|u - u_h\|_{H_0^1(0,1)} \leq C \inf_{v_h \in \mathscr{S}_0^1(\mathscr{T}_h)} \|u - v_h\|_{H_0^1(0,1)} \leq C\|u - \Pi_h u\|_{H_0^1(0,1)},$$

where we used the Lagrange interpolation operator Π_h. Notice that if $u \in H_0^1(0,1)$, then $\Pi_h u(0) = \Pi_h u(1) = 0$, so that $\Pi_h u \in \mathscr{S}_0^{1,0}(\mathscr{T}_h)$.

By the *local* properties of the Lagrange interpolation operator given in Theorem 25.15, we see that

$$\|u - \Pi_h u\|_{H_0^1(0,1)}^2 = \sum_{j=0}^{N} \int_{I_j} |(u - \Pi_h u)'|^2 \, dx \leq C \sum_{j=0}^{N} h_j^2 \int_{I_j} |u''|^2 \, dx.$$

Using that $h = \max_j h_j$ implies the result. □

Notice that the previous result, in conjunction with the Poincaré inequality (D.2), implies that

$$\|u - u_h\|_{L^2(0,1)} \leq C_P \|u - u_h\|_{H_0^1(0,1)} \leq Ch\|u''\|_{L^2(0,1)}.$$

However, Theorem 25.15 shows that the interpolation error is $\mathcal{O}(h^2)$. How can we regain the missing power? For that, we need to study the *dual problem*, i.e., given $g \in L^2(0,1)$, we need to find $z_g \in H_0^1(0,1)$ such that

$$\mathcal{A}(v, z_g) = \int_0^1 gv dx, \quad \forall v \in H_0^1(0,1). \tag{25.5}$$

Notice that the order of the arguments in the bilinear form is switched. This is irrelevant if the bilinear form is symmetric, as in our present case. If it is not symmetric, the order is quite important.

The following result is usually known as *Aubin–Nitsche duality*[5] or *Nitsche's trick*.

Theorem 25.18 ($L^2(\Omega)$-estimate). *Assume that, for every $g \in L^2(0,1)$, there is a unique solution to the dual problem* (25.5)*; furthermore, $z_g'' \in L^2(0,1)$, with the estimate*

$$\|z_g''\|_{L^2(0,1)} \leq C\|g\|_{L^2(0,1)}$$

for some constant $C > 0$. In this case, if $u \in H_0^1(0,1)$ solves (25.4)*, it is such that $u'' \in L^2(0,1)$, and $u_h \in \mathscr{S}_0^{1,0}(\mathscr{T}_h)$ is its finite element approximation, then we have*

$$\|u - u_h\|_{L^2(0,1)} \leq Ch^2\|u''\|_{L^2(0,1)}.$$

[5] Named in honor of the French mathematician Jean-Pierre Aubin (1939–) and the German mathematician Joachim A. Nitsche (1926–1996).

Proof. Define the error $e = u - u_h \in H_0^1(0,1) \subset L^2(0,1)$. Let z_e be the solution to the dual problem (25.5) with data $g = e$. If that is the case, then we have

$$\|e\|_{L^2(0,1)}^2 = \mathcal{A}(e, z_e) = \mathcal{A}(u - u_h, z_e) = \mathcal{A}(u - u_h, z_e - \Pi_h z_e),$$

where the last equality follows from Galerkin orthogonality. Now, using the boundedness of the bilinear form, we obtain

$$\begin{aligned}\|e\|_{L^2(0,1)}^2 &= \mathcal{A}(u - u_h, z_e - \Pi_h z_e)\\ &\leq M\|u - u_h\|_{H_0^1(0,1)}\|z_e - \Pi_h z_e\|_{H_0^1(0,1)}\\ &\leq Ch^2\|u''\|_{L^2(0,1)}\|z_e''\|_{L^2(0,1)},\end{aligned}$$

where we used the convergence estimate of Corollary 25.17 and, since $z_e'' \in L^2(0,1)$, the interpolation estimates of Theorem 25.15. Using the estimate on the second derivatives of z_e then yields the result. □

Remark 25.19 (higher order elements)**.** In our presentation, we chose the space $\mathscr{S}^{1,0}(\mathscr{T}_h)$ to make the discussion as transparent as possible. It is also possible to define finite element spaces of higher order. For $p \in \mathbb{N}$, we define

$$\mathscr{S}^{p,0}(\mathscr{T}_h) = \left\{v_h \in C([0,1]) \;\middle|\; v_{h|I_j} \in \mathbb{P}_p,\ 0 \leq j \leq N\right\},$$
$$\mathscr{S}_0^{p,0}(\mathscr{T}_h) = \mathscr{S}^{p,0}(\mathscr{T}_h) \cap H_0^1(\Omega).$$

The analysis of finite element methods with these spaces follows verbatim what we have done here. The only difference is in the way that the Lagrange interpolation operator is defined. In short, provided that $u^{(p+1)} \in L^2(0,1)$, one can prove that

$$\|u - u_h\|_{H_0^1(0,1)} \leq Ch^p\|u^{(p+1)}\|_{L^2(0,1)}.$$

Remark 25.20 (spaces of variable degree)**.** The spaces $\mathscr{S}^{p,0}(\mathscr{T}_h)$ can be even further generalized and allowed to have a different polynomial degree within each element. To define them, we let $\boldsymbol{p} \in \mathbb{N}^{N+1}$, which is called the *degree vector.* Then

$$\mathscr{S}^{\boldsymbol{p},0}(\mathscr{T}_h) = \left\{v_h \in C([0,1]) \;\middle|\; v_{h|I_j} \in \mathbb{P}_{p_{j+1}},\ 0 \leq j \leq N\right\},$$

with $\mathscr{S}_0^{\boldsymbol{p},0}(\mathscr{T}_h) = \mathscr{S}^{\boldsymbol{p},0}(\mathscr{T}_h) \cap H_0^1(0,1)$. These spaces form the building block of what is known as *hp finite element methods*, where, to increase the accuracy of our numerical approximation, one is allowed to either reduce the local mesh size h_j or increase the polynomial degree p_j. The reader is referred, for instance, to [82] for an account of this methodology.

25.3 The Finite Element Method in Two Dimensions

As we saw in the one-dimensional case of the previous section, the finite element method is a particular version of the Galerkin method. More specifically, the finite element method gives a particular subspace where we seek the approximate solution, and a particular basis for it. In this section, we will present, mostly without

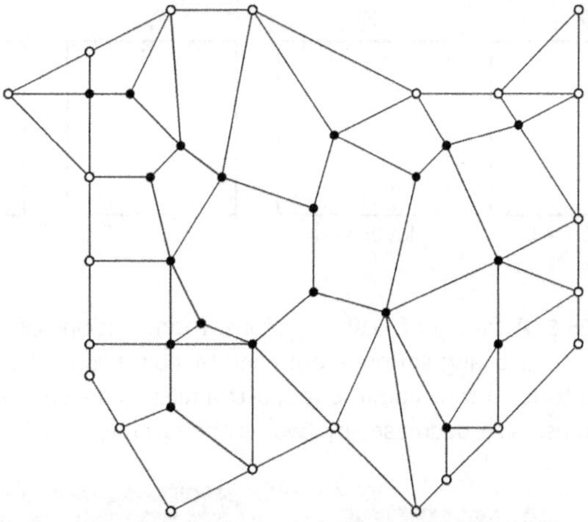

Figure 25.2 A conforming polygonal partition of a polygonal domain Ω. For every pair of elements, only one of the following possibilities holds: either $\overline{K_i} \cap \overline{K_j} = \emptyset$ or $\overline{K_i} \cap \overline{K_j} = e$, a complete edge of both K_i and K_j, or $\overline{K_i} \cap \overline{K_j} = \{x\}$, a shared vertex of K_i and K_j. The filled circles are the interior vertices and the unfilled circles are the boundary vertices.

proofs, the construction and analysis of finite element methods in two dimensions. We refer the reader to [9, 10, 17, 30] for full details, and further developments.

As in the one-dimensional case, we begin with a way to decompose our domain.

Definition 25.21 (polygonal partition)**.** Suppose that $\Omega \subset \mathbb{R}^2$ is an open, bounded, polygonal domain. Let

$$\mathscr{T}_h = \{K_i \mid i = 1, \ldots, N_e\}$$

be a disjoint collection of open subsets of Ω such that

$$\overline{\Omega} = \bigcup_{i=1}^{N_e} \overline{K_i}.$$

The members $K_i \in \mathscr{T}_h$ are called **elements**. $\mathscr{T}_h$ is called a **mesh** or **polygonal partition** of Ω if and only if each K_i is a convex polygon. $\mathscr{T}_h$ is called a **triangulation** of Ω if and only if each element K_i is a triangle. Define the **element diameters** via

$$h_i = \operatorname{diam}(K_i), \quad i = 1, \ldots, N_e.$$

The value

$$h = \max_{1 \le i \le N_e} h_i$$

is called the **global mesh size**. By the set

$$\mathcal{N}_v = \{\boldsymbol{x}_j \mid j = 1, \ldots, N_v\},$$

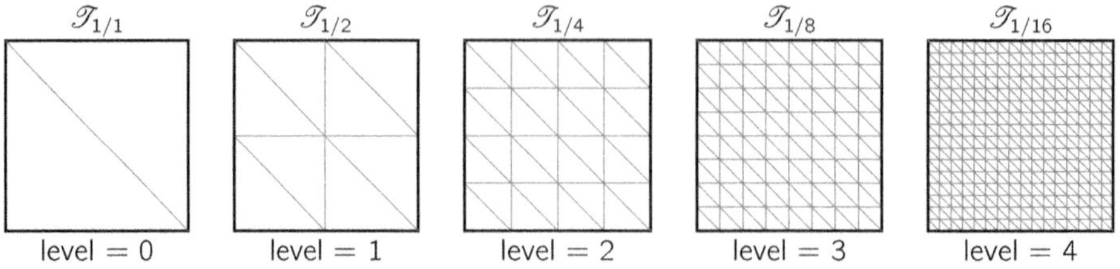

Figure 25.3 A family of nested uniform triangulations of a square. Starting at level = 0, the level = 1 triangulation is obtained by connecting the three midpoints of each triangle to form four congruent sub-triangles. This process can continue indefinitely. The global mesh size decreases by two as the level increases by one.

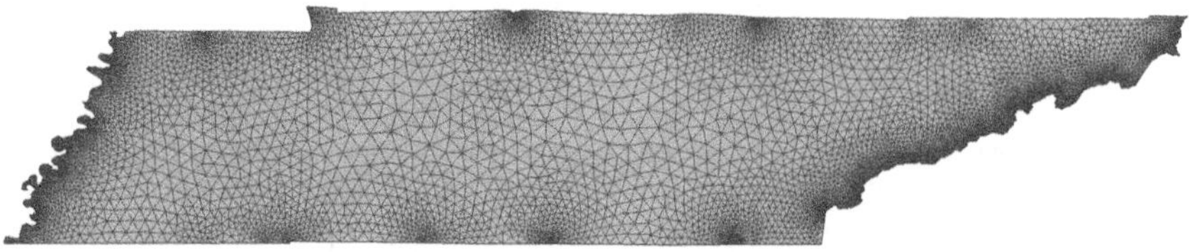

Figure 25.4 A sophisticated triangulation of the state of Tennessee (one of the 50 constituent states of the United States of America).

we denote the set of all **vertices** of $\mathscr{T}_h$, i.e., all the vertex points of the polygons K_i. By

$$\mathcal{N}_v^i = \mathcal{N}_v \cap \Omega = \left\{ \mathbf{x}_j \;\middle|\; j = 1, \ldots, N_v^i \right\},$$

we denote the set of all **interior vertices**. The set $\mathcal{N}_v \setminus \mathcal{N}_v^i$ is the set of **boundary vertices**.

Observe that, from the definition above, a triangulation is a polygonal partition.

Definition 25.22 (conforming partition)**.** A polygonal partition $\mathscr{T}_h$ is called **conforming** if and only if for every pair of elements, only one of the following possibilities holds:

1. $\overline{K_i} \cap \overline{K_j} = \emptyset$,
2. $\overline{K_i} \cap \overline{K_j} = e$, a complete edge of both K_i and K_j, or
3. $\overline{K_i} \cap \overline{K_j} = \{\mathbf{x}\}$, a shared vertex of K_i and K_j.

We show a conforming polygonal partition of a two-dimensional polygonal domain in Figure 25.2. A family of nested triangulations of a square domain is shown in Figure 25.3. In Figure 25.4, we exhibit a sophisticated triangulation of the state of Tennessee, which shows the flexibility of finite element meshing as a means to discretize arbitrary bounded polygonal domains.[6]

We now introduce finite element spaces related to triangulations.

[6] Of course, the state of Tennessee, arguably, does not have a polygonal boundary, especially along the Mississippi River. In such a case, we often approximate the shape with a polygon.

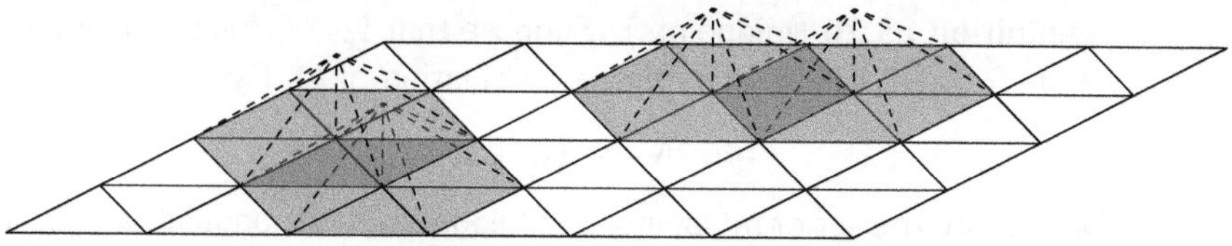

Figure 25.5 Some of the hat basis functions from $\mathscr{S}_0^{1,0}(\mathscr{T}_h)$ subordinate to a uniform conforming triangulation of a rectangle. The supports of the basis functions are the grey shaded triangles in the mesh. The dark grey triangles indicate the regions where the supports intersect.

Definition 25.23 (finite element space). Suppose that $p \in \mathbb{N}$ and $\Omega \subset \mathbb{R}^2$ is an open polygonal domain. Let $\mathscr{T}_h = \{K_i\}$ be a conforming triangulation of Ω. Define

$$\mathscr{S}^{p,0}(\mathscr{T}_h) = \left\{ v \in C(\overline{\Omega}) \middle| v_{|K} \in \mathbb{P}_p, \ \forall K \in \mathscr{T}_h \right\}.$$

The set $\mathscr{S}^{p,0}(\mathscr{T}_h)$ is called the **piecewise polynomial (of degree p) finite element space**. By the set

$$\mathscr{S}_0^{p,0}(\mathscr{T}_h) = \left\{ v \in \mathscr{S}^{p,0}(\mathscr{T}_h) \middle| v_{|\partial\Omega} = 0 \right\},$$

we denote the subspace of functions that vanish on the boundary.

These spaces are suitable to approximate weak solutions to second-order elliptic equations, as the following result shows.

Theorem 25.24 (embedding). *Suppose that $p \in \mathbb{N}$ and $\Omega \subset \mathbb{R}^2$ is an open polygonal domain. Let $\mathscr{T}_h = \{K_i\}$ be a conforming triangulation of Ω. Then $\mathscr{S}^{p,0}(\mathscr{T}_h)$ is a subspace of $H^1(\Omega)$ and $\mathscr{S}_0^{p,0}(\mathscr{T}_h)$ is a subspace of $H_0^1(\Omega)$.*

Proof. See [9, 10]. □

These subspaces are finite dimensional. In fact, we can construct bases for them.

Theorem 25.25 (basis of $\mathscr{S}^{1,0}(\mathscr{T}_h)$). *Suppose that $\Omega \subset \mathbb{R}^2$ is an open polygonal domain and $\mathscr{T}_h = \{K_i\}$ is a conforming triangulation of Ω. Then*

$$\dim(\mathscr{S}^{1,0}(\mathscr{T}_h)) = N_v, \qquad \dim(\mathscr{S}_0^{1,0}(\mathscr{T}_h)) = N_v^i.$$

In particular, defining $\phi_k \in \mathscr{S}_0^{1,0}(\mathscr{T}_h)$ via

$$\phi_k(\boldsymbol{\zeta}_j) = \delta_{j,k}, \quad \forall \boldsymbol{\zeta}_j \in \mathcal{N}_v^i,$$

we see that $\{\phi_1, \dots, \phi_{N_v^i}\}$ is a basis for $\mathscr{S}_0^{1,0}(\mathscr{T}_h)$. The basis for $\mathscr{S}^{1,0}(\mathscr{T}_h)$ is constructed similarly.

Proof. See Problem 25.10. □

The basis functions for $\mathscr{S}_0^{1,0}(\mathscr{T}_h)$ are called hat functions and are illustrated in Figure 25.5. The construction of the bases for higher order spaces, $\mathscr{S}^{p,0}(\mathscr{T}_h)$, with $p \geq 2$, is usually accomplished by adding more nodes.

But, when we make such an approximation, we incur an error, whose estimation is beyond the scope of the text. See, for example, [10] for a discussion of this advanced topic.

Definition 25.26 (midpoints). Suppose that $\Omega \subset \mathbb{R}^2$ is an open polygonal domain and $\mathscr{T}_h = \{K_i\}$ is a conforming triangulation of Ω. By

$$\mathcal{N}_m = \{\boldsymbol{\xi}_j \mid j = 1, \dots, N_m,\},$$

we denote the set of midpoints of all edges in the triangulation, i.e., the **midpoints set**. By

$$\mathcal{N}_m^i = \mathcal{N}_m \cap \Omega = \{\boldsymbol{\xi}_j \mid j = 1, \dots, N_m^i\},$$

we denote the set of all interior edge midpoints, i.e., the **interior midpoints set**. The set $\mathcal{N}_m \backslash \mathcal{N}_m^i$ is the **boundary midpoints set**.

Theorem 25.27 (basis of $\mathscr{S}^{2,0}(\mathscr{T}_h)$). *Suppose that $\Omega \subset \mathbb{R}^2$ is an open polygonal domain and $\mathscr{T}_h = \{K_i\}$ is a conforming triangulation of Ω. Then*

$$\dim(\mathscr{S}^{2,0}(\mathscr{T}_h)) = N_v + N_m, \quad \dim(\mathscr{S}_0^{2,0}(\mathscr{T}_h)) = N_v^i + N_m^i.$$

In particular, defining $\phi_k \in \mathscr{S}_0^{2,0}(\mathscr{T}_h)$ via

$$\phi_k(\boldsymbol{\zeta}_j) = \delta_{j,k}, \quad \forall \boldsymbol{\zeta}_j \in \mathcal{N}_v^i \cup \mathcal{N}_m^i,$$

we see that $\{\phi_1, \dots, \phi_{N_v^i+N_m^i}\}$ is a basis for $\mathscr{S}_0^{2,0}(\mathscr{T}_h)$. The basis for $\mathscr{S}^{2,0}(\mathscr{T}_h)$ is constructed similarly.

Proof. See Problem 25.11. Figure 25.6 gives an illustration of this construction. □

Remark 25.28 (nodal basis). The bases that we constructed in Theorems 25.25 and 25.27 are called *Lagrange nodal bases.* These have the property that the basis elements satisfy $\phi_k(\boldsymbol{\zeta}_j) = \delta_{j,k}$, for $1 \le j, k \le N$, where N is the number of elements in the basis, and $\{\boldsymbol{\zeta}_j\}$ is the *Lagrange nodal set.* We have only defined this basis for $\mathscr{S}_0^{p,0}(\mathscr{T}_h)$ with $p = 1, 2$, where Ω is a polygonal set. But we can construct this type of basis for any $p \in \mathbb{N}$; see, for example, [10].

Having defined finite element spaces, we can use them to approximate weak solutions to elliptic problems. Let us illustrate this in the case of the Poisson problem. Thus, let $\Omega \subset \mathbb{R}^2$ be a bounded polygonal domain, $f \in L^2(\Omega)$, and we seek $u \in H_0^1(\Omega)$ such that

$$\mathcal{A}(u, v) = \int_\Omega \nabla u \cdot \nabla v \, d\boldsymbol{x} = \int_\Omega f v \, d\boldsymbol{x} = F(v), \quad \forall v \in H_0^1(\Omega). \tag{25.6}$$

The finite element method is then a Galerkin method, where we use as a subspace $\mathscr{S}_0^{p,0}(\mathscr{T}_h)$ for some $p \in \mathbb{N}$. Thus, we will seek $u_h \in \mathscr{S}_0^{p,0}(\mathscr{T}_h)$ such that

$$\mathcal{A}(u_h, v_h) = F(v_h), \quad \forall v_h \in \mathscr{S}_0^{p,0}(\mathscr{T}_h). \tag{25.7}$$

Existence and uniqueness of discrete solutions, as well as a quasi-best approximation result

$$\|u - u_h\|_{H_0^1(\Omega)} \le C \inf_{v_h \in \mathscr{S}_0^{p,0}(\mathscr{T}_h)} \|u - v_h\|_{H_0^1(\Omega)},$$

follow from the general theory described in Section 25.1. It remains then to provide error estimates.

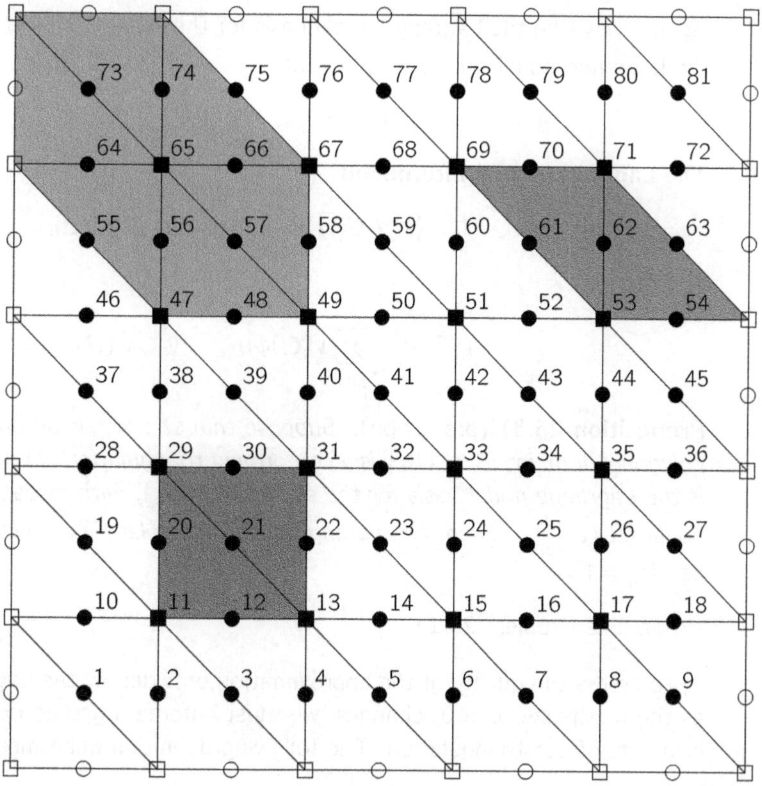

Figure 25.6 A uniform triangulation of a square domain Ω showing the nodes for the piecewise quadratic finite element space $\mathscr{S}_0^{2,0}(\mathscr{T}_h)$. The interior edge midpoint nodes are the filled circles; the unfilled circles are the boundary edge midpoint nodes. The interior vertex nodes are the filled squares; the unfilled squares are the boundary vertex nodes. The supports of the Lagrange nodal basis elements ϕ_{21}, ϕ_{62}, and ϕ_{65} are shown as the shaded regions. There are exactly 81 nodes in the mesh and $\dim(\mathscr{S}_0^{2,0}(\mathscr{T}_h)) = 81$. Observe that $N_v^i = 16$ and $N_m^i = 65$.

Remark 25.29 (sparsity). The supports of the Lagrange nodal basis functions have minimal or no overlap. Consequently, the stiffness matrix A constructed in the abstract Galerkin framework will be sparse, i.e., having only a few nonzero elements.

25.3.1 Basic Error Estimates for the Finite Element Method

We now present some basic error estimates for the finite element method. The idea is to explore the properties of the finite element spaces $\mathscr{S}_0^{p,0}(\mathscr{T}_h)$ to be able to bound the quasi-best approximation provided by Theorem 25.6. To do so, we introduce a so-called *interpolation operator*.

Definition 25.30 (Lagrange interpolant). Let $\Omega \subset \mathbb{R}^2$ be an open, bounded, and polygonal domain. Assume that $\mathscr{T}_h = \{K_i\}$ is a conforming triangulation of Ω. Let

$\{\phi_1, \ldots, \phi_N\}$ be the Lagrange nodal basis for the space $\mathscr{S}_0^{p,0}(\mathscr{T}_h)$, with respect to the Lagrange nodal set $\{\zeta_j\}_{j=1}^N$, so that

$$\phi_j(\zeta_k) = \delta_{j,k}, \quad 1 \le k, j \le N.$$

The **Lagrange nodal interpolant**

$$\Pi_h\colon \left\{ v \in C(\overline{\Omega}) \middle| v|_{\partial\Omega} = 0 \right\} \to \mathscr{S}_0^{p,0}(\mathscr{T}_h)$$

is defined as

$$\Pi_h v = \sum_{j=1}^N v(\zeta_j)\phi_j, \qquad v \in C(\overline{\Omega}).$$

Proposition 25.31 (projection). *Suppose that $\Omega \subset \mathbb{R}^2$ is an open, bounded, and polygonal domain; $\mathscr{T}_h = \{K_i\}$ is a conforming triangulation of Ω; and $\{\phi_1, \ldots, \phi_N\}$ is the Lagrange nodal basis for the space $\mathscr{S}_0^{p,0}(\mathscr{T}_h)$, with respect to the Lagrange nodal set $\{\zeta_j\}_{j=1}^N$. Then the Lagrange nodal interpolant Π_h is a projection operator, i.e., $\Pi_h^2 = \Pi_h$.*

Proof. See Problem 25.12. □

To assess the quality of the approximation provided by the Lagrange interpolant as the mesh size, $h > 0$, changes, we must enforce a geometric condition on the elements of our triangulation. The following definition quantifies this.

Definition 25.32 (shape regularity). Let $\Omega \subset \mathbb{R}^2$ be an open, bounded, and polygonal domain. Let $\{\mathscr{T}_h\}_{h>0}$ be a family of (not necessarily nested) conforming triangulations of Ω, parameterized by $h > 0$; see, for example, Figure 25.3 for the nested case. The family $\{\mathscr{T}_h\}_{h>0}$ is called **shape regular** if and only if there is a constant $C > 0$ such that, for all $h > 0$ and all $K \in \mathscr{T}_h$,

$$1 \le \frac{\rho_{\text{ext}}(K)}{\rho_{\text{int}}(K)} \le C,$$

where $\rho_{\text{ext}}(K)$ is the radius of the smallest circle that circumscribes the triangle K and $\rho_{\text{int}}(K)$ is the radius of the largest circle that is inscribed in K.

Example 25.1 The nested family of triangulations shown in Figure 25.3 is shape regular. All triangles in every triangulation are right and isosceles.

The following result is a cornerstone of piecewise polynomial approximation theory in Sobolev spaces.

Theorem 25.33 (interpolation error estimate). *Suppose that $\Omega \subset \mathbb{R}^2$ is an open, bounded, polygonal domain and $\{\mathscr{T}_h\}_{h>0}$ is a family of (not necessarily nested) conforming, shape regular triangulations of Ω, parameterized by $h > 0$, where*

$h = \max_{K \in \mathscr{T}_h} h_K$. Then there is a constant $C_1 > 0$, independent of h but may depend on p, such that, for any $v \in H^{p+1}(\Omega)$ and all $0 \le m \le p$,

$$\|v - \Pi_h v\|_{H^m(\Omega)} \le C_1 h^{p+1-m} |v|_{H^{p+1}(\Omega)} .$$

Proof. See, for instance, [10]. □

Finally, we can combine Céa's Lemma with this interpolation estimate to provide an error estimate.

Theorem 25.34 (error estimate)**.** *Suppose that $\Omega \subset \mathbb{R}^2$ is an open, bounded, polygonal domain; $f \in L^2(\Omega)$; and $\{\mathscr{T}_h\}_{h>0}$ is a family of (not necessarily nested) conforming, shape regular triangulations of Ω, parameterized by $h > 0$. Suppose that $u \in H_0^1(\Omega) \cap H^{p+1}(\Omega)$ is a weak solution to the Poisson problem (25.6). Suppose that $u_h \in \mathscr{S}_0^{p,0}(\mathscr{T}_h)$ is the finite element approximation defined by (25.7). Then*

$$\|u - u_h\|_{H^1(\Omega)} \le \frac{M}{\alpha} C_1 h^p |u|_{H^{p+1}(\Omega)} ,$$

where $C_1 > 0$ is the constant from Theorem 25.33.

Proof. One needs to use Céa's Lemma, as presented in Theorem 25.6, and the interpolation error estimate from Theorem 25.33. The details are left to the reader as an exercise; see Problem 25.13. □

If we additionally assume that Ω is convex, we can establish a quasi-optimal error estimate in $L^2(\Omega)$ via the Aubin–Nitsche duality technique presented in Theorem 25.18. Convexity is needed because we must invoke the elliptic regularity result presented in Theorem 23.12.

Theorem 25.35 (duality)**.** *Suppose that $\Omega \subset \mathbb{R}^2$ is an open, bounded, polygonal, and convex domain; $f \in L^2(\Omega)$; and $\{\mathscr{T}_h\}_{h>0}$ is a family of (not necessarily nested) conforming, shape regular triangulations of Ω, parameterized by $h > 0$. Suppose that $u \in H_0^1(\Omega)$ is the weak solution of the Poisson problem (25.6) and $u_h \in \mathscr{S}_0^{p,0}(\mathscr{T}_h)$ is the finite element approximation defined by (25.7). Then there is a constant $C_2 > 0$, independent of h and u, such that*

$$\|u - u_h\|_{L^2(\Omega)} \le C_2 h \|u - u_h\|_{H_0^1(\Omega)} .$$

Proof. Set $e = u - u_h \in H_0^1(\Omega)$. Let $z_e \in H_0^1(\Omega)$ be the unique solution of the *dual problem*

$$\mathcal{A}(v, z_e) = \int_\Omega e v \, d\mathbf{x}, \quad \forall v \in H_0^1(\Omega).$$

Notice that, since $\mathcal{A}$ is symmetric, the dual problem is equivalent to the original problem. Since Ω is assumed to be convex, by the elliptic regularity result of Theorem 23.12, we have that $z_e \in H^2(\Omega) \cap H_0^1(\Omega)$ with

$$|z_e|_{H^2(\Omega)} \le C_R \|e\|_{L^2(\Omega)} .$$

Now suppose that $v_h \in \mathscr{S}_0^{p,0}(\mathscr{T}_h)$ is arbitrary and set $v = e$ in the dual problem. Using Galerkin orthogonality, and the boundedness of $\mathcal{A}$, we have that

$$\|e\|_{L^2(\Omega)}^2 = \int_\Omega e^2 \, d\mathbf{x} = \mathcal{A}(e, z_e) = \mathcal{A}(e, z_e - v_h) \leq M \|e\|_{H_0^1(\Omega)} \|z_e - v_h\|_{H_0^1(\Omega)} .$$

Let us choose $v_h = \Pi_h z$, where Π_h is the Lagrange interpolation operator into $\mathscr{S}_0^{1,0}(\mathscr{T}_h)$, the piecewise linear finite element space. Observe that, for any $p \in \mathbb{N}$, we have $\mathscr{S}_0^{1,0}(\mathscr{T}_h) \subseteq \mathscr{S}_0^{p,0}(\mathscr{T}_h)$. Then, by Theorem 25.33,

$$\begin{aligned}\|e\|_{L^2(\Omega)}^2 &\leq M \|e\|_{H_0^1(\Omega)} \|z_e - \Pi_h z_e\|_{H_0^1(\Omega)} \leq C h^{2-1} \|e\|_{H_0^1(\Omega)} |z_e|_{H^2(\Omega)} \\ &\leq C_2 h \|e\|_{H_0^1(\Omega)} \|e\|_{L^2(\Omega)} .\end{aligned}$$

Therefore,

$$\|e\|_{L^2(\Omega)} \leq C_2 h \|e\|_{H_0^1(\Omega)}$$

and the result follows. □

The following is an easy corollary.

Corollary 25.36 ($L^2(\Omega)$ estimate)**.** *Suppose that $\Omega \subset \mathbb{R}^2$ is an open, bounded, convex, polygonal domain; $f \in L^2(\Omega)$; and $\{\mathscr{T}_h\}_{h>0}$ is a family of (not necessarily nested) conforming, shape regular triangulations of Ω, parameterized by $h > 0$. Suppose that, for some $p \in \mathbb{N}$, $u \in H_0^1(\Omega) \cap H^{p+1}(\Omega)$ is a weak solution of the Poisson problem* (25.6)*. Suppose also that $u_h \in \mathscr{S}_0^{p,0}(\mathscr{T}_h)$ is the finite element approximation defined by* (25.7)*. Then there is a constant $C > 0$, independent of h, such that*

$$\|u - u_h\|_{L^2(\Omega)} \leq C h^{p+1} |u|_{H^{p+1}(\Omega)} .$$

Proof. See Problem 25.14. □

Remark 25.37 (linear finite elements)**.** If we set $p = 1$ in Corollary 25.36, i.e., if we suppose that $u \in H_0^1(\Omega) \cap H^2(\Omega)$ is a weak solution of the Poisson problem, then

$$\|u - u_h\|_{L^2(\Omega)} \leq C h^2 |u|_{H^2(\Omega)} ,$$

provided that $\Omega \subset \mathbb{R}^2$ is a convex, bounded, polygonal domain. In other words, we get the expected second-order convergence, as for the finite difference approximation.

Problems

25.1 Prove Theorem 25.5.

25.2 In the setting of Definition 25.10, show that $\{\phi_i\}_{i=0}^{N+1}$ is indeed a basis of $\mathscr{S}^{1,0}(\mathscr{T}_h)$ and, therefore, the dimension of $\mathscr{S}^{1,0}(\mathscr{T}_h)$ is $N+2$.

25.3 In the setting of Definition 25.10, show that $\{\phi_i\}_{i=1}^{N}$ is indeed a basis of $\mathscr{S}_0^{1,0}(\mathscr{T}_h)$ and, therefore, the dimension of $\mathscr{S}_0^{1,0}(\mathscr{T}_h)$ is N.

25.4 Let $L > 0$ and consider the boundary value problem

$$-au'' + cu = f, \quad x \in (0, L), \qquad u(0) = u(L) = 0. \tag{25.8}$$

Assume that a, c, and f are constant; that $a > 0$; and that $c \geq 0$. Write a weak formulation for it. Find an approximate solution to this problem by Galerkin's method over the space $\operatorname{span}\{\varphi_k\}_{k=1}^N$ with $\varphi_k(x) = \sin(\pi k x/L)$.

25.5 Consider the boundary value problem (25.8) with $c = 0$. Find an approximate solution to this problem by Galerkin's method over the space $\operatorname{span}\{\phi_k\}_{k=1}^N$, where the functions ϕ_k are shifted and integrated Legendre's polynomials

$$\phi_k(x) = \int_0^x P_k(y)dy$$

and P_k is the shifted Legendre polynomial of degree k on $[0, L]$.

25.6 Let $L > 0$. Consider the problem

$$-u'' = x, \quad x \in (0, L), \qquad -u'(0) + u(0) = 1, \qquad u'(L) + u(L) = 0.$$

Write down the system of equations that results from Galerkin's method using as a basis:

a) Piecewise linear functions over a uniform mesh.

b) $\varphi_k(x) = \cos(\pi k x/L)$ for $k = 0, \dots, N$.

c) $\psi_k(x) = x^k$ for $k = 0, \dots, N$.

25.7 Consider the two-point boundary value problem

$$-u'' + u = f, \quad \text{in } (0, 1), \qquad -u'(0) = u'(1) = 0.$$

Construct a Galerkin approximation over the subspace $\mathbb{P}_N$ for some $N \in \mathbb{N}$. Since

$$\mathbb{P}_N = \operatorname{span}\{1, x, \dots, x^N\},$$

we can choose the monomials as a computational basis. This, however, turns out to be a terrible idea. Why?

25.8 Show that the finite element and finite difference approximations of

$$-u'' = f, \text{ in } (0, 1), \qquad u(0) = u(1) = 0, \tag{25.9}$$

over a uniform mesh of size h yield the same stiffness matrix.

25.9 The *Green's function* $G(x, y)$ of (25.9) is

$$G(x, y) = \begin{cases} (1 - x)y, & 0 \leq y \leq x \leq 1, \\ x(1 - y), & 0 \leq x \leq y \leq 1. \end{cases}$$

a) Show that the solution of (25.9) is given by

$$u(x) = \int_0^1 G(x, y) f(y) dy.$$

b) Let $\mathscr{S}_0^{1,0}(\mathscr{T}_h)$ be the piecewise linear finite element space over the given mesh $\mathscr{T}_h$. Show that $G(x, x_k) \in \mathscr{S}_0^{1,0}(\mathscr{T}_h)$ if x_k is a node of the mesh.

c) Show that

$$\int_0^1 v'(y)\partial_y G(x, y)dy = v(x), \quad \forall v \in H_0^1(0, 1).$$

d) A numerical method is *interpolant exact* if the numerical solution u_h coincides with the interpolant of the solution, i.e., $u_h = \Pi_h u$. Show that a linear finite element method for (25.9) is interpolant exact.

e) Show that

$$\|u - u_h\|_{L^\infty(0,1)} \leq Ch^2 \|u''\|_{L^\infty(0,1)}$$

for some constant $C > 0$ that is independent of h and the solution u.

25.10 Prove Theorem 25.25.

25.11 Prove Theorem 25.27.

25.12 Prove Proposition 25.31.

25.13 Complete the proof of Theorem 25.34.

25.14 Prove Corollary 25.36.

25.15 Consider the problem

$$\begin{cases} -\Delta u + cu = f, & \text{in } \Omega = (0,1)^2, \\ u = 0, & \text{on } \partial\Omega, \end{cases}$$

where $c > 0$ and $f \in L^2(\Omega)$. Write a weak formulation for this problem, show that it is well posed, and describe a Galerkin approximation method. Prove a Céa-type lemma for this problem. Assuming that the solution to this problem is such that $u \in H^2(\Omega)$, provide an error estimate for a finite element method with $\mathscr{S}_0^{1,0}(\mathscr{T}_h)$.

25.16 Let N be a positive integer. Define $\mathsf{A}_N \in \mathbb{R}^{N \times N}$ via

$$\mathsf{A}_N = \begin{bmatrix} 4 & -1 & 0 & \dots & 0 \\ -1 & 4 & \ddots & & \vdots \\ 0 & \ddots & \ddots & -1 & 0 \\ \vdots & & -1 & 4 & -1 \\ 0 & \dots & 0 & -1 & 4 \end{bmatrix}.$$

Let $\mathsf{O}_N, \mathsf{I}_N \in \mathbb{R}^{N \times N}$ denote the zero and identity matrices, respectively. Now we construct the matrix $\mathsf{A} \in \mathbb{R}^{N^2 \times N^2}$ via

$$\mathsf{A} = \begin{bmatrix} \mathsf{A}_N & -\mathsf{I}_N & \mathsf{O}_N & \dots & \mathsf{O}_N \\ -\mathsf{I}_N & \mathsf{A}_N & \ddots & & \vdots \\ \mathsf{O}_N & \ddots & \ddots & -\mathsf{I}_N & \mathsf{O}_N \\ \vdots & & -\mathsf{I}_N & \mathsf{A}_N & -\mathsf{I}_N \\ \mathsf{O}_N & \dots & \mathsf{O}_N & -\mathsf{I}_N & \mathsf{A}_N \end{bmatrix}.$$

Let $\mathscr{T}_h$ be the uniform triangulation of the domain $\Omega = (0,1)^2$ described in Figure 25.3 such that there are exactly N^2 interior nodes. Number the interior nodes with the lexicographic ordering, i.e., starting in the lower left-hand corner, proceeding by rows to the upper right-hand corner. Show that the stiffness matrix, obtained by using the piecewise linear subspace $\mathscr{S}_0^{1,0}(\mathscr{T}_h)$, is exactly A.

26 Spectral and Pseudo-Spectral Methods for Periodic Elliptic Equations

In this chapter, we will investigate numerical methods for the approximate solution of elliptic equations with periodic boundary conditions in one space dimension. We only consider one spatial dimension for simplicity, but everything in our discussion can be generalized to two- and three-dimensional problems on cuboids as well.

Periodic differential equations, as we will define below, arise in many applications. For instance, they can be used when we want to understand the properties of the solution to a model, without being obfuscated by boundary effects. In addition, many phenomena are intrinsically periodic, in the sense that they exhibit a repeating pattern.

Our discussion of numerical methods begins by illustrating how finite difference and Galerkin methods can be used in this setting. However, the so-called spectral and pseudo-spectral methods are better suited for this type of problems. As we will see, taking advantage of periodicity, these methods can converge much faster than other approaches. The idea here is to use trigonometric functions, the quintessential representation of periodicity, to approximate the solutions.

26.1 Periodic Differential Equations

We refer the reader to Appendix D for the definition and properties of spaces of complex functions, as it is over these spaces that we will introduce the problem we shall be interested in here. Given $c \geq 0$ and $f \in C_{\mathrm{p}}(0,1;\mathbb{C})$, we seek a $[0,1]$-periodic function $u\colon \mathbb{R} \to \mathbb{C}$ that solves the $[0,1]$-*periodic reaction–diffusion problem*,

$$-\frac{\mathrm{d}^2 u(x)}{\mathrm{d}x^2} + cu(x) = f(x), \quad x \in \mathbb{R}. \tag{26.1}$$

If $c = 0$, we call this the $[0,1]$-*periodic Poisson problem.*[1]

Definition 26.1 (classical solution)**.** We say that $u \in C^2_{\mathrm{p}}(0,1;\mathbb{C})$ is a **classical solution** to the $[0,1]$-periodic reaction–diffusion problem if and only if (26.1) holds point-wise.

Remark 26.2 (boundary conditions)**.** One interesting feature of the problem (26.1) is that there are, in some sense, no boundary conditions, since we demand that

[1] Named in honor of the French mathematician, engineer, and physicist Baron Siméon Denis Poisson (1781–1840).

(26.1) holds on the whole real number line. Still, we sometimes say the boundary conditions are of *periodic type*. In any case, for the periodic problem, one must be careful about the value of c.

Proposition 26.3 (compatibility). *Suppose that $f \in C_{\mathrm{p}}(0,1;\mathbb{C})$ and $u \in C^2_{\mathrm{p}}(0,1;\mathbb{C})$ is a classical solution to the $[0,1]$-periodic reaction–diffusion problem with $c \geq 0$. Then*

$$c\int_0^1 u(x)\mathrm{d}x = \int_0^1 f(x)\mathrm{d}x.$$

Consequently, when $c = 0$, a necessary condition for the existence of a solution is

$$\int_0^1 f(x)\mathrm{d}x = 0.$$

Proof. The proof follows using integration by parts, Theorem B.38, and is left to the reader as an exercise; see Problem 26.1. □

The following result provides the existence and uniqueness of solutions.

Theorem 26.4 (existence and uniqueness). *Suppose that $f \in C_{\mathrm{p}}(0,1;\mathbb{C})$. If $c > 0$, there exists a unique classical solution $u \in C^2_{\mathrm{p}}(0,1;\mathbb{C})$ to the $[0,1]$-periodic reaction–diffusion problem (26.1). If $c = 0$ and the right-hand side f satisfies the compatibility condition of Proposition 26.3, then there is a unique classical solution u to (26.1) in the class $\mathring{C}^2_{\mathrm{p}}(0,1;\mathbb{C})$. Finally, if f is real valued, so is u.*

Proof. The existence and uniqueness proof can be obtained using the periodic Green's function, or by using Fourier series techniques; see [31, 49, 99].

To show that u must be real valued when f is, it suffices to take complex conjugates of (26.1) and invoke uniqueness. □

26.2 Finite Difference Approximation

We now describe how to approximate the classical solutions to problem (26.1) using finite difference methods. Many of the techniques that we describe here are similar to those developed in Chapter 24, and so we will follow much of the notation introduced there.

26.2.1 Periodic Grid Functions

We begin with some preliminary definitions. Some of these notions were already introduced in Chapter 13, but we repeat them and adapt them to our interests here.

Definition 26.5 (periodic grid functions). Let $M \in \mathbb{N}$ and $h = \frac{1}{M}$. We define the space of **periodic grid functions** to be

$$\mathcal{V}_{M,\mathrm{p}}(\mathbb{C}) = \left\{ v \in \mathcal{V}(\mathbb{Z}_h;\mathbb{C}) \mid v_{i+mM} = v_i, \ \forall i, m \in \mathbb{Z} \right\}.$$

Notice that the slightly complicated notation is indeed necessary. Periodic functions are defined on the whole $\mathbb{Z}_h$, but they are uniquely characterized by their values in $(0,1]\cap\mathbb{Z}_h$, i.e., the values in ih with $i \in \{1,\dots,M\}$. In addition, since this will be convenient later, we assume from the onset that periodic grid functions are complex valued. We can now introduce norms for the space of periodic functions.

Definition 26.6 (L^p_h-norms)**.** Let $p \in [1,\infty]$. We define, on $\mathcal{V}_{M,\mathrm{p}}(\mathbb{C})$, the norms

$$\|v\|_{L^p_h} = \left(h \sum_{i=1}^{M} |v_i|^p \right)^{1/p}, \quad \forall v \in \mathcal{V}_{M,\mathrm{p}}(\mathbb{C}), \qquad p \in [1,\infty),$$

and

$$\|v\|_{L^\infty_h} = \max_{i=1}^{M} |v_i|, \quad \forall v \in \mathcal{V}_{M,\mathrm{p}}(\mathbb{C}).$$

These are the usual norms on spaces of grid functions. They are just specially adapted to our setting. From this reason, the following result is immediate.

Proposition 26.7 (inner product)**.** *The L^2_h-norm on $\mathcal{V}_{M,\mathrm{p}}(\mathbb{C})$ is induced by the inner product*

$$(v,\varphi)_{L^2_h} = h \sum_{i=1}^{M} v_i \bar{\varphi}_i, \quad \forall v,\varphi \in \mathcal{V}_{M,\mathrm{p}}(\mathbb{C}).$$

Proof. See Problem 26.2. □

Since in the continuous setting, when $c = 0$, spaces of mean-zero, periodic functions become important, we also introduce here the space of mean-zero, periodic grid functions.

Definition 26.8 ($\mathring{\mathcal{V}}_{M,\mathrm{p}}(\mathbb{C})$)**.** Let $M \in \mathbb{N}$. We define

$$\mathring{\mathcal{V}}_{M,\mathrm{p}}(\mathbb{C}) = \left\{ v \in \mathcal{V}_{M,\mathrm{p}}(\mathbb{C}) \;\middle|\; (v,1)_{L^2_h} = 0 \right\},$$

where by 1 we denote the grid function that is constant and has value one at every node.

Finally, we observe that, as in the case of Chapter 24, grid functions can be identified with $\mathbb{C}^M$.

Proposition 26.9 (isomorphism)**.** *For all $M \in \mathbb{N}$, there is a one-to-one correspondence between the space of periodic grid functions $\mathcal{V}_{M,\mathrm{p}}(\mathbb{C})$ and $\mathbb{C}^M$. Therefore, $\dim(\mathcal{V}_{M,\mathrm{p}}(\mathbb{C})) = M$ as a vector space over $\mathbb{C}$.*

Proof. Let $h = \frac{1}{M}$ and $\mathcal{G}_h = (0,1] \cap \mathbb{Z}_h$. Observe that $\#\mathcal{G}_h = M$. Following Proposition 24.3, it is not difficult to establish, see Problem 26.3, the isomorphism between $\mathcal{V}(\mathcal{G}_h;\mathbb{C})$ and $\mathbb{C}^M$. Now if $v \in \mathcal{V}(\mathcal{G}_h;\mathbb{C})$, then we can define its unique periodic extension $v^\uparrow \in \mathcal{V}_{M,\mathrm{p}}(\mathbb{C})$ via

$$v^\uparrow_{i+mM} = v_i, \quad \forall i \in \{1,\dots,N+1\}, \quad \forall m \in \mathbb{Z}.$$

Conversely, for any $v \in \mathcal{V}_{M,\mathrm{p}}(\mathbb{C})$, there is a unique restriction $v^{\downarrow} \in \mathcal{V}(\mathcal{G}_h; \mathbb{C})$ defined via

$$v_i^{\downarrow} = v_i, \quad \forall i \in \{1, \ldots, M\}. \qquad \square$$

Remark 26.10 (notation). As before, we will write

$$\boldsymbol{v} \in \mathbb{C}^M \longleftrightarrow v \in \mathcal{V}(\mathcal{G}_h; \mathbb{C}) \longleftrightarrow v \in \mathcal{V}_{M,\mathrm{p}}(\mathbb{C})$$

to indicate the isomorphism shown above. If we know that the components of the vector are real, we will write $\boldsymbol{v} \in \mathbb{R}^M$ instead. We use the same convention as before, denoting a grid function by a Greek or Roman character and its corresponding canonical vector representative by the boldface of the same character.

26.2.2 Finite Difference Approximation

We are now ready to introduce the finite difference approximation to (26.1).

Definition 26.11 (finite difference approximation). Assume that $c \geq 0$, $f \in C_{\mathrm{p}}(0,1;\mathbb{C})$, and $u \in C_{\mathrm{p}}^2(0,1;\mathbb{C})$ is the classical solution to the one-dimensional periodic reaction–diffusion problem (26.1). Let $M \in \mathbb{N}$ and $h = \frac{1}{M}$. We say that $w \in \mathcal{V}_{M,\mathrm{p}}(\mathbb{C})$ is a **finite difference approximation** to u if and only if

$$-\Delta_h w + cw = f_h, \tag{26.2}$$

where $f_h = \pi_h f$. The **error** is the periodic grid function $e \in \mathcal{V}_{M,\mathrm{p}}(\mathbb{C})$ defined via

$$e = \pi_h u - w.$$

The **consistency error** is the grid function $\mathcal{E}_h[u] \in \mathcal{V}_{M,\mathrm{p}}(\mathbb{C})$,

$$\mathcal{E}_h[u] = -\Delta_h \pi_h u + c\pi_h u - \pi_h f.$$

Above, as usual, π_h denotes the sampling operator, i.e., $(\pi_h v)_i = v(ih)$ for all $i \in \mathbb{Z}$.

The isomorphism of Proposition 26.9 allows us to realize that problem (26.2) can be equivalently written as

$$(\mathsf{A} + ch^2 \mathsf{I}_M)\boldsymbol{w} = h^2 \boldsymbol{f}, \tag{26.3}$$

where $\boldsymbol{w} \in \mathbb{C}^M \longleftrightarrow w \in \mathcal{V}_{M,\mathrm{p}}(\mathbb{C})$, $\boldsymbol{f} \in \mathbb{C}^M \longleftrightarrow f_h \in \mathcal{V}_{M,\mathrm{p}}(\mathbb{C})$, and the matrix $\mathsf{A} \in \mathbb{R}^{M \times M}$ is the so-called *stiffness matrix*, which is defined via

$$\mathsf{A} = \begin{bmatrix} 2 & -1 & 0 & \cdots & -1 \\ -1 & 2 & \ddots & & \vdots \\ 0 & \ddots & \ddots & -1 & 0 \\ \vdots & & -1 & 2 & -1 \\ -1 & \cdots & 0 & -1 & 2 \end{bmatrix}. \tag{26.4}$$

As before, from a practical point of view, computing a finite difference approximation reduces to the solution of a linear system of equations, which, in this case,

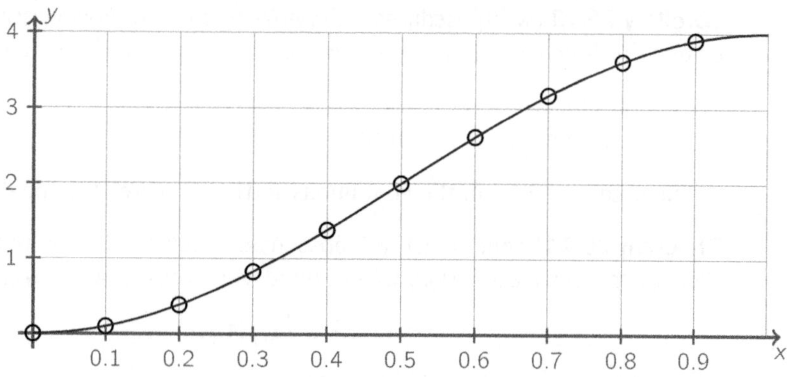

Figure 26.1 The eigenvalues of the periodic stiffness matrix A are shown (open circles) for $M = 10$. Observe that $0 \leq \lambda_m \leq 4$.

is cyclically tridiagonal. In addition, the properties of the matrix A can be used to study the well-posedness of (26.2) and to obtain error estimates.

26.2.3 Well-Posedness and Convergence of the Finite Difference Approximation

As in Theorem 24.25, we study the spectrum of the stiffness matrix A, and this will allow us to establish convergence of the method in the L^2_h-norm.

Theorem 26.12 (spectrum of A). *Let $M \in \mathbb{N}$. Suppose that $\mathsf{A} \in \mathbb{R}^{M \times M}$ is the stiffness matrix defined in (26.4). Consider the set of vectors $S = \{\mathbf{v}_m\}_{m=1}^M \subset \mathbb{C}^M$, where the components of $\mathbf{v}_m$ are*

$$[\mathbf{v}_m]_k = v_{m,k} = e^{2\pi i k m h}, \quad k \in \{1, \ldots, M\}.$$

Then

1. *The set S is an orthonormal set of vectors in the sense that if $\mathbf{v}_j \longleftrightarrow v_j \in \mathcal{V}_{M,\mathrm{p}}(\mathbb{C})$, then*

$$(v_i, v_j)_{L^2_h} = \delta_{i,j}, \quad i, j \in \{1, \ldots, M\}.$$

2. *S is a set of eigenvectors of A and the eigenvalue λ_m corresponding to the eigenvector $\mathbf{v}_m$ is given by*

$$\lambda_m = 4 \sin^2(\pi m h).$$

Since $0 \leq \lambda_m \leq 4$ (see Figure 26.1), for all $m \in \{1, \ldots, M\}$, A is a symmetric positive semi-definite matrix.

3. *A is not invertible. In particular,* $\ker(\mathsf{A}) = \operatorname{span}\{\mathbf{1}\}$.

Proof. The details are similar to those in Theorem 24.25. We leave the details to the reader as an exercise; see Problem 26.4. □

As an immediate consequence, we obtain well-posedness in the case $c > 0$.

Corollary 26.13 (well-posedness). *Suppose that $c > 0$. For every $M \in \mathbb{N}$, there is a unique solution $w \in \mathcal{V}_{M,\mathrm{p}}(\mathbb{C}) \longleftrightarrow \boldsymbol{w} \in \mathbb{C}^M$ to the finite difference problem (26.2).*

Proof. This follows because, when $c > 0$, the system matrix $\mathsf{A} + ch^2\mathsf{I}_M$ for problem (26.3) is symmetric positive denite. □

The case $c = 0$, as in the continuous setting, requires a compatibility condition.

Theorem 26.14 (conditional well-posedness). *Let $c = 0$. For all $M \in \mathbb{N}$, problem (26.2) has a unique solution, which, moreover, satisfies $w \in \mathring{\mathcal{V}}_{M,\mathrm{p}}(\mathbb{C})$ if and only if*

$$f_h \in \mathring{\mathcal{V}}_{M,\mathrm{p}}(\mathbb{C}).$$

Proof. As we observed, $w \in \mathcal{V}_{M,\mathrm{p}}(\mathbb{C}) \longleftrightarrow \boldsymbol{w} \in \mathbb{C}^M$ solves the finite difference problem (26.2) with $c = 0$ if and only if

$$\mathsf{A}\boldsymbol{w} = h^2\boldsymbol{f}.$$

Suppose that $f_h \in \mathring{\mathcal{V}}_{M,\mathrm{p}}(\mathbb{C})$. Recall that

$$\ker(\mathsf{A}) = \operatorname{im}(\mathsf{A}^{\mathsf{H}})^{\perp} = \operatorname{im}(\mathsf{A}^{\mathsf{T}})^{\perp} = \operatorname{im}(\mathsf{A})^{\perp},$$

where we used that A is symmetric. In addition, from Theorem 26.12, we know that $\ker(\mathsf{A}) = \operatorname{span}\{\mathbf{1}\}$. Therefore, the compatibility condition implies that

$$f_h \in \mathring{\mathcal{V}}_{M,\mathrm{p}}(\mathbb{C}) \longleftrightarrow \boldsymbol{f} \in \ker(\mathsf{A})^{\perp} = \operatorname{im}(\mathsf{A}).$$

In other words, there exists $\boldsymbol{w} \in \mathbb{C}^M$ such that $\mathsf{A}\boldsymbol{w} = h^2\boldsymbol{f}$, which shows the existence of a solution. To show uniqueness, assume that we have two solutions $\boldsymbol{w}_1, \boldsymbol{w}_2 \in \mathbb{C}^M$. By linearity,

$$\mathsf{A}(\boldsymbol{w}_1 - \boldsymbol{w}_2) = \mathbf{0} \quad \Longleftrightarrow \quad \boldsymbol{w}_1 - \boldsymbol{w}_2 \in \ker(\mathsf{A}) \quad \Longleftrightarrow \quad (w_1 - w_2, 1)_{L_h^2} = 0.$$

Since $\dim\ker(\mathsf{A}) = 1$, this shows that there is a unique solution with the property $w \in \mathring{\mathcal{V}}_{M,\mathrm{p}}(\mathbb{C})$.

Assume now that the problem has a solution $w \in \mathring{\mathcal{V}}_{M,\mathrm{p}}(\mathbb{C})$. Then

$$h^2(\boldsymbol{f}, \mathbf{1})_2 = (\mathsf{A}\boldsymbol{w}, \mathbf{1})_2 = (\boldsymbol{w}, \mathsf{A}^{\mathsf{T}}\mathbf{1})_2 = (\boldsymbol{w}, \mathsf{A}\mathbf{1})_2 = 0,$$

where, in the last step, we used that A is real and symmetric and $\mathbf{1} \in \ker(\mathsf{A})$. □

Now we prove a convergence result in the L_h^2-norm. For simplicity, we focus on the reaction–diffusion problem, i.e., $c > 0$, and leave the case $c = 0$ for Problem 26.7.

Theorem 26.15 (error estimate). *Let $f \in C_{\mathrm{p}}(0, 1; \mathbb{C})$ and $u \in C_{\mathrm{p}}^4(0, 1; \mathbb{C})$ be a classical solution to the one-dimensional reaction–diffusion problem (26.1) with $c > 0$. Suppose that $M \in \mathbb{N}$ and $w \in \mathcal{V}_{M,\mathrm{p}}(\mathbb{C}) \longleftrightarrow \boldsymbol{w} \in \mathbb{C}^M$ is a solution to the finite difference problem (26.2). There is a constant $C > 0$, independent of h, such that the consistency error satisfies*

$$\|\mathcal{E}_h[u]\|_{L_h^{\infty}} \le Ch^2.$$

In addition, there is a, possibly different, constant $C > 0$, independent of h, for which

$$\|e\|_{L^2_h} \le Ch^2.$$

Proof. The consistency error estimate is a consequence of Taylor's Theorem. The details are left for the reader as an exercise; see Problem 26.6.

To obtain the error estimate, we recall that, as usual,

$$-\Delta_h e + ce = \mathcal{E}_h[u],$$

so that, with the usual identifications,

$$\mathbf{e} = h^2 \left(\mathsf{A} + ch^2 \mathsf{I}_M\right)^{-1} \boldsymbol{\mathcal{E}}_h[u].$$

We leave it to the reader to prove that, if $0 < h < \frac{1}{2}$,

$$\left\| \left(\mathsf{A} + ch^2 \mathsf{I}_M\right)^{-1} \right\|_2 = \frac{1}{ch^2} = Ch^{-2}.$$

If this is the case,

$$\|\mathbf{e}\|_2 \le h^2 \left\| \left(\mathsf{A} + ch^2 \mathsf{I}_M\right)^{-1} \right\|_2 \|\boldsymbol{\mathcal{E}}_h[u]\|_2 \le C \|\boldsymbol{\mathcal{E}}_h[u]\|_2 \le C\sqrt{M}h^2 = Ch^{\frac{3}{2}}.$$

Using the fact that $\|e\|_{L^2_h} = \sqrt{h}\,\|\mathbf{e}\|_2$, the result follows. □

26.3 The Spectral Galerkin Method

In this section, we describe the spectral Galerkin method for solving the periodic Poisson problem, i.e., (26.1) with $c = 0$. This, like the finite element method, is nothing but a Galerkin method where the subspace and its basis are suitably chosen to fit the needs of the problem at hand.

26.3.1 Weak Formulation

Galerkin methods aim to approximate weak solutions, and the correct functional framework for these is Sobolev spaces. The reader is referred to Appendix D for an overview of these spaces. The weak formulation of (26.1) with $c = 0$ reads (see Problem 26.8): Given $f \in \mathring{L}^2(0,1;\mathbb{C})$, find $u \in \mathring{H}^1_{\mathrm{p}}(0,1;\mathbb{C})$ such that

$$\mathcal{A}(u, v) = F(v), \quad \forall v \in \mathring{H}^1_{\mathrm{p}}(0,1;\mathbb{C}), \tag{26.5}$$

where

$$\mathcal{A}(v, w) = \int_0^1 v'(x)\overline{w'(x)}\mathrm{d}x, \qquad F(v) = \int_0^1 f(x)\overline{v(x)}\mathrm{d}x.$$

The analysis of this problem is a slight modification of the theory developed in Section 23.2.3. The first step is the following Poincaré-type inequality for periodic functions.

Theorem 26.16 (Poincaré[2]). *There is a constant $C_P > 0$ such that*

$$\|u\|_{L^2(0,1;\mathbb{C})} \leq C_P |u|_{H^1_p(0,1;\mathbb{C})}$$

for all $u \in \mathring{H}^1_p(0,1;\mathbb{C})$.

Proof. We prove this result for $\mathring{C}^1_p(0,1)$. The more general version can be established using a density argument over the real and imaginary parts. Suppose that $u \in \mathring{C}^1_p(0,1)$. Since u is continuous, periodic, and has zero mean, there is a point $a \in [0,1)$, where $u(a) = 0$. Since the function is periodic, without loss of generality, we may assume that $a = 0$. Then, by the Fundamental Theorem of Calculus, for any $z \in [0,1]$, since $u(0) = 0$,

$$u(z) = u(z) - u(0) = \int_0^z u'(x)\mathrm{d}x.$$

By the Cauchy–Schwarz inequality,

$$\begin{aligned}
|u(z)|^2 &= \left(\int_0^z 1 \cdot u'(x)\mathrm{d}x\right)^2 \\
&\leq \int_0^z 1^2\mathrm{d}x \int_0^z |u'(x)|^2\,\mathrm{d}x \\
&= z\int_0^z |u'(x)|^2\,\mathrm{d}x \\
&\leq \int_0^1 |u'(x)|^2\,\mathrm{d}x.
\end{aligned}$$

Integrating, we have

$$\int_0^1 |u(z)|^2\,\mathrm{d}z \leq \int_0^1 \left[\int_0^1 |u'(x)|^2\,\mathrm{d}x\right]\mathrm{d}z = \int_0^1 |u'(x)|^2\,\mathrm{d}x;$$

in this case, the result holds with $C_P = 1$. □

We can now proceed with the analysis of problem (26.5).

Theorem 26.17 (well-posedness). *Let $f \in \mathring{L}^2(0,1;\mathbb{C})$. Then problem* (26.5) *has a unique solution $u \in \mathring{H}^1_p(0,1;\mathbb{C}) \cap \mathring{H}^2_p(0,1;\mathbb{C})$, which satisfies*

$$\|u\|_{H^1_p(0,1;\mathbb{C})} \leq C\|f\|_{L^2(0,1;\mathbb{C})}$$

for some constant $C > 0$ that is independent of f and u. Furthermore, if, for some $r \in \mathbb{N}$, we have $f \in \mathring{H}^r_p(0,1;\mathbb{C})$, then $u \in \mathring{H}^{r+2}_p(0,1;\mathbb{C})$. Finally, if $f \in \mathring{C}^\infty_p(0,1;\mathbb{C})$, then $u \in \mathring{C}^\infty_p(0,1;\mathbb{C})$.

Proof. We leave the proof of existence and uniqueness to the reader as an exercise; see Problem 26.9.

[2] Named in honor of the French mathematician, theoretical physicist, engineer, and philosopher of science Jules Henri Poincaré (1854–1912).

The further regularity follows, essentially, from the following (formal) argument. Notice, first of all, that, since $f \in L^2(0,1;\mathbb{C})$, the expression

$$v \mapsto G(v) = \int_0^1 f(x)\overline{v'(x)}\mathrm{d}x$$

defines a continuous anti-linear functional in $\mathring{H}^1_\mathrm{p}(0,1;\mathbb{C})$ with

$$\|G\|_{\mathring{H}^1_\mathrm{p}(0,1;\mathbb{C})^*} \leq \|f\|_{L^2(0,1;\mathbb{C})};$$

see again Problem 26.9 for notation and terminology. We can then invoke the well-posedness that we just established to assert that the problem: Find $\tilde{u} \in \mathring{H}^1_\mathrm{p}(0,1;\mathbb{C})$ such that

$$\mathcal{A}(\tilde{u}, v) = G(v), \quad \forall v \in \mathring{H}^1_\mathrm{p}(0,1;\mathbb{C}),$$

has a unique solution which satisfies

$$\|\tilde{u}\|_{H^1_\mathrm{p}(0,1;\mathbb{C})} \leq C\|f\|_{L^2(0,1;\mathbb{C})}.$$

Set now, in (26.5), $v = w'$ with $w \in \mathring{H}^2_\mathrm{p}(0,1;\mathbb{C})$ but otherwise arbitrary and integrate by parts to obtain

$$\mathcal{A}(u', w) = G(w), \quad \forall w \in \mathring{H}^1_\mathrm{p}(0,1;\mathbb{C}).$$

This shows that $\tilde{u} = u'$, so that $u' \in \mathring{H}^1_\mathrm{p}(0,1;\mathbb{C})$, i.e., $u \in \mathring{H}^2_\mathrm{p}(0,1;\mathbb{C})$. □

The construction of the spectral Galerkin method requires knowledge of the eigenvalues and eigenfunctions of the second derivative subject to periodic boundary conditions. We show this in the following example, which should be compared with Theorem 23.14.

Example 26.1 Let us obtain the eigenvalues and eigenfunctions of the differential operator $-\frac{\mathrm{d}^2 v}{\mathrm{d}x^2}$ on the interval $(0,1)$ subject to periodic boundary conditions. Define for $x \in \mathbb{R}$ and $k \in \mathbb{N}$,

$$\psi_k(x) = \sin(2\pi kx), \quad \phi_k(x) = \cos(2\pi kx).$$

Define $\phi_0(x) = 1$ for all $x \in \mathbb{R}$. Then, clearly,

$$-\frac{\mathrm{d}^2\psi_k}{\mathrm{d}x^2} = (2\pi)^2k^2\psi_k = \lambda_k\psi_k, \qquad -\frac{\mathrm{d}^2\phi_k}{\mathrm{d}x^2} = (2\pi)^2k^2\phi_k = \eta_k\phi_k,$$

and $\psi_k, \phi_k \in \mathring{C}^\infty_\mathrm{p}(0,1)$. Observe that the eigenvalues grow unboundedly:

$$\lambda_k = \eta_k = (2\pi)^2k^2 \to \infty, \qquad k \to \infty.$$

The previous computations show that $\{\lambda_k, \phi_k\}_{k<0} \cup \{0, \phi_0\} \cup \{\lambda_k, \psi_k\}_{k>0}$ are indeed eigenvalues and eigenfunctions of this operator.

It is not difficult to see that the eigenfunctions are orthogonal, i.e., for any $k, \ell \in \mathbb{N}$, $k \neq \ell$,

$$(\psi_k, \psi_\ell)_{L^2(0,1)} = 0 = (\phi_k, \phi_\ell)_{L^2(0,1)};$$

for any $k, \ell \in \mathbb{N}$,

$$(\psi_k, \phi_\ell)_{L^2(0,1)} = 0;$$

and, finally, for $k \in \mathbb{N}$,

$$(\psi_k, \phi_0)_{L^2(0,1)} = 0 = (\phi_k, \phi_0)_{L^2(0,1)}.$$

The last orthogonality is a restatement of the mean-zero property. The normalizations are

$$(\psi_k, \psi_k)_{L^2(0,1)} = \frac{1}{2} = (\phi_k, \phi_k)_{L^2(0,1)}, \quad k \in \mathbb{N},$$

and

$$(\phi_0, \phi_0)_{L^2(0,1)} = 1.$$

To simplify the notation and calculations, one can use complex trigonometric functions instead of real ones. Define, for all $x \in \mathbb{R}$, $k \in \mathbb{Z}$,

$$\chi_k(x) = e^{2\pi i k x}.$$

Then

$$-\frac{d^2\chi_k}{dx^2} = (2\pi)^2 k^2 \chi_k = \gamma_k \chi_k,$$

and we observe that

$$\gamma_k = \gamma_{-k} = (2\pi)^2 k^2.$$

In the complex case, the orthogonality property is expressed using the complex L^2-inner product. For the present example,

$$(\chi_k, \chi_\ell)_{L^2(0,1;\mathbb{C})} = \int_0^1 e^{2\pi i k x} e^{-2\pi i \ell x} dx = \begin{cases} 1, & k = \ell, \\ 0, & k \neq \ell. \end{cases}$$

26.3.2 Spectral Galerkin Methods

We are now ready to introduce the subspace that will define the spectral Galerkin method.

Definition 26.18 ($\mathring{\mathcal{S}}_N(0,1;\mathbb{C})$)**.** Let $N \in \mathbb{N}$. The space of complex-valued, mean-zero, trigonometric polynomials of degree at most N is

$$\mathring{\mathcal{S}}_N(0,1;\mathbb{C}) = \left\{ v = \sum_{k=-N}^{N} \alpha_k \chi_k \,\middle|\, \alpha_k \in \mathbb{C}, \alpha_0 = 0 \right\}.$$

Definition 26.19 (spectal Galerkin approximation[3])**.** Suppose that $f \in \mathring{L}^2(0,1;\mathbb{C})$ and $N \in \mathbb{N}$ is given. The **spectral Galerkin approximation** of the periodic Poisson problem, i.e., (26.1) with $c = 0$, is a function $u_N \in \mathring{\mathcal{S}}_N(0,1;\mathbb{C})$ that satisfies

[3] Named in honor of the Russian mathematician and engineer Boris Grigorievich Galerkin (1871–1945).

$$\mathcal{A}(u_N, v_N) = F(v_N), \quad \forall v_N \in \mathring{\mathcal{S}}_N(0,1;\mathbb{C}).$$

Theorem 26.20 (well-posedness). *Suppose that $f \in \mathring{L}^2(0,1;\mathbb{C})$. For all $N \in \mathbb{N}$, there is a unique spectral Galerkin approximation $u_N \in \mathring{\mathcal{S}}_N(0,1;\mathbb{C})$ to the solution of (26.5) in the sense of Definition 26.19. In particular, if*

$$f = \sum_{k\in\mathbb{Z}\setminus\{0\}} \hat{f}_k \chi_k, \qquad \hat{f}_k = (f, \chi_k)_{L^2(0,1;\mathbb{C})},$$

then

$$u_N = \sum_{\substack{k=-N\\k\neq 0}}^{N} \hat{u}_k \chi_k, \qquad \hat{u}_k = \frac{\hat{f}_k}{(2\pi)^2 k^2}. \tag{26.6}$$

Furthermore, if f is real valued, then so is u_N.

Proof. Existence and uniqueness follow after minor modification to the general theory of Galerkin approximations that take into account that functions in our space are complex valued; see Problem 26.10.

To obtain the claimed representation, let $\ell \in \{\pm 1, \pm 2, \ldots, \pm N\}$. Set $v_N = \chi_\ell$ in Definition 26.19. We obtain

$$\begin{aligned}
\hat{f}_\ell &= (f, \chi_\ell)_{L^2(0,1;\mathbb{C})}\\
&= \mathcal{A}(u_N, \chi_\ell)\\
&= \int_0^1 \frac{d}{dx}\left(\sum_{\substack{k=-N\\k\neq 0}}^{N} \hat{u}_k \chi_k(x)\right) \overline{\frac{d\chi_\ell(x)}{dx}} dx\\
&= \sum_{\substack{k=-N\\k\neq 0}}^{N} \hat{u}_k \int_0^1 \frac{d\chi_k(x)}{dx} \overline{\frac{d\chi_\ell(x)}{dx}} dx\\
&= \sum_{\substack{k=-N\\k\neq 0}}^{N} \hat{u}_k (2\pi)^2 k\ell \int_0^1 e^{2\pi i k x} e^{-2\pi i \ell x} dx\\
&= \hat{u}_\ell (2\pi)^2 \ell^2,
\end{aligned}$$

where in the last step we used orthonormality.

Finally, f is real valued if and only if $\overline{\hat{f}_k} = \hat{f}_{-k}$ for all $k \in \mathbb{Z}$. Clearly, u_N inherits this property. □

An error estimate follows from the general theory of Galerkin approximations.

Theorem 26.21 (error estimate). *Suppose that, for some $r \in \mathbb{N}$, we have $f \in \mathring{H}_{\mathrm{p}}^{r-1}(0,1;\mathbb{C})$, so that the weak solution to the periodic Poisson problem (26.5) satisfies $u \in \mathring{H}_{\mathrm{p}}^{r+1}(0,1;\mathbb{C})$. Let $N \in \mathbb{N}$ and $u_N \in \mathring{\mathcal{S}}_N(0,1;\mathbb{C})$ be its spectral*

Galerkin approximation in the sense of Definition 26.19. Then there is a constant $C > 0$, which is independent of f, u, and N, for which we have

$$\|u - u_N\|_{H^1_p(0,1;\mathbb{C})} \le \frac{\gamma}{\alpha} \frac{C}{N^r} |u|_{H^{r+1}_p(0,1;\mathbb{C})}. \tag{26.7}$$

Proof. From Céa's Lemma we obtain the quasi-optimal approximation property

$$\|u - u_N\|_{H^1_p(0,1;\mathbb{C})} \le C \min_{v_N \in \mathring{\mathcal{S}}_N(0,1;\mathbb{C})} \|u - v_N\|_{H^1_p(0,1;\mathbb{C})}.$$

The combination of the last estimate with the approximation result in Theorem 12.31 gives the desired estimate. We leave the remaining details to the reader as an exercise; see Problem 26.11. □

26.4 The Pseudo-Spectral Method

The main motivation for the so-called pseudo-spectral method comes from the following simple example.

Example 26.2 Suppose that $\mathcal{I} \subset \mathbb{Z}$ is an index set of finite cardinality and

$$v(x) = \sum_{k \in \mathcal{I}} \alpha_k e^{2\pi i k x}.$$

Then

$$\frac{dv(x)}{dx} = \sum_{k \in \mathcal{I}} \alpha_k (2\pi i k) e^{2\pi i k x}$$

and

$$\frac{d^2 v(x)}{dx^2} = -\sum_{k \in \mathcal{I}} \alpha_k (2\pi k)^2 e^{2\pi i k x}.$$

In other words, derivatives modify the Fourier coefficients of the series representation of a function in simple ways.

Definition 26.22 (pseudo-spectral derivative). Suppose that $M \in \mathbb{N}$, $v \in \mathcal{V}_{M,p}(\mathbb{C})$, and $\hat{v} \in \mathcal{V}_{M,p}(\mathbb{C})$ is its Discrete Fourier Transform (DFT), in the sense of Definition 13.8. If $M = 2K+1$, $K \in \mathbb{N}$, we define the first- and second-order **pseudo-spectral derivatives**, respectively, of v via

$$\mathcal{D}_h v_j = \sum_{k=-K}^{K} \hat{v}_k (2\pi i k) e^{2\pi i j k h}, \qquad \mathcal{D}_h^2 v_j = -\sum_{k=-K}^{K} \hat{v}_k (2\pi k)^2 e^{2\pi i j k h}.$$

If $M = 2K$, $K \in \mathbb{N}$, the pseudo-spectral derivatives are defined via

$$\mathcal{D}_h v_j = \sum_{k=-K}^{K} \hat{v}_k \omega_k (2\pi i k) e^{2\pi i j k h}, \qquad \mathcal{D}_h^2 v_j = -\sum_{k=-K}^{K} \hat{v}_k \omega_k (2\pi i k)^2 e^{2\pi i j k h},$$

where the weight grid function ω is defined as

$$\omega_k = \begin{cases} \frac{1}{2}, & k = \pm K, \\ 1, & k \neq \pm K. \end{cases}$$

Remark 26.23 (symmetry). Let $M \in \mathbb{N}$. Suppose that $v \in \mathcal{V}_{M,\mathrm{p}}(\mathbb{C})$ and $\hat{v} \in \mathcal{V}_{M,\mathrm{p}}(\mathbb{C})$ is its DFT. Regarding the definitions above, observe that if M is odd, $M = 2K + 1$, $K \in \mathbb{N}$, then

$$v_j = \sum_{k=0}^{M-1} \hat{v}_k e^{2\pi \mathrm{i} jkh} = \sum_{k=-K}^{K} \hat{v}_k e^{2\pi \mathrm{i} jkh},$$

and if M is even, $M = 2K$, $K \in \mathbb{N}$, then

$$v_j = \sum_{k=0}^{M-1} \hat{v}_k e^{2\pi \mathrm{i} jkh} = \sum_{k=-K}^{K} \omega_k \hat{v}_k e^{2\pi \mathrm{i} jkh}.$$

In other words, we do not change anything by shifting our summations. However, note that, when $M = 2K + 1$, $K \in \mathbb{N}$,

$$\sum_{k=0}^{M-1} \hat{v}_k (2\pi \mathrm{i} k) e^{2\pi \mathrm{i} jkh} \neq \sum_{k=-K}^{K} \hat{v}_k (2\pi \mathrm{i} k) e^{2\pi \mathrm{i} jkh},$$

and when $M = 2K$, $K \in \mathbb{N}$,

$$\sum_{k=0}^{M-1} \hat{v}_k (2\pi \mathrm{i} k) e^{2\pi \mathrm{i} jkh} \neq \sum_{k=-K}^{K} \omega_k \hat{v}_k (2\pi \mathrm{i} k) e^{2\pi \mathrm{i} jkh}.$$

The problem is that the grid functions α, γ, defined by $\alpha_k = -2\pi \mathrm{i} k$ and $\gamma_k = (2\pi k)^2$, respectively, are not periodic grid functions. They are, however, odd and even grid functions, respectively. The pseudo-spectral derivatives that we introduced in Definition 26.22 are chosen to be symmetric sums about $k = 0$.

Proposition 26.24 (periodicity). *Suppose that $M \in \mathbb{N}$, $v \in \mathcal{V}_{M,\mathrm{p}}(\mathbb{C})$, and $\hat{v} \in \mathcal{V}_{M,\mathrm{p}}(\mathbb{C})$ is its DFT. The pseudo-spectral derivatives of v are periodic grid functions, i.e., $\mathcal{D}_h v, \mathcal{D}_h^2 v \in \mathcal{V}_{M,\mathrm{p}}(\mathbb{C})$.*

Proof. See Problem 26.12. □

Definition 26.25 (pseudo-spectral method). Let $M \in \mathbb{N}$. Suppose that $f_M \in \mathring{\mathcal{V}}_{m,\mathrm{p}}(\mathbb{C})$ is given, where $\mathring{\mathcal{V}}_{M,\mathrm{p}}(\mathbb{C})$ was introduced in Definition 26.8. Consider the following problem: Find $w \in \mathring{\mathcal{V}}_{M,\mathrm{p}}(\mathbb{C})$ such that

$$-\mathcal{D}_h^2 w = f_M. \tag{26.8}$$

The solution to this problem, if it exists, is called the **pseudo-spectral approximation** of u, where u is the solution of the periodic Poisson problem, (26.1) with $c = 0$.

Let us now show the well-posedness of (26.8).

Theorem 26.26 (well-posedness). *Let $M \in \mathbb{N}$. Suppose that $f_M \in \mathring{\mathcal{V}}_{M,\mathrm{p}}(\mathbb{C})$ is given and $\hat{f} \in \mathcal{V}_{M,\mathrm{p}}(\mathbb{C})$ is its DFT. Problem (26.8) has a unique solution, $w \in \mathring{\mathcal{V}}_{M,\mathrm{p}}(\mathbb{C})$. In particular, if M is odd, i.e., $M = 2K + 1$, then*

$$w_j = \sum_{k=-K}^{K} \hat{w}_k e^{2\pi \mathrm{i} jkh}, \quad \hat{w}_k = \frac{\hat{f}_k}{(2\pi k)^2}, \quad k = \pm 1, \dots, \pm K,$$

with a similar result in the case that M is even.

Proof. Let us, for definiteness, assume that $M \in \mathbb{N}$ is odd, and that $\hat{w}$ and $\hat{f}$ are the DFTs of w and f_M, respectively. Since $(f_M, 1)_{L_h^2} = 0$, $\hat{f}_0 = 0$. We then have that w solves (26.8) if and only if

$$\sum_{k=-K}^{K} \hat{w}_k (2\pi k)^2 e^{2\pi \mathrm{i} jkh} = \sum_{k=-K}^{K} \hat{f}_k e^{2\pi \mathrm{i} jkh},$$

if and only if

$$\hat{w}_k = \frac{\hat{f}_k}{(2\pi k)^2}, \quad k = \pm 1, \dots, \pm K.$$

The case when $M \in \mathbb{N}$ is even is similar; see Problem 26.13. □

As with the spectral Galerkin method, we can prove spectral convergence properties for this approximation. We omit a discussion of this topic for the sake of brevity. The interested reader can consult [13].

Example 26.3 Here, we describe a practical implementation of the pseudo-spectral approximation method. Suppose that $f \in \mathring{C}_{\mathrm{p}}(0, 1; \mathbb{C})$ is given. Let $M \in \mathbb{N}$. Define $\tilde{f} \in \mathcal{V}_{M,\mathrm{p}}(\mathbb{C})$ via

$$\tilde{f}_i = f(ih), \quad \forall i \in \mathbb{Z}.$$

Observe now that, in general, $\tilde{f} \notin \mathring{\mathcal{V}}_{M,\mathrm{p}}(\mathbb{C})$. For this reason, we define the so-called *mean-zero projection*

$$\begin{aligned} P_h \colon \mathcal{V}_{M,\mathrm{p}}(\mathbb{C}) &\to \mathring{\mathcal{V}}_{M,\mathrm{p}}(\mathbb{C}), \\ v &\mapsto P_h v = v - (v, 1)_{L_h^2}. \end{aligned}$$

Let now $\hat{f}$ be the DFT of $P_h \tilde{f}$. Observe that $\hat{f}_0 = 0$, as desired. Assuming that M is odd, the DFT of the solution is then given by

$$\hat{w}_k = \frac{\hat{f}_k}{(2\pi k)^2}, \quad k = \pm 1, \dots, \pm K.$$

Set $\hat{w}_0 = 0$. Finally, we construct the solution via the Inverse Discrete Fourier Transform (IDFT),

$$w = \sum_{k=-K}^{K} \hat{w}_k e^{2\pi \mathrm{i} jkh}.$$

Example 26.4 In this example, we approximate the solution of a problem whose exact solution is known. Suppose that $L > 0$ and u is the $[0, L]$-periodic function

$$u(x) = \exp(\sin(qx)) - C, \quad q = \frac{2\pi}{L},$$

where the constant C can be chosen to satisfy a mean-zero condition, as desired. Suppose that f is such that

$$f = -u'' = q^2 \exp(\sin(qx)) \left[\cos^2(qx) - \sin(qx)\right].$$

It is not difficult to see that f is a mean-zero function, i.e., $\int_0^L f(x)\mathrm{d}x = 0$, and that it is $[0, L]$-periodic. We can apply the finite difference and pseudo-spectral methods to approximate the solutions of u. To do so, we first must adjust f, so that it is mean zero in the discrete sense, using the mean-zero projection defined in Example 26.3.

Listing 26.1 provides the code for computing approximate solutions. Figures 26.2, 26.3, and 26.4 show the output from this code for $M = 8, 16$, and 32, respectively. Observe that the pseudo-spectral approximations are much more accurate than the finite difference approximations. This is illustrated in Figure 26.5.

Remark 26.27. The error analysis of the pseudo-spectral method, which demonstrates the spectral accuracy of the method rigorously, is both interesting and a little bit challenging. We omit it here for the sake of brevity. However, the concepts used in Chapter 27, where we discuss collocation methods at length, can be modified easily to give this analysis. In fact, the pseudo-spectral method is a type of collocation method, as one can easily show. The details are left to the curious reader.

Problems

26.1 Prove Proposition 26.3.

26.2 Prove Proposition 26.7.

26.3 Complete the proof of Proposition 26.9.

26.4 Prove Theorem 26.12.

26.5 Let $M \in \mathbb{N}$ and define $\mathsf{T} \in \mathbb{R}^{M \times M}$ via

$$\mathsf{T} = \begin{bmatrix} \alpha & \beta & 0 & \cdots & 0 \\ \beta & \alpha & \ddots & & \vdots \\ 0 & \ddots & \ddots & \beta & 0 \\ \vdots & & \beta & \alpha & \beta \\ 0 & \cdots & 0 & \beta & \alpha \end{bmatrix}.$$

A matrix of this type is called TST, for *Toeplitz symmetric tridiagonal*. Find the eigenvectors and eigenvalues of this matrix.

26.6 Complete the proof of Theorem 26.15.

26.7 Prove a version of Theorem 26.15 for the case $c = 0$.

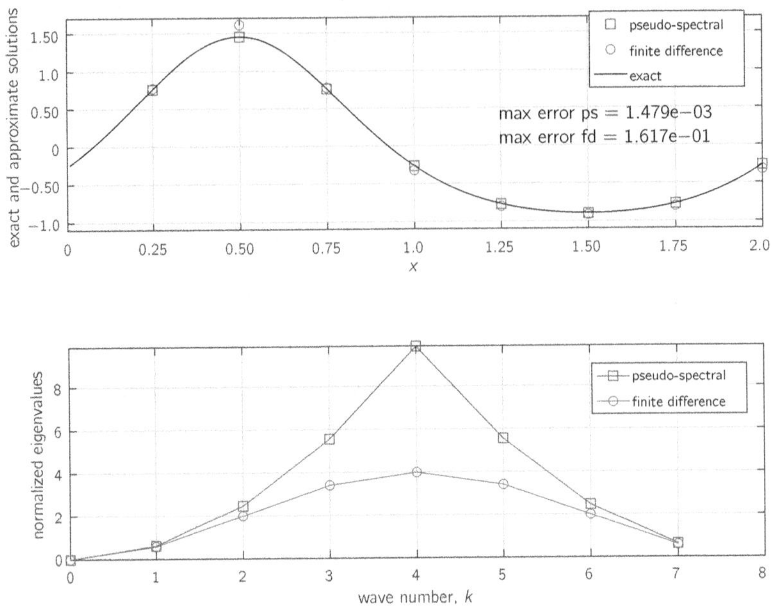

Figure 26.2 Pseudo-spectral and finite difference approximation of the solution to a periodic Poisson problem with eight grid points ($M = 8$); see Example 26.4 for details.

26.8 Derive the weak formulation (26.5) of the periodic Poisson problem, i.e., (26.1) with $c = 0$.

26.9 The purpose of this problem is to complete the proof of Theorem 26.17. To do so, we will consider the following. Let $\mathbb{H}$ be a *complex* Hilbert space with inner product $(\cdot, \cdot)_{\mathbb{H}}$. We say that a functional $F\colon \mathbb{H} \to \mathbb{C}$ is *anti-linear* if

$$F(\alpha v + \beta w) = \bar{\alpha} F(v) + \bar{\beta} F(w), \quad \forall \alpha, \beta \in \mathbb{C}, \quad \forall v, w \in \mathbb{H}.$$

a) Let F be an anti-linear functional. We say that it is *bounded* if there is a constant $C > 0$ such that

$$|F(v)| \le C \|v\|_{\mathbb{H}}, \quad \forall v \in \mathbb{H}.$$

Show that an anti-linear functional is bounded if and only if it is continuous.

b) The *anti-dual space* $\mathbb{H}^*$ is the set of all continuous, anti-linear functionals on $\mathbb{H}$. Show that this is a (complex) vector space, that

$$\|F\|_{\mathbb{H}^*} = \sup_{0 \ne v \in \mathbb{H}} \frac{|F(v)|}{\|v\|_{\mathbb{H}}}$$

defines a norm on $\mathbb{H}^*$, and that $\mathbb{H}^*$ is complete under that norm.
Hint: Revisit Proposition 16.13.

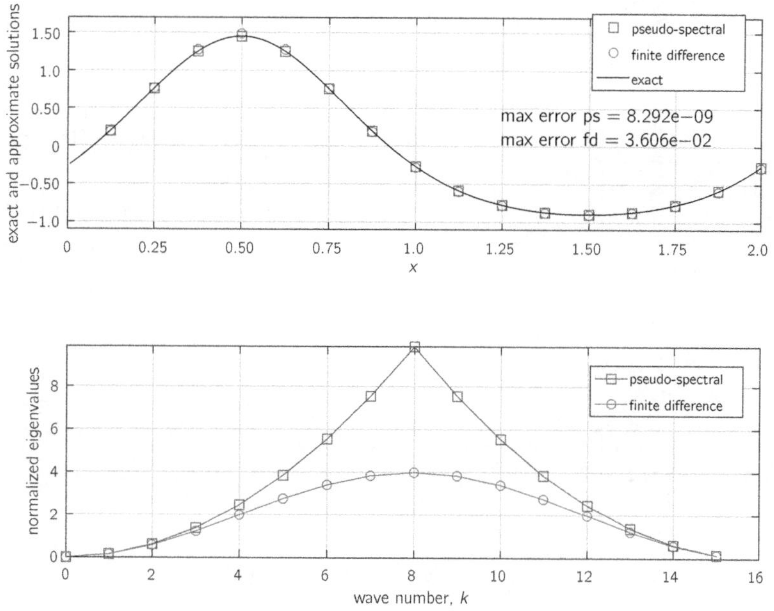

Figure 26.3 Pseudo-spectral and finite difference approximation of the solution to a periodic Poisson problem with 16 grid points ($M = 16$); see Example 26.4 for details.

c) Show that there is a canonical bijection between $\mathbb{H}'$ and $\mathbb{H}^{\star}$. Namely, $F \in \mathbb{H}'$ if and only if $\bar{F} \in \mathbb{H}^{*}$ with equality of norms. Here,

$$\bar{F}\colon v \mapsto \overline{F(v)}, \qquad \forall v \in \mathbb{H}.$$

d) Prove a version of the Riesz Representation Theorem (Theorem 16.14) for the anti-dual space: Let $F \in \mathbb{H}^{*}$. Then there exists a unique element $\mathfrak{R}_* F \in \mathbb{H}$ such that

$$F(v) = (\mathfrak{R}_* F, v)_{\mathbb{H}}, \quad \forall v \in \mathbb{H}.$$

Moreover, $\|\mathfrak{R}_* F\|_{\mathbb{H}} = \|F\|_{\mathbb{H}^*}$.

e) With this construction, show that problem (26.5) is well posed.

26.10 Complete the proof of Theorem 26.20.

26.11 Complete the proof of Theorem 26.21.

26.12 Prove Proposition 26.24.

26.13 Complete the proof of Theorem 26.26.

26.14 Let $M \in \mathbb{N}$.

a) Define the DFT and IDFT for grid functions in $\mathcal{V}_{M,\mathrm{p}}(\mathbb{C})$.

b) Prove that the DFT is bijection from $\mathcal{V}_{M,\mathrm{p}}(\mathbb{C})$ to $\mathcal{V}_{M,\mathrm{p}}(\mathbb{C})$.

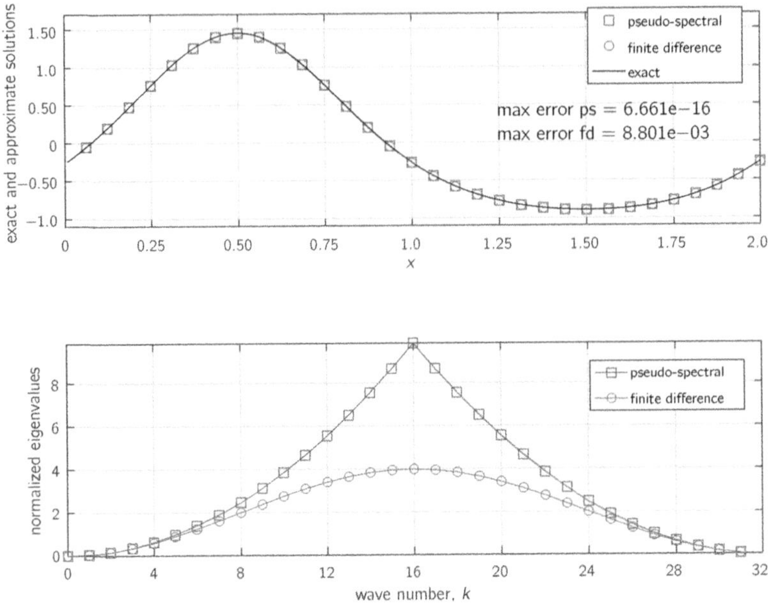

Figure 26.4 Pseudo-spectral and finite difference approximation of the solution to a periodic Poisson problem with 32 grid points ($M = 32$); see Example 26.4 for details.

c) If $w \in \mathcal{V}_{M,\mathrm{p}}(\mathbb{C})$ and $\widehat{w} \in \mathcal{V}_{M,\mathrm{p}}(\mathbb{C})$ is its DFT, prove that

$$h \sum_{k=0}^{M-1} w_k \overline{w_k} = \sum_{k=0}^{M-1} \widehat{w}_k \overline{\widehat{w}_k}.$$

26.15 Consider the periodic Poisson problem on $[0, 1]$ given by (26.1) with $c = 0$. Let $M \in \mathbb{N}$.

a) Describe the pseudo-spectral approximation of this problem quantitatively.

b) Suppose that $f_M \in \mathcal{V}_{M,\mathrm{p}}(\mathbb{C})$. Show that a necessary condition for the solvability of the pseudo-spectral approximation (26.8) is that $(f_M, 1)_{L^2_h} = 0$, i.e., $f_M \in \mathring{\mathcal{V}}_{M,\mathrm{p}}(\mathbb{C})$.

26.16 Let $M \in \mathbb{N}$ and consider the finite difference approximation to the periodic Poisson problem on $[0, 1]$ given in (26.2) with $c = 0$. Recall that the stiffness matrix, $\mathsf{A} \in \mathbb{R}^{M \times M}$, is given in (26.4) and the equivalent matrix problem is

$$\mathsf{A}\boldsymbol{w} = h^2 \boldsymbol{f}.$$

Use the DFT to find a solution to this matrix problem. What assumptions do you need to make about $\boldsymbol{f}$ and $\boldsymbol{w}$ in order to get a unique solution?

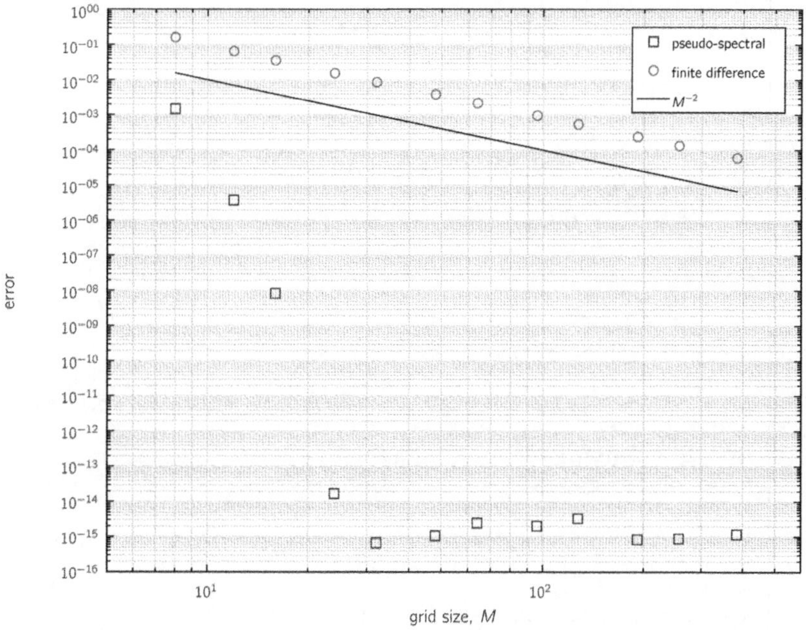

Figure 26.5 Errors in the pseudo-spectral and finite difference approximations of the solution to a periodic Poisson problem; see Example 26.4 for details. The finite difference method exhibits clear second-order convergence, while the error for the pseudo-spectral method fits the classical spectral convergence pattern. Note that the saturation in the pseudo-spectral error, at around 10^{-15}, is due to roundoff error.

Listings

```
function [ePSpec,eFDiff] = PoissonPer(L, N)
%
% This function computes pseudo-spectral and finite difference
% approximations to the periodic Poisson equation:
%
%    -D_{xx} u = f, where u and f are [0,L]-periodic
%
% using the Fast Fourier Transform (FFT) algorithm.
%
% Input
%
%    L : the size of the domain
%    N : the number of points in the periodic grid. Generally
%        this should be a positive, even number.
%
% Output
```

```
%
%   ePSpec : the max norm of the error in the pseudo-spectral
%            computation.
%   eFDiff : the max norm of the error in the finite difference
%            computation.
%
% h is the grid spacing; x holds the nodal values of the grid:
  h = L/N;
  x = (1:N)*h;
%
% Fine grid parameters for representing the exact solution:
  Nf = 256;
  hf = L/Nf;
  xf = (1:Nf)*hf;
%
% q is a useful factor that allows us to easily change the
%   domain size:
  q = 2*pi/L;
%
% Wave numbers, k: This definition is really convenient,
% essentially making k periodic. Note that our array index
% starts at 1, as usual in MATLAB, but the value of k starts
% at 0:
  k = [0:N/2-1 N/2 -N/2+1:-1];
%
% Right-Hand-Side function, f, forced to be discrete mean-zero:
  f = q*q*(cos(q*x).*cos(q*x)-sin(q*x)).*exp(sin(q*x));
  f = -(f-h*sum(f)/L);
%
% Exact solution, forced to be discrete mean-zero:
  uExact = exp(sin(q*x));
  uExactMass = h*sum(uExact)/L;
  uExact = uExact-uExactMass;
%
% Eigenvalues of the pseudo-spectral derivative operator -D^2:
  eigenPSpec = q*q*k.*k;
%
% Modified to avoid dividing by zero below:
  eigenPSpec(1) = 1.0;
%
% Eigenvalues of the finite difference operator -\Delta_h.
% Note that we have applied a shift here to be consistent with
% the pseudo-spectral case:
  eigenFDiff = 4.0*sin(q*(x-h)/2).*sin(q*(x-h)/2)/(h*h);
%
% Modified to avoid dividing by zero below:
  eigenFDiff(1) = 1.0;
%
% Approximations: Note that MATLAB fft's do not respect the fact
% that our data points are real numbers. So we need to get rid
% of any imaginary numbers in the results:
  uPSpec = real((ifft(fft(f)./eigenPSpec)));
  uFDiff = real((ifft(fft(f)./eigenFDiff)));
%
% Error computations:
  ePSpec = max(abs(uPSpec-uExact));
```

```
  eFDiff = max(abs(uFDiff-uExact));
%
% Redefine uExact for plotting:
  uExact = exp(sin(q*xf));
  uExact = uExact-uExactMass;

  hf = figure(1)

  subplot(2,1,1);
  plot(x,uPSpec,'s',x,uFDiff,'o',xf,uExact,'k-')
  grid on, xlabel x, ylabel 'exact and approximate solutions';
  title(['Pseudo-Spectral and Finite Difference ', ...
    'Approximations: N = ', num2str(N)]);
  axis([0,L,-1.1,1.7])
  set(gca,'xTick',0:L/8:L)
  text(1.25,0.425,['max error ps = ' ...
    num2str(ePSpec,'%8.3e')],'fontsize', 12)
  text(1.25,0.125,['max error fd = ' ...
    num2str(eFDiff,'%8.3e')],'fontsize', 12)
  legend('pseudo-spectral','finite difference','exact')

  subplot(2,1,2);
  eigenPSpec(1) = 0.0;
  eigenFDiff(1) = 0.0;
  plot(0:N-1,h*h*eigenPSpec,'-s',0:N-1,h*h*eigenFDiff,'-o')
  grid on, xlabel 'wave number, k', ...
    ylabel 'normalized eigenvalues'
  title(['Normalized Eigenvalues of the Pseudo-Spectral' ...
    ' and Finite Difference Operators']);
  axis([0,N,0,pi*pi])
  set(gca,'xTick',0:N/8:N)
  legend('pseudo-spectral','finite difference')

  s1 = ['000' num2str(N)];
  s2 = s1((length(s1)-3):length(s1));
  s3 = ['OUT/periodicPoisson', s2, '.pdf'];
  exportgraphics(hf, s3)

end
```

Listing 26.1 Approximation of the periodic Poisson equation.

27 Collocation Methods for Elliptic Equations

In this chapter, we will design and analyze collocation methods for solving elliptic boundary value problems (BVPs). We will focus our attention on the following problem: Let $\Omega = (-1, 1)$, $c \geq 0$, and $f : \Omega \to \mathbb{R}$ be given. We seek $u : \bar{\Omega} \to \mathbb{R}$ such that

$$-\frac{\mathrm{d}^2 u(x)}{\mathrm{d}x^2} + cu(x) = f(x), \quad x \in \Omega, \quad u(-1) = u(1) = 0. \tag{27.1}$$

Our approximation strategy will be as follows. Let $\{x_j\}_{j=0}^N \subset [-1, 1]$, which we will call the set of $N + 1$ *collocation points*, with $x_0 = -1$ and $x_N = 1$. We look for an approximate solution of the form

$$w(x) = \sum_{j=0}^{N} w_j L_j(x), \quad x \in [-1, 1],$$

where $L_j \in \mathbb{P}_N$ is the jth Lagrange nodal basis element, defined in (9.3). Thus, it follows that $w(x_j) = w_j$. We can now form $N + 1$ linear equations in $N + 1$ unknowns. The first and last equations are, respectively,

$$w(-1) = w(1) = 0,$$

which come from the boundary conditions. It then follows that $w_0 = 0$ and $w_N = 0$. For the remaining $N - 1$ equations, we require that the approximation w satisfies the differential equation at the interior $N - 1$ collocation points:

$$-w''(x_k) + cw(x_k) = f(x_k), \quad 1 \leq k \leq N - 1.$$

In other words,

$$\sum_{j=1}^{N-1} \left[-L_j''(x_k) + c\delta_{j,k}\right] w_j = f(x_k) = f_k.$$

Let us define the matrix $\mathsf{A} \in \mathbb{R}^{(N-1)\times(N-1)}$ via

$$[\mathsf{A}]_{j,k} = -L_j''(x_k) + c\delta_{j,k}, \quad 1 \leq j, k \leq N - 1, \tag{27.2}$$

and the vectors $\boldsymbol{w}, \boldsymbol{f} \in \mathbb{R}^{N-1}$ as

$$[\boldsymbol{w}]_k = w_k, \quad [\boldsymbol{f}]_k = f_k.$$

Then the approximation w is determined by solving the equation $\mathsf{A}\boldsymbol{w} = \boldsymbol{f}$.

At this stage a numerical analyst must raise several questions. Is A invertible? How does one choose the collocation points? Is the approximation stable? How good is the resulting approximation? How can the error analysis be conducted? We will lean heavily on the excellent books by Shen, Tang, and Wang [85] and Canuto *et al.* [13], as we navigate and answer these questions.

27.1 Weighted Sobolev Spaces and Weak Formulation

To provide a suitable analysis of a collocation method for (27.1), we must define a special type of weak solution; to do so, we need to define a particular weighted Sobolev space. Let

$$\alpha(x) = \frac{1}{\sqrt{1-x^2}}, \quad x \in (-1,1),$$

be the Chebyshev weight function on $(-1,1)$. Observe that we have switched from our usual notation. Since w will be reserved for our approximation, we use α to denote our weight function. To create a weak solution, we will multiply (27.1) by the weight function α and a test function v, and integrate by parts.

Let

$$\mathcal{B}_\alpha(-1,1) = \left\{ g \in C^1([-1,1]) \;\middle|\; \lim_{x\downarrow -1} \alpha(x)g(x) = 0, \; \lim_{x\uparrow 1} \alpha(x)g(x) = 0 \right\}.$$

Suppose that $u \in C^2([-1,1])$ and $v \in \mathcal{B}_\alpha(-1,1)$. Then we find

$$\begin{aligned} -\int_{-1}^{1} \alpha(x)u''(x)v(x)\mathrm{d}x &= -\,\alpha(x)v(x)u'(x)|_{x=-1}^{x=1} + \int_{-1}^{1} (\alpha(x)v(x))'\, u'(x)\mathrm{d}x \\ &= \int_{-1}^{1} (\alpha(x)v(x))'\, u'(x)\mathrm{d}x \\ &= \left(\alpha^{-1}(\alpha v)', u'\right)_{L^2_\alpha(-1,1)}. \end{aligned}$$

Recall that

$$(u,v)_{L^2_\alpha(-1,1)} = \int_{-1}^{1} u(x)v(x)\alpha(x)\mathrm{d}x.$$

In other words, for all $u \in C^2([-1,1])$ and $v \in \mathcal{B}_\alpha(-1,1)$,

$$\left(\alpha^{-1}(\alpha v)', u'\right)_{L^2_\alpha(-1,1)} = -\left(v, u''\right)_{L^2_\alpha(-1,1)}.$$

Remark 27.1 (space $\mathcal{B}_\alpha(-1,1)$)**.** What kind of functions are in $\mathcal{B}_\alpha(-1,1)$? Suppose that $v \in \mathbb{P}_N$, $N \geq 2$ has zeros at $x = \pm 1$. Then

$$v(x) = (x-1)(x+1)r(x), \quad r \in \mathbb{P}_{N-2},$$

and it is clear that

$$\lim_{x\downarrow -1} \alpha(x)v(x) = 0, \; \lim_{x\uparrow 1} \alpha(x)v(x) = 0.$$

Definition 27.2 (weighted Sobolev space). Suppose that α is the Chebyshev weight function on $(-1,1)$ and $m \in \mathbb{N}_0$. We say that $u \in H^m_\alpha(-1,1)$ if and only if $u \in L^2_\alpha(-1,1)$, u is m times weakly differentiable on $(-1,1)$, and

$$|u|^2_{H^\ell_\alpha(-1,1)} = \left\| u^{(\ell)} \right\|^2_{L^2_\alpha(-1,1)} < \infty, \quad \forall \ell \in \{1,\dots,m\}.$$

For $u, v \in H^m_\alpha(-1,1)$, we define

$$(u,v)_{H^m_\alpha(-1,1)} = \sum_{\ell=0}^{m} \left(u^{(\ell)}, v^{(\ell)} \right)_{L^2_\alpha(-1,1)}.$$

The space $H^m_\alpha(-1,1)$ is called a **Chebyshev weighted Sobolev space of order** m.[1]

Theorem 27.3 (properties of $H^m_\alpha(-1,1)$). *Suppose that α is the Chebyshev weight function on $(-1,1)$. Then $H^m_\alpha(-1,1)$ is a Hilbert space when equipped with the inner product $(\cdot,\cdot)_{H^m_\alpha(-1,1)}$. Furthermore, $H^1_\alpha(-1,1) \hookrightarrow C([-1,1])$, meaning that if $u \in H^1_\alpha(-1,1)$, then u coincides, up to a set of measure zero, with a function in $C([-1,1])$ and there is a constant $C > 0$, independent of u, such that*

$$\|u\|_{L^\infty(-1,1)} \le C \|u\|_{H^1_\alpha(-1,1)}.$$

Proof. See [13] or [85]. □

In light of the previous result, we can, in particular, speak of point values of a function in a Chebyshev weighted Sobolev space. For this reason, and to later take into account the boundary conditions, we define a subspace with zero boundary values.

Definition 27.4 ($H^1_{\alpha,0}(-1,1)$). Suppose that α is the Chebyshev weight function on $(-1,1)$. We define

$$H^1_{\alpha,0}(-1,1) = \left\{ u \in H^1_\alpha(-1,1) \;\middle|\; u(-1) = u(1) = 0 \right\}.$$

For all $u, v \in H^1_{\alpha,0}(-1,1)$, set

$$(u,v)_{H^1_{\alpha,0}(-1,1)} = (u', v')_{L^2_\alpha(-1,1)}.$$

Proposition 27.5 (embedding). *Suppose that α is the Chebyshev weight function on $(-1,1)$. Then we have*

$$\mathcal{B}_\alpha(-1,1) \subset H^1_{\alpha,0}(-1,1).$$

Proof. Let $v \in \mathcal{B}_\alpha(-1,1)$. By definition, the function and its first derivative are continuous on $[-1,1]$, and thus bounded there. Consequently,

$$\int_{-1}^{1} \alpha(x) \left(|v(x)|^2 + |v'(x)|^2 \right) dx \le M \int_{-1}^{1} \alpha(x) dx < \infty$$

[1] Named in honor of the Soviet mathematician Sergei Lvovich Sobolev (1908–1989).

and $v \in H^1_\alpha(-1,1)$. To show that the boundary values vanish, we see that, by continuity,

$$v(1) = \lim_{x\uparrow 1} v(x) = \lim_{x\uparrow 1} \alpha(x)v(x)\frac{1}{\alpha(x)} = \lim_{x\uparrow 1} \alpha(x)v(x) \lim_{x\uparrow 1} \frac{1}{\alpha(x)} = 0\cdot 0 = 0,$$

where we used the condition $\alpha(x)v(x) \to 0$ as $x \uparrow 1$. The value at $x = -1$ can be treated similarly; thus, we have the result. □

Theorem 27.6 (Poincaré). *The following Poincaré-type inequality*[2] *is valid: there is a constant $C_{\alpha,P} > 0$ such that*

$$\|u\|_{L^2_\alpha(-1,1)} \le C_{\alpha,P} \|u'\|_{L^2_\alpha(-1,1)} \tag{27.3}$$

for any $u \in H^1_{\alpha,0}(-1,1)$. Consequently, $H^1_{\alpha,0}(-1,1)$ is a Hilbert space with the inner product $(\cdot\,,\cdot)_{H^1_{\alpha,0}(-1,1)}$ and the induced norm

$$\|u\|_{H^1_{\alpha,0}(-1,1)} = \sqrt{(u,u)_{H^1_{\alpha,0}(-1,1)}} = \sqrt{(u',u')_{L^2_\alpha(-1,1)}}, \quad \forall u \in H^1_{\alpha,0}(-1,1).$$

Furthermore, the following norm equivalence is valid:

$$\frac{1}{\sqrt{1+C^2_{\alpha,P}}}\|u\|_{H^1_\alpha(-1,1)} \le \|u\|_{H^1_{\alpha,0}(-1,1)} \le \|u\|_{H^1_\alpha(-1,1)}, \quad \forall u \in H^1_{\alpha,0}(-1,1).$$

Proof. Clearly, the norm equivalence follows from (27.3).

We will prove (27.3) for functions in $\mathcal{B}_\alpha(-1,1)$. A density result, see [13] or [85], closes the argument. Let $u \in \mathcal{B}_\alpha(-1,1)$. As the proof of the previous result shows, we have $u(-1) = 0$. By the Fundamental Theorem of Calculus, we then have

$$\begin{aligned}
u(x) &= u(x) - u(-1) \\
&= \int_{-1}^{x} u'(y)\mathrm{d}y \\
&= \int_{-1}^{x} \alpha^{-1/2}(y)\alpha(y)^{1/2}u'(y)\mathrm{d}y \\
&\le \left(\int_{-1}^{x} \alpha^{-1}(y)\mathrm{d}y\right)^{1/2}\left(\int_{-1}^{x} \alpha(y)|u'(y)|^2\,\mathrm{d}y\right)^{1/2} \\
&\le \left(\int_{-1}^{1} \alpha^{-1}(y)\mathrm{d}y\right)^{1/2}\left(\int_{-1}^{1} \alpha(y)|u'(y)|^2\,\mathrm{d}y\right)^{1/2},
\end{aligned}$$

where, in the last two steps, we applied the Cauchy–Schwarz inequality and the fact that the functions being integrated are nonnegative. With this estimate at hand, we see that

[2] Named in honor of the French mathematician, theoretical physicist, engineer, and philosopher of science Jules Henri Poincaré (1854–1912).

$$\int_{-1}^{1} \alpha(x)|u(x)|^2 \, dx \leq \int_{-1}^{1} \alpha(x) \left(\int_{-1}^{1} \alpha^{-1}(y) dy \right) \left(\int_{-1}^{1} \alpha(y)|u'(y)|^2 \, dy \right) dx$$
$$= \left(\int_{-1}^{1} \alpha(x) dx \right) \left(\int_{-1}^{1} \alpha^{-1}(y) dy \right) \|u'\|^2_{L^2_\alpha(-1,1)} .$$

The estimate of Problem 27.1, part (b) allows us to conclude. In fact, we see that

$$C^2_{\alpha,P} \leq 4C_\alpha .$$

□

Remark 27.7 (embedding)**.** Notice that the proof of the previous result also shows that

$$H^1_{\alpha,0}(-1,1) \subset L^\infty(-1,1),$$

with the corresponding norm estimate.

We are now ready to define and analyze our weak formulation.

Definition 27.8 (weighted weak formulation)**.** Let α be the Chebyshev weight function on $(-1,1)$ and $f \in L^2_\alpha(-1,1)$. The function $u \in H^1_{\alpha,0}(-1,1)$ is called a **Chebyshev weighted weak solution** of (27.1) if and only if

$$\mathcal{A}_\alpha(u,v) = F_\alpha(v), \quad \forall v \in H^1_{\alpha,0}(-1,1), \tag{27.4}$$

where

$$\mathcal{A}_\alpha(u,v) = a_\alpha(u,v) + c(u,v)_{L^2_\alpha(-1,1)},$$
$$a_\alpha(u,v) = \left(u', \alpha^{-1}(\alpha v)'\right)_{L^2_\alpha(-1,1)} = \int_{-1}^{1} u'(x)\left(\alpha(x)v(x)\right)' dx,$$
$$F_\alpha(v) = (f,v)_{L^2_\alpha(-1,1)} .$$

The analysis of this formulation follows the general theory presented in Section 23.2.3.

Theorem 27.9 (coercivity)**.** *The form $a_\alpha(\cdot,\cdot)$, introduced in Definition 27.8, is a nonsymmetric, bounded, and coercive bilinear form. In other words, there are constants $0 < C_{\alpha,1} \leq C_{\alpha,2}$ such that*

$$|a_\alpha(u,v)| \leq C_{\alpha,2} \|u\|_{H^1_{\alpha,0}(-1,1)} \|v\|_{H^1_{\alpha,0}(-1,1)}, \quad \forall u,v \in H^1_{\alpha,0}(-1,1),$$

and

$$C_{\alpha,1} \|u\|^2_{H^1_{\alpha,0}(-1,1)} \leq a_\alpha(u,u), \quad \forall 0 \neq u \in H^1_{\alpha,0}(-1,1).$$

In particular, one can prove that

$$C_{\alpha,1} = \frac{1}{4}, \qquad C_{\alpha,2} = 1 + \sqrt{\frac{8}{3}}.$$

Proof. The fact that this is a nonsymmetric bilinear form is clear from its definition. For the boundedness and coercivity, see, again, [13] or [85]. □

Corollary 27.10 (existence and uniqueness). *Let α be the Chebyshev weight function on $(-1,1)$, $f \in L^2_\alpha(-1,1)$, and $c \geq 0$. Then problem (27.1) has a unique weighted weak solution in the sense of Definition 27.8. Moreover, there is a constant $C > 0$, independent of u, such that*

$$\|u\|_{H^1_{\alpha,0}(-1,1)} \leq C \|f\|_{L^2_\alpha(-1,1)} .$$

Proof. Theorem 27.9 implies that the bilinear form $\mathcal{A}_\alpha$ of Definition 27.8 is bilinear, bounded, and coercive. Clearly, the form F_α is linear and bounded. Thus, an application of the Lax–Milgram Lemma, Theorem 23.10, implies the result. □

27.2 Weighted Spectral Galerkin Approximations

As a natural next step, let us define a spectral Galerkin approximation method for problem (27.1). We will see later how this will connect to collocation approximation methods.

Definition 27.11 (weighted spectral Galerkin approximation). For $N \in \mathbb{N}$, define

$$\mathscr{S}_{N,0}(-1,1) = \{p \in \mathbb{P}_N \mid p(-1) = p(1) = 0\} .$$

Let α be the Chebyshev weight function on $(-1,1)$ and $f \in L^2_\alpha(-1,1)$. The function $u_N \in \mathscr{S}_{N,0}(-1,1)$ is called the **Chebyshev weighted spectral Galerkin approximation**[3] of (27.1) if and only if

$$\mathcal{A}_\alpha(u_N, v_N) = F_\alpha(v_N), \quad \forall v_N \in \mathscr{S}_{N,0}(-1,1), \tag{27.5}$$

where the bilinear form $\mathcal{A}_\alpha$ and the linear form F_α were introduced in Definition 27.8.

The analysis of this method follows the general theory detailed in Section 25.1.

Proposition 27.12 (existence and uniqueness). *Assume that α is the Chebyshev weight function on $(-1,1)$, $f \in L^2_\alpha(-1,1)$, and $c \geq 0$. For every $N \in \mathbb{N}$, the spectral Galerkin approximation (27.5) is well posed. Furthermore, the following stability condition holds:*

$$\|u_N\|_{H^1_\alpha(-1,1)} \leq C \|f\|_{L^2_\alpha(-1,1)}$$

for some $C > 0$ that is independent of N.

Proof. It suffices to, again, apply the Lax–Milgram Lemma, Theorem 23.10. □

The error may be estimated using a Céa-type[4] result.

[3] Named in honor of the Russian mathematician Boris Grigorievich Galerkin (1871–1945).
[4] Named in honor of the French mathematician Jean Céa (1932–).

Theorem 27.13 (weighted Céa). *Suppose that α is the Chebyshev weight function on $(-1,1)$, $f \in L^2_\alpha(-1,1)$, and $c \geq 0$. Let $u \in H^1_{\alpha,0}(-1,1)$ be the unique solution to (27.4) and $u_N \in \mathscr{S}_{N,0}(-1,1)$ be the unique solution to (27.5). Then*

$$\|u - u_N\|_{H^1_{\alpha,0}(-1,1)} \leq \frac{C_{\alpha,2} + \gamma C^2_{\alpha,P}}{C_{\alpha,1}} \inf_{v \in \mathscr{S}_{N,0}(-1,1)} \|u - v\|_{H^1_{\alpha,0}(-1,1)}.$$

Proof. See Problem 27.4 □

Now the weighted Céa's Lemma guarantees that the spectral Galerkin approximation, u_N, is a quasi-best approximation of the weak solution u. We can estimate the norm of the error $u - u_N$, by introducing an optimal order approximation of u from the approximation subspace, $\mathscr{S}_{N,0}$. We will first follow the style of [13], using the weighted elliptic projection operator. Later, we will introduce another method for estimating the error, using a Chebyshev interpolant.

Definition 27.14 (elliptic projection). Let α be the Chebyshev weight function on $(-1,1)$. For $N \in \mathbb{N}$, we define the **Chebyshev weighted elliptic projection** operator

$$P^{N,1}_{\alpha,0} : H^1_{\alpha,0}(-1,1) \to \mathscr{S}_{N,0}(-1,1),$$
$$u \mapsto P_{\alpha,0}[u],$$

where $P^{N,1}_{\alpha,0}[u]$ is the unique solution of

$$\left(P^{N,1}_{\alpha,0}[u], v\right)_{H^1_{\alpha,0}(-1,1)} = (u, v)_{H^1_{\alpha,0}(-1,1)}, \quad \forall v \in \mathscr{S}_{N,0}(-1,1). \tag{27.6}$$

The weighted elliptic projection is nothing but a projection onto a subspace. Thus, it is well defined and stable. The key point is that $P^{N,1}_{\alpha,0}[u]$ approximates u very well. The following optimal-order error estimate can be derived.

Theorem 27.15 (approximation). *Let $m, N \in \mathbb{N}$ with $N+1 > m$ and α be the Chebyshev weight function on $(-1,1)$. Assume that $u \in H^1_{\alpha,0}(-1,1) \cap H^m_\alpha(-1,1)$. Then*

$$\left\|u - P^{N,1}_{\alpha,0}[u]\right\|_{H^1_\alpha(-1,1)} \leq \frac{C}{N^{m-1}} |u|_{H^m_\alpha(-1,1)}$$

for some constant $C > 0$ that is independent of N and u.

Proof. See [13]. □

With the help of this projection, we can then obtain an optimal-order error estimate for (27.5).

Theorem 27.16 (error estimate). *Let $m, N \in \mathbb{N}$ with $N+1 > m$ and α be the Chebyshev weight function on $(-1,1)$. Assume that $u \in H^1_{\alpha,0}(-1,1) \cap H^m_\alpha(-1,1)$ is the unique solution to (27.4) and $u_N \in \mathbb{P}_{N,0}(-1,1)$ is the unique solution to (27.5). Then*

$$\|u - u_N\|_{H^1_\alpha(-1,1)} \leq \frac{C}{N^{m-1}} |u|_{H^m_\alpha(-1,1)},$$

where the constant $C > 0$ is independent of N and u.

Proof. Combine Theorems 27.13 and 27.15. The details are left to the reader as an exercise; see Problem 27.5. □

We have just created and analyzed a numerical method for solving the BVP (27.1). However, there are some practical problems with it; for instance, it requires exact integrations and the stiffness matrix would be dense and nonsymmetric. Moreover, there is no obvious connection between the Galerkin method and the collocation method that is our principal subject of study in this chapter. It turns out that this Galerkin approximation method will be key to defining and analyzing our collocation method.

We must, for the time being, be patient and learn a bit more about Chebyshev's domain of ideas.

27.3 The Chebyshev Projection and the Finite Chebyshev Transform

Let us recall another projection operator; see Proposition 11.8.

Definition 27.17 (Chebyshev projection). Suppose that α is the Chebyshev weight function on $(-1,1)$ and $\{T_j\}_{j=0}^{\infty}$ is the Chebyshev orthogonal polynomial system, which, according to Proposition 10.15, satisfies the orthogonality relation

$$(T_k, T_\ell)_{L^2_\alpha(-1,1)} = \int_{-1}^{1} T_k(x)T_\ell(x)\frac{dx}{\sqrt{1-x^2}} = \frac{\pi}{2\beta_k^\star}\delta_{k,\ell},$$

where

$$\beta_j^\star = \begin{cases} \frac{1}{2}, & j=0, \\ 1, & j\in\mathbb{N}. \end{cases} \tag{27.7}$$

Suppose that $\{\tilde{T}_j\}_{j=0}^{\infty}$ is the normalized Chebyshev polynomial system, defined via

$$\tilde{T}_j = \sqrt{\frac{2\beta_j^\star}{\pi}}T_j, \quad j\in\mathbb{N}_0.$$

For $N\in\mathbb{N}_0$, the projection operator $\mathcal{P}_{\alpha,N}\colon L^2_\alpha(-1,1;\mathbb{C})\to\mathbb{P}_N(\mathbb{C})$, defined via

$$\mathcal{P}_{\alpha,N}[f] = \sum_{j=0}^{N}(f,\tilde{T}_j)_{L^2_\alpha(-1,1;\mathbb{C})}\tilde{T}_j, \quad \forall f\in L^2_\alpha(-1,1;\mathbb{C}),$$

is called the **Chebyshev projection**.

From the fact that this is a projection, its stability immediately follows.

Proposition 27.18 (stability). *Assume that α is the Chebyshev weight function on $(-1,1)$. Then*

$$\|\mathcal{P}_{\alpha,N}[u]\|_{L^2_\alpha(-1,1)} \le \|u\|_{L^2_\alpha(-1,1)} \tag{27.8}$$

for every $u\in L^2_\alpha(-1,1)$.

Proof. See Problem 27.6. □

Remark 27.19 (boundary values). Note that, even if $u \in H^1_{\alpha,0}(-1,1)$, it is not generally true that $\mathcal{P}_{\alpha,N}[f] \in \mathscr{S}_{N,0}(-1,1)$. This is because there is no mechanism to enforce the boundary conditions $\mathcal{P}_{\alpha,N}[f](\pm 1) = 0$. Consequently, this projection is not useful if homogeneous Dirichlet boundary conditions must be maintained. On the other hand, we still have very good approximation properties for the Chebyshev projection.

Theorem 27.20 (error estimate). *Assume that α is the Chebyshev weight function on $(-1,1)$, $\ell \in \{0,1\}$, and $m, N \in \mathbb{N}$ with $N+1 > m > \ell$. If $u \in H^m_\alpha(-1,1)$, then*

$$\|u - \mathcal{P}_{\alpha,N}[u]\|_{H^\ell_\alpha(-1,1)} \leq C N^{\ell - m} |u|_{H^m_\alpha(-1,1)}$$

for some constant $C > 0$ that is independent of N and u.

Proof. See [13]. □

As a consequence of Theorem 11.15, if $f \in L^2_\alpha(-1,1;\mathbb{C})$, then one can show that

$$f = \sum_{j=0}^{\infty} \left[\frac{2\beta_j^\star}{\pi} (f, T_j)_{L^2_\alpha(-1,1;\mathbb{C})} \right] T_j$$

and

$$\begin{aligned} \|f\|^2_{L^2_\alpha(-1,1;\mathbb{C})} &= \sum_{k=0}^{\infty} \left[\frac{2\beta_k^\star}{\pi} (f, T_k)_{L^2_\alpha(-1,1;\mathbb{C})} \right]^2 \frac{\pi}{2\beta_k^\star} \\ &= \sum_{k=0}^{\infty} \frac{2\beta_k^\star}{\pi} \left| (f, T_k)_{L^2_\alpha(-1,1;\mathbb{C})} \right|^2 \\ &= \frac{1}{\pi} \left| (f, T_0)_{L^2_\alpha(-1,1;\mathbb{C})} \right|^2 + \frac{2}{\pi} \sum_{k=1}^{\infty} \left| (f, T_k)_{L^2_\alpha(-1,1;\mathbb{C})} \right|^2 \\ &= \sum_{k=0}^{\infty} \left| (f, \tilde{T}_k)_{L^2_\alpha(-1,1;\mathbb{C})} \right|^2 . \end{aligned}$$

Like for the trigonometric case, we can define a transform via the Chebyshev expansion.

Definition 27.21 (Chebyshev Transform[5]). Suppose that $v \in L^2_\alpha(-1,1)$. Then its **Finite Chebyshev Transform** is defined as

$$\mathcal{F}_\alpha[v]_j = \hat{v}_j = \frac{2\beta_j^\star}{\pi} \int_{-1}^{1} v(x) T_j(x) \alpha(x) \mathrm{d}x.$$

The numbers $\hat{v}_j$ are called the **Chebyshev coefficients**.

Recall that

$$\ell^2(\mathbb{N}_0) = \left\{ a \colon \mathbb{N}_0 \to \mathbb{R} \;\middle|\; \sum_{j=0}^{\infty} |a_j|^2 < \infty \right\}.$$

[5] Named in honor of the Russian mathematician Pafnuty Lvovich Chebyshev (1821–1894).

From Theorem 11.15, we have, for all $v \in L^2_\alpha(-1,1)$,

$$\mathcal{F}_\alpha[v] \in \ell^2(\mathbb{N}_0).$$

Thus, $\mathcal{F}_\alpha \colon L^2_\alpha(-1,1;\mathbb{R}) \to \ell^2(\mathbb{N}_0;\mathbb{R})$. We can then define the inverse Finite Chebyshev Transform.

Definition 27.22 (inverse Chebyshev Transform). The **inverse Finite Chebyshev Transform**, $\mathcal{F}_\alpha^{-1} \colon \ell^2(\mathbb{N}_0) \to L^2_\alpha(-1,1)$, is defined via

$$\mathcal{F}_\alpha^{-1}[c](x) = \sum_{j=0}^{\infty} c_j T_j(x), \quad \forall c \in \ell^2(\mathbb{N}_0).$$

As a consequence of Theorem 11.15, the Finite Chebyshev Transform is an isometric isomorphism from $L^2_\alpha(-1,1)$ onto $\ell^2(\mathbb{N}_0)$, and $\mathcal{F}_\alpha^{-1}$ is the actual inverse of $\mathcal{F}_\alpha$.

27.4 Chebyshev–Gauss–Lobatto Quadrature and Interpolation

One of the drawbacks of the weighted Galerkin approximation that we introduced earlier is that it requires exact integrations. To circumvent these, let us introduce certain numerical quadratures that will make the numerical methods much more practical.

For the next definition, it is a good idea to have a look back at Theorem 10.17.

Definition 27.23 (CGL nodes). Consider the Chebyshev polynomials, $T_N \in \mathbb{P}_N$, $N \in \mathbb{N}_0$, which are defined in (10.12). For $N \in \mathbb{N}$, the zeros of T_N,

$$z_{N-1,k} = \cos\left(\frac{(2k+1)\pi}{2N}\right), \quad k = 0, \dots, N-1,$$

are called the **Chebyshev nodes** of degree $N-1$. The extreme values of T_N,

$$\zeta_{N,k} = \cos\left(\frac{k\pi}{N}\right), \quad k = 0, \dots, N,$$

which satisfy

$$T_N(\zeta_{N,k}) = (-1)^{N+k}, \quad k = 0, \dots, N,$$

are called the **Chebyshev–Gauss–Lobatto (CGL) nodes**[6] of degree N.

Remark 27.24 (ordering). Observe that the Chebyshev and CGL nodes are numbered in decreasing order. There is some utility in this numbering, as we shall see later when we introduce the Cosine Transform.

Remark 27.25 (equivalent characterization). There are other ways to characterize the CGL nodes. Observe that, for $k = 1, \dots, N-1$,

$$T_N'(\zeta_{N,k}) = 0.$$

[6] Named in honor of the Russian mathematician Pafnuty Lvovich Chebyshev (1821–1894), the German mathematician and physicist Johann Carl Friedrich Gauss (1777–1855), and the Dutch mathematician Rehuel Lobatto (1797–1866).

In other words, the interior CGL nodes, $\{\zeta_{N,k}\}_{k=1}^{N-1}$, are the zeros of T_N'. Therefore, the complete set of CGL nodes, $\{\zeta_{N,k}\}_{k=0}^{N}$, is the zero set for the polynomial

$$C_{N+1}(x) = T_N'(x)(1-x^2). \tag{27.9}$$

Since $C_{N+1} \in \mathbb{P}_{N+1}$, it is clear that there are no other zeros besides these.

On the basis of these nodes, we can construct quadrature rules.

Proposition 27.26 (Chebyshev quadrature). *Assume that α is the Chebyshev weight function on $(-1,1)$. Suppose that $\{\zeta_{N,k}\}_{k=0}^{N}$ is the set of Chebyshev nodes of degree N. Consider the simple quadrature rule* (14.4)

$$Q_\alpha^{(-1,1)}[f] = \sum_{j=0}^{N} \beta_j f(\zeta_{N,j}),$$

where the effective weights are chosen to satisfy the exactness condition (14.7). *Then the quadrature weights are precisely*

$$\beta_j = \beta_j^{\mathrm{C}} = \frac{\pi}{N+1}, \quad j = 0, \dots, N. \tag{27.10}$$

Furthermore, the quadrature rule is consistent to exactly order $2N+1$, i.e.,

$$E_Q[q] = \int_{-1}^{1} q(x)\alpha(x)\mathrm{d}x - Q_\alpha^{(-1,1)}[q] = 0, \quad \forall q \in \mathbb{P}_{2N+1},$$

but there is $q \in \mathbb{P}_{2N+2}$ such that

$$E_Q[q] \neq 0.$$

Proof. The simple quadrature rule is precisely a Gaussian quadrature rule of Chebyshev type. Therefore, Theorem 14.46 applies, and the quadrature rule is consistent to exactly order $2N+1$.

We leave to the reader as an exercise the computation of the weights; see Problem 27.7. □

Theorem 27.27 (CGL quadrature). *Assume that α is the Chebyshev weight function on $(-1,1)$. Suppose that $\{\zeta_{N,k}\}_{k=0}^{N}$ is the set of CGL nodes of degree N. Consider the simple quadrature rule* (14.4)

$$Q_\alpha^{(-1,1)}[f] = \sum_{j=0}^{N} \beta_j f(\zeta_{N,j}),$$

where the effective weights are chosen to satisfy the exactness condition (14.7). *Then the quadrature weights are precisely*

$$\beta_j = \beta_j^{\mathrm{CGL}} = \beta_{N,j}^{\star}\frac{\pi}{N}, \tag{27.11}$$

where

$$\beta_{N,j}^{\star} = \begin{cases} \frac{1}{2}, & j = 0, N, \\ 1, & j = 1, \dots, N-1. \end{cases} \tag{27.12}$$

Furthermore, the quadrature rule is consistent to exactly order $2N - 1$*, i.e.,*

$$E_Q[q] = \int_{-1}^{1} q(x)\alpha(x)\mathrm{d}x - Q_\alpha^{(-1,1)}[q] = 0, \quad \forall q \in \mathbb{P}_{2N-1},$$

but there is $q \in \mathbb{P}_{2N}$*, for which*

$$E_Q[q] \neq 0.$$

Proof. The computation of the weights is left to the reader as an exercise; see Problem 27.8.

Notice that, contrary to Proposition 27.26, the quadrature nodes are not the zeros of an orthogonal polynomial. However, the CGL nodes are the zeros of the polynomial C_{N+1}, defined in (27.9). Now one can utilize Corollary 14.47 to prove the result. In particular, we need to show that

$$(C_{N+1}, q)_{L^2_\alpha(-1,1)} = 0, \quad \forall q \in \mathbb{P}_{N-2},$$

but there is $\tilde{q} \in \mathbb{P}_{N-1}$ such that

$$(C_{N+1}, \tilde{q})_{L^2_\alpha(-1,1)} \neq 0.$$

To establish the first result, observe that, for $j \in \{0, \dots, N-2\}$,

$$\begin{aligned}(C_{N+1}, T_j)_{L^2_\alpha(-1,1)} &= \int_{-1}^{1} T_N'(x) T_j(x) \sqrt{1 - x^2}\, \mathrm{d}x \\ &= N \int_0^\pi \sin(N\theta)\cos(j\theta)\sin(\theta)\, \mathrm{d}\theta \\ &= 0,\end{aligned}$$

where we have used the change of variable $x = \cos(\theta)$. On the other hand,

$$(C_{N+1}, T_{N-1})_{L^2_\alpha(-1,1)} = \frac{\pi}{4}.$$

The result is proved, appealing to Corollary 14.47. □

Next, let us define a discrete inner product.

Definition 27.28 (CGL inner product)**.** Suppose that α is the Chebyshev weight function on $(-1, 1)$. Assume that $N \in \mathbb{N}$ and $\{\zeta_{N,k}\}_{k=0}^N$ is the set of CGL nodes of degree N. Consider the simple quadrature rule (14.4)

$$Q_\alpha^{(-1,1)}[f] = \sum_{j=0}^{N} \beta_j^{\mathrm{CGL}} f(\zeta_{N,j}),$$

where the effective weights, β_j^{CGL}, are defined in (27.11). For $p, q \in \mathbb{P}_N$, define

$$\langle p, q \rangle_{\alpha,N} = Q_\alpha^{(-1,1)}[pq] = \sum_{j=0}^{N} p(\zeta_{N,j}) q(\zeta_{N,j}) \beta_j^{\mathrm{CGL}}.$$

Define, for all $p \in \mathbb{P}_N$,

$$\|p\|_{\alpha,N} = \sqrt{\langle p, p \rangle_{\alpha,N}}.$$

The symmetric bilinear form $\langle \cdot, \cdot \rangle_{\alpha,N}$ is called the **CGL lumped-mass inner product**. The object $\| \cdot \|_{\alpha,N}$ is called the **CGL lumped-mass norm**.

With respect to the CGL lumped-mass inner product, we have the following orthogonality condition.

Proposition 27.29 (orthogonality). *For all* $0 \leq k, \ell \leq N$,

$$\langle T_k, T_\ell \rangle_{\alpha,N} = \frac{\pi}{2\beta^\star_{N,k}} \delta_{k,\ell},$$

where $\{T_k\}_{k=0}^{\infty}$ *is the orthogonal system of Chebyshev polynomials defined in* (10.12), α *is the Chebyshev weight function on* $[-1, 1]$, *and* $\beta^\star_{N,k}$ *is defined in* (27.12).

Proof. Recall, from Proposition 10.15, that Chebyshev polynomials satisfy the orthogonality relation

$$(T_k, T_\ell)_{L^2_\alpha(-1,1)} = \frac{\pi}{2\beta^\star_k} \delta_{k,\ell}, \quad k, \ell \in \mathbb{N}_0.$$

If $0 \leq k < \ell \leq N$, then $T_k T_\ell \in \mathbb{P}_{2N-1}$. Consequently,

$$0 = (T_k, T_\ell)_{L^2_\alpha(-1,1)} = \langle T_k, T_\ell \rangle_{\alpha,N}.$$

Suppose that $k = \ell = N$. Then $T_k T_\ell = T_N^2 \in \mathbb{P}_{2N} \backslash \mathbb{P}_{2N-1}$ and

$$(T_N, T_N)_{L^2_\alpha(-1,1)} \neq \langle T_N, T_N \rangle_{\alpha,N}.$$

But, since $T_N^2(\zeta_{N,j}) = 1$,

$$\begin{aligned}
\langle T_N, T_N \rangle_{\alpha,N} &= \frac{\pi}{2N} T_N^2(-1) + \frac{\pi}{2N} T_N^2(1) + \frac{\pi}{N} \sum_{j=1}^{N-1} T_N^2 \left(-\cos\left(\frac{k\pi}{N} \right) \right) \\
&= 2\frac{\pi}{2N} + (N-1)\frac{\pi}{N} \\
&= \pi.
\end{aligned}$$

The other cases are easily established; see Problem 27.10. □

Since $\mathbb{P}_N$ is finite dimensional, all norms are equivalent on it. In particular, the $L^2_\alpha(-1,1)$-norm and the CGL lumped-mass norm. The equivalence constants, however, may depend on N. The following result shows that this is not the case.

Theorem 27.30 (norm equivalence). *Suppose that* α *is the Chebyshev weight function on* $(-1, 1)$. *Then, for every* $N \in \mathbb{N}$ *and all* $p \in \mathbb{P}_N$,

$$\|p\|_{L^2_\alpha(-1,1)} \leq \|p\|_{\alpha,N} \leq \sqrt{2}\, \|p\|_{L^2_\alpha(-1,1)}.$$

Proof. See Problem 27.11. □

Next, let us define a particular type of interpolation.

Definition 27.31 (CGL interpolation operator)**.** Suppose that $N \in \mathbb{N}$ and $\{\zeta_{N,k}\}_{k=0}^{N}$ is the set of CGL nodes of degree N. The **Chebyshev–Gauss–Lobatto (CGL) interpolation operator** is

$$\mathcal{I}_N^{\mathrm{CGL}} : C([-1,1]) \to \mathbb{P}_N,$$
$$v \mapsto \mathcal{I}_N^{\mathrm{CGL}}[v](x) = \sum_{j=0}^{N} v(\zeta_{N,j}) L_j(x),$$

where L_j is the Lagrange nodal basis element, defined in (9.3), subject to the CGL nodes.

There is an interesting alternate formulation for CGL interpolation, which we now give.

Proposition 27.32 (CGL nodes)**.** *Suppose that $N \in \mathbb{N}$ and $\{\zeta_{N,k}\}_{k=0}^{N}$ is the set of CGL nodes of degree N. Then*

$$L_j(x) = \beta_{N,j}^{\star} \frac{(-1)^{j+1} C_{N+1}(x)}{N^2 (x - \zeta_{N,j})},$$

where L_j is the Lagrange nodal basis element, defined in (9.3), *subject to the CGL nodes; C_{N+1} is defined in* (27.9)*; and $\beta_{N,j}^{\star}$ is defined in* (27.12).

Proof. Recall the alternate formulation for the Lagrange nodal basis element given in (9.9),

$$L_j(x) = \frac{\omega_{N+1}(x)}{(x - \zeta_{N,j}) \omega_{N+1}'(\zeta_{N,j})},$$

where $\omega_{N+1} \in \mathbb{P}_{N+1}$ is the nodal polynomial relative to the CGL nodes

$$\omega_{N+1}(x) = \prod_{k=0}^{N} (x - \zeta_{N,k}).$$

It should be clear that, for some constant $c_{N+1} \neq 0$,

$$\omega_{N+1}(x) = c_{N+1} C_{N+1}(x),$$

since the degrees and zeros of the two polynomials coincide. Thus,

$$L_j(x) = \frac{C_{N+1}(x)}{(x - \zeta_{N,j}) C_{N+1}'(\zeta_{N,j})}.$$

The result follows if we can show that

$$C_{N+1}'(\zeta_{N,j}) = \frac{N^2}{\beta_{N,j}^{\star} (-1)^{j+1}}.$$

We leave those details to the reader as an exercise; see Problem 27.12. □

To conclude this section, we state some stability and convergence results for the interpolation operator. First, we should note that, for all $m \in \mathbb{N}$, $H_{\alpha}^{m}(-1,1) \subset C([-1,1])$. Therefore, interpolation on $H_{\alpha}^{m}(-1,1)$ is well defined. See [13] for a proof of the following two results.

Proposition 27.33 (stability). *Assume that α is the Chebyshev weight function on $(-1,1)$ and $N \in \mathbb{N}$. Then there is a constant $C > 0$ such that*

$$\left\|\mathcal{I}_N^{\mathrm{CGL}}[u]\right\|_{H^1_\alpha(-1,1)} \le C \left\|u\right\|_{H^1_\alpha(-1,1)}$$

for every $u \in H^1_\alpha(-1,1)$.

Theorem 27.34 (error estimate). *Suppose that α is the Chebyshev weight function on $(-1,1)$, $\ell \in \{0,1\}$, and $m, N \in \mathbb{N}$ with $N+1 > m > \ell$. If $u \in H^m_\alpha(-1,1)$, then*

$$\left\|u - \mathcal{I}_N^{\mathrm{CGL}}[u]\right\|_{H^\ell_\alpha(-1,1)} \le CN^{\ell-m}|u|_{H^m_\alpha(-1,1)}$$

for some constant $C > 0$ that is independent of N and u.

27.5 The Discrete Cosine Transform

It turns out that, to compute the CGL interpolation operator, the most useful tool is the Discrete Cosine Transform (DCT). Let us define that now and then show how it can be used in the calculation of the CGL interpolant. To do so, we recall the spaces of grid functions that were introduced in Definition 24.2.

Definition 27.35 (DCT). Suppose that $d = 1$, $N \in \mathbb{N}$, and $h = \frac{1}{N}$. For $u \in \mathcal{V}(\bar{\Omega}_h)$, define

$$\hat{u}_k = \frac{2\beta^\star_{N,k}}{N} \sum_{\ell=0}^{N} \beta^\star_{N,\ell} u_\ell \cos\left(\frac{k\pi\ell}{N}\right), \quad k = 0, \dots, N.$$

The grid function $\hat{u} \in \mathcal{V}(\bar{\Omega}_h)$ is called the **Discrete Cosine Transform** (DCT) of u, and we write $\hat{u} = \mathcal{C}_N[u]$. Given the grid function $v \in \mathcal{V}(\bar{\Omega}_h)$, the **inverse Discrete Cosine Transform** (IDCT) is defined as

$$\mathcal{C}_N^{-1}[v]_k = \sum_{\ell=0}^{N} v_\ell \cos\left(\frac{k\pi\ell}{N}\right), \quad k = 0, \dots, N.$$

To explore the properties of the DCT, we will need the following result.

Proposition 27.36 (discrete orthogonality). *Suppose that $N \in \mathbb{N}$. Then, for all $j, k \in \{0, \dots, N\}$,*

$$\sum_{i=0}^{N} \beta^\star_{N,i} \cos\left(\frac{i\pi j}{N}\right) \cos\left(\frac{i\pi k}{N}\right) = \frac{N}{2\beta^\star_{N,j}} \delta_{j,k}.$$

Proof. Use the result of Proposition 27.29. The details are left to the reader as an exercise; see Problem 27.13. ☐

Theorem 27.37 (bijection). *Suppose that $N \in \mathbb{N}$. Then $\mathcal{C}_N[\cdot] \colon \mathcal{V}(\bar{\Omega}_h) \to \mathcal{V}(\bar{\Omega}_h)$ is a linear bijection and $\mathcal{C}_N^{-1}[\cdot] \colon \mathcal{V}(\bar{\Omega}_h) \to \mathcal{V}(\bar{\Omega}_h)$ is its inverse. Thus, for all $u \in \mathcal{V}(\bar{\Omega}_h)$,*

$$u = \mathcal{C}_N^{-1}[\mathcal{C}_N[u]] = \mathcal{C}_N[\mathcal{C}_N^{-1}[u]].$$

Proof. The map $\mathcal{C}_N[\cdot]$ is clearly linear. Let us show that it is also one to one and onto. Suppose that $u, w \in \mathcal{V}(\bar{\Omega}_h)$ have the same DCT, i.e., for every $k \in \{0, \ldots, N\}$,

$$\frac{2\beta^\star_{N,k}}{N}\sum_{\ell=0}^{N}\beta^\star_{N,\ell}u_\ell\cos\left(\frac{k\pi\ell}{N}\right) = \frac{2\beta^\star_{N,k}}{N}\sum_{\ell=0}^{N}\beta^\star_{N,\ell}w_\ell\cos\left(\frac{k\pi\ell}{N}\right).$$

Thus,

$$0 = \sum_{\ell=0}^{N}\beta^\star_{N,\ell}(u_\ell - w_\ell)\cos\left(\frac{k\pi\ell}{N}\right).$$

Using Proposition 27.36, if $1 \le j \le N-1$,

$$\begin{aligned}
0 &= \frac{2}{N}\sum_{k=0}^{N}\beta^\star_{N,k}\sum_{\ell=0}^{N}\beta^\star_{N,\ell}(u_\ell - w_\ell)\cos\left(\frac{k\pi\ell}{N}\right)\cos\left(\frac{k\pi j}{N}\right)\\
&= \frac{2}{N}\sum_{\ell=0}^{N}\beta^\star_{N,\ell}(u_\ell - w_\ell)\sum_{k=0}^{N}\beta^\star_{N,k}\cos\left(\frac{k\pi\ell}{N}\right)\cos\left(\frac{k\pi j}{N}\right)\\
&= \frac{2}{N}\sum_{\ell=0}^{N}\beta^\star_{N,\ell}(u_\ell - w_\ell)\frac{N}{2}\delta_{\ell,j}\\
&= u_j - w_j.
\end{aligned}$$

A similar result will hold if $j = 0$ or $j = N$. Thus, $u_j = w_j$, for all $j \in \{0, \ldots, N\}$, and $\mathcal{C}_N[\cdot]$ is one to one.

To see that $\mathcal{C}_N[\cdot]$ is surjective on $\mathcal{V}(\bar{\Omega}_h)$, let $w \in \mathcal{V}(\bar{\Omega}_h)$ be arbitrary. We will prove that there is an element $u \in \mathcal{V}(\bar{\Omega}_h)$ such that $w = \mathcal{C}_N[u]$. In particular, define

$$u_k = \sum_{\ell=0}^{N} w_\ell\cos\left(\frac{k\pi\ell}{N}\right), \quad k = 0, \ldots, N.$$

Then, if $1 \le j \le N-1$, using Proposition 27.36 again, we find

$$\begin{aligned}
\mathcal{C}_N[u]_j &= \frac{2\beta^\star_{N,j}}{N}\sum_{k=0}^{N}\beta^\star_{N,k}u_k\cos\left(\frac{j\pi k}{N}\right)\\
&= \frac{2\beta^\star_{N,j}}{N}\sum_{k=0}^{N}\beta^\star_{N,k}\sum_{\ell=0}^{N}w_\ell\cos\left(\frac{k\pi\ell}{N}\right)\cos\left(\frac{j\pi k}{N}\right)\\
&= \frac{2\beta^\star_{N,j}}{N}\sum_{\ell=0}^{N}w_\ell\sum_{k=0}^{N}\beta^\star_{N,k}\cos\left(\frac{k\pi\ell}{N}\right)\cos\left(\frac{j\pi k}{N}\right)\\
&= \frac{2\beta^\star_{N,j}}{N}\sum_{\ell=0}^{N}w_\ell\frac{N}{2}\delta_{\ell,j}\\
&= w_j.
\end{aligned}$$

Next, suppose that $j = 0$. Then

$$\begin{aligned}\mathcal{C}_N[u]_0 &= \frac{1}{N}\sum_{\ell=0}^{N} w_\ell \sum_{k=0}^{N} \beta^\star_{N,k}\cos\left(\frac{k\pi\ell}{N}\right)\cos\left(\frac{0\pi k}{N}\right)\\ &= \frac{1}{N}\sum_{\ell=0}^{N} w_\ell N\delta_{\ell,0}\\ &= w_0.\end{aligned}$$

The case $j = N$ is the same. Thus, for $0 \le j \le N$,

$$\mathcal{C}_N[u]_j = w_j$$

and $\mathcal{C}_N[\cdot]$ is onto $\mathcal{V}(\bar{\Omega}_h)$. Our construction also shows that $\mathcal{C}_N^{-1}[\cdot]$ is the inverse of $\mathcal{C}_N[\cdot]$. The proof is complete. □

Next, we will show the link between the DCT and CGL interpolation.

Proposition 27.38 (DCT and CGL interpolation). *Suppose that $N \in \mathbb{N}$, $\{T_k\}_{k=0}^{\infty}$ is the Chebyshev orthogonal system, $\{\zeta_{N,j}\}_{j=0}^{N}$ is the set of CGL nodes, and $v \in C([-1,1])$. Writing the CGL interpolation operator of v in the Chebyshev basis, i.e.,*

$$\mathcal{I}_N^{\mathrm{CGL}}[v] = \sum_{j=0}^{N} c_j T_j,$$

we find, for all $0 \le j \le N$,

$$c_j = \mathcal{C}_N\left[\mathcal{P}_N^{\mathrm{CGL}}[v]\right]_j,$$

where $\mathcal{P}_N^{\mathrm{CGL}}[\cdot]\colon C([-1,1]) \to \mathcal{V}(\bar{\Omega}_h)$ is the CGL grid projection operator, defined by

$$\mathcal{P}_N^{\mathrm{CGL}}[v]_j = v(\zeta_{N,j}), \quad j = 0, \dots, N.$$

Proof. To simplify notation, let us write

$$v_j = \mathcal{P}_N^{\mathrm{CGL}}[v]_j = v(\zeta_{N,j}), \quad j = 0, \dots, N.$$

Then it follows that, for $0 \le k \le N$,

$$\begin{aligned}v_k &= \sum_{j=0}^{N} c_j T_j(\zeta_{N,k})\\ &= \sum_{j=0}^{N} c_j \cos\left(j \cdot \arccos\left[\cos\left(\frac{k\pi}{N}\right)\right]\right)\\ &= \sum_{j=0}^{N} c_j \cos\left(\frac{k\pi j}{N}\right)\\ &= \mathcal{C}_N^{-1}\left[c\right]_k.\end{aligned}$$

Thus, $c = \mathcal{C}_N\left[\mathcal{P}_N^{\mathrm{CGL}}[v]\right]$. □

Let us now introduce a Discrete Chebyshev Transform. Compare the following with Definition 27.21.

Definition 27.39 (Chebyshev–CGL Transform). Suppose α is the Chebyshev weight function on $(-1,1)$, $N \in \mathbb{N}$, and $\{\zeta_{N,j}\}_{j=0}^{N}$ is the set of CGL nodes. For all $u \in \mathcal{V}(\bar{\Omega}_h)$, define

$$\hat{u}_k = \frac{2\beta^\star_{N,k}}{N} \left\langle \mathcal{I}_N^{\mathrm{CGL}}[u], T_k \right\rangle_{\alpha,N}, \quad k = 0, \dots, N,$$

where

$$\mathcal{I}_N^{\mathrm{CGL}}[u](x) = \sum_{j=0}^{N} u_j L_j(x)$$

and L_j is the jth Lagrange nodal basis element, defined in (9.3). The grid function $\hat{u} \in \mathcal{V}(\bar{\Omega}_h)$ is called the **Discrete Chebyshev–CGL Transform**[7] of u, and we write $\hat{u} = \mathcal{F}_{\alpha,N}[u]$. Given the grid function $v \in \mathcal{V}(\bar{\Omega}_h)$, the **inverse Discrete Chebyshev–CGL Transform** is defined as

$$\mathcal{F}_{\alpha,N}^{-1}[v]_k = \sum_{\ell=0}^{N} v_\ell T_\ell(\zeta_{N,k}), \quad k = 0, \dots, N.$$

Proposition 27.40 (equivalence). *The DCT and Discrete Chebyshev–CGL Transforms are identical on the space $\mathcal{V}(\bar{\Omega}_h)$.*

Proof. See Problem 27.14. □

We have defined this Discrete Chebyshev–CGL Transform for grid functions, but we can define it for continuous functions as well.

Definition 27.41 (Chebyshev–CGL transform). Suppose that α is the Chebyshev weight function on $(-1,1)$, $N \in \mathbb{N}$, and $\{\zeta_{N,j}\}_{j=0}^{N}$ is the set of CGL nodes. For all $f \in C([-1,1])$, define

$$\hat{f}_{N,k} = \frac{2\beta^\star_{N,k}}{N} \left\langle \mathcal{I}_N^{\mathrm{CGL}}[f], T_k \right\rangle_{\alpha,N} = \mathcal{C}_N \left[\mathcal{P}_N^{\mathrm{CGL}}[f] \right]_k, \quad k = 0, \dots, N.$$

The grid function $\hat{f}_N \in \mathcal{V}(\bar{\Omega}_h)$ is called the **Discrete Chebyshev–CGL Transform** of f, and we write $\hat{f}_N = \mathcal{F}_{\alpha,N}[f]$. The numbers $\hat{f}_{N,k}$ are called the **discrete Chebyshev coefficients** of f.

Remark 27.42 (notation). Strictly speaking, we should denote these new objects using the superscript CGL, i.e.,

$$\hat{f}_N^{\mathrm{CGL}} = \mathcal{F}_{\alpha,N}^{\mathrm{CGL}}[f],$$

since other discrete transforms can be chosen by using a nodal set other than the CGL set. We caution the reader that there are several Discrete Chebyshev Transforms and there is no universal naming convention.

[7] Named in honor of the Russian mathematician Pafnuty Lvovich Chebyshev (1821–1894), the German mathematician and physicist Johann Carl Friedrich Gauss (1777–1855), and the Dutch mathematician Rehuel Lobatto (1797–1866).

Suppose that $f \in C([-1,1])$. Assume that $k, N \in \mathbb{N}_0$, with $0 \le k < N$. Then

$$\begin{aligned}
\hat{f}_k &= \frac{2\beta_k^\star}{\pi}\int_{-1}^{1} f(x)T_k(k)\alpha(x)\mathrm{d}x \\
&= \frac{2\beta_{N,k}^\star}{\pi}\int_{-1}^{1} f(x)T_k(k)\alpha(x)\mathrm{d}x \\
&\approx \frac{2\beta_{N,k}^\star}{\pi}\left\langle \mathcal{I}_N^{\mathrm{CGL}}[f], T_k\right\rangle_{\alpha,N} \\
&= \frac{2\beta_{N,k}^\star}{\pi}\sum_{\ell=0}^{N} f(\zeta_{N,\ell})T_k(\zeta_{N,\ell})\beta_{N,\ell}^\star\frac{\pi}{N} \\
&= \frac{2\beta_{N,k}^\star}{N}\sum_{\ell=0}^{N} \mathcal{P}_N^{\mathrm{CGL}}[f]_\ell T_k(\zeta_{N,\ell})\beta_{N,\ell}^\star \\
&= \frac{2\beta_{N,k}^\star}{N}\sum_{\ell=0}^{N} \mathcal{P}_N^{\mathrm{CGL}}[f]_\ell \cos\left(\frac{k\pi\ell}{N}\right)\beta_{N,\ell}^\star \\
&= \mathcal{C}_N\left[\mathcal{P}_N^{\mathrm{CGL}}[f]\right]_k \\
&= \hat{f}_{N,k}.
\end{aligned}$$

We now want to estimate the size of the difference,

$$\hat{f}_k - \hat{f}_{N,k},$$

which is called the *aliasing error*, assuming some regularity of f. We first need the following auxiliary result.

Lemma 27.43 (aliasing at CGL nodes). *Suppose that $N \in \mathbb{N}$ and $\{\zeta_{N,j}\}_{j=0}^{N}$ is the set of CGL nodes. Then*

$$T_0(\zeta_{N,j}) = T_{2\ell N}(\zeta_{N,j}), \quad \forall \ell \in \mathbb{N}, \quad \forall j \in \{0,\dots,N\}, \tag{27.13}$$

$$T_N(\zeta_{N,j}) = T_{2\ell N+1}(\zeta_{N,j}), \quad \forall \ell \in \mathbb{N}, \quad \forall j \in \{0,\dots,N\}. \tag{27.14}$$

For all $m \in \{1,\dots,N-1\}$,

$$T_m(\zeta_{N,j}) = T_{2\ell N\pm m}(\zeta_{N,j}), \quad \forall \ell \in \mathbb{N}, \quad \forall j \in \{0,\dots,N\}. \tag{27.15}$$

Proof. See Problem 27.15. □

For an illustration of Lemma 27.43, see Figure 27.1.

Theorem 27.44 (aliasing error). *Let $N \in \mathbb{N}$. Suppose that $f \in C^2([-1,1])$. Then*

$$\hat{f}_{N,0} = \hat{f}_0 + \sum_{k=1}^{\infty} \hat{f}_{2kN}, \tag{27.16}$$

$$\hat{f}_{N,N} = \hat{f}_N + \sum_{k=1}^{\infty} \hat{f}_{2kN+1}, \tag{27.17}$$

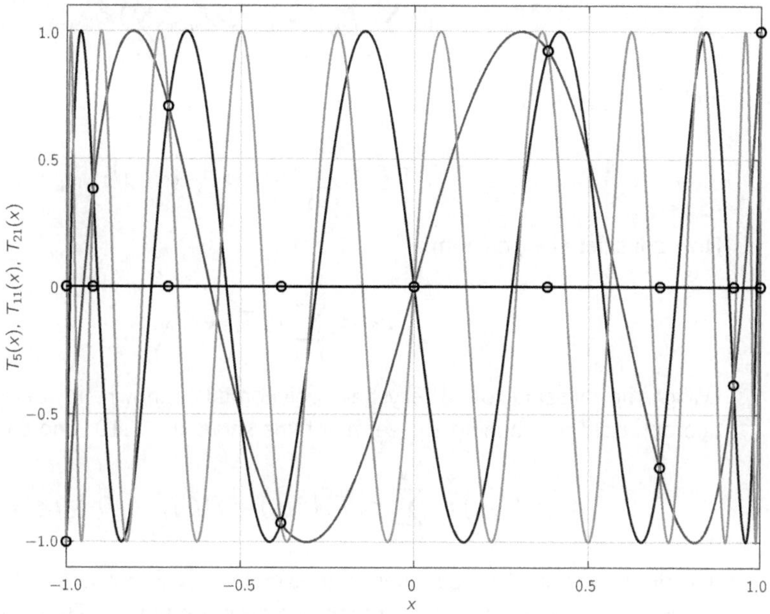

Figure 27.1 Aliasing of the Chebyshev polynomials T_5, T_{11}, and T_{21} on the grid $\{\zeta_{8,j}\}_{j=0}^{8}$, illustrating the results in Lemma 27.43.

and, for $j \in \{1, \ldots, N-1\}$,

$$\hat{f}_{N,j} = \hat{f}_j + \sum_{k=1}^{\infty} \left(\hat{f}_{2kN-j} + \hat{f}_{2kN+j}\right). \tag{27.18}$$

Proof. By Theorem 11.22, the Chebyshev projection of f converges uniformly and absolutely to f on $[-1, 1]$, and we may write

$$f = \sum_{j=0}^{\infty} \hat{f}_j T_j. \tag{27.19}$$

Owing to absolute convergence, we can rearrange the terms of the series, without affecting the result. Furthermore, it follows that

$$\sum_{j=0}^{\infty} |\hat{f}_j| < \infty$$

and, therefore, that the series on the right-hand sides of (27.16)–(27.18) converge absolutely as well. Therefore, we can define the coefficients $\{c_\ell\}_{\ell=0}^{N}$ via

$$c_0 = \hat{f}_0 + \sum_{k=1}^{\infty} \hat{f}_{2kN}, \quad c_N = \hat{f}_N + \sum_{k=1}^{\infty} \hat{f}_{2kN+1},$$

and, for $1 \leq j \leq N-1$,

$$c_j = \hat{f}_j + \sum_{k=1}^{\infty} \left(\hat{f}_{2kN-j} + \hat{f}_{2kN+j} \right).$$

Now consider the polynomial

$$p_N = \sum_{m=0}^{N} c_m T_m \in \mathbb{P}_N.$$

Using the aliasing of Chebyshev polynomials, shown in Lemma 27.43, and appropriately rearranging the terms of the series in 27.19, one can show that

$$p_N(\zeta_{N,j}) = \sum_{m=0}^{N} c_m T_m(\zeta_{N,j}) = f(\zeta_{N,j}), \quad 0 \leq j \leq N.$$

The details are left to the reader as an exercise; see Problem 27.16. This shows that $p_N = \mathcal{I}_N^{\mathrm{CGL}}[f]$, since interpolating polynomials are unique. By Proposition 27.38,

$$\mathcal{I}_N^{\mathrm{CGL}}[f] = \sum_{j=0}^{N} \hat{f}_{N,j} T_j.$$

Therefore,

$$c_j = \hat{f}_{N,j}, \quad 0 \leq j \leq N.$$

The result is proven. □

Remark 27.45 (smoothness). Trefethen [95] shows that one can relax the assumptions for f above. Remarkably, f is only required to be a Lipschitz continuous function on $[-1, 1]$.

27.6 The Chebyshev Collocation Method

Now we are finally ready to define the Chebyshev collocation method.

Definition 27.46 (Chebyshev collocation). Let $N \in \mathbb{N}$ be given. Assume that $f \in L^2_\alpha(-1, 1)$, where α is the Chebyshev weight function on $(-1, 1)$. Set

$$f_N = \mathcal{P}_{\alpha,N}[f] \in \mathbb{P}_N.$$

Suppose that $\{\zeta_{N,k}\}_{k=0}^{N}$ is the set of CGL nodes of degree N. The function

$$w(x) = \sum_{j=0}^{N} w_j L_j(x), \quad L_j(x) = \prod_{\substack{k=0 \\ k \neq j}}^{N} \frac{x - \zeta_{N,k}}{\zeta_{N,j} - \zeta_{N,k}},$$

is called the **Chebyshev collocation approximation** of (27.1) of degree N if and only if

$$w(-1) = w(1) = 0 \tag{27.20}$$

and

$$-w''(\zeta_{N,k}) + cw(\zeta_{N,k}) = f_N(\zeta_{N,k}), \quad 1 \le k \le N-1. \tag{27.21}$$

Notice that (27.20) and (27.21) represent a system of $N+1$ equations in $N+1$ unknowns, the unknowns being the coefficients $w_0, \dots, w_N$. Clearly, (27.20) implies that

$$w_0 = w_N = 0.$$

Now let us use our CGL quadrature rule to define a new bilinear form to replace the one used in the spectral Galerkin method.

Definition 27.47 (discrete bilinear forms). Suppose that $N \in \mathbb{N}$ and α is the Chebyshev weight function on $(-1,1)$. For $u_N, v_N \in \mathscr{S}_{N,0}(-1,1)$, define

$$a_{\alpha,N}(u_N, v_N) = \langle u_N', \alpha^{-1}(\alpha v)' \rangle_{\alpha,N},$$
$$\mathcal{A}_{\alpha,N}(u_N, v_N) = a_{\alpha,N}(u_N, v_N) + c\langle u_N, v_N \rangle_{\alpha,N}.$$

Now let us give, what turns out to be, an alternate formulation of the collocation approximation.

Definition 27.48 (lumped-mass spectral Galerkin). Suppose that α is the Chebyshev weight function on $(-1,1)$, $f \in L^2_\alpha(-1,1)$, and $N \in \mathbb{N}$. Set

$$f_N = \mathcal{P}_{\alpha,N}[f] \in \mathbb{P}_N.$$

We say that $w_N \in \mathscr{S}_{N,0}(-1,1)$ is a **Chebyshev weighted, lumped-mass, spectral Galerkin approximation** of (27.1) if and only if

$$\mathcal{A}_{\alpha,N}(w_N, v_N) = \langle f_N, v \rangle_{\alpha,N}, \quad \forall v_N \in \mathscr{S}_{N,0}(-1,1). \tag{27.22}$$

Remark 27.49 (variational crime). The formulation of our new variational problem in (27.22) — in contrast to the original Galerkin approximation (27.5) — involves what is known as a *variational crime*. This is because the integrals are replaced by quadratures. A standard Galerkin approximation always uses the original bilinear forms restricted to finite-dimensional subspaces. This change means that our method of error analysis for (27.5), facilitated by the weighted Céa's Lemma of Theorem 27.13, no longer applies.

Theorem 27.50 (consistency). *Suppose that α is the Chebyshev weight function on $(-1,1)$ and $N \in \mathbb{N}$. For all $u_N, v_N \in \mathscr{S}_{N,0}(-1,1)$,*

$$a_{\alpha,N}(u_N, v_N) = a_\alpha(u_N, v_N). \tag{27.23}$$

Furthermore, $w_N \in \mathscr{S}_{N,0}(-1,1)$ solves (27.20) *and* (27.21) *if and only if it solves* (27.22).

Proof. Let us begin by showing (27.23). Our first task will be to show that, for $v_N \in \mathscr{S}_{N,0}(-1,1)$,

$$\alpha^{-1}(\alpha v_N)' \in \mathbb{P}_{N-1}.$$

Expanding the expression, we find

$$\alpha^{-1}(x)\left(\alpha(x)v_N(x)\right)' = v_N'(x) + \frac{x}{1-x^2}v_N(x), \quad \forall x \in (-1,1).$$

Since $v_N \in \mathscr{S}_{N,0}(-1,1)$, we have that $v_N' \in \mathbb{P}_{N-1}$, and that there is $r \in \mathbb{P}_{N-2}$ such that

$$v_N(x) = r(x)(x-1)(x+1).$$

Thus, $\frac{x}{1-x^2}v_N(x) = xr(x) \in \mathbb{P}_{N-1}$, and the first task is complete.

Next, it follows that

$$\alpha^{-1}(\alpha v_N)'\, u_N' \in \mathbb{P}_{2N-2} \subset \mathbb{P}_{2N-1},$$

since $u_N \in \mathbb{P}_N$. By Theorem 27.27,

$$a_{\alpha,N}(u_N, v_N) = a_\alpha(u_N, v_N), \quad \forall u_N, v_N \in \mathscr{S}_{N,0}(-1,1),$$

as claimed.

To show the equivalence, notice, first of all, that using integration by parts, the homogeneous Dirichlet boundary conditions, and Theorem 27.27 again, we have

$$\begin{aligned} a_{\alpha,N}(u_N, v_N) &= \langle u_N', \alpha^{-1}(\alpha v_N)'\rangle_{\alpha,N} \\ &= \int_{-1}^{1} \alpha^{-1}(x)\left(\alpha(x)v_N(x)\right)' u_N'(x)\alpha(x)\mathrm{d}x \\ &= \int_{-1}^{1} \left(\alpha(x)v_N(x)\right)' u_N'(x)\mathrm{d}x \\ &= -\int_{-1}^{1} \alpha(x)v_N(x)u_N''(x)\,\mathrm{d}x \\ &= -\langle u_N'', v_N\rangle_{\alpha,N}. \end{aligned}$$

($\Longleftarrow$) Suppose that $w_N \in \mathscr{S}_{N,0}(-1,1)$ solves (27.22). Then the computations above show that

$$\langle -w_N'' + cw_N - f_N, v_N\rangle_{\alpha,N} = 0$$

for all $v_N \in \mathscr{S}_{N,0}(-1,1)$. We can pick $v_N = L_k \in \mathscr{S}_{N,0}(-1,1)$, the kth Lagrange nodal basis function, with $1 \le k \le N-1$. This implies that

$$-w_N''(\zeta_{N,k}) + cw_N(\zeta_{N,k}) - f_N(\zeta_{N,k}) = 0, \quad 1 \le k \le N-1,$$

which proves that w_N is a Chebyshev collocation approximation.

($\Longrightarrow$) This direction is similar and is left to the reader as an exercise; see Problem 27.17. □

The last result makes a remarkable connection between Chebyshev collocation and Chebyshev weighted, lumped-mass, spectral Galerkin methods. Therefore, providing an analysis for one immediately implies providing an analysis for the other. Let us then provide this analysis.

Theorem 27.51 (coercivity). *Suppose that* α *is the Chebyshev weight function on* $(-1,1)$. *Then* $a_{\alpha,N}(\cdot,\cdot)$ *is a nonsymmetric bilinear form and there are constants* $0 < C_{\alpha,3} \leq C_{\alpha,4}$, *which are independent of* N, *such that*

$$|a_{\alpha,N}(u_N, v_N)| \leq C_{\alpha,4} \|u_N\|_{H^1_{\alpha,0}(-1,1)} \|v_N\|_{H^1_{\alpha,0}(-1,1)}, \quad \forall u_N, v_N \in \mathscr{S}_{N,0}(-1,1),$$

and

$$C_{\alpha,3} \|u_N\|^2_{H^1_{\alpha,0}(-1,1)} \leq a_{\alpha,N}(u_N, u_N), \quad \forall u_N \in \mathscr{S}_{N,0}(-1,1).$$

In particular, one can prove that, as for the continuous case,

$$C_{\alpha,3} = C_{\alpha,1} = \frac{1}{4}, \qquad C_{\alpha,4} = C_{\alpha,2} = 1 + \sqrt{\frac{8}{3}}.$$

Proof. This follows immediately from Theorem 27.9 and the fact that

$$a_{\alpha,N}(u_N, v_N) = a_\alpha(u_N, v_N), \quad \forall u_N, v_N \in \mathscr{S}_{N,0}(-1,1). \qquad \square$$

Theorem 27.52 (uniform well-posedness). *Suppose that* α *is the Chebyshev weight function on* $(-1,1)$ *and* $f \in L^2_\alpha(-1,1)$. *For all* $N \in \mathbb{N}$, *there is a unique function* $w_N \in \mathscr{S}_{N,0}(-1,1)$ *that solves (27.22), or, equivalently, (27.20) and (27.21). Furthermore,* w_N *has the stability property*

$$\|w_N\|_{H^1_{\alpha,0}(-1,1)} \leq 8C_{\alpha,P} \|f\|_{L^2_\alpha(-1,1)}.$$

Proof. We will apply the Lax–Milgram Lemma (Theorem 23.10) to show that (27.22) always has a unique solution. By Theorem 27.51, the Cauchy–Schwarz inequality, Theorem 27.30, and the Poincaré inequality, for all $u_N, v_N \in \mathscr{S}_{N,0}(-1,1)$, we have

$$\begin{aligned}
|\mathcal{A}_{\alpha,N}(u_N, v_N)| &\leq |a_{\alpha,N}(u_N, v_N)| + |c\langle u, v\rangle_{\alpha,N}| \\
&\leq C_{\alpha,2} \|u_N\|_{H^1_{\alpha,0}(-1,1)} \|v_N\|_{H^1_{\alpha,0}(-1,1)} + c \|u_N\|_{\alpha,N} \|v_N\|_{\alpha,N} \\
&\leq C_{\alpha,2} \|u_N\|_{H^1_{\alpha,0}(-1,1)} \|v_N\|_{H^1_{\alpha,0}(-1,1)} + 2c \|u_N\|_{L^2_\alpha(-1,1)} \|v_N\|_{L^2_\alpha(-1,1)} \\
&\leq C_{\alpha,2} \|u_N\|_{H^1_{\alpha,0}(-1,1)} \|v_N\|_{H^1_{\alpha,0}(-1,1)} \\
&\quad + 2cC^2_{\alpha,P} \|u_N\|_{H^1_{\alpha,0}(-1,1)} \|v_N\|_{H^1_{\alpha,0}(-1,1)} \\
&= \hat{C}_{\alpha,4} \|u_N\|_{H^1_{\alpha,0}(-1,1)} \|v_N\|_{H^1_{\alpha,0}(-1,1)},
\end{aligned}$$

where

$$\hat{C}_{\alpha,4} = C_{\alpha,2} + 2cC^2_{\alpha,P}.$$

This proves that $\mathcal{A}_{\alpha,N}(\cdot,\cdot)$ is continuous on the subspace $\mathscr{S}_{N,0}(-1,1)$.

The coercivity of $\mathcal{A}_{\alpha,N}(\cdot,\cdot)$ over the subspace $\mathscr{S}_{N,0}(-1,1)$ is simpler: by Theorem 27.51, for all $u_N \in \mathscr{S}_{N,0}(-1,1)$,

$$\mathcal{A}_{\alpha,N}(u_N, u_N) = a_{\alpha,N}(u_N, u_N) + c\langle u_N, u_N\rangle_{\alpha,N} \geq \hat{C}_{\alpha,3} \|u_N\|^2_{H^1_{\alpha,0}(-1,1)},$$

where

$$\hat{C}_{\alpha,3} = C_{\alpha,3} = C_{\alpha,1} = \frac{1}{4}.$$

Next, let us define

$$F_{\alpha,N}(v) = \langle f_N, v_N \rangle_{\alpha,N}, \quad \forall v_N \in \mathscr{S}_{N,0}(-1,1),$$

where $f_N = \mathcal{P}_{\alpha,N}[f]$. Then, using Theorem 27.30, the stability of the Chebyshev projection (27.8), and the weighted Poincaré inequality (27.3), we get

$$\begin{aligned} |F_{\alpha,N}(v_N)| &= |\langle f_N, v_N \rangle_{\alpha,N}| \\ &\leq \|f_N\|_{\alpha,N} \|v_N\|_{\alpha,N} \\ &\leq 2\|f_N\|_{L^2_\alpha(-1,1)} \|v_N\|_{L^2_\alpha(-1,1)} \\ &\leq 2\|f\|_{L^2_\alpha(-1,1)} \|v_N\|_{L^2_\alpha(-1,1)} \\ &\leq 2C_{\alpha,P}\|f\|_{L^2_\alpha(-1,1)} \|v_N\|_{H^1_\alpha(-1,1)} \end{aligned}$$

for all $v \in \mathscr{S}_{N,0}(-1,1)$. This shows that $F_{\alpha,N}$ is a bounded linear functional on $\mathscr{S}_{N,0}(-1,1)$ and

$$\sup_{\substack{v \in \mathscr{S}_{N,0}(-1,1) \\ v \neq 0}} \frac{|F_{\alpha,N}(v)|}{\|v\|_{H^1_{\alpha,0}(-1,1)}} \leq 2C_{\alpha,P}\|f\|_{L^2_\alpha(-1,1)}.$$

By the Lax–Milgram Lemma, Theorem 23.10, problem (27.22) has a unique solution $w_N \in \mathscr{S}_{N,0}(-1,1)$, and the stability estimate follows as well.

The proof is complete. □

27.7 Error Analysis of the Chebyshev Collocation Method

Since we have committed *variational crimes*, we can no longer use weighted Céa's Lemma, Theorem 27.13, for our analysis. Consequently, we need to chart a new route; ours will be based on Strang's First Lemma.[8]

Theorem 27.53 (Strang). *Suppose that $\mathcal{H}$ is real Hilbert space with the inner product $(\cdot,\cdot)_{\mathcal{H}}$ and induced norm $\|\cdot\|_{\mathcal{H}}$. Assume that $\mathcal{A}\colon \mathcal{H}\times\mathcal{H} \to \mathbb{R}$ is a bounded and coercive bilinear form with constants C_2 and C_1, respectively. Let $F \in \mathcal{H}'$ and $u \in \mathcal{H}$ be the (unique) solution to*

$$\mathcal{A}(u, v) = F(v), \quad \forall v \in \mathcal{H}.$$

Assume that, for $n \in \mathbb{N}$, we have $\mathcal{H}_n \leq \mathcal{H}$ with $\dim \mathcal{H}_n = n$ and $\mathcal{A}_n\colon \mathcal{H}_n\times\mathcal{H}_n \to \mathbb{R}$ is a bounded and coercive bilinear form with constants C_4 and C_3. Let $F_n \in \mathcal{H}_n'$. Denote by $u_n \in \mathcal{H}_n$ the solution to

$$\mathcal{A}_n(u_n, v_n) = F_n(v_n), \quad \forall v_n \in \mathcal{H}_n.$$

[8] Named in honor of the American mathematician William Gilbert Strang (1934–).

Then the following error estimate holds:

$$\|u-u_n\|_{\mathcal{H}} \le \inf_{w_n\in\mathcal{H}_n}\left[\left(1+\frac{C_2}{C_3}\right)\|u-w_n\|_{\mathcal{H}} + \frac{1}{C_3}\sup_{\substack{v_n\in\mathcal{H}_n\\ v_n\neq 0}}\frac{|\mathcal{A}(w_n,v_n)-\mathcal{A}_n(w_n,v_n)|}{\|v_n\|_{\mathcal{H}}}\right] + \frac{1}{C_3}\sup_{\substack{v_n\in\mathcal{H}_n\\ v_n\neq 0}}\frac{|F(v_n)-F_n(v_n)|}{\|v_n\|_{\mathcal{H}}}. \tag{27.24}$$

Proof. The fact that $u\in\mathcal{H}$ and $u_n\in\mathcal{H}_n$ exist and are unique is guaranteed by the Lax–Milgram Lemma, Theorem 23.10. Let us then obtain the error estimate.

Suppose that $w_n\in\mathcal{H}_n$ is arbitrary. Set $e_n=u_n-w_n\in\mathcal{H}_n$. Then, by coercivity of the discrete bilinear form,

$$\begin{aligned} C_3\|e_n\|_{\mathcal{H}}^2 &\le \mathcal{A}_n(e_n,e_n)\\ &= \mathcal{A}(u-w_n,e_n)+\mathcal{A}(w_n,e_n)-\mathcal{A}_n(w_n,e_n)+F_n(e_n)-F(e_n). \end{aligned}$$

Assuming that $e_n\neq 0$, it follows that

$$C_3\|e_n\|_{\mathcal{H}} \le C_2\|u-w_n\|_{\mathcal{H}} + \frac{|\mathcal{A}(w_n,e_n)-\mathcal{A}_n(w_n,e_n)|}{\|e_n\|_{\mathcal{H}}} + \frac{|F_n(e_n)-F(e_n)|}{\|e_n\|_{\mathcal{H}}}.$$

Using the triangle inequality,

$$\|u-u_n\|_{\mathcal{H}} \le \|u-w_n\|_{\mathcal{H}} + \|e_n\|_{\mathcal{H}},$$

and we have

$$\|u-u_n\|_{\mathcal{H}} \le \left(1+\frac{C_2}{C_3}\right)\|u-w_n\|_{\mathcal{H}} + \frac{1}{C_3}\frac{|\mathcal{A}(w_n,e_n)-\mathcal{A}_n(w_n,e_n)|}{\|e_n\|_{\mathcal{H}}} + \frac{1}{C_3}\frac{|F(e_n)-F_n(e_n)|}{\|e_n\|_{\mathcal{H}}}.$$

Taking suprema,

$$\|u-u_n\|_{\mathcal{H}} \le \left(1+\frac{C_2}{C_3}\right)\|u-w_n\|_{\mathcal{H}} + \frac{1}{C_3}\sup_{\substack{v_n\in\mathcal{H}_n\\ v_n\neq 0}}\frac{|\mathcal{A}(w_n,v_n)-\mathcal{A}_n(w_n,v_n)|}{\|v_n\|_{\mathcal{H}}} + \sup_{\substack{v_n\in\mathcal{H}_n\\ v_n\neq 0}}\frac{1}{C_3}\frac{|F(v_n)-F_n(v_n)|}{\|v_n\|_{\mathcal{H}}}.$$

Finally, since this last result holds for an arbitrary $w_n\in\mathcal{H}_n$, we can take infimum to arrive at the the error estimate (27.24). □

With the aid of Strang's First Lemma we can provide an error estimate for the solution of (27.22).

Theorem 27.54 (error estimate)**.** *Suppose that α is the Chebyshev weight function on $(-1,1)$, $m\in\mathbb{N}$, $f\in H_\alpha^m(-1,1)$, and $u\in H_{\alpha,0}^1(-1,1)\cap H_\alpha^m(-1,1)$ is the unique* *solution to (27.4). For all $N \in \mathbb{N}$ with $N+1 > m$, we have that $w_N \in \mathscr{S}_{N,0}(-1,1)$, the unique solution to (27.22), satisfies*

$$\|u - w_N\|_{H^1_\alpha(-1,1)} \le C(N-1)^{1-m}|u|_{H^m_\alpha(-1,1)} + C(N-1)^{-m}|f|_{H^m_\alpha(-1,1)}$$

for some constant $C > 0$ that is independent of N.

Proof. Following the notations that define (27.4) and (27.22), we have, for any $v_N \in \mathscr{S}_{N,0}(-1,1)$,

$$\begin{aligned}
|F_N(v_N) - F(v_N)| &= \left|\langle f_N, v_N\rangle_{\alpha,N} - (f, v_N)_{L^2_\alpha(-1,1)}\right| \\
&= \left|\langle f_N - f_{N-1} + f_{N-1}, v_N\rangle_{\alpha,N} - (f, v_N)_{L^2_\alpha(-1,1)}\right| \\
&= \left|\langle f_N - f_{N-1}, v_N\rangle_{\alpha,N} - (f - f_{N-1}, v_N)_{L^2_\alpha(-1,1)}\right| \\
&\le \|f_N - f_{N-1}\|_{\alpha,N}\,\|v_N\|_{\alpha,N} + \|f - f_{N-1}\|_{L^2_\alpha(-1,1)}\,\|v_N\|_{L^2_\alpha(-1,1)} \\
&\le \left(2\,\|f_N - f_{N-1}\|_{L^2_\alpha(-1,1)} + \|f - f_{N-1}\|_{L^2_\alpha(-1,1)}\right)\|v_N\|_{L^2_\alpha(-1,1)} \\
&\le C_{\alpha,P}\left(2\,\|f_N - f_{N-1}\|_{L^2_\alpha(-1,1)} + \|f - f_{N-1}\|_{L^2_\alpha(-1,1)}\right) \\
&\quad \times \|v_N\|_{H^1_{\alpha,0}(-1,1)}\,.
\end{aligned}$$

Using Theorem 27.20,

$$\begin{aligned}
\|f_N - f_{N-1}\|_{L^2_\alpha(-1,1)} &\le \|f - f_N\|_{L^2_\alpha(-1,1)} + \|f - f_{N-1}\|_{L^2_\alpha(-1,1)} \\
&\le CN^{-m}|f|_{H^m_\alpha(-1,1)} + C(N-1)^{-m}|f|_{H^m_\alpha(-1,1)} \\
&\le C(N-1)^{-m}|f|_{H^m_\alpha(-1,1)}
\end{aligned}$$

for some constant $C > 0$ that is independent of N. Therefore,

$$\frac{|F_N(v_N) - F(v_N)|}{\|v_N\|_{H^1_{\alpha,0}(-1,1)}} \le C(N-1)^{-m}|f|_{H^m_\alpha(-1,1)}$$

for all nonzero $v_N \in \mathscr{S}_{N,0}(-1,1)$.

Next, recall that, for $\ell \in \mathbb{N}$, $P^{\ell,1}_{\alpha,0}[u] \in \mathscr{S}_{\ell,0}(-1,1)$ denotes the weighted elliptic projection defined in (27.6). Using the consistency shown in (27.23), we have, for any $v \in \mathscr{S}_{N,0}(-1,1)$,

$$\begin{aligned}
\left|\mathcal{A}_\alpha(P^{N,1}_{\alpha,0}[u], v) - \mathcal{A}_{\alpha,N}(P^{N,1}_{\alpha,0}[u], v)\right| &= c\left|\left(P^{N,1}_{\alpha,0}[u], v\right)_{L^2_\alpha(-1,1)} - \langle P^{N,1}_{\alpha,0}[u], v\rangle_{\alpha,N}\right| \\
&= c\Big|\left(P^{N,1}_{\alpha,0}[u] - P^{N-1,1}_{\alpha,0}[u], v\right)_{L^2_\alpha(-1,1)} \\
&\quad - \langle P^{N,1}_{\alpha,0}[u] - P^{N-1,1}_{\alpha,0}[u], v\rangle_{\alpha,N}\Big| \\
&\le 3c\,\|v\|_{L^2_\alpha(-1,1)} \\
&\quad \times \left\|P^{N,1}_{\alpha,0}[u] - P^{N-1,1}_{\alpha,0}[u]\right\|_{L^2_\alpha(-1,1)} \\
&\le 3cC^2_{\alpha,P}\,\|v\|_{H^1_{\alpha,0}(-1,1)} \\
&\quad \times \left\|P^{N,1}_{\alpha,0}[u] - P^{N-1,1}_{\alpha,0}[u]\right\|_{H^1_{\alpha,0}(-1,1)}\,.
\end{aligned}$$

Using the triangle inequality and Theorem 27.15, we have

$$
\begin{aligned}
\left\|P_{\alpha,0}^{N,1}[u] - P_{\alpha,0}^{N-1,1}[u]\right\|_{H^1_{\alpha,0}(-1,1)} &\leq \left\|P_{\alpha,0}^{N,1}[u] - u\right\|_{H^1_{\alpha,0}(-1,1)} \\
&\quad + \left\|u - P_{\alpha,0}^{N-1,1}[u]\right\|_{H^1_{\alpha,0}(-1,1)} \\
&\leq CN^{1-m}|u|_{H^m_\alpha(-1,1)} + C(N-1)^{1-m}|u|_{H^m_\alpha(-1,1)} \\
&\leq C(N-1)^{1-m}|u|_{H^m_\alpha(-1,1)}.
\end{aligned}
$$

Then, provided that $v \neq 0$,

$$
\frac{\left|\mathcal{A}_\alpha(P_{\alpha,0}^{N,1}[u], v) - \mathcal{A}_{\alpha,N}(P_{\alpha,0}^{N,1}[u], v)\right|}{\|v\|_{H^1_{\alpha,0}(-1,1)}} \leq C(N-1)^{1-m}|u|_{H^m_\alpha(-1,1)}.
$$

Now, by Theorem 27.53, in particular estimate (27.24) and the estimates we have just constructed,

$$
\begin{aligned}
\|u - u_N\|_{H^1_{\alpha,0}(-1,1)} &\leq \left(1 + \frac{C_{\alpha,2} + \gamma C_{\alpha,P}^2}{C_{\alpha,1}}\right)\left\|u - P_{\alpha,0}^{N,1}[u]\right\|_{H^1_{\alpha,0}(-1,1)} \\
&\quad + \frac{1}{C_{\alpha,1}} \sup_{\substack{v_N \in \mathscr{S}_{N,0}(-1,1) \\ v_N \neq 0}} \frac{\left|\mathcal{A}_\alpha(P_{\alpha,0}^{N,1}[u], v_N) - \mathcal{A}_{\alpha,N}(P_{\alpha,0}^{N,1}[u], v_N)\right|}{\|v_N\|_{H^1_{\alpha,0}(-1,1)}} \\
&\quad + \frac{1}{C_{\alpha,1}} \sup_{\substack{v_N \in \mathscr{S}_{N,0}(-1,1) \\ v_N \neq 0}} \frac{|F(v_N) - F_N(v_N)|}{\|v_N\|_{H^1_{\alpha,0}(-1,1)}} \\
&\leq C(N-1)^{1-m}|u|_{H^m_\alpha(-1,1)} + C(N-1)^{-m}|f|_{H^m_\alpha(-1,1)}
\end{aligned}
$$

for some constant $C > 0$ that is independent of N. The proof is complete. □

27.8 Practical Computation of the Collocation Approximation

We have just learned that our collocation approximation $w_N \in \mathscr{S}_{N,0}(-1,1)$ to the solution of (27.1) is spectrally accurate, i.e.,

$$
\|u - w_N\|_{H^1_\alpha(-1,1)} \leq C(N-1)^{1-m}|u|_{H^m_\alpha(-1,1)} + C(N-1)^{-m}|f|_{H^m_\alpha(-1,1)}.
$$

The next question then is this: How do we compute this approximation? Can this be done efficiently? We will not explore those details here, but see the books by Shen *et al.* [85], Canuto *et al.* [13], and Trefethen [94, 95]. The methods described in Trefethen's books, which are used in the next example, are not "fast" methods per se, though they are simple, and not too expensive in one space dimension, since, often, N needs not be too large to resolve the solution to within a very small tolerance. Essentially, one computes the dense matrix A, as in (27.2), and inverts it by Gaussian elimination. There are fast methods that can be employed, as described in the references above, but we will not discuss these for the sake of brevity.

Example 27.1 In this example, we approximate the solution of the following two-point BVP:

$$- u'' + u = -\frac{15\exp(4x) + \sinh(4)x + \cosh(4)}{16}, \quad u(-1) = u(1) = 0. \quad (27.25)$$

The exact solution is

$$u(x) = \frac{\exp(4x) - \sinh(4)x - \cosh(4)}{16}.$$

For the sake of comparison, we compute approximations based on both the finite difference method of Chapter 24 and the Chebyshev collocation method developed in this chapter; see Figures 27.2 ($N = 8$) and 27.3 ($N = 24$). The errors for both methods are reported in Figure 27.4. The method for solving the linear equation $A\boldsymbol{w} = \boldsymbol{f}$ that arises in the approximation of the Chebyshev collocation method is based on the simple, and not so efficient, strategy in Trefethen [94]. In fact, we have modified one of the codes from [94] for our example.

In Figure 27.4, we compare the errors for the Chebyshev collocation and finite difference methods in the approximation of the solution of the BVP (27.25) for the values $N = 8$, 12, 16, 20, 24, 32, 40, 48, 64, and 80. The Chebyshev method is spectrally accurate, as is confirmed in Figure 27.4. It requires a relatively small value of N, $N = 20$, in particular, to obtain a very accurate solution. Roundoff errors dominate the collocation approximations for larger values of N. The finite difference method, as we proved in Chapter 24, is second-order accurate. This is also confirmed in the present example. While, for $N = 24$, the finite difference approximation looks very good to the human eye, its error is, astonishingly, about 10^{12} times larger than that of the collocation method with the same value of N.

Problems

27.1 Let α be the Chebyshev weight function on $[-1, 1]$. Show that:

a) This is a *doubling* weight: There is a constant $C > 0$ such that, for every $a \in (0, \frac{1}{2}]$, we have

$$\int_{-2a}^{2a} \alpha(x)\mathrm{d}x \leq C \int_{-a}^{a} \alpha(x)\mathrm{d}x.$$

b) There is a constant $C_\alpha > 0$ such that, for all $a \in (0, 1]$, we have

$$\frac{1}{4a^2} \int_{-a}^{a} \alpha(x)\mathrm{d}x \int_{-a}^{a} \alpha(x)^{-1}\,\mathrm{d}x < C_\alpha.$$

Hint: Consider the change of variable $x = \cos t$.

c) This is a *strong doubling* weight: There is a constant $C > 0$ such that, for all $a \in (0, 1]$, we have

$$\int_{-1}^{1} \alpha(x)\mathrm{d}x \leq \frac{C}{a^2} \int_{-a}^{a} \alpha(x)\mathrm{d}x.$$

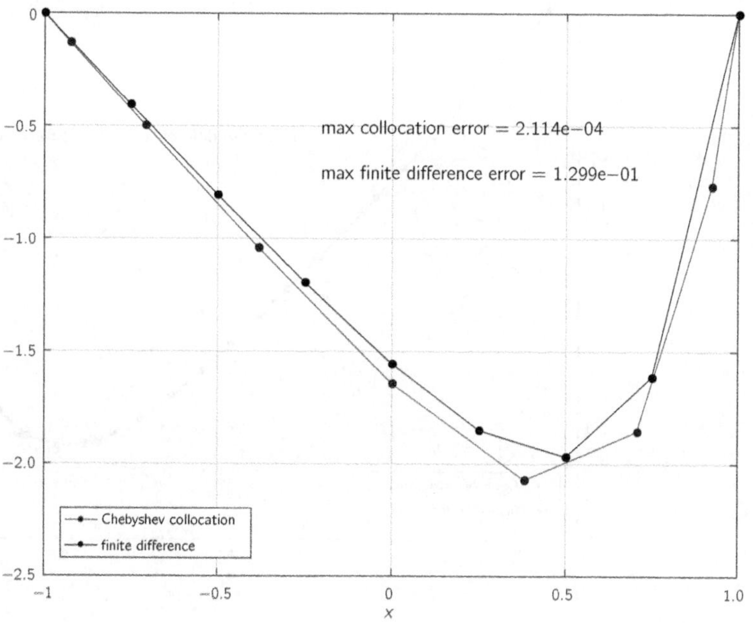

Figure 27.2 Comparison of Chebyshev collocation and finite difference approximations of the BVP (27.25) with $N = 8$.

Hint: Start from $2a = \int_{-a}^{a} 1\,\mathrm{d}x$, apply the Cauchy–Schwarz inequality, and use the previous result.

27.2 Show that, for all $N \in \mathbb{N}$,

$$\mathscr{S}_{N,0}(-1,1) \subset H^1_{\alpha,0}(-1,1).$$

27.3 Let α be the Chebyshev weight function on $[-1,1]$, $q \geq 2$, and ρ be another weight function on $[-1,1]$. Assume that there is a constant $C > 0$ such that, for all $a \in (0,1]$,

$$\int_{-a}^{a} \rho(x)\mathrm{d}x \left(\int_{-a}^{a} \alpha(x)\mathrm{d}x \right)^{-q/2} \leq C.$$

Show that there is a constant $C_{q,\rho} > 0$ such that, for all $v \in H^1_{\alpha,0}(-1,1)$, we have

$$\left(\int_{-1}^{1} |v(x)|^q \rho(x)\mathrm{d}x \right)^{1/q} \leq C_{q,\rho} \|v'\|_{L^2_\alpha(-1,1)} .$$

27.4 Prove Theorem 27.13.

Hint: Revisit Theorem 25.6.

27.5 Prove Theorem 27.16.

27.6 Prove Proposition 27.18.

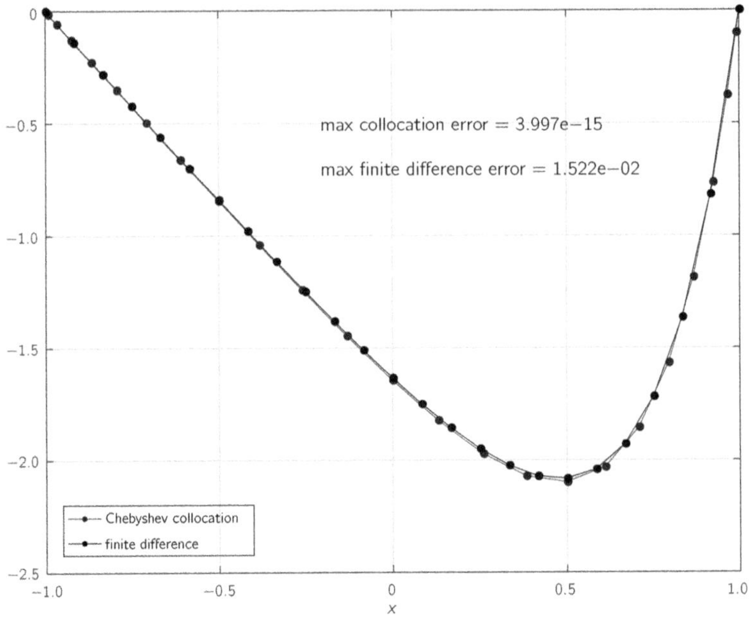

Figure 27.3 Comparison of Chebyshev collocation and finite difference approximations of the BVP (27.25) with $N = 24$. While the finite difference approximation looks very good to the human eye, its error is, astonishingly, about 10^{12} times larger than that of the collocation method.

27.7 Complete the proof of Proposition 27.26.

27.8 Complete the proof of Theorem 27.27.

27.9 Prove that the CGL lumped-mass inner product, introduced in Definition 27.28, is indeed an inner product on $\mathbb{P}_N$.

27.10 Complete the proof of Proposition 27.29.

27.11 Prove Theorem 27.30.

27.12 Complete the proof of Proposition 27.32.

27.13 Complete the proof of Proposition 27.36.

27.14 Prove Proposition 27.40.

27.15 Prove Lemma 27.43.

27.16 Complete the proof of Theorem 27.44.

27.17 Complete the proof of Theorem 27.50.

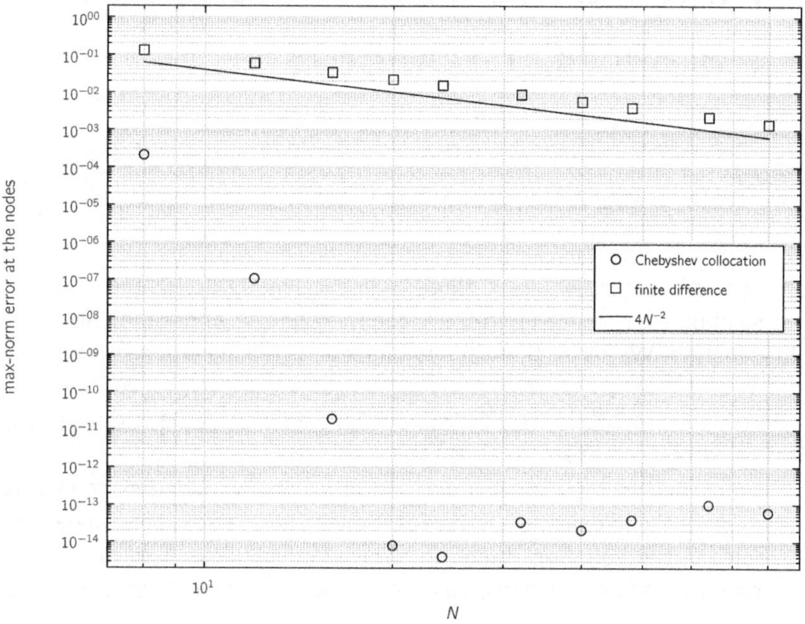

Figure 27.4 Comparison of Chebyshev collocation and finite difference errors in the approximations of the solution of the BVP (27.25) for the values $N = 8$, 12, 16, 20, 24, 32, 40, 48, 64, and 80. The Chebyshev method is spectrally accurate and requires a relatively small value of N, $N = 20$, in particular, to obtain a very accurate solution. Roundoff errors dominate the collocation solutions for larger values of N.

28 Finite Difference Methods for Parabolic Problems

In this chapter, we study finite difference methods for the approximation of the solution to a parabolic problem, focusing on the solution of the heat, or diffusion, equation. We recall that, in one dimension, this equation has the form

$$\frac{\partial u(x,t)}{\partial t} - \frac{\partial^2 u(x,t)}{\partial x^2} = f(x,t), \qquad (x,t) \in \Omega \times (0,T],$$

supplemented by suitable initial and, if necessary, boundary conditions.

Since the independent variables, time t and space x, play fundamentally different roles in the equation, there are two prevalent ideas used in the discretization of time-dependent problems, based on which of the variables is discretized first.

- The *method of lines*: Where one applies any of the methods discussed in Chapters 24–27 to discretize the spatial variable to obtain, for some finite-dimensional space $\mathcal{S}$, the problem: Find $u_{\mathcal{S}} \in C([0,T];\mathcal{S})$ that is a mild solution to

$$u_{\mathcal{S}}'(t) = F_{\mathcal{S}}(t, u_{\mathcal{S}}(t)), \qquad t \in (0,T],$$

  supplemented by suitable initial conditions. This is now an initial value problem (IVP) that can be treated with any of the methods in Part IV. The origin of the name of this approach is that solutions to the IVP in $\mathcal{S}$ draw "lines" on the space–time cylinder $\Omega \times (0,T]$.
- *Rothe's method*[1]: Here, one first applies a finite difference discretization in time to obtain approximations at discrete instances of time which, for instance, read

$$w(t_{k+1}) - \frac{\partial^2 w(x, t_{k+1})}{\partial x^2} = w(t_k) + \tau f(x, t_{k+1}).$$

  This is now an elliptic equation, and we have already discussed several methods for its discretization.

In this chapter, we will only discuss the result of applying a finite difference approximation in space and time and the most rudimentary facts about the analysis of the resulting methods. For more details about this approach, we refer the reader to [56, 50, 79]. The monograph [91] discusses the application of finite element techniques in space to the solution of parabolic problems.

[1] Named in honor of the German mathematician Erich Hans Rothe (1895–1988).

28.1 Space–Time Grid Functions

We begin the discussion with the introduction of space–time grids, and functions on them. We could, in principle, consider a space–time domain merely as a domain in $\mathbb{R}^{d+1}$, but it is sometimes useful to make the difference between space and time, as each plays a different role.

Definition 28.1 (space–time grid). Let $d \in \mathbb{N}$, $\Omega = (0,1)^d$, and $T > 0$. For $K, N \in \mathbb{N}$, we set $\tau = T/K$ and $h = 1/(N+1)$. We define the **space–time grid domain**

$$\bar{C}_h^T = \bar{\Omega}_h \times [0,T]_\tau = \left\{(x, t_k) \mid x \in \bar{\Omega}_h, t_k = k\tau,\ k = 0, \dots, K\right\},$$

where we recall that $\bar{\Omega}_h = \bar{\Omega} \cap \mathbb{Z}_h^d$. We define the **discrete interior** of $\bar{C}_h^T$ to be

$$C_h^T = \Omega_h \times (0,T)_\tau.$$

The **discrete lateral boundary** is

$$\partial_L C_h^T = \partial\Omega_h \times [0,T]_\tau.$$

Finally, the **discrete parabolic boundary** is

$$\partial_p C_h^T = \bar{\Omega}_h \times \{0\} \cup \partial_L C_h^T.$$

Notice that this can be thought of as $K+1$ copies of $\bar{\Omega}_h$, one for each $k \in \{0, \dots, K\}$. We can now define space–time grid functions and norms on them that have the approximation property in suitable spaces of space–time functions.

Definition 28.2 (space–time grid functions). Let C_h^T be a space–time grid domain. We denote by

$$\mathcal{V}(\bar{C}_h^T) = \left\{v \mid \bar{C}_h^T \to \mathbb{R}\right\}$$

the space of **space–time grid functions**. The spaces

$$\mathcal{V}(C_h^T), \qquad \mathcal{V}(\partial_L C_h^T)$$

are defined accordingly. We set

$$\mathcal{V}_0(\bar{C}_h^T) = \left\{v \in \mathcal{V}(\bar{C}_h^T) \mid v(x,t) = 0,\ \forall (x,t) \in \partial_L C_h^T\right\}.$$

If $(ih, k\tau) \in \bar{C}_h^T$ and $v \in \mathcal{V}(\bar{C}_h^T)$, we denote

$$v_i^k = v(ih, k\tau).$$

Finally, if v is a space–time grid function and $k \in \{0, \dots, K\}$, we denote

$$v^k \in \mathcal{V}(\bar{\Omega}_h), \qquad v^k(x) = v(x, k\tau), \quad \forall x \in \bar{\Omega}_h.$$

We comment that spaces of space–time grid functions can also be thought of as a space of *spatial grid function-valued* functions, much as was done with functions in Definition 23.23. In other words,

$$\mathcal{V}(\bar{C}_h^T) = \mathcal{V}\left([0,T]_\tau; \mathcal{V}(\bar{\Omega}_h)\right) = \left(\mathcal{V}(\bar{\Omega}_h)\right)^{K+1}.$$

With this identification, we see that

$$\mathcal{V}_0(\bar{\mathcal{C}}_h^\tau) = \mathcal{V}\left([0,T]_\tau; \mathcal{V}_0(\bar{\Omega}_h)\right).$$

The usefulness of this, at first glance unnecessary and certainly overburdened, notation is that we can introduce norms that have the approximation property with respect to $L^q(0,T;L^p(\Omega))$, for instance.

Definition 28.3 (space–time discrete norms)**.** Let $d \in \{1,2\}$, $p \in [1,\infty]$, and $q \in [1,\infty)$. We define the *space–time norm*

$$\|v\|_{L^q_\tau(L^p_h)} = \left(\tau \sum_{k=1}^{K} \|v^k\|^q_{L^p_h}\right)^{1/q}$$

and

$$\|v\|_{L^\infty_\tau(L^p_h)} = \max_{k=0}^{K} \|v^k\|_{L^p_h}.$$

Proposition 28.4 (approximation property)**.** *Let $d \in \{1,2\}$ and $p,q \in [1,\infty)$, the space–time discrete norms of Definition 28.3, have the approximation property in $L^q(0,T;L^p(\Omega))$. In addition, the space–time discrete $L^\infty_\tau(L^\infty_h)$-norm has the approximation property in $C(\bar{\mathcal{C}})$.*

Proof. See Problem 28.1. □

Having introduced suitable norms, we can discuss the consistency and stability of finite difference methods as before. Since finite difference methods now depend on two parameters, K and N (or τ and h), we usually indicate the order of consistency with respect to each one of them separately.

Definition 28.5 (Courant number[2])**.** Let $\bar{\mathcal{C}}_h^\tau$ be the space–time grid. The **parabolic Courant number** is

$$\mu = \frac{\tau}{h^2}.$$

Definition 28.6 (unconditional stability)**.** We say that a finite difference method is **unconditionally stable** if it is stable, in the sense of Definition 24.16, for all values of the Courant number μ. It is **conditionally stable** if, to attain stability, some condition must be imposed on μ.

28.2 The Initial Boundary Value Problem for the Heat Equation

We are now ready to illustrate the application of the *method of lines* to the solution of parabolic problems. Let $\Omega = (0,1) \subset \mathbb{R}$, $T > 0$, $u_0 \in C(\bar{\Omega})$, $f \in C(\Omega \times (0,T])$. We seek a classical solution to the problem: Find $u \colon \bar{\Omega} \times [0,T] \to \mathbb{R}$ such that

[2] Named in honor of the German–American mathematician Richard Courant (1888–1972).

$$\begin{cases} \dfrac{\partial u(x,t)}{\partial t} - \dfrac{\partial^2 u(x,t)}{\partial x^2} = f(x,t), & (x,t) \in \Omega \times (0,T], \\ u(0,t) = u(1,t) = 0, & \\ u(x,0) = u_0(x), & x \in [0,1]. \end{cases} \tag{28.1}$$

We can semi-discretize this problem in space. Thus, we let $N \in \mathbb{N}$, $h = 1/(N+1)$, and $\bar{\Omega}_h$ be our grid domain. We set $u_{0,h} \in \mathcal{V}(\bar{\Omega}_h)$ as $u_{0,h}(x) = u_0(x)$, and $f_h \in C([0,T]; \mathcal{V}(\Omega_h))$ to be $f_h(x,t) = f(x,t)$. We seek $w\colon [0,T] \to \mathcal{V}_0(\bar{\Omega}_h)$ such that

$$\begin{cases} w'(t) - \Delta_h w(t) = f_h(\cdot, t), & t \in (0,T], \\ w(0) = u_{0,h}. \end{cases} \tag{28.2}$$

We can now apply, for instance, any of the single-step methods in Chapter 18 to obtain a fully discrete method.

Definition 28.7 (fully discrete methods)**.** Let $K, N \in \mathbb{N}$, $\tau = T/K$, $h = 1/(N+1)$, and $\mathcal{C}_h^\tau$ be the discrete space–time cylinder. Let $u \in C(\bar{\mathcal{C}})$ be the classical solution to (28.1). Suppose that $u_{0,h} \in \mathcal{V}(\bar{\Omega}_h)$ and $f_h \in \mathcal{V}(\mathcal{C}_h^\tau)$ are defined as usual. We define the following methods for computing the finite difference approximation $w \in \mathcal{V}_0(\bar{\mathcal{C}}_h^\tau)$ to u.

1. The **backward Euler method**

$$\begin{cases} \bar{\delta}_\tau w^{k+1} - \Delta_h w^{k+1} = f_h^{k+1}, & k = 0, \dots, K-1, \\ w^0 = u_{0,h}, \end{cases} \tag{28.3}$$

   where

$$\bar{\delta}_\tau v^{k+1} = \frac{1}{\tau}\left(v^{k+1} - v^k\right)$$

   is the backward difference in time.

2. The **forward Euler method**

$$\begin{cases} \delta_\tau w^k - \Delta_h w^k = f_h^k, & k = 0, \dots, K-1, \\ w^0 = u_{0,h}, \end{cases} \tag{28.4}$$

   where

$$\delta_\tau v^k = \frac{1}{\tau}\left(v^{k+1} - v^k\right)$$

   is the forward difference in time.

3. The **Crank–Nicolson method**[3]

$$\begin{cases} \bar{\delta}_\tau w^{k+1} - \dfrac{1}{2}\Delta_h \left(w^{k+1} + w^k\right) = \dfrac{1}{2}(f_h^{k+1} + f_h^k), & k = 0, \dots, K-1, \\ w^0 = u_{0,h}. \end{cases} \tag{28.5}$$

[3] Named in honor of the British mathematical physicist John Crank (1916–2006) and the British mathematician and physicist Phyllis Nicolson (1917–1968).

Other single-step, or even multi-step, methods can be used to discretize in time. The consistency of these methods follows by considering, separately, each one of its components. We note that the forward Euler method is explicit, whereas the backward Euler and Crank–Nicolson methods are implicit.

Theorem 28.8 (consistency). *Let $K, N \in \mathbb{N}$, $\tau = T/K$, $h = 1/(N+1)$, and $\bar{C}_h^\tau$ be the discrete space–time cylinder.*

1. *The backward Euler method is consistent, in $C_b(\mathbb{R}^2)$, to order*

$$\tau + h^2.$$

2. *The forward Euler method is consistent, in $C_b(\mathbb{R}^2)$, to order*

$$\tau + h^2.$$

3. *The Crank–Nicolson method is consistent, in $C_b(\mathbb{R}^2)$, to order*

$$\tau^2 + h^2.$$

Proof. See Problem 28.2. □

28.3 Stability and Convergence in the $L^\infty_\tau(L^\infty_h)$-Norm

In this section, we study the stability and convergence in the $L^\infty_\tau(L^\infty_h)$-norm of our three methods. As we will see, the main tool will be a discrete version of the maximum principle of Theorem 23.27.

28.3.1 The Backward Euler Method

Let us now study the stability and convergence of the backward Euler method. We must note that, as part of stability, we must show that this method is well defined, as it is an implicit one. To achieve this, we begin with a discrete maximum principle (DMP). The following result must be compared with Theorem 23.27.

Theorem 28.9 (DMP). *Let $v \in \mathcal{V}(\bar{C}_h^\tau)$ be such that*

$$\bar{\delta}_\tau v^{k+1} - \Delta_h v^{k+1} \leq 0, \quad k = 0, \dots, K-1.$$

Then the function v is either constant or attains its maximum on the discrete parabolic boundary $\partial_p C_h^\tau$.

Proof. We begin by writing the inequality explicitly, i.e.,

$$v_i^{k+1} - v_i^k - \mu\left(v_{i-1}^{k+1} - 2v_i^{k+1} + v_{i+1}^{k+1}\right) \leq 0,$$

where $\mu = \tau/h^2$ is the Courant number. This inequality can be rewritten as

$$v_i^{k+1} \leq \frac{\mu}{1+2\mu} v_{i-1}^{k+1} + \frac{1}{1+2\mu} v_i^k + \frac{\mu}{1+2\mu} v_{i+1}^{k+1}, \tag{28.6}$$

which, since $\frac{1}{1+2\mu}, \frac{\mu}{1+2\mu} \in (0,1)$, shows that v_i^{k+1} is a convex combination of the other values.

Let us assume that the maximum is attained at a point $(ih, (k+1)\tau) \in \Omega_h \times (0, T]_\tau$. This, in particular, implies that

$$v_i^{k+1} \geq v_{i\pm 1}^{k+1}, \qquad v_i^{k+1} \geq v_i^k.$$

The previous inequalities, together with (28.6), imply that

$$v_i^{k+1} = v_{i\pm 1}^{k+1} = v_i^k.$$

If any of the points $((i\pm 1)h, (k+1)\tau)$ or $(ih, k\tau)$ belong to the parabolic boundary, then we have obtained the result. If not, then we can repeat the same argument a finite number of times until we eventually reach the parabolic boundary. □

As an immediate consequence, we have a stability result.

Corollary 28.10 (stability). *Let $w \in \mathcal{V}_0(\bar{C}_h^T)$ be the solution to* (28.3). *Then*

$$\|w\|_{L^\infty_\tau(L^\infty_h)} \leq \|u_{0,h}\|_{L^\infty_h} + C\|f_h\|_{L^\infty_\tau(L^\infty_h)}.$$

As a consequence, the backward Euler method is unconditionally stable.

Proof. We must begin by showing that, for all $k \in \{0, \dots, K-1\}$, the problem

$$(I + \tau\Delta_h)\, w^{k+1} = \tau f_h^{k+1} + w^k = \tilde{f}_h^{k+1},$$

where I is the identity operator, is well posed. We see that this problem can be rewritten as a linear system of equations in $\mathbb{R}^N$ via $w^{k+1} \longleftrightarrow \boldsymbol{w}^{k+1}$

$$\mathsf{B}\boldsymbol{w}^{k+1} = \tilde{\boldsymbol{f}}^{k+1}, \qquad \mathsf{B} = \mathsf{I}_N + \mu\mathsf{A},$$

where the matrix A is the stiffness matrix introduced in (24.7). Clearly, B is invertible for all $\mu > 0$ and so the backward Euler method is well defined.

To obtain stability, as usual, we introduce the comparison function

$$\Phi(x) = \left(x - \frac{1}{2}\right)^2 \geq 0, \quad x \in [0,1],$$

and $\phi \in \mathcal{V}_0(\bar{\Omega}_h)$ is $\phi_i = \Phi(ih)$. As we observed before,

$$-\Phi'' = -2 = -\Delta_h\phi.$$

Define now

$$\psi_\pm = \pm w + \frac{\|f_h\|_{L^\infty_\tau(L^\infty_h)}}{2}\phi \in \mathcal{V}_0(\bar{C}_h^T),$$

and notice that

$$\bar{\delta}_\tau\psi_\pm^{k+1} - \Delta_h\psi_\pm^{k+1} = \pm f_h^{k+1} - \|f_h\|_{L^\infty_\tau(L^\infty_h)} \leq 0.$$

The DMP of Theorem 28.9 then implies that

$$\pm w^{k+1} \leq \psi_\pm^{k+1} \leq \max_{\partial_p C_h^T} \psi_\pm \leq \|u_{0,h}\|_{L^\infty_h} + C\|f_h\|_{L^\infty_\tau(L^\infty_h)},$$

by arguments we have established before. This implies the claimed estimate and the unconditional stability of the method. □

As usual, stability and consistency imply convergence.

Theorem 28.11 (convergence). *Suppose that u is a classical solution to the one-dimensional heat equation (28.1) with the additional regularity*

$$u, \partial_x u, \ldots, \partial_x^4 u, \partial_t u, \partial_t^2 u \in C(\bar{\Omega} \times [0, T]).$$

Let $w \in \mathcal{V}_0(\bar{\mathcal{C}}_h^\tau)$ be the grid function obtained from the backward Euler method (28.3). *Define the error*

$$e = \pi_h^\tau u - w \in \mathcal{V}_0(\bar{\mathcal{C}}_h^\tau), \qquad e_i^k = u(ih, k\tau) - w_i^k.$$

Then we have that there is a constant $C > 0$ for which

$$\|e\|_{L_\tau^\infty(L_h^\infty)} \leq C(\tau + h^2).$$

In other words, the convergence is first order with respect to time and second order with respect to space.

Proof. To prove convergence, we follow the usual approach, i.e., we find an equation that controls the error. Take the difference between the definition of consistency and the method to obtain

$$\bar{\delta}_\tau e^{k+1} - \Delta_h e^{k+1} = \mathcal{E}_h^\tau[u]^{k+1}, \qquad k = 0, \ldots, K-1,$$

where $\mu = \frac{\tau}{h^2}$ is the Courant number; $e_0^k = e_{N+1}^k = 0$ for all $k \in \{0, \ldots, K\}$ and $e_i^0 = 0$ for all $i \in \{0, \ldots, N+1\}$. Moreover, owing to Theorem 28.8,

$$\|\mathcal{E}_h^\tau[u]\|_{L_\tau^\infty(L_h^\infty)} \leq C(\tau + h^2)$$

for some $C > 0$ that is independent of h and τ. The stability of the method, presented in Corollary 28.10, implies that

$$\|e\|_{L_\tau^\infty(L_h^\infty)} \leq C \, \|\mathcal{E}_h^\tau[u]\|_{L_\tau^\infty(L_h^\infty)} \leq C(\tau + h^2),$$

which is what we intended to prove. □

28.3.2 The Forward Euler Method

We will now focus on the forward Euler method (28.4). Notice that this is an *explicit method* so it is automatically well defined. We will now show its (conditional) stability and convergence.

Theorem 28.12 (stability and convergence). *Let u be a classical solution to* (28.1), *which, in addition, satisfies*

$$u, \partial_x u, \ldots, \partial_x^4 u, \partial_t u, \partial_t^2 u \in C(\bar{\Omega} \times [0, T]).$$

Let $w \in \mathcal{V}_0(\bar{\mathcal{C}}_h^\tau)$ be the grid function obtained from the forward Euler method (28.4). *We have:*

1. *If the Courant number satisfies $\mu \le \mu_0 = \frac{1}{2}$, then the method is $L^\infty_\tau(L^\infty_h)$-stable, i.e., there is a constant $C > 0$, independent of h, τ, and w, such that*

$$\|w\|_{L^\infty_\tau(L^\infty_h)} \le \|u_{0,h}\|_{L^\infty_h} + C\|f_h\|_{L^\infty_\tau(L^\infty_h)},$$

so that the forward Euler method is conditionally stable.

2. *Define the error*

$$e = \pi^\tau_h u - w \in \mathcal{V}_0(\bar{C}^\tau_h), \qquad e^k_i = u(ih, k\tau) - w^k_i.$$

Then, provided that $\mu \le \mu_0$, we have that there is a constant $C > 0$ for which

$$\|e\|_{L^\infty_\tau(L^2_h)} \le C(\tau + h^2).$$

In other words, the convergence is first order with respect to time and second order with respect to space.

Proof. See Problem 28.4. □

Remark 28.13 (CFL condition[4]). The restriction on the time and space step sizes in the last result, i.e., $\mu \le \frac{1}{2}$, is called a *CFL-type* condition in honor of R. Courant, K.O. Friedrichs, and H. Lewy, who pioneered the use of finite difference methods in their breakthrough article [21]; see also the translation [22]. It indicates a restriction between the time and space step sizes for stability and/or convergence; namely,

$$\tau \le \frac{h^2}{2},$$

which may be computationally expensive to enforce.

28.3.3 The Crank–Nicolson Method

Let us now study the Crank–Nicolson method, which is implicit. We will obtain that this method is also conditionally stable, but the CFL-type condition is slightly milder.

Theorem 28.14 (stability and convergence). *Let u be a classical solution to* (28.1), *which, in addition, satisfies $u \in C^4(\bar{\Omega} \times [0, T])$. Let $w \in \mathcal{V}_0(\bar{C}^\tau_h)$ be the grid function obtained from the Crank–Nicolson method* (28.5). *We have:*

1. *The method is well defined. In particular, we have that*

$$\mathsf{B}_2 \boldsymbol{w}^{k+1} = \mathsf{A}_1 \boldsymbol{w}^k + \tau \boldsymbol{f}^{k+1},$$

[4] Named in honor of the German–American mathematicians Richard Courant (1888–1972) and Kurt Otto Friedrichs (1901–1982) and the Jewish, German-born, American mathematician Hans Lewy (1904–1988).

where

$$\mathsf{B}_2 = \begin{bmatrix} 1+\mu & -\frac{\mu}{2} & 0 & \cdots & 0 \\ -\frac{\mu}{2} & 1+\mu & \ddots & & \vdots \\ 0 & \ddots & \ddots & -\frac{\mu}{2} & 0 \\ \vdots & & -\frac{\mu}{2} & 1+\mu & -\frac{\mu}{2} \\ 0 & \cdots & 0 & -\frac{\mu}{2} & 1+\mu \end{bmatrix} \in \mathbb{R}^{N\times N}$$

is symmetric positive definite (SPD) and

$$\mathsf{A}_1 = \begin{bmatrix} 1-\mu & \frac{\mu}{2} & 0 & \cdots & 0 \\ \frac{\mu}{2} & 1-\mu & \ddots & & \vdots \\ 0 & \ddots & \ddots & \frac{\mu}{2} & 0 \\ \vdots & & \frac{\mu}{2} & 1-\mu & \frac{\mu}{2} \\ 0 & \cdots & 0 & \frac{\mu}{2} & 1-\mu \end{bmatrix} \in \mathbb{R}^{N\times N}.$$

2. *If the Courant number satisfies $\mu \leq \mu_0 = 1$, then the Crank–Nicolson method is $L^\infty_\tau(L^\infty_h)$-stable, i.e., there is a constant $C > 0$, independent of h, τ, and w, such that*

$$\|w\|_{L^\infty_\tau(L^\infty_h)} \leq \|u_{0,h}\|_{L^\infty_h} + C\|f_h\|_{L^\infty_\tau(L^\infty_h)},$$

so that the Crank–Nicolson method is conditionally stable.

3. *Define the error*

$$e = \pi_h^\tau u - w \in \mathcal{V}_0(\bar{\mathcal{C}}_h^\tau), \qquad e_i^k = u(ih, k\tau) - w_i^k.$$

Then, provided that $\mu \leq \mu_0$, we have that there is a constant $C > 0$ for which

$$\|e\|_{L^\infty_\tau(L^\infty_h)} \leq C\left(\tau^2 + h^2\right).$$

In other words, the convergence is second order with respect to time and space.

Proof. We will prove that the method is well defined and stable, leaving convergence to the reader as an exercise; see Problem 28.5.

To show that the method is well defined, we observe that, with the identification $w^k \longleftrightarrow \boldsymbol{w}^k \in \mathbb{R}^N$, our method can be written as

$$\left(\mathsf{I}_N + \frac{\mu}{2}\mathsf{A}\right)\boldsymbol{w}^{k+1} = \left(\mathsf{I}_N - \frac{\mu}{2}\mathsf{A}\right)\boldsymbol{w}^k + \tau \boldsymbol{f}^{k+1},$$

where A is the stiffness matrix defined in (24.7). Thus, $\mathsf{B}_2 = \mathsf{I}_N + \frac{\mu}{2}\mathsf{A}$ and $\mathsf{A}_1 = \mathsf{I}_N - \frac{\mu}{2}\mathsf{A}$. Observe that B_2 is SPD and strictly diagonally dominant. Therefore, the method is well defined.

Let us now show stability. For simplicity, we consider the case $f_h \equiv 0$. Define $\boldsymbol{z}^k = \mathsf{A}_1\boldsymbol{w}^k$. Then $\mathsf{B}_2\boldsymbol{w}^{k+1} = \boldsymbol{z}^k$. In other words,

$$-\frac{\mu}{2}w_{i-1}^{k+1} + (1+\mu)w_i^{k+1} - \frac{\mu}{2}w_{i+1}^{i+1} = z_i^k,$$

with slight modifications at the boundary nodes where $w_0^k = w_{N+1}^k = 0$ for all $k \in \{0, \ldots, K\}$. Now let $i_0 \in \{1, \ldots, N\}$ satisfy $\left\|\boldsymbol{w}^{k+1}\right\|_\infty = |w_{i_0}^{k+1}|$. Then

$$\left\|\boldsymbol{w}^{k+1}\right\|_\infty = \left|w^{k+1}_{i_0}\right| = \frac{1}{1+\mu}\left|z^k_{i_0} + \frac{\mu}{2}\left(w^{k+1}_{i_0-1} + w^{i_0+1}_{i+1}\right)\right|$$
$$\le \frac{1}{1+\mu}\left[\left|z^k_{i_0}\right| + \frac{\mu}{2}\left(\left|w^{k+1}_{i_0-1}\right| + \left|w^{i_0+1}_{i+1}\right|\right)\right]$$
$$\le \frac{1}{1+\mu}\left[\|\boldsymbol{z}^k\|_\infty + \mu\|\boldsymbol{w}^{k+1}\|_\infty\right].$$

Thus, we have

$$(1+\mu)\left\|\boldsymbol{w}^{k+1}\right\|_\infty \le \left\|\boldsymbol{z}^k\right\|_\infty + \mu\left\|\boldsymbol{w}^{k+1}\right\|_\infty.$$

Now $\boldsymbol{z}^k = \mathsf{A}_1\boldsymbol{w}^k$ can be written in component form as

$$z^k_i = \frac{\mu}{2}w^k_{i-1} + (1-\mu)w^k_i + \frac{\mu}{2}w^k_{i+1}$$

with slight modifications at the boundary nodes. Assuming that $\mu \le 1$, we have that $\frac{\mu}{2}, 1-\mu \in (0,1)$, and z^k_i is a convex combination of w^k_i and $w^k_{i\pm1}$. Suppose that $j_0 \in \{1,\dots,N\}$ satisfies $\|\boldsymbol{z}^k\|_\infty = |z^k_{j_0}|$. Then

$$\|\boldsymbol{z}^k\|_\infty = |z^k_{j_0}| = \left|\frac{\mu}{2}w^k_{j_0-1} + (1-\mu)w^k_{j_0} + \frac{\mu}{2}w^k_{j_0+1}\right|$$
$$\le \frac{\mu}{2}|w^k_{j_0-1}| + (1-\mu)|w^k_{j_0}| + \frac{\mu}{2}|w^k_{j_0+1}|$$
$$\le \|\boldsymbol{w}^k\|_\infty.$$

Putting everything together, if $\mu \le 1$, then we have

$$\left\|w^{k+1}\right\|_{L^\infty_h} = \left\|\boldsymbol{w}^{k+1}\right\|_\infty \le \left\|\boldsymbol{z}^k\right\|_\infty \le \left\|\boldsymbol{w}^k\right\|_\infty = \left\|w^k\right\|_{L^\infty_h}.$$

Consequently,

$$\|w\|_{L^\infty_\tau(L^\infty_h)} \le \left\|w^0\right\|_{L^\infty_h} = \|u_{0,h}\|_{L^\infty_h},$$

and the stability is established. □

Remark 28.15 (alternate proof)**.** The stability of Theorem 28.14 can also be established as follows. Write the method as

$$(1+\mu)w^{k+1}_i = (1-\mu)w^k_i + \frac{\mu}{2}w^{k+1}_{i-1} + \frac{\mu}{2}w^{k+1}_{i+1} + \frac{\mu}{2}w^k_{i-1} + \frac{\mu}{2}w^k_{i+1}.$$

Then, taking absolute values and using the fact that $1-\mu \ge 0$ by assumption, we find

$$(1+\mu)\left|w^{k+1}_i\right| = |1-\mu|\left|w^k_i\right| + \frac{\mu}{2}\left|w^{k+1}_{i-1}\right| + \frac{\mu}{2}\left|w^{k+1}_{i+1}\right| + \frac{\mu}{2}\left|w^k_{i-1}\right| + \frac{\mu}{2}\left|w^k_{i+1}\right|$$
$$= (1-\mu)\left|w^k_i\right| + \frac{\mu}{2}\left|w^{k+1}_{i-1}\right| + \frac{\mu}{2}\left|w^{k+1}_{i+1}\right| + \frac{\mu}{2}\left|w^k_{i-1}\right| + \frac{\mu}{2}\left|w^k_{i+1}\right|$$
$$\le (1-\mu)\left\|\boldsymbol{w}^k\right\|_\infty + \mu\left\|\boldsymbol{w}^k\right\|_\infty + \mu\left\|\boldsymbol{w}^{k+1}\right\|_\infty$$
$$= \left\|\boldsymbol{w}^k\right\|_\infty + \mu\left\|\boldsymbol{w}^{k+1}\right\|_\infty.$$

Since $i \in \{1,2,\dots,N\}$ is arbitrary,

$$(1+\mu)\left\|\boldsymbol{w}^{k+1}\right\|_\infty \le \left\|\boldsymbol{w}^k\right\|_\infty + \mu\left\|\boldsymbol{w}^{k+1}\right\|_\infty.$$

The result follows by cancellation.

28.4 Stability and Convergence in the $L^\infty_\tau(L^2_h)$-Norm

In this section, we examine the stability and convergence of our three methods in the $L^\infty_\tau(L^2_h)$-norm. The technique of analysis changes in this setting to one based on an examination of the eigenvalues of the stiffness matrix.

28.4.1 The Backward Euler Method

Theorem 28.16 (stability and convergence)**.** *Suppose that u is a classical solution to the one-dimensional heat equation (28.1) with the additional regularity*

$$u, \partial_x u, \dots, \partial_x^4 u, \partial_t u, \partial_t^2 u \in C(\bar{\Omega} \times [0, T]).$$

Let $w \in \mathcal{V}_0(\bar{C}_h^\tau)$ be the grid function obtained from the backward Euler method (28.3)*. Then:*

1. *For every $k = 0, \dots, K-1$, one can write $\boldsymbol{w}^{k+1} = \mathsf{A}(\boldsymbol{w}^k + \tau \boldsymbol{f}^{k+1})$ with*

$$\|\mathsf{A}\|_2 \leq 1;$$

consequently, for all τ and h, we have

$$\|w\|_{L^\infty_\tau(L^2_h)} \leq \|u_{0,h}\|_{L^2_h} + \|f_h\|_{L^2_\tau(L^2_h)}.$$

Therefore, the backward Euler method is unconditionally stable in the $L^\infty_\tau(L^2_h)$-norm.

2. *Define the error*

$$e = \pi_h^\tau u - w \in \mathcal{V}_0(\bar{C}_h^\tau).$$

Then there is a constant $C > 0$ for which

$$\|e\|_{L^\infty_\tau(L^2_h)} \leq C(\tau + h^2).$$

Thus, the convergence is first order with respect to time and second order with respect to space.

Proof. See Problem 28.6. □

28.4.2 The Forward Euler Method

As before, it turns out that the forward Euler method is conditionally stable.

Theorem 28.17 (stability and convergence)**.** *Let u be a classical solution to* (28.1)*, which, in addition, satisfies*

$$u, \partial_x u, \dots, \partial_x^4 u, \partial_t u, \partial_t^2 u \in C(\bar{\Omega} \times [0, T]).$$

Let $w \in \mathcal{V}_0(\bar{C}_h^\tau)$ be the grid function obtained from the forward Euler method (28.4)*. We have:*

1. *If the Courant number satisfies* $\mu \leq \mu_0 = \frac{1}{2}$, *then one can write* $\boldsymbol{w}^{k+1} = \mathsf{A}(\boldsymbol{w}^k + \tau \boldsymbol{f}^k)$ *for* $k = 0, \dots, K-1$, *where*

$$\|\mathsf{A}\|_2 \leq 1;$$

consequently, the method is $L^\infty_\tau(L^2_h)$*-stable, i.e., we have*

$$\|w\|_{L^\infty_\tau(L^2_h)} \leq \|u_{0,h}\|_{L^2_h} + \|f_h\|_{L^2_\tau(L^2_h)},$$

so that the forward Euler method is conditionally stable in the $L^\infty_\tau(L^2_h)$*-norm.*

2. *Define the error*

$$e = \pi^\tau_h u - w \in \mathcal{V}_0(\bar{\mathcal{C}}^\tau_h), \qquad e^k_i = u(ih, k\tau) - w^k_i.$$

Then, provided that $\mu \leq \mu_0$, *we have that there is a constant* $C > 0$ *for which*

$$\|e\|_{L^\infty_\tau(L^\infty_h)} \leq C(\tau + h^2).$$

In other words, the (conditional) convergence is first order with respect to time and second order with respect to space.

Proof. We prove stability in the case $f \equiv 0$ and leave the rest to the reader as an exercise; see Problem 28.7. Observe that the forward Euler method is equivalent to $\boldsymbol{w}^{k+1} = \mathsf{A}\boldsymbol{w}^k$, where

$$\mathsf{A} = \begin{bmatrix} 1-2\mu & \mu & 0 & \cdots & 0 \\ \mu & 1-2\mu & \ddots & & \vdots \\ 0 & \ddots & \ddots & \mu & 0 \\ \vdots & & \mu & 1-2\mu & \mu \\ 0 & \cdots & 0 & \mu & 1-2\mu \end{bmatrix}.$$

Observe that the eigenvectors for A are given by

$$S = \{\boldsymbol{v}_1, \boldsymbol{v}_2, \dots, \boldsymbol{v}_N\}, \qquad [\boldsymbol{v}_n]_i = \sin(n\pi i h).$$

The corresponding eigenvalues of A are

$$\lambda_n = 1 - 2\mu\,[1 - \cos(n\pi h)].$$

Now, since $\mathsf{A} \in \mathbb{R}^{N\times N}_{\text{sym}}$, we know that $\|\mathsf{A}\|_2 = \rho(\mathsf{A})$. Observe that if $0 < \mu \leq \mu_0 = \frac{1}{2}$, then

$$0 < 2(1 - \cos(n\pi h)) < 4 \qquad \iff \qquad 1 > \lambda_n > 1 - 4\mu \geq -1,$$

which implies that $|\lambda_n| \leq 1$. Hence, $\|\mathsf{A}\|_2 = \rho(\mathsf{A}) \leq 1$. Therefore,

$$\left\|w^{k+1}\right\|_{L^2_h} = h^{1/2}\left\|\boldsymbol{w}^{k+1}\right\|_2 = h^{1/2}\left\|\mathsf{A}\boldsymbol{w}^k\right\|_2 \leq h^{1/2}\left\|\boldsymbol{w}^k\right\|_2 = \left\|w^k\right\|_{L^2_h}.$$

Consequently, for any $k = 0, \dots, K$,

$$\left\|w^k\right\|_{L^2_h} \leq \|u_{0,h}\|_{L^2_h},$$

as we intended to show. □

28.4.3 The Crank–Nicolson Method

Let us now show that the Crank–Nicolson method is unconditionally stable in the $L^\infty_\tau(L^2_h)$-norm. This is in contrast to the stability in $L^\infty_\tau(L^\infty_h)$, which requires a CFL condition.

Theorem 28.18 (stability and convergence). *Let u be a classical solution to (28.1) with $f \equiv 0$, which, in addition, satisfies $u \in C^4(\bar{\Omega} \times [0, T])$. Let $w \in \mathcal{V}_0(\bar{\mathcal{C}}^\tau_h)$ be the grid function obtained from the Crank–Nicolson method (28.5). We have:*

1. *For $k = 0, \ldots, K-1$, one can write $\boldsymbol{w}^{k+1} = \mathsf{A}\boldsymbol{w}^k$ with*
$$\|\mathsf{A}\|_2 \leq 1;$$
consequently, for any τ and h,
$$\|w\|_{L^\infty_\tau(L^2_h)} \leq \|u_{0,h}\|_{L^2_h}.$$
Thus, the Crank–Nicolson method is unconditionally stable in the $L^\infty_\tau(L^2_h)$-norm.
2. *Define the error*
$$e = \pi^\tau_h u - w \in \mathcal{V}_0(\bar{\mathcal{C}}^\tau_h), \qquad e^k_i = u(ih, k\tau) - w^k_i.$$
Then we have that there is a constant $C > 0$, independent of h and τ, for which
$$\|e\|_{L^\infty_\tau(L^\infty_h)} \leq C(\tau^2 + h^2).$$
In other words, the (unconditional) convergence is second order with respect to time and space.

Proof. We begin by showing stability. Recall that the Crank–Nicolson method is equivalent to $\mathsf{B}_2\boldsymbol{w}^{k+1} = \mathsf{A}_1\boldsymbol{w}^k$, where B_2 and A_1 are given in Theorem 28.14. Since the matrices are of Toeplitz symmetric tridiagonal type, the eigenvectors for B_2 and A_1 are the same and are given by

$$S = \{\boldsymbol{v}_1, \boldsymbol{v}_2, \ldots, \boldsymbol{v}_N\}, \qquad [\boldsymbol{v}_n]_i = \sin(k\pi ih).$$

Calculating $\mathsf{B}_2\boldsymbol{v}_n$ and $\mathsf{A}_1\boldsymbol{v}_n$, one finds

$$\sigma(\mathsf{B}_2) = \{\lambda_n\}_{n=1}^N, \qquad \lambda_n = 1 + \mu\,[1 - \cos(n\pi h)] \in [1, 1 + 2\mu],$$

and

$$\sigma(\mathsf{A}_1) = \{\nu_n\}_{n=1}^N, \qquad \nu_n = 1 - \mu\,[1 - \cos(n\pi h)].$$

This implies that B_2 is SPD and, therefore, invertible. Define $\mathsf{A}_2 = \mathsf{B}_2^{-1}$ and $\mathsf{A} = \mathsf{A}_2\mathsf{A}_1$. Then $\boldsymbol{w}^{k+1} = \mathsf{A}\boldsymbol{w}^k$.

It is an exercise to show that $\mathsf{A} \in \mathbb{R}^{N\times N}$ is symmetric; because of this, we know that $\|\mathsf{A}\|_2 = \rho(\mathsf{A})$. The eigenvalues of A are precisely

$$\sigma_n = \frac{\nu_n}{\lambda_n} = \frac{1 - \mu\,[1 - \cos(n\pi h)]}{1 + \mu\,[1 - \cos(n\pi h)]}, \qquad n = 1, \ldots, N.$$

Therefore, if we can show that $|\sigma_n| \le 1$ for all $n = 1, \dots, N$, then we are able to conclude. Let us do this in an indirect way. The key is to start with

$$1 + \frac{1}{\mu} \ge 1 \ge \cos(n\pi h),$$

which is true for all $\mu > 0$ and all $1 \le n \le N$. This is equivalent to

$$\begin{aligned}
& & -1 - \frac{1}{\mu} \le -1 \le -\cos(n\pi h),\\
&\iff & -\frac{1}{\mu} \le 0 \le 1 - \cos(n\pi h),\\
&\iff & -1 \le 0 \le \mu\,[1 - \cos(n\pi h)],\\
&\iff & -2 \le 0 \le 2\mu\,[1 - \cos(n\pi h)],\\
&\iff & -2 - \mu\,[1 - \cos(n\pi h)] \le -\mu\,[1 - \cos(n\pi h)] \le \mu\,[1 - \cos(n\pi h)],\\
&\iff & -1 - \mu\,[1 - \cos(n\pi h)] \le 1 - \mu\,[1 - \cos(n\pi h)] \le 1 + \mu\,[1 - \cos(n\pi h)],\\
&\iff & -\lambda_n \le \nu_n \le \lambda_n,\\
&\iff & |\nu_n| \le \lambda_n = |\lambda_n|,\\
&\iff & \left|\frac{\nu_n}{\lambda_n}\right| \le 1,\\
&\iff & |\sigma_n| \le 1.
\end{aligned}$$

Hence, $\|\mathsf{A}\|_2 = \rho(\mathsf{A}) \le 1$.

We now show convergence. As usual, the error is defined as $e_i^k = u(x_i, t_k) - w_i^k$. Using the usual techniques, we find the error equation has the form

$$\mathsf{B}_2 \mathbf{e}^{k+1} = \mathsf{A}_1 \mathbf{e}^k + \tau \boldsymbol{\mathcal{E}}_h^\tau[u]^{k+1}.$$

We have already shown that

$$\left|\mathcal{E}_h^\tau[u]_i^k\right| \le C(\tau^2 + h^2).$$

Since B_2 is invertible,

$$\mathbf{e}^{k+1} = \mathsf{A}\mathbf{e}^k + \tau \mathsf{A}_2 \boldsymbol{\mathcal{E}}_h^\tau[u]^{k+1},$$

where $\mathsf{A}_2 = \mathsf{B}_2^{-1}$. Therefore,

$$\begin{aligned}
\left\|\mathbf{e}^{k+1}\right\|_2 &\le \|\mathsf{A}\mathbf{e}^n\|_2 + \tau \left\|\mathsf{A}_2 \boldsymbol{\mathcal{E}}_h^\tau[u]^{k+1}\right\|_2\\
&\le \left\|\mathbf{e}^k\right\|_2 + \tau \|\mathsf{A}_2\|_2 \left\|\boldsymbol{\mathcal{E}}_h^\tau[u]^{k+1}\right\|_2\\
&\le \left\|\mathbf{e}^k\right\|_2 + \tau \|\mathsf{A}_2\|_2 \sqrt{N} \left\|\boldsymbol{\mathcal{E}}_h^\tau[u]^{k+1}\right\|_\infty\\
&\le \left\|\mathbf{e}^k\right\|_2 + \tau \|\mathsf{A}_2\|_2 \sqrt{N} C\left(\tau^2 + h^2\right).
\end{aligned}$$

From stability, we know that $\|\mathsf{A}_2\|_2 = \rho(\mathsf{A}_2) \le 1$; therefore,

$$\left\|e^{k+1}\right\|_{L^2_h} \le \left\|e^k\right\|_{L^2_h} + C\tau h^{1/2}\sqrt{N}\left(\tau^2 + h^2\right) \le \left\|e^k\right\|_{L^2_h} + C\tau\left(\tau^2 + h^2\right).$$

Applying this result recursively, we arrive at

$$\|e\|_{L^\infty(L^2_h)} \le \left\|e^0\right\|_{L^2_h} + C\left(\tau^2 + h^2\right).$$

h	τ	$\epsilon_{h,\tau} = \left\|e^K_{h,\tau}\right\|_{L^\infty_h}$	$\log_2\left(\frac{\epsilon_{2h,2\tau}}{\epsilon_{h,\tau}}\right)$
$\frac{1}{32}$	$\frac{1}{32}$	1.0691e−02	–
$\frac{1}{64}$	$\frac{1}{64}$	2.6595e−03	2.0071
$\frac{1}{128}$	$\frac{1}{128}$	6.6404e−04	2.0018
$\frac{1}{256}$	$\frac{1}{256}$	1.6600e−04	2.0000
$\frac{1}{512}$	$\frac{1}{512}$	4.1502e−05	2.0000

Table 28.1 Convergence of the Crank–Nicolson method for the problem described in Example 28.1. The error is computed at the final time $T = 1.0$. Near-perfect second-order accuracy is observed in the approximation.

The proof is complete. □

Example 28.1 We consider, on the interval $\Omega = (0,1)$, problem (28.1). The initial condition u_0 and forcing term f are constructed, so that the exact solution is

$$u(x,t) = \cos(2\pi t)\left[\exp(\sin(2\pi x)) - 1\right].$$

We approximate this solution using the Crank–Nicolson method (28.5). The code for this approximation is shown in Listing 28.1. Since, in this case, we have access to the exact solution, we can compute the error as τ and h are made smaller; see Table 28.1, where the error is reported at the final time $T = 1.0$. Near-perfect second-order accuracy is observed in the approximation.

28.5 Stability by Energy Techniques

In addition to maximum principles and properties of eigenvalues, there are other techniques to analyze the stability of numerical methods. For instance, a discrete analogue of separation of variables is illustrated in Problem 28.9. In this section, which can be considered the parabolic analogue of Section 24.4.1, we will illustrate the so-called *energy method* to obtain stability. The basic idea is to multiply the method by an appropriate test (grid) function and utilize (a version of) the summation by parts identity given in (24.1).

Theorem 28.19 (stability). *The solution of the backward Euler method* (28.3) *satisfies*

$$\|w\|^2_{L^\infty_\tau(L^2_h)} + \|w\|^2_{L^2_\tau(H^1_h)} \leq \left\|w^0\right\|^2_{L^2_h} + C\,\|f_h\|^2_{L^2_\tau(L^2_h)},$$

where the constant $C > 0$ is independent of h and τ.

Proof. Take the L^2_h-inner product of the method with $2\tau w^{k+1} \in \mathcal{V}_0(\bar{\Omega}_h)$. The summation by parts identity (24.1) yields

$$2\tau\left(\bar{\delta}_\tau w^{k+1}, w^{k+1}\right)_{L^2_h} + 2\tau\|w^{k+1}\|^2_{H^1_h} = 2\tau(f^{k+1}_h, w^{k+1})_{L^2_h}.$$

Now, observe that the so-called *polarization identity*

$$2a(a-b) = a^2 + (a-b)^2 - b^2, \quad \forall a, b \in \mathbb{R}$$

implies that

$$2\tau\left(\bar{\delta}_\tau w^{k+1}, w^{k+1}\right)_{L^2_h} = \|w^{k+1}\|^2_{L^2_h} - \|w^k\|^2_{L^2_h} + \|w^{k+1} - w^k\|^2_{L^2_h}.$$

Now, by the Cauchy–Schwarz inequality, the discrete Poincaré inequality of Theorem 24.34, and Young's inequality, we obtain that

$$2\tau(f_h^{k+1}, w^{k+1})_{L^2_h} \le C\tau\|f_h^{k+1}\|^2_{L^2_h} + \tau\|w^{k+1}\|^2_{H^1_h}.$$

Finally, gathering all the obtained estimates, we infer that

$$\|w^{k+1}\|^2_{L^2_h} + \|w^{k+1} - w^k\|^2_{L^2_h} + \tau\|w^{k+1}\|^2_{H^1_h} \le \|w^k\|^2_{L^2_h} + C\tau\|f_h^{k+1}\|^2_{L^2_h}.$$

Let $k_0 \in \{0, \dots, K\}$ be such that

$$\|w^{k_0}\|_{L^2_h} = \|w\|_{L^\infty_\tau(L^2_h)}.$$

Adding our estimate over $k \in \{0, \dots, k_0 - 1\}$, the stability bound follows. □

To analyze, by a similar technique, the forward Euler method (28.4), we need the inverse inequality of Problem 24.23.

Theorem 28.20 (conditional stability)**.** *Consider the forward Euler method* (28.4)*. Assume that the Courant number satisfies $\mu \le \frac{1}{2C_I}$, where C_I is the constant in the inverse inequality of Problem 24.23. In this setting, there is a constant $C > 0$, independent of τ and h, such that*

$$\|w\|^2_{L^\infty_\tau(L^2_h)} + \|w\|^2_{L^2_\tau(H^1_h)} \le \|w^0\|^2_{L^2_h} + C\,\|f_h\|^2_{L^2_\tau(L^2_h)}.$$

Proof. Let us again take the L^2_h-inner product of the method with $2\tau w^{k+1}$ to obtain, following the derivations of Theorem 28.19, that

$$\begin{aligned}\|w^{k+1}\|^2_{L^2_h} &+ \|w^{k+1} - w^k\|^2_{L^2_h} + 2\tau(\bar{\delta}_h w^{k+1}, \bar{\delta}_h w^k)_{L^2_h} \\ &\le \|w^k\|^2_{L^2_h} + C\tau\|f_h^{k+1}\|^2_{L^2_h} + \frac{\tau}{2}\|w^{k+1}\|^2_{H^1_h}.\end{aligned}$$

Observe now that

$$2\tau(\bar{\delta}_h w^{k+1}, \bar{\delta}_h w^k)_{L^2_h} = 2\tau\|w^{k+1}\|^2_{H^1_h} - 2\tau(\bar{\delta}_h w^{k+1}, \bar{\delta}_h(w^{k+1} - w^k))_{L^2_h},$$

so that

$$\begin{aligned}\|w^{k+1}\|^2_{L^2_h} &+ \|w^{k+1} - w^k\|^2_{L^2_h} + \frac{3}{2}\tau\|w^{k+1}\|^2_{H^1_h} \\ &\le \|w^k\|^2_{L^2_h} + C\tau\|f_h^{k+1}\|^2_{L^2_h} + 2\tau(\bar{\delta}_h w^{k+1}, \bar{\delta}_h(w^{k+1} - w^k))_{L^2_h}.\end{aligned}$$

The Cauchy–Schwarz inequality, Young's inequality, and the inverse inequality of Problem 24.23 then imply that

$$\begin{aligned}2\tau(\bar{\delta}_h w^{k+1}, \bar{\delta}_h(w^{k+1} - w^k))_{L^2_h} &\le \frac{\tau}{2}\|w^{k+1}\|^2_{H^1_h} + 2\tau\|w^{k+1} - w^k\|^2_{H^1_h} \\ &\le \frac{\tau}{2}\|w^{k+1}\|^2_{H^1_h} + 2C_I\tau h^{-2}\|w^{k+1} - w^k\|^2_{L^2_h}.\end{aligned}$$

Since, by assumption, $\mu \leq \frac{1}{2C_I}$, gathering all the obtained bounds and proceeding as in Theorem 28.19 yields the desired result. □

Remark 28.21 (Crank–Nicolson). The stability, using energy techniques, of the Crank–Nicolson method (28.5) is essentially the content of Problem 28.10.

28.6 Advection–Diffusion and Upwinding

Let us, in this section, consider the following problem: Given $u_0 \in C(\bar{\Omega})$ and $b > 0$, find a classical solution to

$$\begin{cases} \partial_t u + b\partial_x u - \partial_{xx}^2 u = 0, & x \in \Omega,\ t \in (0,T], \\ u(0,t) = u(1,t) = 0, & t \in (0,T], \\ u(x,0) = u_0(x), & x \in \Omega, \end{cases} \tag{28.7}$$

with $b > 0$. We assume that the compatibility conditions $u_0(0) = u_0(1) = 0$ hold and that u has regularity

$$u, \partial_x u, \ldots, \partial_x^4 u, \partial_t u, \partial_t^2 u \in C(\bar{\Omega} \times [0,T]).$$

Our goal will be to construct explicit homogeneous methods that are monotone, in the sense of Section 24.4.2. We refer the reader to Section 24.4.2 for some notions that will be relevant in our present discussion. Before proceeding any further, we comment that similar considerations can be done for implicit methods. These will be developed in Problems 28.17 and 28.18.

Let us introduce a *centered* discretization of the problem. Thus, we seek $w \in \mathcal{V}_0(\bar{C}_h^\tau)$ such that

$$w^0 = u_{0,h}, \qquad \delta_\tau w^k + b\mathring{\delta}_h w^k - \Delta_h w^k = 0, \quad k = 1, \ldots, K, \tag{28.8}$$

where $u_{0,h} = \pi_h u_0$ and π_h is the sampling operator.

Theorem 28.22 (conditional stability). *Assume that the mesh Péclet number introduced in Remark 24.39 satisfies*

$$hb \leq 2$$

and the Courant number satisfies

$$\mu \leq \frac{1}{2}.$$

Then the method (28.8) *is stable in the $L_\tau^\infty(L_h^\infty)$-norm.*

Proof. Let k be arbitrary and $i_0 \in \{1, \ldots, N\}$ be such that $\|w^{k+1}\|_{L_h^\infty} = |w_{i_0}^{k+1}|$.

Then we have

$$\begin{aligned}\left\|w^{k+1}\right\|_{L^\infty_h} = |w^{k+1}_{i_0}| &= \left|w^k_{i_0} + \mu\left(w^k_{i_0-1} - 2w^k_{i_0} + w^k_{i_0+1}\right) - \frac{b\mu h}{2}\left(w^k_{i_0+1} - w^k_{i_0-1}\right)\right| \\ &= \left|\mu\left(1+\frac{bh}{2}\right)w^k_{i_0-1} + (1-2\mu)\,w^k_{i_0} + \mu\left(1-\frac{bh}{2}\right)w^k_{i_0+1}\right| \\ &\leq \mu\left|1+\frac{bh}{2}\right||w^k_{i_0-1}| + |1-2\mu|\,|w^k_{i_0}| + \mu\left|1-\frac{bh}{2}\right||w^k_{i_0+1}| \\ &\leq \left(\mu\left|1+\frac{bh}{2}\right| + |1-2\mu| + \mu\left|1-\frac{bh}{2}\right|\right)\left\|w^k\right\|_{L^\infty_h}.\end{aligned}$$

Now the condition on the mesh Péclet number implies that $1-\frac{bh}{2} \geq 0$ and the condition on the Courant number that $1-2\mu \geq 0$. Therefore,

$$\left\|w^{k+1}\right\|_{L^\infty_h} \leq \left\|w^k\right\|_{L^\infty_h},$$

as we intended to show. □

From stability, one can obtain convergence.

Corollary 28.23 (convergence). *Let u be the classical solution to (28.7) and $w \in \mathcal{V}_0(\bar{\mathcal{C}}^\tau_h)$ the solution to (28.8). Define the error*

$$e = \pi^\tau_h u - w, \qquad e^k_i = u(ih, k\tau) - w^k_i.$$

Under the assumption that the mesh Péclet number satisfies

$$hb \leq 2,$$

and the Courant number satisfies

$$\mu \leq \frac{1}{2},$$

we have that there is a constant $C > 0$ such that

$$\|e\|_{L^\infty_\tau(L^\infty_h)} \leq C(\tau + h^2).$$

Proof. We have that, as usual, the error satisfies the equation

$$\delta_\tau e^{k+1} + b\mathring{\delta}_h e^k - \Delta_h e^k = \mathcal{E}^\tau_h[u]^k, \quad k = 1, \ldots, K,$$

where the consistency error satisfies

$$\|\mathcal{E}^\tau_h[u]\|_{L^\infty_\tau(L^\infty_h)} \leq C(\tau + h^2).$$

From stability, we obtain then that

$$\left\|e^{k+1}\right\|_{L^\infty_h} \leq \left\|e^k\right\|_{L^\infty_h} + \tau\left\|\mathcal{E}^\tau_h[u]^k\right\|_{L^\infty_h} \leq \left\|e^k\right\|_{L^\infty_h} + C\tau(\tau + h^2),$$

which, adding over k, implies the claimed convergence result. □

As we see from the previous two results, stability for a centered discretization is obtained under two conditions: a CFL condition, which is expected as the method is explicit, and a condition on the mesh Péclet number. This is due to the fact that we are using a centered discretization. At the expense of a reduced order of convergence, stability can be obtained for any Péclet number using an upwind discretization of the advection term.

We seek $w \in \mathcal{V}^K_{\bar{N},0}(\bar{\mathcal{C}}^\tau_h)$ such that

$$w^0 = u_{0,h}, \qquad \delta_\tau w^k + b\bar{\delta}_h w^k - \Delta_h w^k = 0, \quad k = 1, \dots, K, \tag{28.9}$$

where $u_{0,h} = \pi_h u_0$.

Theorem 28.24 (conditional stability). *Assume that the Courant number satisfies*

$$\mu \leq \frac{1}{2 + bh}.$$

Then the method (28.9) *is stable in the* $L^\infty_\tau(L^\infty_h)$*-norm. Moreover, the error*

$$e = \pi^\tau_h u - w, \qquad e^k_i = u(ih, k\tau) - w^k_i,$$

under the same assumption, satisfies

$$\|e\|_{L^\infty_\tau(L^\infty_h)} \leq C(\tau + h).$$

Proof. See Problem 28.13. □

28.7 The Initial Value Problem for the Heat Equation in One Dimension

This section is the discrete analogue of Section 23.3.1, but for simplicity we will only consider the case $d = 1$. We will approximate the solution of (23.24) with a finite difference method and describe a very general, powerful, and celebrated technique to analyze the stability of difference methods known as *von Neumann stability analysis.*[5]

Let us introduce the finite difference methods that we will use to approximate the solution to (23.24) with $d = 1$.

Definition 28.25 (simple methods). Let $T > 0$, $K, N \in \mathbb{N}$, $\tau = T/K$, and $h = 1/(N+1)$. Let $u \in C_b(\mathbb{R} \times [0, T])$ be the classical solution to (23.24). Suppose that $u_{0,h} \in \mathcal{V}(\mathbb{Z}_h)$ is defined as usual. We defined the following methods for computing the finite difference approximation $w \in \mathcal{V}([0, T]_\tau; \mathcal{V}(\mathbb{Z}_h))u$.

1. The **backward Euler method**

$$\begin{cases} \bar{\delta}_\tau w^{k+1} - \Delta_h w^{k+1} = 0, & k = 0, \dots, K - 1, \\ w^0 = u_{0,h}. \end{cases} \tag{28.10}$$

[5] Named in honor of the Hungarian–American mathematician, physicist, computer scientist, engineer, and polymath John von Neumann (1903–1957).

2. The **forward Euler method**

$$\begin{cases} \delta_{\mathcal{T}} w^k - \Delta_h w^k = 0, & k = 0, \dots, K-1, \\ w^0 = u_{0,h}. \end{cases} \tag{28.11}$$

3. The **Crank–Nicolson method**

$$\begin{cases} \bar{\delta}_{\mathcal{T}} w^{k+1} - \frac{1}{2}\Delta_h \left(w^{k+1} + w^k\right) = 0, & k = 0, \dots, K-1, \\ w^0 = u_{0,h}. \end{cases} \tag{28.12}$$

28.7.1 The Discrete Fourier Transform

The notions of consistency and stability for these, or any other method for (23.24), are taken as usual. Notice, however, that since we are dealing with a problem on the whole space we must deal with grid functions in $\mathcal{V}(\mathbb{Z}_h)$. We need to introduce then a suitable norm on this space.

One possibility for a norm of $\mathcal{V}(\mathbb{Z}_h)$ is

$$\|v\|_{L_h^\infty} = \sup_{j \in \mathbb{Z}} |v_j|,$$

which is the analogue of discrete norms we have seen before. It is important to note that, in this case, this is truly a supremum, as the index j runs over the infinite set $\mathbb{Z}$. Let us, in addition, introduce another one that will be more suited for our purposes.

Definition 28.26 ($L_h^2(\mathbb{Z}_h)$)**.** For $N \in \mathbb{N}$, let $h = 1/(N+1)$. We denote the space of **square summable grid functions** by

$$L_h^2(\mathbb{Z}_h) = \left\{ v \in \mathcal{V}(\mathbb{Z}_h) \;\middle|\; \|v\|_{L_h^2} < \infty \right\},$$

where

$$\|v\|_{L_h^2} = \left(h \sum_{j \in \mathbb{Z}} |v_j|^2 \right)^{1/2}.$$

As usual, if the range of our grid functions is a finite-dimensional vector space $\mathbb{V} \neq \mathbb{R}$, then we denote this by $L_h^2(\mathbb{Z}_h; \mathbb{V})$.

It is not difficult to verify that $L_h^2(\mathbb{Z}_h)$ is a vector space and that $\|\cdot\|_{L_h^2}$ is a norm that makes this a Hilbert space.

Theorem 28.27 (properties of $L_h^2(\mathbb{Z}_h)$)**.** *The space $L_h^2(\mathbb{Z}_h)$ is Hilbert, with the inner product*

$$(v_1, v_2)_{L_h^2} = h \sum_{j \in \mathbb{Z}} v_1(jh) v_2(jh).$$

This inner product induces the L_h^2-norm.

Proof. See Problem 28.19. □

On square summable grid functions, we can define a discrete analogue of the Fourier transform (23.25).

Definition 28.28 (FZT/DFT). The so-called **Fourier-$\mathbb{Z}$ transform** (FZT) or **discrete Fourier transform** (DFT) is the mapping $\mathcal{Z}\colon L^2_h(\mathbb{Z}_h;\mathbb{C}) \to L^2((-\pi,\pi);\mathbb{C})$ defined by

$$\hat{v}(\xi) = \mathcal{Z}[v](\xi) = h\sum_{k\in\mathbb{Z}} v_k e^{-ik\xi}.$$

The properties of the FZT are the content of the following result.

Theorem 28.29 (properties of the FZT). *The FZT is well defined. In particular,*

$$\|\mathcal{Z}[v]\|^2_{L^2(-\pi,\pi;\mathbb{C})} = 2\pi h\|v\|^2_{L^2_h}, \quad \forall v \in L^2_h(\mathbb{Z}_h;\mathbb{C}).$$

The mapping $\mathcal{Z}$ is invertible, and its inverse, for $v \in L^2(-\pi,\pi;\mathbb{C})$, is given by

$$\mathcal{Z}^{-1}[v]_j = \frac{1}{2\pi h}\int_{-\pi}^{\pi} v(\xi)e^{ij\xi}\mathrm{d}\xi.$$

Finally, we have the shift property: if $v \in L^2_h(\mathbb{Z}_h;\mathbb{C})$ and $m \in \mathbb{Z}$, then

$$\mathcal{Z}[v_{\cdot-m}](\xi) = e^{-im\xi}\mathcal{Z}[v](\xi).$$

Proof. First, observe that, for $v \in L^2_h(\mathbb{Z}_h;\mathbb{C})$, we have

$$\begin{aligned}
\|\mathcal{Z}[v]\|^2_{L^2(-\pi,\pi;\mathbb{C})} &= h^2\int_{-\pi}^{\pi}\sum_{k\in\mathbb{Z}} v_k e^{-ik\xi}\cdot\overline{\sum_{k\in\mathbb{Z}} v_m e^{-im\xi}}\mathrm{d}\xi \\
&= h^2\sum_{k,m\in\mathbb{Z}} v_k\overline{v_m}\int_{-\pi}^{\pi} e^{i(m-k)\xi}\mathrm{d}\xi \\
&= 2\pi h^2\sum_{k,m\in\mathbb{Z}} v_k\overline{v_m}\delta_{k,m} \\
&= 2\pi h^2\sum_{k\in\mathbb{Z}}|v_k|^2,
\end{aligned}$$

so that indeed $\mathcal{Z}[v] \in L^2(-\pi,\pi;\mathbb{C})$ with the claimed identity for norms.

To verify the formula for the inverse, consider

$$\begin{aligned}
\mathcal{Z}^{-1}[\mathcal{Z}[v]]_j &= \frac{1}{2\pi h}\int_{-\pi}^{\pi}\mathcal{Z}[v](\xi)e^{ij\xi}\mathrm{d}\xi \\
&= \frac{1}{2\pi}\sum_{k\in\mathbb{Z}} v_k\int_{-\pi}^{\pi} e^{-ik\xi}e^{ij\xi}\mathrm{d}\xi \\
&= \sum_{k\in\mathbb{Z}} v_k\delta_{k,j} = v_j.
\end{aligned}$$

The shift property is immediate

$$\mathcal{Z}[v_{\cdot-m}](\xi) = h\sum_{k\in\mathbb{Z}} v_{k-m}e^{-ik\xi} = he^{-im\xi}\sum_{k\in\mathbb{Z}} v_{k-m}e^{-i(k-m)\xi} = e^{-im\xi}\mathcal{Z}[v](\xi). \quad \square$$

28.7.2 The Symbol of a Finite Difference Method

To motivate the definition we will introduce next, we consider the forward Euler method (28.11). In full analogy to what was done in Section 23.3.1 to reach the solution representation (23.27), we take now the FZT of (28.11) to obtain, using the shift property described in Theorem 28.29, that

$$\widehat{w}^{k+1} - \widehat{w}^k = \mu\left(e^{-i\xi} + e^{i\xi} - 2\right)\widehat{w}^k.$$

In other words,

$$\widehat{w}^{k+1} = (1 - 2\mu + 2\mu\cos\xi)\widehat{w}^k.$$

We now see that all the properties of the method are encoded in the expression

$$\tilde{E}_h^\tau(\xi) = 1 - 2\mu + 2\mu\cos\xi.$$

Definition 28.30 (two-layer method)**.** Let $\mathcal{A}_h, \mathcal{B}_h \colon \mathcal{V}(\mathbb{Z}_h) \to \mathcal{V}(\mathbb{Z}_h)$ be finite difference operators with constant coefficients, i.e.,

$$(\mathcal{A}_h v)_j = \sum_{m\in S_\mathcal{A}} a_m(h,\tau)v_{j-m}, \quad (\mathcal{B}_h v)_j = \sum_{m\in S_\mathcal{B}} b_m(h,\tau)v_{j-m}, \qquad \forall v \in \mathcal{V}(\mathbb{Z}_h),$$

where $S_\mathcal{A}, S_\mathcal{B} \subset \mathbb{Z}$ have finite cardinality and $a_m, b_m \colon \mathbb{R}_+ \times \mathbb{R}_+ \to \mathbb{R}\backslash\{0\}$ are functions. Assume that $\mathcal{A}_h$ is invertible. A **two-layer time stepping finite difference method** with constant coefficients is defined as follows: find $w \in \mathcal{V}([0,T]_\tau; \mathcal{V}(\mathbb{Z}_h))$ such that $w^0 \in \mathcal{V}(\mathbb{Z}_h)$ is given, and

$$\mathcal{A}_h w^{k+1} = \mathcal{B}_h w^k, \quad k = 0, \dots, K-1. \tag{28.13}$$

If $\mathcal{A}_h$ is the identity operator, then we say that the method is **explicit**; otherwise, we say that it is **implicit**.

Definition 28.31 (symbol)**.** The **symbol** or **amplification factor** of the two-layer time stepping finite difference method (28.13) is

$$\tilde{E}_h^\tau(\xi) = \frac{\sum_{m\in S_\mathcal{B}} b_m(h,\tau)e^{-im\xi}}{\sum_{m\in S_\mathcal{A}} a_m(h,\tau)e^{-im\xi}}, \quad \xi \in \mathbb{R}. \tag{28.14}$$

Example 28.2 The symbol of the backward Euler method (28.10) is

$$\tilde{E}_h^\tau(\xi) = \frac{1}{1 + 2\mu - 2\mu\cos\xi};$$

see Problem 28.21.

Example 28.3 The symbol of the forward Euler method (28.11) is

$$\tilde{E}_h^\tau(\xi) = 1 - 2\mu + 2\mu\cos\xi,$$

as we showed at the beginning of our discussion.

Example 28.4 The symbol of the Crank–Nicolson method (28.12) is

$$\tilde{E}_h^{\tau}(\xi) = \frac{1 - \mu + \mu \cos \xi}{1 + \mu - \mu \cos \xi};$$

see Problem 28.22.

The following is nothing but a mere observation.

Proposition 28.32 (properties of the symbol). *Consider the time stepping method* (28.13). *Its symbol $\tilde{E}_h^{\tau}(\xi)$ is, in general, a 2π-periodic rational trigonometric function. If the method is explicit, then the symbol is a trigonometric polynomial.*

Proof. See Problem 28.23. □

Since all the properties of the method (28.13) are encoded in its symbol, it is of no surprise that this is a very useful tool to provide a stability analysis in various norms. For instance, the following is a necessary condition for stability in the $L_\tau^\infty(L_h^\infty)$-norm.

Theorem 28.33 ($L_\tau^\infty(L_h^\infty)$-stability). *Let the time stepping method* (28.13) *with symbol $\tilde{E}_h^{\tau}(\xi)$ be explicit and stable in the $L_\tau^\infty(L_h^\infty)$-norm. Then its symbol satisfies*

$$\left|\tilde{E}_h^{\tau}(\xi)\right| \leq 1, \quad \forall \xi \in \mathbb{R}.$$

Proof. We will argue by contradiction. We assume that the method is stable in the $L_\tau^\infty(L_h^\infty)$-norm, yet there is $\xi_0 \in [-\pi, \pi]$ for which

$$\left|\tilde{E}_h^{\tau}(\xi_0)\right| = q > 1.$$

Let $\varepsilon \in (0, 1)$ and define $v \in \mathcal{V}(\mathbb{Z}_h; \mathbb{C})$ via

$$v_j = \varepsilon e^{\mathrm{i} j \xi_0}, \qquad \|v\|_{L_h^\infty} = \varepsilon.$$

Let $w \in \mathcal{V}([0, T]_\tau; \mathcal{V}(\mathbb{Z}_h))$ be the solution to (28.13) with initial condition $w^0 = v$. Now, since the method is explicit, for any $j \in \mathbb{Z}$, we have

$$w_j^1 = \varepsilon \sum_{m \in S_\mathcal{B}} b_m(h, \tau) e^{\mathrm{i}(j-m)\xi_0} = \varepsilon e^{\mathrm{i} j \xi_0} \sum_{m \in S_\mathcal{B}} b_m(h, \tau) e^{-\mathrm{i} m \xi_0} = \tilde{E}_h^{\tau}(\xi_0) v_j;$$

this implies that, for $k \geq 0$,

$$w^k = \left[\tilde{E}_h^{\tau}(\xi_0)\right]^k v.$$

Consequently,

$$\left\|w^k\right\|_{L_h^\infty} = \varepsilon q^k \to \infty, \qquad k \to \infty,$$

regardless of the value of ε. Thus, the method is not stable in the $L_\tau^\infty(L_h^\infty)$-norm. □

Example 28.5 As we have seen before, for the forward Euler method (28.11),

$$\tilde{E}_h^\tau(\xi) = 1 - 2\mu + 2\mu\cos\xi.$$

Since $\cos\xi \in [-1, 1]$, then

$$\left|\tilde{E}_h^\tau(\xi)\right| \leq 1$$

is possible if and only if

$$-1 \leq 1 - 2\mu - 2\mu \qquad \text{and} \qquad 1 - 2\mu + 2\mu \leq 1.$$

Thus, we require

$$1 - 4\mu \geq -1 \qquad \Longleftrightarrow \qquad \mu \leq \frac{1}{2},$$

which is the CFL-type condition we have obtained before for stability of forward Euler.

28.7.3 The von Neumann Stability Analysis

It turns out that the condition on the modulus of the symbol

$$\left|\tilde{E}_h^\tau(\xi)\right| \leq 1, \quad \forall \xi \in \mathbb{R}, \tag{28.15}$$

which is called the *von Neumann stability condition*, is not only necessary but also sufficient for stability, provided that we study stability in $L_\tau^\infty(L_h^2)$-norms.

Theorem 28.34 (von Neumann[6]). *Consider the time stepping method* (28.13) *with symbol* $\tilde{E}_h^\tau(\xi)$. *This method is stable in the* $L_\tau^\infty(L_h^2)$*-norm if and only if the von Neumann stability condition* (28.15) *holds.*

Proof. Let $w \in \mathcal{V}([0,T]_\tau; \mathcal{V}(\mathbb{Z}_h))$ be the solution to method (28.13) with initial condition $u_{0,h} \in L_h^2(\mathbb{Z}_h)$. Problem 28.20 implies that

$$\hat{w}^{k+1}(\xi) = \tilde{E}_h^\tau(\xi)\hat{w}^k(\xi) \qquad \Longrightarrow \qquad \hat{w}^k(\xi) = \left[\tilde{E}_h^\tau(\xi)\right]^{k+1}\hat{u}_{0,h}(\xi).$$

Assume now that (28.15) holds. Using Theorem 28.29, we have that

$$\begin{aligned}
\|w^k\|_{L_h^2}^2 &= \frac{1}{2\pi h}\|\hat{w}^k\|_{L^2(-\pi,\pi;\mathbb{C})}^2 \\
&= \frac{1}{2\pi h}\left\|\left[\tilde{E}_h^\tau\right]^k \hat{u}_{0,h}\right\|_{L^2(-\pi,\pi;\mathbb{C})}^2 \\
&\leq \frac{1}{2\pi h}\sup_{\xi\in[-\pi,\pi]}\left|\tilde{E}_h^\tau(\xi)\right|^{2k}\|\hat{u}_{0,h}\|_{L^2(-\pi,\pi;\mathbb{C})}^2 \\
&= \sup_{\xi\in[-\pi,\pi]}\left|\tilde{E}_h^\tau(\xi)\right|^{2k}\|u_{0,h}\|_{L_h^2}^2 \\
&\leq \|u_{0,h}\|_{L_h^2}^2,
\end{aligned}$$

which is the desired stability result.

[6] Named in honor of the Hungarian–American mathematician, physicist, computer scientist, engineer, and polymath John von Neumann (1903–1957).

Assume now that (28.15) fails at a point. Owing to Proposition 28.32, the symbol is a rational trigonometric function. Thus, without loss of generality, this is an interior point of $[-\pi, \pi]$, i.e., there are $q > 1$, $\delta > 0$, and $\xi_0 \in (-\pi, \pi)$ such that

$$[\xi_0 - \delta, \xi_0 + \delta] \subseteq [-\pi, \pi]$$

and

$$\left|\tilde{E}_h^{\mathcal{T}}(\xi)\right| \geq q > 1, \quad \forall \xi \in [\xi_0 - \delta, \xi_0 + \delta].$$

Consider now the grid function

$$u_{0,h}(kh) = \frac{e^{ik\xi_0}}{\pi hk} \sin(k\delta).$$

We leave it to the reader as an exercise to show, see Problem 28.24, that

$$u_{0,h} \in L_h^2(\mathbb{Z}_h), \qquad \hat{u}_{0,h}(\xi) = \chi_{[\xi_0-\delta,\xi_0+\delta]}.$$

Using this as an initial condition, we have that

$$\begin{aligned}
\|w^k\|_{L_h^2}^2 &= \frac{1}{2\pi h}\int_{-\pi}^{\pi} \left|\tilde{E}_h^{\mathcal{T}}(\xi)\right|^{2k} |\hat{u}_{0,h}(\xi)|^2 \mathrm{d}\xi \\
&\geq \frac{q^{2k}}{2\pi h}\int_{\xi_0-\delta}^{\xi_0+\delta} |\hat{u}_{0,h}(\xi)|^2 \mathrm{d}\xi \\
&= q^{2k}\|u_{0,h}\|_{L_h^2}^2 \to \infty,
\end{aligned}$$

as $k \to \infty$. In conclusion, the method is not stable. □

Example 28.6 The backward Euler method (28.10) is unconditionally stable in the $L_{\mathcal{T}}^\infty(L_h^2)$-norm. Indeed, we have that

$$\tilde{E}_h^{\mathcal{T}}(\xi) = \frac{1}{1 + 2\mu - 2\mu\cos\xi}.$$

Indeed, since $\cos\xi \in [-1, 1]$,

$$1 = 1 + 2\mu - 2\mu \leq 1 + 2\mu - 2\mu\cos\xi,$$

which implies that

$$0 < \tilde{E}_h^{\mathcal{T}}(\xi) \leq 1, \quad \forall \xi \in [-\pi, \pi].$$

Example 28.7 The forward Euler method (28.11) is stable, in the $L_{\mathcal{T}}^\infty(L_h^2)$-norm, provided that the CFL-type condition

$$\mu \leq \frac{1}{2}$$

holds; see Example 28.5.

Example 28.8 The Crank–Nicolson method (28.12) is unconditionally stable in the $L^\infty_T(L^2_h)$-norm; see Problem 28.25.

Remark 28.35 (periodic boundary conditions)**.** Finally, to end this section, we remark that this powerful von Neumann stability analysis can be used under the assumption of periodic boundary conditions, with very little change in the computations. The details are left to the reader. The essence of the analysis is to use the appropriate eigenfunctions of the finite difference operators in the infinite or periodic settings, where boundary effects can be neglected, to compute the symbol. With this at hand, one can easily assess the stability of the method in the $L^\infty_T(L^2_h)$-norm. Interestingly, the same CFL-type restrictions are required for $L^\infty_T(L^2_h)$-stability in the case of a bounded, periodic, or infinite domain.

28.7.4 Convergence

It turns out that the symbol of a time stepping method can also be used to study the convergence of this method. Here, we briefly describe how this can be achieved. To accomplish this, however, we need to introduce some notions.

Notice that we can identify grid functions with piecewise constant ones. In other words, given $v \in \mathcal{V}(\mathbb{Z}_h)$, we can define $V_h \colon \mathbb{R} \to \mathbb{R}$ by

$$V_h(x) = v_j, \quad x \in \left[\left(j - \frac{1}{2}\right)h, \left(j + \frac{1}{2}\right)h\right).$$

Notice also that

$$\|v\|^2_{L^2_h} = h \sum_{j\in\mathbb{Z}} |v_j|^2 = \sum_{j\in\mathbb{Z}} \int_{(j-\frac{1}{2})h}^{(j+\frac{1}{2})h} |v_j|^2 dx = \sum_{j\in\mathbb{Z}} \int_{(j-\frac{1}{2})h}^{(j+\frac{1}{2})h} |V_h(x)|^2 dx = \|V_h\|^2_{L^2(\mathbb{R};\mathbb{C})},$$

so that $V_h \in L^2(\mathbb{R})$ if and only if $v \in L^2_h(\mathbb{Z}_h)$. Because of this, we can take the Fourier transform of V_h and obtain, see Problem 28.30,

$$\mathcal{F}[V_h](\xi) = \frac{2}{h\xi} \sin\left(\frac{h\xi}{2}\right) \mathcal{Z}[v](h\xi). \tag{28.16}$$

That is, the FZT of a grid function is a dilation of the Fourier transform of its representative times a constant factor.

We can now start to study the convergence of a time stepping method using the symbol. We begin with a definition.

Definition 28.36 (accuracy)**.** Let the time stepping method (28.13) have symbol $\tilde{E}^T_h(\xi)$. We say that the method is **accurate of order** $p \in \mathbb{N}$ if there are constants $C_1, C_2 > 0$ such that whenever

$$|\xi| \le C_2,$$

then

$$\left|\tilde{E}^T_h(\xi) - e^{-\mu\xi^2}\right| \le C_1 |\xi|^{p+2},$$

where $\mu > 0$ is the Courant number.

The meaning of this definition is detailed in Problem 28.31.

Example 28.9 The forward Euler method (28.11) is accurate of order $p = 2$. As we recall,

$$\tilde{E}_h^\tau(\xi) = 1 - 2\mu + 2\mu\cos\xi.$$

Doing a Taylor expansion of this expression about $\xi = 0$, we find there is $\eta \in \mathbb{R}$ for which

$$\begin{aligned}
\tilde{E}_h^\tau(\xi) &= 1 - 2\mu + 2\mu\left(1 - \frac{\xi^2}{2} + \frac{\xi^4}{24} + \eta|\xi|^6\right) \\
&= 1 - \mu\xi^2 + \frac{\mu}{12}\xi^4 + 2\mu\eta|\xi|^6 \\
&= 1 - \mu\xi^2 + \frac{\mu^2}{2}\xi^4 - \frac{\mu}{2}\left(\mu - \frac{1}{6}\right)\xi^4 + 2\mu\eta|\xi|^6.
\end{aligned}$$

If we compare this expression with the Taylor expansion of $e^{-\mu\xi^2}$ around $\xi = 0$, which is

$$e^{-\mu\xi^2} = 1 - \mu\xi^2 + \frac{\mu^2}{2}\xi^4 + C|\xi^6|,$$

we conclude that, if $\mu \neq \frac{1}{6}$, the forward Euler method is accurate of order $p = 2$, and that if $\mu = \frac{1}{6}$, then it is accurate of order $p = 4$.

To gain some intuition into how this notion can be useful, we will look at the error. Usually, to measure the error, we project the exact solution onto the grid. Here, we will go the opposite route, i.e., define

$$e^k(x) = u(x, k\tau) - W_h^k(x),$$

where, for $k \in \{0, \dots, K\}$, the function W_h^k is the piecewise constant representative of the grid function w^k. The advantage of this is that now we are dealing with usual functions. Then we can take the Fourier transform of the error after one step to obtain

$$\begin{aligned}
\hat{e}^1(\xi) &= \hat{u}(\xi, \tau) - \mathcal{F}[W_h^1](\xi) \\
&= \hat{u}(\xi, \tau) - \frac{2}{h\xi}\sin\left(\frac{h\xi}{2}\right)\mathcal{Z}[w^1](h\xi) \\
&= \hat{u}(\xi, \tau) - \frac{2}{h\xi}\sin\left(\frac{h\xi}{2}\right)\tilde{E}_h^\tau(h\xi)\mathcal{Z}[w^0](h\xi) \\
&= \hat{u}(\xi, \tau) - \tilde{E}_h^\tau(h\xi)\mathcal{F}[W_h^0](\xi) \\
&= \left(\hat{u}(\xi, \tau) - \tilde{E}_h^\tau(h\xi)\hat{u}_0(\xi)\right) + \left(\tilde{E}_h^\tau(h\xi)\mathcal{F}\left[u_0 - W_h^0\right](h\xi)\right).
\end{aligned}$$

In other words, the error after one step consists of two terms. The first one

$$\hat{u}(\xi,\tau) - \tilde{E}_h^\tau(h\xi)\hat{u}_0(\xi)$$

encodes the error in the time stepping method. The second term is

$$\tilde{E}_h^\tau(h\xi)\mathcal{F}\left[u_0 - W_h^0\right](h\xi),$$

which depends on how well we approximated the initial condition u_0. Usually, $w^0 = \pi_h u_0$, so that this expression is dependent only on the smoothness of u_0.

Let us focus our attention then on the first term. Using that $\hat{u}(\xi,t) = \hat{u}_0(\xi)e^{-t\xi^2}$, see Section 23.3.1, implies that

$$\hat{u}(\xi,\tau) - \tilde{E}_h^\tau(h\xi)\hat{u}_0(\xi) = \left(e^{-\mu h^2\xi^2} - \tilde{E}_h^\tau(h\xi)\right)\hat{u}_0(\xi),$$

where we also used that $\tau = \mu h^2$, where μ is the Courant number. Let us now assume that $\xi \to 0$ while keeping μ constant and that the method is accurate of order $p \in \mathbb{N}$, then

$$|\hat{u}(\xi,\tau) - \tilde{E}_h^\tau(h\xi)\hat{u}_0(\xi)| \leq Ch^{p+2} = \frac{C}{\mu}\tau h^p.$$

The use of this idea to assert convergence is the content of the following result.

Theorem 28.37 ($L_\tau^\infty(L^2(\mathbb{R}))$-convergence). *Let the time stepping method* (28.13) *be accurate of order $p \in \mathbb{N}$ and stable in $L_\tau^\infty(L_h^2)$. If u_0 is sufficiently smooth, h is sufficiently small, and $\mu = \tau/h^2$ is kept constant, then we have that, for every $k \in \{0,\ldots,K\}$,*

$$\|W_h^k - u(\cdot,t_k)\|_{L^2(\mathbb{R})} \leq Ct_k h^p + \left\|W_h^0 - u_0\right\|_{L^2(\mathbb{R})},$$

where the constant C does not depend on h or τ. It may depend, however, on μ and the smoothness of u_0 via norms of its derivatives.

Proof. We begin with similar computations to those that motivated us. First, by Plancherel's identity of Proposition 23.16,

$$\|W_h^k - u(\cdot,t_k)\|_{L^2(\mathbb{R})} = \|\widehat{W_h^k} - \hat{u}(\cdot,t_k)\|_{L^2(\mathbb{R};\mathbb{C})}.$$

Set $\kappa(h,\xi) = 2\sin(h\xi/2)/(h\xi)$ and observe that

$$\widehat{W_h^k} = \kappa(h,\xi)\mathcal{Z}[w^k](h\xi) = \kappa(h,\xi)\tilde{E}_h^\tau(h\xi)^k\mathcal{Z}[w^0](h\xi) = \tilde{E}_h^\tau(h\xi)^k\widehat{W_h^0}.$$

Therefore, by the triangle inequality,

$$\begin{aligned}\|W_h^k - u(\cdot,t_k)\|_{L^2(\mathbb{R})} &\leq \left\|\tilde{E}_h^\tau(h\,\cdot)^k\hat{u}_0 - \hat{u}(\cdot,t_k)\right\|_{L^2(\mathbb{R};\mathbb{C})} \\ &\quad + \left\|\tilde{E}_h^\tau(h\,\cdot)^k[\widehat{W_h^0} - \hat{u}_0]\right\|_{L^2(\mathbb{R};\mathbb{C})} \\ &\leq \left\|\tilde{E}_h^\tau(h\,\cdot)^k\hat{u}_0 - \hat{u}(\cdot,t_k)\right\|_{L^2(\mathbb{R};\mathbb{C})} + \left\|W_h^0 - u_0\right\|_{L^2(\mathbb{R})},\end{aligned}$$

where, in the last step, we used that the method is $L_\tau^\infty(L_h^2)$-stable; thus, by Theorem 28.34, condition (28.15) holds, followed by, once again, an application of Plancherel's identity. It remains then to bound the first term, for which we observe that

$$\left\|\tilde{E}_h^\tau(h\,\cdot)^k\widehat{u}_0-\widehat{u}(\cdot,t_k)\right\|^2_{L^2(\mathbb{R};\mathbb{C})}=\int_{\mathbb{R}}\left|\tilde{E}_h^\tau(h\xi)^k\widehat{u}_0(\xi)-\widehat{u}(\xi,t_k)\right|^2\mathrm{d}\xi$$
$$=\int_{\mathbb{R}}\left|\tilde{E}_h^\tau(h\xi)^k-e^{-\mu k h^2\xi^2}\right|^2|\widehat{u}_0(\xi)|^2\mathrm{d}\xi. \qquad (28.17)$$

Observe now that, since the method is accurate to order $p\in\mathbb{N}$, for $|\zeta|\le C_2$, we have

$$\left|\tilde{E}_h^\tau(\zeta)^k-e^{-\mu k\zeta^2}\right|=\left|\left(\tilde{E}_h^\tau(\zeta)-e^{-\mu\zeta^2}\right)\sum_{j=0}^{k-1}\tilde{E}_h^\tau(\zeta)^{n-1-j}e^{-\mu j\zeta^2}\right|$$
$$\le\left|\tilde{E}_h^\tau(\zeta)-e^{-\mu\zeta^2}\right|\sum_{j=0}^{k-1}|\tilde{E}_h^\tau(\zeta)^{n-1-j}|$$
$$\le C_1k|\zeta|^{p+2},$$

where we also used the von Neumann stability condition. Using this estimate in (28.17) yields

$$\left\|\tilde{E}_h^\tau(h\,\cdot)^k\widehat{u}_0-\widehat{u}(\cdot,t_k)\right\|^2_{L^2(\mathbb{R};\mathbb{C})}\le C_1^2k^2h^{2(p+2)}\int_{\mathbb{R}}\left|\xi^{p+2}\widehat{u}_0(\xi)\right|^2\mathrm{d}\xi.$$

Use now that $k^2h^4=t_k^2/\mu^2$ and the differentiation property of Proposition 23.16 to conclude then that

$$\left\|\tilde{E}_h^\tau(h\,\cdot)^k\widehat{u}_0-\widehat{u}(\cdot,t_k)\right\|^2_{L^2(\mathbb{R};\mathbb{C})}\le\frac{C_1^2}{\mu^2}t_k^2h^{2p}\left\|u_0^{(p+2)}\right\|^2_{L^2(\mathbb{R})}.$$

Gathering all the obtained bounds, the result now follows. □

Problems

28.1 Prove Proposition 28.4.

28.2 Prove Theorem 28.8.

28.3 Does a maximum principle, such as that of Theorem 28.9, hold for the forward Euler method (28.4)? Perhaps under a CFL-type condition?

28.4 Prove Theorem 28.12.

28.5 Complete the proof of Theorem 28.14.

28.6 Prove Theorem 28.16.

28.7 Complete the proof of Theorem 28.17.

28.8 To approximate the classical solution of (28.1) with $f\equiv 0$, we consider the following method: Find a grid function $w\in\mathcal{V}_0(\bar{C}_h^\tau)$ such that $w^0=\pi_h u_0$ and

$$\bar{\delta}_\tau w^{k+1}-\Delta_h\left(\sigma w^{k+1}+(1-\sigma)w^k\right)=0,\quad k=0,\dots,K-1. \qquad (28.18)$$

Here, $\sigma\in\mathbb{R}$ is a given parameter. Show that if

$$\sigma=\frac{1}{2}-\frac{h^2}{12\tau},$$

then this method is consistent, in $C_b(\mathbb{R}^2)$, with the heat equation (28.1).

Hint: Use $(x_i,t_k+\tau/2)$ as the base point for your Taylor expansions.

28.9 In this problem, we will develop the method of *separation of variables* to study stability, in the $L^\infty_\tau(L^2_h)$-norm, of difference methods.

a) Show that method (28.18) can be equivalently written as

$$\delta_\tau w^k - \sigma\tau\Delta_h\delta_\tau w^k = \Delta_h w^k, \quad k = 1, \dots, K.$$

b) We will seek our solution $w \in \mathcal{V}_0(\bar{\mathcal{C}}_h^\tau)$ as the product of two functions

$$w = X_w T_w, \qquad X_w \in \mathcal{V}_0(\bar{\Omega}_h), \quad T_w \in \mathcal{V}([0, T]_\tau).$$

Show that, for all k,

$$\frac{T_w^{k+1} - T_w^k}{\tau\left(\sigma T_w^{k+1} - (1-\sigma)T_w^k\right)} = \frac{\Delta_h X_w}{X_w} = -\lambda, \tag{28.19}$$

where $\lambda \in \mathbb{R}$.

c) Show that, for any $\lambda > 0$,

$$T_w^{k+1} = qT_w^k, \qquad q = \frac{1-(1-\sigma)\tau\lambda}{1+\sigma\tau\lambda},$$

solves (28.19).

d) Using the results of Theorem 24.25, deduce then that the pairs $\{X_{w,n}, \lambda_n\}_{n=1}^N$

$$\lambda_n = \frac{4}{h^2}\sin^2\left(\frac{n\pi h}{2}\right), \qquad X_{w,n}(x_i) = \sqrt{2}\sin(n\pi i h)$$

are solutions to (28.19). In addition, show that the system $\{X_{w,n}\}_{n=1}^N$ is an orthonormal (in the L^2_h-inner product) basis of $\mathcal{V}_0(\bar{\Omega}_h)$.

e) From the previous two points, show then that the solution to (28.18) has the form

$$w^k = \sum_{n=1}^N u_{0,n} q_n^k X_w, \quad q_n = \frac{1-(1-\sigma)\tau\lambda_n}{1+\sigma\tau\lambda_n},$$

with

$$w^0 = \sum_{n=1}^N u_{0,n} X_w, \qquad \|w^0\|_{L^2_h} = \left(\sum_{n=1}^N |u_{0,n}|^2\right)^{1/2}.$$

f) Show that method (28.18) is stable, in the $L^\infty_\tau(L^2_h)$-norm, provided that

$$\max_{n=1,\dots,N} |q_n| \le 1 \qquad \Longleftrightarrow \qquad \sigma \ge \frac{1}{2} - \frac{h^2}{4\tau}.$$

28.10 Consider the Crank–Nicolson method given in (28.5) with $f_h \equiv 0$. Show that, for every $k \in \{0, \dots, K-1\}$, we have

$$\|w^{k+1}\|^2_{L^2_h} + \frac{\tau}{2}\left\|\bar{\delta}_h(w^{k+1} + w^k)\right\|^2_{L^2_h} = \|w^k\|^2_{L^2_h}.$$

28.11 Consider the so-called *Richardson method*[7] to approximate the classical solution of (28.1) with $f \equiv 0$: Find a grid function $w \in \mathcal{V}_0(\bar{\mathcal{C}}_h^{\tau})$ such that $w^1 = w^0 = \pi_h u_0$ and

$$\frac{1}{2\tau}\left(w^{k+1} - w^{k-1}\right) - \Delta_h w^k = 0, \quad k = 1, \ldots, K-1.$$

Show that this method is *unconditionally unstable* in the $L_\tau^\infty(L_h^2)$-norm. In other words, the method is unstable *regardless* of the choice of τ and h.

28.12 Consider the *Du Fort–Frankel method*[8] to approximate the classical solution of (28.1) with $f \equiv 0$: Find a grid function $w \in \mathcal{V}_0(\bar{\mathcal{C}}_h^{\tau})$ such that $w^1 = w^0 = \pi_h u_0$ and, for $k = 1, \ldots, K-1$ and $i = 1, \ldots, N$,

$$\frac{1}{2\tau}\left(w_i^{k+1} - w_i^{k-1}\right) - \frac{1}{h^2}\left(w_{i-1}^k - w_i^{k+1} - w_i^{k-1} + w_{i+1}^k\right) = 0.$$

Study the consistency and stability, in the $L_\tau^\infty(L_h^2)$-norm, of this method.

28.13 Prove Theorem 28.24.

28.14 Consider the following linear reaction–diffusion problem: Find $u\colon [0,1] \times [0,T] \to \mathbb{R}$ that is a classical solution of

$$\begin{cases} \dfrac{\partial u(x,t)}{\partial t} - \dfrac{\partial^2 u(x,t)}{\partial x^2} + u(x,t) = 0, & (x,t) \in (0,1) \times (0,T], \\ u(0,t) = u(1,t) = 0, & t \in (0,T], \\ u(x,0) = u_0(x), & x \in [0,1]. \end{cases}$$

We propose a Crank–Nicolson-like method for this problem: Find $w \in \mathcal{V}_0(\bar{\mathcal{C}}_h^{\tau})$ that satisfies

$$\bar{\delta}_\tau w^{k+1} - \frac{1}{2}\Delta_h(w^{k+1} + w^k) + \frac{1}{2}(w^{k+1} + w^k) = 0, \quad k = 0, \ldots, K-1.$$

Provide the conditions that guarantee that this method is stable, in the $L_\tau^\infty(L_h^2)$-norm, and convergent with the error estimate

$$\|e\|_{L_\tau^\infty(L_h^2)} \le C(\tau^2 + h^2),$$

where C is independent of h and τ. You may assume, without proof, that the consistency error has the appropriate form you need.

28.15 Consider the heat equation in a periodic domain. That is, we need to find $u\colon \mathbb{R} \times [0,T] \to \mathbb{R}$ that is spatially one-periodic, i.e., for all $t \in [0,T]$ and any $x \in \mathbb{R}$, it satisfies

$$u(x+n,t) = u(x,t), \qquad \forall n \in \mathbb{Z},$$

and is a classical solution of

$$\begin{cases} \dfrac{\partial u(x,t)}{\partial t} = \dfrac{\partial^2 u(x,t)}{\partial x^2}, & (x,t) \in \mathbb{R} \times (0,T], \\ u(x,0) = u_0(x), & x \in \mathbb{R}, \end{cases}$$

[7] Named in honor of the British mathematician, physicist, and meteorologist Lewis Fry Richardson (1881–1953).

[8] Named in honor of E.C. Du Fort and the American computer scientist Stanley Phillips Frankel (1919–1978).

where $u_0 \in C_p^\infty(0,1)$. Devise a Crank–Nicolson approximation to this problem and prove that, for all values of the Courant number $\mu > 0$, this approximation is stable in the $L_\tau^\infty(L_h^2)$-norm. In addition, show that the error satisfies the estimate

$$\|e\|_{L_\tau^\infty(L_h^2)} \leq C(\tau^2 + h^2),$$

where $C > 0$ is a constant that is independent of τ and h.

28.16 Consider the following problem: As usual, $\Omega = (0,1)$ and we seek the solution to the initial boundary value problem

$$\begin{cases} \dfrac{\partial u(x,t)}{\partial t} - \dfrac{\partial}{\partial x}\left(a(x)\dfrac{\partial u(x,t)}{\partial x}\right) = f(x,t), & (x,t) \in \Omega \times (0,T], \\ u(0,t) = u(1,t) = 0, & t \in [0,T], \\ u(x,0) = u_0(x), & x \in \bar{\Omega}. \end{cases}$$

Here, $a \in C(\bar{\Omega})$ is such that there is $\lambda > 0$ for which

$$a(x) \geq \lambda, \quad \forall x \in \bar{\Omega}.$$

In addition, $f \in C(\bar{\mathcal{C}})$ and $u_0 \in C(\bar{\Omega})$ with $u(0) = u(1) = 0$. To approximate the solution to this problem, we propose the following finite difference method: Find $w \in \mathcal{V}_0(\bar{\mathcal{C}}_h^\tau)$ such that $w^0 = \pi_h u_0$ and

$$\bar{\delta}_\tau w^{k+1} + L_h w^{k+1} = f_h^{k+1}, \qquad k = 0, \dots, K-1.$$

Here, π_h denotes the sampling operator, $f_h \in \mathcal{V}(\mathcal{C}_h^\tau)$ is given by $f_h(ih, k\tau) = f(ih, k\tau)$, and L_h is the conservative finite difference method given in (24.14). Show that the solution of this method satisfies, for some constant that is independent of h and τ,

$$\|w\|_{L_\tau^\infty(L_h^2)}^2 + \|w\|_{L_\tau^2(H_h^1)}^2 \leq C\left(\|w^0\|_{L_h^2}^2 + \|f_h\|_{L_\tau^2(L_h^2)}^2\right).$$

Is this method convergent?

28.17 To approximate the classical solution of (28.7), we use an implicit method with a centered discretization of the first derivative

$$\bar{\delta}_\tau w^{k+1} + b\mathring{\delta}_h w^{k+1} - \Delta_h w^{k+1} = 0.$$

Study the stability, in the $L_\tau^\infty(L_h^\infty)$-norm, of this method.

28.18 To approximate the classical solution of (28.7), we use an implicit method with an upwind discretization of the first derivative

$$\bar{\delta}_\tau w^{k+1} + b\bar{\delta}_h w^{k+1} - \Delta_h w^{k+1} = 0.$$

Study the stability, in the $L_\tau^\infty(L_h^\infty)$-norm, of this method.

28.19 Prove Theorem 28.27.

28.20 Show that the definition of the symbol, provided in Definition 28.31, is obtained by taking the discrete Fourier transform of the method (28.13).

28.21 Find the symbol of the backward Euler method (28.10).

28.22 Find the symbol of the Crank–Nicolson method (28.12).

28.23 Prove Proposition 28.32.

28.24 Consider the grid function $v \colon \mathbb{Z}_h \to \mathbb{C}$

$$v_k = \frac{e^{ik\xi_0}}{\pi h k} \sin(k\delta),$$

where $\xi_0 \in (-\pi, \pi)$ and $\delta < \min\{\pi - \xi_0, \xi_0 + \pi\}$. Show that

$$v \in L^2_h(\mathbb{Z}_h), \qquad \hat{v}(\xi) = \chi_{[\xi_0 - \delta, \xi_0 + \delta]}.$$

28.25 Provide all the details for Example 28.8.

28.26 Let $T, \gamma > 0$ and $u_0 \in C(\mathbb{R}) \cap L^2(\mathbb{R})$ be given. Consider the IVP: Find $u \colon \mathbb{R} \times [0, T] \to \mathbb{R}$ such that

$$\begin{cases} \dfrac{\partial u(x,t)}{\partial t} - \dfrac{\partial^2 u(x,t)}{\partial x^2} + \gamma u(x,t) = 0, & (x,t) \in \mathbb{R} \times (0,T], \\ u(x,0) = u_0(x), & x \in \mathbb{R}, \end{cases}$$

which models a diffusion with, since $\gamma > 0$, decay. Let $\theta \in \mathbb{R}$ and consider a family of numerical methods of the form

$$\bar{\delta}_\tau w^{k+1} - \frac{1}{2}\Delta_h \left(w^{k+1} + w^k\right) + \gamma \left(\theta w^{k+1} + (1-\theta) w^k\right) = 0,$$

where τ and h are the time and space step sizes, respectively.

a) By computing the consistency error, show that this method is $O(\tau^p + h^2)$ accurate, where $p = 2$, if $\theta = 1/2$, and $p = 1$, otherwise.

b) Using the von Neumann stability analysis, show that this method is unconditionally stable if $\theta \geq 1/2$.

c) Show that if $\theta = 0$, then the method is stable provided that $\tau \leq 2/\gamma$, independently of $h > 0$.

28.27 Consider the heat equation in a periodic domain. That is, we need to find $u \colon \mathbb{R} \times [0, T] \to \mathbb{R}$ that is spatially one-periodic, i.e., for all $t \in [0, T]$ and any $x \in \mathbb{R}$, it satisfies

$$u(x + n, t) = u(x, t), \qquad \forall n \in \mathbb{Z},$$

and is a classical solution of

$$\begin{cases} \dfrac{\partial u(x,t)}{\partial t} = \dfrac{\partial^2 u(x,t)}{\partial x^2}, & (x,t) \in \mathbb{R} \times (0,T], \\ u(x,0) = u_0(x), & x \in \mathbb{R}, \end{cases}$$

where $u_0 \in C^\infty_{\mathrm{p}}(0,1)$. Which approximation, from the ones given below, is preferable?

a) Leapfrog type:

$$\frac{1}{2\tau}\left(w^{k+1} - w^{k-1}\right) = \Delta_h w^k.$$

b) Skew stencil type:

$$\frac{1}{w\tau}\left(w_j^{k+1} - w_j^{k-1}\right) = \frac{1}{h^2}\left(w_{j-1}^k - \left(w_j^{k+1} + w_j^{k-1}\right) + w_{j+1}^k\right).$$

Justify your choice.

28.28 To approximate the solution to (23.24), we consider the following family of methods: Let $\theta \in [0, 1]$ and

$$\bar{\delta}_\tau w^{k+1} = \Delta_h\left(\theta w^{k+1} + (1-\theta)w^k\right).$$

a) Find the symbol of the method.
b) Find a condition on θ that makes this method unconditionally stable.
c) Is the method convergent?

28.29 Given $T > 0$, consider the following one-dimensional convection–diffusion equation with periodic boundary conditions: Find $u\colon \mathbb{R} \times [0, T]$, a one-periodic function that satisfies

$$\begin{cases} \dfrac{\partial u(x,t)}{\partial t} + b\dfrac{\partial u(x,t)}{\partial x} - \dfrac{\partial^2 u(x,t)}{\partial x^2} = 0, & (x,t) \in \mathbb{R} \times (0,T], \\ u(x,0) = u_0, & x \in \mathbb{R}. \end{cases}$$

Here, $b > 0$ and $u_0 \in C^\infty_{\mathrm{p}}(0,1)$ is one-periodic. Consider the explicit upwind method:

$$\bar{\delta}_\tau w^{k+1} + b\bar{\delta}_h w^k - \Delta_h w^k = 0$$

with periodic boundary conditions and suitable initial conditions. Use a periodic variant of the von Neumann stability analysis to study the stability, in the $L^\infty_\tau(L^2_h)$-norm, of this method.

28.30 Prove identity (28.16).

28.31 Show that the method (28.13), with solution w, is accurate of order $p \in \mathbb{N}$ if and only if there are constants $C_3, C_4 > 0$ such that, whenever

$$h \leq C_3,$$

we have

$$\|\pi_h u(\cdot, \tau) - w^1\|_{L^\infty_h} \leq C_4 \tau h^p.$$

28.32 Show that, with the notation of Theorem 28.37, if $u_0 \in H^1(\mathbb{R})$, then there is a constant, independent of h, for which

$$\|W_h^0 - u_0\|_{L^2(\mathbb{R})} \leq Ch\|u_0'\|_{L^2(\mathbb{R})}.$$

Listings

```
function [error] = DiffusionCrankNic(finalT, N, K, numPlots)
%
% This function computes numerical approximations to solutions
% of the linear diffusion equation
%
% u_t - u_xx = f(x,t)
%
% using the Crank-Nicolson (CN) method on the domain [0,1]. The
% forcing function, f, and the intial conditions are constructed
```

```
% so that the true solution is
%
% u(x,t) = cos(2*pi*t)*(exp(sin(2*pi*x))-1.0).
%
% This can be easily modified.
%
% Input
%   finalT : the final time
%   N : the number of spatial grid points in [0,1]
%   K : the number of time steps in the interval [0,finalT]
%   numPlots : the number of output frames in the time interval
%              [0,finalT]. numPlots must be a divisor of K.
%
% Output
%   error: the max norm error of the CN scheme at t = finalT
%
  error = 0.0;

  if N > 0 && N-floor(N) == 0
    h = 1.0/N;
  else
    display('Error: N must be a positive integer.')
    return
  end

  if (K > 0 && K-floor(K) == 0) && finalT > 0
    tau = finalT/K;
  else
    display('Error: K must be a positive integer, and')
    display('       finalT must be positive.')
    return
  end

  mu = tau/(2.0*h*h);

  x = 0:h:1.0;
  uoCN(1:N+1) = uExact(N,0.0);

  if mod(K,numPlots) == 0
    stepsPerPlot = K/numPlots;
  else
    display('Error: numPlots is not a divisor of K.')
    return
  end
%
% Define the tridiagonal system matrix to be inverted:
%
  a(    1) = 0.0;
  a(2:N-1) = -mu;
  b(1:N-1) = 1.0+2.0*mu;
  c(1:N-2) = -mu;
  c(  N-1) = 0.0;
%
% Main time loop:
%
  for k = 1: numPlots
```

```
    for j = 1: stepsPerPlot
      kk = (k-1)*stepsPerPlot+j;
      currTime = tau*(kk);
      lastTime = tau*(kk-1.0);
      f(1:N-1) = tau*(fForcing(N,currTime) ...
        +fForcing(N,lastTime))/2.0;
      for ell = 1: N-1
        rhs(ell) = f(ell)+uoCN(ell+1)+mu*(uoCN(ell) ...
          -2.0*uoCN(ell+1)+uoCN(ell+2));
      end
      [uCN(2:N),err] = TriDiagonal(a, b, c, rhs);
      uCN(  1) = 0.0;
      uCN(N+1) = 0.0;
      uoCN = uCN;
    end
    hf  = figure(k);
    clf
    plot(x,uExact(N,currTime),'k-',x,uCN,'b-o')
    grid on,
    xlabel("x");
    ylabel('exact and approximate solutions');
    title(['Crank--Nicolson Approximation at T = ', ...
        num2str(currTime), ', h = ', num2str(h), ...
        ', and tau =', num2str(tau)]);
    legend("Exact","Crank--Nicolson")
    set(gca,"xTick",0:0.1:1)

    s1 = ['000', num2str(k)];
    s2 = s1((length(s1)-3):length(s1));
    s3 = ['OUT/diff', s2, '.pdf'];
    exportgraphics(gca, s3)
  end

  error = max(abs(uExact(N,currTime)-uCN));
end

function frc = fForcing(N,t)
%
% Constructs the forcing term:
%
% Input
%   N : to compute the mesh size
%   t : the current time at the time interval midpoint
%
% Output
%   f(1:N-1) : the array of values of the forcing function
%
  frc = zeros(1,N-1);
  h = 1.0/N;
  for i = 1: N-1
    x = i*h;
    tpx = 2.0*pi*x;
    tpt = 2.0*pi*t;
    frc(i) = 2.0*pi*sin(tpt)*(1.0-exp(sin(tpx))) ...
      + 4.0*pi*pi*cos(tpt)*exp(sin(tpx)) ...
      * (sin(tpx)-cos(tpx)*cos(tpx));
```

```
  end
end

function ue = uExact(N,t)
%
% Constructs the exact solution:
%
% Input
%   N : to compute the mesh size
%   t : the current time
%
% Output
%   ue(1:N+1) : the array of values of the exact solution
%
  ue = zeros(1,N+1);
  h = 1.0/N;
  for i = 0: N
    x = i*h;
    tpx = 2.0*pi*x;
    tpt = 2.0*pi*t;
    ue(i+1) = cos(tpt)*(exp(sin(tpx))-1.0);
  end
end
```

Listing 28.1 Approximation of the diffusion equation by the Crank–Nicolson method.

29 Finite Difference Methods for Hyperbolic Problems

In this chapter, we will present fully discrete finite difference methods for hyperbolic problems. We will focus on the Cauchy, or initial value, problem for the transport equation; and the initial, and initial boundary, value problem for the wave equation. Most of the techniques and ideas behind the construction and analysis of these methods have been already presented in Chapters 24 and 28. There is, however, one important feature that must be taken into account here. Namely, since, as observed in Remarks 23.31 and 23.36, the solution of hyperbolic problems has a very specific domain of dependence, this must be respected by the numerical method. This requirement will often manifest itself in two aspects, as follows.

- The spatial grid size h and time step size τ will have to be related in a certain way. This, often, will come in the form of a restriction on the *hyperbolic*[1] *Courant–Friedrichs–Lewy (CFL) number*

$$\mu = \frac{\tau}{h} \leq \mu_0$$

  for some $\mu_0 > 0$ that depends on the problem parameters.
- The solution of a numerical method always reduces to the solution of a linear system of equations

$$\mathsf{A}\boldsymbol{w}^{k+1} = \mathsf{B}\boldsymbol{w}^k + \boldsymbol{f}^{k+1}.$$

  In the case of parabolic equations, the matrices A and B were often symmetric, or at least normal, so that much about the stability of the method could be inferred from the spectrum of A and B. This will *not* be the case for hyperbolic problems. For this reason, looking at the eigenvalues of matrices is often of little use, and other arguments, like energy or von Neumann stability, must be employed.

  The following example illustrates how things can go catastrophically wrong.

Example 29.1 Let us illustrate the aforementioned points with a simple example. Let us consider the transport equation (23.39) with $c = 1$, $f \equiv 0$, and a *downwind* discretization, i.e., we will seek $w \in \mathcal{V}(\mathbb{Z}_h)$ such that $w^0 = \pi_h u_0$ and, for $k = 0, \dots, K-1$,

$$\delta_\tau w^{k+1} + \delta_h w^k = 0.$$

[1] Named in honor of the German-American mathematicians Richard Courant (1888–1972) and Kurt Otto Friedrichs (1901–1982), and the Jewish, German-born, American mathematician Hans Lewy (1904–1988).

In other words, for $i \in \mathbb{Z}$,

$$w_i^{k+1} = w_i^k - \mu(w_{i+1}^k - w_i^k) = (1+\mu)w_i^k - \mu w_{i+1}^k,$$

where $\mu = \frac{\tau}{h}$. We see, first, that the stencil of the method is not symmetric. More importantly, this method can be analyzed with the techniques presented in Section 28.7.3. Its symbol is, see Problem 29.1,

$$\tilde{E}_h^\tau(\xi) = (1+\mu) - \mu e^{-\mathrm{i}\xi};$$

therefore, if $\xi \neq 0$,

$$|\tilde{E}_h^\tau(\xi)|^2 = [1 + \mu(1-\cos\xi)]^2 + \mu^2 \sin^2\xi > 1,$$

meaning that this method is *unconditionally unstable*, i.e., no matter what the value of μ is, the von Neumann stability condition will not be satisfied. This is due to the fact that, to advance in time, we are gathering information from outside the domain of dependence of the solution; see Figure 23.2.

As a last comment before we begin, we mention that we will not cover a very important topic in the approximation of hyperbolic problems: the approximation of hyperbolic systems of conservation laws, as this is far beyond the scope of our elementary discussion. For this topic, we refer the reader to [59, 43, 33, 53].

29.1 The Initial Value Problem for the Transport Equation

In this section, we will construct and analyze finite difference methods for the initial value problem (23.39). We will operate under the notation of Section 28.7.

Let us introduce some methods.

Definition 29.1 (fully discrete methods). Let $T > 0$, $K, N \in \mathbb{N}$, $h = 1/(N+1)$, and $\tau = T/K$. Suppose that $u \in C^1(\mathbb{R}\times[0,T])$ is a classical solution to the Cauchy problem for the transport equation (23.39), with $c > 0$ and $f \equiv 0$. Suppose that $u_{0,h} \in \mathcal{V}(\mathbb{Z}_h)$ is defined as usual. We define the following methods for computing the finite difference approximation $w \in \mathcal{V}([0,T]_\tau; \mathcal{V}(\mathbb{Z}_h))$ to u.

1. The **upwind method**

$$\begin{cases} \delta_\tau w^{k+1} + c\bar{\delta}_h w^k = 0, & k = 0, \dots, K-1, \\ w^0 = u_{0,h}. \end{cases} \tag{29.1}$$

2. The **centered difference method**

$$\begin{cases} \delta_\tau w^{k+1} + c\mathring{\delta}_h w^k = 0, & k = 0, \dots, K-1, \\ w^0 = u_{0,h}. \end{cases} \tag{29.2}$$

3. The **Lax–Friedrichs method**[2]

$$\begin{cases} \dfrac{1}{\tau}\left(w^{k+1} - \dfrac{1}{2}(\mathcal{S}_1 w^k + \mathcal{S}_{-1} w^k)\right) + c\mathring{\delta}_h w^k = 0, \quad k = 0, \ldots, K-1, \\ w^0 = u_{0,h}, \end{cases} \tag{29.3}$$

where $\mathcal{S}_{\pm 1}$ are the shift operators given in Definition 24.5.

4. The **Lax–Wendroff method**[3]

$$\begin{cases} \delta_\tau w^{k+1} + c\mathring{\delta}_h w^k - \dfrac{c^2\tau}{2}\Delta_h w^k = 0, \quad k = 0, \ldots, K-1, \\ w^0 = u_{0,h}. \end{cases} \tag{29.4}$$

5. The **Beam–Warming method**[4]

$$\begin{cases} \delta_\tau w^{k+1} + c\,\mathrm{BDF}_h\, w^k - \dfrac{c^2\tau}{2}\Delta_h w^k = 0, \quad k = 0, \ldots, K-1, \\ w^0 = u_{0,h}, \end{cases} \tag{29.5}$$

where the operator BDF_h was introduced in Proposition 24.20.

6. The **Crank–Nicolson**[5] **method**

$$\begin{cases} \delta_\tau w^{k+1} + \dfrac{c}{2}\mathring{\delta}_h \left(w^{k+1} + w^k\right) = 0, \quad k = 0, \ldots, K-1, \\ w^0 = u_{0,h}. \end{cases} \tag{29.6}$$

Remark 29.2 (explicit methods)**.** We observe that all of the methods above, with the exception of the last, are explicit methods. They can always be solved. The Crank–Nicolson method is, however, implicit; its solvability property is nontrivial.

Let us now show the consistency and stability of some of these methods.

Theorem 29.3 (upwind method)**.** *The upwind method* (29.1) *is stable in the* $L^\infty_\tau(L^2_h)$*-norm provided that the CFL condition*

$$0 < c\mu \leq \mu_0 = 1.$$

Moreover, if u *is a sufficiently smooth classical solution to* (23.39), *then the consistency error satisfies the estimate*

$$|\mathcal{E}_h^\tau[u]_i^k| \leq C\,(\tau + h)\,, \quad i \in \mathbb{Z}, \quad k = 0, \ldots, K,$$

for some $C > 0$ *that is independent of* τ *and* h.

[2] Named in honor of the Hungarian-born American mathematician Peter David Lax (1926–) and the German–American mathematician Kurt Otto Friedrichs (1901–1982).

[3] Named in honor of the Hungarian-born American mathematician Peter David Lax (1926–), the German-American mathematician Kurt Otto Friedrichs (1901–1982), and the American mathematician Burton Wendroff (1930–).

[4] Named in honor of the American mathematicians Richard M. Beam (1935–) and R.F. Warming (1931–).

[5] Named in honor of the British mathematical physicist John Crank (1916–2006) and the British mathematician and physicist Phyllis Nicolson (1917–1968).

Proof. To show stability, we verify the von Neumann stability condition (28.15). It is not difficult to see that the symbol of this method is

$$\tilde{E}_h^\tau(\xi) = 1 - c\mu + c\mu e^{-\mathrm{i}\xi}, \qquad \mu = \frac{\tau}{h}.$$

Thus, if $0 < c\mu \leq 1$,

$$|\tilde{E}_h^\tau(\xi)| = \left|1 - c\mu + c\mu e^{-\mathrm{i}\xi}\right| \leq |1 - c\mu| + c\mu \left|e^{-\mathrm{i}\xi h}\right| = 1 - c\mu + c\mu = 1.$$

Therefore, by Theorem 28.34, the method is $L^\infty_\tau(L^2_h)$-stable provided that the CFL condition holds.

We leave the consistency to the reader as an exercise; see Problem 29.2. □

The upwind method is only first order in time and space. One might want to increase the order of approximation in space by using, for example, the centered difference approximation (29.2). Similarly to Example 29.1, this turns out to be a bad idea.

Theorem 29.4 (centered method)**.** *The centered difference approximation method* (29.2) *is never $L^\infty_\tau(L^2_h)$-stable, regardless of the value of μ.*

Proof. The symbol of this method is

$$\tilde{E}_h^\tau(\xi) = 1 - \frac{c\mu}{2}\left(e^{\mathrm{i}\xi} - e^{-\mathrm{i}\xi}\right) = 1 - \mathrm{i}c\mu \sin(\xi).$$

Consequently,

$$|\tilde{E}_h^\tau(\xi)|^2 = 1 + c^2\mu^2 \sin^2(\xi) \geq 1.$$

□

On the other hand, we have the following result.

Theorem 29.5 (Crank–Nicolson method)**.** *The Crank–Nicolson method* (29.6) *is unconditionally $L^\infty_\tau(L^2_h)$-stable. Furthermore, if u is a sufficiently smooth classical solution to* (23.39)*, the consistency error satisfies*

$$|\mathcal{E}_h^\tau[u]_i^k| \leq C\left(\tau^2 + h^2\right), \quad i \in \mathbb{Z}, \quad k = 0, \ldots, K,$$

for some $C > 0$ that is independent of τ and h.

Proof. See Problem 29.3. Notice that, in this proof, one discovers the curious property that

$$|\tilde{E}_h^\tau(\xi)| = 1.$$

This has an interesting implication. □

Theorem 29.6 (Lax–Friedrichs method)**.** *The Lax–Friedrichs method* (29.3) *is stable in the $L^\infty_\tau(L^2_h)$-norm provided that the CFL condition*

$$0 < c\mu \leq 1$$

holds. Furthermore, if u is a sufficiently smooth classical solution to (23.39)*, then consistency error satisfies*

$$|\mathcal{E}_h^\tau[u]_i^k| \leq C(\tau + h), \quad i \in \mathbb{Z}, \quad k = 0, \ldots, K$$

for some $C > 0$, independent of τ and h, provided that $\tau = \alpha h$ for some constant $\alpha > 0$.

Proof. The symbol of the method is

$$\tilde{E}_h^\tau(\xi) = \cos(\xi) - c\mu i \sin(\xi).$$

We leave the rest of the details to the reader as an exercise; see Problem 29.4.

□

Remark 29.7 (numerical diffusion)**.** Let us give some interpretations of the reason behind the conditional stability of the upwind and Lax–Friedrichs methods. We first observe that the upwind method can be expressed as

$$\delta_\tau w^{k+1} + c\mathring{\delta}_h w^k - \frac{ch}{2}\Delta_h w^k = 0.$$

This looks like an explicit approximation of the advection–diffusion equation

$$\frac{\partial u}{\partial t} + c\frac{\partial u}{\partial x} - \frac{ch}{2}\frac{\partial^2 u}{\partial x^2} = 0, \tag{29.7}$$

where the diffusion is $\mathcal{O}(h)$, i.e., of order h. This small artificial diffusion is called *numerical diffusion*. We can see that added numerical diffusion can stabilize the centered difference method. Indeed, similar to Theorem 28.22, we can show that this explicit method will be stable provided that the mesh Péclet number satisfies

$$\frac{ha}{\frac{ch}{2}} \le 2,$$

which is always true, and the *parabolic* Courant number satisfies

$$\frac{ch}{2}\frac{\tau}{h^2} \le \frac{1}{2} \qquad \Longleftrightarrow \qquad c\frac{\tau}{h} \le 1,$$

which is the stability condition we obtained in Theorem 29.3.

Similarly, the Lax–Friedrichs method can be written as

$$\delta_\tau w^{k+1} + c\mathring{\delta}_h w^k - \frac{h^2}{2\tau}\Delta_h w^k = 0,$$

which is an explicit discretization of the advection–diffusion equation

$$\frac{\partial u}{\partial t} + c\frac{\partial u}{\partial x} - \frac{h^2}{2\tau}\frac{\partial^2 u}{\partial x^2} = 0. \tag{29.8}$$

Here, the numerical diffusion is $\mathcal{O}(\frac{h^2}{2\tau})$. If we recall that, for consistency, the Lax–Friedrichs method requires that $\tau = \alpha h$, then we obtain

$$\frac{\partial u}{\partial t} + c\frac{\partial u}{\partial x} - \frac{h}{2\alpha}\frac{\partial^2 u}{\partial x^2} = 0$$

with numerical diffusion of $\mathcal{O}(h)$. We finally comment that (29.7) and (29.8) are known, in the literature, as the *modified equations* of the upwind and Lax–Friedrichs methods, respectively.

Theorem 29.8 (Lax–Wendroff). *The Lax–Wendroff method* (29.4) *is $L^\infty_\tau(L^2_h)$-stable provided that the condition*

$$0 < c\mu \le 1$$

is satisfied. Furthermore, if u is a sufficiently smooth classical solution to (23.39)*, then, with certain restrictions on the time and space step size, the consistency error satisfies the estimate*

$$|\mathcal{E}_h^\tau[u]_i^k| \le C\left(\tau^2 + h^2\right), \quad i \in \mathbb{Z}, \quad k = 0, \dots, K$$

for some $C > 0$ that is independent of τ and h.

Proof. The symbol of this method is

$$\tilde{E}_h^\tau(\xi) = 1 + c^2\mu^2 (\cos(\xi) - 1) - c\mu \mathrm{i} \sin(\xi).$$

We leave the rest of the details to the reader as an exercise; see Problem 29.5. □

29.2 Positivity and Max-Norm Dissipativity

In Section 23.4.1, we showed that the solution to (23.39) is given by (23.40). From this formula, it is evident that any pointwise bound that is valid for the initial condition will also be valid for the solution at any positive time. For instance, if the initial condition is positive, so will be the exact solution for all positive times.

In this section, we investigate which conditions can guarantee that a numerical method can preserve this property. Let us begin with the upwind method.

Theorem 29.9 (positivity). *Consider the upwind approximation method* (29.1) *to approximate the solution to* (23.39) *with positive velocity $c > 0$. If the CFL condition*

$$0 < c\mu \le \mu_0 = 1$$

holds, then this method will be $L^\infty_\tau(L^\infty_h)$-stable. Moreover, under the same condition, if $w^k \ge 0$, for some $k \in \{0, \dots, K-1\}$, then $w^{k+1} \ge 0$.

Proof. We can write explicitly the method as

$$w_j^{k+1} = w_j^k - c\mu\left(w_j^k - w_{j-1}^k\right), \quad j \in \mathbb{Z}, \quad k = 0, \dots, K-1.$$

Therefore, if $c\mu \in (0, 1]$,

$$\begin{aligned}
|w_j^{k+1}| &= \left|(1 - c\mu) w_j^k + c\mu w_{j-1}^k\right| \\
&\le (1 - c\mu)|w_j^k| + c\mu |w_{j-1}^k| \\
&\le (1 - c\mu)\left\|w^k\right\|_{L_h^\infty} + c\mu \left\|w^k\right\|_{L_h^\infty} \\
&= \left\|w^k\right\|_{L_h^\infty},
\end{aligned}$$

which implies that

$$\left\|w^{k+1}\right\|_{L_h^\infty} \le \left\|w^k\right\|_{L_h^\infty},$$

and this implies the $L^\infty_\tau(L^\infty_h)$-stability.

Suppose now that $w_j^k \geq 0$ for all $j \in \mathbb{Z}$. Since $c\mu$ and $1 - c\mu$ are positive and

$$w_j^{k+1} = (1 - c\mu)w_j^k + c\mu w_{j-1}^k,$$

it is clear that $w_j^{k+1} \geq 0$ for all $j \in \mathbb{Z}$. □

We can generalize this result to more general methods, but first we need to introduce some notions.

Definition 29.10 (positivity). Consider a general two-layer explicit method in the sense of Definition 28.30. We say that this method **reproduces the constant state** if, whenever $w^k \equiv 1$, we obtain that $w^{k+1} \equiv 1$. The method is **max-norm dissipative** if and only if, for every $k = 0, \ldots, K-1$, it satisfies

$$\|w^{k+1}\|_{L_h^\infty} \leq \|w^k\|_{L_h^\infty}.$$

It is **positivity preserving** if and only if whenever $w_j^k \geq 0$, for all $j \in \mathbb{Z}$, then $w_j^{k+1} \geq 0$ for all $j \in \mathbb{Z}$.

The following result gives a general criterion on the coefficients of an explicit finite difference method to be positivity preserving.

Theorem 29.11 (positivity). *Consider*

$$w_j^{k+1} = \sum_{m \in S_{\mathcal{B}}} b_m(h, \tau) w_{j-m}^k,$$

which is an explicit two-layer finite difference method in the sense of Definition 28.30.

1. *If the method reproduces the constant state, then*

$$\sum_{m \in S_{\mathcal{B}}} b_m(h, \tau) = 1.$$

2. *If the method reproduces the constant state and is max-norm dissipative, then the method is positivity preserving.*
3. *If the method reproduces the constant state and is max-norm dissipative, then $b_m(h, \tau) \geq 0$ for all $m \in S_{\mathcal{B}}$.*

Proof. See Problem 29.8. □

The following result, which bears the name *Godunov's Theorem*, limits the consistency order of positivity-preserving methods. For a more detailed presentation, we refer the reader to [100, Chapter 9].

Theorem 29.12 (Godunov[6]). *Consider*

$$w_j^{k+1} = \sum_{m \in S_{\mathcal{B}}} b_m(h, \tau) w_{j-m}^k,$$

[6] Named in honor of the Soviet and Russian mathematician Sergei Konstantinovich Godunov (1929–).

which is an explicit two-layer finite difference method in the sense of Definition 28.30. If the method reproduces the constant state, it is max-norm dissipative, and $\#S_{\mathcal{B}} \geq 2$, then the method is consistent to at most order one.

Proof. (sketch) From the form of the method, we know that its symbol is

$$\tilde{E}_h^\tau(\xi) = \sum_{m \in S_{\mathcal{B}}} b_m(h,\tau) e^{-\mathrm{i}m\xi}.$$

Therefore, as $h \to 0$,

$$\begin{aligned}
|\tilde{E}_h^\tau(h)|^2 &= \sum_{m \in S_{\mathcal{B}}} \sum_{n \in S_{\mathcal{B}}} b_m(h,\tau)\overline{b_n(h,\tau)} e^{-\mathrm{i}(m-n)h} \\
&= \sum_{m,n \in S_{\mathcal{B}}} b_m(h,\tau)\overline{b_n(h,\tau)} \left(1 - \mathrm{i}(m-n)h - \frac{(m-n)^2h^2}{2} + \mathcal{O}(h^3)\right) \\
&= \sum_{m \in S_{\mathcal{B}}} b_m(h,\tau) \overline{\sum_{n \in S_{\mathcal{B}}} b_n(h,\tau)} - \mathrm{i}h \sum_{m,n \in S_{\mathcal{B}}} b_m(h,\tau)\overline{b_n(h,\tau)}(m-n) \\
&\quad - \frac{h^2}{2} \sum_{m,n \in S_{\mathcal{B}}} b_m(h,\tau)\overline{b_n(h,\tau)}(m-n)^2 + \mathcal{O}(h^3).
\end{aligned}$$

From Theorem 29.11, we know that $b_m(h,\tau) \geq 0$, for all $m \in S_{\mathcal{B}}$, and

$$\sum_{m \in S_{\mathcal{B}}} b_m(h,\tau) = 1.$$

This, in turn, implies that

$$\begin{aligned}
\sum_{m \in S_{\mathcal{B}}} b_m(h,\tau) \overline{\sum_{n \in S_{\mathcal{B}}} b_n(h,\tau)} &= 1, \\
\sum_{m,n \in S_{\mathcal{B}}} b_m(h,\tau)\overline{b_n(h,\tau)}(m-n) &= 0, \\
\sum_{m,n \in S_{\mathcal{B}}} b_m(h,\tau)\overline{b_n(h,\tau)}(m-n)^2 &> 0.
\end{aligned}$$

In conclusion, as $h \to 0$,

$$|\tilde{E}_h^\tau(h)|^2 \geq 1 - Ch^2$$

for some $C > 0$.

Consider now, as an initial condition for (23.39), with $f \equiv 0$, the function $u_0(x) = e^{\mathrm{i}x}$. The consistency error then, after one step, is

$$\begin{aligned}
\mathcal{E}_h^\tau[u]_j^1 &= e^{\mathrm{i}(jh - c\tau)} - \sum_{m \in S_{\mathcal{B}}} b_m(h,\tau) e^{\mathrm{i}(j-m)h} \\
&= \left(e^{-\mathrm{i}c\tau} - \tilde{E}_h^\tau(h)\right) e^{\mathrm{i}jh} \\
&= \left(e^{-\mathrm{i}c\mu h} - \tilde{E}_h^\tau(h)\right) e^{\mathrm{i}jh}.
\end{aligned}$$

Therefore,

$$|\mathcal{E}_h^\tau[u]_j^1| = \left|e^{-\mathrm{i}c\mu h} - \tilde{E}_h^\tau(h)\right| \geq C'h,$$

and the method cannot have second-order consistency. □

29.3 The Transport Equation in a Periodic Spatial Domain

We end our discussion of the transport equation with some stability and convergence results for the upwind method in the spatially periodic case. Thus, we consider (23.39) with $c > 0$, $f \equiv 0$, and $u_0 \in C^\infty_{\mathrm{p}}(0,1)$. In this case, the representation formula (23.40) shows then that the solution u will be one-periodic as well.

Theorem 29.13 (upwind method). *Let u be the smooth one-periodic solution to (23.39) with $c > 0$, $f \equiv 0$, and $u_0 \in C^\infty_{\mathrm{p}}(0,1)$. Suppose that $w \in \mathcal{V}([0,T]_\tau; \mathcal{V}_{N+1,\mathrm{p}})$ is the approximation to u computed by the upwind method (29.1). Then, under the CFL condition, $c\mu \leq 1$, the method is $L^\infty_\tau(L^p_h)$-stable for $p \in \{2,\infty\}$, i.e.,*

$$\left\|w^{k+1}\right\|_{L^p_h} \leq \left\|w^k\right\|_{L^p_h}, \quad k = 0, \dots, K-1.$$

Furthermore, under the same CFL condition, the method is convergent in the $L^\infty_\tau(L^p_h)$-norm ($p \in \{2,\infty\}$) with rate

$$\tau + h.$$

Proof. We will prove the convergence for $p = 2$ and leave the other parts to the reader as an exercise; see Problem 29.11. We introduce the error $e \in \mathcal{V}([0,T]_\tau; \mathcal{V}_{N+1,\mathrm{p}})$ and observe that it satisfies $e^0 = 0$ and

$$\delta_\tau e^{k+1} + c\bar{\delta}_h e^k = \mathcal{E}^\tau_h[u]^{k+1}, \qquad k = 0, \dots, K-1,$$

where the consistency error is such that

$$\|\mathcal{E}^\tau_h[u]^{k+1}\|_{L^\infty_h} \leq C(\tau + h)$$

for some $C > 0$ that is independent of τ and h. By Theorem 26.12, the set of vectors

$$S = \left\{\boldsymbol{v}_m \in \mathbb{C}^N \middle| m = 1, \dots, N+1\right\}, \quad [\boldsymbol{v}_m]_n = v_{m,n} = e^{2\pi i m n h}, \quad n = 1, \dots, N+1$$

is orthornormal, in the sense that

$$h(\boldsymbol{v}_m, \boldsymbol{v}_n)_2 = h\boldsymbol{v}_m^{\mathrm{H}}\boldsymbol{v}_n = \delta_{m,n}.$$

The error equation may be expressed in matrix–vector form as

$$e^{k+1} = \mathsf{A}e^k + \tau\mathcal{E}^\tau_h[u]^{k+1},$$

where

$$\mathsf{A} = \begin{bmatrix} 1-c\mu & 0 & 0 & \cdots & c\mu \\ c\mu & 1-c\mu & \ddots & & \vdots \\ 0 & \ddots & \ddots & 0 & 0 \\ \vdots & & c\mu & 1-c\mu & 0 \\ 0 & \cdots & 0 & c\mu & 1-c\mu \end{bmatrix}. \tag{29.9}$$

S is an orthonormal set of eigenvectors for A. In particular,

$$\mathsf{A}\boldsymbol{v}_m = \lambda_m \boldsymbol{v}_m, \qquad \lambda_m = 1 - c\mu + c\mu e^{-2\pi i m h}.$$

Note that the eigenvalues are complex and the matrix A is not symmetric. The computation of the 2-norm of this matrix requires some extra work. In particular, recall that

$$\|\mathsf{A}\|_2 = \max\left\{\sqrt{\nu} \middle| \nu \in \sigma(\mathsf{A}^{\mathsf{H}}\mathsf{A})\right\}.$$

It is a short exercise to show that, in our case, $\nu \in \sigma(\mathsf{A}^{\mathsf{H}}\mathsf{A})$ if and only if $\sqrt{\nu} = |\lambda|$ with $\lambda \in \sigma(\mathsf{A})$. Therefore,

$$\|\mathsf{A}\|_2 = \max\{|\lambda| \; |\lambda \in \sigma(\mathsf{A})\}.$$

Using the CFL condition, observe that, for all $\lambda \in \sigma(\mathsf{A})$,

$$|\lambda| \le |1 - c\mu| + |c\mu| = 1 - c\mu + c\mu = 1.$$

Therefore,

$$\left\|\boldsymbol{e}^{k+1}\right\|_2 \le \|\mathsf{A}\|_2 \left\|\boldsymbol{e}^k\right\|_2 + \tau \left\|\mathcal{E}_h^\tau[u]^{k+1}\right\|_2 \le \left\|\boldsymbol{e}^k\right\|_2 + \tau\sqrt{N}C(\tau + h).$$

Equivalently,

$$\left\|e^{k+1}\right\|_{L_h^2} \le \left\|e^k\right\|_{L_h^2} + C\tau(\tau + h).$$

Therefore, for any $0 \le k \le K$,

$$\left\|e^k\right\|_{L_h^2} \le \left\|e^0\right\|_{L_h^2} + C\tau K(\tau + h) = CT(\tau + h).$$ □

Next, let us analyze the same problem, but using an energy-type argument.

Theorem 29.14 (upwind). *Let u be the smooth one-periodic solution to* (23.39) *with $c > 0$, $f \equiv 0$, and $u_0 \in C_{\mathrm{p}}^\infty(0,1)$. Suppose that $w \in \mathcal{V}([0,T]_\tau; \mathcal{V}_{N+1,\mathrm{p}})$ is the approximation to u computed by the upwind method* (29.1). *Then, under the CFL condition, $c\mu \le 1$, the method is $L_\tau^\infty(L_h^2)$-stable.*

Proof. As mentioned in Remark 29.7, the upwind method can be rewritten as

$$\delta_\tau w^{k+1} + c\mathring{\delta}_h w^k - \frac{ch}{2}\Delta_h w^k = 0.$$

Taking the L_h^2-inner product of the method with $2\tau w^k$ and using summation by parts, the identity

$$(\mathring{\delta}_h v_1, v_2)_{L_h^2} = -(v_1, \mathring{\delta}_h v_2)_{L_h^2}, \quad \forall v_1, v_2 \in \mathcal{V}_{N+1,\mathrm{p}},$$

and the polarization identity

$$2(a - b)b = a^2 - (a - b)^2 - b^2$$

yield

$$\|w^{k+1}\|_{L_h^2}^2 - \|w^k\|_{L_h^2}^2 - \|w^{k+1} - w^k\|_{L_h^2}^2 + ch\tau\|\bar{\delta}_h w^k\|_{L_h^2}^2 = 0.$$

Observe now that, using the method,

$$ch\tau\|\bar{\delta}_h w^k\|_{L_h^2}^2 = ch\tau\left\|\frac{\delta_\tau w^{k+1}}{c}\right\|_{L_h^2}^2 = \frac{1}{c\mu}\|w^{k+1} - w^k\|_{L_h^2}^2.$$

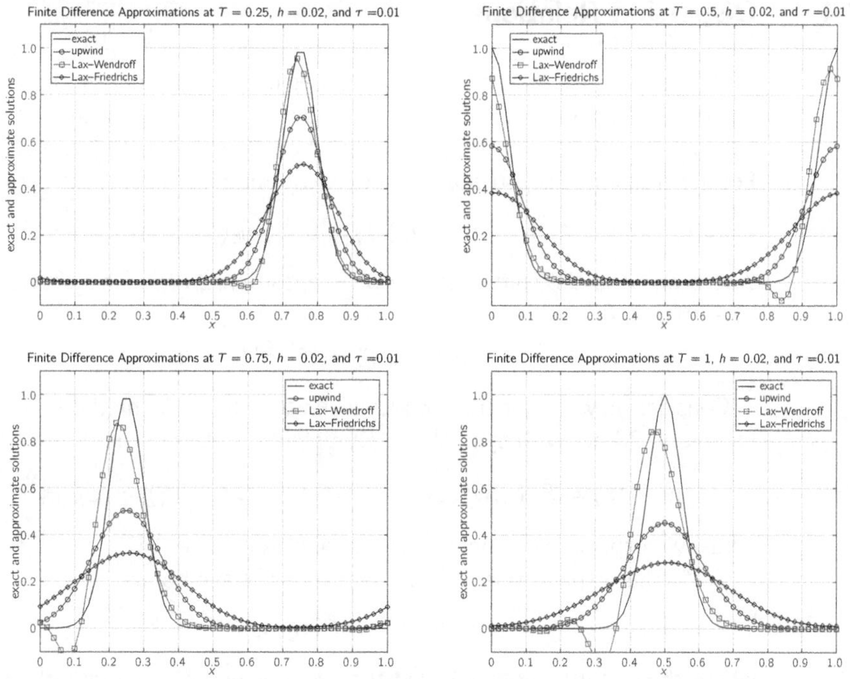

Figure 29.1 Upwind, Lax–Wendroff, and Lax–Friedrichs approximations of the solution to a periodic advection problem at times $T = 0.25, 0.50, 0.75, 1.00$, with $N = 50$, $h = 0.02$, $K = 100$, and $\tau = 0.01$; see Example 29.2 and Listing 29.1.

Rearranging terms, we have

$$c\mu \left\| w^{k+1} \right\|_{L_h^2}^2 + (1 - c\mu) \left\| w^{k+1} - w^k \right\|_{L_h^2}^2 = c\mu \left\| w^k \right\|_{L_h^2}^2 .$$

Clearly, if the CFL condition $\mu \leq 1$ holds,

$$\left\| w^{k+1} \right\|_{L_h^2}^2 \leq \left\| w^k \right\|_{L_h^2}^2 ,$$

from which the claimed $L_\tau^\infty(L_h^2)$-stability immediately follows. □

Example 29.2 Let u be the solution to (23.39) with $c = 1$, $f \equiv 0$, and u_0 being smooth and one-periodic. In fact, it is a Gaussian. Listing 29.1 presents a code that approximates u using the upwind, Lax–Wendroff, and Lax–Friedrich methods. The output at times $T = 0.25, 0.50, 0.75$, and 1.00 are displayed in Figure 29.1. The Lax–Wendroff method seems to give the best approximation. But, curiously, this method is not positivity preserving; see Problem 29.10.

29.4 Dispersion Relations

In this section, we present some properties of second-order methods. To motivate our discussion, we recall that the solution to (23.39) is given by (23.40). Under suitable assumptions on the initial condition, say $u_0 \in L^2(\mathbb{R})$, we can compute its Fourier transform $\hat{u}_0$, so that, in the Fourier domain, (23.40) is given by

$$\hat{u}(\xi, t) = e^{-\mathrm{i}c\xi t}\hat{u}_0(\xi),$$

where we used the shift property of Proposition 23.16. This shows that, for any $m \in \{0, \dots, K-k\}$,

$$\hat{u}(\xi, t_{k+m}) = e^{-\mathrm{i}c\xi m\tau}\hat{u}(\xi, t_k).$$

Consider now a general two-layer finite difference method, in the sense of Definition 28.30, with symbol $\tilde{E}_h^\tau$. By definition of the symbol, we have that

$$\hat{w}^{k+1}(\xi) = \tilde{E}_h^\tau(\xi)\hat{w}^k(\xi) \qquad \Longrightarrow \qquad \hat{w}^{k+m}(\xi) = \left[\tilde{E}_h^\tau(\xi)\right]^m \hat{w}^k(\xi).$$

Thus, we expect the symbol $\tilde{E}_h^\tau(\xi)$ to approximate $e^{-\mathrm{i}c\xi\tau}$.

To further explore this relation, we write the symbol as

$$\tilde{E}_h^\tau(\xi) = |\tilde{E}_h^\tau(\xi)|e^{-\mathrm{i}\omega(\xi)\xi\tau},$$

where the quantity $\omega(\xi)$ is called the *phase speed* of the method. We recall that, according to Theorem 28.34, we must have $|\tilde{E}_h^\tau(\xi)| \leq 1$ for $L_\tau^\infty(L_h^2)$-stability. Thus, at least for the sake of the current illustration, we will assume here that $|\tilde{E}_h^\tau(\xi)| = 1$.

Assume now that the phase speed $\omega(\xi)$ is equal to c. Then, for any initial condition of the form $\hat{u}_0(\xi) = e^{-\mathrm{i}a\xi}$, with $a \in \mathbb{R}$, we would have

$$\hat{u}(\xi, t_k) = e^{-\mathrm{i}(a+ct_k)\xi}$$

and

$$\hat{w}^k(\xi) = e^{-\mathrm{i}(a+ct_k)\xi},$$

which is a correct description of the evolution of the solution. However, in general, $\omega(\xi)$ is only an approximation of c. We then have

$$\hat{u}(\xi, t_k) - \hat{w}^k(\xi) = e^{-\mathrm{i}(a+ct_k)\xi}\left(1 - e^{-\mathrm{i}(\omega(\xi)-c)t_k\xi}\right).$$

The error then is characterized by the quantity $\omega(\xi) - c$, which is known as the *phase error* of a method.

The effect of the phase error, in practice, is manifested by a distortion of the shape of the solution. This is illustrated in Figure 29.2, which was obtained with the code presented in Listing 29.1 and initial condition of the form

$$u_0(x) = \begin{cases} 1, & x \leq \dfrac{1}{2}, \\ 0, & x > \dfrac{1}{2}. \end{cases}$$

We observe that oscillations appear before and after the square wave. This is due to the phase error of the method.

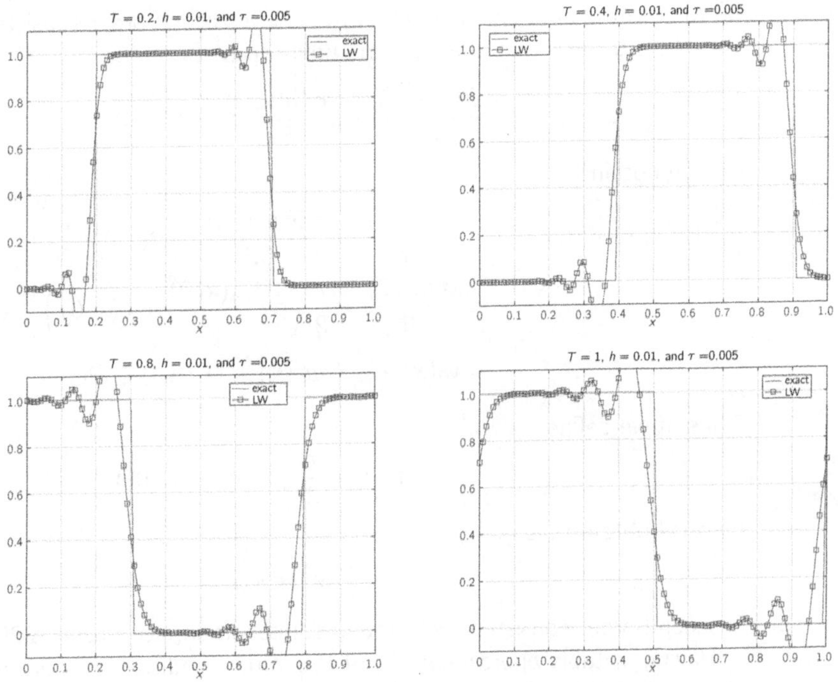

Figure 29.2 Lax–Wendroff (LW) approximation of the solution to a periodic advection problem at times $T = 0.2, 0.4, 0.8, 1.0$, with $N = 100$, $h = 0.01$, $K = 200$, and $\tau = 0.005$. The initial condition is discontinuous. The results were produced with Listing 29.1.

We refer the reader to [88, 59] for a more detailed discussion on dispersion, group velocities, and dispersion relations for finite difference methods.

29.5 The Initial Boundary Value Problem for the Wave Equation

Let us now study a numerical method for the approximation of the initial boundary value problem for the wave equation (23.45) for $d = 1$. Thus, we let $\Omega = (0,1)$ and, as usual, we will seek a space–time grid function $w \in \mathcal{V}_0(\bar{\mathcal{C}}_h^\tau)$ that is defined by a finite difference method that is consistent to order $\tau^2 + h^2$. The first issue we encounter is the approximation of the initial conditions. While the approximation $w^0 = \pi_h u_0$ is clear, the approximation of the initial velocity requires some work. Namely, the easiest and obvious choice,

$$\delta_\tau w^0 = \pi_h v_0,$$

will yield a numerical method that is only consistent to order $\tau + h^2$, due to the fact that we are using a forward difference. Instead, as illustrated in Problem 24.19, we will use the differential equation to achieve second-order consistency (in time). Namely, assuming that the equation holds up to $t = 0$, i.e.,

$$\frac{\partial^2 u(x,0)}{\partial t^2} - a\frac{\partial^2 u(x,0)}{\partial x^2} = f(x,0),$$

we can obtain

$$\begin{aligned}\frac{1}{\tau}\left(u(x,\tau) - u(x,0)\right) &= \frac{\partial u(x,0)}{\partial t} + \frac{\tau}{2}\frac{\partial^2 u(x,0)}{\partial t^2} + \mathcal{O}(\tau^2)\\ &= \frac{\partial u(x,0)}{\partial t} + \frac{\tau}{2}\left(a\frac{\partial^2 u(x,0)}{\partial x^2} + f(x,0)\right) + \mathcal{O}(\tau^2) \qquad (29.10)\\ &= v_0(x) + \frac{\tau}{2}\left(au_0''(x) + f(x,0)\right) + \mathcal{O}(\tau^2).\end{aligned}$$

Thus, if we define

$$v_{0,h} = \pi_h v_0 + \frac{\tau}{2}\pi_h(au_0'' + f(\cdot,0)), \tag{29.11}$$

we shall obtain that

$$\delta_\tau w^0 = v_{0,h} \tag{29.12}$$

is second-order consistent with the initial condition; see Problem 29.13.

Having suitable approximations of the initial conditions, we are ready to define the method. We seek $w \in \mathcal{V}_0(\bar{C}_h^\tau)$ such that

$$\begin{cases}\bar{\delta}_\tau\delta_\tau w^k - a\Delta_h w^k = f_h^k, & k = 1,\dots,K,\\ w^0 = u_{0,h},\\ \delta_\tau w^0 = v_{0,h},\end{cases} \tag{29.13}$$

where $f_h = \pi_h^\tau f$, $u_{0,h} = \pi_h u_0$, and $v_{0,h}$ is given by (29.11).

The following result is an immediate consequence of our construction.

Proposition 29.15 (consistency). *The method* (29.13) *is consistent, in* $C(\mathbb{R}^2)$, *with* (23.45) *up to order* $\tau^2 + h^2$.

Proof. See Problem 29.14. □

Let us now analyze the stability of the method. Before we embark on the technical details, we observe that this is an explicit method. In addition, we recall that the solution to (23.45) has a very well-defined domain of dependency, see Figure 23.3, and that any numerical method that has a chance of success must respect this feature. For this reason, we do not expect the method to be unconditionally stable.

The following results prove the conditional stability of (29.13) in the sense that a discrete analogue of Theorem 23.39 holds. We will do so by means of energy techniques. For a different method of proof, we refer the reader to Problem 29.18.

Let us begin with the case $f_h \equiv 0$.

Theorem 29.16 (conditional stability, $f_h \equiv 0$). *Let* $f_h \equiv 0$. *If*

$$\frac{\sqrt{a}C_I\tau}{2h} \le 1,$$

where C_I is the constant from Problem 24.23, then the method (29.13) *is stable in the sense that the sequence $\{E^k\}_{k=1}^K \subset \mathbb{R}$, defined by*

$$E^{k+1} = \left\|\bar{\delta}_\tau w^{k+1}\right\|_{L^2_h}^2 + \frac{a}{4}\left\|\bar{\delta}_h w^{k+1} + \bar{\delta}_h w^k\right\|_{L^2_h}^2 - \frac{a\tau^2}{4}\left\|\bar{\delta}_\tau\bar{\delta}_h w^{k+1}\right\|_{L^2_h}^2,$$

is independent of k, i.e.,

$$E^{k+1} = E^k.$$

Proof. Take the L^2_h-inner product of the equation that defines the method with $\mathring{\delta}_\tau w^k$ to obtain

$$\left(\bar{\delta}_\tau\delta_\tau w^k, \mathring{\delta}_\tau w^k\right)_{L^2_h} + a\left(\bar{\delta}_h w^k, \bar{\delta}_h\mathring{\delta}_\tau w^k\right)_{L^2_h} = 0.$$

Simple algebraic manipulations allow us to show that, for any grid function $v \in \mathcal{V}(\bar{C}_h^\tau)$,

$$\left(\bar{\delta}_\tau\delta_\tau v^k, \mathring{\delta}_\tau v^k\right)_{L^2_h} = \frac{1}{2\tau}\left[\left\|\bar{\delta}_\tau v^{k+1}\right\|_{L^2_h}^2 - \left\|\bar{\delta}_\tau v^k\right\|_{L^2_h}^2\right] \tag{29.14}$$

and

$$v^k\mathring{\delta}_\tau v^k = \frac{1}{8\tau}\left[(v^{k+1}+v^k)^2 - (v^k+v^{k-1})^2\right] - \frac{\tau}{8}\left[\left(\bar{\delta}_\tau v^{k+1}\right)^2 - \left(\bar{\delta}_\tau v^k\right)^2\right]; \tag{29.15}$$

see Problem 29.16.

From (29.15) with $v = \bar{\delta}_h w$, we then see that

$$\begin{aligned}\left(\bar{\delta}_h w^k, \bar{\delta}_h\mathring{\delta}_\tau w^k\right)_{L^2_h} &= \frac{1}{8\tau}\left[\left\|\bar{\delta}_h w^{k+1} + \bar{\delta}_h w^k\right\|_{L^2_h}^2 - \left\|\bar{\delta}_h w^k + \bar{\delta}_h w^{k-1}\right\|_{L^2_h}^2\right] \\ &\quad - \frac{\tau}{8}\left[\left\|\bar{\delta}_\tau\bar{\delta}_h w^{k+1}\right\|_{L^2_h}^2 - \left\|\bar{\delta}_\tau\bar{\delta}_h w^k\right\|_{L^2_h}^2\right].\end{aligned}$$

The previous considerations then show that $E^{k+1} = E^k$ for all $k = 0, \ldots, K-1$. It thus remains to make sure that $E^k \geq 0$. From the inverse inequality of Problem 24.23, we have that

$$\left\|\bar{\delta}_h\bar{\delta}_\tau w^{k+1}\right\|_{L^2_h} \leq C_I h^{-1}\left\|\bar{\delta}_\tau w^{k+1}\right\|_{L^2_h},$$

so that

$$E^{k+1} \geq \left(\frac{h^2}{C_I^2} - \frac{a\tau^2}{4}\right)\left\|\bar{\delta}_\tau\bar{\delta}_h w^{k+1}\right\|_{L^2_h}^2 + \frac{a}{4}\left\|\bar{\delta}_h w^{k+1} + \bar{\delta}_h w^k\right\|_{L^2_h}^2,$$

which will be nonnegative provided that

$$\frac{\sqrt{a}C_I\tau}{2h} \leq 1,$$

and the claimed stability estimate follows. □

The inhomogeneous case, i.e., $f_h \neq 0$, requires a slightly more restrictive condition. The proof of the following result illustrates a technique known as *negative norm* arguments.

Corollary 29.17 (conditional stability, $f_h \neq 0$). *Assume that there is $\varepsilon \in (0,1)$ for which*

$$\frac{\sqrt{a} C_I \tau}{2h} \leq \sqrt{1-\varepsilon},$$

where C_I is the constant from Problem 24.23, then the method (29.13) is stable in the sense that the sequence $\{E^k\}_{k=1}^K \subset \mathbb{R}$, defined in Theorem 29.16, satisfies

$$\left[E^k\right]^{1/2} \leq \left[E^0\right]^{1/2} + \tau \sum_{m=1}^{k} \|f_h^m\|_{B_h^{-1}}$$

for all $k = 0, \ldots, K$. In the previous expression, we used the energy norm

$$\|z\|_{B_h^{-1}}^2 = \left(B_h^{-1} z, z\right)_{L_h^2}, \qquad B_h = I + \frac{a\tau^2}{4}\Delta_h.$$

Under the given assumptions, the operator B_h^{-1} is self-adjoint with respect to the L_h^2-inner product, positive definite, and the norm of its inverse is uniformly bounded with respect to h.

Proof. Let us provide the main steps of the proof and leave the minute details to the reader as an exercise; see Problem 29.19.

We begin by considering, for $v \in \mathcal{V}_0(\bar{\Omega}_h)$, the expression

$$(v, v)_{L_h^2} - \frac{a\tau^2}{4}\left(\bar{\delta}_h v, \bar{\delta}_h v\right)_{L_h^2} = \left(\left(I + \frac{a\tau^2}{4}\Delta_h\right) v, v\right)_{L_h^2} = (B_h v, v)_{L_h^2}.$$

Clearly, the operator B_h is self-adjoint in the L_h^2-inner product. Using, once again, the inverse inequality of Problem 24.23, we see that

$$(B_h v, v)_{L_h^2} = \left(1 - \frac{a\tau^2 C_I^2}{4h^2}\right)\|v\|_{L_h^2}^2,$$

which, under the assumed restrictions on h and τ, shows that this operator is positive definite and, moreover, that its inverse is uniformly bounded with respect to h and τ in the L_h^2-induced norm.

With the previous observations, we see that we can rewrite the quantity E^{k+1}, defined in Theorem 29.16, as

$$E^{k+1} = \left\|\bar{\delta}_\tau w^{k+1}\right\|_{B_h}^2 + \frac{a}{4}\left\|\bar{\delta}_h w^{k+1} + \bar{\delta}_h w^k\right\|_{L_h^2}^2,$$

where we introduced the B_h-energy norm, which we now know is indeed a norm.

Now, following the proof of Theorem 29.16, we obtain that

$$\begin{aligned}
E^{k+1} - E^k &= 2\tau\left(f_h^k, \mathring{\delta}_\tau w^k\right)_{L_h^2} \\
&= \tau\left(f_h^k, \bar{\delta}_\tau w^{k+1} + \bar{\delta}_\tau w^k\right)_{L_h^2} \\
&\leq \tau\|f_h^k\|_{B_h^{-1}}\left(\left\|\bar{\delta}_\tau w^{k+1}\right\|_{B_h} + \left\|\bar{\delta}_\tau w^k\right\|_{L_h^2}\right) \\
&\leq \tau\|f_h^k\|_{B_h^{-1}}\left(\left[E^{k+1}\right]^{1/2} + \left[E^k\right]^{1/2}\right),
\end{aligned}$$

where the last step makes sense as Theorem 29.16 shows that $E^k \geq 0$. Now we can divide the previously obtained inequality by $\left[E^{k+1}\right]^{1/2} + \left[E^k\right]^{1/2}$ and add over k to obtain the claimed stability estimate. □

29.6 Finite Difference Methods for Hyperbolic Systems

We now present some numerical methods for the initial value problem for a symmetric hyperbolic system (23.46). First, we observe that, as illustrated in Example 23.12, in one dimension the system can be diagonalized and reduced to a system of decoupled transport equations. After this, all of the methods of Section 29.1 can be applied.

Let us here instead present another approach, which does not involve diagonalization and is applicable in more general situations. Our presentation will mostly follow [58, Section 12.2].

Definition 29.18 (finite difference methods). Let $d = 1$, $m \in \mathbb{N}$, and $\boldsymbol{u}$ be the solution of (23.46) with $\boldsymbol{f} \equiv \boldsymbol{0}$. Suppose that $u_{0,h} \in \mathcal{V}(\mathbb{Z}_h; \mathbb{R}^m)$ is defined as usual. We define the following methods for computing the finite difference approximation

$$\boldsymbol{w} \in \mathcal{V}([0,T]_\tau; \mathcal{V}(\mathbb{Z}_h; \mathbb{R}^m))$$

of $\boldsymbol{u}$.

1. **Friedrich's method**

$$\begin{cases} \dfrac{1}{\tau}\left[\boldsymbol{w}^{k+1} - \dfrac{1}{2}\left[\mathcal{S}_1\boldsymbol{w}^k + \mathcal{S}_{-1}\boldsymbol{w}^k\right]\right] + \mathsf{A}\mathring{\delta}_h\boldsymbol{w}^k = \boldsymbol{0}, & k = 0, \dots, K-1, \\ \boldsymbol{w}^0 = u_{0,h}, \end{cases} \tag{29.16}$$

   where $\mathcal{S}_{\pm 1}$ are the shift operators introduced in Definition 24.5.
2. The **Lax–Wendroff method**

$$\begin{cases} \dfrac{1}{\tau}\left(\boldsymbol{w}^{k+1} - \boldsymbol{w}^k\right) + \mathsf{A}\mathring{\delta}_h\boldsymbol{w}^k - \dfrac{\tau}{2}\mathsf{A}^2\Delta_h\boldsymbol{w}^k = \boldsymbol{0}, & k = 0, \dots, K-1, \\ \boldsymbol{w}^0 = u_{0,h}. \end{cases} \tag{29.17}$$

The consistency of these methods can be obtained, as before, by an application of Taylor expansions.

Proposition 29.19 (consistency). *Assume that $\boldsymbol{u}$ is a sufficiently smooth classical solution to (23.46). Then the consistency error for the Friedrichs method (29.16) satisfies*

$$\left\|\mathcal{E}_h^\tau[\boldsymbol{u}]_i^k\right\|_2 \leq C(\tau + h), \quad i \in \mathbb{Z}, \quad k = 0, \dots, K.$$

Under the same assumption, the consistency error for the Lax–Wendroff method satisfies

$$\left\|\mathcal{E}_h^\tau[\boldsymbol{u}]_i^k\right\|_2 \leq C\left(\tau^2 + h^2\right), \quad i \in \mathbb{Z}, \quad k = 0, \dots, K.$$

In both estimates, the constant $C > 0$ is independent of τ and h.

Proof. See Problem 29.20. □

Let us now study the stability of the methods. As expected from setting $m = 1$, these methods cannot be unconditionally stable, as they are the vector-valued version of methods for the transport equation. The approach that we will follow is to extend the von Neumann stability approach of Section 28.7.3 to the vector-valued setting. For simplicity, we confine the discussion to explicit methods.

Definition 29.20 (explicit method)**.** Let $m \in \mathbb{N}$ and consider a **matrix-valued finite difference method** $\mathcal{B}_h \colon \mathcal{V}(\mathbb{Z}_h;\mathbb{R}^m) \to \mathcal{V}_{\bar{N}}(\mathbb{Z}_h;\mathbb{R}^m)$, i.e.,

$$(\mathcal{B}_h \boldsymbol{v})_j = \sum_{m \in S_{\mathcal{B}}} \mathsf{B}_m(h,\tau)\boldsymbol{v}_{j-m}, \qquad \forall \boldsymbol{v} \in \mathcal{V}(\mathbb{Z}_h;\mathbb{R}^m), \tag{29.18}$$

where $S_{\mathcal{B}} \subset \mathbb{Z}$ has finite cardinality and $\mathsf{B}_m \colon \mathbb{R}_+ \times \mathbb{R}_+ \to \mathbb{R}^{m\times m} \setminus \{\mathsf{O}_m\}$. A **two-layer, explicit, matrix-valued, time stepping, finite difference method** with constant coefficients is defined as follows: Find $\boldsymbol{w} \in \mathcal{V}([0,T]_\tau;\mathcal{V}(\mathbb{Z}_h;\mathbb{R}^m))$ such that $\boldsymbol{w}^0$ is given and

$$\boldsymbol{w}^{k+1} = \mathcal{B}_h \boldsymbol{w}^k, \quad k = 0, \dots, K-1.$$

Clearly, all the methods of Definition 29.18 fit the previous definition.

Definition 29.21 (symbol)**.** Consider a two-layer, explicit, matrix-valued, time stepping, finite difference method in the sense of Definition 29.20. Its **matrix-valued symbol** is, for all $\xi \in \mathbb{R}$, the matrix

$$\tilde{\mathsf{E}}_h^\tau(\xi) = \sum_{m \in S_{\mathcal{B}}} \mathsf{B}_m(h,\tau)e^{-\mathrm{i}m\xi} \in \mathbb{C}^{m\times m}.$$

Example 29.3 The matrix-valued symbol of the Friedrichs method (29.16) is

$$\tilde{\mathsf{E}}_h^\tau(\xi) = \mathsf{I}_m \cos\xi - \mu \mathrm{i} \mathsf{A} \sin\xi;$$

see Problem 29.21.

Example 29.4 The matrix-valued symbol of the Friedrichs method (29.17) is

$$\tilde{\mathsf{E}}_h^\tau(\xi) = \mathsf{I}_m + \mu^2\mathsf{A}^2(\cos\xi - 1) - \mu \mathrm{i} \mathsf{A} \sin\xi;$$

see Problem 29.22.

The following result is a vector-valued analogue of Theorem 28.34.

Theorem 29.22 (von Neumann)**.** *Let $m \in \mathbb{N}$ and consider the explicit time stepping method* (29.18) *with symbol $\tilde{\mathsf{E}}_h^\tau(\xi) \in \mathbb{C}^{m\times m}$. This method is stable in the $L_\tau^\infty(L_h^2)$-norm if and only if there exists a constant $C > 0$ such that*

$$\left\|\tilde{\mathsf{E}}_h^\tau(\xi)^k\right\|_2 \le C, \qquad \forall \xi \in \mathbb{R}, \quad k \in \mathbb{N}.$$

We recall that $\|\cdot\|_2$ denotes the induced matrix 2-norm.

Proof. See Problem 29.23. □

We comment that, contrary to the scalar case, the condition of Theorem 29.22 does not imply that

$$\left\|\tilde{\mathsf{E}}_h^{\tau}(\xi)\right\|_2 \leq 1, \tag{29.19}$$

as the following example shows.

Example 29.5 Consider

$$\mathsf{E} = \begin{bmatrix} \frac{1}{2} & 1 \\ 0 & \frac{1}{2} \end{bmatrix}.$$

Then

$$\sigma(\mathsf{E}^{\mathsf{H}}\mathsf{E}) = \left\{\frac{3 \pm 2\sqrt{2}}{4}\right\} \qquad \Longrightarrow \qquad \|\mathsf{E}\|_2 > 1.$$

Yet, for any $k \in \mathbb{N}$,

$$\mathsf{E}^k = \begin{bmatrix} \left(\frac{1}{2}\right)^k & k\left(\frac{1}{2}\right)^{k-1} \\ 0 & \left(\frac{1}{2}\right)^k \end{bmatrix} \qquad \Longrightarrow \qquad \left\|\mathsf{E}^k\right\|_2 \leq C.$$

However, if the condition of Theorem 29.22 holds, then, for every $\lambda(\xi) \in \sigma(\tilde{\mathsf{E}}_h^{\tau}(\xi))$, we must have

$$|\lambda(\xi)|^k \leq C \qquad \Longrightarrow \qquad |\lambda(\xi)| \leq 1.$$

The condition

$$\sigma\left(\tilde{\mathsf{E}}_h^{\tau}(\xi)\right) \subset [-1, 1], \quad \forall \xi \in \mathbb{R} \tag{29.20}$$

is known as the *von Neumann stability condition* for the methods of Definition 29.20. Clearly, this is only a necessary condition. A sufficient stability condition is, obviously, (29.19). The following result shows when this condition is satisfied.

Lemma 29.23 (stability). *Let $m \in \mathbb{N}$ and consider a two-layer, explicit, matrix-valued, finite difference method in the sense of Definition 29.20 with symbol*

$$\tilde{\mathsf{E}}_h^{\tau}(\xi) = \sum_{m \in S_{\mathcal{B}}} \mathsf{B}_m(h, \tau) e^{-\mathrm{i}m\xi} \in \mathbb{C}^{m \times m}.$$

Assume that, for all $m \in S_{\mathcal{B}}$, we have that $\mathsf{B}_m(h, \tau) = \mathsf{B}_m(h, \tau)^{\intercal}$, they are positive semi-definite, and

$$\sum_{m \in S_{\mathcal{B}}} \mathsf{B}_m(h, \tau) = \mathsf{I}_m.$$

Then this method is stable in the $L_{\tau}^{\infty}(L_h^2)$-norm.

Proof. We will prove stability by showing that (29.19) holds. Notice that, for all $\boldsymbol{v}_1, \boldsymbol{v}_2 \in \mathbb{C}^m$, we have

$$\begin{aligned}
\left|\left(\tilde{\mathsf{E}}_h^\tau(\xi)\boldsymbol{v}_1, \boldsymbol{v}_2\right)_2\right| &\leq \sum_{m\in S_\mathcal{B}} |(\mathsf{B}_m(h,\tau)\boldsymbol{v}_1, \boldsymbol{v}_2)_2| \\
&\leq \frac{1}{2}\sum_{m\in S_\mathcal{B}} |(\mathsf{B}_m(h,\tau)\boldsymbol{v}_1, \boldsymbol{v}_1)_2| + \frac{1}{2}\sum_{m\in S_\mathcal{B}} |(\mathsf{B}_m(h,\tau)\boldsymbol{v}_2, \boldsymbol{v}_2)_2| \\
&= \frac{1}{2}\|\boldsymbol{v}_1\|_2^2 + \frac{1}{2}\|\boldsymbol{v}_2\|_2^2,
\end{aligned}$$

where, in the last step, we used that the matrices $\mathsf{B}_m(h,\tau)$ are positive semi-definite and add up to the identity.

Set now, in the previous derivation, $\boldsymbol{v}_2 = \tilde{\mathsf{E}}_h^\tau(\xi)\boldsymbol{v}_1$ while keeping $\boldsymbol{v}_1$ arbitrary. We obtain

$$\left\|\tilde{\mathsf{E}}_h^\tau(\xi)\boldsymbol{v}_1\right\|_2^2 \leq \frac{1}{2}\|\boldsymbol{v}_1\|_2^2 + \frac{1}{2}\left\|\tilde{\mathsf{E}}_h^\tau(\xi)\boldsymbol{v}_1\right\|_2^2,$$

and this clearly implies the result. □

To conclude our discussion, let us provide sufficient conditions for the stability of the Friedrichs method (29.16).

Corollary 29.24 (stability). *Assume that $m \in \mathbb{N}$ and consider the Friedrichs method* (29.16). *If the discretization parameters h and τ are such that*

$$\frac{\tau}{h}\|\mathsf{A}\|_2 \leq 1,$$

then this method is stable in the $L_\tau^\infty(L_h^2)$-norm.

Proof. The symbol for the Friedrichs method is given in Example 29.3:

$$\tilde{\mathsf{E}}_h^\tau(\xi) = \frac{1}{2}(\mathsf{I}_m - \mu\mathsf{A})e^{i\xi} + \frac{1}{2}(\mathsf{I}_m + \mu\mathsf{A})e^{-i\xi}.$$

By assumption, A is symmetric, and thus so is $\mathsf{I}_m \pm \mu\mathsf{A}$. In addition, clearly,

$$\frac{1}{2}(\mathsf{I}_m - \mu\mathsf{A}) + \frac{1}{2}(\mathsf{I}_m + \mu\mathsf{A}) = \mathsf{I}_m.$$

Finally, the condition on $\mu = \frac{\tau}{h}$ guarantees that these matrices are positive semi-definite. Stability then follows from Lemma 29.23. □

Problems

29.1 Complete the details of Example 29.1.

29.2 Complete the proof of Theorem 29.3.

29.3 Prove Theorem 29.5. How does one solve this approximation method?

29.4 Complete the proof of Theorem 29.6.

29.5 Complete the proof of Theorem 29.8.

29.6 Prove that the Beam–Warming method (29.5) is $L_\tau^\infty(L_h^2)$-stable provided that the CFL condition $c\mu \leq 1$ is satisfied.

29.7 Suppose that, in (23.39), we have $c > 0$ and consider the following skewed leapfrog method:

$$w_j^{k+1} = w_{j-2}^{k-1} - \left(\frac{c\tau}{h} - 1\right)(w_j^k - w_{j-2}^k).$$

a) What is the order of consistency of this method?
b) Using the von Neumann stability analysis, for what values of $\frac{c\tau}{h}$ is the method stable?

29.8 In this problem, we will provide a proof of Theorem 29.11.

a) Show that if the method reproduces the constant state, then

$$\sum_{m \in S_{\mathcal{B}}} b_m(h, \tau) = 1.$$

b) Show that if a method reproduces the constant state and is max-norm dissipative, then the method is positivity preserving.
Hint: Suppose that $w_j^k \geq 0$ and $w_m^k = \|w^k\|_{L_h^\infty} = \alpha \geq 0$ for some $m \in \mathbb{Z}$. Define $\eta^k \in \mathcal{V}(\mathbb{Z}_h)$ by

$$\eta_j^k = w_j^k - \frac{\alpha}{2}, \quad j \in \mathbb{Z}.$$

Apply the method to η^k and use the fact that, for all $j \in \mathbb{Z}$,

$$-\frac{\alpha}{2} \leq \eta_j^{k+1} \leq \frac{\alpha}{2}$$

to conclude the result.

c) Show that if a method reproduces the constant state and is max-norm dissipative, then $b_m(h, \tau) \geq 0$ for all $m \in S_{\mathcal{B}}$.
Hint: Use the previous result.

d) Assume that this method is used to approximate the solution to (23.39) with $c = 1$ and $u_0(x) = e^{ix}$. Find an expression for the error in the L_h^2-norm after one step that depends only on the coefficients $\{b_m(h, \tau)\}_{m \in S_{\mathcal{B}}}$, the mesh size $h > 0$, and the Courant number $\mu = \tau/h$, where $\tau > 0$ is the time step size.

29.9 Provide sufficient conditions for the Lax–Friedrichs to be positivity preserving.

29.10 Show that the Lax–Wendroff is not positivity preserving.

29.11 Complete the proof of Theorem 29.13.

29.12 Consider the initial value problem (23.39) for the transport equation with periodic boundary conditions and $c < 0$. Use an energy stability analysis (not the von Newmann analysis) to establish a sufficient condition for stability of methods of the form

$$w_j^{k+1} = \alpha w_{j-1}^k + \beta w_j^k, \qquad j \in \mathbb{Z}, \quad k = 0, \dots, K-1,$$

where $\alpha, \beta \in \mathbb{R}$. Apply your results to the simple method

$$w_j^{k+1} = w_j^k - \frac{c\tau}{h}(w_j^k - w_{j-1}^k), \qquad j \in \mathbb{Z}, \quad k = 0, \dots, K-1.$$

Extend your results to the method

$$w_j^{k+1} = w_j^k - \frac{c\tau}{h}(w_{j+1}^k - w_j^k), \qquad j \in \mathbb{Z}, \quad k = 0, \dots, K-1.$$

29.13 Justify (29.10) and use this to show that (29.12) is consistent up to second order with

$$\frac{\partial u(x,0)}{\partial t} = v_0(x), x \in (0,1).$$

29.14 Prove Proposition 29.15.

29.15 Consider the following variant of (29.13). Let $\sigma \in \mathbb{R}$ be given. Then we seek $w \in \mathcal{V}_0(\bar{C}_h^\tau)$ such that

$$\begin{cases} \bar{\delta}_\tau \delta_\tau w^k - a\Delta_h \left(\sigma w^{k+1} + (1-2\sigma)w^k + \sigma w^{k-1}\right) = f_h^k, & k = 1, \dots, K, \\ w^0 = u_{0,h}, \\ \delta_\tau w^0 = v_{0,h}, \end{cases} \tag{29.21}$$

where $u_{0,h} = \pi_h u_0$ and $v_{0,h}$ is given by (29.11).

a) Show that if $f_h = \pi_h^\tau f$, then, for every $\sigma \in \mathbb{R}$, the method is consistent to order

$$\tau^2 + h^2.$$

b) Show that if

$$f_h = \pi_h^\tau \left(f + \frac{h^2}{12} \frac{\partial^2 f}{\partial x^2} \right)$$

and

$$\sigma \geq \frac{1}{4(1-\varepsilon)} - \frac{h^2}{12 a \tau^2},$$

where $\varepsilon > 0$, then the method is consistent to order

$$\tau^2 + h^4.$$

29.16 Prove identities (29.14) and (29.15).

29.17 Prove the analogue of Theorem 29.16 for (29.21). In other words, show that if $f_h \equiv 0$ and

$$\sigma \geq \frac{1}{4} - \frac{h^2}{a C_I \tau^2},$$

where C_I is the constant of Problem 24.23, then the method (29.21) is stable.

29.18 Prove an analogue of Theorem 29.16 but using the separation of variables technique described in Problem 28.9.

29.19 Provide all the details in the proof of Corollary 29.17.

29.20 Prove Proposition 29.19.

29.21 Provide all the details of Example 29.3.

29.22 Provide all the details of Example 29.4.

29.23 Prove Theorem 29.22.

Hint: Follow Theorem 28.34.

Listings

```
function [errorUW, errorLW, errorLF] = AdvectionPer(finalT, ...
  N, K, numPlots, typeInit)
% This function computes numerical approximations to solutions
% of the linear advection equation
%
% u_t + u_x = 0
%
% using the upwind (UW), Lax-Wendroff (LW), and Lax-Friedrichs
% (LF) methods on the periodic domain [0,1]. The initial data is
% constructed from either a square or a triangular wave.
%
% Input
%   finalT : the final time
%   N : the number of spatial grid points in [0,1]
%   K : the number of time steps in the interval [0,finalT]
%   numPlots : the number of output frames in the time interval
%              [0,finalT]. Must be a divisor of K.
%   typeInit :  is the type of initial condition
%              = 1 for the Gaussian wave
%              = 2 for the triangular wave
%              = 3 for the square wave.
%
% Output
%   errorUW: the max norm error of the UW scheme at t = finalT
%   errorLW: the max norm error of the LW scheme at t = finalT
%   errorLF: the max norm error of the LF scheme at t = finalT
%
  errorUW = 1.0;
  errorLW = 1.0;
  errorLF = 1.0;

  if N > 0 && N-floor(N) == 0
    h = 1.0/N;
  else
    display('Error: N must be a positive integer.')
    return
  end

  if (K > 0 && K-floor(K) == 0) && finalT > 0
    tau = finalT/K;
  else
    display('Error: K must be a positive integer, and')
    display('       finalT must be positive.')
    return
  end

  mu = tau/h;

  x = 0:h:1.0;
  uo = zeros(1,N+1);
  if typeInit == 1
    uo = exp(-20*(sin(pi*(x-0.5))).^2);
```

```
elseif typeInit == 2
  uo = triangle(N,0);
elseif typeInit == 3
  uo = square(N,0);
else
  display('No such initial condition.')
  return
end

uoUW   = zeros(1,N+1); uUW = zeros(1,N+1);
uoLW   = zeros(1,N+2); uLW = zeros(1,N+2);
uoLF   = zeros(1,N+2); uLF = zeros(1,N+2);
uExact = zeros(1,N+1);

uoUW = uo;
uoLW = uo; uoLW(N+2) = uo(2);
uoLF = uo; uoLF(N+2) = uo(2);

if mod(K,numPlots) == 0
  stepsPerPlot = K/numPlots;
else
  display('Error: numPlots is not a divisor of K.')
  return
end

for k = 1: numPlots
  for j = 1: stepsPerPlot
    kk = (k-1)*stepsPerPlot+j;
    for ell = 2: N+1
      uUW(ell) = (1-mu)*uoUW(ell)+mu*uoUW(ell-1);
      uLW(ell) = uoLW(ell)-0.5*mu*(uoLW(ell+1) ...
        -uoLW(ell-1))+0.5*mu*mu*(uoLW(ell+1) ...
        -2.0*uoLW(ell)+uoLW(ell-1));
      uLF(ell) = 0.5*(uoLF(ell+1)+uoLF(ell-1)) ...
        -0.5*mu*(uoLF(ell+1)-uoLF(ell-1));
    end
    uUW(1) = uUW(N+1);
    uoUW = uUW;
    uLW(1) = uLW(N+1); uLW(N+2) = uLW(2);
    uoLW = uLW;
    uLF(1) = uLF(N+1); uLF(N+2) = uLF(2);
    uoLF = uLF;
  end
  currTime = tau*kk;
  if typeInit == 1
    uExact = exp(-20*(sin(pi*(x-0.5-currTime))).^2);
  elseif typeInit == 2
    uExact = triangle(N,currTime);
  elseif typeInit == 3
    uExact = square(N,currTime);
  end
  hf = figure(k);
  clf
  plot(x,uExact,'k-',x,uUW,'b-o',x,uLW(1:N+1),'r-s',x, ...
    uLF(1:N+1),'k-d')
  grid on;
```

```
    xlabel('x');
    ylabel('exact and approximate solutions');
    title(['Finite Difference Approximations at T = ', ...
      num2str(currTime), ', h = ', num2str(h), ...
      ', and tau =', num2str(tau)]);
    legend('Exact','Upwind','Lax--Wendroff','Lax--Friedrichs')
    axis([0,1,-0.1,1.1])
    set(gca,'xTick',0:0.1:1)

    s1 = ['000' num2str(k)];
    s2 = s1((length(s1)-3):length(s1));
    s3 = ['OUT/adv', s2, '.pdf'];
    exportgraphics(gca, s3)
  end

  errorUW = max(abs(uExact-uUW));
  errorLW = max(abs(uExact-uLW(1:N+1)));
  errorLF = max(abs(uExact-uLF(1:N+1)));
end

function u = triangle(N,t)
%
% Constructs a periodic triangle wave traveling with a speed of
% c = 1.
%
% Input
%   N : to compute the mesh size
%   t : the current time
%
% Output
%   u : the array of values of the function
%
  h = 1.0/N;
  for i = 1: N+1
    x = (i-1)*h-t;
    x = x-floor(x);
    if x >= 0.25 && x <= 0.50
      u(i) = 4.0*x-1.0;
    elseif x > 0.50 && x <= 0.75
      u(i) = 3.0-4.0*x;
    else
      u(i) = 0.0;
    end
  end
end

function u = square(N,t)
%
% Constructs a periodic square wave traveling with a speed of
% c = 1.
%
% Input
%   N : to compute the mesh size
%   t : the current time
%
% Output
```

```
%    u : the array of values of the function
  h = 1.0/N;
  for i = 1:N+1
    x = (i-1)*h-t;
    x = x-floor(x);
    if x <= 0.5
      u(i) = 1.0;
    else
      u(i) = 0.0;
    end
  end
end
```

Listing 29.1 Approximation of the periodic linear transport equation.

Appendix A Linear Algebra Review

In this appendix, we review the basic properties of vector spaces, their norms, and inner products.

A.1 The Field of Complex Numbers

We start with some basic definitions.

Definition A.1 (field). A set $\mathbb{K}$ together with two operations $+ \colon \mathbb{K}^2 \to \mathbb{K}$ (addition) and $\cdot \colon \mathbb{K}^2 \to \mathbb{K}$ (multiplication) is called a **field** if the following properties hold.

1. There is an element $0 \in \mathbb{K}$, called **zero**, such that
$$x + 0 = 0 + x = x$$
for all $x \in \mathbb{K}$.
2. For every $x \in \mathbb{K}$, there is an element $-x \in \mathbb{K}$, called the **additive inverse** of x, such that
$$x + (-x) = (-x) + x = 0.$$
3. The **associative property of addition** holds, i.e., for all $x, y, z \in \mathbb{K}$,
$$x + (y + z) = (x + y) + z.$$
4. The **commutative property of addition** holds, i.e., for all $x, y \in \mathbb{K}$,
$$x + y = y + x.$$
5. There is an element $1 \in \mathbb{K}_\star = \mathbb{K}\backslash\{0\}$, called the **multiplicative identity**, such that
$$1 \cdot x = x \cdot 1 = x$$
for all $x \in \mathbb{K}$.
6. For every $x \in \mathbb{K}_\star = \mathbb{K}\backslash\{0\}$, there is an element $x^{-1} \in \mathbb{K}_\star$, called the **multiplicative inverse** of x, such that
$$x \cdot x^{-1} = (x^{-1}) \cdot x = 1.$$

7. The **associative property of multiplication** holds, i.e., for every $x, y, z \in \mathbb{K}$, we have
$$x \cdot (y \cdot z) = (x \cdot y) \cdot z.$$
8. The **commutative property of multiplication** holds, i.e., for all $x, y \in \mathbb{K}$,
$$x \cdot y = y \cdot x.$$
9. The **distributive property** holds, i.e., for all $x, y, z \in \mathbb{K}$,
$$(x + y) \cdot z = x \cdot z + y \cdot z.$$

Remark A.2 (notation)**.** Usually the symbol $\cdot$ is dropped in the multiplication. Thus, from now on, $x \cdot y$ will be replaced by xy, unless the former is needed to avoid ambiguity.

Example A.1 The rational numbers, denoted $\mathbb{Q}$, and the real numbers, denoted $\mathbb{R}$, endowed with the usual rules for addition and multiplication are fields.

Let us now define the field of complex numbers $\mathbb{C}$. This is the collection of all ordered pairs of real numbers, i.e.,
$$\mathbb{C} = \{z = (x, y) \mid x, y \in \mathbb{R}\}.$$
The first component is called the *real* part, $\Re z = x$, while the second one is the *imaginary* part, $\Im z = y$. This way the real numbers can be naturally embedded into $\mathbb{C}$ via $x \mapsto (x, 0) \in \mathbb{C}$. What makes $\mathbb{C}$ different from $\mathbb{R} \times \mathbb{R}$ is how we define operations between elements.

Definition A.3 (complex addition and multiplication)**.** Suppose that $z_1 = (x_1, y_1)$ and $z_2 = (x_2, y_2)$ are arbitrary complex numbers. The standard operations of **complex addition** and **complex multiplication** are defined as
$$z_1 + z_2 = (x_1, y_1) + (x_2, y_2) = (x_1 + x_2, y_1 + y_2)$$
and
$$z_1 \cdot z_2 = (x_1, y_1) \cdot (x_2, y_2) = (x_1 x_2 - y_1 y_2, x_1 y_2 + x_2 y_1).$$

The verification that $\mathbb{C}$ is in fact a field is left to the reader as an exercise. The additive and multiplicative identity elements are clearly $0 = (0, 0)$ and $1 = (1, 0)$, respectively.

Definition A.4 (imaginary unit)**.** The complex number $\mathrm{i} = (0, 1)$ is called the **imaginary unit**. It follows that $\mathrm{i}^2 = -1$, which justifies writing $\mathrm{i} = \sqrt{-1}$.

Having introduced the imaginary unit and the multiplicative identity, we can denote complex numbers in a more familiar way, namely
$$z = (x, y) \in \mathbb{C} \quad \iff \quad z = x + \mathrm{i}y.$$

Definition A.5 (complex conjugate). Suppose that $z = x + iy \in \mathbb{C}$. The **complex conjugate** of z, denoted $\bar{z}$, is the complex number $\bar{z} = x - iy$.

Proposition A.6 (properties of the conjugate). *If $w, z \in \mathbb{C}$, then $\overline{z + w} = \bar{z} + \bar{w}$ and $\overline{zw} = \bar{z}\bar{w}$.*

Recall that, if $x \in \mathbb{R}$, its *absolute value* is defined by

$$|x| = \begin{cases} x, & x \geq 0, \\ -x, & x < 0. \end{cases}$$

Geometrically, $|x|$ represents the length of the segment with endpoints 0 and x; in other words, it tells us how far is x from 0 in distance. There is a similar notion for complex numbers.

Definition A.7 (modulus). Suppose that $z = x + iy \in \mathbb{C}$. The **modulus** of z, denoted $|z|$, is defined by $|z| = \sqrt{x^2 + y^2}$.

A.2 Vector Spaces

We are now ready to define the fundamental concept in linear algebra, the one that will characterize *all* the objects we will be concerned with in this appendix.

Definition A.8 (vector space). Suppose that $\mathbb{K}$ is a field, here called the **field of scalars**, and $\mathbb{V}$ is a set of objects called **vectors**. $\mathbb{V}$ is a **vector space over** $\mathbb{K}$ if and only if there are operations $+ : \mathbb{V}^2 \to \mathbb{V}$ (vector addition) and $\cdot : \mathbb{K} \times \mathbb{V} \to \mathbb{V}$ (scalar multiplication), such that the triple $(\mathbb{V}, +, \cdot)$ satisfies the following.

1. There is an element $0 \in \mathbb{V}$, called a zero vector, such that, for all vectors $x \in \mathbb{V}$,
$$x + 0 = 0 + x = x.$$
2. For every vector $x \in \mathbb{V}$, there is an element $-x \in \mathbb{V}$, called an additive inverse of x, such that
$$x + (-x) = (-x) + x = 0.$$
3. The associative property of vector addition holds, i.e., for all vectors $x, y, z \in \mathbb{V}$, we have
$$x + (y + z) = (x + y) + z.$$
4. The commutative property of vector addition holds, i.e., for all vectors $x, y \in \mathbb{V}$, it holds that
$$x + y = y + x.$$
5. For all scalars $\alpha, \beta \in \mathbb{K}$ and every vector $x \in \mathbb{V}$, we have
$$\alpha \cdot (\beta \cdot x) = (\alpha\beta) \cdot x.$$
6. For every vector $x \in \mathbb{V}$, we have $1 \cdot x = x$, where $1 \in \mathbb{K}$.

7. For all scalars $\alpha, \beta \in \mathbb{K}$ and every vector $x \in \mathbb{V}$, we have

$$(\alpha + \beta) \cdot x = \alpha \cdot x + \beta \cdot x.$$

8. For all $\alpha \in \mathbb{K}$ and $x, y \in \mathbb{V}$, we have

$$\alpha \cdot (x + y) = \alpha \cdot x + \alpha \cdot y.$$

Remark A.9 (notation). Usually the symbol $\cdot$ in the scalar multiplication is suppressed.

Example A.2 Any field is a vector space over itself.

Example A.3 $\mathbb{C}$ is a vector space over $\mathbb{R}$. The subset of $\mathbb{C}$ whose elements have rational real and imaginary parts is a vector space over $\mathbb{Q}$.

Example A.4 The set $\mathbb{P}_n(\mathbb{K})$ of all polynomials of degree no larger than n with coefficients from $\mathbb{K}$ is a vector space over $\mathbb{K}$.

The following is the canonical example of a vector space.

Definition A.10 (n-vectors). Let $\mathbb{K}$ be a field. We define, for any $n \in \mathbb{N}$,

$$\mathbb{K}^n = \left\{ z = \begin{bmatrix} z_1 \\ \vdots \\ z_n \end{bmatrix} \middle| \; z_i \in \mathbb{K}, \; i = 1, \ldots, n \right\}.$$

We call $\mathbb{C}^n$ the set of **complex n-vectors** and $\mathbb{R}^n$ the set of **real n-vectors**.

To extract the ith component of an n-vector $\boldsymbol{z} \in \mathbb{K}^n$, we use the notation $[\boldsymbol{z}]_i = z_i \in \mathbb{K}$. We naturally define n-vector addition and scalar multiplication component-wise via

$$[\boldsymbol{x} + \boldsymbol{y}]_i = x_i + y_i, \quad [\alpha \boldsymbol{x}]_i = \alpha x_i, \quad i = 1, \ldots, n,$$

where $\boldsymbol{x}, \boldsymbol{y} \in \mathbb{K}^n$ are arbitrary n-vectors and $\alpha \in \mathbb{K}$ is an arbitrary scalar.

Proposition A.11 ($\mathbb{K}^n$ is a vector space). *With addition and scalar multiplication defined as above, $\mathbb{K}^n$ is a vector space over $\mathbb{K}$.*

Definition A.12 (linear combination). Let $\mathbb{V}$ be a vector space over $\mathbb{K}$ and

$$S = \{x_1, \ldots, x_k\} \subseteq \mathbb{V}.$$

We call $x \in \mathbb{V}$ a **linear combination** of vectors from S if and only if there are scalars $\alpha_i \in \mathbb{K}$, $i = 1, \ldots, k$ such that

$$x = \sum_{i=1}^{k} \alpha_i x_i \in \mathbb{V}.$$

Definition A.13 (span)**.** Let $S = \{x_1, \ldots, x_k\} \subseteq \mathbb{V}$ be given, where $\mathbb{V}$ is a vector space over the field $\mathbb{K}$. The **span** of the set S is the collection of all linear combinations of elements of S:

$$\operatorname{span}(S) = \langle S \rangle = \left\{ z \in \mathbb{V} \;\middle|\; \exists \alpha_1, \ldots, \alpha_k \in \mathbb{K} \text{ with } z = \sum_{i=1}^{k} \alpha_i x_i \right\}.$$

Definition A.14 (linear dependence)**.** Suppose that $\mathbb{V}$ is a vector space over $\mathbb{K}$. Let $S = \{x_1, \ldots, x_k\} \subseteq \mathbb{V}$ be a given finite set. We say that S is **linearly independent** if and only if

$$\sum_{i=1}^{k} \alpha_i x_i = 0$$

implies that $\alpha_i = 0 \in \mathbb{K}$ for all $i = 1, \ldots, k$. Otherwise, we say that S is **linearly dependent**. If S is an infinite subset of $\mathbb{V}$, we say that S is **linearly independent** if and only if each and every finite subset of S is linearly independent; otherwise, we say that S is **linearly dependent**.

In other words, a set of vectors is linearly independent if the only linear combination that yields the zero vector is the trivial one. This is the same as saying that no element of S can be expressed as a linear combination of the remaining elements of S.

Definition A.15 (subspace)**.** Let $\mathbb{V}$ be a vector space over the field $\mathbb{K}$ with addition $+$ and multiplication $\cdot$. The nonempty subset $\mathbb{W} \subseteq \mathbb{V}$ is called a **subspace** of $\mathbb{V}$ if and only if $(\mathbb{W}, +, \cdot)$ is itself a vector space over $\mathbb{K}$. In general, we write $\mathbb{W} \leq \mathbb{V}$, but, if it is known that $\mathbb{W}$ is a proper subset of $\mathbb{V}$, we write $\mathbb{W} < \mathbb{V}$.

Proposition A.16 (span(S) is a subspace)**.** *Let $S = \{x_1, \ldots, x_k\} \subseteq \mathbb{V}$ be given, where $\mathbb{V}$ is a vector space over the field $\mathbb{K}$. Then* $\operatorname{span}(S)$ *is a subspace of* $\mathbb{V}$.

Proof. See Problem A.5. □

Definition A.17 (dimension)**.** Suppose that $\mathbb{V}$ is a vector space over $\mathbb{K}$. Define

$$J = \{\#(S) \mid S \subseteq \mathbb{V},\ \#(S) < \infty,\ S \text{ is linearly independent}\} \subseteq \mathbb{N},$$

where $\#(S)$ is the cardinality of S, i.e., the number of distinct elements in S. If J is bounded, we say that $\mathbb{V}$ is **finite dimensional**. In this case, we write $\dim(\mathbb{V}) = \max(J)$, and $\dim(\mathbb{V})$ is called the **dimension** of $\mathbb{V}$. If J is unbounded, we say that $\mathbb{V}$ is an **infinite-dimensional** vector space, and we write $\dim(\mathbb{V}) = \infty$. We say that $\dim(\{0\}) = 0$.

Definition A.18 (basis)**.** Suppose that $\mathbb{V}$ is a finite-dimensional vector space with $\dim(\mathbb{V}) = m$. A set $B \subseteq \mathbb{V}$ is called a **basis** of $\mathbb{V}$ if and only if it is linearly independent and $\#(B) = m$.

Remark A.19 (finite dimensions)**.** Observe carefully that we have defined the concept of basis only for finite-dimensional vector spaces. Bases for infinite-dimensional spaces, which are also encountered in our text, require more care.

Example A.5 Define the space of polynomials of arbitrary order,

$$\mathbb{P}(\mathbb{C}) = \bigcup_{k\in\mathbb{N}} \mathbb{P}_{k-1}(\mathbb{C}).$$

The reader can easily show that this is a vector space over the field of complex numbers. The infinite set

$$S = \{1, x, x^2, x^3, x^4, \ldots\}$$

is linearly independent, as the reader can easily show. Therefore, we conclude that $\mathbb{P}$ is of infinite dimensions.

Example A.6 Consider the space of complex-valued continuous functions over the interval $[-1, 1]$, denoted $C([-1, 1]; \mathbb{C})$. If $\alpha, \beta \in \mathbb{C}$ and $f, g \in C([-1, 1]; \mathbb{C})$, then it is easy to see that $\alpha f + \beta g \in C([-1, 1]; \mathbb{C})$. One can easily show that $C([-1, 1]; \mathbb{C})$ is a vector space over the field of complex numbers. The vectors in this case are functions. In fact, $\mathbb{P} \subseteq C([-1, 1]; \mathbb{C})$, as long as we restrict the domain of definition of the polynomials to $[-1, 1]$. This shows that $C([-1, 1]; \mathbb{C})$ is infinite dimensional.

Proposition A.20 (spanning property of a basis). *Suppose that $\mathbb{V}$ is a finite-dimensional vector space and S is a basis. Then* $\operatorname{span}(S) = \mathbb{V}$.

Proof. See Problem A.6. □

Proposition A.21 (generation of subspaces). *Suppose that $\mathbb{V}$ is a finite-dimensional vector space over $\mathbb{K}$ and $S = \{x_1, \ldots, x_k\} \subseteq \mathbb{V}$ is a set of nonzero vectors. Set* $\mathbb{W} = \operatorname{span}(S)$. *Then some subset of S is a basis for $\mathbb{W}$. Moreover,*

$$1 \leq \dim(\mathbb{W}) \leq k \quad \textit{and} \quad \dim(\mathbb{W}) \leq \dim(\mathbb{V}) < \infty.$$

Proof. See Problem A.8. □

Theorem A.22 (basis extension). *Suppose that $\mathbb{V}$ is a finite-dimensional vector space over $\mathbb{K}$ and $S = \{x_1, \ldots, x_k\} \subseteq \mathbb{V}$ is a linearly independent set of vectors. Then* $1 \leq k \leq \dim(\mathbb{V})$. *Moreover, there is a basis T of $\mathbb{V}$ that contains S and*

$$\#(T \backslash S) = \dim(\mathbb{V}) - k.$$

Proof. See Problem A.9. □

Example A.7 Consider the set $B = \{\boldsymbol{e}_1, \ldots, \boldsymbol{e}_n\}$, where

$$[\boldsymbol{e}_i]_j = \delta_{i,j} = \begin{cases} 1, & i = j, \\ 0, & i \neq j. \end{cases}$$

$\delta_{i,j}$ is called the *Kronecker delta function*.[1] Then B is a basis for $\mathbb{C}^n$, and it is called the canonical basis.

Finally, using the basis concept, it can be justified that $\mathbb{C}^n$ is, in some sense, the *only* vector space of dimension n over the field of complex numbers. To see this, we need the following.

Definition A.23 (isomorphism). Suppose that $\mathbb{V}$ and $\mathbb{W}$ are vector spaces over the same field $\mathbb{K}$. A mapping $F\colon \mathbb{V} \to \mathbb{W}$ is called a **vector isomorphism** between $\mathbb{V}$ and $\mathbb{W}$ if and only if F is a bijection (a one-to-one and onto mapping) from $\mathbb{V}$ to $\mathbb{W}$; and F is linear,

$$F(\alpha x + \beta y) = \alpha F(x) + \beta F(y), \quad \forall \alpha, \beta \in \mathbb{K}, \ \forall x, y \in \mathbb{V}.$$

We say that $\mathbb{V}$ and $\mathbb{W}$ are **isomorphic** if and only if there exists a vector isomorphism between them.

Theorem A.24 (isomorphism). *Let $\mathbb{V}$ be a vector space over $\mathbb{C}$ of finite dimension n. Then $\mathbb{V}$ is isomorphic to $\mathbb{C}^n$. In other words, $\mathbb{V}$ can be identified with $\mathbb{C}^n$.*

Proof. The idea is to construct the isomorphism F using a basis for $\mathbb{V}$ and the canonical basis for $\mathbb{C}^n$. □

From this point onward, unless it is indicated otherwise, all vector spaces are assumed to be over the field of complex numbers, $\mathbb{C}$. Such spaces are called *complex vector spaces.*

A.3 Normed Spaces

It is often said that a vector is an object that has a magnitude and a direction. We use norms to characterize lengths (magnitude).

Definition A.25 (norm). Let $\mathbb{V}$ be a complex vector space. A map $\|\cdot\|\colon \mathbb{V} \to \mathbb{R}$ is called a **norm** if and only if it satisfies the following properties.

1. **Positive definiteness:** If $\|x\| = 0$, then $x = 0$.
2. **Nonnegative homogeneity:** For every $\lambda \in \mathbb{C}$ and $x \in \mathbb{V}$, it follows that

$$\|\lambda x\| = |\lambda| \|x\|.$$

3. **Triangle inequality:** For every $x, y \in \mathbb{V}$, it follows that

$$\|x + y\| \leq \|x\| + \|y\|.$$

A vector space equipped with a norm is called a **normed vector space**.

As with the absolute value and modulus, the intuition behind a norm is that we can measure the length of an object.

[1] Named in honor of the German mathematician Leopold Kronecker (1823–1891).

Proposition A.26 (properties of the norm). *Let $\mathbb{V}$ be a complex vector space and suppose that $\|\cdot\|\colon \mathbb{V} \to \mathbb{R}$ is a norm on $\mathbb{V}$. The following are true.*

1. *The norm is nonnegative, i.e., for all $x \in \mathbb{V}$, $\|x\| \geq 0$.*
2. *The norm satisfies the* reverse triangle inequality*: For all $x, y \in \mathbb{V}$, we have*

$$\big|\|x\| - \|y\|\big| \leq \|x - y\|.$$

3. *The norm is a continuous function.*

Proof. See Problem A.10. □

Example A.8 The real number $\mathbb{R}$ equipped with the absolute value $|\cdot|$ is a normed vector space over $\mathbb{R}$.

Example A.9 The complex number $\mathbb{C}$ with the modulus $|\cdot|$ is a normed vector space over $\mathbb{C}$.

Example A.10 Let $\{x_0, \dots, x_n\} \subseteq \mathbb{C}$ be a set of $n+1$ distinct points. The set of polynomials $\mathbb{P}_n(\mathbb{C})$ of degree at most $n \geq 1$, with complex coefficients, equipped with

$$\|p\| = \sum_{i=0}^{n} |p(x_i)|$$

is a normed vector space.

Example A.11 Let $n \in \mathbb{N}$ and $z \in \mathbb{C}^n$. For $p \in [1, \infty)$, we define

$$\|z\|_{\ell^p(\mathbb{C}^n)} = \|z\|_p = \left(\sum_{i=1}^{n} |z_i|^p\right)^{1/p}$$

and, for $p = \infty$,

$$\|z\|_{\ell^\infty(\mathbb{C}^n)} = \|z\|_\infty = \max_{i=1}^{n} |z_i|.$$

Then $\mathbb{C}^n$ equipped with $\|\cdot\|_p$, $1 \leq p \leq \infty$, is a normed vector space. We commonly denote this normed space $\ell^p(\mathbb{C}^n)$.

Proposition A.27 ($\|\cdot\|_p$ norm). *$\|\cdot\|_{\ell^p(\mathbb{C}^n)} : \mathbb{C}^n \to \mathbb{R}$ is a norm.*

Proof. Let us show that the triangle inequality indeed holds for the p-norms with $p \in (1, \infty)$. The other properties are simpler and left to the reader as an exercise; see Problem A.11.

We begin by proving Young's inequality[2] for products: if $a, b \geq 0$ and $p \in (1, \infty)$,

$$ab \leq \frac{a^p}{p} + \frac{b^q}{q},$$

where $q = p/(p-1)$ is "conjugate" to p. Notice that this is trivial if $a = 0$ or $b = 0$. For both a and b positive, we just use the fact that the logarithm function is concave, i.e., for all $t \in (0, 1)$, we have

$$\begin{aligned}\log(ta^p + (1-t)b^q) &\geq t\log(a^p) + (1-t)\log(b^q) \\ &= tp\log(a) + (1-t)q\log(b).\end{aligned}$$

Notice that, since $p > 1$, we can set $t = 1/p \in (0, 1)$ and $(1-t) = 1/q$. With this choice, we then get

$$\log\left(\frac{a^p}{p} + \frac{b^q}{q}\right) \geq \log(a) + \log(b) = \log(ab).$$

We conclude by exponentiating.

The next step is to prove the so-called Hölder inequality,[3] which states that if $\boldsymbol{x}, \boldsymbol{y} \in \mathbb{C}^n$ then we have

$$\sum_{i=1}^{n} |x_i y_i| \leq \|\boldsymbol{x}\|_p \|\boldsymbol{y}\|_q,$$

where q is as before. We begin by noticing that if, for all $i = 1, \ldots, n$, $x_i = 0$ or $y_i = 0$, then the left-hand side of this inequality is zero and there is nothing to prove. Thus, we assume that $\|\boldsymbol{x}\|_p \|\boldsymbol{y}\|_q > 0$. Define $\tilde{x}_i = x_i / \|\boldsymbol{x}\|_p$ and $\tilde{y}_i = y_i / \|\boldsymbol{y}\|_q$. By Young's inequality, we then obtain

$$\sum_{i=1}^{n} |\tilde{x}_i \tilde{y}_i| \leq \frac{1}{p} \sum_{i=1}^{n} |\tilde{x}_i|^p + \frac{1}{q} \sum_{i=1}^{n} |\tilde{y}_i|^q.$$

Notice now that, by construction, $\sum_{i=1}^{n} |\tilde{x}_i|^p = \sum_{i=1}^{n} |\tilde{y}_i|^q = 1$, so that

$$\sum_{i=1}^{n} |\tilde{x}_i \tilde{y}_i| \leq \frac{1}{p} + \frac{1}{q} = 1.$$

Multiply by $\|\boldsymbol{x}\|_p \|\boldsymbol{y}\|_q$ to conclude the Hölder inequality.

We are finally ready to show the so-called Minkowski inequality for sums,[4] which amounts to showing the triangle inequality for $\|\cdot\|_p$. Let $\boldsymbol{x}, \boldsymbol{y} \in \mathbb{C}^n$ be as before. If $\boldsymbol{x} + \boldsymbol{y} = \boldsymbol{0}$, then there is nothing to show; otherwise,

[2] Named in honor of the British mathematician William Henry Young (1863–1942).
[3] Named in honor of the German mathematician Otto Ludwig Hölder (1859–1937).
[4] Named in honor of the German mathematician Hermann Minkowski (1864–1909).

$$
\begin{aligned}
0 < \|\mathbf{x}+\mathbf{y}\|_p^p &= \sum_{i=1}^{n} |x_i + y_i|^p \\
&= \sum_{i=1}^{n} |x_i + y_i||x_i + y_i|^{p-1} \\
&\le \sum_{i=1}^{n} |x_i||x_i + y_i|^{p-1} + \sum_{i=1}^{n} |y_i||x_i + y_i|^{p-1},
\end{aligned}
$$

where we used the triangle inequality for the modulus. The first term on the right-hand side of this inequality, using Hölder, can be rewritten as:

$$
\sum_{i=1}^{n} |x_i||x_i + y_i|^{p-1} \le \left(\sum_{i=1}^{n} |x_i|^p\right)^{1/p} \left(\sum_{i=1}^{n} |x_i + y_i|^{q(p-1)}\right)^{1/q} = \|\mathbf{x}\|_p \|\mathbf{x}+\mathbf{y}\|_p^{p-1},
$$

where we used that $q = p/(p-1)$. A similar treatment of the second term then reveals

$$
0 < \|\mathbf{x}+\mathbf{y}\|_p^p \le (\|\mathbf{x}\|_p + \|\mathbf{y}\|_p) \|\mathbf{x}+\mathbf{y}\|_p^{p-1}.
$$

Divide by $\|\mathbf{x}+\mathbf{y}\|_p^{p-1} > 0$ to conclude the triangle inequality. □

Norms can be used to characterize convergence and continuity. Sometimes it is important and useful to know whether a convergent sequence (with respect to one norm) will also be convergent with respect to another one.

Definition A.28 (norm equivalence)**.** Let $\mathbb{V}$ be a complex vector space and, for $i = 1, 2$, let $\|\cdot\|_{(i)}$ be norms on $\mathbb{V}$. We say that these norms are **equivalent** if there are constants $C_1, C_2 > 0$ such that, for every $x \in \mathbb{V}$, we have

$$
C_1 \|x\|_{(1)} \le \|x\|_{(2)} \le C_2 \|x\|_{(1)} .
$$

As the reader can easily verify, norm equivalence essentially means that a sequence is convergent in one norm if and only if it is convergent in the other. It turns out that, in $\mathbb{C}^n$, all norms are equivalent.

Theorem A.29 (all norms are equivalent)**.** *Suppose that $\|\cdot\| : \mathbb{C}^n \to \mathbb{R}$ is a norm. There exist positive constants $0 < C_1 \le C_2$ such that*

$$
C_1 \|\mathbf{x}\|_\infty \le \|\mathbf{x}\| \le C_2 \|\mathbf{x}\|_\infty , \quad \forall \mathbf{x} \in \mathbb{C}^n.
$$

In other words, in $\mathbb{C}^n$, all norms are equivalent.

Proof. Let $\{\mathbf{e}_i\}_{i=1}^n$ be the canonical basis of $\mathbb{C}^n$. Then, for every $\mathbf{x} \in \mathbb{C}^n$, we have

$$
\mathbf{x} = \sum_{i=1}^{n} x_i \mathbf{e}_i, \qquad x_i \in \mathbb{C}, \qquad \|\mathbf{x}\|_\infty = \max_{i=1}^{n} |x_i|,
$$

and

$$
\|\mathbf{x}\| = \left\| \sum_{i=1}^{n} x_i \mathbf{e}_i \right\| \le \sum_{i=1}^{n} |x_i|\|\mathbf{e}_i\| \le \max_{i=1}^{n} |x_i| \sum_{i=1}^{n} \|\mathbf{e}_i\| = C_2 \|\mathbf{x}\|_\infty
$$

with $C_2 = \sum_{i=1}^n \|\mathbf{e}_i\|$. Since $\mathbf{x}$ is arbitrary, this shows one part of the statement.

To show the other bound, let $F\colon \mathbb{C}^n \to \mathbb{R}$ be defined by $F(\boldsymbol{x}) = \|\boldsymbol{x}\|$. Since, for every $\boldsymbol{x}, \boldsymbol{y} \in \mathbb{C}^n$,

$$|F(\boldsymbol{x}) - F(\boldsymbol{y})| = |\|\boldsymbol{x}\| - \|\boldsymbol{y}\|| \leq \|\boldsymbol{x} - \boldsymbol{y}\| \leq C_2 \|\boldsymbol{x} - \boldsymbol{y}\|_\infty ,$$

the function F is continuous with respect to $\|\cdot\|_\infty$. The set

$$S_\infty^{n-1} = \{\boldsymbol{x} \in \mathbb{C}^n \mid \|\boldsymbol{x}\|_\infty = 1\}$$

is closed and bounded and, therefore, compact; see Corollary B.15. Then, as a consequence of Theorem B.47, there exists $\boldsymbol{x}_0 \in S_\infty^{n-1}$ such that

$$F(\boldsymbol{x}_0) = \inf\left\{F(\boldsymbol{x}) \middle| \boldsymbol{x} \in S_\infty^{n-1}\right\} \iff \|\boldsymbol{x}_0\| = F(\boldsymbol{x}_0) \leq F(\boldsymbol{x}) = \|\boldsymbol{x}\|, \ \forall \boldsymbol{x} \in S_\infty^{n-1}.$$

Notice also that $\boldsymbol{x}_0 \neq \boldsymbol{0}$, so that we can set $C_1 = \|\boldsymbol{x}_0\| > 0$, and the previous statement then reads

$$C_1 \leq \|\boldsymbol{x}\|, \quad \forall \boldsymbol{x} \in S_\infty^{n-1}.$$

Let now $\boldsymbol{y} \in \mathbb{C}_\star^n$ be arbitrary. We can then define $\boldsymbol{x} = \|\boldsymbol{y}\|_\infty^{-1} \boldsymbol{x} \in S_\infty^{n-1}$ and, by the previous reasoning, we find

$$C_1 \leq \|\boldsymbol{x}\| = \left\| \frac{1}{\|\boldsymbol{y}\|_\infty} \boldsymbol{y} \right\| \qquad \iff \qquad C_1 \|\boldsymbol{y}\|_\infty \leq \|\boldsymbol{y}\|.$$

Since $\boldsymbol{y}$ was arbitrary, this provides the claimed bound. □

As the reader may verify, from the result above it follows that *all* norms over $\mathbb{C}^n$ are equivalent; see Problem A.12.

A.4 Inner Product Spaces

In the previous section, we introduced the notion of a norm, which was used to describe the magnitude of a vector. Often, we can also characterize angles between vectors, i.e., their relative directions. Inner products are used for this purpose.

Definition A.30 (inner product)**.** Let $\mathbb{V}$ be a complex vector space. A function $(\cdot,\cdot)\colon \mathbb{V} \times \mathbb{V} \to \mathbb{C}$ is called an **inner product** if and only if:

1. For all $x, y \in \mathbb{V}$, $(x, y) = \overline{(y, x)}$.
2. For all $x, y \in \mathbb{V}$ and $\lambda \in \mathbb{C}$, we have $(\lambda x, y) = \lambda(x, y)$.
3. For all $x, y, z \in \mathbb{V}$, it holds that $(x + y, z) = (x, z) + (y, z)$.
4. For all $x \in \mathbb{V}$, $0 \leq (x, x) \in \mathbb{R}$ and $(x, x) = 0$ implies $x = 0$.

A vector space equipped with an inner product is called a **complex inner product space**. A **real inner product space** is defined analogously.

Example A.12 Let $\mathbf{x}, \mathbf{y} \in \mathbb{C}^n$ be given complex n-vectors. We define

$$(\mathbf{x}, \mathbf{y})_{\ell^2(\mathbb{C}^n)} = (\mathbf{x}, \mathbf{y})_2 = \sum_{i=1}^{n} x_i \bar{y}_i.$$

This is an inner product, a fact that you should prove, and is called the *Euclidean inner product*[5]; see Problem A.14. When $\mathbf{x}, \mathbf{y} \in \mathbb{R}^n$, the object is, after dropping the unnecessary conjugation, a real inner product, which we denote by

$$(\mathbf{x}, \mathbf{y})_{\ell^2(\mathbb{R}^n)} = (\mathbf{x}, \mathbf{y})_2 = \mathbf{x} \cdot \mathbf{y} = \sum_{i=1}^{n} x_i y_i.$$

This inner product makes $\mathbb{R}^n$ a real (as opposed to a complex) inner product space of n dimensions.

Proposition A.31 (Cauchy–Schwarz inequality[6]). *Let* $\mathbb{V}$ *be a complex vector space with inner product* $(\cdot, \cdot)$. *Then, for every* $x, y \in \mathbb{V}$, *we have*

$$|(x, y)| \le (x, x)^{1/2} (y, y)^{1/2},$$

with equality if and only if x *and* y *are linearly dependent.*

Proof. If either $x = 0$ or $y = 0$, there is nothing to prove. It suffices, therefore, to assume that $x \neq 0$ and $y \neq 0$. Define $z = x - \lambda y$, with $\lambda \in \mathbb{C}$ to be chosen. By the properties of the inner product, we have

$$(z, y) = (x - \lambda y, y) = (x, y) - \lambda(y, y).$$

Setting $\lambda = (x, y)/(y, y)$, which is possible since $y \neq 0$, we obtain $(z, y) = 0$. Now, since $x = z + \lambda y$, we find

$$\begin{aligned}
(x, x) &= (z + \lambda y, z + \lambda y) \\
&= (z, z) + \lambda(y, z) + \bar{\lambda}(z, y) + \lambda\bar{\lambda}(y, y) \\
&= (z, z) + |\lambda|^2 (y, y) \\
&\ge |\lambda|^2 (y, y).
\end{aligned}$$

Using the definition of λ allows us to conclude the desired result. □

Corollary A.32 (Euclidean norm). *Let* $\mathbb{V}$ *be a complex vector space with inner product* $(\cdot, \cdot)$. *Then*

$$\|x\| = (x, x)^{1/2} \tag{A.1}$$

defines a norm. In other words, every complex inner product space is a complex normed vector space.

[5] Named in honor of the Greek mathematician widely regarded as the father of geometry, Euclid of Alexandria (300 BCE).

[6] Named in honor of the French mathematician Augustin Louis Cauchy (1789–1857) and the German mathematician Karl Hermann Amandus Schwarz (1843–1921).

Example A.13 For $\mathbb{C}^n$, the natural inner product is the *Euclidean inner product*, defined by

$$(\mathbf{x}, \mathbf{y})_{\ell^2(\mathbb{C}^n)} = (\mathbf{x}, \mathbf{y})_2 = \sum_{i=1}^{n} x_i \bar{y}_i.$$

Observe that

$$(\mathbf{x}, \mathbf{x})_2 = \sum_{i=1}^{n} x_i \bar{x}_i = \sum_{i=1}^{n} |x_i|^2 = \|\mathbf{x}\|_2^2 \in \mathbb{R}.$$

The object on the right is the square of the 2-norm ($p = 2$). Thus, the norm associated with our natural inner product is called the *Euclidean norm*.

With the concept of an inner product comes some notion of perpendicularity, or orthogonality.

Definition A.33 (orthogonality). Let $\mathbb{V}$ be a complex vector space with inner product $(\cdot, \cdot)$. We say that $x, y \in \mathbb{V}$ are **orthogonal**, and we write $x \perp y$, if and only if $(x, y) = 0$. Similarly, we say that the set $S \subseteq \mathbb{V}$ is **orthogonal** if and only if, whenever $x, y \in S$ and $x \neq y$, we have $x \perp y$. Finally, if $V, W \subseteq \mathbb{V}$, we say that V and W are **orthogonal** and write $V \perp W$ if and only if, for all $x \in V$ and $y \in W$, we have $x \perp y$.

Proposition A.34 (property of orthogonal sets). *Let $\mathbb{V}$ be a complex vector space with inner product $(\cdot, \cdot)$. Suppose that $S \subseteq \mathbb{V}$ is an orthogonal set. Then S is linearly independent.*

Proof. See Problem A.16. □

Definition A.35 (orthogonal complement). Let $W \subseteq \mathbb{V}$, where $\mathbb{V}$ is a complex inner product space. The **orthogonal complement** of W is the set

$$W^{\perp} = \{y \in \mathbb{V} \mid x \perp y \ \forall x \in W\}.$$

Proposition A.36 (property of $W^{\perp}$). *Let $\mathbb{V}$ be a complex vector space with inner product $(\cdot, \cdot)$. Suppose that $W \subseteq \mathbb{V}$ is nonempty. Then $W^{\perp} \leq \mathbb{V}$.*

Proof. See Problem A.17. □

A.5 Gram–Schmidt Orthogonalization Process

We now discuss a process by which, from any finite collection of elements of a vector space, we can construct an orthogonal set that spans the original one.

Definition A.37 (Gram–Schmidt process[7]). Let $\mathbb{V}$ be an n-dimensional complex inner product space with inner product $(\cdot, \cdot)$ and norm defined via (A.1). Suppose

[7] Named in honor of the Danish mathematician Jørgen Pedersen Gram (1850–1916) and the German mathematician Erhard Schmidt (1876–1959).

that $\{w_1, \ldots, w_k\} \subseteq \mathbb{V}_\star = \mathbb{V}\backslash\{0\}$, $k \le n$. The **Gram–Schmidt process** is an algorithm for generating the set of vectors $\{q_1, \ldots, q_k\}$ recursively as follows: for $m = 1$,

$$q_1 = \frac{1}{\|w_1\|} w_1.$$

For $m = 2, \ldots, k$, provided that $q_1, \ldots, q_{m-1}$ have already been successfully computed, define

$$r_m = w_m - \sum_{j=1}^{m-1} (w_m, q_j) q_j.$$

If $r_m = 0$, the algorithm terminates without completion; otherwise, the algorithm proceeds with

$$q_m = \frac{1}{\|r_m\|} r_m.$$

Theorem A.38 (properties of Gram–Schmidt). *Let $\mathbb{V}$ be an n-dimensional complex inner product space with inner product $(\cdot, \cdot)$ and norm defined via* (A.1). *Suppose that $\{w_1, \ldots, w_k\} \subseteq \mathbb{V}_\star$, with $k \le n$, is linearly independent. Then the Gram–Schmidt process proceeds to completion to produce an orthonormal set $\{q_1, \ldots, q_k\}$. Furthermore,*

$$\operatorname{span}(\{w_1, \ldots, w_m\}) = \operatorname{span}(\{q_1, \ldots, q_m\})$$

for each $m = 1, \ldots, k$.

Proof. See Problem A.19. □

Theorem A.39 (basis completion theorem). *Suppose that $\mathbb{V}$ is a finite-dimensional inner product space and $\mathbb{W}$ is a proper nontrivial vector subspace. Then there is an orthonormal basis, S, for $\mathbb{W}$. Moreover, any orthonormal basis for $\mathbb{W}$ can be extended to an orthonormal basis, T, for the whole space $\mathbb{V}$ and*

$$\#(T\backslash S) = \dim(\mathbb{V}) - \dim(\mathbb{W}).$$

Proof. See Problem A.20. □

Example A.14 The Gram–Schmidt process can be easily extended to infinite-dimensional inner product spaces. For example, we will see later that the object

$$(f, g)_{L^2(-1,1;\mathbb{C})} = \int_{-1}^{1} f(x)\overline{g(x)}\, dx$$

defines an inner product on the vector space $C([-1, 1]; \mathbb{C})$. In fact, the reader should be able to prove this now. Recall from Example A.5 that $S = \{1, x, x^2, x^3, \ldots\}$ is a linearly independent set in $C([-1, 1]; \mathbb{C})$. Applying the Gram–Schmidt process to S yields the so-called *Legendre polynomials*, which will be used in Chapter 11 to approximate functions and in Chapter 14 to approximate the value of definite integrals.

An algorithmic description of the Gram–Schmidt process, in the case that $\mathbb{V} = \mathbb{R}^n$, is presented in Listing A.1. We must warn the reader, however, *not to use this code.* This algorithm, known as the *classical* Gram–Schmidt process, is numerically unstable. We will see a better algorithm for implementing this process in Section 5.6.

Problems

A.1 Properties of a field. Show that, if $\mathbb{K}$ is a field:

a) The elements $0 \in \mathbb{K}$ and $1 \in \mathbb{K}$ are unique.
b) The additive inverse is unique.
c) The multiplicative inverse, if it exists, is unique.
d) $x0 = 0x = 0$.
e) $xy = 0$ implies that either $x = 0$ or $y = 0$.
f) $-x = (-1)x$.
g) $(-1)(-x) = x$.

A.2 Show that, for every $z \in \mathbb{C}$, $|z| = 0$ if and only if $z = 0$. Show that $|z|^2 = z\bar{z}$ and, therefore, if $z \neq 0$, then

$$z^{-1} = \frac{\bar{z}}{|z|^2}.$$

A.3 Let $z \in \mathbb{C}$. Show that

$$\Re(z) = \frac{1}{2}(z + \bar{z}) \quad \text{and} \quad \Im(z) = \frac{1}{2\mathrm{i}}(z - \bar{z}).$$

A.4 Show that any set of vectors that contains the zero vector is linearly dependent.

A.5 Prove Proposition A.16.

A.6 Prove Proposition A.20.

A.7 Suppose that $\mathbb{V}$ is a finite-dimensional vector space over the field $\mathbb{K}$ and B is a basis for $\mathbb{V}$. Show that basis representations are unique. In other words, if $B = \{v_1, \ldots, v_n\}$ and $x = \sum_{i=1}^{n} \alpha_i v_i$ but also $x = \sum_{i=1}^{n} \beta_i v_i$, then it follows that $\alpha_i = \beta_i \in \mathbb{K}$, $i = 1, \ldots, n$.

A.8 Prove Proposition A.21.

A.9 Prove Theorem A.22.

A.10 Prove Proposition A.26.

A.11 Complete the proof of Proposition A.27.

A.12 Show that all norms in $\mathbb{C}^n$ are equivalent.

Hint: Use Theorem A.29.

A.13 Show that, for every $\boldsymbol{x} \in \mathbb{C}^n$, the following inequalities hold:

$$\|\boldsymbol{x}\|_\infty \le \|\boldsymbol{x}\|_1 \le n\|\boldsymbol{x}\|_\infty,$$
$$\frac{1}{\sqrt{n}}\|\boldsymbol{x}\|_1 \le \|\boldsymbol{x}\|_2 \le \|\boldsymbol{x}\|_1,$$
$$\|\boldsymbol{x}\|_\infty \le \|\boldsymbol{x}\|_2 \le \sqrt{n}\|\boldsymbol{x}\|_\infty.$$

A.14 Complete the details of Example A.12.

A.15 Prove Corollary A.32.

A.16 Prove Proposition A.34.

A.17 Prove Proposition A.36.

A.18 Suppose that $S = \{\mathbf{w}_1, \ldots, \mathbf{w}_k\} \subseteq \mathbb{C}^n$ is an orthonormal set, i.e., $\mathbf{w}_i^{\mathrm{H}}\mathbf{w}_j = \delta_{i,j}$, $i, j = 1, \ldots, k$, with $k < n$. Let $W = \operatorname{span}(S)$ and $\mathbf{v} \in \mathbb{C}^n$ be arbitrary. Show that $\mathbf{v} = \mathbf{w} + \mathbf{p}$, where $\mathbf{p} \in W^{\perp}$ and $\mathbf{w} \in W$.

Hint: Suppose that $\mathbf{w} = \sum_{j=1}^{k} \left(\mathbf{w}_j^{\mathrm{H}}\mathbf{v}\right)\mathbf{w}_j$.

A.19 Prove Theorem A.38.

Hint: Use induction.

A.20 Prove Theorem A.39.

Hint: Use the Gram–Schmidt process.

Listings

```
function [Q, err] = ClassicalGramSchmidt( W )
% The classical Gram-Schmidt orthogonalization process.
%
% Input
%   W(1:n,1:k) : a matrix representing a collection of k column
%                vectors of dimension n
%
% Output
%   Q(1:n,1:k) : a collection of k orthonormal vectors of
%                dimension n
%   err : = 0, if the columns of W are linearly independent
%         = 1, if an error has occurred
%
% WARNING: DO NOT USE THIS CODE!!
% The classical Gram-Schmidt process is numerically unstable
  n = size(W)(1);
  k = size(W)(2);
  Q = zeros(n,k);
  err = 0;
  if k > n
    err = 1;
    return;
  end
  norm_q = norm( W(:,1) );
  if norm_q > eps( norm_q )
    Q(:,1) = W(:,1)/norm_q;
  else
    err = 1;
    return;
  end
  for m=2:k
    r = W(:,m);
    for j = 1:m-1
      r = r - (W(:,m)'*Q(:,j)).*Q(:,j);
    end
    norm_q = norm( r );
    if norm_q > eps( norm_q )
```

```
        Q(:,m) = r/norm_q;
    else
        err = 1;
        return;
    end
  end
end
```

Listing A.1 The classical Gram–Schmidt orthogonalization process.

Appendix B Basic Analysis Review

In this appendix, we review some fundamental results from the analysis of functions of one or more variables. Readers will find the references [6, 76] indispensable.

B.1 Sequences and Compactness in $\mathbb{C}^d$ and $\mathbb{R}^d$

Definition B.1 (sequences and convergence)**.** Let $d \in \mathbb{N}$. A **sequence** is a function $\boldsymbol{x}\colon \mathbb{N} \to \mathbb{C}^d$. We use the notation $\boldsymbol{x}_n = \boldsymbol{x}(n)$. When referring to the complete sequence, we might use different notations $\{\boldsymbol{x}_n\}_{n\in\mathbb{N}} = \{\boldsymbol{x}_n\}_{n=1}^{\infty} \subset \mathbb{C}^d$. We say that the sequence $\{\boldsymbol{x}_n\}_{n\in\mathbb{N}}$ **converges** if and only if there is a point $\boldsymbol{\xi} \in \mathbb{C}^d$ such that, for every $\varepsilon > 0$, there is a positive integer $N \in \mathbb{N}$ such that if n is any integer satisfying $n \geq N$, then

$$\|\boldsymbol{x}_n - \boldsymbol{\xi}\|_2 \leq \varepsilon.$$

For short, we write $\lim_{n\to\infty} \boldsymbol{x}_n = \boldsymbol{\xi}$, or, equivalently, $\boldsymbol{x}_n \to \boldsymbol{\xi}$, as $n \to \infty$ We call $\boldsymbol{\xi}$ the **limit** of the sequence. A sequence $\{\boldsymbol{x}_n\}_{n\in\mathbb{N}}$ is called **bounded** if and only if there is a positive real number M such that, for all $n \in \mathbb{N}$,

$$\|\boldsymbol{x}_n\|_2 \leq M.$$

A sequence $\{\boldsymbol{x}_n\}_{n\in\mathbb{N}}$ is called **Cauchy**[1] if and only if, for every ε, there is a positive integer $N \in \mathbb{N}$ such that if m and n are integers satisfying $m, n \geq N$, then

$$\|\boldsymbol{x}_m - \boldsymbol{x}_n\|_2 \leq \varepsilon.$$

A sequence $\{x_n\}_{n\in\mathbb{N}} \subset \mathbb{R}$ is said to be **increasing** (**decreasing**) if and only if $x_n \leq x_{n+1}$ ($x_n \geq x_{n+1}$) for all $n \in \mathbb{N}$. It is said to be **strictly increasing** (**strictly decreasing**) if and only if $x_n < x_{n+1}$ ($x_n > x_{n+1}$) for all $n \in \mathbb{N}$.

Definition B.2 (subsequence)**.** Let $d \in \mathbb{N}$. Suppose that $\{\boldsymbol{x}_n\}_{n\in\mathbb{N}} \subset \mathbb{C}^d$ is a given sequence. The sequence $\{\boldsymbol{y}_k\}_{k\in\mathbb{N}} \subset \mathbb{C}^d$ is called a **subsequence** of $\{\boldsymbol{x}_n\}_{n\in\mathbb{N}}$ if and only if there is a strictly increasing sequence (of indices) $\{n_k\}_{k\in\mathbb{N}} \subseteq \mathbb{N}$, denoted $n_k = n(k)$, such that $\boldsymbol{y}(k) = \boldsymbol{x}(n(k))$, or, equivalently,

$$\boldsymbol{y}_k = \boldsymbol{x}_{n_k}.$$

For subsequences, we write $\{\boldsymbol{y}_k\}_{k\in\mathbb{N}} = \{\boldsymbol{x}_{n_k}\}_{k\in\mathbb{N}} \subseteq \{\boldsymbol{x}_n\}_{n\in\mathbb{N}}$ for short.

[1] Named in honor of the French mathematician Augustin Louis Cauchy (1789–1857).

Example B.1 Suppose that $x_n = (-1)^n$. This sequence is bounded, but not convergent. Define the sequence of indices $n_k = 2k$, which is strictly increasing. The subsequence $y_k = x_{n_k}$ is convergent, though it is not that interesting:

$$y_k = x_{n_k} = x_{2k} = (-1)^{2k} = 1$$

for all $k \in \mathbb{N}$.

Let us present, mostly without proof, several properties and facts about sequences and their limits.

Theorem B.3 (properties of convergent sequences). *Let $d \in \mathbb{N}$. Suppose that $\{\boldsymbol{x}_n\}_{n\in\mathbb{N}} \subset \mathbb{C}^d$ is a convergent sequence. Then:*

1. *$\{\boldsymbol{x}_n\}_{n\in\mathbb{N}}$ is bounded.*
2. *$\{\boldsymbol{x}_n\}_{n\in\mathbb{N}}$ is Cauchy.*
3. *If, for $\boldsymbol{\xi}_1, \boldsymbol{\xi}_2 \in \mathbb{C}^d$,*

$$\lim_{n\to\infty} \boldsymbol{x}_n = \boldsymbol{\xi}_1, \qquad \lim_{n\to\infty} \boldsymbol{x}_n = \boldsymbol{\xi}_2,$$

then $\boldsymbol{\xi}_1 = \boldsymbol{\xi}_2$. In other words, limits are unique.

When we deal with sequences of real numbers, more refined tools can be invoked.

Definition B.4 (limit superior and inferior). Let $\{a_n\}_{n=1}^{\infty}$ be a bounded sequence of real numbers. Then

$$\liminf_{n\to\infty} a_n = \sup_{n\geq 1} \inf_{m\geq n} a_m$$

and

$$\limsup_{n\to\infty} a_n = \inf_{n\geq 1} \sup_{m\geq n} a_m.$$

The following result is well known.

Proposition B.5 (convergence criterion). *Let $\{a_n\}_{n=1}^{\infty}$ be a bounded sequence of real numbers. Then*

$$\liminf_{n\to\infty} a_n \leq \limsup_{n\to\infty} a_n.$$

Furthermore, $\{a_n\}_{n=1}^{\infty}$ is convergent if and only if

$$\liminf_{n\to\infty} a_n = \limsup_{n\to\infty} a_n.$$

Theorem B.6 (Squeeze Theorem). *Suppose that $\{a_n\}_{n=1}^{\infty} \subset \mathbb{R}$ is a convergent sequence and $\lim_{n\to\infty} a_n = \alpha$. Let $\{b_n\}_{n=1}^{\infty} \subset \mathbb{R}$ be another sequence with the property that*

$$b_n \in \begin{cases} [\alpha, a_n], & a_n \geq \alpha, \\ [a_n, \alpha], & a_n \leq \alpha. \end{cases}$$

Then $b_n \to \alpha$, as $n \to \infty$.

Theorem B.7 (Monotone Convergence Theorem). *Suppose that* $\{x_n\}_{n\in\mathbb{N}} \subset \mathbb{R}$ *is a bounded monotone sequence, i.e., either nondecreasing or nonincreasing. There is a point* $\xi \in \mathbb{R}$ *such that* $\lim_{n\to\infty} x_n = \xi$. *If the sequence is nondecreasing,*

$$\lim_{n\to\infty} x_n = \xi = \sup\{x_n \mid n \in \mathbb{N}\};$$

if nonincreasing,

$$\lim_{n\to\infty} x_n = \xi = \inf\{x_n \mid n \in \mathbb{N}\}.$$

It turns out that, in $\mathbb{C}^d$, a sequence is Cauchy if and only if it converges.

Theorem B.8 (Cauchy criterion). *Let* $d \in \mathbb{N}$. *Suppose that* $\{\boldsymbol{x}_n\}_{n\in\mathbb{N}} \in \mathbb{C}^d$ *is a Cauchy sequence. There is a point* $\boldsymbol{\xi} \in \mathbb{C}^d$ *such that* $\lim_{n\to\infty} \boldsymbol{x}_n = \boldsymbol{\xi}$.

Theorem B.9 (Bolzano–Weierstrass Theorem[2]). *Let* $d \in \mathbb{N}$. *Suppose that the sequence* $\{\boldsymbol{x}_n\}_{n\in\mathbb{N}} \in \mathbb{C}^d$ *is bounded. There is a subsequence* $\{\boldsymbol{x}_{n_k}\}_{k\in\mathbb{N}} \subseteq \{\boldsymbol{x}_n\}_{n\in\mathbb{N}}$ *and a point* $\boldsymbol{\xi} \in \mathbb{C}^d$ *such that* $\lim_{k\to\infty} \boldsymbol{x}_{n_k} = \boldsymbol{\xi}$.

In numerical analysis, it is important not only to know that a sequence converges, but many times we also wish to describe *how fast* the convergence is. The following definition quantifies this.

Definition B.10 (order of convergence). Let $d \in \mathbb{N}$. Suppose that the sequence $\{\boldsymbol{x}_k\}_{k=1}^{\infty} \subset \mathbb{C}^d$ converges to the point $\boldsymbol{\xi} \in \mathbb{C}^d$, i.e., $\boldsymbol{\xi} = \lim_{k\to\infty} \boldsymbol{x}_k$. We say that $\boldsymbol{x}_k$ **converges to $\boldsymbol{\xi}$ at least linearly** if and only if there exists a sequence of positive real numbers $\{\varepsilon_k\}_{k=1}^{\infty}$ that converges to 0 and a real number $\mu \in (0,1)$ such that

$$\|\boldsymbol{x}_k - \boldsymbol{\xi}\|_2 \le \varepsilon_k, \ \forall k \in \mathbb{N}, \qquad \lim_{k\to\infty} \frac{\varepsilon_{k+1}}{\varepsilon_k} = \mu. \tag{B.1}$$

If (B.1) holds with $\|\boldsymbol{x}_k - \boldsymbol{\xi}\|_2 = \varepsilon_k$, for $k = 1, 2, \ldots$, we say that $\boldsymbol{x}_k$ **converges to $\boldsymbol{\xi}$ linearly**, or **exactly linearly**.

We say that $\boldsymbol{x}_k$ **converges to $\boldsymbol{\xi}$ with at least order** q, $q > 1$, if and only if there exists a sequence of positive real numbers $\{\varepsilon_k\}_{k=1}^{\infty}$ that converges to 0 and a real number $\mu > 0$ such that

$$\|\boldsymbol{x}_k - \boldsymbol{\xi}\|_2 \le \varepsilon_k, \ \forall k \in \mathbb{N}, \qquad \lim_{k\to\infty} \frac{\varepsilon_{k+1}}{\varepsilon_k^q} = \mu. \tag{B.2}$$

If (B.2) holds with $\|\boldsymbol{x}_k - \boldsymbol{\xi}\|_2 = \varepsilon_k$, for $k = 1, 2, \ldots$, we say that $\boldsymbol{x}_k$ **converges to $\boldsymbol{\xi}$ with order** q, or with exactly order q. In particular, if $q = 2$, we say that $\boldsymbol{x}_k$ converges to $\boldsymbol{\xi}$ **quadratically**. If $q = 3$, we say that $\boldsymbol{x}_k$ converges to $\boldsymbol{\xi}$ **cubically**.

At this point, it is important to introduce the so-called *Big O notation*, which is one of the so-called Landau symbols.[3] The notation is used to describe the relative order of convergence of two quantities.

[2] Named in honor of the Bohemian mathematician, logician, philosopher, theologian, and Catholic priest of Italian extraction Bernardus Placidus Johann Nepomuk Bolzano (1781–1848) and the German mathematician Karl Theodor Wilhelm Weierstrass (1815–1897).

[3] Named in honor of the German mathematician Edmund Georg Hermann Landau (1877–1938).

Definition B.11 (Big O notation). Let $\{\boldsymbol{a}_k\}_{k\in\mathbb{N}}, \{\boldsymbol{b}_k\}_{k\in\mathbb{N}} \subset \mathbb{C}^d$. If there is a constant $M > 0$ and an integer $K \in \mathbb{N}$ such that, for all $k \geq K$, we have

$$\|\boldsymbol{a}_k\|_2 \leq M\|\boldsymbol{b}_k\|_2,$$

then we write $\boldsymbol{a}_k = \mathcal{O}(\boldsymbol{b}_k)$ and say $\boldsymbol{a}_k$ **is big O of** $\boldsymbol{b}_k$. We use similar notation for functions that depend on a continuous variable.

Example B.2 Suppose that $x_n = 1 - 2^{-n}$. This sequence converges to $\xi = 1$. Let us quantify how fast this convergence is. Observe that

$$|1 - x_n| = |1 - 1 + 2^{-n}| = 2^{-n} = \varepsilon_n.$$

Then

$$\frac{\varepsilon_{n+1}}{\varepsilon_n} = \frac{2^n}{2^{n+1}} = \frac{1}{2} = \mu \in (0, 1).$$

Thus, x_n converges to one exactly linearly. With this example, we see that geometric convergence is just linear convergence. We might say that a sequence x_n converges geometrically if and only if

$$x_n = \xi \pm q^n,$$

where $q \in (-1, 0) \cup (0, 1)$. Then $x_n \to \xi$ exactly linearly and $\mu = |q|$.

A very important notion in analysis is that of compactness, which we now describe.

Definition B.12 (compact). A set $K \subset \mathbb{C}^d$ is called **compact** if and only if, for every sequence $\{\boldsymbol{x}_n\}_{n\in\mathbb{N}}$ in K, there is a subsequence $\{\boldsymbol{y}_k\}_{k\in\mathbb{N}} = \{\boldsymbol{x}_{n_k}\}_{k\in\mathbb{N}} \subseteq \{\boldsymbol{x}_n\}_{n\in\mathbb{N}}$ and a point $\boldsymbol{y} \in K$ such that $\boldsymbol{y}_k \to \boldsymbol{y}$ as $k \to \infty$.

Definition B.13 (closed and bounded). A set $K \subset \mathbb{C}^d$ is called **closed** if and only if whenever $\{\boldsymbol{x}_n\}_{n\in\mathbb{N}} \subset K$ is such that $\boldsymbol{x}_n \to \boldsymbol{x}$ as $n \to \infty$, then it follows that $\boldsymbol{x} \in K$. K is called **bounded** if and only if there exists an $M > 0$ such that $\|\boldsymbol{x}\|_2 \leq M$ for all $\boldsymbol{x} \in K$.

The next theorem establishes a fundamental equivalence.

Theorem B.14 (Heine–Borel Theorem[4]). *A set $K \subset \mathbb{C}$ is compact if and only if it is closed and bounded.*

An immediate corollary of this result is a compactness criterion in $\mathbb{C}^d$.

Corollary B.15 (compactness in $\mathbb{C}^d$). *A set $K \subset \mathbb{C}^d$ is compact if and only if it is closed and bounded with respect to the ∞-norm.*

Proof. Argue component-wise. □

[4] Named in honor of the German mathematician Heinrich Eduard Heine (1821–1881) and the French mathematician and politician Félix Édouard Justin Émile Borel (1871–1956).

Definition B.16 (closure). Suppose that $K \subseteq \mathbb{C}^d$. $\mathbf{x} \in \mathbb{C}$ is called a **limit point of** K if and only if there is a convergent sequence $\{\mathbf{x}_n\}_{n\in\mathbb{N}}$ in K such that $\mathbf{x}_n \to \mathbf{x}$. The set $\overline{K}$ is called the **closure of** K, and is defined by

$$\overline{K} = K \cup \left\{\mathbf{x} \in \mathbb{C}^d \,\middle|\, \exists \{\mathbf{x}_n\}_{n\in\mathbb{N}} \subset K,\ \mathbf{x}_n \to \mathbf{x}\right\}.$$

B.2 Functions of a Single Real Variable

We now list some of the most important properties of functions of a single real variable.

Definition B.17 (interval). Let a and b be extended real numbers, i.e., $a, b \in [-\infty, \infty]$, with $a < b$. A set $I \subseteq \mathbb{R}$ is called an **interval** if and only if $I = (a, b)$, or $I = [a, b)$, or $I = (a, b]$, or $I = [a, b]$. The interval I is called **semi-infinite** if and only if exactly one of its endpoints is infinite.

Definition B.18 (continuity). Let $I \subseteq \mathbb{R}$ be an interval (finite, semi-infinite, or infinite) and $f \colon I \to \mathbb{R}$ be a function. We say that f is **continuous** at $x_0 \in I$ if and only if, for every $\varepsilon > 0$, there is a $\delta > 0$ such that if x is any point of I satisfying $|x - x_0| < \delta$, then it follows that

$$|f(x) - f(x_0)| < \varepsilon.$$

We say that f is **continuous on** I, and we write $f \in C(I)$ if and only if f is continuous at every point of the interval I. We say that f is **uniformly continuous on** I if and only if, for every ε, there exists a $\delta > 0$, only depending upon ε and I, such that if x_1 and x_2 are any two elements of I with $|x_1 - x_2| < \delta$, then it follows that

$$|f(x_1) - f(x_2)| < \varepsilon.$$

A sequential characterization of continuity is as follows.

Theorem B.19 (sequential continuity). *Let $I \subseteq \mathbb{R}$ be an interval (finite, semi-infinite, or infinite) and $f \colon I \to \mathbb{R}$ be a function. The function f is continuous at $x_0 \in I$ if and only if, for every sequence $\{x_n\}_{n\in\mathbb{N}} \subset I$ with $x_n \to x_0$, we have $f(x_n) \to f(x_0)$.*

It turns out that, on compact sets, continuity and uniform continuity are equivalent notions.

Theorem B.20 (uniform continuity). *Suppose that $I = [a, b]$ is a compact interval and $f \colon I \to \mathbb{R}$ is a function. Then if f is continuous on I, it is also uniformly continuous on I. For $I = (a, b)$, a bounded or unbounded open interval, there exist functions that are continuous on I but not uniformly continuous on I.*

Intuitively, continuity means that the function gives values that are "*close to each other*" provided that the arguments are sufficiently close. This closedness can be quantified, for instance, with the following definition.

Definition B.21 (Hölder continuity[5]). Suppose that $I \subseteq \mathbb{R}$ is an interval, $f: I \to \mathbb{R}$ is a function, and $\alpha \in (0, 1]$. We say that f is **Hölder continuous of order** α if and only if there is a constant $C > 0$ such that

$$|f(x) - f(y)| \leq C|x - y|^\alpha$$

for all $x, y \in I$. If $\alpha = 1$, we say that f is **Lipschitz continuous**.[6] For $\alpha \in (0, 1]$, the collection of all functions $f: I \to \mathbb{R}$ that are Hölder continuous of order α is denoted by $C^{0,\alpha}(I)$. Thus, the collection of Lipschitz continuous functions is denoted by $C^{0,1}(I)$.

While Hölder and Lipschitz continuity somewhat characterize the smoothness of a function, a stronger version of this is given by differentiability.

Definition B.22 (differentiability). Suppose that $I \subseteq \mathbb{R}$ is an interval and $f: I \to \mathbb{R}$ is a function. f is said to be **differentiable at** $x_0 \in I$ if and only if the limit

$$L = \lim_{h \to 0} \frac{f(x_0 + h) - f(x_0)}{h}$$

exists. In this case, we write $f'(x_0) = L$. We say that f is **differentiable on** I if and only if f is differentiable at every point in I.

Notice that, if f is differentiable on I, $f': I \to \mathbb{R}$ defines a function. We call this function the *derivative* of f. The properties of the derivative (continuity, etc.) imply several other properties for the function f itself. In addition, we can study the differentiability of f', giving rise to the *second derivative* f''. In an analogous manner, we can talk, for $m \in \mathbb{N}$, about the derivative $f^{(m)}$ of order m of a function, and study its properties.

Definition B.23 (continuously differentiable). Let $I \subseteq \mathbb{R}$ be an interval (finite, semi-infinite, or infinite) and $f: I \to \mathbb{R}$ be a function. Let $m \in \mathbb{N}$. We write $f \in C^m(I)$ if and only if $f \in C(I)$ and all of its derivatives up to and including the mth order derivative exist and are continuous at every point $x \in I$. We write $C^0(I) = C(I)$.

Example B.3 Let $f \in C(a, b)$. We say that $u: [a, b] \to \mathbb{R}$ is a classical solution to the problem

$$u''(x) = f(x), \quad x \in (a, b), \qquad u(a) = u(b) = 0 \tag{B.3}$$

if and only if $u \in C^2(a, b) \cap C([a, b])$ and u satisfies (B.3) pointwise.

---

[5] Named in honor of the German mathematician Otto Ludwig Hölder (1859–1937).
[6] Named in honor of the German mathematician Rudolf Otto Sigismund Lipschitz (1832–1903).

Definition B.24 (boundedness)**.** Let $I \subseteq \mathbb{R}$ be an interval (finite, semi-infinite, or infinite) and $f\colon I \to \mathbb{R}$ be a function. We say that f is **bounded in** I if and only if there is a real number $M \geq 0$ such that, for every $x \in I$,

$$|f(x)| \leq M.$$

We write $f \in C_b^m(I)$ if and only if $f \in C^m(I)$, and f and all of its derivatives up to and including the mth order derivative are bounded on I.

It is of importance to determine the values that a function in $C(I)$, where I is an interval, can take. We now explore this.

Definition B.25 (infimum and supremum of a function)**.** Suppose that $I \subseteq \mathbb{R}$ is an interval and $f : I \to \mathbb{R}$. Set $R = \{f(x) \mid x \in I\} \subset \mathbb{R}$. Then

$$\inf_{x\in I} f(x) = \inf R, \qquad \sup_{x\in I} f(x) = \sup R.$$

Definition B.26 (maximum and minimum of a function)**.** Suppose that $I \subseteq \mathbb{R}$ is an interval and $f\colon I \to \mathbb{R}$. We write

$$f(z) = \min_{x\in I} f(x), \qquad z \in \operatorname*{argmin}_{x\in I} f(x)$$

if and only if $z \in I$ and z satisfies

$$f(z) = \inf_{x\in I} f(x). \tag{B.4}$$

If $z \in I$ is the unique point satisfying (B.4), we write

$$z = \operatorname*{argmin}_{x\in I} f(x).$$

Similarly, we write

$$f(w) = \max_{x\in I} f(x), \qquad w \in \operatorname*{argmax}_{x\in I} f(x)$$

if and only if $w \in I$ and w satisfies

$$f(w) = \sup_{x\in I} f(x). \tag{B.5}$$

If $w \in I$ is the unique point satisfying (B.5), we write

$$w = \operatorname*{argmax}_{x\in I} f(x).$$

The intuition behind continuity is that the graph has "*no gaps*". The following result makes this rigorous.

Theorem B.27 (Intermediate Value Theorem)**.** *Suppose that* $[a, b] \subset \mathbb{R}$ *is compact and* $f \in C([a, b])$. *Set*

$$\alpha = \inf_{x\in[a,b]} f(x), \qquad \beta = \sup_{x\in[a,b]} f(x).$$

There exist $\gamma, \delta \in \mathbb{R}$ *satisfying*

$$-\infty < \gamma \leq f(x) \leq \delta < \infty, \quad \forall x \in [a, b],$$

which implies that

$$\gamma \le \alpha \le \beta \le \delta.$$

Furthermore, for any $y \in [\alpha, \beta]$, there is at least one point $z \in [a, b]$ such that $f(z) = y$. In particular, there are (not necessarily unique) points $x_\alpha, x_\beta \in [a, b]$ such that

$$f(x_\alpha) = \alpha = \min_{x \in [a,b]} f(x), \qquad f(x_\beta) = \beta = \max_{x \in [a,b]} f(x).$$

As a consequence, in compact sets, continuous functions must be bounded.

Corollary B.28 (equality). *Let $[a, b]$ be a compact set and $m \in \mathbb{N}_0$. Then*

$$C^m([a, b]) = C_b^m([a, b]).$$

Let us now investigate the possible values that the derivative of a function can take.

Theorem B.29 (Rolle's Theorem[7]). *Suppose that $-\infty < a < b < \infty$ and $f \in C([a, b])$. Assume that $f'(x)$ exists for all $x \in (a, b)$ and $f(a) = f(b)$. Then there exists at least one point $\xi \in (a, b)$ such that $f'(\xi) = 0$.*

Theorem B.30 (Mean Value Theorem). *Suppose that $-\infty < a < b < \infty$, $f \in C([a, b])$, and $f'(x)$ exists for all $x \in (a, b)$. Then there exists at least one point $\xi \in (a, b)$ such that*

$$f'(\xi) = \frac{f(b) - f(a)}{b - a}.$$

Proof. Define

$$g(x) = f(x) - \frac{x - a}{b - a}\left(f(b) - f(a)\right).$$

Then $g\colon [a, b] \to \mathbb{R}$ satisfies the hypotheses of Rolle's Theorem. In particular, $g(a) = f(a) = g(b)$. By Rolle's Theorem, there is a point $\xi \in (a, b)$ such that $g'(\xi) = 0$. But

$$g'(x) = f'(x) - \frac{f(b) - f(a)}{b - a}$$

for all $x \in (a, b)$. Thus, $f'(\xi) - \frac{f(b)-f(a)}{b-a} = 0$. □

A fundamental notion in numerical analysis, approximation theory, and many applications is that smoothness of a function is equivalent to the fact that this function can be approximated by simpler ones. One way of approximating is via polynomials, and the use of the so-called Taylor polynomial.

Theorem B.31 (Taylor's Theorem I[8]). *Suppose that $-\infty < a < b < \infty$ and n is a nonnegative integer. Assume that $f \in C^n([a, b])$ and $f^{(n+1)}(x)$ exists for all $x \in (a, b)$. For each $x \in (a, b]$, there exists a point $\xi = \xi(x) \in (a, x)$ such that*

$$f(x) = f(a) + f'(a)(x - a) + \cdots + \frac{1}{n!} f^{(n)}(a)(x - a)^n + R_n(x),$$

[7] Named in honor of the French mathematician Michel Rolle (1652–1719).
[8] Named in honor of the British mathematician Brook Taylor (1685–1731).

where the remainder term is given by

$$R_n(x) = \frac{1}{(n+1)!} f^{(n+1)}(\xi)(x-a)^{n+1}.$$

Theorem B.32 (Taylor's Theorem II). *Suppose that n is a nonnegative integer, $[a,b] \subset \mathbb{R}$ is compact, and $c \in [a,b]$. Assume that $f \in C^n([a,b])$ and $f^{(n+1)}(x)$ exists for all $x \in (a,b)$. For each $x \in [a,c) \cup (c,b]$, there exists a point $\xi = \xi(x,c)$, strictly between x and c — i.e., $\xi \in (c,x)$, if $x \in (c,b]$, and $\xi \in (x,c)$, if $x \in [a,c)$ — such that*

$$f(x) = f(c) + f'(c)(x-c) + \cdots + \frac{1}{n!} f^{(n)}(c)(x-c)^n + R_n(x),$$

where the remainder term is

$$R_n(x) = \frac{1}{(n+1)!} f^{(n+1)}(\xi)(x-c)^{n+1}.$$

Remark B.33 (generalization). Taylor's Theorem is an obvious generalization of the Mean Value Theorem B.30. The latter follows from the former upon setting $n = 0$.

Let us now recall some facts about the theory of the Riemann integral.

Definition B.34 (Riemann integrable[9]). Let $I \subseteq \mathbb{R}$ be a finite interval and $f\colon I \to \mathbb{R}$ be a function. We write $f \in \mathcal{R}(I)$ if and only if f is **Riemann integrable** on the interval I, i.e.,

$$0 \le \left| \int_I f(x) \mathrm{d}x \right| \le \int_I |f(x)| \,\mathrm{d}x \le \infty.$$

For $I = [a,b]$, we use the simpler notation $\mathcal{R}(a,b)$ in place of $\mathcal{R}([a,b])$, and similarly for $[a,b)$, etc.

We now study when a function is Riemann integrable, and some properties of the Riemann integral.

Theorem B.35 (integrability condition). *If $f \in C([a,b])$, then $f \in \mathcal{R}(a,b)$.*

The following is a particular case of the so-called Hölder inequality for integrals.

Theorem B.36 (Hölder). *Suppose that $f \in C([a,b])$ and $g \in \mathcal{R}(a,b)$. Then $fg \in \mathcal{R}(a,b)$ and*

$$\left| \int_a^b f(x)g(x)\mathrm{d}x \right| \le \max_{a \le x \le b} |f(x)| \int_a^b |g(x)| \,\mathrm{d}x.$$

The connection between the Riemann integral and the derivative is the content of the so-called *Fundamental Theorem of Calculus* (FToC).

[9] Named in honor of the German mathematician Georg Friedrich Bernhard Riemann (1826–1866).

Theorem B.37 (FToC). *Suppose that $f \in \mathcal{R}(a, b)$ and $F\colon [a, b] \to \mathbb{R}$ is a differentiable function satisfying $F'(x) = f(x)$ for all $x \in [a, b]$. Then*

$$\int_a^b f(x)\mathrm{d}x = F(b) - F(a).$$

The following result, and its generalization to several dimensions, is fundamental in the treatment of boundary value problems.

Theorem B.38 (integration by parts). *Suppose that $[a, b]$ is compact and $f, g \in C([a, b])$. Assume that f and g are differentiable at every point in (a, b) and $f', g' \in \mathcal{R}(a, b)$. Then $fg', f'g \in \mathcal{R}(a, b)$ and*

$$\int_a^b f(x)g'(x)\mathrm{d}x = f(x)g(x)\rvert_{x=a}^{x=b} - \int_a^b f'(x)g(x)\mathrm{d}x.$$

The following result is a consequence of the FToC and integration by parts, and it is known as Taylor's Theorem with integral remainder.

Theorem B.39 (Taylor's Theorem III). *Suppose that $-\infty < a < b < \infty$ and n is a nonnegative integer. Assume that $f \in C^n([a, b])$, $f^{(n+1)}$ exists, and $f^{(n+1)} \in \mathcal{R}(a, b)$. For each $x \in (a, b]$,*

$$f(x) = f(a) + f'(a)(x - a) + \cdots + \frac{1}{n!}f^{(n)}(a)(x - a)^n + R_n(x),$$

where the (integral) remainder is

$$R_n(x) = \frac{1}{n!}\int_a^x f^{(n+1)}(t)(x - t)^n\mathrm{d}t.$$

Proof. The proof is by induction on n.

For $n = 0$, we have, by the FToC,

$$\int_a^x f'(t)\mathrm{d}t = f(x) - f(a),$$

and the $n = 0$ case follows.

For the inductive hypothesis, we assume that the theorem is true for all $n \le k$: for each $x \in (a, b]$,

$$f(x) = f(a) + f'(a)(x - a) + \cdots + \frac{1}{k!}f^{(k)}(a)(x - a)^k + \frac{1}{k!}\int_a^x f^{(k+1)}(t)(x - t)^k\mathrm{d}t,$$

provided that $f \in C^k([a, b])$ and $f^{(k+1)}$ exists and is Riemann integrable on (a, b).

For the inductive step, we assume that $f \in C^{k+1}([a, b])$ and $f^{(k+2)}$ exists and is Riemann integrable on (a, b). Observe that $f \in C^{k+1}([a, b])$ implies that $f^{(k+1)} \in \mathcal{R}(a, b)$. Since the $n = k$ case is valid,

$$f(x) = f(a) + (x - a)f'(a) + \cdots + \frac{1}{k!}f^{(k)}(a)(x - a)^k + \frac{1}{k!}\int_a^x f^{(k+1)}(t)(x - t)^k\mathrm{d}t.$$

Using integration by parts

$$\frac{1}{k!}\int_a^x f^{(k+1)}(t)(x-t)^k \mathrm{d}t = \frac{1}{(k+1)!}f^{(k+1)}(a)(x-a)^{k+1} + \frac{1}{(k+1)!}\int_a^x f^{(k+2)}(t)(x-t)^{k+1}\mathrm{d}t.$$

Thus, the $n = k+1$ case follows, and the result follows by induction. □

The following two theorems are direct consequences of the Intermediate Value Theorem B.27.

Theorem B.40 (Summation Mean Value Theorem). *Suppose that $\{c_i\}_{i=1}^n \subset \mathbb{R}$ is a finite sequence of nonnegative real numbers, $[a,b] \subset \mathbb{R}$ is a compact set, $\{x_i\}_{i=1}^n \subset [a,b]$, and $f \in C([a,b])$. Then there is a point $\xi \in [a,b]$ such that*

$$\sum_{i=1}^n c_i f(x_i) = f(\xi)S, \qquad S = \sum_{i=1}^n c_i.$$

Thus, if $c_i = 1$, for all $i = 1,\ldots,n$, there is a point $\xi \in [a,b]$ such that

$$f(\xi) = \frac{1}{n}\sum_{i=1}^n f(x_i).$$

Proof. By the Intermediate Value Theorem B.27, there are points $x_\alpha, x_\beta \in [a,b]$ such that

$$f(x_\alpha) = \min_{a\le x\le b} f(x) = \alpha, \qquad f(x_\beta) = \max_{a\le x\le b} f(x) = \beta;$$

furthermore, for all $i \in \{1,\ldots,n\}$,

$$\alpha \le f(x_i) \le \beta.$$

Since $c_i \ge 0$, for all $i \in \{1,\ldots,n\}$,

$$\alpha c_i \le c_i f(x_i) \le \beta c_i.$$

Summing over i, we get

$$\alpha S \le \sum_{i=1}^n c_i f(x_i) \le \beta S.$$

The case $S = 0$ is trivial; otherwise, we have

$$\alpha \le \frac{1}{S}\sum_{i=1}^n c_i f(x_i) \le \beta.$$

By the Intermediate Value Theorem B.27, there is a point $\xi \in [a,b]$ such that

$$f(\xi) = \frac{1}{S}\sum_{i=1}^n c_i f(x_i),$$

and the result follows. □

Theorem B.41 (Integral Mean Value Theorem). *Suppose that $-\infty < a < b < \infty$, $f \in C([a,b])$, and $g \in \mathcal{R}(a,b)$. Furthermore, suppose that $g(x) \geq 0$ for all $x \in [a,b]$. Then there exists a point $\xi \in [a,b]$ such that*

$$\int_a^b f(x)g(x)\mathrm{d}x = f(\xi)\int_a^b g(x)\mathrm{d}x.$$

Thus, if $g(x) = 1$, for all $x \in [a,b]$, there exists a point $\xi \in [a,b]$ such that

$$f(\xi) = \frac{1}{b-a}\int_a^b f(x)\mathrm{d}x.$$

Proof. Since f is continuous on $[a,b]$, it is bounded, i.e.,

$$-\infty < \alpha = \min_{x\in[a,b]} f(x) \leq f(x) \leq \max_{x\in[a,b]} f(x) = \beta < \infty$$

for all $x \in [a,b]$. Since $g \geq 0$ on $[a,b]$, it follows that, for all $x \in [a,b]$,

$$\alpha g(x) \leq f(x)g(x) \leq \beta g(x).$$

This implies that

$$\alpha\int_a^b g(x)\mathrm{d}x \leq \int_a^b f(x)g(x)\mathrm{d}x \leq \beta\int_a^b g(x)\mathrm{d}x. \tag{B.6}$$

Since $g \geq 0$ on $[a,b]$ and $g \in \mathcal{R}(a,b)$,

$$0 \leq S = \int_a^b g(x)\mathrm{d}x < \infty.$$

If $S = 0$, then from (B.6) it follows that

$$\int_a^b f(x)g(x)\mathrm{d}x = 0,$$

and the result follows trivially; otherwise, (B.6) implies that

$$\alpha \leq y = \frac{1}{S}\int_a^b f(x)g(x)\mathrm{d}x \leq \beta.$$

By the Intermediate Value Theorem B.27, there is a point $\xi \in [a,b]$ such that $f(\xi) = y$, and the result follows. □

We finally mention that, by arguing component-wise, we can also consider integration of functions whose values lie in a finite-dimensional vector space. The following definition is an example of this approach.

Definition B.42 (Riemann integrable). Let $[a,b] \subset \mathbb{R}$ be a compact integral. We say that the function $f \in \mathcal{R}(a,b;\mathbb{C})$ if $\Re f, \Im f \in \mathcal{R}(a,b)$. For such a function, we define

$$\int_a^b f(x)\mathrm{d}x = \int_a^b \Re f(x)\mathrm{d}x + \mathrm{i}\int_a^b \Im f(x)\mathrm{d}x \in \mathbb{C}.$$

B.3 Functions of Several Variables

We now study functions of several variables.

Definition B.43 (p-ball)**.** Suppose that $d \in \mathbb{N}$ and $\boldsymbol{a} \in \mathbb{C}^d$. For any $r \geq 0$, define, for $1 \leq p \leq \infty$, the **p-ball of radius** r to be

$$B_p(\boldsymbol{a}, r) = \left\{ \boldsymbol{x} \in \mathbb{C}^d \mid \|\boldsymbol{x} - \boldsymbol{a}\|_p < r \right\}.$$

We set $B(\boldsymbol{a}, r) = B_2(\boldsymbol{a}, r)$. We use the notation $\overline{B}_p(\boldsymbol{a}, r) = \overline{B_p(\boldsymbol{a}, r)}$, where the overbar indicates closure with respect to the norm $\|\cdot\|_p$. As before, $\overline{B}(\boldsymbol{a}, r) = \overline{B}_2(\boldsymbol{a}, r)$.

Definition B.44 (convexity)**.** Suppose that $d \in \mathbb{N}$ and $\Omega \subseteq \mathbb{C}^d$. We say that Ω is **convex** if and only if, for all $\boldsymbol{x}, \boldsymbol{y} \in \Omega$ and all $t \in [0, 1]$, we have

$$t\boldsymbol{x} + (1 - t)\boldsymbol{y} \in \Omega.$$

Definition B.45 (domain)**.** Suppose that $d \in \mathbb{N}$ and $\Omega \subseteq \mathbb{C}^d$. We say that Ω is a **domain** if Ω is nonempty, open, and connected.

We define continuity in several variables as before.

Definition B.46 (continuity)**.** Suppose that $d \in \mathbb{N}$, $\Omega \subseteq \mathbb{C}^d$, and $f\colon \Omega \to \mathbb{R}$. We say that $f \in C(\Omega)$ if and only if f is continuous at every point $\boldsymbol{x} \in \Omega$.

As in the one-dimensional case, a continuous function takes all its intermediate values.

Theorem B.47 (Intermediate Value Theorem in $\mathbb{C}^d$)**.** *Suppose that $d \in \mathbb{N}$, $\Omega \subset \mathbb{C}^d$ is compact (closed and bounded), and $f \in C(\Omega; \mathbb{R})$. Set $R = \{f(\boldsymbol{x}) \mid \boldsymbol{x} \in \Omega\}$. Then there are points $\boldsymbol{x}_{\min}, \boldsymbol{x}_{\max} \in \Omega$ such that*

$$f(\boldsymbol{x}_{\min}) = \inf_{\boldsymbol{x} \in \Omega} f(\boldsymbol{x}) = \inf R, \qquad f(\boldsymbol{x}_{\max}) = \sup_{\boldsymbol{x} \in \Omega} f(\boldsymbol{x}) = \sup R,$$

and the extrema are finite,

$$-\infty < f(\boldsymbol{x}_{\min}) \leq f(\boldsymbol{x}_{\max}) < \infty.$$

If Ω is, in addition, convex and $y \in [f(\boldsymbol{x}_{\min}), f(\boldsymbol{x}_{\max})]$, there is at least one point $z \in \Omega$ such that $f(z) = y$.

Definition B.48 (multi-index)**.** Suppose that $d \in \mathbb{N}$. A point $\boldsymbol{\alpha} \in \mathbb{N}_0^d$ is called a **multi-index**. The **order of the multi-index** is denoted $|\boldsymbol{\alpha}|$ and is defined as

$$|\boldsymbol{\alpha}| = \|\boldsymbol{\alpha}\|_1 = \sum_{i=1}^{d} \alpha_i.$$

Let $\boldsymbol{\alpha}$ be a multi-index and $\boldsymbol{x} \in \mathbb{R}^d$. We define the notation

$$\boldsymbol{x}^{\boldsymbol{\alpha}} = x_1^{\alpha_1} x_2^{\alpha_2} \cdots x_d^{\alpha_d}, \qquad \boldsymbol{\alpha}! = \alpha_1! \alpha_2! \cdots \alpha_d!.$$

Multi-indices are useful in giving a shorthand notation for partial derivatives.

Definition B.49 (partial derivatives). Suppose that $d, m \in \mathbb{N}$, $\Omega \subseteq \mathbb{R}^d$, and $f\colon \Omega \to \mathbb{R}$. We say that $f \in C^m(\Omega)$ if and only if $f \in C(\Omega)$ and every partial derivative

$$D^{\boldsymbol{\alpha}} f(\boldsymbol{y}) = \frac{\partial^{|\boldsymbol{\alpha}|} f(\boldsymbol{y})}{\partial \boldsymbol{x}^{\boldsymbol{\alpha}}} = \frac{\partial^{|\boldsymbol{\alpha}|} f(\boldsymbol{y})}{\partial x_1^{\alpha_1} \partial x_2^{\alpha_2} \cdots \partial x_d^{\alpha_d}}$$

of order $|\boldsymbol{\alpha}| \le m$ exists and is continuous at all points $\boldsymbol{y} \in \Omega$.

The first partial derivatives can be gathered into a vector.

Definition B.50 (gradient). Suppose that $d \in \mathbb{N}$, $\Omega \subseteq \mathbb{R}^d$ is open, $\boldsymbol{y} \in \Omega$, and $f \in C(\Omega)$ is such that every partial derivative of order one exists at $\boldsymbol{y}$. The **gradient** of f at $\boldsymbol{y}$ is the vector $\nabla f(\boldsymbol{y}) \in \mathbb{R}^d$ with components

$$[\nabla f(\boldsymbol{y})]_i = \frac{\partial f(\boldsymbol{y})}{\partial x_i}.$$

The following result is a version of Taylor's Theorem for several variables.

Theorem B.51 (Taylor's Theorem I in $\mathbb{R}^d$). *Let $d \in \mathbb{N}$, $\boldsymbol{a} \in \mathbb{R}^d$, and $r > 0$. Suppose that $f \in C^2\left(\overline{B}(\boldsymbol{a}, r); \mathbb{R}\right)$ and $\boldsymbol{x} \in \overline{B}(\boldsymbol{a}, r)$ is fixed. Then there exists $\boldsymbol{\xi} = \boldsymbol{\xi}(\boldsymbol{x}, \boldsymbol{a}) \in B(\boldsymbol{a}, r)$ such that*

$$f(\boldsymbol{x}) = f(\boldsymbol{a}) + \nabla f(\boldsymbol{a})^{\intercal}(\boldsymbol{x} - \boldsymbol{a}) + \frac{1}{2}(\boldsymbol{x} - \boldsymbol{a})^{\intercal} \mathsf{H}_f(\boldsymbol{\xi})(\boldsymbol{x} - \boldsymbol{a}),$$

where $\mathsf{H}_f\colon \overline{B}(\boldsymbol{a}, r) \to \mathbb{R}^{d\times d}$ is the Hessian matrix[10] *of f, which has components*

$$[\mathsf{H}_f(\boldsymbol{y})]_{i,j} = \frac{\partial^2 f(\boldsymbol{y})}{\partial x_i \partial x_j}.$$

Writing $\boldsymbol{\eta} = \boldsymbol{x} - \boldsymbol{a}$, the result may be expressed as

$$f(\boldsymbol{x}) = f(\boldsymbol{a}) + \nabla f(\boldsymbol{a})^{\intercal}\boldsymbol{\eta} + \frac{1}{2}\boldsymbol{\eta}^{\intercal}\mathsf{H}_f(\boldsymbol{\xi})\boldsymbol{\eta}.$$

Proof. Define, for all $t \in [0, 1]$,

$$\phi(t) = f(\boldsymbol{a} + t\boldsymbol{\eta}),$$

and apply Taylor's Theorem B.31. The details are left to the reader as an exercise; see Problem B.12. □

Theorem B.52 (Taylor's Theorem II in $\mathbb{R}^d$). *Let $d, m \in \mathbb{N}$, $\boldsymbol{a} \in \mathbb{R}^d$, and $r > 0$. Suppose that $f \in C^{m+1}\left(\overline{B}(\boldsymbol{a}, r); \mathbb{R}\right)$ and $\boldsymbol{x} \in \overline{B}(\boldsymbol{a}, r)$ is fixed. Assume that, for all $\boldsymbol{x} \in \overline{B}(\boldsymbol{a}, r)$,*

$$|D^{\boldsymbol{\alpha}} f(\boldsymbol{x})| \le A$$

for every multi-index $|\boldsymbol{\alpha}| = m + 1$. Then, for every $\boldsymbol{\eta} \in \overline{B}(\boldsymbol{0}, r)$,

$$f(\boldsymbol{\eta} + \boldsymbol{a}) = f(\boldsymbol{a}) + \sum_{j=1}^{m} \sum_{|\boldsymbol{\alpha}|=j} \frac{1}{\boldsymbol{\alpha}!} D^{\boldsymbol{\alpha}} f(\boldsymbol{a})\boldsymbol{\eta}^{\boldsymbol{\alpha}} + R_m,$$

[10] Named in honor of the German mathematician Ludwig Otto Hesse (1811–1874).

where

$$|R_m| \le \frac{A}{(m+1)!} \|\boldsymbol{\eta}\|_1^{m+1}.$$

Proof. See Problem B.14 □

Remark B.53 (convexity). The previous two results continue to hold when the closed and open balls are replaced by closed and open convex sets, respectively.

The next theorem gives an alternate to Taylor's Theorem that is sufficient for many applications.

Theorem B.54 (Taylor's Theorem III in $\mathbb{R}^d$). *Let $d \in \mathbb{N}$ and $\Omega \subset \mathbb{R}^d$ be a bounded, open, convex set. Suppose that $f \in C^1(\Omega; \mathbb{R})$ with the additional regularity*

$$\|\nabla f(\boldsymbol{x}) - \nabla f(\boldsymbol{y})\|_2 \le \gamma \|\boldsymbol{x} - \boldsymbol{y}\|_2$$

for some $\gamma > 0$ and for all $\boldsymbol{x}, \boldsymbol{y} \in \Omega$. Then, for all $\boldsymbol{x}, \boldsymbol{y} \in \Omega$,

$$|f(\boldsymbol{x}) - f(\boldsymbol{y}) - \nabla f(\boldsymbol{y})^\intercal(\boldsymbol{x} - \boldsymbol{y})| \le \frac{\gamma}{2} \|\boldsymbol{x} - \boldsymbol{y}\|_2^2.$$

Proof. Define, for all $t \in [0, 1]$,

$$\phi(t) = f(\boldsymbol{y} + t(\boldsymbol{x} - \boldsymbol{y})).$$

By the Chain Rule,

$$\phi'(t) = \nabla f(\boldsymbol{y} + t(\boldsymbol{x} - \boldsymbol{y}))^\intercal (\boldsymbol{x} - \boldsymbol{y}).$$

Then, by the Cauchy–Schwarz inequality,

$$|\phi'(t) - \phi'(0)| \le \|\boldsymbol{x} - \boldsymbol{y}\|_2 \|\nabla f(\boldsymbol{y} + t(\boldsymbol{x} - \boldsymbol{y})) - \nabla f(\boldsymbol{y})\|_2 \le \gamma t \|\boldsymbol{x} - \boldsymbol{y}\|_2^2.$$

Define

$$\delta = f(\boldsymbol{x}) - f(\boldsymbol{y}) - \nabla f(\boldsymbol{y})^\intercal(\boldsymbol{x} - \boldsymbol{y}).$$

Then

$$\delta = \phi(1) - \phi(0) - \phi'(0) = \int_0^1 [\phi'(t) - \phi'(0)] \mathrm{d}t.$$

Therefore,

$$\begin{aligned} |\delta| = \left| \int_0^1 [\phi'(t) - \phi'(0)] \mathrm{d}t \right| \le \int_0^1 |\phi'(t) - \phi'(0)| \mathrm{d}t \le \int_0^1 \gamma t \|\boldsymbol{x} - \boldsymbol{y}\|_2^2 \mathrm{d}t \\ \le \frac{\gamma}{2} \|\boldsymbol{x} - \boldsymbol{y}\|_2^2, \end{aligned}$$

as we intended to show. □

The last result can be extended in a natural way for vector-valued functions. To state it, we need a few definitions.

Definition B.55 (Jacobian[11]). Let $m, n \in \mathbb{N}$ and $\Omega \subseteq \mathbb{R}^n$ be open. For $k \in \mathbb{N}_0$, we say that the function $\boldsymbol{f}\colon \Omega \to \mathbb{R}^m$ belongs to $C^k(\Omega; \mathbb{R}^m)$ if and only if, for each $i \in \{1, \ldots, m\}$, we have $f_i \in C^k(\Omega)$. For $\boldsymbol{f} \in C^1(\Omega; \mathbb{R}^m)$ and $\boldsymbol{y} \in \Omega$, the **Jacobian** of $\boldsymbol{f}$ at $\boldsymbol{y}$ is the matrix $\mathsf{J}_{\boldsymbol{f}}(\boldsymbol{y}) \in \mathbb{R}^{m\times n}$ with components

$$[\mathsf{J}_{\boldsymbol{f}}(\boldsymbol{y})]_{i,j} = \frac{\partial f_i(\boldsymbol{y})}{\partial x_j}, \quad i = 1, \ldots, m, \quad j = 1, \ldots, n.$$

Theorem B.56 (Taylor's Theorem IV in $\mathbb{R}^d$). *Let $d \in \mathbb{N}$, $\Omega \subset \mathbb{R}^d$ be a bounded, open, convex set. Assume that $\boldsymbol{f} \in C^1(\Omega; \mathbb{R}^d)$, with the additional regularity that there is $\gamma > 0$ such that, for all $\boldsymbol{x}, \boldsymbol{y} \in \Omega$,*

$$\|\mathsf{J}_{\boldsymbol{f}}(\boldsymbol{x}) - \mathsf{J}_{\boldsymbol{f}}(\boldsymbol{y})\|_2 \leq \gamma \|\boldsymbol{x} - \boldsymbol{y}\|_2 .$$

Then, for all $\boldsymbol{x}, \boldsymbol{y} \in \Omega$,

$$\|\boldsymbol{f}(\boldsymbol{x}) - \boldsymbol{f}(\boldsymbol{y}) - \mathsf{J}_{\boldsymbol{f}}(\boldsymbol{y})(\boldsymbol{x} - \boldsymbol{y})\|_2 \leq \frac{\gamma}{2} \|\boldsymbol{x} - \boldsymbol{y}\|_2^2 .$$

Proof. See Problem B.16. □

For vector-valued functions, a Mean Value Theorem like Theorem B.30 does not hold.

Example B.4 Let $f\colon [0, 2\pi] \to \mathbb{R}^2$ be given by

$$f(t) = \begin{bmatrix}\cos(t) & \sin(t)\end{bmatrix}, \qquad f'(t) = \begin{bmatrix}-\sin(t) & \cos(t)\end{bmatrix}.$$

Notice that $f(2\pi) = f(0)$, yet the derivative never vanishes.

The following result is a generalization of the Mean Value Theorem B.30.

Theorem B.57 (Mean Value Theorem). *Let $m, n \in \mathbb{N}$, $\Omega \subset \mathbb{R}^n$ be a bounded, open, convex set, and $\boldsymbol{a}, \boldsymbol{b} \in \Omega$. Let $\boldsymbol{f} \in C(\Omega; \mathbb{R}^m)$ be differentiable on the open segment with endpoints $\boldsymbol{a}$ and $\boldsymbol{b}$. Then we have*

$$\|\boldsymbol{f}(\boldsymbol{b}) - \boldsymbol{f}(\boldsymbol{a})\|_2 \leq \sup_{t\in[0,1]} \|\mathsf{J}_{\boldsymbol{f}}(t\boldsymbol{b} + (1-t)\boldsymbol{a})\|_2 \|\boldsymbol{b} - \boldsymbol{a}\|_2 .$$

If, in addition, $\boldsymbol{f} \in C^1(\Omega; \mathbb{R}^m)$, then

$$\boldsymbol{f}(\boldsymbol{b}) - \boldsymbol{f}(\boldsymbol{b}) = \int_0^1 \mathsf{J}_{\boldsymbol{f}}(t\boldsymbol{b} + (1-t)\boldsymbol{a})(\boldsymbol{b} - \boldsymbol{a})\mathrm{d}t.$$

We have already defined the gradient. There are other *differential operators*, i.e., expressions involving partial derivatives of a function, that are useful in applications.

[11] Named in honor of the German mathematician Carl Gustav Jacob Jacobi (1804–1851).

Definition B.58 (differential operators). Let $d \in \mathbb{N}$ and $\Omega \subset \mathbb{R}^d$ be a domain. For $v \in C^2(\Omega)$, the **Laplacian**[12] at the point $\boldsymbol{x} \in \Omega$ is

$$\Delta v(\boldsymbol{x}) = \sum_{i=1}^{d} \frac{\partial^2 v(\boldsymbol{x})}{\partial x_i^2} = \operatorname{tr} \mathsf{H}_v(\boldsymbol{x}) = \mathsf{H}_v(\boldsymbol{x}) : \mathsf{I}_d.$$

For $\boldsymbol{w} \in C^1(\Omega; \mathbb{R}^d)$, the **divergence** at the point $\boldsymbol{x} \in \Omega$ is

$$\nabla \cdot \boldsymbol{w}(\boldsymbol{x}) = \sum_{i=1}^{d} \frac{\partial w_i(\boldsymbol{x})}{\partial x_i}.$$

If $d = 2$, the **curl** at the point $\boldsymbol{x} \in \Omega$ is

$$\nabla \times \boldsymbol{w}(\boldsymbol{x}) = \frac{\partial w_2(\boldsymbol{x})}{\partial x_1} - \frac{\partial w_1(\boldsymbol{x})}{\partial x_2}.$$

Finally, if $d = 3$, the **curl** is

$$\begin{aligned}\nabla \times \boldsymbol{w}(\boldsymbol{x}) &= \det \begin{bmatrix} \boldsymbol{e}_1 & \boldsymbol{e}_2 & \boldsymbol{e}_3 \\ \frac{\partial}{\partial x_1} & \frac{\partial}{\partial x_2} & \frac{\partial}{\partial x_3} \\ w_1 & w_2 & w_3 \end{bmatrix} (\boldsymbol{x}) \\ &= \left(\frac{\partial w_3(\boldsymbol{x})}{\partial x_2} - \frac{\partial w_2(\boldsymbol{x})}{\partial x_3} \right) \boldsymbol{e}_1 + \left(\frac{\partial w_1(\boldsymbol{x})}{\partial x_3} - \frac{\partial w_3(\boldsymbol{x})}{\partial x_1} \right) \boldsymbol{e}_2 \\ &\quad + \left(\frac{\partial w_2(\boldsymbol{x})}{\partial x_1} - \frac{\partial w_1(\boldsymbol{x})}{\partial x_2} \right) \boldsymbol{e}_3.\end{aligned}$$

We need to mention a few words about integration of functions of several variables. However, a rigorous presentation will lead us too astray. For this reason, we will assume that the reader is familiar with the fact that, in $\mathbb{R}^d$ with $d > 1$, one cannot integrate over arbitrary subsets, and call those over which integration, in the Riemann sense, can be carried out *admissible*. Much as in the one-dimensional case, if $\Omega \subset \mathbb{R}^d$ is an admissible set, we denote by $f \in \mathcal{R}(\Omega; \mathbb{C})$ the fact that $f : \Omega \to \mathbb{C}$ is Riemann integrable over Ω.

Theorem B.59 (integral identities). *Let $d \in \mathbb{N}$ and $\Omega \subset \mathbb{R}^d$ be a bounded domain with sufficiently smooth boundary and exterior unit boundary normal $\boldsymbol{n} : \partial\Omega \to \mathbb{R}^d$. The following identities hold.*

1. *Integration by parts: If $u, v \in C^1(\bar{\Omega})$,*

$$\int_\Omega \frac{\partial u(\boldsymbol{x})}{\partial x_i} v(\boldsymbol{x}) \mathrm{d}\boldsymbol{x} = \int_{\partial\Omega} u(\boldsymbol{x}) v(\boldsymbol{x}) \boldsymbol{n}(\boldsymbol{x}) \cdot \boldsymbol{e}_i \, \mathrm{d}\boldsymbol{S}(\boldsymbol{x}) - \int_\Omega u(\boldsymbol{x}) \frac{\partial v(\boldsymbol{x})}{\partial x_i} \, \mathrm{d}\boldsymbol{x}.$$

2. *First Green's identity:*[13] *If $v \in C^1(\bar{\Omega})$ and $w \in C^2(\bar{\Omega})$,*

$$\int_\Omega \nabla v(\boldsymbol{x}) \cdot \nabla w(\boldsymbol{x}) \mathrm{d}\boldsymbol{x} = -\int_\Omega v(\boldsymbol{x}) \Delta w(\boldsymbol{x}) \mathrm{d}\boldsymbol{x} + \int_{\partial\Omega} v(\boldsymbol{x}) \frac{\partial w}{\partial n}(\boldsymbol{x}) \mathrm{d}\boldsymbol{S}(\boldsymbol{x}),$$

where $\frac{\partial w}{\partial n}(\boldsymbol{x}) = \boldsymbol{n}(\boldsymbol{x}) \cdot \nabla w(\boldsymbol{x})$.

[12] Named in honor of the French scholar and polymath Pierre-Simon, Marquis de Laplace (1749–1827).

[13] Named in honor of the British mathematical physicist George Green (1793–1841).

3. *Second Green's identity: If* $v, w \in C^2(\bar{\Omega})$,

$$\int_\Omega (v(x)\Delta w(x) - \Delta v(x) w(x))\,\mathrm{d}x = \int_{\partial\Omega} \left(v(x)\frac{\partial w}{\partial n}(x) - \frac{\partial v}{\partial n}(x) w(x) \right) \mathrm{d}S(x).$$

4. *If* $\boldsymbol{v} \in C^1(\bar{\Omega}; \mathbb{R}^d)$ *and* $w \in C^1(\bar{\Omega})$,

$$\int_\Omega \nabla \cdot \boldsymbol{v}(x) w(x) \mathrm{d}x = \int_{\partial\Omega} \boldsymbol{v}(x) \cdot \boldsymbol{n}(x) w(x) \mathrm{d}S(x) - \int_\Omega \boldsymbol{v}(x) \cdot \nabla w(x) \mathrm{d}x.$$

B.4 Sequences of Functions

Finally, we state some facts about sequences of functions, their limits, and the passage to the limit under the derivative and integral.

Definition B.60 (uniform convergence)**.** Suppose that $d \in \mathbb{N}$ and $A \subseteq \mathbb{R}^d$. Let $\{f_n\}_{n\in\mathbb{N}}$ be a sequence of functions $f_n \colon A \to \mathbb{C}$. We say that $\{f_n\}_{n\in\mathbb{N}}$ **converges uniformly** to a function $f \colon A \to \mathbb{C}$ if and only if, for every $\varepsilon > 0$, there is an $N \in \mathbb{N}$ such that if n is any integer satisfying $n \geq N$, then

$$\sup_{x\in A} |f_n(x) - f(x)| \leq \varepsilon.$$

We say that $\{f_n\}_{n\in\mathbb{N}}$ is **uniformly Cauchy** if and only if, given any $\varepsilon > 0$, there is an $N \in \mathbb{N}$ such that if n and m are any two integers satisfying $n, m \geq N$, then

$$\sup_{x\in A} |f_n(x) - f_m(x)| \leq \varepsilon.$$

As in the case of vectors in $\mathbb{C}^d$, sequences of functions converge if and only if they are Cauchy.

Theorem B.61 (uniform Cauchy criterion)**.** *Suppose that* $d \in \mathbb{N}$, $A \subseteq \mathbb{R}^d$, *and* $\{f_n\}_{n\in\mathbb{N}}$ *is a sequence of functions* $f_n \colon A \to \mathbb{C}$. $\{f_n\}_{n\in\mathbb{N}}$ *converges uniformly to a function* $f \colon A \to \mathbb{C}$ *if and only if* $\{f_n\}$ *is uniformly Cauchy. The limit function* f *is unique.*

Theorem B.62 (uniform limit of continuous functions)**.** *Suppose that* $d \in \mathbb{N}$ *and* $A \subseteq \mathbb{R}^d$ *is compact. Let* $\{f_n\}_{n\in\mathbb{N}} \subset C(A; \mathbb{C})$. *If* $\{f_n\}_{n\in\mathbb{N}}$ *converges uniformly to a function* $f \colon A \to \mathbb{C}$, *then* $f \in C(A; \mathbb{C})$.

Theorem B.63 (Weierstrass M-Test[14])**.** *Suppose that* $d \in \mathbb{N}$ *and* $A \subseteq \mathbb{R}^d$ *is compact. Let* $\{f_n\}_{n\in\mathbb{N}} \subset C(A; \mathbb{C})$ *be such that*

$$|f_n(\boldsymbol{x})| \leq M_n, \quad \forall \boldsymbol{x} \in A.$$

If $\sum_{n=1}^\infty M_n < \infty$, *then the series* $\sum_{n=1}^\infty f_n(x)$ *converges uniformly and absolutely on* A. *In other words, the sequence of partial sums*

$$S_m(x) = \sum_{n=1}^m f_n(x), \qquad T_m(x) = \sum_{n=1}^m |f_n(x)|$$

[14] Named in honor of the German mathematician Karl Theodor Wilhelm Weierstrass (1815–1897).

both converge uniformly on A to functions $S \in C(A; \mathbb{C})$ and $T \in C(A)$, respectively.

Theorem B.64 (uniform limits and derivatives). *Suppose that $[a, b] \subseteq \mathbb{R}$ is compact and $\{f_n\}_{n\in\mathbb{N}} \subset C^1([a, b]; \mathbb{C})$. Assume that, for some $x_0 \in [a, b]$, $f_n(x_0)$ converges and $\{f_n'\}_{n\in\mathbb{N}}$ converges uniformly to some function $g\colon [a, b] \to \mathbb{C}$. Then $\{f_n\}_{n\in\mathbb{N}}$ converges uniformly to a differentiable function $f\colon [a, b] \to \mathbb{C}$ and $f' = g$.*

Finally, we discuss uniform limits and integrals.

Theorem B.65 (uniform limits and integrals). *Let $d \in \mathbb{N}$ and $\Omega \subset \mathbb{R}^d$ be bounded and admissible. Assume that $\{f_n\}_{n\in\mathbb{N}} \subset \mathcal{R}(\Omega; \mathbb{C})$ converges uniformly with limit $f\colon \Omega \to \mathbb{C}$. Then $f \in \mathcal{R}(\Omega; \mathbb{C})$ and its integral can be computed as*

$$\lim_{n\to\infty} \int_\Omega f_n(\boldsymbol{x}) \mathrm{d}\boldsymbol{x} = \int_\Omega f(\boldsymbol{x}) \mathrm{d}\boldsymbol{x}.$$

Problems

B.1 Exhibit a sequence that converges exactly quadratically ($q = 2$).
B.2 Exhibit a sequence that converges exactly cubically ($q = 3$).
B.3 Let $f\colon (0, 1) \to \mathbb{R}$ be defined by $f(x) = 1/x$. Show that f is continuous on $(0, 1)$ but not uniformly continuous on $(0, 1)$.
B.4 Show that if $A \subseteq \mathbb{R}$ is not compact, then $C_b(A) \subsetneq C(A)$.
Hint: Consider the previous problem.
B.5 Let $\alpha \in (0, 1)$. Let the function $f_\alpha\colon [-1, 1] \to \mathbb{R}$ be defined by

$$f_\alpha(x) = |x|^\alpha.$$

Show that, for every $\beta \in (0, \alpha]$, we have $f_\alpha \in C^{0,\beta}([-1, 1])$ and that there is no $\varepsilon > 0$ for which $f_\alpha \in C^{0,\alpha+\varepsilon}([-1, 1])$.
B.6 Show that the function $f\colon [-1, 1] \to \mathbb{R}$ defined by $f(x) = |x|$ is such that $f \in C^{0,1}([-1, 1]) \backslash C^1([-1, 1])$.
B.7 Let $f\colon [-1, 1] \to \mathbb{R}$ be defined by

$$f(x) = x \ln |x|.$$

Show that, for every $\alpha \in (0, 1)$, we have $f \in C^{0,\alpha}([-1, 1])$; yet $f \notin C^{0,1}([-1, 1])$.
B.8 Exhibit a function $f \in C^0([-1, 1]) \cap C^1((-1, 1))$, but $f \notin C^1([-1, 1])$.
B.9 Exhibit a function $f \in C^1([a, b]) \cap C^2((a, b))$ such that $f \notin C^2([a, b])$.
B.10 Suppose that $f \in C([a, b])$ and there is a point $c \in (a, b)$ such that $f(c) \neq 0$. Prove that there is a $\delta > 0$ such that $|f(x)| > 0$ for all $x \in (c - \delta, c + \delta)$.
Hint: For this problem, you need to use the definition of continuity.
B.11 Suppose that $f \in C^2([a, b])$. If $f''(x) > 0$, for all $x \in [a, b]$, and there exists a point $c \in (a, b)$ such that $f'(c) = 0$, show that f has a minimum in the interval $[a, b]$ at the point c.
B.12 Prove Theorem B.51.
B.13 Suppose that $f \in C^2(\overline{B}(\boldsymbol{a}, r))$, $\nabla f(\boldsymbol{a}) = \boldsymbol{0}$, and $H_f(\xi)$, the Hessian matrix of f evaluated at $\boldsymbol{\xi}$, is positive definite for all $\boldsymbol{\xi} \in B(\boldsymbol{a}, r)$. Prove that f has a minimum in the closed ball $\overline{B}(\boldsymbol{a}, r)$ at the point $\boldsymbol{a}$.
B.14 Prove Theorem B.52.

B.15 Is it possible to use the Integral Mean Value Theorem B.41 to prove that the remainder terms in the Taylor Theorems B.39 and B.31 are equivalent? If so, give a proof of the equivalence and carefully state your assumptions.

B.16 Prove Theorem B.56.

Appendix C Banach Fixed Point Theorem

In this supplement, we present the well-known Banach contraction mapping principle, both for Banach spaces as well as for complete metric spaces.

C.1 Contractions and Fixed Points in Banach Spaces

Let us first give a very general convergence theory based on the contraction mapping theorem. This section requires some familiarity with Cauchy sequences.

Definition C.1 (contraction). Let X be a normed space with norm $\|\cdot\|_X$. The map $T: X \to X$ is called a **contraction** if and only if there is a real number $q \in (0,1)$, called the **contraction parameter**, such that

$$\|T(x) - T(y)\|_X \le q\|x - y\|_X, \quad \forall x, y \in X.$$

Definition C.2 (fixed point). Let $T: X \to X$. The element $x \in X$ is called a **fixed point** of T if and only if

$$T(x) = x.$$

As is well known, all finite-dimensional normed spaces are complete, i.e., every Cauchy sequence converges. This is not the case in infinite dimensions. As, for instance, Example 16.3 shows, there are infinite-dimensional normed spaces that are not complete. The norm in this case matters.

Definition C.3 (Banach space[1]). A normed space $(X, \|\cdot\|_X)$ that is complete is called **Banach space**.

The reader will see several examples of Banach spaces throughout the text. Their usefulness is particularly evident in Parts II, IV, and V.

We have now arrived at the well-known Banach Contraction Mapping Theorem.

Theorem C.4 (Banach Fixed Point Theorem). *Let X be a complete normed linear space (a Banach space) and $T: X \to X$ a contraction, with contraction parameter $q \in (0,1)$. Then T has a unique fixed point $\bar{x} \in X$. Moreover, if $x_0 \in X$ is any point, then the sequence $\{x_k\}_{k=0}^{\infty} \subset X$ generated by the iteration scheme*

$$x_{k+1} = T(x_k), \quad k \ge 0 \tag{C.1}$$

[1] Named in honor of the Polish mathematician Stefan Banach (1892–1945).

converges to $\bar{x}$*. In addition, we have the following error estimates:*

$$\|x_n - \bar{x}\|_X \leq \frac{q^n}{1-q} \|x_0 - x_1\|_X , \tag{C.2}$$

$$\|x_n - \bar{x}\|_X \leq \frac{q}{1-q} \|x_{n-1} - x_n\|_X , \tag{C.3}$$

$$\|x_n - \bar{x}\|_X \leq q \|x_{n-1} - \bar{x}\|_X . \tag{C.4}$$

Proof. We first observe that, for any x_0, since $T\colon X \to X$, the sequence generated by (C.1) is well defined. In addition, owing to the fact that T is a contraction,

$$\|x_{n+1} - x_n\|_X = \|T(x_n) - T(x_{n-1})\|_X \leq q \|x_n - x_{n-1}\|_X \leq \cdots \leq q^n \|x_1 - x_0\|_X .$$

Thus, for any $m > n \geq 1$, we obtain that

$$\begin{aligned} \|x_m - x_n\|_X &\leq \sum_{j=0}^{m-n-1} \|x_{n+j+1} - x_{n+j}\|_X \leq \sum_{j=0}^{m-n-1} q^{n+j} \|x_1 - x_0\|_X \\ &\leq \frac{q^n}{1-q} \|x_1 - x_0\|_X , \end{aligned}$$

where, since $q \in (0,1)$, we used the convergence of the geometric series. This proves that $\{x_k\}_{k=0}^{\infty}$ is a Cauchy sequence in X. Since X is a Banach space, there is a unique limit point $\bar{x} \in X$. Since T is a contraction, it is also continuous; therefore,

$$\bar{x} = \lim_{n\to\infty} x_{n+1} = \lim_{n\to\infty} T(x_n) = T\left(\lim_{n\to\infty} x_n\right) = T(\bar{x}).$$

This shows that $\bar{x}$ is a fixed point of T; we have proved existence.

To prove the uniqueness of the fixed point, suppose that $\bar{x}_j$, $j = 1, 2$, are fixed points of T. Then

$$\|\bar{x}_1 - \bar{x}_2\|_X \leq q \|\bar{x}_1 - \bar{x}_2\|_X ,$$

which implies that

$$(1-q) \|\bar{x}_1 - \bar{x}_2\|_X \leq 0.$$

The only possibility is that $\|\bar{x}_1 - \bar{x}_2\|_X = 0$, which implies that $\bar{x}_1 = \bar{x}_2$. This shows uniqueness.

Now we prove the estimates. To this end, fix $n \geq 1$ and suppose that $m > n$. Then

$$\|x_n - \bar{x}\|_X \leq \|x_n - x_m\|_X + \|x_m - \bar{x}\|_X \leq \frac{q^n}{1-q} \|x_1 - x_0\|_X + \|x_m - \bar{x}\|_X .$$

Since $x_m \to \bar{x}$, for any $\varepsilon > 0$, there is an $M \in \mathbb{N}$ such that, if $m \geq M$, then

$$\|x_m - \bar{x}\|_X \leq \varepsilon.$$

Thus, for every fixed n, there is an $m \geq \max(n+1, M)$ such that

$$\|x_n - \bar{x}\|_X \leq \frac{q^n}{1-q} \|x_1 - x_0\|_X + \varepsilon.$$

Since $\varepsilon > 0$ is arbitrary, for every fixed n,

$$\|x_n - \bar{x}\|_X \le \frac{q^n}{1-q} \|x_1 - x_0\|_X ,$$

which proves (C.2). To get (C.4), observe that

$$\|x_n - \bar{x}\|_X = \|T(x_n) - T(\bar{x})\|_X \le q \|x_{n-1} - \bar{x}\|_X .$$

Finally, using (C.4), we have

$$\|x_{n-1} - \bar{x}\|_X \le \|x_{n-1} - x_n\|_X + \|x_n - \bar{x}\|_X \le \|x_{n-1} - x_n\|_X + q \|x_{n-1} - \bar{x}\|_X ,$$

which implies that

$$(1-q) \|x_{n-1} - \bar{x}\|_X \le \|x_{n-1} - x_n\|_X \le q \|x_{n-2} - x_{n-1}\|_X .$$

Thus,

$$\|x_{n-1} - \bar{x}\|_X \le \frac{q}{1-q} \|x_{n-2} - x_{n-1}\|_X .$$

Shifting the index, we have (C.4):

$$\|x_n - \bar{x}\|_X \le \frac{q}{1-q} \|x_n - x_{n-1}\|_X .$$

All the statements have been proved. □

C.2 The Contraction Mapping Principle in Metric Spaces

Let us slightly generalize the contraction mapping principle to the case of metric spaces, as this is relevant in one application. We begin by realizing that the notions of convergence, continuity, and many others that we introduced for normed spaces can be naturally generalized. The fact that the space was a vector space was not relevant, only that *distances* between elements of the underlying set can be defined. For this reason, we introduce the following notion.

Definition C.5 (metric space)**.** Let X be a set and $\mathfrak{d} \colon X \times X \to \mathbb{R}$. We say that the pair $(X, \mathfrak{d})$ is a **metric space** if:

1. For every $x, y \in X$, $\mathfrak{d}(x, y) = 0$ if and only if $x = y$.
2. Symmetry: For all $x, y \in X$, we have that $\mathfrak{d}(x, y) = \mathfrak{d}(y, x)$.
3. Triangle inequality: For all $x, y, z \in X$, we have $\mathfrak{d}(x, z) \le \mathfrak{d}(x, y) + \mathfrak{d}(y, z)$.

In this case, the function $\mathfrak{d}$ is called a **metric** or **distance**.

The reader has already seen many examples of metric spaces.

Example C.1 Let $I \subset \mathbb{R}$ be an interval. This is a metric space with

$$\mathfrak{d}(x, y) = |x - y|.$$

Example C.2 Any normed space $(X, \|\cdot\|_X)$ is a metric space with

$$\mathfrak{d}(x, y) = \|x - y\|_X .$$

In fact, any nonempty subset of X is also a metric space with the same distance.

Example C.3 Let $\mathbb{S} \subset \mathbb{C}$ be the unit circle, i.e.,

$$\mathbb{S} = \{z \in \mathbb{C} | |z| = 1\} .$$

While this is a metric space with the metric induced from $\mathbb{C}$, there is another, perhaps more natural, distance. This is the so-called *arcwise* or *intrinsic* distance. Namely, since for every $z \in \mathbb{S}$ we have

$$z = e^{i\theta}, \qquad \theta \in [0, 2\pi),$$

we define

$$\mathfrak{d}_{\mathbb{S}}(z_1, z_2) = \mathfrak{d}_{\mathbb{S}}(e^{i\theta_1}, e^{i\theta_2}) = |\theta_1 - \theta_2|.$$

We leave it to the reader to verify that this is indeed a distance.

All the notions of convergence and continuity can be extended to metric spaces. We only mention those that are relevant for our discussion here.

Definition C.6 (convergence)**.** Let $(X, \mathfrak{d})$ be a metric space. A **sequence** is a mapping $\mathbb{N} \to X$, which we usually denote by $\{x_n\}_{n\in\mathbb{N}} \subset X$. We say that a sequence $\{x_n\}_{n\in\mathbb{N}} \subset X$ **converges** to $x \in X$ if, for every $\varepsilon > 0$, there is $N \in \mathbb{N}$ such that, whenever $n \geq N$,

$$\mathfrak{d}(x_n, x) < \varepsilon.$$

Let $\{x_n\}_{n\in\mathbb{N}} \subset X$ be a sequence. We say that it is **Cauchy** if, for every $\varepsilon > 0$, there is $N \in \mathbb{N}$ such that, whenever $m, n \geq N$, we have

$$\mathfrak{d}(x_n, x_m) < \varepsilon.$$

Proposition C.7 (convergent $\implies$ Cauchy)**.** *Let $(X, \mathfrak{d})$ be a metric space. If the sequence $\{x_n\}_{n\in\mathbb{N}} \subset X$ converges, then it is Cauchy.*

Proof. See Problem C.6. □

The converse is not always true and so we introduce the following.

Definition C.8 (complete space)**.** Let $(X, \mathfrak{d})$ be a metric space. We say that it is **complete** if every Cauchy sequence converges.

Example C.4 Every Banach space is a complete metric space.

Example C.5 The set $(0,1] \subset \mathbb{R}$ with the metric induced from $\mathbb{R}$ is a metric space, but it is not complete. Indeed, for every $n \in \mathbb{N}$, we have that $\frac{1}{n} \in (0,1]$ and the sequence $\{\frac{1}{n}\}_{n\in\mathbb{N}}$ is Cauchy (as it converges in $\mathbb{R}$). However, the limit (which is zero) does not lie in $(0,1]$ and so, as a metric space, this is not complete.

Example C.6 Let $I \subset \mathbb{R}$ be a compact interval and $B \subset \mathbb{R}^d$. Recall that, by

$$C(I; B),$$

we denote the subset of $C(I;\mathbb{R}^d)$ of functions whose values belong to B. If B is compact, then the pair

$$\left(C(I;B), \|\cdot\|_{L^\infty(I;\mathbb{R}^d)}\right)$$

is a complete metric space.

We can also define continuity between metric spaces.

Definition C.9 (continuity). Let $(X, \mathfrak{d}_X)$ and $(Y, \mathfrak{d}_Y)$ be metric spaces. We say that the mapping $T\colon X \to Y$ is **continuous** at $x_0 \in X$ if, for every $\varepsilon > 0$, there is $\delta > 0$ such that, whenever

$$\mathfrak{d}_X(x, x_0) < \delta,$$

we necessarily have

$$\mathfrak{d}_Y(T(x), T(x_0)) < \varepsilon.$$

Let us now extend the notions of the previous sections to the setting of a metric space.

Definition C.10 (contraction). Let $(X, \mathfrak{d})$ be a metric space. We say that the mapping $T\colon X \to X$ is a **contraction** if and only if there is $q \in (0,1)$ such that

$$\mathfrak{d}\left(T(x), T(y)\right) \le q\mathfrak{d}\left(x, y\right), \quad \forall x, y \in X.$$

The number q is usually referred to as the **contraction parameter** of T.

We can now state the contraction mapping principle for metric spaces.

Theorem C.11 (contraction mapping). *Let $(X, \mathfrak{d})$ be a complete metric space and $T\colon X \to X$ a contraction, with contraction parameter $q \in (0,1)$. Then T has a unique fixed point $\bar{x} \in X$. Moreover, if $x_0 \in X$ is any point, then the sequence $\{x_k\}_{k=0}^{\infty} \subset X$ generated by the iteration scheme*

$$x_{k+1} = T(x_k), \quad k \ge 0 \tag{C.5}$$

converges to $\bar{x}$. In addition, we have the following error estimates:

$$\mathfrak{d}(x_n, \bar{x}) \le \frac{q^n}{1-q}\mathfrak{d}(x_0, x_1),$$
$$\mathfrak{d}(x_n, \bar{x}) \le \frac{q}{1-q}\mathfrak{d}(x_{n-1}, x_n),$$
$$\mathfrak{d}(x_n, \bar{x}) \le q\mathfrak{d}(x_{n-1}, \bar{x}).$$

Proof. The proof is a mere exercise in adapting the notation used in Theorem C.4 to this setting; see Problem C.8. □

Problems

C.1 Let X be a finite-dimensional vector space. Assume that the mapping $f: X \to X$ is such that there is an $n \in \mathbb{N}$ for which f^n is a contraction. Show that f has a unique fixed point. Notice that $f^n(x) = \underbrace{f(f(\dots f}_{n \text{ times}}(x)))$.

C.2 Let X be a finite-dimensional vector space and $f_1, f_2: X \to X$ two mappings that *commute*, i.e., $f_1(f_2(x)) = f_2(f_1(x))$ for all $x \in X$. Show that if f_2 has a unique fixed point x_0, then x_0 is a fixed point of f_1.

C.3 Let X be a finite-dimensional vector space and $f: X \to X$ a contraction. Show that, for every $y \in X$, the equation

$$x = f(x) + y$$

has a unique solution. Denote this solution by $x(y)$. Show that $x(y)$ is a continuous function of y.

C.4 The purpose of this problem is to show another proof of the Lax–Milgram Lemma (see Theorem 23.10). Let us recall the setting: $\mathcal{H}$ is a Hilbert space; $\mathcal{A}$ is a bounded and coercive bilinear form on $\mathcal{H}$; and $F \in \mathcal{H}'$. We seek $u \in \mathcal{H}$ such that

$$\mathcal{A}(u, v) = F(v), \quad \forall v \in \mathcal{H}. \tag{C.6}$$

a) Show that, if $x \in \mathcal{H}$ is such that

$$(x, y)_{\mathcal{H}} = 0, \quad \forall y \in \mathcal{H},$$

then $x = 0$.

b) Let $\theta > 0$. Show that problem (C.6) is equivalent to finding $u \in \mathcal{H}$ such that

$$(u, v)_{\mathcal{H}} = (u, v)_{\mathcal{H}} - \theta\left[\mathcal{A}(u, v) - F(v)\right], \quad \forall v \in \mathcal{H}.$$

c) Define $P_\theta: \mathcal{H} \to \mathcal{H}$ via the relation $w = P_\theta(u)$ if

$$(w, v)_{\mathcal{H}} = (u, v)_{\mathcal{H}} - \theta\left[\mathcal{A}(u, v) - F(v)\right], \quad \forall v \in \mathcal{H}.$$

Use the previous two items to show that P_θ is well defined and that (C.6) is equivalent to finding a fixed point of P_θ.

d) Let α, M denote, respectively, the coercivity and boundedness constants of $\mathcal{A}$. Show that, for $\theta \in (0, 2\alpha/M^2)$, the mapping P_θ is a contraction.
Hint: Define $A: \mathcal{H} \to \mathcal{H}$ by $(Av, w)_{\mathcal{H}} = \mathcal{A}(v, w)$. Show that this is a linear mapping. How can you bound $\|Av\|_{\mathcal{H}}$ and $(Av, v)_{\mathcal{H}}$? How do you write $P_\theta(w)$ in terms of A?

C.5 Provide all the details for Example C.3.

C.6 Prove Proposition C.7.

C.7 Let $(X, \mathfrak{d})$ be a metric space and $T: X \to X$ be a contraction. Show that T is continuous at every point $x \in X$.

C.8 Prove Theorem C.11.

Appendix D A (Petting) Zoo of Function Spaces

In this appendix, we collect the (vector) spaces of functions that appear throughout the text. We also present, mostly without proof, their most relevant properties that will be useful to us. A careful and detailed exposition of all these matters is certainly beyond the scope of our text, and there are many useful references for this. We refer the reader, for instance, to [2, 55].

Before we embark on our presentation, we must begin with a remark on notation.

Remark D.1 (notation). Throughout, we will denote spaces of functions by symbols of the form $X(A; B)$. Here, X will possibly be decorated by sub- and super-indices, and others, and this will indicate the property that defines the space of functions. A is the domain of these functions, which will usually be an open or closed subset of $\mathbb{R}^d$ for some $d \in \mathbb{N}$. The range of these functions is B, which is usually a vector space or a subset of it. In an effort to alleviate notation, if $B = \mathbb{R}$, we will suppress this, i.e., $X(A) = X(A; \mathbb{R})$.

D.1 Spaces of Smooth Functions

We refer the reader to Appendix B for the definitions of continuity, Hölder continuity, and differentiability of a function.

Definition D.2 (spaces of smooth functions). Let $d \in \mathbb{N}$ and $\Omega \subset \mathbb{R}^d$ be a bounded domain. For $k \in \mathbb{N}_0$ and $\alpha \in [0, 1]$, we denote by

$$C^{k,\alpha}(\Omega), \qquad C^{k,\alpha}(\overline{\Omega})$$

the **vector spaces of functions that are k-times continuously differentiable** on Ω or $\overline{\Omega}$ respectively, such that the kth derivative is Hölder continuous of order α on Ω or $\overline{\Omega}$ respectively.

In addition, to simplify notation, we set

$$C(\Omega) = C^{0,0}(\Omega), \quad C^k(\Omega) = C^{k,0}(\Omega),$$

and

$$C(\overline{\Omega}) = C^{0,0}(\overline{\Omega}), \quad C^k(\overline{\Omega}) = C^{k,0}(\overline{\Omega}).$$

Finally, we comment that, by arguing component-wise, spaces of vector-valued functions can be defined in a similar manner.

Remark D.3 (notation). In the case $d = 1$, we will usually deal with functions defined on a compact interval $[a, b] \subset \mathbb{R}$ or its interior: the open interval (a, b). In this case, to alleviate notation, we set

$$C^{k,\alpha}(a, b) = C^{k,\alpha}((a, b)).$$

We must immediately remark that whether the set is open or closed makes a substantial difference.

Example D.1 Let $f(x) = x^{-1}$. Then $f \in C(0, 1)$, but $f \notin C([0, 1])$. Notice also that

$$\int_0^1 \frac{1}{x^{1+\varepsilon}} \, dx = +\infty, \quad \forall \varepsilon \geq 0.$$

We now introduce L^p-norms on continuous functions.

Definition D.4 (L^p-norm). Let $d \in \mathbb{N}$ and $\Omega \subset \mathbb{R}^d$ be a bounded domain. For each $p \in [1, \infty)$, define the function $\|\cdot\|_{L^p(\Omega;\mathbb{C})} : C(\bar{\Omega}; \mathbb{C}) \to [0, \infty)$ via

$$\|f\|_{L^p(\Omega;\mathbb{C})} = \left(\int_\Omega |f(\mathbf{x})|^p d\mathbf{x} \right)^{1/p}, \quad \forall f \in C(\bar{\Omega}; \mathbb{C}).$$

This is called the L^p**-norm, or just** p**-norm**. Similarly, the function $\|\cdot\|_{L^\infty(\Omega;\mathbb{C})} : C(\bar{\Omega}; \mathbb{C}) \to [0, \infty)$, defined via

$$\|f\|_{L^\infty(\Omega;\mathbb{C})} = \sup_{\mathbf{x} \in \Omega} |f(\mathbf{x})|, \quad \forall f \in C(\bar{\Omega}; \mathbb{C}),$$

is called the L^∞**-norm, or just** ∞**-norm**.

The definitions above require several remarks. First, we notice that, once again, the fact that the domain of the functions is closed is essential, regardless of the value of p; see Example D.1. In addition, in the case $p = \infty$, it does not matter if the supremum is taken over the open or closed domain. Finally, owing to Theorem B.47, the supremum is actually a maximum.

Proposition D.5 (properties of L^p-norms). *Let $d \in \mathbb{N}$ and $\Omega \subset \mathbb{R}^d$ be a bounded domain. For every $p \in [1, \infty]$, the function $\|\cdot\|_{L^p(\Omega;\mathbb{C})}$ is a norm on $C(\bar{\Omega}; \mathbb{C})$. In particular, in the case $p = 2$, the L^2-norm is generated by the inner product*

$$(f, g)_{L^2(\Omega;\mathbb{C})} = \int_\Omega f(\mathbf{x}) \overline{g(\mathbf{x})} \, d\mathbf{x}, \quad \forall f, g \in C(\bar{\Omega}; \mathbb{C}).$$

Owing to this, the integral Minkowski inequality[1]

$$\|f + g\|_{L^p(\Omega;\mathbb{C})} \leq \|f\|_{L^p(\Omega;\mathbb{C})} + \|g\|_{L^p(\Omega;\mathbb{C})}, \quad \forall f, g \in C(\bar{\Omega}; \mathbb{C})$$

holds. Finally, if $p, q \in [1, \infty]$ are conjugate pairs, i.e.,

$$\frac{1}{p} + \frac{1}{q} = 1,$$

[1] Named in honor of the German mathematician Hermann Minkowski (1864–1909).

with the understanding that $\frac{1}{0} = \infty$ *and* $\frac{1}{\infty} = 0$, *then the* integral Hölder inequality[2]

$$\int_{\Omega} |f(x)||g(x)| \mathrm{d}x \leq \|f\|_{L^p(\Omega;\mathbb{C})} \ \|g\|_{L^q(\Omega;\mathbb{C})}, \quad \forall f, g \in C(\bar{\Omega}; \mathbb{C})$$

holds.

Proof. See Problem D.1 □

Since $C(\bar{\Omega}; \mathbb{C})$ is infinite dimensional, its completeness under a norm is not a trivial matter. The following result shows this.

Theorem D.6 (completeness). *Let* $d \in \mathbb{N}$ *and* $\Omega \subset \mathbb{R}^d$ *be a bounded domain. For every* $1 \leq p < \infty$, *the pair*

$$\left(C(\bar{\Omega}; \mathbb{C}), \|\cdot\|_{L^p(\bar{\Omega};\mathbb{C})}\right)$$

is a normed space, though not a Banach space. On the other hand, the pair

$$\left(C(\bar{\Omega}; \mathbb{C}), \|\cdot\|_{L^\infty(\Omega;\mathbb{C})}\right)$$

is a Banach space.

Proof. The completeness is a consequence of Theorem B.62. The other results are classical and the proofs can be found, for example, in the books by Rudin [76, 77]. □

Now, although this can be done much more generally, we shall only need weighted L^p-norms for functions of one variable. Thus, we introduce them only in this case. We begin with the definition of a weight.

Definition D.7 (weight). Let $[a, b] \subset \mathbb{R}$ be a compact interval. A function $w \colon [a, b] \to \mathbb{R}$ is called a **weight on** $[a, b]$ if and only if it is nonnegative on $[a, b]$, it vanishes only at possibly a finite number of points, $w \in C(a, b)$, and it is Riemann integrable on the interval $[a, b]$, i.e.,

$$0 < \int_a^b w(x) \mathrm{d}x < \infty.$$

Definition D.8 (weighted L^p-norm). Let w be a weight function on the interval $[a, b] \subset \mathbb{R}$ and $p \in [1, \infty)$. The **weighted L^p-norm** of $f \in C([a, b]; \mathbb{C})$ is

$$\|f\|_{L^p_w(a,b;\mathbb{C})} = \left(\int_a^b w(x)|f(x)|^p \mathrm{d}x\right)^{1/p}.$$

The function

$$(\cdot, \cdot)_{L^2_w(a,b;\mathbb{C})} \colon C([a, b]; \mathbb{C}) \times C([a, b]; \mathbb{C}) \to \mathbb{C},$$

$$(f, g)_{L^2_w(a,b;\mathbb{C})} \colon C([a, b]; \mathbb{C}) \mapsto \int_a^b w(x) f(x) \overline{g(x)} \, \mathrm{d}x$$

is called the **weighted L^2-inner product**.

[2] Named in honor of the German mathematician Otto Ludwig Hölder (1859–1937).

Theorem D.9 (properties of weighted L^p-norms). *Let w be a weight function on $[a, b]$. Then, for all $p \in [1, \infty)$, the weighted L^p-norms are norms on $C([a, b]; \mathbb{C})$. In particular, the* weighted Minkowski inequality, *i.e.,*

$$\|f + g\|_{L^p_w(a,b;\mathbb{C})} \leq \|f\|_{L^p_w(a,b;\mathbb{C})} + \|g\|_{L^p_w(a,b;\mathbb{C})}, \quad \forall f, g \in C([a, b]; \mathbb{C}),$$

holds. In addition, if $p, q \in (1, \infty)$ are conjugate exponent pairs, *i.e.,*

$$1 = \frac{1}{p} + \frac{1}{q},$$

then the weighted Hölder inequality *is valid*

$$\int_a^b w(x)|f(x)|\,|g(x)|\mathrm{d}x \leq \|f\|_{L^p_w(a,b;\mathbb{C})} \|g\|_{L^q_w(a,b;\mathbb{C})}.$$

When $p = \infty$, we have the following variant of the weighted Hölder inequality:

$$\int_a^b w(x)|f(x)|\,|g(x)|\mathrm{d}x \leq \|f\|_{L^\infty(a,b;\mathbb{C})} \|g\|_{L^1_w(a,b;\mathbb{C})}.$$

Finally, for any $f \in C([a, b]; \mathbb{C})$,

$$\|f\|_{L^p_w(a,b;\mathbb{C})} \leq \left(\|w\|_{L^1(a,b)}\right)^{1/p} \|f\|_{L^\infty(a,b;\mathbb{C})}. \tag{D.1}$$

Proof. The proof that these are indeed norms is left to the reader as an exercise; see Problem D.1.

Let us prove (D.1). Let $f \in C([a, b]; \mathbb{C})$. Then there is $x_0 \in [a, b]$ such that

$$\|f\|_{L^\infty(a,b;\mathbb{C})} = |f(x_0)| \quad \Longrightarrow \quad |f(x)| \leq |f(x_0)|, \ \forall x \in [a, b].$$

Then, since w is a weight, it is nonnegative and we can estimate

$$\begin{aligned}\|f\|^p_{L^p_w(a,b;\mathbb{C})} &= \int_a^b w(x)|f(x)|^p \mathrm{d}x \\ &\leq |f(x_0)|^p \int_a^b w(x)\mathrm{d}x \\ &= \|f\|^p_{L^\infty(a,b;\mathbb{C})} \|w\|_{L^1(a,b)}.\end{aligned}$$

Taking pth roots, (D.1) follows. □

Clearly, $\|\cdot\|_{L^p_1(a,b)} = \|\cdot\|_{L^p(a,b)}$, so the issue presented in Theorem D.6 remains in this case.

Theorem D.10 (not Banach). *Let w be a weight function on $[a, b]$. For every $1 \leq p < \infty$, the pair*

$$\left(C([a, b]; \mathbb{C}), \|\cdot\|_{L^p_w(a,b;\mathbb{C})}\right)$$

is a normed space, though not a Banach space.

Proof. These facts are well known and the proofs can be found, for example, in the books by Rudin [76, 77]. □

The norm for the case of higher order smoothness is defined similarly.

Definition D.11 ($C^{k,\alpha}$-norm)**.** Let $d \in \mathbb{N}$, $\Omega \subset \mathbb{R}^d$ be a bounded domain, $k \in \mathbb{N}_0$, and $\alpha \in [0,1]$. We define, for all $\in C^{k,\alpha}(\bar{\Omega};\mathbb{C})$,

$$\|f\|_{C^{k,\alpha}(\bar{\Omega};\mathbb{C})} = \max\left\{ \max_{\substack{\beta\in\mathbb{N}_0^d \\ |\beta|\le k}} \left\|D^{\beta} f\right\|_{L^{\infty}(\Omega;\mathbb{C})}, \max_{\substack{\beta\in\mathbb{N}_0^d \\ |\beta|=k}} \sup_{\substack{x,y\in\bar{\Omega} \\ x\neq y}} \frac{|D^{\beta} f(x) - D^{\beta} f(y)|}{\|x-y\|_2^{\alpha}} \right\}.$$

Clearly, $\|\cdot\|_{C^{0,0}(\bar{\Omega};\mathbb{C})} = \|\cdot\|_{L^{\infty}(\Omega;\mathbb{C})}$. These norms make these spaces complete.

Proposition D.12 (completeness)**.** *Let $d \in \mathbb{N}$ and $\Omega \subset \mathbb{R}^d$ be an open bounded domain. For every $k \in \mathbb{N}_0$ and $\alpha \in [0,1]$, the pair*

$$\left(C^{k,\alpha}(\bar{\Omega};\mathbb{C}), \|\cdot\|_{C^{k,\alpha}(\bar{\Omega};\mathbb{C})}\right)$$

is a Banach space.

D.1.1 Periodic Functions

In many problems, periodicity appears as a natural constraint; therefore, we must deal with spaces of periodic functions.

Definition D.13 (periodic function)**.** Let $L > 0$ be given. Suppose that $f\colon \mathbb{R} \to \mathbb{C}$. We say that f is L-**periodic** if and only if

$$f(x+nL) = f(x), \quad \forall n \in \mathbb{Z}, \quad \forall x \in \mathbb{R}.$$

For any $m \in \mathbb{N}_0$, we define $C_{\mathrm{p}}^m(0,L;\mathbb{C})$ as the set of functions

$$C_{\mathrm{p}}^m(0,L;\mathbb{C}) = \{f \in C^m(\mathbb{R};\mathbb{C}) \mid f(x+nL) = f(x),\ \forall n \in \mathbb{Z},\ \forall x \in \mathbb{R}\}.$$

The set $C_{\mathrm{p}}^m(0,L)$ is defined similarly. We also define the set of complex-valued, smooth, mean-zero L-periodic functions as

$$\mathring{C}_{\mathrm{p}}^m(0,L;\mathbb{C}) = \left\{u \in C_{\mathrm{p}}^m(0,L;\mathbb{C}) \middle| \int_0^L u(x)\mathrm{d}x = 0\right\}.$$

Finally, define

$$C_{\mathrm{p}}^{\infty}(0,L;\mathbb{C}) = \bigcap_{m=0}^{\infty} C_{\mathrm{p}}^m(0,L;\mathbb{C}).$$

The set $C_{\mathrm{p}}^{\infty}(0,L)$ is defined similarly.

For simplicity, we will work with one-periodic functions, i.e., with $L = 1$. All of the results can be recast in the case for which $L > 0$ is arbitrary, without much difficulty.

Observe that a one-periodic function is completely characterized by its values on $[0,1]$. Thus, spaces of smooth periodic functions can be normed with any of the norms defined before, assuming that the domain is the unit interval $[0,1]$. On the other hand, it is important to realize that, for any $m \in \mathbb{N}_0$, $C_{\mathrm{p}}^m(0,1;\mathbb{C}) \neq C^m(0,1;\mathbb{C})$. In this situation, the periodicity is vital to its definition. For instance, if $f \in C(0,1;\mathbb{C})$, then it can be periodically extended to all of $\mathbb{R}$, but the extension will not be in $C(\mathbb{R};\mathbb{C})$ unless $f(0) = f(1)$.

D.2 Spaces of Integrable Functions

The motivation for the introduction of the so-called *Lebesgue spaces*[3] is, essentially, the negative results of Theorem D.6 and Theorem D.10. We wish to have spaces of functions that are complete under these norms. To properly develop them and define them would require us to develop *measure theory* [77] or the theory of the *Daniell integral*[4] [75]. We will not do this.

Definition D.14 (L^p space)**.** Let $d \in \mathbb{N}$ and $\Omega \subset \mathbb{R}^d$ be a domain. For $p \in [1, \infty)$, we define

$$L^p(\Omega; \mathbb{C}) = \left\{ f : \Omega \to \mathbb{R} \;\middle|\; \int_\Omega |f(\boldsymbol{x})|^p \mathrm{d}\boldsymbol{x} < \infty \right\}.$$

We also define

$$L^\infty(\Omega; \mathbb{C}) = \left\{ f : \Omega \to \mathbb{R} \;\middle|\; \operatorname*{ess\,sup}_{x \in \Omega} |f(\boldsymbol{x})| < \infty \right\},$$

where ess sup denotes the essential supremum. The L^p-norms are defined as before.

In the one-dimensional case, we can do the same with weighted integrals.

Definition D.15 (w-square integrable)**.** Let $[a, b] \subset \mathbb{R}$ be a compact interval and w be a weight on $[a, b]$. We say that f is w-**square integrable** if and only if

$$\int_a^b w(x)|f(x)|^2 \mathrm{d}x < \infty.$$

The set of all such functions is labeled $L^2_w(a, b; \mathbb{C})$. When the functions are real valued, we write $L^2_w(a, b)$. For two functions $f, g \in L^2_w(a, b; \mathbb{C})$, we define the weighted inner product

$$(f, g)_{L^2_w(a,b;\mathbb{C})} = \int_a^b w(x) f(x) \overline{g(x)} \, \mathrm{d}x,$$

where the overline denotes complex conjugation. The conjugation is dropped if g is real valued.

Remark D.16 (equivalence)**.** The reader familiar with Lebesgue integration will recall that two measurable functions are equivalent in $L^p(\Omega; \mathbb{C})$, or $L^2_w(a, b; \mathbb{C})$, if and only if the set on which they differ has (Lebesgue) measure zero. Strictly speaking, therefore, we think of equivalence classes of functions, rather than functions, as the elements of $L^p(\Omega; \mathbb{C})$, or $L^2_w(a, b; \mathbb{C})$. However, for those not familiar, it is usually fine to think of the elements of the spaces as ordinary functions.

One of the foundational facts of integration theory is the completeness of L^p spaces.

[3] Named in honor of the French mathematician Henri León Lebesgue (1875–1941).
[4] Named in honor of the British mathematician Percy John Daniell (1889–1946).

Theorem D.17 (completeness). *Let $d \in \mathbb{N}$ and $\Omega \subset \mathbb{R}^d$ be a domain. For every $p \in [1, \infty]$, the pair*

$$\left(L^p(\Omega; \mathbb{C}), \|\cdot\|_{L^p(\Omega;\mathbb{C})}\right)$$

is a Banach space. In particular, the pair

$$\left(L^2(\Omega; \mathbb{C}), \|\cdot\|_{L^2(\Omega;\mathbb{C})}\right)$$

is a complex Hilbert space. Similarly, if $[a, b] \subset \mathbb{R}$ is a compact interval and w is a weight function on $[a, b]$, then the pair

$$(L^2_w(a, b), (\cdot, \cdot)_{L^2_w(a,b)})$$

is a real Hilbert space and $(\cdot, \cdot)_{L^2_w(a,b)}$ is a real inner product. The pair

$$(L^2_w(a, b; \mathbb{C}), (\cdot, \cdot)_{L^2_w(a,b;\mathbb{C})})$$

is a complex Hilbert space and $(\cdot, \cdot)_{L^2_w(a,b;\mathbb{C})}$ is a complex inner product. As is standard, the norm is defined as

$$\|f\|_{L^2_w(a,b;\mathbb{C})} = \sqrt{(f, f)_{L^2_w(a,b;\mathbb{C})}}, \quad \forall f \in L^2_w(a, b; \mathbb{C}).$$

Proof. See [77]. □

D.2.1 Periodic Functions

By $L^2_{\mathrm{p}}(0, 1; \mathbb{C})$ we denote the set of all one-periodic, Lebesgue measurable functions $f\colon \mathbb{R} \to \mathbb{C}$ with the property that, for any compact interval $[a, b] \subset \mathbb{R}$,

$$\int_a^b |f(x)|^2 \mathrm{d}x < \infty.$$

This is a vector space over the reals under the usual operations of function addition and scalar multiplication. For two functions $f, g \in L^2_{\mathrm{p}}(0, 1; \mathbb{C})$, we define the inner product

$$(f, g)_{L^2(0,1;\mathbb{C})} = \int_0^1 f(x)\overline{g(x)}\, \mathrm{d}x,$$

where, as usual, the overline denotes complex conjugation. The set $L^2_{\mathrm{p}}(0, 1)$ is defined similarly. Finally, by $\mathring{L}^2(0, 1; \mathbb{C})$, we denote the space of complex-valued, mean-zero square integrable functions on $(0, 1)$, i.e.,

$$\mathring{L}^2(0, 1; \mathbb{C}) = \left\{ u \in L^2(0, 1; \mathbb{C}) \,\middle|\, \int_0^1 u(x)\mathrm{d}x = 0 \right\}.$$

Some standard arguments in integration theory yield the following result.

Theorem D.18 (completeness). *The pair*

$$(L^2_{\mathrm{p}}(0, 1), (\cdot, \cdot)_{L^2(0,1)}) = (L^2(0, 1), (\cdot, \cdot)_{L^2(0,1)})$$

is a real Hilbert space, and the pair

$$\left(L^2_{\mathrm{p}}(0,1;\mathbb{C}), (\cdot,\cdot)_{L^2(0,1;\mathbb{C})}\right) = \left(L^2(0,1;\mathbb{C}), (\cdot,\cdot)_{L^2(0,1;\mathbb{C})}\right)$$

is a complex Hilbert space. As usual, the norm for $L^2_{\mathrm{p}}(0,1;\mathbb{C})$ *is defined from the inner product via*

$$\|f\|_{L^2(0,1;\mathbb{C})} = \sqrt{(f,f)_{L^2(0,1;\mathbb{C})}}.$$

Proof. See [7, 76]. □

Remark D.19 (periodic extension). In case you missed that subtle and curious fact above, here it is again: $L^2_{\mathrm{p}}(0,1;\mathbb{C}) = L^2(0,1;\mathbb{C})$. In other words, periodicity does not matter for merely square integrable functions. This is because, if $f \in L^2(0,1;\mathbb{C})$, it can be easily periodically extended to all of $\mathbb{R}$, and the extension will be in $L^2_{\mathrm{p}}(0,1;\mathbb{C})$.

D.3 Sobolev Spaces

The study of weak solutions of partial differential equations requires one to deal with functions that are differentiable in a sense that is weaker than the classical one. Spaces of functions with such differentiability properties are known as *Sobolev spaces*. There are two approaches to defining these spaces which, in our setting, turn out to be equivalent. We will briefly present both and show the relation between them.

We begin by introducing an auxiliary notion.

Definition D.20 (support). Let $d \in \mathbb{N}$, $\Omega \subset \mathbb{R}^d$ be a domain, and $\phi\colon \Omega \to \mathbb{R}$. The **support** of ϕ is

$$\operatorname{supp}(\phi) = \overline{\{x \in \Omega \mid \phi(x) \neq 0\}}.$$

Next, we restrict our attention to a special class of domains.

Definition D.21 (Lipschitz domain[5]). Let $d \in \mathbb{N}$ and $\Omega \subset \mathbb{R}^d$ be a domain. We say that Ω is a **Lipschitz domain** if and only if its boundary can be locally represented by the graph of a Lipschitz function. We say that Ω is a **uniformly Lipschitz domain** if it is Lipschitz, and the Lipschitz constant of the functions that represent the boundary is uniform.

Every bounded Lipschitz domain is uniformly Lipschitz. From now on, although this can be done much more generally, we will exclusively consider Lipschitz domains.

Definition D.22 (Sobolev space[6]). Let $d \in \mathbb{N}$ and $\Omega \subset \mathbb{R}^d$ be a bounded Lipschitz domain. We say that a function $v\colon \Omega \to \mathbb{R}$ belongs to the *Sobolev space* $H^1(\Omega)$ if there is a sequence $\{\phi_k\}_{k\in\mathbb{N}} \subset C^1(\bar{\Omega})$ such that

$$\|v - \phi_k\|_{L^2(\Omega)} \to 0, \qquad k \to \infty,$$

[5] Named in honor of the German mathematician Rudolf Otto Sigismund Lipschitz (1832–1903).

[6] Named in honor of the Soviet mathematician Sergei Lvovich Sobolev (1908–1989).

and, moreover,

$$\sup_{k\in\mathbb{N}} \|\nabla\phi_k\|_{L^2(\Omega;\mathbb{R}^d)} < \infty.$$

We denote this fact by $v \in H^1(\Omega)$. Finally, if the sequence can be chosen so that there is a compact $K \subset \Omega$ such that

$$\bigcup_{k\in\mathbb{N}} \operatorname{supp}\phi_k \subset K,$$

then we say that the function $v \in H_0^1(\Omega)$.

We immediately observe that $H_0^1(\Omega) \subset H^1(\Omega)$, but that they are not equal. Notice, in addition, that in the previous definition we required that, for any $i = 1, \dots, d$, the sequence of partial derivatives

$$\frac{\partial\phi_k}{\partial x_i}$$

remains bounded in $L^2(\Omega)$. As we have shown before, the space $L^2(\Omega)$ is Hilbert. Owing to Theorem 16.10, we can extract a convergent subsequence. It is possible to show, in addition, that this limit is independent of the subsequence and so we denote it by

$$\frac{\partial v}{\partial x_i},$$

and call it the *strong partial derivative* of the function v. Notice then that, by definition, $v \in H^1(\Omega)$ if and only if $v \in L^2(\Omega)$ and $\nabla v \in L^2(\Omega;\mathbb{R}^d)$, where ∇v is the vector of strong derivatives.

The intuition behind Sobolev spaces is correct. Functions in such spaces may not be smooth, but they can be *approximated* (in the $L^2(\Omega)$-sense) by smooth ones.

Example D.2 The function $v\colon x \mapsto |x|$ is not continuously differentiable, but $v \in H^1(-1,1)$. Indeed, the sequence

$$\phi_k(x) = \begin{cases} |x|, & |x| > \dfrac{1}{k}, \\ \dfrac{k}{2}x^2 + \dfrac{1}{2k}, & |x| \le \dfrac{1}{k} \end{cases}$$

belongs to $C^1([-1,1])$, as the reader can easily verify. In addition, for some $C > 0$,

$$\|v - \phi_k\|_{L^2(-1,1)}^2 = \int_{-1/k}^{1/k} \left[|x| - \frac{k}{2}x^2 + \frac{1}{2k}\right]^2 \mathrm{d}x < \frac{C}{k} \to \infty,$$

as $k \to \infty$. Finally, for every $k \in \mathbb{N}$,

$$\|\phi_k'\|_{L^2(-1,1)}^2 = 2\int_{1/k}^{1} \mathrm{d}x + k^2 \int_{-1/k}^{1/k} x^2 \mathrm{d}x = 2 - \frac{4}{3k} < 2.$$

Let us now state without proof some facts about these spaces.

Proposition D.23 (properties of Sobolev spaces). *Let $d \in \mathbb{N}$ and $\Omega \subset \mathbb{R}^d$ be a bounded Lipschitz domain with sufficiently smooth boundary.*

1. *Completeness: The Sobolev space $H^1(\Omega)$ is Hilbert with respect to the inner product*

$$(v_1, v_2)_{H^1(\Omega)} = \int_\Omega [v_1(\mathbf{x})v_2(\mathbf{x}) + \nabla v_1(\mathbf{x}) \cdot \nabla v_2(\mathbf{x})]\, \mathrm{d}\mathbf{x}, \quad \forall v_1, v_2 \in H^1(\Omega).$$

2. *Subspace: $H^1_0(\Omega)$ is a closed subspace of $H^1(\Omega)$.*
3. *Poincaré inequality:[7] There is a constant $C_P > 0$ that depends on Ω such that, for all $v \in H^1_0(\Omega)$,*

$$\|v\|_{L^2(\Omega)} \le C_P \|\nabla v\|_{L^2(\Omega;\mathbb{R}^d)}. \tag{D.2}$$

As a consequence, the inner product

$$(v_1, v_2)_{H^1_0(\Omega)} = \int_\Omega \nabla v_1(\mathbf{x}) \cdot \nabla v_2(\mathbf{x}) \mathrm{d}\mathbf{x}, \quad \forall v_1, v_2 \in H^1_0(\Omega)$$

induces, on $H^1_0(\Omega)$, an equivalent norm.

4. *Trace inequality: There is a constant that depends on Ω such that*

$$\|v\|_{L^2(\partial\Omega)} \le C\|v\|_{H^1(\Omega)}, \quad \forall v \in H^1(\Omega).$$

5. *Polynomials are dense in $C^1(\bar{\Omega})$, and so they are in $H^1(\Omega)$.*

From the trace inequality, it follows that we can speak of boundary values (sometimes called *traces*) for a function in $H^1(\Omega)$, something that is not feasible for a generic function only in $L^2(\Omega)$. In addition, the meaning of the zero subscript in $H^1_0(\Omega)$ is now clear. These are functions in $H^1(\Omega)$ that "vanish" on the boundary.

Remark D.24 (extension). The Poincaré inequality (D.2) is still true if we assume that the trace of u is zero only on some connected subset of the boundary of positive measure. For example, if $d = 2$, Ω is a polygon, and the boundary trace of u is zero on one of the edges of this polygon, then the inequality is still valid.

In all of the following results, we interpret boundary values through the lenses of boundary trace operators. This is particularly important for boundary integrals. The following result should be compared with Theorem B.59.

Theorem D.25 (integration by parts). *Suppose that $d \in \mathbb{N}$, $\Omega \subset \mathbb{R}^d$ is a bounded Lipschitz domain, and $u, v \in H^1(\Omega)$. Then, for $i = 1, \ldots, d$,*

$$\int_\Omega \frac{\partial u(\mathbf{x})}{\partial x_i} v(\mathbf{x}) \mathrm{d}\mathbf{x} = \int_{\partial\Omega} u(\mathbf{x})v(\mathbf{x})\boldsymbol{n}(\mathbf{x}) \cdot \boldsymbol{e}_i \, \mathrm{d}\boldsymbol{S}(\mathbf{x}) - \int_\Omega u(\mathbf{x}) \frac{\partial v(\mathbf{x})}{\partial x_i} \, \mathrm{d}\mathbf{x},$$

where $\boldsymbol{n}\colon \partial\Omega \to \mathbb{R}^d$ is the outward pointing unit normal vector on the boundary $\partial\Omega$ and $\{\boldsymbol{e}_i\}_{i=1}^d$ is the canonical basis of $\mathbb{R}^d$.

[7] Named in honor of the French mathematician, theoretical physicist, engineer, and philosopher of science Jules Henri Poincaré (1854–1912).

We also need to introduce higher order Sobolev spaces, and this can be done recursively. Essentially, we define $H^m(\Omega)$ as the space of functions whose weak derivatives *up to order m* are in $L^2(\Omega)$.

Definition D.26 (Sobolev space)**.** Let $1 < m \in \mathbb{N}$. Then we define

$$H^m(\Omega) = \left\{ v \in L^2(\Omega) \;\middle|\; \frac{\partial v}{\partial x_j} \in H^{m-1}(\Omega), j = 1, \ldots, d \right\}.$$

For $v \in H^m(\Omega)$ and $j \in \{0, \ldots, m\}$, we define the seminorm

$$|v|_{H^j(\Omega)} = \sqrt{\sum_{\substack{\alpha \in \mathbb{N}_0^d \\ |\alpha| = j}} \|D^\alpha v\|_{L^2(\Omega)}^2}$$

and norm

$$\|v\|_{H^m(\Omega)} = \left(\sum_{j=0}^{m} |v|_{H^j(\Omega)}^2 \right)^{1/2}.$$

It turns out that $H^m(\Omega)$ is Hilbert under this norm.

The following result should be compared with Theorem B.59.

Theorem D.27 (Green's identities[8])**.** *Suppose that $d \in \mathbb{N}$ and $\Omega \subset \mathbb{R}^d$ is a bounded Lipschitz domain. For all $v \in H^1(\Omega)$ and $w \in H^2(\Omega)$,*

$$\int_\Omega \nabla v(\boldsymbol{x}) \cdot \nabla w(\boldsymbol{x}) \mathrm{d}\boldsymbol{x} = -\int_\Omega v(\boldsymbol{x}) \Delta w(\boldsymbol{x}) \mathrm{d}\boldsymbol{x} + \int_{\partial\Omega} v(\boldsymbol{x}) \frac{\partial w}{\partial n}(\boldsymbol{x}) \mathrm{d}\boldsymbol{S}(\boldsymbol{x}),$$

where $\frac{\partial w}{\partial n}(\boldsymbol{x}) = \boldsymbol{n}(x) \cdot \nabla w$. For any $v, w \in H^2(\Omega)$,

$$\int_\Omega (v(\boldsymbol{x})\Delta w(\boldsymbol{x}) - \Delta v(\boldsymbol{x}) w(\boldsymbol{x}))\, \mathrm{d}\boldsymbol{x} = \int_{\partial\Omega} \left(v(\boldsymbol{x}) \frac{\partial w}{\partial n}(\boldsymbol{x}) - \frac{\partial v}{\partial n}(\boldsymbol{x}) w(\boldsymbol{x}) \right) \mathrm{d}\boldsymbol{S}(\boldsymbol{x}).$$

For any $\boldsymbol{v} \in H^1(\Omega; \mathbb{R}^d)$ and $w \in H^1(\Omega)$,

$$\int_\Omega \nabla \cdot \boldsymbol{v}(\boldsymbol{x}) w(\boldsymbol{x}) \mathrm{d}\boldsymbol{x} = \int_{\partial\Omega} \boldsymbol{v}(\boldsymbol{x}) \cdot \boldsymbol{n}(\boldsymbol{x}) w(\boldsymbol{x}) \mathrm{d}\boldsymbol{S}(\boldsymbol{x}) - \int_\Omega \boldsymbol{v}(\boldsymbol{x}) \cdot \nabla w(\boldsymbol{x}) \mathrm{d}\boldsymbol{x}.$$

Another approach to defining Sobolev spaces is via *weak derivatives*. For that, we begin by defining a test function.

Definition D.28 (test function)**.** Let $d \in \mathbb{N}$ and $\Omega \subset \mathbb{R}^d$ be a bounded domain. A function $\phi \colon \Omega \to \mathbb{R}$ is called a **test function** on Ω and we write $\phi \in C_0^\infty(a, b)$ if and only if $\phi \in C^\infty(\Omega)$ and $\mathrm{supp}(\phi) \subset \Omega$.

Remark D.29 (vanishing derivatives)**.** Suppose that $\phi \in C_0^\infty(a, b)$. Since $\mathrm{supp}(\phi)$ is closed and is a proper subset of the open bounded interval (a, b), it follows that ϕ and any of its derivatives vanish on the boundary.

[8] Named in honor of the British mathematical physicist George Green (1793–1841).

Example D.3 Suppose that $a > 1$. The bump function, $\phi\colon (-a, a) \to \mathbb{R}$, defined via

$$\phi(x) = \begin{cases} \exp\left(-\dfrac{1}{1-x^2}\right), & x \in (-1, 1), \\ 0, & x \in (-a, -1] \cup [1, a), \end{cases}$$

is a test function on $(-a, a)$. It is easy to see that $\mathrm{supp}(\phi) = [-1, 1]$.

We can now give another definition of Sobolev spaces.

Definition D.30 (Sobolev space). Let $d \in \mathbb{N}$, $\Omega \subset \mathbb{R}^d$ be a bounded Lipschitz domain and $u \in L^2(\Omega)$. We say that $u \in W^1(\Omega)$ if and only if there exist functions $g_i \in L^2(\Omega)$, $i = 1, \ldots, d$ such that

$$\int_\Omega g_i(\boldsymbol{x})\phi(\boldsymbol{x})\mathrm{d}\boldsymbol{x} = -\int_\Omega u(\boldsymbol{x})\frac{\partial \phi(\boldsymbol{x})}{\partial x_i}\,\mathrm{d}\boldsymbol{x}, \quad \forall \phi \in C_0^\infty(\Omega).$$

The function g_i is called a **weak derivative of** u with respect to x_i. More generally, if $m \in \mathbb{N}$, we say that $u \in W^m(\Omega)$ if and only if there exist

$$\{g_{\boldsymbol{\alpha}}\}_{\substack{\boldsymbol{\alpha} \in \mathbb{N}_0^d \\ |\boldsymbol{\alpha}| \le m}} \subset L^2(\Omega)$$

such that, for every multi-index $\boldsymbol{\alpha} \in \mathbb{N}_0^d$ with $|\boldsymbol{\alpha}| \le m$, we have

$$\int_\Omega g_{\boldsymbol{\alpha}}(\boldsymbol{x})\phi(\boldsymbol{x})\mathrm{d}\boldsymbol{x} = (-1)^{|\boldsymbol{\alpha}|}\int_\Omega u(\boldsymbol{x})D^{\boldsymbol{\alpha}}\phi(\boldsymbol{x})\mathrm{d}\boldsymbol{x}, \quad \forall \phi \in C_0^\infty(\Omega).$$

The function $g_{\boldsymbol{\alpha}}$ is called a **weak derivative of** u of order $\boldsymbol{\alpha}$. We also define the notation $W^0(\Omega) = L^2(\Omega)$. The set $W^m(\Omega)$ is called the **Sobolev space of order** m.

Example D.4 Suppose that $f(x) = |x|$, for $x \in (-2, 2)$. Then $f \in W^1(-2, 2)$, as we will momentarily show. First, recall that, in Example D.3, we showed that f is continuously differentiable and its piecewise derivative was a Heaviside function. In fact, we could take the following as its piecewise derivative:

$$g(x) = \begin{cases} -1, & -2 < x < 0, \\ 0, & x = 0, \\ 1, & 0 < x < 2. \end{cases}$$

Let us show that this also works as a weak derivative of f. Let $\phi \in C_0^\infty(-2,2)$ be arbitrary. Then, using integration by parts,

$$\begin{aligned}\int_{-2}^{2} g(x)\phi(x)\mathrm{d}x &= -\int_{-2}^{0} \phi(x)\mathrm{d}x + \int_{0}^{2} \phi(x)\mathrm{d}x \\ &= -\, x\phi(x)|_{-2}^{0} + \int_{-2}^{0} x\phi'(x)\mathrm{d}x + x\phi(x)|_0^2 - \int_0^2 x\phi'(x)\mathrm{d}x \\ &= \int_{-2}^{0} x\phi'(x)\mathrm{d}x - \int_0^2 x\phi'(x)\mathrm{d}x \\ &= -\int_{-2}^{2} |x|\phi'(x)\mathrm{d}x.\end{aligned}$$

As with the piecewise derivatives defined previously, weak derivatives are, in the pointwise sense, nonunique. They are unique in the L^2-sense.

Remark D.31 (equivalence)**.** Recall that L^2 functions that differ only on a set of measure zero are equivalent.

It is no coincidence that we have called these two sets Sobolev spaces, as the following result shows.

Theorem D.32 (Meyers–Serrin[9])**.** *Let $d \in \mathbb{N}$ and $\Omega \subset \mathbb{R}^d$ be a bounded Lipschitz domain. Then, for every $m \in \mathbb{N}$,*

$$H^m(\Omega) = W^m(\Omega).$$

In particular, the weak and strong derivatives of a function in $H^m(\Omega)$ coincide.

Remark D.33 (vector-valued functions)**.** As always, by arguing component-wise, we can define Sobolev spaces of vector-valued functions.

D.3.1 Periodic Functions

Let us now define Sobolev spaces of periodic functions.

Definition D.34 (periodic Sobolev space)**.** Suppose that $L > 0$ and $m \in \mathbb{N}$. Define

$$H_{\mathrm{p}}^m(0,L;\mathbb{C}) = \{u \in H_{\mathrm{loc}}^m(\mathbb{R};\mathbb{C}) \mid u(x+nL) = u(x), \quad \forall n \in \mathbb{Z}, \quad \forall x \in \mathbb{R}\},$$

where

$$H_{\mathrm{loc}}^m(\mathbb{R};\mathbb{C}) = \left\{ u\colon \mathbb{R} \to \mathbb{C} \;\middle|\; u|_{(c,d)} \in H^m(c,d;\mathbb{C}), \forall c,d \in \mathbb{R}, \ -\infty < c < d < \infty \right\}.$$

The space $H_{\mathrm{p}}^m(0,L;\mathbb{C})$ is called the **periodic Sobolev space of order** m. The subspace of $H_{\mathrm{p}}^m(0,L;\mathbb{C})$, consisting of mean-zero functions, is

$$\mathring{H}_{\mathrm{p}}^m(0,L;\mathbb{C}) = \left\{ f \in H_{\mathrm{p}}^m(0,L;\mathbb{C}) \;\middle|\; \int_0^L f(x)\mathrm{d}x = 0 \right\}.$$

[9] Named in honor of the American mathematicians Norman George Meyers (1930–) and James Burton Serrin (1926–2012), who originally proved a more general version of this result in an article concisely titled $H = W$; see [61].

Remark D.35 (identification). Since an L-periodic function can be uniquely characterized by its values on $[0, L]$, we usually make the identification between a periodic function and its restriction to this interval. With this convention, we could have similarly defined

$$\mathring{H}^m_{\mathrm{p}}(0, L; \mathbb{C}) = H^m_{\mathrm{p}}(0, L; \mathbb{C}) \cap \mathring{L}^2(0, L; \mathbb{C}).$$

As before, we make the simplifying assumption that $L = 1$ from this point on.

We must point out that, for the periodic case, there is an equivalent way to define $H^m_{\mathrm{p}}(0, 1; \mathbb{C})$ using Fourier series. To see how that works, let us first do some calculations. Suppose that $J \subseteq \mathbb{Z}$ is an index set and

$$f(x) = \sum_{j\in J} a_j \exp(2\pi i j x), \quad a_j \in \mathbb{C}, \quad j \in J.$$

If J is finite in cardinality, then f is a smooth, one-periodic function, and we can take any number of derivatives of the sum, term by term, to obtain

$$f^{(m)}(x) = \sum_{j\in J} a_j (2\pi i j)^m \exp(2\pi i j x), \quad \forall x \in \mathbb{R}$$

for any $m \in \mathbb{N}$. Using orthonormality, computing the $L^2(0, 1; \mathbb{C})$ norm of $f^{(m)}$, we find

$$\begin{aligned}
\left\| f^{(m)} \right\|^2_{L^2(0,1;\mathbb{C})} &= \left(\sum_{j\in J} a_j (2\pi i j)^m \exp(2\pi i j \cdot), \sum_{k\in J} a_k (2\pi i k)^m \exp(2\pi i k \cdot) \right)_{L^2(0,1;\mathbb{C})} \\
&= \sum_{j\in J} \sum_{k\in J} a_j (2\pi i j)^m \overline{a_k (2\pi i k)^m} \left(\exp(2\pi i j \cdot), \exp(2\pi i k \cdot)\right)_{L^2(0,1;\mathbb{C})} \\
&= \sum_{j\in J} \sum_{k\in J} a_j (2\pi i j)^m \overline{a_k (2\pi i k)^m}\, \delta_{j,k} \\
&= \sum_{j\in J} |a_j|^2 |2\pi i j|^{2m} \\
&= \sum_{j\in J} |a_j|^2 (4\pi^2 j^2)^m.
\end{aligned}$$

In other words,

$$\left\| f^{(m)} \right\|^2_{L^2(0,1;\mathbb{C})} = \sum_{j\in J} \gamma_j^m |a_j|^2, \qquad \gamma_j = 4\pi^2 j^2, \quad j \in J.$$

Using these calculations as motivation, we can prove the following result.

Theorem D.36 (characterization of $H^m_{\mathrm{p}}(0, 1; \mathbb{C})$). *Suppose that $f \in L^2_{\mathrm{p}}(0, 1; \mathbb{C})$ and $m \in \mathbb{N}$. Let $\Psi = \{\psi_j\}_{j\in\mathbb{Z}}$, with*

$$\psi_j(x) = \exp(2\pi i j x),$$

be the standard orthonormal trigonometric polynomial system on $[0, 1]$ and $\hat{f}_j$ denote the jth Fourier coefficient of f, $\hat{f}_j = (f, \psi_j)_{L^2(0,1;\mathbb{C})}$. Then $f \in H^m_{\mathrm{p}}(0, 1; \mathbb{C})$ if and only if the series

$$\sum_{j=-\infty}^{\infty} \gamma_j^m |\hat{f}_j|^2 = \lim_{n\to\infty} \sum_{j=-n}^{n} \gamma_j^m |\hat{f}_j|^2$$

converges, where

$$\gamma_j = 4\pi^2 j^2, \quad j \in \mathbb{Z}. \tag{D.3}$$

If $f \in H_p^m(0,1;\mathbb{C})$, then, in the L^2-sense, we have, as $n \to \infty$,

$$\sum_{j=-n}^{n} \hat{f}_j (2\pi \mathrm{i} j)^m \psi_j \longrightarrow \frac{\mathrm{d}^m f}{\mathrm{d}x^m},$$

and we are justified in writing

$$\frac{\mathrm{d}^m f}{\mathrm{d}x^m} = \sum_{j=-\infty}^{\infty} \hat{f}_j (2\pi \mathrm{i} j)^m \psi_j.$$

Furthermore, the object

$$|f|_{H^m(0,1;\mathbb{C})}^2 = \left(\frac{\mathrm{d}^m f}{\mathrm{d}x^m}, \frac{\mathrm{d}^m f}{\mathrm{d}x^m} \right)_{L^2(0,1)} = \sum_{j=-\infty}^{\infty} \gamma_j^m |\hat{f}_j|^2 \tag{D.4}$$

is called the Sobolev seminorm of f or order m.

Proof. ($\Longrightarrow$) This direction follows from the Riesz–Fischer Theorem 12.17 and (D.4) follows from Parseval's relation. The details are left to the reader as an exercise; see Problem D.7.

($\Longleftarrow$) This direction is beyond the scope of the text. The interested reader is referred to [97]. □

Problems

D.1 Prove Proposition D.5 and complete the proof of Theorem D.9.
Hint: Revisit the proof of Proposition A.27.

D.2 Prove the following, simplified, version of Poincaré inequality: There is a constant $C > 0$ such that, for any $v \in C^\infty(0,1) \cap C([0,1])$, $v(0) = v(1) = 0$,

$$\int_0^1 v^2 \mathrm{d}x \le C \int_0^1 \left(\frac{\mathrm{d}v}{\mathrm{d}x} \right)^2 \mathrm{d}x.$$

D.3 Prove the *Friedrichs inequality*[10]: Let $d \in \mathbb{N}$ and $\Omega \subset \mathbb{R}^d$ be a bounded domain with sufficiently smooth boundary. There is a constant $C > 0$ such that

$$\|v\|_{L^2(\Omega)} \le C \left(\|\nabla v\|_{L^2(\Omega;\mathbb{R}^d)}^2 + \|v\|_{L^2(\partial\Omega)}^2 \right)^{1/2}, \quad \forall v \in C^1(\bar{\Omega}).$$

Hint: Integrate by parts the identity

$$\int_\Omega |v(\boldsymbol{x})|^2 \mathrm{d}\boldsymbol{x} = \int_\Omega |v(\boldsymbol{x})|^2 \Delta\phi(\boldsymbol{x}) \mathrm{d}\boldsymbol{x},$$

where $\phi(\boldsymbol{x}) = \frac{1}{2d} \|\boldsymbol{x}\|_2^2$.

[10] Named in honor of the German–American mathematician Kurt Otto Friedrichs (1901–1982).

D.4 Let $f(x) = x^\alpha$, for $\alpha \in (0, 1)$. Show that $f \in H^1(0, 1)$ for $\alpha > 1/2$.
D.5 Show that $f(x) = 1 - |x|$ belongs to the Sobolev space $H_0^1(0, 1)$.
D.6 A function $v\colon [0, 1] \to \mathbb{R}$ is said to be Hölder continuous of order $\alpha > 0$, denoted $v \in C^{0,\alpha}([0, 1])$, if and only if

$$\sup_{x,y\in[0,1]:\ x\neq y} \frac{|v(x) - v(y)|}{|x - y|^\alpha} < \infty.$$

a) Show that if a function is Hölder continuous of order $\alpha > 0$, then it is continuous.
b) Show that if a function is Hölder continuous of order $\alpha > 1$, then it is constant.
c) Show that if $v \in H_0^1(0, 1)$, then $v \in C^{0,1/2}([0, 1])$.

D.7 Complete the proof of Theorem D.36.

References

[1] N.I. Achieser. *Theory of Approximation.* Dover Publications, New York, NY, 1992. Translated from the Russian and with a preface by Charles J. Hyman. Reprint of the 1956 English translation.

[2] R.A. Adams and J.J.F. Fournier. *Sobolev Spaces*, 2nd ed. Academic Press, New York, NY, 2003.

[3] L.V. Ahlfors. *Complex Analysis: An Introduction to the Theory of Analytic Functions of One Complex Variable*, 3rd ed. International Series in Pure and Applied Mathematics. McGraw-Hill, New York, NY, 1978.

[4] H. Amann. *Ordinary Differential Equations: An Introduction to Nonlinear Analysis.* Vol. 13 of De Gruyter Studies in Mathematics. Walter de Gruyter, Berlin, 1990. Translated from the German by Gerhard Metzen.

[5] K. Atkinson and W. Han. *Theoretical Numerical Analysis*, 3rd ed. Vol. 39 of Texts in Applied Mathematics Springer-Verlag, New York, NY, 2009.

[6] R.G. Bartle and D.R. Sherbert. *Introduction to Real Analysis*, 4th ed. John Wiley & Sons, Hoboken, NJ, 2011.

[7] R.F. Bass. *Real Analysis for Graduate Students*, 2nd ed. CreateSpace Independent Publishing Platform, Scotts Valley, CA, 2013.

[8] S. Boyd and L. Vandenberghe. *Convex Optimization.* Cambridge University Press, Cambridge, 2004.

[9] D. Braess. *Finite Elements: Theory, Fast Solvers, and Applications in Solid Mechanics*, 3rd ed. Cambridge University Press, Cambridge, 2007.

[10] S.C. Brenner and L.R. Scott. *The Mathematical Theory of Finite Element Methods*, 3rd ed. Vol. 15 of Texts in Applied Mathematics. Springer-Verlag, Berlin, 2007.

[11] J. Burkardt and C. Trenchea. Refactorization of the midpoint rule. *Appl. Math. Lett.*, 107:106438, 2020.

[12] J.C. Butcher. *Numerical Methods for Ordinary Differential Equations*, 2nd ed. John Wiley & Sons, Chichester, 2008.

[13] C. Canuto, M.Y. Hussaini, A. Quarteroni, and T.A. Zang. *Spectral Methods: Fundamentals in Single Domains.* Springer-Verlag, Berlin, 2007.

[14] L. Chen, X. Hu, and S.M. Wise. Convergence analysis of the fast subspace descent methods for convex optimization problems. *Math. Comput.*, 89:2249–2282, 2020.

[15] E.W. Cheney. *Approximation Theory.* McGraw-Hill, New York, NY, 1966.

[16] P.G. Ciarlet. *Introduction to Numerical Linear Algebra and Optimisation.* Cambridge University Press, Cambridge, 1989.

[17] P.G. Ciarlet. *The Finite Element Method for Elliptic Problems.* Vol. 40 of Classics in Applied Mathematics. Society for Industrial and Applied Mathematics, Philadelphia, PA, 2002.

[18] P.G. Ciarlet. *Linear and Nonlinear Functional Analysis with Applications.* Society for Industrial and Applied Mathematics, Philadelphia, PA, 2013.

[19] E.A. Coddington and N. Levinson. *Theory of Ordinary Differential Equations.* McGraw-Hill, New York, NY, 1955.

[20] J.W. Cooley and J.K. Tukey. An algorithm for the machine calculation of complex Fourier series. *Math. Comput.*, 19(2):297–301, 1965.

[21] R. Courant, K. Friedrichs, and H. Lewy. Über die partiellen Differenzengleichungen der mathematischen Physik. *Math. Ann.*, 100(1):32–74, 1928.

[22] R. Courant, K. Friedrichs, and H. Lewy. On the partial difference equations of mathematical physics. *IBM J. Res. Develop.*, 11:215–234, 1967.

[23] R. Dautray and J.-L. Lions. *Evolution Problems II.* Vol. 6 of Mathematical Analysis and Numerical Methods for Science and Technology. Springer-Verlag, Berlin, 1993. Translated from the French by Alan Craig.

[24] P.J. Davis. *Interpolation and Approximation.* Dover Publications, New York, NY, 1975.

[25] P.J. Davis and P. Rabinowitz. *Methods of Numerical Integration.* Academic Press, Orlando, FL, 1984.

[26] J.W. Demmel. *Applied Numerical Linear Algebra.* Society for Industrial and Applied Mathematics, Philadelphia, PA, 1997.

[27] J.E. Dennis, Jr. and R.B. Schnabel. *Numerical Methods for Unconstrained Optimization and Nonlinear Equations.* Vol. 16 of Classics in Applied Mathematics. Society for Industrial and Applied Mathematics, Philadelphia, PA, 1996. Corrected reprint of the 1983 original.

[28] P. Deuflhard. *Newton Methods for Nonlinear Problems: Affine Invariance and Adaptive Algorithms.* Vol. 35 of Springer Series in Computational Mathematics. Springer, Heidelberg, 2011. First softcover printing of the 2006 corrected printing.

[29] D.A. Di Pietro and A. Ern. *Mathematical Aspects of Discontinuous Galerkin Methods.* Vol. 69 of Mathématiques & Applications (Berlin) [Mathematics & Applications]. Springer, Heidelberg, 2012.

[30] A. Ern and J.-L. Guermond. *Theory and Practice of Finite Elements.* Vol. 159 of Applied Mathematical Sciences. Springer-Verlag, New York, NY, 2004.

[31] L.C. Evans. *Partial Differential Equations.* American Mathematical Society, Providence, RI, 2010.

[32] W. Gautschi. *Numerical Analysis*, 2nd ed. Birkhauser-Verlag, New York, NY, 2012.

[33] E. Godlewski and P.-A. Raviart. *Numerical Approximation of Hyperbolic Systems of Conservation Laws.* Vol. 118 of Applied Mathematical Sciences. Springer-Verlag, New York, NY, 1996.

[34] G.H. Golub and C.F. Van Loan. *Matrix Computations*, 4th ed. Johns Hopkins University Press, Baltimore, MD, 2013.

[35] R.E. Greene and S.G. Krantz. *Function Theory of One Complex Variable*, 3rd ed. Vol. 40 of Graduate Studies in Mathematics. American Mathematical Society, Providence, RI, 2006.

[36] L.A. Hageman and D.M. Young. *Applied Iterative Methods.* Academic Press, San Diego, CA, 1981.

[37] E. Hairer, S. Norsett, and G. Wanner. *Solving Ordinary Differential Equations I: Nonstiff Problems*, 2nd ed. Springer-Verlag, Berlin, 1993.

[38] G. Harris and C. Martin. The roots of a polynomial vary continuous as a function of the coefficients. *Proc. Am. Math. Soc.*, 100(2):390–392, 1987.

[39] A. Hatcher. *Algebraic Topology.* Cambridge University Press, Cambridge, 2002.

[40] M.T. Heideman, D.H. Johnson, and C.S. Burrus. Gauss and the history of the fast Fourier transform. *IEEE ASSP Magazine*, 7(7):14–21, 1984.

[41] P. Henrici. *Elements of Numerical Analysis*. John Wiley & Sons, New York, NY, 1964.

[42] E. Hewitt and R.E. Hewitt. The Gibbs-Wilbraham phenomenon: an episode in Fourier analysis. *Arch. Hist. Exact Sci.*, 21(2):129–160, 1979/80.

[43] H. Holden and N.H. Risebro. *Front Tracking for Hyperbolic Conservation Laws*, 2nd ed. Vol. 152 of Applied Mathematical Sciences. Springer, Heidelberg, 2015.

[44] R.A. Horn and C.R. Johnson. *Matrix Analysis*. Cambridge University Press, Cambridge, 1985.

[45] T.W. Hungerford. *Algebra*. Vol. 73 of Graduate Texts in Mathematics. Springer-Verlag, Berlin, 1980. Reprint of the 1974 original.

[46] E. Isaacson and H.B. Keller. *Analysis of Numerical Methods*. John Wiley & Sons, New York, NY, 1966.

[47] A. Iserles. *A First Course in the Numerical Analysis of Differential Equations*, 2nd ed. Cambridge University Press, Cambridge, 2009.

[48] D. Jackson. *The Theory of Approximation*. Vol. 2 of Colloquium Publications. AMS, New York, NY, 1930.

[49] F. John. *Partial Differential Equations*, 4th ed. Applied Mathematical Sciences. Springer-Verlag, Berlin, 1981.

[50] B.S. Jovanović and E. Süli. *Analysis of Finite Difference Schemes: For Linear Partial Differential Equations with Generalized Solutions*. Vol. 46 of Springer Series in Computational Mathematics. Springer, London, 2014.

[51] C.T. Kelley. *Iterative Methods for Linear and Nonlinear Equations*. Vol. 16 of Frontiers in Applied Mathematics. Society for Industrial and Applied Mathematics, Philadelphia, PA, 1995.

[52] R. Kress. *Numerical Analysis*. Springer-Verlag, Berlin, 1998.

[53] D. Kröner. *Numerical Schemes for Conservation Laws*. Wiley-Teubner Series Advances in Numerical Mathematics. John Wiley & Sons, Chichester; B.G. Teubner, Stuttgart, 1997.

[54] V.I. Krylov. *Approximate Calculation of Integrals*. Macmillan, New York, NY, 1962.

[55] A. Kufner, O. John, and S. Fučík. *Function Spaces*. Monographs and Textbooks on Mechanics of Solids and Fluids; Mechanics: Analysis. Noordhoff International Publishing, Leiden; Academia, Prague, 1977.

[56] O.A. Ladyzhenskaya. *The Boundary Value Problems of Mathematical Physics*. Vol. 49 of Applied Mathematical Sciences. Springer-Verlag, New York, 1985. Translated from the Russian by Jack Lohwater [Arthur J. Lohwater].

[57] O. Ladyzhenskaya. *Attractors for Semigroups and Evolution Equations*. Lezioni Lincee. [Lincei Lectures.] Cambridge University Press, Cambridge, 1991.

[58] S. Larsson and V. Thomée. *Partial Differential Equations with Numerical Methods*. Vol. 45 of Texts in Applied Mathematics. Springer-Verlag, Berlin, 2003.

[59] R.J. LeVeque. *Finite Volume Methods for Hyperbolic Problems*. Cambridge University Press, Cambridge, 2002.

[60] J.E. Marsden and M.J. Hoffman. *Basic Complex Analysis*. W.H. Freeman, New York, NY, 1989.

[61] N.G. Meyers and J. Serrin. $H = W$. *Proc. Natl. Acad. Sci. U.S.A.*, 51:1055–1056, 1964.

[62] V.V. Nemytskii and V.V. Stepanov. *Qualitative Theory of Differential Equations*. Princeton Mathematical Series, No. 22. Princeton University Press, Princeton, NJ, 1960.

[63] Y.E. Nesterov. A method for solving the convex programming problem with convergence rate $O(1/k^2)$. *Dokl. Akad. Nauk SSSR*, 269(3):543–547, 1983.

[64] Y.E. Nesterov. *Introductory Lecture Notes on Convex Optimization: A Basic Course*. Vol. 87 of Applied Optimization. Kluwer Academic Publishers, Boston, MA, 2004.

[65] J. Nocedal and S.J. Wright. *Numerical Optimization*, 2nd ed. Springer Series in Operations Research and Financial Engineering. Springer, New York, NY, 2006.

[66] J.J. O'Connor and E.F. Robertson. MacTutor History of Mathematics Archive. Website. `https://mathshistory.st-andrews.ac.uk/`.

[67] J.T. Oden. Finite elements: an introduction. In *Handbook of Numerical Analysis, Vol. II*, pages 3–15. North-Holland, Amsterdam, 1991.

[68] J.M. Ortega and W.C. Rheinboldt. Iterative solution of nonlinear equations in several variables. Vol. 30 of Classics in Applied Mathematics. Society for Industrial and Applied Mathematics, Philadelphia, PA, 2000. Reprint of the 1970 original.

[69] J.-H. Park, A.J. Salgado, and S.M. Wise. Preconditioned accelerated gradient descent methods for locally Lipschitz smooth objectives with applications to the solution of nonlinear PDEs. *J. Sci. Comput.*, 89(1):17, 2021.

[70] B.N. Parlett. *The Symmetric Eigenvalue Problem*. Prentice-Hall, Englewood Cliffs, NJ, 1980.

[71] M.J.D. Powell. *Approximation Theory and Methods*. Cambridge University Press, Cambridge, 1981.

[72] M. Renardy and R.C. Rogers. *An Introduction to Partial Differential Equations*, 2nd ed. Vol. 13 of Texts in Applied Mathematics. Springer-Verlag, Berlin, 2006.

[73] T.J. Rivlin. *An Introduction to the Approximation of Functions*. Dover Books on Advanced Mathematics. Dover Publications, New York, NY, 1981. Corrected reprint of the 1969 original.

[74] R.C. Robinson. *An Introduction to Dynamical Systems—Continuous and Discrete*, 2nd ed. Vol. 19 of Pure and Applied Undergraduate Texts. American Mathematical Society, Providence, RI, 2012.

[75] H.L. Royden. *Real Analysis*, 3rd ed. Macmillan, New York, NY, 1988.

[76] W. Rudin. *Principles of Mathematical Analysis*, 3rd ed. McGraw-Hill, New York, NY, 1976.

[77] W. Rudin. *Real and Complex Analysis*, 3rd ed. McGraw-Hill, New York, NY, 1986.

[78] Y. Saad. *Iterative Methods for Sparse Linear Systems*, 2nd ed. Society for Industrial and Applied Mathematics, Philadelphia, PA, 2003.

[79] A.A. Samarskii. *The Theory of Difference Schemes*, Vol. 240 of Monographs and Textbooks in Pure and Applied Mathematics. Marcel Dekker, New York, NY, 2001.

[80] A.A. Samarskii and A.V. Gulin. *Numerical Methods*. Nauka, Moscow, 1989. [in Russian.]

[81] A.A. Samarskii and E.S. Nikolaev. *Numerical Methods for Grid Equations. Vol. II: Iterative Methods*. Birkhäuser Verlag, Basel, 1989. Translated from the Russian and with a note by Stephen G. Nash.

[82] C. Schwab. *p- and hp-Finite Element Methods. Theory and Applications in Solid and Fluid Mechanics*. Numerical Mathematics and Scientific Computation. Clarendon Press, Oxford University Press, New York, NY, 1998.

[83] L.R. Scott. *Numerical Analysis*. Princeton University Press, Princeton, NJ, 2011.

[84] B. Sendov and A. Andreev. Approximation and interpolation theory. In *Handbook of Numerical Analysis, Vol. III*, pages 223–462. North-Holland, Amsterdam, 1994.

[85] J. Shen, T. Tang, and L.L. Wang. *Spectral Methods: Algorithms, Analysis and Applications*. Springer-Verlag, Berlin, 2011.

[86] J. Stoer and R. Bulirsch. *Introduction to Numerical Analysis*, 3rd ed. Springer-Verlag, Berlin, 2002.

[87] W.A. Strauss. *Partial Differential Equations: An Introduction*, 2nd ed. John Wiley & Sons, Chichester, 2008.

[88] J.C. Strikwerda. *Finite Difference Schemes and Partial Differential Equations*, 2nd ed. Society for Industrial and Applied Mathematics, Philadelphia, PA, 2004.

[89] E. Süli and D.F. Mayers. *An Introduction to Numerical Analysis.* Cambridge University Press, Cambridge, 2003.

[90] R. Temam. *Infinite-Dimensional Dynamical Systems in Mechanics and Physics.* Vol. 68 of Applied Mathematical Sciences. Springer-Verlag, New York, NY, 1988.

[91] V. Thomée. *Galerkin Finite Element Methods for Parabolic Problems.* Vol. 25 of Series in Computational Mathematics. Springer-Verlag, Berlin, 1997.

[92] V. Thomée. From finite differences to finite elements. A short history of numerical analysis of partial differential equations. *J. Comput. Appl. Math.*, 128(1–2):1–54, 2001.

[93] G.P. Tolstov. *Fourier Series.* Prentice-Hall, Englewood Cliffs, NJ, 1989.

[94] L.N. Trefethen. *Spectral Methods in Matlab.* Society for Industrial and Applied Mathematics, Philadelphia, PA, 2001.

[95] L.N. Trefethen. *Interpolation Theory and Interpolation Practice.* Society for Industrial and Applied Mathematics, Philadelphia, PA, 2013.

[96] L.N. Trefethen and D. Bau. *Numerical Linear Algebra.* Society for Industrial and Applied Mathematics, Philadelphia, PA, 1997.

[97] A. Vretblad. *Fourier Analysis and its Applications.* Springer-Verlag, Berlin, 2003.

[98] N.J. Walkington. Nesterov's method for convex optimization. *SIAM Rev.* To appear.

[99] H. Weinberger. *A First Course in Partial Differential Equations.* Xerox, Lexington, MA, 1965.

[100] P. Wesseling. *Principles of Computational Fluid Dynamics.* Vol. 29 of Springer Series in Computational Mathematics. Springer-Verlag, Berlin, 2001.

[101] Wikipedia, The Free Encyclopedia. Website. `https://en.wikipedia.org`

[102] J.H. Wilkinson. *The Algebraic Eigenvalue Problem.* Oxford University Press, Oxford, 1965.

[103] D.M. Young. *Iterative Solution of Large Linear Systems.* Academic Press, Cambridge, MA, 1971.

[104] A. Zygmund. *Trigonometric Series, Vols. I and II*, 3rd ed. Cambridge Mathematical Library. Cambridge University Press, Cambridge, 2002.

Index